# THERMODYNAMICS

EIGHTH EDITION IN SI UNITS

# THERMODYNAMICS

## AN ENGINEERING APPROACH

### EIGHTH EDITION IN SI UNITS

**YUNUS A. ÇENGEL**
*University of Nevada, Reno*

**MICHAEL A. BOLES**
*North Carolina State University*

*Adapted by*

**Mehmet Kanoğlu**
*University of Gaziantep*

## Quotes on Ethics

*Without ethics, everything happens as if we were all five billion passengers on a big machinery and nobody is driving the machinery. And it's going faster and faster, but we don't know where.*

—Jacques Cousteau

*Because you're able to do it and because you have the right to do it doesn't mean it's right to do it.*

—Laura Schlessinger

*A man without ethics is a wild beast loosed upon this world.*

—Manly Hall

*The concern for man and his destiny must always be the chief interest of all technical effort. Never forget it among your diagrams and equations.*

—Albert Einstein

*Cowardice asks the question, 'Is it safe?' Expediency asks the question, 'Is it politic?' Vanity asks the question, 'Is it popular?' But, conscience asks the question, 'Is it right?' And there comes a time when one must take a position that is neither safe, nor politic, nor popular but one must take it because one's conscience tells one that it is right.*

—Martin Luther King, Jr

*To educate a man in mind and not in morals is to educate a menace to society.*

—Theodore Roosevelt

*Politics which revolves around benefit is savagery.*

—Said Nursi

*The true test of civilization is, not the census, nor the size of the cities, nor the crops, but the kind of man that the country turns out.*

—Ralph W. Emerson

*The measure of a man's character is what he would do if he knew he never would be found out.*

—Thomas B. Macaulay

# ABOUT THE AUTHORS

**Yunus A. Çengel** is Professor Emeritus of Mechanical Engineering at the University of Nevada, Reno. He received his B.S. in mechanical engineering from Istanbul Technical University and his M.S. and Ph.D. in mechanical engineering from North Carolina State University. His areas of interest are renewable energy, energy efficiency, energy policies, heat transfer enhancement, and engineering education. He served as the director of the Industrial Assessment Center (IAC) at the University of Nevada, Reno, from 1996 to 2000. He has led teams of engineering students to numerous manufacturing facilities in Northern Nevada and California to perform industrial assessments, and has prepared energy conservation, waste minimization, and productivity enhancement reports for them. He has also served as an advisor for various government organizations and corporations.

Dr. Çengel is also the author or coauthor of the widely adopted textbooks *Heat and Mass Transfer: Fundamentals and Applications* (5th ed., 2015), *Fluid Mechanics:Fundamentals and Applications* (3rd ed., 2014), *Fundamentals of Thermal-Fluid Sciences* (4th ed., 2012), *Introduction to Thermodynamics and Heat Transfer* (2nd ed., 2008), and *Differential Equations for Engineers and Scientists* (1st ed., 2013), all published by McGraw-Hill. Some of his textbooks have been translated into Chinese, Japanese, Korean, Thai, Spanish, Portuguese, Turkish, Italian, Greek, and French.

Dr. Çengel is the recipient of several outstanding teacher awards, and he has received the ASEE Meriam/Wiley Distinguished Author Award for excellence in authorship in 1992 and again in 2000. Dr. Çengel is a registered Professional Engineer in the State of Nevada, and is a member of the American Society of Mechanical Engineers (ASME) and the American Society for Engineering Education (ASEE).

**Michael A. Boles** is Associate Professor of Mechanical and Aerospace Engineering at North Carolina State University, where he earned his Ph.D. in mechanical engineering and is an Alumni Distinguished Professor. Dr. Boles has received numerous awards and citations for excellence as an engineering educator. He is a past recipient of the SAE Ralph R. Teetor Education Award and has been twice elected to the NCSU Academy of Outstanding Teachers. The NCSU ASME student section has consistently recognized him as the outstanding teacher of the year and the faculty member having the most impact on mechanical engineering students.

Dr. Boles specializes in heat transfer and has been involved in the analytical and numerical solution of phase change and drying of porous media. He is a member of the American Society of Mechanical Engineers (ASME), the American Society for Engineering Education (ASEE), and Sigma Xi. Dr. Boles received the ASEE Meriam/Wiley Distinguished Author Award in 1992 for excellence in authorship.

# BRIEF CONTENTS

# CONTENTS

# CHAPTER SIX

## THE SECOND LAW OF THERMODYNAMICS 275

# CHAPTER SEVEN

## ENTROPY 329

# CHAPTER EIGHT

## EXERGY 421

# CHAPTER FIFTEEN
## CHEMICAL REACTIONS 759

# CHAPTER SIXTEEN
## CHEMICAL AND PHASE EQUILIBRIUM 805

# CHAPTER SEVENTEEN
## COMPRESSIBLE FLOW 839

# CHAPTER EIGHTEEN (WEB CHAPTER)
## RENEWABLE ENERGY

# PREFACE

## BACKGROUND

Thermodynamics is an exciting and fascinating subject that deals with energy, and thermodynamics has long been an essential part of engineering curricula all over the world. It has a broad application area ranging from microscopic organisms to common household appliances, transportation vehicles, power generation systems, and even philosophy. This introductory book contains sufficient material for two sequential courses in thermodynamics. Students are assumed to have an adequate background in calculus and physics.

## OBJECTIVES

This book is intended for use as a textbook by undergraduate engineering students in their sophomore or junior year, and as a reference book for practicing engineers. The objectives of this text are

- To cover the *basic principles* of thermodynamics.

- To present a wealth of real-world *engineering examples* to give students a feel for how thermodynamics is applied in engineering practice.

- To develop an *intuitive understanding* of thermodynamics by emphasizing the physics and physical arguments that underpin the theory.

It is our hope that this book, through its careful explanations of concepts and its use of numerous practical examples and figures, helps students develop the necessary skills to bridge the gap between knowledge and the confidence to properly apply knowledge.

## PHILOSOPHY AND GOAL

The philosophy that contributed to the overwhelming popularity of the prior editions of this book has remained unchanged in this edition. Namely, our goal has been to offer an engineering textbook that

- Communicates directly to the minds of tomorrow's engineers in a *simple yet precise* manner.

- Leads students toward a clear understanding and firm grasp of the *basic principles* of thermodynamics.

- Encourages *creative thinking* and development of a *deeper understanding* and *intuitive feel* for thermodynamics.

- Is *read* by students with *interest* and *enthusiasm* rather than being used as an aid to solve problems.

Special effort has been made to appeal to students' natural curiosity and to help them explore the various facets of the exciting subject area of thermodynamics. The enthusiastic responses we have received from users of prior editions—from small colleges to large universities all over the world—and xviii

THERMODYNAMICS

the continued translations into new languages indicate that our objectives have largely been achieved. It is our philosophy that the best way to learn is by practice. Therefore, special effort is made throughout the book to reinforce material that was presented earlier.

Yesterday's engineer spent a major portion of his or her time substituting values into the formulas and obtaining numerical results. However, formula manipulations and number crunching are now being left mainly to computers. Tomorrow's engineer will need a clear understanding and a firm grasp of the *basic principles* so that he or she can understand even the most complex problems, formulate them, and interpret the results. A conscious effort is made to emphasize these basic principles while also providing students with a perspective of how computational tools are used in engineering practice.

The traditional *classical,* or *macroscopic,* approach is used throughout the text, with microscopic arguments serving in a supporting role as appropriate. This approach is more in line with students' intuition and makes learning the subject matter much easier.

## NEW IN THIS EDITION

The primary change in this eighth edition of the text is the effective use of full color to enhance the learning experience of students and to make it more enjoyable. Another significant change is the addition of a new web chapter on Renewable Energy available via the Online Learning Center. The third important change is the update of the R-134a tables to make property values consistent with those from the latest version of EES. All the solved examples and end-of-chapter problems dealing with R-134a are modified to reflect this change. This edition includes numerous new problems with a variety of applications. Problems, whose solutions require parametric investigations and thus the use of a computer, are identified by a computer-EES icon, as before. Some existing problems from previous editions have been removed, and other updates and changes for clarity and readability have been made throughout the text.

The eighth edition also includes **McGraw-Hill's Connect® Engineering.** This online homework management tool allows assignment of algorithmic problems for homework, quizzes and tests. It connects students with the tools and resources they'll need to achieve success. To learn more, visit www.mcgrawhillconnect.com.

**McGraw-Hill LearnSmart™** is also available as an integrated feature of McGraw-Hill Connect® Engineering. It is an adaptive learning system designed to help students learn faster, study more efficiently, and retain more knowledge for greater success. LearnSmart assesses a student's knowledge of course content through a series of adaptive questions. It pinpoints concepts the student does not understand and maps out a personalized study plan for success. Visit the following site for a demonstration: www.mhlearnsmart.com.

# LEARNING TOOLS

## EARLY INTRODUCTION OF THE FIRST LAW OF THERMODYNAMICS

The first law of thermodynamics is introduced early in Chapter 2, "Energy, Energy Transfer, and General Energy Analysis." This introductory chapter sets the framework of establishing a general understanding of various forms of energy, mechanisms of energy transfer, the concept of energy balance, thermo-economics, energy conversion, and conversion efficiency using familiar settings that involve mostly electrical and mechanical forms of energy. It also exposes students to some exciting real-world applications of thermodynamics early in the course, and helps them establish a sense of the monetary value of energy. There is special emphasis on the utilization of renewable energy such as wind power and hydraulic energy, and the efficient use of existing resources.

## EMPHASIS ON PHYSICS

A distinctive feature of this book is its emphasis on the physical aspects of the subject matter in addition to mathematical representations and manipulations. The authors believe that the emphasis in undergraduate education should remain on *developing a sense of underlying physical mechanisms* and a *mastery of solving practical problems* that an engineer is likely to face in the real world. Developing an intuitive understanding should also make the course a more motivating and worthwhile experience for students.

## EFFECTIVE USE OF ASSOCIATION

An observant mind should have no difficulty understanding engineering sciences. After all, the principles of engineering sciences are based on our *everyday experiences* and *experimental observations*. Therefore, a physical, intuitive approach is used throughout this text. Frequently, *parallels are drawn* between the subject matter and students' everyday experiences so that they can relate the subject matter to what they already know. The process of cooking, for example, serves as an excellent vehicle to demonstrate the basic principles of thermodynamics.

## SELF-INSTRUCTING

The material in the text is introduced at a level that an average student can follow comfortably. It speaks *to* students, not *over* students. In fact, it is *self-instructive*. The order of coverage is from *simple* to *general*. That is, it starts with the simplest case and adds complexities gradually. In this way, the basic principles are repeatedly applied to different systems, and students master how to apply the principles instead of how to simplify a general formula. Noting that the principles of sciences are based on experimental observations, all the derivations in this text are based on physical arguments, and thus they are easy to follow and understand.

## EXTENSIVE USE OF ARTWORK

Figures are important learning tools that help students "get the picture," and the text makes very effective use of graphics. This edition of *Thermodynamics: An Engineering Approach,* Eighth Edition features an enhanced art program done in four colors to provide more realism and pedagogical understanding. Further, a large number of figures have been upgraded to become three-dimensional and thus more real-life. Figures attract attention and stimulate curiosity and interest. Most of the figures in this text are intended to serve as a means of emphasizing some key concepts that would otherwise go unnoticed; some serve as page summaries.

## LEARNING OBJECTIVES AND SUMMARIES

Each chapter begins with an *overview* of the material to be covered and chapter-specific *learning objectives*. A *summary* is included at the end of each chapter, providing a quick review of basic concepts and important relations, and pointing out the relevance of the material.

## NUMEROUS WORKED-OUT EXAMPLES WITH A SYSTEMATIC SOLUTIONS PROCEDURE

Each chapter contains several worked-out *examples* that clarify the material and illustrate the use of the basic principles. An *intuitive* and *systematic* approach is used in the solution of the example problems, while maintaining an informal conversational style. The problem is first stated, and the objectives are identified. The assumptions are then stated, together with their justifications. The properties needed to solve the problem are listed separately if appropriate. Numerical values are used together with their units to emphasize that numbers without units are meaningless, and that unit manipulations are as important as manipulating the numerical values with a calculator. The significance of the findings is discussed following the solutions. This approach is also used consistently in the solutions presented in the instructor's solutions manual.

## A WEALTH OF REAL-WORLD END-OF-CHAPTER PROBLEMS

The end-of-chapter problems are grouped under specific topics to make problem selection easier for both instructors and students. Within each group of problems are *Concept Questions,* indicated by "C," to check the students' level of understanding of basic concepts. The problems under *Review Problems* are more comprehensive in nature and are not directly tied to any specific section of a chapter—in some cases they require review of material learned in previous chapters. Problems designated as *Design and Essay* are intended to encourage students to make engineering judgments, to conduct independent exploration of topics of interest, and to communicate their findings in a professional manner. Problems with the ✿ are solved using EES, and complete solutions together with parametric studies are included on the textbook's website. Problems with the 🖳 are comprehensive in nature and are intended to be solved with a computer, possibly using the EES software. Several economics- and safety-related problems are incorporated throughout to promote cost and safety awareness among engineering students. Answers to selected problems are listed immediately following the problem for convenience to students. In addition, to prepare students for the Fundamentals of Engineering Exam (that is becoming more important for the outcome-based ABET 2000 criteria) and to facilitate multiple-choice tests, over 200 *multiple-choice problems* are included in the end-of-chapter problem sets. They are placed under the title *Fundamentals of Engineering (FE) Exam Problems* for easy recognition. These problems are intended to check the understanding of fundamentals and to help readers avoid common pitfalls.

## RELAXED SIGN CONVENTION

The use of a formal sign convention for heat and work is abandoned as it often becomes counterproductive. A physically meaningful and engaging approach is adopted for interactions instead of a mechanical approach. Subscripts "in" and "out," rather than the plus and minus signs, are used to indicate the directions of interactions.

## PHYSICALLY MEANINGFUL FORMULAS

The physically meaningful forms of the balance equations rather than formulas are used to foster deeper understanding and to avoid a cookbook approach. The mass, energy, entropy, and exergy balances for *any system* undergoing *any process* are expressed as

Mass balance:
$$m_{in} - m_{out} = \Delta m_{system}$$

Energy balance:
$$\underbrace{E_{in} - E_{out}}_{\substack{\text{Net energy transfer} \\ \text{by heat, work, and mass}}} = \underbrace{\Delta E_{system}}_{\substack{\text{Change in internal, kinetic,} \\ \text{potential, etc., energies}}}$$

Entropy balance:
$$\underbrace{S_{in} - S_{out}}_{\substack{\text{Net entropy transfer} \\ \text{by heat and mass}}} + \underbrace{S_{gen}}_{\substack{\text{Entropy} \\ \text{generation}}} = \underbrace{\Delta S_{system}}_{\substack{\text{Change} \\ \text{in entropy}}}$$

Exergy balance:
$$\underbrace{X_{in} - X_{out}}_{\substack{\text{Net exergy transfer} \\ \text{by heat, work, and mass}}} - \underbrace{X_{destroyed}}_{\substack{\text{Exergy} \\ \text{destruction}}} = \underbrace{\Delta X_{system}}_{\substack{\text{Change} \\ \text{in exergy}}}$$

These relations reinforce the fundamental principles that during an actual process mass and energy are conserved, entropy is generated, and exergy is destroyed. Students are encouraged to use these forms of balances in early chapters after they specify the system, and to simplify them for the particular problem. A more relaxed approach is used in later chapters as students gain mastery.

## TOPICS OF SPECIAL INTEREST

Most chapters contain a section called "Topic of Special Interest" where interesting aspects of thermodynamics are discussed. Examples include *Thermodynamic Aspects of Biological Systems* in Chapter 4, *Household Refrigerators* in Chapter 6, *Second-Law Aspects of Daily Life* in Chapter 8, and *Saving Fuel and Money by Driving Sensibly* in Chapter 9. The topics selected for these sections provide intriguing extensions to thermodynamics, but they can be ignored if desired without a loss in continuity.

## GLOSSARY OF THERMODYNAMIC TERMS

Throughout the chapters, when an important key term or concept is introduced and defined, it appears in **boldface** type. Fundamental thermodynamic terms and concepts also appear in a glossary located on our accompanying website (www.

mhhe.com/cengel). This unique glossary helps to reinforce key terminology and is an excellent learning and review tool for students as they move forward in their study of thermodynamics. In addition, students can test their knowledge of these fundamental terms by using the flash cards and other interactive resources.

### CONVERSION FACTORS

Frequently used conversion factors and physical constants are listed on the inner cover pages of the text for easy reference.

## SUPPLEMENTS

The following supplements are related to users of this SI edition.

### SOLUTIONS MANUAL

The *Solution Manual* that accompanies this book offers typeset, one-per-page solutions with detail explanations, to end-of-chapter problems.

### POWERPOINT SLIDE

PowerPoint presentation slides for all chapters in the text are available for use in lectures.

### TEXT WEBSITE

Web support is provided for the text specific website at www.mheducation.asia/olc/cengel.

Visit the website for general text information, errata, and author information. The site also includes resources for students including a list of helpful web links. The instructor side of the site includes the solutions manual, the text's images in PowerPoint form, and more!

The following supplements are related to users of this U.S. edition.

### ENGINEERING EQUATION SOLVER (EES)

Developed by Sanford Klein and William Beckman from the University of Wisconsin—Madison, this software combines equation-solving capability and engineering property data. EES can do optimization, parametric analysis, and linear and nonlinear regression, and provides publication-quality plotting capabilities. Thermodynamics and transport properties for air, water, and many other fluids are built in, and EES allows the user to enter property data or functional relationships.

EES is a powerful equation solver with built-in functions and property tables for thermodynamic and transport properties as well as automatic unit checking capability. It requires less time than a calculator for data entry and allows more time for thinking critically about modeling and solving engineering problems. Look for the EES icons in the homework problems sections of the text.

The Limited Academic Version of EES is available for departmental license upon adoption of the Eighth Edition of Thermodynamics: An Engineering Approach (meaning that the text is required for students in the course). You may load this software onto your institution's computer system, for use by students and faculty related to the course, as long as the arrangement between McGraw-Hill Education and F-Chart is in effect. There are minimum order requirements stipulated by F-Chart to qualify.

## PROPERTIES TABLE BOOKLET
### (ISBN 0-07-762477-7)

This booklet provides students with an easy reference to the most important property tables and charts, many of which are found at the back of the textbook in both the SI and English units.

## COSMOS

McGraw-Hill's COSMOS (Complete Online Solutions Manual Organization System) allows instructors to streamline the creation of assignments, quizzes, and tests by using problems and solutions from the textbook, as well as their own custom material. COSMOS is now available online at http://cosmos.mhhe.com/

# ACKNOWLEDGMENTS

The authors would like to acknowledge with appreciation the numerous and valuable comments, suggestions, constructive criticisms, and praise from the following evaluators and reviewers:

**Edward Anderson**
*Texas Tech University*

**John Biddle**
*Cal Poly Pomona University*

**Gianfranco DiGiuseppe**
*Kettering University*

**Shoeleh Di Julio**
*California State University-Northridge*

**Afshin Ghajar**
*Oklahoma State University*

**Harry Hardee**
*New Mexico State University*

**Kevin Lyons**
*North Carolina State University*

**Kevin Macfarlan**
*John Brown University*

**Saeed Manafzadeh**
*University of Illinois-Chicago*

**Alex Moutsoglou**
*South Dakota State University*

**Rishi Raj**
*The City College of New York*

**Maria Sanchez**
*California State University-Fresno*

**Kalyan Srinivasan**
*Mississippi State University*

**Robert Stiger**
*Gonzaga University*

Their suggestions have greatly helped to improve the quality of this text. In particular we would like to express our gratitude to Mehmet Kanoglu of the University of Gaziantep, Turkey, for his valuable contributions, his critical review of the manuscript, and for his special attention to accuracy and detail.

We also would like to thank our students, who provided plenty of feedback from students' perspectives. Finally, we would like to express our appreciation to our wives, Zehra Çengel and Sylvia Boles, and to our children for their continued patience, understanding, and support throughout the preparation of this text.

Yunus A. Çengel
Michael A. Boles

# Online Resources for Students and Instructors

## MCGRAW-HILL CONNECT® ENGINEERING

McGraw-Hill Connect Engineering is a web-based assignment and assessment platform that gives students the means to better connect with their course-work, with their instructors, and with the important concepts that they will need to know for success now and in the future. With Connect Engineering, instructors can deliver assignments, quizzes, and tests easily online. Students can practice important skills at their own pace and on their own schedule.

Connect Engineering for *Thermodynamics: An Engineering Approach*, Eighth Edition is available via the text website at www.mhhe.com/cengel

## COSMOS

McGraw-Hill's COSMOS (Complete Online Solutions Manual Organization System) allows instructors to streamline the creation of assignments, quizzes, and tests by using problems and solutions from the textbook, as well as their own custom material. COSMOS is now available online at http://cosmos.mhhe.com/

## WWW.MHHE.COM/CENGEL

This site offers resources for students and instructors.

The following resources are available for students:

- **Glossary of Key Terms in Thermodynamics**—Bolded terms in the text are defined in this accessible glossary. Organized at the chapter level or available as one large file.

- **Student Study Guide**—This resource outlines the fundamental concepts of the text and is a helpful guide that allows students to focus on the most important concepts. The guide can also serve as a lecture outline for instructors.

- **Learning Objectives**—The chapter learning objectives are outlined here. Organized by chapter and tied to ABET objectives.

- **Self-Quizzing**—Students can test their knowledge using multiple-choice quizzing. These self-tests provide immediate feedback and are an excellent learning tool.

- **Flashcards**—Interactive flashcards test student understanding of the text terms and their definitions. The program also allows students to flag terms that require further understanding.

- **Crossword Puzzles**—An interactive, timed puzzle that provides hints as well as a notes section.

- **Errata**—If errors should be found in the text, they will be reported here.

The following resources are available for instructors under password protection:

- **Instructor Testbank**—Additional problems prepared for instructors to assign to students. Solutions are given, and use of EES is recommended to verify accuracy.
- **Correlation Guide**—New users of this text will appreciate this resource. The guide provides a smooth transition for instructors not currently using the Çengel/Boles text.
- **Image Library**—The electronic version of the figures are supplied for easy integration into course presentations, exams, and assignments.
- **Instructor's Guide**—Provides instructors with helpful tools such as sample syllabi and exams, an ABET conversion guide, a thermodynamics glossary, and chapter objectives.
- **Errata**—If errors should be found in the solutions manual, they will be reported here.
- **Solutions Manual**—The detailed solutions to all text homework problems are provided in PDF form.
- **EES Solutions Manual**—The solutions manual is also available in EES. Any chapter-end problem in the text can be modified and the solution of the modified problem can readily be obtained by changing input parameters in the EES solution and hitting the solve button.
- **PP slides**—Powerpoint presentation slides for all chapters in the text are available for use in lectures
- **Appendices**—These are provided in PDF form for ease of use.

# INTRODUCTION AND BASIC CONCEPTS

**E**very science has a unique vocabulary associated with it, and thermodynamics is no exception. Precise definition of basic concepts forms a sound foundation for the development of a science and prevents possible misunderstandings. We start this chapter with an overview of thermodynamics and the unit systems, and continue with a discussion of some basic concepts such as *system, state, state postulate, equilibrium, and process*. We discuss intensive and extensive properties of a system and define density, specific gravity, and specific weight. We also discuss *temperature* and *temperature scales* with particular emphasis on the International Temperature Scale of 1990. We then present *pressure,* which is the normal force exerted by a fluid per unit area and discuss *absolute* and *gage* pressures, the variation of pressure with depth, and pressure measurement devices, such as manometers and barometers. Careful study of these concepts is essential for a good understanding of the topics in the following chapters. Finally, we present an intuitive systematic *problem-solving technique* that can be used as a model in solving engineering problems.

## OBJECTIVES

The objectives of Chapter 1 are to:

- Identify the unique vocabulary associated with thermodynamics through the precise definition of basic concepts to form a sound foundation for the development of the principles of thermodynamics.
- Review the metric SI and the English unit systems.
- Explain the basic concepts of thermodynamics such as system, state, state postulate, equilibrium, process, and cycle.
- Discuss properties of a system and define density, specific gravity, and specific weight.
- Review concepts of temperature, temperature scales, pressure, and absolute and gage pressure.
- Introduce an intuitive systematic problem-solving technique.

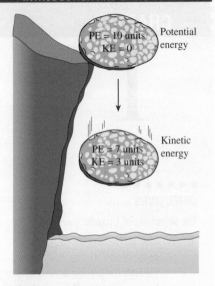

**FIGURE 1–1**

Energy cannot be created or destroyed; it can only change forms (the first law).

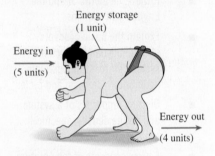

**FIGURE 1–2**

Conservation of energy principle for the human body.

**FIGURE 1–3**

Heat flows in the direction of decreasing temperature.

# 1–1 · THERMODYNAMICS AND ENERGY

Thermodynamics can be defined as the science of *energy*. Although everybody has a feeling of what energy is, it is difficult to give a precise definition for it. Energy can be viewed as the ability to cause changes.

The name *thermodynamics* stems from the Greek words *therme* (heat) and *dynamis* (power), which is most descriptive of the early efforts to convert heat into power. Today the same name is broadly interpreted to include all aspects of energy and energy transformations including power generation, refrigeration, and relationships among the properties of matter.

One of the most fundamental laws of nature is the **conservation of energy principle**. It simply states that during an interaction, energy can change from one form to another but the total amount of energy remains constant. That is, energy cannot be created or destroyed. A rock falling off a cliff, for example, picks up speed as a result of its potential energy being converted to kinetic energy (Fig. 1–1). The conservation of energy principle also forms the backbone of the diet industry: A person who has a greater energy input (food) than energy output (exercise) will gain weight (store energy in the form of fat), and a person who has a smaller energy input than output will lose weight (Fig. 1–2). The change in the energy content of a body or any other system is equal to the difference between the energy input and the energy output, and the energy balance is expressed as $E_{in} - E_{out} = \Delta E$.

The **first law of thermodynamics** is simply an expression of the conservation of energy principle, and it asserts that *energy* is a thermodynamic property. The **second law of thermodynamics** asserts that energy has *quality* as well as *quantity*, and actual processes occur in the direction of decreasing quality of energy. For example, a cup of hot coffee left on a table eventually cools, but a cup of cool coffee in the same room never gets hot by itself (Fig. 1–3). The high-temperature energy of the coffee is degraded (transformed into a less useful form at a lower temperature) once it is transferred to the surrounding air.

Although the principles of thermodynamics have been in existence since the creation of the universe, thermodynamics did not emerge as a science until the construction of the first successful atmospheric steam engines in England by Thomas Savery in 1697 and Thomas Newcomen in 1712. These engines were very slow and inefficient, but they opened the way for the development of a new science.

The first and second laws of thermodynamics emerged simultaneously in the 1850s, primarily out of the works of William Rankine, Rudolph Clausius, and Lord Kelvin (formerly William Thomson). The term *thermodynamics* was first used in a publication by Lord Kelvin in 1849. The first thermodynamics textbook was written in 1859 by William Rankine, a professor at the University of Glasgow.

It is well-known that a substance consists of a large number of particles called *molecules*. The properties of the substance naturally depend on the behavior of these particles. For example, the pressure of a gas in a container is the result of momentum transfer between the molecules and the walls of the container. However, one does not need to know the behavior of the gas particles to determine the pressure in the container. It would be sufficient to attach a pressure gage to the container. This macroscopic approach to the study of thermodynamics that does not require a knowledge of the behavior of individual particles is called **classical thermodynamics**. It provides a direct and easy way to the solution of engineering problems. A more elaborate approach, based on the average behavior of large groups of individual particles, is called **statistical thermodynamics**. This microscopic approach is rather involved and is used in this text only in the supporting role.

## Application Areas of Thermodynamics

All activities in nature involve some interaction between energy and matter; thus, it is hard to imagine an area that does not relate to thermodynamics in some manner. Therefore, developing a good understanding of basic principles of thermodynamics has long been an essential part of engineering education.

Thermodynamics is commonly encountered in many engineering systems and other aspects of life, and one does not need to go very far to see some application areas of it. In fact, one does not need to go anywhere. The heart is constantly pumping blood to all parts of the human body, various energy conversions occur in trillions of body cells, and the body heat generated is constantly rejected to the environment. The human comfort is closely tied to the rate of this metabolic heat rejection. We try to control this heat transfer rate by adjusting our clothing to the environmental conditions.

Other applications of thermodynamics are right where one lives. An ordinary house is, in some respects, an exhibition hall filled with wonders of thermodynamics (Fig. 1–4). Many ordinary household utensils and appliances are designed, in whole or in part, by using the principles of thermodynamics. Some examples include the electric or gas range, the heating and air-conditioning systems, the refrigerator, the humidifier, the pressure cooker, the water heater, the shower, the iron, and even the computer and the TV. On a larger scale, thermodynamics plays a major part in the design and analysis of automotive engines, rockets, jet engines, and conventional or nuclear power plants, solar collectors, and the design of vehicles from ordinary cars to airplanes (Fig. 1–5). The energy-efficient home that you may be living in, for example, is designed on the basis of minimizing heat loss in winter and heat gain in summer. The size, location, and the power input of the fan of your computer is also selected after an analysis that involves thermodynamics.

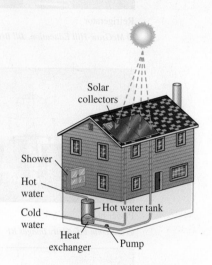

**FIGURE 1–4**
The design of many engineering systems, such as this solar hot water system, involves thermodynamics.

## 1–2 · IMPORTANCE OF DIMENSIONS AND UNITS

Any physical quantity can be characterized by **dimensions**. The magnitudes assigned to the dimensions are called **units**. Some basic dimensions such as mass $m$, length $L$, time $t$, and temperature $T$ are selected as **primary** or **fundamental dimensions**, while others such as velocity $V$, energy $E$, and volume $V$ are expressed in terms of the primary dimensions and are called **secondary dimensions**, or **derived dimensions**.

A number of unit systems have been developed over the years. Despite strong efforts in the scientific and engineering community to unify the world with a single unit system, two sets of units are still in common use today: the **English system**, which is also known as the *United States*

Refrigerator

© McGraw-Hill Education, Jill Braaten

Boats

© Doug Menuez/Getty Images RF

Aircraft and spacecraft

© PhotoLink/Getty Images RF

Power plants

© Malcolm Fife/Getty Images RF

Human body

© Ryan McVay/Getty Images RF

Cars

© Mark Evans/Getty Images RF

Wind turbines

© F. Schussler/PhotoLink/Getty
Images RF

Food processing

Glow Images RF

A piping network in an industrial facility.

Courtesy of UMDE Engineering Contracting
and Trading. Used by permission

**FIGURE 1–5**
Some application areas of thermodynamics.

Customary System (USCS), and the metric **SI** (from *Le Système International d' Unités*), which is also known as the *International System*. The SI is a simple and logical system based on a decimal relationship between the various units, and it is being used for scientific and engineering work in most of the industrialized nations, including England. The English system, however, has no apparent systematic numerical base, and various units in this system are related to each other rather arbitrarily (12 in = 1 ft, 1 mile = 5280 ft, 4 qt = 1 gal, etc.), which makes it confusing and difficult to learn. The United States is the only industrialized country that has not yet fully converted to the metric system.

The systematic efforts to develop a universally acceptable system of units dates back to 1790 when the French National Assembly charged the French Academy of Sciences to come up with such a unit system. An early version of the metric system was soon developed in France, but it did not find universal acceptance until 1875 when *The Metric Convention Treaty* was prepared and signed by 17 nations, including the United States. In this international treaty, meter and gram were established as the metric units for length and mass, respectively, and a *General Conference of Weights and Measures* (CGPM) was established that was to meet every six years. In 1960, the CGPM produced the SI, which was based on six fundamental quantities, and their units were adopted in 1954 at the Tenth General Conference of Weights and Measures: *meter* (m) for length, *kilogram* (kg) for mass, *second* (s) for time, *ampere* (A) for electric current, *degree Kelvin* (°K) for temperature, and *candela* (cd) for luminous intensity (amount of light). In 1971, the CGPM added a seventh fundamental quantity and unit: *mole* (mol) for the amount of matter.

Based on the notational scheme introduced in 1967, the degree symbol was officially dropped from the absolute temperature unit, and all unit names were to be written without capitalization even if they were derived from proper names (Table 1–1). However, the abbreviation of a unit was to be capitalized if the unit was derived from a proper name. For example, the SI unit of force, which is named after Sir Isaac Newton (1647–1723), is *newton* (not Newton), and it is abbreviated as N. Also, the full name of a unit may be pluralized, but its abbreviation cannot. For example, the length of an object can be 5 m or 5 meters, *not* 5 ms or 5 meter. Finally, no period is to be used in unit abbreviations unless they appear at the end of a sentence. For example, the proper abbreviation of meter is m (*not* m.).

The recent move toward the metric system in the United States seems to have started in 1968 when Congress, in response to what was happening in the rest of the world, passed a Metric Study Act. Congress continued to promote a voluntary switch to the metric system by passing the Metric Conversion Act in 1975. A trade bill passed by Congress in 1988 set a September 1992 deadline for all federal agencies to convert to the metric system. However, the deadlines were relaxed later with no clear plans for the future.

The industries that are heavily involved in international trade (such as the automotive, soft drink, and liquor industries) have been quick in converting to the metric system for economic reasons (having a single worldwide design, fewer sizes, smaller inventories, etc.). Today, nearly all the cars manufactured in the United States are metric. Most car owners probably do not realize this until they try an English socket wrench on a metric bolt. Most industries, however, resisted the change, thus slowing down the conversion process.

Presently the United States is a dual-system society, and it will stay that way until the transition to the metric system is completed. This puts an extra burden on today's engineering students, since they are expected to retain their understanding of the English system while learning, thinking, and working in terms of the SI.

As pointed out, the SI is based on a decimal relationship between units. The prefixes used to express the multiples of the various units are listed in Table 1–2. They are standard for all units, and the student is encouraged to memorize them because of their widespread use (Fig. 1–6).

**TABLE 1–1**

The seven fundamental (or primary) dimensions and their units in SI

| Dimension | Unit |
| --- | --- |
| Length | meter (m) |
| Mass | kilogram (kg) |
| Time | second (s) |
| Temperature | kelvin (K) |
| Electric current | ampere (A) |
| Amount of light | candela (cd) |
| Amount of matter | mole (mol) |

**TABLE 1–2**

Standard prefixes in SI units

| Multiple | Prefix |
| --- | --- |
| $10^{24}$ | yotta, Y |
| $10^{21}$ | zetta, Z |
| $10^{18}$ | exa, E |
| $10^{15}$ | peta, P |
| $10^{12}$ | tera, T |
| $10^{9}$ | giga, G |
| $10^{6}$ | mega, M |
| $10^{3}$ | kilo, k |
| $10^{2}$ | hecto, h |
| $10^{1}$ | deka, da |
| $10^{-1}$ | deci, d |
| $10^{-2}$ | centi, c |
| $10^{-3}$ | milli, m |
| $10^{-6}$ | micro, $\mu$ |
| $10^{-9}$ | nano, n |
| $10^{-12}$ | pico, p |
| $10^{-15}$ | femto, f |
| $10^{-18}$ | atto, a |
| $10^{-21}$ | zepto, z |
| $10^{-24}$ | yocto, y |

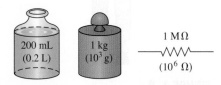

**FIGURE 1–6**

The SI unit prefixes are used in all branches of engineering.

## Some SI and English Units

In SI, the units of mass, length, and time are the kilogram (kg), meter (m), and second (s), respectively. The respective units in the English system are the pound-mass (lbm), foot (ft), and second (s). The pound symbol *lb* is actually the abbreviation of *libra*, which was the ancient Roman unit of weight. The English retained this symbol even after the end of the Roman occupation of Britain in 410. The mass and length units in the two systems are related to each other by

$$1 \text{ lbm} = 0.45359 \text{ kg}$$

$$1 \text{ ft} = 0.3048 \text{ m}$$

In the English system, force is usually considered to be one of the primary dimensions and is assigned a nonderived unit. This is a source of confusion and error that necessitates the use of a dimensional constant ($g_c$) in many formulas. To avoid this nuisance, we consider force to be a secondary dimension whose unit is derived from Newton's second law, that is,

$$\text{Force} = (\text{Mass})(\text{Acceleration})$$

or

$$F = ma \tag{1–1}$$

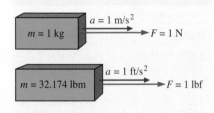

**FIGURE 1–7**
The definition of the force units.

In SI, the force unit is the newton (N), and it is defined as the *force required to accelerate a mass of 1 kg at a rate of 1 m/s²*. In the English system, the force unit is the **pound-force** (lbf) and is defined as the *force required to accelerate a mass of 32.174 lbm (1 slug) at a rate of 1 ft/s²* (Fig. 1–7). That is,

$$1 \text{ N} = 1 \text{ kg·m/s}^2$$

$$1 \text{ lbf} = 32.174 \text{ lbm·ft/s}^2$$

A force of 1 N is roughly equivalent to the weight of a small apple ($m = 102$ g), whereas a force of 1 lbf is roughly equivalent to the weight of four medium apples ($m_{\text{total}} = 454$ g), as shown in Fig. 1–8. Another force unit in common use in many European countries is the *kilogram-force* (kgf), which is the weight of 1 kg mass at sea level (1 kgf = 9.807 N).

The term **weight** is often incorrectly used to express mass, particularly by the "weight watchers." Unlike mass, weight $W$ is a *force*. It is the gravitational force applied to a body, and its magnitude is determined from Newton's second law,

$$W = mg \quad \text{(N)} \tag{1–2}$$

**FIGURE 1–8**
The relative magnitudes of the force units newton (N), kilogram-force (kgf), and pound-force (lbf).

where $m$ is the mass of the body, and $g$ is the local gravitational acceleration ($g$ is 9.807 m/s² or 32.174 ft/s² at sea level and 45° latitude). An ordinary bathroom scale measures the gravitational force acting on a body.

The mass of a body remains the same regardless of its location in the universe. Its weight, however, changes with a change in gravitational acceleration. A body weighs less on top of a mountain since $g$ decreases with altitude. On the surface of the moon, an astronaut weighs about one-sixth of what she or he normally weighs on earth (Fig. 1–9).

At sea level a mass of 1 kg weighs 9.807 N, as illustrated in Fig. 1–10. A mass of 1 lbm, however, weighs 1 lbf, which misleads people to believe that pound-mass and pound-force can be used interchangeably as pound (lb), which is a major source of error in the English system.

It should be noted that the *gravity force* acting on a mass is due to the *attraction* between the masses, and thus it is proportional to the magnitudes of the masses and inversely proportional to the square of the distance between them. Therefore, the gravitational acceleration $g$ at a location depends on the *local density* of the earth's crust, the *distance* to the center of the earth, and to a lesser extent, the positions of the moon and the sun. The value of $g$ varies with location from 9.832 m/s² at the poles (9.789 at the equator) to 7.322 m/s² at 1000 km above sea level. However, at altitudes up to 30 km, the variation of $g$ from the sea-level value of 9.807 m/s² is less than 1 percent. Therefore, for most practical purposes, the gravitational acceleration can be assumed to be *constant* at 9.807 m/s², often rounded to 9.81 m/s². It is interesting to note that at locations below sea level, the value of $g$ increases with distance from the sea level, reaches a maximum at about 4500 m, and then starts decreasing. (What do you think the value of $g$ is at the center of the earth?)

The primary cause of confusion between mass and weight is that mass is usually measured *indirectly* by measuring the *gravity force* it exerts. This approach also assumes that the forces exerted by other effects such as air buoyancy and fluid motion are negligible. This is like measuring the distance to a star by measuring its red shift, or measuring the altitude of an airplane by measuring barometric pressure. Both of these are also indirect measurements. The correct *direct* way of measuring mass is to compare it to a known mass. This is cumbersome, however, and it is mostly used for calibration and measuring precious metals.

*Work,* which is a form of energy, can simply be defined as force times distance; therefore, it has the unit "newton-meter (N·m)," which is called a **joule** (J). That is,

$$1 \text{ J} = 1 \text{ N·m} \tag{1–3}$$

A more common unit for energy in SI is the kilojoule (1 kJ = $10^3$ J). In the English system, the energy unit is the **Btu** (British thermal unit), which is defined as the energy required to raise the temperature of 1 lbm of water at 68°F by 1°F. In the metric system, the amount of energy needed to raise the temperature of 1 g of water at 14.5°C by 1°C is defined as 1 **calorie** (cal), and 1 cal = 4.1868 J. The magnitudes of the kilojoule and Btu are almost identical (1 Btu = 1.0551 kJ). Here is a good way to get a feel for these units: If you light a typical match and let it burn itself out, it yields approximately one Btu (or one kJ) of energy (Fig. 1–11).

The unit for time rate of energy is joule per second (J/s), which is called a **watt** (W). In the case of work, the time rate of energy is called *power*. A commonly used unit of power is horsepower (hp), which is equivalent to 746 W. Electrical energy typically is expressed in the unit kilowatt-hour (kWh), which is equivalent to 3600 kJ. An electric appliance with a rated power of 1 kW consumes 1 kWh of electricity when running continuously

**FIGURE 1–9**
A body weighing 66 kgf on earth will weigh only 11 kgf on the moon.

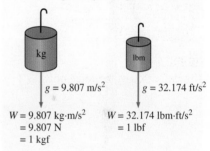

$g$ = 9.807 m/s²   $g$ = 32.174 ft/s²

$W$ = 9.807 kg·m/s²   $W$ = 32.174 lbm·ft/s²
= 9.807 N   = 1 lbf
= 1 kgf

**FIGURE 1–10**
The weight of a unit mass at sea level.

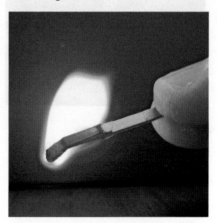

**FIGURE 1–11**
A typical match yields about one Btu (or one kJ) of energy if completely burned.
*Photo by John M. Cimbala*

for one hour. When dealing with electric power generation, the units kW and kWh are often confused. Note that kW or kJ/s is a unit of power, whereas kWh is a unit of energy. Therefore, statements like "the new wind turbine will generate 50 kW of electricity per year" are meaningless and incorrect. A correct statement should be something like "the new wind turbine with a rated power of 50 kW will generate 120,000 kWh of electricity per year."

## Dimensional Homogeneity

We all know that apples and oranges do not add. But we somehow manage to do it (by mistake, of course). In engineering, all equations must be *dimensionally homogeneous*. That is, every term in an equation must have the same unit. If, at some stage of an analysis, we find ourselves in a position to add two quantities that have different units, it is a clear indication that we have made an error at an earlier stage. So checking dimensions can serve as a valuable tool to spot errors.

**FIGURE 1–12**
A wind turbine, as discussed in Example 1–1.
©*Bear Dancer Studios/Mark Dierker RF*

---

**EXAMPLE 1–1    Electric Power Generation by a Wind Turbine**

A school is paying $0.12/kWh for electric power. To reduce its power bill, the school installs a wind turbine (Fig. 1–12) with a rated power of 30 kW. If the turbine operates 2200 hours per year at the rated power, determine the amount of electric power generated by the wind turbine and the money saved by the school per year.

**SOLUTION**   A wind turbine is installed to generate electricity. The amount of electric energy generated and the money saved per year are to be determined.
*Analysis*   The wind turbine generates electric energy at a rate of 30 kW or 30 kJ/s. Then the total amount of electric energy generated per year becomes

$$\text{Total energy} = (\text{Energy per unit time})(\text{Time interval})$$
$$= (30 \text{ kW})(2200 \text{ h})$$
$$= \mathbf{66,000 \text{ kWh}}$$

The money saved per year is the monetary value of this energy determined as

$$\text{Money saved} = (\text{Total energy})(\text{Unit cost of energy})$$
$$= (66,000 \text{ kWh})(\$0.12/\text{kWh})$$
$$= \mathbf{\$7920}$$

*Discussion*   The annual electric energy production also could be determined in kJ by unit manipulations as

$$\text{Total energy} = (30 \text{ kW})(2200 \text{ h})\left(\frac{3600 \text{ s}}{1 \text{ h}}\right)\left(\frac{1 \text{ kJ/s}}{1 \text{ kW}}\right) = 2.38 \times 10^8 \text{ kJ}$$

which is equivalent to 66,000 kWh (1 kWh = 3600 kJ).

---

We all know from experience that units can give terrible headaches if they are not used carefully in solving a problem. However, with some attention and skill, units can be used to our advantage. They can be used to check formulas; sometimes they can even be used to *derive* formulas, as explained in the following example.

## EXAMPLE 1–2    Obtaining Formulas from Unit Considerations

A tank is filled with oil whose density is $\rho$ = 850 kg/m³. If the volume of the tank is $V$ = 2 m³, determine the amount of mass $m$ in the tank.

**SOLUTION** The volume of an oil tank is given. The mass of oil is to be determined.

**Assumptions** Oil is a nearly incompressible substance and thus its density is constant.

**Analysis** A sketch of the system just described is given in Fig. 1–13. Suppose we forgot the formula that relates mass to density and volume. However, we know that mass has the unit of kilograms. That is, whatever calculations we do, we should end up with the unit of kilograms. Putting the given information into perspective, we have

$$\rho = 850 \text{ kg/m}^3 \quad \text{and} \quad V = 2 \text{ m}^3$$

It is obvious that we can eliminate m³ and end up with kg by multiplying these two quantities. Therefore, the formula we are looking for should be

$$m = \rho V$$

Thus,

$$m = (850 \text{ kg/m}^3)(2 \text{ m}^3) = \textbf{1700 kg}$$

**Discussion** Note that this approach may not work for more complicated formulas. Nondimensional constants also may be present in the formulas, and these cannot be derived from unit considerations alone.

**FIGURE 1–13**
Schematic for Example 1–2.

**FIGURE 1–14**
Always check the units in your calculations.

You should keep in mind that a formula that is not dimensionally homogeneous is definitely wrong (Fig. 1–14), but a dimensionally homogeneous formula is not necessarily right.

## Unity Conversion Ratios

Just as all nonprimary dimensions can be formed by suitable combinations of primary dimensions, *all nonprimary units (secondary units) can be formed by combinations of primary units*. Force units, for example, can be expressed as

$$1 \text{ N} = 1 \text{ kg} \frac{\text{m}}{\text{s}^2} \quad \text{and} \quad 1 \text{ lbf} = 32.174 \text{ lbm} \frac{\text{ft}}{\text{s}^2}$$

They can also be expressed more conveniently as **unity conversion ratios** as

$$\frac{1 \text{ N}}{1 \text{ kg·m/s}^2} = 1 \quad \text{and} \quad \frac{1 \text{ lbf}}{32.174 \text{ lbm·ft/s}^2} = 1$$

Unity conversion ratios are identically equal to 1 and are unitless, and thus such ratios (or their inverses) can be inserted conveniently into any calculation to properly convert units (Fig. 1–15). You are encouraged to always use unity conversion ratios such as those given here when converting units. Some textbooks insert the archaic gravitational constant $g_c$ defined as $g_c = 32.174$ lbm·ft/lbf·s² = 1 kg·m/N·s² = 1 into equations in order to force

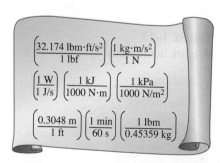

**FIGURE 1–15**
Every unity conversion ratio (as well as its inverse) is exactly equal to one. Shown here are a few commonly used unity conversion ratios.

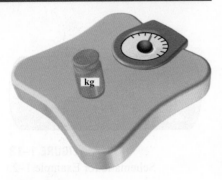

**FIGURE 1–16**
A mass of 1 kg weighs 9.807 N on earth.

Weight?
I thought gram was a unit of mass!

Newton Crunch Cereal
THE BREAKFAST OF ENGINEERS

Net weight:
One pound
(454 grams)

**FIGURE 1–17**
A quirk in the metric system of units.

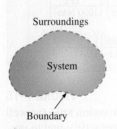

Surroundings

System

Boundary

**FIGURE 1–18**
System, surroundings, and boundary.

units to match. This practice leads to unnecessary confusion and is strongly discouraged by the present authors. We recommend that you instead use unity conversion ratios.

---

**EXAMPLE 1–3    The Weight of One Kilogram**

Using unity conversion ratios, show that 1.00 kg weighs 9.807 N on earth (Fig. 1–16).

**SOLUTION**  A mass of 1.00 kg is subjected to standard earth gravity. Its weight in N is to be determined.
**Assumptions**  Standard sea-level conditions are assumed.
**Properties**  The gravitational constant is $g = 9.807$ m/s$^2$.
**Analysis**  We apply Newton's second law to calculate the weight (force) that corresponds to the known mass and acceleration. The weight of any object is equal to its mass times the local value of gravitational acceleration. Thus,

$$W = mg = (1.00 \text{ kg})(9.807 \text{ m/s}^2)\left(\frac{1 \text{ N}}{1 \text{ kg·m/s}^2}\right) = \mathbf{9.807 \text{ N}}$$

**Discussion**  The quantity in large parentheses in this equation is a unity conversion ratio. Mass is the same regardless of its location. However, on some other planet with a different value of gravitational acceleration, the weight of 1 kg would differ from that calculated here.

---

When you buy a box of breakfast cereal, the printing may say "Net weight: One pound (454 grams)." (See Fig. 1–17.) Technically, this means that the cereal inside the box weighs 1.00 lbf on earth and has a *mass* of 453.6 g (0.4536 kg). Using Newton's second law, the actual weight of the cereal on earth is

$$W = mg = (453.6 \text{ g})(9.81 \text{ m/s}^2)\left(\frac{1 \text{ N}}{1 \text{ kg·m/s}^2}\right)\left(\frac{1 \text{ kg}}{1000 \text{ g}}\right) = 4.49 \text{ N}$$

# 1–3 · SYSTEMS AND CONTROL VOLUMES

A system is defined as a *quantity of matter or a region in space chosen for study*. The mass or region outside the system is called the **surroundings**. The real or imaginary surface that separates the system from its surroundings is called the **boundary** (Fig. 1–18). The boundary of a system can be *fixed* or *movable*. Note that the boundary is the contact surface shared by both the system and the surroundings. Mathematically speaking, the boundary has zero thickness, and thus it can neither contain any mass nor occupy any volume in space.

Systems may be considered to be *closed* or *open,* depending on whether a fixed mass or a fixed volume in space is chosen for study. A **closed system** (also known as a **control mass** or just *system* when the context makes it clear) consists of a fixed amount of mass, and no mass can cross its boundary. That is, no mass can enter or leave a closed system, as shown in

Fig. 1–19. But energy, in the form of heat or work, can cross the boundary; and the volume of a closed system does not have to be fixed. If, as a special case, even energy is not allowed to cross the boundary, that system is called an **isolated system**.

Consider the piston-cylinder device shown in Fig. 1–20. Let us say that we would like to find out what happens to the enclosed gas when it is heated. Since we are focusing our attention on the gas, it is our system. The inner surfaces of the piston and the cylinder form the boundary, and since no mass is crossing this boundary, it is a closed system. Notice that energy may cross the boundary, and part of the boundary (the inner surface of the piston, in this case) may move. Everything outside the gas, including the piston and the cylinder, is the surroundings.

An **open system**, or a **control volume**, as it is often called, is a properly selected region in space. It usually encloses a device that involves mass flow such as a compressor, turbine, or nozzle. Flow through these devices is best studied by selecting the region within the device as the control volume. Both mass and energy can cross the boundary of a control volume.

A large number of engineering problems involve mass flow in and out of a system and, therefore, are modeled as *control volumes*. A water heater, a car radiator, a turbine, and a compressor all involve mass flow and should be analyzed as control volumes (open systems) instead of as control masses (closed systems). In general, *any arbitrary region in space* can be selected as a control volume. There are no concrete rules for the selection of control volumes, but the proper choice certainly makes the analysis much easier. If we were to analyze the flow of air through a nozzle, for example, a good choice for the control volume would be the region within the nozzle.

The boundaries of a control volume are called a *control surface,* and they can be real or imaginary. In the case of a nozzle, the inner surface of the nozzle forms the real part of the boundary, and the entrance and exit areas form the imaginary part, since there are no physical surfaces there (Fig. 1–21a).

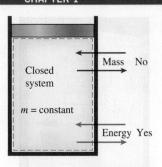

**FIGURE 1–19**
Mass cannot cross the boundaries of a closed system, but energy can.

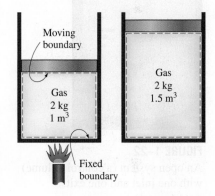

**FIGURE 1–20**
A closed system with a moving boundary.

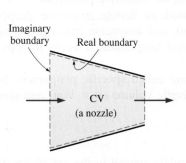

(a) A control volume (CV) with real and imaginary boundaries

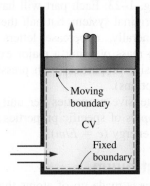

(b) A control volume (CV) with fixed and moving boundaries as well as real and imaginary boundaries

**FIGURE 1–21**
A control volume can involve fixed, moving, real, and imaginary boundaries.

**FIGURE 1–22**
An open system (a control volume)
with one inlet and one exit.

*© McGraw-Hill Education, Christopher Kerrigan*

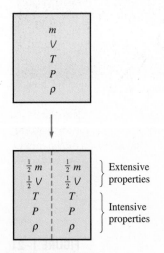

**FIGURE 1–23**
Criterion to differentiate intensive and
extensive properties.

A control volume can be fixed in size and shape, as in the case of a nozzle, or it may involve a moving boundary, as shown in Fig. 1–21*b*. Most control volumes, however, have fixed boundaries and thus do not involve any moving boundaries. A control volume can also involve heat and work interactions just as a closed system, in addition to mass interaction.

As an example of an open system, consider the water heater shown in Fig. 1–22. Let us say that we would like to determine how much heat we must transfer to the water in the tank in order to supply a steady stream of hot water. Since hot water will leave the tank and be replaced by cold water, it is not convenient to choose a fixed mass as our system for the analysis. Instead, we can concentrate our attention on the volume formed by the interior surfaces of the tank and consider the hot and cold water streams as mass leaving and entering the control volume. The interior surfaces of the tank form the control surface for this case, and mass is crossing the control surface at two locations.

In an engineering analysis, the system under study *must* be defined carefully. In most cases, the system investigated is quite simple and obvious, and defining the system may seem like a tedious and unnecessary task. In other cases, however, the system under study may be rather involved, and a proper choice of the system may greatly simplify the analysis.

# 1–4 · PROPERTIES OF A SYSTEM

Any characteristic of a system is called a **property**. Some familiar properties are pressure $P$, temperature $T$, volume $V$, and mass $m$. The list can be extended to include less familiar ones such as viscosity, thermal conductivity, modulus of elasticity, thermal expansion coefficient, electric resistivity, and even velocity and elevation.

Properties are considered to be either *intensive* or *extensive*. **Intensive properties** are those that are independent of the mass of a system, such as temperature, pressure, and density. **Extensive properties** are those whose values depend on the size—or extent—of the system. Total mass, total volume, and total momentum are some examples of extensive properties. An easy way to determine whether a property is intensive or extensive is to divide the system into two equal parts with an imaginary partition, as shown in Fig. 1–23. Each part will have the same value of intensive properties as the original system, but half the value of the extensive properties.

Generally, uppercase letters are used to denote extensive properties (with mass $m$ being a major exception), and lowercase letters are used for intensive properties (with pressure $P$ and temperature $T$ being the obvious exceptions).

Extensive properties per unit mass are called **specific properties**. Some examples of specific properties are specific volume ($v = V/m$) and specific total energy ($e = E/m$).

## Continuum

Matter is made up of atoms that are widely spaced in the gas phase. Yet it is very convenient to disregard the atomic nature of a substance and view it as a continuous, homogeneous matter with no holes, that is, a **continuum**.

The continuum idealization allows us to treat properties as point functions and to assume the properties vary continually in space with no jump discontinuities. This idealization is valid as long as the size of the system we deal with is large relative to the space between the molecules. This is the case in practically all problems, except some specialized ones. The continuum idealization is implicit in many statements we make, such as "the density of water in a glass is the same at any point."

To have a sense of the distance involved at the molecular level, consider a container filled with oxygen at atmospheric conditions. The diameter of the oxygen molecule is about $3 \times 10^{-10}$ m and its mass is $5.3 \times 10^{-26}$ kg. Also, the *mean free path* of oxygen at 1 atm pressure and 20°C is $6.3 \times 10^{-8}$ m. That is, an oxygen molecule travels, on average, a distance of $6.3 \times 10^{-8}$ m (about 200 times of its diameter) before it collides with another molecule.

Also, there are about $3 \times 10^{16}$ molecules of oxygen in the tiny volume of 1 mm³ at 1 atm pressure and 20°C (Fig. 1–24). The continuum model is applicable as long as the characteristic length of the system (such as its diameter) is much larger than the mean free path of the molecules. At very high vacuums or very high elevations, the mean free path may become large (for example, it is about 0.1 m for atmospheric air at an elevation of 100 km). For such cases the **rarefied gas flow theory** should be used, and the impact of individual molecules should be considered. In this text we will limit our consideration to substances that can be modeled as a continuum.

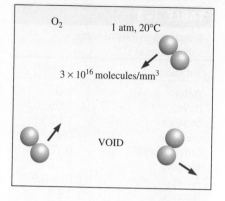

**FIGURE 1–24**
Despite the relatively large gaps between molecules, a gas can usually be treated as a continuum because of the very large number of molecules even in an extremely small volume.

## 1–5 · DENSITY AND SPECIFIC GRAVITY

**Density** is defined as *mass per unit volume* (Fig. 1–25).

*Density:*
$$\rho = \frac{m}{V} \quad (\text{kg/m}^3) \tag{1–4}$$

The reciprocal of density is the **specific volume** $v$, which is defined as *volume per unit mass*. That is,

$$v = \frac{V}{m} = \frac{1}{\rho} \tag{1–5}$$

For a differential volume element of mass $\delta m$ and volume $\delta V$, density can be expressed as $\rho = \delta m/\delta V$.

The density of a substance, in general, depends on temperature and pressure. The density of most gases is proportional to pressure and inversely proportional to temperature. Liquids and solids, on the other hand, are essentially incompressible substances, and the variation of their density with pressure is usually negligible. At 20°C, for example, the density of water changes from 998 kg/m³ at 1 atm to 1003 kg/m³ at 100 atm, a change of just 0.5 percent. The density of liquids and solids depends more strongly on temperature than it does on pressure. At 1 atm, for example, the density of water changes from 998 kg/m³ at 20°C to 975 kg/m³ at 75°C, a change of 2.3 percent, which can still be neglected in many engineering analyses.

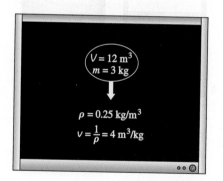

**FIGURE 1–25**
Density is mass per unit volume; specific volume is volume per unit mass.

**TABLE 1-3**

Specific gravities of some substances at 0°C

| Substance | SG |
|-----------|-----|
| Water | 1.0 |
| Blood | 1.05 |
| Seawater | 1.025 |
| Gasoline | 0.7 |
| Ethyl alcohol | 0.79 |
| Mercury | 13.6 |
| Wood | 0.3–0.9 |
| Gold | 19.2 |
| Bones | 1.7–2.0 |
| Ice | 0.92 |
| Air (at 1 atm) | 0.0013 |

Sometimes the density of a substance is given relative to the density of a well-known substance. Then it is called **specific gravity**, or **relative density**, and is defined as *the ratio of the density of a substance to the density of some standard substance at a specified temperature* (usually water at 4°C, for which $\rho_{H_2O} = 1000$ kg/m$^3$). That is,

*Specific gravity:* $$SG = \frac{\rho}{\rho_{H_2O}} \qquad (1-6)$$

Note that the specific gravity of a substance is a dimensionless quantity. However, in SI units, the numerical value of the specific gravity of a substance is exactly equal to its density in g/cm$^3$ or kg/L (or 0.001 times the density in kg/m$^3$) since the density of water at 4°C is 1 g/cm$^3$ = 1 kg/L = 1000 kg/m$^3$. The specific gravity of mercury at 0°C, for example, is 13.6. Therefore, its density at 0°C is 13.6 g/cm$^3$ = 13.6 kg/L = 13,600 kg/m$^3$. The specific gravities of some substances at 0°C are given in Table 1–3. Note that substances with specific gravities less than 1 are lighter than water, and thus they would float on water.

The weight of a unit volume of a substance is called **specific weight** and is expressed as

*Specific weight:* $$\gamma_s = \rho g \quad (N/m^3) \qquad (1-7)$$

where $g$ is the gravitational acceleration.

The densities of liquids are essentially constant, and thus they can often be approximated as being incompressible substances during most processes without sacrificing much in accuracy.

# 1–6 · STATE AND EQUILIBRIUM

Consider a system not undergoing any change. At this point, all the properties can be measured or calculated throughout the entire system, which gives us a set of properties that completely describes the condition, or the **state**, of the system. At a given state, all the properties of a system have fixed values. If the value of even one property changes, the state will change to a different one. In Fig. 1–26 a system is shown at two different states.

Thermodynamics deals with *equilibrium* states. The word **equilibrium** implies a state of balance. In an equilibrium state there are no unbalanced potentials (or driving forces) within the system. A system in equilibrium experiences no changes when it is isolated from its surroundings.

There are many types of equilibrium, and a system is not in thermodynamic equilibrium unless the conditions of all the relevant types of equilibrium are satisfied. For example, a system is in **thermal equilibrium** if the temperature is the same throughout the entire system, as shown in Fig. 1–27. That is, the system involves no temperature differential, which is the driving force for heat flow. **Mechanical equilibrium** is related to pressure, and a system is in mechanical equilibrium if there is no change in pressure at any point of the system with time. However, the pressure may vary within the system with elevation as a result of gravitational effects.

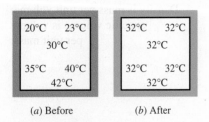

| $m = 2$ kg |
| $T_1 = 20°C$ |
| $V_1 = 1.5$ m$^3$ |

(a) State 1

| $m = 2$ kg |
| $T_2 = 20°C$ |
| $V_2 = 2.5$ m$^3$ |

(b) State 2

**FIGURE 1–26**

A system at two different states.

| 20°C | 23°C |
| | 30°C |
| 35°C | 40°C |
| | 42°C |

(a) Before

| 32°C | 32°C |
| | 32°C |
| 32°C | 32°C |
| | 32°C |

(b) After

**FIGURE 1–27**

A closed system reaching thermal equilibrium.

For example, the higher pressure at a bottom layer is balanced by the extra weight it must carry, and, therefore, there is no imbalance of forces. The variation of pressure as a result of gravity in most thermodynamic systems is relatively small and usually disregarded. If a system involves two phases, it is in **phase equilibrium** when the mass of each phase reaches an equilibrium level and stays there. Finally, a system is in **chemical equilibrium** if its chemical composition does not change with time, that is, no chemical reactions occur. A system will not be in equilibrium unless all the relevant equilibrium criteria are satisfied.

## The State Postulate

As noted earlier, the state of a system is described by its properties. But we know from experience that we do not need to specify all the properties in order to fix a state. Once a sufficient number of properties are specified, the rest of the properties assume certain values automatically. That is, specifying a certain number of properties is sufficient to fix a state. The number of properties required to fix the state of a system is given by the **state postulate**:

> The state of a simple compressible system is completely specified by two independent, intensive properties.

A system is called a **simple compressible system** in the absence of electrical, magnetic, gravitational, motion, and surface tension effects. These effects are due to external force fields and are negligible for most engineering problems. Otherwise, an additional property needs to be specified for each effect that is significant. If the gravitational effects are to be considered, for example, the elevation $z$ needs to be specified in addition to the two properties necessary to fix the state.

The state postulate requires that the two properties specified be independent to fix the state. Two properties are **independent** if one property can be varied while the other one is held constant. Temperature and specific volume, for example, are always independent properties, and together they can fix the state of a simple compressible system (Fig. 1–28). Temperature and pressure, however, are independent properties for single-phase systems, but are dependent properties for multiphase systems. At sea level ($P = 1$ atm), water boils at 100°C, but on a mountaintop where the pressure is lower, water boils at a lower temperature. That is, $T = f(P)$ during a phase-change process; thus, temperature and pressure are not sufficient to fix the state of a two-phase system. Phase-change processes are discussed in detail in Chap. 3.

## 1–7 · PROCESSES AND CYCLES

Any change that a system undergoes from one equilibrium state to another is called a **process**, and the series of states through which a system passes during a process is called the **path** of the process (Fig. 1–29). To describe a process completely, one should specify the initial and final states of the process, as well as the path it follows, and the interactions with the surroundings.

**FIGURE 1–28**
The state of nitrogen is fixed by two independent, intensive properties.

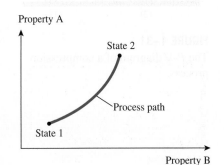

**FIGURE 1–29**
A process between states 1 and 2 and the process path.

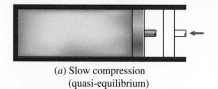

(*a*) Slow compression
(quasi-equilibrium)

(*b*) Very fast compression
(nonquasi-equilibrium)

**FIGURE 1–30**
Quasi-equilibrium and nonquasi-equilibrium compression processes.

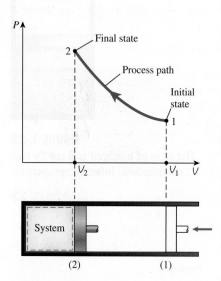

**FIGURE 1–31**
The *P-V* diagram of a compression process.

When a process proceeds in such a manner that the system remains infinitesimally close to an equilibrium state at all times, it is called a **quasi-static**, or **quasi-equilibrium**, **process**. A quasi-equilibrium process can be viewed as a sufficiently slow process that allows the system to adjust itself internally so that properties in one part of the system do not change any faster than those at other parts.

This is illustrated in Fig. 1–30. When a gas in a piston-cylinder device is compressed suddenly, the molecules near the face of the piston will not have enough time to escape and they will have to pile up in a small region in front of the piston, thus creating a high-pressure region there. Because of this pressure difference, the system can no longer be said to be in equilibrium, and this makes the entire process nonquasi-equilibrium. However, if the piston is moved slowly, the molecules will have sufficient time to redistribute and there will not be a molecule pileup in front of the piston. As a result, the pressure inside the cylinder will always be nearly uniform and will rise at the same rate at all locations. Since equilibrium is maintained at all times, this is a quasi-equilibrium process.

It should be pointed out that a quasi-equilibrium process is an idealized process and is not a true representation of an actual process. But many actual processes closely approximate it, and they can be modeled as quasi-equilibrium with negligible error. Engineers are interested in quasi-equilibrium processes for two reasons. First, they are easy to analyze; second, work-producing devices deliver the most work when they operate on quasi-equilibrium processes. Therefore, quasi-equilibrium processes serve as standards to which actual processes can be compared.

Process diagrams plotted by employing thermodynamic properties as coordinates are very useful in visualizing the processes. Some common properties that are used as coordinates are temperature *T*, pressure *P*, and volume *V* (or specific volume *v*). Figure 1–31 shows the *P-V* diagram of a compression process of a gas.

Note that the process path indicates a series of equilibrium states through which the system passes during a process and has significance for quasi-equilibrium processes only. For nonquasi-equilibrium processes, we are not able to characterize the entire system by a single state, and thus we cannot speak of a process path for a system as a whole. A nonquasi-equilibrium process is denoted by a dashed line between the initial and final states instead of a solid line.

The prefix *iso-* is often used to designate a process for which a particular property remains constant. An **isothermal process**, for example, is a process during which the temperature *T* remains constant; an **isobaric process** is a process during which the pressure *P* remains constant; and an **isochoric** (or **isometric**) **process** is a process during which the specific volume *v* remains constant.

A system is said to have undergone a **cycle** if it returns to its initial state at the end of the process. That is, for a cycle the initial and final states are identical.

## The Steady-Flow Process

The terms *steady* and *uniform* are used frequently in engineering, and thus it is important to have a clear understanding of their meanings. The term *steady* implies *no change with time*. The opposite of steady is *unsteady*, or *transient*. The term *uniform*, however, implies *no change with location* over a specified region. These meanings are consistent with their everyday use (steady girlfriend, uniform properties, etc.).

A large number of engineering devices operate for long periods of time under the same conditions, and they are classified as *steady-flow devices*. Processes involving such devices can be represented reasonably well by a somewhat idealized process, called the **steady-flow process**, which can be defined as a *process during which a fluid flows through a control volume steadily* (Fig. 1–32). That is, the fluid properties can change from point to point within the control volume, but at any fixed point they remain the same during the entire process. Therefore, the volume $V$, the mass $m$, and the total energy content $E$ of the control volume remain constant during a steady-flow process (Fig. 1–33).

Steady-flow conditions can be closely approximated by devices that are intended for continuous operation such as turbines, pumps, boilers, condensers, and heat exchangers or power plants or refrigeration systems. Some cyclic devices, such as reciprocating engines or compressors, do not satisfy any of the conditions stated above since the flow at the inlets and the exits will be pulsating and not steady. However, the fluid properties vary with time in a periodic manner, and the flow through these devices can still be analyzed as a steady-flow process by using time-averaged values for the properties.

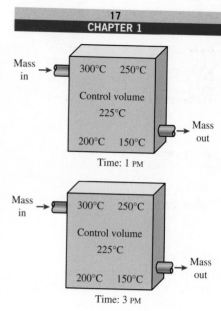

**FIGURE 1–32**
During a steady-flow process, fluid properties within the control volume may change with position but not with time.

# 1–8 ▪ TEMPERATURE AND THE ZEROTH LAW OF THERMODYNAMICS

Although we are familiar with temperature as a measure of "hotness" or "coldness," it is not easy to give an exact definition for it. Based on our physiological sensations, we express the level of temperature qualitatively with words like *freezing cold, cold, warm, hot,* and *red-hot.* However, we cannot assign numerical values to temperatures based on our sensations alone. Furthermore, our senses may be misleading. A metal chair, for example, will feel much colder than a wooden one even when both are at the same temperature.

Fortunately, several properties of materials change with temperature in a *repeatable* and *predictable* way, and this forms the basis for accurate temperature measurement. The commonly used mercury-in-glass thermometer, for example, is based on the expansion of mercury with temperature. Temperature is also measured by using several other temperature-dependent properties.

It is a common experience that a cup of hot coffee left on the table eventually cools off and a cold drink eventually warms up. That is, when a body is brought into contact with another body that is at a different temperature, heat is transferred from the body at higher temperature to the one at lower temperature until both bodies attain the same temperature (Fig. 1–34). At that point, the heat transfer stops, and the two bodies are said to have reached **thermal equilibrium**. The equality of temperature is the only requirement for thermal equilibrium.

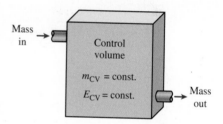

**FIGURE 1–33**
Under steady-flow conditions, the mass and energy contents of a control volume remain constant.

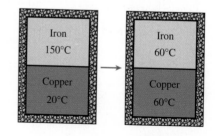

**FIGURE 1–34**
Two bodies reaching thermal equilibrium after being brought into contact in an isolated enclosure.

The **zeroth law of thermodynamics** states that if two bodies are in thermal equilibrium with a third body, they are also in thermal equilibrium with each other. It may seem silly that such an obvious fact is called one of the basic laws of thermodynamics. However, it cannot be concluded from the other laws of thermodynamics, and it serves as a basis for the validity of temperature measurement. By replacing the third body with a thermometer, the zeroth law can be restated as *two bodies are in thermal equilibrium if both have the same temperature reading even if they are not in contact.*

The zeroth law was first formulated and labeled by R. H. Fowler in 1931. As the name suggests, its value as a fundamental physical principle was recognized more than half a century after the formulation of the first and the second laws of thermodynamics. It was named the zeroth law since it should have preceded the first and the second laws of thermodynamics.

## Temperature Scales

Temperature scales enable us to use a common basis for temperature measurements, and several have been introduced throughout history. All temperature scales are based on some easily reproducible states such as the freezing and boiling points of water, which are also called the *ice point* and the *steam point,* respectively. A mixture of ice and water that is in equilibrium with air saturated with vapor at 1 atm pressure is said to be at the ice point, and a mixture of liquid water and water vapor (with no air) in equilibrium at 1 atm pressure is said to be at the steam point.

The temperature scales used in the SI and in the English system today are the **Celsius scale** (formerly called the *centigrade scale;* in 1948 it was renamed after the Swedish astronomer A. Celsius, 1702–1744, who devised it) and the **Fahrenheit scale** (named after the German instrument maker G. Fahrenheit, 1686–1736), respectively. On the Celsius scale, the ice and steam points were originally assigned the values of 0 and 100°C, respectively. The corresponding values on the Fahrenheit scale are 32 and 212°F. These are often referred to as *two-point scales* since temperature values are assigned at two different points.

In thermodynamics, it is very desirable to have a temperature scale that is independent of the properties of any substance or substances. Such a temperature scale is called a **thermodynamic temperature scale**, which is developed later in conjunction with the second law of thermodynamics. The thermodynamic temperature scale in the SI is the **Kelvin scale**, named after Lord Kelvin (1824–1907). The temperature unit on this scale is the **kelvin**, which is designated by K (not °K; the degree symbol was officially dropped from kelvin in 1967). The lowest temperature on the Kelvin scale is absolute zero, or 0 K. Then it follows that only one nonzero reference point needs to be assigned to establish the slope of this linear scale. Using nonconventional refrigeration techniques, scientists have approached absolute zero kelvin (they achieved 0.000000002 K in 1989).

The thermodynamic temperature scale in the English system is the **Rankine scale**, named after William Rankine (1820–1872). The temperature unit on this scale is the **rankine**, which is designated by R.

A temperature scale that turns out to be nearly identical to the Kelvin scale is the **ideal-gas temperature scale**. The temperatures on this scale are measured using a **constant-volume gas thermometer**, which is basically a rigid vessel filled with a gas, usually hydrogen or helium, at low pressure. This thermometer is based on the principle that *at low pressures, the temperature of a gas is proportional to its pressure at constant volume*. That is, the temperature of a gas of fixed volume varies *linearly* with pressure at sufficiently low pressures. Then the relationship between the temperature and the pressure of the gas in the vessel can be expressed as

$$T = a + bP \qquad (1\text{–}8)$$

where the values of the constants $a$ and $b$ for a gas thermometer are determined experimentally. Once $a$ and $b$ are known, the temperature of a medium can be calculated from this relation by immersing the rigid vessel of the gas thermometer into the medium and measuring the gas pressure when thermal equilibrium is established between the medium and the gas in the vessel whose volume is held constant.

An ideal-gas temperature scale can be developed by measuring the pressures of the gas in the vessel at two reproducible points (such as the ice and the steam points) and assigning suitable values to temperatures at those two points. Considering that only one straight line passes through two fixed points on a plane, these two measurements are sufficient to determine the constants $a$ and $b$ in Eq. 1–8. Then the unknown temperature $T$ of a medium corresponding to a pressure reading $P$ can be determined from that equation by a simple calculation. The values of the constants will be different for each thermometer, depending on the type and the amount of the gas in the vessel, and the temperature values assigned at the two reference points. If the ice and steam points are assigned the values 0°C and 100°C, respectively, then the gas temperature scale will be identical to the Celsius scale. In this case the value of the constant $a$ (which corresponds to an absolute pressure of zero) is determined to be −273.15°C regardless of the type and the amount of the gas in the vessel of the gas thermometer. That is, on a *P-T* diagram, all the straight lines passing through the data points in this case will intersect the temperature axis at −273.15°C when extrapolated, as shown in Fig. 1–35. This is the lowest temperature that can be obtained by a gas thermometer, and thus we can obtain an *absolute gas temperature scale* by assigning a value of zero to the constant $a$ in Eq. 1–8. In that case, Eq. 1–8 reduces to $T = bP$, and thus we need to specify the temperature at only *one* point to define an absolute gas temperature scale.

It should be noted that the absolute gas temperature scale is not a thermodynamic temperature scale, since it cannot be used at very low temperatures (due to condensation) and at very high temperatures (due to dissociation and ionization). However, absolute gas temperature is identical to the thermodynamic temperature in the temperature range in which the gas thermometer can be used. Thus, we can view the thermodynamic temperature scale at this point as an absolute gas temperature scale that utilizes an "ideal" or "imaginary" gas that always acts as a low-pressure gas regardless of the temperature. If such a gas thermometer existed, it would read zero kelvin at absolute zero pressure, which corresponds to −273.15°C on the Celsius scale (Fig. 1–36).

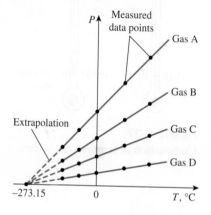

**FIGURE 1–35**

*P* versus *T* plots of the experimental data obtained from a constant-volume gas thermometer using four different gases at different (but low) pressures.

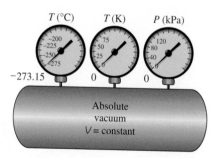

**FIGURE 1–36**

A constant-volume gas thermometer would read −273.15°C at absolute zero pressure.

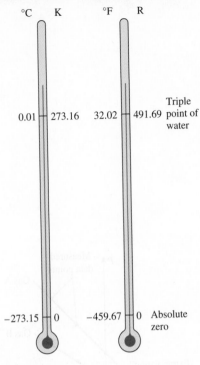

**FIGURE 1–37**
Comparison of temperature scales.

The Kelvin scale is related to the Celsius scale by

$$T(\text{K}) = T(°\text{C}) + 273.15 \qquad (1\text{–}9)$$

The Rankine scale is related to the Fahrenheit scale by

$$T(\text{R}) = T(°\text{F}) + 459.67 \qquad (1\text{–}10)$$

It is common practice to round the constant in Eq. 1–9 to 273 and that in Eq. 1–10 to 460.

The temperature scales in the two unit systems are related by

$$T(\text{R}) = 1.8T(\text{K}) \qquad (1\text{–}11)$$

$$T(°\text{F}) = 1.8T(°\text{C}) + 32 \qquad (1\text{–}12)$$

A comparison of various temperature scales is given in Fig. 1–37.

The reference temperature chosen in the original Kelvin scale was 273.15 K (or 0°C), which is the temperature at which water freezes (or ice melts) and water exists as a solid–liquid mixture in equilibrium under standard atmospheric pressure (the *ice point*). At the Tenth General Conference on Weights and Measures in 1954, the reference point was changed to a much more precisely reproducible point, the *triple point* of water (the state at which all three phases of water coexist in equilibrium), which is assigned the value 273.16 K. The Celsius scale was also redefined at this conference in terms of the ideal-gas temperature scale and a single fixed point, which is again the triple point of water with an assigned value of 0.01°C. The boiling temperature of water (the *steam point*) was experimentally determined to be again 100.00°C, and thus the new and old Celsius scales were in good agreement.

## The International Temperature Scale of 1990 (ITS-90)

The *International Temperature Scale of 1990*, which supersedes the International Practical Temperature Scale of 1968 (IPTS-68), 1948 (ITPS-48), and 1927 (ITS-27), was adopted by the International Committee of Weights and Measures at its meeting in 1989 at the request of the Eighteenth General Conference on Weights and Measures. The ITS-90 is similar to its predecessors except that it is more refined with updated values of fixed temperatures, has an extended range, and conforms more closely to the thermodynamic temperature scale. On this scale, the unit of thermodynamic temperature $T$ is again the kelvin (K), defined as the fraction 1/273.16 of the thermodynamic temperature of the triple point of water, which is sole defining fixed point of both the ITS-90 and the Kelvin scale and is the most important thermometric fixed point used in the calibration of thermometers to ITS-90.

The unit of Celsius temperature is the degree Celsius (°C), which is by definition equal in magnitude to the kelvin (K). A temperature difference may be expressed in kelvins or degrees Celsius. The ice point remains the same at 0°C (273.15 K) in both ITS-90 and ITPS-68, but the steam point is 99.975°C in ITS-90 (with an uncertainty of ±0.005°C) whereas it was 100.000°C in IPTS-68. The change is due to precise measurements made by gas thermometry by paying particular attention to the effect of sorption (the impurities in a gas absorbed by the walls of the bulb at the reference temperature being desorbed at higher temperatures, causing the measured gas pressure to increase).

The ITS-90 extends upward from 0.65 K to the highest temperature practically measurable in terms of the Planck radiation law using monochromatic radiation. It is based on specifying definite temperature values on a number of fixed and easily reproducible points to serve as benchmarks and expressing the variation of temperature in a number of ranges and subranges in functional form.

In ITS-90, the temperature scale is considered in four ranges. In the range of 0.65 to 5 K, the temperature scale is defined in terms of the vapor pressure—temperature relations for $^3$He and $^4$He. Between 3 and 24.5561 K (the triple point of neon), it is defined by means of a properly calibrated helium gas thermometer. From 13.8033 K (the triple point of hydrogen) to 1234.93 K (the freezing point of silver), it is defined by means of platinum resistance thermometers calibrated at specified sets of defining fixed points. Above 1234.93 K, it is defined in terms of the Planck radiation law and a suitable defining fixed point such as the freezing point of gold (1337.33 K).

We emphasize that the magnitudes of each division of 1 K and 1°C are identical (Fig. 1–38). Therefore, when we are dealing with temperature differences $\Delta T$, the temperature interval on both scales is the same. Raising the temperature of a substance by 10°C is the same as raising it by 10 K. That is,

$$\Delta T(K) = \Delta T(°C) \tag{1-13}$$

$$\Delta T(R) = \Delta T(°F) \tag{1-14}$$

Some thermodynamic relations involve the temperature $T$ and often the question arises of whether it is in K or °C. If the relation involves temperature differences (such as $a = b\Delta T$), it makes no difference and either can be used. However, if the relation involves temperatures only instead of temperature differences (such as $a = bT$) then K must be used. When in doubt, it is always safe to use K because there are virtually no situations in which the use of K is incorrect, but there are many thermodynamic relations that will yield an erroneous result if °C is used.

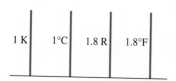

**FIGURE 1–38**
Comparison of magnitudes of various temperature units.

---

■ **EXAMPLE 1–4**    **Expressing Temperature Rise in Different Units**

■ During a heating process, the temperature of a system rises by 10°C. Express
■ this rise in temperature in K, °F, and R.

**SOLUTION**    The temperature rise of a system is to be expressed in different units.

*Analysis* This problem deals with temperature changes, which are identical in Kelvin and Celsius scales. Then,

$$\Delta T(\text{K}) = \Delta T(^\circ\text{C}) = \mathbf{10\ K}$$

The temperature changes in Fahrenheit and Rankine scales are also identical and are related to the changes in Celsius and Kelvin scales through Eqs. 1–11 and 1–14:

$$\Delta T(\text{R}) = 1.8\ \Delta T(\text{K}) = (1.8)(10) = \mathbf{18\ R}$$

and

$$\Delta T(^\circ\text{F}) = \Delta T(\text{R}) = \mathbf{18^\circ F}$$

*Discussion* Note that the units °C and K are interchangeable when dealing with temperature differences.

# 1–9 · PRESSURE

**Pressure** is defined as *a normal force exerted by a fluid per unit area.* Normally, we speak of pressure when we deal with a gas or a liquid. The counterpart of pressure in solids is *normal stress.* Note, however, that pressure is a scaler quantity while stress is a tensor. Since pressure is defined as force per unit area, it has the unit of newtons per square meter ($\text{N/m}^2$), which is called a **pascal** (Pa). That is,

$$1\ \text{Pa} = 1\ \text{N/m}^2$$

The pressure unit pascal is too small for most pressures encountered in practice. Therefore, its multiples *kilopascal* ($1\ \text{kPa} = 10^3\ \text{Pa}$) and *megapascal* ($1\ \text{MPa} = 10^6\ \text{Pa}$) are commonly used. Three other pressure units commonly used in practice, especially in Europe, are *bar, standard atmosphere,* and *kilogram-force per square centimeter:*

$$1\ \text{bar} = 10^5\ \text{Pa} = 0.1\ \text{MPa} = 100\ \text{kPa}$$

$$1\ \text{atm} = 101{,}325\ \text{Pa} = 101.325\ \text{kPa} = 1.01325\ \text{bars}$$

$$1\ \text{kgf/cm}^2 = 9.807\ \text{N/cm}^2 = 9.807 \times 10^4\ \text{N/m}^2 = 9.807 \times 10^4\ \text{Pa}$$

$$= 0.9807\ \text{bar}$$

$$= 0.9679\ \text{atm}$$

Note the pressure units bar, atm, and $\text{kgf/cm}^2$ are almost equivalent to each other. In the English system, the pressure unit is *pound-force per square inch* ($\text{lbf/in}^2$, or psi), and $1\ \text{atm} = 14.696\ \text{psi}$. The pressure units $\text{kgf/cm}^2$ and $\text{lbf/in}^2$ are also denoted by $\text{kg/cm}^2$ and $\text{lb/in}^2$, respectively, and they are commonly used in tire gages. It can be shown that $1\ \text{kgf/cm}^2 = 14.223\ \text{psi}$.

Pressure is also used on solid surfaces as synonymous to *normal stress,* which is the force acting perpendicular to the surface per unit area. For example, a 70-kg person with a total foot imprint area of 280 $\text{cm}^2$ exerts a pressure of 70 kgf/280 cm² = 0.25 kgf/cm² on the floor (Fig. 1–39). If the person stands on one foot, the pressure doubles. If the person gains excessive weight, he or she is likely to encounter foot discomfort because of the increased pressure on the foot (the size of the bottom of the foot does not change with weight gain). This also explains how a person can walk on fresh snow without sinking by wearing large snowshoes, and how a person cuts with little effort when using a sharp knife.

The actual pressure at a given position is called the **absolute pressure**, and it is measured relative to absolute vacuum (i.e., absolute zero pressure). Most pressure-measuring devices, however, are calibrated to read zero in the atmosphere (Fig. 1–40), and so they indicate the difference between the absolute pressure and the local atmospheric pressure. This difference is called the **gage pressure**. $P_{gage}$ can be positive or negative, but pressures below atmospheric pressure are sometimes called **vacuum pressures** and are measured by vacuum gages that indicate the difference between the atmospheric pressure and the absolute pressure. Absolute, gage, and vacuum pressures are related to each other by

$$P_{gage} = P_{abs} - P_{atm} \tag{1–15}$$

$$P_{vac} = P_{atm} - P_{abs} \tag{1–16}$$

This is illustrated in Fig. 1–41.

Like other pressure gages, the gage used to measure the air pressure in an automobile tire reads the gage pressure. Therefore, the common reading of 32.0 psi (2.25 kgf/cm²) indicates a pressure of 32.0 psi above the atmospheric pressure. At a location where the atmospheric pressure is 14.3 psi, for example, the absolute pressure in the tire is 32.0 + 14.3 = 46.3 psi.

In thermodynamic relations and tables, absolute pressure is almost always used. Throughout this text, the pressure $P$ will denote *absolute pressure* unless specified otherwise. Often the letters "a" (for absolute pressure) and "g" (for gage pressure) are added to pressure units (such as psia and psig) to clarify what is meant.

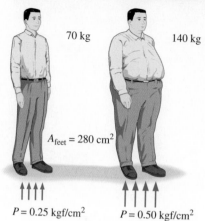

$$P = \sigma_n = \frac{W}{A_{feet}} = \frac{70 \text{ kg}}{280 \text{ cm}^2} = 0.25 \text{ kgf/cm}^2$$

**FIGURE 1–39**
The normal stress (or "pressure") on the feet of a chubby person is much greater than on the feet of a slim person.

**FIGURE 1–40**
Some basic pressure gages.
*Dresser Instruments, Dresser, Inc. Used by permission*

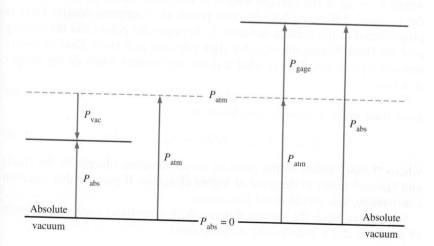

**FIGURE 1–41**
Absolute, gage, and vacuum pressures.

**EXAMPLE 1–5**   **Absolute Pressure of a Vacuum Chamber**

A vacuum gage connected to a chamber reads 40 kPa at a location where the atmospheric pressure is 100 kPa. Determine the absolute pressure in the chamber.

**SOLUTION**   The gage pressure of a vacuum chamber is given. The absolute pressure in the chamber is to be determined.
*Analysis*   The absolute pressure is easily determined from Eq. 1–16 to be

$$P_{abs} = P_{atm} - P_{vac} = 100 - 40 = \textbf{60 kPa}$$

*Discussion*   Note that the *local* value of the atmospheric pressure is used when determining the absolute pressure.

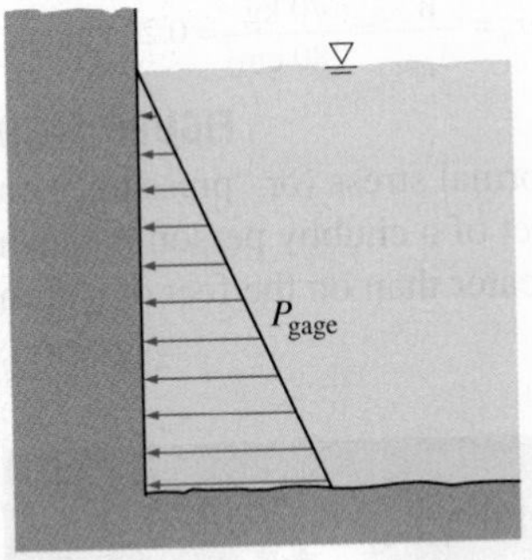

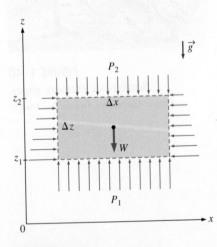

**FIGURE 1–42**
The pressure of a fluid at rest increases with depth (as a result of added weight).

# Variation of Pressure with Depth

It will come as no surprise to you that pressure in a fluid at rest does not change in the horizontal direction. This can be shown easily by considering a thin horizontal layer of fluid and doing a force balance in any horizontal direction. However, this is not the case in the vertical direction in a gravity field. Pressure in a fluid increases with depth because more fluid rests on deeper layers, and the effect of this "extra weight" on a deeper layer is balanced by an increase in pressure (Fig. 1–42).

To obtain a relation for the variation of pressure with depth, consider a rectangular fluid element of height $\Delta z$, length $\Delta x$, and unit depth ($\Delta y = 1$ into the page) in equilibrium, as shown in Fig. 1–43. Assuming the density of the fluid $\rho$ to be constant, a force balance in the vertical $z$-direction gives

$$\sum F_z = ma_z = 0: \quad P_1 \Delta x \Delta y - P_2 \Delta x \Delta y - \rho g \Delta x \Delta y \Delta z = 0$$

where $W = mg = \rho g \Delta x \Delta y \Delta z$ is the weight of the fluid element and $\Delta z = z_2 - z_1$. Dividing by $\Delta x \Delta y$ and rearranging gives

$$\Delta P = P_2 - P_1 = -\rho g \Delta z = -\gamma_s \Delta z \qquad (1\text{–}17)$$

where $\gamma_s = \rho g$ is the *specific weight* of the fluid. Thus, we conclude that the pressure difference between two points in a constant density fluid is proportional to the vertical distance $\Delta z$ between the points and the density $\rho$ of the fluid. Noting the negative sign, *pressure in a static fluid increases linearly with depth*. This is what a diver experiences when diving deeper in a lake.

An easier equation to remember and apply between any two points in the same fluid under hydrostatic conditions is

$$P_{below} = P_{above} + \rho g |\Delta z| = P_{above} + \gamma_s |\Delta z| \qquad (1\text{–}18)$$

where "below" refers to the point at lower elevation (deeper in the fluid) and "above" refers to the point at higher elevation. If you use this equation consistently, you should avoid sign errors.

For a given fluid, the vertical distance $\Delta z$ is sometimes used as a measure of pressure, and it is called the *pressure head*.

**FIGURE 1–43**
Free-body diagram of a rectangular fluid element in equilibrium.

We also conclude from Eq. 1–17 that for small to moderate distances, the variation of pressure with height is negligible for gases because of their low density. The pressure in a tank containing a gas, for example, can be considered to be uniform since the weight of the gas is too small to make a significant difference. Also, the pressure in a room filled with air can be approximated as a constant (Fig. 1–44).

If we take the "above" point to be at the free surface of a liquid open to the atmosphere (Fig. 1–45), where the pressure is the atmospheric pressure $P_{atm}$, then from Eq. 1–18 the pressure at a depth $h$ below the free surface becomes

$$P = P_{atm} + \rho g h \quad \text{or} \quad P_{gage} = \rho g h \quad (1\text{–}19)$$

Liquids are essentially incompressible substances, and thus the variation of density with depth is negligible. This is also the case for gases when the elevation change is not very large. The variation of density of liquids or gases with temperature can be significant, however, and may need to be considered when high accuracy is desired. Also, at great depths such as those encountered in oceans, the change in the density of a liquid can be significant because of the compression by the tremendous amount of liquid weight above.

The gravitational acceleration $g$ varies from 9.807 m/s² at sea level to 9.764 m/s² at an elevation of 14,000 m where large passenger planes cruise. This is a change of just 0.4 percent in this extreme case. Therefore, $g$ can be approximated as a constant with negligible error.

For fluids whose density changes significantly with elevation, a relation for the variation of pressure with elevation can be obtained by dividing Eq. 1–17 by $\Delta z$, and taking the limit as $\Delta z \rightarrow 0$. This yields

$$\frac{dP}{dz} = -\rho g \quad (1\text{–}20)$$

Note that $dP$ is negative when $dz$ is positive since pressure decreases in an upward direction. When the variation of density with elevation is known, the pressure difference between any two points 1 and 2 can be determined by integration to be

$$\Delta P = P_2 - P_1 = -\int_1^2 \rho g \, dz \quad (1\text{–}21)$$

For constant density and constant gravitational acceleration, this relation reduces to Eq. 1–17, as expected.

Pressure in a fluid at rest is independent of the shape or cross section of the container. It changes with the vertical distance, but remains constant in other directions. Therefore, the pressure is the same at all points on a horizontal plane in a given fluid. The Dutch mathematician Simon Stevin (1548–1620) published in 1586 the principle illustrated in Fig. 1–46. Note that the pressures at points $A$, $B$, $C$, $D$, $E$, $F$, and $G$ are the same since they are at the same depth, and they are interconnected by the same static fluid. However, the pressures at points $H$ and $I$ are not the same since these two points cannot be interconnected by the same

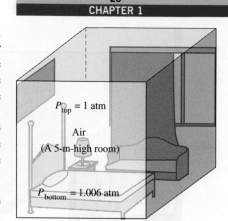

**FIGURE 1–44**

In a room filled with a gas, the variation of pressure with height is negligible.

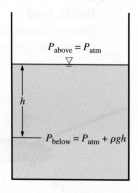

**FIGURE 1–45**

Pressure in a liquid at rest increases linearly with distance from the free surface.

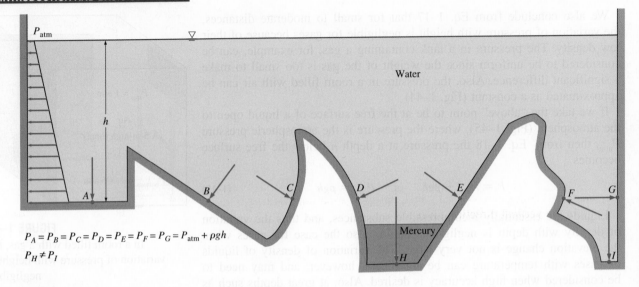

$$P_A = P_B = P_C = P_D = P_E = P_F = P_G = P_{atm} + \rho gh$$
$$P_H \neq P_I$$

**FIGURE 1–46**
Under hydrostatic conditions, the pressure is the same at all points on a horizontal plane in a given fluid regardless of geometry, provided that the points are interconnected by the same fluid.

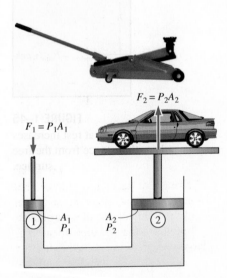

**FIGURE 1–47**
Lifting of a large weight by a small force by the application of Pascal's law. A common example is a hydraulic jack.

*(Top) © Stockbyte/Getty RF*

fluid (i.e., we cannot draw a curve from point $I$ to point $H$ while remaining in the same fluid at all times), although they are at the same depth. (Can you tell at which point the pressure is higher?) Also notice that the pressure force exerted by the fluid is always normal to the surface at the specified points.

A consequence of the pressure in a fluid remaining constant in the horizontal direction is that *the pressure applied to a confined fluid increases the pressure throughout by the same amount*. This is called **Pascal's law**, after Blaise Pascal (1623–1662). Pascal also knew that the force applied by a fluid is proportional to the surface area. He realized that two hydraulic cylinders of different areas could be connected, and the larger could be used to exert a proportionally greater force than that applied to the smaller. "Pascal's machine" has been the source of many inventions that are a part of our daily lives such as hydraulic brakes and lifts. This is what enables us to lift a car easily by one arm, as shown in Fig. 1–47. Noting that $P_1 = P_2$ since both pistons are at the same level (the effect of small height differences is negligible, especially at high pressures), the ratio of output force to input force is determined to be

$$P_1 = P_2 \quad \rightarrow \quad \frac{F_1}{A_1} = \frac{F_2}{A_2} \quad \rightarrow \quad \frac{F_2}{F_1} = \frac{A_2}{A_1} \tag{1–22}$$

The area ratio $A_2/A_1$ is called the *ideal mechanical advantage* of the hydraulic lift. Using a hydraulic car jack with a piston area ratio of $A_2/A_1 = 100$, for example, a person can lift a 1000-kg car by applying a force of just 10 kgf (= 90.8 N).

# 1–10 · PRESSURE MEASUREMENT DEVICES

## The Barometer

Atmospheric pressure is measured by a device called a **barometer**; thus, the atmospheric pressure is often referred to as the *barometric pressure*.

The Italian Evangelista Torricelli (1608–1647) was the first to conclusively prove that the atmospheric pressure can be measured by inverting a mercury-filled tube into a mercury container that is open to the atmosphere, as shown in Fig. 1–48. The pressure at point $B$ is equal to the atmospheric pressure, and the pressure at point $C$ can be taken to be zero since there is only mercury vapor above point $C$ and the pressure is very low relative to $P_{atm}$ and can be neglected to an excellent approximation. Writing a force balance in the vertical direction gives

$$P_{atm} = \rho g h \tag{1–23}$$

where $\rho$ is the density of mercury, $g$ is the local gravitational acceleration, and $h$ is the height of the mercury column above the free surface. Note that the length and the cross-sectional area of the tube have no effect on the height of the fluid column of a barometer (Fig. 1–49).

A frequently used pressure unit is the *standard atmosphere*, which is defined as the pressure produced by a column of mercury 760 mm in height at 0°C ($\rho_{Hg} = 13{,}595$ kg/m³) under standard gravitational acceleration ($g = 9.807$ m/s²). If water instead of mercury were used to measure the standard atmospheric pressure, a water column of about 10.3 m would be needed. Pressure is sometimes expressed (especially by weather forecasters) in terms of the height of the mercury column. The standard atmospheric pressure, for example, is 760 mmHg at 0°C. The unit mmHg is also called the **torr** in honor of Torricelli. Therefore, 1 atm = 760 torr and 1 torr = 133.3 Pa.

Atmospheric pressure $P_{atm}$ changes from 101.325 kPa at sea level to 89.88, 79.50, 54.05, 26.5, and 5.53 kPa at altitudes of 1000, 2000, 5000, 10,000, and 20,000 meters, respectively. The typical atmospheric pressure in Denver (elevation = 1610 m), for example, is 83.4 kPa. Remember that the atmospheric pressure at a location is simply the weight of the air above that location per unit surface area. Therefore, it changes not only with elevation but also with weather conditions.

The decline of atmospheric pressure with elevation has far-reaching ramifications in daily life. For example, cooking takes longer at high altitudes since water boils at a lower temperature at lower atmospheric pressures. Nose bleeding is a common experience at high altitudes since the difference between the blood pressure and the atmospheric pressure is larger in this case, and the delicate walls of veins in the nose are often unable to withstand this extra stress.

For a given temperature, the density of air is lower at high altitudes, and thus a given volume contains less air and less oxygen. So it is no surprise that we tire more easily and experience breathing problems at high altitudes. To compensate for this effect, people living at higher altitudes develop more efficient lungs. Similarly, a 2.0-L car engine will act like a 1.7-L car engine at 1500 m altitude (unless it is turbocharged) because of the 15 percent drop

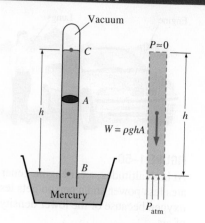

**FIGURE 1–48**
The basic barometer.

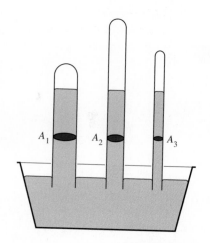

**FIGURE 1–49**
The length or the cross-sectional area of the tube has no effect on the height of the fluid column of a barometer, provided that the tube diameter is large enough to avoid surface tension (capillary) effects.

**FIGURE 1–50**
At high altitudes, a car engine generates less power and a person gets less oxygen because of the lower density of air.

in pressure and thus 15 percent drop in the density of air (Fig. 1–50). A fan or compressor will displace 15 percent less air at that altitude for the same volume displacement rate. Therefore, larger cooling fans may need to be selected for operation at high altitudes to ensure the specified mass flow rate. The lower pressure and thus lower density also affects lift and drag: airplanes need a longer runway at high altitudes to develop the required lift, and they climb to very high altitudes for cruising in order to reduce drag and thus achieve better fuel efficiency.

---

**EXAMPLE 1–6**   **Measuring Atmospheric Pressure with a Barometer**

Determine the atmospheric pressure at a location where the barometric reading is 740 mmHg and the gravitational acceleration is $g = 9.805$ m/s$^2$. Assume the temperature of mercury to be 10°C, at which its density is 13,570 kg/m$^3$.

**SOLUTION**   The barometric reading at a location in height of mercury column is given. The atmospheric pressure is to be determined.
**Assumptions**   The temperature of mercury is assumed to be 10°C.
**Properties**   The density of mercury is given to be 13,570 kg/m$^3$.
**Analysis**   From Eq. 1–23, the atmospheric pressure is determined to be

$$P_{atm} = \rho g h$$

$$= (13{,}570 \text{ kg/m}^3)(9.805 \text{ m/s}^2)(0.740 \text{ m})\left(\frac{1 \text{ N}}{1 \text{ kg·m/s}^2}\right)\left(\frac{1 \text{ kPa}}{1000 \text{ N/m}^2}\right)$$

$$= \textbf{98.5 kPa}$$

**Discussion**   Note that density changes with temperature, and thus this effect should be considered in calculations.

---

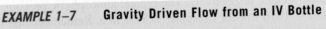

**EXAMPLE 1–7**   **Gravity Driven Flow from an IV Bottle**

Intravenous infusions usually are driven by gravity by hanging the fluid bottle at sufficient height to counteract the blood pressure in the vein and to force the fluid into the body (Fig. 1–51). The higher the bottle is raised, the higher the flow rate of the fluid will be. (*a*) If it is observed that the fluid and the blood pressures balance each other when the bottle is 1.2 m above the arm level, determine the gage pressure of the blood. (*b*) If the gage pressure of the fluid at the arm level needs to be 20 kPa for sufficient flow rate, determine how high the bottle must be placed. Take the density of the fluid to be 1020 kg/m$^3$.

**SOLUTION**   It is given that an IV fluid and the blood pressures balance each other when the bottle is at a certain height. The gage pressure of the blood and elevation of the bottle required to maintain flow at the desired rate are to be determined.
**Assumptions**   1 The IV fluid is incompressible.   2 The IV bottle is open to the atmosphere.

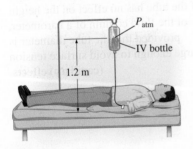

**FIGURE 1–51**
Schematic for Example 1–7.

**Properties** The density of the IV fluid is given to be $\rho = 1020$ kg/m³.

**Analysis** (a) Noting that the IV fluid and the blood pressures balance each other when the bottle is 1.2 m above the arm level, the gage pressure of the blood in the arm is simply equal to the gage pressure of the IV fluid at a depth of 1.2 m,

$$P_{gage,\,arm} = P_{abs} - P_{atm} = \rho g h_{arm-bottle}$$

$$= (1020 \text{ kg/m}^3)(9.81 \text{ m/s}^2)(1.20 \text{ m})\left(\frac{1 \text{ kN}}{1000 \text{ kg·m/s}^2}\right)\left(\frac{1 \text{ kPa}}{1 \text{ kN/m}^2}\right)$$

$$= 12.0 \text{ kPa}$$

(b) To provide a gage pressure of 20 kPa at the arm level, the height of the surface of the IV fluid in the bottle from the arm level is again determined from $P_{gage,\,arm} = \rho g h_{arm-bottle}$ to be

$$h_{arm-bottle} = \frac{P_{gage,\,arm}}{\rho g}$$

$$= \frac{20 \text{ kPa}}{(1020 \text{ kg/m}^3)(9.81 \text{ m/s}^2)}\left(\frac{1000 \text{ kg·m/s}^2}{1 \text{ kN}}\right)\left(\frac{1 \text{ kN/m}^2}{1 \text{ kPa}}\right)$$

$$= 2.00 \text{ m}$$

**Discussion** Note that the height of the reservoir can be used to control flow rates in gravity-driven flows. When there is flow, the pressure drop in the tube due to frictional effects also should be considered. For a specified flow rate, this requires raising the bottle a little higher to overcome the pressure drop.

---

**EXAMPLE 1–8     Hydrostatic Pressure in a Solar Pond with Variable Density**

Solar ponds are small artificial lakes of a few meters deep that are used to store solar energy. The rise of heated (and thus less dense) water to the surface is prevented by adding salt at the pond bottom. In a typical salt gradient solar pond, the density of water increases in the gradient zone, as shown in Fig. 1–52, and the density can be expressed as

$$\rho = \rho_0 \sqrt{1 + \tan^2\left(\frac{\pi}{4}\frac{s}{H}\right)}$$

where $\rho_0$ is the density on the water surface, $s$ is the vertical distance measured downward from the top of the gradient zone ($s = -z$), and $H$ is the thickness of the gradient zone. For $H = 4$ m, $\rho_0 = 1040$ kg/m³, and a thickness of 0.8 m for the surface zone, calculate the gage pressure at the bottom of the gradient zone.

**SOLUTION** The variation of density of saline water in the gradient zone of a solar pond with depth is given. The gage pressure at the bottom of the gradient zone is to be determined.

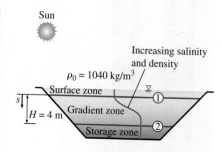

**FIGURE 1–52**
Schematic for Example 1–8.

Sun

Increasing salinity and density

$\rho_0 = 1040$ kg/m³

Surface zone

$s$

$H = 4$ m     Gradient zone

Storage zone

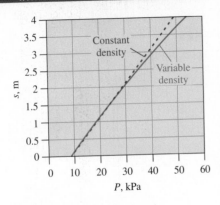

**FIGURE 1–53**
The variation of gage pressure with depth in the gradient zone of the solar pond.

**FIGURE 1–54**
A simple U-tube manometer, with high pressure applied to the right side.

*Photo by John M. Cimbala*

**Assumptions** The density in the surface zone of the pond is constant.

**Properties** The density of brine on the surface is given to be 1040 kg/m³.

**Analysis** We label the top and the bottom of the gradient zone as 1 and 2, respectively. Noting that the density of the surface zone is constant, the gage pressure at the bottom of the surface zone (which is the top of the gradient zone) is

$$P_1 = \rho g h_1 = (1040 \text{ kg/m}^3)(9.81 \text{ m/s}^2)(0.8 \text{ m})\left(\frac{1 \text{ kN}}{1000 \text{ kg·m/s}^2}\right) = 8.16 \text{ kPa}$$

since 1 kN/m² = 1 kPa. Since $s = -z$, the differential change in hydrostatic pressure across a vertical distance of $ds$ is given by

$$dP = \rho g \, ds$$

Integrating from the top of the gradient zone (point 1 where $s = 0$) to any location $s$ in the gradient zone (no subscript) gives

$$P - P_1 = \int_0^s \rho g \, ds \quad \rightarrow \quad P = P_1 + \int_0^s \rho_0 \sqrt{1 + \tan^2\left(\frac{\pi}{4}\frac{s}{H}\right)} g \, ds$$

Performing the integration gives the variation of gage pressure in the gradient zone to be

$$P = P_1 + \rho_0 g \frac{4H}{\pi} \sinh^{-1}\left(\tan \frac{\pi}{4}\frac{s}{H}\right)$$

Then the pressure at the bottom of the gradient zone ($s = H = 4$ m) becomes

$$P_2 = 8.16 \text{ kPa} + (1040 \text{ kg/m}^3)(9.81 \text{ m/s}^2)\frac{4(4 \text{ m})}{\pi} \sinh^{-1}\left(\tan \frac{\pi}{4}\frac{4}{4}\right)\left(\frac{1 \text{ kN}}{1000 \text{ kg·m/s}^2}\right)$$

$$= \mathbf{54.0 \text{ kPa (gage)}}$$

**Discussion** The variation of gage pressure in the gradient zone with depth is plotted in Fig. 1–53. The dashed line indicates the hydrostatic pressure for the case of constant density at 1040 kg/m³ and is given for reference. Note that the variation of pressure with depth is not linear when density varies with depth. That is why integration was required.

## The Manometer

We notice from Eq. 1–17 that an elevation change of $-\Delta z$ in a fluid at rest corresponds to $\Delta P/\rho g$, which suggests that a fluid column can be used to measure pressure differences. A device based on this principle is called a **manometer**, and it is commonly used to measure small and moderate pressure differences. A manometer consists of a glass or plastic U-tube containing one or more fluids such as mercury, water, alcohol, or oil (Fig. 1–54). To keep the size of the manometer to a manageable level, heavy fluids such as mercury are used if large pressure differences are anticipated.

Consider the manometer shown in Fig. 1–55 that is used to measure the pressure in the tank. Since the gravitational effects of gases are negligible, the pressure anywhere in the tank and at position 1 has the same value.

Furthermore, since pressure in a fluid does not vary in the horizontal direction within a fluid, the pressure at point 2 is the same as the pressure at point 1, $P_2 = P_1$.

The differential fluid column of height $h$ is in static equilibrium, and it is open to the atmosphere. Then the pressure at point 2 is determined directly from Eq. 1–18 to be

$$P_2 = P_{atm} + \rho g h \qquad (1\text{–}24)$$

where $\rho$ is the density of the manometer fluid in the tube. Note that the cross-sectional area of the tube has no effect on the differential height $h$, and thus the pressure exerted by the fluid. However, the diameter of the tube should be large enough (more than several millimeters) to ensure that the surface tension effect and thus the capillary rise is negligible.

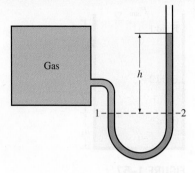

**FIGURE 1–55**
The basic manometer.

---

**■ EXAMPLE 1–9    Measuring Pressure with a Manometer**

A manometer is used to measure the pressure of a gas in a tank. The fluid used has a specific gravity of 0.85, and the manometer column height is 55 cm, as shown in Fig. 1–56. If the local atmospheric pressure is 96 kPa, determine the absolute pressure within the tank.

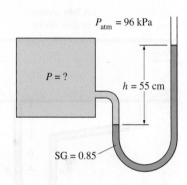

**FIGURE 1–56**
Schematic for Example 1–9.

**SOLUTION**   The reading of a manometer attached to a tank and the atmospheric pressure are given. The absolute pressure in the tank is to be determined.

**Assumptions**   The density of the gas in the tank is much lower than the density of the manometer fluid.

**Properties**   The specific gravity of the manometer fluid is given to be 0.85. We take the standard density of water to be 1000 kg/m³.

**Analysis**   The density of the fluid is obtained by multiplying its specific gravity by the density of water,

$$\rho = SG\,(\rho_{H_2O}) = (0.85)(1000 \text{ kg/m}^3) = 850 \text{ kg/m}^3$$

Then from Eq. 1–24,

$$P = P_{atm} + \rho g h$$

$$= 96 \text{ kPa} + (850 \text{ kg/m}^3)(9.81 \text{ m/s}^2)(0.55 \text{ m})\left(\frac{1 \text{ N}}{1 \text{ kg·m/s}^2}\right)\left(\frac{1 \text{ kPa}}{1000 \text{ N/m}^2}\right)$$

$$= 100.6 \text{ kPa}$$

**Discussion**   Note that the gage pressure in the tank is 4.6 kPa.

---

Some manometers use a slanted or inclined tube in order to increase the resolution (precision) when reading the fluid height. Such devices are called **inclined manometers**.

Many engineering problems and some manometers involve multiple immiscible fluids of different densities stacked on top of each other. Such systems can be analyzed easily by remembering that (1) the pressure change across a fluid column of height $h$ is $\Delta P = \rho g h$, (2) pressure increases downward in a given fluid and decreases upward (i.e., $P_{bottom} > P_{top}$), and

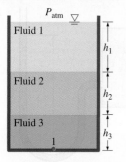

**FIGURE 1–57**

In stacked-up fluid layers at rest, the pressure change across each fluid layer of density $\rho$ and height $h$ is $\rho gh$.

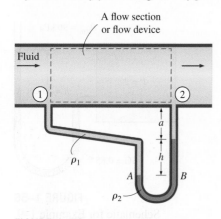

**FIGURE 1–58**

Measuring the pressure drop across a flow section or a flow device by a differential manometer.

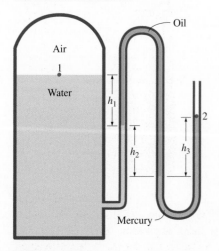

**FIGURE 1–59**

Schematic for Example 1–10; drawing not to scale.

(3) two points at the same elevation in a continuous fluid at rest are at the same pressure.

The last principle, which is a result of *Pascal's law,* allows us to "jump" from one fluid column to the next in manometers without worrying about pressure change as long as we stay in the same continuous fluid and the fluid is at rest. Then the pressure at any point can be determined by starting with a point of known pressure and adding or subtracting $\rho gh$ terms as we advance toward the point of interest. For example, the pressure at the bottom of the tank in Fig. 1–57 can be determined by starting at the free surface where the pressure is $P_{atm}$, moving downward until we reach point 1 at the bottom, and setting the result equal to $P_1$. It gives

$$P_{atm} + \rho_1 g h_1 + \rho_2 g h_2 + \rho_3 g h_3 = P_1$$

In the special case of all fluids having the same density, this relation reduces to $P_{atm} + \rho g(h_1 + h_2 + h_3) = P_1$.

Manometers are particularly well-suited to measure pressure drops across a horizontal flow section between two specified points due to the presence of a device such as a valve or heat exchanger or any resistance to flow. This is done by connecting the two legs of the manometer to these two points, as shown in Fig. 1–58. The working fluid can be either a gas or a liquid whose density is $\rho_1$. The density of the manometer fluid is $\rho_2$, and the differential fluid height is $h$. The two fluids must be immiscible, and $\rho_2$ must be greater than $\rho_1$.

A relation for the pressure difference $P_1 - P_2$ can be obtained by starting at point 1 with $P_1$, moving along the tube by adding or subtracting the $\rho gh$ terms until we reach point 2, and setting the result equal to $P_2$:

$$P_1 + \rho_1 g(a + h) - \rho_2 g h - \rho_1 g a = P_2 \tag{1–25}$$

Note that we jumped from point $A$ horizontally to point $B$ and ignored the part underneath since the pressure at both points is the same. Simplifying,

$$P_1 - P_2 = (\rho_2 - \rho_1)gh \tag{1–26}$$

Note that the distance $a$ must be included in the analysis even though it has no effect on the result. Also, when the fluid flowing in the pipe is a gas, then $\rho_1 \ll \rho_2$ and the relation in Eq. 1–26 simplifies to $P_1 - P_2 \cong \rho_2 gh$.

---

**EXAMPLE 1–10**    **Measuring Pressure with a Multifluid Manometer**

The water in a tank is pressurized by air, and the pressure is measured by a multifluid manometer as shown in Fig. 1–59. The tank is located on a mountain at an altitude of 1400 m where the atmospheric pressure is 85.6 kPa. Determine the air pressure in the tank if $h_1 = 0.1$ m, $h_2 = 0.2$ m, and $h_3 = 0.35$ m. Take the densities of water, oil, and mercury to be 1000 kg/m³, 850 kg/m³, and 13,600 kg/m³, respectively.

**SOLUTION**    The pressure in a pressurized water tank is measured by a multi-fluid manometer. The air pressure in the tank is to be determined.

*Assumption*    The air pressure in the tank is uniform (i.e., its variation with elevation is negligible due to its low density), and thus we can determine the pressure at the air–water interface.

**Properties** The densities of water, oil, and mercury are given to be 1000 kg/m$^3$, 850 kg/m$^3$, and 13,600 kg/m$^3$, respectively.

**Analysis** Starting with the pressure at point 1 at the air–water interface, moving along the tube by adding or subtracting the $\rho gh$ terms until we reach point 2, and setting the result equal to $P_{atm}$ since the tube is open to the atmosphere gives

$$P_1 + \rho_{water}gh_1 + \rho_{oil}gh_2 - \rho_{mercury}gh_3 = P_2 = P_{atm}$$

Solving for $P_1$ and substituting,

$$
\begin{aligned}
P_1 &= P_{atm} - \rho_{water}gh_1 - \rho_{oil}gh_2 + \rho_{mercury}gh_3 \\
&= P_{atm} + g(\rho_{mercury}h_3 - \rho_{water}h_1 - \rho_{oil}h_2) \\
&= 85.6 \text{ kPa} + (9.81 \text{ m/s}^2)[(13{,}600 \text{ kg/m}^3)(0.35 \text{ m}) - (1000 \text{ kg/m}^3)(0.1 \text{ m}) \\
&\quad - (850 \text{ kg/m}^3)(0.2 \text{ m})]\left(\frac{1 \text{ N}}{1 \text{ kg·m/s}^2}\right)\left(\frac{1 \text{ kPa}}{1000 \text{ N/m}^2}\right) \\
&= \mathbf{130 \text{ kPa}}
\end{aligned}
$$

**Discussion** Note that jumping horizontally from one tube to the next and realizing that pressure remains the same in the same fluid simplifies the analysis considerably. Also note that mercury is a toxic fluid, and mercury manometers and thermometers are being replaced by ones with safer fluids because of the risk of exposure to mercury vapor during an accident.

## Other Pressure Measurement Devices

Another type of commonly used mechanical pressure measurement device is the **Bourdon tube**, named after the French engineer and inventor Eugene Bourdon (1808–1884), which consists of a bent, coiled, or twisted hollow metal tube whose end is closed and connected to a dial indicator needle (Fig. 1–60). When the tube is open to the atmosphere, the tube is undeflected, and the needle on the dial at this state is calibrated to read zero (gage pressure). When the fluid inside the tube is pressurized, the tube stretches and moves the needle in proportion to the applied pressure.

Electronics have made their way into every aspect of life, including pressure measurement devices. Modern pressure sensors, called **pressure transducers**, use various techniques to convert the pressure effect to an electrical effect such as a change in voltage, resistance, or capacitance. Pressure transducers are smaller and faster, and they can be more sensitive, reliable, and precise than their mechanical counterparts. They can measure pressures from less than a millionth of 1 atm to several thousands of atm.

A wide variety of pressure transducers is available to measure gage, absolute, and differential pressures in a wide range of applications. *Gage pressure transducers* use the atmospheric pressure as a reference by venting the back side of the pressure-sensing diaphragm to the atmosphere, and they give a zero signal output at atmospheric pressure regardless of altitude. *Absolute pressure transducers* are calibrated to have a zero signal output at full vacuum. *Differential pressure transducers* measure the pressure difference

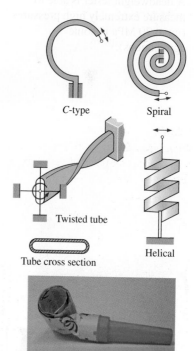

**FIGURE 1–60**

Various types of Bourdon tubes used to measure pressure. They work on the same principle as party noise-makers (bottom photo) due to the flat tube cross section.

*Photo by John M. Cimbala*

between two locations directly instead of using two pressure transducers and taking their difference.

**Strain-gage pressure transducers** work by having a diaphragm deflect between two chambers open to the pressure inputs. As the diaphragm stretches in response to a change in pressure difference across it, the strain gage stretches and a Wheatstone bridge circuit amplifies the output. A capacitance transducer works similarly, but capacitance change is measured instead of resistance change as the diaphragm stretches.

**Piezoelectric transducers**, also called solid-state pressure transducers, work on the principle that an electric potential is generated in a crystalline substance when it is subjected to mechanical pressure. This phenomenon, first discovered by brothers Pierre and Jacques Curie in 1880, is called the piezoelectric (or press-electric) effect. Piezoelectric pressure transducers have a much faster frequency response compared to diaphragm units and are very suitable for high-pressure applications, but they are generally not as sensitive as diaphragm-type transducers, especially at low pressures.

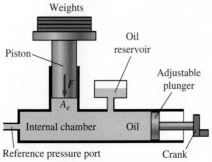

**FIGURE 1–61**
A deadweight tester is able to measure extremely high pressures (up to 70 MPa in some applications).

Another type of mechanical pressure gage called a **deadweight tester** is used primarily for *calibration* and can measure extremely high pressures (Fig. 1–61). As its name implies, a deadweight tester measures pressure *directly* through application of a weight that provides a force per unit area—the fundamental definition of pressure. It is constructed with an internal chamber filled with a fluid (usually oil), along with a tight-fitting piston, cylinder, and plunger. Weights are applied to the top of the piston, which exerts a force on the oil in the chamber. The total force $F$ acting on the oil at the piston–oil interface is the sum of the weight of the piston plus the applied weights. Since the piston cross-sectional area $A_e$ is known, the pressure is calculated as $P = F/A_e$. The only significant source of error is that due to static friction along the interface between the piston and cylinder, but even this error is usually negligibly small. The reference pressure port is connected to either an unknown pressure that is to be measured or to a pressure sensor that is to be calibrated.

## 1–11 · PROBLEM-SOLVING TECHNIQUE

The first step in learning any science is to grasp the fundamentals and to gain a sound knowledge of it. The next step is to master the fundamentals by testing this knowledge. This is done by solving significant real-world problems. Solving such problems, especially complicated ones, requires a systematic approach. By using a step-by-step approach, an engineer can reduce the solution of a complicated problem into the solution of a series of simple problems (Fig. 1–62). When you are solving a problem, we recommend that you use the following steps zealously as applicable. This will help you avoid some of the common pitfalls associated with problem solving.

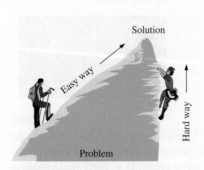

**FIGURE 1–62**
A step-by-step approach can greatly simplify problem solving.

### Step 1: Problem Statement

In your own words, briefly state the problem, the key information given, and the quantities to be found. This is to make sure that you understand the problem and the objectives before you attempt to solve the problem.

## Step 2: Schematic

Draw a realistic sketch of the physical system involved, and list the relevant information on the figure. The sketch does not have to be something elaborate, but it should resemble the actual system and show the key features. Indicate any energy and mass interactions with the surroundings. Listing the given information on the sketch helps one to see the entire problem at once. Also, check for properties that remain constant during a process (such as temperature during an isothermal process), and indicate them on the sketch.

## Step 3: Assumptions and Approximations

State any appropriate assumptions and approximations made to simplify the problem to make it possible to obtain a solution. Justify the questionable assumptions. Assume reasonable values for missing quantities that are necessary. For example, in the absence of specific data for atmospheric pressure, it can be taken to be 1 atm. However, it should be noted in the analysis that the atmospheric pressure decreases with increasing elevation. For example, it drops to 0.83 atm in Denver (elevation 1610 m) (Fig. 1–63).

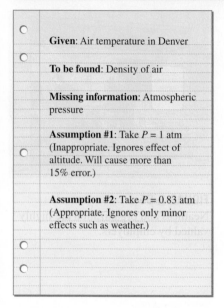

**Given**: Air temperature in Denver

**To be found**: Density of air

**Missing information**: Atmospheric pressure

**Assumption #1**: Take $P = 1$ atm (Inappropriate. Ignores effect of altitude. Will cause more than 15% error.)

**Assumption #2**: Take $P = 0.83$ atm (Appropriate. Ignores only minor effects such as weather.)

**FIGURE 1–63**
The assumptions made while solving an engineering problem must be reasonable and justifiable.

## Step 4: Physical Laws

Apply all the relevant basic physical laws and principles (such as the conservation of mass), and reduce them to their simplest form by utilizing the assumptions made. However, the region to which a physical law is applied must be clearly identified first. For example, the increase in speed of water flowing through a nozzle is analyzed by applying conservation of mass between the inlet and outlet of the nozzle.

## Step 5: Properties

Determine the unknown properties at known states necessary to solve the problem from property relations or tables. List the properties separately, and indicate their source, if applicable.

## Step 6: Calculations

Substitute the known quantities into the simplified relations and perform the calculations to determine the unknowns. Pay particular attention to the units and unit cancellations, and remember that a dimensional quantity without a unit is meaningless. Also, don't give a false implication of high precision by copying all the digits from the screen of the calculator—round the results to an appropriate number of significant digits (see p. 39).

## Step 7: Reasoning, Verification, and Discussion

Check to make sure that the results obtained are reasonable and intuitive, and verify the validity of the questionable assumptions. Repeat the calculations that resulted in unreasonable values. For example, insulating a water heater that uses $80 worth of natural gas a year cannot result in savings of $200 a year (Fig. 1–64).

| | |
|---|---|
| Energy use: | $80/yr |
| Energy saved by insulation: | $200/yr |
| **IMPOSSIBLE!** | |

**FIGURE 1–64**
The results obtained from an engineering analysis must be checked for reasonableness.

**FIGURE 1–65**
Neatness and organization are highly valued by employers.

Also, point out the significance of the results, and discuss their implications. State the conclusions that can be drawn from the results, and any recommendations that can be made from them. Emphasize the limitations under which the results are applicable, and caution against any possible misunderstandings and using the results in situations where the underlying assumptions do not apply. For example, if you determined that wrapping a water heater with a $20 insulation jacket will reduce the energy cost by $30 a year, indicate that the insulation will pay for itself from the energy it saves in less than a year. However, also indicate that the analysis does not consider labor costs, and that this will be the case if you install the insulation yourself.

Keep in mind that the solutions you present to your instructors, and any engineering analysis presented to others, is a form of communication. Therefore neatness, organization, completeness, and visual appearance are of utmost importance for maximum effectiveness (Fig. 1–65). Besides, neatness also serves as a great checking tool since it is very easy to spot errors and inconsistencies in neat work. Carelessness and skipping steps to save time often end up costing more time and unnecessary anxiety.

The approach described here is used in the solved example problems without explicitly stating each step, as well as in the Solutions Manual of this text. For some problems, some of the steps may not be applicable or necessary. For example, often it is not practical to list the properties separately. However, we cannot overemphasize the importance of a logical and orderly approach to problem solving. Most difficulties encountered while solving a problem are not due to a lack of knowledge; rather, they are due to a lack of organization. You are strongly encouraged to follow these steps in problem solving until you develop your own approach that works best for you.

## Engineering Software Packages

You may be wondering why we are about to undertake an in-depth study of the fundamentals of another engineering science. After all, almost all such problems we are likely to encounter in practice can be solved using one of several sophisticated software packages readily available in the market today. These software packages not only give the desired numerical results, but also supply the outputs in colorful graphical form for impressive presentations. It is unthinkable to practice engineering today without using some of these packages. This tremendous computing power available to us at the touch of a button is both a blessing and a curse. It certainly enables engineers to solve problems easily and quickly, but it also opens the door for abuses and misinformation. In the hands of poorly educated people, these software packages are as dangerous as sophisticated powerful weapons in the hands of poorly trained soldiers.

Thinking that a person who can use the engineering software packages without proper training on fundamentals can practice engineering is like thinking that a person who can use a wrench can work as a car mechanic. If it were true that the engineering students do not need all these fundamental courses they are taking because practically everything can be done by computers quickly and easily, then it would also be true that the employers would no longer need high-salaried engineers since any person who knows how to use a word-processing program can also learn how to use those software packages. However, the statistics show that the need for engineers is on the rise, not on the decline, despite the availability of these powerful packages.

We should always remember that all the computing power and the engineering software packages available today are just *tools*, and tools have meaning only in the hands of masters. Having the best word-processing program does not make a person a good writer, but it certainly makes the job of a good writer much easier and makes the writer more productive (Fig. 1–66). Hand calculators did not eliminate the need to teach our children how to add or subtract, and the sophisticated medical software packages did not take the place of medical school training. Neither will engineering software packages replace the traditional engineering education. They will simply cause a shift in emphasis in the courses from mathematics to physics. That is, more time will be spent in the classroom discussing the physical aspects of the problems in greater detail, and less time on the mechanics of solution procedures.

All these marvelous and powerful tools available today put an extra burden on today's engineers. They must still have a thorough understanding of the fundamentals, develop a "feel" of the physical phenomena, be able to put the data into proper perspective, and make sound engineering judgments, just like their predecessors. However, they must do it much better, and much faster, using more realistic models because of the powerful tools available today. The engineers in the past had to rely on hand calculations, slide rules, and later hand calculators and computers. Today they rely on software packages. The easy access to such power and the possibility of a simple misunderstanding or misinterpretation causing great damage make it more important today than ever to have solid training in the fundamentals of engineering. In this text we make an extra effort to put the emphasis on developing an intuitive and physical understanding of natural phenomena instead of on the mathematical details of solution procedures.

**FIGURE 1–66**

An excellent word-processing program does not make a person a good writer; it simply makes a good writer a more efficient writer.

© *Ingram Publishing RF*

## Engineering Equation Solver (EES)

EES is a program that solves systems of linear or nonlinear algebraic or differential equations numerically. It has a large library of built-in thermodynamic property functions as well as mathematical functions, and allows the user to supply additional property data. Unlike some software packages, EES does not solve engineering problems; it only solves the equations supplied by the user. Therefore, the user must understand the problem and formulate it by applying any relevant physical laws and relations. EES saves the user considerable time and effort by simply solving the resulting mathematical equations. This makes it possible to attempt significant engineering problems not suitable for hand calculations, and to conduct parametric studies quickly and conveniently. EES is a very powerful yet intuitive program that is very easy to use, as shown in Examples 1–11 and 1–12. The use and capabilities of EES are explained on the text website.

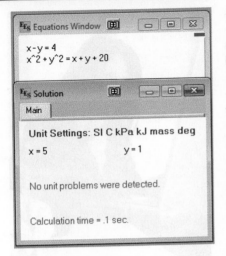

**FIGURE 1–67**

## EXAMPLE 1–11 Solving a System of Equations with EES

The difference of two numbers is 4, and the sum of the squares of these two numbers is equal to the sum of the numbers plus 20. Determine these two numbers. (Fig. 1–67)

**SOLUTION** Relations are given for the difference and the sum of the squares of two numbers. They are to be determined.

**Analysis** We start the EES program by double-clicking on its icon, open a new file, and type the following on the blank screen that appears:

$$x - y = 4$$
$$x^2 + y^2 = x + y + 20$$

which is an exact mathematical expression of the problem statement with $x$ and $y$ denoting the unknown numbers. The solution to this system of two nonlinear equations with two unknowns is obtained by a single click on the "calculator" icon on the taskbar. It gives

$$x = 5 \quad \text{and} \quad y = 1$$

**Discussion** Note that all we did is formulate the problem as we would on paper; EES took care of all the mathematical details of solution. Also note that equations can be linear or nonlinear, and they can be entered in any order with unknowns on either side. Friendly equation solvers such as EES allow the user to concentrate on the physics of the problem without worrying about the mathematical complexities associated with the solution of the resulting system of equations.

## EXAMPLE 1–12 Analyzing a Multifluid Manometer with EES

Reconsider the multifluid manometer discussed in Example 1–10 and replotted in Fig. 1–68. Determine the air pressure in the tank using EES. Also determine what the differential fluid height $h_3$ would be for the same air pressure if the mercury in the last column were replaced by seawater with a density of 1030 kg/m³.

**SOLUTION** The pressure in a water tank is measured by a multifluid manometer. The air pressure in the tank and the differential fluid height $h_3$ if mercury is replaced by seawater are to be determined using EES.

**Analysis** We start the EES program by double-clicking on its icon, open a new file, and type the following on the blank screen that appears (we express the atmospheric pressure in Pa for unit consistency):

g = 9.81
Patm = 85600
h1 = 0.1;    h2 = 0.2;    h3 = 0.35
rw = 1000;   roil = 850;   rm = 13600
P1 + rw*g*h1 + roil*g*h2-rm*g*h3 = Patm

Here $P_1$ is the only unknown, and it is determined by EES to be

$$P_1 = 129647 \text{ Pa} \cong \mathbf{130 \text{ kPa}}$$

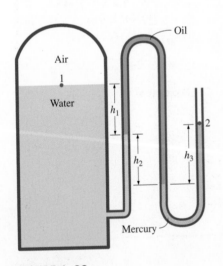

**FIGURE 1–68**
Schematic for Example 1–10; drawing not to scale.

which is identical to the result obtained before. The height of the fluid column $h_3$ when mercury is replaced by seawater is determined easily by replacing "h3 = 0.35" by "P1 = 129647" and "rm = 13600" by "rm = 1030," and clicking on the calculator symbol. It gives

$$h_3 = 4.62 \text{ m}$$

**Discussion** Note that we used the screen like a paper pad and wrote down the relevant information together with the applicable relations in an organized manner. EES did the rest. Equations can be written on separate lines or on the same line by separating them by semicolons, and blank or comment lines can be inserted for readability. EES makes it very easy to ask "what if" questions, and to perform parametric studies, as explained on the text website.

EES also has the capability to check the equations for unit consistency if units are supplied together with numerical values. Units can be specified within brackets [ ] after the specified value. When this feature is utilized, the previous equations would take the following form:

g = 9.81 [m/s^2]
Patm = 85600 [Pa]
h1 = 0.1 [m];      h2 = 0.2 [m];      h3 = 0.35 [m]
rw = 1000 [kg/m^3];   roil = 850 [kg/m^3];   rm = 13600 [kg/m^3]
P1 + rw*g*h1 + roil*g*h2-rm*g*h3 = Patm

# A Remark on Significant Digits

In engineering calculations, the information given is not known to more than a certain number of significant digits, usually three digits. Consequently, the results obtained cannot possibly be accurate to more significant digits. Reporting results in more significant digits implies greater accuracy than exists, and it should be avoided.

For example, consider a 3.75-L container filled with gasoline whose density is 0.845 kg/L, and try to determine its mass. Probably the first thought that comes to your mind is to multiply the volume and density to obtain 3.16875 kg for the mass, which falsely implies that the mass determined is accurate to six significant digits. In reality, however, the mass cannot be more accurate than three significant digits since both the volume and the density are accurate to three significant digits only. Therefore, the result should be rounded to three significant digits, and the mass should be reported to be 3.17 kg instead of what appears in the screen of the calculator. The result 3.16875 kg would be correct only if the volume and density were given to be 3.75000 L and 0.845000 kg/L, respectively. The value 3.75 L implies that we are fairly confident that the volume is accurate within $\pm 0.01$ L, and it cannot be 3.74 or 3.76 L. However, the volume can be 3.746, 3.750, 3.753, etc., since they all round to 3.75 L (Fig. 1–69). It is more appropriate to retain all the digits during intermediate calculations, and to do the rounding in the final step since this is what a computer will normally do.

When solving problems, we will assume the given information to be accurate to at least three significant digits. Therefore, if the length of a pipe is given to be 40 m, we will assume it to be 40.0 m in order to justify using

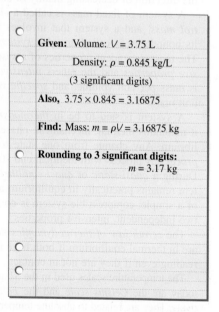

**Given:** Volume: $V = 3.75$ L

Density: $\rho = 0.845$ kg/L

(3 significant digits)

**Also,** $3.75 \times 0.845 = 3.16875$

**Find:** Mass: $m = \rho V = 3.16875$ kg

**Rounding to 3 significant digits:**
$$m = 3.17 \text{ kg}$$

**FIGURE 1–69**
A result with more significant digits than that of given data falsely implies more precision.

three significant digits in the final results. You should also keep in mind that all experimentally determined values are subject to measurement errors and such errors will reflect in the results obtained. For example, if the density of a substance has an uncertainty of 2 percent, then the mass determined using this density value will also have an uncertainty of 2 percent.

You should also be aware that we sometimes knowingly introduce small errors in order to avoid the trouble of searching for more accurate data. For example, when dealing with liquid water, we just use the value of 1000 kg/m$^3$ for density, which is the density value of pure water at 0°C. Using this value at 75°C will result in an error of 2.5 percent since the density at this temperature is 975 kg/m$^3$. The minerals and impurities in the water will introduce additional error. This being the case, you should have no reservation in rounding the final results to a reasonable number of significant digits. Besides, having a few percent uncertainty in the results of engineering analysis is usually the norm, not the exception.

## SUMMARY

In this chapter, the basic concepts of thermodynamics are introduced and discussed. *Thermodynamics* is the science that primarily deals with energy. The *first law of thermodynamics* is simply an expression of the conservation of energy principle, and it asserts that *energy* is a thermodynamic property. The *second law of thermodynamics* asserts that energy has *quality* as well as *quantity,* and actual processes occur in the direction of decreasing quality of energy.

A system of fixed mass is called a *closed system,* or *control mass,* and a system that involves mass transfer across its boundaries is called an *open system,* or *control volume.* The mass-dependent properties of a system are called *extensive properties* and the others *intensive properties. Density* is mass per unit volume, and *specific volume* is volume per unit mass.

A system is said to be in *thermodynamic equilibrium* if it maintains thermal, mechanical, phase, and chemical equilibrium. Any change from one state to another is called a *process.* A process with identical end states is called a *cycle.* During a *quasi-static* or *quasi-equilibrium process,* the system remains practically in equilibrium at all times. The state of a simple, compressible system is completely specified by two independent, intensive properties.

The *zeroth law of thermodynamics* states that two bodies are in thermal equilibrium if both have the same temperature reading even if they are not in contact.

The temperature scales used in the SI and the English system today are the *Celsius scale* and the *Fahrenheit scale,* respectively. They are related to absolute temperature scales by

$$T(K) = T(°C) + 273.15$$

$$T(R) = T(°F) + 459.67$$

The magnitudes of each division of 1 K and 1°C are identical, and so are the magnitudes of each division of 1 R and 1°F. Therefore,

$$\Delta T(K) = \Delta T(°C) \quad \text{and} \quad \Delta T(R) = \Delta T(°F)$$

The normal force exerted by a fluid per unit area is called *pressure,* and its unit is the *pascal,* 1 Pa = 1 N/m$^2$. The pressure relative to absolute vacuum is called the *absolute pressure,* and the difference between the absolute pressure and the local atmospheric pressure is called the *gage pressure.* Pressures below atmospheric pressure are called *vacuum pressures.* The absolute, gage, and vacuum pressures are related by

$$P_{\text{gage}} = P_{\text{abs}} - P_{\text{atm}} \quad \text{(for pressures above } P_{\text{atm}})$$

$$P_{\text{vac}} = P_{\text{atm}} - P_{\text{abs}} \quad \text{(for pressures below } P_{\text{atm}})$$

The pressure at a point in a fluid has the same magnitude in all directions. The variation of pressure with elevation is given by

$$\frac{dP}{dz} = -\rho g$$

where the positive $z$ direction is taken to be upward. When the density of the fluid is constant, the pressure difference across a fluid layer of thickness $\Delta z$ is

$$\Delta P = P_2 - P_1 = \rho g \, \Delta z$$

The absolute and gage pressures in a liquid open to the atmosphere at a depth $h$ from the free surface are

$$P = P_{\text{atm}} + \rho g h \quad \text{or} \quad P_{\text{gage}} = \rho g h$$

Small to moderate pressure differences are measured by a *manometer*. The pressure in a stationary fluid remains constant in the horizontal direction. *Pascal's principle* states that the pressure applied to a confined fluid increases the pressure throughout by the same amount.

The atmospheric pressure is measured by a *barometer* and is given by

$$P_{atm} = \rho g h$$

where $h$ is the height of the liquid column.

## REFERENCES AND SUGGESTED READINGS

1. American Society for Testing and Materials. *Standards for Metric Practice*. ASTM E 380-79, January 1980.

2. A. Bejan. *Advanced Engineering Thermodynamics*. 3rd ed. New York: Wiley, 2006.

3. J. A. Schooley. *Thermometry*. Boca Raton, FL: CRC Press, 1986.

## PROBLEMS*

### Thermodynamics

**1–1C**  What is the difference between the classical and the statistical approaches to thermodynamics?

**1–2C**  Why does a bicyclist pick up speed on a downhill road even when he is not pedaling? Does this violate the conservation of energy principle?

**1–3C**  One of the most amusing things a person can experience is when a car in neutral appears to go uphill when its brakes are released. Can this really happen or is it an optical illusion? How can you verify if a road is pitched uphill or downhill?

**1–4C**  An office worker claims that a cup of cold coffee on his table warmed up to 80°C by picking up energy from the surrounding air, which is at 25°C. Is there any truth to his claim? Does this process violate any thermodynamic laws?

### Mass, Force, and Units

**1–5C**  What is the difference between kg-mass and kg-force?

**1–6C**  Explain why the light-year has the dimension of length.

**1–7C**  What is the net force acting on a car cruising at a constant velocity of 70 km/h (*a*) on a level road and (*b*) on an uphill road?

**1–8**  At 45° latitude, the gravitational acceleration as a function of elevation $z$ above sea level is given by $g = a - bz$, where $a = 9.807$ m/s$^2$ and $b = 3.32 \times 10^{-6}$ s$^{-2}$. Determine the height above sea level where the weight of an object will decrease by 0.3 percent.  *Answer: 8862 m*

**1–9**  What is the weight, in N, of an object with a mass of 200 kg at a location where $g = 9.6$ m/s$^2$?

**1–10**  A 3-kg plastic tank that has a volume of 0.2 m$^3$ is filled with liquid water. Assuming the density of water is 1000 kg/m$^3$, determine the weight of the combined system.

**1–11**  A 3-kg rock is thrown upward with a force of 200 N at a location where the local gravitational acceleration is 9.79 m/s$^2$. Determine the acceleration of the rock, in m/s$^2$.

**1–12**  Solve Prob. 1–11 using EES (or other) software. Print out the entire solution, including the numerical results with proper units.

**1–13**  A 4-kW resistance heater in a water heater runs for 3 hours to raise the water temperature to the desired level. Determine the amount of electric energy used in both kWh and kJ.

**1–14**  A 70-kg astronaut took his bathroom scale (a spring scale) and a beam scale (compares masses) to the moon where the local gravity is $g = 1.67$ m/s$^2$. Determine how much he will weigh (*a*) on the spring scale and (*b*) on the beam scale.  *Answer: (a) 11.9 kgf, (b) 70 kgf*

* Problems designated by a "C" are concept questions, and students are encouraged to answer them all. Problems with the ⊙ icon are solved using EES, and complete solutions together with parametric studies are included on the text website. Problems with the [EES] icon are comprehensive in nature, and are intended to be solved with an equation solver such as EES.

**1–15** The gas tank of a car is filled with a nozzle that discharges gasoline at a constant flow rate. Based on unit considerations of quantities, obtain a relation for the filling time in terms of the volume $V$ of the tank (in L) and the discharge rate of gasoline $V$ (in L/s).

**1–16** A pool of volume $V$ (in $m^3$) is to be filled with water using a hose of diameter $D$ (in m). If the average discharge velocity is $V$ (in m/s) and the filling time is $t$ (in s), obtain a relation for the volume of the pool based on considerations of quantities involved.

## Systems, Properties, State, and Processes

**1–17C** A large fraction of the thermal energy generated in the engine of a car is rejected to the air by the radiator through the circulating water. Should the radiator be analyzed as a closed system or as an open system? Explain.

**FIGURE P1–17C**

© McGraw-Hill Education, Christopher Kerrigan

**1–18C** You are trying to understand how a reciprocating air compressor (a piston-cylinder device) works. What system would you use? What type of system is this?

**1–19C** A can of soft drink at room temperature is put into the refrigerator so that it will cool. Would you model the can of soft drink as a closed system or as an open system? Explain.

**1–20C** What is the difference between intensive and extensive properties?

**1–21C** Is the weight of a system an extensive or intensive property?

**1–22C** Is the state of the air in an isolated room completely specified by the temperature and the pressure? Explain.

**1–23C** The molar specific volume of a system $\bar{v}$ is defined as the ratio of the volume of the system to the number of moles of substance contained in the system. Is this an extensive or intensive property?

**1–24C** What is a quasi-equilibrium process? What is its importance in engineering?

**1–25C** Define the isothermal, isobaric, and isochoric processes.

**1–26C** How would you describe the state of the water in a bathtub? How would you describe the process that this water experiences as it cools?

**1–27C** When analyzing the acceleration of gases as they flow through a nozzle, what would you choose as your system? What type of system is this?

**1–28C** What is specific gravity? How is it related to density?

**1–29** [EES] The density of atmospheric air varies with elevation, decreasing with increasing altitude. (a) Using the data given in the table, obtain a relation for the variation of density with elevation, and calculate the density at an elevation of 7000 m. (b) Calculate the mass of the atmosphere using the correlation you obtained. Assume the earth to be a perfect sphere with a radius of 6377 km, and take the thickness of the atmosphere to be 25 km.

| $z$, km | $\rho$, kg/m$^3$ |
|---------|------------------|
| 6377    | 1.225            |
| 6378    | 1.112            |
| 6379    | 1.007            |
| 6380    | 0.9093           |
| 6381    | 0.8194           |
| 6382    | 0.7364           |
| 6383    | 0.6601           |
| 6385    | 0.5258           |
| 6387    | 0.4135           |
| 6392    | 0.1948           |
| 6397    | 0.08891          |
| 6402    | 0.04008          |

## Temperature

**1–30C** What are the ordinary and absolute temperature scales in the SI and the English system?

**1–31C** Consider an alcohol and a mercury thermometer that read exactly 0°C at the ice point and 100°C at the steam point. The distance between the two points is divided into 100 equal parts in both thermometers. Do you think these thermometers will give exactly the same reading at a temperature of, say, 60°C? Explain.

**1–32C** Consider two closed systems A and B. System A contains 3000 kJ of thermal energy at 20°C, whereas system B contains 200 kJ of thermal energy at 50°C. Now the systems are brought into contact with each other. Determine the direction of any heat transfer between the two systems.

**1–33** The deep body temperature of a healthy person is 37°C. What is it in kelvins?

**1–34** What is the temperature of the heated air at 150°C in °F and R?

**1–35** The temperature of a system rises by 70°C during a heating process. Express this rise in temperature in kelvins.

**1–36** The temperature of ambient air in a certain location is measured to be −40°C. Express this temperature in Fahrenheit (°F), Kelvin (K), and Rankine (R) units.

**1–37** The temperature of a system drops by 45°F during a cooling process. Express this drop in temperature in K, R, and °C.

## Pressure, Manometer, and Barometer

**1–38C** Explain why some people experience nose bleeding and some others experience shortness of breath at high elevations.

**1–39C** A health magazine reported that physicians measured 100 adults' blood pressure using two different arm positions: parallel to the body (along the side) and perpendicular to the body (straight out). Readings in the parallel position were up to 10 percent higher than those in the perpendicular position, regardless of whether the patient was standing, sitting, or lying down. Explain the possible cause for the difference.

**1–40C** Someone claims that the absolute pressure in a liquid of constant density doubles when the depth is doubled. Do you agree? Explain.

**1–41C** Express Pascal's law, and give a real-world example of it.

**1–42C** Consider two identical fans, one at sea level and the other on top of a high mountain, running at identical speeds. How would you compare (a) the volume flow rates and (b) the mass flow rates of these two fans?

**1–43** A vacuum gage connected to a chamber reads 35 kPa at a location where the atmospheric pressure is 92 kPa. Determine the absolute pressure in the chamber.

**1–44** The pressure in a compressed air storage tank is 1200 kPa. What is the tank's pressure in (a) kN and m units; (b) kg, m, and s units; and (c) kg, km, and s units?

**1–45** If the pressure inside a rubber balloon is 1500 mmHg, what is this pressure in pounds-force per square inch (psi)? *Answer:* 29.0 psi

**1–46** A manometer is used to measure the air pressure in a tank. The fluid used has a specific gravity of 1.25, and the differential height between the two arms of the manometer is 72 cm. If the local atmospheric pressure is 87.6 kPa, determine the absolute pressure in the tank for the cases of the manometer arm with the (a) higher and (b) lower fluid level being attached to the tank.

**1–47** The water in a tank is pressurized by air, and the pressure is measured by a multifluid manometer as shown in Fig. P1–47. Determine the gage pressure of air in the tank if $h_1 = 0.2$ m, $h_2 = 0.3$ m, and $h_3 = 0.4$ m. Take the densities of water, oil, and mercury to be 1000 kg/m³, 850 kg/m³, and 13,600 kg/m³, respectively.

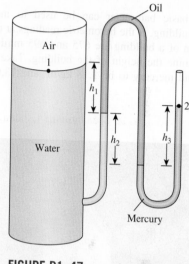

**FIGURE P1–47**

**1–48** Determine the atmospheric pressure at a location where the barometric reading is 750 mmHg. Take the density of mercury to be 13,600 kg/m³.

**1–49** The gage pressure in a liquid at a depth of 3 m is read to be 42 kPa. Determine the gage pressure in the same liquid at a depth of 9 m.

**1–50** The absolute pressure in water at a depth of 9 m is read to be 185 kPa. Determine (a) the local atmospheric pressure, and (b) the absolute pressure at a depth of 5 m in a liquid whose specific gravity is 0.85 at the same location.

**1–51** Determine the pressure exerted on the surface of a submarine cruising 53 m below the free surface of the sea. Assume that the barometric pressure is 101 kPa and the specific gravity of seawater is 1.03.

**1–52** Consider a 70-kg woman who has a total foot imprint area of 400 cm². She wishes to walk on the snow, but the snow cannot withstand pressures greater than 0.5 kPa. Determine the minimum size of the snowshoes needed (imprint area per shoe) to enable her to walk on the snow without sinking.

**1–53** The vacuum pressure of a condenser is given to be 80 kPa. If the atmospheric pressure is 98 kPa, what is the gage pressure and absolute pressure in kPa, kN/m², lbf/in², psi, and mmHg.

**1–54** The barometer of a mountain hiker reads 750 mbars at the beginning of a hiking trip and 650 mbars at the end. Neglecting the effect of altitude on local gravitational acceleration, determine the vertical distance climbed. Assume an average air density of 1.20 kg/m³.  *Answer: 850 m*

**1–55** The basic barometer can be used to measure the height of a building. If the barometric readings at the top and at the bottom of a building are 675 and 695 mmHg, respectively, determine the height of the building. Take the densities of air and mercury to be 1.18 kg/m³ and 13,600 kg/m³, respectively.

**FIGURE P1–55**
© *Royalty-Free/Corbis*

**1–56** ![EES] Solve Prob. 1–55 using EES (or other) software. Print out the entire solution, including the numerical results with proper units.

**1–57** The hydraulic lift in a car repair shop has an output diameter of 30 cm and is to lift cars up to 2000 kg.

Determine the fluid gage pressure that must be maintained in the reservoir.

**1–58** A gas is contained in a vertical, frictionless piston–cylinder device. The piston has a mass of 3.2 kg and a cross-sectional area of 35 cm². A compressed spring above the piston exerts a force of 150 N on the piston. If the atmospheric pressure is 95 kPa, determine the pressure inside the cylinder.  *Answer: 147 kPa*

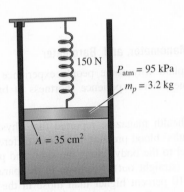

**FIGURE P1–58**

**1–59** ![EES] Reconsider Prob. 1–58. Using EES (or other) software, investigate the effect of the spring force in the range of 0 to 500 N on the pressure inside the cylinder. Plot the pressure against the spring force, and discuss the results.

**1–60** Both a gage and a manometer are attached to a gas tank to measure its pressure. If the reading on the pressure gage is 80 kPa, determine the distance between the two fluid levels of the manometer if the fluid is (*a*) mercury ($\rho = 13{,}600$ kg/m³) or (*b*) water ($\rho = 1000$ kg/m³).

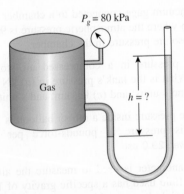

**FIGURE P1–60**

**1–61** ![EES] Reconsider Prob. 1–60. Using EES (or other) software, investigate the effect of the manometer fluid density in the range of 800 to 13,000 kg/m³ on the differential fluid height of the manometer. Plot the differential fluid height against the density, and discuss the results.

**1–62** A manometer containing oil ($\rho = 850$ kg/m³) is attached to a tank filled with air. If the oil-level difference between the two columns is 80 cm and the atmospheric pressure is 98 kPa, determine the absolute pressure of the air in the tank. *Answer:* 105 kPa

**1–63** A mercury manometer ($\rho = 13,600$ kg/m³) is connected to an air duct to measure the pressure inside. The difference in the manometer levels is 15 mm, and the atmospheric pressure is 100 kPa. (a) Judging from Fig. P1–63, determine if the pressure in the duct is above or below the atmospheric pressure. (b) Determine the absolute pressure in the duct.

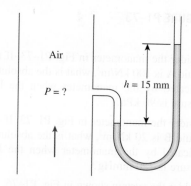

**FIGURE P1–63**

**1–64** Repeat Prob. 1–63 for a differential mercury height of 45 mm.

**1–65** The pressure in a natural gas pipeline is measured by the manometer shown in Fig. P1–65 with one of the arms open to the atmosphere where the local atmospheric pressure is 98 kPa. Determine the absolute pressure in the pipeline.

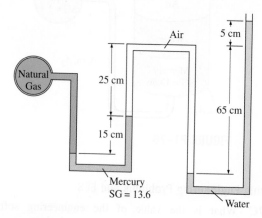

**FIGURE P1–65**

**1–66** Repeat Prob. 1–65 by replacing air by oil with a specific gravity of 0.69.

**1–67** The maximum blood pressure in the upper arm of a healthy person is about 120 mmHg. If a vertical tube open to the atmosphere is connected to the vein in the arm of the person, determine how high the blood will rise in the tube. Take the density of the blood to be 1050 kg/m³.

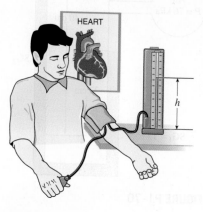

**FIGURE P1–67**

**1–68** Determine the pressure exerted on a diver at 45 m below the free surface of the sea. Assume a barometric pressure of 101 kPa and a specific gravity of 1.03 for seawater. *Answer:* 556 kPa

**1–69** Consider a U-tube whose arms are open to the atmosphere. Now water is poured into the U-tube from one arm, and light oil ($\rho = 790$ kg/m³) from the other. One arm contains 70-cm-high water, while the other arm contains both fluids with an oil-to-water height ratio of 4. Determine the height of each fluid in that arm.

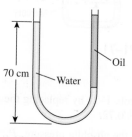

**FIGURE P1–69**

**1–70** Consider a double-fluid manometer attached to an air pipe shown in Fig. P1–70. If the specific gravity of one fluid is 13.55, determine the specific gravity of the other fluid for the indicated absolute pressure of air. Take the atmospheric pressure to be 100 kPa. *Answer:* 5.0

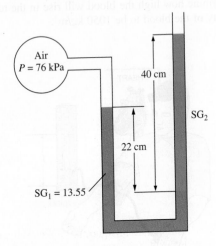

**FIGURE P1–70**

**1–71** Freshwater and seawater flowing in parallel horizontal pipelines are connected to each other by a double U-tube manometer, as shown in Fig. P1–71. Determine the pressure difference between the two pipelines. Take the density of seawater at that location to be $\rho = 1035$ kg/m$^3$. Can the air column be ignored in the analysis?

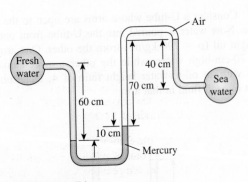

**FIGURE P1–71**

**1–72** Repeat Prob. 1–71 by replacing the air with oil whose specific gravity is 0.72.

**1–73** Calculate the absolute pressure, $P_1$, of the manometer shown in Fig. P1–73 in kPa. The local atmospheric pressure is 758 mmHg.

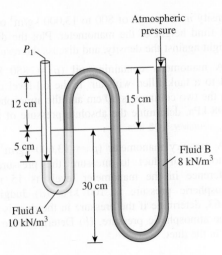

**FIGURE P1–73**

**1–74** Consider the manometer in Fig. P1–73. If the specific weight of fluid A is 100 kN/m$^3$, what is the absolute pressure, in kPa, indicated by the manometer when the local atmospheric pressure is 90 kPa?

**1–75** Consider the manometer in Fig. P1–73. If the specific weight of fluid B is 20 kN/m$^3$, what is the absolute pressure, in kPa, indicated by the manometer when the local atmospheric pressure is 720 mmHg?

**1–76** Consider the system shown in Fig. P1–76. If a change of 0.7 kPa in the pressure of air causes the brine–mercury interface in the right column to drop by 5 mm in the brine level in the right column while the pressure in the brine pipe remains constant, determine the ratio of $A_2/A_1$.

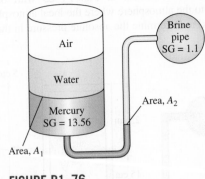

**FIGURE P1–76**

## Solving Engineering Problems and EES

**1–77C** What is the value of the engineering software packages in (*a*) engineering education and (*b*) engineering practice?

**1–78**  Determine a positive real root of this equation using EES:

$$2x^3 - 10x^{0.5} - 3x = -3$$

**1–79**  Solve this system of two equations with two unknowns using EES:

$$x^3 - y^2 = 7.75$$

$$3xy + y = 3.5$$

**1–80**  Solve this system of three equations with three unknowns using EES:

$$x^2y - z = 1$$

$$x - 3y^{0.5} + xz = -2$$

$$x + y - z = 2$$

**1–81** Solve this system of three equations with three unknowns using EES:

$$2x - y + z = 7$$

$$3x^2 + 3y = z + 3$$

$$xy + 2z = 4$$

**1–82** Specific heat is defined as the amount of energy needed to increase the temperature of a unit mass of a substance by one degree. The specific heat of water at room temperature is 4.18 kJ/kg·°C in SI unit system. Using the unit conversion function capability of EES, express the specific heat of water in (a) kJ/kg·K, (b) Btu/lbm·°F, (c) Btu/lbm·R, and (d) kcal/kg·°C units. *Answers:* (a) 4.18, (b) (c) (d) 0.9984

## Review Problems

**1–83** The weight of bodies may change somewhat from one location to another as a result of the variation of the gravitational acceleration $g$ with elevation. Accounting for this variation using the relation in Prob. 1–8, determine the weight of an 80-kg person at sea level ($z = 0$), in Denver ($z = 1610$ m), and on the top of Mount Everest ($z = 8848$ m).

**1–84** A man goes to a traditional market to buy a steak for dinner. He finds a 12-oz steak (1 lbm = 16 oz) for $5.50. He then goes to the adjacent international market and finds a 300-g steak of identical quality for $5.20. Which steak is the better buy?

**1–85** A hydraulic lift is to be used to lift a 2500 kg weight by putting a weight of 25 kg on a piston with a diameter of

10 cm. Determine the diameter of the piston on which the weight is to be placed.

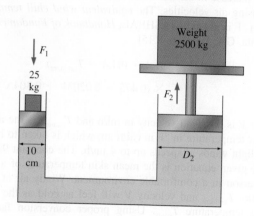

**FIGURE P1–85**

**1–86** The efficiency of a refrigerator increases by 3 percent for each °C rise in the minimum temperature in the device. What is the increase in the efficiency for each (a) K, (b) °F, and (c) R rise in temperature?

**1–87** A house is losing heat at a rate of 1800 kJ/h per °C temperature difference between the indoor and the outdoor temperatures. Express the rate of heat loss from this house per (a) K, (b) °F, and (c) R difference between the indoor and the outdoor temperature.

**1–88** The average temperature of the atmosphere in the world is approximated as a function of altitude by the relation

$$T_{atm} = 288.15 - 6.5z$$

where $T_{atm}$ is the temperature of the atmosphere in K and $z$ is the altitude in km with $z = 0$ at sea level. Determine the average temperature of the atmosphere outside an airplane that is cruising at an altitude of 12,000 m.

**1–89** Joe Smith, an old-fashioned engineering student, believes that the boiling point of water is best suited for use as the reference point on temperature scales. Unhappy that the boiling point corresponds to some odd number in the current absolute temperature scales, he has proposed a new absolute temperature scale that he calls the Smith scale. The temperature unit on this scale is *smith*, denoted by S, and the boiling point of water on this scale is assigned to be 1000 S. From a thermodynamic point of view, discuss if it is an acceptable temperature scale. Also, determine the ice point of water on the Smith scale and obtain a relation between the Smith and Celsius scales.

**1–90** It is well-known that cold air feels much colder in windy weather than what the thermometer reading indicates because of the "chilling effect" of the wind. This effect is due to the increase in the convection heat transfer coefficient with increasing air velocities. The *equivalent wind chill temperature* in °F is given by [ASHRAE, *Handbook of Fundamentals* (Atlanta, GA, 1993), p. 8.15]

$$T_{equiv} = 91.4 - (91.4 - T_{ambient})$$
$$\times (0.475 - 0.0203V + 0.304\sqrt{V})$$

where $V$ is the wind velocity in mi/h and $T_{ambient}$ is the ambient air temperature in °F in calm air, which is taken to be air with light winds at speeds up to 4 mi/h. The constant 91.4°F in the given equation is the mean skin temperature of a resting person in a comfortable environment. Windy air at temperature $T_{ambient}$ and velocity $V$ will feel as cold as the calm air at temperature $T_{equiv}$. Using proper conversion factors, obtain an equivalent relation in SI units where $V$ is the wind velocity in km/h and $T_{ambient}$ is the ambient air temperature in °C.

*Answer:* $T_{equiv} = 33.0 - (33.0 - T_{ambient})$
$$\times (0.475 - 0.0126V + 0.240\sqrt{V})$$

**1–91** Reconsider Prob. 1–90. Using EES (or other) software, plot the equivalent wind chill temperatures in °C as a function of wind velocity in the range of 5 to 60 km/h for the ambient temperatures of –5, 5, and 15°C. Discuss the results.

**1–92** A vertical piston–cylinder device contains a gas at a pressure of 100 kPa. The piston has a mass of 5 kg and a diameter of 12 cm. Pressure of the gas is to be increased by placing some weights on the piston. Determine the local atmospheric pressure and the mass of the weights that will double the pressure of the gas inside the cylinder. *Answers:* 95.7 kPa, 115 kg

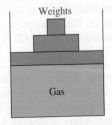

**FIGURE P1–92**

**1–93** An air-conditioning system requires a 35-m-long section of 15-cm diameter duct work to be laid underwater. Determine the upward force the water will exert on the duct. Take the densities of air and water to be 1.3 kg/m³ and 1000 kg/m³, respectively.

**1–94** Balloons are often filled with helium gas because it weighs only about one-seventh of what air weighs under identical conditions. The buoyancy force, which can be expressed as $F_b = \rho_{air} g V_{balloon}$, will push the balloon upward. If the balloon has a diameter of 12 m and carries two people, 85 kg each, determine the acceleration of the balloon when it is first released. Assume the density of air is $\rho = 1.16$ kg/m³, and neglect the weight of the ropes and the cage. *Answer:* 22.4 m/s²

**FIGURE P1–94**

**1–95** Reconsider Prob. 1–94. Using EES (or other) software, investigate the effect of the number of people carried in the balloon on acceleration. Plot the acceleration against the number of people, and discuss the results.

**1–96** Determine the maximum amount of load, in kg, the balloon described in Prob. 1–94 can carry. *Answer:* 900 kg

**1–97** The lower half of a 6-m-high cylindrical container is filled with water ($\rho = 1000$ kg/m³) and the upper half with oil that has a specific gravity of 0.85. Determine the pressure difference between the top and bottom of the cylinder. *Answer:* 54.4 kPa

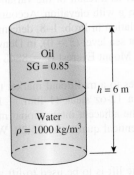

**FIGURE P1–97**

**1–98** A vertical, frictionless piston–cylinder device contains a gas at 180 kPa absolute pressure. The atmospheric pressure outside is 100 kPa, and the piston area is 25 cm². Determine the mass of the piston.

**1–99** A pressure cooker cooks a lot faster than an ordinary pan by maintaining a higher pressure and temperature inside. The lid of a pressure cooker is well sealed, and steam can escape only through an opening in the middle of the lid. A separate metal piece, the petcock, sits on top of this opening and prevents steam from escaping until the pressure force overcomes the weight of the petcock. The periodic escape of the steam in this manner prevents any potentially dangerous pressure buildup and keeps the pressure inside at a constant value. Determine the mass of the petcock of a pressure cooker whose operation pressure is 100 kPa gage and has an opening cross-sectional area of 4 mm². Assume an atmospheric pressure of 101 kPa, and draw the free-body diagram of the petcock. *Answer: 40.8 g*

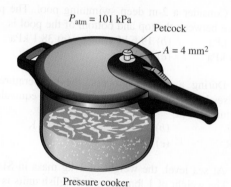

**FIGURE P1–99**

**1–100** A glass tube is attached to a water pipe, as shown in Fig. P1–100. If the water pressure at the bottom of the tube is 110 kPa and the local atmospheric pressure is 99 kPa, determine how high the water will rise in the tube, in m. Take the density of water to be 1000 kg/m³.

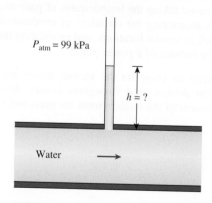

**FIGURE P1–100**

**1–101** Consider a U-tube whose arms are open to the atmosphere. Now equal volumes of water and light oil ($\rho = 790$ kg/m³) are poured from different arms. A person blows from the oil side of the U-tube until the contact surface of the two fluids moves to the bottom of the U-tube, and thus the liquid levels in the two arms are the same. If the fluid height in each arm is 75 cm, determine the gage pressure the person exerts on the oil by blowing.

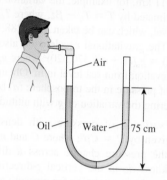

**FIGURE P1–101**

**1–102** The basic barometer can be used as an altitude-measuring device in airplanes. The ground control reports a barometric reading of 753 mmHg while the pilot's reading is 690 mmHg. Estimate the altitude of the plane from ground level if the average air density is 1.20 kg/m³. *Answer: 714 m*

**1–103** A gasoline line is connected to a pressure gage through a double-U manometer, as shown in Fig. P1–103 on the next page. If the reading of the pressure gage is 370 kPa, determine the gage pressure of the gasoline line.

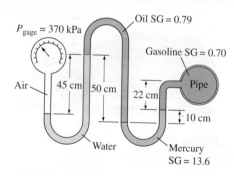

**FIGURE P1–103**

**1–104** Repeat Prob. 1–103 for a pressure gage reading of 180 kPa.

**1–105** The average atmospheric pressure on earth is approximated as a function of altitude by the relation $P_{atm} = 101.325 \ (1 - 0.02256z)^{5.256}$, where $P_{atm}$ is the atmospheric pressure in kPa and $z$ is the altitude in km with $z = 0$ at sea level. Determine the approximate atmospheric pressures at Atlanta ($z = 306$ m), Denver ($z = 1610$ m), Mexico City ($z = 2309$ m), and the top of Mount Everest ($z = 8848$ m).

**1–106** It is well-known that the temperature of the atmosphere varies with altitude. In the troposphere, which extends to an altitude of 11 km, for example, the variation of temperature can be approximated by $T = T_0 - \beta z$, where $T_0$ is the temperature at sea level, which can be taken to be 288.15 K, and $\beta = 0.0065$ K/m. The gravitational acceleration also changes with altitude as $g(z) = g_0/(1 + z/6{,}370{,}320)^2$ where $g_0 = 9.807$ m/s$^2$ and $z$ is the elevation from sea level in m. Obtain a relation for the variation of pressure in the troposphere (*a*) by ignoring and (*b*) by considering the variation of $g$ with altitude.

**1–107** The variation of pressure with density in a thick gas layer is given by $P = C\rho^n$, where $C$ and $n$ are constants. Noting that the pressure change across a differential fluid layer of thickness $dz$ in the vertical $z$-direction is given as $dP = -\rho g\ dz$, obtain a relation for pressure as a function of elevation $z$. Take the pressure and density at $z = 0$ to be $P_0$ and $\rho_0$, respectively.

**1–108** Consider the flow of air through a wind turbine whose blades sweep an area of diameter $D$ (in m). The average air velocity through the swept area is $V$ (in m/s). On the bases of the units of the quantities involved, show that the mass flow rate of air (in kg/s) through the swept area is proportional to air density, the wind velocity, and the square of the diameter of the swept area.

**1–109** The drag force exerted on a car by air depends on a dimensionless drag coefficient, the density of air, the car velocity, and the frontal area of the car. That is, $F_D = $ function ($C_{Drag}$, $A_{front}$, $\rho$, $V$). Based on unit considerations alone, obtain a relation for the drag force.

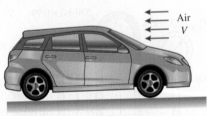

Air
$V$

**FIGURE P1–109**

## Fundamentals of Engineering (FE) Exam Problems

**1–110** An apple loses 4.5 kJ of heat as it cools per °C drop in its temperature. The amount of heat loss from the apple per °F drop in its temperature is
(*a*) 1.25 kJ     (*b*) 2.50 kJ     (*c*) 5.0 kJ
(*d*) 8.1 kJ     (*e*) 4.1 kJ

**1–111** Consider a fish swimming 5 m below the free surface of water. The increase in the pressure exerted on the fish when it dives to a depth of 25 m below the free surface is
(*a*) 196 Pa     (*b*) 5400 Pa     (*c*) 30,000 Pa
(*d*) 196,000 Pa     (*e*) 294,000 Pa

**1–112** The atmospheric pressures at the top and the bottom of a building are read by a barometer to be 96.0 and 98.0 kPa. If the density of air is 1.0 kg/m$^3$, the height of the building is
(*a*) 17 m     (*b*) 20 m     (*c*) 170 m
(*d*) 204 m     (*e*) 252 m

**1–113** Consider a 2-m deep swimming pool. The pressure difference between the top and bottom of the pool is
(*a*) 12.0 kPa     (*b*) 19.6 kPa     (*c*) 38.1 kPa
(*d*) 50.8 kPa     (*e*) 200 kPa

**1–114** During a heating process, the temperature of an object rises by 10°C. This temperature rise is equivalent to a temperature rise of
(*a*) 10°F     (*b*) 42°F     (*c*) 18 K
(*d*) 18 R     (*e*) 283 K

**1–115** At sea level, the weight of 1 kg mass in SI units is 9.81 N. The weight of 1 lbm mass in English units is
(*a*) 1 lbf     (*b*) 9.81 lbf     (*c*) 32.2 lbf
(*d*) 0.1 lbf     (*e*) 0.031 lbf

## Design and Essay Problems

**1–116** Write an essay on different temperature measurement devices. Explain the operational principle of each device, its advantages and disadvantages, its cost, and its range of applicability. Which device would you recommend for use in the following cases: taking the temperatures of patients in a doctor's office, monitoring the variations of temperature of a car engine block at several locations, and monitoring the temperatures in the furnace of a power plant?

**1–117** Write an essay on the various mass- and volume-measurement devices used throughout history. Also, explain the development of the modern units for mass and volume.

# ENERGY, ENERGY TRANSFER, AND GENERAL ENERGY ANALYSIS

Whether we realize it or not, energy is an important part of most aspects of daily life. The quality of life, and even its sustenance, depends on the availability of energy. Therefore, it is important to have a good understanding of the sources of energy, the conversion of energy from one form to another, and the ramifications of these conversions.

Energy exists in numerous forms such as thermal, mechanical, electric, chemical, and nuclear. Even mass can be considered a form of energy. Energy can be transferred to or from a closed system (a fixed mass) in two distinct forms: *heat* and *work*. For control volumes, energy can also be transferred by mass flow. An energy transfer to or from a closed system is *heat* if it is caused by a temperature difference. Otherwise it is *work*, and it is caused by a force acting through a distance.

We start this chapter with a discussion of various forms of energy and energy transfer by heat. We then introduce various forms of work and discuss energy transfer by work. We continue with developing a general intuitive expression for the *first law of thermodynamics*, also known as the *conservation of energy principle*, which is one of the most fundamental principles in nature, and we then demonstrate its use. Finally, we discuss the efficiencies of some familiar energy conversion processes, and examine the impact on energy conversion on the environment. Detailed treatments of the first law of thermodynamics for closed systems and control volumes are given in Chaps. 4 and 5, respectively.

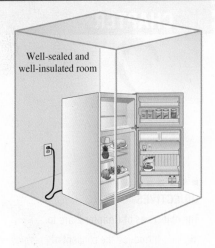

**FIGURE 2–1**
A refrigerator operating with its door open in a well-sealed and well-insulated room.

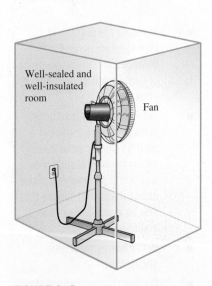

**FIGURE 2–2**
A fan running in a well-sealed and well-insulated room will raise the temperature of air in the room.

# 2–1 · INTRODUCTION

We are familiar with the conservation of energy principle, which is an expression of the first law of thermodynamics, back from our high school years. We are told repeatedly that energy cannot be created or destroyed during a process; it can only change from one form to another. This seems simple enough, but let's test ourselves to see how well we understand and truly believe in this principle.

Consider a room whose door and windows are tightly closed, and whose walls are well-insulated so that heat loss or gain through the walls is negligible. Now let's place a refrigerator in the middle of the room with its door open, and plug it into a wall outlet (Fig. 2–1). You may even use a small fan to circulate the air in order to maintain temperature uniformity in the room. Now, what do you think will happen to the average temperature of air in the room? Will it be increasing or decreasing? Or will it remain constant?

Probably the first thought that comes to mind is that the average air temperature in the room will decrease as the warmer room air mixes with the air cooled by the refrigerator. Some may draw our attention to the heat generated by the motor of the refrigerator, and may argue that the average air temperature may rise if this heating effect is greater than the cooling effect. But they will get confused if it is stated that the motor is made of superconducting materials, and thus there is hardly any heat generation in the motor.

Heated discussion may continue with no end in sight until we remember the conservation of energy principle that we take for granted: If we take the entire room—including the air and the refrigerator—as the system, which is an adiabatic closed system since the room is well-sealed and well-insulated, the only energy interaction involved is the electrical energy crossing the system boundary and entering the room. The conservation of energy requires the energy content of the room to increase by an amount equal to the amount of the electrical energy drawn by the refrigerator, which can be measured by an ordinary electric meter. The refrigerator or its motor does not store this energy. Therefore, this energy must now be in the room air, and it will manifest itself as a rise in the air temperature. The temperature rise of air can be calculated on the basis of the conservation of energy principle using the properties of air and the amount of electrical energy consumed. What do you think would happen if we had a window air conditioning unit instead of a refrigerator placed in the middle of this room? What if we operated a fan in this room instead (Fig. 2–2)?

Note that energy is conserved during the process of operating the refrigerator placed in a room—the electrical energy is converted into an equivalent amount of thermal energy stored in the room air. If energy is already conserved, then what are all those speeches on energy conservation and the measures taken to conserve energy? Actually, by "energy conservation" what is meant is the conservation of the *quality* of energy, not the quantity. Electricity, which is of the highest quality of energy, for example, can always be converted to an equal amount of thermal energy (also called *heat*). But only a small fraction of thermal energy, which is the lowest quality of energy, can be converted back to electricity, as we discuss in Chap. 6. Think about the things that you can do with the electrical energy that the refrigerator has consumed, and the air in the room that is now at a higher temperature.

Now if asked to name the energy transformations associated with the operation of a refrigerator, we may still have a hard time answering because all we see is electrical energy entering the refrigerator and heat dissipated from the refrigerator to the room air. Obviously there is need to study the various forms of energy first, and this is exactly what we do next, followed by a study of the mechanisms of energy transfer.

## 2–2 · FORMS OF ENERGY

Energy can exist in numerous forms such as thermal, mechanical, kinetic, potential, electric, magnetic, chemical, and nuclear (Fig. 2–3), and their sum constitutes the **total energy** $E$ of a system. The total energy of a system on a *unit mass* basis is denoted by $e$ and is expressed as

$$e = \frac{E}{m} \quad \text{(kJ/kg)} \tag{2–1}$$

Thermodynamics provides no information about the absolute value of the total energy. It deals only with the *change* of the total energy, which is what matters in engineering problems. Thus the total energy of a system can be assigned a value of zero ($E = 0$) at some convenient reference point. The change in total energy of a system is independent of the reference point selected. The decrease in the potential energy of a falling rock, for example, depends on only the elevation difference and not the reference level selected.

In thermodynamic analysis, it is often helpful to consider the various forms of energy that make up the total energy of a system in two groups: *macroscopic* and *microscopic*. The **macroscopic** forms of energy are those a system possesses as a whole with respect to some outside reference frame, such as kinetic and potential energies (Fig. 2–4). The **microscopic** forms of energy are those related to the molecular structure of a system and the degree of the molecular activity, and they are independent of outside reference frames. The sum of all the microscopic forms of energy is called the **internal energy** of a system and is denoted by $U$.

The term *energy* was coined in 1807 by Thomas Young, and its use in thermodynamics was proposed in 1852 by Lord Kelvin. The term *internal energy* and its symbol $U$ first appeared in the works of Rudolph Clausius and William Rankine in the second half of the nineteenth century, and it eventually replaced the alternative terms *inner work*, *internal work*, and *intrinsic energy* commonly used at the time.

The macroscopic energy of a system is related to motion and the influence of some external effects such as gravity, magnetism, electricity, and surface tension. The energy that a system possesses as a result of its motion relative to some reference frame is called **kinetic energy** (KE). When all parts of a system move with the same velocity, the kinetic energy is expressed as

$$\text{KE} = m \frac{V^2}{2} \quad \text{(kJ)} \tag{2–2}$$

or, on a unit mass basis,

$$\text{ke} = \frac{V^2}{2} \quad \text{(kJ/kg)} \tag{2–3}$$

(a)

(b)

**FIGURE 2–3**
At least six different forms of energy are encountered in bringing power from a nuclear plant to your home: nuclear, thermal, mechanical, kinetic, magnetic, and electrical.

*(a) ©Creatas/PunchStock RF*
*(b) ©Comstock Images/Jupiterimages RF*

**FIGURE 2–4**
The macroscopic energy of an object changes with velocity and elevation.

where $V$ denotes the velocity of the system relative to some fixed reference frame. The kinetic energy of a rotating solid body is given by $\frac{1}{2}I\omega^2$ where $I$ is the moment of inertia of the body and $\omega$ is the angular velocity.

The energy that a system possesses as a result of its elevation in a gravitational field is called **potential energy** (PE) and is expressed as

$$PE = mgz \quad \text{(kJ)} \tag{2-4}$$

or, on a unit mass basis,

$$pe = gz \quad \text{(kJ/kg)} \tag{2-5}$$

where $g$ is the gravitational acceleration and $z$ is the elevation of the center of gravity of a system relative to some arbitrarily selected reference level.

The magnetic, electric, and surface tension effects are significant in some specialized cases only and are usually ignored. In the absence of such effects, the total energy of a system consists of the kinetic, potential, and internal energies and is expressed as

$$E = U + KE + PE = U + m\frac{V^2}{2} + mgz \quad \text{(kJ)} \tag{2-6}$$

or, on a unit mass basis,

$$e = u + ke + pe = u + \frac{V^2}{2} + gz \quad \text{(kJ/kg)} \tag{2-7}$$

Most closed systems remain stationary during a process and thus experience no change in their kinetic and potential energies. Closed systems whose velocity and elevation of the center of gravity remain constant during a process are frequently referred to as **stationary systems**. The change in the total energy $\Delta E$ of a stationary system is identical to the change in its internal energy $\Delta U$. In this text, a closed system is assumed to be stationary unless stated otherwise.

Control volumes typically involve fluid flow for long periods of time, and it is convenient to express the energy flow associated with a fluid stream in the rate form. This is done by incorporating the **mass flow rate** $\dot{m}$, which is the amount of mass flowing through a cross section per unit time. It is related to the **volume flow rate** $\dot{V}$, which is the volume of a fluid flowing through a cross section per unit time, by

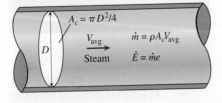

*Mass flow rate:* $\qquad \dot{m} = \rho\dot{V} = \rho A_c V_{avg} \quad \text{(kg/s)} \tag{2-8}$

which is analogous to $m = \rho V$. Here $\rho$ is the fluid density, $A_c$ is the cross-sectional area of flow, and $V_{avg}$ is the average flow velocity normal to $A_c$. The dot over a symbol is used to indicate *time rate* throughout the book. Then the energy flow rate associated with a fluid flowing at a rate of $\dot{m}$ is (Fig. 2–5)

**FIGURE 2–5**
Mass and energy flow rates associated with the flow of steam in a pipe of inner diameter $D$ with an average velocity of $V_{avg}$.

*Energy flow rate:* $\qquad \dot{E} = \dot{m}e \quad \text{(kJ/s or kW)} \tag{2-9}$

which is analogous to $E = me$.

# Some Physical Insight to Internal Energy

Internal energy is defined earlier as the sum of all the *microscopic* forms of energy of a system. It is related to the *molecular structure* and the degree of *molecular activity,* and can be viewed as the sum of the *kinetic* and *potential* energies of the molecules.

To have a better understanding of internal energy, let us examine a system at the molecular level. The molecules of a gas move through space with some velocity, and thus possess some kinetic energy. This is known as the *translational energy.* The atoms of polyatomic molecules rotate about an axis, and the energy associated with this rotation is the *rotational kinetic energy.* The atoms of a polyatomic molecule may also vibrate about their common center of mass, and the energy associated with this back-and-forth motion is the *vibrational kinetic energy.* For gases, the kinetic energy is mostly due to translational and rotational motions, with vibrational motion becoming significant at higher temperatures. The electrons in an atom rotate about the nucleus, and thus possess *rotational kinetic energy.* Electrons at outer orbits have larger kinetic energies. Electrons also spin about their axes, and the energy associated with this motion is the *spin energy.* Other particles in the nucleus of an atom also possess spin energy. The portion of the internal energy of a system associated with the kinetic energies of the molecules is called the **sensible energy** (Fig. 2–6). The average velocity and the degree of activity of the molecules are proportional to the temperature of the gas. Therefore, at higher temperatures, the molecules possess higher kinetic energies, and as a result the system has a higher internal energy.

The internal energy is also associated with various *binding forces* between the molecules of a substance, between the atoms within a molecule, and between the particles within an atom and its nucleus. The forces that bind the *molecules* to each other are, as one would expect, strongest in solids and weakest in gases. If sufficient energy is added to the molecules of a solid or liquid, the molecules overcome these molecular forces and break away, turning the substance into a gas. This is a phase-change process. Because of this added energy, a system in the gas phase is at a higher internal energy level than it is in the solid or the liquid phase. The internal energy associated with the phase of a system is called the **latent energy.** The phase-change process can occur without a change in the chemical composition of a system. Most practical problems fall into this category, and one does not need to pay any attention to the forces binding the atoms in a molecule to each other.

An atom consists of neutrons and positively charged protons bound together by very strong nuclear forces in the nucleus, and negatively charged electrons orbiting around it. The internal energy associated with the atomic bonds in a molecule is called **chemical energy.** During a chemical reaction, such as a combustion process, some chemical bonds are destroyed while others are formed. As a result, the internal energy changes. The nuclear forces are much larger than the forces that bind the electrons to the nucleus. The tremendous amount of energy associated with the strong bonds within the nucleus of the atom itself is called **nuclear energy** (Fig. 2–7). Obviously, we need not be concerned with nuclear energy in thermodynamics unless, of course, we deal with fusion or fission reactions. A chemical reaction involves changes in the structure of the electrons of the atoms, but a

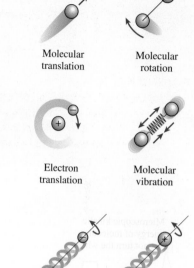

**Molecular translation**    **Molecular rotation**

**Electron translation**    **Molecular vibration**

**Electron spin**    **Nuclear spin**

**FIGURE 2–6**
The various forms of microscopic energies that make up *sensible* energy.

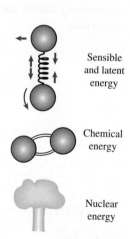

Sensible and latent energy

Chemical energy

Nuclear energy

**FIGURE 2–7**
The internal energy of a system is the sum of all forms of the microscopic energies.

nuclear reaction involves changes in the core or nucleus. Therefore, an atom preserves its identity during a chemical reaction but loses it during a nuclear reaction. Atoms may also possess *electric* and *magnetic dipole-moment energies* when subjected to external electric and magnetic fields due to the twisting of the magnetic dipoles produced by the small electric currents associated with the orbiting electrons.

The forms of energy already discussed, which constitute the total energy of a system, can be *contained* or *stored* in a system, and thus can be viewed as the *static* forms of energy. The forms of energy not stored in a system can be viewed as the *dynamic* forms of energy or as *energy interactions*. The dynamic forms of energy are recognized at the system boundary as they cross it, and they represent the energy gained or lost by a system during a process. The only two forms of energy interactions associated with a closed system are **heat transfer** and **work**. An energy interaction is heat transfer if its driving force is a temperature difference. Otherwise it is work, as explained in the next section. A control volume can also exchange energy via mass transfer since any time mass is transferred into or out of a system, the energy content of the mass is also transferred with it.

In daily life, we frequently refer to the sensible and latent forms of internal energy as *heat*, and we talk about heat content of bodies. In thermodynamics, however, we usually refer to those forms of energy as **thermal energy** to prevent any confusion with *heat transfer*.

Distinction should be made between the macroscopic kinetic energy of an object as a whole and the microscopic kinetic energies of its molecules that constitute the sensible internal energy of the object (Fig. 2–8). The kinetic energy of an object is an *organized* form of energy associated with the orderly motion of all molecules in one direction in a straight path or around an axis. In contrast, the kinetic energies of the molecules are completely *random* and highly *disorganized*. As you will see in later chapters, the organized energy is much more valuable than the disorganized energy, and a major application area of thermodynamics is the conversion of disorganized energy (heat) into organized energy (work). You will also see that the organized energy can be converted to disorganized energy completely, but only a fraction of disorganized energy can be converted to organized energy by specially built devices called *heat engines* (like car engines and power plants). A similar argument can be given for the macroscopic potential energy of an object as a whole and the microscopic potential energies of the molecules.

## More on Nuclear Energy

The best known fission reaction involves the split of the uranium atom (the U-235 isotope) into other elements and is commonly used to generate electricity in nuclear power plants (440 of them in 2004, generating 363,000 MW worldwide), to power nuclear submarines and aircraft carriers, and even to power spacecraft as well as building nuclear bombs.

The percentage of electricity produced by nuclear power is 78 percent in France, 25 percent in Japan, 28 percent in Germany, and 20 percent in the United States. The first nuclear chain reaction was achieved by Enrico Fermi in 1942, and the first large-scale nuclear reactors were built in 1944 for the purpose of producing material for nuclear weapons. When a uranium-235

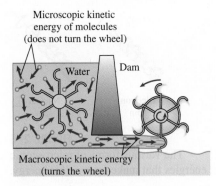

**FIGURE 2–8**
The *macroscopic* kinetic energy is an organized form of energy and is much more useful than the disorganized *microscopic* kinetic energies of the molecules.

atom absorbs a neutron and splits during a fission process, it produces a cesium-140 atom, a rubidium-93 atom, 3 neutrons, and $3.2 \times 10^{-11}$ J of energy. In practical terms, the complete fission of 1 kg of uranium-235 releases $6.73 \times 10^{10}$ kJ of heat, which is more than the heat released when 3000 tons of coal are burned. Therefore, for the same amount of fuel, a nuclear fission reaction releases several million times more energy than a chemical reaction. The safe disposal of used nuclear fuel, however, remains a concern.

Nuclear energy by fusion is released when two small nuclei combine into a larger one. The huge amount of energy radiated by the sun and the other stars originates from such a fusion process that involves the combination of two hydrogen atoms into a helium atom. When two heavy hydrogen (deuterium) nuclei combine during a fusion process, they produce a helium-3 atom, a free neutron, and $5.1 \times 10^{-13}$ J of energy (Fig. 2–9).

Fusion reactions are much more difficult to achieve in practice because of the strong repulsion between the positively charged nuclei, called the *Coulomb repulsion*. To overcome this repulsive force and to enable the two nuclei to fuse together, the energy level of the nuclei must be raised by heating them to about 100 million °C. But such high temperatures are found only in the stars or in exploding atomic bombs (the A-bomb). In fact, the uncontrolled fusion reaction in a hydrogen bomb (the H-bomb) is initiated by a small atomic bomb. The uncontrolled fusion reaction was achieved in the early 1950s, but all the efforts since then to achieve controlled fusion by massive lasers, powerful magnetic fields, and electric currents to generate power have failed.

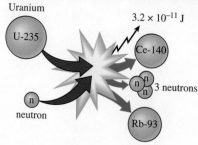

(a) Fission of uranium

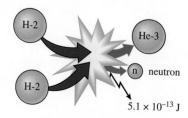

(b) Fusion of hydrogen

**FIGURE 2–9**
The fission of uranium and the fusion of hydrogen during nuclear reactions, and the release of nuclear energy.

---

### ■ EXAMPLE 2–1    A Car Powered by Nuclear Fuel

An average car consumes about 5 L of gasoline a day, and the capacity of the fuel tank of a car is about 50 L. Therefore, a car needs to be refueled once every 10 days. Also, the density of gasoline ranges from 0.68 to 0.78 kg/L, and its lower heating value is about 44,000 kJ/kg (that is, 44,000 kJ of heat is released when 1 kg of gasoline is completely burned). Suppose all the problems associated with the radioactivity and waste disposal of nuclear fuels are resolved, and a car is to be powered by U-235. If a new car comes equipped with 0.1-kg of the nuclear fuel U-235, determine if this car will ever need refueling under average driving conditions (Fig. 2–10).

**SOLUTION**  A car powered by nuclear energy comes equipped with nuclear fuel. It is to be determined if this car will ever need refueling.
*Assumptions*  **1** Gasoline is an incompressible substance with an average density of 0.75 kg/L. **2** Nuclear fuel is completely converted to thermal energy.
*Analysis*  The mass of gasoline used per day by the car is

$$m_{gasoline} = (\rho V)_{gasoline} = (0.75 \text{ kg/L})(5 \text{ L/day}) = 3.75 \text{ kg/day}$$

Noting that the heating value of gasoline is 44,000 kJ/kg, the energy supplied to the car per day is

$$E = (m_{gasoline})(\text{Heating value})$$

$$= (3.75 \text{ kg/day})(44,000 \text{ kJ/kg}) = 165,000 \text{ kJ/day}$$

**FIGURE 2–10**
Schematic for Example 2–1.

The complete fission of 0.1 kg of uranium-235 releases

$$(6.73 \times 10^{10} \text{ kJ/kg})(0.1 \text{ kg}) = 6.73 \times 10^9 \text{ kJ}$$

of heat, which is sufficient to meet the energy needs of the car for

$$\text{No. of days} = \frac{\text{Energy content of fuel}}{\text{Daily energy use}} = \frac{6.73 \times 10^9 \text{ kJ}}{165,000 \text{ kJ/day}} = \textbf{40,790 days}$$

which is equivalent to about 112 years. Considering that no car will last more than 100 years, this car will never need refueling. It appears that nuclear fuel of the size of a cherry is sufficient to power a car during its lifetime.
**Discussion** Note that this problem is not quite realistic since the necessary critical mass cannot be achieved with such a small amount of fuel. Further, all of the uranium cannot be converted in fission, again because of the critical mass problems after partial conversion.

**FIGURE 2–11**
Mechanical energy is a useful concept for flows that do not involve significant heat transfer or energy conversion, such as the flow of gasoline from an underground tank into a car.
©Royalty-Free/Corbis

## Mechanical Energy

Many engineering systems are designed to transport a fluid from one location to another at a specified flow rate, velocity, and elevation difference, and the system may generate mechanical work in a turbine or it may consume mechanical work in a pump or fan during this process (Fig. 2–11). These systems do not involve the conversion of nuclear, chemical, or thermal energy to mechanical energy. Also, they do not involve any heat transfer in any significant amount, and they operate essentially at constant temperature. Such systems can be analyzed conveniently by considering the *mechanical forms of energy* only and the frictional effects that cause the mechanical energy to be lost (i.e., to be converted to thermal energy that usually cannot be used for any useful purpose).

The **mechanical energy** can be defined as *the form of energy that can be converted to mechanical work completely and directly by an ideal mechanical device such as an ideal turbine*. Kinetic and potential energies are the familiar forms of mechanical energy. Thermal energy is not mechanical energy, however, since it cannot be converted to work directly and completely (the second law of thermodynamics).

A pump transfers mechanical energy to a fluid by raising its pressure, and a turbine extracts mechanical energy from a fluid by dropping its pressure. Therefore, the pressure of a flowing fluid is also associated with its mechanical energy. In fact, the pressure unit Pa is equivalent to Pa = $N/m^2$ = $N \cdot m/m^3$ = $J/m^3$, which is energy per unit volume, and the product $Pv$ or its equivalent $P/\rho$ has the unit J/kg, which is energy per unit mass. Note that pressure itself is not a form of energy but a pressure force acting on a fluid through a distance produces work, called *flow work*, in the amount of $P/\rho$ per unit mass. Flow work is expressed in terms of fluid properties, and it is convenient to view it as part of the energy of a flowing fluid and call it *flow energy*. Therefore, the mechanical energy of a flowing fluid can be expressed on a unit mass basis as

$$e_{\text{mech}} = \frac{P}{\rho} + \frac{V^2}{2} + gz \qquad \text{(2–10)}$$

where $P/\rho$ is the *flow energy*, $V^2/2$ is the *kinetic energy*, and $gz$ is the *potential energy* of the fluid, all per unit mass. It can also be expressed in rate form as

$$\dot{E}_{mech} = \dot{m}e_{mech} = \dot{m}\left(\frac{P}{\rho} + \frac{V^2}{2} + gz\right) \qquad (2\text{--}11)$$

where $\dot{m}$ is the mass flow rate of the fluid. Then the mechanical energy change of a fluid during incompressible ($\rho$ = constant) flow becomes

$$\Delta e_{mech} = \frac{P_2 - P_1}{\rho} + \frac{V_2^2 - V_1^2}{2} + g(z_2 - z_1) \qquad (\text{kJ/kg}) \qquad (2\text{--}12)$$

and

$$\Delta \dot{E}_{mech} = \dot{m}\Delta e_{mech} = \dot{m}\left(\frac{P_2 - P_1}{\rho} + \frac{V_2^2 - V_1^2}{2} + g(z_2 - z_1)\right) \qquad (\text{kW}) \qquad (2\text{--}13)$$

Therefore, the mechanical energy of a fluid does not change during flow if its pressure, density, velocity, and elevation remain constant. In the absence of any irreversible losses, the mechanical energy change represents the mechanical work supplied to the fluid (if $\Delta e_{mech} > 0$) or extracted from the fluid (if $\Delta e_{mech} < 0$). The maximum (ideal) power generated by a turbine, for example, is $\dot{W}_{max} = \dot{m}\Delta e_{mech}$, as shown in Fig. 2–12.

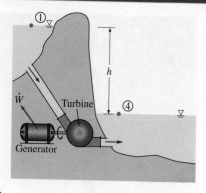

$$\dot{W}_{max} = \dot{m}\Delta e_{mech} = \dot{m}g(z_1 - z_4) = \dot{m}gh$$

since $P_1 \approx P_4 = P_{atm}$ and $V_1 = V_4 \approx 0$

(a)

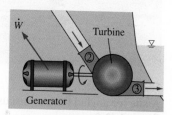

$$\dot{W}_{max} = \dot{m}\Delta e_{mech} = \dot{m}\frac{P_2 - P_3}{\rho} = \dot{m}\frac{\Delta P}{\rho}$$

since $V_2 \approx V_3$ and $z_2 = z_3$

(b)

**FIGURE 2–12**
Mechanical energy is illustrated by an ideal hydraulic turbine coupled with an ideal generator. In the absence of irreversible losses, the maximum produced power is proportional to (a) the change in water surface elevation from the upstream to the downstream reservoir or (b) (close-up view) the drop in water pressure from just upstream to just downstream of the turbine.

■ **EXAMPLE 2–2**    **Wind Energy**

A site evaluated for a wind farm is observed to have steady winds at a speed of 8.5 m/s (Fig. 2–13). Determine the wind energy (a) per unit mass, (b) for a mass of 10 kg, and (c) for a flow rate of 1154 kg/s for air.

**SOLUTION**    A site with a specified wind speed is considered. Wind energy per unit mass, for a specified mass, and for a given mass flow rate of air are to be determined.
*Assumptions*  Wind flows steadily at the specified speed.
*Analysis*  The only harvestable form of energy of atmospheric air is the kinetic energy, which is captured by a wind turbine.
(a) Wind energy per unit mass of air is

$$e = \text{ke} = \frac{V^2}{2} = \frac{(8.5 \text{ m/s})^2}{2}\left(\frac{1 \text{ J/kg}}{1 \text{ m}^2/\text{s}^2}\right) = \textbf{36.1 J/kg}$$

(b) Wind energy for an air mass of 10 kg is

$$E = me = (10 \text{ kg})(36.1 \text{ J/kg}) = \textbf{361 J}$$

(c) Wind energy for a mass flow rate of 1154 kg/s is

$$\dot{E} = \dot{m}e = (1154 \text{ kg/s})(36.1 \text{ J/kg})\left(\frac{1 \text{ kW}}{1000 \text{ J/s}}\right) = \textbf{41.7 kW}$$

*Discussion*  It can be shown that the specified mass flow rate corresponds to a 12-m diameter flow section when the air density is 1.2 kg/m³. Therefore, a wind turbine with a wind span diameter of 12 m has a power generation potential of 41.7 kW. Real wind turbines convert about one-third of this potential to electric power.

**FIGURE 2–13**

A site for a wind farm as discussed in Example 2-2.

*©Ingram Publishing/SuperStock RF*

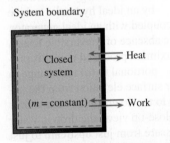

**FIGURE 2–14**

Energy can cross the boundaries of a closed system in the form of heat and work.

## 2–3 · ENERGY TRANSFER BY HEAT

Energy can cross the boundary of a closed system in two distinct forms: *heat* and *work* (Fig. 2–14). It is important to distinguish between these two forms of energy. Therefore, they will be discussed first, to form a sound basis for the development of the laws of thermodynamics.

We know from experience that a can of cold soda left on a table eventually warms up and that a hot baked potato on the same table cools down. When a body is left in a medium that is at a different temperature, energy transfer takes place between the body and the surrounding medium until thermal equilibrium is established, that is, the body and the medium reach the same temperature. The direction of energy transfer is always from the higher temperature body to the lower temperature one. Once the temperature equality is established, energy transfer stops. In the processes described above, energy is said to be transferred in the form of heat.

**Heat** is defined as *the form of energy that is transferred between two systems (or a system and its surroundings) by virtue of a temperature difference* (Fig. 2–15). That is, an energy interaction is heat only if it takes place because of a temperature difference. Then it follows that there cannot be any heat transfer between two systems that are at the same temperature.

Several phrases in common use today—such as heat flow, heat addition, heat rejection, heat absorption, heat removal, heat gain, heat loss, heat storage, heat generation, electrical heating, resistance heating, frictional heating, gas heating, heat of reaction, liberation of heat, specific heat, sensible heat, latent heat, waste heat, body heat, process heat, heat sink, and heat source—are not consistent with the strict thermodynamic meaning of the term *heat*, which limits its use to the *transfer* of thermal energy during a process. However, these phrases are deeply rooted in our vocabulary, and they are used by both ordinary people and scientists without causing any misunderstanding since they are usually interpreted properly instead of being taken literally. (Besides, no acceptable alternatives exist for some of these phrases.) For example, the phrase *body heat* is understood to mean *the thermal energy content* of a body. Likewise, *heat flow* is understood to mean *the transfer of thermal energy*, not the flow of a fluidlike substance called heat, although the latter incorrect interpretation, which is based on the caloric theory, is the origin of this phrase. Also, the transfer of heat into a system is frequently referred to as *heat addition* and the transfer of heat out of a system as *heat rejection*. Perhaps there are thermodynamic reasons for being so reluctant to replace *heat* by *thermal energy*: It takes less time and energy to say, write, and comprehend *heat* than it does *thermal energy*.

Heat is energy in transition. It is recognized only as it crosses the boundary of a system. Consider the hot baked potato one more time. The potato contains energy, but this energy is heat transfer only as it passes through the skin of the potato (the system boundary) to reach the air, as shown in Fig. 2–16. Once in the surroundings, the transferred heat becomes part of the internal energy of the surroundings. Thus, in thermodynamics, the term *heat* simply means *heat transfer*.

A process during which there is no heat transfer is called an **adiabatic process** (Fig. 2–17). The word *adiabatic* comes from the Greek word *adiabatos*, which means *not to be passed*. There are two ways a process can be adiabatic: Either the system is well insulated so that only a negligible amount of heat can pass through the boundary, or both the system and the surroundings are at the same temperature and therefore there is no driving force (temperature difference) for heat transfer. An adiabatic process should not be confused with an isothermal process. Even though there is no heat transfer during an adiabatic process, the energy content and thus the temperature of a system can still be changed by other means such as work.

As a form of energy, heat has energy units, kJ being the most common one. The amount of heat transferred during the process between two states (states 1 and 2) is denoted by $Q_{12}$, or just $Q$. Heat transfer *per unit mass* of a system is denoted $q$ and is determined from

$$q = \frac{Q}{m} \quad \text{(kJ/kg)} \tag{2-14}$$

**FIGURE 2–15**
Temperature difference is the driving force for heat transfer. The larger the temperature difference, the higher is the rate of heat transfer.

Sometimes it is desirable to know the *rate of heat transfer* (the amount of heat transferred per unit time) instead of the total heat transferred over some time interval (Fig. 2–18). The heat transfer rate is denoted $\dot{Q}$, where the overdot stands for the time derivative, or "per unit time." The heat transfer rate $\dot{Q}$ has the unit kJ/s, which is equivalent to kW. When $\dot{Q}$ varies with time, the amount of heat transfer during a process is determined by integrating $\dot{Q}$ over the time interval of the process:

$$Q = \int_{t_1}^{t_2} \dot{Q}\, dt \quad \text{(kJ)} \tag{2-15}$$

When $\dot{Q}$ remains constant during a process, this relation reduces to

$$Q = \dot{Q}\Delta t \quad \text{(kJ)} \tag{2-16}$$

where $\Delta t = t_2 - t_1$ is the time interval during which the process takes place.

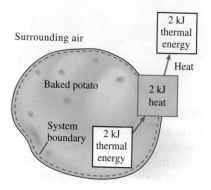

**FIGURE 2–16**
Energy is recognized as heat transfer only as it crosses the system boundary.

## Historical Background on Heat

Heat has always been perceived to be something that produces in us a sensation of warmth, and one would think that the nature of heat is one of the first things understood by mankind. However, it was only in the middle of the nineteenth century that we had a true physical understanding of the nature of heat, thanks to the development at that time of the **kinetic theory**, which treats molecules as tiny balls that are in motion and thus possess kinetic energy. Heat is then defined as the energy associated with the random motion of atoms and molecules. Although it was suggested in the eighteenth and early nineteenth centuries that heat is the manifestation of motion at the molecular level (called the *live force*), the prevailing view of heat until the middle of the nineteenth century was based on the caloric theory proposed by the French chemist Antoine Lavoisier (1744–1794) in 1789. The caloric theory asserts that heat is a fluidlike substance called the **caloric** that is a massless, colorless, odorless, and tasteless substance that can be poured from one body into another (Fig. 2–19). When caloric was added to a body, its temperature increased; and when caloric was removed from a

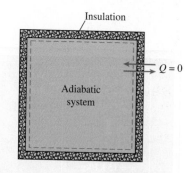

**FIGURE 2–17**
During an adiabatic process, a system exchanges no heat with its surroundings.

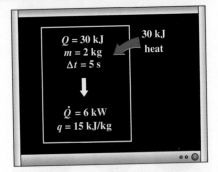

**FIGURE 2–18**

The relationships among $q$, $Q$, and $\dot{Q}$.

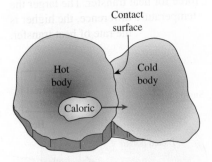

**FIGURE 2–19**

In the early nineteenth century, heat was thought to be an invisible fluid called the *caloric* that flowed from warmer bodies to the cooler ones.

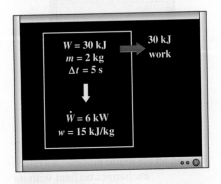

**FIGURE 2–20**

The relationships among $w$, $W$, and $\dot{W}$.

body, its temperature decreased. When a body could not contain any more caloric, much the same way as when a glass of water could not dissolve any more salt or sugar, the body was said to be saturated with caloric. This interpretation gave rise to the terms *saturated liquid* and *saturated vapor* that are still in use today.

The caloric theory came under attack soon after its introduction. It maintained that heat is a substance that could not be created or destroyed. Yet it was known that heat can be generated indefinitely by rubbing one's hands together or rubbing two pieces of wood together. In 1798, the American Benjamin Thompson (Count Rumford) (1754–1814) showed in his papers that heat can be generated continuously through friction. The validity of the caloric theory was also challenged by several others. But it was the careful experiments of the Englishman James P. Joule (1818–1889) published in 1843 that finally convinced the skeptics that heat was not a substance after all, and thus put the caloric theory to rest. Although the caloric theory was totally abandoned in the middle of the nineteenth century, it contributed greatly to the development of thermodynamics and heat transfer.

Heat is transferred by three mechanisms: conduction, convection, and radiation. **Conduction** is the transfer of energy from the more energetic particles of a substance to the adjacent less energetic ones as a result of interaction between particles. **Convection** is the transfer of energy between a solid surface and the adjacent fluid that is in motion, and it involves the combined effects of conduction and fluid motion. **Radiation** is the transfer of energy due to the emission of electromagnetic waves (or photons). An overview of the three mechanisms of heat transfer is given at the end of this chapter as a Topic of Special Interest.

## 2–4 ▪ ENERGY TRANSFER BY WORK

Work, like heat, is an energy interaction between a system and its surroundings. As mentioned earlier, energy can cross the boundary of a closed system in the form of heat or work. Therefore, *if the energy crossing the boundary of a closed system is not heat, it must be work*. Heat is easy to recognize: Its driving force is a temperature difference between the system and its surroundings. Then we can simply say that an energy interaction that is not caused by a temperature difference between a system and its surroundings is work. More specifically, *work is the energy transfer associated with a force acting through a distance*. A rising piston, a rotating shaft, and an electric wire crossing the system boundaries are all associated with work interactions.

Work is also a form of energy transferred like heat and, therefore, has energy units such as kJ. The work done during a process between states 1 and 2 is denoted by $W_{12}$, or simply $W$. The work done *per unit mass* of a system is denoted by $w$ and is expressed as

$$w = \frac{W}{m} \quad \text{(kJ/kg)} \tag{2–17}$$

The work done *per unit time* is called **power** and is denoted $\dot{W}$ (Fig. 2–20). The unit of power is kJ/s, or kW.

Heat and work are *directional quantities*, and thus the complete description of a heat or work interaction requires the specification of both the *magnitude* and *direction*. One way of doing that is to adopt a sign convention. The generally accepted **formal sign convention** for heat and work interactions is as follows: *heat transfer to a system and work done by a system are positive; heat transfer from a system and work done on a system are negative*. Another way is to use the subscripts *in* and *out* to indicate direction (Fig. 2–21). For example, a work input of 5 kJ can be expressed as $W_{in} = 5$ kJ, while a heat loss of 3 kJ can be expressed as $Q_{out} = 3$ kJ. When the direction of a heat or work interaction is not known, we can simply *assume* a direction for the interaction (using the subscript *in* or *out*) and solve for it. A positive result indicates the assumed direction is right. A negative result, on the other hand, indicates that the direction of the interaction is the opposite of the assumed direction. This is just like assuming a direction for an unknown force when solving a statics problem, and reversing the direction when a negative result is obtained for the force. We will use this *intuitive approach* in this book as it eliminates the need to adopt a formal sign convention and the need to carefully assign negative values to some interactions.

Note that a quantity that is transferred to or from a system during an interaction is not a property since the amount of such a quantity depends on more than just the state of the system. Heat and work are *energy transfer mechanisms* between a system and its surroundings, and there are many similarities between them:

1. Both are recognized at the boundaries of a system as they cross the boundaries. That is, both heat and work are *boundary* phenomena.
2. Systems possess energy, but not heat or work.
3. Both are associated with a *process*, not a state. Unlike properties, heat or work has no meaning at a state.
4. Both are *path functions* (i.e., their magnitudes depend on the path followed during a process as well as the end states).

**Path functions** have **inexact differentials** designated by the symbol $\delta$. Therefore, a differential amount of heat or work is represented by $\delta Q$ or $\delta W$, respectively, instead of $dQ$ or $dW$. Properties, however, are **point functions** (i.e., they depend on the state only, and not on how a system reaches that state), and they have **exact differentials** designated by the symbol $d$. A small change in volume, for example, is represented by $dV$, and the total volume change during a process between states 1 and 2 is

$$\int_1^2 dV = V_2 - V_1 = \Delta V$$

That is, the volume change during process 1–2 is always the volume at state 2 minus the volume at state 1, regardless of the path followed (Fig. 2–22). The total work done during process 1–2, however, is

$$\int_1^2 \delta W = W_{12} \quad (not\ \Delta W)$$

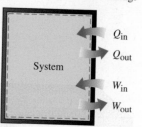

**FIGURE 2–21**
Specifying the directions of heat and work.

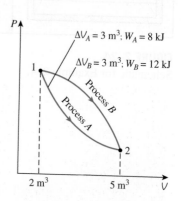

**FIGURE 2–22**
Properties are point functions; but heat and work are path functions (their magnitudes depend on the path followed).

That is, the total work is obtained by following the process path and adding the differential amounts of work ($\delta W$) done along the way. The integral of $\delta W$ *is not* $W_2 - W_1$ (i.e., the work at state 2 minus work at state 1), which is meaningless since work is not a property and systems do not possess work at a state.

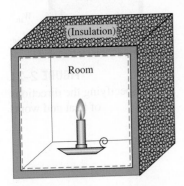

**FIGURE 2–23**
Schematic for Example 2–3.

### EXAMPLE 2–3    Burning of a Candle in an Insulated Room

A candle is burning in a well-insulated room. Taking the room (the air plus the candle) as the system, determine (*a*) if there is any heat transfer during this burning process and (*b*) if there is any change in the internal energy of the system.

**SOLUTION**   A candle burning in a well-insulated room is considered. It is to be determined whether there is any heat transfer and any change in internal energy.
*Analysis*   (*a*) The interior surfaces of the room form the system boundary, as indicated by the dashed lines in Fig. 2–23. As pointed out earlier, heat is recognized as it crosses the boundaries. Since the room is well insulated, we have an adiabatic system and no heat will pass through the boundaries. Therefore, $Q = 0$ for this process.
(*b*) The internal energy involves energies that exist in various forms (sensible, latent, chemical, nuclear). During the process just described, part of the chemical energy is converted to sensible energy. Since there is no increase or decrease in the total internal energy of the system, $\Delta U = 0$ for this process.

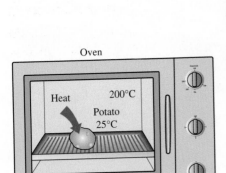

**FIGURE 2–24**
Schematic for Example 2–4.

### EXAMPLE 2–4    Heating of a Potato in an Oven

A potato initially at room temperature (25°C) is being baked in an oven that is maintained at 200°C, as shown in Fig. 2–24. Is there any heat transfer during this baking process?

**SOLUTION**   A potato is being baked in an oven. It is to be determined whether there is any heat transfer during this process.
*Analysis*   This is not a well-defined problem since the system is not specified. Let us assume that we are observing the potato, which will be our system. Then the outer surface of the skin of the potato can be viewed as the system boundary. Part of the energy in the oven will pass through the skin to the potato. Since the driving force for this energy transfer is a temperature difference, this is a heat transfer process.

### EXAMPLE 2–5    Heating of an Oven by Work Transfer

A well-insulated electric oven is being heated through its heating element. If the entire oven, including the heating element, is taken to be the system, determine whether this is a heat or work interaction.

**SOLUTION**   A well-insulated electric oven is being heated by its heating element. It is to be determined whether this is a heat or work interaction.

*Analysis* For this problem, the interior surfaces of the oven form the system boundary, as shown in Fig. 2–25. The energy content of the oven obviously increases during this process, as evidenced by a rise in temperature. This energy transfer to the oven is not caused by a temperature difference between the oven and the surrounding air. Instead, it is caused by *electrons* crossing the system boundary and thus doing work. Therefore, this is a work interaction.

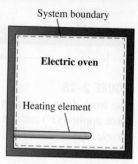

**FIGURE 2–25**
Schematic for Example 2–5.

### ■ EXAMPLE 2–6    Heating of an Oven by Heat Transfer

Answer the question in Example 2–5 if the system is taken as only the air in the oven without the heating element.

**SOLUTION** The question in Example 2–5 is to be reconsidered by taking the system to be only the air in the oven.

*Analysis* This time, the system boundary will include the outer surface of the heating element and will not cut through it, as shown in Fig. 2–26. Therefore, no electrons will be crossing the system boundary at any point. Instead, the energy generated in the interior of the heating element will be transferred to the air around it as a result of the temperature difference between the heating element and the air in the oven. Therefore, this is a heat transfer process.

*Discussion* For both cases, the amount of energy transfer to the air is the same. These two examples show that an energy transfer can be heat or work, depending on how the system is selected.

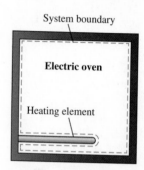

**FIGURE 2–26**
Schematic for Example 2–6.

## Electrical Work

It was pointed out in Example 2–5 that electrons crossing the system boundary do electrical work on the system. In an electric field, electrons in a wire move under the effect of electromotive forces, doing work. When $N$ coulombs of electrical charge move through a potential difference $\mathbf{V}$, the electrical work done is

$$W_e = \mathbf{V}N$$

which can also be expressed in the rate form as

$$\dot{W}_e = \mathbf{V}I \quad (\text{W}) \tag{2–18}$$

where $\dot{W}_e$ is the **electrical power** and $I$ is the number of electrical charges flowing per unit time, that is, the *current* (Fig. 2–27). In general, both $\mathbf{V}$ and $I$ vary with time, and the electrical work done during a time interval $\Delta t$ is expressed as

$$W_e = \int_1^2 \mathbf{V}I \, dt \quad (\text{kJ}) \tag{2–19}$$

When both $\mathbf{V}$ and $I$ remain constant during the time interval $\Delta t$, it reduces to

$$W_e = \mathbf{V}I \, \Delta t \quad (\text{kJ}) \tag{2–20}$$

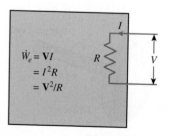

**FIGURE 2–27**
Electrical power in terms of resistance $R$, current $I$, and potential difference $\mathbf{V}$.

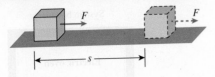

**FIGURE 2–28**
The work done is proportional to the force applied ($F$) and the distance traveled ($s$).

## 2–5 · MECHANICAL FORMS OF WORK

There are several different ways of doing work, each in some way related to a force acting through a distance (Fig. 2–28). In elementary mechanics, the work done by a constant force $F$ on a body displaced a distance $s$ in the direction of the force is given by

$$W = Fs \quad \text{(kJ)} \tag{2–21}$$

If the force $F$ is not constant, the work done is obtained by adding (i.e., integrating) the differential amounts of work,

$$W = \int_1^2 F\,ds \quad \text{(kJ)} \tag{2–22}$$

Obviously, one needs to know how the force varies with displacement to perform this integration. Equations 2–21 and 2–22 give only the magnitude of the work. The sign is easily determined from physical considerations: The work done on a system by an external force acting in the direction of motion is negative, and work done by a system against an external force acting in the opposite direction to motion is positive.

There are two requirements for a work interaction between a system and its surroundings to exist: (1) there must be a *force* acting on the boundary, and (2) the boundary must *move*. Therefore, the presence of forces on the boundary without any displacement of the boundary does not constitute a work interaction. Likewise, the displacement of the boundary without any force to oppose or drive this motion (such as the expansion of a gas into an evacuated space) is not a work interaction since no energy is transferred.

In many thermodynamic problems, mechanical work is the only form of work involved. It is associated with the movement of the boundary of a system or with the movement of the entire system as a whole. Some common forms of mechanical work are discussed next.

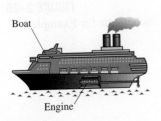

Boat

Engine

**FIGURE 2–29**
Energy transmission through rotating shafts is commonly encountered in practice.

### Shaft Work

Energy transmission with a rotating shaft is very common in engineering practice (Fig. 2–29). Often the torque T applied to the shaft is constant, which means that the force $F$ applied is also constant. For a specified constant torque, the work done during $n$ revolutions is determined as follows: A force $F$ acting through a moment arm $r$ generates a torque T of (Fig. 2–30)

$$T = Fr \quad \rightarrow \quad F = \frac{T}{r} \tag{2–23}$$

This force acts through a distance $s$, which is related to the radius $r$ by

$$s = (2\pi r)n \tag{2–24}$$

Then the shaft work is determined from

$$W_{sh} = Fs = \left(\frac{T}{r}\right)(2\pi rn) = 2\pi n T \quad \text{(kJ)} \tag{2–25}$$

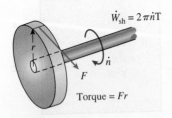

$\dot{W}_{sh} = 2\pi \dot{n} T$

$\dot{n}$

$F$

Torque = $Fr$

**FIGURE 2–30**
Shaft work is proportional to the torque applied and the number of revolutions of the shaft.

The power transmitted through the shaft is the shaft work done per unit time, which can be expressed as

$$\dot{W}_{sh} = 2\pi \dot{n} \text{T} \quad \text{(kW)} \tag{2-26}$$

where $\dot{n}$ is the number of revolutions per unit time.

### EXAMPLE 2–7    Power Transmission by the Shaft of a Car

Determine the power transmitted through the shaft of a car when the torque applied is 200 N·m and the shaft rotates at a rate of 4000 revolutions per minute (rpm).

**SOLUTION** The torque and the rpm for a car engine are given. The power transmitted is to be determined.

**Analysis** A sketch of the car is given in Fig. 2–31. The shaft power is determined directly from

$$\dot{W}_{sh} = 2\pi \dot{n} \text{T} = (2\pi)\left(4000 \, \frac{1}{\text{min}}\right)(200 \, \text{N·m})\left(\frac{1 \, \text{min}}{60 \, \text{s}}\right)\left(\frac{1 \, \text{kJ}}{1000 \, \text{N·m}}\right)$$

$$= \mathbf{83.8 \, kW} \quad \text{(or 112 hp)}$$

**Discussion** Note that power transmitted by a shaft is proportional to torque and the rotational speed.

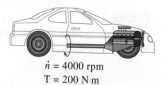

$\dot{n}$ = 4000 rpm
T = 200 N·m

**FIGURE 2–31**
Schematic for Example 2–7.

## Spring Work

It is common knowledge that when a force is applied on a spring, the length of the spring changes (Fig. 2–32). When the length of the spring changes by a differential amount $dx$ under the influence of a force $F$, the work done is

$$\delta W_{spring} = F \, dx \tag{2-27}$$

To determine the total spring work, we need to know a functional relationship between $F$ and $x$. For linear elastic springs, the displacement $x$ is proportional to the force applied (Fig. 2–33). That is,

$$F = kx \quad \text{(kN)} \tag{2-28}$$

where $k$ is the spring constant and has the unit kN/m. The displacement $x$ is measured from the undisturbed position of the spring (that is, $x = 0$ when $F = 0$). Substituting Eq. 2–28 into Eq. 2–27 and integrating yield

$$W_{spring} = \tfrac{1}{2}k(x_2^2 - x_1^2) \quad \text{(kJ)} \tag{2-29}$$

where $x_1$ and $x_2$ are the initial and the final displacements of the spring, respectively, measured from the undisturbed position of the spring.

There are many other forms of mechanical work. Next we introduce some of them briefly.

## Work Done on Elastic Solid Bars

Solids are often modeled as linear springs because under the action of a force they contract or elongate, as shown in Fig. 2–34, and when the force

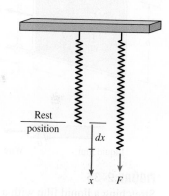

Rest position

$dx$

$x$     $F$

**FIGURE 2–32**
Elongation of a spring under the influence of a force.

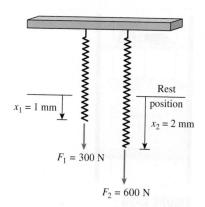

$x_1 = 1$ mm

Rest position

$x_2 = 2$ mm

$F_1 = 300$ N

$F_2 = 600$ N

**FIGURE 2–33**
The displacement of a linear spring doubles when the force is doubled.

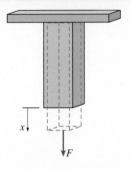

**FIGURE 2–34**
Solid bars behave as springs under the influence of a force.

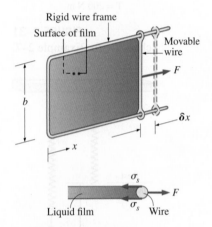

**FIGURE 2–35**
Stretching a liquid film with a U-shaped wire, and the forces acting on the movable wire of length $b$.

**FIGURE 2–36**
The energy transferred to a body while being raised is equal to the change in its potential energy.

is lifted, they return to their original lengths, like a spring. This is true as long as the force is in the elastic range, that is, not large enough to cause permanent (plastic) deformations. Therefore, the equations given for a linear spring can also be used for elastic solid bars. Alternately, we can determine the work associated with the expansion or contraction of an elastic solid bar by replacing pressure $P$ by its counterpart in solids, *normal stress* $\sigma_n = F/A$, in the work expression:

$$W_{elastic} = \int_1^2 F \, dx = \int_1^2 \sigma_n A \, dx \quad (kJ) \qquad (2\text{–}30)$$

where $A$ is the cross-sectional area of the bar. Note that the normal stress has pressure units.

## Work Associated with the Stretching of a Liquid Film

Consider a liquid film such as soap film suspended on a wire frame (Fig. 2–35). We know from experience that it will take some force to stretch this film by the movable portion of the wire frame. This force is used to overcome the microscopic forces between molecules at the liquid–air interfaces. These microscopic forces are perpendicular to any line in the surface, and the force generated by these forces per unit length is called the **surface tension** $\sigma_s$, whose unit is N/m. Therefore, the work associated with the stretching of a film is also called *surface tension work*. It is determined from

$$W_{surface} = \int_1^2 \sigma_s \, dA \quad (kJ) \qquad (2\text{–}31)$$

where $dA = 2b \, dx$ is the change in the surface area of the film. The factor 2 is due to the fact that the film has two surfaces in contact with air. The force acting on the movable wire as a result of surface tension effects is $F = 2b\sigma_s$ where $\sigma_s$ is the surface tension force per unit length.

## Work Done to Raise or to Accelerate a Body

When a body is raised in a gravitational field, its potential energy increases. Likewise, when a body is accelerated, its kinetic energy increases. The conservation of energy principle requires that an equivalent amount of energy must be transferred to the body being raised or accelerated. Remember that energy can be transferred to a given mass by heat and work, and the energy transferred in this case obviously is not heat since it is not driven by a temperature difference. Therefore, it must be work. Then we conclude that (1) the work transfer needed to raise a body is equal to the change in the potential energy of the body, and (2) the work transfer needed to accelerate a body is equal to the change in the kinetic energy of the body (Fig. 2–36). Similarly, the potential or kinetic energy of a body represents the work that can be obtained from the body as it is lowered to the reference level or decelerated to zero velocity.

This discussion together with the consideration for friction and other losses form the basis for determining the required power rating of motors used to drive devices such as elevators, escalators, conveyor belts, and ski lifts. It also plays a primary role in the design of automotive and aircraft engines, and in the determination of the amount of hydroelectric power that can be produced from a given water reservoir, which is simply the potential energy of the water relative to the location of the hydraulic turbine.

■
■ **EXAMPLE 2–8**    **Power Needs of a Car to Climb a Hill**
■

■ Consider a 1200-kg car cruising steadily on a level road at 90 km/h. Now the
■ car starts climbing a hill that is sloped 30° from the horizontal (Fig. 2–37). If
the velocity of the car is to remain constant during climbing, determine the additional power that must be delivered by the engine.

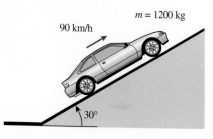

**SOLUTION**    A car is to climb a hill while maintaining a constant velocity. The additional power needed is to be determined.

**Analysis**    The additional power required is simply the work that needs to be done per unit time to raise the elevation of the car, which is equal to the change in the potential energy of the car per unit time:

$$\dot{W}_g = mg\,\Delta z/\Delta t = mgV_{\text{vertical}}$$

$$= (1200 \text{ kg})(9.81 \text{ m/s}^2)(90 \text{ km/h})(\sin 30°)\left(\frac{1 \text{ m/s}}{3.6 \text{ km/h}}\right)\left(\frac{1 \text{ kJ/kg}}{1000 \text{ m}^2/\text{s}^2}\right)$$

$$= 147 \text{ kJ/s} = \mathbf{147 \text{ kW}}    \text{ (or 197 hp)}$$

**Discussion**    Note that the car engine will have to produce almost 200 hp of additional power while climbing the hill if the car is to maintain its velocity.

**FIGURE 2–37**
Schematic for Example 2–8.

■
■ **EXAMPLE 2–9**    **Power Needs of a Car to Accelerate**
■

■ Determine the power required to accelerate a 900-kg car shown in Fig. 2–38
■ from rest to a velocity of 80 km/h in 20 s on a level road.

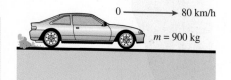

**SOLUTION**    The power required to accelerate a car to a specified velocity is to be determined.

**Analysis**    The work needed to accelerate a body is simply the change in the kinetic energy of the body,

$$W_a = \tfrac{1}{2}m(V_2^2 - V_1^2) = \tfrac{1}{2}(900 \text{ kg})\left[\left(\frac{80,000 \text{ m}}{3600 \text{ s}}\right)^2 - 0^2\right]\left(\frac{1 \text{ kJ/kg}}{1000 \text{ m}^2/\text{s}^2}\right)$$

$$= 222 \text{ kJ}$$

The average power is determined from

$$\dot{W}_a = \frac{W_a}{\Delta t} = \frac{222 \text{ kJ}}{20 \text{ s}} = \mathbf{11.1 \text{ kW}}    \text{ (or 14.9 hp)}$$

**FIGURE 2–38**
Schematic for Example 2–9.

**Discussion**    This is in addition to the power required to overcome friction, rolling resistance, and other imperfections.

## Nonmechanical Forms of Work

The treatment in Section 2–5 represents a fairly comprehensive coverage of mechanical forms of work except the *moving boundary work* that is covered in Chap. 4. Some work modes encountered in practice are not mechanical in nature. However, these nonmechanical work modes can be treated in a similar manner by identifying a *generalized force F* acting in the direction of a *generalized displacement x.* Then the work associated with the differential displacement under the influence of this force is determined from $\delta W = F\,dx$.

Some examples of nonmechanical work modes are **electrical work,** where the generalized force is the *voltage* (the electrical potential) and the generalized displacement is the *electrical charge,* as discussed earlier; **magnetic work,** where the generalized force is the *magnetic field strength* and the generalized displacement is the total *magnetic dipole moment;* and **electrical polarization work,** where the generalized force is the *electric field strength* and the generalized displacement is the *polarization of the medium* (the sum of the electric dipole rotation moments of the molecules). Detailed consideration of these and other nonmechanical work modes can be found in specialized books on these topics.

## 2–6 · THE FIRST LAW OF THERMODYNAMICS

So far, we have considered various forms of energy such as heat $Q$, work $W$, and total energy $E$ individually, and no attempt is made to relate them to each other during a process. The *first law of thermodynamics,* also known as *the conservation of energy principle,* provides a sound basis for studying the relationships among the various forms of energy and energy interactions. Based on experimental observations, the first law of thermodynamics states that *energy can be neither created nor destroyed during a process; it can only change forms.* Therefore, every bit of energy should be accounted for during a process.

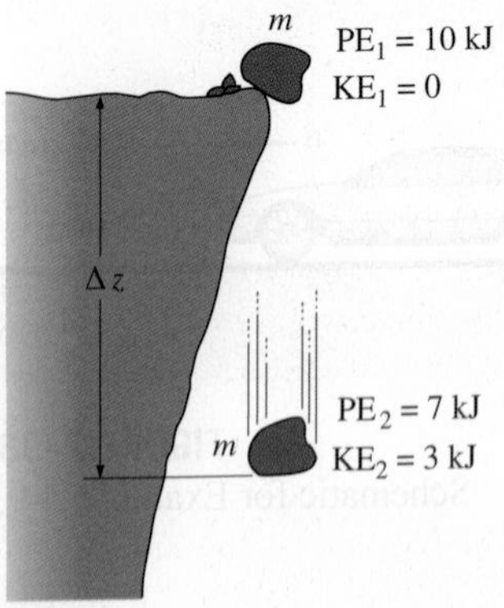

We all know that a rock at some elevation possesses some potential energy, and part of this potential energy is converted to kinetic energy as the rock falls (Fig. 2–39). Experimental data show that the decrease in potential energy $(mg\,\Delta z)$ exactly equals the increase in kinetic energy $[m(V_2^2 - V_1^2)/2]$ when the air resistance is negligible, thus confirming the conservation of energy principle for mechanical energy.

Consider a system undergoing a series of *adiabatic* processes from a specified state 1 to another specified state 2. Being adiabatic, these processes obviously cannot involve any heat transfer, but they may involve several kinds of work interactions. Careful measurements during these experiments indicate the following: *For all adiabatic processes between two specified states of a closed system, the net work done is the same regardless of the nature of the closed system and the details of the process.* Considering that there are an infinite number of ways to perform work interactions under adiabatic conditions, this statement appears to be very powerful, with a potential for far-reaching implications. This statement, which is largely based on the experiments of Joule in the first half of the nineteenth century, cannot be drawn from any other known physical principle and is recognized as a fundamental principle. This principle is called the **first law of thermodynamics** or just the **first law.**

$PE_1 = 10$ kJ
$KE_1 = 0$

$\Delta z$

$PE_2 = 7$ kJ
$KE_2 = 3$ kJ

**FIGURE 2–39**
Energy cannot be created or destroyed; it can only change forms.

A major consequence of the first law is the existence and the definition of the property *total energy E*. Considering that the net work is the same for all adiabatic processes of a closed system between two specified states, the value of the net work must depend on the end states of the system only, and thus it must correspond to a change in a property of the system. This property is the *total energy*. Note that the first law makes no reference to the value of the total energy of a closed system at a state. It simply states that the *change* in the total energy during an adiabatic process must be equal to the net work done. Therefore, any convenient arbitrary value can be assigned to total energy at a specified state to serve as a reference point.

Implicit in the first law statement is the conservation of energy. Although the essence of the first law is the existence of the property *total energy*, the first law is often viewed as a statement of the *conservation of energy* principle. Next, we develop the first law or the conservation of energy relation with the help of some familiar examples using intuitive arguments.

First, we consider some processes that involve heat transfer but no work interactions. The potato baked in the oven is a good example for this case (Fig. 2–40). As a result of heat transfer to the potato, the energy of the potato will increase. If we disregard any mass transfer (moisture loss from the potato), the increase in the total energy of the potato becomes equal to the amount of heat transfer. That is, if 5 kJ of heat is transferred to the potato, the energy increase of the potato will also be 5 kJ.

As another example, consider the heating of water in a pan on top of a range (Fig. 2–41). If 15 kJ of heat is transferred to the water from the heating element and 3 kJ of it is lost from the water to the surrounding air, the increase in energy of the water will be equal to the net heat transfer to water, which is 12 kJ.

Now consider a well-insulated (i.e., adiabatic) room heated by an electric heater as our system (Fig. 2–42). As a result of electrical work done, the energy of the system will increase. Since the system is adiabatic and cannot have any heat transfer to or from the surroundings ($Q = 0$), the conservation of energy principle dictates that the electrical work done on the system must equal the increase in energy of the system.

Next, let us replace the electric heater with a paddle wheel (Fig. 2–43). As a result of the stirring process, the energy of the system will increase. Again, since there is no heat interaction between the system and its surroundings ($Q = 0$), the shaft work done on the system must show up as an increase in the energy of the system.

Many of you have probably noticed that the temperature of air rises when it is compressed (Fig. 2–44). This is because energy is transferred to the air in the form of boundary work. In the absence of any heat transfer ($Q = 0$), the entire boundary work will be stored in the air as part of its total energy. The conservation of energy principle again requires that the increase in the energy of the system be equal to the boundary work done on the system.

We can extend these discussions to systems that involve various heat and work interactions simultaneously. For example, if a system gains 12 kJ of heat during a process while 6 kJ of work is done on it, the increase in the energy of the system during that process is 18 kJ (Fig. 2–45). That is, the change in the energy of a system during a process is simply equal to the net energy transfer to (or from) the system.

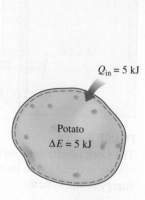

**FIGURE 2–40**
The increase in the energy of a potato in an oven is equal to the amount of heat transferred to it.

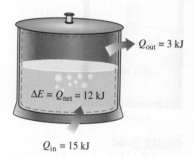

**FIGURE 2–41**
In the absence of any work interactions, the energy change of a system is equal to the net heat transfer.

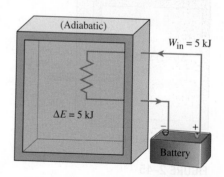

**FIGURE 2–42**
The work (electrical) done on an adiabatic system is equal to the increase in the energy of the system.

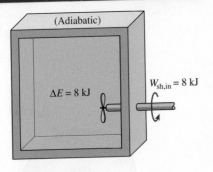

**FIGURE 2–43**
The work (shaft) done on an adiabatic system is equal to the increase in the energy of the system.

**FIGURE 2–44**
The work (boundary) done on an adiabatic system is equal to the increase in the energy of the system.

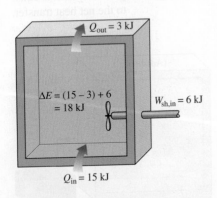

**FIGURE 2–45**
The energy change of a system during a process is equal to the *net* work and heat transfer between the system and its surroundings.

## Energy Balance

In the light of the preceding discussions, the conservation of energy principle can be expressed as follows: *The net change (increase or decrease) in the total energy of the system during a process is equal to the difference between the total energy entering and the total energy leaving the system during that process.* That is,

$$\begin{pmatrix} \text{Total energy} \\ \text{entering the system} \end{pmatrix} - \begin{pmatrix} \text{Total energy} \\ \text{leaving the system} \end{pmatrix} = \begin{pmatrix} \text{Change in the total} \\ \text{energy of the system} \end{pmatrix}$$

or

$$E_{\text{in}} - E_{\text{out}} = \Delta E_{\text{system}}$$

This relation is often referred to as the **energy balance** and is applicable to any kind of system undergoing any kind of process. The successful use of this relation to solve engineering problems depends on understanding the various forms of energy and recognizing the forms of energy transfer.

## Energy Change of a System, $\Delta E_{\text{system}}$

The determination of the energy change of a system during a process involves the evaluation of the energy of the system at the beginning and at the end of the process, and taking their difference. That is,

$$\text{Energy change} = \text{Energy at final state} - \text{Energy at initial state}$$

or

$$\Delta E_{\text{system}} = E_{\text{final}} - E_{\text{initial}} = E_2 - E_1 \tag{2–32}$$

Note that energy is a property, and the value of a property does not change unless the state of the system changes. Therefore, the energy change of a system is zero if the state of the system does not change during the process. Also, energy can exist in numerous forms such as internal (sensible, latent, chemical, and nuclear), kinetic, potential, electric, and magnetic, and their sum constitutes the *total energy E* of a system. In the absence of electric, magnetic, and surface tension effects (i.e., for simple compressible systems), the change in the total energy of a system during a process is the sum of the changes in its internal, kinetic, and potential energies and can be expressed as

$$\Delta E = \Delta U + \Delta KE + \Delta PE \tag{2–33}$$

where

$$\Delta U = m(u_2 - u_1)$$

$$\Delta KE = \tfrac{1}{2}m(V_2^2 - V_1^2)$$

$$\Delta PE = mg(z_2 - z_1)$$

When the initial and final states are specified, the values of the specific internal energies $u_1$ and $u_2$ can be determined directly from the property tables or thermodynamic property relations.

Most systems encountered in practice are stationary, that is, they do not involve any changes in their velocity or elevation during a process (Fig. 2–46).

Thus, for **stationary systems**, the changes in kinetic and potential energies are zero (that is, $\Delta KE = \Delta PE = 0$), and the total energy change relation in Eq. 2–33 reduces to $\Delta E = \Delta U$ for such systems. Also, the energy of a system during a process will change even if only one form of its energy changes while the other forms of energy remain unchanged.

# Mechanisms of Energy Transfer, $E_{in}$ and $E_{out}$

Energy can be transferred to or from a system in three forms: *heat, work,* and *mass flow.* Energy interactions are recognized at the system boundary as they cross it, and they represent the energy gained or lost by a system during a process. The only two forms of energy interactions associated with a fixed mass or closed system are *heat transfer* and *work.*

1. **Heat Transfer**, $Q$ Heat transfer to a system (heat gain) increases the energy of the molecules and thus the internal energy of the system, and heat transfer from a system (heat loss) decreases it since the energy transferred out as heat comes from the energy of the molecules of the system.

2. **Work Transfer**, $W$ An energy interaction that is not caused by a temperature difference between a system and its surroundings is work. A rising piston, a rotating shaft, and an electrical wire crossing the system boundaries are all associated with work interactions. Work transfer to a system (i.e., work done on a system) increases the energy of the system, and work transfer from a system (i.e., work done by the system) decreases it since the energy transferred out as work comes from the energy contained in the system. Car engines and hydraulic, steam, or gas turbines produce work while compressors, pumps, and mixers consume work.

3. **Mass Flow**, $m$ Mass flow in and out of the system serves as an additional mechanism of energy transfer. When mass enters a system, the energy of the system increases because mass carries energy with it (in fact, mass is energy). Likewise, when some mass leaves the system, the energy contained within the system decreases because the leaving mass takes out some energy with it. For example, when some hot water is taken out of a water heater and is replaced by the same amount of cold water, the energy content of the hot-water tank (the control volume) decreases as a result of this mass interaction (Fig. 2–47).

Noting that energy can be transferred in the forms of heat, work, and mass, and that the net transfer of a quantity is equal to the difference between the amounts transferred in and out, the energy balance can be written more explicitly as

$$E_{in} - E_{out} = (Q_{in} - Q_{out}) + (W_{in} - W_{out}) + (E_{mass,in} - E_{mass,out}) = \Delta E_{system} \quad \text{(2–34)}$$

where the subscripts "in" and "out" denote quantities that enter and leave the system, respectively. All six quantities on the right side of the equation represent "amounts," and thus they are *positive* quantities. The direction of any energy transfer is described by the subscripts "in" and "out."

The heat transfer $Q$ is zero for adiabatic systems, the work transfer $W$ is zero for systems that involve no work interactions, and the energy transport

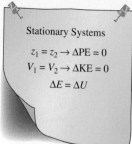

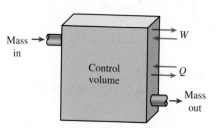

**FIGURE 2–46**
For stationary systems, $\Delta KE = \Delta PE = 0$; thus $\Delta E = \Delta U$.

Stationary Systems
$$z_1 = z_2 \rightarrow \Delta PE = 0$$
$$V_1 = V_2 \rightarrow \Delta KE = 0$$
$$\Delta E = \Delta U$$

**FIGURE 2–47**
The energy content of a control volume can be changed by mass flow as well as heat and work interactions.

with mass $E_{mass}$ is zero for systems that involve no mass flow across their boundaries (i.e., closed systems).

Energy balance for any system undergoing any kind of process can be expressed more compactly as

$$\underbrace{E_{in} - E_{out}}_{\substack{\text{Net energy transfer} \\ \text{by heat, work, and mass}}} = \underbrace{\Delta E_{system}}_{\substack{\text{Change in internal, kinetic,} \\ \text{potential, etc., energies}}} \quad \text{(kJ)} \qquad \text{(2–35)}$$

or, in the **rate form**, as

$$\underbrace{\dot{E}_{in} - \dot{E}_{out}}_{\substack{\text{Rate of net energy transfer} \\ \text{by heat, work, and mass}}} = \underbrace{dE_{system}/dt}_{\substack{\text{Rate of change in internal,} \\ \text{kinetic, potential, etc., energies}}} \quad \text{(kW)} \qquad \text{(2–36)}$$

For constant rates, the total quantities during a time interval $\Delta t$ are related to the quantities per unit time as

$$Q = \dot{Q} \, \Delta t, \quad W = \dot{W} \, \Delta t, \quad \text{and} \quad \Delta E = (dE/dt) \, \Delta t \quad \text{(kJ)} \qquad \text{(2–37)}$$

The energy balance can be expressed on a **per unit mass** basis as

$$e_{in} - e_{out} = \Delta e_{system} \quad \text{(kJ/kg)} \qquad \text{(2–38)}$$

which is obtained by dividing all the quantities in Eq. 2–35 by the mass $m$ of the system. Energy balance can also be expressed in the differential form as

$$\delta E_{in} - \delta E_{out} = dE_{system} \quad \text{or} \quad \delta e_{in} - \delta e_{out} = de_{system} \qquad \text{(2–39)}$$

For a closed system undergoing a **cycle**, the initial and final states are identical, and thus $\Delta E_{system} = E_2 - E_1 = 0$. Then the energy balance for a cycle simplifies to $E_{in} - E_{out} = 0$ or $E_{in} = E_{out}$. Noting that a closed system does not involve any mass flow across its boundaries, the energy balance for a cycle can be expressed in terms of heat and work interactions as

$$W_{net,out} = Q_{net,in} \quad \text{or} \quad \dot{W}_{net,out} = \dot{Q}_{net,in} \quad \text{(for a cycle)} \qquad \text{(2–40)}$$

That is, the net work output during a cycle is equal to net heat input (Fig. 2–48).

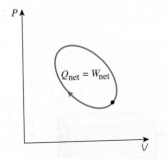

**FIGURE 2–48**
For a cycle $\Delta E = 0$, thus $Q = W$.

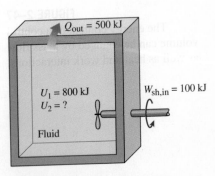

**FIGURE 2–49**
Schematic for Example 2–10.

---

**EXAMPLE 2–10**    **Cooling of a Hot Fluid in a Tank**

A rigid tank contains a hot fluid that is cooled while being stirred by a paddle wheel. Initially, the internal energy of the fluid is 800 kJ. During the cooling process, the fluid loses 500 kJ of heat, and the paddle wheel does 100 kJ of work on the fluid. Determine the final internal energy of the fluid. Neglect the energy stored in the paddle wheel.

**SOLUTION**   A fluid in a rigid tank looses heat while being stirred. The final internal energy of the fluid is to be determined.

**Assumptions**   **1** The tank is stationary and thus the kinetic and potential energy changes are zero, $\Delta KE = \Delta PE = 0$. Therefore, $\Delta E = \Delta U$ and internal energy is the only form of the system's energy that may change during this process. **2** Energy stored in the paddle wheel is negligible.

**Analysis**   Take the contents of the tank as the system (Fig. 2–49). This is a *closed system* since no mass crosses the boundary during the process.

We observe that the volume of a rigid tank is constant, and thus there is no moving boundary work. Also, heat is lost from the system and shaft work is done on the system. Applying the energy balance on the system gives

$$\underbrace{E_{in} - E_{out}}_{\substack{\text{Net energy transfer} \\ \text{by heat, work, and mass}}} = \underbrace{\Delta E_{system}}_{\substack{\text{Change in internal, kinetic,} \\ \text{potential, etc., energies}}}$$

$$W_{sh,in} - Q_{out} = \Delta U = U_2 - U_1$$

$$100 \text{ kJ} - 500 \text{ kJ} = U_2 - 800 \text{ kJ}$$

$$U_2 = \textbf{400 kJ}$$

Therefore, the final internal energy of the system is 400 kJ.

---

■ **EXAMPLE 2–11**    **Acceleration of Air by a Fan**

■ A fan that consumes 20 W of electric power when operating is claimed to ■ discharge air from a ventilated room at a rate of 1.0 kg/s at a discharge ■ velocity of 8 m/s (Fig. 2–50). Determine if this claim is reasonable.

**SOLUTION**   A fan is claimed to increase the velocity of air to a specified value while consuming electric power at a specified rate. The validity of this claim is to be investigated.

***Assumptions***   The ventilating room is relatively calm, and air velocity in it is negligible.

***Analysis***   First, let's examine the energy conversions involved: The motor of the fan converts part of the electrical power it consumes to mechanical (shaft) power, which is used to rotate the fan blades in air. The blades are shaped such that they impart a large fraction of the mechanical power of the shaft to air by mobilizing it. In the limiting ideal case of no losses (no conversion of electrical and mechanical energy to thermal energy) in steady operation, the electric power input will be equal to the rate of increase of the kinetic energy of air. Therefore, for a control volume that encloses the fan-motor unit, the energy balance can be written as

$$\underbrace{\dot{E}_{in} - \dot{E}_{out}}_{\substack{\text{Rate of net energy transfer} \\ \text{by heat, work, and mass}}} = \underbrace{dE_{system}/dt}_{\substack{\text{Rate of change in internal, kinetic,} \\ \text{potential, etc., energies}}}^{\nearrow \, 0 \, (\text{steady})} = 0 \rightarrow \dot{E}_{in} = \dot{E}_{out}$$

$$\dot{W}_{elect,\,in} = \dot{m}_{air}\, ke_{out} = \dot{m}_{air}\, \frac{V_{out}^2}{2}$$

Solving for $V_{out}$ and substituting gives the maximum air outlet velocity to be

$$V_{out} = \sqrt{\frac{2\dot{W}_{elect,in}}{\dot{m}_{air}}} = \sqrt{\frac{2(20 \text{ J/s})}{1.0 \text{ kg/s}} \left(\frac{1 \text{ m}^2/\text{s}^2}{1 \text{ J/kg}}\right)} = 6.3 \text{ m/s}$$

which is less than 8 m/s. Therefore, the claim is **false**.

***Discussion***   The conservation of energy principle requires the energy to be preserved as it is converted from one form to another, and it does not allow any energy to be created or destroyed during a process. From the first law point of view, there is nothing wrong with the conversion of the entire electrical energy

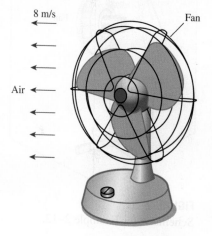

8 m/s

Fan

Air

**FIGURE 2–50**
Schematic for Example 2–11.

into kinetic energy. Therefore, the first law has no objection to air velocity reaching 6.3 m/s—but this is the upper limit. Any claim of higher velocity is in violation of the first law, and thus impossible. In reality, the air velocity will be considerably lower than 6.3 m/s because of the losses associated with the conversion of electrical energy to mechanical shaft energy, and the conversion of mechanical shaft energy to kinetic energy or air.

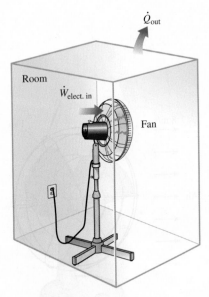

**FIGURE 2–51**
Schematic for Example 2–12.

### EXAMPLE 2–12    Heating Effect of a Fan

A room is initially at the outdoor temperature of 25°C. Now a large fan that consumes 200 W of electricity when running is turned on (Fig. 2–51). The heat transfer rate between the room and the outdoor air is given as $\dot{Q} = UA(T_i - T_o)$ where $U = 6$ W/m²·°C is the overall heat transfer coefficient, $A = 30$ m² is the exposed surface area of the room, and $T_i$ and $T_o$ are the indoor and outdoor air temperatures, respectively. Determine the indoor air temperature when steady operating conditions are established.

**SOLUTION**  A large fan is turned on and kept on in a room that looses heat to the outdoors. The indoor air temperature is to be determined when steady operation is reached.

**Assumptions**  **1** Heat transfer through the floor is negligible. **2** There are no other energy interactions involved.

**Analysis**  The electricity consumed by the fan is energy input for the room, and thus the room gains energy at a rate of 200 W. As a result, the room air temperature tends to rise. But as the room air temperature rises, the rate of heat loss from the room increases until the rate of heat loss equals the electric power consumption. At that point, the temperature of the room air, and thus the energy content of the room, remains constant, and the conservation of energy for the room becomes

$$\underbrace{\dot{E}_{in} - \dot{E}_{out}}_{\substack{\text{Rate of net energy transfer} \\ \text{by heat, work, and mass}}} = \underbrace{dE_{system}/dt}_{\substack{\text{Rate of change in internal, kinetic,} \\ \text{potential, etc., energies}}}^{\nearrow 0(\text{steady})} = 0 \rightarrow \dot{E}_{in} = \dot{E}_{out}$$

$$\dot{W}_{\text{elect,in}} = \dot{Q}_{out} = UA(T_i - T_o)$$

Substituting,

$$200 \text{ W} = (6 \text{ W/m}^2\cdot°\text{C})(30 \text{ m}^2)(T_i - 25°\text{C})$$

It gives

$$T_i = \textbf{26.1°C}$$

Therefore, the room air temperature will remain constant after it reaches 26.1°C.

**Discussion**  Note that a 200-W fan heats a room just like a 200-W resistance heater. In the case of a fan, the motor converts part of the electric energy it draws to mechanical energy in the form of a rotating shaft while the remaining part is dissipated as heat to the room air because of the motor inefficiency (no motor converts 100 percent of the electric energy it receives to mechanical energy, although some large motors come close with a conversion efficiency of over 97 percent). Part of the mechanical energy of the shaft is converted to kinetic energy of air through the blades, which is then converted to thermal energy as air molecules slow down because of friction. At the end, the entire electric energy drawn by the fan motor is converted to thermal energy of air, which manifests itself as a rise in temperature.

■ **EXAMPLE 2–13    Annual Lighting Cost of a Classroom**

The lighting needs of a classroom are met by 30 fluorescent lamps, each consuming 80 W of electricity (Fig. 2–52). The lights in the classroom are kept on for 12 hours a day and 250 days a year. For a unit electricity cost of 11 cents per kWh, determine annual energy cost of lighting for this classroom. Also, discuss the effect of lighting on the heating and air-conditioning requirements of the room.

**SOLUTION**    The lighting of a classroom by fluorescent lamps is considered. The annual electricity cost of lighting for this classroom is to be determined, and the lighting's effect on the heating and air-conditioning requirements is to be discussed.

*Assumptions*    The effect of voltage fluctuations is negligible so that each fluorescent lamp consumes its rated power.

*Analysis*    The electric power consumed by the lamps when all are on and the number of hours they are kept on per year are

$$\text{Lighting power} = (\text{Power consumed per lamp}) \times (\text{No. of lamps})$$
$$= (80 \text{ W/lamp})(30 \text{ lamps})$$
$$= 2400 \text{ W} = 2.4 \text{ kW}$$

$$\text{Operating hours} = (12 \text{ h/day})(250 \text{ days/year}) = 3000 \text{ h/year}$$

Then the amount and cost of electricity used per year become

$$\text{Lighting energy} = (\text{Lighting power})(\text{Operating hours})$$
$$= (2.4 \text{ kW})(3000 \text{ h/year}) = 7200 \text{ kWh/year}$$

$$\text{Lighting cost} = (\text{Lighting energy})(\text{Unit cost})$$
$$= (7200 \text{ kWh/year})(\$0.11/\text{kWh}) = \mathbf{\$792/year}$$

Light is absorbed by the surfaces it strikes and is converted to thermal energy. Disregarding the light that escapes through the windows, the entire 2.4 kW of electric power consumed by the lamps eventually becomes part of thermal energy of the classroom. Therefore, the lighting system in this room reduces the heating requirements by 2.4 kW, but increases the air-conditioning load by 2.4 kW.

*Discussion*    Note that the annual lighting cost of this classroom alone is close to $800. This shows the importance of energy conservation measures. If incandescent light bulbs were used instead of fluorescent tubes, the lighting costs would be four times as much since incandescent lamps use four times as much power for the same amount of light produced.

**FIGURE 2–52**
Fluorescent lamps lighting a classroom as discussed in Example 2–13.

Water heater

| Type | Efficiency |
| --- | --- |
| Gas, conventional | 55% |
| Gas, high-efficiency | 62% |
| Electric, conventional | 90% |
| Electric, high-efficiency | 94% |

**FIGURE 2–53**

Typical efficiencies of conventional and high-efficiency electric and natural gas water heaters.

©*The McGraw-Hill Companies, Inc./Christopher Kerrigan, photographer*

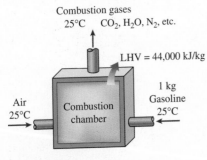

Combustion gases
25°C    $CO_2$, $H_2O$, $N_2$, etc.

LHV = 44,000 kJ/kg

Air
25°C

Combustion chamber

1 kg
Gasoline
25°C

**FIGURE 2–54**

The definition of the heating value of gasoline.

# 2–7 · ENERGY CONVERSION EFFICIENCIES

*Efficiency* is one of the most frequently used terms in thermodynamics, and it indicates how well an energy conversion or transfer process is accomplished. Efficiency is also one of the most frequently misused terms in thermodynamics and a source of misunderstandings. This is because efficiency is often used without being properly defined first. Next, we will clarify this further and define some efficiencies commonly used in practice.

Efficiency, in general, can be expressed in terms of the desired output and the required input as

$$\text{Efficiency} = \frac{\text{Desired output}}{\text{Required input}} \qquad (2\text{–}41)$$

If you are shopping for a water heater, a knowledgeable salesperson will tell you that the efficiency of a conventional electric water heater is about 90 percent (Fig. 2–53). You may find this confusing, since the heating elements of electric water heaters are resistance heaters, and the efficiency of all resistance heaters is 100 percent as they convert all the electrical energy they consume into thermal energy. A knowledgeable salesperson will clarify this by explaining that the heat losses from the hot-water tank to the surrounding air amount to 10 percent of the electrical energy consumed, and the **efficiency of a water heater** is defined as the ratio of the *energy delivered to the house by hot water* to the *energy supplied to the water heater*. A clever salesperson may even talk you into buying a more expensive water heater with thicker insulation that has an efficiency of 94 percent. If you are a knowledgeable consumer and have access to natural gas, you will probably purchase a gas water heater whose efficiency is only 55 percent since a gas unit costs about the same as an electric unit to purchase and install, but the annual energy cost of a gas unit will be much less than that of an electric unit.

Perhaps you are wondering how the efficiency for a gas water heater is defined, and why it is much lower than the efficiency of an electric heater. As a general rule, the efficiency of equipment that involves the combustion of a fuel is based on the **heating value of the fuel**, which is *the amount of heat released when a unit amount of fuel at room temperature is completely burned and the combustion products are cooled to the room temperature* (Fig. 2–54). Then the performance of combustion equipment can be characterized by **combustion efficiency**, defined as

$$\eta_{\text{combustion}} = \frac{Q}{\text{HV}} = \frac{\text{Amount of heat released during combustion}}{\text{Heating value of the fuel burned}} \qquad (2\text{–}42)$$

A combustion efficiency of 100 percent indicates that the fuel is burned completely and the stack gases leave the combustion chamber at room temperature, and thus the amount of heat released during a combustion process is equal to the heating value of the fuel.

Most fuels contain hydrogen, which forms water when burned, and the heating value of a fuel will be different, depending on whether the water in combustion products is in the liquid or vapor form. The heating value is called the *lower heating value,* or LHV, when the water leaves as a vapor, and the *higher heating value,* or HHV, when the water in the combustion gases is completely condensed and thus the heat of vaporization is also recovered. The difference between these two heating values is equal to the product of the amount of water and the enthalpy of vaporization of water at room temperature. For example, the lower and higher heating values of gasoline are 44,000 kJ/kg and 47,300 kJ/kg, respectively. An efficiency definition should make it clear whether it is based on the higher or lower heating value of the fuel. Efficiencies of cars and jet engines are normally based on *lower heating values* since water normally leaves as a vapor in the exhaust gases, and it is not practical to try to recuperate the heat of vaporization. Efficiencies of furnaces, on the other hand, are based on *higher heating values.*

The efficiency of space heating systems of residential and commercial buildings is usually expressed in terms of the **annual fuel utilization efficiency**, or AFUE, which accounts for the combustion efficiency as well as other losses such as heat losses to unheated areas and start-up and cool-down losses. The AFUE of most new heating systems is about 85 percent, although the AFUE of some old heating systems is under 60 percent. The AFUE of some new high-efficiency furnaces exceeds 96 percent, but the high cost of such furnaces cannot be justified for locations with mild to moderate winters. Such high efficiencies are achieved by reclaiming most of the heat in the flue gases, condensing the water vapor, and discharging the flue gases at temperatures as low as 38°C instead of about 200°C for the conventional models.

For *car engines,* the work output is understood to be the power delivered by the crankshaft. But for power plants, the work output can be the mechanical power at the turbine exit, or the electrical power output of the generator. A generator is a device that converts mechanical energy to electrical energy, and the effectiveness of a generator is characterized by the **generator efficiency**, which is the ratio of the *electrical power output* to the *mechanical power input.* The *thermal efficiency* of a power plant, which is of primary interest in thermodynamics, is usually defined as the ratio of the net shaft work output of the turbine to the heat input to the working fluid. The effects of other factors are incorporated by defining an **overall efficiency** for the power plant as the ratio of the *net electrical power output* to the *rate of fuel energy input.* That is,

$$\eta_{overall} = \eta_{combustion}\, \eta_{thermal}\, \eta_{generator} = \frac{\dot{W}_{net,electric}}{HHV \times \dot{m}_{fuel}} \tag{2–43}$$

The overall efficiencies are about 26–30 percent for gasoline automotive engines, 34–40 percent for diesel engines, and up to 60 percent for large power plants.

We are all familiar with the conversion of electrical energy to *light* by incandescent lightbulbs, fluorescent tubes, and high-intensity discharge lamps. The efficiency for the conversion of electricity to light can be defined as the ratio of the energy converted to light to the electrical energy consumed. For example, common incandescent lightbulbs convert about 5 percent of the electrical energy they consume to light; the rest of the energy consumed is dissipated as heat, which adds to the cooling load of the air conditioner in summer. However, it is more common to express the effectiveness of this

**TABLE 2–1**

The efficacy of different lighting systems

| Type of lighting | Efficacy, lumens/W |
|---|---|
| *Combustion* | |
| Candle | 0.3 |
| Kerosene lamp | 1–2 |
| *Incandescent* | |
| Ordinary | 6–20 |
| Halogen | 15–35 |
| *Fluorescent* | |
| Compact | 40–87 |
| Tube | 60–120 |
| *High-intensity discharge* | |
| Mercury vapor | 40–60 |
| Metal halide | 65–118 |
| High-pressure sodium | 85–140 |
| Low-pressure sodium | 70–200 |
| *Solid-State* | |
| LED | 20–160 |
| OLED | 15–60 |
| Theoretical limit | 300* |

*This value depends on the spectral distribution of the assumed ideal light source. For white light sources, the upper limit is about 300 lm/W for metal halide, 350 lm/W for fluorescents, and 400 lm/W for LEDs. Spectral maximum occurs at a wavelength of 555 nm (green) with a light output of 683 lm/W.

**FIGURE 2–55**

A 15-W compact fluorescent lamp provides as much light as a 60-W incandescent lamp.

15 W      60 W

conversion process by **lighting efficacy**, which is defined as the *amount of light output in lumens per W of electricity consumed.*

The efficacy of different lighting systems is given in Table 2–1. Note that a compact fluorescent lightbulb produces about four times as much light as an incandescent lightbulb per W, and thus a 15-W fluorescent bulb can replace a 60-W incandescent lightbulb (Fig. 2–55). Also, a compact fluorescent bulb lasts about 10,000 h, which is 10 times as long as an incandescent bulb, and it plugs directly into the socket of an incandescent lamp. Therefore, despite their higher initial cost, compact fluorescents reduce the lighting costs considerably through reduced electricity consumption. Sodium-filled high-intensity discharge lamps provide the most efficient lighting, but their use is limited to outdoor use because of their yellowish light.

We can also define efficiency for cooking appliances since they convert electrical or chemical energy to heat for cooking. The **efficiency of a cooking appliance** can be defined as the ratio of the *useful energy transferred to the food to the energy consumed by the appliance* (Fig. 2–56). Electric ranges are more efficient than gas ranges, but it is much cheaper to cook with natural gas than with electricity because of the lower unit cost of natural gas (Table 2–2).

The cooking efficiency depends on user habits as well as the individual appliances. Convection and microwave ovens are inherently more efficient than conventional ovens. On average, convection ovens save about *one-third* and microwave ovens save about *two-thirds* of the energy used by conventional ovens. The cooking efficiency can be increased by using the smallest oven for baking, using a pressure cooker, using an electric slow cooker for stews and soups, using the smallest pan that will do the job, using the smaller heating element for small pans on electric ranges, using flat-bottomed pans on electric burners to assure good contact, keeping burner drip pans clean and shiny, defrosting frozen foods in the refrigerator before cooking, avoiding preheating unless it is necessary, keeping the pans covered during cooking, using timers and thermometers to avoid overcooking, using the self-cleaning feature of ovens right after cooking, and keeping inside surfaces of microwave ovens clean.

**TABLE 2–2**

Energy costs of cooking a casserole with different appliances*

[From J. T. Amann, A. Wilson, and K. Ackerly, *Consumer Guide to Home Energy Savings*, 9th ed., American Council for an Energy-Efficient Economy, Washington, D.C., 2007, p. 163.]

| Cooking appliance | Cooking temperature | Cooking time | Energy used | Cost of energy |
|---|---|---|---|---|
| Electric oven | 350°F (177°C) | 1 h | 2.0 kWh | $0.19 |
| Convection oven (elect.) | 325°F (163°C) | 45 min | 1.39 kWh | $0.13 |
| Gas oven | 350°F (177°C) | 1 h | 0.112 therm | $0.13 |
| Frying pan | 420°F (216°C) | 1 h | 0.9 kWh | $0.09 |
| Toaster oven | 425°F (218°C) | 50 min | 0.95 kWh | $0.09 |
| Crockpot | 200°F (93°C) | 7 h | 0.7 kWh | $0.07 |
| Microwave oven | "High" | 15 min | 0.36 kWh | $0.03 |

*Assumes a unit cost of $0.095/kWh for electricity and $1.20/therm (1 therm = 105,500 kJ) for gas.

Using energy-efficient appliances and practicing energy conservation measures help our pocketbooks by reducing our utility bills. It also helps the **environment** by reducing the amount of pollutants emitted to the atmosphere during the combustion of fuel at home or at the power plants where electricity is generated. The combustion of *each therm of natural gas* produces 6.4 kg of carbon dioxide, which causes global climate change; 4.7 g of nitrogen oxides and 0.54 g of hydrocarbons, which cause smog; 2.0 g of carbon monoxide, which is toxic; and 0.030 g of sulfur dioxide, which causes acid rain. Each therm of natural gas saved eliminates the emission of these pollutants while saving $0.60 for the average consumer in the United States. Each kWh of electricity conserved saves 0.4 kg of coal and 1.0 kg of $CO_2$ and 15 g of $SO_2$ from a coal power plant.

$$\text{Efficiency} = \frac{\text{Energy utilized}}{\text{Energy supplied to appliance}}$$

$$= \frac{3 \text{ kWh}}{5 \text{ kWh}} = 0.60$$

**FIGURE 2–56**
The efficiency of a cooking appliance represents the fraction of the energy supplied to the appliance that is transferred to the food.

■ **EXAMPLE 2–14**   **Cost of Cooking with Electric and Gas Ranges**

■ The efficiency of cooking appliances affects the internal heat gain from them since an inefficient appliance consumes a greater amount of energy for the same task, and the excess energy consumed shows up as heat in the living space. The efficiency of open burners is determined to be 73 percent for electric units and 38 percent for gas units (Fig. 2–57). Consider a 2-kW electric burner at a location where the unit costs of electricity and natural gas are $0.09/kWh and $1.20/therm (1 therm = 105,500 kJ), respectively. Determine the rate of energy consumption by the burner and the unit cost of utilized energy for both electric and gas burners.

**SOLUTION**   The operation of electric and gas ranges is considered. The rate of energy consumption and the unit cost of utilized energy are to be determined.
*Analysis*   The efficiency of the electric heater is given to be 73 percent. Therefore, a burner that consumes 2 kW of electrical energy will supply

$$\dot{Q}_{\text{utilized}} = (\text{Energy input}) \times (\text{Efficiency}) = (2 \text{ kW})(0.73) = \mathbf{1.46 \text{ kW}}$$

of useful energy. The unit cost of utilized energy is inversely proportional to the efficiency, and is determined from

$$\text{Cost of utilized energy} = \frac{\text{Cost of energy input}}{\text{Efficiency}} = \frac{\$0.09/\text{kWh}}{0.73} = \mathbf{\$0.123/kWh}$$

Noting that the efficiency of a gas burner is 38 percent, the energy input to a gas burner that supplies utilized energy at the same rate (1.46 kW) is

$$\dot{Q}_{\text{input, gas}} = \frac{\dot{Q}_{\text{utilized}}}{\text{Efficiency}} = \frac{1.46 \text{ kW}}{0.38} = \mathbf{3.84 \text{ kW}}$$

Therefore, a gas burner should have a rating of at least 3.84 kW to perform as well as the electric unit.
Noting that 1 therm = 29.3 kWh, the unit cost of utilized energy in the case of a gas burner is determined to be

$$\text{Cost of utilized energy} = \frac{\text{Cost of energy input}}{\text{Efficiency}} = \frac{\$1.20/29.3 \text{ kWh}}{0.38}$$

$$= \mathbf{\$0.108/kWh}$$

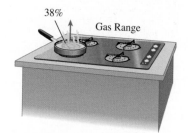

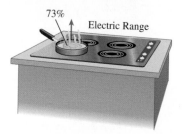

**FIGURE 2–57**
Schematic of the 73 percent efficient electric heating unit and 38 percent efficient gas burner discussed in Example 2–14.

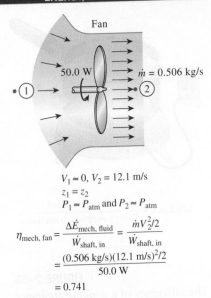

$V_1 \approx 0$, $V_2 = 12.1$ m/s
$z_1 = z_2$
$P_1 \approx P_{atm}$ and $P_2 \approx P_{atm}$

$$\eta_{mech, fan} = \frac{\Delta \dot{E}_{mech, fluid}}{\dot{W}_{shaft, in}} = \frac{\dot{m} V_2^2/2}{\dot{W}_{shaft, in}}$$
$$= \frac{(0.506 \text{ kg/s})(12.1 \text{ m/s})^2/2}{50.0 \text{ W}}$$
$$= 0.741$$

**FIGURE 2–58**
The mechanical efficiency
of a fan is the ratio of the rate of
increase of the mechanical energy
of air to the mechanical power input.

**Discussion** The cost of utilized gas is less than that of utilized electricity. Therefore, despite its higher efficiency, cooking with an electric burner will cost about 14 percent more compared to a gas burner in this case. This explains why cost-conscious consumers always ask for gas appliances, and it is not wise to use electricity for heating purposes.

# Efficiencies of Mechanical and Electrical Devices

The transfer of mechanical energy is usually accomplished by a rotating shaft, and thus mechanical work is often referred to as *shaft work*. A pump or a fan receives shaft work (usually from an electric motor) and transfers it to the fluid as mechanical energy (less frictional losses). A turbine, on the other hand, converts the mechanical energy of a fluid to shaft work. In the absence of any irreversibilities such as friction, mechanical energy can be converted entirely from one mechanical form to another, and the **mechanical efficiency** of a device or process can be defined as (Fig. 2–58)

$$\eta_{mech} = \frac{\text{Mechanical energy output}}{\text{Mechanical energy input}} = \frac{E_{mech, out}}{E_{mech, in}} = 1 - \frac{E_{mech, loss}}{E_{mech, in}} \quad (2\text{--}44)$$

A conversion efficiency of less than 100 percent indicates that conversion is less than perfect and some losses have occurred during conversion. A mechanical efficiency of 97 percent indicates that 3 percent of the mechanical energy input is converted to thermal energy as a result of frictional heating, and this will manifest itself as a slight rise in the temperature of the fluid.

In fluid systems, we are usually interested in increasing the pressure, velocity, and/or elevation of a fluid. This is done by *supplying mechanical energy* to the fluid by a pump, a fan, or a compressor (we will refer to all of them as pumps). Or we are interested in the reverse process of *extracting mechanical energy* from a fluid by a turbine and producing mechanical power in the form of a rotating shaft that can drive a generator or any other rotary device. The degree of perfection of the conversion process between the mechanical work supplied or extracted and the mechanical energy of the fluid is expressed by the **pump efficiency** and **turbine efficiency**, defined as

$$\eta_{pump} = \frac{\text{Mechanical energy increase of the fluid}}{\text{Mechanical energy input}} = \frac{\Delta \dot{E}_{mech, fluid}}{\dot{W}_{shaft, in}} = \frac{\dot{W}_{pump, u}}{\dot{W}_{pump}} \quad (2\text{--}45)$$

where $\Delta \dot{E}_{mech, fluid} = \dot{E}_{mech, out} - \dot{E}_{mech, in}$ is the rate of increase in the mechanical energy of the fluid, which is equivalent to the **useful pumping power** $\dot{W}_{pump, u}$ supplied to the fluid, and

$$\eta_{turbine} = \frac{\text{Mechanical energy output}}{\text{Mechanical energy decrease of the fluid}} = \frac{\dot{W}_{shaft, out}}{|\Delta \dot{E}_{mech, fluid}|} = \frac{\dot{W}_{turbine}}{\dot{W}_{turbine, e}} \quad (2\text{--}46)$$

where $|\Delta \dot{E}_{mech, fluid}| = \dot{E}_{mech, in} - \dot{E}_{mech, out}$ is the rate of decrease in the mechanical energy of the fluid, which is equivalent to the mechanical power extracted from the fluid by the turbine $\dot{W}_{turbine, e}$, and we use the absolute value sign to avoid negative values for efficiencies. A pump or turbine efficiency of 100 percent indicates perfect conversion between the shaft work and the mechanical energy of the fluid, and this value can be approached (but never attained) as the frictional effects are minimized.

Electrical energy is commonly converted to *rotating mechanical energy* by electric motors to drive fans, compressors, robot arms, car starters, and so forth. The effectiveness of this conversion process is characterized by the *motor efficiency* $\eta_{motor}$, which is the ratio of the *mechanical energy output* of the motor to the *electrical energy input*. The full-load motor efficiencies range from about 35 percent for small motors to over 97 percent for large high-efficiency motors. The difference between the electrical energy consumed and the mechanical energy delivered is dissipated as waste heat.

The mechanical efficiency should not be confused with the **motor efficiency** and the **generator efficiency**, which are defined as

*Motor:*
$$\eta_{motor} = \frac{\text{Mechanical power output}}{\text{Electric power input}} = \frac{\dot{W}_{shaft,out}}{\dot{W}_{elect,in}} \tag{2–47}$$

and

*Generator:*
$$\eta_{generator} = \frac{\text{Electric power output}}{\text{Mechanical power input}} = \frac{\dot{W}_{elect,out}}{\dot{W}_{shaft,in}} \tag{2–48}$$

A pump is usually packaged together with its motor, and a turbine with its generator. Therefore, we are usually interested in the **combined** or **overall efficiency** of pump–motor and turbine–generator combinations (Fig. 2–59), which are defined as

$$\eta_{pump-motor} = \eta_{pump}\eta_{motor} = \frac{\dot{W}_{pump,u}}{\dot{W}_{elect,in}} = \frac{\Delta\dot{E}_{mech,fluid}}{\dot{W}_{elect,in}} \tag{2–49}$$

and

$$\eta_{turbine-gen} = \eta_{turbine}\eta_{generator} = \frac{\dot{W}_{elect,out}}{\dot{W}_{turbine,e}} = \frac{\dot{W}_{elect,out}}{|\Delta\dot{E}_{mech,fluid}|} \tag{2–50}$$

All the efficiencies just defined range between 0 and 100 percent. The lower limit of 0 percent corresponds to the conversion of the entire mechanical or electric energy input to thermal energy, and the device in this case functions like a resistance heater. The upper limit of 100 percent corresponds to the case of perfect conversion with no friction or other irreversibilities, and thus no conversion of mechanical or electric energy to thermal energy.

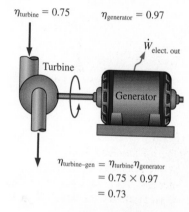

$\eta_{turbine} = 0.75 \qquad \eta_{generator} = 0.97$

$\dot{W}_{elect.\,out}$

Turbine

Generator

$$\eta_{turbine-gen} = \eta_{turbine}\eta_{generator}$$
$$= 0.75 \times 0.97$$
$$= 0.73$$

**FIGURE 2–59**
The overall efficiency of a turbine–generator is the product of the efficiency of the turbine and the efficiency of the generator, and represents the fraction of the mechanical power of the fluid converted to electrical power.

---

■ **EXAMPLE 2–15**  **Power Generation from a Hydroelectric Plant**

Electric power is to be generated by installing a hydraulic turbine–generator at a site 70 m below the free surface of a large water reservoir that can supply water at a rate of 1500 kg/s steadily (Fig. 2–60). If the mechanical power output of the turbine is 800 kW and the electric power generation is 750 kW, determine the turbine efficiency and the combined turbine–generator efficiency of this plant. Neglect losses in the pipes.

**SOLUTION**  A hydraulic turbine-generator installed at a large reservoir is to generate electricity. The combined turbine–generator efficiency and the turbine efficiency are to be determined.

*Assumptions*  **1** The water elevation in the reservoir remains constant. **2** The mechanical energy of water at the turbine exit is negligible.

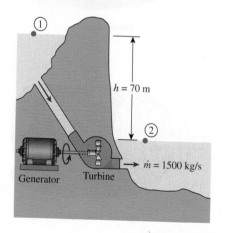

$h = 70$ m

$\dot{m} = 1500$ kg/s

Generator     Turbine

**FIGURE 2–60**
Schematic for Example 2–15.

*Analysis* We take the free surface of water in the reservoir to be point 1 and the turbine exit to be point 2. We also take the turbine exit as the reference level ($z_2 = 0$) so that the potential energies at 1 and 2 are $pe_1 = gz_1$ and $pe_2 = 0$. The flow energy $P/\rho$ at both points is zero since both 1 and 2 are open to the atmosphere ($P_1 = P_2 = P_{atm}$). Further, the kinetic energy at both points is zero ($ke_1 = ke_2 = 0$) since the water at point 1 is essentially motionless, and the kinetic energy of water at turbine exit is assumed to be negligible. The potential energy of water at point 1 is

$$pe_1 = gz_1 = (9.81 \text{ m/s}^2)(70 \text{ m})\left(\frac{1 \text{ kJ/kg}}{1000 \text{ m}^2/\text{s}^2}\right) = 0.687 \text{ kJ/kg}$$

Then the rate at which the mechanical energy of water is supplied to the turbine becomes

$$|\Delta \dot{E}_{mech,fluid}| = \dot{m}(e_{mech,in} - e_{mech,out}) = \dot{m}(pe_1 - 0) = \dot{m}pe_1$$
$$= (1500 \text{ kg/s})(0.687 \text{ kJ/kg})$$
$$= 1031 \text{ kW}$$

The combined turbine–generator and the turbine efficiency are determined from their definitions to be

$$\eta_{turbine-gen} = \frac{\dot{W}_{elect,out}}{|\Delta \dot{E}_{mech,fluid}|} = \frac{750 \text{ kW}}{1031 \text{ kW}} = 0.727 \text{ or } 72.7\%$$

$$\eta_{turbine} = \frac{\dot{W}_{elect,out}}{|\dot{E}_{mech,fluid}|} = \frac{800 \text{kW}}{1031 \text{kW}} = 0.776 \text{ or } 77.6\%$$

Therefore, the reservoir supplies 1031 kW of mechanical energy to the turbine, which converts 800 kW of it to shaft work that drives the generator, which then generates 750 kW of electric power.

*Discussion* This problem can also be solved by taking point 1 to be at the turbine inlet, and using flow energy instead of potential energy. It would give the same result since the flow energy at the turbine inlet is equal to the potential energy at the free surface of the reservoir.

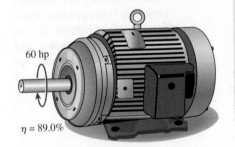

60 hp

$\eta = 89.0\%$

Standard motor

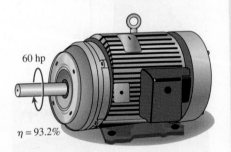

60 hp

$\eta = 93.2\%$

High-efficiency motor

**FIGURE 2–61**
Schematic for Example 2–16.

**EXAMPLE 2–16** **Cost Savings Associated with High-Efficiency Motors**

A 60-hp electric motor (a motor that delivers 60 hp of shaft power at full load) that has an efficiency of 89.0 percent is worn out and is to be replaced by a 93.2 percent efficient high-efficiency motor (Fig. 2–61). The motor operates 3500 hours a year at full load. Taking the unit cost of electricity to be $0.08/kWh, determine the amount of energy and money saved as a result of installing the high-efficiency motor instead of the standard motor. Also, determine the simple payback period if the purchase prices of the standard and high-efficiency motors are $4520 and $5160, respectively.

**SOLUTION** A worn-out standard motor is to be replaced by a high-efficiency one. The amount of electrical energy and money saved as well as the simple payback period are to be determined.

*Assumptions* The load factor of the motor remains constant at 1 (full load) when operating.

*Analysis* The electric power drawn by each motor and their difference can be expressed as

$$\dot{W}_{\text{electric in, standard}} = \dot{W}_{\text{shaft}}/\eta_{\text{st}} = (\text{Rated power})(\text{Load factor})/\eta_{\text{st}}$$

$$\dot{W}_{\text{electric in, efficient}} = \dot{W}_{\text{shaft}}/\eta_{\text{eff}} = (\text{Rated power})(\text{Load factor})/\eta_{\text{eff}}$$

$$\text{Power savings} = \dot{W}_{\text{electric in, standard}} - \dot{W}_{\text{electric in, efficient}}$$

$$= (\text{Rated power})(\text{Load factor})(1/\eta_{\text{st}} - 1/\eta_{\text{eff}})$$

where $\eta_{\text{st}}$ is the efficiency of the standard motor, and $\eta_{\text{eff}}$ is the efficiency of the comparable high-efficiency motor. Then the annual energy and cost savings associated with the installation of the high-efficiency motor become

$$\begin{aligned}\text{Energy savings} &= (\text{Power savings})(\text{Operating hours})\\ &= (\text{Rated power})(\text{Operating hours})(\text{Load factor})(1/\eta_{\text{st}} - 1\eta_{\text{eff}})\\ &= (60 \text{ hp})(0.7457 \text{ kW/hp})(3500 \text{ h/year})(1)(1/0.89 - 1/0.93.2)\\ &= \textbf{7929 kWh/year}\end{aligned}$$

$$\begin{aligned}\text{Cost savings} &= (\text{Energy savings})(\text{Unit cost of energy})\\ &= (7929 \text{ kWh/year})(\$0.08/\text{kWh})\\ &= \textbf{\$634/year}\end{aligned}$$

Also,

$$\text{Excess initial cost} = \text{Purchase price differential} = \$5160 = \$4520 = \$640$$

This gives a simple payback period of

$$\text{Simple payback period} = \frac{\text{Excess initial cost}}{\text{Annual cost savings}} = \frac{\$640}{\$634/\text{year}} = \textbf{1.01 year}$$

*Discussion* Note that the high-efficiency motor pays for its price differential within about one year from the electrical energy it saves. Considering that the service life of electric motors is several years, the purchase of the higher efficiency motor is definitely indicated in this case.

## 2–8 · ENERGY AND ENVIRONMENT

The conversion of energy from one form to another often affects the environment and the air we breathe in many ways, and thus the study of energy is not complete without considering its impact on the environment (Fig. 2–62). Fossil fuels such as coal, oil, and natural gas have been powering the industrial development and the amenities of modern life that we enjoy since the 1700s, but this has not been without any undesirable side effects. From the soil we farm and the water we drink to the air we breathe, the environment has been paying a heavy toll for it. Pollutants emitted during the combustion of fossil fuels are responsible for smog, acid rain, global warming, and climate change. The environmental pollution has reached such high levels that

**FIGURE 2–62**

Energy conversion processes are often accompanied by environmental pollution.

©Comstock Images/Alamy RF

**FIGURE 2–63**
Motor vehicles are the largest
source of air pollution.

it became a serious threat to vegetation, wild life, and human health. Air pollution has been the cause of numerous health problems including asthma and cancer. It is estimated that over 60,000 people in the United States alone die each year due to heart and lung diseases related to air pollution.

Hundreds of elements and compounds such as benzene and formaldehyde are known to be emitted during the combustion of coal, oil, natural gas, and wood in electric power plants, engines of vehicles, furnaces, and even fireplaces. Some compounds are added to liquid fuels for various reasons (such as MTBE to raise the octane number of the fuel and also to oxygenate the fuel in winter months to reduce urban smog). The largest source of air pollution is the motor vehicles, and the pollutants released by the vehicles are usually grouped as hydrocarbons (HC), nitrogen oxides ($NO_x$), and carbon monoxide (CO) (Fig. 2–63). The HC emissions are a large component of volatile organic compounds (VOCs) emissions, and the two terms are generally used interchangeably for motor vehicle emissions. A significant portion of the VOC or HC emissions are caused by the evaporation of fuels during refueling or spillage during spitback or by evaporation from gas tanks with faulty caps that do not close tightly. The solvents, propellants, and household cleaning products that contain benzene, butane, or other HC products are also significant sources of HC emissions.

The increase of environmental pollution at alarming rates and the rising awareness of its dangers made it necessary to control it by legislation and international treaties. In the United States, the Clean Air Act of 1970 (whose passage was aided by the 14-day smog alert in Washington that year) set limits on pollutants emitted by large plants and vehicles. These early standards focused on emissions of hydrocarbons, nitrogen oxides, and carbon monoxide. The new cars were required to have catalytic converters in their exhaust systems to reduce HC and CO emissions. As a side benefit, the removal of lead from gasoline to permit the use of catalytic converters led to a significant reduction in toxic lead emissions.

Emission limits for HC, $NO_x$, and CO from cars have been declining steadily since 1970. The Clean Air Act of 1990 made the requirements on emissions even tougher, primarily for ozone, CO, nitrogen dioxide, and particulate matter (PM). As a result, today's industrial facilities and vehicles emit a fraction of the pollutants they used to emit a few decades ago. The HC emissions of cars, for example, decreased from about 5 g/km (grams per km) in 1970 to 0.25 g/km in 1980 and about 0.06 g/km in 1999. This is a significant reduction since many of the gaseous toxics from motor vehicles and liquid fuels are hydrocarbons.

Children are most susceptible to the damages caused by air pollutants since their organs are still developing. They are also exposed to more pollution since they are more active, and thus they breathe faster. People with heart and lung problems, especially those with asthma, are most affected by air pollutants. This becomes apparent when the air pollution levels in their neighborhoods rise to high levels.

## Ozone and Smog

If you live in a metropolitan area such as Los Angeles, you are probably familiar with urban smog—the dark yellow or brown haze that builds up in a large stagnant air mass and hangs over populated areas on calm hot summer days. *Smog* is made up mostly of ground-level ozone ($O_3$), but it also contains numerous other chemicals, including carbon monoxide (CO), particulate matter such as soot and dust, volatile organic compounds (VOCs) such as benzene, butane, and other hydrocarbons. The harmful ground-level ozone should not be confused with the useful ozone layer high in the stratosphere that protects the earth from the sun's harmful ultraviolet rays. Ozone at ground level is a pollutant with several adverse health effects.

The primary source of both nitrogen oxides and hydrocarbons is the motor vehicles. Hydrocarbons and nitrogen oxides react in the presence of sunlight on hot calm days to form ground-level ozone, which is the primary component of smog (Fig. 2–64). The smog formation usually peaks in late afternoons when the temperatures are highest and there is plenty of sunlight. Although ground-level smog and ozone form in urban areas with heavy traffic or industry, the prevailing winds can transport them several hundred miles to other cities. This shows that pollution knows of no boundaries, and it is a global problem.

*Ozone* irritates eyes and damages the air sacs in the lungs where oxygen and carbon dioxide are exchanged, causing eventual hardening of this soft and spongy tissue. It also causes shortness of breath, wheezing, fatigue, headaches, and nausea, and aggravates respiratory problems such as asthma. Every exposure to ozone does a little damage to the lungs, just like cigarette smoke, eventually reducing the individual's lung capacity. Staying indoors and minimizing physical activity during heavy smog minimizes damage. Ozone also harms vegetation by damaging leaf tissues. To improve the air quality in areas with the worst ozone problems, reformulated gasoline (RFG) that contains at least 2 percent oxygen was introduced. The use of RFG has resulted in significant reduction in the emission of ozone and other pollutants, and its use is mandatory in many smog-prone areas.

The other serious pollutant in smog is *carbon monoxide,* which is a colorless, odorless, poisonous gas. It is mostly emitted by motor vehicles, and it can build to dangerous levels in areas with heavy congested traffic. It deprives the body's organs from getting enough oxygen by binding with the red blood cells that would otherwise carry oxygen. At low levels, carbon monoxide decreases the amount of oxygen supplied to the brain and other organs and muscles, slows body reactions and reflexes, and impairs judgment. It poses a serious threat to people with heart disease because of the fragile condition of the circulatory system and to fetuses because of the oxygen needs of the developing brain. At high levels, it can be fatal, as evidenced by numerous deaths caused by cars that are warmed up in closed garages or by exhaust gases leaking into the cars.

Smog also contains suspended particulate matter such as dust and soot emitted by vehicles and industrial facilities. Such particles irritate the eyes and the lungs since they may carry compounds such as acids and metals.

## Acid Rain

Fossil fuels are mixtures of various chemicals, including small amounts of sulfur. The sulfur in the fuel reacts with oxygen to form sulfur dioxide ($SO_2$), which is an air pollutant. The main source of $SO_2$ is the electric power plants that burn high-sulfur coal. The Clean Air Act of 1970 has limited the $SO_2$ emissions severely, which forced the plants to install $SO_2$

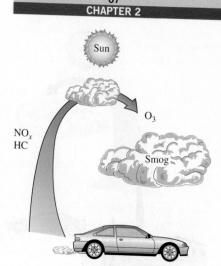

**FIGURE 2–64**
Ground-level ozone, which is the primary component of smog, forms when HC and $NO_x$ react in the presence of sunlight in hot calm days.

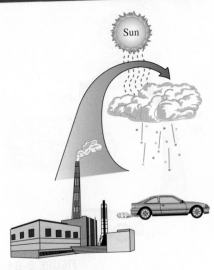

**FIGURE 2–65**
Sulfuric acid and nitric acid are formed when sulfur oxides and nitric oxides react with water vapor and other chemicals high in the atmosphere in the presence of sunlight.

scrubbers, to switch to low-sulfur coal, or to gasify the coal and recover the sulfur. Motor vehicles also contribute to $SO_2$ emissions since gasoline and diesel fuel also contain small amounts of sulfur. Volcanic eruptions and hot springs also release sulfur oxides (the cause of the rotten egg smell).

The sulfur oxides and nitric oxides react with water vapor and other chemicals high in the atmosphere in the presence of sunlight to form sulfuric and nitric acids (Fig. 2–65). The acids formed usually dissolve in the suspended water droplets in clouds or fog. These acid-laden droplets, which can be as acidic as lemon juice, are washed from the air on to the soil by rain or snow. This is known as **acid rain**. The soil is capable of neutralizing a certain amount of acid, but the amounts produced by the power plants using inexpensive high-sulfur coal has exceeded this capability, and as a result many lakes and rivers in industrial areas such as New York, Pennsylvania, and Michigan have become too acidic for fish to grow. Forests in those areas also experience a slow death due to absorbing the acids through their leaves, needles, and roots. Even marble structures deteriorate due to acid rain. The magnitude of the problem was not recognized until the early 1970s, and serious measures have been taken since then to reduce the sulfur dioxide emissions drastically by installing scrubbers in plants and by desulfurizing coal before combustion.

## The Greenhouse Effect:
## Global Warming and Climate Change

You have probably noticed that when you leave your car under direct sunlight on a sunny day, the interior of the car gets much warmer than the air outside, and you may have wondered why the car acts like a heat trap. This is because glass at thicknesses encountered in practice transmits over 90 percent of radiation in the visible range and is practically opaque (nontransparent) to radiation in the longer wavelength infrared regions. Therefore, glass allows the solar radiation to enter freely but blocks the infrared radiation emitted by the interior surfaces. This causes a rise in the interior temperature as a result of the thermal energy buildup in the car. This heating effect is known as the **greenhouse effect**, since it is utilized primarily in greenhouses.

The greenhouse effect is also experienced on a larger scale on earth. The surface of the earth, which warms up during the day as a result of the absorption of solar energy, cools down at night by radiating part of its energy into deep space as infrared radiation. Carbon dioxide ($CO_2$), water vapor, and trace amounts of some other gases such as methane and nitrogen oxides act like a blanket and keep the earth warm at night by blocking the heat radiated from the earth (Fig. 2–66). Therefore, they are called "greenhouse gases," with $CO_2$ being the primary component. Water vapor is usually taken out of this list since it comes down as rain or snow as part of the water cycle and human activities in producing water (such as the burning of fossil fuels) do not make much difference on its concentration in the atmosphere (which is mostly due to evaporation from rivers, lakes, oceans, etc.). $CO_2$ is different, however, in that people's activities do make a difference in $CO_2$ concentration in the atmosphere.

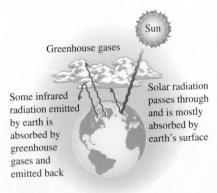

Greenhouse gases

Some infrared radiation emitted by earth is absorbed by greenhouse gases and emitted back

Solar radiation passes through and is mostly absorbed by earth's surface

**FIGURE 2–66**
The greenhouse effect on earth.

The greenhouse effect makes life on earth possible by keeping the earth warm (about 30°C warmer). However, excessive amounts of these gases disturb the delicate balance by trapping too much energy, which causes the average temperature of the earth to rise and the climate at some localities to change. These undesirable consequences of the greenhouse effect are referred to as **global warming** or **global climate change**.

The global climate change is due to the excessive use of fossil fuels such as coal, petroleum products, and natural gas in electric power generation, transportation, buildings, and manufacturing, and it has been a concern in recent decades. In 1995, a total of 6.5 billion tons of carbon was released to the atmosphere as $CO_2$. The current concentration of $CO_2$ in the atmosphere is about 360 ppm (or 0.36 percent). This is 20 percent higher than the level a century ago, and it is projected to increase to over 700 ppm by the year 2100. Under normal conditions, vegetation consumes $CO_2$ and releases $O_2$ during the photosynthesis process, and thus keeps the $CO_2$ concentration in the atmosphere in check. A mature, growing tree consumes about 12 kg of $CO_2$ a year and exhales enough oxygen to support a family of four. However, deforestation and the huge increase in the $CO_2$ production in recent decades disturbed this balance.

In a 1995 report, the world's leading climate scientists concluded that the earth has already warmed about 0.5°C during the last century, and they estimate that the earth's temperature will rise another 2°C by the year 2100. A rise of this magnitude is feared to cause severe changes in weather patterns with storms and heavy rains and flooding at some parts and drought in others, major floods due to the melting of ice at the poles, loss of wetlands and coastal areas due to rising sea levels, variations in water supply, changes in the ecosystem due to the inability of some animal and plant species to adjust to the changes, increases in epidemic diseases due to the warmer temperatures, and adverse side effects on human health and socioeconomic conditions in some areas.

The seriousness of these threats has moved the United Nations to establish a committee on climate change. A world summit in 1992 in Rio de Janeiro, Brazil, attracted world attention to the problem. The agreement prepared by the committee in 1992 to control greenhouse gas emissions was signed by 162 nations. In the 1997 meeting in Kyoto (Japan), the world's industrialized countries adopted the Kyoto protocol and committed to reduce their $CO_2$ and other greenhouse gas emissions by 5 percent below the 1990 levels by 2008 to 2012. In December 2011, countries agreed in Durban, South Africa to forge a new deal forcing the biggest polluting countries to limit greenhouse gas emissions. The Kyoto protocol is extended allowing five more years to finalize a wider agreement. It was agreed to work in a new, legally binding accord to cut greenhouse gases. This should be decided by 2015 and come into force by 2020. Greenhouse gas emissions can be reduced by increasing conservation efforts and improving conversion efficiencies, while meeting new energy demands by the use of renewable energy (such as hydroelectric, solar, wind, and geothermal energy) rather than by fossil fuels.

The United States is the largest contributor of greenhouse gases, with over 5 tons of carbon emissions per person per year. Major sources of greenhouse gas emissions are industrial sector and transportation. Each kilowatt-hour of electricity produced by a fossil-fuelled power plant produces 0.6

**FIGURE 2–67**

The average car produces several times its weight in $CO_2$ every year (it is driven 22,000 km a year, consumes 2300 L of gasoline, and produces 2.5 kg of $CO_2$ per liter).

©*Emma Lee/Life File/Getty Images RF*

**FIGURE 2–68**

Renewable energies such as wind are called "green energy" since they emit no pollutants or greenhouse gases.

©*J. Luke/PhotoLink/Getty Images RF*

to 1.0 kg carbon dioxide. Each liter of gasoline burned by a vehicle produces about 2.5 kg of $CO_2$. An average car in the United States is driven about 22,000 km a year, and it consumes about 2300 L of gasoline. Therefore, a car emits about 5500 kg of $CO_2$ to the atmosphere a year, which is about four times the weight of a typical car (Fig. 2–67). This and other emissions can be reduced significantly by buying an energy-efficient car that burns less fuel over the same distance, and by driving sensibly. Saving fuel also saves money and the environment. For example, choosing a vehicle that consumes 8 L/100 km rather than 12 L/100 km will prevent 2 tons of $CO_2$ from being released to the atmosphere every year while reducing the fuel cost by $900 per year (under average driving conditions of 22,000 km a year and at a fuel cost of $1.06/L).

It is clear from these discussions that considerable amounts of pollutants are emitted as the chemical energy in fossil fuels is converted to thermal, mechanical, or electrical energy via combustion, and thus power plants, motor vehicles, and even stoves take the blame for air pollution. In contrast, no pollution is emitted as electricity is converted to thermal, chemical, or mechanical energy, and thus electric cars are often touted as "zero emission" vehicles and their widespread use is seen by some as the ultimate solution to the air pollution problem. It should be remembered, however, that the electricity used by the electric cars is generated somewhere else mostly by burning fuel and thus emitting pollution. Therefore, each time an electric car consumes 1 kWh of electricity, it bears the responsibility for the pollutions emitted as 1 kWh of electricity (plus the conversion and transmission losses) is generated elsewhere. The electric cars can be claimed to be zero emission vehicles only when the electricity they consume is generated by emission-free renewable resources such as hydroelectric, solar, wind, and geothermal energy (Fig. 2–68). Therefore, the use of renewable energy should be encouraged worldwide, with incentives, as necessary, to make the earth a better place to live in. The advancements in thermodynamics have contributed greatly in recent decades to improve conversion efficiencies (in some cases doubling them) and thus to reduce pollution. As individuals, we can also help by practicing energy conservation measures and by making energy efficiency a high priority in our purchases.

**EXAMPLE 2–17    Reducing Air Pollution by Geothermal Heating**

A geothermal power plant in Nevada is generating electricity using geothermal water extracted at 180°C, and reinjected back to the ground at 85°C. It is proposed to utilize the reinjected brine for heating the residential and commercial buildings in the area, and calculations show that the geothermal heating system can save 18 million therms (1 therm = 105,500 kJ) of natural gas a year. Determine the amount of $NO_x$ and $CO_2$ emissions the geothermal system will save a year. Take the average $NO_x$ and $CO_2$ emissions of gas furnaces to be 0.0047 kg/therm and 6.4 kg/therm, respectively.

**SOLUTION**    The gas heating systems in an area are being replaced by a geothermal district heating system. The amounts of $NO_x$ and $CO_2$ emissions saved per year are to be determined.

*Analysis* The amounts of emissions saved per year are equivalent to the amounts emitted by furnaces when 18 million therms of natural gas are burned,

$$NO_x \text{ savings} = (NO_x \text{ emission per therm})(\text{No. of therms per year})$$
$$= (0.0047 \text{ kg/therm})(18 \times 10^6 \text{ therm/year})$$
$$= \mathbf{8.5 \times 10^4 \text{ kg/year}}$$

$$CO_2 \text{ savings} = (CO_2 \text{ emission per therm})(\text{No. of therms per year})$$
$$= (6.4 \text{ kg/therm})(18 \times 10^6 \text{ therm/year})$$
$$= \mathbf{1.2 \times 10^8 \text{ kg/year}}$$

*Discussion* A typical car on the road generates about 8.5 kg of $NO_x$ and 6000 kg of $CO_2$ a year. Therefore the environmental impact of replacing the gas heating systems in the area by the geothermal heating system is equivalent to taking 10,000 cars off the road for $NO_x$ emission and taking 20,000 cars off the road for $CO_2$ emission. The proposed system should have a significant effect on reducing smog in the area.

## TOPIC OF SPECIAL INTEREST*  Mechanisms of Heat Transfer

Heat can be transferred in three different ways: *conduction, convection,* and *radiation.* We will give a brief description of each mode to familiarize the reader with the basic mechanisms of heat transfer. All modes of heat transfer require the existence of a temperature difference, and all modes of heat transfer are from the high-temperature medium to a lower temperature one.

**Conduction** is the transfer of energy from the more energetic particles of a substance to the adjacent less energetic ones as a result of interactions between the particles. Conduction can take place in solids, liquids, or gases. In gases and liquids, conduction is due to the collisions of the molecules during their random motion. In solids, it is due to the combination of vibrations of molecules in a lattice and the energy transport by free electrons. A cold canned drink in a warm room, for example, eventually warms up to the room temperature as a result of heat transfer from the room to the drink through the aluminum can by conduction (Fig. 2–69).

It is observed that the rate of heat conduction $\dot{Q}_{cond}$ through a layer of constant thickness $\Delta x$ is proportional to the temperature difference $\Delta T$ across the layer and the area $A$ normal to the direction of heat transfer, and is inversely proportional to the thickness of the layer. Therefore,

$$\dot{Q}_{cond} = k_t A \frac{\Delta T}{\Delta x} \quad \text{(W)} \tag{2-51}$$

where the constant of proportionality $k_t$ is the **thermal conductivity** of the material, which is a measure of the ability of a material to conduct heat (Table 2–3). Materials such as copper and silver, which are good electric conductors, are also good heat conductors, and therefore have high $k_t$ values.

*This section can be skipped without a loss in continuity

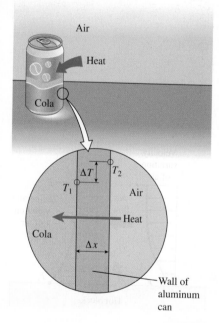

**FIGURE 2–69**
Heat conduction from warm air to a cold canned drink through the wall of the aluminum can.

## TABLE 2–3

Thermal conductivities of some materials at room conditions

| Material | Thermal conductivity, W/m·K |
|---|---|
| Diamond | 2300 |
| Silver | 429 |
| Copper | 401 |
| Gold | 317 |
| Aluminum | 237 |
| Iron | 80.2 |
| Mercury ($\ell$) | 8.54 |
| Glass | 1.4 |
| Brick | 0.72 |
| Water ($\ell$) | 0.613 |
| Human skin | 0.37 |
| Wood (oak) | 0.17 |
| Helium ($g$) | 0.152 |
| Soft rubber | 0.13 |
| Glass fiber | 0.043 |
| Air ($g$) | 0.026 |
| Urethane, rigid foam | 0.026 |

Materials such as rubber, wood, and styrofoam are poor conductors of heat, and therefore have low $k_t$ values.

In the limiting case of $\Delta x \to 0$, the equation above reduces to the differential form

$$\dot{Q}_{cond} = -k_t A \frac{dT}{dx} \quad \text{(W)} \tag{2–52}$$

which is known as **Fourier's law** of heat conduction. It indicates that the rate of heat conduction in a direction is proportional to the *temperature gradient* in that direction. Heat is conducted in the direction of decreasing temperature, and the temperature gradient becomes negative when temperature decreases with increasing $x$. Therefore, a negative sign is added in Eq. 2–52 to make heat transfer in the positive $x$ direction a positive quantity.

Temperature is a measure of the kinetic energies of the molecules. In a liquid or gas, the kinetic energy of the molecules is due to the random motion of the molecules as well as the vibrational and rotational motions. When two molecules possessing different kinetic energies collide, part of the kinetic energy of the more energetic (higher temperature) molecule is transferred to the less energetic (lower temperature) particle, in much the same way as when two elastic balls of the same mass at different velocities collide, part of the kinetic energy of the faster ball is transferred to the slower one.

In solids, heat conduction is due to two effects: the lattice vibrational waves induced by the vibrational motions of the molecules positioned at relatively fixed position in a periodic manner called a *lattice*, and the energy transported via the free flow of electrons in the solid. The thermal conductivity of a solid is obtained by adding the lattice and the electronic components. The thermal conductivity of pure metals is primarily due to the electronic component, whereas the thermal conductivity of nonmetals is primarily due to the lattice component. The lattice component of thermal conductivity strongly depends on the way the molecules are arranged. For example, the thermal conductivity of diamond, which is a highly ordered crystalline solid, is much higher than the thermal conductivities of pure metals, as can be seen from Table 2–3.

**Convection** is the mode of energy transfer between a solid surface and the adjacent liquid or gas that is in motion, and it involves the combined effects of *conduction* and *fluid motion*. The faster the fluid motion, the greater the convection heat transfer. In the absence of any bulk fluid motion, heat transfer between a solid surface and the adjacent fluid is by pure conduction. The presence of bulk motion of the fluid enhances the heat transfer between the solid surface and the fluid, but it also complicates the determination of heat transfer rates.

Consider the cooling of a hot block by blowing of cool air over its top surface (Fig. 2–70). Energy is first transferred to the air layer adjacent to the surface of the block by conduction. This energy is then carried away from the surface by convection; that is, by the combined effects of conduction within the air, which is due to random motion of air molecules, and the bulk or macroscopic motion of the air, which removes the heated air near the surface and replaces it by the cooler air.

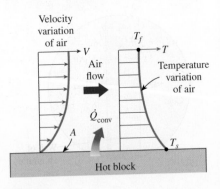

**FIGURE 2–70**

Heat transfer from a hot surface to air by convection.

Convection is called **forced convection** if the fluid is *forced* to flow in a tube or over a surface by external means such as a fan, pump, or the wind. In contrast, convection is called **free** (or **natural**) **convection** if the fluid motion is caused by buoyancy forces induced by density differences due to the variation of temperature in the fluid (Fig. 2–71). For example, in the absence of a fan, heat transfer from the surface of the hot block in Fig. 2–70 will be by natural convection since any motion in the air in this case will be due to the rise of the warmer (and thus lighter) air near the surface and the fall of the cooler (and thus heavier) air to fill its place. Heat transfer between the block and surrounding air will be by conduction if the temperature difference between the air and the block is not large enough to overcome the resistance of air to move and thus to initiate natural convection currents.

Heat transfer processes that involve *change of phase* of a fluid are also considered to be convection because of the fluid motion induced during the process such as the rise of the vapor bubbles during *boiling* or the fall of the liquid droplets during *condensation*.

The rate of heat transfer by convection $\dot{Q}_{conv}$ is determined from **Newton's law of cooling**, expressed as

$$\dot{Q}_{conv} = hA(T_s - T_f) \quad \text{(W)} \quad (2\text{–}53)$$

where $h$ is the **convection heat transfer coefficient**, $A$ is the surface area through which heat transfer takes place, $T_s$ is the surface temperature, and $T_f$ is bulk fluid temperature away from the surface. (At the surface, the fluid temperature equals the surface temperature of the solid.)

The convection heat transfer coefficient $h$ is not a property of the fluid. It is an experimentally determined parameter whose value depends on all the variables that influence convection such as the surface geometry, the nature of fluid motion, the properties of the fluid, and the bulk fluid velocity. Typical values of $h$, in $W/m^2 \cdot K$, are in the range of 2–25 for the free convection of gases, 50–1000 for the free convection of liquids, 25–250 for the forced convection of gases, 50–20,000 for the forced convection of liquids, and 2500–100,000 for convection in boiling and condensation processes.

**Radiation** is the energy emitted by matter in the form of electromagnetic waves (or photons) as a result of the changes in the electronic configurations of the atoms or molecules. Unlike conduction and convection, the transfer of energy by radiation does not require the presence of an intervening medium (Fig. 2–72). In fact, energy transfer by radiation is fastest (at the speed of light) and it suffers no attenuation in a vacuum. This is exactly how the energy of the sun reaches the earth.

In heat transfer studies, we are interested in *thermal radiation,* which is the form of radiation emitted by bodies because of their temperature. It differs from other forms of electromagnetic radiation such as X-rays, gamma rays, microwaves, radio waves, and television waves that are not related to temperature. All bodies at a temperature above absolute zero emit thermal radiation.

Radiation is a *volumetric phenomenon,* and all solids, liquids, and gases emit, absorb, or transmit radiation of varying degrees. However, radiation is usually considered to be a *surface phenomenon* for solids that are opaque to

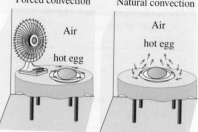

**FIGURE 2–71**
The cooling of a boiled egg by forced and natural convection.

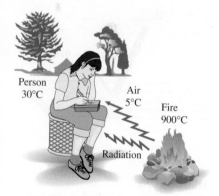

**FIGURE 2–72**
Unlike conduction and convection, heat transfer by radiation can occur between two bodies, even when they are separated by a medium colder than both of them.

## TABLE 2–4

Emissivity of some materials at 300 K

| Material | Emissivity |
| --- | --- |
| Aluminum foil | 0.07 |
| Anodized aluminum | 0.82 |
| Polished copper | 0.03 |
| Polished gold | 0.03 |
| Polished silver | 0.02 |
| Polished stainless steel | 0.17 |
| Black paint | 0.98 |
| White paint | 0.90 |
| White paper | 0.92–0.97 |
| Asphalt pavement | 0.85–0.93 |
| Red brick | 0.93–0.96 |
| Human skin | 0.95 |
| Wood | 0.82–0.92 |
| Soil | 0.93–0.96 |
| Water | 0.96 |
| Vegetation | 0.92–0.96 |

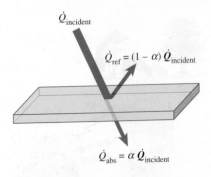

### FIGURE 2–73

The absorption of radiation incident on an opaque surface of absorptivity $\alpha$.

thermal radiation such as metals, wood, and rocks since the radiation emitted by the interior regions of such material can never reach the surface, and the radiation incident on such bodies is usually absorbed within a few microns from the surface.

The maximum rate of radiation that can be emitted from a surface at an *absolute* temperature $T_s$ is given by the *Stefan–Boltzmann law* as

$$\dot{Q}_{emit,max} = \sigma A T_s^4 \quad (W) \qquad (2\text{--}54)$$

where A is the surface area and $\sigma = 5.67 \times 10^{-8}$ W/m²·K⁴ is the **Stefan–Boltzmann constant**. The idealized surface that emits radiation at this maximum rate is called a **blackbody**, and the radiation emitted by a blackbody is called **blackbody radiation**. The radiation emitted by all *real* surfaces is less than the radiation emitted by a blackbody at the same temperatures and is expressed as

$$\dot{Q}_{emit} = \varepsilon \sigma A T_s^4 \quad (W) \qquad (2\text{--}55)$$

where $\varepsilon$ is the **emissivity** of the surface. The property emissivity, whose value is in the range $0 \leq \varepsilon \leq 1$, is a measure of how closely a surface approximates a blackbody for which $\varepsilon = 1$. The emissivities of some surfaces are given in Table 2–4.

Another important radiation property of a surface is its **absorptivity**, $\alpha$, which is the fraction of the radiation energy incident on a surface that is absorbed by the surface. Like emissivity, its value is in the range $0 \leq \alpha \leq 1$. A blackbody absorbs the entire radiation incident on it. That is, a blackbody is a perfect absorber ($\alpha = 1$) as well as a perfect emitter.

In general, both $\varepsilon$ and $\alpha$ of a surface depend on the temperature and the wavelength of the radiation. **Kirchhoff's law** of radiation states that the emissivity and the absorptivity of a surface are equal at the same temperature and wavelength. In most practical applications, the dependence of $\varepsilon$ and $\alpha$ on the temperature and wavelength is ignored, and the average absorptivity of a surface is taken to be equal to its average emissivity. The rate at which a surface absorbs radiation is determined from (Fig. 2–73)

$$\dot{Q}_{abs} = \alpha \dot{Q}_{incident} \quad (W) \qquad (2\text{--}56)$$

where $\dot{Q}_{incident}$ is the rate at which radiation is incident on the surface and $\alpha$ is the absorptivity of the surface. For opaque (nontransparent) surfaces, the portion of incident radiation that is not absorbed by the surface is reflected back.

The difference between the rates of radiation emitted by the surface and the radiation absorbed is the *net* radiation heat transfer. If the rate of radiation absorption is greater than the rate of radiation emission, the surface is said to be *gaining* energy by radiation. Otherwise, the surface is said to be *losing* energy by radiation. In general, the determination of the net rate of heat transfer by radiation between two surfaces is a complicated matter since it depends on the properties of the surfaces, their orientation relative to each other, and the interaction of the medium between the surfaces with radiation. However, in the special case of a relatively small surface of emissivity $\varepsilon$ and surface area A at *absolute* temperature $T_s$ that is completely enclosed by a much larger surface at *absolute* temperature $T_{surr}$

separated by a gas (such as air) that does not intervene with radiation (i.e., the amount of radiation emitted, absorbed, or scattered by the medium is negligible), the net rate of radiation heat transfer between these two surfaces is determined from (Fig. 2–74)

$$\dot{Q}_{rad} = \varepsilon\sigma A(T_s^4 - T_{surr}^4) \quad (W) \tag{2–57}$$

In this special case, the emissivity and the surface area of the surrounding surface do not have any effect on the net radiation heat transfer.

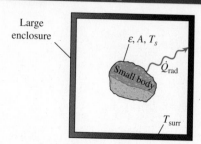

**FIGURE 2–74**
Radiation heat transfer between a body and the inner surfaces of a much larger enclosure that completely surrounds it.

### EXAMPLE 2–18    Heat Transfer from a Person

Consider a person standing in a breezy room at 20°C. Determine the total rate of heat transfer from this person if the exposed surface area and the average outer surface temperature of the person are 1.6 m$^2$ and 29°C, respectively, and the convection heat transfer coefficient is 6 W/m$^2$·°C (Fig. 2–75).

**SOLUTION** A person is standing in a breezy room. The total rate of heat loss from the person is to be determined.

**Assumptions** **1** The emissivity and heat transfer coefficient are constant and uniform. **2** Heat conduction through the feet is negligible. **3** Heat loss by evaporation is disregarded.

**Analysis** The heat transfer between the person and the air in the room will be by convection (instead of conduction) since it is conceivable that the air in the vicinity of the skin or clothing will warm up and rise as a result of heat transfer from the body, initiating natural convection currents. It appears that the experimentally determined value for the rate of convection heat transfer in this case is 6 W per unit surface area (m$^2$) per unit temperature difference (in K or °C) between the person and the air away from the person. Thus, the rate of convection heat transfer from the person to the air in the room is, from Eq. 2–53,

$$\dot{Q}_{conv} = hA(T_s - T_f)$$
$$= (6 \text{ W/m}^2\text{·°C})(1.6 \text{ m}^2)(29 - 20) \text{ °C}$$
$$= 86.4 \text{ W}$$

The person will also lose heat by radiation to the surrounding wall surfaces. We take the temperature of the surfaces of the walls, ceiling, and the floor to be equal to the air temperature in this case for simplicity, but we recognize that this does not need to be the case. These surfaces may be at a higher or lower temperature than the average temperature of the room air, depending on the outdoor conditions and the structure of the walls. Considering that air does not intervene with radiation and the person is completely enclosed by the surrounding surfaces, the net rate of radiation heat transfer from the person to the surrounding walls, ceiling, and the floor is, from Eq. 2–57,

$$\dot{Q}_{rad} = \varepsilon\sigma A(T_s^4 - T_{surr}^4)$$
$$= (0.95)(5.67 \times 10^{-8} \text{ W/m}^2\text{·K}^4)(1.6 \text{ m}^2) \times [(29 + 273)^4 - (20 + 273)^4]\text{K}^4$$
$$= 81.7 \text{ W}$$

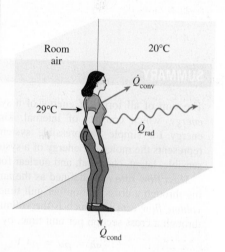

**FIGURE 2–75**
Heat transfer from the person described in Example 2–18.

Note that we must use *absolute* temperatures in radiation calculations. Also note that we used the emissivity value for the skin and clothing at room temperature since the emissivity is not expected to change significantly at a slightly higher temperature.

Then the rate of total heat transfer from the body is determined by adding these two quantities to be

$$\dot{Q}_{total} = \dot{Q}_{conv} + \dot{Q}_{rad} = 86.4 + 81.7 = \textbf{168.1 W}$$

The heat transfer would be much higher if the person were not dressed since the exposed surface temperature would be higher. Thus, an important function of the clothes is to serve as a barrier against heat transfer.

**Discussion** In the above calculations, heat transfer through the feet to the floor by conduction, which is usually very small, is neglected. Heat transfer from the skin by perspiration, which is the dominant mode of heat transfer in hot environments, is not considered here.

# SUMMARY

The sum of all forms of energy of a system is called *total energy,* which consists of internal, kinetic, and potential energy for simple compressible systems. *Internal energy* represents the molecular energy of a system and may exist in sensible, latent, chemical, and nuclear forms.

*Mass flow rate* $\dot{m}$ is defined as the amount of mass flowing through a cross section per unit time. It is related to the *volume flow rate* $\dot{V}$, which is the volume of a fluid flowing through a cross section per unit time, by

$$\dot{m} = \rho\dot{V} = \rho A_c V_{avg}$$

The energy flow rate associated with a fluid flowing at a rate of $\dot{m}$ is

$$\dot{E} = \dot{m}e$$

which is analogous to $E = me$.

The *mechanical energy* is defined as *the form of energy that can be converted to mechanical work completely and directly by a mechanical device such as an ideal turbine.* It is expressed on a unit mass basis and rate form as

$$e_{mech} = \frac{P}{\rho} + \frac{V^2}{2} + gz$$

and

$$\dot{E}_{mech} = \dot{m}e_{mech} = \dot{m}\left(\frac{P}{\rho} + \frac{V^2}{2} + gz\right)$$

where $P/\rho$ is the *flow energy*, $V^2/2$ is the *kinetic energy*, and $gz$ is the *potential energy* of the fluid per unit mass.

Energy can cross the boundaries of a closed system in the form of heat or work. For control volumes, energy can also be transported by mass. If the energy transfer is due to a temperature difference between a closed system and its surroundings, it is *heat;* otherwise, it is *work.*

Work is the energy transferred as a force acts on a system through a distance. Various forms of work are expressed as follows:

*Electrical work:* $W_e = \mathbf{V}I\Delta t$

*Shaft work:* $W_{sh} = 2\pi nT$

*Spring work:* $W_{spring} = \frac{1}{2}k(x_2^2 - x_1^2)$

The *first law of thermodynamics* is essentially an expression of the conservation of energy principle, also called the *energy balance.* The general mass and energy balances for *any system* undergoing *any process* can be expressed as

$$\underbrace{E_{in} - E_{out}}_{\substack{\text{Net energy transfer} \\ \text{by heat, work, and mass}}} = \underbrace{\Delta E_{system}}_{\substack{\text{Change in internal, kinetic,} \\ \text{potential, etc., energies}}} \quad \text{(kJ)}$$

It can also be expressed in the *rate form* as

$$\underbrace{\dot{E}_{in} - \dot{E}_{out}}_{\substack{\text{Rate of net energy transfer} \\ \text{by heat, work, and mass}}} = \underbrace{dE_{system}/dt}_{\substack{\text{Rate of change in internal,} \\ \text{kinetic, potential, etc., energies}}} \quad \text{(kW)}$$

The efficiencies of various devices are defined as

$$\eta_{pump} = \frac{\Delta\dot{E}_{mech,fluid}}{\dot{W}_{shaft,in}} = \frac{\dot{W}_{pump,u}}{\dot{W}_{pump}}$$

$$\eta_{turbine} = \frac{\dot{W}_{shaft,out}}{|\Delta \dot{E}_{mech,fluid}|} = \frac{\dot{W}_{turbine}}{\dot{W}_{turbine,e}}$$

$$\eta_{motor} = \frac{\text{Mechanical power output}}{\text{Electric power input}} = \frac{\dot{W}_{shaft,out}}{\dot{W}_{elect,in}}$$

$$\eta_{generator} = \frac{\text{Electric power output}}{\text{Mechanical power input}} = \frac{\dot{W}_{elect,out}}{\dot{W}_{shaft,in}}$$

$$\eta_{pump-motor} = \eta_{pump}\eta_{motor} = \frac{\Delta \dot{E}_{mech,fluid}}{\dot{W}_{elect,in}}$$

$$\eta_{turbine-gen} = \eta_{turbine}\eta_{generator} = \frac{\dot{W}_{elect,out}}{|\Delta \dot{E}_{mech,fluid}|}$$

The conversion of energy from one form to another is often associated with adverse effects on the environment, and environmental impact should be an important consideration in the conversion and utilization of energy.

## REFERENCES AND SUGGESTED READINGS

1. ASHRAE *Handbook of Fundamentals*. SI version. Atlanta, GA: American Society of Heating, Refrigerating, and Air-Conditioning Engineers, Inc., 1993.

2. Y. A. Çengel. "An Intuitive and Unified Approach to Teaching Thermodynamics." ASME International Mechanical Engineering Congress and Exposition, Atlanta, Georgia, AES-Vol. 36, pp. 251–260, November 17–22, 1996.

## PROBLEMS*

### Forms of Energy

**2–1C**  What is total energy? Identify the different forms of energy that constitute the total energy.

**2–2C**  List the forms of energy that contribute to the internal energy of a system.

**2–3C**  How are heat, internal energy, and thermal energy related to each other?

**2–4C**  What is mechanical energy? How does it differ from thermal energy? What are the forms of mechanical energy of a fluid stream?

**2–5C**  Natural gas, which is mostly methane $CH_4$, is a fuel and a major energy source. Can we say the same about hydrogen gas, $H_2$?

**2–6C**  Portable electric heaters are commonly used to heat small rooms. Explain the energy transformation involved during this heating process.

**2–7C**  Consider the process of heating water on top of an electric range. What are the forms of energy involved during this process? What are the energy transformations that take place?

**2–8**  Calculate the total potential energy, in kJ, of an object with a mass of 90 kg when it is 3 m above a datum level at a location where standard gravitational acceleration exists.

**2–9**  A person gets into an elevator at the lobby level of a hotel together with his 30-kg suitcase, and gets out at the 10th floor 35 m above. Determine the amount of energy consumed by the motor of the elevator that is now stored in the suitcase.

**2–10**  Electric power is to be generated by installing a hydraulic turbine–generator at a site 120 m below the free surface of a large water reservoir that can supply water at a rate of 2400 kg/s steadily. Determine the power generation potential.

---

* Problems designated by a "C" are concept questions, and students are encouraged to answer them all. Problems with the 🔘 icon are solved using EES, and complete solutions together with parametric studies are included on the text website. Problems with the 📘 icon are comprehensive in nature, and are intended to be solved with an equation solver such as EES.

**2–11** At a certain location, wind is blowing steadily at 10 m/s. Determine the mechanical energy of air per unit mass and the power generation potential of a wind turbine with 60-m-diameter blades at that location. Take the air density to be 1.25 kg/m³.

**2–12** A water jet that leaves a nozzle at 60 m/s at a flow rate of 120 kg/s is to be used to generate power by striking the buckets located on the perimeter of a wheel. Determine the power generation potential of this water jet.

**2–13** Two sites are being considered for wind power generation. In the first site, the wind blows steadily at 7 m/s for 3000 hours per year, whereas in the second site the wind blows at 10 m/s for 1500 hours per year. Assuming the wind velocity is negligible at other times for simplicity, determine which is a better site for wind power generation. *Hint:* Note that the mass flow rate of air is proportional to wind velocity.

**2–14** A river flowing steadily at a rate of 175 m³/s is considered for hydroelectric power generation. It is determined that a dam can be built to collect water and release it from an elevation difference of 80 m to generate power. Determine how much power can be generated from this river water after the dam is filled.

**2–15** Consider a river flowing toward a lake at an average velocity of 3 m/s at a rate of 500 m³/s at a location 90 m above the lake surface. Determine the total mechanical energy of the river water per unit mass and the power generation potential of the entire river at that location.

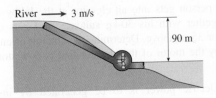

**FIGURE P2–15**

### Energy Transfer by Heat and Work

**2–16C** When is the energy crossing the boundaries of a closed system heat and when is it work?

**2–17C** Consider an automobile traveling at a constant speed along a road. Determine the direction of the heat and work interactions, taking the following as the system: (a) the car radiator, (b) the car engine, (c) the car wheels, (d) the road, and (e) the air surrounding the car.

**2–18C** Consider an electric refrigerator located in a room. Determine the direction of the work and heat interactions (in or out) when the following are taken as the system: (a) the contents of the refrigerator, (b) all parts of the refrigerator including the contents, and (c) everything contained within the room during a winter day.

**FIGURE P2–18C**

**2–19C** A gas in a piston-cylinder device is compressed, and as a result its temperature rises. Is this a heat or work interaction?

**2–20C** A room is heated by an iron that is left plugged in. Is this a heat or work interaction? Take the entire room, including the iron, as the system.

**2–21C** A room is heated as a result of solar radiation coming in through the windows. Is this a heat or work interaction for the room?

**2–22C** An insulated room is heated by burning candles. Is this a heat or work interaction? Take the entire room, including the candles, as the system.

**2–23** A small electrical motor produces 5 W of mechanical power. What is this power in (a) N, m, and s units; and (b) kg, m, and s units? *Answers:* (a) 5 N·m/s, (b) 5 kg·m²/s³

**2–24** A model aircraft internal-combustion engine produces 10 W of power. How much power is this in (a) lbf·ft/s and (b) hp?

## Mechanical Forms of Work

**2–25C** Lifting a weight to a height of 20 m takes 20 s for one crane and 10 s for another. Is there any difference in the amount of work done on the weight by each crane?

**2–26** An 80-kg man is pushing a 45-kg cart with its contents up a ramp that is inclined at an angle of 10° from the horizontal. Determine the work needed to move along this ramp a distance of 30 m considering (a) the man and (b) the cart and its contents as the system. Express your answers in both N·m and kJ.

**FIGURE P2–26**

©McGraw-Hill Education/Lars A.Niki

**2–27** The force F required to compress a spring a distance x is given by $F - F_0 = kx$ where k is the spring constant and $F_0$ is the preload. Determine the work required to compress a spring whose spring constant is $k = 3.5$ kN/cm a distance of one cm starting from its free length where $F_0 = 0$ kN. Express your answer in both kN·m and kJ.

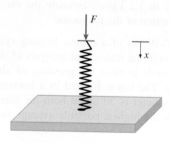

**FIGURE P2–27**

**2–28** Determine the energy required to accelerate a 1300-kg car from 10 to 60 km/h on an uphill road with a vertical rise of 40 m.

**2–29** Determine the torque applied to the shaft of a car that transmits 335 kW and rotates at a rate of 3000 rpm.

**2–30** Determine the work required to deflect a linear spring with a spring constant of 70 kN/m by 20 cm from its rest position.

**2–31** A ski lift has a one-way length of 1 km and a vertical rise of 200 m. The chairs are spaced 20 m apart, and each chair can seat three people. The lift is operating at a steady speed of 10 km/h. Neglecting friction and air drag and assuming that the average mass of each loaded chair is 250 kg, determine the power required to operate this ski lift. Also estimate the power required to accelerate this ski lift in 5 s to its operating speed when it is first turned on.

**2–32** The engine of a 1500-kg automobile has a power rating of 75 kW. Determine the time required to accelerate this car from rest to a speed of 100 km/h at full power on a level road. Is your answer realistic?

**2–33** Determine the power required for a 1150-kg car to climb a 100-m-long uphill road with a slope of 30° (from horizontal) in 12 s (a) at a constant velocity, (b) from rest to a final velocity of 30 m/s, and (c) from 35 m/s to a final velocity of 5 m/s. Disregard friction, air drag, and rolling resistance. *Answers:* (a) 47.0 kW, (b) 90.1 kW, (c) −10.5 kW

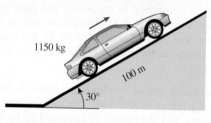

**FIGURE P2–33**

## The First Law of Thermodynamics

**2–34C** What are the different mechanisms for transferring energy to or from a control volume?

**2–35C** On a hot summer day, a student turns his fan on when he leaves his room in the morning. When he returns in the evening, will the room be warmer or cooler than the neighboring rooms? Why? Assume all the doors and windows are kept closed.

**2–36** Water is being heated in a closed pan on top of a range while being stirred by a paddle wheel. During the process, 30 kJ of heat is transferred to the water, and 5 kJ of heat is lost to the surrounding air. The paddle-wheel work amounts to 500 N·m. Determine the final energy of the system if its initial energy is 10 kJ. *Answer: 35.5 kJ*

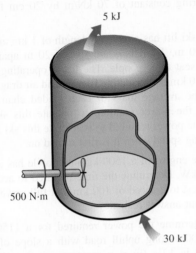

5 kJ

500 N·m

30 kJ

**FIGURE P2–36**

**2–37** A vertical piston-cylinder device contains water and is being heated on top of a range. During the process, 65 kJ of heat is transferred to the water, and heat losses from the side walls amount to 8 kJ. The piston rises as a result of evaporation, and 5 kJ of work is done by the vapor. Determine the change in the energy of the water for this process. *Answer: 52 kJ*

**2–38** At winter design conditions, a house is projected to lose heat at a rate of 60,000 kJ/h. The internal heat gain from people, lights, and appliances is estimated to be 6000 kJ/h. If this house is to be heated by electric resistance heaters, determine the required rated power of these heaters in kW to maintain the house at constant temperature.

**2–39** A water pump that consumes 2 kW of electric power when operating is claimed to take in water from a lake and pump it to a pool whose free surface is 30 m above the free surface of the lake at a rate of 50 L/s. Determine if this claim is reasonable.

**2–40** A classroom that normally contains 40 people is to be air-conditioned with window air-conditioning units of 5-kW cooling capacity. A person at rest may be assumed to dissipate heat at a rate of about 360 kJ/h. There are 10 lightbulbs in the room, each with a rating of 100 W. The rate of heat transfer to the classroom through the walls and the windows is estimated to be 15,000 kJ/h. If the room air is to be maintained at a constant temperature of 21°C, determine the number of window air-conditioning units required. *Answer: 2 units*

**2–41** A university campus has 200 classrooms and 400 faculty offices. The classrooms are equipped with 12 fluorescent tubes, each consuming 110 W, including the electricity used by the ballasts. The faculty offices, on average, have half as many tubes. The campus is open 240 days a year. The classrooms and faculty offices are not occupied an average of 4 h a day, but the lights are kept on. If the unit cost of electricity is $0.11/kWh, determine how much the campus will save a year if the lights in the classrooms and faculty offices are turned off during unoccupied periods.

**2–42** The lighting requirements of an industrial facility are being met by 700 40-W standard fluorescent lamps. The lamps are close to completing their service life and are to be replaced by their 34-W high-efficiency counterparts that operate on the existing standard ballasts. The standard and high-efficiency fluorescent lamps can be purchased in quantity at a cost of $1.77 and $2.26 each, respectively. The facility operates 2800 hours a year, and all of the lamps are kept on during operating hours. Taking the unit cost of electricity to be $0.105/kWh and the ballast factor to be 1.1 (i.e., ballasts consume 10 percent of the rated power of the lamps), determine how much energy and money will be saved per year as a result of switching to the high-efficiency fluorescent lamps. Also, determine the simple payback period.

**2–43** Consider a room that is initially at the outdoor temperature of 20°C. The room contains a 40-W lightbulb, a 110-W TV set, a 300-W refrigerator, and a 1200-W iron. Assuming no heat transfer through the walls, determine the rate of increase of the energy content of the room when all of these electric devices are on.

**2–44** Consider a fan located in a 1 m × 1 m square duct. Velocities at various points at the outlet are measured, and the average flow velocity is determined to be 7 m/s. Taking the air density to 1.2 kg/m³, estimate the minimum electric power consumption of the fan motor.

**2–45** The 60-W fan of a central heating system is to circulate air through the ducts. The analysis of the flow shows that the fan needs to raise the pressure of air by 50 Pa to maintain flow. The fan is located in a horizontal flow section whose diameter is 30 cm at both the inlet and the outlet. Determine the highest possible average flow velocity in the duct.

**2–46** The driving force for fluid flow is the pressure difference, and a pump operates by raising the pressure of a fluid (by converting the mechanical shaft work to flow energy). A gasoline pump is measured to consume 3.8 kW of electric power when operating. If the pressure differential between the outlet and inlet of the pump is measured to be 7 kPa and the changes in velocity and elevation are negligible, determine the maximum possible volume flow rate of gasoline.

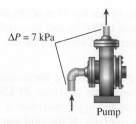

$\Delta P = 7$ kPa

Pump

**FIGURE P2–46**

**2–47** An escalator in a shopping center is designed to move 50 people, 75 kg each, at a constant speed of 0.6 m/s at 45° slope. Determine the minimum power input needed to drive this escalator. What would your answer be if the escalator velocity were to be doubled?

**2–48** Consider a 1400-kg car cruising at constant speed of 70 km/s. Now the car starts to pass another car, by accelerating to 110 km/h in 5 s. Determine the additional power needed to achieve this acceleration. What would your answer be if the total mass of the car were only 700 kg? *Answers:* 77.8 kW, 38.9 kW

## Energy Conversion Efficiencies

**2–49C** How is the combined pump–motor efficiency of a pump and motor system defined? Can the combined pump–motor efficiency be greater than either the pump or the motor efficiency?

**2–50C** Define turbine efficiency, generator efficiency, and combined turbine–generator efficiency.

**2–51C** Can the combined turbine-generator efficiency be greater than either the turbine efficiency or the generator efficiency? Explain.

**2–52** Consider a 24-kW hooded electric open burner in an area where the unit costs of electricity and natural gas are $0.10/kWh and $1.20/therm (1 therm = 105,500 kJ), respectively. The efficiency of open burners can be taken to be 73 percent for electric burners and 38 percent for gas burners. Determine the rate of energy consumption and the unit cost of utilized energy for both electric and gas burners.

**2–53** A 75-hp (shaft output) motor that has an efficiency of 91.0 percent is worn out and is to be replaced by a high-efficiency motor that has an efficiency of 95.4 percent. The motor operates 4368 hours a year at a load factor of 0.75. Taking the cost of electricity to be $0.12/kWh, determine the amount of energy and money saved as a result of installing the high-efficiency motor instead of the standard motor. Also, determine the simple payback period if the purchase prices of the standard and high-efficiency motors are $5449 and $5520, respectively.

**2–54** Consider an electric motor with a shaft power output of 20 kW and an efficiency of 88 percent. Determine the rate at which the motor dissipates heat to the room it is in when the motor operates at full load. In winter, this room is normally heated by a 2-kW resistance heater. Determine if it is necessary to turn the heater on when the motor runs at full load.

**2–55** The steam requirements of a manufacturing facility are being met by a boiler whose rated heat input is $5.2 \times 10^6$ kJ/h. The combustion efficiency of the boiler is measured to be 0.7 by a hand-held flue gas analyzer. After tuning up the boiler, the combustion efficiency rises to 0.8. The boiler operates 4200 hours a year intermittently. Taking the unit cost of energy to be $4.12/$10^6$ kJ, determine the annual energy and cost savings as a result of tuning up the boiler.

**2–56** Reconsider Prob. 2–55. Using EES (or other) software, study the effects of the unit cost of energy, the new combustion efficiency on the annual energy, and cost savings. Let the efficiency vary from 0.7 to 0.9, and the unit cost to vary from $4 to $6 per million kJ. Plot the annual energy and cost savings against the efficiency for unit costs of $4, $5, and $6 per million kJ, and discuss the results.

**2–57** A geothermal pump is used to pump brine whose density is 1050 kg/m³ at a rate of 0.3 m³/s from a depth of 200 m. For a pump efficiency of 74 percent, determine the required power input to the pump. Disregard frictional losses in the pipes, and assume the geothermal water at 200 m depth to be exposed to the atmosphere.

**2–58** An exercise room has 6 weight-lifting machines that have no motors and 7 treadmills each equipped with a 2.5-hp (shaft output) motor. The motors operate at an average load factor of 0.7, at which their efficiency is 0.77. During peak evening hours, all 12 pieces of exercising equipment are used continuously, and there are also two people doing light exercises while waiting in line for one piece of the equipment. Assuming the average rate of heat dissipation from people in an exercise room is 600 W, determine the rate of heat gain of the exercise room from people and the equipment at peak load conditions.

**2–59** A room is cooled by circulating chilled water through a heat exchanger located in a room. The air is circulated through the heat exchanger by a 0.25-hp (shaft output) fan. Typical efficiency of small electric motors driving 0.25-hp equipment is 54 percent. Determine the rate of heat supply by the fan–motor assembly to the room.

**2–60** The water in a large lake is to be used to generate electricity by the installation of a hydraulic turbine-generator at a location where the depth of the water is 50 m. Water is to be supplied at a rate of 5000 kg/s. If the electric power generated is measured to be 1862 kW and the generator efficiency is 95 percent, determine (*a*) the overall efficiency of the turbine—generator, (*b*) the mechanical efficiency of the turbine, and (*c*) the shaft power supplied by the turbine to the generator.

**2–61** A 7-hp (shaft) pump is used to raise water to an elevation of 15 m. If the mechanical efficiency of the pump is 82 percent, determine the maximum volume flow rate of water.

**2–62** At a certain location, wind is blowing steadily at 7 m/s. Determine the mechanical energy of air per unit mass and the power generation potential of a wind turbine with 80-m-diameter blades at that location. Also determine the actual electric power generation assuming an overall efficiency of 30 percent. Take the air density to be 1.25 kg/m³.

**2–63** ![EES] Reconsider Prob. 2–62. Using EES (or other) software, investigate the effect of wind velocity and the blade span diameter on wind power generation. Let the velocity vary from 5 to 20 m/s in increments of 5 m/s, and the diameter vary from 20 to 120 m in increments of 20 m. Tabulate the results, and discuss their significance.

**2–64** Water is pumped from a lake to a storage tank 15 m above at a rate of 70 L/s while consuming 15.4 kW of electric power. Disregarding any frictional losses in the pipes and any changes in kinetic energy, determine (a) the overall efficiency of the pump–motor unit and (b) the pressure difference between the inlet and the exit of the pump.

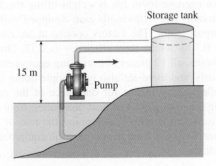

**FIGURE P2–64**

**2–65** Large wind turbines with blade span diameters of over 100 m are available for electric power generation. Consider a wind turbine with a blade span diameter of 100 m installed at a site subjected to steady winds at 8 m/s. Taking the overall efficiency of the wind turbine to be 32 percent and the air density to be 1.25 kg/m³, determine the electric power generated by this wind turbine. Also, assuming steady winds of 8 m/s during a 24-hour period, determine the amount of electric energy and the revenue generated per day for a unit price of $0.09/kWh for electricity.

**2–66** A hydraulic turbine has 85 m of elevation difference available at a flow rate of 0.25 m³/s, and its overall turbine–generator efficiency is 91 percent. Determine the electric power output of this turbine.

**2–67** Water is pumped from a lower reservoir to a higher reservoir by a pump that provides 20 kW of shaft power. The free surface of the upper reservoir is 45 m higher than that of the lower reservoir. If the flow rate of water is measured to be 0.03 m³/s, determine mechanical power that is converted to thermal energy during this process due to frictional effects.

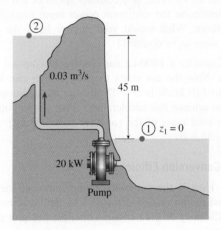

**FIGURE P2–67**

**2–68** The water behind Hoover Dam in Nevada is 206 m higher than the Colorado River below it. At what rate must water pass through the hydraulic turbines of this dam to produce 100 MW of power if the turbines are 100 percent efficient?

**FIGURE P2–68**

*Photo by Lynn Betts, USDA Natural Resources Conservation Society*

**2–69** An oil pump is drawing 44 kW of electric power while pumping oil with $\rho = 860$ kg/m³ at a rate of 0.1 m³/s.

The inlet and outlet diameters of the pipe are 8 cm and 12 cm, respectively. If the pressure rise of oil in the pump is measured to be 500 kPa and the motor efficiency is 90 percent, determine the mechanical efficiency of the pump.

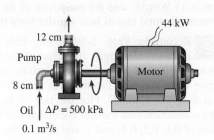

**FIGURE P2–69**

## Energy and Environment

**2–70C** How does energy conversion affect the environment? What are the primary chemicals that pollute the air? What is the primary source of these pollutants?

**2–71C** What is acid rain? Why is it called a "rain"? How do the acids form in the atmosphere? What are the adverse effects of acid rain on the environment?

**2–72C** Why is carbon monoxide a dangerous air pollutant? How does it affect human health at low and at high levels?

**2–73C** What is the greenhouse effect? How does the excess $CO_2$ gas in the atmosphere cause the greenhouse effect? What are the potential long-term consequences of greenhouse effect? How can we combat this problem?

**2–74C** What is smog? What does it consist of? How does ground-level ozone form? What are the adverse effects of ozone on human health?

**2–75** A Ford Taurus driven 20,000 km a year will use about 2500 L of gasoline compared to a Ford Explorer that would use 3200 L. About 2.4 kg of $CO_2$, which causes global warming, is released to the atmosphere when a liter of gasoline is burned. Determine the extra amount of $CO_2$ production a man is responsible for during a 5-year period if he trades his Taurus for an Explorer.

**2–76** Consider a household that uses 14,000 kWh of electricity per year and 3400 L of fuel oil during a heating season. The average amount of $CO_2$ produced is 3.2 kg/L of fuel oil and 0.70 kg/kWh of electricity. If this household reduces its oil and electricity usage by 15 percent as a result of implementing some energy conservation measures, determine the reduction in the amount of $CO_2$ emissions by that household per year.

**2–77** When a hydrocarbon fuel is burned, almost all of the carbon in the fuel burns completely to form $CO_2$ (carbon dioxide), which is the principal gas causing the greenhouse effect and thus global climate change. On average, 0.59 kg of $CO_2$ is produced for each kWh of electricity generated from a power plant that burns natural gas. A typical new household refrigerator uses about 700 kWh of electricity per year. Determine the amount of $CO_2$ production that is due to the refrigerators in a city with 300,000 households.

**2–78** Repeat Prob. 2–77 assuming the electricity is produced by a power plant that burns coal. The average production of $CO_2$ in this case is 1.1 kg per kWh.

**2–79** A typical car driven 20,000 km a year emits to the atmosphere about 11 kg per year of $NO_x$ (nitrogen oxides), which cause smog in major population areas. Natural gas burned in the furnace emits about 4.3 g of $NO_x$ per therm (1 therm = 105,500 kJ), and the electric power plants emit about 7.1 g of $NO_x$ per kWh of electricity produced. Consider a household that has two cars and consumes 9000 kWh of electricity and 1200 therms of natural gas. Determine the amount of $NO_x$ emission to the atmosphere per year for which this household is responsible.

**FIGURE P2–79**

## Special Topic: Mechanisms of Heat Transfer

**2–80C** What are the mechanisms of heat transfer?

**2–81C** Which is a better heat conductor, diamond or silver?

**2–82C** How does forced convection differ from natural convection?

**2–83C** What is a blackbody? How do real bodies differ from a blackbody?

**2–84C** Define emissivity and absorptivity. What is Kirchhoff's law of radiation?

**2–85C** Does any of the energy of the sun reach the earth by conduction or convection?

**2–86** The inner and outer surfaces of a 5-m × 6-m brick wall of thickness 30 cm and thermal conductivity 0.69 W/m·°C are maintained at temperatures of 20°C and 5°C, respectively. Determine the rate of heat transfer through the wall, in W.

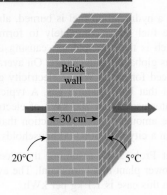

**FIGURE P2–86**

**2–87** The inner and outer surfaces of a 0.5-cm-thick 2-m $\times$ 2-m window glass in winter are 15°C and 6°C, respectively. If the thermal conductivity of the glass is 0.78 W/m·°C, determine the amount of heat loss, in kJ, through the glass over a period of 10 h. What would your answer be if the glass were 1-cm thick?

**2–88** Reconsider Prob. 2–87. Using EES (or other) software, investigate the effect of glass thickness on heat loss for the specified glass surface temperatures. Let the glass thickness vary from 0.2 to 2 cm. Plot the heat loss versus the glass thickness, and discuss the results.

**2–89** An aluminum pan whose thermal conductivity is 237 W/m·°C has a flat bottom whose diameter is 20 cm and thickness 0.6 cm. Heat is transferred steadily to boiling water in the pan through its bottom at a rate of 700 W. If the inner surface of the bottom of the pan is 105°C, determine the temperature of the outer surface of the bottom of the pan.

**2–90** The inner and outer glasses of a 2-m $\times$ 2-m double pane window are at 18°C and 6°C, respectively. If the 1-cm space between the two glasses is filled with still air, determine the rate of heat transfer through the air layer by conduction, in kW.

**2–91** Two surfaces of a 2-cm-thick plate are maintained at 0°C and 100°C, respectively. If it is determined that heat is transferred through the plate at a rate of 500 W/m$^2$, determine its thermal conductivity.

**2–92** Hot air at 80°C is blown over a 2-m $\times$ 4-m flat surface at 30°C. If the convection heat transfer coefficient is 55 W/m$^2$·°C, determine the rate of heat transfer from the air to the plate, in kW.

**2–93** For heat transfer purposes, a standing man can be modeled as a 30-cm diameter, 175-cm long vertical cylinder with both the top and bottom surfaces insulated and with the side surface at an average temperature of 34°C. For a convection heat transfer coefficient of 10 W/m$^2$·°C, determine the

rate of heat loss from this man by convection in an environment at 20°C. *Answer:* 231 W

**2–94** A 9-cm-diameter spherical ball whose surface is maintained at a temperature of 110°C is suspended in the middle of a room at 20°C. If the convection heat transfer coefficient is 15 W/m$^2$·C and the emissivity of the surface is 0.8, determine the total rate of heat transfer from the ball.

**2–95** Reconsider Prob. 2–94. Using EES (or other) software, investigate the effect of the convection heat transfer coefficient and surface emissivity on the heat transfer rate from the ball. Let the heat transfer coefficient vary from 5 to 30 W/m$^2$·°C. Plot the rate of heat transfer against the convection heat transfer coefficient for the surface emissivities of 0.1, 0.5, 0.8, and 1, and discuss the results.

**2–96** A 1000-W iron is left on the ironing board with its base exposed to the air at 23°C. The convection heat transfer coefficient between the base surface and the surrounding air is 20 W/m$^2$·°C. If the base has an emissivity of 0.4 and a surface area of 0.02 m$^2$, determine the temperature of the base of the iron.

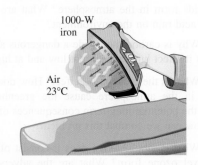

**FIGURE P2–96**

**2–97** A 7-cm-external-diameter, 18-m-long hot-water pipe at 80°C is losing heat to the surrounding air at 5°C by natural convection with a heat transfer coefficient of 25 W/m$^2$·°C. Determine the rate of heat loss from the pipe by natural convection, in kW.

**2–98** A thin metal plate is insulated on the back and exposed to solar radiation on the front surface. The exposed surface of the plate has an absorptivity of 0.8 for solar radiation. If solar radiation is incident on the plate at a rate of 450 W/m$^2$ and the surrounding air temperature is 25°C, determine the surface temperature of the plate when the heat loss by convection equals the solar energy absorbed by the plate. Assume the convection heat transfer coefficient to be 50 W/m$^2$·°C, and disregard heat loss by radiation.

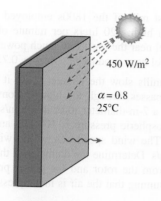

$450 \text{ W/m}^2$

$\alpha = 0.8$
$25°C$

**FIGURE P2–98**

**2–99** 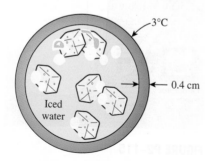 Reconsider Prob. 2–98. Using EES (or other) software, investigate the effect of the convection heat transfer coefficient on the surface temperature of the plate. Let the heat transfer coefficient vary from 10 to 90 W/$m^2\cdot°C$. Plot the surface temperature against the convection heat transfer coefficient, and discuss the results.

**2–100** The outer surface of a spacecraft in space has an emissivity of 0.6 and an absorptivity of 0.2 for solar radiation. If solar radiation is incident on the spacecraft at a rate of 1000 W/$m^2$, determine the surface temperature of the spacecraft when the radiation emitted equals the solar energy absorbed.

**2–101** Reconsider Prob. 2–100. Using EES (or other) software, investigate the effect of the surface emissivity and absorptivity of the spacecraft on the equilibrium surface temperature. Plot the surface temperature against emissivity for solar absorptivities of 0.1, 0.5, 0.8, and 1, and discuss the results.

**2–102** A hollow spherical iron container whose outer diameter is 40 cm and thickness is 0.4 cm is filled with iced water at 0°C. If the outer surface temperature is 3°C, determine the approximate rate of heat loss from the sphere, and the rate at which ice melts in the container.

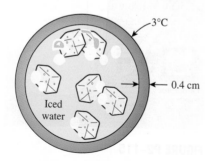

3°C

0.4 cm

Iced water

**FIGURE P2–102**

## Review Problems

**2–103** Consider a vertical elevator whose cabin has a total mass of 800 kg when fully loaded and 150 kg when empty. The weight of the elevator cabin is partially balanced by a 400-kg counterweight that is connected to the top of the cabin by cables that pass through a pulley located on top of the elevator well. Neglecting the weight of the cables and assuming the guide rails and the pulleys to be frictionless, determine (a) the power required while the fully loaded cabin is rising at a constant speed of 1.2 m/s and (b) the power required while the empty cabin is descending at a constant speed of 1.2 m/s.

What would your answer be to (a) if no counterweight were used? What would your answer be to (b) if a friction force of 800 N has developed between the cabin and the guide rails?

**2–104** Consider a homeowner who is replacing his 25-year-old natural gas furnace that has an efficiency of 55 percent. The homeowner is considering a conventional furnace that has an efficiency of 82 percent and costs $1600 and a high-efficiency furnace that has an efficiency of 95 percent and costs $2700. The homeowner would like to buy the high-efficiency furnace if the savings from the natural gas pay for the additional cost in less than 8 years. If the homeowner presently pays $1200 a year for heating, determine if he should buy the conventional or high-efficiency model.

**2–105** A homeowner is considering these heating systems for heating his house: Electric resistance heating with $0.12/kWh and 1 kWh = 3600 kJ, gas heating with $1.24/therm and 1 therm = 105,500 kJ, and oil heating with $2.3/gal and 1 gal of oil = 138,500 kJ. Assuming efficiencies of 100 percent for the electric furnace and 87 percent for the gas and oil furnaces, determine the heating system with the lowest energy cost.

**2–106** The U.S. Department of Energy estimates that 570,000 barrels of oil would be saved per day if every household in the United States lowered the thermostat setting in winter by 6°F (3.3°C). Assuming the average heating season to be 180 days and the cost of oil to be $110/barrel, determine how much money would be saved per year.

**2–107** The U.S. Department of Energy estimates that up to 10 percent of the energy use of a house can be saved by caulking and weatherstripping doors and windows to reduce air leaks at a cost of about $90 for materials for an average home with 12 windows and 2 doors. Caulking and weatherstripping every gas-heated home properly would save enough energy to heat about 4 million homes. The savings can be increased by installing storm windows. Determine how long it will take for the caulking and weatherstripping to pay for itself from the energy they save for a house whose annual energy use is $1500.

**2–108** The force required to compress the gas in a gas spring a distance $x$ is given by

$$F = \frac{\text{Constant}}{x^k}$$

where the constant is determined by the geometry of this device and $k$ is determined by the gas used in the device. One such device has a constant of 5.2 N·m$^{1.4}$ and $k = 1.4$. Determine the work, in kJ, required to compress this device from 5 cm to 18 cm.  *Answer:* 0.0173 kJ

**2–109** A man weighing 800 N pushes a block weighing 450 N along a horizontal plane. The dynamic coefficient of friction between the block and plane is 0.2. Assuming that the block is moving at constant speed, calculate the work required to move the block a distance of 30 m considering (*a*) the man and (*b*) the block as the system. Express your answers in both N·m and kJ.

**2–110** A diesel engine with an engine volume of 4.0 L and an engine speed of 2500 rpm operates on an air–fuel ratio of 18 kg air/kg fuel. The engine uses light diesel fuel that contains 750 ppm (parts per million) of sulfur by mass. All of this sulfur is exhausted to the environment where the sulfur is converted to sulfurous acid ($H_2SO_3$). If the rate of the air entering the engine is 336 kg/h, determine the mass flow rate of sulfur in the exhaust. Also, determine the mass flow rate of sulfurous acid added to the environment if for each kmol of sulfur in the exhaust, one kmol sulfurous acid will be added to the environment.

**2–111** Leaded gasoline contains lead that ends up in the engine exhaust. Lead is a very toxic engine emission. The use of leaded gasoline in the United States has been unlawful for most vehicles since the 1980s. However, leaded gasoline is still used in some parts of the world. Consider a city with 70,000 cars using leaded gasoline. The gasoline contains 0.15 g/L of lead and 50 percent of lead is exhausted to the environment. Assuming that an average car travels 15,000 km per year with a gasoline consumption of 8.5 L/100 km, determine the amount of lead put into the atmosphere per year in that city. *Answer:* 6694 kg

**2–112** Consider a TV set that consumes 120 W of electric power when it is on and is kept on for an average of 6 hours per day. For a unit electricity cost of 12 cents per kWh, determine the cost of electricity this TV consumes per month (30 days).

**2–113** A grist mill of the 1800s employed a water wheel that was 14 m high; 320 liters per minute of water flowed on to the wheel near the top. How much power, in kW, could this water wheel have produced?  *Answer:* 0.732 kW

**2–114** Windmills slow the air and cause it to fill a larger channel as it passes through the blades. Consider a circular windmill with a 7-m-diameter rotor in a 8 m/s wind on a day when the atmospheric pressure is 100 kPa and the temperature is 20°C. The wind speed behind the windmill is measured at 6.5 m/s. Determine the diameter of the wind channel downstream from the rotor and the power produced by this windmill, presuming that the air is incompressible.

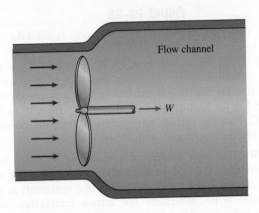

**FIGURE P2–114**

**2–115** In a hydroelectric power plant, 65 m$^3$/s of water flows from an elevation of 90 m to a turbine, where electric power is generated. The overall efficiency of the turbine–generator is 84 percent. Disregarding frictional losses in piping, estimate the electric power output of this plant.  *Answer:* 48.2 MW

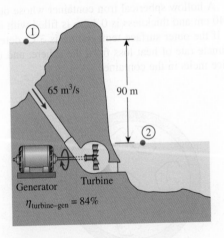

**FIGURE P2–115**

**2–116** The demand for electric power is usually much higher during the day than it is at night, and utility companies often sell power at night at much lower prices to encourage consumers to use the available power generation capacity and to avoid building new expensive power plants that will be used only a short time during peak periods. Utilities are also willing to purchase power produced during the day from private parties at a high price.

Suppose a utility company is selling electric power for $0.05/kWh at night and is willing to pay $0.12/kWh for power produced during the day. To take advantage of this opportunity, an entrepreneur is considering building a large reservoir 40 m above the lake level, pumping water from the lake to the reservoir at night using cheap power, and letting the water flow from the reservoir back to the lake during the day, producing power as the pump–motor operates as a turbine–generator during reverse flow. Preliminary analysis shows that a water flow rate of 2 m³/s can be used in either direction. The combined pump–motor and turbine–generator efficiencies are expected to be 75 percent each. Disregarding the frictional losses in piping and assuming the system operates for 10 h each in the pump and turbine modes during a typical day, determine the potential revenue this pump–turbine system can generate per year.

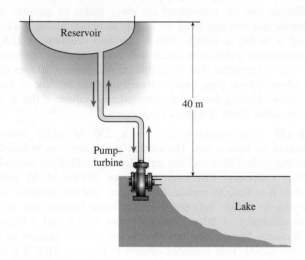

**FIGURE P2–116**

**2–117** The pump of a water distribution system is powered by a 15-kW electric motor whose efficiency is 90 percent. The water flow rate through the pump is 50 L/s. The diameters of the inlet and outlet pipes are the same, and the elevation difference across the pump is negligible. If the pressures at the inlet and outlet of the pump are measured to be 100 kPa and 300 kPa (absolute), respectively, determine the mechanical efficiency of the pump. *Answer:* 74.1 percent

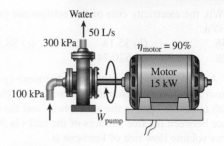

**FIGURE P2–117**

## Fundamentals of Engineering (FE) Exam Problems

**2–118** On a hot summer day, the air in a well-sealed room is circulated by a 0.50-hp fan driven by a 65 percent efficient motor. (Note that the motor delivers 0.50 hp of net shaft power to the fan.) The rate of energy supply from the fan-motor assembly to the room is
(a) 0.769 kJ/s     (b) 0.325 kJ/s     (c) 0.574 kJ/s
(d) 0.373 kJ/s     (e) 0.242 kJ/s

**2–119** A fan is to accelerate quiescent air to a velocity of 12 m/s at a rate of 3 m³/s. If the density of air is 1.15 kg/m³, the minimum power that must be supplied to the fan is
(a) 248 W     (b) 72 W     (c) 497 W
(d) 216 W     (e) 162 W

**2–120** A 2-kW electric resistance heater in a room is turned on and kept on for 50 min. The amount of energy transferred to the room by the heater is
(a) 2 kJ     (b) 100 kJ     (c) 3000 kJ
(d) 6000 kJ     (e) 12,000 kJ

**2–121** A 900-kg car cruising at a constant speed of 60 km/s is to accelerate to 100 km/h in 4 s. The additional power needed to achieve this acceleration is
(a) 56 kW     (b) 222 kW     (c) 2.5 kW
(d) 62 kW     (e) 90 kW

**2–122** The elevator of a large building is to raise a net mass of 400 kg at a constant speed of 12 m/s using an electric motor. Minimum power rating of the motor should be
(a) 0 kW     (b) 4.8 kW     (c) 47 kW
(d) 12 kW     (e) 36 kW

**2–123** Electric power is to be generated in a hydroelectric power plant that receives water at a rate of 70 m³/s from an elevation of 65 m using a turbine–generator with an efficiency of 85 percent. When frictional losses in piping are disregarded, the electric power output of this plant is
(a) 3.9 MW     (b) 38 MW     (c) 45 MW
(d) 53 MW     (e) 65 MW

**2–124** Consider a refrigerator that consumes 320 W of electric power when it is running. If the refrigerator runs only one quarter of the time and the unit cost of electricity is

$0.09/kWh, the electricity cost of this refrigerator per month (30 days) is

(a) $3.56          (b) $5.18          (c) $8.54
(d) $9.28          (e) $20.74

**2–125**  A 2-kW pump is used to pump kerosene ($\rho = 0.820$ kg/L) from a tank on the ground to a tank at a higher elevation. Both tanks are open to the atmosphere, and the elevation difference between the free surfaces of the tanks is 30 m. The maximum volume flow rate of kerosene is

(a) 8.3 L/s          (b) 7.2 L/s          (c) 6.8 L/s
(d) 12.1 L/s          (e) 17.8 L/s

**2–126**  A glycerin pump is powered by a 5-kW electric motor. The pressure differential between the outlet and the inlet of the pump at full load is measured to be 211 kPa. If the flow rate through the pump is 18 L/s and the changes in elevation and the flow velocity across the pump are negligible, the overall efficiency of the pump is

(a) 69 percent          (b) 72 percent          (c) 76 percent
(d) 79 percent          (e) 82 percent

**2–127**  A 75-hp compressor in a facility that operates at full load for 2500 h a year is powered by an electric motor that has an efficiency of 93 percent. If the unit cost of electricity is $0.06/kWh, the annual electricity cost of this compressor is

(a) $7802          (b) $9021          (c) $12,100
(d) $8389          (e) $10,460

### The Following Problems Are Based on the Optional Special Topic of Heat Transfer

**2–128**  A 10-cm high and 20-cm wide circuit board houses on its surface 100 closely spaced chips, each generating heat at a rate of 0.08 W and transferring it by convection to the surrounding air at 25°C. Heat transfer from the back surface of the board is negligible. If the convection heat transfer coefficient on the surface of the board is 10 W/m²·°C and radiation heat transfer is negligible, the average surface temperature of the chips is

(a) 26°C          (b) 45°C          (c) 15°C
(d) 80°C          (e) 65°C

**2–129**  A 50-cm-long, 0.2-cm-diameter electric resistance wire submerged in water is used to determine the boiling heat transfer coefficient in water at 1 atm experimentally. The surface temperature of the wire is measured to be 130°C when a wattmeter indicates the electric power consumption to be 4.1 kW. Then the heat transfer coefficient is

(a) 43,500 W/m²·°C          (b) 137 W/m²·°C
(c) 68,330 W/m²·°C          (d) 10,038 W/m²·°C
(e) 37,540 W/m²·°C

**2–130**  A 3-m² hot black surface at 80°C is losing heat to the surrounding air at 25°C by convection with a convection heat transfer coefficient of 12 W/m²·°C, and by radiation to

the surrounding surfaces at 15°C. The total rate of heat loss from the surface is

(a) 1987 W          (b) 2239 W          (c) 2348 W
(d) 3451 W          (e) 3811 W

**2–131**  Heat is transferred steadily through a 0.2-m thick 8 m × 4 m wall at a rate of 2.4 kW. The inner and outer surface temperatures of the wall are measured to be 15°C and 5°C. The average thermal conductivity of the wall is

(a) 0.002 W/m·°C          (b) 0.75 W/m·°C          (c) 1.0 W/m·°C
(d) 1.5 W/m·°C          (e) 3.0 W/m·°C

**2–132**  The roof of an electrically heated house is 7-m long, 10-m wide, and 0.25-m thick. It is made of a flat layer of concrete whose thermal conductivity is 0.92 W/m·°C. During a certain winter night, the temperatures of the inner and outer surfaces of the roof are measured to be 15°C and 4°C, respectively. The average rate of heat loss through the roof that night was

(a) 41 W          (b) 177 W          (c) 4894 W
(d) 5567 W          (e) 2834 W

### Design and Essay Problems

**2–133**  Conduct a literature survey that reviews that concepts of thermal pollution and its current state of the art.

**2–134**  An average vehicle puts out nearly 2.4 kg of carbon dioxide into the atmosphere for every gallon of gasoline it burns, and thus one thing we can do to reduce global warming is to buy a vehicle with higher fuel economy. A U.S. government publication states that a vehicle that consumes 8 L/100 km rather than 10 L/100 km will prevent 10 tons of carbon dioxide from being released over the lifetime of the vehicle. Making reasonable assumptions, evaluate if this is a reasonable claim or a gross exaggeration.

**2–135**  Your neighbor lives in a 250 m² older house heated by natural gas. The current gas heater was installed in the early 1980s and has an efficiency (called the Annual Fuel Utilization Efficiency rating, or AFUE) of 65 percent. It is time to replace the furnace, and the neighbor is trying to decide between a conventional furnace that has an efficiency of 80 percent and costs $1500 and a high-efficiency furnace that has an efficiency of 95 percent and costs $2500. Your neighbor offered to pay you $100 if you help him make the right decision. Considering the weather data, typical heating loads, and the price of natural gas in your area, make a recommendation to your neighbor based on a convincing economic analysis.

**2–136**  Solar energy reaching the earth is about 1350 W/m² outside the earth's atmosphere, and 950 W/m² on earth's surface normal to the sun on a clear day. Someone is marketing 2 m × 3 m photovoltaic cell panels with the claim that a single panel can meet the electricity needs of a house. How do you evaluate this claim? Photovoltaic cells have a conversion efficiency of about 15 percent.

**2–137** Find out the prices of heating oil, natural gas, and electricity in your area, and determine the cost of each per kWh of energy supplied to the house as heat. Go through your utility bills and determine how much money you spent for heating last January. Also determine how much your January heating bill would be for each of the heating systems if you had the latest and most efficient system installed.

**2–138** Prepare a report on the heating systems available in your area for residential buildings. Discuss the advantages and disadvantages of each system and compare their initial and operating costs. What are the important factors in the selection of a heating system? Give some guidelines. Identify the conditions under which each heating system would be the best choice in your area.

**2–139** An electrical-generation utility sometimes pumps liquid water into an elevated reservoir during periods of low electrical consumption. This water is used to generate electricity during periods when the demand for electricity exceeds the utility's ability to produce electricity. Discuss this energy-storage scheme from a conversion efficiency perspective as compared to storing a compressed phase-changing substance.

**2–140** The roofs of many homes in the United States are covered with photovoltaic (PV) solar cells that resemble roof tiles, generating electricity quietly from solar energy. An article stated that over its projected 30-year service life, a 4-kW roof PV system in California will reduce the production of $CO_2$ that causes global warming by 196,000 kg, sulfates that cause acid rain by 1300 kg, and nitrates that cause smog by 750 kg. The article also claims that a PV roof will save 115,000 kg of coal, 80,000 liters of oil, and 760,000 $m^3$ of natural gas. Making reasonable assumptions for incident solar radiation, efficiency, and emissions, evaluate these claims and make corrections if necessary.

# PROPERTIES OF PURE SUBSTANCES

**W**e start this chapter with the introduction of the concept of a *pure substance* and a discussion of the physics of phase-change processes. We then illustrate the various property diagrams and *P-v-T* surfaces of pure substances. After demonstrating the use of the property tables, the hypothetical substance *ideal gas* and the *ideal-gas equation of state* are discussed. The *compressibility factor,* which accounts for the deviation of real gases from ideal-gas behavior, is introduced, and some of the best-known equations of state such as the van der Waals, Beattie-Bridgeman, and Benedict-Webb-Rubin equations are presented.

■ ■ ■ ■ ■ ■ ■

**OBJECTIVES**

The objectives of Chapter 3 are to:

■ Introduce the concept of a pure substance.

■ Discuss the physics of phase-change processes.

■ Illustrate the *P-v*, *T-v*, and *P-T* property diagrams and *P-v-T* surfaces of pure substances.

■ Demonstrate the procedures for determining thermodynamic properties of pure substances from tables of property data.

■ Describe the hypothetical substance "ideal gas" and the ideal-gas equation of state.

■ Apply the ideal-gas equation of state in the solution of typical problems.

■ Introduce the compressibility factor, which accounts for the deviation of real gases from ideal-gas behavior.

■ Present some of the best-known equations of state.

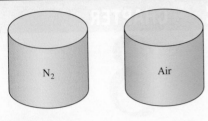

**FIGURE 3–1**
Nitrogen and gaseous air are
pure substances.

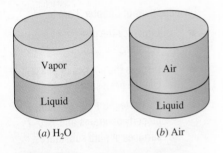

(a) $H_2O$          (b) Air

**FIGURE 3–2**
A mixture of liquid and gaseous water
is a pure substance, but a mixture of
liquid and gaseous air is not.

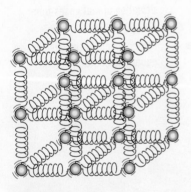

**FIGURE 3–3**
The molecules in a solid are kept at
their positions by the large springlike
intermolecular forces.

# 3–1 · PURE SUBSTANCE

A substance that has a fixed chemical composition throughout is called a
**pure substance**. Water, nitrogen, helium, and carbon dioxide, for example,
are all pure substances.

A pure substance does not have to be of a single chemical element or
compound, however. A mixture of various chemical elements or compounds
also qualifies as a pure substance as long as the mixture is homogeneous.
Air, for example, is a mixture of several gases, but it is often considered
to be a pure substance because it has a uniform chemical composition
(Fig. 3–1). However, a mixture of oil and water is not a pure substance.
Since oil is not soluble in water, it will collect on top of the water, forming
two chemically dissimilar regions.

A mixture of two or more phases of a pure substance is still a pure substance
as long as the chemical composition of all phases is the same (Fig. 3–2). A
mixture of ice and liquid water, for example, is a pure substance because both
phases have the same chemical composition. A mixture of liquid air and gaseous air, however, is not a pure substance since the composition of liquid air
is different from the composition of gaseous air, and thus the mixture is no
longer chemically homogeneous. This is due to different components in air
condensing at different temperatures at a specified pressure.

# 3–2 · PHASES OF A PURE SUBSTANCE

We all know from experience that substances exist in different phases.
At room temperature and pressure, copper is a solid, mercury is a liquid,
and nitrogen is a gas. Under different conditions, each may appear in a different phase. Even though there are three principal phases—solid, liquid,
and gas—a substance may have several phases within a principal phase,
each with a different molecular structure. Carbon, for example, may exist as
graphite or diamond in the solid phase. Helium has two liquid phases; iron
has three solid phases. Ice may exist at seven different phases at high pressures. A phase is identified as having a distinct molecular arrangement that
is homogeneous throughout and separated from the others by easily identifi-
able boundary surfaces. The two phases of $H_2O$ in iced water represent a
good example of this.

When studying phases or phase changes in thermodynamics, one does
not need to be concerned with the molecular structure and behavior of different phases. However, it is very helpful to have some understanding of
the molecular phenomena involved in each phase, and a brief discussion of
phase transformations follows.

Intermolecular bonds are strongest in solids and weakest in gases. One
reason is that molecules in solids are closely packed together, whereas in
gases they are separated by relatively large distances.

The molecules in a **solid** are arranged in a three-dimensional pattern
(lattice) that is repeated throughout (Fig. 3–3). Because of the small distances between molecules in a solid, the attractive forces of molecules on
each other are large and keep the molecules at fixed positions. Note that the
attractive forces between molecules turn to repulsive forces as the distance
between the molecules approaches zero, thus preventing the molecules from

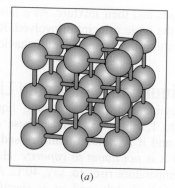 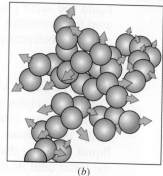 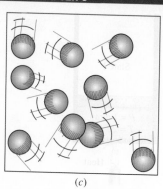

(a)　　　　　　　　　　(b)　　　　　　　　　　(c)

**FIGURE 3–4**

The arrangement of atoms in different phases: (a) molecules are at relatively fixed positions in a solid, (b) groups of molecules move about each other in the liquid phase, and (c) molecules move about at random in the gas phase.

piling up on top of each other. Even though the molecules in a solid cannot move relative to each other, they continually oscillate about their equilibrium positions. The velocity of the molecules during these oscillations depends on the temperature. At sufficiently high temperatures, the velocity (and thus the momentum) of the molecules may reach a point where the intermolecular forces are partially overcome and groups of molecules break away (Fig. 3–4). This is the beginning of the melting process.

The molecular spacing in the **liquid** phase is not much different from that of the solid phase, except the molecules are no longer at fixed positions relative to each other and they can rotate and translate freely. In a liquid, the intermolecular forces are weaker relative to solids, but still relatively strong compared with gases. The distances between molecules generally experience a slight increase as a solid turns liquid, with water being a notable exception.

In the **gas** phase, the molecules are far apart from each other, and a molecular order is nonexistent. Gas molecules move about at random, continually colliding with each other and the walls of the container they are in. Particularly at low densities, the intermolecular forces are very small, and collisions are the only mode of interaction between the molecules. Molecules in the gas phase are at a considerably higher energy level than they are in the liquid or solid phases. Therefore, the gas must release a large amount of its energy before it can condense or freeze.

# 3–3 · PHASE-CHANGE PROCESSES OF PURE SUBSTANCES

There are many practical situations where two phases of a pure substance coexist in equilibrium. Water exists as a mixture of liquid and vapor in the boiler and the condenser of a steam power plant. The refrigerant turns from liquid to vapor in the freezer of a refrigerator. Even though many home owners consider the freezing of water in underground pipes as the most important phase-change process, attention in this section is focused on the

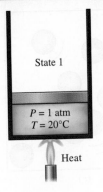

**FIGURE 3–5**
At 1 atm and 20°C, water exists in the liquid phase (*compressed liquid*).

**FIGURE 3–6**
At 1 atm pressure and 100°C, water exists as a liquid that is ready to vaporize (*saturated liquid*).

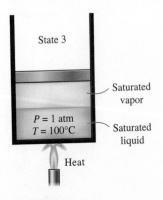

**FIGURE 3–7**
As more heat is transferred, part of the saturated liquid vaporizes (*saturated liquid–vapor mixture*).

liquid and vapor phases and their mixture. As a familiar substance, water is used to demonstrate the basic principles involved. Remember, however, that all pure substances exhibit the same general behavior.

## Compressed Liquid and Saturated Liquid

Consider a piston–cylinder device containing liquid water at 20°C and 1 atm pressure (state 1, Fig. 3–5). Under these conditions, water exists in the liquid phase, and it is called a **compressed liquid**, or a **subcooled liquid**, meaning that it is *not about to vaporize*. Heat is now transferred to the water until its temperature rises to, say, 40°C. As the temperature rises, the liquid water expands slightly, and so its specific volume increases. To accommodate this expansion, the piston moves up slightly. The pressure in the cylinder remains constant at 1 atm during this process since it depends on the outside barometric pressure and the weight of the piston, both of which are constant. Water is still a compressed liquid at this state since it has not started to vaporize.

As more heat is transferred, the temperature keeps rising until it reaches 100°C (state 2, Fig. 3–6). At this point water is still a liquid, but any heat addition will cause some of the liquid to vaporize. That is, a phase-change process from liquid to vapor is about to take place. A liquid that is *about to vaporize* is called a **saturated liquid**. Therefore, state 2 is a saturated liquid state.

## Saturated Vapor and Superheated Vapor

Once boiling starts, the temperature stops rising until the liquid is completely vaporized. That is, the temperature will remain constant during the entire phase-change process if the pressure is held constant. This can easily be verified by placing a thermometer into boiling pure water on top of a stove. At sea level ($P = 1$ atm), the thermometer will always read 100°C if the pan is uncovered or covered with a light lid. During a boiling process, the only change we will observe is a large increase in the volume and a steady decline in the liquid level as a result of more liquid turning to vapor.

Midway about the vaporization line (state 3, Fig. 3–7), the cylinder contains equal amounts of liquid and vapor. As we continue transferring heat, the vaporization process continues until the last drop of liquid is vaporized (state 4, Fig. 3–8). At this point, the entire cylinder is filled with vapor that is on the borderline of the liquid phase. Any heat loss from this vapor will cause some of the vapor to condense (phase change from vapor to liquid). A vapor that is *about to condense* is called a **saturated vapor**. Therefore, state 4 is a saturated vapor state. A substance at states between 2 and 4 is referred to as a **saturated liquid–vapor mixture** since the *liquid and vapor phases coexist* in equilibrium at these states.

One the phase-change process is completed, we are back to a single-phase region again (this time vapor), and further transfer of heat results in an increase in both the temperature and the specific volume (Fig. 3–9). At state 5, the temperature of the vapor is, let us say, 300°C; and if we transfer some heat from the vapor, the temperature may drop somewhat but no condensation will take place as long as the temperature remains above 100°C (for $P = 1$ atm). A vapor that is *not about to condense* (i.e., not a saturated vapor) is called a **superheated vapor**. Therefore, water at state 5 is a superheated vapor. This constant-pressure phase-change process is illustrated on a $T$-$v$ diagram in Fig. 3–10.

If the entire process described here is reversed by cooling the water while maintaining the pressure at the same value, the water will go back to state 1, retracing the same path, and in so doing, the amount of heat released will exactly match the amount of heat added during the heating process.

In our daily life, water implies liquid water and steam implies water vapor. In thermodynamics, however, both water and steam usually mean only one thing: $H_2O$.

## Saturation Temperature and Saturation Pressure

It probably came as no surprise to you that water started to boil at 100°C. Strictly speaking, the statement "water boils at 100°C" is incorrect. The correct statement is "water boils at 100°C at 1 atm pressure." The only reason water started boiling at 100°C was because we held the pressure constant at 1 atm (101.325 kPa). If the pressure inside the cylinder were raised to 500 kPa by adding weights on top of the piston, water would start boiling at 151.8°C. That is, *the temperature at which water starts boiling depends on the pressure; therefore, if the pressure is fixed, so is the boiling temperature.*

At a given pressure, the temperature at which a pure substance changes phase is called the **saturation temperature** $T_{sat}$. Likewise, at a given temperature, the pressure at which a pure substance changes phase is called the **saturation pressure** $P_{sat}$. At a pressure of 101.325 kPa, $T_{sat}$ is 99.97°C. Conversely, at a temperature of 99.97°C, $P_{sat}$ is 101.325 kPa. (At 100.00°C, $P_{sat}$ is 101.42 kPa in the ITS-90 discussed in Chap. 1.)

Saturation tables that list the saturation pressure against the temperature (or the saturation temperature against the pressure) are available for practically

**FIGURE 3–8**
At 1 atm pressure, the temperature remains constant at 100°C until the last drop of liquid is vaporized (*saturated vapor*).

**FIGURE 3–9**
As more heat is transferred, the temperature of the vapor starts to rise (*superheated vapor*).

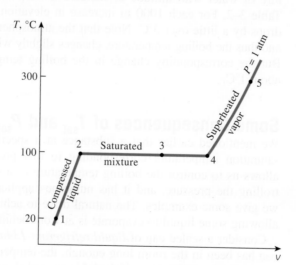

**FIGURE 3–10**
$T$-$v$ diagram for the heating process of water at constant pressure.

## TABLE 3-1

Saturation (or vapor) pressure of water at various temperatures

| Temperature $T$, °C | Saturation Pressure $P_{sat}$, kPa |
|---|---|
| −10 | 0.260 |
| −5 | 0.403 |
| 0 | 0.611 |
| 5 | 0.872 |
| 10 | 1.23 |
| 15 | 1.71 |
| 20 | 2.34 |
| 25 | 3.17 |
| 30 | 4.25 |
| 40 | 7.38 |
| 50 | 12.35 |
| 100 | 101.3 (1 atm) |
| 150 | 475.8 |
| 200 | 1554 |
| 250 | 3973 |
| 300 | 8581 |

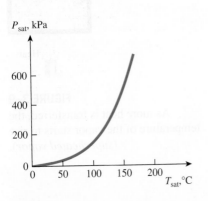

**FIGURE 3–11**
The liquid–vapor saturation curve of a pure substance (numerical values are for water).

all substances. A partial listing of such a table is given in Table 3–1 for water. This table indicates that the pressure of water changing phase (boiling or condensing) at 25°C must be 3.17 kPa, and the pressure of water must be maintained at 3976 kPa (about 40 atm) to have it boil at 250°C. Also, water can be frozen by dropping its pressure below 0.61 kPa.

It takes a large amount of energy to melt a solid or vaporize a liquid. The amount of energy absorbed or released during a phase-change process is called the **latent heat**. More specifically, the amount of energy absorbed during melting is called the **latent heat of fusion** and is equivalent to the amount of energy released during freezing. Similarly, the amount of energy absorbed during vaporization is called the **latent heat of vaporization** and is equivalent to the energy released during condensation. The magnitudes of the latent heats depend on the temperature or pressure at which the phase change occurs. At 1 atm pressure, the latent heat of fusion of water is 333.7 kJ/kg and the latent heat of vaporization is 2256.5 kJ/kg.

During a phase-change process, pressure and temperature are obviously dependent properties, and there is a definite relation between them, that is, $T_{sat} = f(P_{sat})$. A plot of $T_{sat}$ versus $P_{sat}$, such as the one given for water in Fig. 3–11, is called a **liquid–vapor saturation curve**. A curve of this kind is characteristic of all pure substances.

It is clear from Fig. 3–11 that $T_{sat}$ increases with $P_{sat}$. Thus, a substance at higher pressures boils at higher temperatures. In the kitchen, higher boiling temperatures mean shorter cooking times and energy savings. A beef stew, for example, may take 1 to 2 h to cook in a regular pan that operates at 1 atm pressure, but only 20 min in a pressure cooker operating at 3 atm absolute pressure (corresponding boiling temperature: 134°C).

The atmospheric pressure, and thus the boiling temperature of water, decreases with elevation. Therefore, it takes longer to cook at higher altitudes than it does at sea level (unless a pressure cooker is used). For example, the standard atmospheric pressure at an elevation of 2000 m is 79.50 kPa, which corresponds to a boiling temperature of 93.3°C as opposed to 100°C at sea level (zero elevation). The variation of the boiling temperature of water with altitude at standard atmospheric conditions is given in Table 3–2. For each 1000 m increase in elevation, the boiling temperature drops by a little over 3°C. Note that the atmospheric pressure at a location, and thus the boiling temperature, changes slightly with the weather conditions. But the corresponding change in the boiling temperature is no more than about 1°C.

## Some Consequences of $T_{sat}$ and $P_{sat}$ Dependence

We mentioned earlier that a substance at a specified pressure boils at the saturation temperature corresponding to that pressure. This phenomenon allows us to control the boiling temperature of a substance by simply controlling the pressure, and it has numerous applications in practice. Below we give some examples. The natural drive to achieve phase equilibrium by allowing some liquid to evaporate is at work behind the scenes.

Consider a sealed can of *liquid refrigerant-134a* in a room at 25°C. If the can has been in the room long enough, the temperature of the refrigerant in the can is also 25°C. Now, if the lid is opened slowly and some refrigerant is allowed to escape, the pressure in the can will start dropping until it reaches the atmospheric pressure. If you are holding the can, you will notice its temperature dropping rapidly, and even ice forming outside the can if the air is humid. A thermometer inserted in the can will register $-26°C$ when the pressure drops to 1 atm, which is the saturation temperature of refrigerant-134a at that pressure. The temperature of the liquid refrigerant will remain at $-26°C$ until the last drop of it vaporizes.

Another aspect of this interesting physical phenomenon is that a liquid cannot vaporize unless it absorbs energy in the amount of the latent heat of vaporization, which is 217 kJ/kg for refrigerant-134a at 1 atm. Therefore, the rate of vaporization of the refrigerant depends on the rate of heat transfer to the can: the larger the rate of heat transfer, the higher the rate of vaporization. The rate of heat transfer to the can and thus the rate of vaporization of the refrigerant can be minimized by insulating the can heavily. In the limiting case of no heat transfer, the refrigerant will remain in the can as a liquid at $-26°C$ indefinitely.

The boiling temperature of *nitrogen* at atmospheric pressure is $-196°C$ (see Table A–3a). This means the temperature of liquid nitrogen exposed to the atmosphere must be $-196°C$ since some nitrogen will be evaporating. The temperature of liquid nitrogen remains constant at $-196°C$ until it is depleted. For this reason, nitrogen is commonly used in low-temperature scientific studies (such as superconductivity) and cryogenic applications to maintain a test chamber at a constant temperature of $-196°C$. This is done by placing the test chamber into a liquid nitrogen bath that is open to the atmosphere. Any heat transfer from the environment to the test section is absorbed by the nitrogen, which evaporates isothermally and keeps the test chamber temperature constant at $-196°C$ (Fig. 3–12). The entire test section must be insulated heavily to minimize heat transfer and thus liquid nitrogen consumption. Liquid nitrogen is also used for medical purposes to burn off unsightly spots on the skin. This is done by soaking a cotton swap in liquid nitrogen and wetting the target area with it. As the nitrogen evaporates, it freezes the affected skin by rapidly absorbing heat from it.

A practical way of cooling leafy vegetables is **vacuum cooling**, which is based on *reducing the pressure* of the sealed cooling chamber to the saturation pressure at the desired low temperature, and evaporating some water from the products to be cooled. The heat of vaporization during evaporation is absorbed from the products, which lowers the product temperature. The saturation pressure of water at 0°C is 0.61 kPa, and the products can be cooled to 0°C by lowering the pressure to this level. The cooling rate can be increased by lowering the pressure below 0.61 kPa, but this is not desirable because of the danger of freezing and the added cost.

In vacuum cooling, there are two distinct stages. In the first stage, the products at ambient temperature, say at 25°C, are loaded into the chamber, and the operation begins. The temperature in the chamber remains constant until the *saturation pressure* is reached, which is 3.17 kPa at 25°C. In the second stage that follows, saturation conditions are maintained inside at progressively *lower pressures* and the corresponding *lower temperatures* until the desired temperature is reached (Fig. 3–13).

**TABLE 3–2**

Variation of the standard atmospheric pressure and the boiling (saturation) temperature of water with altitude

| Elevation, m | Atmospheric pressure, kPa | Boiling temperature, °C |
|---|---|---|
| 0 | 101.33 | 100.0 |
| 1,000 | 89.55 | 96.5 |
| 2,000 | 79.50 | 93.3 |
| 5,000 | 54.05 | 83.3 |
| 10,000 | 26.50 | 66.3 |
| 20,000 | 5.53 | 34.7 |

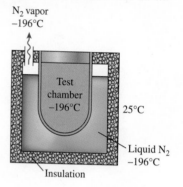

**FIGURE 3–12**

The temperature of liquid nitrogen exposed to the atmosphere remains constant at $-196°C$, and thus it maintains the test chamber at $-196°C$.

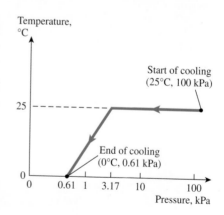

**FIGURE 3–13**

The variation of the temperature of fruits and vegetables with pressure during vacuum cooling from 25°C to 0°C.

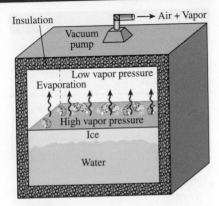

**FIGURE 3–14**
In 1775, ice was made by evacuating
the air space in a water tank.

Vacuum cooling is usually more expensive than the conventional refrigerated cooling, and its use is limited to applications that result in much faster cooling. Products with large surface area per unit mass and a high tendency to release moisture such as lettuce and spinach are well-suited for vacuum cooling. Products with low surface area to mass ratio are not suitable, especially those that have relatively impervious peels such as tomatoes and cucumbers. Some products such as mushrooms and green peas can be vacuum cooled successfully by wetting them first.

The vacuum cooling just described becomes **vacuum freezing** if the vapor pressure in the vacuum chamber is dropped below 0.61 kPa, the saturation pressure of water at 0°C. The idea of making ice by using a vacuum pump is nothing new. Dr. William Cullen actually made ice in Scotland in 1775 by evacuating the air in a water tank (Fig. 3–14).

**Package icing** is commonly used in small-scale cooling applications to remove heat and keep the products cool during transit by taking advantage of the large latent heat of fusion of water, but its use is limited to products that are not harmed by contact with ice. Also, ice provides *moisture* as well as *refrigeration*.

## 3–4 ▪ PROPERTY DIAGRAMS FOR PHASE-CHANGE PROCESSES

The variations of properties during phase-change processes are best studied and understood with the help of property diagrams. Next, we develop and discuss the *T-v*, *P-v*, and *P-T* diagrams for pure substances.

### 1 The *T-v* Diagram

The phase-change process of water at 1 atm pressure was described in detail in the last section and plotted on a *T-v* diagram in Fig. 3–10. Now we repeat this process at different pressures to develop the *T-v* diagram.

Let us add weights on top of the piston until the pressure inside the cylinder reaches 1 MPa. At this pressure, water has a somewhat smaller specific volume than it does at 1 atm pressure. As heat is transferred to the water at this new pressure, the process follows a path that looks very much like the process path at 1 atm pressure, as shown in Fig. 3–15, but there are some noticeable differences. First, water starts boiling at a much higher temperature (179.9°C) at this pressure. Second, the specific volume of the saturated liquid is larger and the specific volume of the saturated vapor is smaller than the corresponding values at 1 atm pressure. That is, the horizontal line that connects the saturated liquid and saturated vapor states is much shorter.

As the pressure is increased further, this saturation line continues to shrink, as shown in Fig. 3–15, and it becomes a point when the pressure reaches 22.06 MPa for the case of water. This point is called the **critical point**, and it is defined as *the point at which the saturated liquid and saturated vapor states are identical.*

The temperature, pressure, and specific volume of a substance at the critical point are called, respectively, the *critical temperature* $T_{cr}$, *critical*

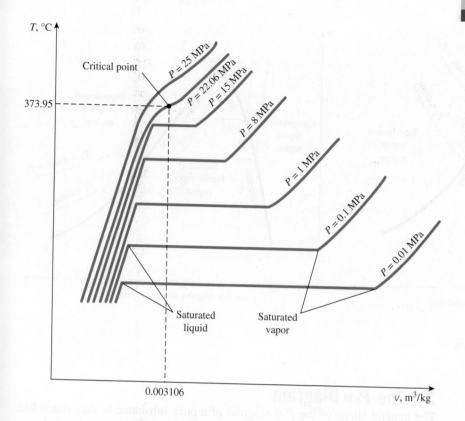

FIGURE 3–15

$T$-$v$ diagram of constant-pressure
phase-change processes of a pure
substance at various pressures
(numerical values are for water).

*pressure* $P_{cr}$, and *critical specific volume* $v_{cr}$. The critical-point properties of water are $P_{cr}$ = 22.06 MPa, $T_{cr}$ = 373.95°C, and $v_{cr}$ = 0.003106 m³/kg. For helium, they are 0.23 MPa, −267.85°C, and 0.01444 m³/kg. The critical properties for various substances are given in Table A–1 in the appendix.

At pressures above the critical pressure, there is not a distinct phase-change process (Fig. 3–16). Instead, the specific volume of the substance continually increases, and at all times there is only one phase present. Eventually, it resembles a vapor, but we can never tell when the change has occurred. Above the critical state, there is no line that separates the compressed liquid region and the superheated vapor region. However, it is customary to refer to the substance as superheated vapor at temperatures above the critical temperature and as compressed liquid at temperatures below the critical temperature.

The saturated liquid states in Fig. 3–15 can be connected by a line called the **saturated liquid line**, and saturated vapor states in the same figure can be connected by another line, called the **saturated vapor line**. These two lines meet at the critical point, forming a dome as shown in Fig. 3–17a. All the compressed liquid states are located in the region to the left of the saturated liquid line, called the **compressed liquid region**. All the superheated vapor states are located to the right of the saturated vapor line, called the **superheated vapor region**. In these two regions, the substance exists in a single phase, a liquid or a vapor. All the states that involve both phases in equilibrium are located under the dome, called the **saturated liquid–vapor mixture region**, or the **wet region**.

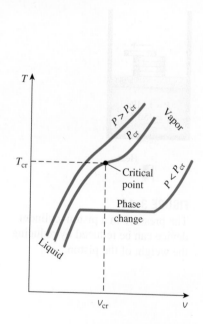

FIGURE 3–16

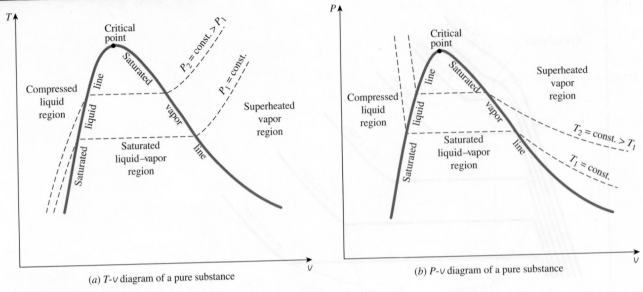

(a) T-v diagram of a pure substance

(b) P-v diagram of a pure substance

**FIGURE 3–17**
Property diagrams of a pure substance.

**FIGURE 3–18**
The pressure in a piston–cylinder device can be reduced by reducing the weight of the piston.

## 2   The *P-v* Diagram

The general shape of the *P-v* diagram of a pure substance is very much like the *T-v* diagram, but the *T* = constant lines on this diagram have a downward trend, as shown in Fig. 3–17*b*.

Consider again a piston–cylinder device that contains liquid water at 1 MPa and 150°C. Water at this state exists as a compressed liquid. Now the weights on top of the piston are removed one by one so that the pressure inside the cylinder decreases gradually (Fig. 3–18). The water is allowed to exchange heat with the surroundings so its temperature remains constant. As the pressure decreases, the volume of the water increases slightly. When the pressure reaches the saturation-pressure value at the specified temperature (0.4762 MPa), the water starts to boil. During this vaporization process, both the temperature and the pressure remain constant, but the specific volume increases. Once the last drop of liquid is vaporized, further reduction in pressure results in a further increase in specific volume. Notice that during the phase-change process, we did not remove any weights. Doing so would cause the pressure and therefore the temperature to drop [since $T_{sat} = f(P_{sat})$], and the process would no longer be isothermal.

When the process is repeated for other temperatures, similar paths are obtained for the phase-change processes. Connecting the saturated liquid and the saturated vapor states by a curve, we obtain the *P-v* diagram of a pure substance, as shown in Fig. 3–17*b*.

## Extending the Diagrams to Include the Solid Phase

The two equilibrium diagrams developed so far represent the equilibrium states involving the liquid and the vapor phases only. However, these diagrams can easily be extended to include the solid phase as well as the solid–liquid and the

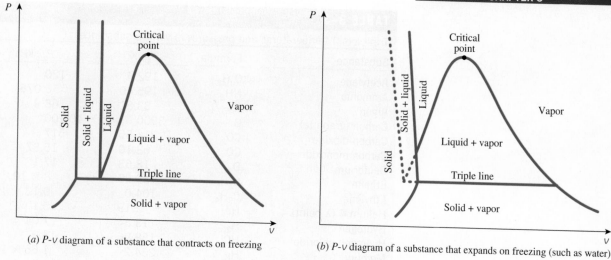

(a) P-v diagram of a substance that contracts on freezing

(b) P-v diagram of a substance that expands on freezing (such as water)

**FIGURE 3–19**
P-v diagrams of different substances.

solid–vapor saturation regions. The basic principles discussed in conjunction with the liquid–vapor phase-change process apply equally to the solid–liquid and solid–vapor phase-change processes. Most substances contract during a solidification (i.e., freezing) process. Others, like water, expand as they freeze. The P-v diagrams for both groups of substances are given in Figs. 3–19a and 3–19b. These two diagrams differ only in the solid–liquid saturation region. The T-v diagrams look very much like the P-v diagrams, especially for substances that contract on freezing.

The fact that water expands upon freezing has vital consequences in nature. If water contracted on freezing as most other substances do, the ice formed would be heavier than the liquid water, and it would settle to the bottom of rivers, lakes, and oceans instead of floating at the top. The sun's rays would never reach these ice layers, and the bottoms of many rivers, lakes, and oceans would be covered with ice at times, seriously disrupting marine life.

We are all familiar with two phases being in equilibrium, but under some conditions all three phases of a pure substance coexist in equilibrium (Fig. 3–20). On P-v or T-v diagrams, these triple-phase states form a line called the **triple line**. The states on the triple line of a substance have the same pressure and temperature but different specific volumes. The triple line appears as a point on the P-T diagrams and, therefore, is often called the **triple point**. The triple-point temperatures and pressures of various substances are given in Table 3–3. For water, the triple-point temperature and pressure are 0.01°C and 0.6117 kPa, respectively. That is, all three phases of water coexist in equilibrium only if the temperature and pressure have precisely these values. No substance can exist in the liquid phase in stable equilibrium at pressures below the triple-point pressure. The same can be said for temperature for substances that contract on freezing. However, substances at high pressures can exist in the liquid phase at temperatures below the triple-point temperature. For example, water cannot exist in liquid form

**FIGURE 3–20**
At triple-point pressure and temperature, a substance exists in three phases in equilibrium.

**TABLE 3–3**

Triple-point temperatures and pressures of various substances

| Substance | Formula | $T_{tp}$, K | $P_{tp}$, kPa |
|---|---|---|---|
| Acetylene | $C_2H_2$ | 192.4 | 120 |
| Ammonia | $NH_3$ | 195.40 | 6.076 |
| Argon | A | 83.81 | 68.9 |
| Carbon (graphite) | C | 3900 | 10,100 |
| Carbon dioxide | $CO_2$ | 216.55 | 517 |
| Carbon monoxide | CO | 68.10 | 15.37 |
| Deuterium | $D_2$ | 18.63 | 17.1 |
| Ethane | $C_2H_6$ | 89.89 | $8 \times 10^{-4}$ |
| Ethylene | $C_2H_4$ | 104.0 | 0.12 |
| Helium 4 ($\lambda$ point) | He | 2.19 | 5.1 |
| Hydrogen | $H_2$ | 13.84 | 7.04 |
| Hydrogen chloride | HCl | 158.96 | 13.9 |
| Mercury | Hg | 234.2 | $1.65 \times 10^{-7}$ |
| Methane | $CH_4$ | 90.68 | 11.7 |
| Neon | Ne | 24.57 | 43.2 |
| Nitric oxide | NO | 109.50 | 21.92 |
| Nitrogen | $N_2$ | 63.18 | 12.6 |
| Nitrous oxide | $N_2O$ | 182.34 | 87.85 |
| Oxygen | $O_2$ | 54.36 | 0.152 |
| Palladium | Pd | 1825 | $3.5 \times 10^{-3}$ |
| Platinum | Pt | 2045 | $2.0 \times 10^{-4}$ |
| Sulfur dioxide | $SO_2$ | 197.69 | 1.67 |
| Titanium | Ti | 1941 | $5.3 \times 10^{-3}$ |
| Uranium hexafluoride | $UF_6$ | 337.17 | 151.7 |
| Water | $H_2O$ | 273.16 | 0.61 |
| Xenon | Xe | 161.3 | 81.5 |
| Zinc | Zn | 692.65 | 0.065 |

*Source:* Data from National Bureau of Standards (U.S.) Circ., 500 (1952).

in equilibrium at atmospheric pressure at temperatures below 0°C, but it can exist as a liquid at −20°C at 200 MPa pressure. Also, ice exists at seven different solid phases at pressures above 100 MPa.

There are two ways a substance can pass from the solid to vapor phase: either it melts first into a liquid and subsequently evaporates, or it evaporates directly without melting first. The latter occurs at pressures below the triple-point value, since a pure substance cannot exist in the liquid phase at those pressures (Fig. 3–21). Passing from the solid phase directly into the vapor phase is called **sublimation**. For substances that have a triple-point pressure above the atmospheric pressure such as solid $CO_2$ (dry ice), sublimation is the only way to change from the solid to vapor phase at atmospheric conditions.

**FIGURE 3–21**
At low pressures (below the triple-point value), solids evaporate without melting first (*sublimation*).

## 3 ■ The *P-T* Diagram

Figure 3–22 shows the *P-T* diagram of a pure substance. This diagram is often called the **phase diagram** since all three phases are separated from

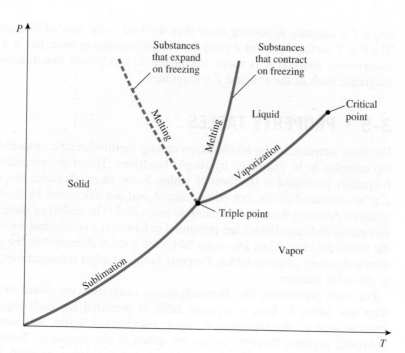

**FIGURE 3–22**
*P-T* diagram of pure substances.

each other by three lines. The sublimation line separates the solid and vapor regions, the vaporization line separates the liquid and vapor regions, and the melting (or fusion) line separates the solid and liquid regions. These three lines meet at the triple point, where all three phases coexist in equilibrium. The vaporization line ends at the critical point because no distinction can be made between liquid and vapor phases above the critical point. Substances that expand and contract on freezing differ only in the melting line on the *P-T* diagram.

## The *P-v-T* Surface

The state of a simple compressible substance is fixed by any two independent, intensive properties. Once the two appropriate properties are fixed, all the other properties become dependent properties. Remembering that any equation with two independent variables in the form $z = z(x, y)$ represents a surface in space, we can represent the *P-v-T* behavior of a substance as a surface in space, as shown in Figs. 3–23 and 3–24. Here $T$ and $v$ may be viewed as the independent variables (the base) and $P$ as the dependent variable (the height).

All the points on the surface represent equilibrium states. All states along the path of a quasi-equilibrium process lie on the *P-v-T* surface since such a process must pass through equilibrium states. The single-phase regions appear as curved surfaces on the *P-v-T* surface, and the two-phase regions as surfaces perpendicular to the *P-T* plane. This is expected since the projections of two-phase regions on the *P-T* plane are lines.

All the two-dimensional diagrams we have discussed so far are merely projections of this three-dimensional surface onto the appropriate planes. A *P-v* diagram is just a projection of the *P-v-T* surface on the *P-v* plane,

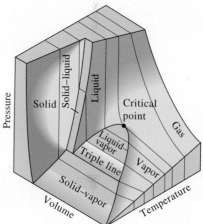

**FIGURE 3–23**
*P-v-T* surface of a substance that *contracts* on freezing.

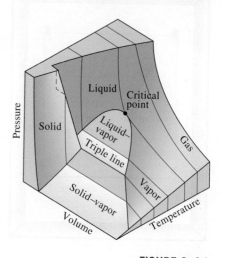

**FIGURE 3–24**
*P-v-T* surface of a substance that *expands* on freezing (like water).

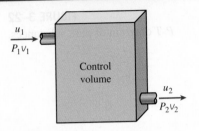

**FIGURE 3–25**

The combination $u + Pv$ is frequently encountered in the analysis of control volumes.

and a $T$-$v$ diagram is nothing more than the bird's-eye view of this surface. The $P$-$v$-$T$ surfaces present a great deal of information at once, but in a thermodynamic analysis it is more convenient to work with two-dimensional diagrams, such as the $P$-$v$ and $T$-$v$ diagrams.

## 3–5 · PROPERTY TABLES

For most substances, the relationships among thermodynamic properties are too complex to be expressed by simple equations. Therefore, properties are frequently presented in the form of tables. Some thermodynamic properties can be measured easily, but others cannot and are calculated by using the relations between them and measurable properties. The results of these measurements and calculations are presented in tables in a convenient format. In the following discussion, the steam tables are used to demonstrate the use of thermodynamic property tables. Property tables of other substances are used in the same manner.

For each substance, the thermodynamic properties are listed in more than one table. In fact, a separate table is prepared for each region of interest such as the superheated vapor, compressed liquid, and saturated (mixture) regions. Property tables are given in the appendix. Before we get into the discussion of property tables, we define a new property called *enthalpy*.

### Enthalpy—A Combination Property

A person looking at the tables will notice two new properties: enthalpy $h$ and entropy $s$. Entropy is a property associated with the second law of thermodynamics, and we will not use it until it is properly defined in Chap. 7. However, it is appropriate to introduce enthalpy at this point.

In the analysis of certain types of processes, particularly in power generation and refrigeration (Fig. 3–25), we frequently encounter the combination of properties $u + Pv$. For the sake of simplicity and convenience, this combination is defined as a new property, **enthalpy**, and given the symbol $h$:

$$h = u + Pv \quad \text{(kJ/kg)} \tag{3–1}$$

or,

$$H = U + PV \quad \text{(kJ)} \tag{3–2}$$

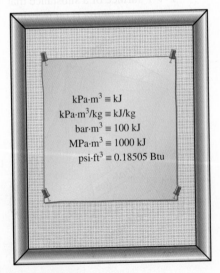

**FIGURE 3–26**

The product *pressure* × *volume* has energy units.

Both the total enthalpy $H$ and specific enthalpy $h$ are simply referred to as enthalpy since the context clarifies which one is meant. Notice that the equations given above are dimensionally homogeneous. That is, the unit of the pressure–volume product may differ from the unit of the internal energy by only a factor (Fig. 3–26). For example, it can be easily shown that 1 kPa·m³ = 1 kJ. In some tables encountered in practice, the internal energy $u$ is frequently not listed, but it can always be determined from $u = h - Pv$.

The widespread use of the property enthalpy is due to Professor Richard Mollier, who recognized the importance of the group $u + Pv$ in the analysis of steam turbines and in the representation of the properties of steam in tabular and graphical form (as in the famous Mollier chart). Mollier referred to the group $u + Pv$ as *heat content* and *total heat*. These terms were not quite consistent with the modern thermodynamic terminology and were replaced in the 1930s by the term *enthalpy* (from the Greek word *enthalpien*, which means *to heat*).

## 1a  Saturated Liquid and Saturated Vapor States

The properties of saturated liquid and saturated vapor for water are listed in Tables A–4 and A–5. Both tables give the same information. The only difference is that in Table A–4 properties are listed under temperature and in Table A–5 under pressure. Therefore, it is more convenient to use Table A–4 when *temperature* is given and Table A–5 when *pressure* is given. The use of Table A–4 is illustrated in Fig. 3–27.

The subscript $f$ is used to denote properties of a saturated liquid, and the subscript $g$ to denote the properties of saturated vapor. These symbols are commonly used in thermodynamics and originated from German. Another subscript commonly used is $fg$, which denotes the difference between the saturated vapor and saturated liquid values of the same property. For example,

$v_f$ = specific volume of saturated liquid

$v_g$ = specific volume of saturated vapor

$v_{fg}$ = difference between $v_g$ and $v_f$ (that is $v_{fg} = v_g - v_f$)

The quantity $h_{fg}$ is called the **enthalpy of vaporization** (or latent heat of vaporization). It represents the amount of energy needed to vaporize a unit mass of saturated liquid at a given temperature or pressure. It decreases as the temperature or pressure increases and becomes zero at the critical point.

| Temp. °C $T$ | Sat. press. kPa $P_{sat}$ | Specific volume m³/kg | |
|---|---|---|---|
| | | Sat. liquid $v_f$ | Sat. vapor $v_g$ |
| 85 | 57.868 | 0.001032 | 2.8261 |
| 90 | 70.183 | 0.001036 | 2.3593 |
| 95 | 84.609 | 0.001040 | 1.9808 |

Specific temperature

Corresponding saturation pressure

Specific volume of saturated liquid

Specific volume of saturated vapor

**FIGURE 3–27**
A partial list of Table A–4.

---

### EXAMPLE 3–1    Pressure of Saturated Liquid in a Tank

A rigid tank contains 50 kg of saturated liquid water at 90°C. Determine the pressure in the tank and the volume of the tank.

**SOLUTION**  A rigid tank contains saturated liquid water. The pressure and volume of the tank are to be determined.
**Analysis**  The state of the saturated liquid water is shown on a *T-v* diagram in Fig. 3–28. Since saturation conditions exist in the tank, the pressure must be the saturation pressure at 90°C:

$$P = P_{sat\ @\ 90°C} = 70.183\ \text{kPa}\quad\text{(Table A-4)}$$

The specific volume of the saturated liquid at 90°C is

$$v = v_{f\ @\ 90°C} = 0.001036\ \text{m}^3/\text{kg}\quad\text{(Table A-4)}$$

Then the total volume of the tank becomes

$$V = mv = (50\ \text{kg})(0.001036\ \text{m}^3/\text{kg}) = \mathbf{0.0518\ m^3}$$

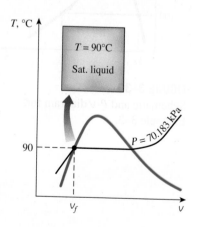

**FIGURE 3–28**
Schematic and *T-v* diagram for Example 3–1.

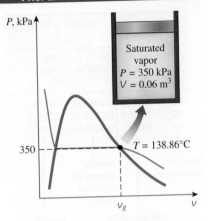

**FIGURE 3–29**
Schematic and $P$-$v$ diagram for
Example 3–2.

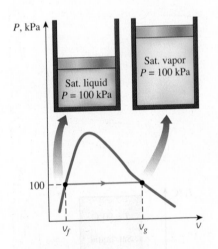

**FIGURE 3–30**
Schematic and $P$-$v$ diagram for
Example 3–3.

**EXAMPLE 3–2    Temperature of Saturated Vapor in a Cylinder**

A piston–cylinder device contains 0.06 m³ of saturated water vapor at 350-kPa
pressure. Determine the temperature and the mass of the vapor inside the cylinder.

**SOLUTION**  A cylinder contains saturated water vapor. The temperature and
the mass of vapor are to be determined.
**Analysis**  The state of the saturated water vapor is shown on a $P$-$v$ diagram
in Fig. 3–29. Since the cylinder contains saturated vapor at 350 kPa, the
temperature inside must be the saturation temperature at this pressure:

$$T = T_{\text{sat @ 350 kPa}} = \mathbf{138.86°C} \quad \text{(Table A–5)}$$

The specific volume of the saturated vapor at 350 kPa is

$$v = v_{g \text{ @ 350 kPa}} = 0.52422 \text{ m}^3/\text{kg} \quad \text{(Table A–5)}$$

Then the mass of water vapor inside the cylinder becomes

$$m = \frac{V}{v} = \frac{0.06 \text{ m}^3}{0.52422 \text{ m}^3/\text{kg}} = \mathbf{0.114 \text{ kg}}$$

**EXAMPLE 3–3    Volume and Energy Change during Evaporation**

A mass of 200 g of saturated liquid water is completely vaporized at a constant pressure of 100 kPa. Determine (a) the volume change and (b) the
amount of energy transferred to the water.

**SOLUTION**  Saturated liquid water is vaporized at constant pressure. The
volume change and the energy transferred are to be determined.
**Analysis**  (a) The process described is illustrated on a $P$-$v$ diagram in
Fig. 3–30. The volume change per unit mass during a vaporization process
is $v_{fg}$, which is the difference between $v_g$ and $v_f$. Reading these values from
Table A–5 at 100 kPa and substituting yield

$$v_{fg} = v_g - v_f = 1.6941 - 0.001043 = 1.6931 \text{ m}^3/\text{kg}$$

Thus,

$$\Delta V = m v_{fg} = (0.2 \text{ kg})(1.6931 \text{ m}^3/\text{kg}) = \mathbf{0.3386 \text{ m}^3}$$

(b) The amount of energy needed to vaporize a unit mass of a substance at a
given pressure is the enthalpy of vaporization at that pressure, which is $h_{fg} =$
2257.5 kJ/kg for water at 100 kPa. Thus, the amount of energy transferred is

$$m h_{fg} = (0.2 \text{ kg})(2257.5 \text{ kJ/kg}) = \mathbf{451.5 \text{ kJ}}$$

**Discussion**  Note that we have considered the first four decimal digits of $v_{fg}$
and disregarded the rest. This is because $v_g$ has significant numbers to the
first four decimal places only, and we do not know the numbers in the other
decimal places. Copying all the digits from the calculator would mean that
we are assuming $v_g = 1.694100$, which is not necessarily the case. It could
very well be that $v_g = 1.694138$ since this number, too, would truncate
to 1.6941. All the digits in our result (1.6931) are significant. But if we
did not truncate the result, we would obtain $v_{fg} = 1.693057$, which falsely
implies that our result is accurate to the sixth decimal place.

## 1b Saturated Liquid–Vapor Mixture

During a vaporization process, a substance exists as part liquid and part vapor. That is, it is a mixture of saturated liquid and saturated vapor (Fig. 3–31). To analyze this mixture properly, we need to know the proportions of the liquid and vapor phases in the mixture. This is done by defining a new property called the **quality** $x$ as the ratio of the mass of vapor to the total mass of the mixture:

$$x = \frac{m_{\text{vapor}}}{m_{\text{total}}} \qquad (3\text{–}3)$$

where

$$m_{\text{total}} = m_{\text{liquid}} + m_{\text{vapor}} = m_f + m_g$$

Quality has significance for *saturated mixtures* only. It has no meaning in the compressed liquid or superheated vapor regions. Its value is between 0 and 1. The quality of a system that consists of *saturated liquid* is 0 (or 0 percent), and the quality of a system consisting of *saturated vapor* is 1 (or 100 percent). In saturated mixtures, quality can serve as one of the two independent intensive properties needed to describe a state. Note that *the properties of the saturated liquid are the same whether it exists alone or in a mixture with saturated vapor*. During the vaporization process, only the amount of saturated liquid changes, not its properties. The same can be said about a saturated vapor.

A saturated mixture can be treated as a combination of two subsystems: the saturated liquid and the saturated vapor. However, the amount of mass for each phase is usually not known. Therefore, it is often more convenient to imagine that the two phases are mixed well, forming a homogeneous mixture (Fig. 3–32). Then the properties of this "mixture" will simply be the average properties of the saturated liquid–vapor mixture under consideration. Here is how it is done.

Consider a tank that contains a saturated liquid–vapor mixture. The volume occupied by saturated liquid is $V_f$, and the volume occupied by saturated vapor is $V_g$. The total volume $V$ is the sum of the two:

$$V = V_f + V_g$$

$$V = mv \longrightarrow m_t v_{\text{avg}} = m_f v_f + m_g v_g$$

$$m_f = m_t - m_g \longrightarrow m_t v_{\text{avg}} = (m_t - m_g)v_f + m_g v_g$$

Dividing by $m_t$ yields

$$v_{\text{avg}} = (1 - x)v_f + xv_g$$

since $x = m_g/m_t$. This relation can also be expressed as

$$v_{\text{avg}} = v_f + xv_{fg} \quad (\text{m}^3/\text{kg}) \qquad (3\text{–}4)$$

where $v_{fg} = v_g - v_f$. Solving for quality, we obtain

$$x = \frac{v_{\text{avg}} - v_f}{v_{fg}} \qquad (3\text{–}5)$$

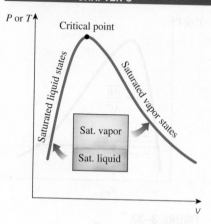

**FIGURE 3–31**

The relative amounts of liquid and vapor phases in a saturated mixture are specified by the *quality x.*

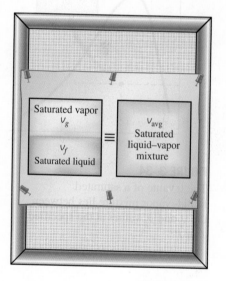

**FIGURE 3–32**

A two-phase system can be treated as a homogeneous mixture for convenience.

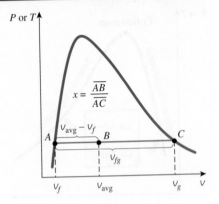

**FIGURE 3–33**
Quality is related to the horizontal distances on P-$v$ and T-$v$ diagrams.

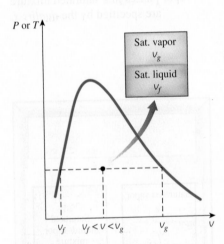

**FIGURE 3–34**
The $v$ value of a saturated liquid–vapor mixture lies between the $v_f$ and $v_g$ values at the specified T or P.

Based on this equation, quality can be related to the horizontal distances on a P-$v$ or T-$v$ diagram (Fig. 3–33). At a given temperature or pressure, the numerator of Eq. 3–5 is the distance between the actual state and the saturated liquid state, and the denominator is the length of the entire horizontal line that connects the saturated liquid and saturated vapor states. A state of 50 percent quality lies in the middle of this horizontal line.

The analysis given above can be repeated for internal energy and enthalpy with the following results:

$$u_{avg} = u_f + xu_{fg} \quad (\text{kJ/kg}) \tag{3–6}$$

$$h_{avg} = h_f + xh_{fg} \quad (\text{kJ/kg}) \tag{3–7}$$

All the results are of the same format, and they can be summarized in a single equation as

$$y_{avg} = y_f + xy_{fg}$$

where $y$ is $v$, $u$, or $h$. The subscript "avg" (for "average") is usually dropped for simplicity. The values of the average properties of the mixtures are always *between* the values of the saturated liquid and the saturated vapor properties (Fig. 3–34). That is,

$$y_f \le y_{avg} \le y_g$$

Finally, all the saturated-mixture states are located under the saturation curve, and to analyze saturated mixtures, all we need are saturated liquid and saturated vapor data (Tables A–4 and A–5 in the case of water).

**EXAMPLE 3–4    Pressure and Volume of a Saturated Mixture**

A rigid tank contains 10 kg of water at 90°C. If 8 kg of the water is in the liquid form and the rest is in the vapor form, determine (a) the pressure in the tank and (b) the volume of the tank.

**SOLUTION** A rigid tank contains saturated mixture. The pressure and the volume of the tank are to be determined.
**Analysis** (a) The state of the saturated liquid–vapor mixture is shown in Fig. 3–35. Since the two phases coexist in equilibrium, we have a saturated mixture, and the pressure must be the saturation pressure at the given temperature:

$$P = P_{\text{sat @ 90°C}} = 70.183 \text{ kPa} \quad (\text{Table A–4})$$

(b) At 90°C, we have $v_f = 0.001036$ m³/kg and $v_g = 2.3593$ m³/kg (Table A–4). One way of finding the volume of the tank is to determine the volume occupied by each phase and then add them:

$$V = V_f + V_g = m_f v_f + m_g v_g$$

$$= (8 \text{ kg})(0.001036 \text{ m}^3/\text{kg}) + (2 \text{ kg})(2.3593 \text{ m}^3/\text{kg})$$

$$= 4.73 \text{ m}^3$$

Another way is to first determine the quality $x$, then the average specific volume $v$, and finally the total volume:

$$x = \frac{m_g}{m_t} = \frac{2 \text{ kg}}{10 \text{ kg}} = 0.2$$

$$v = v_f + xv_{fg}$$

$$= 0.001036 \text{ m}^3/\text{kg} + (0.2)[(2.3593 - 0.001036) \text{ m}^3/\text{kg}]$$

$$= 0.473 \text{ m}^3/\text{kg}$$

and

$$V = mv = (10 \text{ kg})(0.473 \text{ m}^3/\text{kg}) = 4.73 \text{ m}^3$$

**Discussion** The first method appears to be easier in this case since the masses of each phase are given. In most cases, however, the masses of each phase are not available, and the second method becomes more convenient.

■ **EXAMPLE 3–5** **Properties of Saturated Liquid–Vapor Mixture**

An 80-L vessel contains 4 kg of refrigerant-134a at a pressure of 160 kPa. Determine (a) the temperature, (b) the quality, (c) the enthalpy of the refrigerant, and (d) the volume occupied by the vapor phase.

**SOLUTION** A vessel is filled with refrigerant-134a. Some properties of the refrigerant are to be determined.

**Analysis** (a) The state of the saturated liquid–vapor mixture is shown in Fig. 3–36. At this point we do not know whether the refrigerant is in the compressed liquid, superheated vapor, or saturated mixture region. This can be determined by comparing a suitable property to the saturated liquid and saturated vapor values. From the information given, we can determine the specific volume:

$$v = \frac{V}{m} = \frac{0.080 \text{ m}^3}{4 \text{ kg}} = 0.02 \text{ m}^3/\text{kg}$$

At 160 kPa, we read

$$v_f = 0.0007435 \text{ m}^3/\text{kg}$$

$$v_g = 0.12355 \text{ m}^3/\text{kg} \qquad \text{(Table A–12)}$$

Obviously, $v_f < v < v_g$, and, the refrigerant is in the saturated mixture region. Thus, the temperature must be the saturation temperature at the specified pressure:

$$T = T_{\text{sat @ 160 kPa}} = -15.60°C$$

(b) Quality can be determined from

$$x = \frac{v - v_f}{v_{fg}} = \frac{0.02 - 0.0007435}{0.12355 - 0.0007435} = 0.157$$

(c) At 160 kPa, we also read from Table A–12 that $h_f = 31.18$ kJ/kg and $h_{fg} = 209.96$ kJ/kg. Then,

$$h = h_f + xh_{fg}$$

$$= 31.18 \text{ kJ/kg} + (0.157)(209.96 \text{ kJ/kg})$$

$$= 64.1 \text{ kJ/kg}$$

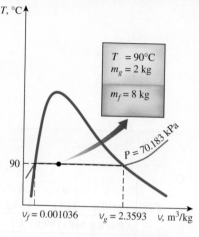

**FIGURE 3–35**
Schematic and $T$-$v$ diagram for Example 3–4.

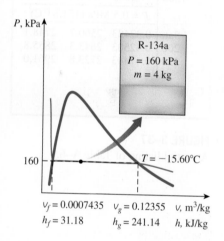

**FIGURE 3–36**
Schematic and $P$-$v$ diagram for Example 3–5.

(d) The mass of the vapor is

$$m_g = xm_t = (0.157)(4 \text{ kg}) = 0.628 \text{ kg}$$

and the volume occupied by the vapor phase is

$$V_g = m_g v_g = (0.628 \text{ kg})(0.12355 \text{ m}^3/\text{kg}) = \mathbf{0.0776 \text{ m}^3} \text{ (or 77.6 L)}$$

The rest of the volume (2.4 L) is occupied by the liquid.

Property tables are also available for saturated solid–vapor mixtures. Properties of saturated ice–water vapor mixtures, for example, are listed in Table A–8. Saturated solid–vapor mixtures can be handled just as saturated liquid–vapor mixtures.

| $T$,°C | $v$ m³/kg | $u$ kJ/kg | $h$ kJ/kg |
|---|---|---|---|
| | $P = 0.1$ MPa (99.61°C) | | |
| Sat. | 1.6941 | 2505.6 | 2675.0 |
| 100 | 1.6959 | 2506.2 | 2675.8 |
| 150 | 1.9367 | 2582.9 | 2776.6 |
| ⋮ | ⋮ | ⋮ | ⋮ |
| 1300 | 7.2605 | 4687.2 | 5413.3 |
| | $P = 0.5$ MPa (151.83°C) | | |
| Sat. | 0.37483 | 2560.7 | 2748.1 |
| 200 | 0.42503 | 2643.3 | 2855.8 |
| 250 | 0.47443 | 2723.8 | 2961.0 |

**FIGURE 3–37**
A partial listing of Table A–6.

## 2 Superheated Vapor

In the region to the right of the saturated vapor line and at temperatures above the critical point temperature, a substance exists as superheated vapor. Since the superheated region is a single-phase region (vapor phase only), temperature and pressure are no longer dependent properties and they can conveniently be used as the two independent properties in the tables. The format of the superheated vapor tables is illustrated in Fig. 3–37.

In these tables, the properties are listed against temperature for selected pressures starting with the saturated vapor data. The saturation temperature is given in parentheses following the pressure value.

Compared to saturated vapor, superheated vapor is characterized by

Lower pressures ($P < P_{sat}$ at a given $T$)
Higher tempreatures ($T > T_{sat}$ at a given $P$)
Higher specific volumes ($v > v_g$ at a given $P$ or $T$)
Higher internal energies ($u > u_g$ at a given $P$ or $T$)
Higher enthalpies ($h > h_g$ at a given $P$ or $T$)

---

**EXAMPLE 3–6    Internal Energy of Superheated Vapor**

Determine the internal energy of water at 200 kPa and 300°C.

**SOLUTION**  The internal energy of water at a specified state is to be determined.
**Analysis**  At 200 kPa, the saturation temperature is 120.21°C. Since $T > T_{sat}$, the water is in the superheated vapor region. Then the internal energy at the given temperature and pressure is determined from the superheated vapor table (Table A–6) to be

$$u = \mathbf{2808.8 \text{ kJ/kg}}$$

■ *EXAMPLE 3–7*    **Temperature of Superheated Vapor**

Determine the temperature of water at a state of $P = 0.5$ MPa and $h = 2890$ kJ/kg.

**SOLUTION**    The temperature of water at a specified state is to be determined.
*Analysis*    At 0.5 MPa, the enthalpy of saturated water vapor is $h_g = 2748.1$ kJ/kg.
Since $h > h_g$, as shown in Fig. 3–38, we again have superheated vapor.
Under 0.5 MPa in Table A–6 we read

| $T$, °C | $h$, kJ/kg |
|---------|-----------|
| 200     | 2855.8    |
| 250     | 2961.0    |

Obviously, the temperature is between 200 and 250°C. By linear interpolation it is determined to be

$$T = 216.3°C$$

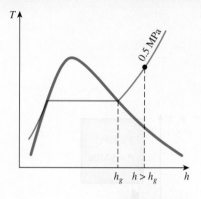

**FIGURE 3–38**
At a specified $P$, superheated vapor exists at a higher $h$ than the saturated vapor (Example 3–7).

## 3  Compressed Liquid

Compressed liquid tables are not as commonly available, and Table A–7 is the only compressed liquid table in this text. The format of Table A–7 is very much like the format of the superheated vapor tables. One reason for the lack of compressed liquid data is the relative independence of compressed liquid properties from pressure. Variation of properties of compressed liquid with pressure is very mild. Increasing the pressure 100 times often causes properties to change less than 1 percent.

In the absence of compressed liquid data, a general approximation is *to treat compressed liquid as saturated liquid at the given temperature* (Fig. 3–39). This is because the compressed liquid properties depend on temperature much more strongly than they do on pressure. Thus,

$$y \cong y_{f @ T} \tag{3–8}$$

for compressed liquids, where $y$ is $v$, $u$, or $h$. Of these three properties, the property whose value is most sensitive to variations in the pressure is the enthalpy $h$. Although the above approximation results in negligible error in $v$ and $u$, the error in $h$ may reach undesirable levels. However, the error in $h$ at low to moderate pressures and temperatures can be reduced significantly by evaluating it from

$$h \cong h_{f @ T} + v_{f @ T}(P - P_{sat @ T}) \tag{3–9}$$

instead of taking it to be just $h_f$. Note, however, that the approximation in Eq. 3–9 does not yield any significant improvement at moderate to high temperatures and pressures, and it may even backfire and result in greater error due to overcorrection at very high temperatures and pressures (*see* Kostic, 2006).

In general, a compressed liquid is characterized by

Higher pressures ($P > P_{sat}$ at a given $T$)

Lower tempreatures ($T < T_{sat}$ at a given $P$)

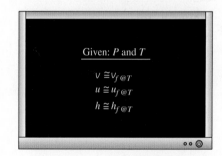

Given: $P$ and $T$

$v \cong v_{f @ T}$

$u \cong u_{f @ T}$

$h \cong h_{f @ T}$

**FIGURE 3–39**
A compressed liquid may be approximated as a saturated liquid at the given temperature.

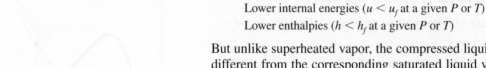

Lower specific volumes ($v < v_f$ at a given $P$ or $T$)
Lower internal energies ($u < u_f$ at a given $P$ or $T$)
Lower enthalpies ($h < h_f$ at a given $P$ or $T$)

But unlike superheated vapor, the compressed liquid properties are not much different from the corresponding saturated liquid values.

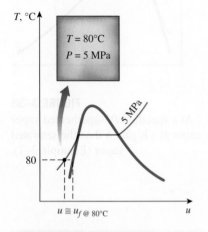

**FIGURE 3–40**
Schematic and $T$-$u$ diagram
for Example 3–8.

---

### EXAMPLE 3–8    Approximating Compressed Liquid as Saturated Liquid

Determine the internal energy of compressed liquid water at 80°C and 5 MPa, using (a) data from the compressed liquid table and (b) saturated liquid data. What is the error involved in the second case?

**SOLUTION**   The exact and approximate values of the internal energy of liquid water are to be determined.

**Analysis**   At 80°C, the saturation pressure of water is 47.416 kPa, and since 5 MPa > $P_{sat}$, we obviously have compressed liquid, as shown in Fig. 3–40.

(a) From the compressed liquid table (Table A–7)

$$\left.\begin{array}{r} P = 5 \text{ MPa} \\ T = 80°C \end{array}\right\} \quad u = 333.82 \text{ kJ/kg}$$

(b) From the saturation table (Table A–4), we read

$$u \cong u_{f @ 80°C} = 334.97 \text{ kJ/kg}$$

The error involved is

$$\frac{334.97 - 333.82}{333.82} \times 100 = 0.34\%$$

which is less than 1 percent.

## Reference State and Reference Values

The values of $u$, $h$, and $s$ cannot be measured directly, and they are calculated from measurable properties using the relations between thermodynamic properties. However, those relations give the *changes* in properties, not the values of properties at specified states. Therefore, we need to choose a convenient *reference state* and assign a value of *zero* for a convenient property or properties at that state. For water, the state of saturated liquid at 0.01°C is taken as the reference state, and the internal energy and entropy are assigned zero values at that state. For refrigerant-134a, the state of saturated liquid at −40°C is taken as the reference state, and the enthalpy and entropy are assigned zero values at that state. Note that some properties may have negative values as a result of the reference state chosen.

It should be mentioned that sometimes different tables list different values for some properties at the same state as a result of using a different reference state. However, in thermodynamics we are concerned with the *changes* in properties, and the reference state chosen is of no consequence in calculations as long as we use values from a single consistent set of tables or charts.

## ■ EXAMPLE 3–9  The Use of Steam Tables to Determine Properties

Determine the missing properties and the phase descriptions in the following table for water:

|     | $T$, °C | $P$, kPa | $u$, kJ/kg | $x$ | Phase description |
|-----|---------|----------|------------|-----|-------------------|
| (a) |         | 200      |            | 0.6 |                   |
| (b) | 125     |          | 1600       |     |                   |
| (c) |         | 1000     | 2950       |     |                   |
| (d) | 75      | 500      |            |     |                   |
| (e) |         | 850      |            | 0.0 |                   |

**SOLUTION** Properties and phase descriptions of water are to be determined at various states.

**Analysis** (a) The quality is given to be $x = 0.6$, which implies that 60 percent of the mass is in the vapor phase and the remaining 40 percent is in the liquid phase. Therefore, we have saturated liquid–vapor mixture at a pressure of 200 kPa. Then the temperature must be the saturation temperature at the given pressure:

$$T = T_{\text{sat @ 200 kPa}} = \mathbf{120.21°C} \quad \text{(Table A-5)}$$

At 200 kPa, we also read from Table A–5 that $u_f = 504.50$ kJ/kg and $u_{fg} = 2024.6$ kJ/kg. Then the average internal energy of the mixture is

$$
\begin{aligned}
u &= u_f + xu_{fg} \\
&= 504.50 \text{ kJ/kg} + (0.6)(2024.6 \text{ kJ/kg}) \\
&= \mathbf{1719.26 \text{ kJ/kg}}
\end{aligned}
$$

(b) This time the temperature and the internal energy are given, but we do not know which table to use to determine the missing properties because we have no clue as to whether we have saturated mixture, compressed liquid, or superheated vapor. To determine the region we are in, we first go to the saturation table (Table A–4) and determine the $u_f$ and $u_g$ values at the given temperature. At 125°C, we read $u_f = 524.83$ kJ/kg and $u_g = 2534.3$ kJ/kg. Next we compare the given $u$ value to these $u_f$ and $u_g$ values, keeping in mind that

$$
\begin{array}{lll}
\text{if} & u < u_f & \text{we have } \textit{compressed liquid} \\
\text{if} & u_f \leq u \leq u_g & \text{we have } \textit{saturated mixture} \\
\text{if} & u > u_g & \text{we have } \textit{superheated vapor}
\end{array}
$$

In our case the given $u$ value is 1600, which falls between the $u_f$ and $u_g$ values at 125°C. Therefore, we have saturated liquid–vapor mixture. Then the pressure must be the saturation pressure at the given temperature:

$$P = P_{\text{sat @ 125°C}} = \mathbf{232.23 \text{ kPa}} \quad \text{(Table A–4)}$$

The quality is determined from

$$x = \frac{u - u_f}{u_{fg}} = \frac{1600 - 524.83}{2009.5} = \mathbf{0.535}$$

The criteria above for determining whether we have compressed liquid, saturated mixture, or superheated vapor can also be used when enthalpy $h$ or specific volume $v$ is given instead of internal energy $u$, or when pressure is given instead of temperature.

(c) This is similar to case (b), except pressure is given instead of temperature. Following the argument given above, we read the $u_f$ and $u_g$ values at the specified pressure. At 1 MPa, we have $u_f = 761.39$ kJ/kg and $u_g = 2582.8$ kJ/kg. The specified $u$ value is 2950 kJ/kg, which is greater than the $u_g$ value at 1 MPa. Therefore, we have superheated vapor, and the temperature at this state is determined from the superheated vapor table by interpolation to be

$$T = 395.2°C \quad \text{(Table A-6)}$$

We would leave the quality column blank in this case since quality has no meaning for a superheated vapor.

(d) In this case the temperature and pressure are given, but again we cannot tell which table to use to determine the missing properties because we do not know whether we have saturated mixture, compressed liquid, or superheated vapor. To determine the region we are in, we go to the saturation table (Table A-5) and determine the saturation temperature value at the given pressure. At 500 kPa, we have $T_{sat} = 151.83°C$. We then compare the given $T$ value to this $T_{sat}$ value, keeping in mind that

if     $T < T_{\text{sat @ given } P}$     we have *compressed liquid*

if     $T = T_{\text{sat @ given } P}$     we have *saturated mixture*

if     $T > T_{\text{sat @ given } P}$     we have *superheated vapor*

In our case, the given $T$ value is 75°C, which is less than the $T_{sat}$ value at the specified pressure. Therefore, we have compressed liquid (Fig. 3–41), and normally we would determine the internal energy value from the compressed liquid table. But in this case the given pressure is much lower than the lowest pressure value in the compressed liquid table (which is 5 MPa), and therefore we are justified to treat the compressed liquid as saturated liquid at the given temperature (*not* pressure):

$$u \cong u_{f\text{ @ 75°C}} = 313.99 \text{ kJ/kg} \quad \text{(Table A-4)}$$

We would leave the quality column blank in this case since quality has no meaning in the compressed liquid region.

(e) The quality is given to be $x = 0$, and thus we have saturated liquid at the specified pressure of 850 kPa. Then the temperature must be the saturation temperature at the given pressure, and the internal energy must have the saturated liquid value:

$$T = T_{\text{sat @ 850 kPa}} = 172.94°C$$

$$u = u_{f\text{ @ 850 kPa}} = 731.00 \text{ kJ/kg} \quad \text{(Table A-5)}$$

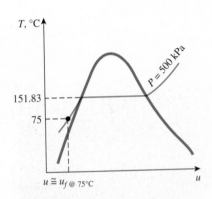

**FIGURE 3–41**
At a given $P$ and $T$, a pure substance will exist as a compressed liquid if $T < T_{\text{sat @ } P}$.

# 3–6 · THE IDEAL-GAS EQUATION OF STATE

Property tables provide very accurate information about the properties, but they are bulky and vulnerable to typographical errors. A more practical and desirable approach would be to have some simple relations among the properties that are sufficiently general and accurate.

Any equation that relates the pressure, temperature, and specific volume of a substance is called an **equation of state**. Property relations that involve other properties of a substance at equilibrium states are also referred to as equations of state. There are several equations of state, some simple and others very complex. The simplest and best-known equation of state for substances in the gas phase is the ideal-gas equation of state. This equation predicts the $P$-$v$-$T$ behavior of a gas quite accurately within some properly selected region.

*Gas* and *vapor* are often used as synonymous words. The vapor phase of a substance is customarily called a *gas* when it is above the critical temperature. *Vapor* usually implies a gas that is not far from a state of condensation.

In 1662, Robert Boyle, an Englishman, observed during his experiments with a vacuum chamber that the pressure of gases is inversely proportional to their volume. In 1802, J. Charles and J. Gay-Lussac, Frenchmen, experimentally determined that at low pressures the volume of a gas is proportional to its temperature. That is,

$$P = R\left(\frac{T}{v}\right)$$

or

$$Pv = RT \tag{3-10}$$

where the constant of proportionality $R$ is called the **gas constant**. Equation 3–10 is called the **ideal-gas equation of state**, or simply the **ideal-gas relation**, and a gas that obeys this relation is called an **ideal gas**. In this equation, $P$ is the absolute pressure, $T$ is the absolute temperature, and $v$ is the specific volume.

The gas constant $R$ is different for each gas (Fig. 3–42) and is determined from

$$R = \frac{R_u}{M} \quad \text{(kJ/kg·K or kPa·m}^3\text{/kg·K)}$$

where $R_u$ is the **universal gas constant** and $M$ is the molar mass (also called *molecular weight*) of the gas. The constant $R_u$ is the same for all substances, and its value is

$$R_u = \begin{cases} 8.31447 \text{ kJ/kmol·K} \\ 8.31447 \text{ kPa·m}^3\text{/kmol·K} \\ 0.0831447 \text{ bar·m}^3\text{/kmol·K} \\ 1.98588 \text{ Btu/lbmol·R} \\ 10.7316 \text{ psia·ft}^3\text{/lbmol·R} \\ 1545.37 \text{ ft·lbf/lbmol·R} \end{cases} \tag{3-11}$$

| Substance | $R$, kJ/kg·K |
|-----------|--------------|
| Air       | 0.2870       |
| Helium    | 2.0769       |
| Argon     | 0.2081       |
| Nitrogen  | 0.2968       |

**FIGURE 3–42**
Different substances have different gas constants.

The **molar mass** $M$ can simply be defined as *the mass of one mole* (also called a *gram-mole*, abbreviated gmol) *of a substance in grams*, or *the mass of one kmol* (also called a *kilogram-mole*, abbreviated kgmol) *in kilograms*. In English units, it is the mass of 1 lbmol in lbm. Notice that the molar mass of a substance has the same numerical value in both unit systems because of the way it is defined. When we say the molar mass of nitrogen is 28, it simply means the mass of 1 kmol of nitrogen is 28 kg, or the mass of 1 lbmol of nitrogen is 28 lbm. That is, $M = 28$ kg/kmol $= 28$ lbm/lbmol.

The mass of a system is equal to the product of its molar mass $M$ and the mole number $N$:

$$m = MN \quad \text{(kg)} \tag{3–12}$$

The values of $R$ and $M$ for several substances are given in Table A–1.

The ideal-gas equation of state can be written in several different forms:

$$V = mv \longrightarrow PV = mRT \tag{3–13}$$

$$mR = (MN)R = NR_u \longrightarrow PV = NR_uT \tag{3–14}$$

$$V = N\bar{v} \longrightarrow P\bar{v} = R_uT \tag{3–15}$$

where $\bar{v}$ is the molar specific volume, that is, the volume per unit mole (in $m^3/kmol$). A bar above a property denotes values on a *unit-mole basis* throughout this text (Fig. 3–43).

By writing Eq. 3–13 twice for a fixed mass and simplifying, the properties of an ideal gas at two different states are related to each other by

$$\frac{P_1 V_1}{T_1} = \frac{P_2 V_2}{T_2} \tag{3-16}$$

An ideal gas is an *imaginary* substance that obeys the relation $Pv = RT$. It has been experimentally observed that the ideal-gas relation given closely approximates the $P$-$v$-$T$ behavior of real gases at low densities. At low pressures and high temperatures, the density of a gas decreases, and the gas behaves as an ideal gas under these conditions. What constitutes low pressure and high temperature is explained later.

In the range of practical interest, many familiar gases such as air, nitrogen, oxygen, hydrogen, helium, argon, neon, krypton, and even heavier gases such as carbon dioxide can be treated as ideal gases with negligible error (often less than 1 percent). Dense gases such as water vapor in steam power plants and refrigerant vapor in refrigerators, however, should not be treated as ideal gases. Instead, the property tables should be used for these substances.

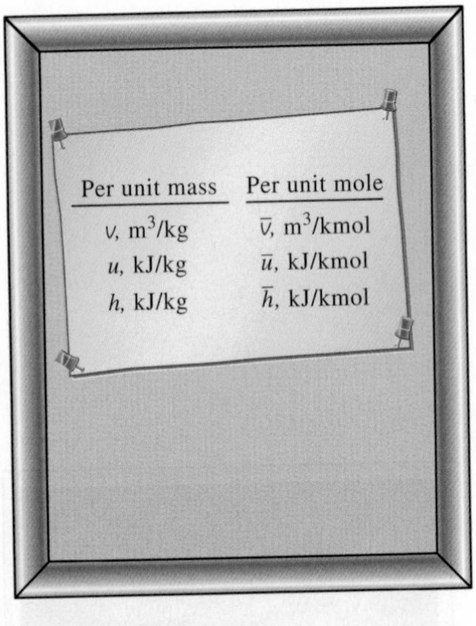

| Per unit mass | Per unit mole |
|---|---|
| $v$, $m^3/kg$ | $\bar{v}$, $m^3/kmol$ |
| $u$, kJ/kg | $\bar{u}$, kJ/kmol |
| $h$, kJ/kg | $\bar{h}$, kJ/kmol |

**FIGURE 3–43**
Properties per unit mole are denoted with a bar on the top.

**FIGURE 3–44**
©*Stockbyte/Getty Images RF*

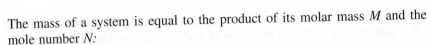

*EXAMPLE 3-10*     **Temperature Rise of Air in a Tire During a Trip**

The gage pressure of an automobile tire is measured to be 210 kPa before a trip and 220 kPa after the trip at a location where the atmospheric pressure is 95 kPa (Fig. 3–44). Assuming the volume of the tire remains constant and the air temperature before the trip is 25°C, determine air temperature in the tire after the trip.

**SOLUTION** The pressure in an automobile tire is measured before and after a trip. The temperature of air in the tire after the trip is to be determined.

**Assumptions** 1 The volume of the tire remains constant. 2 Air is an ideal gas.
**Properties** The local atmospheric pressure is 95 kPa.
**Analysis** The absolute pressures in the tire before and after the trip are

$$P_1 = P_{\text{gage,1}} + P_{\text{atm}} = 210 + 95 = 305 \text{ kPa}$$
$$P_2 = P_{\text{gage,2}} + P_{\text{atm}} = 220 + 95 = 315 \text{ kPa}$$

Note that air is an ideal gas and the volume is constant, the air temperatures after the trip is determined to be

$$\frac{P_1 V_1}{T_1} = \frac{P_2 V_2}{T_2} \longrightarrow T_2 = \frac{P_2}{P_1} T_1 = \frac{315 \text{ kPa}}{305 \text{ kPa}}(25 + 273 \text{ K}) = 307.8 \text{ K} = \mathbf{34.8°C}$$

Therefore, the absolute temperature of air in the tire will increase by 6.9% during this trip.

**Discussion** Note that the air temperature has risen nearly 10°C during this trip. This shows the importance of measuring the tire pressures before long trips to avoid errors due to temperature rise of air in tire. Also note that the unit Kelvin is used for temperature in the ideal gas relation.

## Is Water Vapor an Ideal Gas?

This question cannot be answered with a simple yes or no. The error involved in treating water vapor as an ideal gas is calculated and plotted in Fig. 3–45. It is clear from this figure that at pressures below 10 kPa, water

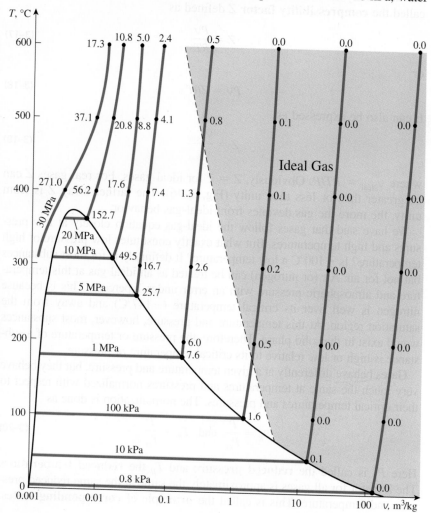

**FIGURE 3–45**
Percentage of error
$([|v_{\text{table}} - v_{\text{ideal}}|/v_{\text{table}}] \times 100)$
involved in assuming steam to be an ideal gas, and the region where steam can be treated as an ideal gas with less than 1 percent error.

vapor can be treated as an ideal gas, regardless of its temperature, with negligible error (less than 0.1 percent). At higher pressures, however, the ideal-gas assumption yields unacceptable errors, particularly in the vicinity of the critical point and the saturated vapor line (over 100 percent). Therefore, in air-conditioning applications, the water vapor in the air can be treated as an ideal gas with essentially no error since the pressure of the water vapor is very low. In steam power plant applications, however, the pressures involved are usually very high; therefore, ideal-gas relations should not be used.

## 3–7 ▪ COMPRESSIBILITY FACTOR—A MEASURE OF DEVIATION FROM IDEAL-GAS BEHAVIOR

The ideal-gas equation is very simple and thus very convenient to use. However, as illustrated in Fig. 3–45, gases deviate from ideal-gas behavior significantly at states near the saturation region and the critical point. This deviation from ideal-gas behavior at a given temperature and pressure can accurately be accounted for by the introduction of a correction factor called the **compressibility factor** $Z$ defined as

$$Z = \frac{Pv}{RT} \tag{3–17}$$

or

$$Pv = ZRT \tag{3–18}$$

It can also be expressed as

$$Z = \frac{v_{actual}}{v_{ideal}} \tag{3–19}$$

where $v_{ideal} = RT/P$. Obviously, $Z = 1$ for ideal gases. For real gases $Z$ can be greater than or less than unity (Fig. 3–46). The farther away $Z$ is from unity, the more the gas deviates from ideal-gas behavior.

We have said that gases follow the ideal-gas equation closely at low pressures and high temperatures. But what exactly constitutes low pressure or high temperature? Is $-100°C$ a low temperature? It definitely is for most substances but not for air. Air (or nitrogen) can be treated as an ideal gas at this temperature and atmospheric pressure with an error under 1 percent. This is because nitrogen is well over its critical temperature ($-147°C$) and away from the saturation region. At this temperature and pressure, however, most substances would exist in the solid phase. Therefore, the pressure or temperature of a substance is high or low relative to its critical temperature or pressure.

Gases behave differently at a given temperature and pressure, but they behave very much the same at temperatures and pressures normalized with respect to their critical temperatures and pressures. The normalization is done as

$$P_R = \frac{P}{P_{cr}} \quad \text{and} \quad T_R = \frac{T}{T_{cr}} \tag{3–20}$$

Here $P_R$ is called the **reduced pressure** and $T_R$ the **reduced temperature**. The $Z$ factor for all gases is approximately the same at the same reduced pressure and temperature. This is called the **principle of corresponding states**.

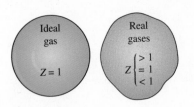

**FIGURE 3–46**
The compressibility factor is unity for ideal gases.

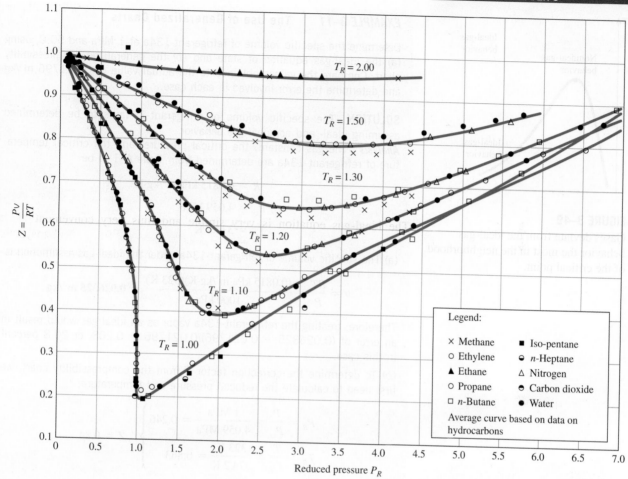

**FIGURE 3–47**

Comparison of $Z$ factors for various gases.

*Source: Gour-Jen Su, "Modified Law of Corresponding States," Ind. Eng. Chem. (international ed.) 38 (1946), p. 803.*

In Fig. 3–47, the experimentally determined $Z$ values are plotted against $P_R$ and $T_R$ for several gases. The gases seem to obey the principle of corresponding states reasonably well. By curve-fitting all the data, we obtain the **generalized compressibility chart** that can be used for all gases (Fig. A–15).

The following observations can be made from the generalized compressibility chart:

1. At very low pressures ($P_R \ll 1$), gases behave as an ideal gas regardless of temperature (Fig. 3–48).
2. At high temperatures ($T_R > 2$), ideal-gas behavior can be assumed with good accuracy regardless of pressure (except when $P_R \gg 1$).
3. The deviation of a gas from ideal-gas behavior is greatest in the vicinity of the critical point (Fig. 3–49).

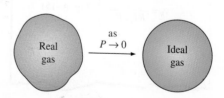

**FIGURE 3–48**

At very low pressures, all gases approach ideal-gas behavior (regardless of their temperature).

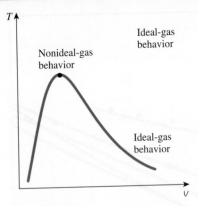

**FIGURE 3–49**
Gases deviate from the ideal-gas behavior the most in the neighborhood of the critical point.

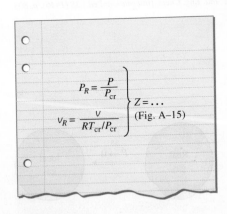

$$P_R = \frac{P}{P_{cr}}$$
$$v_R = \frac{v}{RT_{cr}/P_{cr}}$$
$$Z = \dots$$
(Fig. A–15)

**FIGURE 3–50**
The compressibility factor can also be determined from a knowledge of $P_R$ and $v_R$.

**EXAMPLE 3–11    The Use of Generalized Charts**

Determine the specific volume of refrigerant-134a at 1 MPa and 50°C, using (a) the ideal-gas equation of state and (b) the generalized compressibility chart. Compare the values obtained to the actual value of 0.021796 m³/kg and determine the error involved in each case.

**SOLUTION**  The specific volume of refrigerant-134a is to be determined assuming ideal- and nonideal-gas behavior.

**Analysis**  The gas constant, the critical pressure, and the critical temperature of refrigerant-134a are determined from Table A–1 to be

$$R = 0.0815 \text{ kPa·m}^3/\text{kg·K}$$

$$P_{cr} = 4.059 \text{ MPa}$$

$$T_{cr} = 374.2 \text{ K}$$

(a) The specific volume of refrigerant-134a under the ideal-gas assumption is

$$v = \frac{RT}{P} = \frac{(0.0815 \text{ kPa·m}^3/\text{kg·K})(323 \text{ K})}{1000 \text{ kPa}} = 0.026325 \text{ m}^3/\text{kg}$$

Therefore, treating the refrigerant-134a vapor as an ideal gas would result in an error of $(0.026325 - 0.021796)/0.021796 = 0.208$, or 20.8 percent in this case.

(b) To determine the correction factor $Z$ from the compressibility chart, we first need to calculate the reduced pressure and temperature:

$$\left. \begin{array}{l} P_R = \dfrac{P}{P_{cr}} = \dfrac{1 \text{ MPa}}{4.059 \text{ MPa}} = 0.246 \\[2mm] T_R = \dfrac{T}{T_{cr}} = \dfrac{323 \text{ K}}{374.2 \text{ K}} = 0.863 \end{array} \right\} \quad Z = 0.84$$

Thus

$$v = Zv_{\text{ideal}} = (0.84)(0.026325 \text{ m}^3/\text{kg}) = 0.022113 \text{ m}^3/\text{kg}$$

**Discussion**  The error in this result is less than **2 percent**. Therefore, in the absence of tabulated data, the generalized compressibility chart can be used with confidence.

When $P$ and $v$, or $T$ and $v$, are given instead of $P$ and $T$, the generalized compressibility chart can still be used to determine the third property, but it would involve tedious trial and error. Therefore, it is necessary to define one more reduced property called the **pseudo-reduced specific volume** $v_R$ as

$$v_R = \frac{v_{\text{actual}}}{RT_{cr}/P_{cr}} \tag{3–21}$$

Note that $v_R$ is defined differently from $P_R$ and $T_R$. It is related to $T_{cr}$ and $P_{cr}$ instead of $v_{cr}$. Lines of constant $v_R$ are also added to the compressibility charts, and this enables one to determine $T$ or $P$ without having to resort to time-consuming iterations (Fig. 3–50).

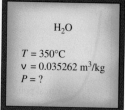

**EXAMPLE 3–12**   **Using Generalized Charts to Determine Pressure**

Determine the pressure of water vapor at 350°C and 0.035262 m³/kg, using (a) the steam tables, (b) the ideal-gas equation, and (c) the generalized compressibility chart.

**SOLUTION**   The pressure of water vapor is to be determined in three different ways.

**Analysis**   A sketch of the system is given in Fig. 3–51. The gas constant, the critical pressure, and the critical temperature of steam are determined from Table A–1 to be

$$R = 0.4615 \text{ kPa·m}^3/\text{kg·K}$$

$$P_{cr} = 22.06 \text{ MPa}$$

$$T_{cr} = 647.1 \text{ K}$$

(a) The pressure at the specified state is determined from Table A–6 to be

$$\left. \begin{array}{l} v = 0.035262 \text{ m}^3/\text{kg} \\ T = 350°C \end{array} \right\} \quad P = 7.0 \text{ MPa}$$

This is the experimentally determined value, and thus it is the most accurate.

(b) The pressure of steam under the ideal-gas assumption is determined from the ideal-gas relation to be

$$P = \frac{RT}{v} = \frac{(0.4615 \text{ kPa·m}^3/\text{kg·K})(623 \text{ K})}{0.035262 \text{ m}^3/\text{kg}} = 8.15 \text{ MPa}$$

Therefore, treating the steam as an ideal gas would result in an error of (8.15 − 7.0)/7.0 = 0.164, or 16.4 percent in this case.

(c) To determine the correction factor Z from the compressibility chart (Fig. A–15), we first need to calculate the pseudo-reduced specific volume and the reduced temperature:

$$\left. \begin{array}{l} v_R = \dfrac{v_{actual}}{RT_{cr}/P_{cr}} = \dfrac{(0.035262 \text{ m}^3/\text{kg})(22,060 \text{ kPa})}{(0.4615 \text{ kPa·m}^3/\text{kg·K})(647.1 \text{ K})} = 2.605 \\[3mm] T_R = \dfrac{T}{T_{cr}} = \dfrac{623 \text{ K}}{647.1 \text{ K}} = 0.96 \end{array} \right\} P_R = 0.31$$

Thus,

$$P = P_R P_{cr} = (0.31)(22.06 \text{ MPa}) = 6.84 \text{ MPa}$$

**Discussion**   Using the compressibility chart reduced the error from 16.4 to 2.3 percent, which is acceptable for most engineering purposes (Fig. 3–52). A bigger chart, of course, would give better resolution and reduce the reading errors. Notice that we did not have to determine Z in this problem since we could read $P_R$ directly from the chart.

**FIGURE 3–51**
Schematic for Example 3–12.

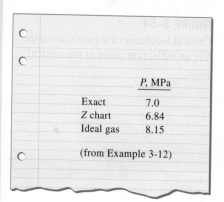

| | P, MPa |
|---|---|
| Exact | 7.0 |
| Z chart | 6.84 |
| Ideal gas | 8.15 |
| (from Example 3-12) | |

**FIGURE 3–52**
Results obtained by using the compressibility chart are usually within a few percent of actual values.

**FIGURE 3–53**
Several equations of state have been proposed throughout history.

# 3–8 · OTHER EQUATIONS OF STATE

The ideal-gas equation of state is very simple, but its range of applicability is limited. It is desirable to have equations of state that represent the P-v-T behavior of substances accurately over a larger region with no limitations. Such equations are naturally more complicated. Several equations have been proposed for this purpose (Fig. 3–53), but we shall discuss only three: the *van der Waals* equation

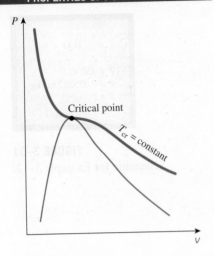

**FIGURE 3–54**
Critical isotherm of a pure substance has an inflection point at the critical state.

because it is one of the earliest, the *Beattie-Bridgeman* equation of state because it is one of the best known and is reasonably accurate, and the *Benedict-Webb-Rubin* equation because it is one of the more recent and is very accurate.

## van der Waals Equation of State

The van der Waals equation of state was proposed in 1873, and it has two constants that are determined from the behavior of a substance at the critical point. It is given by

$$\left(P + \frac{a}{v^2}\right)(v - b) = RT \tag{3–22}$$

Van der Waals intended to improve the ideal-gas equation of state by including two of the effects not considered in the ideal-gas model: the *intermolecular attraction forces* and the *volume occupied by the molecules themselves*. The term $a/v^2$ accounts for the intermolecular forces, and $b$ accounts for the volume occupied by the gas molecules. In a room at atmospheric pressure and temperature, the volume actually occupied by molecules is only about one-thousandth of the volume of the room. As the pressure increases, the volume occupied by the molecules becomes an increasingly significant part of the total volume. Van der Waals proposed to correct this by replacing $v$ in the ideal-gas relation with the quantity $v - b$, where $b$ represents the volume occupied by the gas molecules per unit mass.

The determination of the two constants appearing in this equation is based on the observation that the critical isotherm on a $P$-$v$ diagram has a horizontal inflection point at the critical point (Fig. 3–54). Thus, the first and the second derivatives of $P$ with respect to $v$ at the critical point must be zero. That is,

$$\left(\frac{\partial P}{\partial v}\right)_{T=T_{cr}=\text{const}} = 0 \quad \text{and} \quad \left(\frac{\partial^2 P}{\partial v^2}\right)_{T=T_{cr}=\text{const}} = 0$$

By performing the differentiations and eliminating $v_{cr}$, the constants $a$ and $b$ are determined to be

$$a = \frac{27R^2T_{cr}^2}{64P_{cr}} \quad \text{and} \quad b = \frac{RT_{cr}}{8P_{cr}} \tag{3–23}$$

The constants $a$ and $b$ can be determined for any substance from the critical-point data alone (Table A–1).

The accuracy of the van der Waals equation of state is often inadequate, but it can be improved by using values of $a$ and $b$ that are based on the actual behavior of the gas over a wider range instead of a single point. Despite its limitations, the van der Waals equation of state has a historical value in that it was one of the first attempts to model the behavior of real gases. The van der Waals equation of state can also be expressed on a unit-mole basis by replacing the $v$ in Eq. 3–22 by $\bar{v}$ and the $R$ in Eqs. 3–22 and 3–23 by $R_u$.

## Beattie-Bridgeman Equation of State

The Beattie-Bridgeman equation, proposed in 1928, is an equation of state based on five experimentally determined constants. It is expressed as

$$P = \frac{R_u T}{\bar{v}^2}\left(1 - \frac{c}{\bar{v}T^3}\right)(\bar{v} + B) - \frac{A}{\bar{v}^2} \tag{3–24}$$

**TABLE 3–4**

Constants that appear in the Beattie-Bridgeman and the Benedict-Webb-Rubin equations of state

(a) When $P$ is in kPa, $\bar{v}$ is in m³/kmol, $T$ is in K, and $R_u$ = 8.314 kPa·m³/kmol·K, the five constants in the Beattie-Bridgeman equation are as follows:

| Gas | $A_0$ | $a$ | $B_0$ | $b$ | $c$ |
|---|---|---|---|---|---|
| Air | 131.8441 | 0.01931 | 0.04611 | −0.001101 | $4.34 \times 10^4$ |
| Argon, Ar | 130.7802 | 0.02328 | 0.03931 | 0.0 | $5.99 \times 10^4$ |
| Carbon dioxide, $CO_2$ | 507.2836 | 0.07132 | 0.10476 | 0.07235 | $6.60 \times 10^5$ |
| Helium, He | 2.1886 | 0.05984 | 0.01400 | 0.0 | 40 |
| Hydrogen, $H_2$ | 20.0117 | −0.00506 | 0.02096 | −0.04359 | 504 |
| Nitrogen, $N_2$ | 136.2315 | 0.02617 | 0.05046 | −0.00691 | $4.20 \times 10^4$ |
| Oxygen, $O_2$ | 151.0857 | 0.02562 | 0.04624 | 0.004208 | $4.80 \times 10^4$ |

*Source:* Gordon J. Van Wylen and Richard E. Sonntag, *Fundamentals of Classical Thermodynamics,* English/SI Version, 3rd ed. (New York: John Wiley & Sons, 1986), p. 46, table 3.3.

(b) When $P$ is in kPa, $\bar{v}$ is in m³/kmol, $T$ is in K, and $R_u$ = 8.314 kPa·m³/kmol·K, the eight constants in the Benedict-Webb-Rubin equation are as follows:

| Gas | $a$ | $A_0$ | $b$ | $B_0$ | $c$ | $C_0$ | $\alpha$ | $\gamma$ |
|---|---|---|---|---|---|---|---|---|
| n-Butane, $C_4H_{10}$ | 190.68 | 1021.6 | 0.039998 | 0.12436 | $3.205 \times 10^7$ | $1.006 \times 10^8$ | $1.101 \times 10^{-3}$ | 0.0340 |
| Carbon dioxide, $CO_2$ | 13.86 | 277.30 | 0.007210 | 0.04991 | $1.511 \times 10^6$ | $1.404 \times 10^7$ | $8.470 \times 10^{-5}$ | 0.00539 |
| Carbon monoxide, CO | 3.71 | 135.87 | 0.002632 | 0.05454 | $1.054 \times 10^5$ | $8.673 \times 10^5$ | $1.350 \times 10^{-4}$ | 0.0060 |
| Methane, $CH_4$ | 5.00 | 187.91 | 0.003380 | 0.04260 | $2.578 \times 10^5$ | $2.286 \times 10^6$ | $1.244 \times 10^{-4}$ | 0.0060 |
| Nitrogen, $N_2$ | 2.54 | 106.73 | 0.002328 | 0.04074 | $7.379 \times 10^4$ | $8.164 \times 10^5$ | $1.272 \times 10^{-4}$ | 0.0053 |

*Source:* Kenneth Wark, *Thermodynamics,* 4th ed. (New York: McGraw-Hill, 1983), p. 815, table A-21M. Originally published in H. W. Cooper and J. C. Goldfrank, *Hydrocarbon Processing* 46, no. 12 (1967), p. 141.

where

$$A = A_0\left(1 - \frac{a}{\bar{v}}\right) \quad \text{and} \quad B = B_0\left(1 - \frac{b}{\bar{v}}\right) \tag{3–25}$$

The constants appearing in the above equation are given in Table 3–4 for various substances. The Beattie-Bridgeman equation is known to be reasonably accurate for densities up to about $0.8\rho_{cr}$, where $\rho_{cr}$ is the density of the substance at the critical point.

## Benedict-Webb-Rubin Equation of State

Benedict, Webb, and Rubin extended the Beattie-Bridgeman equation in 1940 by raising the number of constants to eight. It is expressed as

$$P = \frac{R_u T}{\bar{v}} + \left(B_0 R_u T - A_0 - \frac{C_0}{T^2}\right)\frac{1}{\bar{v}^2} + \frac{bR_u T - a}{\bar{v}^3} + \frac{a\alpha}{\bar{v}^6} + \frac{c}{\bar{v}^3 T^2}\left(1 + \frac{\gamma}{\bar{v}^2}\right)e^{-\gamma/\bar{v}^2}$$

$$\tag{3–26}$$

The values of the constants appearing in this equation are given in Table 3–4. This equation can handle substances at densities up to about $2.5\rho_{cr}$. In 1962, Strobridge further extended this equation by raising the number of constants to 16 (Fig. 3–55).

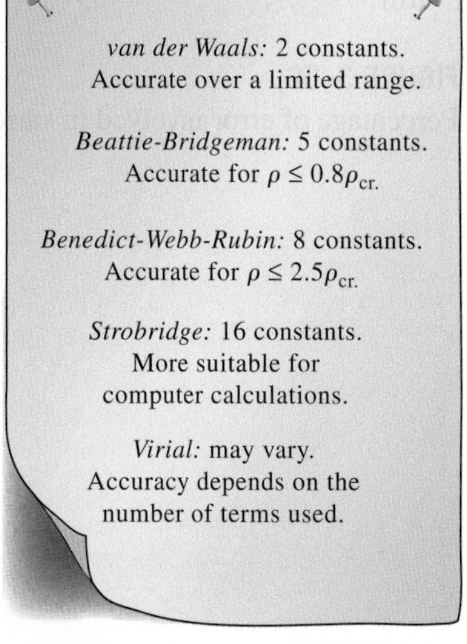

*van der Waals:* 2 constants. Accurate over a limited range.

*Beattie-Bridgeman:* 5 constants. Accurate for $\rho \le 0.8\rho_{cr}$.

*Benedict-Webb-Rubin:* 8 constants. Accurate for $\rho \le 2.5\rho_{cr}$.

*Strobridge:* 16 constants. More suitable for computer calculations.

*Virial:* may vary. Accuracy depends on the number of terms used.

**FIGURE 3–55**

Complex equations of state represent the $P$-$v$-$T$ behavior of gases more accurately over a wider range.

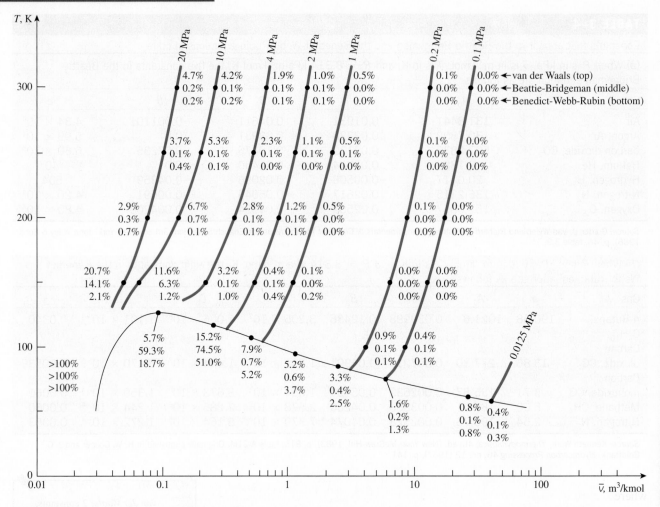

**FIGURE 3–56**
Percentage of error involved in various equations of state for nitrogen (% error = $[(|v_{table} - v_{equation}|)/v_{table}] \times 100$).

## Virial Equation of State

The equation of state of a substance can also be expressed in a series form as

$$P = \frac{RT}{v} + \frac{a(T)}{v^2} + \frac{b(T)}{v^3} + \frac{c(T)}{v^4} + \frac{d(T)}{v^5} + \ldots \qquad (3\text{-}27)$$

This and similar equations are called the *virial equations of state,* and the coefficients $a(T)$, $b(T)$, $c(T)$, and so on, that are functions of temperature alone are called *virial coefficients*. These coefficients can be determined experimentally or theoretically from statistical mechanics. Obviously, as the pressure approaches zero, all the virial coefficients will vanish and the equation will reduce to the ideal-gas equation of state. The $P$-$v$-$T$ behavior of a substance can be represented accurately with the virial equation of state over a wider range by including a sufficient number of terms. The equations of state discussed here are applicable to the gas phase of the substances only, and thus should not be used for liquids or liquid–vapor mixtures.

Complex equations represent the $P$-$v$-$T$ behavior of substances reasonably well and are very suitable for digital computer applications. For hand calculations, however, it is suggested that the reader use the property tables or the simpler equations of state for convenience. This is particularly true for specific-volume calculations since all the earlier equations are implicit in $v$ and require a trial-and-error approach. The accuracy of the van der Waals, Beattie-Bridgeman, and Benedict-Webb-Rubin equations of state is illustrated in Fig. 3–56. It is apparent from this figure that the Benedict-Webb-Rubin equation of state is usually the most accurate.

■ **EXAMPLE 3–13** **Different Methods of Evaluating Gas Pressure**

Predict the pressure of nitrogen gas at $T = 175$ K and $v = 0.00375$ m$^3$/kg on the basis of (a) the ideal-gas equation of state, (b) the van der Waals equation of state, (c) the Beattie-Bridgeman equation of state, and (d) the Benedict-Webb-Rubin equation of state. Compare the values obtained to the experimentally determined value of 10,000 kPa.

**SOLUTION** The pressure of nitrogen gas is to be determined using four different equations of state.

**Properties** The gas constant of nitrogen gas is 0.2968 kPa·m$^3$/kg·K (Table A–1).

**Analysis** (a) Using the ideal-gas equation of state, the pressure is found to be

$$P = \frac{RT}{v} = \frac{(0.2968 \text{ kPa·m}^3/\text{kg·K})(175 \text{ K})}{0.00375 \text{ m}^3/\text{kg}} = 13{,}851 \text{ kPa}$$

which is in error by 38.5 percent.

(b) The van der Waals constants for nitrogen are determined from Eq. 3–23 to be

$$a = 0.175 \text{ m}^6\text{·kPa/kg}^2$$

$$b = 0.00138 \text{ m}^3/\text{kg}$$

From Eq. 3–22,

$$P = \frac{RT}{v - b} - \frac{a}{v^2} = 9471 \text{ kPa}$$

which is in error by 5.3 percent.

(c) The constants in the Beattie-Bridgeman equation are determined from Table 3–4 to be

$$A = 102.29$$

$$B = 0.05378$$

$$c = 4.2 \times 10^4$$

Also, $\bar{v} = Mv = (28.013 \text{ kg/mol})(0.00375 \text{ m}^3/\text{kg}) = 0.10505$ m$^3$/kmol. Substituting these values into Eq. 3–24, we obtain

$$P = \frac{R_u T}{\bar{v}^2}\left(1 - \frac{c}{\bar{v}T^3}\right)(\bar{v} + B) - \frac{A}{\bar{v}^2} = 10{,}110 \text{ kPa}$$

which is in error by 1.1 percent.

(d) The constants in the Benedict-Webb-Rubin equation are determined from Table 3–4 to be

$$a = 2.54 \qquad A_0 = 106.73$$
$$b = 0.002328 \qquad B_0 = 0.04074$$
$$c = 7.379 \times 10^4 \qquad C_0 = 8.164 \times 10^5$$
$$\alpha = 1.272 \times 10^{-4} \qquad \gamma = 0.0053$$

Substituting these values into Eq. 3–26 gives

$$P = \frac{R_u T}{\bar{v}} + \left(B_0 R_u T - A_0 - \frac{C_0}{T^2}\right)\frac{1}{\bar{v}^2} + \frac{b R_u T - a}{\bar{v}^3}$$
$$+ \frac{a\alpha}{\bar{v}^6} + \frac{c}{\bar{v}^3 T^2}\left(1 + \frac{\gamma}{\bar{v}^2}\right)e^{-\gamma/\bar{v}^2}$$
$$= \mathbf{10{,}009 \ kPa}$$

which is in error by only 0.09 percent. Thus, the accuracy of the Benedict-Webb-Rubin equation of state is rather impressive in this case.

---

## TOPIC OF SPECIAL INTEREST*    Vapor Pressure and Phase Equilibrium

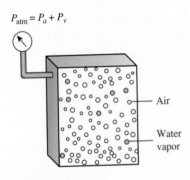

$P_{atm} = P_a + P_v$

— Air

— Water vapor

**FIGURE 3–57**
Atmospheric pressure is the sum of the dry air pressure $P_a$ and the vapor pressure $P_v$.

The pressure in a gas container is due to the individual molecules striking the wall of the container and exerting a force on it. This force is proportional to the average velocity of the molecules and the number of molecules per unit volume of the container (i.e., molar density). Therefore, the pressure exerted by a gas is a strong function of the density and the temperature of the gas. For a gas mixture, the pressure measured by a sensor such as a transducer is the sum of the pressures exerted by the individual gas species, called the *partial pressure*. It can be shown (see Chap. 13) that the partial pressure of a gas in a mixture is proportional to the number of moles (or the mole fraction) of that gas.

Atmospheric air can be viewed as a mixture of dry air (air with zero moisture content) and water vapor (also referred to as moisture), and the atmospheric pressure is the sum of the pressure of dry air $P_a$ and the pressure of water vapor, called the **vapor pressure** $P_v$ (Fig. 3–57). That is,

$$P_{atm} = P_a + P_v \tag{3–28}$$

(Note that in some applications, the phrase "vapor pressure" is used to indicate saturation pressure.) The vapor pressure constitutes a small fraction (usually under 3 percent) of the atmospheric pressure since air is mostly nitrogen and oxygen, and the water molecules constitute a small fraction (usually under 3 percent) of the total molecules in the air. However, the amount of water vapor in the air has a major impact on thermal comfort and many processes such as drying.

---

*This section can be skipped without a loss in continuity.

Air can hold a certain amount of moisture only, and the ratio of the actual amount of moisture in the air at a given temperature to the maximum amount air can hold at that temperature is called the **relative humidity** $\phi$. The relative humidity ranges from 0 for dry air to 100 percent for **saturated air** (air that cannot hold any more moisture). The vapor pressure of saturated air at a given temperature is equal to the saturation pressure of water at that temperature. For example, the vapor pressure of saturated air at 25°C is 3.17 kPa.

The amount of moisture in the air is completely specified by the temperature and the relative humidity, and the vapor pressure is related to relative humidity $\phi$ by

$$P_v = \phi P_{sat @ T} \qquad (3-29)$$

where $P_{sat @ T}$ is the saturation pressure of water at the specified temperature. For example, the vapor pressure of air at 25°C and 60 percent relative humidity is

$$P_v = \phi P_{sat @ 25°C} = 0.6 \times (3.17 \text{ kPa}) = 1.90 \text{ kPa}$$

The desirable range of relative humidity for thermal comfort is 40 to 60 percent.

Note that the amount of moisture air can hold is proportional to the saturation pressure, which increases with temperature. Therefore, air can hold more moisture at higher temperatures. Dropping the temperature of moist air reduces its moisture capacity and may result in the condensation of some of the moisture in the air as suspended water droplets (fog) or as a liquid film on cold surfaces (dew). So it is no surprise that fog and dew are common occurrences at humid locations especially in the early morning hours when the temperatures are the lowest. Both fog and dew disappear (evaporate) as the air temperature rises shortly after sunrise. You also may have noticed that electronic devices such as camcorders come with warnings against bringing them into moist indoors when the devices are cold to avoid moisture condensation on the sensitive electronics of the devices.

It is a common observation that whenever there is an imbalance of a commodity in a medium, nature tends to redistribute it until a "balance" or "equality" is established. This tendency is often referred to as the *driving force,* which is the mechanism behind many naturally occurring transport phenomena such as heat transfer, fluid flow, electric current, and mass transfer. If we define the amount of a commodity per unit volume as the *concentration* of that commodity, we can say that the flow of a commodity is always in the direction of decreasing concentration, that is, from the region of high concentration to the region of low concentration (Fig. 3–58). The commodity simply creeps away during redistribution, and thus the flow is a *diffusion process.*

We know from experience that a wet T-shirt hanging in an open area eventually dries, a small amount of water left in a glass evaporates, and the aftershave in an open bottle quickly disappears. These and many other similar examples suggest that there is a driving force between the two phases of a substance that forces the mass to transform from one phase to another. The magnitude of this force depends on the relative concentrations of the two phases. A wet T-shirt dries much faster in dry air than it would in humid air. In fact, it does not dry at all if the relative humidity of the environment is 100 percent and thus the air is saturated. In this case, there is no transformation from the liquid phase to the vapor phase, and the two phases are in

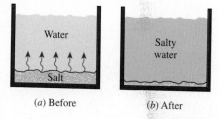

(a) Before   (b) After

**FIGURE 3–58**

Whenever there is a concentration difference of a physical quantity in a medium, nature tends to equalize things by forcing a flow from the high to the low concentration region.

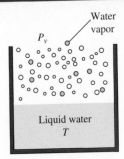

**FIGURE 3–59**
When open to the atmosphere, water is in phase equilibrium with the vapor in the air if the vapor pressure is equal to the saturation pressure of water.

**phase equilibrium.** For liquid water that is open to the atmosphere, the criterion for phase equilibrium can be expressed as follows: *The vapor pressure in the air must be equal to the saturation pressure of water at the water temperature.* That is (Fig. 3–59),

Phase equilibrium criterion for water exposed to air:   $P_v = P_{sat\,@\,T}$   (3–30)

Therefore, if the vapor pressure in the air is less than the saturation pressure of water at the water temperature, some liquid will evaporate. The larger the difference between the vapor and saturation pressures, the higher the rate of evaporation. The evaporation has a cooling effect on water, and thus reduces its temperature. This, in turn, reduces the saturation pressure of water and thus the rate of evaporation until some kind of quasi-steady operation is reached. This explains why water is usually at a considerably lower temperature than the surrounding air, especially in dry climates. It also suggests that the rate of evaporation of water can be increased by increasing the water temperature and thus the saturation pressure of water.

Note that the air at the water surface is always saturated because of the direct contact with water, and thus the vapor pressure. Therefore, the vapor pressure at the lake surface is the saturation pressure of water at the temperature of the water at the surface. If the air is not saturated, then the vapor pressure decreases to the value in the air at some distance from the water surface, and the difference between these two vapor pressures is the driving force for the evaporation of water.

The natural tendency of water to evaporate in order to achieve phase equilibrium with the water vapor in the surrounding air forms the basis for the operation of the **evaporative coolers** (also called the *swamp coolers*). In such coolers, hot and dry outdoor air is forced to flow through a wet cloth before entering a building. Some of the water evaporates by absorbing heat from the air, and thus cooling it. Evaporative coolers are commonly used in dry climates and provide effective cooling. They are much cheaper to run than air conditioners since they are inexpensive to buy, and the fan of an evaporative cooler consumes much less power than the compressor of an air conditioner.

Boiling and evaporation are often used interchangeably to indicate *phase change from liquid to vapor.* Although they refer to the same physical process, they differ in some aspects. **Evaporation** occurs at the *liquid–vapor interface* when the vapor pressure is less than the saturation pressure of the liquid at a given temperature. Water in a lake at 20°C, for example, evaporates to air at 20°C and 60 percent relative humidity since the saturation pressure of water at 20°C is 2.34 kPa, and the vapor pressure of air at 20°C and 60 percent relative humidity is 1.4 kPa. Other examples of evaporation are the drying of clothes, fruits, and vegetables; the evaporation of sweat to cool the human body; and the rejection of waste heat in wet cooling towers. Note that evaporation involves no bubble formation or bubble motion (Fig. 3–60).

**Boiling,** on the other hand, occurs at the *solid–liquid interface* when a liquid is brought into contact with a surface maintained at a temperature $T_s$ sufficiently above the saturation temperature $T_{sat}$ of the liquid. At 1 atm, for example, liquid water in contact with a solid surface at 110°C boils since the saturation temperature of water at 1 atm is 100°C. The boiling process is characterized by the rapid motion of *vapor bubbles* that form at the solid–liquid interface, detach from the surface when they reach a certain size, and attempt to rise to the free surface of the liquid. When cooking, we do not say water is boiling unless we see the bubbles rising to the top.

©John A Rizzo/Getty Images RF    ©David Chasey/Getty Images RF

**FIGURE 3–60**
A liquid-to-vapor phase change process is called *evaporation* if it occurs at a liquid–vapor interface, and *boiling* if it occurs at a solid–liquid interface.

■ **EXAMPLE 3–14**    **Temperature Drop of a Lake Due to Evaporation**

On a summer day, the air temperature over a lake is measured to be 25°C. Determine water temperature of the lake when phase equilibrium conditions are established between the water in the lake and the vapor in the air for relative humidities of 10, 80, and 100 percent for the air (Fig. 3–61).

**SOLUTION** Air at a specified temperature is blowing over a lake. The equilibrium temperatures of water for three different cases are to be determined.
*Analysis* The saturation pressure of water at 25°C, from Table 3–1, is 3.17 kPa. Then the vapor pressures at relative humidities of 10, 80, and 100 percent are determined from Eq. 3–29 to be

Relative humidity = 10%:  $P_{v1} = \phi_1 P_{sat\ @\ 25°C} = 0.1 \times (3.17\ \text{kPa})$
$$= 0.317\ \text{kPa}$$

Relative humidity = 80%:  $P_{v2} = \phi_2 P_{sat\ @\ 25°C} = 0.8 \times (3.17\ \text{kPa})$
$$= 2.536\ \text{kPa}$$

Relative humidity = 100%  $P_{v3} = \phi_3 P_{sat\ @25°C} = 1.0 \times (3.17\ \text{kPa})$
$$= 3.17\ \text{kPa}$$

The saturation temperatures corresponding to these pressures are determined from Table 3–1 (or Table A–5) by interpolation to be

$$T_1 = -8.0°C \quad T_2 = 21.2°C \quad \text{and} \quad T_3 = 25°C$$

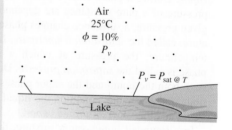

**FIGURE 3–61**
Schematic for Example 3–14.

Therefore, water will freeze in the first case even though the surrounding air is hot. In the last case the water temperature will be the same as the surrounding air temperature.

**Discussion** You are probably skeptical about the lake freezing when the air is at 25°C, and you are right. The water temperature drops to −8°C in the limiting case of no heat transfer to the water surface. In practice the water temperature drops below the air temperature, but it does not drop to −8°C because (1) it is very unlikely for the air over the lake to be so dry (a relative humidity of just 10 percent) and (2) as the water temperature near the surface drops, heat transfer from the air and the lower parts of the water body will tend to make up for this heat loss and prevent the water temperature from dropping too much. The water temperature stabilizes when the heat gain from the surrounding air and the water body equals the heat loss by evaporation, that is, when a *dynamic balance* is established between heat and mass transfer instead of phase equilibrium. If you try this experiment using a shallow layer of water in a well-insulated pan, you can actually freeze the water if the air is very dry and relatively cool.

# SUMMARY

A substance that has a fixed chemical composition throughout is called a *pure substance*. A pure substance exists in different phases depending on its energy level. In the liquid phase, a substance that is not about to vaporize is called a *compressed* or *subcooled liquid*. In the gas phase, a substance that is not about to condense is called a *superheated vapor*. During a phase-change process, the temperature and pressure of a pure substance are dependent properties. At a given pressure, a substance changes phase at a fixed temperature, called the *saturation temperature*. Likewise, at a given temperature, the pressure at which a substance changes phase is called the *saturation pressure*. During a boiling process, both the liquid and the vapor phases coexist in equilibrium, and under this condition the liquid is called *saturated liquid* and the vapor *saturated vapor*.

In a saturated liquid–vapor mixture, the mass fraction of vapor is called the *quality* and is expressed as

$$x = \frac{m_{vapor}}{m_{total}}$$

Quality may have values between 0 (saturated liquid) and 1 (saturated vapor). It has no meaning in the compressed liquid or superheated vapor regions. In the saturated mixture region, the average value of any intensive property $y$ is determined from

$$y = y_f + xy_{fg}$$

where $f$ stands for saturated liquid and $g$ for saturated vapor.

In the absence of compressed liquid data, a general approximation is to treat a compressed liquid as a saturated liquid at the given *temperature*,

$$y \cong y_{f@T}$$

where $y$ stands for $v$, $u$, or $h$.

The state beyond which there is no distinct vaporization process is called the *critical point*. At supercritical pressures, a substance gradually and uniformly expands from the liquid to vapor phase. All three phases of a substance coexist in equilibrium at states along the *triple line* characterized by triple-line temperature and pressure. The compressed liquid has lower $v$, $u$, and $h$ values than the saturated liquid at the same $T$ or $P$. Likewise, superheated vapor has higher $v$, $u$, and $h$ values than the saturated vapor at the same $T$ or $P$.

Any relation among the pressure, temperature, and specific volume of a substance is called an *equation of state*. The simplest and best-known equation of state is the *ideal-gas equation of state*, given as

$$Pv = RT$$

where $R$ is the gas constant. Caution should be exercised in using this relation since an ideal gas is a fictitious substance. Real gases exhibit ideal-gas behavior at relatively low pressures and high temperatures.

The deviation from ideal-gas behavior can be properly accounted for by using the *compressibility factor* $Z$, defined as

$$Z = \frac{Pv}{RT} \quad \text{or} \quad Z = \frac{v_{actual}}{v_{ideal}}$$

The $Z$ factor is approximately the same for all gases at the same *reduced temperature* and *reduced pressure*, which are defined as

$$T_R = \frac{T}{T_{cr}} \quad \text{and} \quad P_R = \frac{P}{P_{cr}}$$

where $P_{cr}$ and $T_{cr}$ are the critical pressure and temperature, respectively. This is known as the *principle of corresponding states*. When either $P$ or $T$ is unknown, it can be determined from the compressibility chart with the help of the *pseudo-reduced specific volume*, defined as

$$v_R = \frac{v_{actual}}{RT_{cr}/P_{cr}}$$

The $P$-$v$-$T$ behavior of substances can be represented more accurately by more complex equations of state. Three of the best known are

van der Waals: $\quad \left(P + \dfrac{a}{v^2}\right)(v - b) = RT$

where

$$a = \frac{27R^2 T_{cr}^2}{64P_{cr}} \quad \text{and} \quad b = \frac{RT_{cr}}{8P_{cr}}$$

Beattie-Bridgeman: $\quad P = \dfrac{R_u T}{\overline{v}^2}\left(1 - \dfrac{c}{\overline{v}T^3}\right)(\overline{v} + B) - \dfrac{A}{\overline{v}^2}$

where

$$A = A_0\left(1 - \frac{a}{\overline{v}}\right) \quad \text{and} \quad B = B_0\left(1 - \frac{b}{\overline{v}}\right)$$

Benedict-Webb-Rubin:

$$P = \frac{R_u T}{\overline{v}} + \left(B_0 R_u T - A_0 - \frac{C_0}{T^2}\right)\frac{1}{\overline{v}^2} + \frac{bR_u T - a}{\overline{v}^3} + \frac{a\alpha}{\overline{v}^6}$$
$$+ \frac{c}{\overline{v}^3 T^2}\left(1 + \frac{\gamma}{\overline{v}^2}\right)e^{-\gamma/\overline{v}^2}$$

where $R_u$ is the universal gas constant and $\overline{v}$ is the molar specific volume.

## REFERENCES AND SUGGESTED READINGS

1. ASHRAE *Handbook of Fundamentals*. SI version. Atlanta, GA: American Society of Heating, Refrigerating, and Air-Conditioning Engineers, Inc., 1993.

2. ASHRAE *Handbook of Refrigeration*. SI version. Atlanta, GA: American Society of Heating, Refrigerating, and Air-Conditioning Engineers, Inc., 1994.

3. A. Bejan. *Advanced Engineering Thermodynamics*. 3rd ed. New York: Wiley, 2006.

4. M. Kostic. *Analysis of Enthalpy Approximation for Compressed Liquid Water*. ASME J. Heat Transfer, Vol. 128, pp. 421–426, 2006.

## PROBLEMS*

### Pure Substances, Phase-Change Processes, Property Diagrams

**3–1C** Is iced water a pure substance? Why?

**3–2C** What is the difference between saturated vapor and superheated vapor?

**3–3C** Is there any difference between the intensive properties of saturated vapor at a given temperature and the vapor of a saturated mixture at the same temperature?

**3–4C** Why are the temperature and pressure dependent properties in the saturated mixture region?

**3–5C** Is it true that water boils at higher temperature at higher pressure? Explain

**3–6C** What is the difference between the critical point and the triple point?

**3–7C** Is it possible to have water vapor at $-10°C$?

**3–8C** A househusband is cooking beef stew for his family in a pan that is (a) uncovered, (b) covered with a light

* Problems designated by a "C" are concept questions, and students are encouraged to answer them all. Problems with the 🌀 icon are solved using EES, and complete solutions together with parametric studies are included on the text website. Problems with the 📘 icon are comprehensive in nature, and are intended to be solved with an equation solver such as EES.

lid, and (c) covered with a heavy lid. For which case will the cooking time be the shortest? Why?

## Property Tables

**3–9C** In what kind of pot will a given volume of water boil at a higher temperature: a tall and narrow one or a short and wide one? Explain.

**3–10C** It is well known that warm air in a cooler environment rises. Now consider a warm mixture of air and gasoline on top of an open gasoline can. Do you think this gas mixture will rise in a cooler environment?

**3–11C** Does the amount of heat absorbed as 1 kg of saturated liquid water boils at 100°C have to be equal to the amount of heat released as 1 kg of saturated water vapor condenses at 100°C?

**3–12C** Does the reference point selected for the properties of a substance have any effect on thermodynamic analysis? Why?

**3–13C** What is the physical significance of $h_{fg}$? Can it be obtained from a knowledge of $h_f$ and $h_g$? How?

**3–14C** Does $h_{fg}$ change with pressure? How?

**3–15C** Is it true that it takes more energy to vaporize 1 kg of saturated liquid water at 100°C than it would at 120°C?

**3–16C** What is quality? Does it have any meaning in the superheated vapor region?

**3–17C** Which process requires more energy: completely vaporizing 1 kg of saturated liquid water at 1 atm pressure or completely vaporizing 1 kg of saturated liquid water at 8 atm pressure?

**3–18C** In the absence of compressed liquid tables, how is the specific volume of a compressed liquid at a given P and T determined?

**3–19C** In 1775, Dr. William Cullen made ice in Scotland by evacuating the air in a water tank. Explain how that device works, and discuss how the process can be made more efficient.

**3–20** Complete this table for $H_2O$:

| T, °C | P, kPa | u, kJ/kg | Phase description |
|-------|--------|----------|-------------------|
|       | 400    | 1450     |                   |
| 220   |        |          | Saturated vapor   |
| 190   | 2500   |          |                   |
|       | 4000   | 3040     |                   |

**3–21** Complete this table for $H_2O$:

| T, °C | P, kPa | v, m³/kg | Phase description |
|-------|--------|----------|-------------------|
| 50    |        | 7.72     |                   |
|       | 400    |          | Saturated vapor   |
| 250   | 500    |          |                   |
| 110   | 350    |          |                   |

**3–22** Reconsider Prob. 3–21. Using EES (or other) software, determine the missing properties of water. Repeat the solution for refrigerant-134a, refrigerant-22, and ammonia.

**3–23** Complete this table for $H_2O$:

| T, °C | P, kPa | v, m³/kg | Phase description |
|-------|--------|----------|-------------------|
| 140   |        | 0.05     |                   |
|       | 550    |          | Saturated liquid  |
| 125   | 750    |          |                   |
| 500   |        | 0.140    |                   |

**3–24** Complete this table for refrigerant-134a:

| T, °C | P, kPa | v, m³/kg | Phase description |
|-------|--------|----------|-------------------|
| −4    | 320    |          |                   |
| 10    |        | 0.0065   |                   |
|       | 850    |          | Saturated vapor   |
| 90    | 600    |          |                   |

**3–25** Complete this table for refrigerant-134a:

| T, °C | P, kPa | h, kJ/kg | x | Phase description |
|-------|--------|----------|---|-------------------|
|       | 600    | 180      |   |                   |
| −10   |        |          | 0.6 |                 |
| −14   | 500    |          |   |                   |
|       | 1200   | 300.63   |   |                   |
| 44    |        |          | 1.0 |                 |

**3–26** A 1.8-m³ rigid tank contains steam at 220°C. One-third of the volume is in the liquid phase and the rest is in the vapor form. Determine (a) the pressure of the steam, (b) the quality of the saturated mixture, and (c) the density of the mixture.

Steam
1.8 m³
220°C

**FIGURE P3–26**

**3–27** A piston–cylinder device contains 0.85 kg of refrigerant- 134a at −10°C. The piston that is free to move has a mass of 12 kg and a diameter of 25 cm. The local atmospheric pressure is 88 kPa. Now, heat is transferred to refrigerant-134a until the temperature is 15°C. Determine (a) the final pressure, (b) the change in the volume of the cylinder, and (c) the change in the enthalpy of the refrigerant-134a.

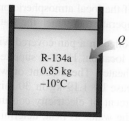

R-134a
0.85 kg
−10°C

Q

**FIGURE P3–27**

**3–28** R-134a, whose specific volume is 0.04471 m³/kg, flows through a tube at 600 kPa. What is the temperature in the tube?

**3–29** 10-kg of R-134a fill a 1.348-m³ rigid container at an initial temperature of −40°C. The container is then heated until the pressure is 200 kPa. Determine the final temperature and the initial pressure. *Answers:* 66.3°C, 51.25 kPa

**3–30** A 9-m³ container is filled with 300 kg of R-134a at 10°C. What is the specific enthalpy of the R-134a in the container?

**3–31** Refrigerant-134a at 200 kPa and 25°C flows through a refrigeration line. Determine its specific volume.

**3–32** The average atmospheric pressure in Denver (elevation = 1610 m) is 83.4 kPa. Determine the temperature at which water in an uncovered pan boils in Denver. *Answer:* 94.6°C

**3–33** A spring-loaded piston-cylinder device is initially filled with 0.1 kg of an R-134a liquid-vapor mixture whose temperature is −34°C and whose quality is 80 percent. The spring constant in the spring force relation $F = kx$ is 6.6 kN/m, and the piston diameter is 30 cm. The R-134a undergoes a process that increases its volume by 40 percent. Calculate the final temperature and enthalpy of the R-134a. *Answers:* −30.9°C, 13.7 kJ/kg

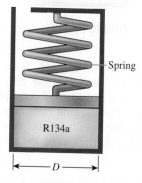

Spring

R134a

D

**FIGURE P3–33**

**3–34** One kilogram of water fills a 0.1546-m³ weighted piston-cylinder device at a temperature of 350°C. The piston-cylinder device is now cooled until its temperature is 100°C. Determine the final pressure of water, in MPa, and the volume, in m³. *Answers:* 1.8 MPa, 0.001043 m³

**3–35** Three kilograms of water in a container have a pressure of 100 kPa and temperature of 150°C. What is the volume of this container?

**3–36** Water is to be boiled at sea level in a 30-cm-diameter stainless steel pan placed on top of a 3-kW electric burner. If 60 percent of the heat generated by the burner is transferred to the water during boiling, determine the rate of evaporation of water.

Vapor

60%     40%

3 kW

**FIGURE P3–36**

**3–37** Repeat Prob, 3–36 for a location at an elevation of 1500 m where the atmospheric pressure is 84.5 kPa and thus the boiling temperature of water is 95°C.

**3–38** 10-kg of R-134a at 300 kPa fills a rigid container whose volume is 14 L. Determine the temperature and total enthalpy in the container. The container is now heated until the pressure is 600 kPa. Determine the temperature and total enthalpy when the heating is completed.

R-134a
300 kPa
10 kg
14 L

Q

**FIGURE P3–38**

**3–39** 100-kg of R-134a at 200 kPa are contained in a piston-cylinder device whose volume is 12.322 m³. The piston is now moved until the volume is one-half its original size. This is done such that the pressure of the R-134a does not change. Determine the final temperature and the change in the total internal energy of the R-134a.

**3–40** Water initially at 200 kPa and 300°C is contained in a piston-cylinder device fitted with stops. The water is allowed to cool at constant pressure until it exists as a saturated vapor and the piston rests on the stops. Then the water continues to cool until the pressure is 100 kPa. On the $T$-$v$ diagrams sketch, with respect to the saturation lines, the process curves passing through both the initial, intermediate, and final states of the water. Label the $T$, $P$ and $v$ values for end states on the process curves. Find the overall change in internal energy between the initial and final states per unit mass of water.

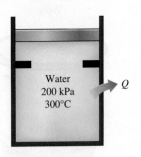

Water
200 kPa
300°C

$Q$

**FIGURE P3–40**

**3–41** Saturated steam coming off the turbine of a steam power plant at 40°C condenses on the outside of a 3-cm-outer-diameter, 35-m-long tube at a rate of 130 kg/h. Determine the rate of heat transfer from the steam to the cooling water flowing through the pipe.

**3–42** Water in a 5-cm-deep pan is observed to boil at 98°C. At what temperature will the water in a 40-cm-deep pan boil? Assume both pans are full of water.

**3–43** A cooking pan whose inner diameter is 20 cm is filled with water and covered with a 4-kg lid. If the local atmospheric pressure is 101 kPa, determine the temperature at which the water starts boiling when it is heated. *Answer:* 100.2°C

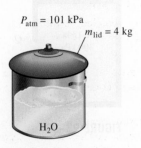

$P_{atm} = 101$ kPa

$m_{lid} = 4$ kg

$H_2O$

**FIGURE P3–43**

**3–44** Reconsider Prob. 3–43. Using EES (or other) software, investigate the effect of the mass of the lid on the boiling temperature of water in the pan. Let the mass vary from 1 kg to 10 kg. Plot the boiling temperature against the mass of the lid, and discuss the results.

**3–45** Water is being heated in a vertical piston–cylinder device. The piston has a mass of 40 kg and a cross-sectional area of 150 cm². If the local atmospheric pressure is 100 kPa, determine the temperature at which the water starts boiling.

**3–46** Water is boiled in a pan covered with a poorly fitting lid at a specified location. Heat is supplied to the pan by a 2-kW resistance heater. The amount of water in the pan is observed to decrease by 1.19 kg in 30 minutes. If it is estimated that 75 percent of electricity consumed by the heater is transferred to the water as heat, determine the local atmospheric pressure in that location. *Answer:* 85.4 kPa

**3–47** A rigid tank with a volume of 1.8 m³ contains 15 kg of saturated liquid–vapor mixture of water at 90°C. Now the water is slowly heated. Determine the temperature at which the liquid in the tank is completely vaporized. Also, show the process on a $T$-$v$ diagram with respect to saturation lines. *Answer:* 202.9°C

**3–48** A piston–cylinder device contains 0.005 m³ of liquid water and 0.9 m³ of water vapor in equilibrium at 600 kPa. Heat is transferred at constant pressure until the temperature reaches 200°C.

(a) What is the initial temperature of the water?
(b) Determine the total mass of the water.
(c) Calculate the final volume.
(d) Show the process on a $P$-$v$ diagram with respect to saturation lines.

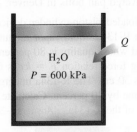

$H_2O$

$P = 600$ kPa

$Q$

**FIGURE P3–48**

**3–49** Reconsider Prob. 3–48. Using EES (or other) software, investigate the effect of pressure on the total mass of water in the tank. Let the pressure vary from 0.1 MPa to 1 MPa. Plot the total mass of water against pressure, and discuss the results. Also, show the process in Prob. 3–51 on a $P$-$v$ diagram using the property plot feature of EES.

**3–50** A 0.14-m³ rigid tank contains a saturated mixture of refrigerant-34a at 400 kPa. If the saturated liquid occupies 20 percent of the volume, determine the quality and the total mass of the refrigerant in the tank.

**3–51** Superheated water vapor at 1.4 MPa and 250°C is allowed to cool at constant volume until the temperature drops to 120°C. At the final state, determine (*a*) the pressure, (*b*) the quality, and (*c*) the enthalpy. Also, show the process on a *T-v* diagram with respect to saturation lines. *Answers:* (*a*) 198.7 kPa, (*b*) 0.1825, (*c*) 905.7 kJ/kg

**3–52** [EES] Reconsider Prob. 3–51. Using EES (or other) software, investigate the effect of initial pressure on the quality of water at the final state. Let the pressure vary from 700 kPa to 2000 kPa. Plot the quality against initial pressure, and discuss the results. Also, show the process in Prob. 3–51 on a *T-v* diagram using the property plot feature of EES.

**3–53** A piston–cylinder device contains 0.6 kg of steam at 200°C and 0.5 MPa. Steam is cooled at constant pressure until one-half of the mass condenses.

(*a*) Show the process on a *T-v* diagram.
(*b*) Find the final temperature.
(*c*) Determine the volume change.

**3–54** A rigid tank contains water vapor at 250°C and an unknown pressure. When the tank is cooled to 124°C, the vapor starts condensing. Estimate the initial pressure in the tank. *Answer:* 0.30 MPa

**3–55** A piston-cylinder device initially contains 1.4-kg saturated liquid water at 200°C. Now heat is transferred to the water until the volume quadruples and the cylinder contains saturated vapor only. Determine (*a*) the volume of the tank, (*b*) the final temperature and pressure, and (*c*) the internal energy change of the water.

**FIGURE P3–55**

Water
1.4 kg
200°C
*Q*

**3–56** 100 grams of R-134a initially fill a weighted piston-cylinder device at 60 kPa and −20°C. The device is then heated until the temperature is 100°C. Determine the change in the device's volume as a result of the heating. *Answer:* 0.0168 m³

R-134a
60 kPa
−20°C
100 g
*Q*

**FIGURE P3–56**

**3–57** A rigid vessel contains 8 kg of refrigerant-134a at 500 kPa and 120°C. Determine the volume of the vessel and the total internal energy. *Answers:* 0.494 m³, 2639 kJ

**3–58** A rigid tank initially contains 1.4-kg saturated liquid water at 200°C. At this state, 25 percent of the volume is occupied by water and the rest by air. Now heat is supplied to the water until the tank contains saturated vapor only. Determine (*a*) the volume of the tank, (*b*) the final temperature and pressure, and (*c*) the internal energy change of the water.

Water
1.4 kg
200°C
*Q*

**FIGURE P3–58**

**3–59** A piston-cylinder device initially contains 50 L of liquid water at 40°C and 200 kPa. Heat is transferred to the water at constant pressure until the entire liquid is vaporized.

(*a*) What is the mass of the water?
(*b*) What is the final temperature?
(*c*) Determine the total enthalpy change.
(*d*) Show the process on a *T-v* diagram with respect to saturation lines.

*Answers:* (*a*) 49.61 kg, (*b*) 120.21°C, (*c*) 125,950 kJ

## Ideal Gas

**3–60C** Under what conditions is the ideal-gas assumption suitable for real gases?

**3–61C** What is the difference between $R$ and $R_u$? How are these two related?

**3–62C** Propane and methane are commonly used for heating in winter, and the leakage of these fuels, even for short periods, poses a fire danger for homes. Which gas leakage do you think poses a greater risk for fire? Explain.

**3–63** A 400-L rigid tank contains 5 kg of air at 25°C. Determine the reading on the pressure gage if the atmospheric pressure is 97 kPa.

**3–64** A 0.09-m³ container is filled with 0.9 kg of oxygen at a pressure of 600 kPa. What is the temperature of the oxygen?

**3–65** A 2-kg mass of helium is maintained at 300 kPa and 27°C in a rigid container. How large is the container, in m³?

**3–66** The pressure gage on a 2.5-m³ oxygen tank reads 500 kPa. Determine the amount of oxygen in the tank if the temperature is 28°C and the atmospheric pressure is 97 kPa.

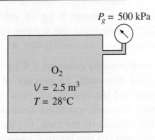

**FIGURE P3–66**

**3–67** A spherical balloon with a diameter of 9 m is filled with helium at 27°C and 200 kPa. Determine the mole number and the mass of the helium in the balloon.
*Answers:* 30.6 kmol, 123 kg

**3–68** Reconsider Prob. 3–67. Using EES (or other) software, investigate the effect of the balloon diameter on the mass of helium contained in the balloon for the pressures of (a) 100 kPa and (b) 200 kPa. Let the diameter vary from 5 m to 15 m. Plot the mass of helium against the diameter for both cases.

**3–69** The air in an automobile tire with a volume of 0.015 m$^3$ is at 32°C and 135 kPa (gage). Determine the amount of air that must be added to raise the pressure to the recommended value of 225 kPa (gage). Assume the atmospheric pressure to be 100 kPa and the temperature and the volume to remain constant. *Answer:* 0.0154 kg

**3–70** A 1-m$^3$ tank containing air at 10°C and 350 kPa is connected through a valve to another tank containing 3 kg of air at 35°C and 200 kPa. Now the valve is opened, and the entire system is allowed to reach thermal equilibrium with the surroundings, which are at 20°C. Determine the volume of the second tank and the final equilibrium pressure of air. *Answers:* 1.33 m$^3$, 264 kPa

**3–71** A rigid tank whose volume is unknown is divided into two parts by a partition. One side of the tank contains an ideal gas at 927°C. The other side is evacuated and has a volume twice the size of the part containing the gas. The partition is now removed and the gas expands to fill the entire tank. Heat is now applied to the gas until the pressure equals the initial pressure. Determine the final temperature of the gas. *Answer:* 3327°C

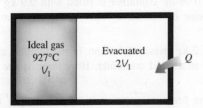

**FIGURE P3–71**

**3–72** Argon in the amount of 1.5 kg fills a 0.04-m$^3$ piston-cylinder device at 550 kPa. The piston is now moved by changing the weights until the volume is twice its original size. During this process, argon's temperature is maintained constant. Determine the final pressure in the device.

**3–73** A rigid tank contains 10 kg of air at 150 kPa and 20°C. More air is added to the tank until the pressure and temperature rise to 250 kPa and 33°C, respectively. Determine the amount of air added to the tank. *Answer:* 5.96 kg

### Compressibility Factor

**3–74C** What is the principle of corresponding states?

**3–75C** How are the reduced pressure and reduced temperature defined?

**3–76** Determine the specific volume of superheated water vapor at 15 MPa and 350°C, using (a) the ideal-gas equation, (b) the generalized compressibility chart, and (c) the steam tables. Also determine the error involved in the first two cases. *Answers:* (a) 0.01917 m$^3$/kg, 67.0 percent, (b) 0.01246 m$^3$/kg, 8.5 percent, (c) 0.01148 m$^3$/kg

**3–77** Reconsider Prob. 3–76. Solve the problem using the generalized compressibility factor feature of the EES software. Again using EES, compare the specific volume of water for the three cases at 15 MPa over the temperature range of 350 to 600°C in 25°C intervals. Plot the percent error involved in the ideal-gas approximation against temperature, and discuss the results.

**3–78** Determine the specific volume of superheated water vapor at 3.5 MPa and 450°C based on (a) the ideal-gas equation, (b) the generalized compressibility chart, and (c) the steam tables. Determine the error involved in the first two cases.

**3–79** Somebody claims that oxygen gas at 160 K and 3 MPa can be treated as an ideal gas with an error of less than 10 percent. Is this claim valid?

**3–80** Ethane in a rigid vessel is to be heated from 550 kPa and 40°C until its temperature is 280°C. What is the final pressure of the ethane as predicted by the compressibility chart?

**3–81** Ethylene is heated at constant pressure from 5 MPa and 20°C to 200°C. Using the compressibility chart, determine the change in the ethylene's specific volume as a result of this heating. *Answer:* 0.0172 m$^3$/kg

**3–82** What is the percentage of error involved in treating carbon dioxide at 7 MPa and 380 K as an ideal gas?

**3–83** Saturated water vapor at 350°C is heated at constant pressure until its volume has doubled. Determine the final temperature using the ideal gas equation of state, the compressibility charts, and the steam tables.

**3–84** Methane at 10 MPa and 300 K is heated at constant pressure until its volume has increased by 80 percent. Determine the final temperature using the ideal gas equation of state and the compressibility factor. Which of these two results is more accurate?

**3–85** Carbon dioxide gas enters a pipe at 3 MPa and 500 K at a rate of 2 kg/s. $CO_2$ is cooled at constant pressure as it flows in the pipe and the temperature of $CO_2$ drops to 450 K at the exit. Determine the volume flow rate and the density of carbon dioxide at the inlet and the volume flow rate at the exit of the pipe using (a) the ideal-gas equation and (b) the generalized compressibility chart. Also, determine (c) the error involved in the first case.

**FIGURE P3–85**

3 MPa
500 K
2 kg/s
$CO_2 \longrightarrow$
450 K

**3–86** A 0.016773-$m^3$ tank contains 1 kg of refrigerant-134a at 110°C. Determine the pressure of the refrigerant, using (a) the ideal-gas equation, (b) the generalized compressibility chart, and (c) the refrigerant tables. *Answers:* (a) 1.861 MPa, (b) 1.583 MPa, (c) 1.6 MPa

## Other Equations of State

**3–87C** What is the physical significance of the two constants that appear in the van der Waals equation of state? On what basis are they determined?

**3–88** A 3.27-$m^3$ tank contains 100 kg of nitrogen at 175 K. Determine the pressure in the tank, using (a) the ideal-gas equation, (b) the van der Waals equation, and (c) the Beattie-Bridgeman equation. Compare your results with the actual value of 1505 kPa.

**3–89** Methane is heated in a rigid container from 80 kPa and 20°C to 300°C. Determine the final pressure of the methane treating it as (a) an ideal gas and (b) a Benedict-Webb-Rubin gas.

**3–90** Refrigerant-134a at 1.6 MPa has a specific volume of 0.01343 $m^3$/kg. Determine the temperature of the refrigerant based on (a) the ideal-gas equation, (b) the van der Waals equation, and (c) the refrigerant tables.

**3–91** Nitrogen at 150 K has a specific volume of 0.041884 $m^3$/kg. Determine the pressure of the nitrogen, using (a) the ideal-gas equation and (b) the Beattie-Bridgeman equation. Compare your results to the experimental value of 1000 kPa. *Answers:* (a) 1063 kPa, (b) 1000.4 kPa

**3–92** Reconsider Prob. 3–91. Using EES (or other) software, compare the pressure results of the ideal-gas and Beattie-Bridgeman equations with nitrogen data supplied by EES. Plot temperature versus specific volume

for a pressure of 1000 kPa with respect to the saturated liquid and saturated vapor lines of nitrogen over the range of 110 K < T < 150 K.

**3–93** 1-kg of carbon dioxide is compressed from 1 MPa and 200°C to 3 MPa in a piston-cylinder device arranged to execute a polytropic process for which $PV^{1.2}$ = constant. Determine the final temperature treating the carbon dioxide as (a) an ideal gas and (b) a van der Waals gas.

**3–94** A 1-$m^3$ tank contains 2.841 kg of steam at 0.6 MPa. Determine the temperature of the steam, using (a) the ideal-gas equation, (b) the van der Waals equation, and (c) the steam tables. *Answers:* (a) 457.6 K, (b) 465.9 K, (c) 473 K

**3–95** Reconsider Prob. 3-94. Solve the problem using EES (or other) software. Again using the EES, compare the temperature of water for the three cases at constant specific volume over the pressure range of 0.1 MPa to 1 MPa in 0.1 MPa increments. Plot the percent error involved in the ideal-gas approximation against pressure, and discuss the results.

## Special Topic: Vapor Pressure and Phase Equilibrium

**3–96** During a hot summer day at the beach when the air temperature is 30°C, someone claims the vapor pressure in the air to be 5.2 kPa. Is this claim reasonable?

**3–97** Consider a glass of water in a room that is at 20°C and 40 percent relative humidity. If the water temperature is 15°C, determine the vapor pressure (a) at the free surface of the water and (b) at a location in the room far from the glass.

**3–98** On a certain day, the temperature and relative humidity of air over a large swimming pool are measured to be 25°C and 60 percent, respectively. Determine the water temperature of the pool when phase equilibrium conditions are established between the water in the pool and the vapor in the air.

**3–99** During a hot summer day when the air temperature is 35°C and the relative humidity is 70 percent, you buy a supposedly "cold" canned drink from a store. The store owner claims that the temperature of the drink is below 10°C. Yet the drink does not feel so cold and you are skeptical since you notice no condensation forming outside the can. Can the store owner be telling the truth?

**3–100** Consider two rooms that are identical except that one is maintained at 25°C and 40 percent relative humidity while the other is maintained at 20°C and 55 percent relative humidity. Noting that the amount of moisture is proportional to the vapor pressure, determine which room contains more moisture.

**3–101** A thermos bottle is half-filled with water and is left open to the atmospheric air at 20°C and 35 percent relative humidity. If heat transfer to the water through the thermos walls and the free surface is negligible, determine the temperature of water when phase equilibrium is established.

## Review Problems

**3–102** Carbon-dioxide gas at 3 MPa and 500 K flows steadily in a pipe at a rate of 0.4 kmol/s. Determine (a) the volume and mass flow rates and the density of carbon dioxide at this state. If $CO_2$ is cooled at constant pressure as it flows in the pipe so that the temperature of $CO_2$ drops to 450 K at the exit of the pipe, determine (b) the volume flow rate at the exit of the pipe.

3 MPa
500 K
0.4 kmol/s

$CO_2 \longrightarrow$

450 K

**FIGURE P3–102**

**3–103** A tank contains argon at 600°C and 200 kPa gage. The argon is cooled in a process by heat transfer to the surroundings such that the argon reaches a final equilibrium state at 300°C. Determine the final gage pressure of the argon. Assume atmospheric pressure is 100 kPa.

**3–104** The combustion in a gasoline engine may be approximated by a constant volume heat addition process. There exists the air–fuel mixture in the cylinder before the combustion and the combustion gases after it, and both may be approximated as air, an ideal gas. In a gasoline engine, the cylinder conditions are 1.2 MPa and 450°C before the combustion and 1750°C after it. Determine the pressure at the end of the combustion process. *Answer:* 3.36 MPa

Combustion
chamber
1.2 MPa
450°C

**FIGURE P3–104**

**3–105** One kilogram of R-134a fills a 0.090 m³ rigid container at an initial temperature of −40°C. The container is then heated until the pressure is 280 kPa. Determine the initial pressure and final temperature. *Answers:* 51.25 kPa, 50°C

**3–106** A rigid tank with a volume of 0.117 m³ contains 1 kg of refrigerant-134a vapor at 240 kPa. The refrigerant is now allowed to cool. Determine the pressure when the refrigerant first starts condensing. Also, show the process on a P-v diagram with respect to saturation lines.

**3–107** One kilogram of water fills a 0.2825 m³ weighted piston-cylinder device at a temperature of 350°C. The piston-cylinder device is now cooled until its temperature is 50°C. Determine the final pressure and volume of the water.

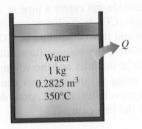

Water
1 kg
0.2825 m³
350°C

$Q$

**FIGURE P3–107**

**3–108** Ethane at 10 MPa and 100°C is heated at constant pressure until its volume has increased by 60 percent. Determine the final temperature using (a) the ideal gas equation of state and (b) the compressibility factor. Which of these two results is the more accurate?

**3–109** A 13-m³ tank contains nitrogen at 17°C and 600 kPa. Some nitrogen is allowed to escape until the pressure in the tank drops to 400 kPa. If the temperature at this point is 15°C, determine the amount of nitrogen that has escaped. *Answer:* 29.8 kg

**3–110** A 10-kg mass of superheated refrigerant-134a at 1.2 MPa and 70°C is cooled at constant pressure until it exists as a compressed liquid at 20°C.

(a) Show the process on a T-v diagram with respect to saturation lines.
(b) Determine the change in volume.
(c) Find the change in total internal energy.
*Answers:* (b) −0.187 m³ (c) −1984 kJ

**3–111** A 4-L rigid tank contains 2 kg of saturated liquid–vapor mixture of water at 50°C. The water is now slowly heated until it exists in a single phase. At the final state, will the water be in the liquid phase or the vapor phase? What would your answer be if the volume of the tank were 400 L instead of 4 L?

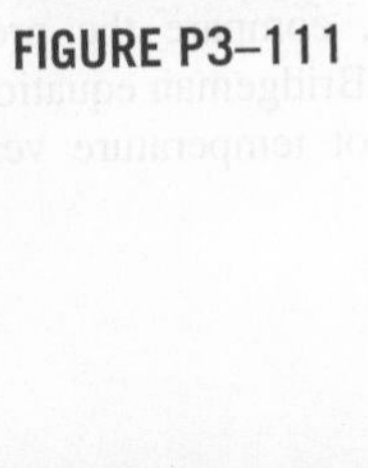

$H_2O$
V = 4 L
m = 2 kg
T = 50°C

**FIGURE P3–111**

**3–112** The gage pressure of an automobile tire is measured to be 200 kPa before a trip and 220 kPa after the trip at a location where the atmospheric pressure is 90 kPa. Assuming the volume of the tire remains constant at 0.035 m³, determine the percent increase in the absolute temperature of the air in the tire.

**3–113** A piston-cylinder device initially contains 0.2 kg of steam at 200 kPa and 300°C. Now, the steam is cooled at constant pressure until it is at 150°C. Determine the volume change of the cylinder during this process using the compressibility factor and compare the result to the actual value.

**FIGURE P3–113**

**3–114** Steam at 400°C has a specific volume of 0.02 m³/kg. Determine the pressure of the steam based on (a) the ideal-gas equation, (b) the generalized compressibility chart, and (c) the steam tables. *Answers:* (a) 15,529 kPa, (b) 12,574 kPa, (c) 12,515 kPa

**3–115** A tank whose volume is unknown is divided into two parts by a partition. One side of the tank contains 0.03 m³ of refrigerant-134a that is a saturated liquid at 0.9 MPa, while the other side is evacuated. The partition is now removed, and the refrigerant fills the entire tank. If the final state of the refrigerant is 20°C and 280 kPa, determine the volume of the tank.

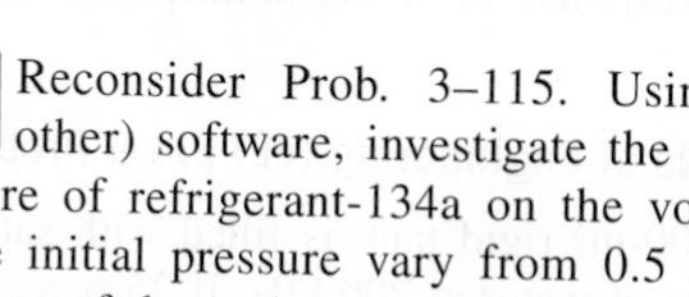

| R-134a<br>$V = 0.03$ m³<br>$P = 0.9$ MPa | Evacuated |
|---|---|

**FIGURE P3–115**

**3–116** [EES] Reconsider Prob. 3–115. Using EES (or other) software, investigate the effect of the initial pressure of refrigerant-134a on the volume of the tank. Let the initial pressure vary from 0.5 to 1.5 MPa. Plot the volume of the tank versus the initial pressure, and discuss the results.

**3–117** Liquid propane is commonly used as a fuel for heating homes, powering vehicles such as forklifts, and filling portable picnic tanks. Consider a propane tank that initially contains 5 L of liquid propane at the environment temperature of 20°C. If a hole develops in the connecting tube of a propane tank and the propane starts to leak out, determine the temperature of propane when the pressure in the tank drops to 1 atm. Also, determine the total amount of heat transfer from the environment to the tank to vaporize the entire propane in the tank.

**FIGURE P3–117**

**3–118** Repeat Prob. 3–117 for isobutane.

**3–119** A tank contains helium at 37°C and 140 kPa gage. The helium is heated in a process by heat transfer from the surroundings such that the helium reaches a final equilibrium state at 200°C. Determine the final gage pressure of the helium. Assume atmospheric pressure is 100 kPa.

**3–120** If sufficient data are provided, complete the blank cells in the following table of properties of water. In the last column describe the condition of water as compressed liquid, saturated mixture, superheated vapor, or insufficient information; and, if applicable, give the quality.

| P, kPa | T, °C | v, m³/kg | u, kJ/kg | Phase description |
|---|---|---|---|---|
| | 250 | | 2728.9 | |
| 300 | | | 1560.0 | |
| 101.42 | 100 | | | |
| 3000 | 180 | | | |

**3-121** Water initially at 300 kPa and 0.5 m³/kg is contained in a piston-cylinder device fitted with stops so that the water supports the weight of the piston and the force of the atmosphere. The water is heated until it reaches the saturated vapor state and the piston rests against the stops. With the piston against the stops, the water is further heated until the pressure is 600 kPa. On the P-v and T-v diagrams sketch, with respect to the saturation lines, the process curves passing through both the initial and final states of the water. Label the states on the process as 1, 2, and 3. On both the P-v and T-v diagrams, sketch the isotherms passing through the states and show their values, in °C, on the isotherms.

$Q$ → Water
300 kPa
0.5 m³/kg

**FIGURE P3–121**

**3–122** 0.2 kg of argon is compressed from 6895 kPa and 149°C to 13,790 kPa in a piston-cylinder device which executes a polytropic process for which $PV^{1.6}$ = constant. Determine the final temperature treating the argon as (*a*) an ideal gas and (*b*) a Beattie-Bridgeman gas.

**3–123** Although balloons have been around since 1783 when the first balloon took to the skies in France, a real breakthrough in ballooning occurred in 1960 with the design of the modern hot-air balloon fueled by inexpensive propane and constructed of lightweight nylon fabric. Over the years, ballooning has become a sport and a hobby for many people around the world. Unlike balloons filled with the light helium gas, hot-air balloons are open to the atmosphere. Therefore, the pressure in the balloon is always the same as the local atmospheric pressure, and the balloon is never in danger of exploding.

Hot-air balloons range from about 15 to 25 m in diameter. The air in the balloon cavity is heated by a propane burner located at the top of the passenger cage. The flames from the burner that shoot into the balloon heat the air in the balloon cavity, raising the air temperature at the top of the balloon from 65°C to over 120°C. The air temperature is maintained at the desired levels by periodically firing the propane burner.

The buoyancy force that pushes the balloon upward is proportional to the density of the cooler air outside the balloon and the volume of the balloon, and can be expressed as

$$F_B = \rho_{\text{cool air}}\, g V_{\text{balloon}}$$

where $g$ is the gravitational acceleration. When air resistance is negligible, the buoyancy force is opposed by (1) the weight of the hot air in the balloon, (2) the weight of the cage, the ropes, and the balloon material, and (3) the weight of the people and other load in the cage. The operator of the balloon can control the height and the vertical motion of the balloon by firing the burner or by letting some hot air in the balloon escape, to be replaced by cooler air. The forward motion of the balloon is provided by the winds.

Consider a 20-m-diameter hot-air balloon that, together with its cage, has a mass of 80 kg when empty. This balloon is hanging still in the air at a location where the atmospheric pressure and temperature are 90 kPa and 15°C, respectively,

while carrying three 65-kg people. Determine the average temperature of the air in the balloon. What would your response be if the atmospheric air temperature were 30°C?

**FIGURE P3–123**
©PhotoLink/Getty Images RF

**3–124** Reconsider Prob. 3-123. Using EES (or other) software, investigate the effect of the environment temperature on the average air temperature in the balloon when the balloon is suspended in the air. Assume the environment temperature varies from −10 to 30°C. Plot the average air temperature in the balloon versus the environment temperature, and discuss the results. Investigate how the number of people carried affects the temperature of the air in the balloon.

**3–125** Consider an 18-m-diameter hot-air balloon that, together with its cage, has a mass of 120 kg when empty. The air in the balloon, which is now carrying two 70-kg people, is heated by propane burners at a location where the atmospheric pressure and temperature are 93 kPa and 12°C, respectively. Determine the average temperature of the air in the balloon when the balloon first starts rising. What would your response be if the atmospheric air temperature were 25°C?

## Fundamentals of Engineering (FE) Exam Problems

**3–126** A 300-m³ rigid tank is filled with saturated liquid–vapor mixture of water at 200 kPa. If 25 percent of the mass is liquid and 75 percent of the mass is vapor, the total mass in the tank is

(a) 451 kg     (b) 556 kg     (c) 300 kg
(d) 331 kg     (e) 195 kg

**3–127** Water is boiled at 1 atm pressure in a coffee maker equipped with an immersion-type electric heating element. The coffee maker initially contains 1 kg of water. Once boiling started, it is observed that half of the water in the coffee maker evaporated in 10 minutes. If the heat loss from the coffee maker is negligible, the power rating of the heating element is

(a) 3.8 kW     (b) 2.2 kW     (c) 1.9 kW
(d) 1.6 kW     (e) 0.8 kW

**3–128** A 1-$m^3$ rigid tank contains 10 kg of water (in any phase or phases) at 160°C. The pressure in the tank is

(a) 738 kPa     (b) 618 kPa     (c) 370 kPa
(d) 2000 kPa     (e) 1618 kPa

**3–129** Water is boiling at 1 atm pressure in a stainless steel pan on an electric range. It is observed that 2 kg of liquid water evaporates in 30 min. The rate of heat transfer to the water is

(a) 2.51 kW     (b) 2.32 kW     (c) 2.97 kW
(d) 0.47 kW     (e) 3.12 kW

**3–130** Water is boiled in a pan on a stove at sea level. During 10 min of boiling, it is observed that 200 g of water has evaporated. Then the rate of heat transfer to the water is

(a) 0.84 kJ/min     (b) 45.1 kJ/min     (c) 41.8 kJ/min
(d) 53.5 kJ/min     (e) 225.7 kJ/min

**3–131** A 3-$m^3$ rigid vessel contains steam at 4 MPa and 500°C. The mass of the steam is

(a) 3 kg     (b) 9 kg     (c) 26 kg
(d) 35 kg     (e) 52 kg

**3–132** Consider a sealed can that is filled with refrigerant-134a. The contents of the can are at the room temperature of 25°C. Now a leak develops, and the pressure in the can drops to the local atmospheric pressure of 90 kPa. The temperature of the refrigerant in the can is expected to drop to (rounded to the nearest integer)

(a) 0°C     (b) −29°C     (c) −16°C
(d) 5°C     (e) 25°C

**3–133** A rigid tank contains 2 kg of an ideal gas at 4 atm and 40°C. Now a valve is opened, and half of mass of the gas is allowed to escape. If the final pressure in the tank is 2.2 atm, the final temperature in the tank is

(a) 71°C     (b) 44°C     (c) −100°C
(d) 20°C     (e) 172°C

**3–134** The pressure of an automobile tire is measured to be 190 kPa (gage) before a trip and 215 kPa (gage) after the trip at a location where the atmospheric pressure is 95 kPa. If the temperature of air in the tire before the trip is 25°C, the air temperature after the trip is

(a) 51.1°C     (b) 64.2°C     (c) 27.2°C
(d) 28.3°C     (e) 25.0°C

**Design and Essay Problems**

**3–135** In an article on tire maintenance, it is stated that tires lose air over time, and pressure losses as high as 90 kPa per year are measured. The article recommends checking tire pressure at least once a month to avoid low tire pressure that hurts fuel efficiency and causes uneven thread wear on tires. Taking the beginning tire pressure to be 220 kPa (gage) and the atmospheric pressure to be 100 kPa, determine the fraction of air that can be lost from a tire per year.

**3–136** It is well known that water freezes at 0°C at atmospheric pressure. The mixture of liquid water and ice at 0°C is said to be at stable equilibrium since it cannot undergo any changes when it is isolated from its surroundings. However, when water is free of impurities and the inner surfaces of the container are smooth, the temperature of water can be lowered to −2°C or even lower without any formation of ice at atmospheric pressure. But at that state even a small disturbance can initiate the formation of ice abruptly, and the water temperature stabilizes at 0°C following this sudden change. The water at −2°C is said to be in a *metastable state*. Write an essay on metastable states and discuss how they differ from stable equilibrium states.

**3–137** A solid normally absorbs heat as it melts, but there is a known exception at temperatures close to absolute zero. Find out which solid it is and give a physical explanation for it.

# ENERGY ANALYSIS OF CLOSED SYSTEMS

I n Chap. 2, we considered various forms of energy and energy transfer, and we developed a general relation for the conservation of energy principle or energy balance. Then in Chap. 3, we learned how to determine the thermodynamics properties of substances. In this chapter, we apply the energy balance relation to systems that do not involve any mass flow across their boundaries; that is, closed systems.

We start this chapter with a discussion of the *moving boundary work* or *P dV work* commonly encountered in reciprocating devices such as automotive engines and compressors. We continue by applying the *general energy balance* relation, which is simply expressed as $E_{in} - E_{out} = \Delta E_{system}$, to systems that involve pure substance. Then we define *specific heats*, obtain relations for the internal energy and enthalpy of *ideal gases* in terms of specific heats and temperature changes, and perform energy balances on various systems that involve ideal gases. We repeat this for systems that involve solids and liquids, which are approximated as *incompressible substances*.

## OBJECTIVES

The objectives of Chapter 4 are to:

- Examine the moving boundary work or $P\,dV$ work commonly encountered in reciprocating devices such as automotive engines and compressors.

- Identify the first law of thermodynamics as simply a statement of the conservation of energy principle for closed (fixed mass) systems.

- Develop the general energy balance applied to closed systems.

- Define the specific heat at constant volume and the specific heat at constant pressure.

- Relate the specific heats to the calculation of the changes in internal energy and enthalpy of ideal gases.

- Describe incompressible substances and determine the changes in their internal energy and enthalpy.

- Solve energy balance problems for closed (fixed mass) systems that involve heat and work interactions for general pure substances, ideal gases, and incompressible substances.

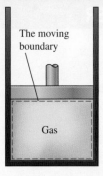

**FIGURE 4–1**

The work associated with a moving boundary is called *boundary work*.

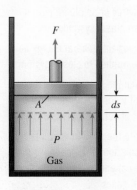

**FIGURE 4–2**

A gas does a differential amount of work $\delta W_b$ as it forces the piston to move by a differential amount $ds$.

# 4–1 · MOVING BOUNDARY WORK

One form of mechanical work frequently encountered in practice is associated with the expansion or compression of a gas in a piston–cylinder device. During this process, part of the boundary (the inner face of the piston) moves back and forth. Therefore, the expansion and compression work is often called **moving boundary work**, or simply **boundary work** (Fig. 4–1). Some call it the $P\,dV$ work for reasons explained later. Moving boundary work is the primary form of work involved in *automobile engines*. During their expansion, the combustion gases force the piston to move, which in turn forces the crankshaft to rotate.

The moving boundary work associated with real engines or compressors cannot be determined exactly from a thermodynamic analysis alone because the piston usually moves at very high speeds, making it difficult for the gas inside to maintain equilibrium. Then the states through which the system passes during the process cannot be specified, and no process path can be drawn. Work, being a path function, cannot be determined analytically without a knowledge of the path. Therefore, the boundary work in real engines or compressors is determined by direct measurements.

In this section, we analyze the moving boundary work for a *quasi-equilibrium process*, a process during which the system remains nearly in equilibrium at all times. A quasi-equilibrium process, also called a *quasistatic process*, is closely approximated by real engines, especially when the piston moves at low velocities. Under identical conditions, the work output of the engines is found to be a maximum, and the work input to the compressors to be a minimum when quasi-equilibrium processes are used in place of nonquasi-equilibrium processes. Below, the work associated with a moving boundary is evaluated for a quasi-equilibrium process.

Consider the gas enclosed in the piston–cylinder device shown in Fig. 4–2. The initial pressure of the gas is $P$, the total volume is $V$, and the cross-sectional area of the piston is $A$. If the piston is allowed to move a distance $ds$ in a quasi-equilibrium manner, the differential work done during this process is

$$\delta W_b = F\,ds = PA\,ds = P\,dV \qquad (4\text{–}1)$$

That is, the boundary work in the differential form is equal to the product of the absolute pressure $P$ and the differential change in the volume $dV$ of the system. This expression also explains why the moving boundary work is sometimes called the $P\,dV$ work.

Note in Eq. 4–1 that $P$ is the absolute pressure, which is always positive. However, the volume change $dV$ is positive during an expansion process (volume increasing) and negative during a compression process (volume decreasing). Thus, the boundary work is positive during an expansion process and negative during a compression process. Therefore, Eq. 4–1 can be viewed as an expression for boundary work output, $W_{b,\text{out}}$. A negative result indicates boundary work input (compression).

The total boundary work done during the entire process as the piston moves is obtained by adding all the differential works from the initial state to the final state:

$$W_b = \int_1^2 P\,dV \qquad (\text{kJ}) \qquad (4\text{–}2)$$

This integral can be evaluated only if we know the functional relationship between $P$ and $V$ during the process. That is, $P = f(V)$ should be available. Note that $P = f(V)$ is simply the equation of the process path on a $P$-$V$ diagram.

The quasi-equilibrium expansion process described is shown on a $P$-$V$ diagram in Fig. 4–3. On this diagram, the differential area $dA$ is equal to $P\, dV$, which is the differential work. The total area $A$ under the process curve 1–2 is obtained by adding these differential areas:

$$\text{Area} = A = \int_1^2 dA = \int_1^2 P\, dV \tag{4–3}$$

A comparison of this equation with Eq. 4–2 reveals that *the area under the process curve on a P-V diagram is equal, in magnitude, to the work done during a quasi-equilibrium expansion or compression process of a closed system.* (On the $P$-$v$ diagram, it represents the boundary work done per unit mass.)

A gas can follow several different paths as it expands from state 1 to state 2. In general, each path will have a different area underneath it, and since this area represents the magnitude of the work, the work done will be different for each process (Fig. 4–4). This is expected, since work is a path function (i.e., it depends on the path followed as well as the end states). If work were not a path function, no cyclic devices (car engines, power plants) could operate as work-producing devices. The work produced by these devices during one part of the cycle would have to be consumed during another part, and there would be no net work output. The cycle shown in Fig. 4–5 produces a net work output because the work done by the system during the expansion process (area under path $A$) is greater than the work done on the system during the compression part of the cycle (area under path $B$), and the difference between these two is the net work done during the cycle (the colored area).

If the relationship between $P$ and $V$ during an expansion or a compression process is given in terms of experimental data instead of in a functional form, obviously we cannot perform the integration analytically. We can, however, plot the $P$-$V$ diagram of the process using these data points and calculate the area underneath graphically to determine the work done.

Strictly speaking, the pressure $P$ in Eq. 4–2 is the pressure at the inner surface of the piston. It becomes equal to the pressure of the gas in the cylinder only if the process is quasi-equilibrium and thus the entire gas in the cylinder is at the same pressure at any given time. Equation 4–2 can also be used for nonquasi-equilibrium processes provided that the pressure *at the inner face of the piston* is used for $P$. (Besides, we cannot speak of the pressure of a *system* during a nonquasi-equilibrium process since properties are defined for equilibrium states only.) Therefore, we can generalize the boundary work relation by expressing it as

$$W_b = \int_1^2 P_i\, dV \tag{4–4}$$

where $P_i$ is the pressure at the inner face of the piston.

Note that work is a mechanism for energy interaction between a system and its surroundings, and $W_b$ represents the amount of energy transferred

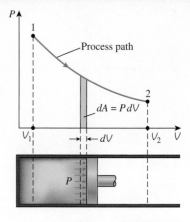

**FIGURE 4–3**
The area under the process curve on a $P$-$V$ diagram represents the boundary work.

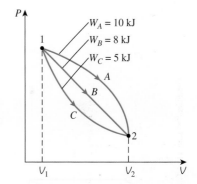

**FIGURE 4–4**
The boundary work done during a process depends on the path followed as well as the end states.

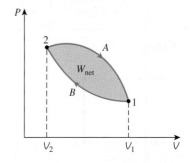

**FIGURE 4–5**
The net work done during a cycle is the difference between the work done by the system and the work done on the system.

from the system during an expansion process (or to the system during a compression process). Therefore, it has to appear somewhere else and we must be able to account for it since energy is conserved. In a car engine, for example, the boundary work done by the expanding hot gases is used to overcome friction between the piston and the cylinder, to push atmospheric air out of the way, and to rotate the crankshaft. Therefore,

$$W_b = W_{\text{friction}} + W_{\text{atm}} + W_{\text{crank}} = \int_1^2 (F_{\text{friction}} + P_{\text{atm}}A + F_{\text{crank}})\,dx \qquad \textbf{(4–5)}$$

Of course the work used to overcome friction appears as frictional heat and the energy transmitted through the crankshaft is transmitted to other components (such as the wheels) to perform certain functions. But note that the energy transferred by the system as work must equal the energy received by the crankshaft, the atmosphere, and the energy used to overcome friction. The use of the boundary work relation is not limited to the quasi-equilibrium processes of gases only. It can also be used for solids and liquids.

---

**EXAMPLE 4–1    Boundary Work for a Constant-Volume Process**

A rigid tank contains air at 500 kPa and 150°C. As a result of heat transfer to the surroundings, the temperature and pressure inside the tank drop to 65°C and 400 kPa, respectively. Determine the boundary work done during this process.

**SOLUTION**  Air in a rigid tank is cooled, and both the pressure and temperature drop. The boundary work done is to be determined.
**Analysis**  A sketch of the system and the P-V diagram of the process are shown in Fig. 4–6. The boundary work can be determined from Eq. 4–2 to be

$$W_b = \int_1^2 P \, d\cancelto{0}{V} = 0$$

**Discussion**  This is expected since a rigid tank has a constant volume and $dV = 0$ in this equation. Therefore, there is no boundary work done during this process. That is, the boundary work done during a constant-volume process is always zero. This is also evident from the P-V diagram of the process (the area under the process curve is zero).

---

**EXAMPLE 4–2    Boundary Work for a Constant-Pressure Process**

A frictionless piston–cylinder device contains 5 kg of steam at 400 kPa and 200°C. Heat is now transferred to the steam until the temperature reaches 250°C. If the piston is not attached to a shaft and its mass is constant, determine the work done by the steam during this process.

**SOLUTION**  Steam in a piston cylinder device is heated and the temperature rises at constant pressure. The boundary work done is to be determined.

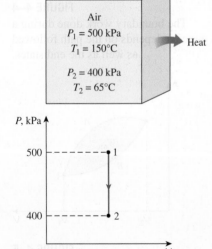

**FIGURE 4–6**
Schematic and P-V diagram for Example 4–1.

Air
$P_1 = 500$ kPa
$T_1 = 150°C$
Heat
$P_2 = 400$ kPa
$T_2 = 65°C$

$P$, kPa

500 ------●1

400 ------●2

$V$

*Analysis* A sketch of the system and the *P-v* diagram of the process are shown in Fig. 4–7.

*Assumption* The expansion process is quasi-equilibrium.

*Analysis* Even though it is not explicitly stated, the pressure of the steam within the cylinder remains constant during this process since both the atmospheric pressure and the weight of the piston remain constant. Therefore, this is a constant-pressure process, and, from Eq. 4–2

$$ W_b = \int_1^2 P\,dV = P_0 \int_1^2 dV = P_0(V_2 - V_1) \tag{4–6} $$

or

$$ W_b = mP_0(v_2 - v_1) $$

since $V = mv$. From the superheated vapor table (Table A–6), the specific volumes are determined to be $v_1 = 0.53434$ m³/kg at state 1 (400 kPa, 200°C) and $v_2 = 0.59520$ m³/kg at state 2 (400 kPa, 250°C). Substituting these values yields

$$ W_b = (5\text{ kg})(400\text{ kPa})[(0.59520 - 0.53434)\text{ m}^3/\text{kg}]\left(\frac{1\text{ kJ}}{1\text{ kPa·m}^3}\right) $$

$$ = \mathbf{122\ kJ} $$

*Discussion* The positive sign indicates that the work is done by the system. That is, the steam used 122 kJ of its energy to do this work. The magnitude of this work could also be determined by calculating the area under the process curve on the *P-V* diagram, which is simply $P_0\,\Delta V$ for this case.

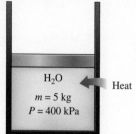

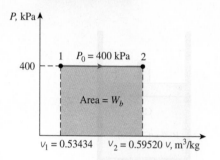

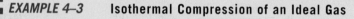

**FIGURE 4–7**
Schematic and *P-v* diagram for Example 4–2.

---

■ **EXAMPLE 4–3**   **Isothermal Compression of an Ideal Gas**

A piston–cylinder device initially contains 0.4 m³ of air at 100 kPa and 80°C. The air is now compressed to 0.1 m³ in such a way that the temperature inside the cylinder remains constant. Determine the work done during this process.

**SOLUTION** Air in a piston–cylinder device is compressed isothermally. The boundary work done is to be determined.

*Analysis* A sketch of the system and the *P-V* diagram of the process are shown in Fig. 4–8.

*Assumptions* **1** The compression process is quasi-equilibrium. **2** At specified conditions, air can be considered to be an ideal gas since it is at a high temperature and low pressure relative to its critical-point values.

*Analysis* For an ideal gas at constant temperature $T_0$,

$$ PV = mRT_0 = C \quad \text{or} \quad P = \frac{C}{V} $$

where $C$ is a constant. Substituting this into Eq. 4–2, we have

$$ W_b = \int_1^2 P\,dV = \int_1^2 \frac{C}{V}\,dV = C\int_1^2 \frac{dV}{V} = C\ln\frac{V_2}{V_1} = P_1V_1\ln\frac{V_2}{V_1} \tag{4–7} $$

In Eq. 4–7, $P_1V_1$ can be replaced by $P_2V_2$ or $mRT_0$. Also, $V_2/V_1$ can be replaced by $P_1/P_2$ for this case since $P_1V_1 = P_2V_2$.

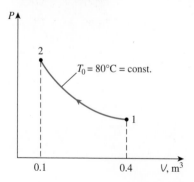

**FIGURE 4–8**
Schematic and *P-V* diagram for Example 4–3.

Substituting the numerical values into Eq. 4–7 yields

$$W_b = (100 \text{ kPa})(0.4 \text{ m}^3)\left(\ln\frac{0.1}{0.4}\right)\left(\frac{1 \text{ kJ}}{1 \text{ kPa·m}^3}\right)$$

$$= -55.5 \text{ kJ}$$

**Discussion** The negative sign indicates that this work is done on the system (a work input), which is always the case for compression processes.

## Polytropic Process

During actual expansion and compression processes of gases, pressure and volume are often related by $PV^n = C$, where $n$ and $C$ are constants. A process of this kind is called a **polytropic process** (Fig. 4–9). Below we develop a general expression for the work done during a polytropic process. The pressure for a polytropic process can be expressed as

$$P = CV^{-n} \tag{4–8}$$

Substituting this relation into Eq. 4–2, we obtain

$$W_b = \int_1^2 P\,dV = \int_1^2 CV^{-n}\,dV = C\frac{V_2^{-n+1} - V_1^{-n+1}}{-n + 1} = \frac{P_2 V_2 - P_1 V_1}{1 - n} \tag{4–9}$$

since $C = P_1 V_1^n = P_2 V_2^n$. For an ideal gas ($PV = mRT$), this equation can also be written as

$$W_b = \frac{mR(T_2 - T_1)}{1 - n} \quad n \neq 1 \quad \text{(kJ)} \tag{4–10}$$

For the special case of $n = 1$ the boundary work becomes

$$W_b = \int_1^2 P\,dV = \int_1^2 CV^{-1}\,dV = PV \ln\left(\frac{V_2}{V_1}\right)$$

For an ideal gas this result is equivalent to the isothermal process discussed in the previous example.

Gas

$PV^n = C = \text{const.}$

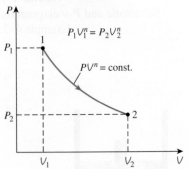

$P_1 V_1^n = P_2 V_2^n$

$PV^n = \text{const.}$

**FIGURE 4–9**
Schematic and $P\text{-}V$ diagram for a polytropic process.

---

**EXAMPLE 4–4** **Expansion of a Gas against a Spring**

A piston–cylinder device contains 0.05 m³ of a gas initially at 200 kPa. At this state, a linear spring that has a spring constant of 150 kN/m is touching the piston but exerting no force on it. Now heat is transferred to the gas, causing the piston to rise and to compress the spring until the volume inside the cylinder doubles. If the cross-sectional area of the piston is 0.25 m², determine (*a*) the final pressure inside the cylinder, (*b*) the total work done by the gas, and (*c*) the fraction of this work done against the spring to compress it.

**SOLUTION** A gas in a piston–cylinder device equipped with a linear spring expands as a result of heating. The final gas pressure, the total work done, and the fraction of the work done to compress the spring are to be determined.

**Assumptions** **1** The expansion process is quasi-equilibrium. **2** The spring is linear in the range of interest.

**Analysis** A sketch of the system and the *P-V* diagram of the process are shown in Fig. 4–10.

(*a*) The enclosed volume at the final state is

$$V_2 = 2V_1 = (2)(0.05 \text{ m}^3) = 0.1 \text{ m}^3$$

Then the displacement of the piston (and of the spring) becomes

$$x = \frac{\Delta V}{A} = \frac{(0.1 - 0.05) \text{ m}^3}{0.25 \text{ m}^2} = 0.2 \text{ m}$$

The force applied by the linear spring at the final state is

$$F = kx = (150 \text{ kN/m})(0.2 \text{ m}) = 30 \text{ kN}$$

The additional pressure applied by the spring on the gas at this state is

$$P = \frac{F}{A} = \frac{30 \text{ kN}}{0.25 \text{ m}^2} = 120 \text{ kPa}$$

Without the spring, the pressure of the gas would remain constant at 200 kPa while the piston is rising. But under the effect of the spring, the pressure rises linearly from 200 kPa to

$$200 + 120 = \mathbf{320 \text{ kPa}}$$

at the final state.

(*b*) An easy way of finding the work done is to plot the process on a *P-V* diagram and find the area under the process curve. From Fig. 4–10 the area under the process curve (a trapezoid) is determined to be

$$W = \text{area} = \frac{(200 + 320) \text{ kPa}}{2} [(0.1 - 0.05) \text{ m}^3] \left( \frac{1 \text{ kJ}}{1 \text{ kPa·m}^3} \right) = \mathbf{13 \text{ kJ}}$$

Note that the work is done by the system.

(*c*) The work represented by the rectangular area (region I) is done against the piston and the atmosphere, and the work represented by the triangular area (region II) is done against the spring. Thus,

$$W_{\text{spring}} = \tfrac{1}{2}[(320 - 200) \text{ kPa}](0.05 \text{ m}^3) \left( \frac{1 \text{ kJ}}{1 \text{ kPa·m}^3} \right) = \mathbf{3 \text{ kJ}}$$

**Discussion** This result could also be obtained from

$$W_{\text{spring}} = \tfrac{1}{2}k(x_2^2 - x_1^2) = \tfrac{1}{2}(150 \text{ kN/m})[(0.2 \text{ m})^2 - 0^2] \left( \frac{1 \text{ kJ}}{1 \text{ kN·m}} \right) = 3 \text{ kJ}$$

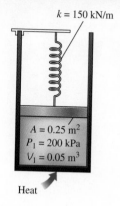

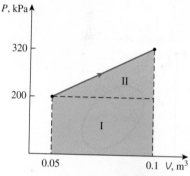

**FIGURE 4–10**
Schematic and *P-V* diagram for Example 4–4.

# 4–2 · ENERGY BALANCE FOR CLOSED SYSTEMS

Energy balance for any system undergoing any kind of process was expressed as (see Chap. 2)

$$\underbrace{E_{\text{in}} - E_{\text{out}}}_{\substack{\text{Net energy transfer} \\ \text{by heat, work, and mass}}} = \underbrace{\Delta E_{\text{system}}}_{\substack{\text{Change in internal, kinetic,} \\ \text{potential, etc., energies}}} \quad \text{(kJ)} \qquad (4\text{–}11)$$

or, in the **rate form**, as

$$\underbrace{\dot{E}_{\text{in}} - \dot{E}_{\text{out}}}_{\substack{\text{Rate of net energy transfer} \\ \text{by heat, work, and mass}}} = \underbrace{dE_{\text{system}}/dt}_{\substack{\text{Rate of change in internal,} \\ \text{kinetic, potential, etc., energies}}} \quad \text{(kW)} \qquad \text{(4–12)}$$

For constant rates, the total quantities during a time interval $\Delta t$ are related to the quantities per unit time as

$$Q = \dot{Q}\,\Delta t, \quad W = \dot{W}\,\Delta t, \quad \text{and} \quad \Delta E = (dE/dt)\Delta t \quad \text{(kJ)} \qquad \text{(4–13)}$$

The energy balance can be expressed on a **per unit mass** basis as

$$e_{\text{in}} - e_{\text{out}} = \Delta e_{\text{system}} \quad \text{(kJ/kg)} \qquad \text{(4–14)}$$

which is obtained by dividing all the quantities in Eq. 4–11 by the mass $m$ of the system. Energy balance can also be expressed in the differential form as

$$\delta E_{\text{in}} - \delta E_{\text{out}} = dE_{\text{system}} \quad \text{or} \quad \delta e_{\text{in}} - \delta e_{\text{out}} = de_{\text{system}} \qquad \text{(4–15)}$$

For a closed system undergoing a **cycle**, the initial and final states are identical, and thus $\Delta E_{\text{system}} = E_2 - E_1 = 0$. Then, the energy balance for a cycle simplifies to $E_{\text{in}} - E_{\text{out}} = 0$ or $E_{\text{in}} = E_{\text{out}}$. Noting that a closed system does not involve any mass flow across its boundaries, the energy balance for a cycle can be expressed in terms of heat and work interactions as

$$W_{\text{net,out}} = Q_{\text{net,in}} \quad \text{or} \quad \dot{W}_{\text{net,out}} = \dot{Q}_{\text{net,in}} \quad \text{(for a cycle)} \qquad \text{(4–16)}$$

That is, the net work output during a cycle is equal to net heat input (Fig. 4–11).

The energy balance (or the first-law) relations already given are intuitive in nature and are easy to use when the magnitudes and directions of heat and work transfers are known. However, when performing a general analytical study or solving a problem that involves an unknown heat or work interaction, we need to assume a direction for the heat or work interactions. In such cases, it is common practice to use the classical thermodynamics sign convention and to assume heat to be transferred *into the system* (heat input) in the amount of $Q$ and work to be done *by the system* (work output) in the amount of $W$, and then to solve the problem. The energy balance relation in that case for a closed system becomes

$$Q_{\text{net,in}} - W_{\text{net,out}} = \Delta E_{\text{system}} \quad \text{or} \quad Q - W = \Delta E \qquad \text{(4–17)}$$

where $Q = Q_{\text{net,in}} = Q_{\text{in}} - Q_{\text{out}}$ is the *net heat input* and $W = W_{\text{net,out}} = W_{\text{out}} - W_{\text{in}}$ is the *net work output*. Obtaining a negative quantity for $Q$ or $W$ simply means that the assumed direction for that quantity is wrong and should be reversed. Various forms of this "traditional" first-law relation for closed systems are given in Fig. 4–12.

The first law cannot be proven mathematically, but no process in nature is known to have violated the first law, and this should be taken as sufficient proof. Note that if it were possible to prove the first law on the basis of other physical principles, the first law then would be a consequence of those principles instead of being a fundamental physical law itself.

As energy quantities, heat and work are not that different, and you probably wonder why we keep distinguishing them. After all, the change in the

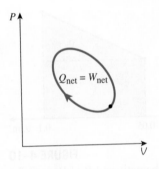

**FIGURE 4–11**
For a cycle $\Delta E = 0$, thus $Q = W$.

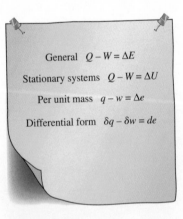

General $\quad Q - W = \Delta E$

Stationary systems $\quad Q - W = \Delta U$

Per unit mass $\quad q - w = \Delta e$

Differential form $\quad \delta q - \delta w = de$

**FIGURE 4–12**
Various forms of the first-law relation for closed systems.

energy content of a system is equal to the amount of energy that crosses the system boundaries, and it makes no difference whether the energy crosses the boundary as heat or work. It seems as if the first-law relations would be much simpler if we had just one quantity that we could call *energy interaction* to represent both heat and work. Well, from the first-law point of view, heat and work are not different at all. From the second-law point of view, however, heat and work are very different, as is discussed in later chapters.

### EXAMPLE 4–5 Electric Heating of a Gas at Constant Pressure

A piston–cylinder device contains 25 g of saturated water vapor that is maintained at a constant pressure of 300 kPa. A resistance heater within the cylinder is turned on and passes a current of 0.2 A for 5 min from a 120-V source. At the same time, a heat loss of 3.7 kJ occurs. (*a*) Show that for a closed system the boundary work $W_b$ and the change in internal energy $\Delta U$ in the first-law relation can be combined into one term, $\Delta H$, for a constant-pressure process. (*b*) Determine the final temperature of the steam.

**SOLUTION** Saturated water vapor in a piston–cylinder device expands at constant pressure as a result of heating. It is to be shown that $\Delta U + W_b = \Delta H$, and the final temperature is to be determined.

*Assumptions* **1** The tank is stationary and thus the kinetic and potential energy changes are zero, $\Delta KE = \Delta PE = 0$. Therefore, $\Delta E = \Delta U$ and internal energy is the only form of energy of the system that may change during this process. **2** Electrical wires constitute a very small part of the system, and thus the energy change of the wires can be neglected.

*Analysis* We take the contents of the cylinder, including the resistance wires, as the *system* (Fig. 4–13). This is a *closed system* since no mass crosses the system boundary during the process. We observe that a piston–cylinder device typically involves a moving boundary and thus boundary work $W_b$. The pressure remains constant during the process and thus $P_2 = P_1$. Also, heat is lost from the system and electrical work $W_e$ is done on the system.

(*a*) This part of the solution involves a general analysis for a closed system undergoing a quasi-equilibrium constant-pressure process, and thus we consider a general closed system. We take the direction of heat transfer $Q$ to be to the system and the work $W$ to be done by the system. We also express the work as the sum of boundary and other forms of work (such as electrical and shaft). Then, the energy balance can be expressed as

$$\underbrace{E_{\text{in}} - E_{\text{out}}}_{\substack{\text{Net energy transfer} \\ \text{by heat, work, and mass}}} = \underbrace{\Delta E_{\text{system}}}_{\substack{\text{Change in internal, kinetic,} \\ \text{potential, etc., energies}}}$$

$$Q - W = \Delta U + \cancelto{0}{\Delta KE} + \cancelto{0}{\Delta PE}$$

$$Q - W_{\text{other}} - W_b = U_2 - U_1$$

For a constant-pressure process, the boundary work is given as $W_b = P_0(V_2 - V_1)$. Substituting this into the preceding relation gives

$$Q - W_{\text{other}} - P_0(V_2 - V_1) = U_2 - U_1$$

However,

$$P_0 = P_2 = P_1 \rightarrow Q - W_{\text{other}} = (U_2 + P_2 V_2) - (U_1 + P_1 V_1)$$

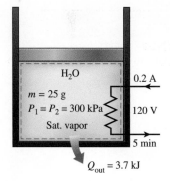

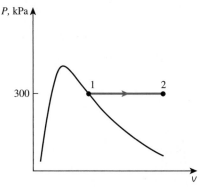

**FIGURE 4–13**
Schematic and *P*-*v* diagram for Example 4–5.

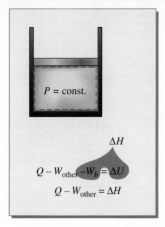

**FIGURE 4–14**
For a closed system undergoing a quasi-equilibrium, $P = $ constant process, $\Delta U + W_b = \Delta H$. Note that this relation is NOT valid for closed systems processes during which pressure DOES NOT remain constant.

Also $H = U + PV$, and thus

$$Q - W_{\text{other}} = H_2 - H_1 \quad \text{(kJ)} \qquad \text{(4–18)}$$

which is the desired relation (Fig. 4–14). This equation is very convenient to use in the analysis of closed systems undergoing a **constant-pressure quasi-equilibrium process** since the boundary work is automatically taken care of by the enthalpy terms, and one no longer needs to determine it separately.

(b) The only other form of work in this case is the electrical work, which can be determined from

$$W_e = \mathbf{V}I\Delta t = (120 \text{ V})(0.2 \text{ A})(300 \text{ s})\left(\frac{1 \text{ kJ/s}}{1000 \text{ VA}}\right) = 7.2 \text{ kJ}$$

$$\textit{State 1:} \quad \left.\begin{array}{l} P_1 = 300 \text{ kPa} \\ \text{sat. vapor} \end{array}\right\} \; h_1 = h_{g \text{ @ } 300 \text{ kPa}} = 2724.9 \text{ kJ/kg} \quad \text{(Table A–5)}$$

The enthalpy at the final state can be determined directly from Eq. 4–18 by expressing heat transfer from the system and work done on the system as negative quantities (since their directions are opposite to the assumed directions). Alternately, we can use the general energy balance relation with the simplification that the boundary work is considered automatically by replacing $\Delta U$ by $\Delta H$ for a constant-pressure expansion or compression process:

$$\underbrace{E_{\text{in}} - E_{\text{out}}}_{\substack{\text{Net energy transfer} \\ \text{by heat, work, and mass}}} = \underbrace{\Delta E_{\text{system}}}_{\substack{\text{Change in internal, kinetic,} \\ \text{potential, etc., energies}}}$$

$$W_{e,\text{in}} - Q_{\text{out}} - W_b = \Delta U$$

$$W_{e,\text{in}} - Q_{\text{out}} = \Delta H = m(h_2 - h_1) \quad \text{(since } P = \text{constant)}$$

$$7.2 \text{ kJ} - 3.7 \text{ kJ} = (0.025 \text{ kg})(h_2 - 2724.9) \text{ kJ/kg}$$

$$h_2 = 2864.9 \text{ kJ/kg}$$

Now the final state is completely specified since we know both the pressure and the enthalpy. The temperature at this state is

$$\textit{State 2:} \quad \left.\begin{array}{l} P_2 = 300 \text{ kPa} \\ h_2 = 2864.9 \text{ kJ/kg} \end{array}\right\} \; T_2 = \mathbf{200°C} \quad \text{(Table A–6)}$$

Therefore, the steam will be at 200°C at the end of this process.

**Discussion** Strictly speaking, the potential energy change of the steam is not zero for this process since the center of gravity of the steam rose somewhat. Assuming an elevation change of 1 m (which is rather unlikely), the change in the potential energy of the steam would be 0.0002 kJ, which is very small compared to the other terms in the first-law relation. Therefore, in problems of this kind, the potential energy term is always neglected.

---

### EXAMPLE 4–6　Unrestrained Expansion of Water

A rigid tank is divided into two equal parts by a partition. Initially, one side of the tank contains 5 kg of water at 200 kPa and 25°C, and the other side is evacuated. The partition is then removed, and the water expands into the entire tank. The water is allowed to exchange heat with its surroundings until the temperature in the tank returns to the initial value of 25°C. Determine (a) the volume of the tank, (b) the final pressure, and (c) the heat transfer for this process.

**SOLUTION**  One half of a rigid tank is filled with liquid water while the other side is evacuated. The partition between the two parts is removed and water is allowed to expand and fill the entire tank while the temperature is maintained constant. The volume of the tank, the final pressure, and the heat transfer are to be to determined.

**Assumptions**  **1** The system is stationary and thus the kinetic and potential energy changes are zero, $\Delta KE = \Delta PE = 0$ and $\Delta E = \Delta U$. **2** The direction of heat transfer is to the system (heat gain, $Q_{in}$). A negative result for $Q_{in}$ indicates the assumed direction is wrong and thus it is a heat loss. **3** The volume of the rigid tank is constant, and thus there is no energy transfer as boundary work. **4** There is no electrical, shaft, or any other kind of work involved.

**Analysis**  We take the contents of the tank, including the evacuated space, as the *system* (Fig. 4–15). This is a *closed system* since no mass crosses the system boundary during the process. We observe that the water fills the entire tank when the partition is removed (possibly as a liquid–vapor mixture).

(a) Initially the water in the tank exists as a compressed liquid since its pressure (200 kPa) is greater than the saturation pressure at 25°C (3.1698 kPa). Approximating the compressed liquid as a saturated liquid at the given temperature, we find

$$v_1 \cong v_{f@\,25°C} = 0.001003 \text{ m}^3/\text{kg} \cong 0.001 \text{ m}^3/\text{kg}\quad \text{(Table A–4)}$$

Then the initial volume of the water is

$$V_1 = mv_1 = (5 \text{ kg})(0.001 \text{ m}^3/\text{kg}) = 0.005 \text{ m}^3$$

The total volume of the tank is twice this amount:

$$V_{tank} = (2)(0.005 \text{ m}^3) = \textbf{0.01 m}^3$$

(b) At the final state, the specific volume of the water is

$$v_2 = \frac{V_2}{m} = \frac{0.01 \text{ m}^3}{5 \text{ kg}} = 0.002 \text{ m}^3/\text{kg}$$

which is twice the initial value of the specific volume. This result is expected since the volume doubles while the amount of mass remains constant.

$$\text{At } 25°C:\quad v_f = 0.001003 \text{ m}^3/\text{kg} \quad \text{and} \quad v_g = 43.340 \text{ m}^3/\text{kg}\quad \text{(Table A–4)}$$

Since $v_f < v_2 < v_g$, the water is a saturated liquid–vapor mixture at the final state, and thus the pressure is the saturation pressure at 25°C:

$$P_2 = P_{sat\,@\,25°C} = \textbf{3.1698 kPa}\quad \text{(Table A–4)}$$

(c) Under stated assumptions and observations, the energy balance on the system can be expressed as

$$\underbrace{E_{in} - E_{out}}_{\substack{\text{Net energy transfer} \\ \text{by heat, work, and mass}}} = \underbrace{\Delta E_{system}}_{\substack{\text{Change in internal, kinetic,} \\ \text{potential, etc., energies}}}$$

$$Q_{in} = \Delta U = m(u_2 - u_1)$$

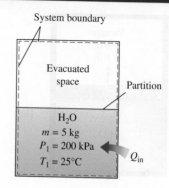

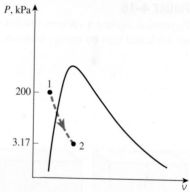

**FIGURE 4–15**
Schematic and P-v diagram for Example 4–6.

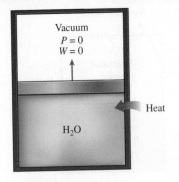

**FIGURE 4–16**
Expansion against a vacuum involves
no work and thus no energy transfer.

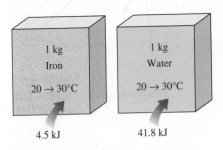

**FIGURE 4–17**
It takes different amounts of energy
to raise the temperature of different
substances by the same amount.

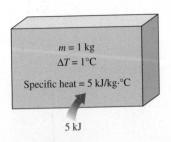

**FIGURE 4–18**
Specific heat is the energy required
to raise the temperature of a unit
mass of a substance by one degree
in a specified way.

Notice that even though the water is expanding during this process, the system chosen involves fixed boundaries only (the dashed lines) and therefore the moving boundary work is zero (Fig. 4–16). Then $W = 0$ since the system does not involve any other forms of work. (Can you reach the same conclusion by choosing the water as our system?) Initially,

$$u_1 \cong u_{f\,@\,25°C} = 104.83 \text{ kJ/kg}$$

The quality at the final state is determined from the specific volume information:

$$x_2 = \frac{v_2 - v_f}{v_{fg}} = \frac{0.002 - 0.001}{43.34 - 0.001} = 2.3 \times 10^{-5}$$

Then

$$u_2 = u_f + x_2 u_{fg}$$
$$= 104.83 \text{ kJ/kg} + (2.3 \times 10^{-5})(2304.3 \text{ kJ/kg})$$
$$= 104.88 \text{ kJ/kg}$$

Substituting yields

$$Q_{\text{in}} = (5 \text{ kg})[(104.88 - 104.83) \text{ kJ/kg}] = \textbf{0.25 kJ}$$

***Discussion*** The positive sign indicates that the assumed direction is correct, and heat is transferred to the water.

# 4–3 · SPECIFIC HEATS

We know from experience that it takes different amounts of energy to raise the temperature of identical masses of different substances by one degree. For example, we need about 4.5 kJ of energy to raise the temperature of 1 kg of iron from 20 to 30°C, whereas it takes about 9 times this energy (41.8 kJ to be exact) to raise the temperature of 1 kg of liquid water by the same amount (Fig. 4–17). Therefore, it is desirable to have a property that will enable us to compare the energy storage capabilities of various substances. This property is the specific heat.

The **specific heat** is defined as *the energy required to raise the temperature of a unit mass of a substance by one degree* (Fig. 4–18). In general, this energy depends on how the process is executed. In thermodynamics, we are interested in two kinds of specific heats: **specific heat at constant volume** $c_v$ and **specific heat at constant pressure** $c_p$.

Physically, the specific heat at constant volume $c_v$ can be viewed as *the energy required to raise the temperature of the unit mass of a substance by one degree as the volume is maintained constant.* The energy required to do the same as the pressure is maintained constant is the specific heat at constant pressure $c_p$. This is illustrated in Fig. 4–19. The specific heat at constant pressure $c_p$ is always greater than $c_v$ because at constant pressure the system is allowed to expand and the energy for this expansion work must also be supplied to the system.

Now we attempt to express the specific heats in terms of other thermodynamic properties. First, consider a fixed mass in a stationary closed system undergoing a constant-volume process (and thus no expansion or compression work is involved). The conservation of energy principle $e_{in} - e_{out} = \Delta e_{system}$ for this process can be expressed in the differential form as

$$\delta e_{in} - \delta e_{out} = du$$

The left-hand side of this equation represents the net amount of energy transferred to the system. From the definition of $c_v$, this energy must be equal to $c_v \, dT$, where $dT$ is the differential change in temperature. Thus,

$$c_v \, dT = du \quad \text{at constant volume}$$

or

$$c_v = \left(\frac{\partial u}{\partial T}\right)_v \qquad (4\text{-}19)$$

Similarly, an expression for the specific heat at constant pressure $c_p$ can be obtained by considering a constant-pressure expansion or compression process. It yields

$$c_p = \left(\frac{\partial h}{\partial T}\right)_p \qquad (4\text{-}20)$$

Equations 4–19 and 4–20 are the defining equations for $c_v$ and $c_p$, and their interpretation is given in Fig. 4–20.

Note that $c_v$ and $c_p$ are expressed in terms of other properties; thus, they must be properties themselves. Like any other property, the specific heats of a substance depend on the state that, in general, is specified by two independent, intensive properties. That is, the energy required to raise the temperature of a substance by one degree is different at different temperatures and pressures (Fig. 4–21). But this difference is usually not very large.

A few observations can be made from Eqs. 4–19 and 4–20. First, these equations are *property relations* and as such *are independent of the type of processes*. They are valid for *any* substance undergoing *any* process. The only relevance $c_v$ has to a constant-volume process is that $c_v$ happens to be the energy transferred to a system during a constant-volume process per unit mass, per unit degree rise in temperature. This is how the values of $c_v$ are determined. This is also how the name *specific heat at constant volume* originated. Likewise, the energy transferred to a system per unit mass per unit temperature rise during a constant-pressure process happens to be equal to $c_p$. This is how the values of $c_p$ can be determined and also explains the origin of the name *specific heat at constant pressure*.

Another observation that can be made from Eqs. 4–19 and 4–20 is that $c_v$ is related to the changes in *internal energy* and $c_p$ to the changes in *enthalpy*. In fact, it would be more proper to define $c_v$ as *the change in the internal energy of a substance per unit change in temperature at constant volume*. Likewise, $c_p$ can be defined as *the change in the enthalpy of a substance per unit change in temperature at constant pressure*. In other words, $c_v$ is a measure of the variation of internal energy of a substance with temperature, and $c_p$ is a measure of the variation of enthalpy of a substance with temperature.

Both the internal energy and enthalpy of a substance can be changed by the transfer of *energy* in any form, with heat being only one of them. Therefore,

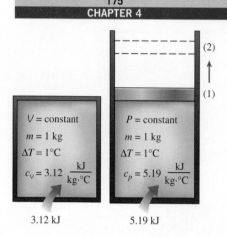

**FIGURE 4–19**

Constant-volume and constant-pressure specific heats $c_v$ and $c_p$ (values given are for helium gas).

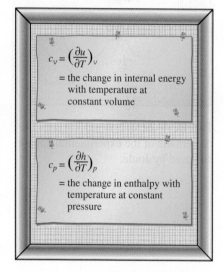

**FIGURE 4–20**

Formal definitions of $c_v$ and $c_p$.

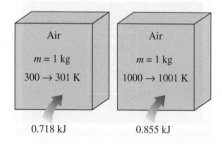

**FIGURE 4–21**

The specific heat of a substance changes with temperature.

the term *specific energy* is probably more appropriate than the term *specific heat*, which implies that energy is transferred (and stored) in the form of heat.

A common unit for specific heats is kJ/kg·°C or kJ/kg·K. Notice that these two units are *identical* since $\Delta T(°C) = \Delta T(K)$, and 1°C change in temperature is equivalent to a change of 1 K. The specific heats are sometimes given on a *molar basis*. They are then denoted by $\bar{c}_v$ and $\bar{c}_p$ and have the unit kJ/kmol·°C or kJ/kmol·K.

## 4–4 · INTERNAL ENERGY, ENTHALPY, AND SPECIFIC HEATS OF IDEAL GASES

We defined an ideal gas as a gas whose temperature, pressure, and specific volume are related by

$$P v = RT$$

It has been demonstrated mathematically (Chap. 12) and experimentally (Joule, 1843) that for an idea gas the internal energy is a function of the temperature only. That is,

$$u = u(T) \tag{4–21}$$

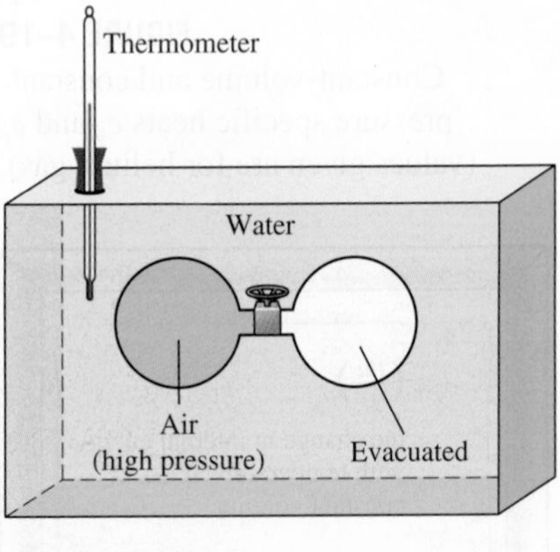

In his classical experiment, Joule submerged two tanks connected with a pipe and a valve in a water bath, as shown in Fig. 4–22. Initially, one tank contained air at a high pressure and the other tank was evacuated. When thermal equilibrium was attained, he opened the valve to let air pass from one tank to the other until the pressures equalized. Joule observed no change in the temperature of the water bath and assumed that no heat was transferred to or from the air. Since there was also no work done, he concluded that the internal energy of the air did not change even though the volume and the pressure changed. Therefore, he reasoned, the internal energy is a function of temperature only and not a function of pressure or specific volume. (Joule later showed that for gases that deviate significantly from ideal gas behavior, the internal energy is not a function of temperature alone.)

Using the definition of enthalpy and the equation of state of an ideal gas, we have

$$\left.\begin{array}{r} h = u + P v \\ P v = RT \end{array}\right\} \quad h = u + RT$$

Since $R$ is constant and $u = u(T)$, it follows that the enthalpy of an ideal gas is also a function of temperature only:

$$h = h(T) \tag{4–22}$$

Since $u$ and $h$ depend only on temperature for an ideal gas, the specific heats $c_v$ and $c_p$ also depend, at most, on temperature only. Therefore, at a given temperature, $u$, $h$, $c_v$, and $c_p$ of an ideal gas have fixed values regardless of the specific volume or pressure (Fig. 4–23). Thus, for ideal gases, the partial derivatives in Eqs. 4–19 and 4–20 can be replaced by ordinary derivatives. Then, the differential changes in the internal energy and enthalpy of an ideal gas can be expressed as

$$du = c_v(T)\, dT \tag{4–23}$$

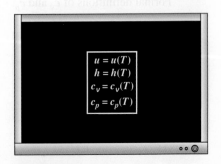

**FIGURE 4–22**
Schematic of the experimental apparatus used by Joule.

$$u = u(T)$$
$$h = h(T)$$
$$c_v = c_v(T)$$
$$c_p = c_p(T)$$

**FIGURE 4–23**
For ideal gases, $u$, $h$, $c_v$, and $c_p$ vary with temperature only.

and

$$dh = c_p(T)\, dT \tag{4-24}$$

The change in internal energy or enthalpy for an ideal gas during a process from state 1 to state 2 is determined by integrating these equations:

$$\Delta u = u_2 - u_1 = \int_1^2 c_v(T)\, dT \quad \text{(kJ/kg)} \tag{4-25}$$

and

$$\Delta h = h_2 - h_1 = \int_1^2 c_p(T)\, dT \quad \text{(kJ/kg)} \tag{4-26}$$

To carry out these integrations, we need to have relations for $c_v$ and $c_p$ as functions of temperature.

At low pressures, all real gases approach ideal-gas behavior, and therefore their specific heats depend on temperature only. The specific heats of real gases at low pressures are called *ideal-gas specific heats*, or *zero-pressure specific heats*, and are often denoted $c_{p0}$ and $c_{v0}$. Accurate analytical expressions for ideal-gas specific heats, based on direct measurements or calculations from statistical behavior of molecules, are available and are given as third-degree polynomials in the appendix (Table A–2c) for several gases. A plot of $\bar{c}_{p0}(T)$ data for some common gases is given in Fig. 4–24.

The use of ideal-gas specific heat data is limited to low pressures, but these data can also be used at moderately high pressures with reasonable accuracy as long as the gas does not deviate from ideal-gas behavior significantly.

The integrations in Eqs. 4–25 and 4–26 are straightforward but rather time-consuming and thus impractical. To avoid these laborious calculations, $u$ and $h$ data for a number of gases have been tabulated over small temperature intervals. These tables are obtained by choosing an arbitrary reference point and performing the integrations in Eqs. 4–25 and 4–26 by treating state 1 as the reference state. In the ideal-gas tables given in the appendix, zero kelvin is chosen as the reference state, and both the enthalpy and the internal energy are assigned zero values at that state (Fig. 4–25). The choice of the reference state has no effect on $\Delta u$ or $\Delta h$ calculations. The $u$ and $h$ data are given in kJ/kg for air (Table A–17) and usually in kJ/kmol for other gases. The unit kJ/kmol is very convenient in the thermodynamic analysis of chemical reactions.

Some observations can be made from Fig. 4–24. First, the specific heats of gases with complex molecules (molecules with two or more atoms) are higher and increase with temperature. Also, the variation of specific heats with temperature is smooth and may be approximated as linear over small temperature intervals (a few hundred degrees or less). Therefore, the specific heat functions in Eqs. 4–25 and 4–26 can be replaced by the constant average specific heat values. Then, the integrations in these equations can be performed, yielding

$$u_2 - u_1 = c_{v,\text{avg}}(T_2 - T_1) \quad \text{(kJ/kg)} \tag{4-27}$$

and

$$h_2 - h_1 = c_{p,\text{avg}}(T_2 - T_1) \quad \text{(kJ/kg)} \tag{4-28}$$

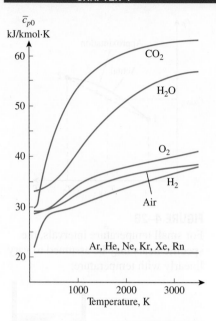

**FIGURE 4–24**

Ideal-gas constant-pressure specific heats for some gases (see Table A–2c for $c_p$ equations).

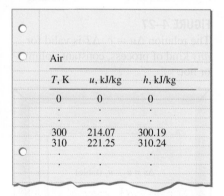

| Air | | |
|---|---|---|
| $T$, K | $u$, kJ/kg | $h$, kJ/kg |
| 0 | 0 | 0 |
| . | . | . |
| . | . | . |
| . | . | . |
| 300 | 214.07 | 300.19 |
| 310 | 221.25 | 310.24 |
| . | . | . |
| . | . | . |

**FIGURE 4–25**

In the preparation of ideal-gas tables, 0 K is chosen as the reference temperature.

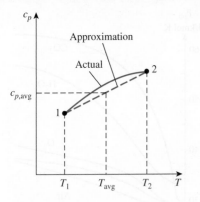

**FIGURE 4–26**
For small temperature intervals, the specific heats may be assumed to vary linearly with temperature.

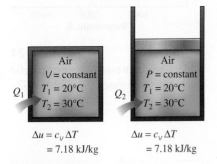

**FIGURE 4–27**
The relation $\Delta u = c_v \Delta T$ is valid for *any* kind of process, constant-volume or not.

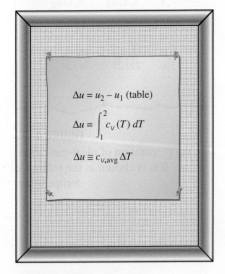

**FIGURE 4–28**
Three ways of calculating $\Delta u$.

The specific heat values for some common gases are listed as a function of temperature in Table A–2b. The average specific heats $c_{p,\text{avg}}$ and $c_{v,\text{avg}}$ are evaluated from this table at the average temperature $(T_1 + T_2)/2$, as shown in Fig. 4–26. If the final temperature $T_2$ is not known, the specific heats may be evaluated at $T_1$ or at the anticipated average temperature. Then $T_2$ can be determined by using these specific heat values. The value of $T_2$ can be refined, if necessary, by evaluating the specific heats at the new average temperature.

Another way of determining the average specific heats is to evaluate them at $T_1$ and $T_2$ and then take their average. Usually both methods give reasonably good results, and one is not necessarily better than the other.

Another observation that can be made from Fig. 4–24 is that the ideal-gas specific heats of *monatomic gases* such as argon, neon, and helium remain constant over the entire temperature range. Thus, $\Delta u$ and $\Delta h$ of monatomic gases can easily be evaluated from Eqs. 4–27 and 4–28.

Note that the $\Delta u$ and $\Delta h$ relations given previously are not restricted to any kind of process. They are valid for all processes. The presence of the constant-volume specific heat $c_v$ in an equation should not lead one to believe that this equation is valid for a constant-volume process only. On the contrary, the relation $\Delta u = c_{v,\text{avg}} \Delta T$ is valid for *any* ideal gas undergoing *any* process (Fig. 4–27). A similar argument can be given for $c_p$ and $\Delta h$.

To summarize, there are three ways to determine the internal energy and enthalpy changes of ideal gases (Fig. 4–28):

1. By using the tabulated $u$ and $h$ data. This is the easiest and most accurate way when tables are readily available.

2. By using the $c_v$ or $c_p$ relations as a function of temperature and performing the integrations. This is very inconvenient for hand calculations but quite desirable for computerized calculations. The results obtained are very accurate.

3. By using average specific heats. This is very simple and certainly very convenient when property tables are not available. The results obtained are reasonably accurate if the temperature interval is not very large.

## Specific Heat Relations of Ideal Gases

A special relationship between $c_p$ and $c_v$ for ideal gases can be obtained by differentiating the relation $h = u + RT$, which yields

$$dh = du + R\, dT$$

Replacing $dh$ by $c_p dT$ and $du$ by $c_v dT$ and dividing the resulting expression by $dT$, we obtain

$$c_p = c_v + R \quad \text{(kJ/kg·K)} \tag{4–29}$$

This is an important relationship for ideal gases since it enables us to determine $c_v$ from a knowledge of $c_p$ and the gas constant $R$.

When the specific heats are given on a molar basis, $R$ in the above equation should be replaced by the universal gas constant $R_u$ (Fig. 4–29).

$$\bar{c}_p = \bar{c}_v + R_u \quad \text{(kJ/kmol·K)} \tag{4–30}$$

At this point, we introduce another ideal-gas property called the **specific heat ratio** $k$, defined as

$$k = \frac{c_p}{c_v} \tag{4-31}$$

The specific ratio also varies with temperature, but this variation is very mild. For monatomic gases, its value is essentially constant at 1.667. Many diatomic gases, including air, have a specific heat ratio of about 1.4 at room temperature.

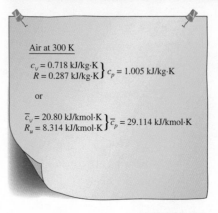

Air at 300 K

$\left.\begin{array}{l} c_v = 0.718 \text{ kJ/kg·K} \\ R = 0.287 \text{ kJ/kg·K} \end{array}\right\}\ c_p = 1.005 \text{ kJ/kg·K}$

or

$\left.\begin{array}{l} \bar{c}_v = 20.80 \text{ kJ/kmol·K} \\ R_u = 8.314 \text{ kJ/kmol·K} \end{array}\right\}\ \bar{c}_p = 29.114 \text{ kJ/kmol·K}$

**FIGURE 4–29**
The $c_p$ of an ideal gas can be determined from a knowledge of $c_v$ and $R$.

■ **EXAMPLE 4–7**    **Evaluation of the $\Delta u$ of an Ideal Gas**

Air at 300 K and 200 kPa is heated at constant pressure to 600 K. Determine the change in internal energy of air per unit mass, using (a) data from the air table (Table A–17), (b) the functional form of the specific heat (Table A–2c), and (c) the average specific heat value (Table A–2b).

**SOLUTION**  The internal energy change of air is to be determined in three different ways.
**Assumptions**  At specified conditions, air can be considered to be an ideal gas since it is at a high temperature and low pressure relative to its critical-point values.
**Analysis**  The internal energy change $\Delta u$ of ideal gases depends on the initial and final temperatures only, and not on the type of process. Thus, the following solution is valid for any kind of process.
(a) One way of determining the change in internal energy of air is to read the $u$ values at $T_1$ and $T_2$ from Table A–17 and take the difference:

$$u_1 = u_{@\,300\,K} = 214.07 \text{ kJ/kg}$$

$$u_2 = u_{@\,600\,K} = 434.78 \text{ kJ/kg}$$

Thus,

$$\Delta u = u_2 - u_1 = (434.78 - 214.07)\text{ kJ/kg} = \textbf{220.71 kJ/kg}$$

(b) The $\bar{c}_p(T)$ of air is given in Table A–2c in the form of a third-degree polynomial expressed as

$$\bar{c}_p(T) = a + bT + cT^2 + dT^3$$

where $a = 28.11$, $b = 0.1967 \times 10^{-2}$, $c = 0.4802 \times 10^{-5}$, and $d = -1.966 \times 10^{-9}$. From Eq. 4–30,

$$\bar{c}_v(T) = \bar{c}_p - R_u = (a - R_u) + bT + cT^2 + dT^3$$

From Eq. 4–25,

$$\Delta\bar{u} = \int_1^2 \bar{c}_v(T)\,dT = \int_{T_1}^{T_2} [(a - R_u) + bT + cT^2 + dT^3]\,dT$$

Performing the integration and substituting the values, we obtain

$$\Delta\bar{u} = 6447 \text{ kJ/kmol}$$

The change in the internal energy on a unit-mass basis is determined by dividing this value by the molar mass of air (Table A–1):

$$\Delta u = \frac{\Delta \bar{u}}{M} = \frac{6447 \text{ kJ/kmol}}{28.97 \text{ kg/kmol}} = \textbf{222.5 kJ/kg}$$

which differs from the tabulated value by 0.8 percent.

(c) The average value of the constant-volume specific heat $c_{v,\text{avg}}$ is determined from Table A–2b at the average temperature of $(T_1 + T_2)/2 = 450$ K to be

$$c_{v,\text{avg}} = c_{v \, @ \, 450 \text{ K}} = 0.733 \text{ kJ/kg·K}$$

Thus,

$$\Delta u = c_{v,\text{avg}}(T_2 - T_1) = (0.733 \text{ kJ/kg·K})[(600 - 300)\text{K}]$$
$$= 220 \text{ kJ/kg}$$

**Discussion** This answer differs from the tabulated value (220.71 kJ/kg) by only 0.4 percent. This close agreement is not surprising since the assumption that $c_v$ varies linearly with temperature is a reasonable one at temperature intervals of only a few hundred degrees. If we had used the $c_v$ value at $T_1 = 300$ K instead of at $T_{\text{avg}}$, the result would be 215.4 kJ/kg, which is in error by about 2 percent. Errors of this magnitude are acceptable for most engineering purposes.

---

*EXAMPLE 4–8*     **Heating of a Gas in a Tank by Stirring**

An insulated rigid tank initially contains 0.7 kg of helium at 27°C and 350 kPa. A paddle wheel with a power rating of 0.015 kW is operated within the tank for 30 min. Determine (a) the final temperature and (b) the final pressure of the helium gas.

**SOLUTION** Helium gas in an insulated rigid tank is stirred by a paddle wheel. The final temperature and pressure of helium are to be determined.
*Assumptions* **1** Helium is an ideal gas since it is at a very high temperature relative to its critical-point value of −268°C. **2** Constant specific heats can be used for helium. **3** The system is stationary and thus the kinetic and potential energy changes are zero, $\Delta \text{KE} = \Delta \text{PE} = 0$ and $\Delta E = \Delta U$. **4** The volume of the tank is constant, and thus there is no boundary work. **5** The system is adiabatic and thus there is no heat transfer.
*Analysis* We take the contents of the tank as the *system* (Fig. 4–30). This is a *closed system* since no mass crosses the system boundary during the process. We observe that there is shaft work done on the system.

(a) The amount of paddle-wheel work done on the system is

$$W_{\text{sh}} = \dot{W}_{\text{sh}} \Delta t = (0.015 \text{ kW})(30 \text{ min}) \left(\frac{60 \text{ s}}{1 \text{ min}}\right) = 27 \text{ kJ}$$

Under the stated assumptions and observations, the energy balance on the system can be expressed as

$$\underbrace{E_{\text{in}} - E_{\text{out}}}_{\substack{\text{Net energy transfer} \\ \text{by heat, work, and mass}}} = \underbrace{\Delta E_{\text{system}}}_{\substack{\text{Change in internal, kinetic,} \\ \text{potential, etc., energies}}}$$

$$W_{\text{sh,in}} = \Delta U = m(u_2 - u_1) = mc_{v,\text{avg}}(T_2 - T_1)$$

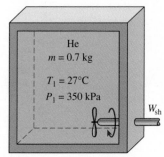

He
$m = 0.7$ kg
$T_1 = 27°C$
$P_1 = 350$ kPa

$W_{\text{sh}}$

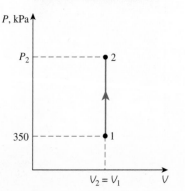

**FIGURE 4–30**
Schematic and *P-V* diagram for Example 4–8.

As we pointed out earlier, the ideal-gas specific heats of monatomic gases (helium being one of them) are constant. The $c_v$ value of helium is determined from Table A–2a to be $c_v = 3.1156$ kJ/kg·°C. Substituting this and other known quantities into the above equation, we obtain

$$27 \text{ kJ} = (0.7 \text{ kg})(3.1156 \text{ kJ/kg·°C})(T_2 - 27\text{°C})$$

$$T_2 = 39.4\text{°C}$$

(b) The final pressure is determined from the ideal-gas relation

$$\frac{P_1 V_1}{T_1} = \frac{P_2 V_2}{T_2}$$

where $V_1$ and $V_2$ are identical and cancel out. Then the final pressure becomes

$$\frac{350 \text{ kPa}}{(27 + 273) \text{ K}} = \frac{P_2}{(39.4 + 273)\text{K}}$$

$$P_2 = \mathbf{364 \text{ kPa}}$$

**Discussion** Note that the pressure in the ideal-gas relation is always the absolute pressure.

---

### ■ EXAMPLE 4–9　　Heating of a Gas by a Resistance Heater

A piston–cylinder device initially contains 0.5 m³ of nitrogen gas at 400 kPa and 27°C. An electric heater within the device is turned on and is allowed to pass a current of 2 A for 5 min from a 120-V source. Nitrogen expands at constant pressure, and a heat loss of 2800 J occurs during the process. Determine the final temperature of nitrogen.

**SOLUTION** Nitrogen gas in a piston–cylinder device is heated by an electric resistance heater. Nitrogen expands at constant pressure while some heat is lost. The final temperature of nitrogen is to be determined.
**Assumptions** **1** Nitrogen is an ideal gas since it is at a high temperature and low pressure relative to its critical-point values of −147°C, and 3.39 MPa. **2** The system is stationary and thus the kinetic and potential energy changes are zero, $\Delta KE = \Delta PE = 0$ and $\Delta E = \Delta U$. **3** The pressure remains constant during the process and thus $P_2 = P_1$. **4** Nitrogen has constant specific heats at room temperature.
**Analysis** We take the contents of the cylinder as the *system* (Fig. 4–31). This is a *closed system* since no mass crosses the system boundary during the process. We observe that a piston–cylinder device typically involves a moving boundary and thus boundary work, $W_b$. Also, heat is lost from the system and electrical work $W_e$ is done on the system.
First, let us determine the electrical work done on the nitrogen:

$$W_e = \mathbf{V}I \Delta t = (120 \text{ V})(2 \text{ A})(5 \times 60 \text{ s})\left(\frac{1 \text{ kJ/s}}{1000 \text{ VA}}\right) = 72 \text{ kJ}$$

The mass of nitrogen is determined from the ideal-gas relation:

$$m = \frac{P_1 V_1}{RT_1} = \frac{(400 \text{ kPa})(0.5 \text{ m}^3)}{(0.297 \text{ kPa·m}^3/\text{kg·K})(300 \text{ K})} = 2.245 \text{ kg}$$

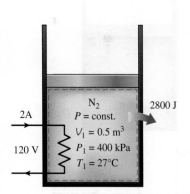

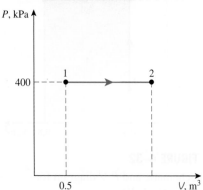

**FIGURE 4–31**
Schematic and *P-V* diagram for Example 4–9.

Under the stated assumptions and observations, the energy balance on the system can be expressed as

$$\underbrace{E_{in} - E_{out}}_{\substack{\text{Net energy transfer} \\ \text{by heat, work, and mass}}} = \underbrace{\Delta E_{system}}_{\substack{\text{Change in internal, kinetic,} \\ \text{potential, etc., energies}}}$$

$$W_{e,in} - Q_{out} - W_{b,out} = \Delta U$$

$$W_{e,in} - Q_{out} = \Delta H = m(h_2 - h_1) = mc_p(T_2 - T_1)$$

since $\Delta U + W_b = \Delta H$ for a closed system undergoing a quasi-equilibrium expansion or compression process at constant pressure. From Table A–2a, $c_p = 1.039$ kJ/kg·K for nitrogen at room temperature. The only unknown quantity in the previous equation is $T_2$, and it is found to be

$$72 \text{ kJ} - 2.8 \text{ kJ} = (2.245 \text{ kg})(1.039 \text{ kJ/kg·K})(T_2 - 27°C)$$

$$T_2 = \mathbf{56.7°C}$$

***Discussion*** Note that we could also solve this problem by determining the boundary work and the internal energy change rather than the enthalpy change.

---

### EXAMPLE 4–10    Heating of a Gas at Constant Pressure

A piston–cylinder device initially contains air at 150 kPa and 27°C. At this state, the piston is resting on a pair of stops, as shown in Fig. 4–32, and the enclosed volume is 400 L. The mass of the piston is such that a 350-kPa pressure is required to move it. The air is now heated until its volume has doubled. Determine (*a*) the final temperature, (*b*) the work done by the air, and (*c*) the total heat transferred to the air.

**SOLUTION** Air in a piston–cylinder device with a set of stops is heated until its volume is doubled. The final temperature, work done, and the total heat transfer are to be determined.

***Assumptions*** **1** Air is an ideal gas since it is at a high temperature and low pressure relative to its critical-point values. **2** The system is stationary and thus the kinetic and potential energy changes are zero, $\Delta KE = \Delta PE = 0$ and $\Delta E = \Delta U$. **3** The volume remains constant until the piston starts moving, and the pressure remains constant afterwards. **4** There are no electrical, shaft, or other forms of work involved.

***Analysis*** We take the contents of the cylinder as the *system* (Fig. 4–32). This is a *closed system* since no mass crosses the system boundary during the process. We observe that a piston-cylinder device typically involves a moving boundary and thus boundary work, $W_b$. Also, the boundary work is done by the system, and heat is transferred to the system.

(*a*) The final temperature can be determined easily by using the ideal-gas relation between states 1 and 3 in the following form:

$$\frac{P_1 V_1}{T_1} = \frac{P_3 V_3}{T_3} \longrightarrow \frac{(150 \text{ kPa})(V_1)}{300 \text{ K}} = \frac{(350 \text{ kPa})(2V_1)}{T_3}$$

$$T_3 = \mathbf{1400 \text{ K}}$$

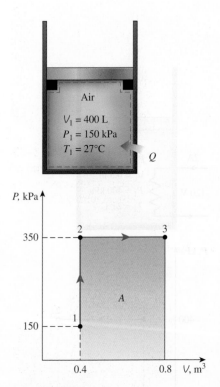

**FIGURE 4–32**

Schematic and *P-V* diagram for Example 4–10.

(b) The work done could be determined by integration, but for this case it is much easier to find it from the area under the process curve on a P-V diagram, shown in Fig. 4–32:

$$A = (V_2 - V_1)P_2 = (0.4 \text{ m}^3)(350 \text{ kPa}) = 140 \text{ m}^3 \cdot \text{kPa}$$

Therefore,

$$W_{13} = 140 \text{ kJ}$$

The work is done by the system (to raise the piston and to push the atmospheric air out of the way), and thus it is work output.

(c) Under the stated assumptions and observations, the energy balance on the system between the initial and final states (process 1–3) can be expressed as

$$\underbrace{E_{\text{in}} - E_{\text{out}}}_{\substack{\text{Net energy transfer} \\ \text{by heat, work, and mass}}} = \underbrace{\Delta E_{\text{system}}}_{\substack{\text{Change in internal, kinetic,} \\ \text{potential, etc., energies}}}$$

$$Q_{\text{in}} - W_{b,\text{out}} = \Delta U = m(u_3 - u_1)$$

The mass of the system can be determined from the ideal-gas relation:

$$m = \frac{P_1 V_1}{RT_1} = \frac{(150 \text{ kPa}) (0.4 \text{ m}^3)}{(0.287 \text{ kPa} \cdot \text{m}^3/\text{kg} \cdot \text{K}) (300 \text{ K})} = 0.697 \text{ kg}$$

The internal energies are determined from the air table (Table A–17) to be

$$u_1 = u_{@\,300\,\text{K}} = 214.07 \text{ kJ/kg}$$

$$u_3 = u_{@\,1400\,\text{K}} = 1113.52 \text{ kJ/kg}$$

Thus,

$$Q_{\text{in}} - 140 \text{ kJ} = (0.697 \text{ kg})[(1113.52 - 214.07) \text{ kJ/kg}]$$

$$Q_{\text{in}} = 767 \text{ kJ}$$

**Discussion** The positive sign verifies that heat is transferred to the system.

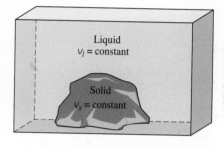

**FIGURE 4–33**
The specific volumes of incompressible substances remain constant during a process.

# 4–5 · INTERNAL ENERGY, ENTHALPY, AND SPECIFIC HEATS OF SOLIDS AND LIQUIDS

A substance whose specific volume (or density) is constant is called an **incompressible substance**. The specific volumes of solids and liquids essentially remain constant during a process (Fig. 4–33). Therefore, liquids and solids can be approximated as incompressible substances without sacrificing much in accuracy. The constant-volume assumption should be taken to imply that the energy associated with the volume change is negligible compared with other forms of energy. Otherwise, this assumption would be ridiculous for studying the thermal stresses in solids (caused by volume change with temperature) or analyzing liquid-in-glass thermometers.

It can be mathematically shown that (see Chap. 12) the constant-volume and constant-pressure specific heats are identical for incompressible substances (Fig. 4–34). Therefore, for solids and liquids, the subscripts on $c_p$

**FIGURE 4–34**
The $c_v$ and $c_p$ values of incompressible substances are identical and are denoted by c.

and $c_v$ can be dropped, and both specific heats can be represented by a single symbol $c$. That is,

$$c_p = c_v = c \tag{4–32}$$

This result could also be deduced from the physical definitions of constant-volume and constant-pressure specific heats. Specific heat values for several common liquids and solids are given in Table A–3.

## Internal Energy Changes

Like those of ideal gases, the specific heats of incompressible substances depend on temperature only. Thus, the partial differentials in the defining equation of $c_v$ can be replaced by ordinary differentials, which yield

$$du = c_v\,dT = c(T)\,dT \tag{4–33}$$

The change in internal energy between states 1 and 2 is then obtained by integration:

$$\Delta u = u_2 - u_1 = \int_1^2 c(T)\,dT \quad \text{(kJ/kg)} \tag{4–34}$$

The variation of specific heat $c$ with temperature should be known before this integration can be carried out. For small temperature intervals, a $c$ value at the average temperature can be used and treated as a constant, yielding

$$\Delta u \cong c_{\text{avg}}(T_2 - T_1) \quad \text{(kJ/kg)} \tag{4–35}$$

## Enthalpy Changes

Using the definition of enthalpy $h = u + Pv$ and noting that $v = \text{constant}$, the differential form of the enthalpy change of incompressible substances can be determined by differentiation to be

$$dh = du + v\,dP + P\,dv^{\nearrow 0} = du + v\,dP \tag{4–36}$$

Integrating,

$$\Delta h = \Delta u + v\,\Delta P \cong c_{\text{avg}}\,\Delta T + v\,\Delta P \quad \text{(kJ/kg)} \tag{4–37}$$

For *solids*, the term $v\,\Delta P$ is insignificant and thus $\Delta h = \Delta u \cong c_{\text{avg}}\Delta T$. For *liquids*, two special cases are commonly encountered:

1. *Constant-pressure processes*, as in heaters ($\Delta P = 0$): $\Delta h = \Delta u \cong c_{\text{avg}}\Delta T$
2. *Constant-temperature processes*, as in pumps ($\Delta T = 0$): $\Delta h = v\,\Delta P$

For a process between states 1 and 2, the last relation can be expressed as $h_2 - h_1 = v(P_2 - P_1)$. By taking state 2 to be the compressed liquid state at a given $T$ and $P$ and state 1 to be the saturated liquid state at the same temperature, the enthalpy of the compressed liquid can be expressed as

$$h_{@P,T} \cong h_{f@T} + v_{f@T}(P - P_{\text{sat}@T}) \tag{4–38}$$

as discussed in Chap. 3. This is an improvement over the assumption that the enthalpy of the compressed liquid could be taken as $h_f$ at the given temperature (that is, $h_{@P,T} \cong h_{f@T}$). However, the contribution of the last term is often very small, and is neglected. (Note that at high temperature and pressures, Eq. 4–38 may overcorrect the enthalpy and result in a larger error than the approximation $h \cong h_{f @ T}$.)

■
■ **EXAMPLE 4–11    Enthalpy of Compressed Liquid**
■
■ Determine the enthalpy of liquid water at 100°C and 15 MPa (a) by using
■ compressed liquid tables, (b) by approximating it as a saturated liquid, and
■ (c) by using the correction given by Eq. 4–38.

**SOLUTION**  The enthalpy of liquid water is to be determined exactly and approximately.

**Analysis**  At 100°C, the saturation pressure of water is 101.42 kPa, and since $P > P_{sat}$, the water exists as a compressed liquid at the specified state.

(a) From compressed liquid tables, we read

$$\left. \begin{array}{l} P = 15 \text{ MPa} \\ T = 100°C \end{array} \right\} \quad h = 430.39 \text{ kJ/kg} \quad \text{(Table A–7)}$$

This is the exact value.

(b) Approximating the compressed liquid as a saturated liquid at 100°C, as is commonly done, we obtain

$$h \cong h_{f @ 100°C} = 419.17 \text{ kJ/kg}$$

This value is in error by about 2.6 percent.

(c) From Eq. 4–38,

$$h_{@P,T} \cong h_{f @ T} + v_{f @ T}(P - P_{sat @ T})$$

$$= (419.17 \text{ kJ/kg}) + (0.001 \text{ m}^3 \text{ kg})[(15,000 - 101.42) \text{ kPa}]\left(\frac{1 \text{ kJ}}{1 \text{ kPa·m}^3}\right)$$

$$= 434.07 \text{ kJ/kg}$$

**Discussion**  Note that the correction term reduced the error from 2.6 to about 1 percent in this case. However, this improvement in accuracy is often not worth the extra effort involved.

■
■ **EXAMPLE 4–12    Cooling of an Iron Block by Water**
■
■ A 50-kg iron block at 80°C is dropped into an insulated tank that contains
■ 0.5 m³ of liquid water at 25°C. Determine the temperature when thermal equilibrium is reached.

**SOLUTION**  An iron block is dropped into water in an insulated tank. The final temperature when thermal equilibrium is reached is to be determined.

**Assumptions**  **1** Both water and the iron block are incompressible substances. **2** Constant specific heats at room temperature can be used for water and the iron. **3** The system is stationary and thus the kinetic and potential energy changes are zero, $\Delta KE = \Delta PE = 0$ and $\Delta E = \Delta U$. **4** There are no electrical, shaft, or other forms of work involved. **5** The system is well-insulated and thus there is no heat transfer.

**FIGURE 4–35**
Schematic for Example 4–12.

*Analysis* We take the entire contents of the tank as the *system* (Fig. 4–35). This is a *closed system* since no mass crosses the system boundary during the process. We observe that the volume of a rigid tank is constant, and thus there is no boundary work. The energy balance on the system can be expressed as

$$\underbrace{E_{in} - E_{out}}_{\substack{\text{Net energy transfer} \\ \text{by heat, work, and mass}}} = \underbrace{\Delta E_{system}}_{\substack{\text{Change in internal, kinetic,} \\ \text{potential, etc., energies}}}$$

$$0 = \Delta U$$

The total internal energy $U$ is an extensive property, and therefore it can be expressed as the sum of the internal energies of the parts of the system. Then the total internal energy change of the system becomes

$$\Delta U_{sys} = \Delta U_{iron} + \Delta U_{water} = 0$$

$$[mc(T_2 - T_1)]_{iron} + [mc(T_2 - T_1)]_{water} = 0$$

The specific volume of liquid water at or about room temperature can be taken to be 0.001 m³/kg. Then the mass of the water is

$$m_{water} = \frac{V}{v} = \frac{0.5 \text{ m}^3}{0.001 \text{ m}^3/\text{kg}} = 500 \text{ kg}$$

The specific heats of iron and liquid water are determined from Table A–3 to be $c_{iron}$ = 0.45 kJ/kg·°C and $c_{water}$ = 4.18 kJ/kg·°C. Substituting these values into the energy equation, we obtain

$$(50 \text{ kg})(0.45 \text{ kJ/kg·°C})(T_2 - 80°C) + (500 \text{ kg})(4.18 \text{ kJ/kg·°C})(T_2 - 25°C) = 0$$

$$T_2 = \textbf{25.6°C}$$

Therefore, when thermal equilibrium is established, both the water and iron will be at 25.6°C.
*Discussion* The small rise in water temperature is due to its large mass and large specific heat.

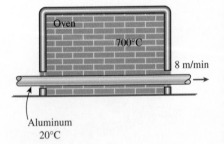

**FIGURE 4–36**
Schematic for Example 4–13

***EXAMPLE 4–13***     **Heating of Aluminum Rods in a Furnace**

Long cylindrical aluminum rods ($\rho$ = 2700 kg/m³ and $c_p$ = 0.973 kJ/kg·K) of 5-cm diameter are heat treated from 20°C to an average temperature of 400°C by drawing them at a velocity of 8 m/min through a long oven. Determine the rate of heat transfer to the rods in the oven.

**SOLUTION** Aluminum rods are to be heated in an oven to a specified average temperature. The rate of heat transfer to the rods is to be determined.
*Assumptions* **1** The thermal properties of the rods are constant. **2** There are no changes in kinetic and potential energies. **3** The balls are at a uniform temperature when they leave the oven.
*Analysis* Aluminum rods pass through the oven at a constant speed of 8 m/min. That is, an external observer will see that an 8-m long section of cold rods enters and an 8-m long section of hot rods leaves the oven

every minute. We take the 8-m long section of the rod as the system. The energy balance for this closed system can be expressed as

$$\underbrace{E_{\text{in}} - E_{\text{out}}}_{\substack{\text{Net energy transfer} \\ \text{by heat, work, and mass}}} = \underbrace{\Delta E_{\text{system}}}_{\substack{\text{Change in internal, kinetic,} \\ \text{potential, etc., energies}}}$$

$$Q_{\text{in}} = \Delta U_{\text{rod}} = m(u_2 - u_1)$$
$$Q_{\text{in}} = mc(T_2 - T_1)$$

The density and specific heat of the rods are given to be $\rho = 2700$ kg/m$^3$ and $c = 0.973$ kJ/kg·K $= 0.973$ kJ/kg·°C. The amount of heat transfer to an 8-m long section of the rod as it is heated to the specified temperature is determined to be

$$m = \rho V = \rho \frac{\pi D^2}{4} L = (2700 \text{ kg/m}^3) \frac{\pi(0.05 \text{ m})^2}{4} (8 \text{ m}) = 42.41 \text{ kg}$$

$$Q_{\text{in}} = mc\,(T_2 - T_1) = (42.41 \text{ kg}) (0.973 \text{ kJ/kg·°C}) (400 - 20)°\text{C}$$
$$= 15,680 \text{ kJ (per 8-m section)}$$

Considering that an 8-m long section of the rods is heated every minute, the rate of heat transfer to the rods in the oven becomes

$$\dot{Q}_{\text{in}} = Q_{\text{in}}/\Delta t = 15,680 \text{ kJ/min} = \textbf{261 kJ/s}$$

*Discussion* This problem can also be solved by working with the *rate* form of the equations as

$$\dot{m} = \rho \dot{V} = \rho \frac{\pi D^2}{4} L/\Delta t = \rho \frac{\pi D^2}{4} V = (2700 \text{ kg/m}^3) \frac{\pi(0.05 \text{ m})^2}{4} (8 \text{ m/min})$$
$$= 42.41 \text{ kg/min}$$

$$\dot{Q}_{\text{in}} = \dot{m}c(T_2 - T_1) = (42.41 \text{ kg/min}) (0.973 \text{ kJ/kg·°C}) (400 - 20)°\text{C}$$
$$= 15,680 \text{ kJ/min}$$

which is identical to the result obtained before.

## TOPIC OF SPECIAL INTEREST*    Thermodynamic Aspects of Biological Systems

An important and exciting application area of thermodynamics is biological systems, which are the sites of rather complex and intriguing energy transfer and transformation processes. Biological systems are not in thermodynamic equilibrium, and thus they are not easy to analyze. Despite their complexity, biological systems are primarily made up of four simple elements: hydrogen, oxygen, carbon, and nitrogen. In the human body, hydrogen accounts for 63 percent, oxygen 25.5 percent, carbon 9.5 percent, and nitrogen 1.4 percent of all the atoms. The remaining 0.6 percent of the atoms comes from 20 other elements essential for life. By mass, about 72 percent of the human body is water.

The building blocks of living organisms are *cells*, which resemble miniature factories performing functions that are vital for the survival of organisms.

*This section can be skipped without a loss in continuity.

**FIGURE 4–37**
An average person dissipates energy to the surroundings at a rate of 84 W when resting.

©*Janis Christie/Getty Images RF*

**FIGURE 4–38**
Two fast-dancing people supply more energy to a room than a 1-kW electric resistance heater.

A biological system can be as simple as a single cell. The human body contains about 100 trillion cells with an average diameter of 0.01 mm. The membrane of the cell is a semipermeable wall that allows some substances to pass through it while excluding others.

In a typical cell, thousands of chemical reactions occur every second during which some molecules are broken down and energy is released and some new molecules are formed. This high level of chemical activity in the cells, which maintains the human body at a temperature of 37°C while performing the necessary bodily tasks, is called **metabolism**. In simple terms, metabolism refers to the burning of foods such as carbohydrates, fat, and protein. The rate of metabolism in the resting state is called the *basal metabolic rate*, which is the rate of metabolism required to keep a body performing the necessary functions (such as breathing and blood circulation) at zero external activity level. The metabolic rate can also be interpreted as the energy consumption rate for a body. For an average male (30 years old, 70 kg, 1.8-m² body surface area), the basal metabolic rate is 84 W. That is, the body dissipates energy to the environment at a rate of 84 W, which means that the body is converting chemical energy of the food (or of the body fat if the person has not eaten) into thermal energy at a rate of 84 W (Fig. 4–37). The metabolic rate increases with the level of activity, and it may exceed 10 times the basal metabolic rate when a body is doing strenuous exercise. That is, two people doing heavy exercising in a room may be supplying more energy to the room than a 1-kW electrical resistance heater (Fig. 4–38). The fraction of sensible heat varies from about 40 percent in the case of heavy work to about 70 percent in the case of light work. The rest of the energy is rejected from the body by perspiration in the form of latent heat.

The basal metabolic rate varies with sex, body size, general health conditions, and so forth, and decreases considerably with age. This is one of the reasons people tend to put on weight in their late twenties and thirties even though they do not increase their food intake. The brain and the liver are the major sites of metabolic activity. These two organs are responsible for almost 50 percent of the basal metabolic rate of an adult human body although they constitute only about 4 percent of the body mass. In small children, it is remarkable that about half of the basal metabolic activity occurs in the brain alone.

The biological reactions in cells occur essentially at constant temperature, pressure, and volume. The temperature of the cell tends to rise when some chemical energy is converted to heat, but this energy is quickly transferred to the circulatory system, which transports it to outer parts of the body and eventually to the environment through the skin.

The muscle cells function very much like an engine, converting the chemical energy into mechanical energy (work) with a conversion efficiency of close to 20 percent. When the body does no net work on the environment (such as moving some furniture upstairs), the entire work is also converted to heat. In that case, the entire chemical energy in the food released during metabolism in the body is eventually transferred to the environment. A TV set that consumes electricity at a rate of 300 W must reject heat to its environment at a rate of 300 W in steady operation regardless of what goes on in the set. That is, turning on a 300-W TV set or three 100-W light bulbs will produce the same heating effect in a room as a 300-W resistance heater

(Fig. 4–39). This is a consequence of the conservation of energy principle, which requires that the energy input into a system must equal the energy output when the total energy content of a system remains constant during a process.

## Food and Exercise

The energy requirements of a body are met by the food we eat. The nutrients in the food are considered in three major groups: carbohydrates, proteins, and fats. *Carbohydrates* are characterized by having hydrogen and oxygen atoms in a 2:1 ratio in their molecules. The molecules of carbohydrates range from very simple (as in plain sugar) to very complex or large (as in starch). Bread and plain sugar are the major sources of carbohydrates. *Proteins* are very large molecules that contain carbon, hydrogen, oxygen, and nitrogen, and they are essential for the building and repairing of the body tissues. Proteins are made up of smaller building blocks called *amino acids*. Complete proteins such as meat, milk, and eggs have all the amino acids needed to build body tissues. Plant source proteins such as those in fruits, vegetables, and grains lack one or more amino acids, and are called incomplete proteins. *Fats* are relatively small molecules that consist of carbon, hydrogen, and oxygen. Vegetable oils and animal fats are major sources of fats. Most foods we eat contain all three nutrition groups at varying amounts. The typical average American diet consists of 45 percent carbohydrate, 40 percent fat, and 15 percent protein, although it is recommended that in a healthy diet less than 30 percent of the calories should come from fat.

The energy content of a given food is determined by burning a small sample of the food in a device called a *bomb calorimeter*, which is basically a well-insulated rigid tank (Fig. 4–40). The tank contains a small combustion chamber surrounded by water. The food is ignited and burned in the combustion chamber in the presence of excess oxygen, and the energy released is transferred to the surrounding water. The energy content of the food is calculated on the basis of the conservation of energy principle by measuring the temperature rise of the water. The carbon in the food is converted into $CO_2$ and hydrogen into $H_2O$ as the food burns. The same chemical reactions occur in the body, and thus the same amount of energy is released.

Using dry (free of water) samples, the average energy contents of the three basic food groups are determined by bomb calorimeter measurements to be 18.0 MJ/kg for carbohydrates, 22.2 MJ/kg for proteins, and 39.8 MJ/kg for fats. These food groups are not entirely metabolized in the human body, however. The fraction of metabolizable energy contents are 95.5 percent for carbohydrates, 77.5 percent for proteins, and 97.7 percent for fats. That is, the fats we eat are almost entirely metabolized in the body, but close to one quarter of the protein we eat is discarded from the body unburned. This corresponds to 4.1 Calories/g for proteins and carbohydrates and 9.3 Calories/g for fats (Fig. 4–41) commonly seen in nutrition books and on food labels. The energy contents of the foods we normally eat are much lower than the values above because of the large water content (water adds bulk to the food but it cannot be metabolized or burned, and thus it has no energy value). Most vegetables, fruits, and meats, for example, are mostly water. The average metabolizable energy

**FIGURE 4–39**
Some arrangements that supply a room the same amount of energy as a 300-W electric resistance heater.

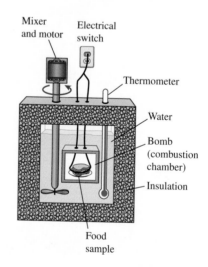

**FIGURE 4–40**
Schematic of a bomb calorimeter used to determine the energy content of food samples.

3 cookies (32 g)

Fat: (8 g)(9.3 Cal/g) = 74.4 Cal
Protein: (2 g)(4.1 Cal/g) = 8.2 Cal
Carbohydrates: (21 g)(4.1 Cal/g) = 86.1 Cal
Other: (1 g)(0 Cal/g) = 0

TOTAL (for 32 g): 169 Cal

**FIGURE 4–41**
Evaluating the calorie content of one serving of chocolate chip cookies (values are for Chips Ahoy cookies made by Nabisco).

©Comstock/Punchstock RF

contents of the three basic food groups are 4.2 MJ/kg for carbohydrates, 8.4 MJ/kg for proteins, and 33.1 MJ/kg for fats. Note that 1 kg of natural fat contains almost 8 times the metabolizable energy of 1 kg of natural carbohydrates. Thus, a person who fills his stomach with fatty foods is consuming much more energy than a person who fills his stomach with carbohydrates such as bread or rice.

The metabolizable energy content of foods is usually expressed by nutritionists in terms of the capitalized *Calories*. One Calorie is equivalent to one *kilocalorie* (1000 calories), which is equivalent to 4.1868 kJ. That is,

$$1 \text{ Cal (Calorie)} = 1000 \text{ calories} = 1 \text{ kcal (kilocalorie)} = 4.1868 \text{ kJ}$$

The calorie notation often causes confusion since it is not always followed in the tables or articles on nutrition. When the topic is food or fitness, a calorie normally means a kilocalorie whether it is capitalized or not.

The **daily calorie needs** of people vary greatly with age, gender, the state of health, the activity level, the body weight, and the composition of the body as well as other factors. A small person needs fewer calories than a larger person of the same sex and age. An average man needs about 2400 to 2700 Calories a day. The daily need of an average woman varies from 1800 to 2200 Calories. The daily calorie needs are about 1600 for sedentary women and some older adults; 2000 for sedentary men and most older adults; 2200 for most children, teenage girls, and active women; 2800 for teenage boys, active men, and some very active women; and above 3000 for very active men. The *average* value of calorie intake is usually taken to be 2000 Calories per day. The daily calorie needs of a person can be determined by multiplying the body weight in pounds (which is 2.205 times the body weight in kg) by 11 for a sedentary person, 13 for a moderately active person, 15 for a moderate exerciser or physical laborer, and 18 for an extremely active exerciser or physical laborer. The extra calories a body consumes are usually stored as fat, which serves as the spare energy of the body for use when the energy intake of the body is less than the needed amount.

Like other natural fat, 1 kg of human body fat contains about 33.1 MJ of metabolizable energy. Therefore, a starving person (zero energy intake) who uses up 2200 Calories (9211 kJ) a day can meet his daily energy intake requirements by burning only 9211/33,100 = 0.28 kg of body fat. So it is no surprise that people are known to survive over 100 days without eating. (They still need to drink water, however, to replenish the water lost through the lungs and the skin to avoid the dehydration that may occur in just a few days.) Although the desire to get rid of the excess fat in a thin world may be overwhelming at times, starvation diets are not recommended because the body soon starts to consume its own muscle tissue in addition to fat. A healthy diet should involve regular exercise while allowing a reasonable amount of calorie intake.

The average metabolizable energy contents of various foods and the energy consumption during various activities are given in Tables 4–1 and 4–2. Considering that no two hamburgers are alike, and that no two people walk exactly the same way, there is some uncertainty in these values, as you would expect. Therefore, you may encounter somewhat different values in other books or magazines for the same items.

**TABLE 4–1**

Approximate metabolizable energy content of some common foods
(1 Calorie = 4.1868 kJ)

| Food | Calories | Food | Calories | Food | Calories |
|---|---|---|---|---|---|
| Apple (one, medium) | 70 | Fish sandwich | 450 | Milk (skim, 200 ml) | 76 |
| Baked potato (plain) | 250 | French fries (regular) | 250 | Milk (whole, 200 ml) | 136 |
| Baked potato with cheese | 550 | Hamburger | 275 | Peach (one, medium) | 65 |
| Bread (white, one slice) | 70 | Hot dog | 300 | Pie (one $\frac{1}{8}$ slice, 23 cm diameter) | 300 |
| Butter (one teaspoon) | 35 | Ice cream (100 ml, 10% fat) | 110 | | |
| Cheeseburger | 325 | | | Pizza (large, cheese, one $\frac{1}{8}$ slice) | 350 |
| Chocolate candy bar (20 g) | 105 | Lettuce salad with French dressing | 150 | | |
| Cola (200 ml) | 87 | | | | |
| Egg (one) | 80 | | | | |

The rates of energy consumption listed in Table 4–2 during some activities are for a 68-kg adult. The energy consumed for smaller or larger adults can be determined using the proportionality of the metabolism rate and the body size. For example, the rate of energy consumption by a 68-kg bicyclist is listed in Table 4–2 to be 639 Calories/h. Then the rate of energy consumption by a 50-kg bicyclist is

$$(50 \text{ kg})\frac{639 \text{ Cal/h}}{68 \text{ kg}} = 470 \text{ Cal/h}$$

For a 100-kg person, it would be 940 Cal/h.

The thermodynamic analysis of the human body is rather complicated since it involves mass transfer (during breathing, perspiring, etc.) as well as energy transfer. As such, it should be treated as an open system. However, the energy transfer with mass is difficult to quantify. Therefore, the human body is often modeled as a closed system for simplicity by treating energy transported with mass as just energy transfer. For example, eating is modeled as the transfer of energy into the human body in the amount of the metabolizable energy content of the food.

## Dieting

Most diets are based on *calorie counting*; that is, the conservation of energy principle: a person who consumes more calories than his or her body burns will gain weight whereas a person who consumes less calories than his or her body burns will lose weight. Yet, people who eat whatever they want whenever they want without gaining any weight are living proof that the calorie-counting technique alone does not work in dieting. Obviously there is more to dieting than keeping track of calories. It should be noted that the phrases *weight gain* and *weight loss* are misnomers. The correct phrases should be *mass gain* and *mass loss*. A man who goes to space loses practically all of his weight but none of his mass. When the topic is food and fitness, *weight* is understood to mean *mass*, and weight is expressed in mass units.

Researchers on nutrition proposed several theories on dieting. One theory suggests that some people have very "food efficient" bodies. These people need fewer calories than other people do for the same activity, just like

**TABLE 4–2**

Approximate energy consumption of a 68-kg adult during some activities
(1 Calorie = 4.1868 kJ)

| Activity | Calories/h |
|---|---|
| Basal metabolism | 72 |
| Basketball | 550 |
| Bicycling (21 km/h) | 639 |
| Cross-country skiing (13 km/h) | 936 |
| Driving a car | 180 |
| Eating | 99 |
| Fast dancing | 600 |
| Fast running (13 km/h) | 936 |
| Jogging (8 km/h) | 540 |
| Swimming (fast) | 860 |
| Swimming (slow) | 288 |
| Tennis (advanced) | 480 |
| Tennis (beginner) | 288 |
| Walking (7.2 km/h) | 432 |
| Watching TV | 72 |

a fuel-efficient car needing less fuel for traveling a given distance. It is interesting that we want our cars to be fuel efficient but we do not want the same high efficiency for our bodies. One thing that frustrates the dieters is that the body interprets dieting as *starvation* and starts using the energy reserves of the body more stringently. Shifting from a normal 2000-Calorie daily diet to an 800-Calorie diet without exercise is observed to lower the basal metabolic rate by 10 to 20 percent. Although the metabolic rate returns to normal once the dieting stops, extended periods of low-calorie dieting without adequate exercise may result in the loss of considerable muscle tissue together with fat. With less muscle tissue to burn calories, the metabolic rate of the body declines and stays below normal even after a person starts eating normally. As a result, the person regains the weight he or she has lost in the form of fat, plus more. The basal metabolic rate remains about the same in people who exercise while dieting.

Regular moderate exercise is part of any healthy dieting program for good reason: it builds or preserves muscle tissue that burns calories much faster than the fat tissue does. It is interesting that aerobic exercise continues burning calories for several hours after the workout, raising the overall metabolic rate considerably.

Another theory suggests that people with *too many fat cells* developed during childhood or adolescence are much more likely to gain weight. Some people believe that the fat content of the bodies is controlled by the setting of a "fat control" mechanism, much like the temperature of a house is controlled by the thermostat setting.

Some people put the blame for weight problems simply on the *genes*. Considering that 80 percent of the children of overweight parents are also overweight, heredity may indeed play an important role in the way a body stores fat. Researchers from the University of Washington and the Rockefeller University have identified a gene, called the RIIbeta, that seems to control the rate of metabolism. The body tries to keep the body fat at a particular level, called the **set point**, that differs from person to person (Fig. 4–42). This is done by *speeding up* the metabolism and thus burning extra calories much faster when a person tends to gain weight and by *slowing down* the metabolism and thus burning calories at a slower rate when a person tends to lose weight. Therefore, a person who just became slim burns fewer calories than does a person of the same size who has always been slim. Even exercise does not seem to change that. Then to keep the weight off, the newly slim person should consume no more calories than he or she can burn. Note that in people with high metabolic rates, the body dissipates the extra calories as body heat instead of storing them as fat, and thus there is no violation of the conservation of energy principle.

In some people, a *genetic flaw* is believed to be responsible for the extremely low rates of metabolism. Several studies concluded that losing weight for such people is nearly impossible. That is, obesity is a biological phenomenon. However, even such people will not gain weight unless they eat more than their body can burn. They just must learn to be content with little food to remain slim, and forget about ever having a normal "eating" life. For most people, genetics determine the range of normal weights. A person may end up at the high or low end of that range, depending on

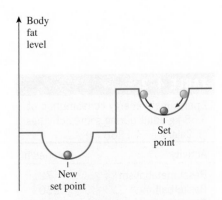

**FIGURE 4–42**
The body tends to keep the body fat level at a *set point* by speeding up metabolism when a person splurges and by slowing it down when the person starves.

eating and exercise habits. This also explains why some genetically identical twins are not so identical when it comes to body weight. *Hormone imbalance* is also believed to cause excessive weight gain or loss.

Based on his experience, the first author of this book has also developed a diet called the *"sensible diet."* It consists of two simple rules: eat *whatever* you want *whenever* you want *as much as* you want provided that (1) you do not eat unless you are hungry and (2) you stop eating before you get stuffed. In other words, *listen to your body and don't impose on it.* Don't expect to see this unscientific diet advertised anywhere since there is nothing to be sold and thus no money to be made. Also, it is not as easy as it sounds since food is at the center stage of most leisure activities in social life, and eating and drinking have become synonymous with having a good time. However, it is comforting to know that the human body is quite forgiving of occasional impositions.

Being *overweight* is associated with a long list of health risks from high blood pressure to some forms of cancer, especially for people who have a weight-related medical condition such as diabetes, hypertension, and heart disease. Therefore, people often wonder if their weight is in the proper range. Well, the answer to this question is not written in stone, but if you cannot see your toes or you can pinch your love handles more than an inch, you don't need an expert to tell you that you went over your range. On the other hand, some people who are obsessed with the weight issue try to lose more weight even though they are actually underweight. Therefore, it is useful to have a scientific criterion to determine physical fitness. The range of healthy weight for adults is usually expressed in terms of the **body mass index** (BMI), defined, in SI units, as

$$\text{BMI} = \frac{W\,(\text{kg})}{H^2\,(\text{m}^2)} \quad \text{with} \quad \begin{array}{l} \text{BMI} < 19 \quad \text{underweight} \\ 19 \leq \text{BMI} \leq 25 \quad \text{healthy weight} \\ \text{BMI} > 25 \quad \text{overweight} \end{array} \quad (4\text{--}39)$$

where $W$ is the weight (actually, the mass) of the person in kg and $H$ is the height in m. Therefore, a BMI of 25 is the upper limit for the healthy weight and a person with a BMI of 27 is 8 percent overweight. It can be shown that the formula above is equivalent in English units to BMI $= 705\,W/H^2$ where $W$ is in pounds and $H$ is in inches. The proper range of weight for adults of various heights is given in Table 4–3 in both SI and English units.

### TABLE 4–3

The range of healthy weight for adults of various heights (Source: National Institute of Health)

| English Units | | SI Units | |
|---|---|---|---|
| Height, in | Healthy Weight, lbm* | Height, m | Healthy weight, kg* |
| 58 | 91–119 | 1.45 | 40–53 |
| 60 | 97–127 | 1.50 | 43–56 |
| 62 | 103–136 | 1.55 | 46–60 |
| 64 | 111–146 | 1.60 | 49–64 |
| 66 | 118–156 | 1.65 | 52–68 |
| 68 | 125–165 | 1.70 | 55–72 |
| 70 | 133–175 | 1.75 | 58–77 |
| 72 | 140–185 | 1.80 | 62–81 |
| 74 | 148–195 | 1.85 | 65–86 |
| 76 | 156–205 | 1.90 | 69–90 |

*The upper and lower limits of healthy range correspond to mass body indexes of 19 and 25, respectively.

### EXAMPLE 4–14    Burning Off Lunch Calories

A 90-kg man had two hamburgers, a regular serving of french fries, and a 200-ml Coke for lunch (Fig. 4–43). Determine how long it will take for him to burn the lunch calories off (*a*) by watching TV and (*b*) by fast swimming. What would your answers be for a 45-kg man?

**SOLUTION**  A man had lunch at a restaurant. The time it will take for him to burn the lunch calories by watching TV and by fast swimming are to be determined.

*Assumptions*  The values in Tables 4–1 and 4–2 are applicable for food and exercise.

**FIGURE 4–43**
A typical lunch discussed in Example 4–14.
©*John A. Rizzo/Getty Images RF*

*Analysis* (a) We take the human body as our *system* and treat it as *a closed system* whose energy content remains unchanged during the process. Then the conservation of energy principle requires that the energy input into the body must be equal to the energy output. The net energy input in this case is the metabolizable energy content of the food eaten. It is determined from Table 4–1 to be

$$E_{\text{in}} = 2 \times E_{\text{hamburger}} + E_{\text{fries}} + E_{\text{cola}}$$
$$= 2 \times 275 + 250 + 87$$
$$= 887 \text{ Cal}$$

The rate of energy output for a 68-kg man watching TV is given in Table 4–2 to be 72 Calories/h. For a 90-kg man it becomes

$$E_{\text{out}} = (90 \text{ kg})\frac{72 \text{ Cal/h}}{68 \text{ kg}} = 95.3 \text{ Cal/h}$$

Therefore, it will take

$$\Delta t = \frac{887 \text{ Cal}}{95.3 \text{ Cal/h}} = 9.3 \text{ h}$$

to burn the lunch calories off by watching TV.

(b) It can be shown in a similar manner that it takes only **47 min** to burn the lunch calories off by fast swimming.

*Discussion* The 45-kg man is half as large as the 90-kg man. Therefore, expending the same amount of energy takes twice as long in each case: **18.6 h** by watching TV and **94 min** by fast swimming.

---

### EXAMPLE 4–15    Losing Weight by Switching to Fat-Free Chips

The fake fat olestra passes through the body undigested, and thus adds zero calorie to the diet. Although foods cooked with olestra taste pretty good, they may cause abdominal discomfort and the long-term effects are unknown. A 1-oz (28.3-g) serving of regular potato chips has 10 g of fat and 150 Calories, whereas 1 oz of the so-called fat-free chips fried in olestra has only 75 Calories. Consider a person who eats 1 oz of regular potato chips every day at lunch without gaining or losing any weight. Determine how much weight this person will lose in one year if he or she switches to fat-free chips (Fig. 4–44).

**SOLUTION** A person switches from regular potato chips to fat-free ones. The weight the person loses in one year is to be determined.
*Assumptions* Exercising and other eating habits remain the same.
*Analysis* The person who switches to the fat-free chips consumes 75 fewer Calories a day. Then the annual reduction in calories consumed becomes

$$E_{\text{reduced}} = (75 \text{ Cal/day})(365 \text{ day/year}) = 27,375 \text{ Cal/year}$$

The metabolizable energy content of 1 kg of body fat is 33,100 kJ. Therefore, assuming the deficit in the calorie intake is made up by burning body fat, the person who switches to fat-free chips will lose

$$m_{\text{fat}} = \frac{E_{\text{reduced}}}{\text{Energy content of fat}} = \frac{27,375 \text{ Cal}}{33,100 \text{ kJ/kg}}\left(\frac{4.1868 \text{ kJ}}{1 \text{ Cal}}\right) = \mathbf{3.46 \text{ kg}}$$

of body fat that year.

**FIGURE 4–44**
Schematic for Example 4–15.

# SUMMARY

Work is the energy transferred as a force acts on a system through a distance. The most common form of mechanical work is the *boundary work*, which is the work associated with the expansion and compression of substances. On a $P$-$V$ diagram, the area under the process curve represents the boundary work for a quasi-equilibrium process. Various forms of boundary work are expressed as follows:

(1) General
$$W_b = \int_1^2 P \, dV$$

(2) Isobaric process
$$W_b = P_0(V_2 - V_1) \qquad (P_1 = P_2 = P_0 = \text{constant})$$

(3) Polytropic process
$$W_b = \frac{P_2 V_2 - P_1 V_1}{1 - n} \quad (n \neq 1) \qquad (PV^n = \text{constant})$$

(4) Isothernal process of an ideal gas
$$W_b = P_1 V_1 \ln \frac{V_2}{V_1}$$
$$= mRT_0 \ln \frac{V_2}{V_1} \qquad (PV = mRT_0 = \text{constant})$$

The first law of thermodynamics is essentially an expression of the conservation of energy principle, also called the energy balance. The general energy balances for *any system* undergoing *any process* can be expressed as

$$\underbrace{E_{\text{in}} - E_{\text{out}}}_{\substack{\text{Net energy transfer} \\ \text{by heat, work, and mass}}} = \underbrace{\Delta E_{\text{system}}}_{\substack{\text{Change in internal, kinetic,} \\ \text{potential, etc., energies}}}$$

It can also be expressed in the *rate form* as

$$\underbrace{\dot{E}_{\text{in}} - \dot{E}_{\text{out}}}_{\substack{\text{Rate of net energy transfer} \\ \text{by heat, work, and mass}}} = \underbrace{dE_{\text{system}}/dt}_{\substack{\text{Rate of change in internal,} \\ \text{kinetic, potential, etc., energies}}}$$

Taking heat transfer *to* the system and work done *by* the system to be positive quantities, the energy balance for a closed system can also be expressed as

$$Q - W = \Delta U + \Delta KE + \Delta PE$$

where

$$W = W_{\text{other}} + W_b$$
$$\Delta U = m(u_2 - u_1)$$
$$\Delta KE = \tfrac{1}{2}m(V_2^2 - V_1^2)$$
$$\Delta PE = mg(z_2 - z_1)$$

For a *constant-pressure process*, $W_b + \Delta U = \Delta H$. Thus,

$$Q - W_{\text{other}} = \Delta H + \Delta KE + \Delta PE$$

Note that the relation above is limited to constant pressure processes of closed system, and is NOT valid for processes during which pressure varies.

The amount of energy needed to raise the temperature of a unit mass of a substance by one degree is called the *specific heat at constant volume* $c_v$ for a constant-volume process and the *specific heat at constant pressure* $c_p$ for a constant-pressure process. They are defined as

$$c_v = \left(\frac{\partial u}{\partial T}\right)_v \quad \text{and} \quad c_p = \left(\frac{\partial h}{\partial T}\right)_p$$

For ideal gases $u$, $h$, $c_v$, and $c_p$ are functions of temperature alone. The $\Delta u$ and $\Delta h$ of ideal gases are expressed as

$$\Delta u = u_2 - u_1 = \int_1^2 c_v(T) \, dT \cong c_{v,\text{avg}}(T_2 - T_1)$$

$$\Delta h = h_2 - h_1 = \int_1^2 c_p(T) \, dT \cong c_{p,\text{avg}}(T_2 - T_1)$$

For ideal gases, $c_v$ and $c_p$ are related by

$$c_p = c_v + R$$

where $R$ is the gas constant. The *specific heat ratio* $k$ is defined as

$$k = \frac{c_p}{c_v}$$

For *incompressible substances* (liquids and solids), both the constant-pressure and constant-volume specific heats are identical and denoted by $c$:

$$c_p = c_v = c$$

The $\Delta u$ and $\Delta h$ of imcompressible substances are given by

$$\Delta u = \int_1^2 c(T) \, dT \cong c_{\text{avg}}(T_2 - T_1)$$

$$\Delta h = \Delta u + v\Delta P$$

# REFERENCES AND SUGGESTED READINGS

**1.** ASHRAE *Handbook of Fundamentals*. SI version. Atlanta, GA: American Society of Heating, Refrigerating, and Air-Conditioning Engineers, Inc., 1993.

**2.** ASHRAE *Handbook of Refrigeration*. SI version. Atlanta, GA: American Society of Heating, Refrigerating, and Air-Conditioning Engineers, Inc., 1994.

# PROBLEMS*

## Moving Boundary Work

**4–1C** An ideal gas at a given state expands to a fixed final volume first at constant pressure and then at constant temperature. For which case is the work done greater?

**4–2** Nitrogen at an initial state of 300 K, 150 kPa, and 0.2 m³ is compressed slowly in an isothermal process to a final pressure of 800 kPa. Determine the work done during this process.

**4–3** The volume of 1 kg of helium in a piston-cylinder device is initially 5 m³. Now helium is compressed to 2 m³ while its pressure is maintained constant at 180 kPa. Determine the initial and final temperatures of helium as well as the work required to compress it, in kJ.

**4–4** A piston–cylinder device initially contains 0.07 m³ of nitrogen gas at 130 kPa and 120°C. The nitrogen is now expanded polytropically to a state of 100 kPa and 100°C. Determine the boundary work done during this process.

**4–5** A piston–cylinder device with a set of stops initially contains 0.6 kg of steam at 1.0 MPa and 400°C. The location of the stops corresponds to 40 percent of the initial volume. Now the steam is cooled. Determine the compression work if the final state is (a) 1.0 MPa and 250°C and (b) 500 kPa. (c) Also determine the temperature at the final state in part (b).

**FIGURE P4–5**

**4–6** A piston–cylinder device initially contains 0.07 m³ of nitrogen gas at 130 kPa and 180°C. The nitrogen is now expanded to a pressure of 80 kPa polytropically with a polytropic exponent whose value is equal to the specific heat ratio

(called *isentropic expansion*). Determine the final temperature and the boundary work done during this process.

**4–7** A mass of 5 kg of saturated water vapor at 300 kPa is heated at constant pressure until the temperature reaches 200°C. Calculate the work done by the steam during this process. *Answer:* 166 kJ

**4–8** 1-m³ of saturated liquid water at 200°C is expanded isothermally in a closed system until its quality is 80 percent. Determine the total work produced by this expansion, in kJ.

**4–9** A gas is compressed from an initial volume of 0.42 m³ to a final volume of 0.12 m³. During the quasi-equilibrium process, the pressure changes with volume according to the relation $P = aV + b$, where $a = -1200$ kPa/m³ and $b = 600$ kPa. Calculate the work done during this process (a) by plotting the process on a P-V diagram and finding the area under the process curve and (b) by performing the necessary integrations.

Gas

$P = aV + b$

**FIGURE P4–9**

**4–10** A mass of 1.5 kg of air at 120 kPa and 24°C is contained in a gas-tight, frictionless piston–cylinder device. The air is now compressed to a final pressure of 600 kPa. During the process, heat is transferred from the air such that the temperature inside the cylinder remains constant. Calculate the work input during this process. *Answer:* 206 kJ

**4–11** During some actual expansion and compression processes in piston–cylinder devices, the gases have been observed to satisfy the relationship $PV^n = C$, where $n$ and $C$ are constants. Calculate the work done when a gas expands from 350 kPa and 0.03 m³ to a final volume of 0.2 m³ for the case of $n = 1.5$.

**4–12** Reconsider Prob. 4–11. Using the EES (or other) software, plot the process described in the problem on a P-V diagram, and investigate the effect of the polytropic exponent $n$ on the boundary work. Let the polytropic exponent vary from 1.1 to 1.6. Plot the boundary work versus the polytropic exponent, and discuss the results.

**4–13** A frictionless piston–cylinder device contains 5 kg of nitrogen at 100 kPa and 250 K. Nitrogen is now compressed slowly according to the relation $PV^{1.4} = $ constant until it

---

\* Problems designated by a "C" are concept questions, and students are encouraged to answer them all. Problems with the ⚙ icon are solved using EES, and complete solutions together with parametric studies are included on the text website. Problems with the 💾 icon are comprehensive in nature, and are intended to be solved with an equation solver such as EES.

reaches a final temperature of 360 K. Calculate the work input during this process. *Answer: 408 kJ*

$N_2$

$PV^{1.4} = \text{const.}$

**FIGURE P4–13**

**4–14** The equation of state of a gas is given as $\overline{v}(P + 10/\overline{v}^2) = R_u T$, where the units of $\overline{v}$ and $P$ are m³/kmol and kPa, respectively. Now 0.2 kmol of this gas is expanded in a quasi-equilibrium manner from 2 to 4 m³ at a constant temperature of 350 K. Determine (a) the unit of the quantity 10 in the equation and (b) the work done during this isothermal expansion process.

**4–15** Reconsider Prob. 4–14. Using the integration feature of the EES software, calculate the work done, and compare your result with the "hand-calculated" result obtained in Prob. 4–14. Plot the process described in the problem on a P-v diagram.

**4–16** During an expansion process, the pressure of a gas changes from 100 to 700 kPa according to the relation $P = aV + b$, where $a = 1220$ kPa/m³ and $b$ is a constant. If the initial volume of the gas is 0.2 m³, calculate the work done during the process. *Answer: 197 kJ*

**4–17** A piston–cylinder device initially contains 0.4 kg of nitrogen gas at 160 kPa and 140°C. The nitrogen is now expanded isothermally to a pressure of 100 kPa. Determine the boundary work done during this process. *Answer: 23.0 kJ*

$N_2$
160 kPa
140°C

**FIGURE P4–17**

**4–18** Hydrogen is contained in a piston–cylinder device at 100 kPa and 0.4 m³. At this state, a linear spring ($F \propto x$) with a spring constant of 20,300 N/m is touching the piston but exerts no force on it. The cross-sectional area of the piston is 0.3 m². Heat is transferred to the hydrogen, causing it to expand until its

volume doubles. Determine (a) the final pressure, (b) the total work done by the hydrogen, and (c) the fraction of this work done against the spring. Also, show the process on a P-V diagram.

**4–19** A piston–cylinder device contains 0.15 kg of air initially at 2 MPa and 350°C. The air is first expanded isothermally to 500 kPa, then compressed polytropically with a polytropic exponent of 1.2 to the initial pressure, and finally compressed at the constant pressure to the initial state. Determine the boundary work for each process and the net work of the cycle.

**4–20** 1-kg of water that is initially at 90°C with a quality of 10 percent occupies a spring-loaded piston–cylinder device, such as that in Fig. P4–20. This device is now heated until the pressure rises to 800 kPa and the temperature is 250°C. Determine the total work produced during this process, in kJ. *Answer: 24.5 kJ*

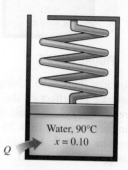

Water, 90°C
$x = 0.10$

$Q$

**FIGURE P4–20**

**4–21** 0.75-kg water that is initially at 0.5 MPa and 30 percent quality occupies a spring-loaded piston–cylinder device. This device is now cooled until the water is a saturated liquid at 100°C. Calculate the total work produced during this process, in kJ.

**4–22** An ideal gas undergoes two processes in a piston-cylinder device as follows:

1-2 Polytropic compression from $T_1$ and $P_1$ with a polytropic exponent $n$ and a compression ratio of $r = V_1/V_2$.

2-3 Constant pressure expansion at $P_3 = P_2$ until $V_3 = V_1$.

   (a) Sketch the processes on a single P-V diagram.

   (b) Obtain an expression for the ratio of the compression-to-expansion work as a function of $n$ and $r$.

   (c) Find the value of this ratio for values of $n = 1.4$ and $r = 6$.

*Answers: (b)          (c) 0.256*

**4–23** A piston–cylinder device contains 50 kg of water at 250 kPa and 25°C. The cross-sectional area of the piston is 0.1 m². Heat is now transferred to the water, causing part of it to evaporate and expand. When the volume reaches 0.2 m³, the piston reaches a linear spring whose spring constant is 100 kN/m. More heat is transferred to the water until the piston rises 20 cm more. Determine (a) the final pressure and temperature and (b) the work done during this process. Also, show the process on a P-V diagram. *Answers: (a) 450 kPa, 147.9°C, (b) 44.5 kJ*

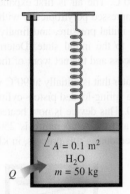

**FIGURE P4–23**

**4–24** Reconsider Prob. 4–23. Using the EES software, investigate the effect of the spring constant on the final pressure in the cylinder and the boundary work done. Let the spring constant vary from 50 kN/m to 500 kN/m. Plot the final pressure and the boundary work against the spring constant, and discuss the results.

### Closed System Energy Analysis

**4–25** Complete the table below on the basis of the conservation of energy principle for a closed system.

| $Q_{in}$ kJ | $W_{out}$ kJ | $E_1$ kJ | $E_2$ kJ | $m$ kg | $e_2 - e_1$ kJ/kg |
|---|---|---|---|---|---|
| 350 | — | 1020 | 860 | 3 | — |
| 350 | 130 | 550 | — | 5 | — |
| — | 260 | 600 | — | 2 | 150 |
| −500 | — | 1400 | 900 | 7 | — |
| — | −50 | 1000 | — | 3 | −200 |

**4–26** A rigid container equipped with a stirring device contains 2.5 kg of motor oil. Determine the rate of specific energy increase when heat is transferred to the oil at a rate of 1 W, and 1.5 W of power is applied to the stirring device.

**4–27** A 0.5-m³ rigid tank contains refrigerant-134a initially at 160 kPa and 40 percent quality. Heat is now transferred to the refrigerant until the pressure reaches 700 kPa. Determine (a) the mass of the refrigerant in the tank and (b) the amount of heat transferred. Also, show the process on a P-v diagram with respect to saturation lines.

**4–28** A 0.6-m³ rigid tank initially contains saturated refrigerant-134a vapor at 1200 kPa. As a result of heat transfer from the refrigerant, the pressure drops to 400 kPa. Show the process on a P-v diagram with respect to saturation lines, and

determine (a) the final temperature, (b) the amount of refrigerant that has condensed, and (c) the heat transfer.

**4–29** A rigid 10-L vessel initially contains a mixture of liquid water and vapor at 100°C with 12.3 percent quality. The mixture is then heated until its temperature is 150°C. Calculate the heat transfer required for this process. *Answer: 46.9 kJ*

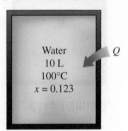

**FIGURE P4–29**

**4–30** A fixed mass of saturated water vapor at 400 kPa is isothermally cooled until it is a saturated liquid. Calculate the amount of heat rejected during this process, in kJ/kg.

**4–31** A piston–cylinder device contains steam initially at 1 MPa, 450°C, and 2.5 m³. Steam is allowed to cool at constant pressure until it first starts condensing. Show the process on a T-v diagram with respect to saturation lines and determine (a) the mass of the steam, (b) the final temperature, and (c) the amount of heat transfer.

**4–32** An insulated piston–cylinder device contains 5 L of saturated liquid water at a constant pressure of 175 kPa. Water is stirred by a paddle wheel while a current of 8 A flows for 45 min through a resistor placed in the water. If one-half of the liquid is evaporated during this constant-pressure process and the paddle-wheel work amounts to 400 kJ, determine the voltage of the source. Also, show the process on a P-v diagram with respect to saturation lines. *Answer: 224 V*

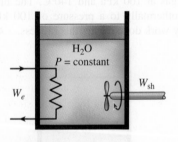

**FIGURE P4–32**

**4–33** A piston–cylinder device initially contains steam at 200 kPa, 200°C, and 0.4 m³. At this state, a linear spring ($F \propto x$) is touching the piston but exerts no force on it. Heat is now slowly transferred to the steam, causing the pressure and the volume to rise to 250 kPa and 0.6 m³, respectively. Show the process on a P-v diagram with respect to saturation lines and determine (a) the final temperature, *(b)* the work done by the steam, and *(c)* the total heat transferred. *Answers: (a)* 606°C, *(b)* 45 kJ, *(c)* 288 kJ

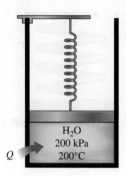

**FIGURE P4–33**

**4–34** Reconsider Prob. 4–33. Using EES (or other) software, investigate the effect of the initial temperature of steam on the final temperature, the work done, and the total heat transfer. Let the initial temperature vary from 150 to 250°C. Plot the final results against the initial temperature, and discuss the results.

**4–35** A piston–cylinder device initially contains 0.8 m³ of saturated water vapor at 250 kPa. At this state, the piston is resting on a set of stops, and the mass of the piston is such that a pressure of 300 kPa is required to move it. Heat is now slowly transferred to the steam until the volume doubles. Show the process on a *P-v* diagram with respect to saturation lines and determine *(a)* the final temperature, *(b)* the work done during this process, and *(c)* the total heat transfer. *Answers: (a)* 662°C, *(b)* 240 kJ, *(c)* 1213 kJ

**4–36** A 40-L electrical radiator containing heating oil is placed in a 50-m³ room. Both the room and the oil in the radiator are initially at 10°C. The radiator with a rating of 2.4 kW is now turned on. At the same time, heat is lost from the room at an average rate of 0.35 kJ/s. After some time, the average temperature is measured to be 20°C for the air in the room, and 50°C for the oil in the radiator. Taking the density and the specific heat of the oil to be 950 kg/m³ and 2.2 kJ/kg·°C, respectively, determine how long the heater is kept on. Assume the room is well-sealed so that there are no air leaks.

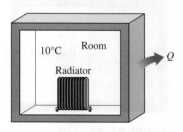

**FIGURE P4–36**

**4–37** Steam at 75 kPa and 8 percent quality is contained in a spring-loaded piston–cylinder device, as shown in Fig. P4–37, with an initial volume of 2 m³. Steam is now heated until its volume is 5 m³ and its pressure is 225 kPa. Determine the heat transferred to and the work produced by the steam during this process.

**FIGURE P4–37**

**4–38** Saturated R-134a vapor at 40°C is condensed at constant pressure to a saturated liquid in a closed piston–cylinder system. Calculate the heat transfer and work done during this process, in kJ/kg.

**4–39** An insulated tank is divided into two parts by a partition. One part of the tank contains 2.5 kg of compressed liquid water at 60°C and 600 kPa while the other part is evacuated. The partition is now removed, and the water expands to fill the entire tank. Determine the final temperature of the water and the volume of the tank for a final pressure of 10 kPa.

**FIGURE P4–39**

**4–40** Reconsider Prob. 4–39. Using EES (or other) software, investigate the effect of the initial pressure of water on the final temperature in the tank. Let the initial pressure vary from 100 to 600 kPa. Plot the final temperature against the initial pressure, and discuss the results.

**Specific Heats, Δ*u*, and Δ*h* of Ideal Gases**

**4–41C** Is the relation Δ*u* = *mc*$_{v,avg}$Δ*T* restricted to constant-volume processes only, or can it be used for any kind of process of an ideal gas?

**4–42C** Is the relation $\Delta h = mc_{p,avg}\Delta T$ restricted to constant-pressure processes only, or can it be used for any kind of process of an ideal gas?

**4–43C** Is the energy required to heat air from 295 to 305 K the same as the energy required to heat it from 345 to 355 K? Assume the pressure remains constant in both cases.

**4–44C** A fixed mass of an ideal gas is heated from 50 to 80°C at a constant pressure of (a) 1 atm and (b) 3 atm. For which case do you think the energy required will be greater? Why?

**4–45C** A fixed mass of an ideal gas is heated from 50 to 80°C at a constant volume of (a) 1 m³ and (b) 3 m³. For which case do you think the energy required will be greater? Why?

**4–46C** A fixed mass of an ideal gas is heated from 50 to 80°C (a) at constant volume and (b) at constant pressure. For which case do you think the energy required will be greater? Why?

**4–47** Show that for an ideal gas $\bar{c}_p = \bar{c}_v + R_u$.

**4–48** What is the change in the enthalpy, in kJ/kg, of oxygen as its temperature changes from 150 to 250°C? Is there any difference if the temperature change were from 0 to 100°C? Does the pressure at the beginning and end of this process have any effect on the enthalpy change?

**4–49** The temperature of 2 kg of neon is increased from 20 to 180°C. Calculate the change in the total internal energy of the neon, in kJ. Would the internal energy change be any different if the neon were replaced with argon?

**4–50** Calculate the change in the enthalpy of argon, in kJ/kg, when it is cooled from 75 to 25°C. If neon had undergone this same change of temperature, would its enthalpy change have been any different?

**4–51** Determine the internal energy change $\Delta u$ of hydrogen, in kJ/kg, as it is heated from 200 to 800 K, using (a) the empirical specific heat equation as a function of temperature (Table A–2c), (b) the $c_v$ value at the average temperature (Table A–2b), and (c) the $c_v$ value at room temperature (Table A–2a).

**4–52** Determine the enthalpy change $\Delta h$ of nitrogen, in kJ/kg, as it is heated from 600 to 1000 K, using (a) the empirical specific heat equation as a function of temperature (Table A–2c), (b) the $c_p$ value at the average temperature (Table A–2b), and (c) the $c_p$ value at room temperature (Table A–2a).
Answers: (a) 447.8 kJ/kg, (b) 448.4 kJ/kg, (c) 415.6 kJ/kg

**4–53** 0.03-m³ of air is contained in the spring-loaded piston-cylinder device shown in Fig. P4–53. The spring constant is 875 N/m, and the piston diameter is 25 cm. When no force is exerted by the spring on the piston, the state of the air is 2250 kPa and 240°C. This device is now cooled until the volume is one-half its original size. Determine the change in the

specific internal energy and enthalpy of the air. *Answers:* 185 kJ/kg, 258 kJ/kg

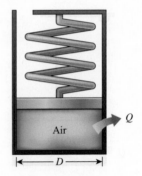

**FIGURE P4–53**

### Closed-System Energy Analysis: Ideal Gases

**4–54C** Is it possible to compress an ideal gas isothermally in an adiabatic piston–cylinder device? Explain.

**4–55** A 3-m³ rigid tank contains hydrogen at 250 kPa and 550 K. The gas is now cooled until its temperature drops to 350 K. Determine (a) the final pressure in the tank and (b) the amount of heat transfer.

**4–56** A 0.285-m³ tank contains oxygen initially at 101 kPa and 27°C. A paddle wheel within the tank is rotated until the pressure inside rises to 140 kPa. During the process 20 kJ of heat is lost to the surroundings. Determine the paddle-wheel work done. Neglect the energy stored in the paddle wheel.

**4–57** Nitrogen gas to 150 kPa and 40°C initially occupies a volume of 0.025 m³ in a rigid container equipped with a stirring paddle wheel. After 6 kJ of paddle wheel work is done on nitrogen, what is its final temperature? *Answer:* 240°C

**4–58** An insulated rigid tank is divided into two equal parts by a partition. Initially, one part contains 4 kg of an ideal gas at 800 kPa and 50°C, and the other part is evacuated. The partition is now removed, and the gas expands into the entire tank. Determine the final temperature and pressure in the tank.

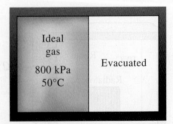

**FIGURE P4–58**

**4–59** A 4-m × 5-m × 6-m room is to be heated by a baseboard resistance heater. It is desired that the resistance heater be able to raise the air temperature in the room from 5 to 25°C within 11 min. Assuming no heat losses from the room and an atmospheric pressure of 100 kPa, determine the required power of the resistance heater. Assume constant specific heats at room temperature. *Answer: 3.28 kW*

**4–60** A student living in a 3-m × 4-m × 4-m dormitory room turns on her 100-W fan before she leaves the room on a summer day, hoping that the room will be cooler when she comes back in the evening. Assuming all the doors and windows are tightly closed and disregarding any heat trans-fer through the walls and the windows, determine the temperature in the room when she comes back 8 h later. Use specific heat values at room temperature, and assume the room to be at 100 kPa and 20°C in the morning when she leaves. *Answer: 90.3°C*

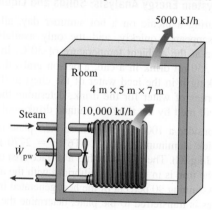

**FIGURE P4–60**

**4–61** A 4-m × 5-m × 7-m room is heated by the radiator of a steam-heating system. The steam radiator transfers heat at a rate of 10,000 kJ/h, and a 100-W fan is used to distribute the warm air in the room. The rate of heat loss from the room is estimated to be about 5000 kJ/h. If the initial temperature of the room air is 10°C, determine how long it will take for the air temperature to rise to 20°C. Assume constant specific heats at room temperature.

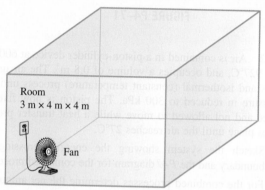

**FIGURE P4–61**

**4–62** Argon is compressed in a polytropic process with $n = 1.2$ from 120 kPa and 10°C to 800 kPa in a piston–cylinder device. Determine the work produced and heat transferred during this compression process, in kJ/kg.

**FIGURE P4–62**

**4–63** An insulated piston–cylinder device contains 100 L of air at 400 kPa and 25°C. A paddle wheel within the cylinder is rotated until 15 kJ of work is done on the air while the pressure is held constant. Determine the final temperature of the air. Neglect the energy stored in the paddle wheel.

**4–64** A spring-loaded piston-cylinder device contains 1 kg of carbon dioxide. This system is heated from 100 kPa and 25°C to 1000 kPa and 300°C. Determine the total heat transfer to and work produced by this system.

**FIGURE P4–64**

**4–65** A piston–cylinder device contains 0.7 m³ of nitrogen at 280 kPa and 370°C. Nitrogen is now allowed to cool at constant pressure until the temperature drops to 90°C. Using specific heats at the average temperature, determine the amount of heat loss.

**4–66** Air is contained in a variable-load piston-cylinder device equipped with a paddle wheel. Initially, air is at 400 kPa and 17°C. The paddle wheel is now turned by an external electric motor until 75 kJ/kg of work has been transferred to air. During this process, heat is transferred to maintain a constant air temperature while allowing the gas volume to triple. Calculate the required amount of heat transfer, in kJ/kg. *Answer: 16.4 kJ/kg*

**FIGURE P4–66**

**4–67** A mass of 15 kg of air in a piston–cylinder device is heated from 25 to 77°C by passing current through a resistance heater inside the cylinder. The pressure inside the cylinder is held constant at 300 kPa during the process, and a heat loss of 60 kJ occurs. Determine the electric energy supplied, in kWh. *Answer:* 0.235 kWh

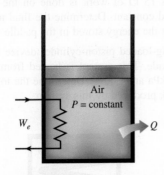

**FIGURE P4–67**

**4–68** A piston–cylinder device contains 2.2 kg of nitrogen initially at 100 kPa and 25°C. The nitrogen is now compressed slowly in a polytropic process during which $PV^{1.3} =$ constant until the volume is reduced by one-half. Determine the work done and the heat transfer for this process.

**4–69** Reconsider Prob. 4–68. Using EES (or other) software, plot the process described in the problem on a $P$-$V$ diagram, and investigate the effect of the polytropic exponent $n$ on the boundary work and heat transfer. Let the polytropic exponent vary from 1.0 to 1.4. Plot the boundary work and the heat transfer versus the polytropic exponent, and discuss the results.

**4–70** A piston–cylinder device, with a set of stops on the top, initially contains 3 kg of air at 200 kPa and 27°C. Heat is now transferred to the air, and the piston rises until it hits the stops, at which point the volume is twice the initial volume. More heat is transferred until the pressure inside the cylinder also doubles. Determine the work done and the amount of heat transfer for this process. Also, show the process on a $P$-$v$ diagram.

**4–71** Air is contained in a cylinder device fitted with a piston-cylinder. The piston initially rests on a set of stops, and a pressure of 200 kPa is required to move the piston. Initially, the air is at 100 kPa and 23°C and occupies a volume of 0.25 m³. Determine the amount of heat transferred to the air, in kJ, while increasing the temperature to 700 K. Assume air has constant specific heats evaluated at 300 K. *Answer:* 94.5 kJ

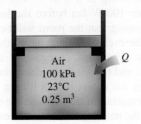

**FIGURE P4–71**

**4–72** Air is contained in a piston-cylinder device at 600 kPa and 927°C, and occupies a volume of 0.8 m³. The air undergoes and isothermal (constant temperature) process until the pressure in reduced to 300 kPa. The piston is now fixed in place and not allowed to move while a heat transfer process takes place until the air reaches 27°C.

(a) Sketch the system showing the energies crossing the boundary and the $P$-$V$ diagram for the combined processes.

(b) For the combined processes determine the net amount of heat transfer, in kJ, and its direction.

Assume air has constant specific heats evaluated at 300 K.

**4–73** A piston–cylinder device contains 4 kg of argon at 250 kPa and 35°C. During a quasi-equilibrium, isothermal expansion process, 15 kJ of boundary work is done by the system, and 3 kJ of paddle-wheel work is done on the system. Determine the heat transfer for this process.

### Closed-System Energy Analysis: Solids and Liquids

**4–74** During a picnic on a hot summer day, all the cold drinks disappeared quickly, and the only available drinks were those at the ambient temperature of 30°C. In an effort to cool a 350-ml drink in a can, a person grabs the can and starts shaking it in the iced water of the chest at 0°C. Using the properties of water for the drink, determine the mass of ice that will melt by the time the canned drink cools to 3°C.

**4–75** Consider a 1000-W iron whose base plate is made of 0.5-cm-thick aluminum alloy 2024-T6 ($\rho = 2770$ kg/m³ and $c_p = 875$ J/kg·°C). The base plate has a surface area of 0.03 m². Initially, the iron is in thermal equilibrium with the ambient air at 22°C. Assuming 90 percent of the heat generated in the resistance wires is transferred to the plate, determine the minimum time needed for the plate temperature to reach 200°C.

**FIGURE P4–75**

**4–76** Stainless steel ball bearings ($\rho$ = 8085 kg/m$^3$ and $c_p$ = 0.480 kJ/kg·°C) having a diameter of 1.2 cm are to be quenched in water at a rate of 800 per minute. The balls leave the oven at a uniform temperature of 900°C and are exposed to air at 25°C for a while before they are dropped into the water. If the temperature of the balls drops to 850°C prior to quenching, determine the rate of heat transfer from the balls to the air.

**4–77** In a production facility, 4-cm-thick 0.6-m × 0.6-m square brass plates ($\rho$ = 8530 kg/m$^3$ and $c_p$ = 0.38 kJ/kg·°C) that are initially at a uniform temperature of 24°C are heated by passing them through an oven at 800°C at a rate of 300 per minute. If the plates remain in the oven until their average temperature rises to 500°C, determine the rate of heat transfer to the plates in the furnace.

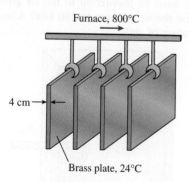

Furnace, 800°C

4 cm→

Brass plate, 24°C

**FIGURE P4–77**

**4–78** Long cylindrical steel rods ($\rho$ = 7833 kg/m$^3$ and $c_p$ = 0.465 kJ/kg·°C) of 8-cm diameter are heat-treated by drawing them at a velocity of 2 m/min through an oven maintained at 900°C. If the rods enter the oven at 30°C and leave at a mean temperature of 700°C, determine the rate of heat transfer to the rods in the oven.

**4–79** An electronic device dissipating 25 W has a mass of 20 g and a specific heat of 850 J/kg·°C. The device is lightly

used, and it is on for 5 min and then off for several hours, during which it cools to the ambient temperature of 25°C. Determine the highest possible temperature of the device at the end of the 5-min operating period. What would your answer be if the device were attached to a 0.5-kg aluminum heat sink? Assume the device and the heat sink to be nearly isothermal.

**4–80** [EES] Reconsider Prob. 4–79. Using EES (or other) software, investigate the effect of the mass of the heat sink on the maximum device temperature. Let the mass of heat sink vary from 0 to 1 kg. Plot the maximum temperature against the mass of heat sink, and discuss the results.

**4–81** If you ever slapped someone or got slapped yourself, you probably remember the burning sensation. Imagine you had the unfortunate occasion of being slapped by an angry person, which caused the temperature of the affected area of your face to rise by 2.4°C (ouch!). Assuming the slapping hand has a mass of 0.9 kg and about 0.150 kg of the tissue on the face and the hand is affected by the incident, estimate the velocity of the hand just before impact. Take the specific heat of the tissue to be 3.8 kJ/kg·K.

**4–82** In a manufacturing facility, 5-cm-diameter brass balls ($\rho$ = 8522 kg/m$^3$ and $c_p$ = 0.385 kJ/kg·°C) initially at 120°C are quenched in a water bath at 50°C for a period of 2 min at a rate of 100 balls per minute. If the temperature of the balls after quenching is 74°C, determine the rate at which heat needs to be removed from the water in order to keep its temperature constant at 50°C.

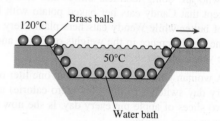

120°C        Brass balls

50°C

Water bath

**FIGURE P4–82**

**4–83** Repeat Prob. 4–82 for aluminum balls.

## Special Topic: Biological Systems

**4–84C** Is the metabolizable energy content of a food the same as the energy released when it is burned in a bomb calorimeter? If not, how does it differ?

**4–85C** Is the number of prospective occupants an important consideration in the design of heating and cooling systems of classrooms? Explain.

**4–86C** What do you think of a diet program that allows for generous amounts of bread and rice provided that no butter or margarine is added?

**4–87** The average specific heat of the human body is 3.6 kJ/kg·°C. If the body temperature of an 80-kg man rises from 37°C to 39°C during strenuous exercise, determine the increase in the thermal energy of the body as a result of this rise in body temperature.

**4–88** Consider two identical 80-kg men who are eating identical meals and doing identical things except that one of them jogs for 30 min every day while the other watches TV. Determine the weight difference between the two in a month.
*Answer:* 1.04 kg

**4–89** A 68-kg woman is planning to bicycle for an hour. If she is to meet her entire energy needs while bicycling by eating 30-g chocolate candy bars, determine how many candy bars she needs to take with her.

**4–90** A 90-kg man gives in to temptation and eats an entire 1-L box of ice cream. How long does this man need to jog to burn off the calories he consumed from the ice cream?
*Answer:* 1.54 h

**4–91** A 60-kg man used to have an apple every day after dinner without losing or gaining any weight. He now eats a 200-ml serving of ice cream instead of an apple and walks 20 min every day. On this new diet, how much weight will he lose or gain per month?  *Answer:* 0.087-kg gain

**4–92** Consider a man who has 20 kg of body fat when he goes on a hunger strike. Determine how long he can survive on his body fat alone.

**4–93** Consider two identical 50-kg women, Candy and Wendy, who are doing identical things and eating identical food except that Candy eats her baked potato with four teaspoons of butter while Wendy eats hers plain every evening. Determine the difference in the weights of Candy and Wendy after one year.  *Answer:* 6.5 kg

**4–94** A woman who used to drink about one liter of regular cola every day switches to diet cola (zero calorie) and starts eating two slices of apple pie every day. Is she now consuming fewer or more calories?

**4–95** A person eats a McDonald's Big Mac sandwich (530 Cal), a second person eats a Burger King Whopper sandwich (640 Cal), and a third person eats 50 olives with regular french fries (350 Cal) for lunch. Determine who consumes the most calories. An olive contains about 5 Calories.

**4–96** A 75-kg man decides to lose 5 kg without cutting down his intake of 4000 Calories a day. Instead, he starts fast swimming, fast dancing, jogging, and biking each for an hour every day. He sleeps or relaxes the rest of the day. Determine how long it will take him to lose 5 kg.

**4–97** The range of healthy weight for adults is usually expressed in terms of the *body mass index* (BMI), defined, in SI units, as

$$BMI = \frac{W\,(\text{kg})}{H^2\,(\text{m}^2)}$$

where $W$ is the weight (actually, the mass) of the person in kg and $H$ is the height in m, and the range of healthy weight is $19 \leq BMI \geq 25$. Convert the previous formula to English units such that the weight is in pounds and the height in inches. Also, calculate your own BMI, and if it is not in the healthy range, determine how many pounds (or kg) you need to gain or lose to be fit.

**4–98** The body mass index (BMI) of a 1.6-m tall woman who normally has 3 large slices of cheese pizza and a 400-ml Coke for lunch is 30. She now decides to change her lunch to 2 slices of pizza and a 200-ml Coke. Assuming that the deficit in the calorie intake is made up by burning body fat, determine how long it will take for the BMI of this person to drop to 20. Use the data in the text for calories and take the metabolizable energy content of 1 kg of body fat to be 33,100 kJ.  *Answer:* 463 days

## Review Problems

**4–99** The temperature of air changes from 0 to 10°C while its velocity changes from zero to a final velocity, and its elevation changes from zero to a final elevation. At which values of final air velocity and final elevation will the internal, kinetic, and potential energy changes be equal?
*Answers:* 120 m/s, 732 m

**4–100** Consider a piston–cylinder device that contains 0.5 kg air. Now, heat is transferred to the air at constant pressure and the air temperature increases by 5°C. Determine the expansion work done during this process.

**4–101** Air in the amount of 1 kg is contained in a well-insulated, rigid vessel equipped with a stirring paddle wheel. The initial state of this air is 210 kPa and 15°C. How much work, in kJ, must be transferred to the air with the paddle wheel to raise the air pressure to 280 kPa? Also, what is the final temperature of air?

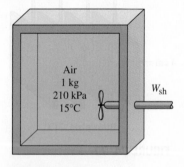

**FIGURE P4–101**

**4–102** Air is expanded in a polytropic process with $n = 1.2$ from 1 MPa and 400°C to 110 kPa in a piston-cylinder device. Determine the final temperature of the air.

**4–103** Nitrogen at 100 kPa and 25°C in a rigid vessel is heated until its pressure is 300 kPa. Calculate the work done and the heat transferred during this process, in kJ/kg.

**4–104** A well-insulated rigid vessel contains 3 kg of saturated liquid water at 40°C. The vessel also contains an electrical resistor that draws 10 amperes when 50 volts are applied. Determine the final temperature in the vessel after the resistor has been operating for 30 minutes. *Answer:* 119°C

**4–105** In order to cool 1 ton of water at 20°C in an insulated tank, a person pours 80 kg of ice at −5°C into the water. Determine the final equilibrium temperature in the tank. The melting temperature and the heat of fusion of ice at atmospheric pressure are 0°C and 333.7 kJ/kg, respectively. *Answer:* 12.4°C

**4–106** A mass of 3 kg of saturated liquid–vapor mixture of water is contained in a piston–cylinder device at 160 kPa. Initially, 1 kg of the water is in the liquid phase and the rest is in the vapor phase. Heat is now transferred to the water, and the piston, which is resting on a set of stops, starts moving when the pressure inside reaches 500 kPa. Heat transfer continues until the total volume increases by 20 percent. Determine (a) the initial and final temperatures, (b) the mass of liquid water when the piston first starts moving, and (c) the work done during this process. Also, show the process on a P-v diagram.

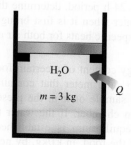

**FIGURE P4–106**

**4–107** A mass of 12 kg of saturated refrigerant-134a vapor is contained in a piston–cylinder device at 240 kPa. Now 300 kJ of heat is transferred to the refrigerant at constant pressure while a 110-V source supplies current to a resistor within the cylinder for 6 min. Determine the current supplied if the final temperature is 70°C. Also, show the process on a T-v diagram with respect to the saturation lines. *Answer:* 12.8 A

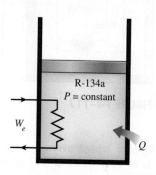

**FIGURE P4–107**

**4–108** Saturated water vapor at 200°C is condensed to a saturated liquid at 50°C in a spring-loaded piston-cylinder device. Determine the heat transfer for this process, in kJ/kg.

**4–109** A piston–cylinder device contains 0.8 kg of an ideal gas. Now, the gas is cooled at constant pressure until its temperature decreases by 10°C. If 16.6 kJ of compression work is done during this process, determine the gas constant and the molar mass of the gas. Also, determine the constant-volume and constant-pressure specific heats of the gas if its specific heat ratio is 1.667.

**FIGURE P4–109**

**4–110** A piston–cylinder device contains helium gas initially at 100 kPa, 10°C, and 0.2 m³. The helium is now compressed in a polytropic process ($PV^n$ = constant) to 700 kPa and 290°C. Determine the heat loss or gain during this process. *Answer:* 6.51 kJ loss

**FIGURE P4–110**

**4–111** An insulated piston–cylinder device initially contains 0.01 m³ of saturated liquid–vapor mixture with a quality of 0.2 at 120°C. Now some ice at 0°C is added to the cylinder. If the cylinder contains saturated liquid at 120°C when thermal equilibrium is established, determine the amount of ice added. The melting temperature and the heat of fusion of ice at atmospheric pressure are 0°C and 333.7 kJ/kg, respectively.

**4–112** Nitrogen gas is expanded in a polytropic process with $n = 1.25$ from 2 MPa and 1200 K to 200 kPa in a piston–cylinder device. How much work is produced and heat is transferred during this expansion process, in kJ/kg?

**4–113** A passive solar house that is losing heat to the outdoors at an average rate of 50,000 kJ/h is maintained at 22°C at all times during a winter night for 10 h. The house is to be heated by 50 glass containers each containing 20 L of water that is heated to 80°C during the day by absorbing solar energy. A thermostat-controlled 15-kW back-up electric resistance heater turns on whenever necessary to keep the house at 22°C. (a) How long did the electric heating system run that night? (b) How long would the electric heater run that night if the house incorporated no solar heating? *Answers:* (a) 4.77 h, (b) 9.26 h

**FIGURE P4–113**

**4–114** One ton (1000 kg) of liquid water at 50°C is brought into a well-insulated and well-sealed 4-m × 5-m × 6-m room initially at 15°C and 95 kPa. Assuming constant specific heats for both air and water at room temperature, determine the final equilibrium temperature in the room. *Answer:* 49.2°C

**4–115** Water is boiled at sea level in a coffee maker equipped with an immersion-type electric heating element. The coffee maker contains 1 L of water when full. Once boiling starts, it is observed that half of the water in the coffee maker evaporates in 25 min. Determine the power rating of the electric heating element immersed in water. Also, determine how long it will take for this heater to raise the temperature of 1 L of cold water from 18°C to the boiling temperature.

**FIGURE P4–115**

**4–116** A 3-m × 4-m × 5-m room is to be heated by one ton (1000 kg) of liquid water contained in a tank that is placed in the room. The room is losing heat to the outside at an average rate of 6000 kJ/h. The room is initially at 20°C and 100 kPa and is maintained at an average temperature of 20°C at all times. If the hot water is to meet the heating requirements of this room for a 24-h period, determine the minimum temperature of the water when it is first brought into the room. Assume constant specific heats for both air and water at room temperature.

**4–117** The energy content of a certain food is to be determined in a bomb calorimeter that contains 3 kg of water by burning a 2-g sample of it in the presence of 100 g of air in the reaction chamber. If the water temperature rises by 3.2°C when equilibrium is established, determine the energy content of the food, in kJ/kg, by neglecting the thermal energy stored in the reaction chamber and the energy supplied by the mixer. What is a rough estimate of the error involved in neglecting the thermal energy stored in the reaction chamber? *Answer:* 20,060 kJ/kg

**FIGURE P4–117**

**4–118** A 68-kg man whose average body temperature is 39°C drinks 1 L of cold water at 3°C in an effort to cool down. Taking the average specific heat of the human body to be 3.6 kJ/kg·°C, determine the drop in the average body temperature of this person under the influence of this cold water.

**4–119** An insulated rigid tank initially contains 1.4-kg saturated liquid water at 200°C and air. At this state, 25 percent of the volume is occupied by liquid water and the rest by air. Now an electric resistor placed in the tank is turned on, and the tank is observed to contain saturated water vapor after 20 min. Determine (a) the volume of the tank, (b) the final temperature, and (c) the electric power rating of the resistor. Neglect energy added to the air. *Answers:* (a) 0.00648 m³, (b) 371°C, (c) 1.58 kW

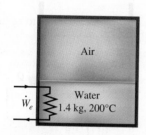

**FIGURE P4–119**

**4–120** A 0.3-L glass of water at 20°C is to be cooled with ice to 5°C. Determine how much ice needs to be added to the water, in grams, if the ice is at (a) 0°C and (b) −20°C. Also determine how much water would be needed if the cooling is to be done with cold water at 0°C. The melting temperature and the heat of fusion of ice at atmospheric pressure are 0°C and 333.7 kJ/kg, respectively, and the density of water is 1 kg/L.

**4–121** [EES] Reconsider Prob. 4–120. Using EES (or other) software, investigate the effect of the initial temperature of the ice on the final mass required. Let the ice temperature vary from −26 to 0°C. Plot the mass of ice against the initial temperature of ice, and discuss the results.

**4–122** A well-insulated 3-m × 4-m × 6-m room initially at 7°C is heated by the radiator of a steam heating system. The radiator has a volume of 15 L and is filled with superheated vapor at 200 kPa and 200°C. At this moment both the inlet and the exit valves to the radiator are closed. A 120-W fan is used to distribute the air in the room. The pressure of the steam is observed to drop to 100 kPa after 45 min as a result of heat transfer to the room. Assuming constant specific heats for air at room temperature, determine the average temperature of air in 45 min. Assume the air pressure in the room remains constant at 100 kPa.

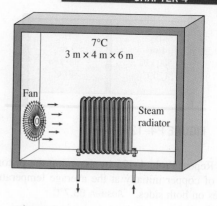

**FIGURE P4–122**

**4–123** Two rigid tanks are connected by a valve. Tank A contains 0.2 m³ of water at 400 kPa and 80 percent quality. Tank B contains 0.5 m³ of water at 200 kPa and 250°C. The valve is now opened, and the two tanks eventually come to the same state. Determine the pressure and the amount of heat transfer when the system reaches thermal equilibrium with the surroundings at 25°C. *Answers:* 3.17 kPa, 2170 kJ

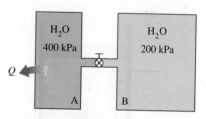

**FIGURE P4–123**

**4–124** [EES] Reconsider Prob. 4–123. Using EES (or other) software, investigate the effect of the environment temperature on the final pressure and the heat transfer. Let the environment temperature vary from 0 to 50°C. Plot the final results against the environment temperature, and discuss the results.

**4–125** Consider a well-insulated horizontal rigid cylinder that is divided into two compartments by a piston that is free to move but does not allow either gas to leak into the other side. Initially, one side of the piston contains 1 m³ of $N_2$ gas at 500 kPa and 120°C while the other side contains 1 m³ of He gas at 500 kPa and 40°C. Now thermal equilibrium is established in the cylinder as a result of heat transfer through the piston. Using constant specific heats at room temperature, determine the final equilibrium temperature in the cylinder. What would your answer be if the piston were not free to move?

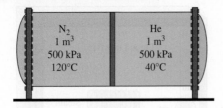

**FIGURE P4–125**

**4–126** Repeat Prob. 4–125 by assuming the piston is made of 8 kg of copper initially at the average temperature of the two gases on both sides. *Answer:* 83.7°C

**4–127** Reconsider Prob. 4–126. Using EES (or other) software, investigate the effect of the mass of the copper piston on the final equilibrium temperature. Let the mass of piston vary from 1 to 10 kg. Plot the final temperature against the mass of piston, and discuss the results.

**4–128** An insulated piston-cylinder device initially contains 1.8-kg saturated liquid water at 120°C. Now an electric resistor placed in the tank is turned on for 10 min until the volume quadruples. Determine (*a*) the volume of the tank, (*b*) the final temperature, and (*c*) the electrical power rating of the resistor. *Answers:* (*a*) 0.00763 m³, (*b*) 120°C, (*c*) 0.0236 kW

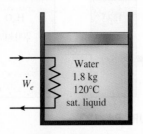

**FIGURE P4–128**

**4–129** A vertical 12-cm diameter piston–cylinder device contains an ideal gas at the ambient conditions of 1 bar and 24°C. Initially, the inner face of the piston is 20 cm from the base of the cylinder. Now an external shaft connected to the piston exerts a force corresponding to a boundary work input of 0.1 kJ. The temperature of the gas remains constant during the process. Determine (*a*) the amount of heat transfer, (*b*) the final pressure in the cylinder, and (*c*) the distance that the piston is displaced.

**4–130** A vertical 12-cm diameter piston–cylinder device contains an ideal gas at the ambient conditions of 1 bar and 24°C. Initially, the inner face of the piston is 20 cm from the base of the cylinder. Now an external shaft connected to the piston exerts a force corresponding to a boundary work input of 0.1 kJ. The temperature of the gas remains constant during the process. Determine (*a*) the amount of heat transfer, (*b*) the final pressure in the cylinder, and (*c*) the distance that the piston is displaced.

**4–131** A piston–cylinder device initially contains 0.35-kg steam at 3.5 MPa, superheated by 7.4°C. Now the steam loses heat to the surroundings and the piston moves down, hitting a set of stops at which point the cylinder contains saturated liquid water. The cooling continues until the cylinder contains water at 200°C. Determine (*a*) the final pressure and the quality (if mixture), (*b*) the boundary work, (*c*) the amount of heat transfer when the piston first hits the stops, (*d*) and the total heat transfer.

**FIGURE P4–131**

**4–132** An insulated rigid tank is divided into two compartments of different volumes. Initially, each compartment contains the same ideal gas at identical pressure but at different temperatures and masses. The wall separating the two compartments is removed and the two gases are allowed to mix. Assuming constant specific heats, find the simplest expression for the mixture temperature written in the form

$$T_3 = f\left(\frac{m_1}{m_3}, \frac{m_2}{m_3}, T_1, T_2\right)$$

where $m_3$ and $T_3$ are the mass and temperature of the final mixture, respectively.

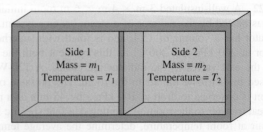

**FIGURE P4–132**

**4–133** One kilogram of carbon dioxide is compressed from 0.5 MPa and 200°C to 3 MPa in a piston-cylinder device arranged to execute a polytropic process with $n = 1.3$. Use the compressibility factor to determine the final temperature.

**4–134** In solar-heated buildings, energy is often stored as sensible heat in rocks, concrete, or water during the day for use at night. To minimize the storage space, it is desirable to use a material that can store a large amount of heat while experiencing a small temperature change. A large amount of heat can be stored essentially at constant temperature during a phase change process, and thus materials that change phase at about room temperature such as glaubers salt (sodium sulfate decahydrate), which has a melting point of 32°C and a heat of fusion of 329 kJ/L, are very suitable for this purpose. Determine how much heat can be stored in a 5-m³ storage space using (*a*) glaubers salt undergoing a phase change, (*b*) granite rocks with a heat capacity of 2.32 kJ/kg · °C and a temperature change of 20°C, and (*c*) water with a heat capacity of 4.00 kJ/k · °C and a temperature change of 20°C.

**4–135** The early steam engines were driven by the atmospheric pressure acting on the piston fitted into a cylinder filled with saturated steam. A vacuum was created in the cylinder by cooling the cylinder externally with cold water, and thus condensing the steam.

Consider a piston–cylinder device with a piston surface area of 0.1 m² initially filled with 0.05 m³ of saturated water vapor at the atmospheric pressure of 100 kPa. Now cold water is poured outside the cylinder, and the steam inside starts condensing as a result of heat transfer to the cooling water outside. If the piston is stuck at its initial position, determine the friction force acting on the piston and the amount of heat transfer when the temperature inside the cylinder drops to 30°C.

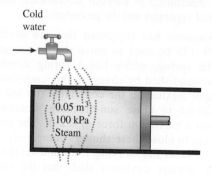

Cold
water

0.05 m³
100 kPa
Steam

**FIGURE P4–135**

## Fundamentals of Engineering (FE) Exam Problems

**4–136** The specific heat of a material is given in a strange unit to be $c = 3.60$ kJ/kg·°F. The specific heat of this material in the SI units of kJ/kg·°C is

(*a*) 2.00 kJ/kg·°C    (*b*) 3.20 kJ/kg·°C    (*c*) 3.60 kJ/kg·°C
(*d*) 4.80 kJ/kg·°C    (*e*) 6.48 kJ/kg·°C

**4–137** A 3-m³ rigid tank contains nitrogen gas at 500 kPa and 300 K. Now heat is transferred to the nitrogen in the tank and the pressure of nitrogen rises to 800 kPa. The work done during this process is
(*a*) 500 kJ          (*b*) 1500 kJ       (*c*) 0 kJ
(*d*) 900 kJ          (*e*) 2400 kJ

**4–138** A 0.5-m³ rigid tank contains nitrogen gas at 600 kPa and 300 K. Now the gas is compressed isothermally to a volume of 0.1 m³. The work done on the gas during this compression process is
(*a*) 720 kJ          (*b*) 483 kJ        (*c*) 240 kJ
(*d*) 175 kJ          (*e*) 143 kJ

**4–139** A well-sealed room contains 60 kg of air at 200 kPa and 25°C. Now solar energy enters the room at an average rate of 0.8 kJ/s while a 120-W fan is turned on to circulate the air in the room. If heat transfer through the walls is negligible, the air temperature in the room in 30 min will be
(*a*) 25.6°C        (*b*) 49.8°C       (*c*) 53.4°C
(*d*) 52.5°C        (*e*) 63.4°C

**4–140** A 2-kW baseboard electric resistance heater in a vacant room is turned on and kept on for 15 min. The mass of the air in the room is 75 kg, and the room is tightly sealed so that no air can leak in or out. The temperature rise of air at the end of 15 min is
(*a*) 8.5°C         (*b*) 12.4°C       (*c*) 24.0°C
(*d*) 33.4°C        (*e*) 54.8°C

**4–141** A room contains 75 kg of air at 100 kPa and 15°C. The room has a 250-W refrigerator (the refrigerator consumes 250 W of electricity when running), a 120-W TV, a 1.8-kW electric resistance heater, and a 50-W fan. During a cold winter day, it is observed that the refrigerator, the TV, the fan, and the electric resistance heater are running continuously but the air temperature in the room remains constant. The rate of heat loss from the room that day is
(*a*) 5832 kJ/h      (*b*) 6192 kJ/h     (*c*) 7560 kJ/h
(*d*) 7632 kJ/h      (*e*) 7992 kJ/h

**4–142** A piston–cylinder device contains 5 kg of air at 400 kPa and 30°C. During a quasi-equilibrium isothermal expansion process, 15 kJ of boundary work is done by the system, and 3 kJ of paddle-wheel work is done on the system. The heat transfer during this process is
(*a*) 12 kJ          (*b*) 18 kJ         (*c*) 2.4 kJ
(*d*) 3.5 kJ          (*e*) 60 kJ

**4–143** A 6-pack canned drink is to be cooled from 18°C to 3°C. The mass of each canned drink is 0.355 kg. The drinks can be treated as water, and the energy stored in the aluminum can itself is negligible. The amount of heat transfer from the 6 canned drinks is

(a) 22 kJ     (b) 32 kJ     (c) 134 kJ
(d) 187 kJ     (e) 223 kJ

**4–144** A glass of water with a mass of 0.45 kg at 20°C is to be cooled to 0°C by dropping ice cubes at 0°C into it. The latent heat of fusion of ice is 334 kJ/kg, and the specific heat of water is 4.18 kJ/kg·°C. The amount of ice that needs to be added is
(a) 56 g     (b) 113 g     (c) 124 g
(d) 224 g     (e) 450 g

**4–145** A 2-kW electric resistance heater submerged in 5-kg water is turned on and kept on for 10 min. During the process, 300 kJ of heat is lost from the water. The temperature rise of water is
(a) 0.4°C     (b) 43.1°C     (c) 57.4°C
(d) 71.8°C     (e) 180°C

**4–146** 1.5 kg of liquid water initially at 12°C is to be heated at 95°C in a teapot equipped with a 800-W electric heating element inside. The specific heat of water can be taken to be 4.18 kJ/kg·°C, and the heat loss from the water during heating can be neglected. The time it takes to heat water to the desired temperature is
(a) 5.9 min     (b) 7.3 min     (c) 10.8 min
(d) 14.0 min     (e) 17.0 min

**4–147** An ordinary egg with a mass of 0.1 kg and a specific heat of 3.32 kJ/kg·°C is dropped into boiling water at 95°C. If the initial temperature of the egg is 5°C, the maximum amount of heat transfer to the egg is
(a) 12 kJ     (b) 30 kJ     (c) 24 kJ
(d) 18 kJ     (e) infinity

**4–148** An apple with an average mass of 0.18 kg and average specific heat of 3.65 kJ/kg·°C is cooled from 22°C to 5°C. The amount of heat transferred from the apple is
(a) 0.85 kJ     (b) 62.1 kJ     (c) 17.7 kJ
(d) 11.2 kJ     (e) 7.1 kJ

**4–149** The specific heat at constant volume for an ideal gas is given by $c_v = 0.7 + (2.7 \times 10^{-4})T$ (kJ/kg·K) where $T$ is in kelvin. The change in the internal energy for this ideal gas undergoing a process in which the temperature changes from 27 to 127°C is most nearly
(a) 70 kJ/kg     (b) 72.1 kJ/kg     (c) 79.5 kJ/kg
(d) 82.1 kJ/kg     (e) 84.0 kJ/kg

**4–150** An ideal gas has a gas constant $R = 0.3$ kJ/kg·K and a constant-volume specific heat $c_v = 0.7$ kJ/kg·K. If the gas has a temperature change of 100°C, choose the correct answer for each of the following:

1. The change in enthalpy is, in kJ/kg
(a) 30     (b) 70     (c) 100
(d) insufficient information to determine

2. The change in internal energy is, in kJ/kg
(a) 30     (b) 70     (c) 100
(d) insufficient information to determine

3. The work done is, in kJ/kg
(a) 30     (b) 70     (c) 100
(d) insufficient information to determine

4. The heat transfer is, in kJ/kg
(a) 30     (b) 70     (c) 100
(d) insufficient information to determine

5. The change in the pressure-volume product is, in kJ/kg
(a) 30     (b) 70     (c) 100
(d) insufficient information to determine

**4–151** An ideal gas undergoes a constant temperature (isothermal) process in a closed system. The heat transfer and work are, respectively
(a) $0, -c_v\Delta T$     (b) $c_v\Delta T, 0$
(c) $c_p\Delta T, R\Delta T$     (d) $R \ln(T_2/T_1), R \ln(T_2/T_1)$

**4–152** An ideal gas undergoes a constant volume (isochoric) process in a closed system. The heat transfer and work are, respectively
(a) $0, -c_v\Delta T$     (b) $c_v\Delta T, 0$
(c) $c_p\Delta T, R\Delta T$     (d) $R \ln(T_2/T_1), R \ln(T_2/T_1)$

**4–153** An ideal gas undergoes a constant pressure (isobaric) process in a closed system. The heat transfer and work are, respectively
(a) $0, -c_v\Delta T$     (b) $c_v\Delta T, 0$
(c) $c_p\Delta T, R\Delta T$     (d) $R \ln(T_2/T_1), R \ln(T_2/T_1)$

## Design and Essay Problems

**4–154** Find out how the specific heats of gases, liquids, and solids are determined in national laboratories. Describe the experimental apparatus and the procedures used.

**4–155** Someone has suggested that the device shown in Fig. P4–155 be used to move the maximum force $F$ against the spring, which has a spring constant of $k$. This is accomplished by changing the temperature of the liquid–vapor mixture in the container. You are to design such a device to close sun-blocking window shutters that require a maximum force of 2.2 N. The piston must move 15 cm to close these shutters completely. You elect to use R-134a as the working fluid and arrange the liquid–vapor mixture container such that the temperature changes from 20°C when shaded from the sun to 40°C when exposed to the full sun. Select the sizes of the various components in this system to do this task. Also select the necessary spring constant and the amount of R-134a to be used.

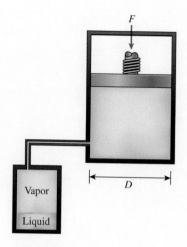

**FIGURE P4–155**

**4–156** You are asked to design a heating system for a swimming pool that is 2 m deep, 25 m long, and 25 m wide. Your client desires that the heating system be large enough to raise the water temperature from 20 to 30°C in 3 h. The rate of heat loss from the water to the air at the outdoor design conditions is determined to be 960 W/m², and the heater must also be able to maintain the pool at 30°C at those conditions. Heat losses to the ground are expected to be small and can be disregarded. The heater considered is a natural gas furnace whose efficiency is 80 percent. What heater size (in kW input) would you recommend to your client?

**4–157** It is claimed that fruits and vegetables are cooled by 6°C for each percentage point of weight loss as moisture during vacuum cooling. Using calculations, demonstrate if this claim is reasonable.

**4–158** Using a thermometer, measure the boiling temperature of water and calculate the corresponding saturation pressure. From this information, estimate the altitude of your town and compare it with the actual altitude value.

**4–159** Design an experiment complete with instrumentation to determine the specific heats of a gas using a resistance heater. Discuss how the experiment will be conducted, what measurements need to be taken, and how the specific heats will be determined. What are the sources of error in your system? How can you minimize the experimental error?

# MASS AND ENERGY ANALYSIS OF CONTROL VOLUMES

In Chap. 4, we applied the general energy balance relation expressed as $E_{in} - E_{out} = \Delta E_{system}$ to closed systems. In this chapter, we extend the energy analysis to systems that involve mass flow across their boundaries i.e., control volumes, with particular emphasis to steady-flow systems.

We start this chapter with the development of the general *conservation of mass* relation for control volumes, and we continue with a discussion of flow work and the energy of fluid streams. We then apply the energy balance to systems that involve *steady-flow processes* and analyze the common steady-flow devices such as nozzles, diffusers, compressors, turbines, throttling devices, mixing chambers, and heat exchangers. Finally, we apply the energy balance to general *unsteady-flow processes* such as the charging and discharging of vessels.

## OBJECTIVES

The objectives of Chapter 5 are to:

- Develop the conservation of mass principle.

- Apply the conservation of mass principle to various systems including steady- and unsteady-flow control volumes.

- Apply the first law of thermodynamics as the statement of the conservation of energy principle to control volumes.

- Identify the energy carried by a fluid stream crossing a control surface as the sum of internal energy, flow work, kinetic energy, and potential energy of the fluid and to relate the combination of the internal energy and the flow work to the property enthalpy.

- Solve energy balance problems for common steady-flow devices such as nozzles, compressors, turbines, throttling valves, mixers, heaters, and heat exchangers.

- Apply the energy balance to general unsteady-flow processes with particular emphasis on the uniform-flow process as the model for commonly encountered charging and discharging processes.

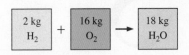

**FIGURE 5–1**
Mass is conserved even during chemical reactions.

# 5–1 · CONSERVATION OF MASS

The conservation of mass principle is one of the most fundamental principles in nature. We are all familiar with this principle, and it is not difficult to understand. A person does not have to be a rocket scientist to figure out how much vinegar-and-oil dressing will be obtained by mixing 100 g of oil with 25 g of vinegar. Even chemical equations are balanced on the basis of the conservation of mass principle. When 16 kg of oxygen reacts with 2 kg of hydrogen, 18 kg of water is formed (Fig. 5–1). In an electrolysis process, the water separates back to 2 kg of hydrogen and 16 kg of oxygen.

Technically, mass is not exactly conserved. It turns out that mass $m$ and energy $E$ can be converted to each other according to the well-known formula proposed by Albert Einstein (1879–1955):

$$E = mc^2 \tag{5–1}$$

where $c$ is the speed of light in a vacuum, which is $c = 2.9979 \times 10^8$ m/s. This equation suggests that there is equivalence between mass and energy. All physical and chemical systems exhibit energy interactions with their surroundings, but the amount of energy involved is equivalent to an extremely small mass compared to the system's total mass. For example, when 1 kg of liquid water is formed from oxygen and hydrogen at normal atmospheric conditions, the amount of energy released is 15.8 MJ, which corresponds to a mass of only $1.76 \times 10^{-10}$ kg. However, in nuclear reactions, the mass equivalence of the amount of energy interacted is a significant fraction of the total mass involved. Therefore, in most engineering analyses, we consider both mass and energy as conserved quantities.

For *closed systems,* the conservation of mass principle is implicitly used by requiring that the mass of the system remain constant during a process. For *control volumes,* however, mass can cross the boundaries, and so we must keep track of the amount of mass entering and leaving the control volume.

## Mass and Volume Flow Rates

The amount of mass flowing through a cross section per unit time is called the **mass flow rate** and is denoted by $\dot{m}$. The dot over a symbol is used to indicate *time rate of change*.

A fluid flows into or out of a control volume, usually through pipes or ducts. The differential mass flow rate of fluid flowing across a small area element $dA_c$ in a cross section of a pipe is proportional to $dA_c$ itself, the fluid density $\rho$, and the component of the flow velocity normal to $dA_c$, which we denote as $V_n$, and is expressed as (Fig. 5–2)

$$\delta\dot{m} = \rho V_n dA_c \tag{5–2}$$

Note that both $\delta$ and $d$ are used to indicate differential quantities, but $\delta$ is typically used for quantities (such as heat, work, and mass transfer) that are *path functions* and have *inexact differentials,* while $d$ is used for quantities (such as properties) that are *point functions* and have *exact differentials.* For flow through an annulus of inner radius $r_1$ and outer radius $r_2$, for example,

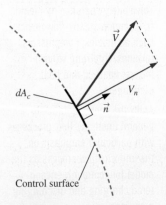

Control surface

**FIGURE 5–2**
The normal velocity $V_n$ for a surface is the component of velocity perpendicular to the surface.

$$\int_1^2 dA_c = A_{c2} - A_{c1} = \pi(r_2^2 - r_1^2) \text{ but } \int_1^2 \delta\dot{m} = \dot{m}_{\text{total}} \text{ (total mass flow rate through}$$

the annulus), not $\dot{m}_2 - \dot{m}_1$. For specified values of $r_1$ and $r_2$, the value of the integral of $dA_c$ is fixed (thus the names point function and exact differential), but this is not the case for the integral of $\delta\dot{m}$ (thus the names path function and inexact differential).

The mass flow rate through the entire cross-sectional area of a pipe or duct is obtained by integration:

$$\dot{m} = \int_{A_c} \delta\dot{m} = \int_{A_c} \rho V_n \, dA_c \quad \text{(kg/s)} \tag{5–3}$$

While Eq. 5–3 is always valid (in fact it is *exact*), it is not always practical for engineering analyses because of the integral. We would like instead to express mass flow rate in terms of average values over a cross section of the pipe. In a general compressible flow, both $\rho$ and $V_n$ vary across the pipe. In many practical applications, however, the density is essentially uniform over the pipe cross section, and we can take $\rho$ outside the integral of Eq. 5–3. Velocity, however, is *never* uniform over a cross section of a pipe because of the no-slip condition at the walls. Rather, the velocity varies from zero at the walls to some maximum value at or near the centerline of the pipe. We define the **average velocity** $V_{\text{avg}}$ as the average value of $V_n$ across the entire cross section of the pipe (Fig. 5–3),

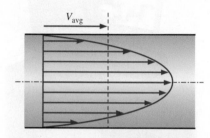

**FIGURE 5–3**
The average velocity $V_{\text{avg}}$ is defined as the average speed through a cross section.

*Average velocity:*
$$V_{\text{avg}} = \frac{1}{A_c} \int_{A_c} V_n \, dA_c \tag{5–4}$$

where $A_c$ is the area of the cross section normal to the flow direction. Note that if the speed were $V_{\text{avg}}$ all through the cross section, the mass flow rate would be identical to that obtained by integrating the actual velocity profile. Thus for incompressible flow or even for compressible flow where $\rho$ is approximated as uniform across $A_c$, Eq. 5–3 becomes

$$\dot{m} = \rho V_{\text{avg}} A_c \quad \text{(kg/s)} \tag{5–5}$$

For compressible flow, we can think of $\rho$ as the bulk average density over the cross section, and then Eq. 5–5 can be used as a reasonable approximation. For simplicity, we drop the subscript on the average velocity. Unless otherwise stated, $V$ denotes the average velocity in the flow direction. Also, $A_c$ denotes the cross-sectional area normal to the flow direction.

The volume of the fluid flowing through a cross section per unit time is called the **volume flow rate** $\dot{V}$ (Fig. 5–4) and is given by

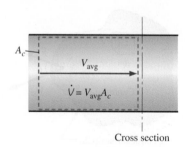

**FIGURE 5–4**
The volume flow rate is the volume of fluid flowing through a cross section per unit time.

$$\dot{V} = \int_{A_c} V_n \, dA_c = V_{\text{avg}} A_c = V A_c \quad \text{(m}^3\text{/s)} \tag{5–6}$$

An early form of Eq. 5–6 was published in 1628 by the Italian monk Benedetto Castelli (circa 1577–1644). Note that many fluid mechanics textbooks use $Q$ instead of $\dot{V}$ for volume flow rate. We use $\dot{V}$ to avoid confusion with heat transfer.

The mass and volume flow rates are related by

$$\dot{m} = \rho\dot{V} = \frac{\dot{V}}{v} \qquad (5\text{-}7)$$

where $v$ is the specific volume. This relation is analogous to $m = \rho V = V/v$, which is the relation between the mass and the volume of a fluid in a container.

## Conservation of Mass Principle

The **conservation of mass principle** for a control volume can be expressed as: *The net mass transfer to or from a control volume during a time interval $\Delta t$ is equal to the net change (increase or decrease) of the total mass within the control volume during $\Delta t$.* That is,

$$\left( \begin{array}{c} \text{Total mass entering} \\ \text{the CV during } \Delta t \end{array} \right) - \left( \begin{array}{c} \text{Total mass leaving} \\ \text{the CV during } \Delta t \end{array} \right) = \left( \begin{array}{c} \text{Net change of mass} \\ \text{within the CV during } \Delta t \end{array} \right)$$

or

$$m_{\text{in}} - m_{\text{out}} = \Delta m_{\text{CV}} \quad (\text{kg}) \qquad (5\text{-}8)$$

where $\Delta m_{\text{CV}} = m_{\text{final}} - m_{\text{initial}}$ is the change in the mass of the control volume during the process (Fig. 5–5). It can also be expressed in *rate form* as

$$\dot{m}_{\text{in}} - \dot{m}_{\text{out}} = dm_{\text{CV}}/dt \quad (\text{kg/s}) \qquad (5\text{-}9)$$

where $\dot{m}_{\text{in}}$ and $\dot{m}_{\text{out}}$ are the total rates of mass flow into and out of the control volume, and $dm_{\text{CV}}/dt$ is the rate of change of mass within the control volume boundaries. Equations 5–8 and 5–9 are often referred to as the **mass balance** and are applicable to any control volume undergoing any kind of process.

Consider a control volume of arbitrary shape, as shown in Fig. 5–6. The mass of a differential volume $dV$ within the control volume is $dm = \rho \, dV$. The total mass within the control volume at any instant in time $t$ is determined by integration to be

Total mass within the CV: $\qquad m_{\text{CV}} = \displaystyle\int_{\text{CV}} \rho \, dV \qquad (5\text{-}10)$

Then the time rate of change of the amount of mass within the control volume is expressed as

Rate of change of mass within the CV: $\qquad \dfrac{dm_{\text{CV}}}{dt} = \dfrac{d}{dt}\displaystyle\int_{\text{CV}} \rho \, dV \qquad (5\text{-}11)$

For the special case of no mass crossing the control surface (i.e., the control volume is a closed system), the conservation of mass principle reduces to $dm_{\text{CV}}/dt = 0$. This relation is valid whether the control volume is fixed, moving, or deforming.

**FIGURE 5–5**

Conservation of mass principle for an ordinary bathtub.

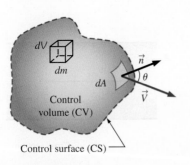

**FIGURE 5–6**

The differential control volume $dV$ and the differential control surface $dA$ used in the derivation of the conservation of mass relation.

Now consider mass flow into or out of the control volume through a differential area $dA$ on the control surface of a fixed control volume. Let $\vec{n}$ be the outward unit vector of $dA$ normal to $dA$ and $\vec{V}$ be the flow velocity at $dA$ relative to a fixed coordinate system, as shown in Fig. 5–6. In general, the velocity may cross $dA$ at an angle $\theta$ off the normal of $dA$, and the mass flow rate is proportional to the normal component of velocity $V_n = V\cos\theta$ ranging from a maximum outflow of $\vec{V}$ for $\theta = 0$ (flow is normal to $dA$) to a minimum of zero for $\theta = 90°$ (flow is tangent to $dA$) to a maximum *inflow* of $\vec{V}$ for $\theta = 180°$ (flow is normal to $dA$ but in the opposite direction). Making use of the concept of dot product of two vectors, the magnitude of the normal component of velocity is

*Normal component of velocity:*
$$V_n = V\cos\theta = \vec{V}\cdot\vec{n} \tag{5–12}$$

The mass flow rate through $dA$ is proportional to the fluid density $\rho$, normal velocity $V_n$, and the flow area $dA$, and is expressed as

*Differential mass flow rate:*
$$\delta\dot{m} = \rho V_n\, dA = \rho(V\cos\theta)\, dA = \rho(\vec{V}\cdot\vec{n})\, dA \tag{5–13}$$

The net flow rate into or out of the control volume through the entire control surface is obtained by integrating $\delta\dot{m}$ over the entire control surface,

*Net mass flow rate:*
$$\dot{m}_{net} = \int_{CS}\delta\dot{m} = \int_{CS}\rho V_n\, dA = \int_{CS}\rho(\vec{V}\cdot\vec{n})\, dA \tag{5–14}$$

Note that $V_n = \vec{V}\cdot\vec{n} = V\cos\theta$ is positive for $\theta < 90°$ (outflow) and negative for $\theta > 90°$ (inflow). Therefore, the direction of flow is automatically accounted for, and the surface integral in Eq. 5–14 directly gives the *net* mass flow rate. A positive value for $\dot{m}_{net}$ indicates a net outflow of mass and a negative value indicates a net inflow of mass.

Rearranging Eq. 5–9 as $dm_{CV}/dt + \dot{m}_{out} - \dot{m}_{in} = 0$, the conservation of mass relation for a fixed control volume is then expressed as

*General conservation of mass:*
$$\frac{d}{dt}\int_{CV}\rho\, dV + \int_{CS}\rho(\vec{V}\cdot\vec{n})\, dA = 0 \tag{5–15}$$

It states that *the time rate of change of mass within the control volume plus the net mass flow rate through the control surface is equal to zero.*

Splitting the surface integral in Eq. 5–15 into two parts—one for the outgoing flow streams (positive) and one for the incoming flow streams (negative)—the general conservation of mass relation can also be expressed as

$$\frac{d}{dt}\int_{CV}\rho\, dV + \sum_{out}\rho|V_n|A - \sum_{in}\rho|V_n|A = 0 \tag{5–16}$$

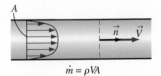

$$\dot{m} = \rho(V\cos\theta)(A/\cos\theta) = \rho VA$$

(a) Control surface *at an angle* to the flow

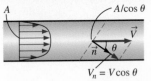

$$\dot{m} = \rho VA$$

(b) Control surface *normal* to the flow

**FIGURE 5–7**

A control surface should always be selected *normal to the flow* at all locations where it crosses the fluid flow to avoid complications, even though the result is the same.

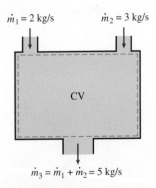

**FIGURE 5–8**

Conservation of mass principle for a two-inlet–one-outlet steady-flow system.

where $A$ represents the area for an inlet or outlet, and the summation signs are used to emphasize that *all* the inlets and outlets are to be considered. Using the definition of mass flow rate, Eq. 5–16 can also be expressed as

$$\frac{d}{dt}\int_{CV}\rho\, dV = \sum_{in}\dot{m} - \sum_{out}\dot{m} \quad\text{or}\quad \frac{dm_{CV}}{dt} = \sum_{in}\dot{m} - \sum_{out}\dot{m} \qquad (5\text{--}17)$$

There is considerable flexibility in the selection of a control volume when solving a problem. Many control volume choices are available, but some are more convenient to work with. A control volume should not introduce any unnecessary complications. A wise choice of a control volume can make the solution of a seemingly complicated problem rather easy. A simple rule in selecting a control volume is to make the control surface *normal to the flow* at all locations where it crosses the fluid flow, whenever possible. This way the dot product $\vec{V}\cdot\vec{n}$ simply becomes the magnitude of the velocity, and the integral $\int_A \rho(\vec{V}\cdot\vec{n})\, dA$ becomes simply $\rho VA$ (Fig. 5–7).

Equations 5–15 and 5–16 are also valid for moving or deforming control volumes provided that the *absolute velocity* $\vec{V}$ is replaced by the *relative velocity* $\vec{V}_r$, which is the fluid velocity relative to the control surface.

## Mass Balance for Steady-Flow Processes

During a steady-flow process, the total amount of mass contained within a control volume does not change with time ($m_{CV}$ = constant). Then the conservation of mass principle requires that the total amount of mass entering a control volume equal the total amount of mass leaving it. For a garden hose nozzle in steady operation, for example, the amount of water entering the nozzle per unit time is equal to the amount of water leaving it per unit time.

When dealing with steady-flow processes, we are not interested in the amount of mass that flows in or out of a device over time; instead, we are interested in the amount of mass flowing per unit time, that is, *the mass flow rate $\dot{m}$. The conservation of mass principle* for a general steady-flow system with multiple inlets and outlets is expressed in rate form as (Fig. 5–8)

*Steady flow:* $$\sum_{in}\dot{m} = \sum_{out}\dot{m} \quad\text{(kg/s)} \qquad (5\text{--}18)$$

It states that *the total rate of mass entering a control volume is equal to the total rate of mass leaving it.*

Many engineering devices such as nozzles, diffusers, turbines, compressors, and pumps involve a single stream (only one inlet and one outlet). For these cases, we typically denote the inlet state by the subscript 1 and the outlet state by the subscript 2, and drop the summation signs. Then Eq. 5–18 reduces, for *single-stream steady-flow systems,* to

*Steady flow (single stream):* $\quad \dot{m}_1 = \dot{m}_2 \quad \rightarrow \quad \rho_1 V_1 A_1 = \rho_2 V_2 A_2$ $\qquad$ (5–19)

## Special Case: Incompressible Flow

The conservation of mass relations can be simplified even further when the fluid is incompressible, which is usually the case for liquids. Canceling the density from both sides of the general steady-flow relation gives

*Steady, incompressible flow:* $\quad \displaystyle\sum_{in} \dot{V} = \sum_{out} \dot{V} \quad$ (m$^3$/s) $\qquad$ (5–20)

For single-stream steady-flow systems Eq. 5–20 becomes

*Steady, incompressible flow (single stream):* $\quad \dot{V}_1 = \dot{V}_2 \rightarrow V_1 A_1 = V_2 A_2 \qquad$ (5–21)

It should always be kept in mind that there is no such thing as a "conservation of volume" principle. Therefore, the volume flow rates into and out of a steady-flow device may be different. The volume flow rate at the outlet of an air compressor is much less than that at the inlet even though the mass flow rate of air through the compressor is constant (Fig. 5–9). This is due to the higher density of air at the compressor exit. For steady flow of liquids, however, the volume flow rates remain nearly constant since liquids are essentially incompressible (constant-density) substances. Water flow through the nozzle of a garden hose is an example of the latter case.

The conservation of mass principle requires every bit of mass to be accounted for during a process. If you can balance your checkbook (by keeping track of deposits and withdrawals, or by simply observing the "conservation of money" principle), you should have no difficulty applying the conservation of mass principle to engineering systems.

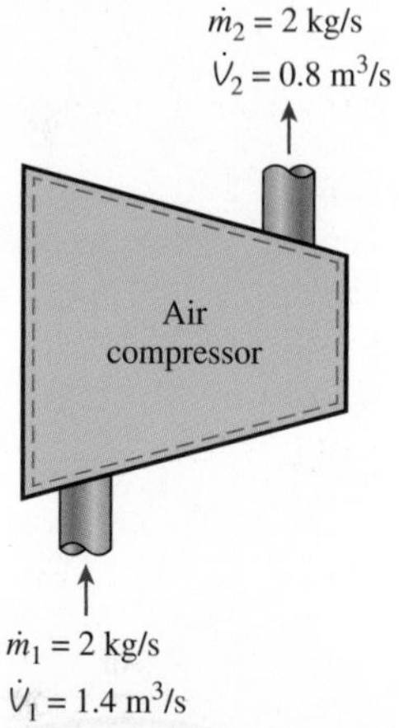

$\dot{m}_2 = 2$ kg/s
$\dot{V}_2 = 0.8$ m$^3$/s

Air compressor

$\dot{m}_1 = 2$ kg/s
$\dot{V}_1 = 1.4$ m$^3$/s

**FIGURE 5–9**
During a steady-flow process, volume flow rates are not necessarily conserved although mass flow rates are.

---

### ■ *EXAMPLE 5–1*  Water Flow through a Garden Hose Nozzle

A garden hose attached with a nozzle is used to fill a 40-L bucket. The inner diameter of the hose is 2 cm, and it reduces to 0.8 cm at the nozzle exit (Fig. 5–10). If it takes 50 s to fill the bucket with water, determine (*a*) the volume and mass flow rates of water through the hose, and (*b*) the average velocity of water at the nozzle exit.

**SOLUTION**  A garden hose is used to fill a water bucket. The volume and mass flow rates of water and the exit velocity are to be determined.
*Assumptions*  **1** Water is a nearly incompressible substance. **2** Flow through the hose is steady. **3** There is no waste of water by splashing.
*Properties*  We take the density of water to be 1000 kg/m$^3$ = 1 kg/L.

**FIGURE 5–10**
Schematic for Example 5–1.
*Photo by John M. Cimbala*

*Analysis* (*a*) Noting that 40 L of water are discharged in 50 s, the volume and mass flow rates of water are

$$\dot{V} = \frac{V}{\Delta t} = \frac{40 \text{ L}}{50 \text{ s}} = \textbf{0.8 L/s}$$

$$\dot{m} = \rho\dot{V} = (1 \text{ kg/L})(0.8 \text{ L/s}) = \textbf{0.8 kg/s}$$

(*b*) The cross-sectional area of the nozzle exit is

$$A_e = \pi r_e^2 = \pi(0.4 \text{ cm})^2 = 0.5027 \text{ cm}^2 = 0.5027 \times 10^{-4} \text{ m}^2$$

The volume flow rate through the hose and the nozzle is constant. Then the average velocity of water at the nozzle exit becomes

$$V_e = \frac{\dot{V}}{A_e} = \frac{0.8 \text{ L/s}}{0.5027 \times 10^{-4} \text{ m}^2}\left(\frac{1 \text{ m}^3}{1000 \text{ L}}\right) = \textbf{15.9 m/s}$$

*Discussion* It can be shown that the average velocity in the hose is 2.5 m/s. Therefore, the nozzle increases the water velocity by over six times.

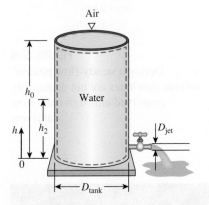

Air

$h_0$

Water

$h$   $h_2$   $D_{jet}$

0

$D_{tank}$

**FIGURE 5–11**
Schematic for Example 5–2.

---

### EXAMPLE 5–2   Discharge of Water from a Tank

A 1.2-m-high, 0.9-m-diameter cylindrical water tank whose top is open to the atmosphere is initially filled with water. Now the discharge plug near the bottom of the tank is pulled out, and a water jet whose diameter is 1.3 cm streams out (Fig. 5–11). The average velocity of the jet is approximated as $V = \sqrt{2gh}$, where $h$ is the height of water in the tank measured from the center of the hole (a variable) and $g$ is the gravitational acceleration. Determine how long it takes for the water level in the tank to drop to 0.6 m from the bottom.

**SOLUTION** The plug near the bottom of a water tank is pulled out. The time it takes for half of the water in the tank to empty is to be determined.
*Assumptions* 1 Water is a nearly incompressible substance. 2 The distance between the bottom of the tank and the center of the hole is negligible compared to the total water height. 3 The gravitational acceleration is 9.81 m/s$^2$.
*Analysis* We take the volume occupied by water as the control volume. The size of the control volume decreases in this case as the water level drops, and thus this is a variable control volume. (We could also treat this as a fixed control volume that consists of the interior volume of the tank by disregarding the air that replaces the space vacated by the water.) This is obviously an unsteady-flow problem since the properties (such as the amount of mass) within the control volume change with time.

The conservation of mass relation for a control volume undergoing any process is given in rate form as

$$\dot{m}_{\text{in}} - \dot{m}_{\text{out}} = \frac{dm_{\text{CV}}}{dt} \tag{1}$$

During this process no mass enters the control volume ($\dot{m}_{in} = 0$), and the mass flow rate of discharged water is

$$\dot{m}_{out} = (\rho V A)_{out} = \rho\sqrt{2gh}\,A_{jet} \qquad (2)$$

where $A_{jet} = \pi D_{jet}^2/4$ is the cross-sectional area of the jet, which is constant. Noting that the density of water is constant, the mass of water in the tank at any time is

$$m_{CV} = \rho V = \rho A_{tank}h \qquad (3)$$

where $A_{tank} = D_{tank}^2/4$ is the base area of the cylindrical tank. Substituting Eqs. 2 and 3 into the mass balance relation (Eq. 1) gives

$$-\rho\sqrt{2gh}\,A_{jet} = \frac{d(\rho A_{tank}h)}{dt} \rightarrow -\rho\sqrt{2gh}(\pi D_{jet}^2/4) = \frac{\rho(\pi D_{tank}^2/4)dh}{dt}$$

Canceling the densities and other common terms and separating the variables give

$$dt = -\frac{D_{tank}^2}{D_{jet}^2}\frac{dh}{\sqrt{2gh}}$$

Integrating from $t = 0$ at which $h = h_0$ to $t = t$ at which $h = h_2$ gives

$$\int_0^t dt = -\frac{D_{tank}^2}{D_{jet}^2\sqrt{2g}}\int_{h_0}^{h_2}\frac{dh}{\sqrt{h}} \rightarrow t = \frac{\sqrt{h_0}-\sqrt{h_2}}{\sqrt{g/2}}\left(\frac{D_{tank}}{D_{jet}}\right)^2$$

Substituting, the time of discharge is determined to be

$$t = \frac{\sqrt{1.2\text{ m}}-\sqrt{0.6\text{ m}}}{\sqrt{9.81/2\text{ m/s}^2}}\left(\frac{0.9\text{ m}}{0.013\text{ m}}\right)^2 = 694\text{ s} = \textbf{11.6 min}$$

Therefore, it takes 11.6 min after the discharge hole is unplugged for half of the tank to be emptied.

**Discussion** Using the same relation with $h_2 = 0$ gives $t = 39.5$ min for the discharge of the entire amount of water in the tank. Therefore, emptying the bottom half of the tank takes much longer than emptying the top half. This is due to the decrease in the average discharge velocity of water with decreasing $h$.

# 5–2 · FLOW WORK AND THE ENERGY OF A FLOWING FLUID

Unlike closed systems, control volumes involve mass flow across their boundaries, and some work is required to push the mass into or out of the control volume. This work is known as the **flow work**, or **flow energy**, and is necessary for maintaining a continuous flow through a control volume.

To obtain a relation for flow work, consider a fluid element of volume $V$ as shown in Fig. 5–12. The fluid immediately upstream forces this fluid element to enter the control volume; thus, it can be regarded as an imaginary piston. The fluid element can be chosen to be sufficiently small so that it has uniform properties throughout.

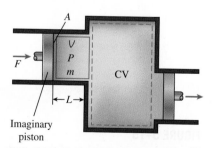

**FIGURE 5–12**
Schematic for flow work.

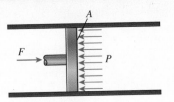

**FIGURE 5–13**

In the absence of acceleration, the force applied on a fluid by a piston is equal to the force applied on the piston by the fluid.

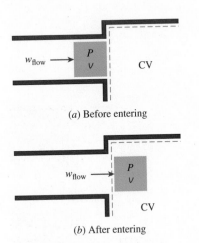

(a) Before entering

(b) After entering

**FIGURE 5–14**

Flow work is the energy needed to push a fluid into or out of a control volume, and it is equal to $Pv$.

If the fluid pressure is $P$ and the cross-sectional area of the fluid element is $A$ (Fig. 5–13), the force applied on the fluid element by the imaginary piston is

$$F = PA \qquad (5\text{-}22)$$

To push the entire fluid element into the control volume, this force must act through a distance $L$. Thus, the work done in pushing the fluid element across the boundary (i.e., the flow work) is

$$W_{\text{flow}} = FL = PAL = PV \quad (\text{kJ}) \qquad (5\text{-}23)$$

The flow work per unit mass is obtained by dividing both sides of this equation by the mass of the fluid element:

$$w_{\text{flow}} = Pv \quad (\text{kJ/kg}) \qquad (5\text{-}24)$$

The flow work relation is the same whether the fluid is pushed into or out of the control volume (Fig. 5–14).

It is interesting that unlike other work quantities, flow work is expressed in terms of properties. In fact, it is the product of two properties of the fluid. For that reason, some people view it as a *combination property* (like enthalpy) and refer to it as *flow energy, convected energy,* or *transport energy* instead of flow work. Others, however, argue rightfully that the product $Pv$ represents energy for flowing fluids only and does not represent any form of energy for nonflow (closed) systems. Therefore, it should be treated as work. This controversy is not likely to end, but it is comforting to know that both arguments yield the same result for the energy balance equation. In the discussions that follow, we consider the flow energy to be part of the energy of a flowing fluid, since this greatly simplifies the energy analysis of control volumes.

## Total Energy of a Flowing Fluid

As we discussed in Chap. 2, the total energy of a simple compressible system consists of three parts: internal, kinetic, and potential energies (Fig. 5–15). On a unit-mass basis, it is expressed as

$$e = u + \text{ke} + \text{pe} = u + \frac{V^2}{2} + gz \quad (\text{kJ/kg}) \qquad (5\text{-}25)$$

where $V$ is the velocity and $z$ is the elevation of the system relative to some external reference point.

**FIGURE 5–15**

The total energy consists of three parts for a nonflowing fluid and four parts for a flowing fluid.

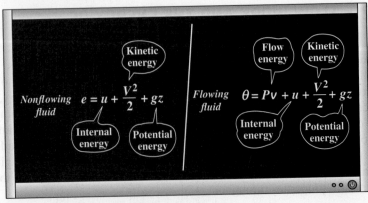

Nonflowing fluid $\quad e = u + \dfrac{V^2}{2} + gz$

Flowing fluid $\quad \theta = Pv + u + \dfrac{V^2}{2} + gz$

The fluid entering or leaving a control volume possesses an additional form of energy—the *flow energy* $P_v$, as already discussed. Then the total energy of a **flowing fluid** on a unit-mass basis (denoted by $\theta$) becomes

$$\theta = P_v + e = P_v + (u + \text{ke} + \text{pe}) \qquad \text{(5–26)}$$

But the combination $P_v + u$ has been previously defined as the enthalpy $h$. So the relation in Eq. 5–26 reduces to

$$\theta = h + \text{ke} + \text{pe} = h + \frac{V^2}{2} + gz \quad \text{(kJ/kg)} \qquad \text{(5–27)}$$

By using the enthalpy instead of the internal energy to represent the energy of a flowing fluid, one does not need to be concerned about the flow work. The energy associated with pushing the fluid into or out of the control volume is automatically taken care of by enthalpy. In fact, this is the main reason for defining the property enthalpy. From now on, the energy of a fluid stream flowing into or out of a control volume is represented by Eq. 5–27, and no reference will be made to flow work or flow energy.

## Energy Transport by Mass

Noting that $\theta$ is total energy per unit mass, the total energy of a flowing fluid of mass $m$ is simply $m\theta$, provided that the properties of the mass $m$ are uniform. Also, when a fluid stream with uniform properties is flowing at a mass flow rate of $\dot{m}$, the rate of energy flow with that stream is $\dot{m}\theta$ (Fig. 5–16). That is,

*Amount of energy transport:* $\quad E_{\text{mass}} = m\theta = m\left(h + \dfrac{V^2}{2} + gz\right) \quad \text{(kJ)} \qquad \text{(5–28)}$

*Rate of energy transport:* $\quad \dot{E}_{\text{mass}} = \dot{m}\theta = \dot{m}\left(h + \dfrac{V^2}{2} + gz\right) \quad \text{(kW)} \qquad \text{(5–29)}$

When the kinetic and potential energies of a fluid stream are negligible, as is often the case, these relations simplify to $E_{\text{mass}} = mh$ and $\dot{E}_{\text{mass}} = \dot{m}h$.

In general, the total energy transported by mass into or out of the control volume is not easy to determine since the properties of the mass at each inlet or exit may be changing with time as well as over the cross section. Thus, the only way to determine the energy transport through an opening as a result of mass flow is to consider sufficiently small differential masses $dm$ that have uniform properties and to add their total energies during flow.

Again noting that $\theta$ is total energy per unit mass, the total energy of a flowing fluid of mass $\delta m$ is $\theta\,\delta m$. Then the total energy transported by mass through an inlet or exit ($m_i\theta_i$ and $m_e\theta_e$) is obtained by integration. At an inlet, for example, it becomes

$$E_{\text{in,mass}} = \int_{m_i} \theta_i\,\delta m_i = \int_{m_i} \left(h_i + \frac{V_i^2}{2} + gz_i\right)\delta m_i \qquad \text{(5–30)}$$

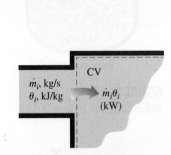

**FIGURE 5–16**
The product $\dot{m}_i\theta_i$ is the energy transported into control volume by mass per unit time.

Most flows encountered in practice can be approximated as being steady and one-dimensional, and thus the simple relations in Eqs. 5–28 and 5–29 can be used to represent the energy transported by a fluid stream.

Steam

150 kPa

Pressure cooker

**FIGURE 5–17**
Schematic for Example 5–3.

---

### EXAMPLE 5–3    Energy Transport by Mass

Steam is leaving a 4-L pressure cooker whose operating pressure is 150 kPa (Fig. 5–17). It is observed that the amount of liquid in the cooker has decreased by 0.6 L in 40 min after the steady operating conditions are established, and the cross-sectional area of the exit opening is 8 mm². Determine (a) the mass flow rate of the steam and the exit velocity, (b) the total and flow energies of the steam per unit mass, and (c) the rate at which energy leaves the cooker by steam.

**SOLUTION** Steam leaves a pressure cooker at a specified pressure. The velocity, flow rate, the total and flow energies, and the rate of energy transfer by mass are to be determined.

**Assumptions**   **1** The flow is steady, and the initial start-up period is disregarded. **2** The kinetic and potential energies are negligible, and thus they are not considered. **3** Saturation conditions exist within the cooker at all times so that steam leaves the cooker as a saturated vapor at the cooker pressure.

**Properties**   The properties of saturated liquid water and water vapor at 150 kPa are $v_f = 0.001053$ m³/kg, $v_g = 1.1594$ m³/kg, $u_g = 2519.2$ kJ/kg, and $h_g = 2693.1$ kJ/kg (Table A–5).

**Analysis**   (a) Saturation conditions exist in a pressure cooker at all times after the steady operating conditions are established. Therefore, the liquid has the properties of saturated liquid and the exiting steam has the properties of saturated vapor at the operating pressure. The amount of liquid that has evaporated, the mass flow rate of the exiting steam, and the exit velocity are

$$m = \frac{\Delta V_{\text{liquid}}}{v_f} = \frac{0.6\ \text{L}}{0.001053\ \text{m}^3/\text{kg}}\left(\frac{1\ \text{m}^3}{1000\ \text{L}}\right) = 0.570\ \text{kg}$$

$$\dot{m} = \frac{m}{\Delta t} = \frac{0.570\ \text{kg}}{40\ \text{min}} = 0.0142\ \text{kg/min} = \mathbf{2.37 \times 10^{-4}\ kg/s}$$

$$V = \frac{\dot{m}}{\rho_g A_c} = \frac{\dot{m} v_g}{A_c} = \frac{(2.37 \times 10^{-4}\ \text{kg/s})(1.1594\ \text{m}^3/\text{kg})}{8 \times 10^{-6}\ \text{m}^2} = \mathbf{34.3\ m/s}$$

(b) Noting that $h = u + Pv$ and that the kinetic and potential energies are disregarded, the flow and total energies of the exiting steam are

$$e_{\text{flow}} = Pv = h - u = 2693.1 - 2519.2 = \mathbf{173.9\ kJ/kg}$$

$$\theta = h + \text{ke} + \text{pe} \cong h = \mathbf{2693.1\ kJ/kg}$$

Note that the kinetic energy in this case is ke $= V^2/2 = (34.3\ \text{m/s})^2/2 = 588\ \text{m}^2/\text{s}^2 = 0.588$ kJ/kg, which is small compared to enthalpy.

(c) The rate at which energy is leaving the cooker by mass is simply the product of the mass flow rate and the total energy of the exiting steam per unit mass,

$$\dot{E}_{\text{mass}} = \dot{m}\theta = (2.37 \times 10^{-4}\ \text{kg/s})(2693.1\ \text{kJ/kg}) = 0.638\ \text{kJ/s} = \mathbf{0.638\ kW}$$

*Discussion* The numerical value of the energy leaving the cooker with steam alone does not mean much since this value depends on the reference point selected for enthalpy (it could even be negative). The significant quantity is the difference between the enthalpies of the exiting vapor and the liquid inside (which is $h_{fg}$) since it relates directly to the amount of energy supplied to the cooker.

# 5–3 · ENERGY ANALYSIS OF STEADY-FLOW SYSTEMS

A large number of engineering devices such as turbines, compressors, and nozzles operate for long periods of time under the same conditions once the transient start-up period is completed and steady operation is established, and they are classified as *steady-flow devices* (Fig. 5–18). Processes involving such devices can be represented reasonably well by a somewhat idealized process, called the **steady-flow process**, which was defined in Chap. 1 as *a process during which a fluid flows through a control volume steadily*. That is, the fluid properties can change from point to point within the control volume, but at any point, they remain constant during the entire process. (Remember, *steady* means *no change with time*.)

During a steady-flow process, no intensive or extensive properties *within the control volume* change with time. Thus, the volume $V$, the mass $m$, and the total energy content $E$ of the control volume remain constant (Fig. 5–19). As a result, the boundary work is zero for steady-flow systems (since $V_{CV}$ = constant), and the total mass or energy entering the control volume must be equal to the total mass or energy leaving it (since $m_{CV}$ = constant and $E_{CV}$ = constant). These observations greatly simplify the analysis.

The fluid properties at an inlet or exit remain constant during a steady-flow process. The properties may, however, be different at different inlets and exits. They may even vary over the cross section of an inlet or an exit. However, all properties, including the velocity and elevation, must remain constant with time at a fixed point at an inlet or exit. It follows that the mass flow rate of the fluid at an opening must remain constant during a steady-flow process (Fig. 5–20). As an added simplification, the fluid properties at an opening are usually considered to be uniform (at some average value) over the cross section. Thus, the fluid properties at an inlet or exit may be specified by the average single values. Also, the *heat* and *work* interactions between a steady-flow system and its surroundings do not change with time. Thus, the power delivered by a system and the rate of heat transfer to or from a system remain constant during a steady-flow process.

The *mass balance* for a general steady-flow system was given in Sec. 5–1 as

$$\sum_{in} \dot{m} = \sum_{out} \dot{m} \quad \text{(kg/s)} \tag{5–31}$$

The mass balance for a single-stream (one-inlet and one-outlet) steady-flow system was given as

$$\dot{m}_1 = \dot{m}_2 \quad \longrightarrow \quad \rho_1 V_1 A_1 = \rho_2 V_2 A_2 \tag{5–32}$$

**FIGURE 5–18**
Many engineering systems such as power plants operate under steady conditions.
©*Malcolm Fife /Getty Images RF*

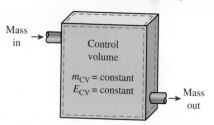

**FIGURE 5–19**
Under steady-flow conditions, the mass and energy contents of a control volume remain constant.

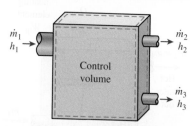

**FIGURE 5–20**
Under steady-flow conditions, the fluid properties at an inlet or exit remain constant (do not change with time).

where the subscripts 1 and 2 denote the inlet and the exit states, respectively, $\rho$ is density, $V$ is the average flow velocity in the flow direction, and $A$ is the cross-sectional area normal to flow direction.

During a steady-flow process, the total energy content of a control volume remains constant ($E_{CV} =$ constant), and thus the change in the total energy of the control volume is zero ($\Delta E_{CV} = 0$). Therefore, the amount of energy entering a control volume in all forms (by heat, work, and mass) must be equal to the amount of energy leaving it. Then the rate form of the general energy balance reduces for a steady-flow process to

$$\underbrace{\dot{E}_{in} - \dot{E}_{out}}_{\substack{\text{Rate of net energy transfer} \\ \text{by heat, work, and mass}}} = \underbrace{dE_{system}/dt}_{\substack{\text{Rate of change in internal, kinetic,} \\ \text{potential, etc., energies}}}^{\nearrow \, 0 \text{ (steady)}} = 0 \tag{5-33}$$

or

$$\textit{Energy balance:} \qquad \underbrace{\dot{E}_{in}}_{\substack{\text{Rate of net energy transfer in} \\ \text{by heat, work, and mass}}} = \underbrace{\dot{E}_{out}}_{\substack{\text{Rate of net energy transfer out} \\ \text{by heat, work, and mass}}} \quad \text{(kW)} \tag{5-34}$$

Noting that energy can be transferred by heat, work, and mass only, the energy balance in Eq. 5–34 for a general steady-flow system can also be written more explicitly as

$$\dot{Q}_{in} + \dot{W}_{in} + \sum_{in} \dot{m}\theta = \dot{Q}_{out} + \dot{W}_{out} + \sum_{out} \dot{m}\theta \tag{5-35}$$

or

$$\dot{Q}_{in} + \dot{W}_{in} + \underbrace{\sum_{in} \dot{m}\left(h + \frac{V^2}{2} + gz\right)}_{\text{for each inlet}} = \dot{Q}_{out} + \dot{W}_{out} + \underbrace{\sum_{out} \dot{m}\left(h + \frac{V^2}{2} + gz\right)}_{\text{for each exit}} \tag{5-36}$$

since the energy of a flowing fluid per unit mass is $\theta = h + ke + pe = h + V^2/2 + gz$. The energy balance relation for steady-flow systems first appeared in 1859 in a German thermodynamics book written by Gustav Zeuner.

Consider, for example, an ordinary electric hot-water heater under steady operation, as shown in Fig. 5–21. A cold-water stream with a mass flow rate $\dot{m}$ is continuously flowing into the water heater, and a hot-water stream of the same mass flow rate is continuously flowing out of it. The water heater (the control volume) is losing heat to the surrounding air at a rate of $\dot{Q}_{out}$, and the electric heating element is supplying electrical work (heating) to the water at a rate of $\dot{W}_{in}$. On the basis of the conservation of energy principle, we can say that the water stream experiences an increase in its total energy as it flows through the water heater that is equal to the electric energy supplied to the water minus the heat losses.

The energy balance relation just given is intuitive in nature and is easy to use when the magnitudes and directions of heat and work transfers are known. When performing a general analytical study or solving a problem that involves an unknown heat or work interaction, however, we need to assume a direction for the heat or work interactions. In such cases, it is

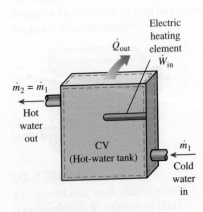

**FIGURE 5–21**
A water heater in steady operation.

common practice to assume heat to be transferred *into the system* (heat input) at a rate of $\dot{Q}$, and work produced *by the system* (work output) at a rate of $\dot{W}$, and then solve the problem. The first-law or energy balance relation in that case for a general steady-flow system becomes

$$\dot{Q} - \dot{W} = \sum_{out} \underbrace{\dot{m}\left(h + \frac{V^2}{2} + gz\right)}_{\text{for each exit}} - \sum_{in} \underbrace{\dot{m}\left(h + \frac{V^2}{2} + gz\right)}_{\text{for each inlet}} \qquad \textbf{(5–37)}$$

Obtaining a negative quantity for $\dot{Q}$ or $\dot{W}$ simply means that the assumed direction is wrong and should be reversed. For single-stream devices, the steady-flow energy balance equation becomes

$$\dot{Q} - \dot{W} = \dot{m}\left[h_2 - h_1 + \frac{V_2^2 - V_1^2}{2} + g(z_2 - z_1)\right] \qquad \textbf{(5–38)}$$

Dividing Eq. 5–38 by $\dot{m}$ gives the energy balance on a unit-mass basis as

$$q - w = h_2 - h_1 + \frac{V_2^2 - V_1^2}{2} + g(z_2 - z_1) \qquad \textbf{(5–39)}$$

where $q = \dot{Q}/\dot{m}$ and $w = \dot{W}/\dot{m}$ are the heat transfer and work done per unit mass of the working fluid, respectively. When the fluid experiences negligible changes in its kinetic and potential energies (that is, $\Delta ke \cong 0$, $\Delta pe \cong 0$), the energy balance equation is reduced further to

$$q - w = h_2 - h_1 \qquad \textbf{(5–40)}$$

The various terms appearing in the above equations are as follows:

$\dot{Q}$ = **rate of heat transfer between the control volume and its surroundings**. When the control volume is losing heat (as in the case of the water heater), $\dot{Q}$ is negative. If the control volume is well insulated (i.e., adiabatic), then $\dot{Q} = 0$.

$\dot{W}$ = **power**. For steady-flow devices, the control volume is constant; thus, there is no boundary work involved. The work required to push mass into and out of the control volume is also taken care of by using enthalpies for the energy of fluid streams instead of internal energies. Then $\dot{W}$ represents the remaining forms of work done per unit time (Fig. 5–22). Many steady-flow devices, such as turbines, compressors, and pumps, transmit power through a shaft, and $\dot{W}$ simply becomes the shaft power for those devices. If the control surface is crossed by electric wires (as in the case of an electric water heater), $\dot{W}$ represents the electrical work done per unit time. If neither is present, then $\dot{W} = 0$.

$\Delta h = h_2 - h_1$. The enthalpy change of a fluid can easily be determined by reading the enthalpy values at the exit and inlet states from the tables. For ideal gases, it can be approximated by $\Delta h = c_{p,\text{avg}}(T_2 - T_1)$. Note that $(kg/s)(kJ/kg) \equiv kW$.

$\Delta ke = (V_2^2 - V_1^2)/2$. The unit of kinetic energy is $m^2/s^2$, which is equivalent to J/kg (Fig. 5–23). The enthalpy is usually given in kJ/kg. To add these two quantities, the kinetic energy should be expressed in kJ/kg. This is easily accomplished by dividing it by 1000. A velocity of

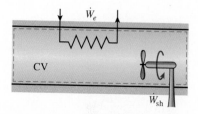

**FIGURE 5–22**
Under steady operation, shaft work and electrical work are the only forms of work a simple compressible system may involve.

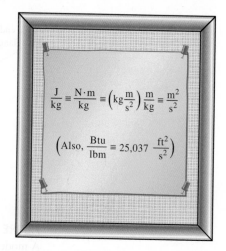

**FIGURE 5–23**
The units $m^2/s^2$ and J/kg are equivalent.

| $V_1$ m/s | $V_2$ m/s | $\Delta ke$ kJ/kg |
|-----------|-----------|-------------------|
| 0 | 45 | 1 |
| 50 | 67 | 1 |
| 100 | 110 | 1 |
| 200 | 205 | 1 |
| 500 | 502 | 1 |

**FIGURE 5–24**

At very high velocities, even small changes in velocities can cause significant changes in the kinetic energy of the fluid.

45 m/s corresponds to a kinetic energy of only 1 kJ/kg, which is a very small value compared with the enthalpy values encountered in practice. Thus, the kinetic energy term at low velocities can be neglected. When a fluid stream enters and leaves a steady-flow device at about the same velocity ($V_1 \cong V_2$), the change in the kinetic energy is close to zero regardless of the velocity. Caution should be exercised at high velocities, however, since small changes in velocities may cause significant changes in kinetic energy (Fig. 5–24).

$\Delta pe = g(z_2 - z_1)$. A similar argument can be given for the potential energy term. A potential energy change of 1 kJ/kg corresponds to an elevation difference of 102 m. The elevation difference between the inlet and exit of most industrial devices such as turbines and compressors is well below this value, and the potential energy term is always neglected for these devices. The only time the potential energy term is significant is when a process involves pumping a fluid to high elevations and we are interested in the required pumping power.

# 5–4 · SOME STEADY-FLOW ENGINEERING DEVICES

Many engineering devices operate essentially under the same conditions for long periods of time. The components of a steam power plant (turbines, compressors, heat exchangers, and pumps), for example, operate nonstop for months before the system is shut down for maintenance (Fig. 5–25). Therefore, these devices can be conveniently analyzed as steady-flow devices.

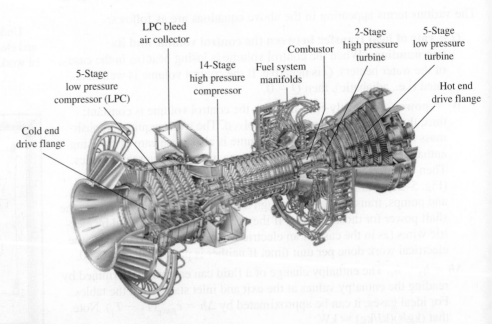

**FIGURE 5–25**

A modern land-based gas turbine used for electric power production. This is a General Electric LM5000 turbine. It has a length of 6.2 m, it weighs 12.5 tons, and produces 55.2 MW at 3600 rpm with steam injection.

*Courtesy of GE Power Systems.*

In this section, some common steady-flow devices are described, and the thermodynamic aspects of the flow through them are analyzed. The conservation of mass and the conservation of energy principles for these devices are illustrated with examples.

# 1 Nozzles and Diffusers

Nozzles and diffusers are commonly utilized in jet engines, rockets, spacecraft, and even garden hoses. A **nozzle** is a device that *increases the velocity of a fluid* at the expense of pressure. A **diffuser** is a device that *increases the pressure of a fluid* by slowing it down. That is, nozzles and diffusers perform opposite tasks. The cross-sectional area of a nozzle decreases in the flow direction for subsonic flows and increases for supersonic flows. The reverse is true for diffusers.

The rate of heat transfer between the fluid flowing through a nozzle or a diffuser and the surroundings is usually very small ($\dot{Q} \approx 0$) since the fluid has high velocities, and thus it does not spend enough time in the device for any significant heat transfer to take place. Nozzles and diffusers typically involve no work ($\dot{W} = 0$) and any change in potential energy is negligible ($\Delta pe \cong 0$). But nozzles and diffusers usually involve very high velocities, and as a fluid passes through a nozzle or diffuser, it experiences large changes in its velocity (Fig. 5–26). Therefore, the kinetic energy changes must be accounted for in analyzing the flow through these devices ($\Delta ke \neq 0$).

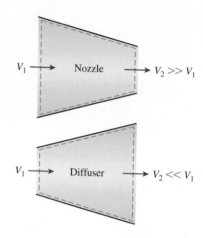

**FIGURE 5–26**
Nozzles and diffusers are shaped so that they cause large changes in fluid velocities and thus kinetic energies.

---

■ **EXAMPLE 5–4**    **Deceleration of Air in a Diffuser**

■ Air at 10°C and 80 kPa enters the diffuser of a jet engine steadily with a velocity of 200 m/s. The inlet area of the diffuser is 0.4 m². The air leaves the diffuser with a velocity that is very small compared with the inlet velocity. Determine (a) the mass flow rate of the air and (b) the temperature of the air leaving the diffuser.

**SOLUTION**  Air enters the diffuser of a jet engine steadily at a specified velocity. The mass flow rate of air and the temperature at the diffuser exit are to be determined.

**Assumptions**  **1** This is a steady-flow process since there is no change with time at any point and thus $\Delta m_{CV} = 0$ and $\Delta E_{CV} = 0$. **2** Air is an ideal gas since it is at a high temperature and low pressure relative to its critical-point values. **3** The potential energy change is zero, $\Delta pe = 0$. **4** Heat transfer is negligible. **5** Kinetic energy at the diffuser exit is negligible. **6** There are no work interactions.

**Analysis**  We take the *diffuser* as the system (Fig. 5–27). This is a *control volume* since mass crosses the system boundary during the process. We observe that there is only one inlet and one exit and thus $\dot{m}_1 = \dot{m}_2 = \dot{m}$.

(a) To determine the mass flow rate, we need to find the specific volume of the air first. This is determined from the ideal-gas relation at the inlet conditions:

$$v_1 = \frac{RT_1}{P_1} = \frac{(0.287 \text{ kPa·m}^3/\text{kg·K})(283 \text{ K})}{80 \text{ kPa}} = 1.015 \text{ m}^3/\text{kg}$$

**FIGURE 5–27**
The diffuser of a jet engine discussed in Example 5–4.

*Photo by Yunus Çengel*

Then,

$$\dot{m} = \frac{1}{v_1} V_1 A_1 = \frac{1}{1.015 \text{ m}^3/\text{kg}} (200 \text{ m/s})(0.4 \text{ m}^2) = 78.8 \text{ kg/s}$$

Since the flow is steady, the mass flow rate through the entire diffuser remains constant at this value.

(b) Under stated assumptions and observations, the energy balance for this steady-flow system can be expressed in the rate form as

$$\underbrace{\dot{E}_{in} - \dot{E}_{out}}_{\substack{\text{Rate of net energy transfer} \\ \text{by heat, work, and mass}}} = \underbrace{dE_{system}/dt}_{\substack{\text{Rate of change in internal, kinetic,} \\ \text{potential, etc., energies}}} \nearrow^{0 \text{ (steady)}} = 0$$

$$\dot{E}_{in} = \dot{E}_{out}$$

$$\dot{m}\left(h_1 + \frac{V_1^2}{2}\right) = \dot{m}\left(h_2 + \frac{V_2^2}{2}\right) \qquad (\text{since } \dot{Q} \cong 0, \dot{W} = 0, \text{ and } \Delta pe \cong 0)$$

$$h_2 = h_1 - \frac{V_2^2 - V_1^2}{2}$$

The exit velocity of a diffuser is usually small compared with the inlet velocity ($V_2 << V_1$); thus, the kinetic energy at the exit can be neglected. The enthalpy of air at the diffuser inlet is determined from the air table (Table A–17) to be

$$h_1 = h_{@ 283 \text{ K}} = 283.14 \text{ kJ/kg}$$

Substituting, we get

$$h_2 = 283.14 \text{ kJ/kg} - \frac{0 - (200 \text{ m/s})^2}{2}\left(\frac{1 \text{ kJ/kg}}{1000 \text{ m}^2/\text{s}^2}\right)$$

$$= 303.14 \text{ kJ/kg}$$

From Table A–17, the temperature corresponding to this enthalpy value is

$$T_2 = 303 \text{ K}$$

**Discussion** This result shows that the temperature of the air increases by about 20°C as it is slowed down in the diffuser. The temperature rise of the air is mainly due to the conversion of kinetic energy to internal energy.

---

**EXAMPLE 5–5    Acceleration of Steam in a Nozzle**

Steam at 1.8 MPa and 400°C steadily enters a nozzle whose inlet area is 0.02 m². The mass flow rate of steam through the nozzle is 5 kg/s. Steam leaves the nozzle at 1.4 MPa with a velocity of 275 m/s. Heat losses from the nozzle per unit mass of the steam are estimated to be 2.8 kJ/kg. Determine (a) the inlet velocity and (b) the exit temperature of the steam.

**SOLUTION** Steam enters a nozzle steadily at a specified flow rate and velocity. The inlet velocity of steam and the exit temperature are to be determined.

**Assumptions** **1** This is a steady-flow process since there is no change with time at any point and thus $\Delta m_{CV} = 0$ and $\Delta E_{CV} = 0$. **2** There are no work interactions. **3** The potential energy change is zero, $\Delta pe = 0$.

**Analysis** We take the *nozzle* as the system (Fig. 5–28). This is a *control volume* since mass crosses the system boundary during the process. We observe that there is only one inlet and one exit and thus $\dot{m}_1 = \dot{m}_2 = \dot{m}$.

(a) The specific volume and enthalpy of steam at the nozzle inlet are

$$\left. \begin{array}{l} P_1 = 1.8 \text{ MPa} \\ T_1 = 400°C \end{array} \right\} \begin{array}{l} v_1 = 0.16849 \text{ m}^3/\text{kg} \\ h_1 = 3251.6 \text{ kJ/kg} \end{array} \quad \text{(Table A–6)}$$

Then,

$$\dot{m} = \frac{1}{v_1} V_1 A_1$$

$$5 \text{ kg/s} = \frac{1}{0.16849 \text{ m}^3/\text{kg}} (V_1)(0.02 \text{ m}^2)$$

$$V_1 = \textbf{42.1 m/s}$$

(b) Under stated assumptions and observations, the energy balance for this steady-flow system can be expressed in the rate form as

$$\underbrace{\dot{E}_{in} - \dot{E}_{out}}_{\substack{\text{Rate of net energy transfer} \\ \text{by heat, work, and mass}}} = \underbrace{dE_{system}/dt}_{\substack{\text{Rate of change in internal, kinetic,} \\ \text{potential, etc., energies}}} \nearrow^{0 \text{ (steady)}} = 0$$

$$\dot{E}_{in} = \dot{E}_{out}$$

$$\dot{m}\left(h_1 + \frac{V_1^2}{2}\right) = \dot{Q}_{out} + \dot{m}\left(h_2 + \frac{V_2^2}{2}\right) \quad \text{(since } \dot{W} = 0, \text{ and } \Delta pe \cong 0)$$

Dividing by the mass flow rate $\dot{m}$ and substituting, $h_2$ is determined to be

$$h_2 = h_1 - q_{out} - \frac{V_2^2 - V_1^2}{2}$$

$$= (3251.6 - 2.8) \text{ kJ/kg} - \frac{(275 \text{ m/s})^2 - (42.1 \text{ m/s})^2}{2} \left(\frac{1 \text{ kJ/kg}}{1000 \text{ m}^2/\text{s}^2}\right)$$

$$= 3211.9 \text{ kJ/kg}$$

Then,

$$\left. \begin{array}{l} P_2 = 1.4 \text{ MPa} \\ h_2 = 3211.9 \text{ kJ/kg} \end{array} \right\} \quad T_2 = \textbf{378.6°C} \quad \text{(Table A–6)}$$

**Discussion** Note that the temperature of steam drops by 21.4°C as it flows through the nozzle. This drop in temperature is mainly due to the conversion of internal energy to kinetic energy. (The heat loss is too small to cause any significant effect in this case.)

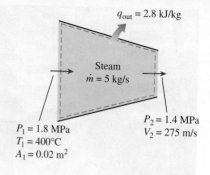

$q_{out} = 2.8 \text{ kJ/kg}$

Steam
$\dot{m} = 5 \text{ kg/s}$

$P_1 = 1.8 \text{ MPa}$
$T_1 = 400°C$
$A_1 = 0.02 \text{ m}^2$

$P_2 = 1.4 \text{ MPa}$
$V_2 = 275 \text{ m/s}$

**FIGURE 5–28**
Schematic for Example 5–5.

**FIGURE 5–29**
Turbine blades attached to the turbine shaft.
*©Royalty-Free/Corbis*

## 2 Turbines and Compressors

In steam, gas, or hydroelectric power plants, the device that drives the electric generator is the turbine. As the fluid passes through the turbine, work is done against the blades, which are attached to the shaft. As a result, the shaft rotates, and the turbine produces work (Fig. 5–29).

Compressors, as well as pumps and fans, are devices used to increase the pressure of a fluid. Work is supplied to these devices from an external source through a rotating shaft. Therefore, compressors involve work inputs. Even though these three devices function similarly, they do differ in the tasks they perform. A *fan* increases the pressure of a gas slightly and is mainly used to mobilize a gas. A *compressor* is capable of compressing the gas to very high pressures. *Pumps* work very much like compressors except that they handle liquids instead of gases.

Note that turbines produce power output whereas compressors, pumps, and fans require power input. Heat transfer from turbines is usually negligible ($\dot{Q} \approx 0$) since they are typically well insulated. Heat transfer is also negligible for compressors unless there is intentional cooling. Potential energy changes are negligible for all of these devices ($\Delta pe \cong 0$). The velocities involved in these devices, with the exception of turbines and fans, are usually too low to cause any significant change in the kinetic energy ($\Delta ke \cong 0$). The fluid velocities encountered in most turbines are very high, and the fluid experiences a significant change in its kinetic energy. However, this change is usually very small relative to the change in enthalpy, and thus it is often disregarded.

---

**EXAMPLE 5–6**   **Compressing Air by a Compressor**

Air at 100 kPa and 280 K is compressed steadily to 600 kPa and 400 K. The mass flow rate of the air is 0.02 kg/s, and a heat loss of 16 kJ/kg occurs during the process. Assuming the changes in kinetic and potential energies are negligible, determine the necessary power input to the compressor.

**SOLUTION**   Air is compressed steadily by a compressor to a specified temperature and pressure. The power input to the compressor is to be determined.

**Assumptions   1** This is a steady-flow process since there is no change with time at any point and thus $\Delta m_{CV} = 0$ and $\Delta E_{CV} = 0$. **2** Air is an ideal gas since it is at a high temperature and low pressure relative to its critical-point values. **3** The kinetic and potential energy changes are zero, $\Delta ke = \Delta pe = 0$.

**Analysis**   We take the *compressor* as the system (Fig. 5–30). This is a *control volume* since mass crosses the system boundary during the process. We observe that there is only one inlet and one exit and thus $\dot{m}_1 = \dot{m}_2 = \dot{m}$. Also, heat is lost from the system and work is supplied to the system.

Under stated assumptions and observations, the energy balance for this steady-flow system can be expressed in the rate form as

$$\underbrace{\dot{E}_{in} - \dot{E}_{out}}_{\substack{\text{Rate of net energy transfer} \\ \text{by heat, work, and mass}}} = \underbrace{dE_{system}/dt}_{\substack{\text{Rate of change in internal, kinetic,} \\ \text{potential, etc., energies}}} \overset{0 \text{ (steady)}}{\nearrow} = 0$$

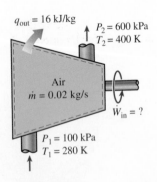

$q_{out} = 16$ kJ/kg

$P_2 = 600$ kPa
$T_2 = 400$ K

Air
$\dot{m} = 0.02$ kg/s

$\dot{W}_{in} = ?$

$P_1 = 100$ kPa
$T_1 = 280$ K

**FIGURE 5–30**
Schematic for Example 5–6.

$$\dot{E}_{in} = \dot{E}_{out}$$

$$\dot{W}_{in} + \dot{m}h_1 = \dot{Q}_{out} + \dot{m}h_2 \quad (\text{since } \Delta\text{ke} = \Delta\text{pe} \cong 0)$$

$$\dot{W}_{in} = \dot{m}q_{out} + \dot{m}(h_2 - h_1)$$

The enthalpy of an ideal gas depends on temperature only, and the enthalpies of the air at the specified temperatures are determined from the air table (Table A–17) to be

$$h_1 = h_{@\,280\,K} = 280.13 \text{ kJ/kg}$$

$$h_2 = h_{@\,400\,K} = 400.98 \text{ kJ/kg}$$

Substituting, the power input to the compressor is determined to be

$$\dot{W}_{in} = (0.02 \text{ kg/s})(16 \text{ kJ/kg}) + (0.02 \text{ kg/s})(400.98 - 280.13) \text{ kJ/kg}$$

$$= 2.74 \text{ kW}$$

**Discussion** Note that the mechanical energy input to the compressor manifests itself as a rise in enthalpy of air and heat loss from the compressor.

■
■ **EXAMPLE 5–7**  **Power Generation by a Steam Turbine**
■
■ The power output of an adiabatic steam turbine is 5 MW, and the inlet and
■ the exit conditions of the steam are as indicated in Fig. 5–31.

(a) Compare the magnitudes of $\Delta h$, $\Delta$ke, and $\Delta$pe.
(b) Determine the work done per unit mass of the steam flowing through the turbine.
(c) Calculate the mass flow rate of the steam.

**SOLUTION** The inlet and exit conditions of a steam turbine and its power output are given. The changes in kinetic energy, potential energy, and enthalpy of steam, as well as the work done per unit mass and the mass flow rate of steam are to be determined.
**Assumptions** 1 This is a steady-flow process since there is no change with time at any point and thus $\Delta m_{CV} = 0$ and $\Delta E_{CV} = 0$. 2 The system is adiabatic and thus there is no heat transfer.
**Analysis** We take the *turbine* as the system. This is a *control volume* since mass crosses the system boundary during the process. We observe that there is only one inlet and one exit and thus $\dot{m}_1 = \dot{m}_2 = \dot{m}$. Also, work is done by the system. The inlet and exit velocities and elevations are given, and thus the kinetic and potential energies are to be considered.

(a) At the inlet, steam is in a superheated vapor state, and its enthalpy is

$$\left. \begin{array}{r} P_1 = 2 \text{ MPa} \\ T_1 = 400°C \end{array} \right\} \quad h_1 = 3248.4 \text{ kJ/kg} \quad \text{(Table A–6)}$$

At the turbine exit, we obviously have a saturated liquid–vapor mixture at 15-kPa pressure. The enthalpy at this state is

$$h_2 = h_f + x_2 h_{fg} = [225.94 + (0.9)(2372.3)] \text{ kJ/kg} = 2361.01 \text{ kJ/kg}$$

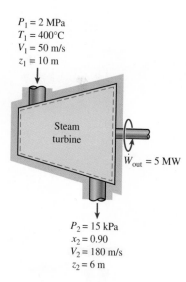

$P_1 = 2$ MPa
$T_1 = 400°C$
$V_1 = 50$ m/s
$z_1 = 10$ m

Steam turbine

$\dot{W}_{out} = 5$ MW

$P_2 = 15$ kPa
$x_2 = 0.90$
$V_2 = 180$ m/s
$z_2 = 6$ m

**FIGURE 5–31**
Schematic for Example 5–7.

Then

$$\Delta h = h_2 - h_1 = (2361.01 - 3248.4) \text{ kJ/kg} = -887.39 \text{ kJ/kg}$$

$$\Delta \text{ke} = \frac{V_2^2 - V_1^2}{2} = \frac{(180 \text{ m/s})^2 - (50 \text{ m/s})^2}{2}\left(\frac{1 \text{ kJ/kg}}{1000 \text{ m}^2/\text{s}^2}\right) = 14.95 \text{ kJ/kg}$$

$$\Delta \text{pe} = g(z_2 - z_1) = (9.81 \text{ m/s}^2)[(6 - 10) \text{ m}]\left(\frac{1 \text{ kJ/kg}}{1000 \text{ m}^2/\text{s}^2}\right) = -0.04 \text{ kJ/kg}$$

(b) The energy balance for this steady-flow system can be expressed in the rate form as

$$\underbrace{\dot{E}_{\text{in}} - \dot{E}_{\text{out}}}_{\substack{\text{Rate of net energy transfer} \\ \text{by heat, work, and mass}}} = \underbrace{dE_{\text{system}}/dt}_{\substack{\text{Rate of change in internal, kinetic,} \\ \text{potential, etc., energies}}}^{\nearrow 0 \text{ (steady)}} = 0$$

$$\dot{E}_{\text{in}} = \dot{E}_{\text{out}}$$

$$\dot{m}\left(h_1 + \frac{V_1^2}{2} + gz_1\right) = \dot{W}_{\text{out}} + \dot{m}\left(h_2 + \frac{V_2^2}{2} + gz_2\right) \quad (\text{since } \dot{Q} = 0)$$

Dividing by the mass flow rate $\dot{m}$ and substituting, the work done by the turbine per unit mass of the steam is determined to be

$$w_{\text{out}} = -\left[(h_2 - h_1) + \frac{V_2^2 - V_1^2}{2} + g(z_2 - z_1)\right] = -(\Delta h + \Delta \text{ke} + \Delta \text{pe})$$

$$= -[-887.39 + 14.95 - 0.04] \text{ kJ/kg} = 872.48 \text{ kJ/kg}$$

(c) The required mass flow rate for a 5-MW power output is

$$\dot{m} = \frac{\dot{W}_{\text{out}}}{w_{\text{out}}} = \frac{5000 \text{ kJ/s}}{872.48 \text{ kJ/kg}} = 5.73 \text{ kg/s}$$

**Discussion** Two observations can be made from these results. First, the change in potential energy is insignificant in comparison to the changes in enthalpy and kinetic energy. This is typical for most engineering devices. Second, as a result of low pressure and thus high specific volume, the steam velocity at the turbine exit can be very high. Yet the change in kinetic energy is a small fraction of the change in enthalpy (less than 2 percent in our case) and is therefore often neglected.

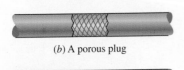

(a) An adjustable valve

(b) A porous plug

(c) A capillary tube

**FIGURE 5–32**
Throttling valves are devices that cause large pressure drops in the fluid.

# 3 Throttling Valves

Throttling valves are *any kind of flow-restricting devices* that cause a significant pressure drop in the fluid. Some familiar examples are ordinary adjustable valves, capillary tubes, and porous plugs (Fig. 5–32). Unlike turbines, they produce a pressure drop without involving any work. The pressure drop in the fluid is often accompanied by a *large drop in temperature*, and for that reason throttling devices are commonly used in refrigeration and air-conditioning applications. The magnitude of the temperature drop (or, sometimes, the temperature rise) during a throttling process is governed by a property called the *Joule-Thomson coefficient,* discussed in Chap. 12.

Throttling valves are usually small devices, and the flow through them may be assumed to be adiabatic ($q \cong 0$) since there is neither sufficient time nor large enough area for any effective heat transfer to take place. Also, there is no work done ($w = 0$), and the change in potential energy, if any, is very small ($\Delta\text{pe} \cong 0$). Even though the exit velocity is often considerably higher than the inlet velocity, in many cases, the increase in kinetic energy is insignificant ($\Delta\text{ke} \cong 0$). Then the conservation of energy equation for this single-stream steady-flow device reduces to

$$h_2 \cong h_1 \quad \text{(kJ/kg)} \tag{5-41}$$

That is, enthalpy values at the inlet and exit of a throttling valve are the same. For this reason, a throttling valve is sometimes called an *isenthalpic device.* Note, however, that for throttling devices with large exposed surface areas such as capillary tubes, heat transfer may be significant.

To gain some insight into how throttling affects fluid properties, let us express Eq. 5–41 as follows:

$$u_1 + P_1 v_1 = u_2 + P_2 v_2$$

or

$$\text{Internal energy} + \text{Flow energy} = \text{Constant}$$

Thus the final outcome of a throttling process depends on which of the two quantities increases during the process. If the flow energy increases during the process ($P_2 v_2 > P_1 v_1$), it can do so at the expense of the internal energy. As a result, internal energy decreases, which is usually accompanied by a drop in temperature. If the product $Pv$ decreases, the internal energy and the temperature of a fluid will increase during a throttling process. In the case of an ideal gas, $h = h(T)$, and thus the temperature has to remain constant during a throttling process (Fig. 5–33).

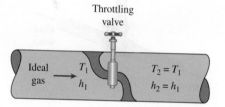

**FIGURE 5–33**
The temperature of an ideal gas does not change during a throttling ($h = \text{constant}$) process since $h = h(T)$.

---

■ **EXAMPLE 5–8**    **Expansion of Refrigerant-134a in a Refrigerator**

■ Refrigerant-134a enters the capillary tube of a refrigerator as saturated liq-
■ uid at 0.8 MPa and is throttled to a pressure of 0.12 MPa. Determine the
■ quality of the refrigerant at the final state and the temperature drop during
■ this process.

**SOLUTION**    Refrigerant-134a that enters a capillary tube as saturated liquid is throttled to a specified pressure. The exit quality of the refrigerant and the temperature drop are to be determined.
*Assumptions*  **1** Heat transfer from the tube is negligible. **2** Kinetic energy change of the refrigerant is negligible.
*Analysis*  A capillary tube is a simple flow-restricting device that is commonly used in refrigeration applications to cause a large pressure drop in the

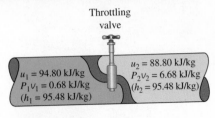

Throttling
valve

$u_1 = 94.80$ kJ/kg
$P_1 v_1 = 0.68$ kJ/kg
$(h_1 = 95.48$ kJ/kg)

$u_2 = 88.80$ kJ/kg
$P_2 v_2 = 6.68$ kJ/kg
$(h_2 = 95.48$ kJ/kg)

**FIGURE 5–34**
During a throttling process, the enthalpy (flow energy + internal energy) of a fluid remains constant. But internal and flow energies may be converted to each other.

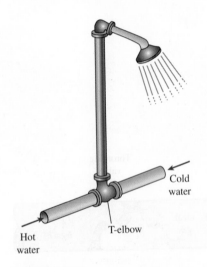

Cold
water

T-elbow

Hot
water

**FIGURE 5–35**
The T-elbow of an ordinary shower serves as the mixing chamber for the hot- and the cold-water streams.

refrigerant. Flow through a capillary tube is a throttling process; thus, the enthalpy of the refrigerant remains constant (Fig. 5–34).

$$\text{At inlet:} \quad \left.\begin{array}{l} P_1 = 0.8 \text{ MPa} \\ \text{sat. liquid} \end{array}\right\} \quad \begin{array}{l} T_1 = T_{\text{sat @ 0.8 MPa}} = 31.31°C \\ h_1 = h_{f\text{ @ 0.8 MPa}} = 95.48 \text{ kJ/kg} \end{array} \quad \text{(Table A–12)}$$

$$\text{At exit:} \quad \begin{array}{l} P_2 = 0.12 \text{ MPa} \\ (h_2 = h_1) \end{array} \longrightarrow \begin{array}{l} h_f = 22.47 \text{ kJ/kg} \quad T_{\text{sat}} = -22.32°C \\ h_g = 236.99 \text{ kJ/kg} \end{array}$$

Obviously $h_f < h_2 < h_g$; thus, the refrigerant exists as a saturated mixture at the exit state. The quality at this state is

$$x_2 = \frac{h_2 - h_f}{h_{fg}} = \frac{95.48 - 22.47}{236.99 - 22.47} = \mathbf{0.340}$$

Since the exit state is a saturated mixture at 0.12 MPa, the exit temperature must be the saturation temperature at this pressure, which is −22.32°C. Then the temperature change for this process becomes

$$\Delta T = T_2 - T_1 = (-22.32 - 31.31)°C = \mathbf{-53.63°C}$$

***Discussion*** Note that the temperature of the refrigerant drops by 53.63°C during this throttling process. Also note that 34.0 percent of the refrigerant vaporizes during this throttling process, and the energy needed to vaporize this refrigerant is absorbed from the refrigerant itself.

## 4a   Mixing Chambers

In engineering applications, mixing two streams of fluids is not a rare occurrence. The section where the mixing process takes place is commonly referred to as a **mixing chamber**. The mixing chamber does not have to be a distinct "chamber." An ordinary T-elbow or a Y-elbow in a shower, for example, serves as the mixing chamber for the cold- and hot-water streams (Fig. 5–35).

The conservation of mass principle for a mixing chamber requires that the sum of the incoming mass flow rates equal the mass flow rate of the outgoing mixture.

Mixing chambers are usually well insulated ($q \cong 0$) and usually do not involve any kind of work ($w = 0$). Also, the kinetic and potential energies of the fluid streams are usually negligible (ke $\cong$ 0, pe $\cong$ 0). Then all there is left in the energy equation is the total energies of the incoming streams and the outgoing mixture. The conservation of energy principle requires that these two equal each other. Therefore, the conservation of energy equation becomes analogous to the conservation of mass equation for this case.

***EXAMPLE 5–9***     **Mixing of Hot and Cold Waters in a Shower**

Consider an ordinary shower where hot water at 60°C is mixed with cold water at 10°C. If it is desired that a steady stream of warm water at 45°C be supplied, determine the ratio of the mass flow rates of the hot to cold water. Assume the heat losses from the mixing chamber to be negligible and the mixing to take place at a pressure of 150 kPa.

**SOLUTION** In a shower, cold water is mixed with hot water at a specified temperature. For a specified mixture temperature, the ratio of the mass flow rates of the hot to cold water is to be determined.

**Assumptions** **1** This is a steady-flow process since there is no change with time at any point and thus $\Delta m_{CV} = 0$ and $\Delta E_{CV} = 0$. **2** The kinetic and potential energies are negligible, ke $\cong$ pe $\cong$ 0. **3** Heat losses from the system are negligible and thus $\dot{Q} \cong 0$. **4** There is no work interaction involved.

**Analysis** We take the *mixing chamber* as the system (Fig. 5–36). This is a *control volume* since mass crosses the system boundary during the process. We observe that there are two inlets and one exit.

Under the stated assumptions and observations, the mass and energy balances for this steady-flow system can be expressed in the rate form as follows:

Mass balance: $\quad \dot{m}_{in} - \dot{m}_{out} = \underbrace{dm_{system}/dt}_{\nearrow\ 0\ (steady)} = 0$

$$\dot{m}_{in} = \dot{m}_{out} \rightarrow \dot{m}_1 + \dot{m}_2 = \dot{m}_3$$

Energy balance: $\quad \underbrace{\dot{E}_{in} - \dot{E}_{out}}_{\substack{\text{Rate of net energy transfer} \\ \text{by heat, work, and mass}}} = \underbrace{dE_{system}/dt}_{\substack{\text{Rate of change in internal, kinetic,} \\ \text{potential, etc., energies}}}^{\nearrow\ 0\ (steady)} = 0$

$$\dot{E}_{in} = \dot{E}_{out}$$

$$\dot{m}_1 h_1 + \dot{m}_2 h_2 = \dot{m}_3 h_3 \quad (\text{since } \dot{Q} \cong 0, \dot{W} = 0, \text{ke} \cong \text{pe} \cong 0)$$

Combining the mass and energy balances,

$$\dot{m}_1 h_1 + \dot{m}_2 h_2 = (\dot{m}_1 + \dot{m}_2)h_3$$

Dividing this equation by $\dot{m}_2$ yields

$$y h_1 + h_2 = (y + 1)h_3$$

where $y = \dot{m}_1/\dot{m}_2$ is the desired mass flow rate ratio.

The saturation temperature of water at 150 kPa is 111.35°C. Since the temperatures of all three streams are below this value ($T < T_{sat}$), the water in all three streams exists as a compressed liquid (Fig. 5–37). A compressed liquid can be approximated as a saturated liquid at the given temperature. Thus,

$$h_1 \cong h_{f\,@\,60°C} = 251.18 \text{ kJ/kg}$$

$$h_2 \cong h_{f\,@\,10°C} = 42.022 \text{ kJ/kg}$$

$$h_3 \cong h_{f\,@\,45°C} = 188.44 \text{ kJ/kg}$$

Solving for $y$ and substituting yields

$$y = \frac{h_3 - h_2}{h_1 - h_3} = \frac{188.44 - 42.022}{251.18 - 188.44} = \mathbf{2.33}$$

**Discussion** Note that the mass flow rate of the hot water must be 2.33 times the mass flow rate of the cold water for the mixture to leave at 45°C.

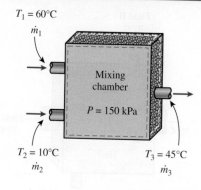

$T_1 = 60°C$
$\dot{m}_1$

Mixing chamber

$P = 150$ kPa

$T_2 = 10°C$
$\dot{m}_2$

$T_3 = 45°C$
$\dot{m}_3$

**FIGURE 5–36**
Schematic for Example 5–9.

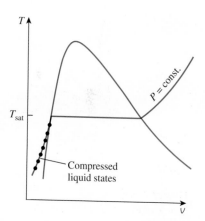

**FIGURE 5–37**
A substance exists as a compressed liquid at temperatures below the saturation temperatures at the given pressure.

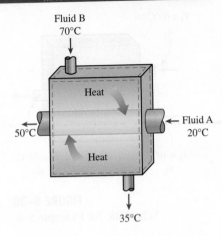

**FIGURE 5–38**
A heat exchanger can be as simple as
two concentric pipes.

## 4b  Heat Exchangers

As the name implies, **heat exchangers** are devices where two moving fluid
streams exchange heat without mixing. Heat exchangers are widely used in
various industries, and they come in various designs.

The simplest form of a heat exchanger is a *double-tube* (also called
*tube-and-shell*) *heat exchanger,* shown in Fig. 5–38. It is composed of
two concentric pipes of different diameters. One fluid flows in the inner
pipe, and the other in the annular space between the two pipes. Heat is
transferred from the hot fluid to the cold one through the wall separating
them. Sometimes the inner tube makes a couple of turns inside the shell to
increase the heat transfer area, and thus the rate of heat transfer. The mixing chambers discussed earlier are sometimes classified as *direct-contact*
heat exchangers.

The conservation of mass principle for a heat exchanger in steady operation requires that the sum of the inbound mass flow rates equal the sum of
the outbound mass flow rates. This principle can also be expressed as follows: *Under steady operation, the mass flow rate of each fluid stream flowing through a heat exchanger remains constant.*

Heat exchangers typically involve no work interactions ($w = 0$) and
negligible kinetic and potential energy changes ($\Delta\text{ke} \cong 0$, $\Delta\text{pe} \cong 0$) for
each fluid stream. The heat transfer rate associated with heat exchang-
ers depends on how the control volume is selected. Heat exchangers are
intended for heat transfer between two fluids *within* the device, and the
outer shell is usually well insulated to prevent any heat loss to the surrounding medium.

When the entire heat exchanger is selected as the control volume, $\dot{Q}$
becomes zero, since the boundary for this case lies just beneath the insulation and little or no heat crosses the boundary (Fig. 5–39). If, however,
only one of the fluids is selected as the control volume, then heat will
cross this boundary as it flows from one fluid to the other and $\dot{Q}$ will not
be zero. In fact, $\dot{Q}$ in this case will be the rate of heat transfer between the
two fluids.

**FIGURE 5–39**
The heat transfer associated with
a heat exchanger may be zero or
nonzero depending on how the control
volume is selected.

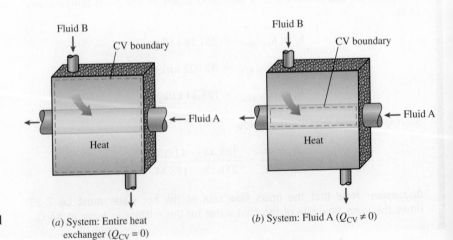

(a) System: Entire heat
exchanger ($Q_{CV} = 0$)

(b) System: Fluid A ($Q_{CV} \neq 0$)

### ■ EXAMPLE 5–10    Cooling of Refrigerant-134a by Water

Refrigerant-134a is to be cooled by water in a condenser. The refrigerant enters the condenser with a mass flow rate of 6 kg/min at 1 MPa and 70°C and leaves at 35°C. The cooling water enters at 300 kPa and 15°C and leaves at 25°C. Neglecting any pressure drops, determine (a) the mass flow rate of the cooling water required and (b) the heat transfer rate from the refrigerant to water.

**SOLUTION**  Refrigerant-134a is cooled by water in a condenser. The mass flow rate of the cooling water and the rate of heat transfer from the refrigerant to the water are to be determined.

*Assumptions*  **1** This is a steady-flow process since there is no change with time at any point and thus $\Delta m_{CV} = 0$ and $\Delta E_{CV} = 0$. **2** The kinetic and potential energies are negligible, ke $\cong$ pe $\cong$ 0. **3** Heat losses from the system are negligible and thus $\dot{Q} \cong 0$. **4** There is no work interaction.

*Analysis*  We take the *entire heat exchanger* as the system (Fig. 5–40). This is a *control volume* since mass crosses the system boundary during the process. In general, there are several possibilities for selecting the control volume for multiple-stream steady-flow devices, and the proper choice depends on the situation at hand. We observe that there are two fluid streams (and thus two inlets and two exits) but no mixing.

(a) Under the stated assumptions and observations, the mass and energy balances for this steady-flow system can be expressed in the rate form as follows:

*Mass balance:*
$$\dot{m}_{in} = \dot{m}_{out}$$

for each fluid stream since there is no mixing. Thus,

$$\dot{m}_1 = \dot{m}_2 = \dot{m}_w$$

$$\dot{m}_3 = \dot{m}_4 = \dot{m}_R$$

*Energy balance:*

$$\underbrace{\dot{E}_{in} - \dot{E}_{out}}_{\substack{\text{Rate of net energy transfer} \\ \text{by heat, work, and mass}}} = \underbrace{dE_{system}/dt}_{\substack{\text{Rate of change in internal, kinetic,} \\ \text{potential, etc., energies}}}^{\nearrow 0 \text{ (steady)}} = 0$$

$$\dot{E}_{in} = \dot{E}_{out}$$

$$\dot{m}_1 h_1 + \dot{m}_3 h_3 = \dot{m}_2 h_2 + \dot{m}_4 h_4 \quad (\text{since } \dot{Q} \cong 0, \dot{W} = 0, \text{ke} \cong \text{pe} \cong 0)$$

Combining the mass and energy balances and rearranging give

$$\dot{m}_w(h_1 - h_2) = \dot{m}_R(h_4 - h_3)$$

Now we need to determine the enthalpies at all four states. Water exists as a compressed liquid at both the inlet and the exit since the temperatures at both locations are below the saturation temperature of water at 300 kPa (133.52°C). Approximating the compressed liquid as a saturated liquid at the given temperatures, we have

$$h_1 \cong h_{f\,@\,15°C} = 62.982 \text{ kJ/kg}$$

$$h_2 \cong h_{f\,@\,25°C} = 104.83 \text{ kJ/kg}$$

(Table A–4)

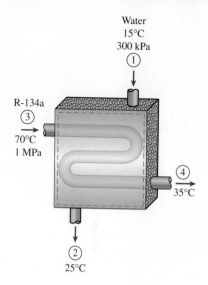

Water
15°C
300 kPa
①

R-134a
③
⟶
70°C
1 MPa

④
⟶
35°C

②
↓
25°C

**FIGURE 5–40**
Schematic for Example 5–10.

The refrigerant enters the condenser as a superheated vapor and leaves as a compressed liquid at 35°C. From refrigerant-134a tables,

$$\left.\begin{array}{l} P_3 = 1 \text{ MPa} \\ T_3 = 70°C \end{array}\right\} \quad h_3 = 303.87 \text{ kJ/kg} \quad \text{(Table A–13)}$$

$$\left.\begin{array}{l} P_4 = 1 \text{ MPa} \\ T_4 = 35°C \end{array}\right\} \quad h_4 \cong h_{f\,@\,35°C} = 100.88 \text{ kJ/kg} \quad \text{(Table A–11)}$$

Substituting, we find

$$\dot{m}_w(62.982 - 104.83) \text{ kJ/kg} = (6 \text{ kg/min})[(100.88 - 303.87) \text{ kJ/kg}]$$

$$\dot{m}_w = \textbf{29.1 kg/min}$$

(b) To determine the heat transfer from the refrigerant to the water, we have to choose a control volume whose boundary lies on the path of heat transfer. We can choose the volume occupied by either fluid as our control volume. For no particular reason, we choose the volume occupied by the water. All the assumptions stated earlier apply, except that the heat transfer is no longer zero. Then assuming heat to be transferred to water, the energy balance for this single-stream steady-flow system reduces to

$$\underbrace{\dot{E}_{in} - \dot{E}_{out}}_{\substack{\text{Rate of net energy transfer} \\ \text{by heat, work, and mass}}} = \underbrace{dE_{system}/dt}_{\substack{\text{Rate of change in internal, kinetic,} \\ \text{potential, etc., energies}}}^{\nearrow 0 \text{ (steady)}} = 0$$

$$\dot{E}_{in} = \dot{E}_{out}$$

$$\dot{Q}_{w,\,in} + \dot{m}_w h_1 = \dot{m}_w h_2$$

Rearranging and substituting,

$$\dot{Q}_{w,in} = \dot{m}_w(h_2 - h_1) = (29.1 \text{ kg/min})[(104.83 - 62.982) \text{ kJ/kg}]$$

$$= \textbf{1218 kJ/min}$$

**Discussion** Had we chosen the volume occupied by the refrigerant as the control volume (Fig. 5–41), we would have obtained the same result for $\dot{Q}_{R,out}$ since the heat gained by the water is equal to the heat lost by the refrigerant.

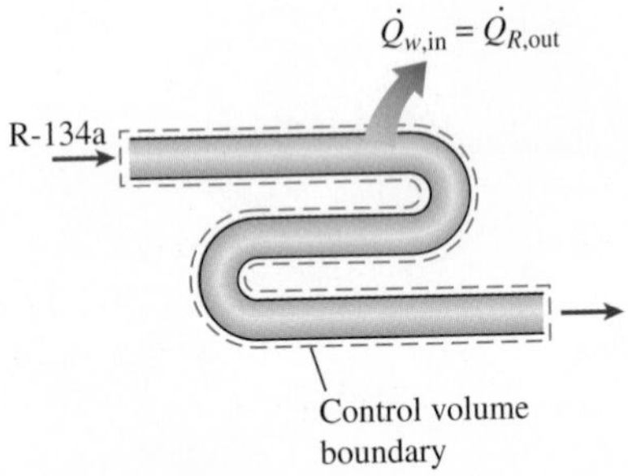

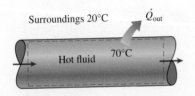

**FIGURE 5–41**

In a heat exchanger, the heat transfer depends on the choice of the control volume.

Surroundings 20°C  $\dot{Q}_{out}$

Hot fluid  70°C

**FIGURE 5–42**

Heat losses from a hot fluid flowing through an uninsulated pipe or duct to the cooler environment may be very significant.

## 5 Pipe and Duct Flow

The transport of liquids or gases in pipes and ducts is of great importance in many engineering applications. Flow through a pipe or a duct usually satisfies the steady-flow conditions and thus can be analyzed as a steady-flow process. This, of course, excludes the transient start-up and shut-down periods. The control volume can be selected to coincide with the interior surfaces of the portion of the pipe or the duct that we are interested in analyzing.

Under normal operating conditions, the amount of heat gained or lost by the fluid may be very significant, particularly if the pipe or duct is long (Fig. 5–42). Sometimes heat transfer is desirable and is the sole purpose of

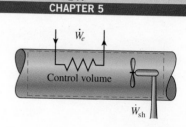

**FIGURE 5–43**
Pipe or duct flow may involve more than one form of work at the same time.

the flow. Water flow through the pipes in the furnace of a power plant, the flow of refrigerant in a freezer, and the flow in heat exchangers are some examples of this case. At other times, heat transfer is undesirable, and the pipes or ducts are insulated to prevent any heat loss or gain, particularly when the temperature difference between the flowing fluid and the surroundings is large. Heat transfer in this case is negligible.

If the control volume involves a heating section (electric wires), a fan, or a pump (shaft), the work interactions should be considered (Fig. 5–43). Of these, fan work is usually small and often neglected in energy analysis.

The velocities involved in pipe and duct flow are relatively low, and the kinetic energy changes are usually insignificant. This is particularly true when the pipe or duct diameter is constant and the heating effects are negligible. Kinetic energy changes may be significant, however, for gas flow in ducts with variable cross-sectional areas especially when the compressibility effects are significant. The potential energy term may also be significant when the fluid undergoes a considerable elevation change as it flows in a pipe or duct.

■ **EXAMPLE 5–11**   **Electric Heating of Air in a House**

The electric heating systems used in many houses consist of a simple duct with resistance heaters. Air is heated as it flows over resistance wires. Consider a 15-kW electric heating system. Air enters the heating section at 100 kPa and 17°C with a volume flow rate of 150 m³/min. If heat is lost from the air in the duct to the surroundings at a rate of 200 W, determine the exit temperature of air.

**SOLUTION**   The electric heating system of a house is considered. For specified electric power consumption and air flow rate, the air exit temperature is to be determined.
**Assumptions**   **1** This is a steady-flow process since there is no change with time at any point and thus $\Delta m_{CV} = 0$ and $\Delta E_{CV} = 0$. **2** Air is an ideal gas since it is at a high temperature and low pressure relative to its critical-point values. **3** The kinetic and potential energy changes are negligible, $\Delta ke \cong \Delta pe \cong 0$. **4** Constant specific heats at room temperature can be used for air.
**Analysis**   We take the *heating section portion of the duct* as the system (Fig. 5–44). This is a *control volume* since mass crosses the system boundary during the process. We observe that there is only one inlet and one exit and thus $\dot{m}_1 = \dot{m}_2 = \dot{m}$. Also, heat is lost from the system and electrical work is supplied to the system.

At temperatures encountered in heating and air-conditioning applications, $\Delta h$ can be replaced by $c_p \Delta T$ where $c_p = 1.005$ kJ/kg·°C—the value at room temperature—with negligible error (Fig. 5–45). Then the energy balance for this steady-flow system can be expressed in the rate form as

$$\underbrace{\dot{E}_{in} - \dot{E}_{out}}_{\substack{\text{Rate of net energy transfer} \\ \text{by heat, work, and mass}}} = \underbrace{dE_{system}/dt}_{\substack{\text{Rate of change in internal, kinetic,} \\ \text{potential, etc., energies}}}^{\nearrow 0 \text{ (steady)}} = 0$$

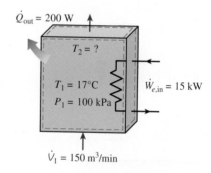

**FIGURE 5–44**
Schematic for Example 5–11.

$\Delta h = 1.005 \, \Delta T$ (kJ/kg)

**FIGURE 5–45**
The error involved in $\Delta h = c_p \, \Delta T$, where $c_p = 1.005$ kJ/kg·°C, is less than 0.5 percent for air in the temperature range $-20$ to $70$°C.

$$\dot{E}_{in} = \dot{E}_{out}$$

$$\dot{W}_{e,in} + \dot{m}h_1 = \dot{Q}_{out} + \dot{m}h_2 \quad (\text{since } \Delta ke \cong \Delta pe \cong 0)$$

$$\dot{W}_{e,in} - \dot{Q}_{out} = \dot{m}c_p(T_2 - T_1)$$

From the ideal-gas relation, the specific volume of air at the inlet of the duct is

$$v_1 = \frac{RT_1}{P_1} = \frac{(0.287 \text{ kPa·m}^3/\text{kg·K})(290 \text{ K})}{100 \text{ kPa}} = 0.832 \text{ m}^3/\text{kg}$$

The mass flow rate of the air through the duct is determined from

$$\dot{m} = \frac{\dot{V}_1}{v_1} = \frac{150 \text{ m}^3/\text{min}}{0.832 \text{ m}^3/\text{kg}} \left(\frac{1 \text{ min}}{60 \text{ s}}\right) = 3.0 \text{ kg/s}$$

Substituting the known quantities, the exit temperature of the air is determined to be

$$(15 \text{ kJ/s}) - (0.2 \text{ kJ/s}) = (3 \text{ kg/s})(1.005 \text{ kJ/kg·°C})(T_2 - 17)\text{°C}$$

$$T_2 = \mathbf{21.9°C}$$

**Discussion** Note that heat loss from the duct reduces the exit temperature of air.

## 5–5 · ENERGY ANALYSIS OF UNSTEADY-FLOW PROCESSES

During a steady-flow process, no changes occur within the control volume; thus, one does not need to be concerned about what is going on within the boundaries. Not having to worry about any changes within the control volume with time greatly simplifies the analysis.

Many processes of interest, however, involve *changes* within the control volume with time. Such processes are called *unsteady-flow,* or *transient-flow,* processes. The steady-flow relations developed earlier are obviously not applicable to these processes. When an unsteady-flow process is analyzed, it is important to keep track of the mass and energy contents of the control volume as well as the energy interactions across the boundary.

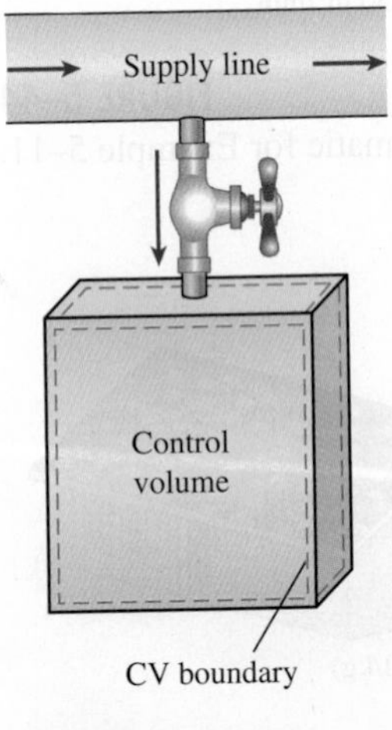

Some familiar unsteady-flow processes are the charging of rigid vessels from supply lines (Fig. 5–46), discharging a fluid from a pressurized vessel, driving a gas turbine with pressurized air stored in a large container, inflating tires or balloons, and even cooking with an ordinary pressure cooker.

Unlike steady-flow processes, unsteady-flow processes start and end over some finite time period instead of continuing indefinitely. Therefore in this section, we deal with changes that occur over some time interval $\Delta t$ instead of with the rate of changes (changes per unit time). An unsteady-flow system, in some respects, is similar to a closed system, except that the mass within the system boundaries does not remain constant during a process.

Another difference between steady- and unsteady-flow systems is that steady-flow systems are fixed in space, size, and shape. Unsteady-flow systems,

**FIGURE 5–46**
Charging of a rigid tank from a supply line is an unsteady-flow process since it involves changes within the control volume.

however, are not (Fig. 5–47). They are usually stationary; that is, they are fixed in space, but they may involve moving boundaries and thus boundary work.

The *mass balance* for any system undergoing any process can be expressed as (*see* Sec. 5–1)

$$m_{in} - m_{out} = \Delta m_{system} \quad (kg) \qquad (5\text{–}42)$$

where $\Delta m_{system} = m_{final} - m_{initial}$ is the change in the mass of the system. For control volumes, it can also be expressed more explicitly as

$$m_i - m_e = (m_2 - m_1)_{CV} \qquad (5\text{–}43)$$

where $i$ = inlet, $e$ = exit, 1 = initial state, and 2 = final state of the control volume. Often one or more terms in the equation above are zero. For example, $m_i = 0$ if no mass enters the control volume during the process, $m_e = 0$ if no mass leaves, and $m_1 = 0$ if the control volume is initially evacuated.

The energy content of a control volume changes with time during an unsteady-flow process. The magnitude of change depends on the amount of energy transfer across the system boundaries as heat and work as well as on the amount of energy transported into and out of the control volume by mass during the process. When analyzing an unsteady-flow process, we must keep track of the energy content of the control volume as well as the energies of the incoming and outgoing flow streams.

The general energy balance was given earlier as

*Energy balance:*
$$\underbrace{E_{in} - E_{out}}_{\substack{\text{Net energy transfer} \\ \text{by heat, work, and mass}}} = \underbrace{\Delta E_{system}}_{\substack{\text{Change in internal, kinetic,} \\ \text{potential, etc., energies}}} \quad (kJ) \qquad (5\text{–}44)$$

The general unsteady-flow process, in general, is difficult to analyze because the properties of the mass at the inlets and exits may change during a process. Most unsteady-flow processes, however, can be represented reasonably well by the **uniform-flow process**, which involves the following idealization: *The fluid flow at any inlet or exit is uniform and steady, and thus the fluid properties do not change with time or position over the cross section of an inlet or exit. If they do, they are averaged and treated as constants for the entire process.*

Note that unlike the steady-flow systems, the state of an unsteady-flow system may change with time, and that the state of the mass leaving the control volume at any instant is the same as the state of the mass in the control volume at that instant. The initial and final properties of the control volume can be determined from the knowledge of the initial and final states, which are completely specified by two independent intensive properties for simple compressible systems.

Then the energy balance for a uniform-flow system can be expressed explicitly as

$$\left( Q_{in} + W_{in} + \sum_{in} m\theta \right) - \left( Q_{out} + W_{out} + \sum_{out} m\theta \right) = (m_2 e_2 - m_1 e_1)_{system} \qquad (5\text{–}45)$$

where $\theta = h + ke + pe$ is the energy of a fluid stream at any inlet or exit per unit mass, and $e = u + ke + pe$ is the energy of the nonflowing fluid within the control volume per unit mass. When the kinetic and potential

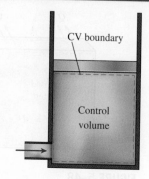

**FIGURE 5–47**
The shape and size of a control volume may change during an unsteady-flow process.

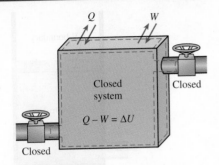

**FIGURE 5–48**
The energy equation of a uniform-flow system reduces to that of a closed system when all the inlets and exits are closed.

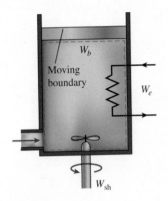

**FIGURE 5–49**
A uniform-flow system may involve electrical, shaft, and boundary work all at once.

energy changes associated with the control volume and fluid streams are negligible, as is usually the case, the energy balance above simplifies to

$$Q - W = \sum_{out} mh - \sum_{in} mh + (m_2 u_2 - m_1 u_1)_{system} \qquad (5\text{–}46)$$

where $Q = Q_{net,in} = Q_{in} - Q_{out}$ is the net heat input and $W = W_{net,out} = W_{out} - W_{in}$ is the net work output. Note that if no mass enters or leaves the control volume during a process ($m_i = m_e = 0$, and $m_1 = m_2 = m$), this equation reduces to the energy balance relation for closed systems (Fig. 5–48). Also note that an unsteady-flow system may involve boundary work as well as electrical and shaft work (Fig. 5–49).

Although both the steady-flow and uniform-flow processes are somewhat idealized, many actual processes can be approximated reasonably well by one of these with satisfactory results. The degree of satisfaction depends on the desired accuracy and the degree of validity of the assumptions made.

---

**EXAMPLE 5–12    Charging of a Rigid Tank by Steam**

A rigid, insulated tank that is initially evacuated is connected through a valve to a supply line that carries steam at 1 MPa and 300°C. Now the valve is opened, and steam is allowed to flow slowly into the tank until the pressure reaches 1 MPa, at which point the valve is closed. Determine the final temperature of the steam in the tank.

**SOLUTION**   A valve connecting an initially evacuated tank to a steam line is opened, and steam flows in until the pressure inside rises to the line level. The final temperature in the tank is to be determined.

**Assumptions**  **1** This process can be analyzed as a *uniform-flow process* since the properties of the steam entering the control volume remain constant during the entire process. **2** The kinetic and potential energies of the streams are negligible, ke $\cong$ pe $\cong$ 0. **3** The tank is stationary and thus its kinetic and potential energy changes are zero; that is, $\Delta KE = \Delta PE = 0$ and $\Delta E_{system} = \Delta U_{system}$. **4** There are no boundary, electrical, or shaft work interactions involved. **5** The tank is well insulated and thus there is no heat transfer.

**Analysis**   We take the *tank* as the system (Fig. 5–50). This is a *control volume* since mass crosses the system boundary during the process. We observe that this is an unsteady-flow process since changes occur within the control volume. The control volume is initially evacuated and thus $m_1 = 0$ and $m_1 u_1 = 0$. Also, there is one inlet and no exits for mass flow.

Noting that microscopic energies of flowing and nonflowing fluids are represented by enthalpy $h$ and internal energy $u$, respectively, the mass and energy balances for this uniform-flow system can be expressed as

*Mass balance:*    $m_{in} - m_{out} = \Delta m_{system} \rightarrow m_i = m_2 - m_1 \overset{0}{=} m_2$

*Energy balance:*

$$\underbrace{E_{in} - E_{out}}_{\substack{\text{Net energy transfer} \\ \text{by heat, work, and mass}}} = \underbrace{\Delta E_{system}}_{\substack{\text{Change in internal, kinetic,} \\ \text{potential, etc., energies}}}$$

$$m_i h_i = m_2 u_2 \quad (\text{since } W = Q = 0, \text{ ke} \cong \text{pe} \cong 0, m_1 = 0)$$

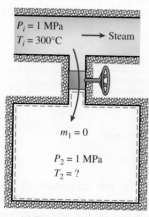

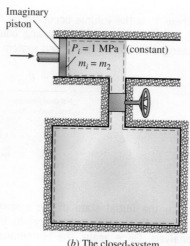

(a) Flow of steam into an evacuated tank

(b) The closed-system equivalence

**FIGURE 5–50**
Schematic for Example 5–12.

Combining the mass and energy balances gives

$$u_2 = h_i$$

That is, the final internal energy of the steam in the tank is equal to the enthalpy of the steam entering the tank. The enthalpy of the steam at the inlet state is

$$\left. \begin{array}{l} P_i = 1 \text{ MPa} \\ T_i = 300°C \end{array} \right\} \quad h_i = 3051.6 \text{ kJ/kg} \quad \text{(Table A–6)}$$

which is equal to $u_2$. Since we now know two properties at the final state, it is fixed and the temperature at this state is determined from the same table to be

$$\left. \begin{array}{l} P_2 = 1 \text{ MPa} \\ u_2 = 3051.6 \text{ kJ/kg} \end{array} \right\} \quad T_2 = \textbf{456.1°C}$$

***Discussion*** Note that the temperature of the steam in the tank has increased by 156.1°C. This result may be surprising at first, and you may be wondering where the energy to raise the temperature of the steam came from. The answer lies in the enthalpy term $h = u + Pv$. Part of the energy represented by enthalpy is the flow energy $Pv$, and this flow energy is converted to sensible internal energy once the flow ceases to exist in the control volume, and it shows up as an increase in temperature (Fig. 5–51).

***Alternative solution*** This problem can also be solved by considering the region within the tank and the mass that is destined to enter the tank as a closed system, as shown in Fig. 5–50b. Since no mass crosses the boundaries, viewing this as a closed system is appropriate.

During the process, the steam upstream (the imaginary piston) will push the enclosed steam in the supply line into the tank at a constant pressure of 1 MPa. Then the boundary work done during this process is

$$W_{b,\text{in}} = -\int_1^2 P_i \, dV = -P_i(V_2 - V_1) = -P_i[V_{\text{tank}} - (V_{\text{tank}} + V_i)] = P_i V_i$$

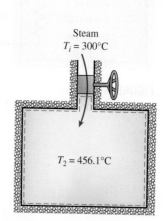

**FIGURE 5–51**
The temperature of steam rises from 300 to 456.1°C as it enters a tank as a result of flow energy being converted to internal energy.

where $V_i$ is the volume occupied by the steam before it enters the tank and $P_i$ is the pressure at the moving boundary (the imaginary piston face). The energy balance for the closed system gives

$$\underbrace{E_{\text{in}} - E_{\text{out}}}_{\substack{\text{Net energy transfer} \\ \text{by heat, work, and mass}}} = \underbrace{\Delta E_{\text{system}}}_{\substack{\text{Change in internal, kinetic,} \\ \text{potential, etc., energies}}}$$

$$W_{b,\text{in}} = \Delta U$$

$$m_i P_i V_i = m_2 u_2 - m_i u_i$$

$$u_2 = u_i + P_i V_i = h_i$$

since the initial state of the system is simply the line conditions of the steam. This result is identical to the one obtained with the uniform-flow analysis. Once again, the temperature rise is caused by the so-called flow energy or flow work, which is the energy required to move the fluid during flow.

---

### EXAMPLE 5–13    Discharge of Heated Air at Constant Temperature

An insulated 8-m³ rigid tank contains air at 600 kPa and 400 K. A valve connected to the tank is now opened, and air is allowed to escape until the pressure inside drops to 200 kPa. The air temperature during the process is maintained constant by an electric resistance heater placed in the tank. Determine the electrical energy supplied to air during this process.

**SOLUTION** Pressurized air in an insulated rigid tank equipped with an electric heater is allowed to escape at constant temperature until the pressure inside drops to a specified value. The amount of electrical energy supplied to air is to be determined.

**Assumptions** **1** This is an unsteady process since the conditions within the device are changing during the process, but it can be analyzed as a uniform-flow process since the exit conditions remain constant. **2** Kinetic and potential energies are negligible. **3** The tank is insulated and thus heat transfer is negligible. **4** Air is an ideal gas with variable specific heats.

**Analysis** We take the contents of the tank as the system, which is a control volume since mass crosses the boundary (Fig. 5–52). Noting that the microscopic energies of flowing and nonflowing fluids are represented by enthalpy $h$ and internal energy $u$, respectively, the mass and energy balances for this uniform-flow system can be expressed as

*Mass balance:*    $m_{\text{in}} - m_{\text{out}} = \Delta m_{\text{system}} \quad \rightarrow \quad m_e = m_1 - m_2$

*Energy balance:*    $\underbrace{E_{\text{in}} - E_{\text{out}}}_{\substack{\text{Net energy transfer} \\ \text{by heat, work, and mass}}} = \underbrace{\Delta E_{\text{system}}}_{\substack{\text{Change in internal, kinetic,} \\ \text{potential, etc., energies}}}$

$$W_{e,\text{in}} - m_e h_e = m_2 u_2 - m_1 u_1 \quad (\text{since } Q \cong \text{ke} \cong \text{pe} \cong 0)$$

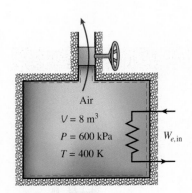

Air
$V = 8 \text{ m}^3$
$P = 600 \text{ kPa}$
$T = 400 \text{ K}$
$W_{e,\text{in}}$

**FIGURE 5–52**
Schematic for Example 5–13.

The gas constant of air is $R = 0.287$ kPa·m³/kg·K (Table A-1). The initial and final masses of air in the tank and the discharged amount are determined from the ideal gas relation to be

$$m_1 = \frac{P_1 V_1}{RT_1} = \frac{(600 \text{ kPa})(8 \text{ m}^3)}{(0.287 \text{ kPa·m}^3/\text{kg·K})(400 \text{ K})} = 41.81 \text{ kg}$$

$$m_2 = \frac{P_2 V_2}{RT_2} = \frac{(200 \text{ kPa})(8 \text{ m}^3)}{(0.287 \text{ kPa·m}^3/\text{kg·K})(400 \text{ K})} = 13.94 \text{ kg}$$

$$m_e = m_1 - m_2 = 41.81 - 13.94 = 27.87 \text{ kg}$$

The enthalpy and internal energy of air at 400 K are $h_e = 400.98$ kJ/kg and $u_1 = u_2 = 286.16$ kJ/kg (Table A-17). The the electrical energy supplied to air is determined from the energy balance to be

$$W_{e,\text{in}} = m_e h_e + m_2 u_2 - m_1 u_1$$
$$= (27.87 \text{ kg})(400.98 \text{ kJ/kg}) + (13.94 \text{ kg})(286.16 \text{ kJ/kg})$$
$$- (41.81 \text{ kg})(286.16 \text{ kJ/kg})$$
$$= 3200 \text{ kJ} = \textbf{0.889 kWh}$$

since 1 kWh = 3600 kJ.

**Discussion** If the temperature of discharged air changes during the process, the problem can be solved with reasonable accuracy by evaluating $h_e$ at the average discharge temperature $T_e = (T_2 + T_1)/2$, and treating it as constant.

## TOPIC OF SPECIAL INTEREST*        General Energy Equation

One of the most fundamental laws in nature is the **first law of thermodynamics**, also known as the **conservation of energy principle**, which provides a sound basis for studying the relationships among the various forms of energy and energy interactions. It states that *energy can be neither created nor destroyed during a process; it can only change forms.*

The energy content of a fixed quantity of mass (a closed system) can be changed by two mechanisms: *heat transfer Q* and *work transfer W.* Then the conservation of energy for a fixed quantity of mass can be expressed in rate form as

$$\dot{Q} - \dot{W} = \frac{dE_{\text{sys}}}{dt} \quad \text{or} \quad \dot{Q} - \dot{W} = \frac{d}{dt} \int_{\text{sys}} \rho e\, dV \qquad \text{(5–47)}$$

where $\dot{Q} = \dot{Q}_{\text{net,in}} = \dot{Q}_{\text{in}} - \dot{Q}_{\text{out}}$ is the net rate of heat transfer to the system (negative, if from the system), $\dot{W} = \dot{W}_{\text{net,out}} = \dot{W}_{\text{out}} - \dot{W}_{\text{in}}$ is the net power output from the system in all forms (negative, if power input) and $dE_{\text{sys}}/dt$

*This section can be skipped without a loss in continuity.

is the rate of change of the total energy content of the system. The overdot stands for time rate. For simple compressible systems, total energy consists of internal, kinetic, and potential energies, and it is expressed on a unit-mass basis as

$$e = u + \text{ke} + \text{pe} = u + \frac{V^2}{2} + gz \tag{5–48}$$

Note that total energy is a property, and its value does not change unless the state of the system changes.

An energy interaction is *heat* if its driving force is a temperature difference, and it is *work* if it is associated with a force acting through a distance, as explained in Chap. 2. A system may involve numerous forms of work, and the total work can be expressed as

$$W_{\text{total}} = W_{\text{shaft}} + W_{\text{pressure}} + W_{\text{viscous}} + W_{\text{other}} \tag{5–49}$$

where $W_{\text{shaft}}$ is the work transmitted by a rotating shaft, $W_{\text{pressure}}$ is the work done by the pressure forces on the control surface, $W_{\text{viscous}}$ is the work done by the normal and shear components of viscous forces on the control surface, and $W_{\text{other}}$ is the work done by other forces such as electric, magnetic, and surface tension, which are insignificant for simple compressible systems and are not considered in this text. We do not consider $W_{\text{viscous}}$ either since it is usually small relative to other terms in control volume analysis. But it should be kept in mind that the work done by shear forces as the blades shear through the fluid may need to be considered in a refined analysis of turbomachinery.

## Work Done by Pressure Forces

Consider a gas being compressed in the piston-cylinder device shown in Fig. 5–53a. When the piston moves down a differential distance $ds$ under the influence of the pressure force $PA$, where $A$ is the cross-sectional area of the piston, the boundary work done *on* the system is $\delta W_{\text{boundary}} = PA\,ds$. Dividing both sides of this relation by the differential time interval $dt$ gives the time rate of boundary work (i.e., *power*),

$$\delta \dot{W}_{\text{pressure}} = \delta \dot{W}_{\text{boundary}} = PAV_{\text{piston}}$$

where $V_{\text{piston}} = ds/dt$ is the piston velocity, which is the velocity of the moving boundary at the piston face.

Now consider a material chunk of fluid (a system) of arbitrary shape, which moves with the flow and is free to deform under the influence of pressure, as shown in Fig. 5–53b. Pressure always acts inward and normal to the surface, and the pressure force acting on a differential area $dA$ is $P\,dA$. Again noting that work is force times distance and distance traveled per unit time is velocity, the time rate at which work is done by pressure forces on this differential part of the system is

$$\delta \dot{W}_{\text{pressure}} = P\,dA\,V_n = P\,dA(\vec{V} \cdot \vec{n}) \tag{5–50}$$

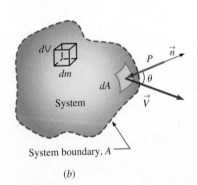

**FIGURE 5–53**
The pressure force acting on (a) the moving boundary of a system in a piston–cylinder device, and (b) the differential surface area of a system of arbitrary shape.

since the normal component of velocity through the differential area $dA$ is $V_n = V \cos \theta = \vec{V} \cdot \vec{n}$. Note that $\vec{n}$ is the outer normal of $dA$, and thus the quantity $\vec{V} \cdot \vec{n}$ is positive for expansion and negative for compression. The total rate of work done by pressure forces is obtained by integrating $\delta \dot{W}_{pressure}$ over the entire surface $A$,

$$\dot{W}_{pressure,net out} = \int_A P(\vec{V} \cdot \vec{n}) \, dA = \int_A \frac{P}{\rho} \rho (\vec{V} \cdot \vec{n}) \, dA \quad (5\text{-}51)$$

In light of these discussions, the net power transfer can be expressed as

$$\dot{W}_{net,out} = \dot{W}_{shaft,net out} + \dot{W}_{pressure,net out} = \dot{W}_{shaft,net out} + \int_A (\vec{V} \cdot \vec{n}) \, dA \quad (5\text{-}52)$$

Then the rate form of the conservation of energy relation for a closed system becomes

$$\dot{Q}_{net,in} - \dot{W}_{shaft,net out} - \dot{W}_{pressure,net out} = \frac{dE_{sys}}{dt} \quad (5\text{-}53)$$

To obtain a relation for the conservation of energy for a *control volume*, we apply the Reynolds transport theorem by replacing the extensive property $B$ with total energy $E$, and its associated intensive property $b$ with total energy per unit mass $e$, which is $e = u + \text{ke} + \text{pe} = u + V^2/2 + gz$ (Fig. 5–54). This yields

$$\frac{dE_{sys}}{dt} = \frac{d}{dt} \int_{CV} e\rho \, dV + \int_{CS} e\rho (\vec{V} \cdot \vec{n}) \, A \quad (5\text{-}54)$$

Substituting the left-hand side of Eq. 5–53 into Eq. 5–54, the general form of the energy equation that applies to fixed, moving, or deforming control volumes becomes

$$\dot{Q}_{net,in} - \dot{W}_{shaft,net out} - \dot{W}_{pressure,net out} = \frac{d}{dt} \int_{CV} e\rho \, dV + \int_{CS} e\rho (\vec{V}_r \cdot \vec{n}) \, dA \quad (5\text{-}55)$$

which can be stated as

$$\begin{pmatrix} \text{The net rate of energy} \\ \text{transfer into a CV by} \\ \text{heat and work transfer} \end{pmatrix} = \begin{pmatrix} \text{The time rate of} \\ \text{change of the energy} \\ \text{content of the CV} \end{pmatrix} + \begin{pmatrix} \text{The net flow rate of} \\ \text{energy out of the control} \\ \text{surface by mass flow} \end{pmatrix}$$

Here $\vec{V}_r = \vec{V} - \vec{V}_{CS}$ is the fluid velocity relative to the control surface, and the product $\rho(\vec{V}_r \cdot \vec{n}) \, dA$ represents the mass flow rate through area element $dA$ into or out of the control volume. Again noting that $\vec{n}$ is the outer normal of $dA$, the quantity $\vec{V}_r \cdot \vec{n}$ and thus mass flow is positive for outflow and negative for inflow.

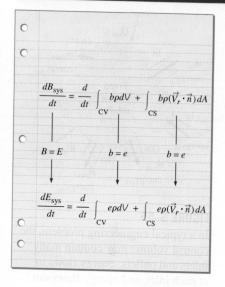

**FIGURE 5–54**
The conservation of energy equation is obtained by replacing an extensive property $B$ in the Reynolds transport theorem by energy $E$ and its associated intensive property $b$ by $e$ (Ref. 3).

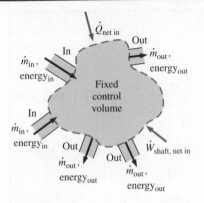

**FIGURE 5–55**
In a typical engineering problem, the control volume may contain many inlets and outlets; energy flows in at each inlet, and energy flows out at each outlet. Energy also enters the control volume through net heat transfer and net shaft work.

Substituting the surface integral for the rate of pressure work from Eq. 5–51 into Eq. 5–55 and combining it with the surface integral on the right give

$$\dot{Q}_{net,in} - \dot{W}_{shaft,net\ out} = \frac{d}{dt}\int_{CV} e\rho\, dV + \int_{CS}\left(\frac{P}{\rho} + e\right)\rho(\vec{V}_r\cdot\vec{n})\, dA \quad \text{(5–56)}$$

This is a very convenient form for the energy equation since pressure work is now combined with the energy of the fluid crossing the control surface and we no longer have to deal with pressure work.

The term $P/\rho = Pv = w_{flow}$ is the *flow work*, which is the work associated with pushing a fluid into or out of a control volume per unit mass. Note that the fluid velocity at a solid surface is equal to the velocity of the solid surface because of the no-slip condition and is zero for nonmoving surfaces. As a result, the pressure work along the portions of the control surface that coincide with nonmoving solid surfaces is zero. Therefore, pressure work for fixed control volumes can exist only along the imaginary part of the control surface where the fluid enters and leaves the control volume (i.e., inlets and outlets).

This equation is not in a convenient form for solving practical engineering problems because of the integrals, and thus it is desirable to rewrite it in terms of average velocities and mass flow rates through inlets and outlets. If $P/\rho + e$ is nearly uniform across an inlet or outlet, we can simply take it outside the integral. Noting that $\dot{m} = \int_{A_c} \rho(\vec{V}_r\cdot\vec{n})\, dA_c$ is the mass flow rate across an inlet or outlet, the rate of inflow or outflow of energy through the inlet or outlet can be approximated as $\dot{m}(P/\rho + e)$. Then the energy equation becomes (Fig. 5–55)

$$\dot{Q}_{net,in} - \dot{W}_{shaft,net\ out} = \frac{d}{dt}\int_{CV} e\rho\, dV + \sum_{out}\dot{m}\left(\frac{P}{\rho} + e\right) - \sum_{in}\dot{m}\left(\frac{P}{\rho} + e\right) \quad \text{(5–57)}$$

where $e = u + V^2/2 + gz$ is the total energy per unit mass for both the control volume and flow streams. Then,

$$\dot{Q}_{net,in} - \dot{W}_{shaft,net\ out} = \frac{d}{dt}\int_{CV} e\rho\, dV + \sum_{out}\dot{m}\left(\frac{P}{\rho} + u + \frac{V^2}{2} + gz\right) - \sum_{in}\dot{m}\left(\frac{P}{\rho} + u + \frac{V^2}{2} + gz\right)$$

$$\text{(5–58)}$$

or

$$\dot{Q}_{net,in} - \dot{W}_{shaft,net\ out} = \frac{d}{dt}\int_{CV} e\rho\, dV + \sum_{out}\dot{m}\left(h + \frac{V^2}{2} + gz\right) - \sum_{in}\dot{m}\left(h + \frac{V^2}{2} + gz\right)$$

$$\text{(5–59)}$$

where we used the definition of enthalpy $h = u + Pv = u + P/\rho$. The last two equations are fairly general expressions of conservation of energy, but their use is still limited to uniform flow at inlets and outlets and negligible work due to viscous forces and other effects. Also, the subscript "net,in" stands for "net input," and thus any heat or work transfer is positive if *to* the system and negative if *from* the system.

# SUMMARY

The *conservation of mass principle* states that the net mass transfer to or from a system during a process is equal to the net change (increase or decrease) in the total mass of the system during that process, and is expressed as

$$m_{in} - m_{out} = \Delta m_{system} \quad \text{and} \quad \dot{m}_{in} - \dot{m}_{out} = dm_{system}/dt$$

where $\Delta m_{system} = m_{final} - m_{initial}$ is the change in the mass of the system during the process, $\dot{m}_{in}$ and $\dot{m}_{out}$ are the total rates of mass flow into and out of the system, and $dm_{system}/dt$ is the rate of change of mass within the system boundaries. The relations above are also referred to as the mass balance and are applicable to any system undergoing any kind of process.

The amount of mass flowing through a cross section per unit time is called the *mass flow rate,* and is expressed as

$$\dot{m} = \rho V A$$

where $\rho$ = density of fluid, $V$ = average fluid velocity normal to $A$, and $A$ = cross-sectional area normal to flow direction. The volume of the fluid flowing through a cross section per unit time is called the *volume flow rate* and is expressed as

$$\dot{V} = VA = \dot{m}/\rho$$

The work required to push a unit mass of fluid into or out of a control volume is called *flow work* or *flow energy,* and is expressed as $w_{flow} = Pv$. In the analysis of control volumes, it is convenient to combine the flow energy and internal energy into *enthalpy.* Then the total energy of a flowing fluid is expressed as

$$\theta = h + ke + pe = h + \frac{V^2}{2} + gz$$

The total energy transported by a flowing fluid of mass $m$ with uniform properties is $m\theta$. The rate of energy transport by a fluid with a mass flow rate of $\dot{m}$ is $\dot{m}\theta$. When the kinetic and potential energies of a fluid stream are negligible, the amount and rate of energy transport become $E_{mass} = mh$ and $\dot{E}_{mass} = \dot{m}h$, respectively.

The *first law of thermodynamics* is essentially an expression of the conservation of energy principle, also called the *energy balance.* The general mass and energy balances for *any system* undergoing *any process* can be expressed as

$$\underbrace{E_{in} - E_{out}}_{\substack{\text{Net energy transfer} \\ \text{by heat, work, and mass}}} = \underbrace{\Delta E_{system}}_{\substack{\text{Change in internal, kinetic,} \\ \text{potential, etc., energies}}}$$

It can also be expressed in the *rate form* as

$$\underbrace{E_{in} - E_{out}}_{\substack{\text{Rate of net energy transfer} \\ \text{by heat, work, and mass}}} = \underbrace{dE_{system}/dt}_{\substack{\text{Rate of change in internal, kinetic,} \\ \text{potential, etc., energies}}}$$

Thermodynamic processes involving control volumes can be considered in two groups: steady-flow processes and unsteady-flow processes. During a *steady-flow process,* the fluid flows through the control volume steadily, experiencing no change with time at a fixed position. The mass and energy content of the control volume remain constant during a steady-flow process. Taking heat transfer *to* the system and work done *by* the system to be positive quantities, the conservation of mass and energy equations for steady-flow processes are expressed as

$$\sum_{in} \dot{m} = \sum_{out} \dot{m}$$

$$\dot{Q} - \dot{W} = \sum_{out} \dot{m}\underbrace{\left(h + \frac{V^2}{2} + gz\right)}_{\text{for each exit}} - \sum_{in} \dot{m}\underbrace{\left(h + \frac{V^2}{2} + gz\right)}_{\text{for each inlet}}$$

These are the most general forms of the equations for steady-flow processes. For single-stream (one-inlet–one-exit) systems such as nozzles, diffusers, turbines, compressors, and pumps, they simplify to

$$\dot{m}_1 = \dot{m}_2 \longrightarrow \frac{1}{v_1}V_1A_1 = \frac{1}{v_2}V_2A_2$$

$$\dot{Q} - \dot{W} = \dot{m}\left[h_2 - h_1 + \frac{V_2^2 - V_1^2}{2} + g(z_2 - z_1)\right]$$

In these relations, subscripts 1 and 2 denote the inlet and exit states, respectively.

Most unsteady-flow processes can be modeled as a *uniform-flow process,* which requires that the fluid flow at any inlet or exit is uniform and steady, and thus the fluid properties do not change with time or position over the cross section of an inlet or exit. If they do, they are averaged and treated as constants for the entire process. When kinetic and potential energy changes associated with the control volume and the fluid streams are negligible, the mass and energy balance relations for a uniform-flow system are expressed as

$$m_{in} - m_{out} = \Delta m_{system}$$

$$Q - W = \sum_{out} mh - \sum_{in} mh + (m_2 u_2 - m_1 u_1)_{system}$$

where $Q = Q_{net,in} = Q_{in} - Q_{out}$ is the net heat input and $W = W_{net,out} = W_{out} - W_{in}$ is the net work output.

When solving thermodynamic problems, it is recommended that the general form of the energy balance $E_{in} - E_{out} = \Delta E_{system}$ be used for all problems, and simplify it for the particular problem instead of using the specific relations given here for different processes.

# REFERENCES AND SUGGESTED READINGS

1. ASHRAE Handbook of Fundamentals. SI version. Atlanta, GA: American Society of Heating, Refrigerating, and Air-Conditioning Engineers, Inc., 1993.

2. ASHRAE Handbook of Refrigeration. SI version. Atlanta, GA: American Society of Heating, Refrigerating, and Air-Conditioning Engineers, Inc., 1994.

3. Y. A. Çengel and J. M. Cimbala, *Fluid Mechanics: Fundamentals and Applications,* 3rd ed. New York: McGraw-Hill, 2014.

# PROBLEMS*

## Conservation of Mass

**5–1C**  When is the flow through a control volume steady?

**5–2C**  Define mass and volume flow rates. How are they related to each other?

**5–3C**  Does the amount of mass entering a control volume have to be equal to the amount of mass leaving during an unsteady-flow process?

**5–4C**  Consider a device with one inlet and one outlet. If the volume flow rates at the inlet and at the outlet are the same, is the flow through this device necessarily steady? Why?

**5–5**  The ventilating fan of the bathroom of a building has a volume flow rate of 30 L/s and runs continuously. If the density of air inside is 1.20 kg/m³, determine the mass of air vented out in one day.

**5–6**  Air enters a 28-cm diameter pipe steadily at 200 kPa and 20°C with a velocity of 5 m/s. Air is heated as it flows, and leaves the pipe at 180 kPa and 40°C. Determine (a) the volume flow rate of air at the inlet, (b) the mass flow rate of air, and (c) the velocity and volume flow rate at the exit.

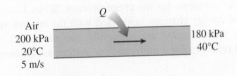

**FIGURE P5–6**

**5–7**  A steady-flow compressor is used to compress helium from 100 kPa and 20°C at the inlet to 1400 kPa and 315°C at the outlet. The outlet area and velocity are 0.001 m² and 30 m/s, respectively, and the inlet velocity is 15 m/s. Determine the mass flow rate and the inlet area.  *Answers:* 0.0344 kg/s, 0.0140 m²

**5–8**  A 2-m³ rigid tank initially contains air whose density is 1.18 kg/m³. The tank is connected to a high-pressure supply line through a valve. The valve is opened, and air is allowed to enter the tank until the density in the tank rises to 5.30 kg/m³. Determine the mass of air that has entered the tank.  *Answer:* 8.24 kg

**5–9**  A cyclone separator like that in Fig. P5–9 is used to remove fine solid particles, such as fly ash, that are suspended in a gas stream. In the flue-gas system of an electrical power plant, the weight fraction of fly ash in the exhaust gases is approximately 0.001. Determine the mass flow rates at the two outlets (flue gas and fly ash) when 10 kg/s of flue gas and ash mixture enters this unit. Also determine the amount of fly ash collected per year.

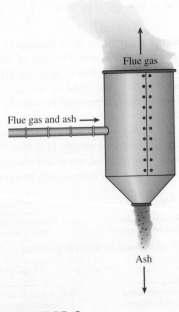

**FIGURE P5–9**

---

* Problems designated by a "C" are concept questions, and students are encouraged to answer them all. Problems with the 🌀 icon are solved using EES, and complete solutions together with parametric studies are included on the text website. Problems with the 🔲 icon are comprehensive in nature, and are intended to be solved with an equation solver such as EES.

**5–10** A spherical hot-air balloon is initially filled with air at 120 kPa and 20°C with an initial diameter of 5 m. Air enters this balloon at 120 kPa and 20°C with a velocity of 3 m/s through a 1-m diameter opening. How many minutes will it take to inflate this balloon to a 15-m diameter when the pressure and temperature of the air in the balloon remain the same as the air entering the balloon? *Answer:* 12.0 min

**FIGURE P5–10**

*©Photo Link/Getty Images RF*

**5–11** A desktop computer is to be cooled by a fan whose flow rate is 0.34 m³/min. Determine the mass flow rate of air through the fan at an elevation of 3400 m where the air density is 0.7 kg/m³. Also, if the average velocity of air is not to exceed 110 m/min, determine the diameter of the casing of the fan. *Answers:* 0.238 kg/min, 0.063 m

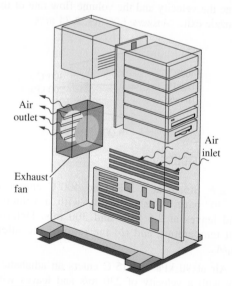

**FIGURE P5–11**

**5–12** A pump increases the water pressure from 100 kPa at the inlet to 900 kPa at the outlet. Water enters this pump at 15°C through a 1-cm-diameter opening and exits through a 1.5-cm-diameter opening. Determine the velocity of the water at the inlet and outlet when the mass flow rate through the pump is 0.5 kg/s. Will these velocities change significantly if the inlet temperature is raised to 40°C?

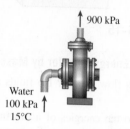

900 kPa

Water
100 kPa
15°C

**FIGURE P5–12**

**5–13** Refrigerant-134a enters a 28-cm-diameter pipe steadily at 200 kPa and 20°C with a velocity of 5 m/s. The refrigerant gains heat as it flows and leaves the pipe at 180 kPa and 40°C. Determine (*a*) the volume flow rate of the refrigerant at the inlet, (*b*) the mass flow rate of the refrigerant, and (*c*) the velocity and volume flow rate at the exit.

**5–14** A smoking lounge is to accommodate 15 heavy smokers. The minimum fresh air requirement for smoking lounges is specified to be 30 L/s per person (ASHRAE, Standard 62, 1989). Determine the minimum required flow rate of fresh air that needs to be supplied to the lounge, and the diameter of the duct if the air velocity is not to exceed 8 m/s.

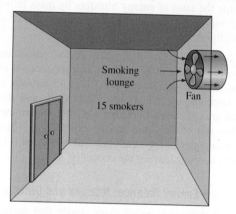

Smoking lounge

15 smokers

Fan

**FIGURE P5–14**

**5–15** Consider a 300-L storage tank of a solar water heating system initially filled with warm water at 45°C. Warm water is withdrawn from the tank through a 2-cm diameter hose at an average velocity of 0.5 m/s while cold water enters the tank at 20°C at a rate of 15 L/min. Determine the amount of water in the tank after a 20-minute period. Assume the pressure in the tank remains constant at 1 atm. *Answer:* 189 kg

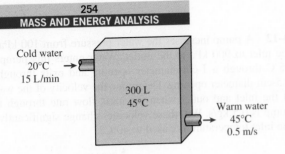

**FIGURE P5-15**

## Flow Work and Energy Transfer by Mass

**5-16C** What is flow energy? Do fluids at rest possess any flow energy?

**5-17C** How do the energies of a flowing fluid and a fluid at rest compare? Name the specific forms of energy associated with each case.

**5-18** A house is maintained at 1 atm and 24°C, and warm air inside a house is forced to leave the house at a rate of 150 m³/h as a result of outdoor air at 5°C infiltrating into the house through the cracks. Determine the rate of net energy loss of the house due to mass transfer. *Answer:* 0.945 kW

**5-19** A water pump increases the water pressure from 100 kPa to 600 kPa. Determine the flow work, in kJ/kg, required by the pump.

**5-20** Refrigerant-134a enters the compressor of a refrigeration system as saturated vapor at 0.14 MPa, and leaves as superheated vapor at 0.8 MPa and 60°C at a rate of 0.06 kg/s. Determine the rates of energy transfers by mass into and out of the compressor. Assume the kinetic and potential energies to be negligible.

**5-21** Steam is leaving a pressure cooker whose operating pressure is 150 kPa. It is observed that the amount of liquid in the cooker has decreased by 2.3 L in 45 minutes after the steady operating conditions are established, and the cross-sectional area of the exit opening is 1 cm². Determine (*a*) the mass flow rate of the steam and the exit velocity, (*b*) the total and flow energies of the steam per unit mass, and (*c*) the rate at which energy is leaving the cooker by steam.

## Steady-Flow Energy Balance: Nozzles and Diffusers

**5-22C** A diffuser is an adiabatic device that decreases the kinetic energy of the fluid by slowing it down. What happens to this *lost* kinetic energy?

**5-23C** The kinetic energy of a fluid increases as it is accelerated in an adiabatic nozzle. Where does this energy come from?

**5-24C** Is heat transfer to or from the fluid desirable as it flows through a nozzle? How will heat transfer affect the fluid velocity at the nozzle exit?

**5-25** The stators in a gas turbine are designed to increase the kinetic energy of the gas passing through them adiabatically. Air enters a set of these nozzles at 2100 kPa and 370°C with a velocity of 25 m/s and exits at 1750 kPa and 340°C. Calculate the velocity at the exit of the nozzles.

**5-26** The diffuser in a jet engine is designed to decrease the kinetic energy of the air entering the engine compressor without any work or heat interactions. Calculate the velocity at the exit of a diffuser when air at 100 kPa and 30°C enters it with a velocity of 350 m/s and the exit state is 200 kPa and 90°C.

**FIGURE P5-26**

*©Stockbyte/Punchstock RF*

**5-27** Air at 600 kPa and 500 K enters an adiabatic nozzle that has an inlet-to-exit area ratio of 2:1 with a velocity of 120 m/s and leaves with a velocity of 380 m/s. Determine (*a*) the exit temperature and (*b*) the exit pressure of the air. *Answers:* (*a*) 437 K, (*b*) 331 kPa

**5-28** Steam enters a nozzle at 400°C and 800 kPa with a velocity of 10 m/s, and leaves at 300°C and 200 kPa while losing heat at a rate of 25 kW. For an inlet area of 800 cm², determine the velocity and the volume flow rate of the steam at the nozzle exit. *Answers:* 606 m/s, 2.74 m³/s

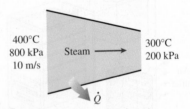

**FIGURE P5-28**

**5-29** Steam at 3 MPa and 400°C enters an adiabatic nozzle steadily with a velocity of 40 m/s and leaves at 2.5 MPa and 300 m/s. Determine (*a*) the exit temperature and (*b*) the ratio of the inlet to exit area $A_1/A_2$.

**5-30** Air at 90 kPa and 15°C enters an adiabatic diffuser steadily with a velocity of 230 m/s and leaves with a low velocity at a pressure of 100 kPa. The exit area of the diffuser is 3 times the inlet area. Determine (a) the exit temperature and (b) the exit velocity of the air.

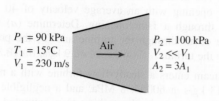

$P_1 = 90$ kPa
$T_1 = 15°C$
$V_1 = 230$ m/s

Air

$P_2 = 100$ kPa
$V_2 \ll V_1$
$A_2 = 3A_1$

**FIGURE P5–30**

**5–31** Carbon dioxide enters an adiabatic nozzle steadily at 1 MPa and 500°C with a mass flow rate of 6000 kg/h and leaves at 100 kPa and 450 m/s. The inlet area of the nozzle is 40 cm². Determine (a) the inlet velocity and (b) the exit temperature. *Answers:* (a) 60.8 m/s, (b) 686 K

**5–32** Refrigerant-134a at 700 kPa and 120°C enters an adiabatic nozzle steadily with a velocity of 20 m/s and leaves at 400 kPa and 30°C. Determine (a) the exit velocity and (b) the ratio of the inlet to exit area $A_1/A_2$.

**5–33** Nitrogen gas at 60 kPa and 7°C enters an adiabatic diffuser steadily with a velocity of 275 m/s and leaves at 85 kPa and 27°C. Determine (a) the exit velocity of the nitrogen and (b) the ratio of the inlet to exit area $A_1/A_2$.

**5–34** [EES] Reconsider Prob. 5–33. Using EES (or other) software, investigate the effect of the inlet velocity on the exit velocity and the ratio of the inlet-to-exit area. Let the inlet velocity vary from 210 to 350 m/s. Plot the final results against the inlet velocity, and discuss the results.

**5–35** Refrigerant-134a enters a diffuser steadily as saturated vapor at 600 kPa with a velocity of 160 m/s, and it leaves at 700 kPa and 40°C. The refrigerant is gaining heat at a rate of 2 kJ/s as it passes through the diffuser. If the exit area is 80 percent greater than the inlet area, determine (a) the exit velocity and (b) the mass flow rate of the refrigerant. *Answers:* (a) 82.1 m/s, (b) 0.298 kg/s

**5–36** Steam at 4 MPa and 400°C enters a nozzle steadily with a velocity of 60 m/s, and it leaves at 2 MPa and 300°C. The inlet area of the nozzle is 50 cm², and heat is being lost at a rate of 75 kJ/s. Determine (a) the mass flow rate of the steam, (b) the exit velocity of the steam, and (c) the exit area of the nozzle.

**5–37** Air at 80 kPa, 27°C, and 220 m/s enters a diffuser at a rate of 2.5 kg/s and leaves at 42°C. The exit area of the diffuser is 400 cm². The air is estimated to lose heat at a rate of 18 kJ/s during this process. Determine (a) the exit velocity and (b) the exit pressure of the air. *Answers:* (a) 62.0 m/s, (b) 91.1 kPa

## Turbines and Compressors

**5–38C** Consider an air compressor operating steadily. How would you compare the volume flow rates of the air at the compressor inlet and exit?

**5–39C** Will the temperature of air rise as it is compressed by an adiabatic compressor? Why?

**5–40C** Somebody proposes the following system to cool a house in the summer: Compress the regular outdoor air, let it cool back to the outdoor temperature, pass it through a turbine, and discharge the cold air leaving the turbine into the house. From a thermodynamic point of view, is the proposed system sound?

**5–41** Refrigerant-134a enters an adiabatic compressor as saturated vapor at −24°C and leaves at 0.8 MPa and 60°C. The mass flow rate of the refrigerant is 1.2 kg/s. Determine (a) the power input to the compressor and (b) the volume flow rate of the refrigerant at the compressor inlet.

**5–42** Refrigerant-134a enters a compressor at 180 kPa as a saturated vapor with a flow rate of 0.35 m³/min and leaves at 700 kPa. The power supplied to the refrigerant during compression process is 2.35 kW. What is the temperature of R-134a at the exit of the compressor? *Answer:* 48.9°C

**5–43** Steam flows steadily through an adiabatic turbine. The inlet conditions of the steam are 4 MPa, 500°C, and 80 m/s, and the exit conditions are 30 kPa, 92 percent quality, and 50 m/s. The mass flow rate of the steam is 12 kg/s. Determine (a) the change in kinetic energy, (b) the power output, and (c) the turbine inlet area. *Answers:* (a) −1.95 kJ/kg, (b) 12.1 MW, (c) 0.0130 m²

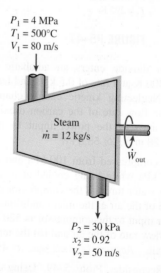

$P_1 = 4$ MPa
$T_1 = 500°C$
$V_1 = 80$ m/s

Steam
$\dot{m} = 12$ kg/s

$\dot{W}_{out}$

$P_2 = 30$ kPa
$x_2 = 0.92$
$V_2 = 50$ m/s

**FIGURE P5–43**

**5–44** Reconsider Prob. 5–43. Using EES (or other) software, investigate the effect of the turbine exit pressure on the power output of the turbine. Let the exit pressure vary from 10 to 200 kPa. Plot the power output against the exit pressure, and discuss the results.

**5–45** Steam enters an adiabatic turbine at 10 MPa and 500°C and leaves at 10 kPa with a quality of 90 percent. Neglecting the changes in kinetic and potential energies, determine the mass flow rate required for a power output of 5 MW. *Answer:* 4.852 kg/s

**5–46** Steam flows steadily through a turbine at a rate of 20,000 kg/h, entering at 7 MPa and 500°C and leaving at 40 kPa as saturated vapor. If the power generated by the turbine is 4 MW, determine the rate of heat loss from the steam.

**5–47** Helium is to be compressed from 105 kPa and 295 K to 700 kPa and 460 K. A heat loss of 15 kJ/kg occurs during the compression process. Neglecting kinetic energy changes, determine the power input required for a mass flow rate of 60 kg/min.

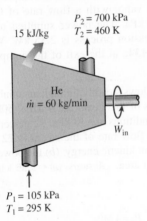

$P_2 = 700$ kPa
$T_2 = 460$ K

15 kJ/kg

He
$\dot{m} = 60$ kg/min

$\dot{W}_{in}$

$P_1 = 105$ kPa
$T_1 = 295$ K

**FIGURE P5–47**

**5–48** Carbon dioxide enters an adiabatic compressor at 100 kPa and 300 K at a rate of 0.5 kg/s and leaves at 600 kPa and 450 K. Neglecting kinetic energy changes, determine (*a*) the volume flow rate of the carbon dioxide at the compressor inlet and (*b*) the power input to the compressor. *Answers:* (*a*) 0.28 m³/s, (*b*) 68.8 kW

**5–49** Air is compressed from 100 kPa and 15°C to a pressure of 1000 kPa while being cooled at a rate of 20 kJ/kg by circulating water through the compressor casing. The volume flow rate of the air at the inlet conditions is 140 m³/min, and the power input to the compressor is 520 kW. Determine (*a*) the mass flow rate of the air and (*b*) the temperature at the compressor exit. *Answers:* (*a*) 2.82 kg/s, (*b*) 451 K

**5–50** Reconsider Prob. 5–49. Using EES (or other) software, investigate the effect of the rate of cooling of the compressor on the exit temperature of air. Let the

cooling rate vary from 0 to 200 kJ/kg. Plot the air exit temperature against the rate of cooling, and discuss the results.

**5–51** An adiabatic gas turbine expands air at 1300 kPa and 500°C to 100 kPa and 127°C. Air enters the turbine through a 0.2-m² opening with an average velocity of 40 m/s, and exhausts through a 1-m² opening. Determine (*a*) the mass flow rate of air through the turbine and (*b*) the power produced by the turbine. *Answers:* (*a*) 46.9 kg/s, (*b*) 18.3 MW

**5–52** Steam enters a steady-flow turbine with a mass flow rate of 13 kg/s at 600°C, 8 MPa, and a negligible velocity. The steam expands in the turbine to a saturated vapor at 300 kPa where 10 percent of the steam is removed for some other use. The remainder of the steam continues to expand to the turbine exit where the pressure is 10 kPa and quality is 85 percent. If the turbine is adiabatic, determine the rate of work done by the steam during this process. *Answer:* 17.8 MW

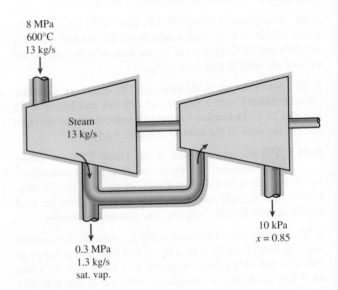

8 MPa
600°C
13 kg/s

Steam
13 kg/s

10 kPa
x = 0.85

0.3 MPa
1.3 kg/s
sat. vap.

**FIGURE P5–52**

**5–53** Steam flows steadily into a turbine with a mass flow rate of 26 kg/s and a negligible velocity at 6 MPa and 600°C. The steam leaves the turbine at 0.5 MPa and 200°C with a velocity of 180 m/s. The rate of work done by the steam in the turbine is measured to be 20 MW. If the elevation change between the turbine inlet and exit is negligible, determine the rate of heat transfer associated with this process. *Answer:* 455 kW

**5–54** Air enters the compressor of a gas-turbine plant at ambient conditions of 100 kPa and 25°C with a low velocity and exits at 1 MPa and 347°C with a velocity of 90 m/s. The compressor is cooled at a rate of 1500 kJ/min, and the power input to the compressor is 250 kW. Determine the mass flow rate of air through the compressor.

## Throttling Valves

**5–55C** Why are throttling devices commonly used in refrigeration and air-conditioning applications?

**5–56C** Would you expect the temperature of air to drop as it undergoes a steady-flow throttling process? Explain.

**5–57C** Would you expect the temperature of a liquid to change as it is throttled? Explain.

**5–58C** During a throttling process, the temperature of a fluid drops from 30 to −20°C. Can this process occur adiabatically?

**5–59** Refrigerant-134a is throttled from the saturated liquid state at 700 kPa to a pressure of 160 kPa. Determine the temperature drop during this process and the final specific volume of the refrigerant. *Answers: 42.3°C, 0.0345 m³/kg*

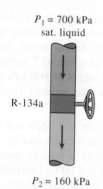

$P_1$ = 700 kPa
sat. liquid

R-134a

$P_2$ = 160 kPa

**FIGURE P5–59**

**5–60** Saturated liquid-vapor mixture of water, called wet steam, in a steam line at 1500 kPa is throttled to 50 kPa and 100°C. What is the quality in the steam line? *Answer: 0.944*

Throttling valve

Steam
1.5 MPa

50 kPa
100°C

**FIGURE P5–60**

**5–61** Refrigerant-134a at 800 kPa and 25°C is throttled to a temperature of −20°C. Determine the pressure and the internal energy of the refrigerant at the final state. *Answers: 133 kPa, 80.7 kJ/kg*

**5–62** A well-insulated valve is used to throttle steam from 8 MPa and 350°C to 2 MPa. Determine the final temperature of the steam. *Answer: 285°C*

**5–63** Reconsider Prob. 5–62. Using EES (or other) software, investigate the effect of the exit pressure of steam on the exit temperature after throttling. Let

the exit pressure vary from 6 to 1 MPa. Plot the exit temperature of steam against the exit pressure, and discuss the results.

## Mixing Chambers and Heat Exchangers

**5–64C** Consider a steady-flow mixing process. Under what conditions will the energy transported into the control volume by the incoming streams be equal to the energy transported out of it by the outgoing stream?

**5–65C** Consider a steady-flow heat exchanger involving two different fluid streams. Under what conditions will the amount of heat lost by one fluid be equal to the amount of heat gained by the other?

**5–66C** When two fluid streams are mixed in a mixing chamber, can the mixture temperature be lower than the temperature of both streams? Explain.

**5–67** Liquid water at 300 kPa and 20°C is heated in a chamber by mixing it with superheated steam at 300 kPa and 300°C. Cold water enters the chamber at a rate of 1.8 kg/s. If the mixture leaves the mixing chamber at 60°C, determine the mass flow rate of the superheated steam required. *Answer: 0.107 kg/s*

**5–68** In steam power plants, open feedwater heaters are frequently utilized to heat the feedwater by mixing it with steam bled off the turbine at some intermediate stage. Consider an open feedwater heater that operates at a pressure of 1000 kPa. Feedwater at 50°C and 1000 kPa is to be heated with superheated steam at 200°C and 1000 kPa. In an ideal feedwater heater, the mixture leaves the heater as saturated liquid at the feedwater pressure. Determine the ratio of the mass flow rates of the feedwater and the superheated vapor for this case. *Answer: 3.73*

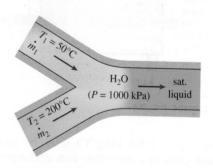

$T_1$ = 50°C
$\dot{m}_1$

$H_2O$
(P = 1000 kPa)

sat.
liquid

$T_2$ = 200°C
$\dot{m}_2$

**FIGURE P5–68**

**5–69** Water at 15°C and 150 kPa is heated in a chamber by mixing it with saturated water vapor at 150 kPa. If both streams enter the mixing chamber at the same mass flow rate, determine the temperature and the quality of the exiting stream. *Answers: 111°C, 0.409*

**5–70**  A stream of refrigerant-134a at 1 MPa and 20°C is mixed with another stream at 1 MPa and 80°C. If the mass flow rate of the cold stream is twice that of the hot one, determine the temperature and the quality of the exit stream.

**5–71**  Reconsider Prob. 5–70. Using EES (or other) software, investigate the effect of the mass flow rate of the cold stream of R-134a on the temperature and the quality of the exit stream. Let the ratio of the mass flow rate of the cold stream to that of the hot stream vary from 1 to 4. Plot the mixture temperature and quality against the cold-to-hot mass flow rate ratio, and discuss the results.

**5–72**  A heat exchanger is to heat water ($c_p$ = 4.18 kJ/kg·°C) from 25 to 60°C at a rate of 0.2 kg/s. The heating is to be accomplished by geothermal water ($c_p$ = 4.31 kJ/kg·°C) available at 140°C at a mass flow rate of 0.3 kg/s. Determine the rate of heat transfer in the heat exchanger and the exit temperature of geothermal water.

**5–73**  Steam is to be condensed on the shell side of a heat exchanger at 25°C. Cooling water enters the tubes at 10°C at a rate of 20 kg/s and leaves at 20°C. Assuming the heat exchanger to be well-insulated, determine the rate of heat transfer in the heat exchanger and the rate of condensation of the steam.

**5–74**  A thin-walled double-pipe counter-flow heat exchanger is used to cool oil ($c_p$ = 2.20 kJ/kg·°C) from 150 to 40°C at a rate of 2 kg/s by water ($c_p$ = 4.18 kJ/kg·°C) that enters at 22°C at a rate of 1.5 kg/s. Determine the rate of heat transfer in the heat exchanger and the exit temperature of water.

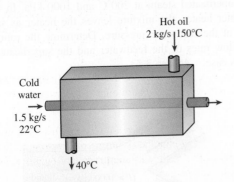

Hot oil
2 kg/s 150°C

Cold
water

1.5 kg/s
22°C

40°C

**FIGURE P5–74**

**5–75**  Air ($c_p$ = 1.005 kJ/kg·°C) is to be preheated by hot exhaust gases in a cross-flow heat exchanger before it enters the furnace. Air enters the heat exchanger at 95 kPa and 20°C at a rate of 0.6 m³/s. The combustion gases ($c_p$ = 1.10 kJ/kg·°C) enter at 160°C at a rate of 0.95 kg/s and leave at 95°C. Determine the rate of heat transfer to the air and its outlet temperature.

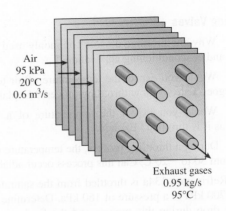

Air
95 kPa
20°C
0.6 m³/s

Exhaust gases
0.95 kg/s
95°C

**FIGURE P5–75**

**5–76**  In a steam heating system, air is heated by being passed over some tubes through which steam flows steadily. Steam enters the heat exchanger at 200 kPa and 200°C at a rate of 7 kg/min and leaves at 175 kPa and 100°C. Air enters at 100 kPa and 27°C and leaves at 55°C. Determine the volume flow rate of air at the inlet.

**5–77**  Refrigerant-134a at 1 MPa and 90°C is to be cooled to 1 MPa and 30°C in a condenser by air. The air enters at 100 kPa and 27°C with a volume flow rate of 600 m³/min and leaves at 95 kPa and 60°C. Determine the mass flow rate of the refrigerant.  *Answer:* 100 kg/min

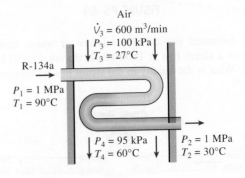

Air
$\dot{V}_3$ = 600 m³/min
$P_3$ = 100 kPa
$T_3$ = 27°C

R-134a

$P_1$ = 1 MPa
$T_1$ = 90°C

$P_4$ = 95 kPa
$T_4$ = 60°C

$P_2$ = 1 MPa
$T_2$ = 30°C

**FIGURE P5–77**

**5–78**  An air-conditioning system involves the mixing of cold air and warm outdoor air before the mixture is routed to the conditioned room in steady operation. Cold air enters the mixing chamber at 7°C and 105 kPa at a rate of 0.55 m³/s while warm air enters at 34°C and 105 kPa. The air leaves the room at 24°C. The ratio of the mass flow rates of the hot to cold air streams is 1.6. Using variable specific heats, determine (a) the mixture temperature at the inlet of the room and (b) the rate of heat gain of the room.

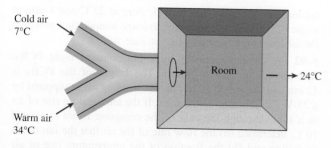

Cold air
7°C

Room

→ 24°C

Warm air
34°C

**FIGURE P5–78**

**5–79** Hot exhaust gases of an internal combustion engine are to be used to produce saturated water vapor at 2 MPa pressure. The exhaust gases enter the heat exchanger at 400°C at a rate of 32 kg/min while water enters at 15°C. The heat exchanger is not well insulated, and it is estimated that 10 percent of heat given up by the exhaust gases is lost to the surroundings. If the mass flow rate of the exhaust gases is 15 times that of the water, determine (a) the temperature of the exhaust gases at the heat exchanger exit and (b) the rate of heat transfer to the water. Use the constant specific heat properties of air for the exhaust gases.

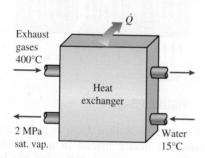

$\dot{Q}$

Exhaust
gases
400°C

Heat
exchanger

2 MPa
sat. vap.

Water
15°C

**FIGURE P5–79**

**5–80** The evaporator of a refrigeration cycle is basically a heat exchanger in which a refrigerant is evaporated by absorbing heat from a fluid. Refrigerant-22 enters an evaporator at 200 kPa with a quality of 22 percent and a flow rate of 2.65 L/h. R-22 leaves the evaporator at the same pressure superheated by 5°C. The refrigerant is evaporated by absorbing heat from air whose flow rate is 0.75 kg/s. Determine (a) the rate of heat absorbed from the air and (b) the temperature change of air. The properties of R-22 at the inlet and exit of the condenser are $h_1 = 220.2$ kJ/kg, $v_1 = 0.0253$ m³/kg, and $h_2 = 398.0$ kJ/kg.

**5–81** Steam is to be condensed in the condenser of a steam power plant at a temperature of 50°C with cooling water from a nearby lake, which enters the tubes of the condenser at 18°C at a rate of 101 kg/s and leaves at 27°C.

Determine the rate of condensation of the steam in the condenser. *Answer:* 1.60 kg/s

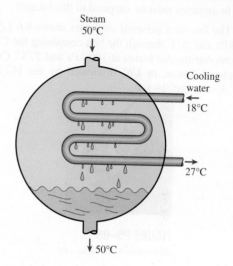

Steam
50°C

Cooling
water
18°C

27°C

50°C

**FIGURE P5–81**

**5–82** Reconsider Prob. 5–81. Using EES (or other) software, investigate the effect of the inlet temperature of cooling water on the rate of condensation of steam. Let the inlet temperature vary from 10 to 20°C, and assume the exit temperature to remain constant. Plot the rate of condensation of steam against the inlet temperature of the cooling water, and discuss the results.

**5–83** Two mass streams of the same ideal gas are mixed in a steady-flow chamber while receiving energy by heat transfer from the surroundings. The mixing process takes place at constant pressure with no work and negligible changes in kinetic and potential energies. Assume the gas has constant specific heats.

(a) Determine the expression for the final temperature of the mixture in terms of the rate of heat transfer to the mixing chamber and the inlet and exit mass flow rates.

(b) Obtain an expression for the volume flow rate at the exit of the mixing chamber in terms of the volume flow rates of the two inlet streams and the rate of heat transfer to the mixing chamber.

(c) For the special case of adiabatic mixing, show that the exit volume flow rate is the sum of the two inlet volume flow rates.

**Pipe and Duct Flow**

**5–84** Water enters a boiler at 3500 kPa as a saturated liquid and leaves at 300°C at the same pressure. Calculate the heat transfer per unit mass of water.

**5–85** A 110-volt electrical heater is used to warm 8.6 L/s of air at 100 kPa and 15°C to 100 kPa and 30°C. How much current in amperes must be supplied to this heater?

**5–86** The fan on a personal computer draws 8.6 L/s of air at 100 kPa and 20°C through the box containing the CPU and other components. Air leaves at 100 kPa and 27°C. Calculate the electrical power, in kW, dissipated by the PC components. *Answer:* 0.0719 kW

**FIGURE P5–86**

©*PhotoDisc/Getty Images RF*

**5–87** A sealed electronic box is to be cooled by tap water flowing through the channels on two of its sides. It is specified that the temperature rise of the water not exceed 4°C. The power dissipation of the box is 2 kW, which is removed entirely by water. If the box operates 24 hours a day, 365 days a year, determine the mass flow rate of water flowing through the box and the amount of cooling water used per year.

**5–88** Repeat Prob. 5–87 for a power dissipation of 4 kW.

**5–89** The components of an electronic system dissipating 180 W are located in a 1.4-m-long horizontal duct whose cross section is 20 cm × 20 cm. The components in the duct are cooled by forced air that enters the duct at 30°C and 1 atm at a rate of 0.6 m³/min and leaves at 40°C. Determine the rate of heat transfer from the outer surfaces of the duct to the ambient. *Answer:* 63 W

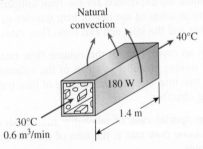

**FIGURE P5–89**

**5–90** Repeat Prob. 5–89 for a circular horizontal duct of diameter 20 cm.

**5–91** Consider a hollow-core printed circuit board 9 cm high and 18 cm long, dissipating a total of 15 W. The width of the air gap in the middle of the PCB is 0.25 cm. If the

cooling air enters the 12-cm-wide core at 25°C and 1 atm at a rate of 0.8 L/s, determine the average temperature at which the air leaves the hollow core. *Answer:* 46.0°C

**5–92** A computer cooled by a fan contains eight PCBs, each dissipating 10 W power. The height of the PCBs is 12 cm and the length is 18 cm. The cooling air is supplied by a 25-W fan mounted at the inlet. If the temperature rise of air as it flows through the case of the computer is not to exceed 10°C, determine (a) the flow rate of the air that the fan needs to deliver and (b) the fraction of the temperature rise of air that is due to the heat generated by the fan and its motor. *Answers:* (a) 0.0104 kg/s, (b) 24 percent

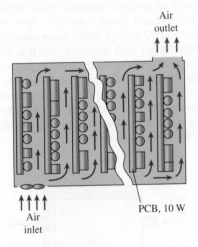

**FIGURE P5–92**

**5–93** A 4-m × 5-m × 6-m room is to be heated by an electric resistance heater placed in a short duct in the room. Initially, the room is at 15°C, and the local atmospheric pressure is 98 kPa. The room is losing heat steadily to the outside at a rate of 150 kJ/min. A 200-W fan circulates the air steadily through the duct and the electric heater at an average mass flow rate of 40 kg/min. The duct can be assumed to be adiabatic, and there is no air leaking in or out of the room. If it takes 20 min for the room air to reach an average temperature of 25°C, find (a) the power rating of the electric heater and (b) the temperature rise that the air experiences each time it passes through the heater.

**5–94** A long roll of 2-m-wide and 0.5-cm-thick 1-Mn manganese steel plate ($\rho = 7854$ kg/m³ and $c_p = 0.434$ kJ/kg·°C) coming off a furnace at 820°C is to be quenched in an oil bath at 45°C to a temperature of 51.1°C. If the metal sheet is moving at a steady velocity of 10 m/min, determine the required rate of heat removal from the oil to keep its temperature constant at 45°C. *Answer:* 4368 kW

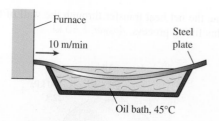

**FIGURE P5–94**

**5–95** Reconsider Prob. 5–94. Using EES (or other) software, investigate the effect of the moving velocity of the steel plate on the rate of heat transfer from the oil bath. Let the velocity vary from 5 to 50 m/min. Plot the rate of heat transfer against the plate velocity, and discuss the results.

**5–96** The hot-water needs of a household are to be met by heating water at 13°C to 82°C by a parabolic solar collector at a rate of 1.8 kg/s. Water flows through a 3.2-cm-diameter thin aluminum tube whose outer surface is black-anodized in order to maximize its solar absorption ability. The centerline of the tube coincides with the focal line of the collector, and a glass sleeve is placed outside the tube to minimize the heat losses. If solar energy is transferred to water at a net rate of 350 W per m length of the tube, determine the required length of the parabolic collector to meet the hot-water requirements of this house.

**5–97** A house has an electric heating system that consists of a 300-W fan and an electric resistance heating element placed in a duct. Air flows steadily through the duct at a rate of 0.6 kg/s and experiences a temperature rise of 7°C. The rate of heat loss from the air in the duct is estimated to be 300 W. Determine the power rating of the electric resistance heating element. *Answer:* 4.22 kW

**5–98** Steam enters a long, horizontal pipe with an inlet diameter of $D_1 = 16$ cm at 2 MPa and 300°C with a velocity of 2.5 m/s. Farther downstream, the conditions are 1.8 MPa and 250°C, and the diameter is $D_2 = 14$ cm. Determine (*a*) the mass flow rate of the steam and (*b*) the rate of heat transfer. *Answers:* (*a*) 0.401 kg/s, (*b*) 45.1 kJ/s

**5–99** Refrigerant-134a enters the condenser of a refrigerator at 900 kPa and 60°C, and leaves as a saturated liquid at the same pressure. Determine the heat transfer from the refrigerant per unit mass.

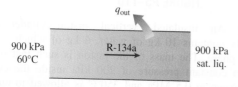

**FIGURE P5–99**

**5–100** Saturated liquid water is heated at constant pressure in a steady-flow device until it is a saturated vapor. Calculate the heat transfer, in kJ/kg, when the vaporization is done at a pressure of 500 kPa.

**5–101** Water is heated in an insulated, constant-diameter tube by a 7-kW electric resistance heater. If the water enters the heater steadily at 20°C and leaves at 75°C, determine the mass flow rate of water.

**5–102** Air at 300 K and 100 kPa steadily flows into a hair dryer having electrical work input of 1500 W. Because of the size of the air intake, the inlet velocity of the air is negligible. The air temperature and velocity at the hair dryer exit are 80°C and 21 m/s, respectively. The flow process is both constant pressure and adiabatic. Assume air has constant specific heats evaluated at 300 K. (*a*) Determine the air mass flow rate into the hair dryer, in kg/s. (*b*) Determine the air volume flow rate at the hair dryer exit, in m³/s. *Answers:* (*a*) 0.0280 kg/s, (*b*) 0.0284 m³/s

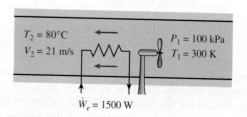

**FIGURE P5–102**

**5–103** Reconsider Prob. 5–102. Using EES (or other) software, investigate the effect of the exit velocity on the mass flow rate and the exit volume flow rate. Let the exit velocity vary from 5 to 25 m/s. Plot the mass flow rate and exit volume flow rate against the exit velocity, and discuss the results.

**5–104** Air enters the duct of an air-conditioning system at 100 kPa and 10°C at a volume flow rate of 13 m³/min. The diameter of the duct is 25 cm, and heat is transferred to the air in the duct from the surroundings at a rate of 2 kJ/s. Determine (*a*) the velocity of the air at the duct inlet and (*b*) the temperature of the air at the exit.

## Charging and Discharging Processes

**5–105** A rigid, insulated tank that is initially evacuated is connected through a valve to a supply line that carries steam at 4 MPa. Now the valve is opened, and steam is allowed to flow into the tank until the pressure reaches 4 MPa, at which point the valve is closed. If the final temperature of the steam in the tank is 550°C, determine the temperature of the steam in the supply line and the flow work per unit mass of the steam.

**5–106** A 2-m³ rigid insulated tank initially containing saturated water vapor at 1 MPa is connected through a valve to a supply line that carries steam at 400°C. Now the valve is opened, and steam is allowed to flow slowly into the tank until the pressure in the tank rises to 2 MPa. At this instant the tank temperature is measured to be 300°C. Determine the mass of the steam that has entered and the pressure of the steam in the supply line.

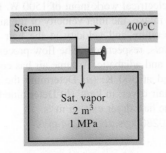

**FIGURE P5–106**

**5–107** A rigid, insulated tank that is initially evacuated is connected through a valve to a supply line that carries helium at 200 kPa and 120°C. Now the valve is opened, and helium is allowed to flow into the tank until the pressure reaches 200 kPa, at which point the valve is closed. Determine the flow work of the helium in the supply line and the final temperature of the helium in the tank. *Answers:* 816 kJ/kg, 655 K

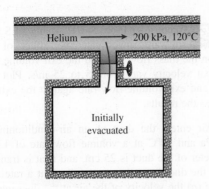

**FIGURE P5–107**

**5–108** Consider a 35-L evacuated rigid bottle that is surrounded by the atmosphere at 100 kPa and 22°C. A valve at the neck of the bottle is now opened and the atmospheric air is allowed to flow into the bottle. The air trapped in the bottle eventually reaches thermal equilibrium with the atmosphere as a result of heat transfer through the wall of the bottle. The valve remains open during the process so that the trapped air also reaches mechanical equilibrium with the atmosphere.

Determine the net heat transfer through the wall of the bottle during this filling process. *Answer:* 3.50 kJ

**FIGURE P5–108**

**5–109** A 0.2-m³ rigid tank equipped with a pressure regulator contains steam at 2 MPa and 300°C. The steam in the tank is now heated. The regulator keeps the steam pressure constant by letting out some steam, but the temperature inside rises. Determine the amount of heat transferred when the steam temperature reaches 500°C.

**5–110** A 4-L pressure cooker has an operating pressure of 175 kPa. Initially, one-half of the volume is filled with liquid and the other half with vapor. If it is desired that the pressure cooker not run out of liquid water for 1 h, determine the highest rate of heat transfer allowed.

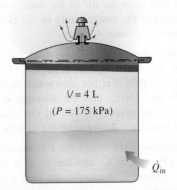

**FIGURE P5–110**

**5–111** An insulated, vertical piston–cylinder device initially contains 10 kg of water, 6 kg of which is in the vapor phase. The mass of the piston is such that it maintains a constant pressure of 200 kPa inside the cylinder. Now steam at 0.5 MPa and 350°C is allowed to enter the cylinder from a supply line until all the liquid in the cylinder has vaporized. Determine (*a*) the final temperature in

the cylinder and (b) the mass of the steam that has entered.
*Answers:* (a) 120.2°C, (b) 19.07 kg

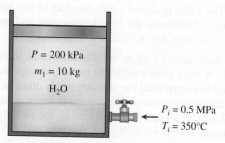

**FIGURE P5–111**

**5–112**   A scuba diver's 0.057-m³ air tank is to be filled with air from a compressed air line at 850 kPa and 25°C. Initially, the air in this tank is at 140 kPa and 15°C. Presuming that the tank is well insulated, determine the temperature and mass in the tank when it is filled to 850 kPa.

**5–113**   An air-conditioning system is to be filled from a rigid container that initially contains 5 kg of liquid R-134a at 24°C. The valve connecting this container to the air-conditioning system is now opened until the mass in the container is 0.25 kg, at which time the valve is closed. During this time, only liquid R-134a flows from the container. Presuming that the process is isothermal while the valve is open, determine the final quality of the R-134a in the container and the total heat transfer.   *Answers:* 0.506, 22.6 kJ

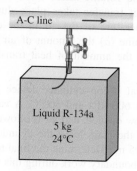

**FIGURE P5–113**

**5–114**   Oxygen is supplied to a medical facility from ten 43-L compressed oxygen tanks. Initially, these tanks are at 10 MPa and 27°C. The oxygen is removed from these tanks slowly enough that the temperature in the tanks remains at 27°C. After two weeks, the pressure in the tanks is 2 MPa. Determine the mass of oxygen used and the total heat transfer to the tanks.

**5–115**   A 0.06-m³ rigid tank initially contains refrigerant-134a at 0.8 MPa and 100 percent quality. The tank is

connected by a valve to a supply line that carries refrigerant-134a at 1.2 MPa and 36°C. Now the valve is opened, and the refrigerant is allowed to enter the tank. The valve is closed when it is observed that the tank contains saturated liquid at 1.2 MPa. Determine (a) the mass of the refrigerant that has entered the tank and (b) the amount of heat transfer.   *Answers:* (a) 64.8 kg, (b) 627 kJ

**5–116**   A 0.3-m³ rigid tank is filled with saturated liquid water at 200°C. A valve at the bottom of the tank is opened, and liquid is withdrawn from the tank. Heat is transferred to the water such that the temperature in the tank remains constant. Determine the amount of heat that must be transferred by the time one-half of the total mass has been withdrawn.

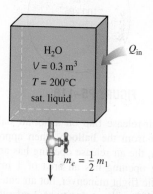

**FIGURE P5–116**

**5–117**   A 0.3-m³ rigid tank initially contains refrigerant-134a at 14°C. At this state, 55 percent of the mass is in the vapor phase, and the rest is in the liquid phase. The tank is connected by a valve to a supply line where refrigerant at 1.4 MPa and 100°C flows steadily. Now the valve is opened slightly, and the refrigerant is allowed to enter the tank. When the pressure in the tank reaches 1 MPa, the entire refrigerant in the tank exists in the vapor phase only. At this point the valve is closed. Determine (a) the final temperature in the tank, (b) the mass of refrigerant that has entered the tank, and (c) the heat transfer between the system and the surroundings.

**5–118**   A balloon that initially contains 50 m³ of steam at 100 kPa and 150°C is connected by a valve to a large reservoir that supplies steam at 150 kPa and 200°C. Now the valve is opened, and steam is allowed to enter the balloon until the pressure equilibrium with the steam at the supply line is reached. The material of the balloon is such that its volume increases linearly with pressure. Heat transfer also takes place between the balloon and the surroundings, and the mass of the steam in the balloon doubles at the end of the process. Determine the final temperature and the boundary work during this process.

Steam
150 kPa
200°C

Steam
50 m³
100 kPa
150°C

**FIGURE P5–118**

**5–119** The air-release flap on a hot-air balloon is used to release hot air from the balloon when appropriate. On one hot-air balloon, the air release opening has an area of 0.5 m², and the filling opening has an area of 1 m². During a two minute adiabatic flight maneuver, hot air enters the balloon at 100 kPa and 35°C with a velocity of 2 m/s; the air in the balloon remains at 100 kPa and 35°C; and air leaves the balloon through the air-release flap at velocity 1 m/s. At the start of this maneuver, the volume of the balloon is 75 m³. Determine the final volume of the balloon and work produced by the air inside the balloon as it expands the balloon skin.

**FIGURE P5–119**
©*Photo Link/Getty Images RF*

**5–120** An insulated 0.15-m³ tank contains helium at 3 MPa and 130°C. A valve is now opened, allowing some helium to escape. The valve is closed when one-half of the initial mass has escaped. Determine the final temperature and pressure in the tank. *Answers:* 257 K, 956 kPa

**5–121** An insulated 1.15-m³ rigid tank contains air at 350 kPa and 50°C. A valve connected to the tank is now opened, and air is allowed to escape until the pressure inside drops to 175 kPa. The air temperature during this process is maintained constant by an electric resistance heater placed in the tank. Determine the electrical work done during this process.

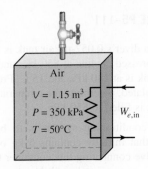

Air

$V = 1.15$ m³
$P = 350$ kPa
$T = 50°C$

$W_{e,\text{in}}$

**FIGURE P5–121**

**5–122** A vertical piston–cylinder device initially contains 0.2 m³ of air at 20°C. The mass of the piston is such that it maintains a constant pressure of 300 kPa inside. Now a valve connected to the cylinder is opened, and air is allowed to escape until the volume inside the cylinder is decreased by one-half. Heat transfer takes place during the process so that the temperature of the air in the cylinder remains constant. Determine (*a*) the amount of air that has left the cylinder and (*b*) the amount of heat transfer. *Answers:* (*a*) 0.357 kg, (*b*) 0

**5–123** A vertical piston–cylinder device initially contains 0.25 m³ of air at 600 kPa and 300°C. A valve connected to the cylinder is now opened, and air is allowed to escape until three-quarters of the mass leave the cylinder at which point the volume is 0.05 m³. Determine the final temperature in the cylinder and the boundary work during this process.

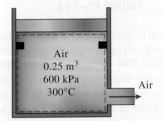

Air
0.25 m³
600 kPa
300°C

Air

**FIGURE P5–123**

**5–124** A vertical piston–cylinder device initially contains 0.01 m³ of steam at 200°C. The mass of the frictionless piston is such that it maintains a constant pressure of 500 kPa inside. Now steam at 1 MPa and 350°C is allowed to enter the cylinder from a supply line until the volume inside doubles. Neglecting any heat transfer that may have taken place during the process, determine (*a*) the final temperature of the steam in the cylinder and (*b*) the amount of mass that has entered. *Answers: (a) 261.7°C, (b) 0.0176 kg*

**5–125** The air in an insulated, rigid compressed-air tank whose volume is 0.5 m³ is initially at 4000 kPa and 20°C. Enough air is now released from the tank to reduce the pressure to 2000 kPa. Following this release, what is the temperature of the remaining air in the tank?

**FIGURE P5–125**

*©C Squared Studios/Getty Images RF*

**5–126** An insulated vertical piston–cylinder device initially contains 0.8 m³ of refrigerant-134a at 1.4 MPa and 120°C. A linear spring at this point applies full force to the piston. A valve connected to the cylinder is now opened, and refrigerant is allowed to escape. The spring unwinds as the piston moves down, and the pressure and volume drop to 0.7 MPa and 0.5 m³ at the end of the process. Determine (*a*) the amount of refrigerant that has escaped and (*b*) the final temperature of the refrigerant.

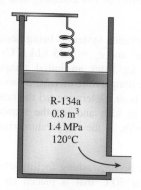

R-134a
0.8 m³
1.4 MPa
120°C

**FIGURE P5–126**

## Review Problems

**5–127** The air in a 6-m × 5-m × 4-m hospital room is to be completely replaced by conditioned air every 15 min. If the

average air velocity in the circular air duct leading to the room is not to exceed 5 m/s, determine the minimum diameter of the duct.

**5–128** A long roll of 1-m-wide and 0.5-cm-thick 1-Mn manganese steel plate ($\rho = 7854$ kg/m³) coming off a furnace is to be quenched in an oil bath to a specified temperature. If the metal sheet is moving at a steady velocity of 10 m/min, determine the mass flow rate of the steel plate through the oil bath.

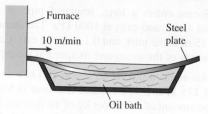

Furnace

10 m/min

Steel plate

Oil bath

**FIGURE P5–128**

**5–129** Air at 4.18 kg/m³ enters a nozzle that has an inlet-to-exit area ratio of 2:1 with a velocity of 120 m/s and leaves with a velocity of 380 m/s. Determine the density of air at the exit. *Answer: 2.64 kg/m³*

**5–130** An air compressor compresses 15 L/s of air at 120 kPa and 20°C to 800 kPa and 300°C while consuming 6.2 kW of power. How much of this power is being used to increase the pressure of the air versus the power needed to move the fluid through the compressor? *Answers: 4.48 kW, 1.72 kW*

**5–131** Saturated refrigerant-134a vapor at 34°C is to be condensed as it flows in a 1-cm-diameter tube at a rate of 0.1 kg/min. Determine the rate of heat transfer from the refrigerant. What would your answer be if the condensed refrigerant is cooled to 20°C?

**5–132** A steam turbine operates with 1.6 MPa and 350°C steam at its inlet and saturated vapor at 30°C at its exit. The mass flow rate of the steam is 22 kg/s, and the turbine produces 12,350 kW of power. Determine the rate at which heat is lost through the casing of this turbine.

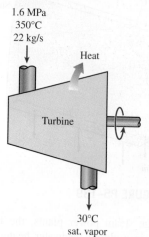

1.6 MPa
350°C
22 kg/s

Heat

Turbine

30°C
sat. vapor

**FIGURE P5–132**

**5–133** Nitrogen gas flows through a long, constant-diameter adiabatic pipe. It enters at 700 kPa and 50°C and leaves at 350 kPa and 20°C. Calculate the velocity of the nitrogen at the pipe's inlet and outlet.

**5–134** A 110-V electric hot-water heater warms 0.1 L/s of water from 18 to 30°C. Calculate the current in amperes that must be supplied to this heater. *Answer:* 45.6 A

**5–135** Steam enters a long, insulated pipe at 1200 kPa, 250°C, and 4 m/s, and exits at 1000 kPa. The diameter of the pipe is 0.15 m at the inlet, and 0.1 m at the exit. Calculate the mass flow rate of the steam and its speed at the pipe outlet.

**5–136** Air enters a pipe at 65°C and 200 kPa and leaves at 60°C and 175 kPa. It is estimated that heat is lost from the pipe in the amount of 3.3 kJ per kg of air flowing in the pipe. The diameter ratio for the pipe is $D_1/D_2 = 1.4$. Using constant specific heats for air, determine the inlet and exit velocities of the air. *Answers:* 29.9 m/s, 66.1 m/s

**5–137** Steam enters a nozzle with a low velocity at 150°C and 200 kPa, and leaves as a saturated vapor at 75 kPa. There is a heat transfer from the nozzle to the surroundings in the amount of 26 kJ for every kilogram of steam flowing through the nozzle. Determine (*a*) the exit velocity of the steam and (*b*) the mass flow rate of the steam at the nozzle entrance if the nozzle exit area is 0.001 m².

**5–138** In a gas-fired boiler, water is boiled at 180°C by hot gases flowing through a stainless steel pipe submerged in water. If the rate of heat transfer from the hot gases to water is 48 kJ/s, determine the rate of evaporation of water.

**5–139** Saturated steam at 1 atm condenses on a vertical plate that is maintained at 90°C by circulating cooling water through the other side. If the rate of heat transfer by condensation to the plate is 180 kJ/s, determine the rate at which the condensate drips off the plate at the bottom.

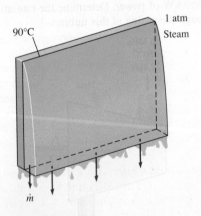

90°C

1 atm
Steam

$\dot{m}$

**FIGURE P5–139**

**5–140** In large steam power plants, the feedwater is frequently heated in a closed feedwater heater by using steam extracted from the turbine at some stage. Steam enters the feedwater heater at 1 MPa and 200°C and leaves as saturated liquid at the same pressure. Feedwater enters the heater at 2.5 MPa and 50°C and leaves at 10°C below the exit temperature of the steam. Determine the ratio of the mass flow rates of the extracted steam and the feedwater.

**5–141** Cold water enters a steam generator at 20°C and leaves as saturated vapor at 200°C. Determine the fraction of heat used in the steam generator to preheat the liquid water from 20°C to the saturation temperature of 200°C.

**5–142** Cold water enters a steam generator at 20°C and leaves as saturated vapor at the boiler pressure. At what pressure will the amount of heat needed to preheat the water to saturation temperature be equal to the heat needed to vaporize the liquid at the boiler pressure?

**5–143** An ideal gas expands in an adiabatic turbine from 1200 K and 900 kPa to 800 K. Determine the turbine inlet volume flow rate of the gas, in m³/s, required to produce turbine work output at the rate of 650 kW. The average values of the specific heats for this gas over the temperature range and the gas constant are $c_p = 1.13$ kJ/kg·K, $c_v = 0.83$ kJ/kg·K, and $R = 0.30$ kJ/kg·K.

**5–144** Chickens with an average mass of 2.2 kg and average specific heat of 3.54 kJ/kg·°C are to be cooled by chilled water that enters a continuous-flow-type immersion chiller at 0.5°C. Chickens are dropped into the chiller at a uniform temperature of 15°C at a rate of 500 chickens per hour and are cooled to an average temperature of 3°C before they are taken out. The chiller gains heat from the surroundings at a rate of 200 kJ/h. Determine (*a*) the rate of heat removal from the chickens, in kW, and (*b*) the mass flow rate of water, in kg/s, if the temperature rise of water is not to exceed 2°C.

**5–145** Repeat Prob. 5–144 assuming heat gain of the chiller is negligible.

**5–146** A refrigeration system is being designed to cool eggs ($\rho = 1080$ kg/m³ and $c_p = 3.35$ kJ/kg·°C) with an average mass of 0.065 kg from an initial temperature of 30°C to a final average temperature of 10°C by air at 1°C at a rate of 10,000 eggs per hour. Determine (*a*) the rate of heat removal from the eggs, in kJ/h and (*b*) the required volume flow rate of air, in m³/h, if the temperature rise of air is not to exceed 6°C.

**5–147** A glass bottle washing facility uses a well-agitated hot-water bath at 50°C that is placed on the ground. The bottles enter at a rate of 450 per minute at an ambient temperature of 20°C and leave at the water temperature. Each bottle has a mass of 150 g and removes 0.2 g of water as it leaves the bath wet. Make-up water is supplied at 15°C.

Disregarding any heat losses from the outer surfaces of the bath, determine the rate at which (*a*) water and (*b*) heat must be supplied to maintain steady operation.

**5–148** The heat of hydration of dough, which is 15 kJ/kg, will raise its temperature to undesirable levels unless some cooling mechanism is utilized. A practical way of absorbing the heat of hydration is to use refrigerated water when kneading the dough. If a recipe calls for mixing 2 kg of flour with 1 kg of water, and the temperature of the city water is 15°C, determine the temperature to which the city water must be cooled before mixing in order for the water to absorb the entire heat of hydration when the water temperature rises to 15°C. Take the specific heats of the flour and the water to be 1.76 and 4.18 kJ/kg·°C, respectively. *Answer:* 4.2°C

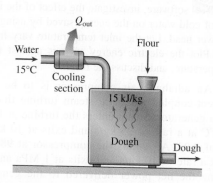

**FIGURE P5–148**

**5–149** Long aluminum wires of diameter 5 mm ($\rho = 2702$ kg/m³ and $c_p = 0.896$ kJ/kg·°C) are extruded at a temperature of 350°C and are cooled to 50°C in atmospheric air at 25°C. If the wire is extruded at a velocity of 8 m/min, determine the rate of heat transfer from the wire to the extrusion room.

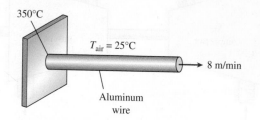

**FIGURE P5–149**

**5–150** Repeat Prob. 5–149 for a copper wire ($\rho = 8950$ kg/m³ and $c_p = 0.383$ kJ/kg·°C).

**5–151** Steam at 550 kPa and 200°C is mixed with water at 15°C and 550 kPa steadily in an adiabatic device. Steam enters the device at a rate of 0.02 kg/s, while the water enters at 0.45 kg/s. Determine the temperature of the mixture leaving this device when the outlet pressure is 550 kPa. *Answer:* 43.4°C

**5–152** A constant-pressure R-134a vapor separation unit separates the liquid and vapor portions of a saturated mixture into two separate outlet streams. Determine the flow power needed to pass 6 L/s of R-134a at 320 kPa and 55 percent quality through this unit. What is the mass flow rate, in kg/s, of the two outlet streams?

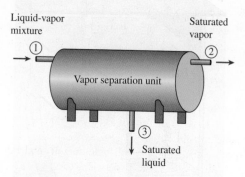

**FIGURE P5–152**

**5–153** Consider two identical buildings: one in Los Angeles, California, where the atmospheric pressure is 101 kPa and the other in Denver, Colorado, where the atmospheric pressure is 83 kPa. Both buildings are maintained at 21°C, and the infiltration rate for both buildings is 1.2 air changes per hour (ACH). That is, the entire air in the building is replaced completely by the outdoor air 1.2 times per hour on a day when the outdoor temperature at both locations is 10°C. Disregarding latent heat, determine the ratio of the heat losses by infiltration at the two cities.

**5–154** It is well established that indoor air quality (IAQ) has a significant effect on general health and productivity of employees at a workplace. A study showed that enhancing IAQ by increasing the building ventilation from 140 m³/min to 560 m³/min increased the productivity by 0.25 percent, valued at $90 per person per year, and decreased the respiratory illnesses by 10 percent for an average annual savings of $39 per person while increasing the annual energy consumption by $6 and the equipment cost by about $4 per person per year (*ASHRAE Journal,* December 1998). For a workplace with 120 employees, determine the net monetary benefit of installing an enhanced IAQ system to the employer per year. *Answer:* $14,280/yr

**5–155** The ventilating fan of the bathroom of a building has a volume flow rate of 30 L/s and runs continuously. The building is located in San Francisco, California, where the average winter temperature is 12.2°C, and is maintained at 22°C at all times. The building is heated by electricity whose unit cost is $0.12/kWh. Determine the amount and cost of the heat "vented out" per month in winter.

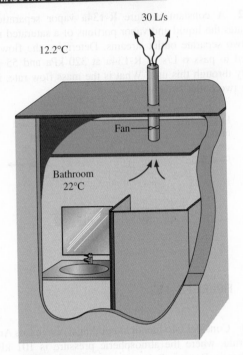

30 L/s

12.2°C

Fan

Bathroom
22°C

**FIGURE P5–155**

**5–156** Determine the rate of sensible heat loss from a building due to infiltration if the outdoor air at −5°C and 95 kPa enters the building at a rate of 60 L/s when the indoors is maintained at 25°C.

**5–157** An air-conditioning system requires airflow at the main supply duct at a rate of 130 m³/min. The average velocity of air in the circular duct is not to exceed 8 m/s to avoid excessive vibration and pressure drops. Assuming the fan converts 80 percent of the electrical energy it consumes into kinetic energy of air, determine the size of the electric motor needed to drive the fan and the diameter of the main duct. Take the density of air to be 1.20 kg/m³.

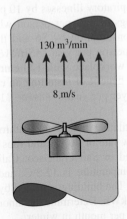

130 m³/min

8 m/s

**FIGURE P5–157**

**5–158** The maximum flow rate of standard shower heads is about 3.5 gpm (13.3 L/min) and can be reduced to 2.75 gpm (10.5 L/min) by switching to low-flow shower heads that are equipped with flow controllers. Consider a family of four, with each person taking a 5-min shower every morning. City water at 15°C is heated to 55°C in an electric water heater and tempered to 42°C by cold water at the T-elbow of the shower before being routed to the shower heads. Assuming a constant specific heat of 4.18 kJ/kg·°C for water, determine (a) the ratio of the flow rates of the hot and cold water as they enter the T-elbow and (b) the amount of electricity that will be saved per year, in kWh, by replacing the standard shower heads by the low-flow ones.

**5–159** 🖳 Reconsider Prob. 5–158. Using EES (or other) software, investigate the effect of the inlet temperature of cold water on the energy saved by using the low-flow shower head. Let the inlet temperature vary from 10°C to 20°C. Plot the electric energy savings against the water inlet temperature, and discuss the results.

**5–160** An adiabatic air compressor is to be powered by a direct-coupled adiabatic steam turbine that is also driving a generator. Steam enters the turbine at 12.5 MPa and 500°C at a rate of 25 kg/s and exits at 10 kPa and a quality of 0.92. Air enters the compressor at 98 kPa and 295 K at a rate of 10 kg/s and exits at 1 MPa and 620 K. Determine the net power delivered to the generator by the turbine.

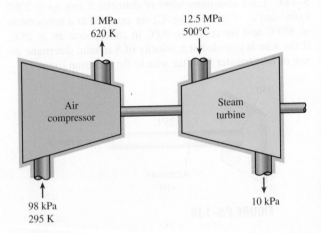

1 MPa
620 K

12.5 MPa
500°C

Air
compressor

Steam
turbine

98 kPa
295 K

10 kPa

**FIGURE P5–160**

**5–161** Determine the power input for a compressor that compresses helium from 110 kPa and 20°C to 400 kPa and 200°C. Helium enters this compressor through a 0.1-m² pipe at a velocity of 9 m/s.

**5–162** Refrigerant 134a enters a compressor with a mass flow rate of 5 kg/s and a negligible velocity. The refrigerant enters the compressor as a saturated vapor at 10°C and leaves the compressor at 1400 kPa with an enthalpy of 281.39 kJ/kg and a velocity of 50 m/s. The rate of work done on the refrigerant is measured to be 132.4 kW. If the elevation change between the compressor inlet and exit is negligible, determine the rate of heat transfer associated with this process, in kW.

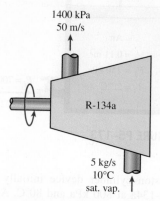

1400 kPa
50 m/s

R-134a

5 kg/s
10°C
sat. vap.

FIGURE P5–162

**5–163** Submarines change their depth by adding or removing air from rigid ballast tanks, thereby displacing seawater in the tanks. Consider a submarine that has a 700 m³ air-ballast tank originally partially filled with 100 m³ of air at 1500 kPa and 15°C. For the submarine to surface, air at 1500 kPa and 20°C is pumped into the ballast tank, until it is entirely filled with air. The tank is filled so quickly that the process is adiabatic and the seawater leaves the tank at 15°C. Determine the final temperature and mass of the air in the ballast tank.

**5–164** In Prob. 5-163, presume that air is added to the tank in such a way that the temperature and pressure of the air in the tank remain constant. Determine the final mass of the air in the ballast tank under this condition. Also determine the total heat transfer while the tank is being filled in this manner.

**5–165** Water flows through a shower head steadily at a rate of 10 L/min. An electric resistance heater placed in the water pipe heats the water from 16 to 43°C. Taking the density of water to be 1 kg/L, determine the electric power input to the heater, in kW.

In an effort to conserve energy, it is proposed to pass the drained warm water at a temperature of 39°C through a heat exchanger to preheat the incoming cold water. If the heat exchanger has an effectiveness of 0.50 (that is, it recovers only half of the energy that can possibly be transferred from the drained water to incoming cold water), determine the electric power input required in this case. If the price of the electric energy is 11.5 ¢/kWh, determine how much money is saved during a 10-min shower as a result of installing this heat exchanger.

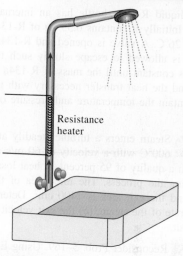

Resistance
heater

FIGURE P5–165

**5–166** Reconsider Prob. 5–165. Using EES (or other) software, investigate the effect of the heat exchanger effectiveness on the money saved. Let effectiveness range from 20 to 90 percent. Plot the money saved against the effectiveness, and discuss the results.

**5–167** A tank with an internal volume of 1 m³ contains air at 800 kPa and 25°C. A valve on the tank is opened allowing air to escape and the pressure inside quickly drops to 150 kPa, at which point the valve is closed. Assume there is negligible heat transfer from the tank to the air left in the tank.

(a) Using the approximation $h_e \approx$ constant $= h_{e,avg} = 0.5(h_1 + h_2)$, calculate the mass withdrawn during the process.

(b) Consider the same process but broken into two parts. That is, consider an intermediate state at $P_2 = 400$ kPa, calculate the mass removed during the process from $P_1 = 800$ kPa to $P_2$ and then the mass removed during the process from $P_2$ to $P_3 = 150$ kPa, using the type of approximation used in part (a), and add the two to get the total mass removed.

(c) Calculate the mass removed if the variation of $h_e$ is accounted for.

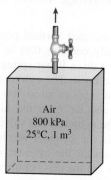

Air
800 kPa
25°C, 1 m³

FIGURE P5–167

**5–168** A liquid R-134a bottle has an internal volume of 0.0015 m³. Initially it contains 0.55 kg of R-134a (saturated mixture) at 26°C. A valve is opened and R-134a vapor only (no liquid) is allowed to escape slowly such that temperature remains constant until the mass of R-134a remaining is 0.15 kg. Find the heat transfer necessary with the surroundings to maintain the temperature and pressure of the R-134a constant.

**5–169** Steam enters a turbine steadily at 7 MPa and 600°C with a velocity of 60 m/s and leaves at 25 kPa with a quality of 95 percent. A heat loss of 20 kJ/kg occurs during the process. The inlet area of the turbine is 150 cm², and the exit area is 1400 cm². Determine (a) the mass flow rate of the steam, (b) the exit velocity, and (c) the power output.

**5–170** Reconsider Prob. 5–169. Using EES (or other) software, investigate the effects of turbine exit area and turbine exit pressure on the exit velocity and power output of the turbine. Let the exit pressure vary from 10 to 50 kPa (with the same quality), and the exit area to vary from 1000 to 3000 cm². Plot the exit velocity and the power outlet against the exit pressure for the exit areas of 1000, 2000, and 3000 cm², and discuss the results.

**5–171** In large gas-turbine power plants, air is preheated by the exhaust gases in a heat exchanger called the *regenerator* before it enters the combustion chamber. Air enters the regenerator at 1 MPa and 550 K at a mass flow rate of 800 kg/min. Heat is transferred to the air at a rate of 3200 kJ/s. Exhaust gases enter the regenerator at 140 kPa and 800 K and leave at 130 kPa and 600 K. Treating the exhaust gases as air, determine (a) the exit temperature of the air and (b) the mass flow rate of exhaust gases.
*Answers:* (a) 775 K, (b) 14.9 kg/s

**5–172** It is proposed to have a water heater that consists of an insulated pipe of 7.5-cm diameter and an electric resistor inside. Cold water at 20°C enters the heating section steadily at a rate of 24 L/min. If water is to be heated to 48°C, determine (a) the power rating of the resistance heater and (b) the average velocity of the water in the pipe.

**5–173** An insulated vertical piston–cylinder device initially contains 0.11 m³ of air at 150 kPa and 22°C. At this state, a linear spring touches the piston but exerts no force on it. The cylinder is connected by a valve to a line that supplies air at 700 kPa and 22°C. The valve is opened, and air from the high-pressure line is allowed to enter the cylinder. The valve is turned off when the pressure inside the cylinder reaches 600 kPa. If the enclosed volume inside the cylinder doubles during this process, determine (a) the mass of air that entered the cylinder, and (b) the final temperature of the air inside the cylinder.

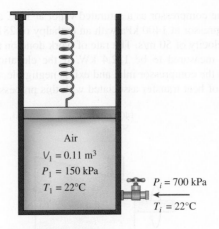

Air
$V_1 = 0.11$ m³
$P_1 = 150$ kPa
$T_1 = 22°C$

$P_i = 700$ kPa
$T_i = 22°C$

**FIGURE P5–173**

**5–174** A piston–cylinder device initially contains 2 kg of refrigerant-134a at 800 kPa and 80°C. At this state, the piston is touching on a pair of stops at the top. The mass of the piston is such that a 500-kPa pressure is required to move it. A valve at the bottom of the tank is opened, and R-134a is withdrawn from the cylinder. After a while, the piston is observed to move and the valve is closed when half of the refrigerant is withdrawn from the tank and the temperature in the tank drops to 20°C. Determine (a) the work done and (b) the heat transfer. *Answers:* (a) 11.6 kJ, (b) 60.7 kJ

**5–175** A piston–cylinder device initially contains 1.2 kg of air at 700 kPa and 200°C. At this state, the piston is touching on a pair of stops. The mass of the piston is such that 600-kPa pressure is required to move it. A valve at the bottom of the tank is opened, and air is withdrawn from the cylinder. The valve is closed when the volume of the cylinder decreases to 80 percent of the initial volume. If it is estimated that 40 kJ of heat is lost from the cylinder, determine (a) the final temperature of the air in the cylinder, (b) the amount of mass that has escaped from the cylinder, and (c) the work done. Use constant specific heats at the average temperature.

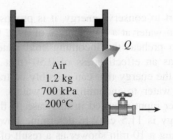

$Q$

Air
1.2 kg
700 kPa
200°C

**FIGURE P5–175**

**5–176** In a single-flash geothermal power plant, geothermal water enters the flash chamber (a throttling valve) at 230°C as a saturated liquid at a rate of 50 kg/s. The steam resulting from the flashing process enters a turbine and leaves at 20 kPa with a moisture content of 5 percent. Determine the temperature of the steam after the flashing process and the power output from the turbine if the pressure of the steam at the exit of the flash chamber is (a) 1 MPa, (b) 500 kPa, (c) 100 kPa, (d) 50 kPa.

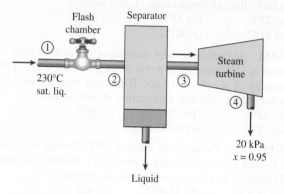

**FIGURE P5–176**

**5–177** The turbocharger of an internal combustion engine consists of a turbine and a compressor. Hot exhaust gases flow through the turbine to produce work and the work output from the turbine is used as the work input to the compressor. The pressure of ambient air is increased as it flows through the compressor before it enters the engine cylinders. Thus, the purpose of a turbocharger is to increase the pressure of air so that more air gets into the cylinder. Consequently, more fuel can be burned and more power can be produced by the engine.

In a turbocharger, exhaust gases enter the turbine at 400°C and 120 kPa at a rate of 0.02 kg/s and leave at 350°C. Air enters the compressor at 50°C and 100 kPa and leaves at 130 kPa at a rate of 0.018 kg/s. The compressor increases the air pressure with a side effect: It also increases the air temperature, which increases the possibility of a gasoline engine to experience an engine knock. To avoid this, an aftercooler is placed after the compressor to cool the warm air by cold ambient air before it enters the engine cylinders. It is estimated that the aftercooler must decrease the air temperature below 80°C if knock is to be avoided. The cold ambient air enters the aftercooler at 30°C and leaves at 40°C. Disregarding any frictional losses in the turbine and the compressor and treating the exhaust gases as air, determine (a) the temperature of the air at the compressor outlet and (b) the minimum volume flow rate of ambient air required to avoid knock.

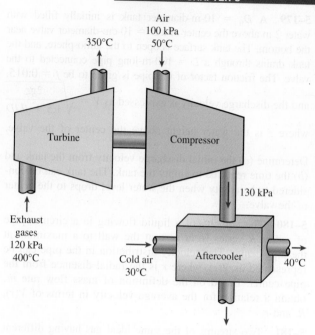

**FIGURE P5–177**

**5–178** A building with an internal volume of 400 m³ is to be heated by a 30-kW electric resistance heater placed in the duct inside the building. Initially, the air in the building is at 14°C, and the local atmospheric pressure is 95 kPa. The building is losing heat to the surroundings at a steady rate of 450 kJ/min. Air is forced to flow through the duct and the heater steadily by a 250-W fan, and it experiences a temperature rise of 5°C each time it passes through the duct, which may be assumed to be adiabatic.

(a) How long will it take for the air inside the building to reach an average temperature of 24°C?

(b) Determine the average mass flow rate of air through the duct.

*Answers:* (a) 146 s, (b) 6.02 kg/s

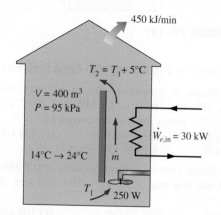

**FIGURE P5–178**

**5–179** A $D_0$ = 10-m-diameter tank is initially filled with water 2 m above the center of a $D$ = 10-cm-diameter valve near the bottom. The tank surface is open to the atmosphere, and the tank drains through a $L$ = 100-m-long pipe connected to the valve. The friction factor of the pipe is given to be $f$ = 0.015, and the discharge velocity is expressed as $V = \sqrt{\dfrac{2gz}{1.5 + fL/D}}$

where $z$ is the water height above the center of the valve.

Determine (a) the initial discharge velocity from the tank and (b) the time required to empty the tank. The tank can be considered to be empty when the water level drops to the center of the valve.

**5–180** The velocity of a liquid flowing in a circular pipe of radius $R$ varies from zero at the wall to a maximum at the pipe center. The velocity distribution in the pipe can be represented as $V(r)$, where $r$ is the radial distance from the pipe center. Based on the definition of mass flow rate $\dot{m}$, obtain a relation for the average velocity in terms of $V(r)$, $R$, and $r$.

**5–181** Two streams of the same ideal gas having different mass flow rates and temperatures are mixed in a steady-flow, adiabatic mixing device. Assuming constant specific heats, find the simplest expression for the mixture temperature written in the form

$$T_3 = f\left(\frac{\dot{m}_1}{\dot{m}_3}, \frac{\dot{m}_2}{\dot{m}_3}, T_1, T_2\right)$$

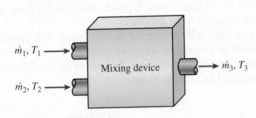

**FIGURE P5–181**

## Fundamentals of Engineering (FE) Exam Problems

**5–182** Steam is compressed by an adiabatic compressor from 0.2 MPa and 150°C to 2.5 MPa and 250°C at a rate of 1.30 kg/s. The power input to the compressor is
(a) 144 kW   (b) 234 kW   (c) 438 kW
(d) 717 kW   (e) 901 kW

**5–183** Steam enters a diffuser steadily at 0.5 MPa, 300°C, and 122 m/s at a rate of 3.5 kg/s. The inlet area of the diffuser is
(a) 15 cm²   (b) 50 cm²   (c) 105 cm²
(d) 150 cm²   (e) 190 cm²

**5–184** An adiabatic heat exchanger is used to heat cold water at 15°C entering at a rate of 5 kg/s by hot air at 90°C entering also at a rate of 5 kg/s. If the exit temperature of hot air is 20°C, the exit temperature of cold water is
(a) 27°C   (b) 32°C   (c) 52°C
(d) 85°C   (e) 90°C

**5–185** A heat exchanger is used to heat cold water at 15°C entering at a rate of 2 kg/s by hot air at 85°C entering at a rate of 3 kg/s. The heat exchanger is not insulated and is losing heat at a rate of 25 kJ/s. If the exit temperature of hot air is 20°C, the exit temperature of cold water is
(a) 28°C   (b) 35°C   (c) 38°C
(d) 41°C   (e) 80°C

**5–186** An adiabatic heat exchanger is used to heat cold water at 15°C entering at a rate of 5 kg/s by hot water at 90°C entering at a rate of 4 kg/s. If the exit temperature of hot water is 50°C, the exit temperature of cold water is
(a) 42°C   (b) 47°C   (c) 55°C
(d) 78°C   (e) 90°C

**5–187** In a shower, cold water at 10°C flowing at a rate of 5 kg/min is mixed with hot water at 60°C flowing at a rate of 2 kg/min. The exit temperature of the mixture is
(a) 24.3°C   (b) 35.0°C   (c) 40.0°C
(d) 44.3°C   (e) 55.2°C

**5–188** In a heating system, cold outdoor air at 7°C flowing at a rate of 4 kg/min is mixed adiabatically with heated air at 70°C flowing at a rate of 3 kg/min. The exit temperature of the mixture is
(a) 34°C   (b) 39°C   (c) 45°C
(d) 63°C   (e) 77°C

**5–189** Hot combustion gases (assumed to have the properties of air at room temperature) enter a gas turbine at 1 MPa and 1500 K at a rate of 0.1 kg/s, and exit at 0.2 MPa and 900 K. If heat is lost from the turbine to the surroundings at a rate of 15 kJ/s, the power output of the gas turbine is
(a) 15 kW   (b) 30 kW   (c) 45 kW
(d) 60 kW   (e) 75 kW

**5–190** Steam expands in a turbine from 4 MPa and 500°C to 0.5 MPa and 250°C at a rate of 1350 kg/h. Heat is lost from the turbine at a rate of 25 kJ/s during the process. The power output of the turbine is
(a) 157 kW   (b) 207 kW   (c) 182 kW
(d) 287 kW   (e) 246 kW

**5–191** Steam is compressed by an adiabatic compressor from 0.2 MPa and 150°C to 0.8 MPa and 350°C at a rate of 1.30 kg/s. The power input to the compressor is
(a) 511 kW   (b) 393 kW   (c) 302 kW
(d) 717 kW   (e) 901 kW

**5–192** Refrigerant-134a is compressed by a compressor from the saturated vapor state at 0.14 MPa to 0.9 MPa and

60°C at a rate of 0.108 kg/s. The refrigerant is cooled at a rate of 1.10 kJ/s during compression. The power input to the compressor is

(a) 4.94 kW  (b) 6.04 kW  (c) 7.14 kW
(d) 7.50 kW  (e) 8.13 kW

**5–193** Refrigerant-134a expands in an adiabatic turbine from 1.2 MPa and 100°C to 0.18 MPa and 50°C at a rate of 1.25 kg/s. The power output of the turbine is

(a) 44.7 kW  (b) 66.4 kW  (c) 72.7 kW
(d) 89.2 kW  (e) 112.0 kW

**5–194** Refrigerant-134a at 1.4 MPa and 90°C is throttled to a pressure of 0.6 MPa. The temperature of the refrigerant after throttling is

(a) 22°C  (b) 56°C  (c) 82°C
(d) 80°C  (e) 90°C

**5–195** Air at 27°C and 5 atm is throttled by a valve to 1 atm. If the valve is adiabatic and the change in kinetic energy is negligible, the exit temperature of air will be

(a) 10°C  (b) 15°C  (c) 20°C
(d) 23°C  (e) 27°C

**5–196** Steam at 1 MPa and 300°C is throttled adiabatically to a pressure of 0.4 MPa. If the change in kinetic energy is negligible, the specific volume of the steam after throttling is

(a) 0.358 m³/kg  (b) 0.233 m³/kg  (c) 0.375 m³/kg
(d) 0.646 m³/kg  (e) 0.655 m³/kg

**5–197** Air is to be heated steadily by an 8-kW electric resistance heater as it flows through an insulated duct. If the air enters at 50°C at a rate of 2 kg/s, the exit temperature of air is

(a) 46.0°C  (b) 50.0°C  (c) 54.0°C
(d) 55.4°C  (e) 58.0°C

## Design and Essay Problems

**5–198** You have been given the responsibility of picking a steam turbine for an electrical-generation station that is to produce 300 MW of electrical power that will sell for $0.05 per kilowatt-hour. The boiler will produce steam at 5 MPa and 400°C, and the condenser is planned to operate at 25°C. The cost of generating and condensing the steam is $0.01 per kilowatt-hour of electricity produced. You have narrowed your selection to the three turbines in the table below. Your criterion for selection is to pay for the equipment as quickly as possible. Which turbine should you choose?

| Turbine | Capacity (MW) | $\eta$ | Cost ($Million) | Operating Cost ($/kWh) |
|---------|---------------|--------|-----------------|------------------------|
| A | 50 | 0.9 | 5 | 0.01 |
| B | 100 | 0.92 | 11 | 0.01 |
| C | 100 | 0.93 | 10.5 | 0.015 |

**5–199** You are to design a small, directional control rocket to operate in space by providing as many as 100 bursts of 5 seconds each with a mass flow rate of 0.2 kg/s at a velocity of 120 m/s. Storage tanks that will contain up to 20 MPa are available, and the tanks will be located in an environment whose temperature is 5°C. Your design criterion is to minimize the volume of the storage tank. Should you use a compressed-air or an R-134a system?

**5–200** An air cannon uses compressed air to propel a projectile from rest to a final velocity. Consider an air cannon that is to accelerate a 10-gram projectile to a speed of 300 m/s using compressed air, whose temperature cannot exceed 20°C. The volume of the storage tank is not to exceed 0.1 m³. Select the storage volume size and maximum storage pressure that requires the minimum amount of energy to fill the tank.

**5–201** Design a 1200-W electric hair dryer such that the air temperature and velocity in the dryer will not exceed 50°C and 3 m/s, respectively.

**5–202** Design a scalding unit for slaughtered chickens to loosen their feathers before they are routed to feather-picking machines with a capacity of 1200 chickens per hour under the following conditions:

The unit will be of an immersion type filled with hot water at an average temperature of 53°C at all times. Chicken with an average mass of 2.2 kg and an average temperature of 36°C will be dipped into the tank, held in the water for 1.5 min, and taken out by a slow-moving conveyor. The chicken is expected to leave the tank 15 percent heavier as a result of the water that sticks to its surface. The center-to-center distance between chickens in any direction will be at least 30 cm. The tank can be as wide as 3 m and as high as 60 cm. The water is to be circulated through and heated by a natural gas furnace, but the temperature rise of water will not exceed 5°C as it passes through the furnace. The water loss is to be made up by the city water at an average temperature of 16°C. The walls and the floor of the tank are well-insulated. The unit operates 24 h a day and 6 days a week. Assuming reasonable values for the average properties, recommend reasonable values for (a) the mass flow rate of the makeup water that must be supplied to the tank, (b) the rate of heat transfer from the water to the chicken, in kW, (c) the size of the heating system in kJ/h, and (d) the operating cost of the scalding unit per month for a unit cost of $1.12/therm of natural gas.

# THE SECOND LAW OF THERMODYNAMICS

To this point, we have focused our attention on the first law of thermodynamics, which requires that energy be conserved during a process. In this chapter, we introduce the second law of thermodynamics, which asserts that processes occur in a certain direction and that energy has quality as well as quantity. A process cannot take place unless it satisfies both the first and second laws of thermodynamics. In this chapter, the thermal energy reservoirs, reversible and irreversible processes, heat engines, refrigerators, and heat pumps are introduced first. Various statements of the second law are followed by a discussion of perpetual-motion machines and the thermodynamic temperature scale. The Carnot cycle is introduced next, and the Carnot principles are discussed. Finally, the idealized Carnot heat engines, refrigerators, and heat pumps are examined.

■ ■ ■ ■ ■ ■ ■
## OBJECTIVES
The objectives of Chapter 6 are to:

■ Introduce the second law of thermodynamics.

■ Identify valid processes as those that satisfy both the first and second laws of thermodynamics.

■ Discuss thermal energy reservoirs, reversible and irreversible processes, heat engines, refrigerators, and heat pumps.

■ Describe the Kelvin–Planck and Clausius statements of the second law of thermodynamics.

■ Discuss the concepts of perpetual-motion machines.

■ Apply the second law of thermodynamics to cycles and cyclic devices.

■ Apply the second law to develop the absolute thermodynamic temperature scale.

■ Describe the Carnot cycle.

■ Examine the Carnot principles, idealized Carnot heat engines, refrigerators, and heat pumps.

■ Determine the expressions for the thermal efficiencies and coefficients of performance for reversible heat engines, heat pumps, and refrigerators.

**FIGURE 6–1**
A cup of hot coffee does not get hotter in a cooler room.

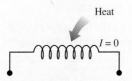

**FIGURE 6–2**
Transferring heat to a wire will not generate electricity.

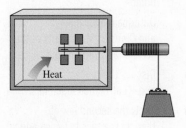

**FIGURE 6–3**
Transferring heat to a paddle wheel will not cause it to rotate.

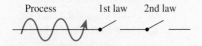

**FIGURE 6–4**
Processes occur in a certain direction, and not in the reverse direction.

**FIGURE 6–5**
A process must satisfy both the first and second laws of thermodynamics to proceed.

# 6–1 · INTRODUCTION TO THE SECOND LAW

In Chaps. 4 and 5, we applied the *first law of thermodynamics,* or the *conservation of energy principle,* to processes involving closed and open systems. As pointed out repeatedly in those chapters, energy is a conserved property, and no process is known to have taken place in violation of the first law of thermodynamics. Therefore, it is reasonable to conclude that a process must satisfy the first law to occur. However, as explained here, satisfying the first law alone does not ensure that the process will actually take place.

It is common experience that a cup of hot coffee left in a cooler room eventually cools off (Fig. 6–1). This process satisfies the first law of thermodynamics since the amount of energy lost by the coffee is equal to the amount gained by the surrounding air. Now let us consider the reverse process—the hot coffee getting even hotter in a cooler room as a result of heat transfer from the room air. We all know that this process never takes place. Yet, doing so would not violate the first law as long as the amount of energy lost by the air is equal to the amount gained by the coffee.

As another familiar example, consider the heating of a room by the passage of electric current through a resistor (Fig. 6–2). Again, the first law dictates that the amount of electric energy supplied to the resistance wires be equal to the amount of energy transferred to the room air as heat. Now let us attempt to reverse this process. It will come as no surprise that transferring some heat to the wires does not cause an equivalent amount of electric energy to be generated in the wires.

Finally, consider a paddle-wheel mechanism that is operated by the fall of a mass (Fig. 6–3). The paddle wheel rotates as the mass falls and stirs a fluid within an insulated container. As a result, the potential energy of the mass decreases, and the internal energy of the fluid increases in accordance with the conservation of energy principle. However, the reverse process, raising the mass by transferring heat from the fluid to the paddle wheel, does not occur in nature, although doing so would not violate the first law of thermodynamics.

It is clear from these arguments that processes proceed in a *certain direction* and not in the reverse direction (Fig. 6–4). The first law places no restriction on the direction of a process, but satisfying the first law does not ensure that the process can actually occur. This inadequacy of the first law to identify whether a process can take place is remedied by introducing another general principle, the *second law of thermodynamics.* We show later in this chapter that the reverse processes discussed above violate the second law of thermodynamics. This violation is easily detected with the help of a property, called *entropy,* defined in Chap. 7. A process cannot occur unless it satisfies both the first and the second laws of thermodynamics (Fig. 6–5).

There are numerous valid statements of the second law of thermodynamics. Two such statements are presented and discussed later in this chapter in relation to some engineering devices that operate on cycles.

The use of the second law of thermodynamics is not limited to identifying the direction of processes. The second law also asserts that energy has *quality* as well as quantity. The first law is concerned with the quantity of energy and the transformations of energy from one form to another with no regard to its quality. Preserving the quality of energy is a major concern to engineers, and the second law provides the necessary means to determine the quality as well as the degree of degradation of energy during a process. As discussed later in this chapter, more of high-temperature energy can be converted to work, and thus it has a higher quality than the same amount of energy at a lower temperature.

The second law of thermodynamics is also used in determining the *theoretical limits* for the performance of commonly used engineering systems, such as heat engines and refrigerators, as well as predicting the *degree of completion* of chemical reactions. The second law is also closely associated with the concept of *perfection*. In fact, the second law *defines* perfection for thermodynamic processes. It can be used to quantify the level of perfection of a process, and point the direction to eliminate imperfections effectively.

# 6–2 · THERMAL ENERGY RESERVOIRS

In the development of the second law of thermodynamics, it is very convenient to have a hypothetical body with a relatively large *thermal energy capacity* (mass × specific heat) that can supply or absorb finite amounts of heat without undergoing any change in temperature. Such a body is called a **thermal energy reservoir**, or just a reservoir. In practice, large bodies of water such as oceans, lakes, and rivers as well as the atmospheric air can be modeled accurately as thermal energy reservoirs because of their large thermal energy storage capabilities or thermal masses (Fig. 6–6). The *atmosphere*, for example, does not warm up as a result of heat losses from residential buildings in winter. Likewise, megajoules of waste energy dumped in large rivers by power plants do not cause any significant change in water temperature.

A *two-phase system* can also be modeled as a reservoir since it can absorb and release large quantities of heat while remaining at constant temperature. Another familiar example of a thermal energy reservoir is the *industrial furnace*. The temperatures of most furnaces are carefully controlled, and they are capable of supplying large quantities of thermal energy as heat in an essentially isothermal manner. Therefore, they can be modeled as reservoirs.

A body does not actually have to be very large to be considered a reservoir. Any physical body whose thermal energy capacity is large relative to the amount of energy it supplies or absorbs can be modeled as one. The air in a room, for example, can be treated as a reservoir in the analysis of the heat dissipation from a TV set in the room, since the amount of heat transfer from the TV set to the room air is not large enough to have a noticeable effect on the room air temperature.

A reservoir that supplies energy in the form of heat is called a **source**, and one that absorbs energy in the form of heat is called a **sink** (Fig. 6–7). Thermal energy reservoirs are often referred to as **heat reservoirs** since they supply or absorb energy in the form of heat.

Heat transfer from industrial sources to the environment is of major concern to environmentalists as well as to engineers. Irresponsible management of waste energy can significantly increase the temperature of portions of the environment, causing what is called *thermal pollution*. If it is not carefully controlled, thermal pollution can seriously disrupt marine life in lakes

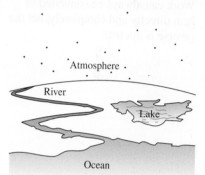

**FIGURE 6–6**
Bodies with relatively large thermal masses can be modeled as thermal energy reservoirs.

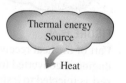

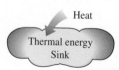

**FIGURE 6–7**
A source supplies energy in the form of heat, and a sink absorbs it.

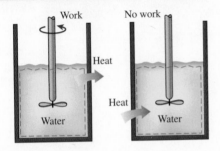

**FIGURE 6–8**

Work can always be converted to heat directly and completely, but the reverse is not true.

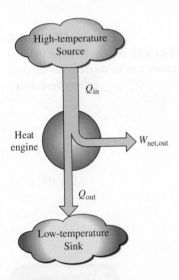

**FIGURE 6–9**

Part of the heat received by a heat engine is converted to work, while the rest is rejected to a sink.

and rivers. However, by careful design and management, the waste energy dumped into large bodies of water can be used to improve the quality of marine life by keeping the local temperature increases within safe and desirable levels.

## 6–3 · HEAT ENGINES

As pointed out earlier, work can easily be converted to other forms of energy, but converting other forms of energy to work is not that easy. The mechanical work done by the shaft shown in Fig. 6–8, for example, is first converted to the internal energy of the water. This energy may then leave the water as heat. We know from experience that any attempt to reverse this process will fail. That is, transferring heat to the water does not cause the shaft to rotate. From this and other observations, we conclude that work can be converted to heat directly and completely, but converting heat to work requires the use of some special devices. These devices are called **heat engines**.

Heat engines differ considerably from one another, but all can be characterized by the following (Fig. 6–9):

1. They receive heat from a high-temperature source (solar energy, oil furnace, nuclear reactor, etc.).
2. They convert part of this heat to work (usually in the form of a rotating shaft).
3. They reject the remaining waste heat to a low-temperature sink (the atmosphere, rivers, etc.).
4. They operate on a cycle.

Heat engines and other cyclic devices usually involve a fluid to and from which heat is transferred while undergoing a cycle. This fluid is called the **working fluid**.

The term *heat engine* is often used in a broader sense to include work-producing devices that do not operate in a thermodynamic cycle. Engines that involve internal combustion such as gas turbines and car engines fall into this category. These devices operate in a mechanical cycle but not in a thermodynamic cycle since the working fluid (the combustion gases) does not undergo a complete cycle. Instead of being cooled to the initial temperature, the exhaust gases are purged and replaced by fresh air-and-fuel mixture at the end of the cycle.

The work-producing device that best fits into the definition of a heat engine is the *steam power plant*, which is an external-combustion engine. That is, combustion takes place outside the engine, and the thermal energy released during this process is transferred to the steam as heat. The schematic of a basic steam power plant is shown in Fig. 6–10. This is a rather simplified diagram, and the discussion of actual steam power plants is given in later chapters. The various quantities shown on this figure are as follows:

$Q_{in}$ = amount of heat supplied to steam in boiler from a high-temperature source (furnace)

$Q_{out}$ = amount of heat rejected from steam in condenser to a low-temperature sink (the atmosphere, a river, etc.)

$W_{out}$ = amount of work delivered by steam as it expands in turbine

$W_{in}$ = amount of work required to compress water to boiler pressure

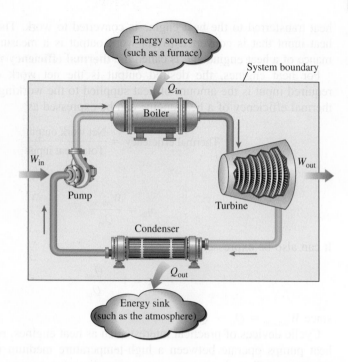

**FIGURE 6–10**
Schematic of a steam power plant.

Notice that the directions of the heat and work interactions are indicated by the subscripts *in* and *out*. Therefore, all four of the described quantities are always *positive*.

The net work output of this power plant is simply the difference between the total work output of the plant and the total work input (Fig. 6–11):

$$W_{net,out} = W_{out} - W_{in} \quad (kJ) \qquad (6\text{–}1)$$

The net work can also be determined from the heat transfer data alone. The four components of the steam power plant involve mass flow in and out, and therefore should be treated as open systems. These components, together with the connecting pipes, however, always contain the same fluid (not counting the steam that may leak out, of course). No mass enters or leaves this combination system, which is indicated by the shaded area on Fig. 6–10; thus, it can be analyzed as a closed system. Recall that for a closed system undergoing a cycle, the change in internal energy $\Delta U$ is zero, and therefore the net work output of the system is also equal to the net heat transfer to the system:

$$W_{net,out} = Q_{in} - Q_{out} \quad (kJ) \qquad (6\text{–}2)$$

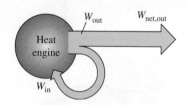

**FIGURE 6–11**
A portion of the work output of a heat engine is consumed internally to maintain continuous operation.

## Thermal Efficiency

In Eq. 6–2, $Q_{out}$ represents the magnitude of the energy wasted in order to complete the cycle. But $Q_{out}$ is never zero; thus, the net work output of a heat engine is always less than the amount of heat input. That is, only part of the

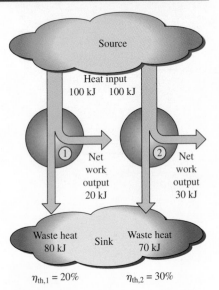

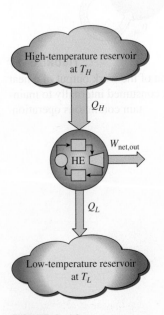

**FIGURE 6–12**
Some heat engines perform better than others (convert more of the heat they receive to work).

**FIGURE 6–13**
Schematic of a heat engine.

heat transferred to the heat engine is converted to work. The fraction of the heat input that is converted to net work output is a measure of the performance of a heat engine and is called the **thermal efficiency** $\eta_{th}$ (Fig. 6–12).

For heat engines, the desired output is the net work output, and the required input is the amount of heat supplied to the working fluid. Then the thermal efficiency of a heat engine can be expressed as

$$\text{Thermal efficiency} = \frac{\text{Net work output}}{\text{Total heat input}} \tag{6-3}$$

or

$$\eta_{th} = \frac{W_{net,out}}{Q_{in}} \tag{6-4}$$

It can also be expressed as

$$\eta_{th} = 1 - \frac{Q_{out}}{Q_{in}} \tag{6-5}$$

since $W_{net,out} = Q_{in} - Q_{out}$.

Cyclic devices of practical interest such as heat engines, refrigerators, and heat pumps operate between a high-temperature medium (or reservoir) at temperature $T_H$ and a low-temperature medium (or reservoir) at temperature $T_L$. To bring uniformity to the treatment of heat engines, refrigerators, and heat pumps, we define these two quantities:

$Q_H$ = magnitude of heat transfer between the cyclic device and the high-temperature medium at temperature $T_H$

$Q_L$ = magnitude of heat transfer between the cyclic device and the low-temperature medium at temperature $T_L$

Notice that both $Q_L$ and $Q_H$ are defined as *magnitudes* and therefore are positive quantities. The direction of $Q_H$ and $Q_L$ is easily determined by inspection. Then, the net work output and thermal efficiency relations for any heat engine (shown in Fig. 6–13) can also be expressed as

$$W_{net,out} = Q_H - Q_L$$

and

$$\eta_{th} = \frac{W_{net,out}}{Q_H} \quad \text{or} \quad \eta_{th} = 1 - \frac{Q_L}{Q_H} \tag{6-6}$$

The thermal efficiency of a heat engine is always less than unity since both $Q_L$ and $Q_H$ are defined as positive quantities.

Thermal efficiency is a measure of how efficiently a heat engine converts the heat that it receives to work. Heat engines are built for the purpose of converting heat to work, and engineers are constantly trying to improve the efficiencies of these devices since increased efficiency means less fuel consumption and thus lower fuel bills and less pollution.

The thermal efficiencies of work-producing devices are relatively low. Ordinary spark-ignition automobile engines have a thermal efficiency of about 25 percent. That is, an automobile engine converts about 25 percent of the chemical energy of the gasoline to mechanical work. This number is as high as 40 percent for diesel engines and large gas-turbine plants and as high as 60 percent for large combined gas-steam power plants. Thus, even with the most efficient heat engines available today, almost one-half of the energy supplied ends up in the rivers, lakes, or the atmosphere as waste or useless energy (Fig. 6–14).

## Can We Save $Q_{out}$?

In a steam power plant, the condenser is the device where large quantities of waste heat is rejected to rivers, lakes, or the atmosphere. Then one may ask, can we not just take the condenser out of the plant and save all that waste energy? The answer to this question is, unfortunately, a firm *no* for the simple reason that without a heat rejection process in a condenser, the cycle cannot be completed. (Cyclic devices such as steam power plants cannot run continuously unless the cycle is completed.) This is demonstrated next with the help of a simple heat engine.

Consider the simple heat engine shown in Fig. 6–15 that is used to lift weights. It consists of a piston–cylinder device with two sets of stops. The working fluid is the gas contained within the cylinder. Initially, the gas temperature is 30°C. The piston, which is loaded with the weights, is resting on top of the lower stops. Now 100 kJ of heat is transferred to the gas in the cylinder from a source at 100°C, causing it to expand and to raise the loaded piston until the piston reaches the upper stops, as shown in the figure. At this point, the load is removed, and the gas temperature is observed to be 90°C.

The work done on the load during this expansion process is equal to the increase in its potential energy, say 15 kJ. Even under ideal conditions (weightless piston, no friction, no heat losses, and quasi-equilibrium expansion), the amount of heat supplied to the gas is greater than the work done since part of the heat supplied is used to raise the temperature of the gas.

Now let us try to answer this question: *Is it possible to transfer the 85 kJ of excess heat at 90°C back to the reservoir at 100°C for later use?* If it is, then we will have a heat engine that can have a thermal efficiency of 100 percent under ideal conditions. The answer to this question is again

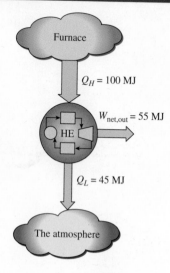

**FIGURE 6–14**
Even the most efficient heat engines reject almost one-half of the energy they receive as waste heat.

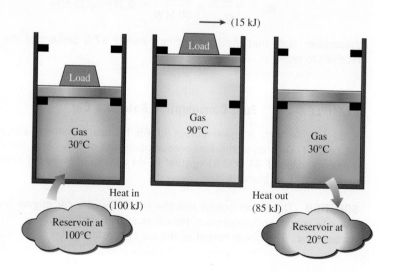

**FIGURE 6–15**
A heat-engine cycle cannot be completed without rejecting some heat to a low-temperature sink.

*no*, for the very simple reason that heat is always transferred from a high-temperature medium to a low-temperature one, and never the other way around. Therefore, we cannot cool this gas from 90 to 30°C by transferring heat to a reservoir at 100°C. Instead, we have to bring the system into contact with a low-temperature reservoir, say at 20°C, so that the gas can return to its initial state by rejecting its 85 kJ of excess energy as heat to this reservoir. This energy cannot be recycled, and it is properly called *waste energy*.

We conclude from this discussion that every heat engine must *waste* some energy by transferring it to a low-temperature reservoir in order to complete the cycle, even under idealized conditions. The requirement that a heat engine exchange heat with at least two reservoirs for continuous operation forms the basis for the Kelvin–Planck expression of the second law of thermodynamics discussed later in this section.

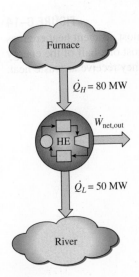

**FIGURE 6–16**
Schematic for Example 6–1.

---

### EXAMPLE 6–1  Net Power Production of a Heat Engine

Heat is transferred to a heat engine from a furnace at a rate of 80 MW. If the rate of waste heat rejection to a nearby river is 50 MW, determine the net power output and the thermal efficiency for this heat engine.

**SOLUTION** The rates of heat transfer to and from a heat engine are given. The net power output and the thermal efficiency are to be determined.
**Assumptions** Heat losses through the pipes and other components are negligible.
**Analysis** A schematic of the heat engine is given in Fig. 6–16. The furnace serves as the high-temperature reservoir for this heat engine and the river as the low-temperature reservoir. The given quantities can be expressed as

$$\dot{Q}_H = 80 \text{ MW} \quad \text{and} \quad \dot{Q}_L = 50 \text{ MW}$$

The net power output of this heat engine is

$$\dot{W}_{net,out} = \dot{Q}_H - \dot{Q}_L = (80 - 50) \text{ MW} = \mathbf{30 \ MW}$$

Then the thermal efficiency is easily determined to be

$$\eta_{th} = \frac{\dot{W}_{net,out}}{\dot{Q}_H} = \frac{30 \text{ MW}}{80 \text{ MW}} = \mathbf{0.375} \text{ (or 37.5%)}$$

**Discussion** Note that the heat engine converts 37.5 percent of the heat it receives to work.

---

### EXAMPLE 6–2  Fuel Consumption Rate of a Car

A car engine with a power output of 65 hp has a thermal efficiency of 24 percent. Determine the fuel consumption rate of this car if the fuel has a heating value of 44,000 kJ/kg (that is, 44,000 kJ of energy is released for each lbm of fuel burned).

**SOLUTION** The power output and the efficiency of a car engine are given. The rate of fuel consumption of the car is to be determined.
**Assumptions** The power output of the car is constant.

*Analysis*  A schematic of the car engine is given in Fig. 6–17. The car engine is powered by converting 24 percent of the chemical energy released during the combustion process to work. The amount of energy input required to produce a power output of 65 hp is determined from the definition of thermal efficiency to be

$$\dot{Q}_H = \frac{\dot{W}_{net,out}}{\eta_{th}} = \frac{65 \text{ hp}}{0.24}\left(\frac{0.7457 \text{ kW}}{1 \text{ hp}}\right) = 202 \text{ kW}$$

To supply energy at this rate, the engine must burn fuel at a rate of

$$\dot{m}_{fuel} = \frac{202 \text{ kJ/s}}{44{,}000 \text{ kJ/kg}} = \textbf{0.00459 kg/s} = \textbf{16.5 kg/h}$$

since 44,000 kJ of thermal energy is released for each kg of fuel burned.

*Discussion*  Note that if the thermal efficiency of the car could be doubled, the rate of fuel consumption would be reduced by half.

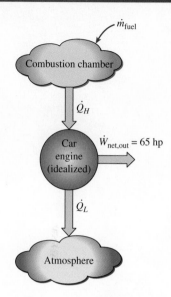

**FIGURE 6–17**
Schematic for Example 6–2.

# The Second Law of Thermodynamics: Kelvin–Planck Statement

We have demonstrated earlier with reference to the heat engine shown in Fig. 6–15 that, even under ideal conditions, a heat engine must reject some heat to a low-temperature reservoir in order to complete the cycle. That is, no heat engine can convert all the heat it receives to useful work. This limitation on the thermal efficiency of heat engines forms the basis for the Kelvin–Planck statement of the second law of thermodynamics, which is expressed as follows:

> It is impossible for any device that operates on a cycle to receive heat from a single reservoir and produce a net amount of work.

That is, a heat engine must exchange heat with a low-temperature sink as well as a high-temperature source to keep operating. The Kelvin–Planck statement can also be expressed as *no heat engine can have a thermal efficiency of 100 percent* (Fig. 6–18), or as *for a power plant to operate, the working fluid must exchange heat with the environment as well as the furnace.*

Note that the impossibility of having a 100 percent efficient heat engine is not due to friction or other dissipative effects. It is a limitation that applies to both the idealized and the actual heat engines. Later in this chapter, we develop a relation for the maximum thermal efficiency of a heat engine. We also demonstrate that this maximum value depends on the reservoir temperatures only.

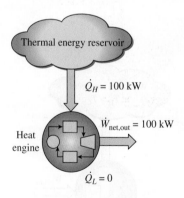

**FIGURE 6–18**
A heat engine that violates the Kelvin–Planck statement of the second law.

## 6–4 · REFRIGERATORS AND HEAT PUMPS

We all know from experience that heat is transferred in the direction of decreasing temperature, that is, from high-temperature mediums to low-temperature ones. This heat transfer process occurs in nature without requiring any devices. The reverse process, however, cannot occur by itself.

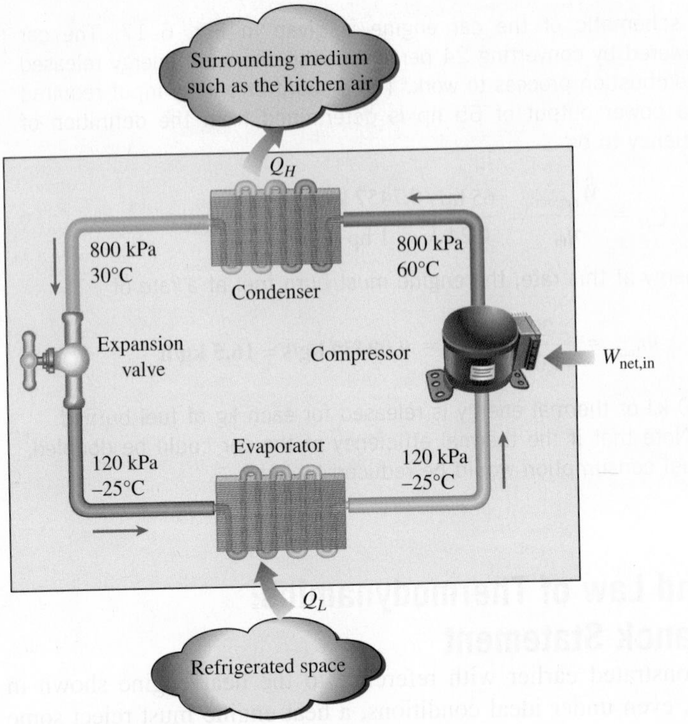

**FIGURE 6–19**
Basic components of a refrigeration system and typical operating conditions.

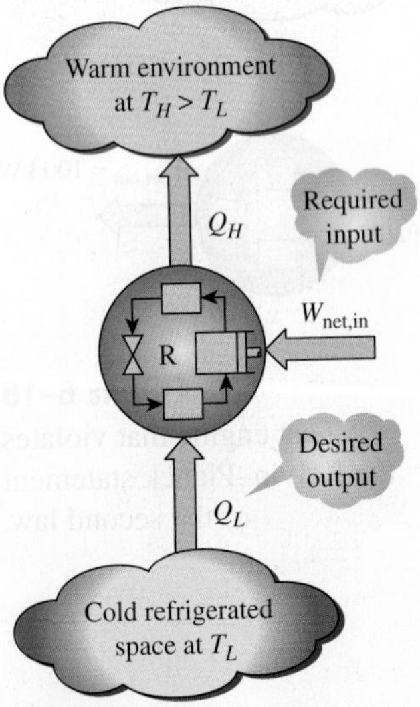

**FIGURE 6–20**
The objective of a refrigerator is to remove $Q_L$ from the cooled space.

The transfer of heat from a low-temperature medium to a high-temperature one requires special devices called **refrigerators**.

Refrigerators, like heat engines, are cyclic devices. The working fluid used in the refrigeration cycle is called a **refrigerant**. The most frequently used refrigeration cycle is the *vapor-compression refrigeration cycle*, which involves four main components: a compressor, a condenser, an expansion valve, and an evaporator, as shown in Fig. 6–19.

The refrigerant enters the compressor as a vapor and is compressed to the condenser pressure. It leaves the compressor at a relatively high temperature and cools down and condenses as it flows through the coils of the condenser by rejecting heat to the surrounding medium. It then enters a capillary tube where its pressure and temperature drop drastically due to the throttling effect. The low-temperature refrigerant then enters the evaporator, where it evaporates by absorbing heat from the refrigerated space. The cycle is completed as the refrigerant leaves the evaporator and reenters the compressor.

In a household refrigerator, the freezer compartment where heat is absorbed by the refrigerant serves as the evaporator, and the coils, usually behind the refrigerator where heat is dissipated to the kitchen air, serve as the condenser.

A refrigerator is shown schematically in Fig. 6–20. Here $Q_L$ is the magnitude of the heat removed from the refrigerated space at temperature $T_L$, $Q_H$ is the magnitude of the heat rejected to the warm environment at temperature $T_H$, and $W_{net,in}$ is the net work input to the refrigerator. As discussed before, $Q_L$ and $Q_H$ represent magnitudes and thus are positive quantities.

## Coefficient of Performance

The *efficiency* of a refrigerator is expressed in terms of the **coefficient of performance** (COP), denoted by $COP_R$. The objective of a refrigerator is to remove heat ($Q_L$) from the refrigerated space. To accomplish this objective, it requires a work input of $W_{net,in}$. Then the COP of a refrigerator can be expressed as

$$COP_R = \frac{\text{Desired output}}{\text{Required input}} = \frac{Q_L}{W_{net,in}} \qquad (6\text{–}7)$$

This relation can also be expressed in rate form by replacing $Q_L$ by $\dot{Q}_L$ and $W_{net,in}$ by $\dot{W}_{net,in}$.

The conservation of energy principle for a cyclic device requires that

$$W_{net,in} = Q_H - Q_L \quad (kJ) \qquad (6\text{–}8)$$

Then the COP relation becomes

$$COP_R = \frac{Q_L}{Q_H - Q_L} = \frac{1}{Q_H/Q_L - 1} \qquad (6\text{–}9)$$

Notice that the value of $COP_R$ can be *greater than unity*. That is, the amount of heat removed from the refrigerated space can be greater than the amount of work input. This is in contrast to the thermal efficiency, which can never be greater than 1. In fact, one reason for expressing the efficiency of a refrigerator by another term—the coefficient of performance—is the desire to avoid the oddity of having efficiencies greater than unity.

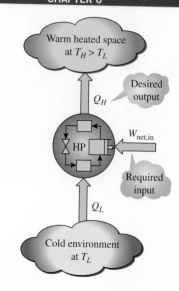

**FIGURE 6–21**
The objective of a heat pump is to supply heat $Q_H$ into the warmer space.

## Heat Pumps

Another device that transfers heat from a low-temperature medium to a high-temperature one is the **heat pump**, shown schematically in Fig. 6–21. Refrigerators and heat pumps operate on the same cycle but differ in their objectives. The objective of a refrigerator is to maintain the refrigerated space at a low emperature by removing heat from it. Discharging this heat to a higher-temperature medium is merely a necessary part of the operation, not the purpose. The objective of a heat pump, however, is to maintain a heated space at a high temperature. This is accomplished by absorbing heat from a low-temperature source, such as well water or cold outside air in winter, and supplying this heat to the high-temperature medium such as a house (Fig. 6–22).

An ordinary refrigerator that is placed in the window of a house with its door open to the cold outside air in winter will function as a heat pump since it will try to cool the outside by absorbing heat from it and rejecting this heat into the house through the coils behind it (Fig. 6–23).

The measure of performance of a heat pump is also expressed in terms of the **coefficient of performance** $COP_{HP}$, defined as

$$COP_{HP} = \frac{\text{Desired output}}{\text{Required input}} = \frac{Q_H}{W_{net,in}} \qquad (6\text{–}10)$$

which can also be expressed as

$$COP_{HP} = \frac{Q_H}{Q_H - Q_L} = \frac{1}{1 - Q_L/Q_H} \qquad (6\text{–}11)$$

A comparison of Eqs. 6–7 and 6–10 reveals that

$$COP_{HP} = COP_R + 1 \qquad (6\text{–}12)$$

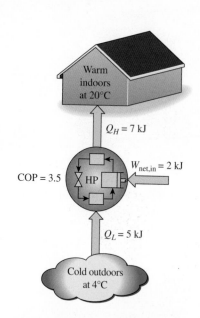

**FIGURE 6–22**
The work supplied to a heat pump is used to extract energy from the cold outdoors and carry it into the warm indoors.

for fixed values of $Q_L$ and $Q_H$. This relation implies that the coefficient of performance of a heat pump is always greater than unity since $COP_R$ is a positive quantity. That is, a heat pump will function, at worst, as a resistance heater, supplying as much energy to the house as it consumes. In reality, however, part of $Q_H$ is lost to the outside air through piping and other devices, and $COP_{HP}$ may drop below unity when the outside air temperature is too low. When this happens, the system usually switches to a resistance heating mode. Most heat pumps in operation today have a seasonally averaged COP of 2 to 3.

Most existing heat pumps use the cold outside air as the heat source in winter, and they are referred to as *air-source heat pumps*. The COP of such heat pumps is about 3.0 at design conditions. Air-source heat pumps are not appropriate for cold climates since their efficiency drops considerably when temperatures are below the freezing point. In such cases, geothermal (also called ground-source) heat pumps that use the ground as the heat source can be used. Geothermal heat pumps require the burial of pipes in the ground 1 to 2 m deep. Such heat pumps are more expensive to install, but they are also more efficient (up to 45 percent more efficient than air-source heat pumps). The COP of ground-source heat pumps can be as high as 6 in the cooling mode.

**Air conditioners** are basically refrigerators whose refrigerated space is a room or a building instead of the food compartment. A window air-conditioning unit cools a room by absorbing heat from the room air and discharging it to the outside. The same air-conditioning unit can be used as a heat pump in winter by installing it backwards. In this mode, the unit absorbs heat from the cold outside and delivers it to the room. Air-conditioning systems that are equipped with proper controls and a reversing valve operate as air conditioners in summer and as heat pumps in winter.

## Performance of Refrigerators, Air-Conditioners, and Heat Pumps

The performance of air conditioners and heat pumps is often expressed in terms of the **energy efficiency ratio** (EER) or **seasonal energy efficiency ratio** (SEER) determined by following certain testing standards. SEER is the ratio the total amount of heat removed by an air conditioner or heat pump during a normal cooling season (in Btu) to the total amount of electricity consumed (in watt-hours, Wh), and it is a measure of seasonal performance of cooling equipment. EER, on the other hand, is a measure of the instantaneous energy efficiency, and is defined as the ratio of the rate of heat removal from the cooled space by the cooling equipment to the rate of electricity consumption in steady operation. Therefore, both EER and SEER have the unit Btu/Wh. Considering that 1 kWh = 3412 Btu and thus 1 Wh = 3.412 Btu, a device that removes 1 kWh of heat from the cooled space for each kWh of electricity it consumes (COP = 1) will have an EER of 3.412. Therefore, the relation between EER (or SEER) and COP is

$$\text{EER} = 3.412 \, COP_R$$

To promote the efficient use of energy, governments worldwide have mandated minimum standards for the performance of energy consuming equipment. Most air conditioners or heat pumps in the market have SEER values from 13 to 21, which correspond to COP values of 3.8 to 6.2. Best performance is achieved using units equipped with variable-speed drives (also called inverters). Variable-speed compressors and fans allow the unit to operate at maximum efficiency for varying heating/cooling needs and weather conditions as determined by a microprocessor. In the air-conditioning mode, for example, they operate at higher speeds on hot days and at lower speeds on cooler days, enhancing both efficiency and comfort.

The EER or COP of a refrigerator decreases with decreasing refrigeration temperature. Therefore, it is not economical to refrigerate to a lower temperature than needed. The COPs of refrigerators are in the range of 2.6–3.0 for cutting and preparation rooms; 2.3–2.6 for meat, deli, dairy, and produce; 1.2–1.5 for frozen foods; and 1.0–1.2 for ice cream units. Note that the COP of freezers is about half of the COP of meat refrigerators, and thus it costs twice as much to cool the meat products with refrigerated air that is cold enough to cool frozen foods. It is good energy conservation practice to use separate refrigeration systems to meet different refrigeration needs.

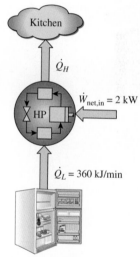

**FIGURE 6–23**
Schematic for Example 6–3.

■ **EXAMPLE 6–3    Heat Rejection by a Refrigerator**

The food compartment of a refrigerator, shown in Fig. 6–23, is maintained at 4°C by removing heat from it at a rate of 360 kJ/min. If the required power input to the refrigerator is 2 kW, determine (a) the coefficient of performance of the refrigerator and (b) the rate of heat rejection to the room that houses the refrigerator.

**SOLUTION** The power consumption of a refrigerator is given. The COP and the rate of heat rejection are to be determined.
**Assumptions** Steady operating conditions exist.
**Analysis** (a) The coefficient of performance of the refrigerator is

$$\text{COP}_R = \frac{\dot{Q}_L}{\dot{W}_{net,in}} = \frac{360 \text{ kJ/min}}{2 \text{ kW}}\left(\frac{1 \text{ kW}}{60 \text{ kJ/min}}\right) = 3$$

That is, 3 kJ of heat is removed from the refrigerated space for each kJ of work supplied.
(b) The rate at which heat is rejected to the room that houses the refrigerator is determined from the conservation of energy relation for cyclic devices,

$$\dot{Q}_H = \dot{Q}_L + \dot{W}_{net,in} = 360 \text{ kJ/min} + (2 \text{ kW})\left(\frac{60 \text{ kJ/min}}{1 \text{ kW}}\right) = 480 \text{ kJ/min}$$

**Discussion** Notice that both the energy removed from the refrigerated space as heat and the energy supplied to the refrigerator as electrical work eventually show up in the room air and become part of the internal energy of the air. This demonstrates that energy can change from one form to another, can move from one place to another, but is never destroyed during a process.

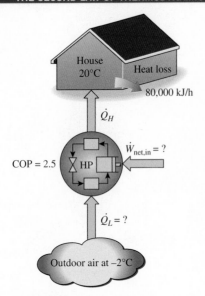

**FIGURE 6–24**
Schematic for Example 6–4.

*EXAMPLE 6–4*  **Heating a House by a Heat Pump**

A heat pump is used to meet the heating requirements of a house and main-tain it at 20°C. On a day when the outdoor air temperature drops to −2°C, the house is estimated to lose heat at a rate of 80,000 kJ/h. If the heat pump under these conditions has a COP of 2.5, determine (*a*) the power consumed by the heat pump and (*b*) the rate at which heat is absorbed from the cold outdoor air.

**SOLUTION**   The COP of a heat pump is given. The power consumption and the rate of heat absorption are to be determined.
**Assumptions**   Steady operating conditions exist.
**Analysis**   (*a*) The power consumed by this heat pump, shown in Fig. 6–24, is determined from the definition of the coefficient of performance to be

$$\dot{W}_{net,in} = \frac{\dot{Q}_H}{COP_{HP}} = \frac{80{,}000 \text{ kJ/h}}{2.5} = \textbf{32{,}000 kJ/h} \text{ (or 8.9 kW)}$$

(*b*) The house is losing heat at a rate of 80,000 kJ/h. If the house is to be maintained at a constant temperature of 20°C, the heat pump must deliver heat to the house at the same rate, that is, at a rate of 80,000 kJ/h. Then the rate of heat transfer from the outdoor becomes

$$\dot{Q}_L = \dot{Q}_H - \dot{W}_{net,in} = (80{,}000 - 32{,}000) \text{ kJ/h} = \textbf{48{,}000 kJ/h}$$

**Discussion**   Note that 48,000 of the 80,000 kJ/h heat delivered to the house is actually extracted from the cold outdoor air. Therefore, we are pay-ing only for the 32,000-kJ/h energy that is supplied as electrical work to the heat pump. If we were to use an electric resistance heater instead, we would have to supply the entire 80,000 kJ/h to the resistance heater as electric energy. This would mean a heating bill that is 2.5 times higher. This explains the popularity of heat pumps as heating systems and why they are preferred to simple electric resistance heaters despite their considerably higher initial cost.

# The Second Law of Thermodynamics: Clausius Statement

There are two classical statements of the second law—the Kelvin–Planck statement, which is related to heat engines and discussed in the preceding section, and the Clausius statement, which is related to refrigerators or heat pumps. The Clausius statement is expressed as follows:

> It is impossible to construct a device that operates in a cycle and produces no effect other than the transfer of heat from a lower-temperature body to a higher-temperature body.

It is common knowledge that heat does not, of its own volition, trans-fer from a cold medium to a warmer one. The Clausius statement does not imply that a cyclic device that transfers heat from a cold medium to a warmer one is impossible to construct. In fact, this is precisely what a com-mon household refrigerator does. It simply states that a refrigerator cannot operate unless its compressor is driven by an external power source, such as an electric motor (Fig. 6–25). This way, the net effect on the surroundings involves the consumption of some energy in the form of work, in addition to the transfer of heat from a colder body to a warmer one. That is, it leaves a trace in the surroundings. Therefore, a household refrigerator is in complete compliance with the Clausius statement of the second law.

Both the Kelvin–Planck and the Clausius statements of the second law are negative statements, and a negative statement cannot be proved. Like any other physical law, the second law of thermodynamics is based on experimental observations. To date, no experiment has been conducted that contradicts the second law, and this should be taken as sufficient proof of its validity.

## Equivalence of the Two Statements

The Kelvin–Planck and the Clausius statements are equivalent in their consequences, and either statement can be used as the expression of the second law of thermodynamics. Any device that violates the Kelvin–Planck statement also violates the Clausius statement, and vice versa. This can be demonstrated as follows.

Consider the heat-engine-refrigerator combination shown in Fig. 6–26a, operating between the same two reservoirs. The heat engine is assumed to have, in violation of the Kelvin–Planck statement, a thermal efficiency of 100 percent, and therefore it converts all the heat $Q_H$ it receives to work $W$. This work is now supplied to a refrigerator that removes heat in the amount of $Q_L$ from the low-temperature reservoir and rejects heat in the amount of $Q_L + Q_H$ to the high-temperature reservoir. During this process, the high-temperature reservoir receives a net amount of heat $Q_L$ (the difference between $Q_L + Q_H$ and $Q_H$). Thus, the combination of these two devices can be viewed as a refrigerator, as shown in Fig. 6–26b, that transfers heat in

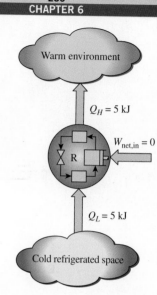

**FIGURE 6–25**
A refrigerator that violates the Clausius statement of the second law.

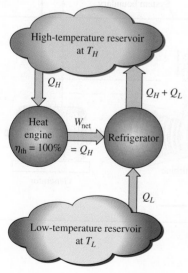

(a) A refrigerator that is powered by a 100 percent efficient heat engine

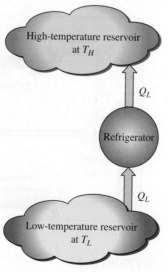

(b) The equivalent refrigerator

**FIGURE 6–26**
Proof that the violation of the Kelvin–Planck statement leads to the violation of the Clausius statement.

an amount of $Q_L$ from a cooler body to a warmer one without requiring any input from outside. This is clearly a violation of the Clausius statement. Therefore, a violation of the Kelvin–Planck statement results in the violation of the Clausius statement.

It can also be shown in a similar manner that a violation of the Clausius statement leads to the violation of the Kelvin–Planck statement. Therefore, the Clausius and the Kelvin–Planck statements are two equivalent expressions of the second law of thermodynamics.

# 6–5 · PERPETUAL-MOTION MACHINES

We have repeatedly stated that a process cannot take place unless it satisfies both the first and second laws of thermodynamics. Any device that violates either law is called a **perpetual-motion machine**, and despite numerous attempts, no perpetual-motion machine is known to have worked. But this has not stopped inventors from trying to create new ones.

A device that violates the first law of thermodynamics (by *creating* energy) is called a **perpetual-motion machine of the first kind** (PMM1), and a device that violates the second law of thermodynamics is called a **perpetual-motion machine of the second kind** (PMM2).

Consider the steam power plant shown in Fig. 6–27. It is proposed to heat the steam by resistance heaters placed inside the boiler, instead of by the energy supplied from fossil or nuclear fuels. Part of the electricity generated by the plant is to be used to power the resistors as well as the pump. The rest of the electric energy is to be supplied to the electric network as the net work output. The inventor claims that once the system is started, this power plant will produce electricity indefinitely without requiring any energy input from the outside.

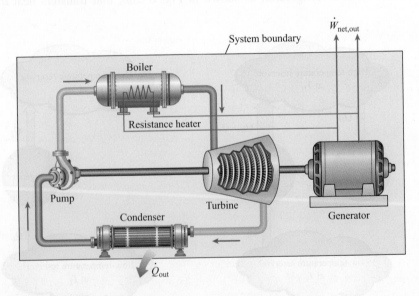

**FIGURE 6–27**

A perpetual-motion machine that violates the first law of thermodynamics (PMM1).

Well, here is an invention that could solve the world's energy problem—if it works, of course. A careful examination of this invention reveals that the system enclosed by the shaded area is continuously supplying energy to the outside at a rate of $\dot{Q}_{out} + \dot{W}_{net,out}$ without receiving any energy. That is, this system is creating energy at a rate of $\dot{Q}_{out} + \dot{W}_{net,out}$, which is clearly a violation of the first law. Therefore, this wonderful device is nothing more than a PMM1 and does not warrant any further consideration.

Now let us consider another novel idea by the same inventor. Convinced that energy cannot be created, the inventor suggests the following modification that will greatly improve the thermal efficiency of that power plant without violating the first law. Aware that more than one-half of the heat transferred to the steam in the furnace is discarded in the condenser to the environment, the inventor suggests getting rid of this wasteful component and sending the steam to the pump as soon as it leaves the turbine, as shown in Fig. 6–28. This way, all the heat transferred to the steam in the boiler will be converted to work, and thus the power plant will have a theoretical efficiency of 100 percent. The inventor realizes that some heat losses and friction between the moving components are unavoidable and that these effects will hurt the efficiency somewhat, but still expects the efficiency to be no less than 80 percent (as opposed to 40 percent in most actual power plants) for a carefully designed system.

Well, the possibility of doubling the efficiency would certainly be very tempting to plant managers and, if not properly trained, they would probably give this idea a chance, since intuitively they see nothing wrong with it. A student of thermodynamics, however, will immediately label this device as a PMM2, since it works on a cycle and does a net amount of work while exchanging heat with a single reservoir (the furnace) only. It satisfies the first law but violates the second law, and therefore it will not work.

Countless perpetual-motion machines have been proposed throughout history, with many more still being proposed. Some proposers have even gone so far as to patent their inventions, only to find out that what they actually have in their hands is a worthless piece of paper.

Some perpetual-motion machine inventors were very successful in fund-raising. For example, a Philadelphia carpenter named J. W. Kelly collected millions of dollars between 1874 and 1898 from investors in his *hydropneumatic-pulsating-vacu-engine,* which supposedly could push a railroad train 3000 miles on 1 L of water. Of course, it never did. After his death in 1898, the investigators discovered that the demonstration machine was powered by a hidden motor. Recently, a group of investors was set to invest $2.5 million into a mysterious *energy augmentor,* which multiplied whatever power it took in, but their lawyer wanted an expert opinion first. Confronted by the scientists, the "inventor" fled the scene without even attempting to run his demo machine.

Tired of applications for perpetual-motion machines, the U.S. Patent Office decreed in 1918 that it would no longer consider any perpetual-motion machine applications. However, several such patent applications were still filed, and some made it through the patent office undetected. Some applicants whose patent applications were denied sought legal action. For example, in 1982 the U.S. Patent Office dismissed as just another perpetual-motion machine a huge device that involves several hundred kilograms of

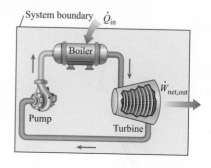

**FIGURE 6–28**

A perpetual-motion machine that violates the second law of thermodynamics (PMM2).

rotating magnets and kilometers of copper wire that is supposed to be generating more electricity than it is consuming from a battery pack. However, the inventor challenged the decision, and in 1985 the National Bureau of Standards finally tested the machine just to certify that it is battery-operated. However, it did not convince the inventor that his machine will not work.

The proposers of perpetual-motion machines generally have innovative minds, but they usually lack formal engineering training, which is very unfortunate. No one is immune from being deceived by an innovative perpetual-motion machine. As the saying goes, however, if something sounds too good to be true, it probably is.

# 6–6 ▪ REVERSIBLE AND IRREVERSIBLE PROCESSES

The second law of thermodynamics states that no heat engine can have an efficiency of 100 percent. Then one may ask, what is the highest efficiency that a heat engine can possibly have? Before we can answer this question, we need to define an idealized process first, which is called the *reversible process*.

The processes that were discussed at the beginning of this chapter occurred in a certain direction. Once having taken place, these processes cannot reverse themselves spontaneously and restore the system to its initial state. For this reason, they are classified as *irreversible processes*. Once a cup of hot coffee cools, it will not heat up by retrieving the heat it lost from the surroundings. If it could, the surroundings, as well as the system (coffee), would be restored to their original condition, and this would be a reversible process.

A **reversible process** is defined as a *process that can be reversed without leaving any trace on the surroundings* (Fig. 6–29). That is, both the system *and* the surroundings are returned to their initial states at the end of the reverse process. This is possible only if the net heat *and* net work exchange between the system and the surroundings is zero for the combined (original and reverse) process. Processes that are not reversible are called **irreversible processes**.

It should be pointed out that a system can be restored to its initial state following a process, regardless of whether the process is reversible or irreversible. But for reversible processes, this restoration is made without leaving any net change on the surroundings, whereas for irreversible processes, the surroundings usually do some work on the system and therefore does not return to their original state.

Reversible processes actually do not occur in nature. They are merely *idealizations* of actual processes. Reversible processes can be approximated by actual devices, but they can never be achieved. That is, all the processes occurring in nature are irreversible. You may be wondering, then, *why* we are bothering with such fictitious processes. There are two reasons. First, they are easy to analyze, since a system passes through a series of equilibrium states during a reversible process. Second, they serve as idealized models to which actual processes can be compared.

In daily life, the concepts of Mr. Right and Ms. Right are also idealizations, just like the concept of a reversible (perfect) process. People who insist on finding Mr. or Ms. Right to settle down are bound to remain Mr. or Ms. Single for the rest of their lives. The possibility of finding the perfect prospective mate is no higher than the possibility of finding a perfect

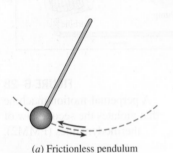

(*a*) Frictionless pendulum

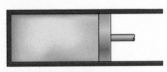

(*b*) Quasi-equilibrium expansion and compression of a gas

**FIGURE 6–29**
Two familiar reversible processes.

(reversible) process. Likewise, a person who insists on perfection in friends is bound to have no friends.

Engineers are interested in reversible processes because work-producing devices such as car engines and gas or steam turbines *deliver the most work,* and work-consuming devices such as compressors, fans, and pumps *consume the least work* when reversible processes are used instead of irreversible ones (Fig. 6–30).

Reversible processes can be viewed as *theoretical limits* for the corresponding irreversible ones. Some processes are more irreversible than others. We may never be able to have a reversible process, but we can certainly approach it. The more closely we approximate a reversible process, the more work delivered by a work-producing device or the less work required by a work-consuming device.

The concept of reversible processes leads to the definition of the **second-law efficiency** for actual processes, which is the degree of approximation to the corresponding reversible processes. This enables us to compare the performance of different devices that are designed to do the same task on the basis of their efficiencies. The better the design, the lower the irreversibilities and the higher the second-law efficiency.

## Irreversibilities

The factors that cause a process to be irreversible are called **irreversibilities**. They include friction, unrestrained expansion, mixing of two fluids, heat transfer across a finite temperature difference, electric resistance, inelastic deformation of solids, and chemical reactions. The presence of any of these effects renders a process irreversible. A reversible process involves none of these. Some of the frequently encountered irreversibilities are discussed briefly below.

**Friction** is a familiar form of irreversibility associated with bodies in motion. When two bodies in contact are forced to move relative to each other (a piston in a cylinder, for example, as shown in Fig. 6–31), a friction force that opposes the motion develops at the interface of these two bodies, and some work is needed to overcome this friction force. The energy supplied as work is eventually converted to heat during the process and is transferred to the bodies in contact, as evidenced by a temperature rise at the interface. When the direction of the motion is reversed, the bodies are restored to their original position, but the interface does not cool, and heat is not converted back to work. Instead, more of the work is converted to heat while overcoming the friction forces that also oppose the reverse motion. Since the system (the moving bodies) and the surroundings cannot be returned to their original states, this process is irreversible. Therefore, any process that involves friction is irreversible. The larger the friction forces involved, the more irreversible the process is.

Friction does not always involve two solid bodies in contact. It is also encountered between a fluid and solid and even between the layers of a fluid moving at different velocities. A considerable fraction of the power produced by a car engine is used to overcome the friction (the drag force) between the air and the external surfaces of the car, and it eventually becomes part of the internal energy of the air. It is not possible to reverse this process and recover that lost power, even though doing so would not violate the conservation of energy principle.

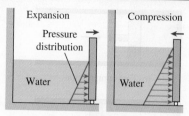

*(a)* Slow (reversible) process

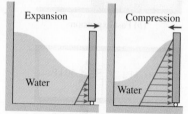

*(b)* Fast (irreversible) process

**FIGURE 6–30**

Reversible processes deliver the most and consume the least work.

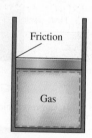

**FIGURE 6–31**

Friction renders a process irreversible.

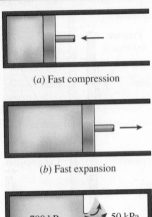

(a) Fast compression

(b) Fast expansion

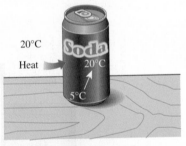

700 kPa | 50 kPa

(c) Unrestrained expansion

**FIGURE 6–32**
Irreversible compression and expansion processes.

20°C
Heat
20°C
5°C

(a) An irreversible heat transfer process

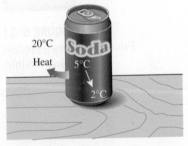

20°C
Heat
5°C
2°C

(b) An impossible heat transfer process

**FIGURE 6–33**
(a) Heat transfer through a temperature difference is irreversible, and (b) the reverse process is impossible.

Another example of irreversibility is the **unrestrained expansion of a gas** separated from a vacuum by a membrane, as shown in Fig. 6–32. When the membrane is ruptured, the gas fills the entire tank. The only way to restore the system to its original state is to compress it to its initial volume, while transferring heat from the gas until it reaches its initial temperature. From the conservation of energy considerations, it can easily be shown that the amount of heat transferred from the gas equals the amount of work done on the gas by the surroundings. The restoration of the surroundings involves conversion of this heat completely to work, which would violate the second law. Therefore, unrestrained expansion of a gas is an irreversible process.

A third form of irreversibility familiar to us all is **heat transfer** through a finite temperature difference. Consider a can of cold soda left in a warm room (Fig. 6–33). Heat is transferred from the warmer room air to the cooler soda. The only way this process can be reversed and the soda restored to its original temperature is to provide refrigeration, which requires some work input. At the end of the reverse process, the soda will be restored to its initial state, but the surroundings will not be. The internal energy of the surroundings will increase by an amount equal in magnitude to the work supplied to the refrigerator. The restoration of the surroundings to the initial state can be done only by converting this excess internal energy completely to work, which is impossible to do without violating the second law. Since only the system, not both the system and the surroundings, can be restored to its initial condition, heat transfer through a finite temperature difference is an irreversible process.

Heat transfer can occur only when there is a temperature difference between a system and its surroundings. Therefore, it is physically impossible to have a reversible heat transfer process. But a heat transfer process becomes less and less irreversible as the temperature difference between the two bodies approaches zero. Then, heat transfer through a differential temperature difference $dT$ can be considered to be reversible. As $dT$ approaches zero, the process can be reversed in direction (at least theoretically) without requiring any refrigeration. Notice that reversible heat transfer is a conceptual process and cannot be duplicated in the real world.

The smaller the temperature difference between two bodies, the smaller the heat transfer rate will be. Any significant heat transfer through a small temperature difference requires a very large surface area and a very long time. Therefore, even though approaching reversible heat transfer is desirable from a thermodynamic point of view, it is impractical and not economically feasible.

## Internally and Externally Reversible Processes

A typical process involves interactions between a system and its surroundings, and a reversible process involves no irreversibilities associated with either of them.

A process is called **internally reversible** if no irreversibilities occur within the boundaries of the system during the process. During an internally reversible process, a system proceeds through a series of equilibrium states, and when the process is reversed, the system passes through exactly the same equilibrium states while returning to its initial state. That is, the paths of the forward and reverse processes coincide for an internally reversible process. The quasi-equilibrium process is an example of an internally reversible process.

A process is called **externally reversible** if no irreversibilities occur outside the system boundaries during the process. Heat transfer between a reservoir and a system is an externally reversible process if the outer surface of the system is at the temperature of the reservoir.

A process is called **totally reversible**, or simply **reversible**, if it involves no irreversibilities within the system or its surroundings (Fig. 6–34). A totally reversible process involves no heat transfer through a finite temperature difference, no nonquasi-equilibrium changes, and no friction or other dissipative effects.

As an example, consider the transfer of heat to two identical systems that are undergoing a constant-pressure (thus constant-temperature) phase-change process, as shown in Fig. 6–35. Both processes are internally reversible, since both take place isothermally and both pass through exactly the same equilibrium states. The first process shown is externally reversible also, since heat transfer for this process takes place through an infinitesimal temperature difference $dT$. The second process, however, is externally irreversible, since it involves heat transfer through a finite temperature difference $\Delta T$.

**FIGURE 6–34**
A reversible process involves no internal and external irreversibilities.

# 6–7 ■ THE CARNOT CYCLE

We mentioned earlier that heat engines are cyclic devices and that the working fluid of a heat engine returns to its initial state at the end of each cycle. Work is done by the working fluid during one part of the cycle and on the working fluid during another part. The difference between these two is the net work delivered by the heat engine. The efficiency of a heat-engine cycle greatly depends on how the individual processes that make up the cycle are executed. The net work, thus the cycle efficiency, can be maximized by using processes that require the least amount of work and deliver the most, that is, by using *reversible processes*. Therefore, it is no surprise that the most efficient cycles are reversible cycles, that is, cycles that consist entirely of reversible processes.

Reversible cycles cannot be achieved in practice because the irreversibilities associated with each process cannot be eliminated. However, reversible cycles provide upper limits on the performance of real cycles. Heat engines and refrigerators that work on reversible cycles serve as models to which actual heat engines and refrigerators can be compared. Reversible cycles also serve as starting points in the development of actual cycles and are modified as needed to meet certain requirements.

Probably the best known reversible cycle is the **Carnot cycle**, first proposed in 1824 by French engineer Sadi Carnot. The theoretical heat engine that operates on the Carnot cycle is called the **Carnot heat engine**. The Carnot cycle is composed of four reversible processes—two isothermal and two adiabatic—and it can be executed either in a closed or a steady-flow system.

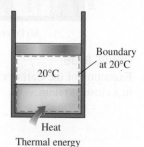

**FIGURE 6–35**
Totally and internally reversible heat transfer processes.

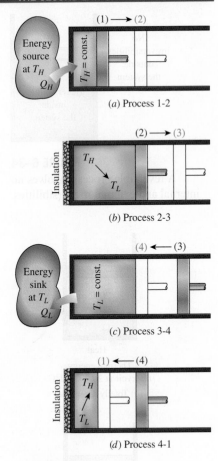

(a) Process 1-2

(b) Process 2-3

(c) Process 3-4

(d) Process 4-1

**FIGURE 6–36**
Execution of the Carnot cycle
in a closed system.

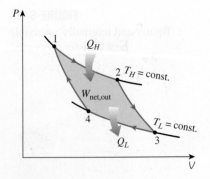

**FIGURE 6–37**
P-V diagram of the Carnot cycle.

Consider a closed system that consists of a gas contained in an adiabatic piston–cylinder device, as shown in Fig. 6–36. The insulation of the cylinder head is such that it may be removed to bring the cylinder into contact with reservoirs to provide heat transfer. The four reversible processes that make up the Carnot cycle are as follows:

**Reversible Isothermal Expansion** (process 1-2, $T_H$ = constant). Initially (state 1), the temperature of the gas is $T_H$ and the cylinder head is in close contact with a source at temperature $T_H$. The gas is allowed to expand slowly, doing work on the surroundings. As the gas expands, the temperature of the gas tends to decrease. But as soon as the temperature drops by an infinitesimal amount $dT$, some heat is transferred from the reservoir into the gas, raising the gas temperature to $T_H$. Thus, the gas temperature is kept constant at $T_H$. Since the temperature difference between the gas and the reservoir never exceeds a differential amount $dT$, this is a reversible heat transfer process. It continues until the piston reaches position 2. The amount of total heat transferred to the gas during this process is $Q_H$.

**Reversible Adiabatic Expansion** (process 2-3, temperature drops from $T_H$ to $T_L$). At state 2, the reservoir that was in contact with the cylinder head is removed and replaced by insulation so that the system becomes adiabatic. The gas continues to expand slowly, doing work on the surroundings until its temperature drops from $T_H$ to $T_L$ (state 3). The piston is assumed to be frictionless and the process to be quasi-equilibrium, so the process is reversible as well as adiabatic.

**Reversible Isothermal Compression** (process 3-4, $T_L$ = constant). At state 3, the insulation at the cylinder head is removed, and the cylinder is brought into contact with a sink at temperature $T_L$. Now the piston is pushed inward by an external force, doing work on the gas. As the gas is compressed, its temperature tends to rise. But as soon as it rises by an infinitesimal amount $dT$, heat is transferred from the gas to the sink, causing the gas temperature to drop to $T_L$. Thus, the gas temperature remains constant at $T_L$. Since the temperature difference between the gas and the sink never exceeds a differential amount $dT$, this is a reversible heat transfer process. It continues until the piston reaches state 4. The amount of heat rejected from the gas during this process is $Q_L$.

**Reversible Adiabatic Compression** (process 4-1, temperature rises from $T_L$ to $T_H$). State 4 is such that when the low-temperature reservoir is removed, the insulation is put back on the cylinder head, and the gas is compressed in a reversible manner, the gas returns to its initial state (state 1). The temperature rises from $T_L$ to $T_H$ during this reversible adiabatic compression process, which completes the cycle.

The P-V diagram of this cycle is shown in Fig. 6–37. Remembering that on a P-V diagram the area under the process curve represents the boundary work for quasi-equilibrium (internally reversible) processes, we see that the area under curve 1-2-3 is the work done by the gas during the expansion part of the cycle, and the area under curve 3-4-1 is the work done on the gas during the compression part of the cycle. The area enclosed by the path of the cycle (area 1-2-3-4-1) is the difference between these two and represents the net work done during the cycle.

Notice that if we acted stingily and compressed the gas at state 3 adiabatically instead of isothermally in an effort *to save* $Q_L$, we would end up back at state 2, retracing the process path 3-2. By doing so we would save $Q_L$, but we would not be able to obtain any net work output from this engine. This illustrates once more the necessity of a heat engine exchanging heat with at least two reservoirs at different temperatures to operate in a cycle and produce a net amount of work.

The Carnot cycle can also be executed in a steady-flow system. It is discussed in later chapters in conjunction with other power cycles.

Being a reversible cycle, the Carnot cycle is the most efficient cycle operating between two specified temperature limits. Even though the Carnot cycle cannot be achieved in reality, the efficiency of actual cycles can be improved by attempting to approximate the Carnot cycle more closely.

## The Reversed Carnot Cycle

The Carnot heat-engine cycle just described is a totally reversible cycle. Therefore, all the processes that comprise it can be *reversed,* in which case it becomes the **Carnot refrigeration cycle**. This time, the cycle remains exactly the same, except that the directions of any heat and work interactions are reversed: Heat in the amount of $Q_L$ is absorbed from the low-temperature reservoir, heat in the amount of $Q_H$ is rejected to a high-temperature reservoir, and a work input of $W_{net,in}$ is required to accomplish all this.

The *P-V* diagram of the reversed Carnot cycle is the same as the one given for the Carnot cycle, except that the directions of the processes are reversed, as shown in Fig. 6–38.

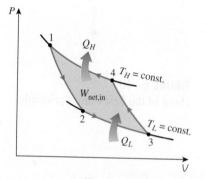

**FIGURE 6–38**
*P-V* diagram of the reversed Carnot cycle.

## 6–8 · THE CARNOT PRINCIPLES

The second law of thermodynamics puts limits on the operation of cyclic devices as expressed by the Kelvin–Planck and Clausius statements. A heat engine cannot operate by exchanging heat with a single reservoir, and a refrigerator cannot operate without a net energy input from an external source.

We can draw valuable conclusions from these statements. Two conclusions pertain to the thermal efficiency of reversible and irreversible (i.e., actual) heat engines, and they are known as the **Carnot principles** (Fig. 6–39), expressed as follows:

1. The efficiency of an irreversible heat engine is always less than the efficiency of a reversible one operating between the same two reservoirs.
2. The efficiencies of all reversible heat engines operating between the same two reservoirs are the same.

These two statements can be proved by demonstrating that the violation of either statement results in the violation of the second law of thermodynamics.

To prove the first statement, consider two heat engines operating between the same reservoirs, as shown in Fig. 6–40. One engine is reversible and the other is irreversible. Now each engine is supplied with the same amount of heat $Q_H$. The amount of work produced by the reversible heat engine is $W_{rev}$, and the amount produced by the irreversible one is $W_{irrev}$.

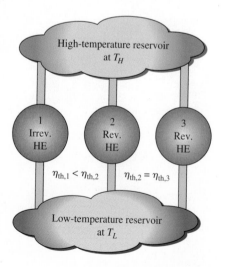

**FIGURE 6–39**
The Carnot principles.

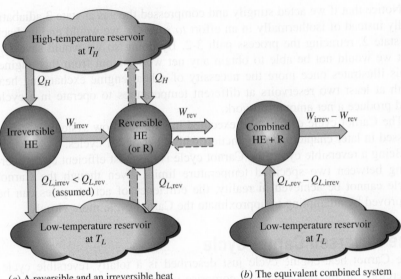

(a) A reversible and an irreversible heat engine operating between the same two reservoirs (the reversible heat engine is then reversed to run as a refrigerator)

(b) The equivalent combined system

**FIGURE 6–40**
Proof of the first Carnot principle.

In violation of the first Carnot principle, we assume that the irreversible heat engine is more efficient than the reversible one (that is, $\eta_{th,irrev} > \eta_{th,rev}$) and thus delivers more work than the reversible one. Now let the reversible heat engine be reversed and operate as a refrigerator. This refrigerator will receive a work input of $W_{rev}$ and reject heat to the high-temperature reservoir. Since the refrigerator is rejecting heat in the amount of $Q_H$ to the high-temperature reservoir and the irreversible heat engine is receiving the same amount of heat from this reservoir, the net heat exchange for this reservoir is zero. Thus, it could be eliminated by having the refrigerator discharge $Q_H$ directly into the irreversible heat engine.

Now considering the refrigerator and the irreversible engine together, we have an engine that produces a net work in the amount of $W_{irrev} - W_{rev}$ while exchanging heat with a single reservoir—a violation of the Kelvin–Planck statement of the second law. Therefore, our initial assumption that $\eta_{th,irrev} > \eta_{th,rev}$ is incorrect. Then we conclude that no heat engine can be more efficient than a reversible heat engine operating between the same reservoirs.

The second Carnot principle can also be proved in a similar manner. This time, let us replace the irreversible engine by another reversible engine that is more efficient and thus delivers more work than the first reversible engine. By following through the same reasoning, we end up having an engine that produces a net amount of work while exchanging heat with a single reservoir, which is a violation of the second law. Therefore, we conclude that no reversible heat engine can be more efficient than a reversible one operating between the same two reservoirs, regardless of how the cycle is completed or the kind of working fluid used.

# 6–9 · THE THERMODYNAMIC TEMPERATURE SCALE

A temperature scale that is independent of the properties of the substances that are used to measure temperature is called a **thermodynamic temperature scale**. Such a temperature scale offers great conveniences in thermodynamic calculations, and its derivation is given below using some reversible heat engines.

The second Carnot principle discussed in Section 6–8 states that all reversible heat engines have the same thermal efficiency when operating between the same two reservoirs (Fig. 6–41). That is, the efficiency of a reversible engine is independent of the working fluid employed and its properties, the way the cycle is executed, or the type of reversible engine used. Since energy reservoirs are characterized by their temperatures, the thermal efficiency of reversible heat engines is a function of the reservoir temperatures only. That is,

$$\eta_{th,rev} = g(T_H, T_L)$$

or

$$\frac{Q_H}{Q_L} = f(T_H, T_L) \qquad \text{(6–13)}$$

since $\eta_{th} = 1 - Q_L/Q_H$. In these relations $T_H$ and $T_L$ are the temperatures of the high- and low-temperature reservoirs, respectively.

The functional form of $f(T_H, T_L)$ can be developed with the help of the three reversible heat engines shown in Fig. 6–42. Engines A and C are supplied with the same amount of heat $Q_1$ from the high-temperature reservoir at $T_1$. Engine C rejects $Q_3$ to the low-temperature reservoir at $T_3$. Engine B receives the heat $Q_2$ rejected by engine A at temperature $T_2$ and rejects heat in the amount of $Q_3$ to a reservoir at $T_3$.

The amounts of heat rejected by engines B and C must be the same since engines A and B can be combined into one reversible engine operating between the same reservoirs as engine C and thus the combined engine will have the same efficiency as engine C. Since the heat input to engine C is the same as the heat input to the combined engines A and B, both systems must reject the same amount of heat.

Applying Eq. 6–13 to all three engines separately, we obtain

$$\frac{Q_1}{Q_2} = f(T_1, T_2), \quad \frac{Q_2}{Q_3} = f(T_2, T_3), \quad \text{and} \quad \frac{Q_1}{Q_3} = f(T_1, T_3)$$

Now consider the identity

$$\frac{Q_1}{Q_3} = \frac{Q_1}{Q_2}\frac{Q_2}{Q_3}$$

which corresponds to

$$f(T_1, T_3) = f(T_1, T_2) \cdot f(T_2, T_3)$$

A careful examination of this equation reveals that the left-hand side is a function of $T_1$ and $T_3$, and therefore the right-hand side must also be a function of

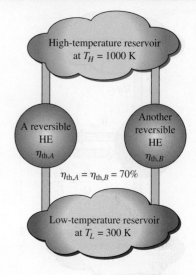

**FIGURE 6–41**
All reversible heat engines operating between the same two reservoirs have the same efficiency (the second Carnot principle).

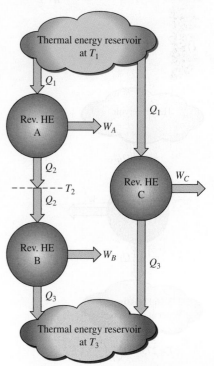

**FIGURE 6–42**
The arrangement of heat engines used to develop the thermodynamic temperature scale.

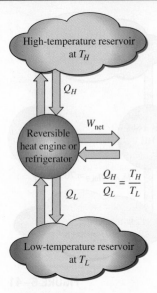

**FIGURE 6–43**
For reversible cycles, the heat transfer ratio $Q_H/Q_L$ can be replaced by the absolute temperature ratio $T_H/T_L$.

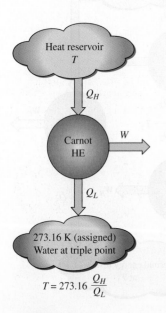

**FIGURE 6–44**
A conceptual experimental setup to determine thermodynamic temperatures on the Kelvin scale by measuring heat transfers $Q_H$ and $Q_L$.

$T_1$ and $T_3$ only, and not $T_2$. That is, the value of the product on the right-hand side of this equation is independent of the value of $T_2$. This condition will be satisfied only if the function $f$ has the following form:

$$f(T_1, T_2) = \frac{\phi(T_1)}{\phi(T_2)} \quad \text{and} \quad f(T_2, T_3) = \frac{\phi(T_2)}{\phi(T_3)}$$

so that $\phi(T_2)$ will cancel from the product of $f(T_1, T_2)$ and $f(T_2, T_3)$, yielding

$$\frac{Q_1}{Q_3} = f(T_1, T_3) = \frac{\phi(T_1)}{\phi(T_3)} \tag{6–14}$$

This relation is much more specific than Eq. 6–13 for the functional form of $Q_1/Q_3$ in terms of $T_1$ and $T_3$.

For a reversible heat engine operating between two reservoirs at temperatures $T_H$ and $T_L$, Eq. 6–14 can be written as

$$\frac{Q_H}{Q_L} = \frac{\phi(T_H)}{\phi(T_L)} \tag{6–15}$$

This is the only requirement that the second law places on the ratio of heat transfers to and from the reversible heat engines. Several functions $\phi(T)$ satisfy this equation, and the choice is completely arbitrary. Lord Kelvin first proposed taking $\phi(T) = T$ to define a thermodynamic temperature scale as (Fig. 6–43)

$$\left(\frac{Q_H}{Q_L}\right)_{\text{rev}} = \frac{T_H}{T_L} \tag{6–16}$$

This temperature scale is called the **Kelvin scale**, and the temperatures on this scale are called **absolute temperatures**. On the Kelvin scale, the temperature ratios depend on the ratios of heat transfer between a reversible heat engine and the reservoirs and are independent of the physical properties of any substance. On this scale, temperatures vary between zero and infinity.

The thermodynamic temperature scale is not completely defined by Eq. 6–16 since it gives us only a ratio of absolute temperatures. We also need to know the magnitude of a kelvin. At the International Conference on Weights and Measures held in 1954, the triple point of water (the state at which all three phases of water exist in equilibrium) was assigned the value 273.16 K (Fig. 6–44). The *magnitude of a kelvin* is defined as 1/273.16 of the temperature interval between absolute zero and the triple-point temperature of water. The magnitudes of temperature units on the Kelvin and Celsius scales are identical (1 K ≡ 1°C). The temperatures on these two scales differ by a constant 273.15:

$$T(°C) = T(K) - 273.15 \tag{6–17}$$

Even though the thermodynamic temperature scale is defined with the help of the reversible heat engines, it is not possible, nor is it practical, to actually operate such an engine to determine numerical values on the absolute temperature scale. Absolute temperatures can be measured accurately by other means, such as the constant-volume ideal-gas thermometer together with extrapolation techniques as discussed in Chap. 1. The validity of Eq. 6–16 can be demonstrated from physical considerations for a reversible cycle using an ideal gas as the working fluid.

## 6–10 · THE CARNOT HEAT ENGINE

The hypothetical heat engine that operates on the reversible Carnot cycle is called the **Carnot heat engine**. The thermal efficiency of any heat engine, reversible or irreversible, is given by Eq. 6–6 as

$$\eta_{th} = 1 - \frac{Q_L}{Q_H}$$

where $Q_H$ is heat transferred to the heat engine from a high-temperature reservoir at $T_H$, and $Q_L$ is heat rejected to a low-temperature reservoir at $T_L$. For reversible heat engines, the heat transfer ratio in the above relation can be replaced by the ratio of the absolute temperatures of the two reservoirs, as given by Eq. 6–16. Then the efficiency of a Carnot engine, or any reversible heat engine, becomes

$$\eta_{th,rev} = 1 - \frac{T_L}{T_H} \qquad (6\text{–}18)$$

This relation is often referred to as the **Carnot efficiency**, since the Carnot heat engine is the best known reversible engine. *This is the highest efficiency a heat engine operating between the two thermal energy reservoirs at temperatures $T_L$ and $T_H$ can have* (Fig. 6–45). All irreversible (i.e., actual) heat engines operating between these temperature limits ($T_L$ and $T_H$) have lower efficiencies. An actual heat engine cannot reach this maximum theoretical efficiency value because it is impossible to completely eliminate all the irreversibilities associated with the actual cycle.

Note that $T_L$ and $T_H$ in Eq. 6–18 are *absolute temperatures*. Using °C for temperatures in this relation gives results grossly in error.

The thermal efficiencies of actual and reversible heat engines operating between the same temperature limits compare as follows (Fig. 6–46):

$$\eta_{th} \begin{cases} < & \eta_{th,rev} \quad \text{irreversible heat engine} \\ = & \eta_{th,rev} \quad \text{reversible heat engine} \\ > & \eta_{th,rev} \quad \text{impossible heat engine} \end{cases} \qquad (6\text{–}19)$$

Most work-producing devices (heat engines) in operation today have efficiencies under 40 percent, which appear low relative to 100 percent. However, when the performance of actual heat engines is assessed, the efficiencies should not be compared to 100 percent; instead, they should be compared to the efficiency of a reversible heat engine operating between the same temperature limits—because this is the true theoretical upper limit for the efficiency, not 100 percent.

The maximum efficiency of a steam power plant operating between $T_H = 1000$ K and $T_L = 300$ K is 70 percent, as determined from Eq. 6–18. Compared with this value, an actual efficiency of 40 percent does not seem so bad, even though there is still plenty of room for improvement.

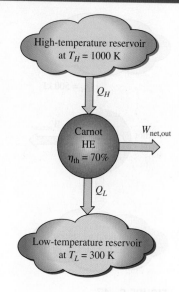

**FIGURE 6–45**
The Carnot heat engine is the most efficient of all heat engines operating between the same high- and low-temperature reservoirs.

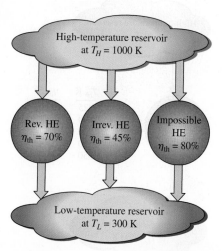

**FIGURE 6–46**
No heat engine can have a higher efficiency than a reversible heat engine operating between the same high- and low-temperature reservoirs.

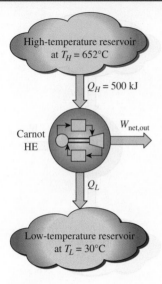

**FIGURE 6–47**
Schematic for Example 6–5.

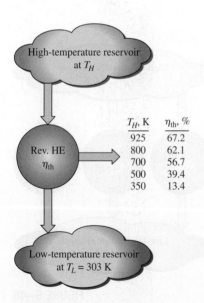

| $T_H$, K | $\eta_{th}$, % |
|---|---|
| 925 | 67.2 |
| 800 | 62.1 |
| 700 | 56.7 |
| 500 | 39.4 |
| 350 | 13.4 |

**FIGURE 6–48**
The fraction of heat that can be converted to work as a function of source temperature (for $T_L = 303$ K).

It is obvious from Eq. 6–18 that the efficiency of a Carnot heat engine increases as $T_H$ is increased, or as $T_L$ is decreased. This is to be expected since as $T_L$ decreases, so does the amount of heat rejected, and as $T_L$ approaches zero, the Carnot efficiency approaches unity. This is also true for actual heat engines. *The thermal efficiency of actual heat engines can be maximized by supplying heat to the engine at the highest possible temperature* (limited by material strength) *and rejecting heat from the engine at the lowest possible temperature* (limited by the temperature of the cooling medium such as rivers, lakes, or the atmosphere).

---

**EXAMPLE 6–5**     **Analysis of a Carnot Heat Engine**

A Carnot heat engine, shown in Fig. 6–47, receives 500 kJ of heat per cycle from a high-temperature source at 652°C and rejects heat to a low-temperature sink at 30°C. Determine (a) the thermal efficiency of this Carnot engine and (b) the amount of heat rejected to the sink per cycle.

**SOLUTION** The heat supplied to a Carnot heat engine is given. The thermal efficiency and the heat rejected are to be determined.
**Analysis** (a) The Carnot heat engine is a reversible heat engine, and so its efficiency can be determined from Eq. 6–18 to be

$$\eta_{th,rev} = 1 - \frac{T_L}{T_H} = 1 - \frac{(30 + 273)\ \text{K}}{(652 + 273)\ \text{K}} = 0.672$$

That is, this Carnot heat engine converts 67.2 percent of the heat it receives to work.

(b) The amount of heat rejected $Q_L$ by this reversible heat engine is easily determined from Eq. 6–16 to be

$$Q_{L,rev} = \frac{T_L}{T_H} Q_{H,rev} = \frac{(30 + 273)\ \text{K}}{(652 + 273)\ \text{K}} (500\ \text{kJ}) = 164\ \text{kJ}$$

**Discussion** Note that this Carnot heat engine rejects to a low-temperature sink 164 kJ of the 500 kJ of heat it receives during each cycle.

---

## The Quality of Energy

The Carnot heat engine in Example 6–5 receives heat from a source at 925 K and converts 67.2 percent of it to work while rejecting the rest (32.8 percent) to a sink at 303 K. Now let us examine how the thermal efficiency varies with the source temperature when the sink temperature is held constant.

The thermal efficiency of a Carnot heat engine that rejects heat to a sink at 303 K is evaluated at various source temperatures using Eq. 6–18 and is listed in Fig. 6–49. Clearly, the thermal efficiency decreases as the source temperature is lowered. When heat is supplied to the heat engine at 500 instead of 925 K, for example, the thermal efficiency drops from 67.2 to 39.4 percent. That is, the fraction of heat that can be converted to work drops to 39.4 percent when the temperature of the source drops to 500 K. When the source temperature is 350 K, this fraction becomes a mere 13.4 percent.

These efficiency values show that energy has **quality** as well as quantity. It is clear from the thermal efficiency values in Fig. 6–48 that *more of the* *high-temperature thermal energy can be converted to work. Therefore, the higher the temperature, the higher the quality of the energy* (Fig. 6–49).

Large quantities of solar energy, for example, can be stored in large bodies of water called *solar ponds* at about 350 K. This stored energy can then be supplied to a heat engine to produce work (electricity). However, the efficiency of solar pond power plants is very low (under 5 percent) because of the low quality of the energy stored in the source, and the construction and maintenance costs are relatively high. Therefore, they are not competitive even though the energy supply of such plants is free. The temperature (and thus the quality) of the solar energy stored could be raised by utilizing concentrating collectors, but the equipment cost in that case becomes very high.

Work is a more valuable form of energy than heat since 100 percent of work can be converted to heat, but only a fraction of heat can be converted to work. When heat is transferred from a high-temperature body to a lower-temperature one, it is degraded since less of it now can be converted to work. For example, if 100 kJ of heat is transferred from a body at 1000 K to a body at 300 K, at the end we will have 100 kJ of thermal energy stored at 300 K, which has no practical value. But if this conversion is made through a heat engine, up to $1 - 300/1000 = 70$ percent of it could be converted to work, which is a more valuable form of energy. Thus 70 kJ of work potential is wasted as a result of this heat transfer, and energy is degraded.

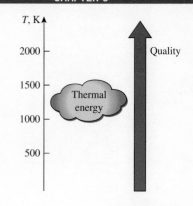

**FIGURE 6–49**
The higher the temperature of the thermal energy, the higher its quality.

## Quantity versus Quality in Daily Life

At times of energy crisis, we are bombarded with speeches and articles on how to "conserve" energy. Yet we all know that the *quantity* of energy is already conserved. What is not conserved is the *quality* of energy, or the work potential of energy. Wasting energy is synonymous to converting it to a less useful form. One unit of high-quality energy can be more valuable than three units of lower-quality energy. For example, a finite amount of thermal energy at high temperature is more attractive to power plant engineers than a vast amount of thermal energy at low temperature, such as the energy stored in the upper layers of the oceans at tropical climates.

As part of our culture, we seem to be fascinated by quantity, and little attention is given to quality. However, quantity alone cannot give the whole picture, and we need to consider quality as well. That is, we need to look at something from both the first- and second-law points of view when evaluating something, even in nontechnical areas. Below we present some ordinary events and show their relevance to the second law of thermodynamics.

Consider two students Andy and Wendy. Andy has 10 friends who never miss his parties and are always around during fun times. However, they seem to be busy when Andy needs their help. Wendy, on the other hand, has five friends. They are never too busy for her, and she can count on them at times of need. Let us now try to answer the question, *Who has more friends?* From the first-law point of view, which considers quantity only, it is obvious that Andy has more friends. However, from the second-law point of view, which considers quality as well, there is no doubt that Wendy is the one with more friends.

Another example with which most people will identify is the multibillion-dollar diet industry, which is primarily based on the first law of thermodynamics. However, considering that 90 percent of the people who lose weight gain it back quickly, with interest, suggests that the first law alone does not give the whole picture. People who seem to be eating whatever they want, whenever they want, without gaining weight are living proof that the calorie-counting technique (the first law) leaves many questions on dieting unanswered. Obviously, more research focused on the second-law effects of dieting is needed before we can fully understand the weight-gain and weight-loss process.

It is tempting to judge things on the basis of their *quantity* instead of their *quality* since assessing quality is much more difficult than assessing quantity. However, assessments made on the basis of quantity only (the first law) may be grossly inadequate and misleading.

## 6–11 · THE CARNOT REFRIGERATOR AND HEAT PUMP

A refrigerator or a heat pump that operates on the reversed Carnot cycle is called a **Carnot refrigerator**, or a **Carnot heat pump**. The coefficient of performance of any refrigerator or heat pump, reversible or irreversible, is given by Eqs. 6–9 and 6–11 as

$$\text{COP}_\text{R} = \frac{1}{Q_H/Q_L - 1} \quad \text{and} \quad \text{COP}_\text{HP} = \frac{1}{1 - Q_L/Q_H}$$

where $Q_L$ is the amount of heat absorbed from the low-temperature medium and $Q_H$ is the amount of heat rejected to the high-temperature medium. The COPs of all reversible refrigerators or heat pumps can be determined by replacing the heat transfer ratios in the above relations by the ratios of the absolute temperatures of the high- and low-temperature reservoirs, as expressed by Eq. 6–16. Then the COP relations for reversible refrigerators and heat pumps become

$$\text{COP}_\text{R,rev} = \frac{1}{T_H/T_L - 1} \quad (6\text{–}20)$$

and

$$\text{COP}_\text{HP,rev} = \frac{1}{1 - T_L/T_H} \quad (6\text{–}21)$$

*These are the highest coefficients of performance that a refrigerator or a heat pump operating between the temperature limits of $T_L$ and $T_H$ can have.* All actual refrigerators or heat pumps operating between these temperature limits ($T_L$ and $T_H$) have lower coefficients of performance (Fig. 6–50).

The coefficients of performance of actual and reversible refrigerators operating between the same temperature limits can be compared as follows:

$$\text{COP}_\text{R} \begin{cases} < \text{COP}_\text{R,rev} & \text{irreversible refrigerator} \\ = \text{COP}_\text{R,rev} & \text{reversible refrigerator} \\ > \text{COP}_\text{R,rev} & \text{impossible refrigerator} \end{cases} \quad (6\text{–}22)$$

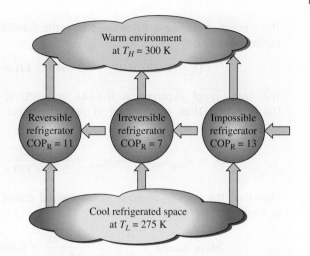

**FIGURE 6–50**
No refrigerator can have a higher COP
than a reversible refrigerator operating
between the same temperature limits.

A similar relation can be obtained for heat pumps by replacing all $COP_R$'s in Eq. 6–22 by $COP_{HP}$.

The COP of a reversible refrigerator or heat pump is the maximum theoretical value for the specified temperature limits. Actual refrigerators or heat pumps may approach these values as their designs are improved, but they can never reach them.

As a final note, the COPs of both the refrigerators and the heat pumps decrease as $T_L$ decreases. That is, it requires more work to absorb heat from lower-temperature media. As the temperature of the refrigerated space approaches zero, the amount of work required to produce a finite amount of refrigeration approaches infinity and $COP_R$ approaches zero.

■
■ **EXAMPLE 6–6    A Carnot Refrigeration Cycle Operating in the**
■                 **Saturation Dome**
■

■ A Carnot refrigeration cycle is executed in a closed system in the saturated liquid–vapor mixture region using 0.8 kg of refrigerant-134a as the working fluid (Fig. 6–51). The maximum and the minimum temperatures in the cycle are 20 and −8°C, respectively. It is known that the refrigerant is saturated liquid at the end of the heat rejection process, and the net work input to the cycle is 15 kJ. Determine the fraction of the mass of the refrigerant that vaporizes during the heat addition process, and the pressure at the end of the heat rejection process.

**SOLUTION** A Carnot refrigeration cycle is executed in a closed system. The mass fraction of the refrigerant that vaporizes during the heat addition process and the pressure at the end of the heat rejection process are to be determined.
**Assumptions** The refrigerator operates on the ideal Carnot cycle.
**Analysis** Knowing the high and low temperatures, the coefficient of performance of the cycle is

$$COP_R = \frac{1}{T_H/T_L - 1} = \frac{1}{(20 + 273 \text{ K})/(-8 + 273 \text{ K}) - 1} = 9.464$$

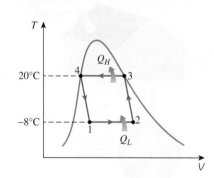

**FIGURE 6–51**
Schematic for Example 6–6.

The amount of cooling is determined from the definition of the coefficient of performance to be

$$Q_L = \text{COP}_R \times W_{in} = (9.464)(15 \text{ kJ}) = 142 \text{ kJ}$$

The enthalpy of vaporization R-134a at $-8°C$ is $h_{fg} = 204.59 \text{ kJ/kg}$ (Table A-11). Then the amount of refrigerant that vaporizes during heat absorption becomes

$$Q_L = m_{evap}h_{fg@-8°C} \rightarrow m_{evap} = \frac{142 \text{ kJ}}{204.59 \text{ kJ/kg}} = 0.694 \text{ kg}$$

Therefore, the fraction of mass that vaporized during heat addition process to the refrigerant is

$$\text{Mass fraction} = \frac{m_{evap}}{m_{total}} = \frac{0.694 \text{ kg}}{0.8 \text{ kg}} = \textbf{0.868} \text{ or } \textbf{86.8\%}$$

The pressure at the end of heat rejection process is simply the saturation pressure at heat rejection temperature,

$$P_4 = P_{sat@20°C} = \textbf{572.1 kPa}$$

**Discussion** Carnot cycle is an idealized refrigeration cycle, thus it cannot be achieved in practice. Practical refrigeration cycles are analyzed in Chap. 11.

---

### EXAMPLE 6–7    Heating a House by a Carnot Heat Pump

A heat pump is to be used to heat a house during the winter, as shown in Fig. 6–52. The house is to be maintained at 21°C at all times. The house is estimated to be losing heat at a rate of 135,000 kJ/h when the outside temperature drops to −5°C. Determine the minimum power required to drive this heat pump.

**SOLUTION** A heat pump maintains a house at a constant temperature. The required minimum power input to the heat pump is to be determined.
**Assumptions** Steady operating conditions exist.
**Analysis** The heat pump must supply heat to the house at a rate of $\dot{Q}_H = 135,000$ kJ/h = 37.5 kW. The power requirements are minimum when a reversible heat pump is used to do the job. The COP of a reversible heat pump operating between the house and the outside air is

$$\text{COP}_{HP,rev} = \frac{1}{1 - T_L/T_H} = \frac{1}{1 - (-5 + 273 \text{ K})/(21 + 273 \text{ K})} = 11.3$$

Then, the required power input to this reversible heat pump becomes

$$\dot{W}_{net,in} = \frac{\dot{Q}_H}{\text{COP}_{HP}} = \frac{37.5 \text{ kW}}{11.3} = \textbf{3.32 kW}$$

**Discussion** This reversible heat pump can meet the heating requirements of this house by consuming electric power at a rate of 3.32 kW only. If this

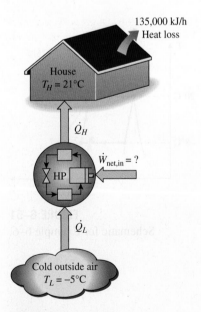

135,000 kJ/h
Heat loss

House
$T_H = 21°C$

$\dot{Q}_H$

$\dot{W}_{net,in} = ?$

HP

$\dot{Q}_L$

Cold outside air
$T_L = -5°C$

**FIGURE 6–52**
Schematic for Example 6–7.

house were to be heated by electric resistance heaters instead, the power consumption would jump up 11.3 times to 37.5 kW. This is because in resistance heaters the electric energy is converted to heat at a one-to-one ratio. With a heat pump, however, energy is absorbed from the outside and carried to the inside using a refrigeration cycle that consumes only 3.32 kW. Notice that the heat pump does not create energy. It merely transports it from one medium (the cold outdoors) to another (the warm indoors).

**TOPIC OF SPECIAL INTEREST\***   Household Refrigerators

Refrigerators to preserve perishable foods have long been one of the essential appliances in a household. They have proven to be highly durable and reliable, providing satisfactory service for over 15 years. A typical household refrigerator is actually a combination refrigerator-freezer since it has a freezer compartment to make ice and to store frozen food.

Today's refrigerators use much less energy as a result of using *smaller* and *higher-efficiency* motors and compressors, *better insulation materials, larger coil surface areas,* and *better door seals* (Fig. 6–53). At an average electricity rate of 8.3 cents per kWh, an average refrigerator costs about \$72 a year to run, which is half the annual operating cost of a refrigerator 25 years ago. Replacing a 25-year-old, 0.5-m³ refrigerator with a new energy-efficient model will save over 1000 kWh of electricity per year. For the environment, this means a reduction of over 1 ton of $CO_2$, which causes global climate change, and over 10 kg of $SO_2$, which causes acid rain.

Despite the improvements made in several areas during the past 100 years in household refrigerators, the basic *vapor-compression refrigeration cycle* has remained unchanged. The alternative *absorption refrigeration* and *thermoelectric refrigeration* systems are currently more expensive and less efficient, and they have found limited use in some specialized applications (Table 6–1).

A household refrigerator is designed to maintain the freezer section at −18°C and the refrigerator section at 3°C. Lower freezer temperatures increase energy consumption without improving the storage life of frozen foods significantly. Different temperatures for the storage of specific foods can be maintained in the refrigerator section by using *special-purpose* compartments.

Practically all full-size refrigerators have a large *air-tight* drawer for leafy vegetables and fresh fruits to seal in moisture and to protect them from the drying effect of cool air circulating in the refrigerator. A covered *egg compartment* in the lid extends the life of eggs by slowing down the moisture loss from the eggs. It is common for refrigerators to have a special warmer compartment for *butter* in the door to maintain butter at spreading temperature. The compartment also isolates butter and prevents it from absorbing *odors* and *tastes* from other food items. Some upscale models have a temperature-controlled *meat compartment* maintained at −0.5°C, which keeps meat at the lowest safe temperature without freezing it, and thus extending its

\*This section can be skipped without a loss in continuity.

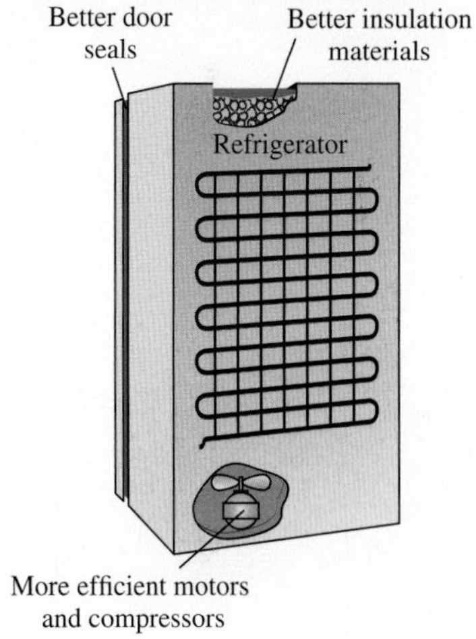

**FIGURE 6–53**
Today's refrigerators are much more efficient because of the improvements in technology and manufacturing.

**TABLE 6–1**
Typical operating efficiencies of some refrigeration systems for a freezer temperature of −18°C and ambient temperature of 32°C

| Type of refrigeration system | Coefficient of performance |
| --- | --- |
| Vapor-compression | 1.3 |
| Absorption refrigeration | 0.4 |
| Thermoelectric refrigeration | 0.1 |

storage life. The more expensive models come with an automatic *icemaker* located in the freezer section that is connected to the water line, as well as automatic ice and chilled-water dispensers. A typical icemaker can produce 2 to 3 kg of ice per day and store 3 to 5 kg of ice in a removable ice storage container.

Household refrigerators consume from about 90 to 600 W of electrical energy when running and are designed to perform satisfactorily in environments at up to 43°C. Refrigerators run intermittently, as you may have noticed, running about 30 percent of the time under normal use in a house at 25°C.

For specified external dimensions, a refrigerator is desired to have *maximum* food storage volume, *minimum* energy consumption, and the *lowest* possible cost to the consumer. The total food storage volume has been increased over the years without an increase in the external dimensions by using thinner but more effective insulation and minimizing the space occupied by the compressor and the condenser. Switching from the fiber-glass insulation (thermal conductivity $k = 0.032$–$0.040$ W/m·°C) to expanded-in-place urethane foam insulation ($k = 0.019$ W/m·°C) made it possible to reduce the wall thickness of the refrigerator by almost half, from about 90 to 48 mm for the freezer section and from about 70 to 40 mm for the refrigerator section. The rigidity and bonding action of the foam also provide additional structural support. However, the entire shell of the refrigerator must be carefully sealed to prevent any water leakage or moisture migration into the insulation since moisture degrades the effectiveness of insulation.

The size of the compressor and the other components of a refrigeration system are determined on the basis of the anticipated heat load (or refrigeration load), which is the rate of heat flow into the refrigerator. The heat load consists of the *predictable part*, such as heat transfer through the walls and door gaskets of the refrigerator, fan motors, and defrost heaters (Fig. 6–54), and the *unpredictable part*, which depends on the user habits such as opening the door, making ice, and loading the refrigerator. The amount of *energy*

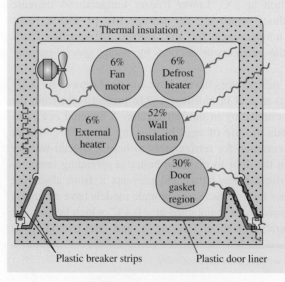

**FIGURE 6–54**

The cross section of a refrigerator showing the relative magnitudes of various effects that constitute the predictable heat load.

consumed by the refrigerator can be minimized by practicing good *conservation measures* as discussed below.

1. *Open the refrigerator door the fewest times possible* for the shortest duration possible. Each time the refrigerator door is opened, the cool air inside is replaced by the warmer air outside, which needs to be cooled. Keeping the refrigerator or freezer full will save energy by reducing the amount of cold air that can escape each time the door is opened.

2. *Cool the hot foods* to room temperature first before putting them into the refrigerator. Moving a hot pan from the oven directly into the refrigerator not only wastes energy by making the refrigerator work longer, but it also causes the nearby perishable foods to spoil by creating a warm environment in its immediate surroundings (Fig. 6–55).

3. *Clean the condenser coils* located behind or beneath the refrigerator. The dust and grime that collect on the coils act as insulation that slows down heat dissipation through them. Cleaning the coils a couple of times a year with a damp cloth or a vacuum cleaner will improve cooling ability of the refrigerator while cutting down the power consumption by a few percent. Sometimes a fan is used to force-cool the condensers of large or built-in refrigerators, and the strong air motion keeps the coils clean.

4. *Check the door gasket* for air leaks. This can be done by placing a flashlight into the refrigerator, turning off the kitchen lights, and looking for light leaks. Heat transfer through the door gasket region accounts for almost one-third of the regular heat load of the refrigerators, and thus any defective door gaskets must be repaired immediately.

5. *Avoid unnecessarily low temperature settings.* The recommended temperatures for freezers and refrigerators are $-18°C$ and $3°C$, respectively. Setting the freezer temperature below $-18°C$ adds significantly to the energy consumption but does not add much to the storage life of frozen foods. Keeping temperatures $6°C$ below recommended levels can increase the energy use by as much as 25 percent.

6. *Avoid excessive ice build-up* on the interior surfaces of the evaporator. The ice layer on the surface acts as insulation and slows down heat transfer from the freezer section to the refrigerant. The refrigerator should be defrosted by manually turning off the temperature control switch when the ice thickness exceeds a few millimeters.

Defrosting is done automatically in no-frost refrigerators by supplying heat to the evaporator by a 300-W to 1000-W resistance heater or by hot refrigerant gas, periodically for short periods. The water is then drained to a pan outside where it is evaporated using the heat dissipated by the condenser. The no-frost evaporators are basically finned tubes subjected to air flow circulated by a fan. Practically all the frost collects on fins, which are the coldest surfaces, leaving the exposed surfaces of the freezer section and the frozen food frost-free.

7. *Use the power-saver switch* that controls the heating coils and prevents condensation on the outside surfaces in humid environments. The low-wattage heaters are used to raise the temperature of the outer surfaces

Warm air
30°C

Hot food
80°C

5°C

**FIGURE 6–55**
Putting hot foods into the refrigerator without cooling them first not only wastes energy but also could spoil the foods nearby.

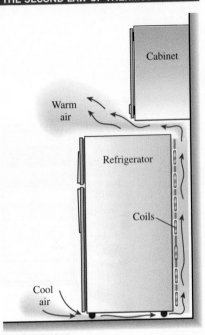

of the refrigerator at critical locations above the dew point in order to avoid water droplets forming on the surfaces and sliding down. Condensation is most likely to occur in summer in hot and humid climates in homes without air-conditioning. The moisture formation on the surfaces is undesirable since it may cause the painted finish of the outer surface to deteriorate and it may wet the kitchen floor. About 10 percent of the total energy consumed by the refrigerator can be saved by turning this heater off and keeping it off unless there is visible condensation on the outer surfaces.

8. *Do not block the air flow passages* to and from the condenser coils of the refrigerator. The heat dissipated by the condenser to the air is carried away by air that enters through the bottom and sides of the refrigerator and leaves through the top. Any blockage of this natural convection air circulation path by large objects such as several cereal boxes on top of the refrigerator will degrade the performance of the condenser and thus the refrigerator (Fig. 6–56).

**FIGURE 6–56**
The condenser coils of a refrigerator must be cleaned periodically, and the airflow passages must not be blocked to maintain high performance.

These and other commonsense conservation measures will result in a reduction in the energy and maintenance costs of a refrigerator as well as an extended trouble-free life of the device.

---

**EXAMPLE 6–8**  **Malfunction of a Refrigerator Light Switch**

The interior lighting of refrigerators is provided by incandescent lamps whose switches are actuated by the opening of the refrigerator door. Consider a refrigerator whose 40-W lightbulb remains on continuously as a result of a malfunction of the switch (Fig. 6–57). If the refrigerator has a coefficient of performance of 1.3 and the cost of electricity is 12 cents per kWh, determine the increase in the energy consumption of the refrigerator and its cost per year if the switch is not fixed.

**SOLUTION**  The lightbulb of a refrigerator malfunctions and remains on. The increases in the electricity consumption and cost are to be determined.
**Assumptions**  The life of the lightbulb is more than 1 year.
**Analysis**  The lightbulb consumes 40 W of power when it is on, and thus adds 40 W to the heat load of the refrigerator. Noting that the COP of the refrigerator is 1.3, the power consumed by the refrigerator to remove the heat generated by the lightbulb is

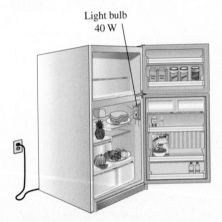

**FIGURE 6–57**
Schematic for Example 6–8.

$$\dot{W}_{\text{refrig}} = \frac{\dot{Q}_{\text{refrig}}}{\text{COP}_R} = \frac{40 \text{ W}}{1.3} = 30.8 \text{ W}$$

Therefore, the total additional power consumed by the refrigerator is

$$\dot{W}_{\text{total,additional}} = \dot{W}_{\text{light}} + \dot{W}_{\text{refrig}} = 40 + 30.8 = 70.8 \text{ W}$$

The total number of hours in a year is

$$\text{Annual hours} = (365 \text{ days/yr})(24 \text{ h/day}) = 8760 \text{ h/yr}$$

Assuming the refrigerator is opened 20 times a day for an average of 30 s, the light would normally be on for

Normal operating hours = (20 times/day)(30 s/time)(1 h/3600 s)(365 days/yr)

= 61 h/yr

Then the additional hours the light remains on as a result of the malfunction becomes

Additional operating hours = Annual hours − Normal operating hours

= 8760 − 61 = 8699 h/yr

Therefore, the additional electric power consumption and its cost per year are

Additional power consumption = $\dot{W}_{\text{total,additional}}$ × (Additional operating hours)

= (0.0708 kW)(8699 h/yr) = **616 kWh/yr**

and

Additional power cost = (Additional power consumption)(Unit cost)

= (616 kWh/yr)($0.12/kWh) = **$73.9/yr**

**Discussion**  Note that not repairing the switch will cost the homeowner about $75 a year. This is alarming when we consider that at $0.12/kWh, a typical refrigerator consumes about $100 worth of electricity a year.

## SUMMARY

The *second law of thermodynamics* states that processes occur in a certain direction, not in any direction. A process does not occur unless it satisfies both the first and the second laws of thermodynamics. Bodies that can absorb or reject finite amounts of heat isothermally are called *thermal energy reservoirs* or *heat reservoirs*.

Work can be converted to heat directly, but heat can be converted to work only by some devices called *heat engines*. The *thermal efficiency* of a heat engine is defined as

$$\eta_{\text{th}} = \frac{W_{\text{net,out}}}{Q_H} = 1 - \frac{Q_L}{Q_H}$$

where $W_{\text{net,out}}$ is the net work output of the heat engine, $Q_H$ is the amount of heat supplied to the engine, and $Q_L$ is the amount of heat rejected by the engine.

Refrigerators and heat pumps are devices that absorb heat from low-temperature media and reject it to higher-temperature ones. The performance of a refrigerator or a heat pump is expressed in terms of the *coefficient of performance*, which is defined as

$$\text{COP}_R = \frac{Q_L}{W_{\text{net,in}}} = \frac{1}{Q_H/Q_L - 1}$$

$$\text{COP}_{\text{HP}} = \frac{Q_H}{W_{\text{net,in}}} = \frac{1}{1 - Q_L/Q_H}$$

The *Kelvin–Planck statement* of the second law of thermodynamics states that no heat engine can produce a net amount of work while exchanging heat with a single reservoir only. The *Clausius statement* of the second law states that no device can transfer heat from a cooler body to a warmer one without leaving an effect on the surroundings.

Any device that violates the first or the second law of thermodynamics is called a *perpetual-motion machine*.

A process is said to be *reversible* if both the system and the surroundings can be restored to their original conditions. Any other process is *irreversible*. The effects such as friction, non-quasi-equilibrium expansion or compression, and heat transfer through a finite temperature difference render a process irreversible and are called *irreversibilities*.

The *Carnot cycle* is a reversible cycle that is composed of four reversible processes, two isothermal and two adiabatic. The *Carnot principles* state that the thermal efficiencies of all reversible heat engines operating between the same two reservoirs are the same, and that no heat engine is more efficient than a reversible one operating between the same two reservoirs. These statements form the basis for establishing a *thermodynamic temperature scale* related to the heat transfers between a reversible device and the high- and low-temperature reservoirs by

$$\left(\frac{Q_H}{Q_L}\right)_{rev} = \frac{T_H}{T_L}$$

Therefore, the $Q_H/Q_L$ ratio can be replaced by $T_H/T_L$ for reversible devices, where $T_H$ and $T_L$ are the absolute temperatures of the high- and low-temperature reservoirs, respectively.

A heat engine that operates on the reversible Carnot cycle is called a *Carnot heat engine*. The thermal efficiency of a Carnot heat engine, as well as all other reversible heat engines, is given by

$$\eta_{th,rev} = 1 - \frac{T_L}{T_H}$$

This is the maximum efficiency a heat engine operating between two reservoirs at temperatures $T_H$ and $T_L$ can have.

The COPs of reversible refrigerators and heat pumps are given in a similar manner as

$$COP_{R,rev} = \frac{1}{T_H/T_L - 1}$$

and

$$COP_{HP,rev} = \frac{1}{1 - T_L/T_H}$$

Again, these are the highest COPs a refrigerator or a heat pump operating between the temperature limits of $T_H$ and $T_L$ can have.

## REFERENCES AND SUGGESTED READINGS

1. *ASHRAE Handbook of Refrigeration,* SI version. Atlanta, GA: American Society of Heating, Refrigerating, and Air-Conditioning Engineers, Inc. 1994.

2. D. Stewart. "Wheels Go Round and Round, but Always Run Down." November 1986, *Smithsonian,* pp. 193–208.

3. J. T. Amann, A. Wilson, and K. Ackerly, *Consumer Guide to Home Energy Saving,* 9th ed., American Council for an Energy-Efficient Economy, Washington, D. C., 2007.

## PROBLEMS*

### Second Law of Thermodynamics and Thermal Energy Reservoirs

**6–1C** Describe an imaginary process that violates both the first and the second laws of thermodynamics.

**6–2C** Describe an imaginary process that satisfies the first law but violates the second law of thermodynamics.

**6–3C** Describe an imaginary process that satisfies the second law but violates the first law of thermodynamics.

---

* Problems designated by a "C" are concept questions, and students are encouraged to answer them all. Problems with the 🌐 icon are solved using EES, and complete solutions together with parametric studies are included on the text website. Problems with the [ᴱᴱˢ] icon are comprehensive in nature, and are intended to be solved with an equation solver such as EES.

**6–4C** An experimentalist claims to have raised the temperature of a small amount of water to 150°C by transferring heat from high-pressure steam at 120°C. Is this a reasonable claim? Why? Assume no refrigerator or heat pump is used in the process.

**6–5C** What is a thermal energy reservoir? Give some examples.

**6–6C** Consider the process of baking potatoes in a conventional oven. Can the hot air in the oven be treated as a thermal energy reservoir? Explain.

### Heat Engines and Thermal Efficiency

**6–7C** What are the characteristics of all heat engines?

**6–8C** What is the Kelvin–Planck expression of the second law of thermodynamics?

**6–9C** Is it possible for a heat engine to operate without rejecting any waste heat to a low-temperature reservoir? Explain.

**6–10C** Baseboard heaters are basically electric resistance heaters and are frequently used in space heating. A home owner claims that her 5-year-old baseboard heaters have a conversion efficiency of 100 percent. Is this claim in violation of any thermodynamic laws? Explain.

**6–11C** Does a heat engine that has a thermal efficiency of 100 percent necessarily violate (a) the first law and (b) the second law of thermodynamics? Explain.

**6–12C** In the absence of any friction and other irreversibilities, can a heat engine have an efficiency of 100 percent? Explain.

**6–13C** Are the efficiencies of all the work-producing devices, including the hydroelectric power plants, limited by the Kelvin–Planck statement of the second law? Explain.

**6–14C** Consider a pan of water being heated (a) by placing it on an electric range and (b) by placing a heating element in the water. Which method is a more efficient way of heating water? Explain.

**6–15** A steam power plant receives heat from a furnace at a rate of 280 GJ/h. Heat losses to the surrounding air from the steam as it passes through the pipes and other components are estimated to be about 8 GJ/h. If the waste heat is transferred to the cooling water at a rate of 145 GJ/h, determine (a) net power output and (b) the thermal efficiency of this power plant. *Answers:* (a) 35.3 MW, (b) 45.4 percent

**6–16** A heat engine has a heat input of $3 \times 10^4$ kJ/h and a thermal efficiency of 40 percent. Calculate the power it will produce, in kW.

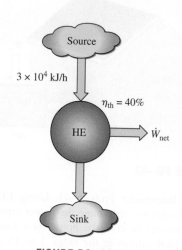

**FIGURE P6–16**

**6–17** The thermal efficiency of a general heat engine is 35 percent, and it produces 60 hp. At what rate is heat transferred to this engine, in kJ/s?

**6–18** A 600-MW steam power plant, which is cooled by a nearby river, has a thermal efficiency of 40 percent. Determine

the rate of heat transfer to the river water. Will the actual heat transfer rate be higher or lower than this value? Why?

**6–19** A heat engine that pumps water out of an underground mine accepts 700 kJ of heat and produces 250 kJ of work. How much heat does it reject, in kJ?

**6–20** A heat engine with a thermal efficiency of 45 percent rejects 500 kJ/kg of heat. How much heat does it receive? *Answer:* 909 kJ/kg

**6–21** A steam power plant with a power output of 150 MW consumes coal at a rate of 60 tons/h. If the heating value of the coal is 30,000 kJ/kg, determine the overall efficiency of this plant. *Answer:* 30.0 percent

**6–22** An automobile engine consumes fuel at a rate of 22 L/h and delivers 55 kW of power to the wheels. If the fuel has a heating value of 44,000 kJ/kg and a density of 0.8 g/cm$^3$, determine the efficiency of this engine. *Answer:* 25.6 percent

**6–23** In 2001, the United States produced 51 percent of its electricity in the amount of $1.878 \times 10^{12}$ kWh from coal-fired power plants. Taking the average thermal efficiency to be 34 percent, determine the amount of thermal energy rejected by the coal-fired power plants in the United States that year.

**6–24** Solar energy stored in large bodies of water, called solar pounds, is being used to generate electricity. If such a solar power plant has an efficiency of 3 percent and a net power output of 180 kW, determine the average value of the required solar energy collection rate, in kJ/h.

**6–25** A coal-burning steam power plant produces a net power of 300 MW with an overall thermal efficiency of 32 percent. The actual gravimetric air–fuel ratio in the furnace is calculated to be 12 kg air/kg fuel. The heating value of the coal is 28,000 kJ/kg. Determine (a) the amount of coal consumed during a 24-hour period and (b) the rate of air flowing through the furnace. *Answers:* (a) $2.89 \times 10^6$ kg, (b) 402 kg/s

**6–26** An Ocean Thermal Energy Conversion (OTEC) power plant built in Hawaii in 1987 was designed to operate between the temperature limits of 30°C at the ocean surface and 5°C at a depth of 640 m. About 50 m$^3$/min of cold seawater was to be pumped from deep ocean through a 1-m-diameter pipe to serve as the cooling medium or heat sink. If the cooling water experiences a temperature rise of 3.3°C and the thermal efficiency is 2.5 percent, determine the amount of power generated. Take the density of seawater to be 1025 kg/m$^3$.

## Refrigerators and Heat Pumps

**6–27** What is the difference between a refrigerator and a heat pump?

**6–28** What is the difference between a refrigerator and an air conditioner?

**6–29C** In a refrigerator, heat is transferred from a lower-temperature medium (the refrigerated space) to a higher-temperature one (the kitchen air). Is this a violation of the second law of thermodynamics? Explain.

**6–30C** A heat pump is a device that absorbs energy from the cold outdoor air and transfers it to the warmer indoors. Is this a violation of the second law of thermodynamics? Explain.

**6–31C** Define the coefficient of performance of a refrigerator in words. Can it be greater than unity?

**6–32C** Define the coefficient of performance of a heat pump in words. Can it be greater than unity?

**6–33C** A heat pump that is used to heat a house has a COP of 2.5. That is, the heat pump delivers 2.5 kWh of energy to the house for each 1 kWh of electricity it consumes. Is this a violation of the first law of thermodynamics? Explain.

**6–34C** A refrigerator has a COP of 1.5. That is, the refrigerator removes 1.5 kWh of energy from the refrigerated space for each 1 kWh of electricity it consumes. Is this a violation of the first law of thermodynamics? Explain.

**6–35C** What is the Clausius expression of the second law of thermodynamics?

**6–36C** Show that the Kelvin–Planck and the Clausius expressions of the second law are equivalent.

**6–37** Determine the COP of a refrigerator that removes heat from the food compartment at a rate of 5040 kJ/h for each kW of power it consumes. Also, determine the rate of heat rejection to the outside air.

**6–38** Determine the COP of a heat pump that supplies energy to a house at a rate of 8000 kJ/h for each kW of electric power it draws. Also, determine the rate of energy absorption from the outdoor air. *Answers:* 2.22, 4400 kJ/h

**6–39** A refrigerator used to cool a computer requires 1.2 kW of electrical power and has a COP of 1.8. Calculate the cooling effect of this refrigerator, in kW.

**6–40** An air conditioner removes heat steadily from a house at a rate of 750 kJ/min while drawing electric power at a rate of 6 kW. Determine (*a*) the COP of this air conditioner and (*b*) the rate of heat transfer to the outside air. *Answers:* (*a*) 2.08, (*b*) 1110 kJ/min

**6–41** A food department is kept at −12°C by a refrigerator in an environment at 30°C. The total heat gain to the food department is estimated to be 3300 kJ/h and the heat rejection in the condenser is 4800 kJ/h. Determine the power input to the compressor, in kW and the COP of the refrigerator.

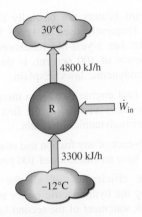

**FIGURE P6–41**

**6–42** A household refrigerator that has a power input of 450 W and a COP of 1.5 is to cool 5 large watermelons, 10 kg each, to 8°C. If the watermelons are initially at 28°C, determine how long it will take for the refrigerator to cool them. The watermelons can be treated as water whose specific heat is 4.2 kJ/kg·°C. Is your answer realistic or optimistic? Explain. *Answer:* 104 min

**6–43** When a man returns to his well-sealed house on a summer day, he finds that the house is at 35°C. He turns on the air conditioner, which cools the entire house to 20°C in 30 min. If the COP of the air-conditioning system is 2.8, determine the power drawn by the air conditioner. Assume the entire mass within the house is equivalent to 800 kg of air for which $c_v = 0.72$ kJ/kg·°C and $c_p = 1.0$ kJ/kg·°C.

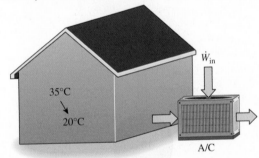

**FIGURE P6–43**

**6–44** Reconsider Prob. 6–43. Using EES (or other) software, determine the power input required by the air conditioner to cool the house as a function for air-conditioner EER ratings in the range 5 to 15. Discuss your results and include representative costs of air-conditioning units in the EER rating range.

**6–45** A heat pump with a COP of 2.5 supplies energy to a house at a rate of 60,000 kJ/h. Determine (*a*) the electric power drawn by the heat pump and (*b*) the rate of heat absorption from the outside air. *Answers:* (*a*) 6.67 kW, (*b*) 36,000 kJ/h

**6–46** Bananas are to be cooled from 24 to 13°C at a rate of 215 kg/h by a refrigeration system. The power input to the refrigerator is 1.4 kW. Determine the rate of cooling, in kJ/min, and the COP of the refrigerator. The specific heat of banana above freezing is 3.35 kJ/kg·°C.

**6–47** A heat pump is used to maintain a house at a constant temperature of 23°C. The house is losing heat to the outside air through the walls and the windows at a rate of 85,000 kJ/h while the energy generated within the house from people, lights, and appliances amounts to 4000 kJ/h. For a COP of 3.2, determine the required power input to the heat pump. *Answer:* 7.03 kW

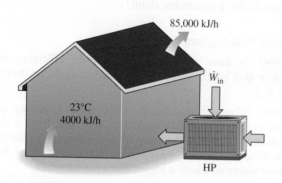

FIGURE P6–47

**6–48** Water enters an ice machine at 13°C and leaves as ice at −4°C. If the COP of the ice machine is 2.4 during this operation, determine the required power input for an ice production rate of 12 kg/h. (393 kJ of energy needs to be removed from each kg of water at 13°C to turn it into ice at −4°C.)

**6–49** A household refrigerator runs one-fourth of the time and removes heat from the food compartment at an average rate of 800 kJ/h. If the COP of the refrigerator is 2.2, determine the power the refrigerator draws when running.

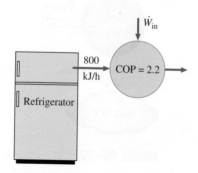

FIGURE P6–49

**6–50** A heat pump used to heat a house runs about one-third of the time. The house is losing heat at an average rate of 22,000 kJ/h. If the COP of the heat pump is 2.8, determine the power the heat pump draws when running.

**6–51** Consider a building whose annual air-conditioning load is estimated to be 40,000 kWh in an area where the unit cost of electricity is $0.10/kWh. Two air conditioners are considered for the building. Air conditioner A has a seasonal average COP of 2.3 and costs $5500 to purchase and install. Air conditioner B has a seasonal average COP of 3.6 and costs $7000 to purchase and install. All else being equal, determine which air conditioner is a better buy.

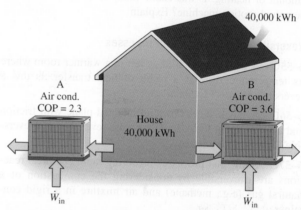

FIGURE P6–51

**6–52** Refrigerant-134a enters the condenser of a residential heat pump at 800 kPa and 35°C at a rate of 0.018 kg/s and leaves at 800 kPa as a saturated liquid. If the compressor consumes 1.2 kW of power, determine (*a*) the COP of the heat pump and (*b*) the rate of heat absorption from the outside air.

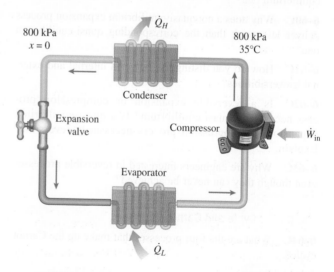

FIGURE P6–52

## Perpetual-Motion Machines

**6–53** An inventor claims to have developed a resistance heater that supplies 1.2 kWh of energy to a room for each kWh of electricity it consumes. Is this a reasonable claim, or has the inventor developed a perpetual-motion machine? Explain.

**6–54** It is common knowledge that the temperature of air rises as it is compressed. An inventor thought about using this high-temperature air to heat buildings. He used a compressor driven by an electric motor. The inventor claims that the compressed hot-air system is 25 percent more efficient than a resistance heating system that provides an equivalent amount of heating. Is this claim valid, or is this just another perpetual-motion machine? Explain.

## Reversible and Irreversible Processes

**6–55C** A cold canned drink is left in a warmer room where its temperature rises as a result of heat transfer. Is this a reversible process? Explain.

**6–56C** A block slides down an inclined plane with friction and no restraining force. Is this process reversible or irreversible? Justify your answer.

**6–57C** Show that processes involving rapid chemical reactions are irreversible by considering the combustion of a natural gas (e.g., methane) and air mixture in a rigid container.

**6–58C** Show that processes that use work for mixing are irreversible by considering an adiabatic system whose contents are stirred by turning a paddle wheel inside the system (e.g., stirring a cake mix with an electric mixer).

**6–59C** Why does a nonquasi-equilibrium compression process require a larger work input than the corresponding quasi-equilibrium one?

**6–60C** Why does a nonquasi-equilibrium expansion process deliver less work than the corresponding quasi-equilibrium one?

**6–61C** How do you distinguish between internal and external irreversibilities?

**6–62C** Is a reversible expansion or compression process necessarily quasi-equilibrium? Is a quasi-equilibrium expansion or compression process necessarily reversible? Explain.

**6–63C** Why are engineers interested in reversible processes even though they can never be achieved?

## The Carnot Cycle and Carnot Principles

**6–64C** What are the four processes that make up the Carnot cycle?

**6–65C** What are the two statements known as the Carnot principles?

**6–66C** Is it possible to develop (*a*) an actual and (*b*) a reversible heat-engine cycle that is more efficient than a Carnot cycle operating between the same temperature limits? Explain.

**6–67C** Somebody claims to have developed a new reversible heat-engine cycle that has a higher theoretical efficiency than the Carnot cycle operating between the same temperature limits. How do you evaluate this claim?

**6–68C** Somebody claims to have developed a new reversible heat-engine cycle that has the same theoretical efficiency as the Carnot cycle operating between the same temperature limits. Is this a reasonable claim?

## Carnot Heat Engines

**6–69C** Is there any way to increase the efficiency of a Carnot heat engine other than by increasing $T_H$ or decreasing $T_L$?

**6–70C** Consider two actual power plants operating with solar energy. Energy is supplied to one plant from a solar pond at 80°C and to the other from concentrating collectors that raise the water temperature to 600°C. Which of these power plants will have a higher efficiency? Explain.

**6–71** From a work-production perspective, which is more valuable: (*a*) thermal energy reservoirs at 675 K and 325 K or (*b*) thermal energy reservoirs at 625 K and 275 K?

**6–72** A heat engine is operating on a Carnot cycle and has a thermal efficiency of 55 percent. The waste heat from this engine is rejected to a nearby lake at 15°C at a rate of 800 kJ/min. Determine (*a*) the power output of the engine and (*b*) the temperature of the source.    *Answers:* (*a*) 16.3 kW, (*b*) 640 K

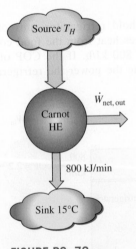

**FIGURE P6–72**

**6–73** A Carnot heat engine receives 650 kJ of heat from a source of unknown temperature and rejects 250 kJ of it to a sink at 24°C. Determine (a) the temperature of the source and (b) the thermal efficiency of the heat engine.

**6–74** A Carnot heat engine operates between a source at 1000 K and a sink at 300 K. If the heat engine is supplied with heat at a rate of 800 kJ/min, determine (a) the thermal efficiency and (b) the power output of this heat engine. *Answers:* (a) 70 percent, (b) 9.33 kW

**6–75** A heat engine operates between a source at 477°C and a sink at 25°C. If heat is supplied to the heat engine at a steady rate of 65,000 kJ/min, determine the maximum power output of this heat engine.

**6–76** Reconsider Prob. 6–75. Using EES (or other) software, study the effects of the temperatures of the heat source and the heat sink on the power produced and the cycle thermal efficiency. Let the source temperature vary from 300 to 1000°C, and the sink temperature to vary from 0 to 50°C. Plot the power produced and the cycle efficiency against the source temperature for sink temperatures of 0°C, 25°C, and 50°C, and discuss the results.

**6–77** An inventor claims to have devised a cyclical engine for use in space vehicles that operates with a nuclear-fuel-generated energy source whose temperature is 510 K and a sink at 270 K that radiates waste heat to deep space. He also claims that this engine produces 4.1 kW while rejecting heat at a rate of 15,000 kJ/h. Is this claim valid?

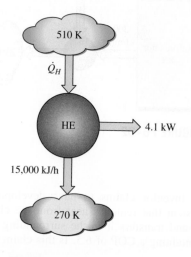

**FIGURE P6–77**

**6–78** A heat engine receives heat from a heat source at 1200°C and has a thermal efficiency of 40 percent. The heat engine does maximum work equal to 500 kJ. Determine the heat supplied to the heat engine by the heat source, the heat rejected to the heat sink, and the temperature of the heat sink.

**6–79** In tropical climates, the water near the surface of the ocean remains warm throughout the year as a result of solar energy absorption. In the deeper parts of the ocean, however, the water remains at a relatively low temperature since the sun's rays cannot penetrate very far. It is proposed to take advantage of this temperature difference and construct a power plant that will absorb heat from the warm water near the surface and reject the waste heat to the cold water a few hundred meters below. Determine the maximum thermal efficiency of such a plant if the water temperatures at the two respective locations are 24 and 3°C.

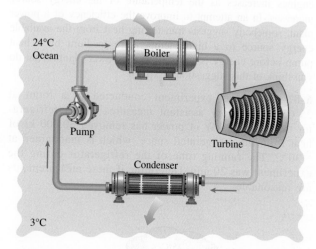

**FIGURE P6–79**

**6–80** A well-established way of power generation involves the utilization of geothermal energy—the energy of hot water that exists naturally underground—as the heat source. If a supply of hot water at 140°C is discovered at a location where the environmental temperature is 20°C, determine the maximum thermal efficiency a geothermal power plant built at that location can have. *Answer:* 29.1 percent

## Carnot Refrigerators and Heat Pumps

**6–81C** A homeowner buys a new refrigerator and a new air conditioner. Which one of these devices would you expect to have a higher COP? Why?

**6–82C** A homeowner buys a new refrigerator with no freezer compartment and a deep freezer for the new kitchen. Which of these devices would you expect to have a lower COP? Why?

**6–83C** How can we increase the COP of a Carnot refrigerator?

**6–84C** In an effort to conserve energy in a heat-engine cycle, somebody suggests incorporating a refrigerator that will absorb some of the waste energy $Q_L$ and transfer it to the energy source of the heat engine. Is this a smart idea? Explain.

**6–85C** It is well established that the thermal efficiency of a heat engine increases as the temperature $T_L$ at which heat is rejected from the heat engine decreases. In an effort to increase the efficiency of a power plant, somebody suggests refrigerating the cooling water before it enters the condenser, where heat rejection takes place. Would you be in favor of this idea? Why?

**6–86C** It is well known that the thermal efficiency of heat engines increases as the temperature of the energy source increases. In an attempt to improve the efficiency of a power plant, somebody suggests transferring heat from the available energy source to a higher-temperature medium by a heat pump before energy is supplied to the power plant. What do you think of this suggestion? Explain.

**6–87** During an experiment conducted in a room at 25°C, a laboratory assistant measures that a refrigerator that draws 2 kW of power has removed 30,000 kJ of heat from the refrigerated space, which is maintained at −30°C. The running time of the refrigerator during the experiment was 20 min. Determine if these measurements are reasonable.

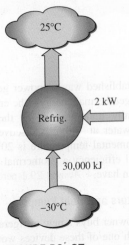

**FIGURE P6–87**

**6–88** A Carnot refrigerator operates in a room in which the temperature is 22°C and consumes 2 kW of power when operating. If the food compartment of the refrigerator is to be maintained at 3°C, determine the rate of heat removal from the food compartment.

**6–89** An air-conditioning system operating on the reversed Carnot cycle is required to transfer heat from a house at a rate of 750 kJ/min to maintain its temperature at 24°C. If the outdoor air temperature is 35°C, determine the power required to operate this air-conditioning system. *Answer: 0.46 kW*

**6–90** An inventor claims to have developed a heat pump that produces a 200-kW heating effect for a 293 K heated zone while only using 75 kW of power and a heat source at 273 K. Justify the validity of this claim.

**6–91** A heat pump operates on a Carnot heat pump cycle with a COP of 8.7. It keeps a space at 24°C by consuming 2.15 kW of power. Determine the temperature of the reservoir from which the heat is absorbed and the heating load provided by the heat pump. *Answers: 263 K, 18.7 kW*

**6–92** A refrigerator is to remove heat from the cooled space at a rate of 300 kJ/min to maintain its temperature at −8°C. If the air surrounding the refrigerator is at 25°C, determine the minimum power input required for this refrigerator. *Answer: 0.623 kW*

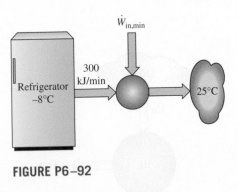

**FIGURE P6–92**

**6–93** An inventor claims to have developed a refrigeration system that removes heat from the closed region at −12°C and transfers it to the surrounding air at 25°C while maintaining a COP of 6.5. Is this claim reasonable? Why?

**6–94** A heat pump is used to maintain a house at 25°C by extracting heat from the outside air on a day when the outside air temperature is 4°C. The house is estimated to lose heat at a rate of 110,000 kJ/h, and the heat pump consumes 4.75 kW of electric power when running. Is this heat pump powerful enough to do the job?

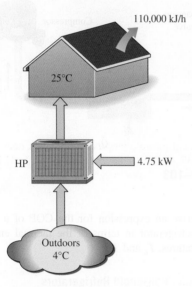

**FIGURE P6–94**

**6–95** An air-conditioning system is used to maintain a house at 22°C when the temperature outside is 32°C. If this air-conditioning system draws 3.7 kW of power when operating, determine the maximum rate of heat removal from the house that it can accomplish.

**6–96** A refrigerator operating on the reversed Carnot cycle has a measured work input of 200 kW and heat rejection of 2000 kW to a heat reservoir at 27°C. Determine the cooling load supplied to the refrigerator, in kW, and the temperature of the heat source, in °C. *Answers:* 1800 kW, −3°C

**6–97** A commercial refrigerator with refrigerant-134a as the working fluid is used to keep the refrigerated space at −35°C by rejecting waste heat to cooling water that enters the condenser at 18°C at a rate of 0.25 kg/s and leaves at 26°C. The refrigerant enters the condenser at 1.2 MPa and 50°C and leaves at the same pressure subcooled by 5°C. If the compressor consumes 3.3 kW of power, determine (*a*) the mass flow rate of the refrigerant, (*b*) the refrigeration load, (*c*) the COP, and (*d*) the

minimum power input to the compressor for the same refrigeration load.

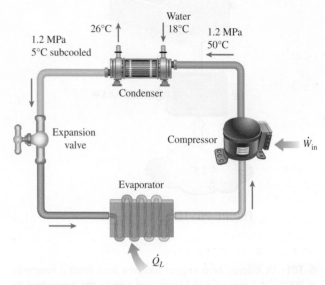

**FIGURE P6–97**

**6–98** The performance of a heat pump degrades (i.e., its COP decreases) as the temperature of the heat source decreases. This makes using heat pumps at locations with severe weather conditions unattractive. Consider a house that is heated and maintained at 20°C by a heat pump during the winter. What is the maximum COP for this heat pump if heat is extracted from the outdoor air at (*a*) 10°C, (*b*) −5°C, and (*c*) −30°C?

**6–99** A heat pump is to be used for heating a house in winter. The house is to be maintained at 26°C at all times. When the temperature outdoors drops to −4°C, the heat losses from the house are estimated to be 70,000 kJ/h. Determine the minimum power required to run this heat pump if heat is extracted from (*a*) the outdoor air at −4°C and (*b*) the well water at 10°C.

**6–100** A Carnot heat pump is to be used to heat a house and maintain it at 25°C in winter. On a day when the average outdoor temperature remains at about 2°C, the house is estimated to lose heat at a rate of 55,000 kJ/h. If the heat pump consumes 4.8 kW of power while operating, determine (*a*) how long the heat pump ran on that day; (*b*) the total heating costs, assuming an average price of 11¢/kWh for electricity; and (*c*) the heating cost for the same day if resistance heating is used instead of a heat pump. *Answers:* (*a*) 5.90 h, (*b*) $3.11, (*c*) $40.3

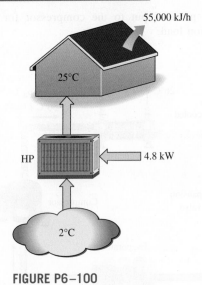

**FIGURE P6-100**

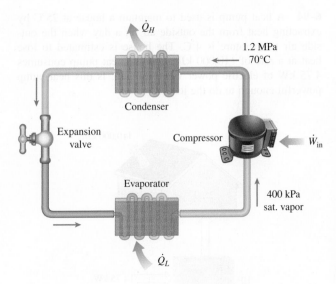

**FIGURE P6-103**

**6-101** A Carnot heat engine receives heat from a reservoir at 900°C at a rate of 800 kJ/min and rejects the waste heat to the ambient air at 27°C. The entire work output of the heat engine is used to drive a refrigerator that removes heat from the refrigerated space at −5°C and transfers it to the same ambient air at 27°C. Determine (a) the maximum rate of heat removal from the refrigerated space and (b) the total rate of heat rejection to the ambient air. *Answers:* (a) 4982 kJ/min, (b) 5782 kJ/min

**6-102** The structure of a house is such that it loses heat at a rate of 3800 kJ/h per °C difference between the indoors and outdoors. A heat pump that requires a power input of 4 kW is used to maintain this house at 24°C. Determine the lowest outdoor temperature for which the heat pump can meet the heating requirements of this house. *Answer:* −9.5°C

**6-103** An air-conditioner with refrigerant-134a as the working fluid is used to keep a room at 23°C by rejecting the waste heat to the outdoor air at 34°C. The room gains heat through the walls and the windows at a rate of 250 kJ/min while the heat generated by the computer, TV, and lights amounts to 900 W. The refrigerant enters the compressor at 400 kPa as a saturated vapor at a rate of 80 L/min and leaves at 1200 kPa and 70°C. Determine (a) the actual COP, (b) the maximum COP, and (c) the minimum volume flow rate of the refrigerant at the compressor inlet for the same compressor inlet and exit conditions. *Answers:* (a) 4.33, (b) 26.9, (c) 12.9 L/min

**6-104** Derive an expression for the COP of a completely reversible refrigerator in terms of the thermal energy reservoir temperatures, $T_L$ and $T_H$.

**Special Topic: Household Refrigerators**

**6-105C** Why are today's refrigerators much more efficient than those built in the past?

**6-106C** Explain how you can reduce the energy consumption of your household refrigerator.

**6-107C** Why is it important to clean the condenser coils of a household refrigerator a few times a year? Also, why is it important not to block airflow through the condenser coils?

**6-108C** Someone proposes that the refrigeration system of a supermarket be overdesigned so that the entire air-conditioning needs of the store can be met by refrigerated air without installing any air-conditioning system. What do you think of this proposal?

**6-109C** Someone proposes that the entire refrigerator/freezer requirements of a store be met using a large freezer that supplies sufficient cold air at −20°C instead of installing separate refrigerators and freezers. What do you think of this proposal?

**6-110** The "Energy Guide" label of a refrigerator states that the refrigerator will consume $170 worth of electricity per year under normal use if the cost of electricity is $0.125/kWh. If the electricity consumed by the lightbulb is negligible and the refrigerator consumes 400 W when running, determine the fraction of the time the refrigerator will run.

**6–111** The interior lighting of refrigerators is usually provided by incandescent lamps whose switches are actuated by the opening of the refrigerator door. Consider a refrigerator whose 40-W lightbulb remains on about 60 h per year. It is proposed to replace the lightbulb by an energy-efficient bulb that consumes only 18 W but costs $25 to purchase and install. If the refrigerator has a coefficient of performance of 1.3 and the cost of electricity is 8 cents per kWh, determine if the energy savings of the proposed lightbulb justify its cost.

**6–112** It is commonly recommended that hot foods be cooled first to room temperature by simply waiting a while before they are put into the refrigerator to save energy. Despite this commonsense recommendation, a person keeps cooking a large pan of stew three times a week and putting the pan into the refrigerator while it is still hot, thinking that the money saved is probably too little. But he says he can be convinced if you can show that the money saved is significant. The average mass of the pan and its contents is 5 kg. The average temperature of the kitchen is 23°C, and the average temperature of the food is 95°C when it is taken off the stove. The refrigerated space is maintained at 3°C, and the average specific heat of the food and the pan can be taken to be 3.9 kJ/kg·°C. If the refrigerator has a coefficient of performance of 1.5 and the cost of electricity is 10 cents per kWh, determine how much this person will save a year by waiting for the food to cool to room temperature before putting it into the refrigerator.

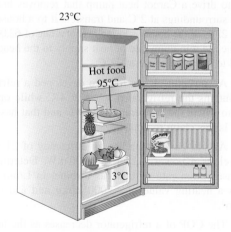

**FIGURE P6–112**

**6–113** It is often stated that the refrigerator door should be opened as few times as possible for the shortest duration of time to save energy. Consider a household refrigerator whose interior volume is 0.9 m³ and average internal temperature is 4°C. At any given time, one-third of the refrigerated space is occupied by food items, and the remaining 0.6 m³ is filled with air. The average temperature and pressure in the kitchen are 20°C and 95 kPa, respectively. Also, the moisture contents of the air in the kitchen and the refrigerator are 0.010 and 0.004 kg per kg of air, respectively, and thus 0.006 kg of water vapor is condensed and removed for each kg of air that enters. The refrigerator door is opened an average of 20 times a day, and each time half of the air volume in the refrigerator is replaced by the warmer kitchen air. If the refrigerator has a coefficient of performance of 1.4 and the cost of electricity is 11.5 cents per kWh, determine the cost of the energy wasted per year as a result of opening the refrigerator door? What would your answer be if the kitchen air were very dry and thus a negligible amount of water vapor condensed in the refrigerator?

## Review Problems

**6–114** An air-conditioning system is used to maintain a house at a constant temperature of 20°C. The house is gaining heat from outdoors at a rate of 20,000 kJ/h, and the heat generated in the house from the people, lights, and appliances amounts to 8000 kJ/h. For a COP of 2.5, determine the required power input to this air-conditioning system. *Answer:* 3.11 kW

**6–115** A Carnot heat pump is used to heat and maintain a residential building at 24°C. An energy analysis of the house reveals that it loses heat at a rate of 4500 kJ/h per °C temperature difference between the indoors and the outdoors. For an outdoor temperature of 2°C, determine (*a*) the coefficient of performance and (*b*) the required power input to the heat pump. *Answers:* (*a*) 13.5, (*b*) 2.04 kW

**6–116** A heat engine receives heat from a heat source at 1200°C and rejects heat to a heat sink at 50°C. The heat engine does maximum work equal to 500 kJ. Determine the heat supplied to the heat engine by the heat source and the heat rejected to the heat sink.

**6–117** A heat pump creates a heating effect of 32,000 kJ/h for a space maintained at 295 K while using 1.8 kW of electrical power. What is the minimum temperature of the source that satisfies the second law of thermodynamics? *Answer:* 235 K

**6–118** A heat pump with a COP of 2.8 is used to heat an air-tight house. When running, the heat pump consumes 5 kW of power. If the temperature in the house is 7°C when the heat pump is turned on, how long will it take for the heat pump to raise the temperature of the house to 22°C? Is this answer realistic or optimistic? Explain. Assume the entire mass within the house (air, furniture, etc.) is equivalent to 1500 kg of air. *Answer:* 19.2 min

**6–119** A promising method of power generation involves collecting and storing solar energy in large artificial lakes a few meters deep, called solar ponds. Solar energy is absorbed by all parts of the pond, and the water temperature rises everywhere. The top part of the pond, however, loses to the atmosphere much of the heat it absorbs, and as a result, its temperature drops. This cool water serves as insulation for the bottom part of the pond and helps trap the energy there. Usually, salt is planted at the bottom of the pond to prevent the rise of this hot water to the top. A power plant that uses an organic fluid, such as alcohol, as the working fluid can be operated between the top and the bottom portions of the pond. If the water temperature is 35°C near the surface and 80°C near the bottom of the pond, determine the maximum thermal efficiency that this power plant can have. Is it realistic to use 35 and 80°C for temperatures in the calculations? Explain. *Answer:* 12.7 percent

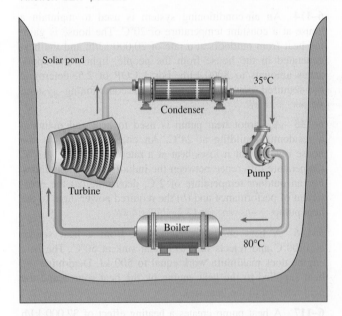

**FIGURE P6–119**

**6–120** Consider a Carnot refrigeration cycle executed in a closed system in the saturated liquid–vapor mixture region using 0.96 kg of refrigerant-134a as the working fluid. It is known that the maximum absolute temperature in the cycle is 1.2 times the minimum absolute temperature, and the net work input to the cycle is 22 kJ. If the refrigerant changes from saturated vapor to saturated liquid during the heat rejection process, determine the minimum pressure in the cycle.

**6–121** [EES] Reconsider Prob. 6–120. Using EES (or other) software, investigate the effect of the net work input on the minimum pressure. Let the work input vary from

10 to 30 kJ. Plot the minimum pressure in the refrigeration cycle as a function of net work input, and discuss the results.

**6–122** Consider two Carnot heat engines operating in series. The first engine receives heat from the reservoir at 1400 K and rejects the waste heat to another reservoir at temperature $T$. The second engine receives this energy rejected by the first one, converts some of it to work, and rejects the rest to a reservoir at 300 K. If the thermal efficiencies of both engines are the same, determine the temperature $T$. *Answer:* 648 K

**6–123** A Carnot heat engine receives heat at 900 K and rejects the waste heat to the environment at 300 K. The entire work output of the heat engine is used to drive a Carnot refrigerator that removes heat from the cooled space at −15°C at a rate of 250 kJ/min and rejects it to the same environment at 300 K. Determine (*a*) the rate of heat supplied to the heat engine and (*b*) the total rate of heat rejection to the environment.

**6–124** [EES] Reconsider Prob. 6–123. Using EES (or other) software, investigate the effects of the heat engine source temperature, the environment temperature, and the cooled space temperature on the required heat supply to the heat engine and the total rate of heat rejection to the environment. Let the source temperature vary from 500 to 1000 K, the environment temperature vary from 275 to 325 K, and the cooled space temperature vary from −20 to 0°C. Plot the required heat supply against the source temperature for the cooled space temperature of −15°C and environment temperatures of 275, 300, and 325 K, and discuss the results.

**6–125** A heat engine operates between two reservoirs at 800 and 20°C. One-half of the work output of the heat engine is used to drive a Carnot heat pump that removes heat from the cold surroundings at 2°C and transfers it to a house maintained at 22°C. If the house is losing heat at a rate of 62,000 kJ/h, determine the minimum rate of heat supply to the heat engine required to keep the house at 22°C.

**6–126** An inventor claims to have developed a refrigerator that maintains the refrigerated space at 4°C while operating in a room where the temperature is 29°C and that has a COP of 13.5. Is this claim reasonable?

**6–127** An old gas turbine has an efficiency of 21 percent and develops a power output of 6000 kW. Determine the fuel consumption rate of this gas turbine, in L/min, if the fuel has a heating value of 42,000 kJ/kg and a density of 0.8 g/cm³.

**6–128** The COP of a refrigerator decreases as the temperature of the refrigerated space is decreased. That is, removing heat from a medium at a very low temperature will require a large work input. Determine the minimum work input required to remove 1 kJ of heat from liquid helium at 3 K when the outside temperature is 300 K. *Answer:* 99 kJ

**6–129** Consider a Carnot heat-pump cycle executed in a steady-flow system in the saturated liquid–vapor mixture region using refrigerant-134a flowing at a rate of 0.22 kg/s as the working fluid. It is known that the maximum absolute temperature in the cycle is 1.2 times the minimum absolute temperature, and the net power input to the cycle is 5 kW. If the refrigerant changes from saturated vapor to saturated liquid during the heat rejection process, determine the ratio of the maximum to minimum pressures in the cycle.

**6–130** Replacing incandescent lights with energy-efficient fluorescent lights can reduce the lighting energy consumption to one-fourth of what it was before. The energy consumed by the lamps is eventually converted to heat, and thus switching to energy-efficient lighting also reduces the cooling load in summer but increases the heating load in winter. Consider a building that is heated by a natural gas furnace with an efficiency of 80 percent and cooled by an air conditioner with a COP of 3.5. If electricity costs $0.12/kWh and natural gas costs $1.40/therm (1 therm = 105,500 kJ), determine if efficient lighting will increase or decrease the total energy cost of the building (*a*) in summer and (*b*) in winter.

**6–131** A heat pump supplies heat energy to a house at the rate of 140,000 kJ/h when the house is maintained at 25°C. Over a period of one month, the heat pump operates for 100 hours to transfer energy from a heat source outside the house to inside the house. Consider a heat pump receiving heat from two different outside energy sources. In one application the heat pump receives heat from the outside air at 0°C. In a second application the heat pump receives heat from a lake having a water temperature of 10°C. If electricity costs $0.105/kWh, determine the maximum money saved by using the lake water rather than the outside air as the outside energy source.

**6–132** The cargo space of a refrigerated truck whose inner dimensions are 12 m × 2.3 m × 3.5 m is to be precooled from 25°C to an average temperature of 5°C. The construction of the truck is such that a transmission heat gain occurs at a rate of 120 W/°C. If the ambient temperature is 25°C, determine how long it will take for a system with a refrigeration capacity of 11 kW to precool this truck.

**6–133** The maximum flow rate of a standard shower head is about 3.5 gpm (13.3 L/min) and can be reduced to 2.75 gpm (10.5 L/min) by switching to a low-flow shower head that is equipped with flow controllers. Consider a family of four, with each person taking a 6-minute shower every morning. City water at 15°C is heated to 55°C in an oil water heater whose efficiency is 65 percent and then tempered to 42°C by cold water at the T-elbow of the shower before being routed to the shower head. The price of heating oil is $2.80/gal and its heating value is 146,300 kJ/gal. Assuming a constant specific heat of 4.18 kJ/kg·°C for water, determine the amount of oil and money saved per year by replacing the standard shower heads by the low-flow ones.

**6–134** [EES] Using EES (or other) software, determine the maximum work that can be extracted from a pond containing $10^5$ kg of water at 350 K when the temperature of the surroundings is 300 K. Notice that the temperature of water in the pond will be gradually decreasing as energy is extracted from it; therefore, the efficiency of the engine will be decreasing. Use temperature intervals of (*a*) 5 K, (*b*) 2 K, and (*c*) 1 K until the pond temperature drops to 300 K. Also solve this problem exactly by integration and compare the results.

**6–135** A refrigeration system is to cool bread loaves with an average mass of 350 g from 30 to −10°C at a rate of 1200 loaves per hour by refrigerated air at −30°C. Taking the average specific and latent heats of bread to be 2.93 kJ/kg·°C and 109.3 kJ/kg, respectively, determine (*a*) the rate of heat removal from the breads, in kJ/h; (*b*) the required volume flow rate of air, in m³/h, if the temperature rise of air is not to exceed 8°C; and (*c*) the size of the compressor of the refrigeration system, in kW, for a COP of 1.2 for the refrigeration system.

**6–136** The drinking water needs of a production facility with 20 employees is to be met by a bubbler type water fountain. The refrigerated water fountain is to cool water from 22 to 8°C and supply cold water at a rate of 0.4 L per hour per person. Heat is transferred to the reservoir from the surroundings at 25°C at a rate of 45 W. If the COP of the refrigeration system is 2.9, determine the size of the compressor, in W, that will be suitable for the refrigeration system of this water cooler.

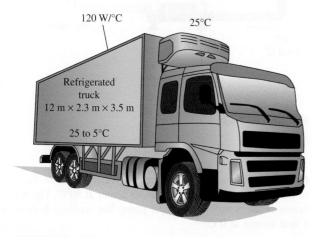

**FIGURE P6–132**

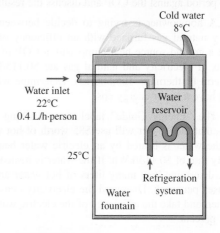

**FIGURE P6–136**

**6–137** A typical electric water heater has an efficiency of 95 percent and costs $350 a year to operate at a unit cost of electricity of $0.11/kWh. A typical heat pump–powered water heater has a COP of 3.3 but costs about $800 more to install. Determine how many years it will take for the heat pump water heater to pay for its cost differential from the energy it saves.

Water heater

**FIGURE P6–137**

*©McGraw-Hill Education//Christopher Kerrigan*

**6–138** Reconsider Prob. 6–137. Using EES (or other) software, investigate the effect of the heat pump COP on the yearly operation costs and the number of years required to break even. Let the COP vary from 2 to 5. Plot the payback period against the COP and discuss the results.

**6–139** A homeowner is trying to decide between a high-efficiency natural gas furnace with an efficiency of 97 percent and a ground-source heat pump with a COP of 3.5. The unit costs of electricity and natural gas are $0.115/kWh and $1.42/therm (1 therm = 105,500 kJ). Determine which system will have a lower energy cost.

**6–140** The "Energy Guide" label on a washing machine indicates that the washer will use $85 worth of hot water per year if the water is heated by an electric water heater at an electricity rate of $0.113/kWh. If the water is heated from 12 to 55°C, determine how many liters of hot water an average family uses per week. Disregard the electricity consumed by the washer, and take the efficiency of the electric water heater to be 91 percent.

**6–141** The kitchen, bath, and other ventilation fans in a house should be used sparingly since these fans can discharge a houseful of warmed or cooled air in just one hour. Consider a 200-m² house whose ceiling height is 2.8 m. The house is heated by a 96 percent efficient gas heater and is maintained at 22°C and 92 kPa. If the unit cost of natural gas is $1.20/therm (1 therm = 105,500 kJ), determine the cost of energy "vented out" by the fans in 1 h. Assume the average outdoor temperature during the heating season to be 5°C.

**6–142** Repeat Prob. 6–141 for the air-conditioning cost in a dry climate for an outdoor temperature of 33°C. Assume the COP of the air-conditioning system to be 2.1, and the unit cost of electricity to be $0.12/kWh.

**6–143** A heat pump with refrigerant-134a as the working fluid is used to keep a space at 25°C by absorbing heat from geothermal water that enters the evaporator at 60°C at a rate of 0.065 kg/s and leaves at 40°C. Refrigerant enters the evaporator at 12°C with a quality of 15 percent and leaves at the same pressure as saturated vapor. If the compressor consumes 1.6 kW of power, determine (a) the mass flow rate of the refrigerant, (b) the rate of heat supply, (c) the COP, and (d) the minimum power input to the compressor for the same rate of heat supply. *Answers:* (a) 0.0338 kg/s, (b) 7.04 kW, (c) 4.40, (d) 0.740 kW

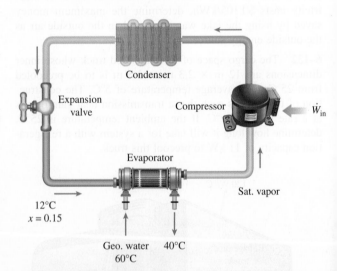

**FIGURE P6–143**

**6–144** Cold water at 10°C enters a water heater at the rate of 0.02 m³/min and leaves the water heater at 50°C. The water heater receives heat from a heat pump that receives heat from a heat source at 0°C.

(a) Assuming the water to be an incompressible liquid that does not change phase during heat addition, determine the rate of heat supplied to the water, in kJ/s.

(b) Assuming the water heater acts as a heat sink having an average temperature of 30°C, determine the minimum power supplied to the heat pump, in kW.

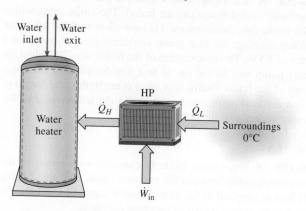

**FIGURE P6–144**

**6–145** A heat pump receives heat from a lake that has an average winter time temperature of 6°C and supplies heat into a house having an average temperature of 23°C.

(a) If the house loses heat to the atmosphere at the rate of 52,000 kJ/h, determine the minimum power supplied to the heat pump, in kW.

(b) A heat exchanger is used to transfer the energy from the lake water to the heat pump. If the lake water temperature decreases by 5°C as it flows through the lake water-to-heat pump heat exchanger, determine the minimum mass flow rate of lake water, in kg/s. Neglect the effect of the lake water pump.

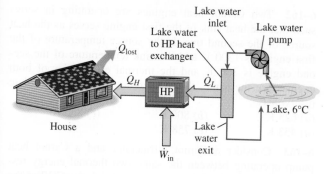

**FIGURE P6–145**

**6–146** Prove that the COP of all completely reversible refrigerators must be the same when the reservoir temperatures are the same.

**6–147** A Carnot heat engine is operating between a source at $T_H$ and a sink at $T_L$. If it is desired to double the thermal efficiency of this engine, what should the new source temperature be? Assume the sink temperature is held constant.

**6–148** When discussing Carnot engines, it is assumed that the engine is in thermal equilibrium with the source and the sink during the heat addition and heat rejection processes, respectively. That is, it is assumed that $T_H^* = T_H$ and $T_L^* = T_L$ so that there is no external irreversibility. In that case, the thermal efficiency of the Carnot engine is $\eta_C = 1 - T_L/T_H$.

In reality, however, we must maintain a reasonable temperature difference between the two heat transfer media in order to have an acceptable heat transfer rate through a finite heat exchanger surface area. The heat transfer rates in that case can be expressed as

$$\dot{Q}_H = (h_A)_H(T_H - T_H^*)$$
$$\dot{Q}_L = (hA)_L(T_L^* - T_L)$$

where $h$ and $A$ are the heat transfer coefficient and heat transfer surface area, respectively. When the values of $h$, $A$, $T_H$, and $T_L$ are fixed, show that the power output will be a maximum when

$$\frac{T_L^*}{T_H^*} = \left(\frac{T_L}{T_H}\right)^{1/2}$$

Also, show that the maximum net power output in this case is

$$\dot{W}_{C,max} = \frac{(hA)_H T_H}{1 + (hA)_H/(hA)_L}\left[1 - \left(\frac{T_L}{T_H}\right)^{1/2}\right]^2$$

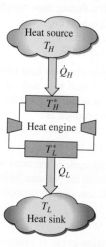

**FIGURE P6–148**

**6–149**  Show that $COP_{HP} = COP_R + 1$ when both the heat pump and the refrigerator have the same $Q_L$ and $Q_H$ values.

## Fundamentals of Engineering (FE) Exam Problems

**6–150**  A 2.4-m high 200-m² house is maintained at 22°C by an air-conditioning system whose COP is 3.2. It is estimated that the kitchen, bath, and other ventilating fans of the house discharge a houseful of conditioned air once every hour. If the average outdoor temperature is 32°C, the density of air is 1.20 kg/m³, and the unit cost of electricity is $0.10/kWh, the amount of money "vented out" by the fans in 10 hours is
(a) $0.50    (b) $1.60    (c) $5.00
(d) $11.00   (e) $16.00

**6–151**  The drinking water needs of an office are met by cooling tab water in a refrigerated water fountain from 23 to 6°C at an average rate of 10 kg/h. If the COP of this refrigerator is 3.1, the required power input to this refrigerator is
(a) 197 W    (b) 612 W    (c) 64 W
(d) 109 W    (e) 403 W

**6–152**  The label on a washing machine indicates that the washer will use $85 worth of hot water if the water is heated by a 90 percent efficient electric heater at an electricity rate of $0.09/kWh. If the water is heated from 18 to 45°C, the amount of hot water an average family uses per year is
(a) 11.6 tons    (b) 15.8 tons    (c) 27.1 tons
(d) 30.1 tons    (e) 33.5 tons

**6–153**  A heat pump is absorbing heat from the cold outdoors at 5°C and supplying heat to a house at 25°C at a rate of 18,000 kJ/h. If the power consumed by the heat pump is 1.9 kW, the coefficient of performance of the heat pump is
(a) 1.3    (b) 2.6    (c) 3.0
(d) 3.8    (e) 13.9

**6–154**  A heat engine cycle is executed with steam in the saturation dome. The pressure of steam is 1 MPa during heat addition, and 0.4 MPa during heat rejection. The highest possible efficiency of this heat engine is
(a) 8.0%    (b) 15.6%    (c) 20.2%
(d) 79.8%   (e) 100%

**6–155**  A heat engine receives heat from a source at 1000°C and rejects the waste heat to a sink at 50°C. If heat is supplied to this engine at a rate of 100 kJ/s, the maximum power this heat engine can produce is
(a) 25.4 kW    (b) 55.4 kW    (c) 74.6 kW
(d) 95.0 kW    (e) 100 kW

**6–156**  A heat pump cycle is executed with R–134a under the saturation dome between the pressure limits of 1.4 and 0.16 MPa. The maximum coefficient of performance of this heat pump is
(a) 1.1    (b) 3.8    (c) 4.8
(d) 5.3    (e) 2.9

**6–157**  A refrigeration cycle is executed with R-134a under the saturation dome between the pressure limits of 1.6 and 0.2 MPa. If the power consumption of the refrigerator is 3 kW, the maximum rate of heat removal from the cooled space of this refrigerator is
(a) 0.45 kJ/s    (b) 0.78 kJ/s    (c) 3.0 kJ/s
(d) 11.6 kJ/s    (e) 14.6 kJ/s

**6–158**  A heat pump with a COP of 3.2 is used to heat a perfectly sealed house (no air leaks). The entire mass within the house (air, furniture, etc.) is equivalent to 1200 kg of air. When running, the heat pump consumes electric power at a rate of 5 kW. The temperature of the house was 7°C when the heat pump was turned on. If heat transfer through the envelope of the house (walls, roof, etc.) is negligible, the length of time the heat pump must run to raise the temperature of the entire contents of the house to 22°C is
(a) 13.5 min    (b) 43.1 min    (c) 138 min
(d) 18.8 min    (e) 808 min

**6–159**  A heat engine cycle is executed with steam in the saturation dome between the pressure limits of 7 and 2 MPa. If heat is supplied to the heat engine at a rate of 150 kJ/s, the maximum power output of this heat engine is
(a) 8.1 kW    (b) 19.7 kW    (c) 38.6 kW
(d) 107 kW    (e) 130 kW

**6–160**  An air-conditioning system operating on the reversed Carnot cycle is required to remove heat from the house at a rate of 32 kJ/s to maintain its temperature constant at 20°C. If the temperature of the outdoors is 35°C, the power required to operate this air-conditioning system is
(a) 0.58 kW    (b) 3.20 kW    (c) 1.56 kW
(d) 2.26 kW    (e) 1.64 kW

**6–161**  A refrigerator is removing heat from a cold medium at 3°C at a rate of 7200 kJ/h and rejecting the waste heat to a medium at 30°C. If the coefficient of performance of the refrigerator is 2, the power consumed by the refrigerator is
(a) 0.1 kW    (b) 0.5 kW    (c) 1.0 kW
(d) 2.0 kW    (e) 5.0 kW

**6–162**  Two Carnot heat engines are operating in series such that the heat sink of the first engine serves as the heat source of the second one. If the source temperature of the first engine is 1300 K and the sink temperature of the second engine is 300 K and the thermal efficiencies of both engines are the same, the temperature of the intermediate reservoir is
(a) 625 K    (b) 800 K    (c) 860 K
(d) 453 K    (e) 758 K

**6–163**  Consider a Carnot refrigerator and a Carnot heat pump operating between the same two thermal energy reservoirs. If the COP of the refrigerator is 3.4, the COP of the heat pump is
(a) 1.7    (b) 2.4    (c) 3.4
(d) 4.4    (e) 5.0

**6–164**  A typical new household refrigerator consumes about 680 kWh of electricity per year and has a coefficient of performance of 1.4. The amount of heat removed by this refrigerator from the refrigerated space per year is

(*a*) 952 MJ/yr   (*b*) 1749 MJ/yr   (*c*) 2448 MJ/yr
(*d*) 3427 MJ/yr   (*e*) 4048 MJ/yr

**6–165**   A window air conditioner that consumes 1 kW of electricity when running and has a coefficient of performance of 3 is placed in the middle of a room, and is plugged in. The rate of cooling or heating this air conditioner will provide to the air in the room when running is

(*a*) 3 kJ/s, cooling   (*b*) 1 kJ/s, cooling   (*c*) 0.33 kJ/s, heating
(*d*) 1 kJ/s, heating   (*e*) 3 kJ/s, heating

## Design and Essay Problems

**6–166**   Devise a Carnot heat engine using steady-flow components, and describe how the Carnot cycle is executed in that engine. What happens when the directions of heat and work interactions are reversed?

**6–167**   When was the concept of the heat pump conceived and by whom? When was the first heat pump built, and when were the heat pumps first mass-produced?

**6–168**   The sun supplies electromagnetic energy to the earth. It appears to have an effective temperature of approximately 5800 K. On a clear summer day in North America, the energy incident on a surface facing the sun is approximately 0.95 kW/m². The electromagnetic solar energy can be converted into thermal energy by being absorbed on a darkened surface. How might you characterize the work potential of the sun's energy when it is to be used to produce work?

**6–169**   In the search to reduce thermal pollution and take advantage of renewable energy sources, some people have proposed that we take advantage of such sources as discharges from electrical power plants, geothermal energy, and ocean thermal energy. Although many of these sources contain an enormous amount of energy, the amount of work they are capable of producing is limited. How might you use the work potential to assign an "energy quality" to these proposed sources? Test your proposed "energy quality" measure by applying it to the ocean thermal source, where the temperature 30 m below the surface is perhaps 5°C lower than at the surface. Apply it also to the geothermal water source, where the temperature 2 to 3 km below the surface is perhaps 150°C hotter than at the surface.

**6–170**   Using a thermometer, measure the temperature of the main food compartment of your refrigerator, and check if it is between 1 and 4°C. Also, measure the temperature of the freezer compartment, and check if it is at the recommended value of −18°C.

**6–171**   Using a timer (or watch) and a thermometer, conduct the following experiment to determine the rate of heat gain of your refrigerator. First make sure that the door of the refrigerator is not opened for at least a few hours so that steady operating conditions are established. Start the timer when the refrigerator stops running and measure the time $\Delta t_1$ it stays off before it kicks in. Then, measure the time $\Delta t_2$ it stays on. Noting that the heat removed during $\Delta t_2$ is equal to the heat gain of the refrigerator during $\Delta t_1 + \Delta t_2$ and using the power consumed by the refrigerator when it is running, determine the average rate of heat gain for your refrigerator, in W. Take the COP (coefficient of performance) of your refrigerator to be 1.3 if it is not available.

**6–172**   Design a hydrocooling unit that can cool fruits and vegetables from 30 to 5°C at a rate of 20,000 kg/h under the following conditions:

The unit will be of flood type, which will cool the products as they are conveyed into the channel filled with water. The products will be dropped into the channel filled with water at one end and be picked up at the other end. The channel can be as wide as 3 m and as high as 90 cm. The water is to be circulated and cooled by the evaporator section of a refrigeration system. The refrigerant temperature inside the coils is to be −2°C, and the water temperature is not to drop below 1°C and not to exceed 6°C.

Assuming reasonable values for the average product density, specific heat, and porosity (the fraction of air volume in a box), recommend reasonable values for (*a*) the water velocity through the channel and (*b*) the refrigeration capacity of the refrigeration system.

# ENTROPY

I n Chap. 6, we introduced the second law of thermodynamics and applied it to cycles and cyclic devices. In this chapter, we apply the second law to processes. The first law of thermodynamics deals with the property *energy* and the conservation of it. The second law leads to the definition of a new property called *entropy*. Entropy is a somewhat abstract property, and it is difficult to give a physical description of it without considering the microscopic state of the system. Entropy is best understood and appreciated by studying its uses in commonly encountered engineering processes, and this is what we intend to do.

This chapter starts with a discussion of the Clausius inequality, which forms the basis for the definition of entropy, and continues with the increase of entropy principle. Unlike energy, entropy is a nonconserved property, and there is no such thing as *conservation of entropy*. Next, the entropy changes that take place during processes for pure substances, incompressible substances, and ideal gases are discussed, and a special class of idealized processes, called *isentropic processes*, is examined. Then, the reversible steady-flow work and the isentropic efficiencies of various engineering devices such as turbines and compressors are considered. Finally, entropy balance is introduced and applied to various systems.

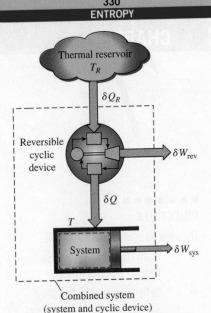

**FIGURE 7–1**
The system considered in the development of the Clausius inequality.

# 7–1 · ENTROPY

The second law of thermodynamics often leads to expressions that involve inequalities. An irreversible (i.e., actual) heat engine, for example, is less efficient than a reversible one operating between the same two thermal energy reservoirs. Likewise, an irreversible refrigerator or a heat pump has a lower coefficient of performance (COP) than a reversible one operating between the same temperature limits. Another important inequality that has major consequences in thermodynamics is the **Clausius inequality**. It was first stated by the German physicist R. J. E. Clausius (1822–1888), one of the founders of thermodynamics, and is expressed in 1865 as

$$\oint \frac{\delta Q}{T} \leq 0$$

That is, *the cyclic integral of $\delta Q/T$ is always less than or equal to zero.* This inequality is valid for all cycles, reversible or irreversible. The symbol $\oint$ (integral symbol with a circle in the middle) is used to indicate that the integration is to be performed over the entire cycle. Any heat transfer to or from a system can be considered to consist of differential amounts of heat transfer. Then the cyclic integral of $\delta Q/T$ can be viewed as the sum of all these differential amounts of heat transfer divided by the temperature at the boundary.

To demonstrate the validity of the Clausius inequality, consider a system connected to a thermal energy reservoir at a constant thermodynamic (i.e., absolute) temperature of $T_R$ through a *reversible* cyclic device (Fig. 7–1). The cyclic device receives heat $\delta Q_R$ from the reservoir and supplies heat $\delta Q$ to the system whose temperature at that part of the boundary is $T$ (a variable) while producing work $\delta W_{rev}$. The system produces work $\delta W_{sys}$ as a result of this heat transfer. Applying the energy balance to the combined system identified by dashed lines yields

$$\delta W_C = \delta Q_R - dE_C$$

where $\delta W_C$ is the total work of the combined system ($\delta W_{rev} + \delta W_{sys}$) and $dE_C$ is the change in the total energy of the combined system. Considering that the cyclic device is a *reversible* one, we have

$$\frac{\delta Q_R}{T_R} = \frac{\delta Q}{T}$$

where the sign of $\delta Q$ is determined with respect to the system (positive if *to* the system and negative if *from* the system) and the sign of $\delta Q_R$ is determined with respect to the reversible cyclic device. Eliminating $\delta Q_R$ from the two relations above yields

$$\delta W_C = T_R \frac{\delta Q}{T} - dE_C$$

We now let the system undergo a cycle while the cyclic device undergoes an integral number of cycles. Then the preceding relation becomes

$$W_C = T_R \oint \frac{\delta Q}{T}$$

since the cyclic integral of energy (the net change in the energy, which is a property, during a cycle) is zero. Here $W_C$ is the cyclic integral of $\delta W_C$, and it represents the net work for the combined cycle.

It appears that the combined system is exchanging heat with a single thermal energy reservoir while involving (producing or consuming) work $W_C$ during a cycle. On the basis of the Kelvin–Planck statement of the second law, which states that *no system can produce a net amount of work while operating in a cycle and exchanging heat with a single thermal energy reservoir,* we reason that $W_C$ cannot be a work output, and thus it cannot be a positive quantity. Considering that $T_R$ is the thermodynamic temperature and thus a positive quantity, we must have

$$\oint \frac{\delta Q}{T} \leq 0 \tag{7-1}$$

which is the *Clausius inequality.* This inequality is valid for all thermodynamic cycles, reversible or irreversible, including the refrigeration cycles.

If no irreversibilities occur within the system as well as the reversible cyclic device, then the cycle undergone by the combined system is internally reversible. As such, it can be reversed. In the reversed cycle case, all the quantities have the same magnitude but the opposite sign. Therefore, the work $W_C$, which could not be a positive quantity in the regular case, cannot be a negative quantity in the reversed case. Then it follows that $W_{C,\text{int rev}} = 0$ since it cannot be a positive or negative quantity, and therefore

$$\oint \left(\frac{\delta Q}{T}\right)_{\text{int rev}} = 0 \tag{7-2}$$

for internally reversible cycles. Thus, we conclude that the equality in the Clausius inequality holds for totally or just internally reversible cycles and the inequality for the irreversible ones.

To develop a relation for the definition of entropy, let us examine Eq. 7–2 more closely. Here we have a quantity whose cyclic integral is zero. Let us think for a moment what kind of quantities can have this characteristic. We know that the cyclic integral of *work* is not zero. (It is a good thing that it is not. Otherwise, heat engines that work on a cycle such as steam power plants would produce zero net work.) Neither is the cyclic integral of heat.

Now consider the volume occupied by a gas in a piston–cylinder device undergoing a cycle, as shown in Fig. 7–2. When the piston returns to its initial position at the end of a cycle, the volume of the gas also returns to its initial value. Thus, the net change in volume during a cycle is zero. This is also expressed as

$$\oint dV = 0 \tag{7-3}$$

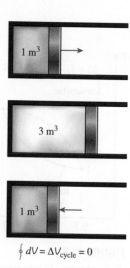

$$\oint dV = \Delta V_{\text{cycle}} = 0$$

**FIGURE 7–2**

The net change in volume (a property) during a cycle is always zero.

That is, the cyclic integral of volume (or any other property) is zero. Conversely, a quantity whose cyclic integral is zero depends on the *state* only and not the process path, and thus it is a property. Therefore, the quantity $(\delta Q/T)_{\text{int rev}}$ must represent a property in the differential form.

Clausius realized in 1865 that he had discovered a new thermodynamic property, and he chose to name this property **entropy**. It is designated $S$ and is defined as

$$dS = \left(\frac{\delta Q}{T}\right)_{\text{int rev}} \quad \text{(kJ/K)} \tag{7–4}$$

Entropy is an extensive property of a system and sometimes is referred to as *total entropy*. Entropy per unit mass, designated $s$, is an intensive property and has the unit kJ/kg·K. The term *entropy* is generally used to refer to both total entropy and entropy per unit mass since the context usually clarifies which one is meant.

The entropy change of a system during a process can be determined by integrating Eq. 7–4 between the initial and the final states:

$$\Delta S = S_2 - S_1 = \int_1^2 \left(\frac{\delta Q}{T}\right)_{\text{int rev}} \quad \text{(kJ/K)} \tag{7–5}$$

Notice that we have actually defined the *change* in entropy instead of entropy itself, just as we defined the change in energy instead of the energy itself when we developed the first-law relation. Absolute values of entropy are determined on the basis of the third law of thermodynamics, which is discussed later in this chapter. Engineers are usually concerned with the *changes* in entropy. Therefore, the entropy of a substance can be assigned a zero value at some arbitrarily selected reference state, and the entropy values at other states can be determined from Eq. 7–5 by choosing state 1 to be the reference state ($S = 0$) and state 2 to be the state at which entropy is to be determined.

To perform the integration in Eq. 7–5, one needs to know the relation between $Q$ and $T$ during a process. This relation is often not available, and the integral in Eq. 7–5 can be performed for a few cases only. For the majority of cases we have to rely on tabulated data for entropy.

Note that entropy is a property, and like all other properties, it has fixed values at fixed states. Therefore, the entropy change $\Delta S$ between two specified states is the same no matter what path, reversible or irreversible, is followed during a process (Fig. 7–3).

Also note that the integral of $\delta Q/T$ gives us the value of entropy change *only if* the integration is carried out along an *internally reversible* path between the two states. The integral of $\delta Q/T$ along an irreversible path is not a property, and in general, different values will be obtained when the integration is carried out along different irreversible paths. Therefore, even for irreversible processes, the entropy change should be determined by carrying out this integration along some convenient *imaginary* internally reversible path between the specified states.

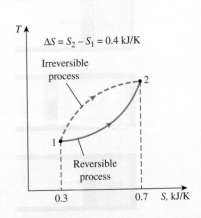

**FIGURE 7–3**
The entropy change between two specified states is the same whether the process is reversible or irreversible.

# A Special Case: Internally Reversible Isothermal Heat Transfer Processes

Recall that isothermal heat transfer processes are internally reversible. Therefore, the entropy change of a system during an internally reversible isothermal heat transfer process can be determined by performing the integration in Eq. 7–5:

$$\Delta S = \int_1^2 \left(\frac{\delta Q}{T}\right)_{\text{int rev}} = \int_1^2 \left(\frac{\delta Q}{T_0}\right)_{\text{int rev}} = \frac{1}{T_0} \int_1^2 (\delta Q)_{\text{int rev}}$$

which reduces to

$$\Delta S = \frac{Q}{T_0} \quad \text{(kJ/K)} \tag{7–6}$$

where $T_0$ is the constant temperature of the system and $Q$ is the heat transfer for the internally reversible process. Equation 7–6 is particularly useful for determining the entropy changes of thermal energy reservoirs that can absorb or supply heat indefinitely at a constant temperature.

Notice that the entropy change of a system during an internally reversible isothermal process can be positive or negative, depending on the direction of heat transfer. Heat transfer to a system increases the entropy of a system, whereas heat transfer from a system decreases it. In fact, losing heat is the only way the entropy of a system can be decreased.

---

### ■ EXAMPLE 7–1    Entropy Change during an Isothermal Process

A piston–cylinder device contains a liquid–vapor mixture of water at 300 K. During a constant-pressure process, 750 kJ of heat is transferred to the water. As a result, part of the liquid in the cylinder vaporizes. Determine the entropy change of the water during this process.

**SOLUTION**  Heat is transferred to a liquid–vapor mixture of water in a piston–cylinder device at constant pressure. The entropy change of water is to be determined.

**Assumptions**  No irreversibilities occur within the system boundaries during the process.

**Analysis**  We take the *entire water* (liquid + vapor) in the cylinder as the system (Fig. 7–4). This is a *closed system* since no mass crosses the system boundary during the process. We note that the temperature of the system remains constant at 300 K during this process since the temperature of a pure substance remains constant at the saturation value during a phase-change process at constant pressure.

The system undergoes an internally reversible, isothermal process, and thus its entropy change can be determined directly from Eq. 7–6 to be

$$\Delta S_{\text{sys,isothermal}} = \frac{Q}{T_{\text{sys}}} = \frac{750 \text{ kJ}}{300 \text{ K}} = \textbf{2.5 kJ/K}$$

**Discussion**  Note that the entropy change of the system is positive, as expected, since heat transfer is to the system.

$T = 300 \text{ K} = \text{const.}$

$\Delta S_{\text{sys}} = \dfrac{Q}{T} = 2.5 \,\dfrac{\text{kJ}}{\text{K}}$

$Q = 750 \text{ kJ}$

**FIGURE 7–4**

Schematic for Example 7–1.

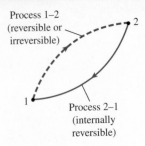

Process 1–2
(reversible or
irreversible)

2

1

Process 2–1
(internally
reversible)

**FIGURE 7–5**
A cycle composed of a reversible and
an irreversible process.

# 7–2 · THE INCREASE OF ENTROPY PRINCIPLE

Consider a cycle that is made up of two processes: process 1-2, which is arbitrary (reversible or irreversible), and process 2-1, which is internally reversible, as shown in Figure 7–5. From the Clausius inequality,

$$\oint \frac{\delta Q}{T} \leq 0$$

or

$$\int_1^2 \frac{\delta Q}{T} + \int_2^1 \left(\frac{\delta Q}{T}\right)_{\text{int rev}} \leq 0$$

The second integral in the previous relation is recognized as the entropy change $S_1 - S_2$. Therefore,

$$\int_1^2 \frac{\delta Q}{T} + S_1 - S_2 \leq 0$$

which can be rearranged as

$$S_2 - S_1 \geq \int_1^2 \frac{\delta Q}{T} \tag{7-7}$$

It can also be expressed in differential form as

$$dS \geq \frac{\delta Q}{T} \tag{7-8}$$

where the equality holds for an internally reversible process and the inequality for an irreversible process. We may conclude from these equations that the entropy change of a closed system during an irreversible process is greater than the integral of $\delta Q/T$ evaluated for that process. In the limiting case of a reversible process, these two quantities become equal. We again emphasize that $T$ in these relations is the *thermodynamic temperature* at the *boundary* where the differential heat $\delta Q$ is transferred between the system and the surroundings.

The quantity $\Delta S = S_2 - S_1$ represents the *entropy change* of the system. For a reversible process, it becomes equal to $\int_1^2 \delta Q/T$, which represents the *entropy transfer* with heat.

The inequality sign in the preceding relations is a constant reminder that the entropy change of a closed system during an irreversible process is always greater than the entropy transfer. That is, some entropy is *generated* or *created* during an irreversible process, and this generation is due entirely to the presence of irreversibilities. The entropy generated during a process is called **entropy generation** and is denoted by $S_{\text{gen}}$. Noting that the difference between the entropy change of a closed system and the entropy transfer is equal to entropy generation, Eq. 7-7 can be rewritten as an equality as

$$\Delta S_{\text{sys}} = S_2 - S_1 = \int_1^2 \frac{\delta Q}{T} + S_{\text{gen}} \tag{7-9}$$

Note that the entropy generation $S_{gen}$ is always a *positive* quantity or zero. Its value depends on the process, and thus it is *not* a property of the system. Also, in the absence of any entropy transfer, the entropy change of a system is equal to the entropy generation.

Equation 7–7 has far-reaching implications in thermodynamics. For an isolated system (or simply an adiabatic closed system), the heat transfer is zero, and Eq. 7–7 reduces to

$$\Delta S_{isolated} \geq 0 \qquad (7\text{–}10)$$

This equation can be expressed as *the entropy of an isolated system during a process always increases or, in the limiting case of a reversible process, remains constant.* In other words, it *never* decreases. This is known as the **increase of entropy principle**. Note that in the absence of any heat transfer, entropy change is due to irreversibilities only, and their effect is always to increase entropy.

Entropy is an extensive property, thus the total entropy of a system is equal to the sum of the entropies of the parts of the system. An isolated system may consist of any number of subsystems (Fig. 7–6). A system and its surroundings, for example, constitute an isolated system since both can be enclosed by a sufficiently large arbitrary boundary across which there is no heat, work, or mass transfer (Fig. 7–7). Therefore, a system and its surroundings can be viewed as the two subsystems of an isolated system, and the entropy change of this isolated system during a process is the sum of the entropy changes of the system and its surroundings, which is equal to the entropy generation since an isolated system involves no entropy transfer. That is,

$$S_{gen} = \Delta S_{total} = \Delta S_{sys} + \Delta S_{surr} \geq 0 \qquad (7\text{–}11)$$

where the equality holds for reversible processes and the inequality for irreversible ones. Note that $\Delta S_{surr}$ refers to the change in the entropy of the surroundings as a result of the occurrence of the process under consideration.

Since no actual process is truly reversible, we can conclude that some entropy is generated during a process, and therefore the entropy of the universe, which can be considered to be an isolated system, is continuously increasing. The more irreversible a process, the larger the entropy generated during that process. No entropy is generated during reversible processes ($S_{gen} = 0$).

Entropy increase of the universe is a major concern not only to engineers but also to philosophers, theologians, economists, and environmentalists since entropy is viewed as a measure of the disorder (or "mixed-up-ness") in the universe.

The increase of entropy principle does not imply that the entropy of a system cannot decrease. The entropy change of a system *can* be negative during a process (Fig. 7–8), but entropy generation cannot. The increase of entropy principle can be summarized as follows:

$$S_{gen} \begin{cases} > 0 & \text{Irreversible process} \\ = 0 & \text{Reversible process} \\ < 0 & \text{Impossible process} \end{cases}$$

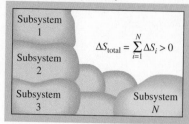

**FIGURE 7–6**
The entropy change of an isolated system is the sum of the entropy changes of its components, and is never less than zero.

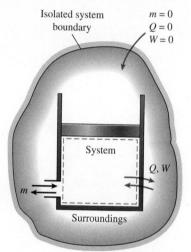

**FIGURE 7–7**
A system and its surroundings form an isolated system.

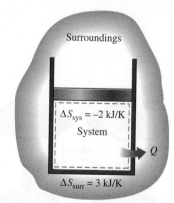

$$S_{gen} = \Delta S_{total} = \Delta S_{sys} + \Delta S_{surr} = 1 \text{ kJ/K}$$

**FIGURE 7–8**
The entropy change of a system can be negative, but the entropy generation cannot.

This relation serves as a criterion in determining whether a process is reversible, irreversible, or impossible.

Things in nature have a tendency to change until they attain a state of equilibrium. The increase of entropy principle dictates that the entropy of an isolated system increases until the entropy of the system reaches a *maximum* value. At that point, the system is said to have reached an equilibrium state since the increase of entropy principle prohibits the system from undergoing any change of state that results in a decrease in entropy.

## Some Remarks about Entropy

In light of the preceding discussions, we draw the following conclusions:

1. Processes can occur in a *certain* direction only, not in *any* direction. A process must proceed in the direction that complies with the increase of entropy principle, that is, $S_{gen} \geq 0$. A process that violates this principle is impossible. This principle often forces chemical reactions to come to a halt before reaching completion.

2. Entropy is a *nonconserved property,* and there is *no* such thing as the *conservation of entropy principle*. Entropy is conserved during the idealized reversible processes only and increases during *all* actual processes.

3. The performance of engineering systems is degraded by the presence of irreversibilities, and *entropy generation* is a measure of the magnitudes of the irreversibilities present during that process. The greater the extent of irreversibilities, the greater the entropy generation. Therefore, entropy generation can be used as a quantitative measure of irreversibilities associated with a process. It is also used to establish criteria for the performance of engineering devices. This point is illustrated further in Example 7–2.

---

**EXAMPLE 7–2**    **Entropy Generation during Heat Transfer Processes**

A heat source at 800 K loses 2000 kJ of heat to a sink at (a) 500 K and (b) 750 K. Determine which heat transfer process is more irreversible.

**SOLUTION**   Heat is transferred from a heat source to two heat sinks at different temperatures. The heat transfer process that is more irreversible is to be determined.

**Analysis**   A sketch of the reservoirs is shown in Fig. 7–9. Both cases involve heat transfer through a finite temperature difference, and therefore both are irreversible. The magnitude of the irreversibility associated with each process can be determined by calculating the total entropy change for each case. The total entropy change for a heat transfer process involving two reservoirs (a source and a sink) is the sum of the entropy changes of each reservoir since the two reservoirs form an adiabatic system.

Or do they? The problem statement gives the impression that the two reservoirs are in direct contact during the heat transfer process. But this cannot be the case since the temperature at a point can have only one value, and thus it cannot be 800 K on one side of the point of contact and 500 K on the other side. In other words, the temperature function cannot have a jump discontinuity.

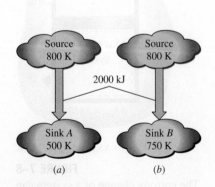

**FIGURE 7–9**
Schematic for Example 7–2.

Therefore, it is reasonable to assume that the two reservoirs are separated by a partition through which the temperature drops from 800 K on one side to 500 K (or 750 K) on the other. In that case, the entropy change of the partition should also be considered when evaluating the total entropy change for this process. However, considering that entropy is a property and the values of properties depend on the state of a system, we can argue that the entropy change of the partition is zero since the partition appears to have undergone a *steady* process and thus experienced no change in its properties at any point. We base this argument on the fact that the temperature on both sides of the partition and thus throughout remains constant during this process. Therefore, we are justified to assume that $\Delta S_{\text{partition}} = 0$ since the entropy (as well as the energy) content of the partition remains constant during this process.

The entropy change for each reservoir can be determined from Eq. 7–6 since each reservoir undergoes an internally reversible, isothermal process.

(a) For the heat transfer process to a sink at 500 K:

$$\Delta S_{\text{source}} = \frac{Q_{\text{source}}}{T_{\text{source}}} = \frac{-2000 \text{ kJ}}{800 \text{ K}} = -2.5 \text{ kJ/K}$$

$$\Delta S_{\text{sink}} = \frac{Q_{\text{sink}}}{T_{\text{sink}}} = \frac{2000 \text{ kJ}}{500 \text{ K}} = +4.0 \text{ kJ/K}$$

and

$$S_{\text{gen}} = \Delta S_{\text{total}} = \Delta S_{\text{source}} + \Delta S_{\text{sink}} = (-2.5 + 4.0) \text{ kJ/K} = \mathbf{1.5 \text{ kJ/K}}$$

Therefore, 1.5 kJ/K of entropy is generated during this process. Noting that both reservoirs have undergone internally reversible processes, the entire entropy generation took place in the partition.

(b) Repeating the calculations in part (a) for a sink temperature of 750 K, we obtain

$$\Delta S_{\text{source}} = -2.5 \text{ kJ/k}$$

$$\Delta S_{\text{sink}} = +2.7 \text{ kJ/K}$$

and

$$S_{\text{gen}} = \Delta S_{\text{total}} = (-2.5 + 2.7) \text{ kJ/K} = \mathbf{0.2 \text{ kJ/K}}$$

The total entropy change for the process in part (b) is smaller, and therefore it is less irreversible. This is expected since the process in (b) involves a smaller temperature difference and thus a smaller irreversibility.

**Discussion** The irreversibilities associated with both processes could be eliminated by operating a Carnot heat engine between the source and the sink. For this case it can be shown that $\Delta S_{\text{total}} = 0$.

# 7–3 · ENTROPY CHANGE OF PURE SUBSTANCES

Entropy is a property, and thus the value of entropy of a system is fixed once the state of the system is fixed. Specifying two intensive independent properties fixes the state of a simple compressible system, and thus the value of entropy and of other properties at that state. Starting with its defining relation, the entropy change of a substance can be expressed in terms of other properties (see Sec. 7–7). But in general, these relations are too complicated and are not practical to use for hand calculations. Therefore, using a suitable

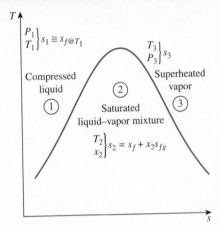

**FIGURE 7–10**

The entropy of a pure substance is determined from the tables (like other properties).

reference state, the entropies of substances are evaluated from measurable property data following rather involved computations, and the results are tabulated in the same manner as the other properties such as $v$, $u$, and $h$ (Fig. 7–10).

The entropy values in the property tables are given relative to an arbitrary reference state. In steam tables the entropy of saturated liquid $s_f$ at 0.01°C is assigned the value of zero. For refrigerant-134a, the zero value is assigned to saturated liquid at −40°C. The entropy values become negative at temperatures below the reference value.

The value of entropy at a specified state is determined just like any other property. In the compressed liquid and superheated vapor regions, it can be obtained directly from the tables at the specified state. In the saturated mixture region, it is determined from

$$s = s_f + xs_{fg} \quad (\text{kJ/kg·K})$$

where $x$ is the quality and $s_f$ and $s_{fg}$ values are listed in the saturation tables. In the absence of compressed liquid data, the entropy of the compressed liquid can be approximated by the entropy of the saturated liquid at the given temperature:

$$s_{@ T,P} \cong s_{f @ T} \quad (\text{kJ/kg·K})$$

The entropy change of a specified mass $m$ (a closed system) during a process is simply

$$\Delta S = m\Delta s = m(s_2 - s_1) \quad (\text{kJ/K}) \qquad (7–12)$$

which is the difference between the entropy values at the final and initial states.

When studying the second-law aspects of processes, entropy is commonly used as a coordinate on diagrams such as the $T$-$s$ and $h$-$s$ diagrams. The general characteristics of the $T$-$s$ diagram of pure substances are shown in Fig. 7–11 using data for water. Notice from this diagram that the constant-volume lines are steeper than the constant-pressure lines and the constant-pressure lines are

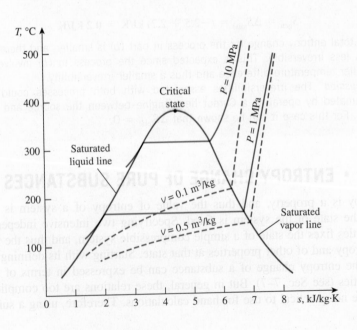

**FIGURE 7–11**

Schematic of the $T$-$s$ diagram for water.

parallel to the constant-temperature lines in the saturated liquid–vapor mixture region. Also, the constant-pressure lines almost coincide with the saturated liquid line in the compressed liquid region.

### ■ EXAMPLE 7–3    Entropy Change of a Substance in a Tank

A rigid tank contains 5 kg of refrigerant-134a initially at 20°C and 140 kPa. The refrigerant is now cooled while being stirred until its pressure drops to 100 kPa. Determine the entropy change of the refrigerant during this process.

**SOLUTION**   The refrigerant in a rigid tank is cooled while being stirred. The entropy change of the refrigerant is to be determined.

**Assumptions**   The volume of the tank is constant and thus $v_2 = v_1$.

**Analysis**   We take the refrigerant in the tank as the system (Fig. 7–12). This is a *closed system* since no mass crosses the system boundary during the process. We note that the change in entropy of a substance during a process is simply the difference between the entropy values at the final and initial states. The initial state of the refrigerant is completely specified.

Recognizing that the specific volume remains constant during this process, the properties of the refrigerant at both states are

State 1:    $\left. \begin{array}{l} P_1 = 140 \text{ kPa} \\ T_1 = 20°C \end{array} \right\}$    $\begin{array}{l} s_1 = 1.0625 \text{ kJ/kg·K} \\ v_1 = 0.16544 \text{ m}^3/\text{kg} \end{array}$

State 2:    $\left. \begin{array}{l} P_2 = 100 \text{ kPa} \\ (v_2 = v_1) \end{array} \right\}$    $\begin{array}{l} v_f = 0.0007258 \text{ m}^3/\text{kg} \\ v_g = 0.19255 \text{ m}^3/\text{kg} \end{array}$

The refrigerant is a saturated liquid–vapor mixture at the final state since $v_f < v_2 < v_g$ at 100 kPa pressure. Therefore, we need to determine the quality first:

$$x_2 = \frac{v_2 - v_f}{v_{fg}} = \frac{0.16544 - 0.0007258}{0.19255 - 0.0007258} = 0.859$$

Thus,

$$s_2 = s_f + x_2 s_{fg} = 0.07182 + (0.859)(0.88008) = 0.8278 \text{ kJ/kg·K}$$

Then, the entropy change of the refrigerant during this process is

$$\Delta S = m(s_2 - s_1) = (5 \text{ kg})(0.8278 - 1.0625) \text{ kJ/kg·K}$$

$$= -1.173 \text{ kJ/K}$$

**Discussion**   The negative sign indicates that the entropy of the system is decreasing during this process. This is not a violation of the second law, however, since it is the *entropy generation* $S_{gen}$ that cannot be negative.

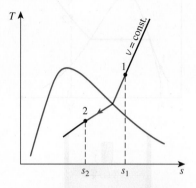

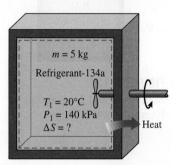

**FIGURE 7–12**
Schematic and *T-s* diagram for Example 7–3.

### ■ EXAMPLE 7–4    Entropy Change during a Constant-Pressure Process

A piston–cylinder device initially contains 1.5 kg of liquid water at 150 kPa and 20°C. The water is now heated at constant pressure by the addition of 4000 kJ of heat. Determine the entropy change of the water during this process.

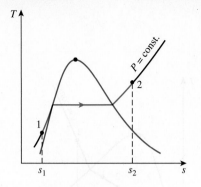

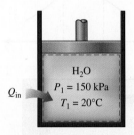

**FIGURE 7–13**
Schematic and *T-s* diagram for Example 7–4.

**SOLUTION** Liquid water in a piston–cylinder device is heated at constant pressure. The entropy change of water is to be determined.

**Assumptions** **1** The tank is stationary and thus the kinetic and potential energy changes are zero, $\Delta KE = \Delta PE = 0$. **2** The process is quasi-equilibrium. **3** The pressure remains constant during the process and thus $P_2 = P_1$.

**Analysis** We take the water in the cylinder as the *system* (Fig. 7–13). This is a *closed system* since no mass crosses the system boundary during the process. We note that a piston–cylinder device typically involves a moving boundary and thus boundary work $W_b$. Also, heat is transferred to the system.

Water exists as a compressed liquid at the initial state since its pressure is greater than the saturation pressure of 2.3392 kPa at 20°C. By approximating the compressed liquid as a saturated liquid at the given temperature, the properties at the initial state are

$$\text{State 1:} \quad \left. \begin{array}{l} P_1 = 150 \text{ kPa} \\ T_1 = 20°C \end{array} \right\} \quad \begin{array}{l} s_1 \cong s_{f@\,20°C} = 0.2965 \text{ kJ/kg·K} \\ h_1 \cong h_{f@\,20°C} = 83.915 \text{ kJ/kg} \end{array}$$

At the final state, the pressure is still 150 kPa, but we need one more property to fix the state. This property is determined from the energy balance,

$$\underbrace{E_{\text{in}} - E_{\text{out}}}_{\substack{\text{Net energy transfer} \\ \text{by heat, work, and mass}}} = \underbrace{\Delta E_{\text{system}}}_{\substack{\text{Change in internal, kinetic,} \\ \text{potential, etc., energies}}}$$

$$Q_{\text{in}} - W_b = \Delta U$$

$$Q_{\text{in}} = \Delta H = m(h_2 - h_1)$$

$$4000 \text{ kJ} = (1.5 \text{ kg})(h_2 - 83.915 \text{ kJ/kg})$$

$$h_2 = 2750.6 \text{ kJ/kg}$$

since $\Delta U + W_b = \Delta H$ for a constant-pressure quasi-equilibrium process. Then,

$$\text{State 2:} \quad \left. \begin{array}{l} P_2 = 150 \text{ kPa} \\ h_2 = 2750.6 \text{ kJ/kg} \end{array} \right\} \quad \begin{array}{l} s_2 = 7.3674 \text{ kJ/kg·K} \\ \text{(Table A-6, interpolation)} \end{array}$$

Therefore, the entropy change of water during this process is

$$\Delta S = m(s_2 - s_1) = (1.5 \text{ kg})(7.3674 - 0.2965) \text{ kJ/kg·K}$$
$$= \textbf{10.61 kJ/K}$$

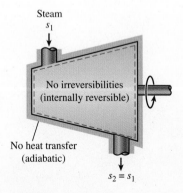

**FIGURE 7–14**
During an internally reversible, adiabatic (isentropic) process, the entropy remains constant.

# 7–4 · ISENTROPIC PROCESSES

We mentioned earlier that the entropy of a fixed mass can be changed by (1) heat transfer and (2) irreversibilities. Then it follows that the entropy of a fixed mass does not change during a process that is *internally reversible* and *adiabatic* (Fig. 7–14). A process during which the entropy remains constant is called an **isentropic process**. It is characterized by

*Isentropic process:* $\qquad \Delta s = 0 \quad \text{or} \quad s_2 = s_1 \quad$ (kJ/kg·K) $\qquad$ **(7–13)**

That is, a substance will have the same entropy value at the end of the process as it does at the beginning if the process is carried out in an isentropic manner.

Many engineering systems or devices such as pumps, turbines, nozzles, and diffusers are essentially adiabatic in their operation, and they perform best when the irreversibilities, such as the friction associated with the process, are minimized. Therefore, an isentropic process can serve as an appropriate model for actual processes. Also, isentropic processes enable us to define efficiencies for processes to compare the actual performance of these devices to the performance under idealized conditions.

It should be recognized that a *reversible adiabatic* process is necessarily isentropic ($s_2 = s_1$), but an *isentropic* process is not necessarily a reversible adiabatic process. (The entropy increase of a substance during a process as a result of irreversibilities may be offset by a decrease in entropy as a result of heat losses, for example.) However, the term *isentropic process* is customarily used in thermodynamics to imply an *internally reversible, adiabatic process*.

## EXAMPLE 7–5    Isentropic Expansion of Steam in a Turbine

Steam enters an adiabatic turbine at 5 MPa and 450°C and leaves at a pressure of 1.4 MPa. Determine the work output of the turbine per unit mass of steam if the process is reversible.

**SOLUTION**    Steam is expanded in an adiabatic turbine to a specified pressure in a reversible manner. The work output of the turbine is to be determined.
**Assumptions**    1 This is a steady-flow process since there is no change with time at any point and thus $\Delta m_{CV} = 0$, $\Delta E_{CV} = 0$, and $\Delta S_{CV} = 0$. 2 The process is reversible. 3 Kinetic and potential energies are negligible. 4 The turbine is adiabatic and thus there is no heat transfer.
**Analysis**    We take the *turbine* as the system (Fig. 7–15). This is a *control volume* since mass crosses the system boundary during the process. We note that there is only one inlet and one exit, and thus $\dot{m}_1 = \dot{m}_2 = \dot{m}$.

The power output of the turbine is determined from the rate form of the energy balance,

$$\underbrace{\dot{E}_{in} - \dot{E}_{out}}_{\substack{\text{Rate of net energy transfer} \\ \text{by heat, work, and mass}}} = \underbrace{dE_{system}/dt}_{\substack{\text{Rate of change in internal, kinetic,} \\ \text{potential, etc., energies}}}^{\nearrow^{0 \text{ (steady)}}} = 0$$

$$\dot{E}_{in} = \dot{E}_{out}$$

$$\dot{m}h_1 = \dot{W}_{out} + \dot{m}h_2 \quad (\text{since } \dot{Q} = 0, \text{ke} \cong \text{pe} \cong 0)$$

$$\dot{W}_{out} = \dot{m}(h_1 - h_2)$$

The inlet state is completely specified since two properties are given. But only one property (pressure) is given at the final state, and we need one more property to fix it. The second property comes from the observation that the process is reversible and adiabatic, and thus isentropic. Therefore, $s_2 = s_1$, and

*State 1:* $\left. \begin{array}{l} P_1 = 5 \text{ MPa} \\ T_1 = 450°C \end{array} \right\}$ $\begin{array}{l} h_1 = 3317.2 \text{ kJ/kg} \\ s_1 = 6.8210 \text{ kJ/kg·K} \end{array}$

*State 2:* $\left. \begin{array}{l} P_2 = 1.4 \text{ MPa} \\ s_2 = s_1 \end{array} \right\}$ $h_2 = 2967.4 \text{ kJ/kg}$

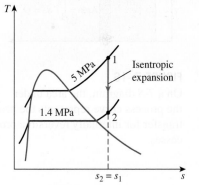

$P_1 = 5 \text{ MPa}$
$T_1 = 450°C$

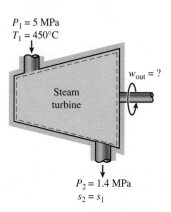

$w_{out} = ?$

Steam turbine

$P_2 = 1.4 \text{ MPa}$
$s_2 = s_1$

**FIGURE 7–15**
Schematic and *T-s* diagram for Example 7–5.

Then, the work output of the turbine per unit mass of the steam becomes

$$w_{out} = h_1 - h_2 = 3317.2 - 2967.4 = \textbf{349.8 kJ/kg}$$

## 7–5 · PROPERTY DIAGRAMS INVOLVING ENTROPY

Property diagrams serve as great visual aids in the thermodynamic analysis of processes. We have used $P$-$v$ and $T$-$v$ diagrams extensively in previous chapters in conjunction with the first law of thermodynamics. In the second-law analysis, it is very helpful to plot the processes on diagrams for which one of the coordinates is entropy. The two diagrams commonly used in the second-law analysis are the *temperature-entropy* and the *enthalpy-entropy* diagrams.

Consider the defining equation of entropy (Eq. 7–4). It can be rearranged as

$$\delta Q_{\text{int rev}} = T\,dS \quad \text{(kJ)} \tag{7–14}$$

As shown in Fig. 7–16, $\delta Q_{\text{rev int}}$ corresponds to a differential area on a $T$-$S$ diagram. The total heat transfer during an internally reversible process is determined by integration to be

$$Q_{\text{int rev}} = \int_1^2 T\,dS \quad \text{(kJ)} \tag{7–15}$$

which corresponds to the area under the process curve on a $T$-$S$ diagram. Therefore, we conclude that *the area under the process curve on a $T$-$S$ diagram represents heat transfer during an internally reversible process.* This is somewhat analogous to reversible boundary work being represented by the area under the process curve on a $P$-$V$ diagram. Note that the area under the process curve represents heat transfer for processes that are internally (or totally) reversible. The area has no meaning for irreversible processes.

Equations 7–14 and 7–15 can also be expressed on a unit-mass basis as

$$\delta q_{\text{int rev}} = T\,ds \quad \text{(kJ/kg)} \tag{7–16}$$

and

$$q_{\text{int rev}} = \int_1^2 T\,ds \quad \text{(kJ/kg)} \tag{7–17}$$

To perform the integrations in Eqs. 7–15 and 7–17, one needs to know the relationship between $T$ and $s$ during a process. One special case for which these integrations can be performed easily is the *internally reversible isothermal process.* It yields

$$Q_{\text{int rev}} = T_0\,\Delta S \quad \text{(kJ)} \tag{7–18}$$

or

$$q_{\text{int rev}} = T_0\,\Delta s \quad \text{(kJ/kg)} \tag{7–19}$$

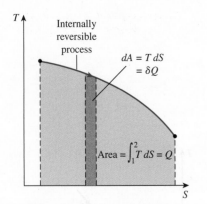

**FIGURE 7–16**
On a $T$-$S$ diagram, the area under the process curve represents the heat transfer for internally reversible processes.

where $T_0$ is the constant temperature and $\Delta S$ is the entropy change of the system during the process.

An isentropic process on a $T$-$s$ diagram is easily recognized as a *vertical-line segment*. This is expected since an isentropic process involves no heat transfer, and therefore the area under the process path must be zero (Fig. 7–17). The $T$-$s$ diagrams serve as valuable tools for visualizing the second-law aspects of processes and cycles, and thus they are frequently used in thermodynamics. The $T$-$s$ diagram of water is given in the appendix in Fig. A–9.

Another diagram commonly used in engineering is the enthalpy-entropy diagram, which is quite valuable in the analysis of steady-flow devices such as turbines, compressors, and nozzles. The coordinates of an $h$-$s$ diagram represent two properties of major interest: enthalpy, which is a primary property in the first-law analysis of the steady-flow devices, and entropy, which is the property that accounts for irreversibilities during adiabatic processes. In analyzing the steady flow of steam through an adiabatic turbine, for example, the vertical distance between the inlet and the exit states $\Delta h$ is a measure of the work output of the turbine, and the horizontal distance $\Delta s$ is a measure of the irreversibilities associated with the process (Fig. 7–18).

The $h$-$s$ diagram is also called a **Mollier diagram** after the German scientist R. Mollier (1863–1935). An $h$-$s$ diagram is given in the appendix for steam in Fig. A–10.

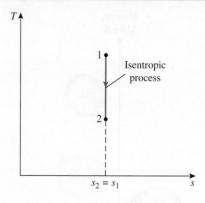

**FIGURE 7–17**
The isentropic process appears as a vertical line segment on a $T$-$s$ diagram.

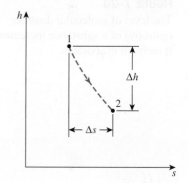

**FIGURE 7–18**
For adiabatic steady-flow devices, the vertical distance $\Delta h$ on an $h$-$s$ diagram is a measure of work, and the horizontal distance $\Delta s$ is a measure of irreversibilities.

---

■ **EXAMPLE 7–6**    **The $T$-$S$ Diagram of the Carnot Cycle**

■ Show the Carnot cycle on a $T$-$S$ diagram and indicate the areas that repre-
■ sent the heat supplied $Q_H$, heat rejected $Q_L$, and the net work output $W_{net,out}$
on this diagram.

**SOLUTION**   The Carnot cycle is to be shown on a $T$-$S$ diagram, and the areas that represent $Q_H$, $Q_L$, and $W_{net,out}$ are to be indicated.
*Analysis*   Recall that the Carnot cycle is made up of two reversible isothermal ($T$ = constant) processes and two isentropic ($s$ = constant) processes. These four processes form a rectangle on a $T$-$S$ diagram, as shown in Fig. 7–19.

On a $T$-$S$ diagram, the area under the process curve represents the heat transfer for that process. Thus the area $A12B$ represents $Q_H$, the area $A43B$ represents $Q_L$, and the difference between these two (the area in color) represents the net work since

$$W_{net,out} = Q_H - Q_L$$

Therefore, the area enclosed by the path of a cycle (area 1234) on a $T$-$S$ diagram represents the net work. Recall that the area enclosed by the path of a cycle also represents the net work on a $P$-$V$ diagram.

---

# 7–6 ■ WHAT IS ENTROPY?

It is clear from the previous discussion that entropy is a useful property and serves as a valuable tool in the second-law analysis of engineering devices. But this does not mean that we know and understand entropy well. Because we do not. In fact, we cannot even give an adequate answer to the question,

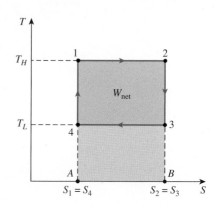

**FIGURE 7–19**
The $T$-$S$ diagram of a Carnot cycle
(Example 7–6).

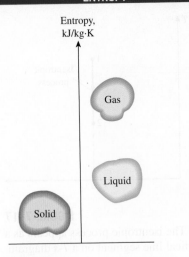

Entropy,
kJ/kg·K

Gas

Liquid

Solid

**FIGURE 7–20**
The level of molecular disorder
(entropy) of a substance increases as
it melts or evaporates.

what is entropy? Not being able to describe entropy fully, however, does not take anything away from its usefulness. We could not define *energy* either, but it did not interfere with our understanding of energy transformations and the conservation of energy principle. Granted, entropy is not a household word like energy. But with continued use, our understanding of entropy will deepen, and our appreciation of it will grow. The next discussion should shed some light on the physical meaning of entropy by considering the microscopic nature of matter.

Entropy can be viewed as a measure of *molecular disorder,* or *molecular randomness.* As a system becomes more disordered, the positions of the molecules become less predictable and the entropy increases. Thus, it is not surprising that the entropy of a substance is lowest in the solid phase and highest in the gas phase (Fig. 7–20). In the solid phase, the molecules of a substance continually oscillate about their equilibrium positions, but they cannot move relative to each other, and their position at any instant can be predicted with good certainty. In the gas phase, however, the molecules move about at random, collide with each other, and change direction, making it extremely difficult to predict accurately the microscopic state of a system at any instant. Associated with this molecular chaos is a high value of entropy.

When viewed microscopically from a statistical thermodynamics point of view, an isolated system that appears to be at a state of equilibrium actually exhibits a high level of activity because of the continual motion of the molecules. To each state of macroscopic equilibrium, there corresponds a large number of molecular microscopic states or molecular configurations. Boltzmann first hypothesized that the entropy of a system at a specified macrostate is related to the total number of possible relevant microstates of that system, $W$ (from *Wahrscheinlichkeit,* the German word for 'probability'). This thought was formulated later by Plank using a constant $k$ with the entropy unit of J/K named after Boltzmann (and incribed on Boltzmann's tombstone) as

$$S = k \ln W \tag{7–20a}$$

which is known as the **Boltzmann relation.** The thermal motion randomness or disorder, as related to entropy is later generalized by Gibbs as a measure of the sum of all microstates' uncertainties, i.e., probabilities, as

$$S = -k \sum p_i \log p_i \tag{7-20b}$$

**Gibbs' formulation** is more general since it allows for non-uniform probability, $p_i$, of microstates. With an increase of particle momenta or thermal disorder and volume occupied, more information is required for the characterization of the system, relative to more ordered systems. Gibbs' formulation reduces to Boltzmann relation for equi-probable, uniform probability of all $W$ microstates since $p_i = 1/W = $ constant $<< 1$.

From a microscopic point of view, the entropy of a system increases whenever the thermal randomness or disorder (i.e., the number of possible relevant molecular microstates corresponding to a given bulk macrostate) of a system increases. Thus, entropy can be viewed as a measure of thermal randomness or molecular disorder, which increases anytime an isolated system undergoes a process.

As mentioned earlier, the molecules of a substance in solid phase continually oscillate, creating an uncertainty about their position. These oscillations, however, fade as the temperature is decreased, and the molecules supposedly become motionless at absolute zero. This represents a state of ultimate molecular order (and minimum energy). Therefore, *the entropy of a pure crystalline substance at absolute zero temperature is zero* since there is no uncertainty about the state of the molecules at that instant (Fig. 7–21). This statement is known as the **third law of thermodynamics**. The third law of thermodynamics provides an absolute reference point for the determination of entropy. The entropy determined relative to this point is called **absolute entropy**, and it is extremely useful in the thermodynamic analysis of chemical reactions. Notice that the entropy of a substance that is not pure crystalline (such as a solid solution) is not zero at absolute zero temperature. This is because more than one molecular configuration exists for such substances, which introduces some uncertainty about the microscopic state of the substance.

Molecules in the gas phase possess a considerable amount of kinetic energy. However, we know that no matter how large their kinetic energies are, the gas molecules do not rotate a paddle wheel inserted into the container and produce work. This is because the gas molecules, and the energy they possess, are disorganized. Probably the number of molecules trying to rotate the wheel in one direction at any instant is equal to the number of molecules that are trying to rotate it in the opposite direction, causing the wheel to remain motionless. Therefore, we cannot extract any useful work directly from disorganized energy (Fig. 7–22).

Now consider a rotating shaft shown in Fig. 7–23. This time the energy of the molecules is completely organized since the molecules of the shaft are rotating in the same direction together. This organized energy can readily be used to perform useful tasks such as raising a weight or generating electricity. Being an organized form of energy, work is free of disorder or randomness and thus free of entropy. *There is no entropy transfer associated with energy transfer as work.* Therefore, in the absence of any friction, the process of raising a weight by a rotating shaft (or a flywheel) does not produce any entropy. Any process that does not produce a net entropy is reversible, and thus the process just described can be reversed by lowering the weight. Therefore, energy is not degraded during this process, and no potential to do work is lost.

Instead of raising a weight, let us operate the paddle wheel in a container filled with a gas, as shown in Fig. 7–24. The paddle-wheel work in this case is converted to the internal energy of the gas, as evidenced by a rise in gas temperature, creating a higher level of molecular disorder in the container. This process is quite different from raising a weight since the organized paddle-wheel energy is now converted to a highly disorganized form of energy, which cannot be converted back to the paddle wheel as the rotational kinetic energy. Only a portion of this energy can be converted to work by partially reorganizing it through the use of a heat engine. Therefore, energy is degraded during this process, the ability to do work is

Pure crystal
$T = 0$ K
Entropy = 0

**FIGURE 7–21**
A pure crystalline substance at absolute zero temperature is in perfect order, and its entropy is zero (the third law of thermodynamics).

**FIGURE 7–22**
Disorganized energy does not create much useful effect, no matter how large it is.

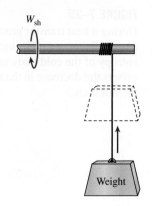

$W_{sh}$

Weight

**FIGURE 7–23**
In the absence of friction, raising a weight by a rotating shaft does not create any disorder (entropy), and thus energy is not degraded during this process.

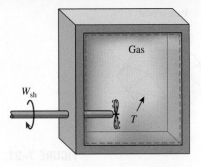

**FIGURE 7–24**
The paddle-wheel work done on a gas increases the level of disorder (entropy) of the gas, and thus energy is degraded during this process.

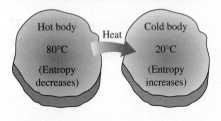

**FIGURE 7–25**
During a heat transfer process, the net entropy increases. (The increase in the entropy of the cold body more than offsets the decrease in the entropy of the hot body.)

reduced, molecular disorder is produced, and associated with all this is an increase in entropy.

The *quantity* of energy is always preserved during an actual process (the first law), but the *quality* is bound to decrease (the second law). This decrease in quality is always accompanied by an increase in entropy. As an example, consider the transfer of 10 kJ of energy as heat from a hot medium to a cold one. At the end of the process, we still have the 10 kJ of energy, but at a lower temperature and thus at a lower quality.

Heat is, in essence, a form of *disorganized energy*, and some disorganization (entropy) flows with heat (Fig. 7–25). As a result, the entropy and the level of molecular disorder or randomness of the hot body decreases with the entropy and the level of molecular disorder of the cold body increases. The second law requires that the increase in entropy of the cold body be greater than the decrease in entropy of the hot body, and thus the net entropy of the combined system (the cold body and the hot body) increases. That is, the combined system is at a state of greater disorder at the final state. Thus we can conclude that processes can occur only in the direction of increased overall entropy or molecular disorder. That is, the entire universe is getting more and more chaotic every day.

## Entropy and Entropy Generation in Daily Life

The concept of entropy can also be applied to other areas. Entropy can be viewed as a measure of disorder or disorganization in a system. Likewise, entropy generation can be viewed as a measure of disorder or disorganization generated during a process. The concept of entropy is not used in daily life nearly as extensively as the concept of energy, even though entropy is readily applicable to various aspects of daily life. The extension of the entropy concept to nontechnical fields is not a novel idea. It has been the topic of several articles, and even some books. Next, we present several ordinary events and show their relevance to the concept of entropy and entropy generation.

Efficient people lead low-entropy (highly organized) lives. They have a place for everything (minimum uncertainty), and it takes minimum energy for them to locate something. Inefficient people, on the other hand, are disorganized and lead high-entropy lives. It takes them minutes (if not hours) to find something they need, and they are likely to create a bigger disorder as they are searching since they will probably conduct the search in a disorganized manner. People leading high-entropy lifestyles are always on the run, and never seem to catch up.

Consider two identical buildings, each containing one million books. In the first building, the books are *piled* on top of each other, whereas in the second building they are *highly organized, shelved, and indexed* for easy reference. There is no doubt about which building a student will prefer to go to for checking out a certain book. Yet, some may argue from the first-law point of view that these two buildings are equivalent since the mass and knowledge content of the two buildings are identical, despite the high level of disorganization (entropy) in the first building. This example illustrates that any realistic comparisons should involve the second-law point of view.

Two *textbooks* that seem to be identical because both cover basically the same topics and present the same information may actually be *very* different depending on *how* they cover the topics. After all, two seemingly identical cars are not so identical if one goes only half as many miles as the other one on the same amount of fuel. Likewise, two seemingly identical books are not so identical if it takes twice as long to learn a topic from one of them as it does from the other. Thus, comparisons made on the basis of the first law only may be highly misleading.

Having a disorganized (high-entropy) *army* is like having no army at all. It is no coincidence that the command centers of any armed forces are among the primary targets during a war. One army that consists of 10 divisions is 10 times more powerful than 10 armies each consisting of a single division. Likewise, one country that consists of 10 states is more powerful than 10 countries, each consisting of a single state. The *United States* would not be such a powerful country if there were 50 independent countries in its place instead of a single country with 50 states. The European Union has the potential to be a new economic and political superpower. The old cliché "divide and conquer" can be rephrased as "increase the entropy and conquer."

We know that mechanical friction is always accompanied by entropy generation, and thus reduced performance. We can generalize this to daily life: *friction in the workplace* with fellow workers is bound to generate entropy, and thus adversely affect performance (Fig. 7–26). It results in reduced productivity.

We also know that *unrestrained expansion* (or explosion) and uncontrolled electron exchange (chemical reactions) generate entropy and are highly irreversible. Likewise, unrestrained opening of the mouth to scatter angry words is highly irreversible since this generates entropy, and it can cause considerable damage. A person who gets up in anger is bound to sit down at a loss. Hopefully, someday we will be able to come up with some procedures to quantify entropy generated during nontechnical activities, and maybe even pinpoint its primary sources and magnitude.

**FIGURE 7–26**
As in mechanical systems, friction in the workplace is bound to generate entropy and reduce performance.
© *PhotoLink/Getty Images RF*

# 7–7 · THE *T ds* RELATIONS

Recall that the quantity $(\delta Q/T)_{\text{int rev}}$ corresponds to a differential change in the property *entropy*. The entropy change for a process, then, can be evaluated by integrating $\delta Q/T$ along some imaginary internally reversible path between the actual end states. For isothermal internally reversible processes, this integration is straightforward. But when the temperature varies during the process, we have to have a relation between $\delta Q$ and $T$ to perform this integration. Finding such relations is what we intend to do in this section.

The differential form of the conservation of energy equation for a closed stationary system (a fixed mass) containing a simple compressible substance can be expressed for an internally reversible process as

$$\delta Q_{\text{int rev}} - \delta W_{\text{int rev,out}} = dU \qquad (7\text{–}21)$$

But

$$\delta Q_{\text{int rev}} = T\,dS$$

$$\delta W_{\text{int rev,out}} = P\,dV$$

Thus,

$$T\,dS = dU + P\,dV \quad \text{(kJ)} \tag{7-22}$$

or

$$T\,ds = du + P\,dv \quad \text{(kJ/kg)} \tag{7-23}$$

This equation is known as the first $T\,ds$, or *Gibbs, equation*. Notice that the only type of work interaction a simple compressible system may involve as it undergoes an internally reversible process is the boundary work.

The second $T\,ds$ equation is obtained by eliminating $du$ from Eq. 7–23 by using the definition of enthalpy ($h = u + Pv$):

$$\left. \begin{array}{l} h = u + Pv \quad \longrightarrow \quad dh = du + P\,dv + v\,dP \\ \text{(Eq. 7--23)} \quad \longrightarrow \quad T\,ds = du + P\,dv \end{array} \right\} \; T\,ds = dh - v\,dP \tag{7-24}$$

Equations 7–23 and 7–24 are extremely valuable since they relate entropy changes of a system to the changes in other properties. Unlike Eq. 7–4, they are property relations and therefore are independent of the type of the processes.

These $T\,ds$ relations are developed with an internally reversible process in mind since the entropy change between two states must be evaluated along a reversible path. However, the results obtained are valid for both reversible and irreversible processes since entropy is a property and the change in a property between two states is independent of the type of process the system undergoes. Equations 7–23 and 7–24 are relations between the properties of a unit mass of a simple compressible system as it undergoes a change of state, and they are applicable whether the change occurs in a closed or an open system (Fig. 7–27).

Explicit relations for differential changes in entropy are obtained by solving for $ds$ in Eqs. 7–23 and 7–24:

$$ds = \frac{du}{T} + \frac{P\,dv}{T} \tag{7-25}$$

and

$$ds = \frac{dh}{T} - \frac{v\,dP}{T} \tag{7-26}$$

The entropy change during a process can be determined by integrating either of these equations between the initial and the final states. To perform these integrations, however, we must know the relationship between $du$ or $dh$ and the temperature (such as $du = c_v\,dT$ and $dh = c_p\,dT$ for ideal gases) as well as the equation of state for the substance (such as the ideal-gas equation of state $Pv = RT$). For substances for which such relations exist, the integration of Eq. 7–25 or 7–26 is straightforward. For other substances, we have to rely on tabulated data.

The $T\,ds$ relations for nonsimple systems, that is, systems that involve more than one mode of quasi-equilibrium work, can be obtained in a similar manner by including all the relevant quasi-equilibrium work modes.

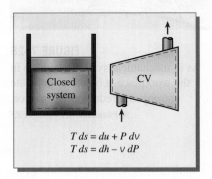

**FIGURE 7–27**

The $T\,ds$ relations are valid for both reversible and irreversible processes and for both closed and open systems.

# 7–8 · ENTROPY CHANGE OF LIQUIDS AND SOLIDS

Recall that liquids and solids can be approximated as *incompressible substances* since their specific volumes remain nearly constant during a process. Thus, $dv \cong 0$ for liquids and solids, and Eq. 7–25 for this case reduces to

$$ds = \frac{du}{T} = \frac{c\,dT}{T} \tag{7–27}$$

since $c_p = c_v = c$ and $du = c\,dT$ for incompressible substances. Then the entropy change during a process is determined by integration to be

*Liquids, solids:*     $s_2 - s_1 = \displaystyle\int_1^2 c(T)\frac{dT}{T} \cong c_{avg} \ln\frac{T_2}{T_1}$   (kJ/kg·K)     (7–28)

where $c_{avg}$ is the *average* specific heat of the substance over the given temperature interval. Note that the entropy change of a truly incompressible substance depends on temperature only and is independent of pressure.

Equation 7–28 can be used to determine the entropy changes of solids and liquids with reasonable accuracy. However, for liquids that expand considerably with temperature, it may be necessary to consider the effects of volume change in calculations. This is especially the case when the temperature change is large.

A relation for isentropic processes of liquids and solids is obtained by setting the entropy change relation above equal to zero. It gives

*Isentropic:*     $s_2 - s_1 = c_{avg} \ln\dfrac{T_2}{T_1} = 0 \quad\longrightarrow\quad T_2 = T_1$     (7–29)

That is, the temperature of a truly incompressible substance remains constant during an isentropic process. Therefore, the isentropic process of an incompressible substance is also isothermal. This behavior is closely approximated by liquids and solids.

---

■ **EXAMPLE 7–7     Effect of Density of a Liquid on Entropy**

■ Liquid methane is commonly used in various cryogenic applications. The critical temperature of methane is 191 K (or −82°C), and thus methane must be maintained below 191 K to keep it in liquid phase. The properties of liquid methane at various temperatures and pressures are given in Table 7–1. Determine the entropy change of liquid methane as it undergoes a process from 110 K and 1 MPa to 120 K and 5 MPa (a) using tabulated properties and (b) approximating liquid methane as an incompressible substance. What is the error involved in the latter case?

**SOLUTION** Liquid methane undergoes a process between two specified states. The entropy change of methane is to be determined by using actual data and by assuming methane to be incompressible.

*Analysis* (a) We consider a unit mass of liquid methane (Fig. 7–28). The properties of the methane at the initial and final states are

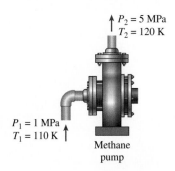

$P_2 = 5$ MPa
$T_2 = 120$ K

$P_1 = 1$ MPa
$T_1 = 110$ K

Methane pump

**FIGURE 7–28**
Schematic for Example 7–7.

## TABLE 7–1

Properties of liquid methane

| Temp., $T$, K | Pressure, $P$, MPa | Density, $\rho$, kg/m³ | Enthalpy, $h$, kJ/kg | Entropy, $s$, kJ/kg·K | Specific heat, $c_p$, kJ/kg·K |
|---|---|---|---|---|---|
| 110 | 0.5 | 425.3 | 208.3 | 4.878 | 3.476 |
|  | 1.0 | 425.8 | 209.0 | 4.875 | 3.471 |
|  | 2.0 | 426.6 | 210.5 | 4.867 | 3.460 |
|  | 5.0 | 429.1 | 215.0 | 4.844 | 3.432 |
| 120 | 0.5 | 410.4 | 243.4 | 5.185 | 3.551 |
|  | 1.0 | 411.0 | 244.1 | 5.180 | 3.543 |
|  | 2.0 | 412.0 | 245.4 | 5.171 | 3.528 |
|  | 5.0 | 415.2 | 249.6 | 5.145 | 3.486 |

*State 1:* $\left.\begin{array}{l} P_1 = 1 \text{ MPa} \\ T_1 = 110 \text{ K} \end{array}\right\}$ $\begin{array}{l} s_1 = 4.875 \text{ kJ/kg·K} \\ c_{p1} = 3.471 \text{ kJ/kg·K} \end{array}$

*State 2:* $\left.\begin{array}{l} P_2 = 5 \text{ MPa} \\ T_2 = 120 \text{ K} \end{array}\right\}$ $\begin{array}{l} s_2 = 5.145 \text{ kJ/kg·K} \\ c_{p2} = 3.486 \text{ kJ/kg·K} \end{array}$

Therefore,

$$\Delta s = s_2 - s_1 = 5.145 - 4.875 = \mathbf{0.270 \ kJ/kg \cdot K}$$

(b) Approximating liquid methane as an incompressible substance, its entropy change is determined to be

$$\Delta s = c_{\text{avg}} \ln \frac{T_2}{T_1} = (3.4785 \text{ kJ/kg·K}) \ln \frac{120 \text{ K}}{110 \text{ K}} = \mathbf{0.303 \ kJ/kg \cdot K}$$

since

$$c_{\text{avg}} = \frac{c_{p1} + c_{p2}}{2} = \frac{3.471 + 3.486}{2} = 3.4785 \text{ kJ/kg·K}$$

Therefore, the error involved in approximating liquid methane as an incompressible substance is

$$\text{Error} = \frac{|\Delta s_{\text{actual}} - \Delta s_{\text{ideal}}|}{\Delta s_{\text{actual}}} = \frac{|0.270 - 0.303|}{0.270} = \mathbf{0.122} \text{ (or 12.2\%)}$$

*Discussion* This result is not surprising since the density of liquid methane changes during this process from 425.8 to 415.2 kg/m³ (about 3 percent), which makes us question the validity of the incompressible substance assumption. Still, this assumption enables us to obtain reasonably accurate results with less effort, which proves to be very convenient in the absence of compressed liquid data.

## EXAMPLE 7–8    Economics of Replacing a Valve by a Turbine

A cryogenic manufacturing facility handles liquid methane at 115 K and 5 MPa at a rate of 0.280 m³/s . A process requires dropping the pressure of liquid methane to 1 MPa, which is done by throttling the liquid methane by passing it through a flow resistance such as a valve. A recently hired engineer proposes to replace the throttling valve by a turbine in order to produce power while dropping the pressure to 1 MPa. Using data from Table 7–1, determine the maximum amount of power that can be produced by such a turbine. Also, determine how much this turbine will save the facility from electricity usage costs per year if the turbine operates continuously (8760 h/yr) and the facility pays $0.075/kWh for electricity.

**SOLUTION**   Liquid methane is expanded in a turbine to a specified pressure at a specified rate. The maximum power that this turbine can produce and the amount of money it can save per year are to be determined.
**Assumptions**   **1** This is a steady-flow process since there is no change with time at any point and thus $\Delta m_{CV} = 0$, $\Delta E_{CV} = 0$, and $\Delta S_{CV} = 0$. **2** The turbine is adiabatic and thus there is no heat transfer. **3** The process is reversible. **4** Kinetic and potential energies are negligible.
**Analysis**   We take the *turbine* as the system (Fig. 7–29). This is a *control volume* since mass crosses the system boundary during the process. We note that there is only one inlet and one exit and thus $\dot{m}_1 = \dot{m}_2 = \dot{m}$.

The assumptions above are reasonable since a turbine is normally well insulated and it must involve no irreversibilities for best performance and thus *maximum* power production. Therefore, the process through the turbine must be *reversible adiabatic* or *isentropic*. Then, $s_2 = s_1$ and

State 1: $\left. \begin{array}{l} P_1 = 5 \text{ MPa} \\ T_1 = 115 \text{ K} \end{array} \right\}$ $\begin{array}{l} h_1 = 232.3 \text{ kJ/kg} \\ s_1 = 4.9945 \text{ kJ/kg·K} \\ \rho_1 = 422.15 \text{ kg/m}^3 \end{array}$

State 2: $\left. \begin{array}{l} P_2 = 1 \text{ MPa} \\ s_2 = s_1 \end{array} \right\}$ $h_2 = 222.8 \text{ kJ/kg}$

Also, the mass flow rate of liquid methane is

$$\dot{m} = \rho_1 \dot{V}_1 = (422.15 \text{ kg/m}^3)(0.280 \text{ m}^3/\text{s}) = 118.2 \text{ kg/s}$$

Then the power output of the turbine is determined from the rate form of the energy balance to be

$$\underbrace{\dot{E}_{in} - \dot{E}_{out}}_{\substack{\text{Rate of net energy transfer} \\ \text{by heat, work, and mass}}} = \underbrace{dE_{system}/dt}_{\substack{\text{Rate of change in internal,} \\ \text{kinetic, potential, etc., energies}}}^{\nearrow 0 \text{ (steady)}} = 0$$

$$\dot{E}_{in} = \dot{E}_{out}$$
$$\dot{m}h_1 = \dot{W}_{out} + \dot{m}h_2 \quad (\text{since } \dot{Q} = 0, \text{ ke} \cong \text{ pe} \cong 0)$$
$$\dot{W}_{out} = \dot{m}(h_1 - h_2)$$
$$= (118.2 \text{ kg/s})(232.3 - 222.8) \text{ kJ/kg}$$
$$= \mathbf{1123 \text{ kW}}$$

For continuous operation (365 × 24 = 8760 h), the amount of power produced per year is

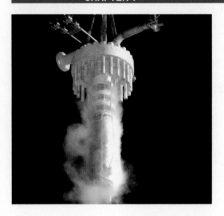

**FIGURE 7–29**
Liquefied Natural Gas (LNG)
Turbine after being removed
from an LNG tank
*Courtesy of Ebara International Corp.,*
*Cryodynamics Division, Sparks, Nevada*

Annual power production $= \dot{W}_{out} \times \Delta t = (1123 \text{ kW})(8760 \text{ h/yr})$
$$= 0.9837 \times 10^7 \text{ kWh/yr}$$

At \$0.075/kWh, the amount of money this turbine can save the facility is

Annual money savings $=$ (Annual power production)(Unit cost of power)
$$= (0.9837 \times 10^7 \text{ kWh/yr})(\$0.075/\text{kWh})$$
$$= \textbf{\$737,800/yr}$$

That is, this turbine can save the facility \$737,800 a year by simply taking advantage of the potential that is currently being wasted by a throttling valve, and the engineer who made this observation should be rewarded.

*Discussion* This example shows the importance of the property entropy since it enabled us to quantify the work potential that is being wasted. In practice, the turbine will not be isentropic, and thus the power produced will be less. The analysis above gave us the upper limit. An actual turbine-generator assembly can utilize about 80 percent of the potential and produce more than 900 kW of power while saving the facility more than \$600,000 a year.

It can also be shown that the temperature of methane drops to 113.9 K (a drop of 1.1 K) during the isentropic expansion process in the turbine instead of remaining constant at 115 K as would be the case if methane were assumed to be an incompressible substance. The temperature of methane would rise to 116.6 K (a rise of 1.6 K) during the throttling process.

# 7–9 · THE ENTROPY CHANGE OF IDEAL GASES

An expression for the entropy change of an ideal gas can be obtained from Eq. 7–25 or 7–26 by employing the property relations for ideal gases (Fig. 7–30). By substituting $du = c_v \, dT$ and $P = RT/v$ into Eq. 7–25, the differential entropy change of an ideal gas becomes

$$ds = c_v \frac{dT}{T} + R \frac{dv}{v} \tag{7–30}$$

The entropy change for a process is obtained by integrating this relation between the end states:

$$s_2 - s_1 = \int_1^2 c_v(T) \frac{dT}{T} + R \ln \frac{v_2}{v_1} \tag{7–31}$$

A second relation for the entropy change of an ideal gas is obtained in a similar manner by substituting $dh = c_p \, dT$ and $v = RT/P$ into Eq. 7–26 and integrating. The result is

$$s_2 - s_1 = \int_1^2 c_p(T) \frac{dT}{T} - R \ln \frac{P_2}{P_1} \tag{7–32}$$

The specific heats of ideal gases, with the exception of monatomic gases, depend on temperature, and the integrals in Eqs. 7–31 and 7–32 cannot be performed unless the dependence of $c_v$ and $c_p$ on temperature is known.

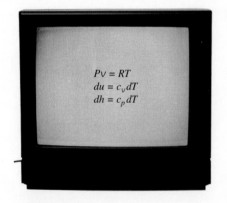

$Pv = RT$
$du = c_v \, dT$
$dh = c_p \, dT$

**FIGURE 7–30**
A broadcast from channel IG.
*© Photodisc/Getty Images RF*

Even when the $c_v(T)$ and $c_p(T)$ functions are available, performing long integrations every time entropy change is calculated is not practical. Then two reasonable choices are left: either perform these integrations by simply assuming constant specific heats or evaluate those integrals once and tabulate the results. Both approaches are presented next.

## Constant Specific Heats (Approximate Analysis)

Assuming constant specific heats for ideal gases is a common approximation, and we used this assumption before on several occasions. It usually simplifies the analysis greatly, and the price we pay for this convenience is some loss in accuracy. The magnitude of the error introduced by this assumption depends on the situation at hand. For example, for monatomic ideal gases such as helium, the specific heats are independent of temperature, and therefore the constant-specific-heat assumption introduces no error. For ideal gases whose specific heats vary almost linearly in the temperature range of interest, the possible error is minimized by using specific heat values evaluated at the average temperature (Fig. 7–31). The results obtained in this way usually are sufficiently accurate if the temperature range is not greater than a few hundred degrees.

The entropy-change relations for ideal gases under the constant-specific-heat assumption are easily obtained by replacing $c_v(T)$ and $c_p(T)$ in Eqs. 7–31 and 7–32 by $c_{v,avg}$ and $c_{p,avg}$, respectively, and performing the integrations. We obtain

$$s_2 - s_1 = c_{v,avg} \ln\frac{T_2}{T_1} + R \ln\frac{v_2}{v_1} \quad \text{(kJ/kg·K)} \qquad \text{(7–33)}$$

and

$$s_2 - s_1 = c_{p,avg} \ln\frac{T_2}{T_1} - R \ln\frac{P_2}{P_1} \quad \text{(kJ/kg·K)} \qquad \text{(7–34)}$$

Entropy changes can also be expressed on a unit-mole basis by multiplying these relations by molar mass:

$$\bar{s}_2 - \bar{s}_1 = \bar{c}_{v,avg} \ln\frac{T_2}{T_1} + R_u \ln\frac{v_2}{v_1} \quad \text{(kJ/kmol·K)} \qquad \text{(7–35)}$$

and

$$\bar{s}_2 - \bar{s}_1 = \bar{c}_{p,avg} \ln\frac{T_2}{T_1} - R_u \ln\frac{P_2}{P_1} \quad \text{(kJ/kmol·K)} \qquad \text{(7–36)}$$

## Variable Specific Heats (Exact Analysis)

When the temperature change during a process is large and the specific heats of the ideal gas vary nonlinearly within the temperature range, the assumption of constant specific heats may lead to considerable errors in entropy-change calculations. For those cases, the variation of specific heats

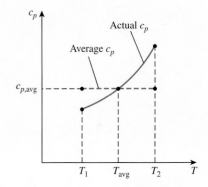

**FIGURE 7–31**
Under the constant-specific-heat assumption, the specific heat is assumed to be constant at some average value.

with temperature should be properly accounted for by utilizing accurate relations for the specific heats as a function of temperature. The entropy change during a process is then determined by substituting these $c_v(T)$ or $c_p(T)$ relations into Eq. 7–31 or 7–32 and performing the integrations.

Instead of performing these laborious integrals each time we have a new process, it is convenient to perform these integrals once and tabulate the results. For this purpose, we choose absolute zero as the reference temperature and define a function $s°$ as

$$s° = \int_0^T c_p(T) \frac{dT}{T} \qquad (7\text{–}37)$$

Obviously, $s°$ is a function of temperature alone, and its value is zero at absolute zero temperature. The values of $s°$ are calculated at various temperatures, and the results are tabulated in the appendix as a function of temperature for air. Given this definition, the integral in Eq. 7–32 becomes

$$\int_1^2 c_p(T) \frac{dT}{T} = s_2° - s_1° \qquad (7\text{–}38)$$

where $s_2°$ is the value of $s°$ at $T_2$ and $s_1°$ is the value at $T_1$. Thus,

$$s_2 - s_1 = s_2° - s_1° - R \ln \frac{P_2}{P_1} \quad \text{(kJ/kg·K)} \qquad (7\text{–}39)$$

It can also be expressed on a unit-mole basis as

$$\bar{s}_2 - \bar{s}_1 = \bar{s}_2° - \bar{s}_1° - R_u \ln \frac{P_2}{P_1} \quad \text{(kJ/kmol·K)} \qquad (7\text{–}40)$$

Note that unlike internal energy and enthalpy, the entropy of an ideal gas varies with specific volume or pressure as well as the temperature. Therefore, entropy cannot be tabulated as a function of temperature alone. The $s°$ values in the tables account for the temperature dependence of entropy (Fig. 7–32). The variation of entropy with pressure is accounted for by the last term in Eq. 7–39. Another relation for entropy change can be developed based on Eq. 7–31, but this would require the definition of another function and tabulation of its values, which is not practical.

| $T$, K | $s°$, kJ/kg·K |
|---|---|
| . | . |
| . | . |
| . | . |
| 300 | 1.70203 |
| 310 | 1.73498 |
| 320 | 1.76690 |
| . | . |
| . | . |
| . | . |

(Table A-17)

**FIGURE 7–32**

The entropy of an ideal gas depends on both $T$ and $P$. The function $s°$ represents only the temperature-dependent part of entropy.

---

**EXAMPLE 7–9    Entropy Change of an Ideal Gas**

Air is compressed from an initial state of 100 kPa and 17°C to a final state of 600 kPa and 57°C. Determine the entropy change of air during this compression process by using (a) property values from the air table and (b) average specific heats.

**SOLUTION** Air is compressed between two specified states. The entropy change of air is to be determined by using tabulated property values and also by using average specific heats.

**Assumptions** Air is an ideal gas since it is at a high temperature and low pressure relative to its critical-point values. Therefore, entropy change relations developed under the ideal-gas assumption are applicable.

**Analysis** A sketch of the system and the *T-s* diagram for the process are given in Fig. 7–33. We note that both the initial and the final states of air are completely specified.

(*a*) The properties of air are given in the air table (Table A–17). Reading $s°$ values at given temperatures and substituting, we find

$$s_2 - s_1 = s_2° - s_1° - R \ln\frac{P_2}{P_1}$$

$$= [(1.79783 - 1.66802) \text{ kJ/kg·K}] - (0.287 \text{ kJ/kg·K}) \ln\frac{600 \text{ kPa}}{100 \text{ kPa}}$$

$$= \mathbf{-0.3844 \text{ kJ/kg·K}}$$

(*b*) The entropy change of air during this process can also be determined approximately from Eq. 7–34 by using a $c_p$ value at the average temperature of 37°C (Table A–2*b*) and treating it as a constant:

$$s_2 - s_1 = c_{p,\text{avg}} \ln\frac{T_2}{T_1} - R \ln\frac{P_2}{P_1}$$

$$= (1.006 \text{ kJ/kg·K}) \ln\frac{330 \text{ K}}{290 \text{ K}} - (0.287 \text{ kJ/kg·K}) \ln\frac{600 \text{ kPa}}{100 \text{ kPa}}$$

$$= \mathbf{-0.3842 \text{ kJ/kg·K}}$$

**Discussion** The two results above are almost identical since the change in temperature during this process is relatively small (Fig. 7–34). When the temperature change is large, however, they may differ significantly. For those cases, Eq. 7–39 should be used instead of Eq. 7–34 since it accounts for the variation of specific heats with temperature.

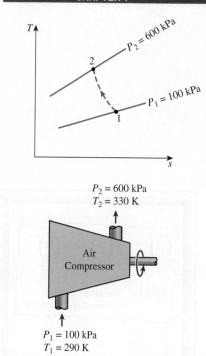

**FIGURE 7–33**

Schematic and *T-s* diagram for Example 7–9.

# Isentropic Processes of Ideal Gases

Several relations for the isentropic processes of ideal gases can be obtained by setting the entropy-change relations developed previously equal to zero. Again, this is done first for the case of constant specific heats and then for the case of variable specific heats.

## Constant Specific Heats (Approximate Analysis)

When the constant-specific-heat assumption is valid, the isentropic relations for ideal gases are obtained by setting Eqs. 7–33 and 7–34 equal to zero. From Eq. 7–33,

$$\ln\frac{T_2}{T_1} = -\frac{R}{c_v} \ln\frac{v_2}{v_1}$$

which can be rearranged as

$$\ln\frac{T_2}{T_1} = \ln\left(\frac{v_1}{v_2}\right)^{R/c_v} \tag{7–41}$$

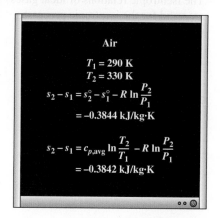

**FIGURE 7–34**

For small temperature differences, the exact and approximate relations for entropy changes of ideal gases give almost identical results.

or

$$\left(\frac{T_2}{T_1}\right)_{s=\text{const.}} = \left(\frac{v_1}{v_2}\right)^{k-1} \quad \text{(ideal gas)} \tag{7-42}$$

since $R = c_p - c_v$, $k = c_p/c_v$, and thus $R/c_v = k - 1$.

Equation 7–42 is the *first isentropic relation* for ideal gases under the constant-specific-heat assumption. The *second isentropic relation* is obtained in a similar manner from Eq. 7–34 with the following result:

$$\left(\frac{T_2}{T_1}\right)_{s=\text{const.}} = \left(\frac{P_2}{P_1}\right)^{(k-1)/k} \quad \text{(ideal gas)} \tag{7-43}$$

The *third isentropic relation* is obtained by substituting Eq. 7–43 into Eq. 7–42 and simplifying:

$$\left(\frac{P_2}{P_1}\right)_{s=\text{const.}} = \left(\frac{v_1}{v_2}\right)^{k} \quad \text{(ideal gas)} \tag{7-44}$$

Equations 7–42 through 7–44 can also be expressed in a compact form as

$$Tv^{k-1} = \text{constant} \tag{7-45}$$

$$TP^{(1-k)/k} = \text{constant} \quad \text{(ideal gas)} \tag{7-46}$$

$$Pv^{k} = \text{constant} \tag{7-47}$$

The specific heat ratio $k$, in general, varies with temperature, and thus an average $k$ value for the given temperature range should be used.

Note that the ideal-gas isentropic relations above, as the name implies, are strictly valid for isentropic processes only when the constant-specific-heat assumption is appropriate (Fig. 7–35).

## Variable Specific Heats (Exact Analysis)

When the constant-specific-heat assumption is not appropriate, the isentropic relations developed previously yields results that are not quite accurate. For such cases, we should use an isentropic relation obtained from Eq. 7–39 that accounts for the variation of specific heats with temperature. Setting this equation equal to zero gives

$$0 = s_2^\circ - s_1^\circ - R\ln\frac{P_2}{P_1}$$

or

$$s_2^\circ = s_1^\circ + R\ln\frac{P_2}{P_1} \tag{7-48}$$

where $s_2^\circ$ is the $s^\circ$ value at the end of the isentropic process.

## Relative Pressure and Relative Specific Volume

Equation 7–48 provides an accurate way of evaluating property changes of ideal gases during isentropic processes since it accounts for the variation

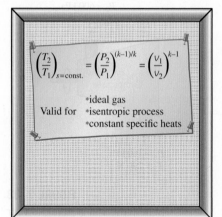

$$\left(\frac{T_2}{T_1}\right)_{s=\text{const.}} = \left(\frac{P_2}{P_1}\right)^{(k-1)/k} = \left(\frac{v_1}{v_2}\right)^{k-1}$$

Valid for
*ideal gas
*isentropic process
*constant specific heats

**FIGURE 7–35**
The isentropic relations of ideal gases are valid for the isentropic processes of ideal gases only.

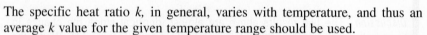

of specific heats with temperature. However, it involves tedious iterations when the volume ratio is given instead of the pressure ratio. This is quite an inconvenience in optimization studies, which usually require numerous repetitive calculations. To remedy this deficiency, we define two new dimensionless quantities associated with isentropic processes.

The definition of the first is based on Eq. 7–48, which can be rearranged as

$$\frac{P_2}{P_1} = \exp\frac{s_2^\circ - s_1^\circ}{R}$$

or

$$\frac{P_2}{P_1} = \frac{\exp(s_2^\circ/R)}{\exp(s_1^\circ/R)}$$

The quantity $\exp(s^\circ/R)$ is defined as the **relative pressure** $P_r$. With this definition, the last relation becomes

$$\left(\frac{P_2}{P_1}\right)_{s=\text{const.}} = \frac{P_{r2}}{P_{r1}} \tag{7–49}$$

Note that the relative pressure $P_r$ is a *dimensionless* quantity that is a function of temperature only since $s^\circ$ depends on temperature alone. Therefore, values of $P_r$ can be tabulated against temperature. This is done for air in Table A–17. The use of $P_r$ data is illustrated in Fig. 7–36.

Sometimes specific volume ratios are given instead of pressure ratios. This is particularly the case when automotive engines are analyzed. In such cases, one needs to work with volume ratios. Therefore, we define another quantity related to specific volume ratios for isentropic processes. This is done by utilizing the ideal-gas relation and Eq. 7–49:

$$\frac{P_1 v_1}{T_1} = \frac{P_2 v_2}{T_2} \quad \rightarrow \quad \frac{v_2}{v_1} = \frac{T_2 P_1}{T_1 P_2} = \frac{T_2 P_{r1}}{T_1 P_{r2}} = \frac{T_2/P_{r2}}{T_1/P_{r1}}$$

The quantity $T/P_r$ is a function of temperature only and is defined as **relative specific volume** $v_r$. Thus,

$$\left(\frac{v_2}{v_1}\right)_{s=\text{const.}} = \frac{v_{r2}}{v_{r1}} \tag{7–50}$$

Equations 7–49 and 7–50 are strictly valid for isentropic processes of ideal gases only. They account for the variation of specific heats with temperature and therefore give more accurate results than Eqs. 7–42 through 7–47. The values of $P_r$ and $v_r$ are listed for air in Table A–17.

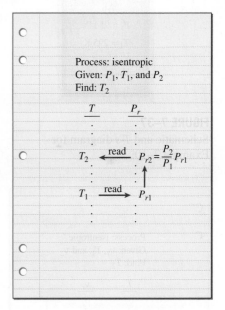

Process: isentropic
Given: $P_1$, $T_1$, and $P_2$
Find: $T_2$

$$T \qquad P_r$$

$$T_2 \xleftarrow{\text{read}} P_{r2} = \frac{P_2}{P_1} P_{r1}$$

$$T_1 \xrightarrow{\text{read}} P_{r1}$$

**FIGURE 7–36**
The use of $P_r$ data for calculating the final temperature during an isentropic process.

■ **EXAMPLE 7–10**　**Isentropic Compression of Air in a Car Engine**

■ Air is compressed in a car engine from 22°C and 95 kPa in a reversible
■ and adiabatic manner. If the compression ratio $V_1/V_2$ of this engine is 8,
■ determine the final temperature of the air.

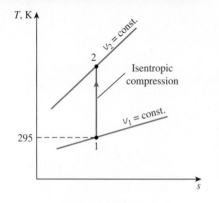

**FIGURE 7–37**

Schematic and *T-s* diagram for
Example 7–10.

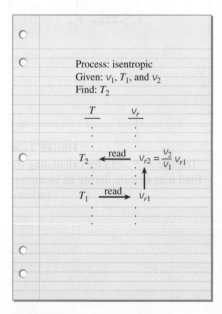

**FIGURE 7–38**

The use of $v_r$ data for calculating the
final temperature during an isentropic
process (Example 7–10).

**SOLUTION**  Air is compressed in a car engine isentropically. For a given
compression ratio, the final air temperature is to be determined.

**Assumptions**  At specified conditions, air can be treated as an ideal gas.
Therefore, the isentropic relations for ideal gases are applicable.

**Analysis**  A sketch of the system and the *T-s* diagram for the process are
given in Fig. 7–37.

This process is easily recognized as being isentropic since it is both reversible and adiabatic. The final temperature for this isentropic process can be
determined from Eq. 7–50 with the help of relative specific volume data
(Table A–17), as illustrated in Fig. 7–38.

For closed systems:
$$\frac{V_2}{V_1} = \frac{v_2}{v_1}$$

At $T_1 = 295$ K:
$$v_{r1} = 647.9$$

From Eq. 7–50:  $v_{r2} = v_{r1}\left(\frac{v_2}{v_1}\right) = (647.9)\left(\frac{1}{8}\right) = 80.99 \rightarrow T_2 = \mathbf{662.7\ K}$

Therefore, the temperature of air will increase by 367.7°C during this
process.

**ALTERNATIVE SOLUTION**  The final temperature could also be determined
from Eq. 7–42 by assuming constant specific heats for air:

$$\left(\frac{T_2}{T_1}\right)_{s=\text{const.}} = \left(\frac{v_1}{v_2}\right)^{k-1}$$

The specific heat ratio *k* also varies with temperature, and we need to use
the value of *k* corresponding to the average temperature. However, the
final temperature is not given, and so we cannot determine the average
temperature in advance. For such cases, calculations can be started with
a *k* value at the initial or the anticipated average temperature. This value
could be refined later, if necessary, and the calculations can be repeated.
We know that the temperature of the air will rise considerably during this
adiabatic compression process, so we *guess* the average temperature to
be about 450 K. The *k* value at this anticipated average temperature is
determined from Table A–2*b* to be 1.391. Then, the final temperature of
air becomes

$$T_2 = (295\ \text{K})(8)^{1.391-1} = 665.2\ \text{K}$$

This gives an average temperature value of 480.1 K, which is sufficiently
close to the assumed value of 450 K. Therefore, it is not necessary to repeat
the calculations by using the *k* value at this average temperature.

The result obtained by assuming constant specific heats for this case is
in error by about 0.4 percent, which is rather small. This is not surprising
since the temperature change of air is relatively small (only a few hundred
degrees) and the specific heats of air vary almost linearly with temperature
in this temperature range.

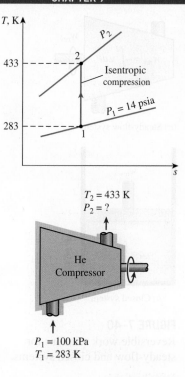

**FIGURE 7–39**
Schematic and *T-s* diagram for
Example 7–11.

■ *EXAMPLE 7–11*     **Isentropic Compression of an Ideal Gas**
■
■ Helium gas is compressed by an adiabatic compressor from an initial state
■ of 100 kPa and 10°C to a final temperature of 160°C in a reversible manner.
■ Determine the exit pressure of helium.

**SOLUTION**   Helium is compressed from a given state to a specified pressure
isentropically. The exit pressure of helium is to be determined.
*Assumptions*   At specified conditions, helium can be treated as an ideal
gas. Therefore, the isentropic relations developed earlier for ideal gases are
applicable.
*Analysis*   A sketch of the system and the *T-s* diagram for the process are
given in Fig. 7–39.
  The specific heat ratio *k* of helium is 1.667 and is independent of temperature in the region where it behaves as an ideal gas. Thus the final pressure
of helium can be determined from Eq. 7–43:

$$P_2 = P_1\left(\frac{T_2}{T_1}\right)^{k/(k-1)} = (100 \text{ kPa})\left(\frac{433 \text{ K}}{283 \text{ K}}\right)^{1.667/0.667} = \textbf{289 kPa}$$

# 7–10 · REVERSIBLE STEADY-FLOW WORK

The work done during a process depends on the path followed as well as
on the properties at the end states. Recall that reversible (quasi-equilibrium)
moving boundary work associated with closed systems is expressed in terms
of the fluid properties as

$$W_b = \int_1^2 P\,dV$$

We mentioned that the quasi-equilibrium work interactions lead to the maximum work output for work-producing devices and the minimum work input
for work-consuming devices.
  It would also be very insightful to express the work associated with
steady-flow devices in terms of fluid properties.
  Taking the positive direction of work to be from the system (work output),
the energy balance for a steady-flow device undergoing an internally
reversible process can be expressed in differential form as

$$\delta q_{\text{rev}} - \delta w_{\text{rev}} = dh + d\text{ke} + d\text{pe}$$

But

$$\left.\begin{array}{ll} \delta q_{\text{rev}} = T\,ds & (\text{Eq. 7–16}) \\ T\,ds = dh - v\,dP & (\text{Eq. 7–24}) \end{array}\right\} \quad \delta q_{\text{rev}} = dh - v\,dP$$

Substituting this into the relation above and canceling *dh* yield

$$-\delta w_{\text{rev}} = v\,dP + d\text{ke} + d\text{pe}$$

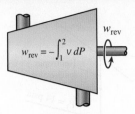

$$w_{rev} = -\int_1^2 v\,dP$$

(a) Steady-flow system

$$w_{rev} = \int_1^2 P\,dv$$

(b) Closed system

**FIGURE 7–40**
Reversible work relations for steady-flow and closed systems.

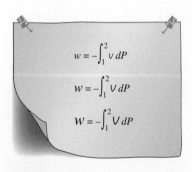

$$w = -\int_1^2 v\,dP$$

$$w = -\int_1^2 v\,dP$$

$$w = -\int_1^2 v\,dP$$

**FIGURE 7–41**
The larger the specific volume, the greater the work produced (or consumed) by a steady-flow device.

Integrating, we find

$$w_{rev} = -\int_1^2 v\,dP - \Delta ke - \Delta pe \quad (kJ/kg) \tag{7–51}$$

When the changes in kinetic and potential energies are negligible, this equation reduces to

$$w_{rev} = -\int_1^2 v\,dP \quad (kJ/kg) \tag{7–52}$$

Equations 7–51 and 7–52 are relations for the *reversible work output* associated with an internally reversible process in a steady-flow device. They will give a negative result when work is done on the system. To avoid the negative sign, Eq. 7–51 can be written for work input to steady-flow devices such as compressors and pumps as

$$w_{rev,in} = \int_1^2 v\,dP + \Delta ke + \Delta pe \tag{7–53}$$

The resemblance between the $v\,dP$ in these relations and $P\,dv$ is striking. They should not be confused with each other, however, since $P\,dv$ is associated with reversible boundary work in closed systems (Fig. 7–40).

Obviously, one needs to know $v$ as a function of $P$ for the given process to perform the integration. When the working fluid is *incompressible,* the specific volume $v$ remains constant during the process and can be taken out of the integration. Then Eq. 7–51 simplifies to

$$w_{rev} = -v(P_2 - P_1) - \Delta ke - \Delta pe \quad (kJ/kg) \tag{7–54}$$

For the steady flow of a liquid through a device that involves no work interactions (such as a nozzle or a pipe section), the work term is zero, and the equation above can be expressed as

$$v(P_2 - P_1) + \frac{V_2^2 - V_1^2}{2} + g(z_2 - z_1) = 0 \tag{7–55}$$

which is known as the **Bernoulli equation** in fluid mechanics. It is developed for an internally reversible process and thus is applicable to incompressible fluids that involve no irreversibilities such as friction or shock waves. This equation can be modified, however, to incorporate these effects.

Equation 7–52 has far-reaching implications in engineering regarding devices that produce or consume work steadily such as turbines, compressors, and pumps. It is obvious from this equation that the reversible steady-flow work is closely associated with the specific volume of the fluid flowing through the device. *The larger the specific volume, the larger the reversible work produced or consumed by the steady-flow device* (Fig. 7–41). This conclusion is equally valid for actual steady-flow devices. Therefore, every effort should be made to keep the specific volume of a fluid as small as possible during a compression process to minimize the work input and as large as possible during an expansion process to maximize the work output.

In steam or gas power plants, the pressure rise in the pump or compressor is equal to the pressure drop in the turbine if we disregard the pressure losses in various other components. In steam power plants, the pump handles liquid, which has a very small specific volume, and the turbine handles vapor, whose specific volume is many times larger. Therefore, the work output of the turbine is much larger than the work input to the pump. This is one of the reasons for the wide-spread use of steam power plants in electric power generation.

If we were to compress the steam exiting the turbine back to the turbine inlet pressure before cooling it first in the condenser in order to "save" the heat rejected, we would have to supply all the work produced by the turbine back to the compressor. In reality, the required work input would be even greater than the work output of the turbine because of the irreversibilities present in both processes.

In gas power plants, the working fluid (typically air) is compressed in the gas phase, and a considerable portion of the work output of the turbine is consumed by the compressor. As a result, a gas power plant delivers less net work per unit mass of the working fluid.

■ **EXAMPLE 7–12**   **Compressing a Substance in the Liquid versus Gas Phases**

■ Determine the compressor work input required to compress steam isentropically from 100 kPa to 1 MPa, assuming that the steam exists as (a) saturated liquid and (b) saturated vapor at the inlet state.

**SOLUTION**   Steam is to be compressed from a given pressure to a specified pressure isentropically. The work input is to be determined for the cases of steam being a saturated liquid and saturated vapor at the inlet.
*Assumptions* **1** Steady operating conditions exist. **2** Kinetic and potential energy changes are negligible. **3** The process is given to be isentropic.
*Analysis* We take first the turbine and then the pump as the *system*. Both are *control volumes* since mass crosses the boundary. Sketches of the pump and the turbine together with the *T-s* diagram are given in Fig. 7–42.
(a) In this case, steam is a saturated liquid initially, and its specific volume is

$$v_1 = v_{f\,@\,100\,kPa} = 0.001043\ m^3/kg \qquad (Table\ A–5)$$

which remains essentially constant during the process. Thus,

$$w_{rev,in} = \int_1^2 v\,dP \cong v_1(P_2 - P_1)$$

$$= (0.001043\ m^3/kg)[(1000 - 100)\ kPa]\left(\frac{1\ kJ}{1\ kPa\cdot m^3}\right)$$

$$= \mathbf{0.94\ kJ/kg}$$

(b) This time, steam is a saturated vapor initially and remains a vapor during the entire compression process. Since the specific volume of a gas changes considerably during a compression process, we need to know how $v$ varies

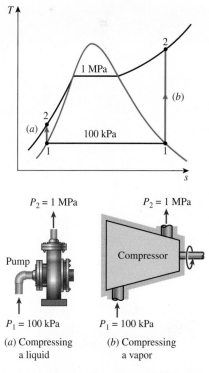

**FIGURE 7–42**
Schematic and *T-s* diagram for Example 7–12.

with $P$ to perform the integration in Eq. 7–53. This relation, in general, is not readily available. But for an isentropic process, it is easily obtained from the second $T \, ds$ relation by setting $ds = 0$:

$$\left. \begin{array}{l} T \, ds = dh - v \, dP \quad \text{(Eq. 7-24)} \\ ds = 0 \quad \text{(isentropic process)} \end{array} \right\} \quad v \, dP = dh$$

Thus,

$$w_{\text{rev,in}} = \int_1^2 v \, dP = \int_1^2 dh = h_2 - h_1$$

This result could also be obtained from the energy balance relation for an isentropic steady-flow process. Next we determine the enthalpies:

State 1: $\left. \begin{array}{l} P_1 = 100 \text{ kPa} \\ \text{(sat. vapor)} \end{array} \right\}$  $\begin{array}{l} h_1 = 2675.0 \text{ kJ/kg} \\ s_1 = 7.3589 \text{ kJ/kg·K} \end{array}$  (Table A–5)

State 2: $\left. \begin{array}{l} P_2 = 1 \text{ MPa} \\ s_2 = s_1 \end{array} \right\}$  $h_2 = 3194.5 \text{ kJ/kg}$   (Table A–6)

Thus,

$$w_{\text{rev,in}} = (3194.5 - 2675.0) \text{ kJ/kg} = \mathbf{519.5 \text{ kJ/kg}}$$

*Discussion*  Note that compressing steam in the vapor form would require over 500 times more work than compressing it in the liquid form between the same pressure limits.

# Proof that Steady-Flow Devices Deliver the Most and Consume the Least Work When the Process is Reversible

We have shown in Chap. 6 that cyclic devices (heat engines, refrigerators, and heat pumps) deliver the most work and consume the least when reversible processes are used. Now we demonstrate that this is also the case for individual devices such as turbines and compressors in steady operation.

Consider two steady-flow devices, one reversible and the other irreversible, operating between the same inlet and exit states. Again taking heat transfer to the system and work done by the system to be positive quantities, the energy balance for each of these devices can be expressed in the differential form as

Actual:  $\delta q_{\text{act}} - \delta w_{\text{act}} = dh + d\text{ke} + d\text{pe}$

Reversible:  $\delta q_{\text{rev}} - \delta w_{\text{rev}} = dh + d\text{ke} + d\text{pe}$

The right-hand sides of these two equations are identical since both devices are operating between the same end states. Thus,

$$\delta q_{\text{act}} - \delta w_{\text{act}} = \delta q_{\text{rev}} - \delta w_{\text{rev}}$$

or

$$\delta w_{\text{rev}} - \delta w_{\text{act}} = \delta q_{\text{rev}} - \delta q_{\text{act}}$$

However,

$$\delta q_{\text{rev}} = T \, ds$$

Substituting this relation into the preceding equation and dividing each term by $T$, we obtain

$$\frac{\delta w_{\text{rev}} - \delta w_{\text{act}}}{T} = ds - \frac{\delta q_{\text{act}}}{T} \geq 0$$

since

$$ds \geq \frac{\delta q_{\text{act}}}{T}$$

Also, $T$ is the absolute temperature, which is always positive. Thus,

$$\delta w_{\text{rev}} \geq \delta w_{\text{act}}$$

or

$$w_{\text{rev}} \geq w_{\text{act}}$$

Therefore, work-producing devices such as turbines ($w$ is positive) deliver more work, and work-consuming devices such as pumps and compressors ($w$ is negative) require less work when they operate reversibly (Fig. 7–43).

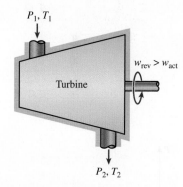

**FIGURE 7–43**
A reversible turbine delivers more work than an irreversible one if both operate between the same end states.

## 7–11 · MINIMIZING THE COMPRESSOR WORK

We have just shown that the work input to a compressor is minimized when the compression process is executed in an internally reversible manner. When the changes in kinetic and potential energies are negligible, the compressor work is given by (Eq. 7–53)

$$w_{\text{rev,in}} = \int_1^2 v \, dP \tag{7–56}$$

Obviously one way of minimizing the compressor work is to approximate an internally reversible process as much as possible by minimizing the irreversibilities such as friction, turbulence, and nonquasi-equilibrium compression. The extent to which this can be accomplished is limited by economic considerations. A second (and more practical) way of reducing the compressor work is to keep the specific volume of the gas as small as possible during the compression process. This is done by maintaining the temperature of the gas as low as possible during compression since the specific volume of a gas is proportional to temperature. Therefore, reducing the work input to a compressor requires that the gas be cooled as it is compressed.

To have a better understanding of the effect of cooling during the compression process, we compare the work input requirements for three kinds of processes: *an isentropic process* (involves no cooling), *a polytropic process* (involves some cooling), and *an isothermal process* (involves maximum cooling). Assuming all three processes are executed between the same pressure levels ($P_1$ and $P_2$) in an internally reversible manner and the gas behaves as an ideal gas ($Pv = RT$) with constant specific heats, we see that the compression work is determined by performing the integration in Eq. 7–56 for each case, with the following results:

Isentropic ($Pv^k$ = constant):

$$w_{\text{comp,in}} = \frac{kR(T_2 - T_1)}{k - 1} = \frac{kRT_1}{k - 1}\left[\left(\frac{P_2}{P_1}\right)^{(k-1)/k} - 1\right] \qquad (7\text{–}57a)$$

Polytropic ($Pv^n$ = constant):

$$w_{\text{comp,in}} = \frac{nR(T_2 - T_1)}{n - 1} = \frac{nRT_1}{n - 1}\left[\left(\frac{P_2}{P_1}\right)^{(n-1)/n} - 1\right] \qquad (7\text{–}57b)$$

Isothermal ($Pv$ = constant):

$$w_{\text{comp,in}} = RT \ln\frac{P_2}{P_1} \qquad (7\text{–}57c)$$

The three processes are plotted on a $P$-$v$ diagram in Fig. 7–44 for the same inlet state and exit pressure. On a $P$-$v$ diagram, the area to the left of the process curve is the integral of $v\,dP$. Thus it is a measure of the steady-flow compression work. It is interesting to observe from this diagram that of the three internally reversible cases considered, the adiabatic compression ($Pv^k$ = constant) requires the maximum work and the isothermal compression ($T$ = constant or $Pv$ = constant) requires the minimum. The work input requirement for the polytropic case ($Pv^n$ = constant) is between these two and decreases as the polytropic exponent $n$ is decreased, by increasing the heat rejection during the compression process. If sufficient heat is removed, the value of $n$ approaches unity and the process becomes isothermal. One common way of cooling the gas during compression is to use cooling jackets around the casing of the compressors.

## Multistage Compression with Intercooling

It is clear from these arguments that cooling a gas as it is compressed is desirable since this reduces the required work input to the compressor. However, often it is not possible to have adequate cooling through the casing of the compressor, and it becomes necessary to use other techniques to achieve effective cooling. One such technique is **multistage compression with intercooling**, where the gas is compressed in stages and cooled between each stage by passing it through a heat exchanger called an *intercooler*. Ideally, the cooling process takes place at constant pressure, and the gas is cooled to the initial temperature $T_1$ at each intercooler. Multistage

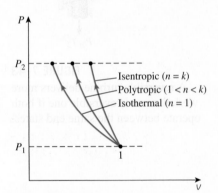

**FIGURE 7–44**

*P*-*v* diagrams of isentropic, polytropic, and isothermal compression processes between the same pressure limits.

compression with intercooling is especially attractive when a gas is to be compressed to very high pressures.

The effect of intercooling on compressor work is graphically illustrated on $P$-$v$ and $T$-$s$ diagrams in Fig. 7–45 for a two-stage compressor. The gas is compressed in the first stage from $P_1$ to an intermediate pressure $P_x$, cooled at constant pressure to the initial temperature $T_1$, and compressed in the second stage to the final pressure $P_2$. The compression processes, in general, can be modeled as polytropic ($Pv^n = $ constant) where the value of $n$ varies between $k$ and 1. The colored area on the $P$-$v$ diagram represents the work saved as a result of two-stage compression with intercooling. The process paths for single-stage isothermal and polytropic processes are also shown for comparison.

The size of the colored area (the saved work input) varies with the value of the intermediate pressure $P_x$, and it is of practical interest to determine the conditions under which this area is maximized. The total work input for a two-stage compressor is the sum of the work inputs for each stage of compression, as determined from Eq. 7–57b:

$$w_{comp,in} = w_{comp\,I,in} + w_{comp\,II,in} \tag{7–58}$$

$$= \frac{nRT_1}{n-1}\left[\left(\frac{P_x}{P_1}\right)^{(n-1)/n} - 1\right] + \frac{nRT_1}{n-1}\left[\left(\frac{P_2}{P_x}\right)^{(n-1)/n} - 1\right]$$

The only variable in this equation is $P_x$. The $P_x$ value that minimizes the total work is determined by differentiating this expression with respect to $P_x$ and setting the resulting expression equal to zero. It yields

$$P_x = (P_1 P_2)^{1/2} \quad \text{or} \quad \frac{P_x}{P_1} = \frac{P_2}{P_x} \tag{7–59}$$

That is, *to minimize compression work during two-stage compression, the pressure ratio across each stage of the compressor must be the same.* When this condition is satisfied, the compression work at each stage becomes identical, that is, $w_{comp\,I,in} = w_{comp\,II,in}$.

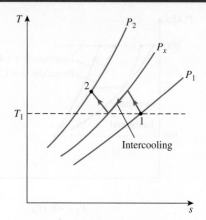

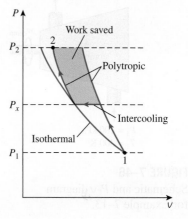

**FIGURE 7–45**
$P$-$v$ and $T$-$s$ diagrams for a two-stage steady-flow compression process.

---

### EXAMPLE 7–13    Work Input for Various Compression Processes

Air is compressed steadily by a reversible compressor from an inlet state of 100 kPa and 300 K to an exit pressure of 900 kPa. Determine the compressor work per unit mass for (*a*) isentropic compression with $k = 1.4$, (*b*) polytropic compression with $n = 1.3$, (*c*) isothermal compression, and (*d*) ideal two-stage compression with intercooling with a polytropic exponent of 1.3.

**SOLUTION** Air is compressed reversibly from a specified state to a specified pressure. The compressor work is to be determined for the cases of isentropic, polytropic, isothermal, and two-stage compression.
*Assumptions* **1** Steady operating conditions exist. **2** At specified conditions, air can be treated as an ideal gas. **3** Kinetic and potential energy changes are negligible.

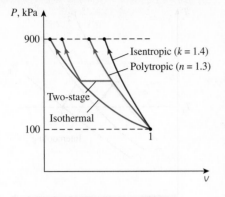

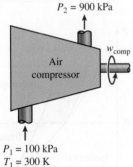

**FIGURE 7–46**
Schematic and P-v diagram
for Example 7–13.

*Analysis* We take the compressor to be the system. This is a control volume since mass crosses the boundary. A sketch of the system and the *T-s* diagram for the process are given in Fig. 7–46.

The steady-flow compression work for all these four cases is determined by using the relations developed earlier in this section:

(a) Isentropic compression with $k = 1.4$:

$$w_{\text{comp,in}} = \frac{kRT_1}{k-1}\left[\left(\frac{P_2}{P_1}\right)^{(k-1)/k} - 1\right]$$

$$= \frac{(1.4)(0.287 \text{ kJ/kg·K})(300 \text{ K})}{1.4 - 1}\left[\left(\frac{900 \text{ kPa}}{100 \text{ kPa}}\right)^{(1.4-1)/1.4} - 1\right]$$

$$= \textbf{263.2 kJ/kg}$$

(b) Polytropic compression with $n = 1.3$:

$$w_{\text{comp,in}} = \frac{nRT_1}{n-1}\left[\left(\frac{P_2}{P_1}\right)^{(n-1)/n} - 1\right]$$

$$= \frac{(1.3)(0.287 \text{ kJ/kg·K})(300 \text{ K})}{1.3 - 1}\left[\left(\frac{900 \text{ kPa}}{100 \text{ kPa}}\right)^{(1.3-1)/1.3} - 1\right]$$

$$= \textbf{246.4 kJ/kg}$$

(c) Isothermal compression:

$$w_{\text{comp,in}} = RT \ln\frac{P_2}{P_1} = (0.287 \text{ kJ/kg·K})(300 \text{ K}) \ln\frac{900 \text{ kPa}}{100 \text{ kPa}}$$

$$= \textbf{189.2 kJ/kg}$$

(d) Ideal two-stage compression with intercooling ($n = 1.3$): In this case, the pressure ratio across each stage is the same, and its value is

$$P_x = (P_1P_2)^{1/2} = [(100 \text{ kPa})(900 \text{ kPa})]^{1/2} = 300 \text{ kPa}$$

The compressor work across each stage is also the same. Thus the total compressor work is twice the compression work for a single stage:

$$w_{\text{comp,in}} = 2w_{\text{comp I,in}} = 2\frac{nRT_1}{n-1}\left[\left(\frac{P_x}{P_1}\right)^{(n-1)/n} - 1\right]$$

$$= \frac{2(1.3)(0.287 \text{ kJ/kg·K})(300 \text{ K})}{1.3 - 1}\left[\left(\frac{300 \text{ kPa}}{100 \text{ kPa}}\right)^{(1.3-1)/1.3} - 1\right]$$

$$= \textbf{215.3 kJ/kg}$$

*Discussion* Of all four cases considered, the isothermal compression requires the minimum work and the isentropic compression the maximum. The compressor work is decreased when two stages of polytropic compression are utilized instead of just one. As the number of compressor stages is increased, the compressor work approaches the value obtained for the isothermal case.

# 7–12 · ISENTROPIC EFFICIENCIES OF STEADY-FLOW DEVICES

We mentioned repeatedly that irreversibilities inherently accompany all actual processes and that their effect is always to downgrade the performance of devices. In engineering analysis, it would be very desirable to have some parameters that would enable us to quantify the degree of degradation of energy in these devices. In the last chapter we did this for cyclic devices, such as heat engines and refrigerators, by comparing the actual cycles to the idealized ones, such as the Carnot cycle. A cycle that was composed entirely of reversible processes served as the *model cycle* to which the actual cycles could be compared. This idealized model cycle enabled us to determine the theoretical limits of performance for cyclic devices under specified conditions and to examine how the performance of actual devices suffered as a result of irreversibilities.

Now we extend the analysis to discrete engineering devices working under steady-flow conditions, such as turbines, compressors, and nozzles, and we examine the degree of degradation of energy in these devices as a result of irreversibilities. However, first we need to define an ideal process that serves as a model for the actual processes.

Although some heat transfer between these devices and the surrounding medium is unavoidable, many steady-flow devices are intended to operate under adiabatic conditions. Therefore, the model process for these devices should be an adiabatic one. Furthermore, an ideal process should involve no irreversibilities since the effect of irreversibilities is always to downgrade the performance of engineering devices. Thus, the ideal process that can serve as a suitable model for adiabatic steady-flow devices is the *isentropic* process (Fig. 7–47).

The more closely the actual process approximates the idealized isentropic process, the better the device performs. Thus, it would be desirable to have a parameter that expresses quantitatively how efficiently an actual device approximates an idealized one. This parameter is the **isentropic** or **adiabatic efficiency**, which is a measure of the deviation of actual processes from the corresponding idealized ones.

Isentropic efficiencies are defined differently for different devices since each device is set up to perform different tasks. Next, we define the isentropic efficiencies of turbines, compressors, and nozzles by comparing the actual performance of these devices to their performance under isentropic conditions for the same inlet state and exit pressure.

## Isentropic Efficiency of Turbines

For a turbine under steady operation, the inlet state of the working fluid and the exhaust pressure are fixed. Therefore, the ideal process for an adiabatic turbine is an isentropic process between the inlet state and the exhaust pressure. The desired output of a turbine is the work produced, and the **isentropic efficiency of a turbine** is defined as *the ratio of the actual work output of the turbine to the work output that would be achieved if the process between the inlet state and the exit pressure were isentropic:*

$$\eta_T = \frac{\text{Actual turbine work}}{\text{Isentropic turbine work}} = \frac{w_a}{w_s} \qquad (7\text{–}60)$$

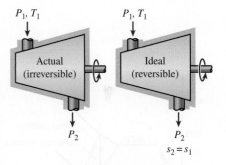

**FIGURE 7–47**
The isentropic process involves no irreversibilities and serves as the ideal process for adiabatic devices.

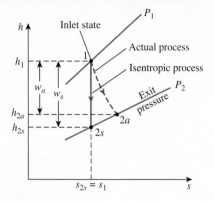

**FIGURE 7–48**
The *h-s* diagram for the actual
and isentropic processes of
an adiabatic turbine.

Usually the changes in kinetic and potential energies associated with a fluid stream flowing through a turbine are small relative to the change in enthalpy and can be neglected. Then, the work output of an adiabatic turbine simply becomes the change in enthalpy, and Eq. 7–60 becomes

$$\eta_T \cong \frac{h_1 - h_{2a}}{h_1 - h_{2s}} \qquad (7\text{–}61)$$

where $h_{2a}$ and $h_{2s}$ are the enthalpy values at the exit state for actual and isentropic processes, respectively (Fig. 7–48).

The value of $\varepsilon_T$ greatly depends on the design of the individual components that make up the turbine. Well-designed, large turbines have isentropic efficiencies above 90 percent. For small turbines, however, it may drop even below 70 percent. The value of the isentropic efficiency of a turbine is determined by measuring the actual work output of the turbine and by calculating the isentropic work output for the measured inlet conditions and the exit pressure. This value can then be used conveniently in the design of power plants.

---

**EXAMPLE 7–14    Isentropic Efficiency of a Steam Turbine**

Steam enters an adiabatic turbine steadily at 3 MPa and 400°C and leaves at 50 kPa and 100°C. If the power output of the turbine is 2 MW, determine (*a*) the isentropic efficiency of the turbine and (*b*) the mass flow rate of the steam flowing through the turbine.

**SOLUTION**    Steam flows steadily in a turbine between inlet and exit states. For a specified power output, the isentropic efficiency and the mass flow rate are to be determined.
**Assumptions**    **1** Steady operating conditions exist. **2** The changes in kinetic and potential energies are negligible.
**Analysis**    A sketch of the system and the *T-s* diagram of the process are given in Fig. 7–49.
(*a*) The enthalpies at various states are

*State 1:* $\quad \left. \begin{array}{l} P_1 = 3 \text{ MPA} \\ T_1 = 400°C \end{array} \right\} \quad \begin{array}{l} h_1 = 3231.7 \text{ kJ/kg} \\ s_1 = 6.9235 \text{ kJ/kg·K} \end{array} \quad$ (Table A–6)

*State 2a:* $\quad \left. \begin{array}{l} P_{2a} = 50 \text{ kPa} \\ T_{2a} = 100°C \end{array} \right\} \quad h_{2a} = 2682.4 \text{ kJ/kg} \quad$ (Table A–6)

The exit enthalpy of the steam for the isentropic process $h_{2s}$ is determined from the requirement that the entropy of the steam remain constant ($s_{2s} = s_1$):

*State 2s:* $\quad \left. \begin{array}{l} P_{2s} = 50 \text{ kPa} \\ (s_{2s} = s_1) \end{array} \right. \longrightarrow \quad \begin{array}{l} s_f = 1.0912 \text{ kJ/kg·K} \\ s_g = 7.5931 \text{ kJ/kg·K} \end{array} \quad$ (Table A–5)

Obviously, at the end of the isentropic process steam exists as a saturated mixture since $s_f < s_{2s} < s_g$. Thus, we need to find the quality at state 2s first:

$$x_{2s} = \frac{s_{2s} - s_f}{s_{fg}} = \frac{6.9235 - 1.0912}{6.5019} = 0.897$$

---

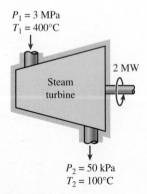

**FIGURE 7–49**
Schematic and *T-s* diagram for
Example 7–14.

and

$$h_{2s} = h_f + x_{2s}h_{fg} = 340.54 + 0.897(2304.7) = 2407.9 \text{ kJ/kg}$$

By substituting these enthalpy values into Eq. 7–61, the isentropic efficiency of this turbine is determined to be

$$\eta_T \cong \frac{h_1 - h_{2a}}{h_1 - h_{2s}} = \frac{3231.7 - 2682.4}{3231.7 - 2407.9} = \textbf{0.667} \text{ (or 66.7%)}$$

(b) The mass flow rate of steam through this turbine is determined from the energy balance for steady-flow systems:

$$\dot{E}_{in} = \dot{E}_{out}$$

$$\dot{m}h_1 = \dot{W}_{a,out} + \dot{m}h_{2a}$$

$$\dot{W}_{a,out} = \dot{m}(h_1 - h_{2a})$$

$$2 \text{ MW}\left(\frac{1000 \text{ kJ/s}}{1 \text{ MW}}\right) = \dot{m}(3231.7 - 2682.4) \text{ kJ/kg}$$

$$\dot{m} = \textbf{3.64 kg/s}$$

## Isentropic Efficiencies of Compressors and Pumps

The **isentropic efficiency of a compressor** is defined as *the ratio of the work input required to raise the pressure of a gas to a specified value in an isentropic manner to the actual work input:*

$$\eta_C = \frac{\text{Isentropic compressor work}}{\text{Actual compressor work}} = \frac{w_s}{w_a} \qquad (7\text{–}62)$$

Notice that the isentropic compressor efficiency is defined with the *isentropic work input in the numerator* instead of in the denominator. This is because $w_s$ is a smaller quantity than $w_a$, and this definition prevents $\eta_C$ from becoming greater than 100 percent, which would falsely imply that the actual compressors performed better than the isentropic ones. Also notice that the inlet conditions and the exit pressure of the gas are the same for both the actual and the isentropic compressor.

When the changes in kinetic and potential energies of the gas being compressed are negligible, the work input to an adiabatic compressor becomes equal to the change in enthalpy, and Eq. 7–62 for this case becomes

$$\eta_C \cong \frac{h_{2s} - h_1}{h_{2a} - h_1} \qquad (7\text{–}63)$$

where $h_{2a}$ and $h_{2s}$ are the enthalpy values at the exit state for actual and isentropic compression processes, respectively, as illustrated in Fig. 7–50. Again, the value of $\eta_C$ greatly depends on the design of the compressor. Well-designed compressors have isentropic efficiencies that range from 80 to 90 percent.

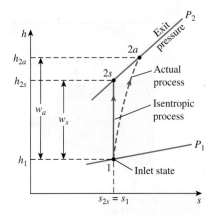

**FIGURE 7–50**

The *h-s* diagram of the actual and isentropic processes of an adiabatic compressor.

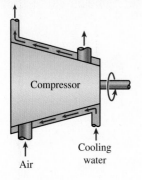

**FIGURE 7–51**
Compressors are sometimes intentionally cooled to minimize the work input.

When the changes in potential and kinetic energies of a liquid are negligible, the isentropic efficiency of a pump is defined similarly as

$$\eta_P = \frac{w_s}{w_a} = \frac{v(P_2 - P_1)}{h_{2a} - h_1} \quad (7\text{–}64)$$

When no attempt is made to cool the gas as it is compressed, the actual compression process is nearly adiabatic and the reversible adiabatic (i.e., isentropic) process serves well as the ideal process. However, sometimes *compressors are cooled intentionally* by utilizing fins or a water jacket placed around the casing to reduce the work input requirements (Fig. 7–51). In this case, the isentropic process is not suitable as the model process since the device is no longer adiabatic and the isentropic compressor efficiency defined above is meaningless. A realistic model process for compressors that are intentionally cooled during the compression process is the *reversible isothermal process*. Then we can conveniently define an **isothermal efficiency** for such cases by comparing the actual process to a reversible isothermal one:

$$\eta_C = \frac{w_t}{w_a} \quad (7\text{–}65)$$

where $w_t$ and $w_a$ are the required work inputs to the compressor for the reversible isothermal and actual cases, respectively.

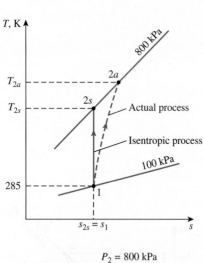

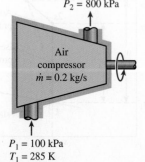

**FIGURE 7–52**
Schematic and *T-s* diagram for Example 7–15.

**EXAMPLE 7–15**    **Effect of Efficiency on Compressor Power Input**

Air is compressed by an adiabatic compressor from 100 kPa and 12°C to a pressure of 800 kPa at a steady rate of 0.2 kg/s. If the isentropic efficiency of the compressor is 80 percent, determine (*a*) the exit temperature of air and (*b*) the required power input to the compressor.

**SOLUTION**   Air is compressed to a specified pressure at a specified rate. For a given isentropic efficiency, the exit temperature and the power input are to be determined.
**Assumptions**   **1** Steady operating conditions exist. **2** Air is an ideal gas. **3** The changes in kinetic and potential energies are negligible.
**Analysis**   A sketch of the system and the *T-s* diagram of the process are given in Fig. 7–52.
(*a*) We know only one property (pressure) at the exit state, and we need to know one more to fix the state and thus determine the exit temperature. The property that can be determined with minimal effort in this case is $h_{2a}$ since the isentropic efficiency of the compressor is given. At the compressor inlet,

$$T_1 = 285 \text{ K} \quad \rightarrow \quad h_1 = 285.14 \text{ kJ/kg} \quad \text{(Table A–17)}$$
$$(P_{r1} = 1.1584)$$

The enthalpy of the air at the end of the isentropic compression process is determined by using one of the isentropic relations of ideal gases,

$$P_{r2} = P_{r1}\left(\frac{P_2}{P_1}\right) = 1.1584\left(\frac{800 \text{ kPa}}{100 \text{ kPa}}\right) = 9.2672$$

and

$$P_{r2} = 9.2672 \quad \rightarrow \quad h_{2s} = 517.05 \text{ kJ/kg}$$

Substituting the known quantities into the isentropic efficiency relation, we have

$$\eta_C \cong \frac{h_{2s} - h_1}{h_{2a} - h_1} \quad \rightarrow \quad 0.80 = \frac{(517.05 - 285.14) \text{ kJ/kg}}{(h_{2a} - 285.14) \text{ kJ/kg}}$$

Thus,

$$h_{2a} = 575.03 \text{ kJ/kg} \quad \rightarrow \quad T_{2a} = 569.5 \text{ K}$$

(b) The required power input to the compressor is determined from the energy balance for steady-flow devices,

$$\dot{E}_{in} = \dot{E}_{out}$$

$$\dot{m}h_1 + \dot{W}_{a,in} = \dot{m}h_{2a}$$

$$\dot{W}_{a,in} = \dot{m}(h_{2a} - h_1)$$

$$= (0.2 \text{ kg/s})[(575.03 - 285.14) \text{ kJ/kg}]$$

$$= 58.0 \text{ kW}$$

**Discussion** Notice that in determining the power input to the compressor, we used $h_{2a}$ instead of $h_{2s}$ since $h_{2a}$ is the actual enthalpy of the air as it exits the compressor. The quantity $h_{2s}$ is a hypothetical enthalpy value that the air would have if the process were isentropic.

## Isentropic Efficiency of Nozzles

Nozzles are essentially adiabatic devices and are used to accelerate a fluid. Therefore, the isentropic process serves as a suitable model for nozzles. The **isentropic efficiency of a nozzle** is defined as *the ratio of the actual kinetic energy of the fluid at the nozzle exit to the kinetic energy value at the exit of an isentropic nozzle for the same inlet state and exit pressure.* That is,

$$\eta_N = \frac{\text{Actual KE at nozzle exit}}{\text{Isentropic KE at nozzle exit}} = \frac{V_{2a}^2}{V_{2s}^2} \tag{7-66}$$

Note that the exit pressure is the same for both the actual and isentropic processes, but the exit state is different.

Nozzles involve no work interactions, and the fluid experiences little or no change in its potential energy as it flows through the device. If, in addition, the inlet velocity of the fluid is small relative to the exit velocity, the energy balance for this steady-flow device reduces to

$$h_1 = h_{2a} + \frac{V_{2a}^2}{2}$$

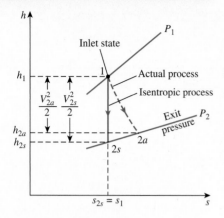

**FIGURE 7–53**
The h-s diagram of the actual and isentropic processes of an adiabatic nozzle.

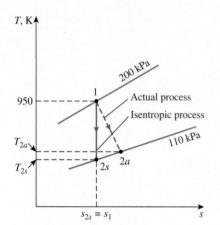

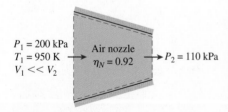

**FIGURE 7–54**
Schematic and T-s diagram for Example 7–16.

Then the isentropic efficiency of the nozzle can be expressed in terms of enthalpies as

$$\eta_N \cong \frac{h_1 - h_{2a}}{h_1 - h_{2s}} \qquad (7\text{–}67)$$

where $h_{2a}$ and $h_{2s}$ are the enthalpy values at the nozzle exit for the actual and isentropic processes, respectively (Fig. 7–53). Isentropic efficiencies of nozzles are typically above 90 percent, and nozzle efficiencies above 95 percent are not uncommon.

---

**EXAMPLE 7–16**    **Effect of Efficiency on Nozzle Exit Velocity**

Air at 200 kPa and 950 K enters an adiabatic nozzle at low velocity and is discharged at a pressure of 110 kPa. If the isentropic efficiency of the nozzle is 92 percent, determine (a) the maximum possible exit velocity, (b) the exit temperature, and (c) the actual exit velocity of the air. Assume constant specific heats for air.

**SOLUTION**    The acceleration of air in a nozzle is considered. For specified exit pressure and isentropic efficiency, the maximum and actual exit velocities and the exit temperature are to be determined.
**Assumptions**    1 Steady operating conditions exist. 2 Air is an ideal gas. 3 The inlet kinetic energy is negligible.
**Analysis**    A sketch of the system and the T-s diagram of the process are given in Fig. 7–54.

The temperature of air will drop during this acceleration process because some of its internal energy is converted to kinetic energy. This problem can be solved accurately by using property data from the air table. But we will assume constant specific heats (thus sacrifice some accuracy) to demonstrate their use. Let us guess the average temperature of the air to be about 850 K. Then, the average values of $c_p$ and $k$ at this anticipated average temperature are determined from Table A–2b to be $c_p$ = 1.11 kJ/kg·K and $k$ = 1.349.

(a) The exit velocity of the air will be a maximum when the process in the nozzle involves no irreversibilities. The exit velocity in this case is determined from the steady-flow energy equation. However, first we need to determine the exit temperature. For the isentropic process of an ideal gas we have:

$$\frac{T_{2s}}{T_1} = \left(\frac{P_{2s}}{P_1}\right)^{(k-1)/k}$$

or

$$T_{2s} = T_1\left(\frac{P_{2s}}{P_1}\right)^{(k-1)/k} = (950 \text{ K})\left(\frac{110 \text{ kPa}}{200 \text{ kPa}}\right)^{0.349/1.349} = 814 \text{ K}$$

This gives an average temperature of 882 K, which is somewhat higher than the assumed average temperature (850 K). This result could be refined by reevaluating the $k$ value at 882 K and repeating the calculations, however, it is not warranted since the two average temperatures are sufficiently close (doing so would change the temperature by only 0.6 K, which is not significant).

Now we can determine the isentropic exit velocity of the air from the energy balance for this isentropic steady-flow process:

$$e_{in} = e_{out}$$

$$h_1 + \frac{V_1^2}{2} = h_{2s} + \frac{V_{2s}^2}{2}$$

or

$$V_{2s} = \sqrt{2(h_1 - h_{2s})} = \sqrt{2c_{p,avg}(T_1 - T_{2s})}$$

$$= \sqrt{2(1.11 \text{ kJ/kg·K})[(950 - 814) \text{ K}]\left(\frac{1000 \text{ m}^2/\text{s}^2}{1 \text{ kJ/kg}}\right)}$$

$$= \textbf{549 m/s}$$

(b) The actual exit temperature of the air is higher than the isentropic exit temperature evaluated above and is determined from

$$\eta_N \cong \frac{h_1 - h_{2a}}{h_1 - h_{2s}} = \frac{c_{p,avg}(T_1 - T_{2a})}{c_{p,avg}(T_1 - T_{2s})}$$

or

$$0.92 = \frac{950 - T_{2a}}{950 - 814} \rightarrow T_{2a} = \textbf{825 K}$$

That is, the temperature is 11 K higher at the exit of the actual nozzle as a result of irreversibilities such as friction. It represents a loss since this rise in temperature comes at the expense of kinetic energy (Fig. 7–55).

(c) The actual exit velocity of air can be determined from the definition of isentropic efficiency of a nozzle,

$$\eta_N = \frac{V_{2a}^2}{V_{2s}^2} \rightarrow V_{2a} = \sqrt{\eta_N V_{2s}^2} = \sqrt{0.92(549 \text{ m/s})^2} = \textbf{527 m/s}$$

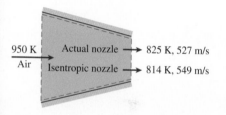

**FIGURE 7–55**
A substance leaves actual nozzles at a higher temperature (thus a lower velocity) as a result of friction.

# 7–13 · ENTROPY BALANCE

The property *entropy* is a measure of molecular disorder or randomness of a system, and the second law of thermodynamics states that entropy can be created but it cannot be destroyed. Therefore, the entropy change of a system during a process is greater than the entropy transfer by an amount equal to the entropy generated during the process within the system, and the *increase of entropy principle* for any system is expressed as (Fig. 7–56)

$$\begin{pmatrix} \text{Total} \\ \text{entropy} \\ \text{entering} \end{pmatrix} - \begin{pmatrix} \text{Total} \\ \text{entropy} \\ \text{leaving} \end{pmatrix} + \begin{pmatrix} \text{Total} \\ \text{entropy} \\ \text{generated} \end{pmatrix} = \begin{pmatrix} \text{Change in the} \\ \text{total entropy} \\ \text{of the system} \end{pmatrix}$$

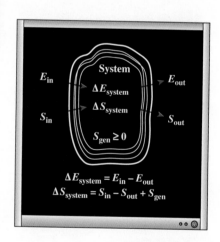

**FIGURE 7–56**
Energy and entropy balances for a system.

or

$$S_{in} - S_{out} + S_{gen} = \Delta S_{system} \qquad (7\text{--}68)$$

which is a verbal statement of Eq. 7–9. This relation is often referred to as the **entropy balance** and is applicable to any system undergoing any process. The entropy balance relation above can be stated as: *the entropy change of a system during a process is equal to the net entropy transfer through the system boundary and the entropy generated within the system.* Next, we discuss the various terms in that relation.

## Entropy Change of a System, $\Delta S_{system}$

Despite the reputation of entropy as being vague and abstract and the intimidation associated with it, entropy balance is actually easier to deal with than energy balance since, unlike energy, entropy does not exist in various forms. Therefore, the determination of entropy change of a system during a process involves evaluating entropy of the system at the beginning and at the end of the process and taking their difference. That is,

Entropy change = Entropy at final state − Entropy at initial state

or

$$\Delta S_{system} = S_{final} - S_{initial} = S_2 - S_1 \qquad (7\text{--}69)$$

Note that entropy is a property, and the value of a property does not change unless the state of the system changes. Therefore, the entropy change of a system is zero if the state of the system does not change during the process. For example, the entropy change of steady-flow devices such as nozzles, compressors, turbines, pumps, and heat exchangers is zero during steady operation.

When the properties of the system are not uniform, the entropy of the system can be determined by integration from

$$S_{system} = \int s\, \delta m = \int_V s \rho\, dV \qquad (7\text{--}70)$$

where $V$ is the volume of the system and $\rho$ is density.

## Mechanisms of Entropy Transfer, $S_{in}$ and $S_{out}$

Entropy can be transferred to or from a system by two mechanisms: *heat transfer* and *mass flow* (in contrast, energy is transferred by work also). Entropy transfer is recognized at the system boundary as it crosses the boundary, and it represents the entropy gained or lost by a system during a process. The only form of entropy interaction associated with a fixed mass or closed system is *heat transfer,* and thus the entropy transfer for an adiabatic closed system is zero.

## 1 Heat Transfer

Heat is, in essence, a form of disorganized energy, and some disorganization (entropy) will flow with heat. Heat transfer to a system increases the entropy of that system and thus the level of molecular disorder or randomness, and heat transfer from a system decreases it. In fact, heat rejection is the only way the entropy of a fixed mass can be decreased. The ratio of the heat transfer $Q$ at a location to the absolute temperature $T$ at that location is called the *entropy flow* or *entropy transfer* and is expressed as (Fig. 7–57)

*Entropy transfer by heat transfer:*  $\qquad S_{\text{heat}} = \dfrac{Q}{T}$   $(T = \text{constant})$  (7–71)

The quantity $Q/T$ represents the entropy transfer accompanied by heat transfer, and the direction of entropy transfer is the same as the direction of heat transfer since thermodynamic temperature $T$ is always a positive quantity.

When the temperature $T$ is not constant, the entropy transfer during a process 1-2 can be determined by integration (or by summation if appropriate) as

$$S_{\text{heat}} = \int_1^2 \frac{\delta Q}{T} \cong \sum \frac{Q_k}{T_k} \qquad (7\text{–}72)$$

where $Q_k$ is the heat transfer through the boundary at temperature $T_k$ at location $k$.

When two systems are in contact, the entropy transfer from the warmer system is equal to the entropy transfer into the cooler one at the point of contact. That is, no entropy can be created or destroyed at the boundary since the boundary has no thickness and occupies no volume.

Note that **work** is entropy-free, and no entropy is transferred by work. Energy is transferred by both heat and work, whereas entropy is transferred only by heat. That is,

*Entropy transfer by work:*  $\qquad S_{\text{work}} = 0$  (7–73)

The first law of thermodynamics makes no distinction between heat transfer and work; it considers them as *equals*. The distinction between heat transfer and work is brought out by the second law: *an energy interaction that is accompanied by entropy transfer is heat transfer, and an energy interaction that is not accompanied by entropy transfer is work.* That is, no entropy is exchanged during a work interaction between a system and its surroundings. Thus, only *energy* is exchanged during work interaction whereas both *energy* and *entropy* are exchanged during heat transfer (Fig. 7–58).

## 2 Mass Flow

Mass contains entropy as well as energy, and the entropy and energy contents of a system are proportional to the mass. (When the mass of a system is doubled, so are the entropy and energy contents of the system.) Both entropy and energy are carried into or out of a system by streams of matter, and the rates of entropy and energy transport into or out of a system are proportional to the mass flow rate. Closed systems do not involve any mass flow and thus any entropy transfer by mass. When a mass in the amount of $m$ enters or leaves a system, entropy in the amount of $ms$, where $s$ is the specific entropy (entropy per unit mass entering or leaving), accompanies it (Fig. 7–59). That is,

*Entropy transfer by mass flow:*  $\qquad S_{\text{mass}} = ms$  (7–74)

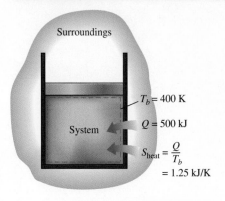

**FIGURE 7–57**
Heat transfer is always accompanied by entropy transfer in the amount of $Q/T$, where $T$ is the boundary temperature.

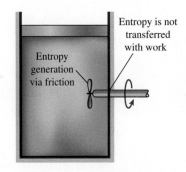

**FIGURE 7–58**
No entropy accompanies work as it crosses the system boundary. But entropy may be generated within the system as work is dissipated into a less useful form of energy.

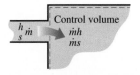

**FIGURE 7–59**
Mass contains entropy as well as energy, and thus mass flow into or out of system is always accompanied by energy and entropy transfer.

Therefore, the entropy of a system increases by $ms$ when mass in the amount of $m$ enters and decreases by the same amount when the same amount of mass at the same state leaves the system. When the properties of the mass change during the process, the entropy transfer by mass flow can be determined by integration from

$$\dot{S}_{mass} = \int_{A_c} s\rho V_n \, dA_c \quad \text{and} \quad S_{mass} = \int s \, \delta m = \int_{\Delta t} \dot{S}_{mass} \, dt \qquad \text{(7-75)}$$

where $A_c$ is the cross-sectional area of the flow and $V_n$ is the local velocity normal to $dA_c$.

## Entropy Generation, $S_{gen}$

Irreversibilities such as friction, mixing, chemical reactions, heat transfer through a finite temperature difference, unrestrained expansion, nonquasi-equilibrium compression, or expansion always cause the entropy of a system to increase, and entropy generation is a measure of the entropy created by such effects during a process.

For a *reversible process* (a process that involves no irreversibilities), the entropy generation is zero and thus the *entropy change* of a system is equal to the *entropy transfer*. Therefore, the entropy balance relation in the reversible case becomes analogous to the energy balance relation, which states that *energy change* of a system during a process is equal to the *energy transfer* during that process. However, note that the energy change of a system equals the energy transfer for *any* process, but the entropy change of a system equals the entropy transfer only for a *reversible* process.

The entropy transfer by heat $Q/T$ is zero for adiabatic systems, and the entropy transfer by mass $ms$ is zero for systems that involve no mass flow across their boundary (i.e., closed systems).

Entropy balance for *any system* undergoing *any process* can be expressed more explicitly as

$$\underbrace{S_{in} - S_{out}}_{\substack{\text{Net entropy transfer} \\ \text{by heat and mass}}} + \underbrace{S_{gen}}_{\substack{\text{Entropy} \\ \text{generation}}} = \underbrace{\Delta S_{system}}_{\substack{\text{Change} \\ \text{in entropy}}} \quad \text{(kJ/K)} \qquad \text{(7-76)}$$

or, in the **rate form**, as

$$\underbrace{\dot{S}_{in} - \dot{S}_{out}}_{\substack{\text{Rate of net entropy} \\ \text{transfer by heat} \\ \text{and mass}}} + \underbrace{\dot{S}_{gen}}_{\substack{\text{Rate of entropy} \\ \text{generation}}} = \underbrace{dS_{system}/dt}_{\substack{\text{Rate of change} \\ \text{in entropy}}} \quad \text{(kW/K)} \qquad \text{(7-77)}$$

where the rates of entropy transfer by heat transferred at a rate of $\dot{Q}$ and mass flowing at a rate of $\dot{m}$ are $\dot{S}_{heat} = \dot{Q}/T$ and $\dot{S}_{mass} = \dot{m}s$. The entropy balance can also be expressed on a **unit-mass basis** as

$$(s_{in} - s_{out}) + s_{gen} = \Delta s_{system} \quad \text{(kJ/kg·K)} \qquad \text{(7-78)}$$

where all the quantities are expressed per unit mass of the system. Note that for a *reversible process,* the entropy generation term $S_{gen}$ drops out from all of the relations above.

The term $S_{gen}$ represents the entropy generation *within the system boundary* only (Fig. 7–60), and not the entropy generation that may occur outside the system boundary during the process as a result of external irreversibilities. Therefore, a process for which $S_{gen} = 0$ is *internally reversible,* but not necessarily *totally* reversible. The *total* entropy generated during a process can be determined by applying the entropy balance to an *extended system* that includes the system itself and its immediate surroundings where external irreversibilities might be occurring (Fig. 7–61). Also, the entropy change in this case is equal to the sum of the entropy change of the system and the entropy change of the immediate surroundings. Note that under steady conditions, the state and thus the entropy of the immediate surroundings (let us call it the "buffer zone") at any point does not change during the process, and the entropy change of the buffer zone is zero. The entropy change of the buffer zone, if any, is usually small relative to the entropy change of the system, and thus it is usually disregarded.

When evaluating the entropy transfer between an extended system and the surroundings, the boundary temperature of the extended system is simply taken to be the *environment temperature.*

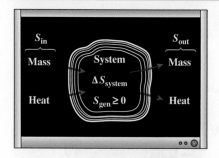

**FIGURE 7–60**
Mechanisms of entropy transfer for a general system.

## Closed Systems

A closed system involves *no mass flow* across its boundaries, and its entropy change is simply the difference between the initial and final entropies of the system. The *entropy change* of a closed system is due to the *entropy transfer* accompanying heat transfer and the *entropy generation* within the system boundaries. Taking the positive direction of heat transfer to be *to* the system, the general entropy balance relation (Eq. 7–76) can be expressed for a closed system as

*Closed system:* $\qquad \sum \dfrac{Q_k}{T_k} + S_{gen} = \Delta S_{system} = S_2 - S_1 \quad$ (kJ/K) $\qquad$ **(7–79)**

The entropy balance relation above can be stated as:

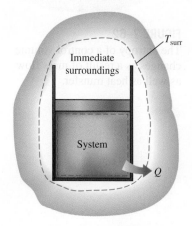

**FIGURE 7–61**
Entropy generation outside system boundaries can be accounted for by writing an entropy balance on an extended system that includes the system and its immediate surroundings.

> The entropy change of a closed system during a process is equal to the sum of the net entropy transferred through the system boundary by heat transfer and the entropy generated within the system boundaries.

For an *adiabatic process* ($Q = 0$), the entropy transfer term in the above relation drops out and the entropy change of the closed system becomes equal to the entropy generation within the system boundaries. That is,

*Adiabatic closed system:* $\qquad S_{gen} = \Delta S_{adiabatic\ system}$ $\qquad$ **(7–80)**

Noting that any closed system and its surroundings can be treated as an adiabatic system and the total entropy change of a system is equal to the sum of the entropy changes of its parts, the entropy balance for a closed system and its surroundings can be written as

*System + Surroundings:* $\qquad S_{gen} = \sum \Delta S = \Delta S_{system} + \Delta S_{surroundings}$ $\qquad$ **(7–81)**

where $\Delta S_{system} = m(s_2 - s_1)$ and the entropy change of the surroundings can be determined from $\Delta S_{surr} = Q_{surr}/T_{surr}$ if its temperature is constant. At initial

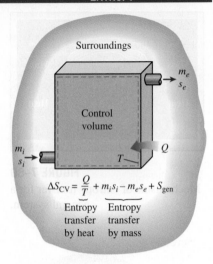

Surroundings

Control
volume

$m_i$
$s_i$

$m_e$
$s_e$

$Q$

$T$

$$\Delta S_{CV} = \frac{Q}{T} + m_i s_i - m_e s_e + S_{gen}$$

Entropy          Entropy
transfer         transfer
by heat          by mass

**FIGURE 7–62**
The entropy of a control volume
changes as a result of mass flow
as well as heat transfer.

stages of studying entropy and entropy transfer, it is more instructive to start with the general form of the entropy balance (Eq. 7–76) and to simplify it for the problem under consideration. The specific relations above are convenient to use after a certain degree of intuitive understanding of the material is achieved.

## Control Volumes

The entropy balance relations for control volumes differ from those for closed systems in that they involve one more mechanism of entropy exchange: *mass flow across the boundaries.* As mentioned earlier, mass possesses entropy as well as energy, and the amounts of these two extensive properties are proportional to the amount of mass (Fig. 7–62).

Taking the positive direction of heat transfer to be *to* the system, the general entropy balance relations (Eqs. 7–76 and 7–77) can be expressed for control volumes as

$$\sum \frac{Q_k}{T_k} + \sum m_i s_i - \sum m_e s_e + S_{gen} = (S_2 - S_1)_{CV} \quad \text{(kJ/K)} \quad \text{(7–82)}$$

or, in the rate form, as

$$\sum \frac{\dot{Q}_k}{T_k} + \sum \dot{m}_i s_i - \sum \dot{m}_e s_e + \dot{S}_{gen} = dS_{CV}/dt \quad \text{(kW/K)} \quad \text{(7–83)}$$

This entropy balance relation can be stated as:

> The rate of entropy change within the control volume during a process is equal to the sum of the rate of entropy transfer through the control volume boundary by heat transfer, the net rate of entropy transfer into the control volume by mass flow, and the rate of entropy generation within the boundaries of the control volume as a result of irreversibilities.

Most control volumes encountered in practice such as turbines, compressors, nozzles, diffusers, heat exchangers, pipes, and ducts operate steadily, and thus they experience no change in their entropy. Therefore, the entropy balance relation for a general **steady-flow process** can be obtained from Eq. 7–83 by setting $dS_{CV}/dt = 0$ and rearranging to give

*Steady-flow:* $\quad \dot{S}_{gen} = \sum \dot{m}_e s_e - \sum \dot{m}_i s_i - \sum \frac{\dot{Q}_k}{T_k}$ $\quad$ (7–84)

For *single-stream* (one inlet and one exit) steady-flow devices, the entropy balance relation simplifies to

*Steady-flow, single-stream:* $\quad \dot{S}_{gen} = \dot{m}(s_e - s_i) - \sum \frac{\dot{Q}_k}{T_k}$ $\quad$ (7–85)

For the case of an *adiabatic* single-stream device, the entropy balance relation further simplifies to

*Steady-flow, single-stream, adiabatic:* $\quad \dot{S}_{gen} = \dot{m}(s_e - s_i)$ $\quad$ (7–86)

which indicates that the specific entropy of the fluid must increase as it flows through an adiabatic device since $\dot{S}_{gen} \geq 0$ (Fig. 7–63). If the flow through the device is *reversible* and *adiabatic,* then the entropy remains constant, $s_e = s_i$, regardless of the changes in other properties.

### EXAMPLE 7–17    Entropy Generation in a Wall

Consider steady heat transfer through a 5-m × 7-m brick wall of a house of thickness 30 cm. On a day when the temperature of the outdoors is 0°C, the house is maintained at 27°C. The temperatures of the inner and outer surfaces of the brick wall are measured to be 20°C and 5°C, respectively, and the rate of heat transfer through the wall is 1035 W. Determine the rate of entropy generation in the wall, and the rate of total entropy generation associated with this heat transfer process.

**SOLUTION**    Steady heat transfer through a wall is considered. For specified heat transfer rate, wall temperatures, and environment temperatures, the entropy generation rate within the wall and the total entropy generation rate are to be determined.

**Assumptions**  **1** The process is steady, and thus the rate of heat transfer through the wall is constant. **2** Heat transfer through the wall is one-dimensional.

**Analysis**  We first take the *wall* as the system (Fig. 7–64). This is a *closed system* since no mass crosses the system boundary during the process. We note that the entropy change of the wall is zero during this process since the state and thus the entropy of the wall do not change anywhere in the wall. Heat and entropy are entering from one side of the wall and leaving from the other side.

The rate form of the entropy balance for the wall simplifies to

$$\underbrace{\dot{S}_{in} - \dot{S}_{out}}_{\substack{\text{Rate of net entropy} \\ \text{transfer by heat} \\ \text{and mass}}} + \underbrace{\dot{S}_{gen}}_{\substack{\text{Rate of entropy} \\ \text{generation}}} = \underbrace{dS_{system}/dt}_{\substack{\text{Rate of change} \\ \text{in entropy}}}^{\;0\ (steady)}$$

$$\left(\frac{\dot{Q}}{T}\right)_{in} - \left(\frac{\dot{Q}}{T}\right)_{out} + \dot{S}_{gen} = 0$$

$$\frac{1035\ \text{W}}{293\ \text{K}} - \frac{1035\ \text{W}}{278\ \text{K}} + \dot{S}_{gen} = 0$$

Therefore, the rate of entropy generation in the wall is

$$\dot{S}_{gen} = \mathbf{0.191\ W/K}$$

Note that entropy transfer by heat at any location is $Q/T$ at that location, and the direction of entropy transfer is the same as the direction of heat transfer.

To determine the rate of total entropy generation during this heat transfer process, we extend the system to include the regions on both sides of the wall that experience a temperature change. Then, one side of the system boundary becomes room temperature while the other side becomes the temperature of the outdoors. The entropy balance for this *extended*

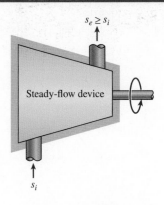

**FIGURE 7–63**
The entropy of a substance always increases (or remains constant in the case of a reversible process) as it flows through a single-stream, adiabatic, steady-flow device.

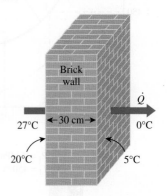

**FIGURE 7–64**
Schematic for Example 7–17.

*system* (system + immediate surroundings) is the same as that given above, except the two boundary temperatures are now 300 and 273 K instead of 293 and 278 K, respectively. Then the rate of total entropy generation becomes

$$\frac{1035 \text{ W}}{300 \text{ K}} - \frac{1035 \text{ W}}{273 \text{ K}} + \dot{S}_{\text{gen,total}} = 0 \quad \rightarrow \quad \dot{S}_{\text{gen,total}} = \textbf{0.341 W/K}$$

**Discussion** Note that the entropy change of this extended system is also zero since the state of air does not change at any point during the process. The differences between the two entropy generations is 0.150 W/K, and it represents the entropy generated in the air layers on both sides of the wall. The entropy generation in this case is entirely due to irreversible heat transfer through a finite temperature difference.

***EXAMPLE 7–18***     **Entropy Generation During a Throttling Process**

Steam at 7 MPa and 450°C is throttled in a valve to a pressure of 3 MPa during a steady-flow process. Determine the entropy generated during this process and check if the increase of entropy principle is satisfied.

**SOLUTION** Steam is throttled to a specified pressure. The entropy generated during this process is to be determined, and the validity of the increase of entropy principle is to be verified.
***Assumptions*** **1** This is a steady-flow process since there is no change with time at any point and thus $\Delta m_{\text{CV}} = 0$, $\Delta E_{\text{CV}} = 0$, and $\Delta S_{\text{CV}} = 0$. **2** Heat transfer to or from the valve is negligible. **3** The kinetic and potential energy changes are negligible, $\Delta \text{ke} = \Delta \text{pe} = 0$.
***Analysis*** We take the throttling valve as the *system* (Fig. 7–65). This is a *control volume* since mass crosses the system boundary during the process. We note that there is only one inlet and one exit and thus $\dot{m}_1 = \dot{m}_2 = \dot{m}$. Also, the enthalpy of a fluid remains nearly constant during a throttling process and thus $h_2 \cong h_1$.

The entropy of the steam at the inlet and the exit states is determined from the steam tables to be

*State 1:*    $\left. \begin{array}{l} P_1 = 7 \text{ MPa} \\ T_1 = 450°C \end{array} \right\}$   $\begin{array}{l} h_1 = 3288.3 \text{ kJ/kg} \\ s_1 = 6.6353 \text{ kJ/kg·K} \end{array}$

*State 2:*    $\left. \begin{array}{l} P_2 = 3 \text{ MPa} \\ h_2 = h_1 \end{array} \right\}$   $s_2 = 7.0046 \text{ kJ/kg·K}$

Then, the entropy generation per unit mass of the steam is determined from the entropy balance applied to the throttling valve,

$$\underbrace{\dot{S}_{\text{in}} - \dot{S}_{\text{out}}}_{\substack{\text{Rate of net entropy} \\ \text{transfer by heat} \\ \text{and mass}}} + \underbrace{\dot{S}_{\text{gen}}}_{\substack{\text{Rate of entropy} \\ \text{generation}}} = \underbrace{dS_{\text{system}}/dt}_{\substack{\text{Rate of change} \\ \text{in entropy}}} \overset{\text{0 (steady)}}{\nearrow}$$

$$\dot{m}s_1 - \dot{m}s_2 + \dot{S}_{\text{gen}} = 0$$

$$\dot{S}_{\text{gen}} = \dot{m}(s_2 - s_1)$$

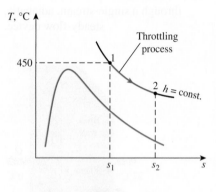

$T$, °C

Throttling process

450

1

2   $h = \text{const.}$

$s_1$    $s_2$    $s$

$P_1 = 7$ MPa
$T_1 = 450$ °C

$P_2 = 3$ MPa

**FIGURE 7–65**
Schematic and *T-s* diagram for Example 7–18.

Dividing by mass flow rate and substituting gives

$$s_{gen} = s_2 - s_1 = 7.0046 - 6.6353 = 0.3693 \text{ kJ/kg·K}$$

This is the amount of entropy generated per unit mass of steam as it is throttled from the inlet state to the final pressure, and it is caused by unrestrained expansion. The increase of entropy principle is obviously satisfied during this process since the entropy generation is positive.

■ **EXAMPLE 7–19**　**Entropy Generated when a Hot Block Is Dropped in a Lake**

A 50-kg block of iron casting at 500 K is thrown into a large lake that is at a temperature of 285 K. The iron block eventually reaches thermal equilibrium with the lake water. Assuming an average specific heat of 0.45 kJ/kg·K for the iron, determine (a) the entropy change of the iron block, (b) the entropy change of the lake water, and (c) the entropy generated during this process.

**SOLUTION** A hot iron block is thrown into a lake, and cools to the lake temperature. The entropy changes of the iron and of the lake as well as the entropy generated during this process are to be determined.
**Assumptions** 1 Both the water and the iron block are incompressible substances. 2 Constant specific heats can be used for the water and the iron. 3 The kinetic and potential energy changes of the iron are negligible, $\Delta KE = \Delta PE = 0$ and thus $\Delta E = \Delta U$.
**Properties** The specific heat of the iron is 0.45 kJ/kg·K (Table A–3).
**Analysis** We take the *iron casting* as the system (Fig. 7–66). This is a *closed system* since no mass crosses the system boundary during the process.

To determine the entropy change for the iron block and for the lake, first we need to know the final equilibrium temperature. Given that the thermal energy capacity of the lake is very large relative to that of the iron block, the lake will absorb all the heat rejected by the iron block without experiencing any change in its temperature. Therefore, the iron block will cool to 285 K during this process while the lake temperature remains constant at 285 K.
(a) The entropy change of the iron block can be determined from

$$\Delta S_{iron} = m(s_2 - s_1) = mc_{avg} \ln\frac{T_2}{T_1}$$
$$= (50 \text{ kg})(0.45 \text{ kJ/kg·K}) \ln\frac{285 \text{ K}}{500 \text{ K}}$$
$$= -12.65 \text{ kJ/K}$$

(b) The temperature of the lake water remains constant during this process at 285 K. Also, the amount of heat transfer from the iron block to the lake is determined from an energy balance on the iron block to be

$$\underbrace{E_{in} - E_{out}}_{\substack{\text{Net energy transfer by} \\ \text{heat, work, and mass}}} = \underbrace{\Delta E_{system}}_{\substack{\text{Change in internal, kinetic,} \\ \text{potential, etc., energies}}}$$

$$-Q_{out} = \Delta U = mc_{avg}(T_2 - T_1)$$

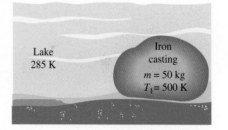

Lake
285 K

Iron casting
$m = 50$ kg
$T_1 = 500$ K

**FIGURE 7–66**
Schematic for Example 7–19.

or

$$Q_{\text{out}} = mc_{\text{avg}}(T_1 - T_2) = (50 \text{ kg})(0.45 \text{ kJ/kg·K})(500 - 285) \text{ K} = 4838 \text{ kJ}$$

Then, the entropy change of the lake becomes

$$\Delta S_{\text{lake}} = \frac{Q_{\text{lake}}}{T_{\text{lake}}} = \frac{+4838 \text{ kJ}}{285 \text{ K}} = \mathbf{16.97 \text{ kJ/K}}$$

(*c*) The entropy generated during this process can be determined by applying an entropy balance on an *extended system* that includes the iron block and its immediate surroundings so that the boundary temperature of the extended system is at 285 K at all times:

$$\underbrace{S_{\text{in}} - S_{\text{out}}}_{\substack{\text{Net entropy transfer} \\ \text{by heat and mass}}} + \underbrace{S_{\text{gen}}}_{\substack{\text{Entropy} \\ \text{generation}}} = \underbrace{\Delta S_{\text{system}}}_{\substack{\text{Change} \\ \text{in entropy}}}$$

$$-\frac{Q_{\text{out}}}{T_b} + S_{\text{gen}} = \Delta S_{\text{system}}$$

or

$$S_{\text{gen}} = \frac{Q_{\text{out}}}{T_b} + \Delta S_{\text{system}} = \frac{4838 \text{ kJ}}{285 \text{ K}} - (12.65 \text{ kJ/K}) = \mathbf{4.32 \text{ kJ/K}}$$

**Discussion** The entropy generated can also be determined by taking the iron block and the entire lake as the system, which is an isolated system, and applying an entropy balance. An isolated system involves no heat or entropy transfer, and thus the entropy generation in this case becomes equal to the total entropy change,

$$S_{\text{gen}} = \Delta S_{\text{total}} = \Delta S_{\text{system}} + \Delta S_{\text{lake}} = -12.65 + 16.97 = 4.32 \text{ kJ/K}$$

which is the same result obtained above.

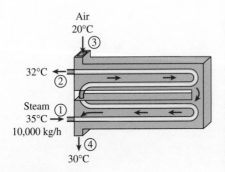

**FIGURE 7–67**
Schematic for Example 7–20

### EXAMPLE 7–20    Entropy Generation in a Heat Exchanger

Air in a large building is kept warm by heating it with steam in a heat exchanger (Fig. 7–67). Saturated water vapor enters this unit at 35°C at a rate of 10,000 kg/h and leaves as saturated liquid at 32°C. Air at 1-atm pressure enters the unit at 20°C and leaves at 30°C at about the same pressure. Determine the rate of entropy generation associated with this process.

**SOLUTION** Air is heated by steam in a heat exchanger. The rate of entropy generation associated with this process is to be determined.
**Assumptions** 1 Steady operating conditions exist. 2 The heat exchanger is well-insulated so that heat loss to the surroundings is negligible and thus heat transfer from the hot fluid is equal to the heat transfer to the cold fluid. 3 Changes in the kinetic and potential energies of fluid streams are negligible. 4 Air is an ideal gas with constant specific heats at room temperature. 5 The pressure of air remains constant.

*Analysis* The rate of entropy generation within the heat exchanger is determined by applying the rate form of the entropy balance on the entire heat exchanger:

$$\underbrace{\dot{S}_{in} - \dot{S}_{out}}_{\substack{\text{Rate of net entropy} \\ \text{transfer by heat} \\ \text{and mass}}} + \underbrace{\dot{S}_{gen}}_{\substack{\text{Rate of entropy} \\ \text{generation}}} = \underbrace{dS_{system}/dt}_{\substack{\text{Rate of change} \\ \text{in entropy}}} \nearrow^{0 \text{ (steady)}}$$

$$\dot{m}_{steam} s_1 + \dot{m}_{air} s_3 - \dot{m}_{steam} s_2 - \dot{m}_{air} s_4 + \dot{S}_{gen} = 0$$

$$\dot{S}_{gen} = \dot{m}_{steam}(s_2 - s_1) + \dot{m}_{air}(s_4 - s_3)$$

The specific heat of air at room temperature is $c_p = 1.005$ kJ/kg·°C (Table A–2a). The properties of the steam at the inlet and exit states are

$$\left. \begin{array}{l} T_1 = 35°C \\ x_1 = 1 \end{array} \right\} \quad \begin{array}{l} h_1 = 2564.6 \text{ kJ/kg} \\ s_1 = 8.3517 \text{ kJ/kg·K} \end{array} \quad \text{(Table A–4)}$$

$$\left. \begin{array}{l} T_2 = 32°C \\ x_2 = 0 \end{array} \right\} \quad \begin{array}{l} h_2 = 134.10 \text{ kJ/kg} \\ s_2 = 0.4641 \text{ kJ/kg·K} \end{array} \quad \text{(Table A–4)}$$

From an energy balance the heat transferred from steam is equal to the heat transferred to the air. Then, the mass flow rate of air is determined to be

$$\dot{Q} = \dot{m}_{steam}(h_1 - h_2) = (10,000/3600 \text{ kg/s})(2564.6 - 134.10) \text{ kJ/kg} = 6751 \text{ kW}$$

$$\dot{m}_{air} = \frac{\dot{Q}}{c_p(T_4 - T_3)} = \frac{6751 \text{ kW}}{(1.005 \text{ kJ/kg·°C})(30 - 20)°C} = 671.7 \text{ kg/s}$$

Substituting into the entropy balance relation, the rate of entropy generation becomes

$$\dot{S}_{gen} = \dot{m}_{steam}(s_2 - s_1) + \dot{m}_{air}(s_4 - s_3)$$
$$= \dot{m}_{steam}(s_2 - s_1) + \dot{m}_{air} c_p \ln\frac{T_4}{T_3}$$
$$= (10,000/3600 \text{ kg/s})(0.4641 - 8.3517) \text{ kJ/kg·K}$$
$$+ (671.7 \text{ kg/s})(1.005 \text{ kJ/kg·K}) \ln\frac{303 \text{ K}}{293 \text{ K}}$$
$$= 0.745 \text{ kW/K}$$

*Discussion* Note that the pressure of air remains nearly constant as it flows through the heat exchanger, and thus the pressure term is not included in the entropy change expression for air.

## EXAMPLE 7–21    Entropy Generation Associated with Heat Transfer

A frictionless piston–cylinder device contains a saturated liquid–vapor mixture of water at 100°C. During a constant-pressure process, 600 kJ of heat is transferred to the surrounding air at 25°C. As a result, part of the water vapor contained in the cylinder condenses. Determine (*a*) the entropy

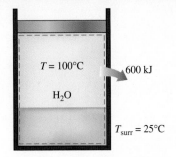

$T = 100°C$

$H_2O$

600 kJ

$T_{surr} = 25°C$

**FIGURE 7–68**
Schematic for Example 7–21.

change of the water and (*b*) the total entropy generation during this heat transfer process.

**SOLUTION** Saturated liquid–vapor mixture of water loses heat to its surroundings, and some of the vapor condenses. The entropy change of the water and the total entropy generation are to be determined.

**Assumptions** **1** There are no irreversibilities involved within the system boundaries, and thus the process is internally reversible. **2** The water temperature remains constant at 100°C everywhere, including the boundaries.

**Analysis** We first take the *water in the cylinder* as the system (Fig. 7–68). This is a *closed system* since no mass crosses the system boundary during the process. We note that the pressure and thus the temperature of water in the cylinder remain constant during this process. Also, the entropy of the system decreases during the process because of heat loss.

(*a*) Noting that water undergoes an internally reversible isothermal process, its entropy change can be determined from

$$\Delta S_{system} = \frac{Q}{T_{system}} = \frac{-600 \text{ kJ}}{(100 + 273 \text{ K})} = -1.61 \text{ kJ/K}$$

(*b*) To determine the total entropy generation during this process, we consider the *extended system*, which includes the water, the piston–cylinder device, and the region immediately outside the system that experiences a temperature change so that the entire boundary of the extended system is at the surrounding temperature of 25°C. The entropy balance for this *extended system* (system + immediate surroundings) yields

$$\underbrace{S_{in} - S_{out}}_{\substack{\text{Net entropy transfer} \\ \text{by heat and mass}}} + \underbrace{S_{gen}}_{\substack{\text{Entropy} \\ \text{generation}}} = \underbrace{\Delta S_{system}}_{\substack{\text{Change} \\ \text{in entropy}}}$$

$$-\frac{Q_{out}}{T_b} + S_{gen} = \Delta S_{sysem}$$

or

$$S_{gen} = \frac{Q_{out}}{T_b} + \Delta S_{system} = \frac{600 \text{ kJ}}{(25 + 273) \text{ K}} + (-1.61 \text{ kJ/K}) = 0.40 \text{ kJ/K}$$

The entropy generation in this case is entirely due to irreversible heat transfer through a finite temperature difference.

Note that the entropy change of this extended system is equivalent to the entropy change of water since the piston–cylinder device and the immediate surroundings do not experience any change of state at any point, and thus any change in any property, including entropy.

**Discussion** For the sake of argument, consider the reverse process (i.e., the transfer of 600 kJ of heat from the surrounding air at 25°C to saturated water at 100°C) and see if the increase of entropy principle can detect the impossibility of this process. This time, heat transfer will be to the water (heat gain instead of heat loss), and thus the entropy change of water will be +1.61 kJ/K. Also, the entropy transfer at the boundary of the extended system will have the same magnitude but opposite direction. This will result in an entropy generation of −0.4 kJ/K. The negative sign for the entropy generation indicates that the reverse process is *impossible*.

To complete the discussion, let us consider the case where the surrounding air temperature is a differential amount below 100°C (say 99.999 ... 9°C) instead of being 25°C. This time, heat transfer from the saturated water to the surrounding air will take place through a differential temperature difference rendering this process *reversible*. It can be shown that $S_{gen} = 0$ for this process.

Remember that reversible processes are idealized processes, and they can be approached but never reached in reality.

# Entropy Generation Associated with a Heat Transfer Process

In Example 7–21 it is determined that 0.4 kJ/K of entropy is generated during the heat transfer process, but it is not clear where exactly the entropy generation takes place, and how. To pinpoint the location of entropy generation, we need to be more precise about the description of the system, its surroundings, and the system boundary.

In that example, we assumed both the system and the surrounding air to be isothermal at 100 and 25°C, respectively. This assumption is reasonable if both fluids are well mixed. The inner surface of the wall must also be at 100°C while the outer surface is at 25°C since two bodies in physical contact must have the same temperature at the point of contact. Considering that entropy transfer with heat transfer $Q$ through a surface at constant temperature $T$ is $Q/T$, the entropy transfer from the water into the wall is $Q/T_{sys} = 1.61$ kJ/K. Likewise, entropy transfer from the outer surface of the wall into the surrounding air is $Q/T_{surr} = 2.01$ kJ/K. Obviously, entropy in the amount of $2.01 - 1.61 = 0.4$ kJ/K is generated in the wall, as illustrated in Fig. 7–69b.

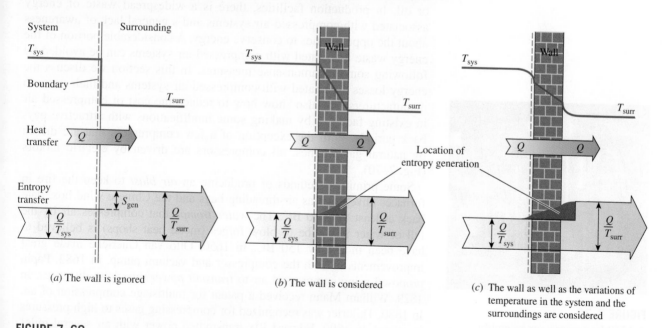

(a) The wall is ignored

(b) The wall is considered

(c) The wall as well as the variations of temperature in the system and the surroundings are considered

**FIGURE 7–69**

Graphical representation of entropy generation during a heat transfer process through a finite temperature difference.

Identifying the location of entropy generation enables us to determine whether a process is internally reversible. A process is internally reversible if no entropy is generated within the system boundaries. Therefore, the heat transfer process discussed in Example 7–21 is internally reversible if the inner surface of the wall is taken as the system boundary, and thus the system excludes the container wall. If the system boundary is taken to be the outer surface of the container wall, then the process is no longer internally reversible since the wall, which is the site of entropy generation, is now part of the system.

For thin walls, it is very tempting to ignore the mass of the wall and to regard the wall as the boundary between the system and the surroundings. This seemingly harmless choice hides the site of the entropy generation from view and is a source of confusion. The temperature in this case drops suddenly from $T_{sys}$ to $T_{surr}$ at the boundary surface, and confusion arises as to which temperature to use in the relation $Q/T$ for entropy transfer at the boundary.

Note that if the system and the surrounding air are not isothermal as a result of insufficient mixing, then part of the entropy generation will occur in both the system and the surrounding air in the vicinity of the wall, as shown in Fig. 7–69c.

## TOPIC OF SPECIAL INTEREST*     Reducing the Cost of Compressed Air

Compressed air at gage pressures of 550 to 1000 kPa is commonly used in industrial facilities to perform a wide variety of tasks such as *cleaning, operating pneumatic equipment,* and even *refrigeration.* It is often referred to as the *fourth utility* after electricity, water, and natural gas or oil. In production facilities, there is a widespread waste of energy associated with compressed-air systems and a general lack of awareness about the opportunities to conserve energy. A considerable portion of the energy waste associated with compressed-air systems can be avoided by following some commonsense measures. In this section we discuss the energy losses associated with compressed-air systems and their costs to manufacturers. We also show how to reduce the cost of compressed air in existing facilities by making some modifications with attractive payback periods. With the exception of a few compressors that are driven by natural gas engines, all compressors are driven by electric motors (Fig. 7–70).

Some primitive methods of producing an *air blast* to keep the fire in furnaces alive, such as air-threading bags and the Chinese wind box, date back at least to 2000 BC. The *water trompe* that compresses air by the fall of water in a tube to blow forges (metal heat shops) is believed to have been in use by 150 BC. In 1650, Otto van Guericke made great improvements in both the compressor and vacuum pump. In 1683, Papin proposed using compressed air to *transmit power* over long distances. In 1829, William Mann received a patent for multistage compression of air. In 1830, Thilorier was recognized for compressing gases to high pressures in *stages.* In 1890, Edward Rix transmitted power with air several miles to operate lifting machines in the North Star mine near Grass Valley,

**FIGURE 7–70**

A 1250-hp compressor assembly.

*Courtesy of Dresser Rand Company, Painted Post, NY*

California, by using a compressor driven by Pelton wheels. In 1872, *cooling* was adapted to increase efficiency by spraying water directly into the cylinder through the air inlet valves. This "wet compression" was abandoned later because of the problems it caused. The cooling then was accomplished externally by *water jacketing* the cylinders. The first large-scale compressor used in the United States was a four-cylinder unit built in 1866 for use in the Hoosac tunnel. The cooling was first accomplished by water injection into the cylinder, and later by running a stream of water over the cylinder. Major advances in recent compressor technology are due to Burleigh, Ingersoll, Sergeant, Rand, and Clayton, among others.

The compressors used range from a few horsepower to more than 10,000 hp in size, and they are among the major energy-consuming equipment in most manufacturing facilities. Manufacturers are quick to identify energy (and thus money) losses from *hot surfaces* and to insulate those surfaces. However, somehow they are not so sensitive when it comes to saving *compressed air* since they view air as being free, and the only time the air leaks and dirty air filters get some attention is when the air and pressure losses interfere with the normal operation of the plant. However, paying attention to the compressed-air system and practicing some simple conservation measures can result in considerable energy and cost savings for the plants.

The hissing of *air leaks* can sometimes be heard even in high-noise manufacturing facilities. *Pressure drops* at end-use points in the order of 40 percent of the compressor-discharged pressure are not uncommon. Yet a common response to such a problem is the installation of a larger compressor instead of checking the system and finding out what the problem is. The latter corrective action is usually taken only after the larger compressor also fails to eliminate the problem. The energy wasted in compressed-air systems because of poor installation and maintenance can account for up to 50 percent of the energy consumed by the compressor, and about half of this amount can be saved by simple measures.

The cost of electricity to operate a compressor for one year can exceed the purchase price of the compressor. This is especially the case for larger compressors operating two or three shifts. For example, operating a 125-hp compressor powered by a 90-percent efficient electric motor at full load for 6000 hours a year at $0.085/kWh will cost $52,820 a year in electricity cost, which greatly exceeds the purchase and installation cost of a typical unit (Fig. 7–71).

Below we describe some procedures to reduce the cost of compressed air in industrial facilities and quantify the energy and cost savings associated with them. Once the compressor power wasted is determined, the *annual energy* (usually electricity) and *cost savings* can be determined from

$$\text{Energy savings} = (\text{Power saved})(\text{Operating hours})/\eta_{\text{motor}} \qquad \textbf{(7–87)}$$

and

$$\text{Cost savings} = (\text{Energy savings})(\text{Unit cost of energy}) \qquad \textbf{(7–88)}$$

where $\eta_{\text{motor}}$ is the efficiency of the motor driving the compressor and the unit cost of energy is usually expressed in dollars per kilowatt hour (1 kWh = 3600 kJ).

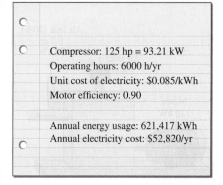

Compressor: 125 hp = 93.21 kW
Operating hours: 6000 h/yr
Unit cost of electricity: $0.085/kWh
Motor efficiency: 0.90

Annual energy usage: 621,417 kWh
Annual electricity cost: $52,820/yr

**FIGURE 7–71**
The cost of electricity to operate a compressor for one year can exceed the purchase price of the compressor.

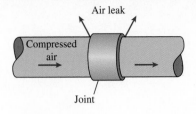

**FIGURE 7–72**
Air leaks commonly occur at joints and connections.

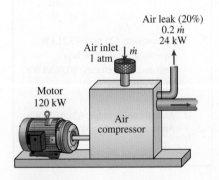

**FIGURE 7–73**
The energy wasted as compressed air escapes through the leaks is equivalent to the energy it takes to compress it.

# 1   Repairing Air Leaks on Compressed-Air Lines

Air leaks are the greatest single cause of energy loss in manufacturing facilities associated with compressed-air systems. It takes energy to compress the air, and thus the loss of compressed air is a loss of energy for the facility. A compressor must work harder and longer to make up for the lost air and must use more energy in the process. Several studies at plants have revealed that up to 40 percent of the compressed air is lost through leaks. Eliminating the air leaks totally is impractical, and a leakage rate of 10 percent is considered acceptable.

Air leaks, in general, occur at the *joints, flange connections, elbows, reducing bushes, sudden expansions, valve systems, filters, hoses, check valves, relief valves, extensions,* and the *equipment* connected to the compressed-air lines (Fig. 7–72). Expansion and contraction as a result of thermal cycling and vibration are common causes of loosening at the joints, and thus air leaks. Therefore, it is a good practice to *check* the joints for tightness and to *tighten* them periodically. Air leaks also commonly occur at the points of end use or where the compressed-air lines are connected to the equipment that operates on compressed air. Because of the frequent opening and closing of the compressed-air lines at these points, the gaskets wear out quickly, and they need to be replaced periodically.

There are many ways of detecting air leaks in a compressed-air system. Perhaps the simplest way of detecting a large air leak is to listen for it. The high velocity of the air escaping the line produces a *hissing sound* that is difficult not to notice except in environments with a high noise level. Another way of detecting air leaks, especially small ones, is to test the suspected area with *soap water* and to watch for soap bubbles. This method is obviously not practical for a large system with many connections. A modern way of checking for air leaks is to use an acoustic leak detector, which consists of a directional microphone, amplifiers, audio filters, and digital indicators.

A practical way of quantifying the air leaks in a production facility in its entirety is to conduct a *pressure drop test*. The test is conducted by stopping all the operations that use compressed air and by shutting down the compressors and closing the pressure relief valve, which relieves pressure automatically if the compressor is equipped with one. This way, any pressure drop in the compressed-air lines is due to the cumulative effects of air leaks. The drop in pressure in the system with time is observed, and the test is conducted until the pressure drops by an amount that can be measured accurately, usually 0.5 atm. The time it takes for the pressure to drop by this amount is measured, and the decay of pressure as a function of time is recorded. The total volume of the compressed-air system, including the compressed-air tanks, the headers, accumulators, and the primary compressed-air lines, is calculated. Ignoring the small lines will make the job easier and will cause the result to be more conservative. The rate of air leak can be determined using the ideal gas equation of state.

The amount of *mechanical energy wasted* as a unit mass of air escapes through the leaks is equivalent to the actual amount of energy it takes to compress it, and is determined from Eq. 7–57, modified as (Fig. 7–73)

$$w_{comp,in} = \frac{w_{reversible\ comp,in}}{\eta_{comp}} = \frac{nRT_1}{\eta_{comp}(n-1)}\left[\left(\frac{P_2}{P_1}\right)^{(n-1)/n} - 1\right] \tag{7–89}$$

where $n$ is the polytropic compression exponent ($n = 1.4$ when the compression is isentropic and $1 < n < 1.4$ when there is intercooling) and $\eta_{\text{comp}}$ is the compressor efficiency, whose value usually ranges between 0.7 and 0.9.

Using compressible-flow theory (see Chap. 17), it can be shown that whenever the line pressure is above 2 atm, which is usually the case, the velocity of air at the leak site must be equal to the local *speed of sound*. Then the **mass flow rate of air** through a leak of minimum cross-sectional area $A$ becomes

$$\dot{m}_{\text{air}} = C_{\text{discharge}} \left( \frac{2}{k+1} \right)^{1/(k-1)} \frac{P_{\text{line}}}{RT_{\text{line}}} A \sqrt{kR \left( \frac{2}{k+1} \right) T_{\text{line}}} \qquad \textbf{(7–90)}$$

where $k$ is the specific heat ratio ($k = 1.4$ for air) and $C_{\text{discharge}}$ is a discharge (or loss) coefficient that accounts for imperfections in flow at the leak site. Its value ranges from about 0.60 for an orifice with sharp edges to 0.97 for a well-rounded circular hole. The air-leak sites are imperfect in shape, and thus the discharge coefficient can be taken to be 0.65 in the absence of actual data. Also, $T_{\text{line}}$ and $P_{\text{line}}$ are the temperature and pressure in the compressed-air line, respectively.

Once $\dot{m}_{\text{air}}$ and $w_{\text{comp,in}}$ are available, the **power wasted** by the leaking compressed air (or the power saved by repairing the leak) is determined from

$$\text{Power saved} = \text{Power wasted} = \dot{m}_{\text{air}} w_{\text{comp,in}} \qquad \textbf{(7–91)}$$

### EXAMPLE 7–22   Energy and Cost Savings by Fixing Air Leaks

The compressors of a production facility maintain the compressed-air lines at a (gauge) pressure of 700 kPa at sea level where the atmospheric pressure is 101 kPa (Fig. 7–74). The average temperature of air is 20°C at the compressor inlet and 24°C in the compressed-air lines. The facility operates 4200 hours a year, and the average price of electricity is $0.078/kWh. Taking the compressor efficiency to be 0.8, the motor efficiency to be 0.92, and the discharge coefficient to be 0.65, determine the energy and money saved per year by sealing a leak equivalent to a 3-mm-diameter hole on the compressed-air line.

**SOLUTION** An air leak in the compressed air lines of a facility is considered. The energy and money saved per year by sealing the leak are to be determined.
**Assumptions** 1 Steady operating conditions exist. 2 Air is an ideal gas. 3 Pressure losses in the compressed air lines are negligible.
**Analysis** We note that the absolute pressure is the sum of the gauge and atmospheric pressures.

The work needed to compress a unit mass of air at 20°C from the atmospheric pressure of 101 kPa to $700 + 101 = 801$ kPa is

$$\begin{aligned} w_{\text{comp,in}} &= \frac{nRT_1}{\eta_{\text{comp}}(n-1)} \left[ \left( \frac{P_2}{P_1} \right)^{(n-1)/n} - 1 \right] \\ &= \frac{(1.4)(0.287 \text{ kJ/kg·K})(293 \text{ K})}{(0.8)(1.4-1)} \left[ \left( \frac{801 \text{ kPa}}{101 \text{ kPa}} \right)^{0.4/1.4} - 1 \right] = 296.9 \text{ kJ/kg} \end{aligned}$$

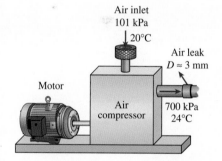

**FIGURE 7–74**
Schematic for Example 7–22.

Air inlet
101 kPa
20°C
Air leak
$D \approx 3$ mm
Motor
Air compressor
700 kPa
24°C

The cross-sectional area of the 3-mm-diameter hole is

$$A = \pi D^2/4 = \pi(3 \times 10^{-3} \text{ m})^2/4 = 7.069 \times 10^{-6} \text{ m}^2$$

Noting that the line conditions are 297 K and 801 kPa, the mass flow rate of the air leaking through the hole is determined to be

$$\dot{m}_{air} = C_{discharge}\left(\frac{2}{k+1}\right)^{1/(k-1)}\frac{P_{line}}{RT_{line}}A\sqrt{kR\left(\frac{2}{k+1}\right)T_{line}}$$

$$= (0.65)\left(\frac{2}{1.4+1}\right)^{1/(1.4-1)}\frac{801 \text{ kPa}}{(0.287 \text{ kPa·m}^3/\text{kg·K})(297 \text{ K})}(7.069 \times 10^{-6} \text{ m}^2)$$

$$\times \sqrt{(1.4)(0.287 \text{ kJ/kg·K})\left(\frac{1000 \text{ m}^2/\text{s}^2}{1 \text{ kJ/kg}}\right)\left(\frac{2}{1.4+1}\right)(297 \text{ K})}$$

$$= 0.008632 \text{ kg/s}$$

Then, the power wasted by the leaking compressed air becomes

$$\text{Power wasted} = \dot{m}_{air}w_{comp,in}$$

$$= (0.008632 \text{ kg/s})(296.9 \text{ kJ/kg})$$

$$= 2.563 \text{ kW}$$

The compressor operates 4200 h/yr, and the motor efficiency is 0.92. Then the annual energy and cost savings resulting from repairing this leak are determined to be

$$\text{Energy savings} = (\text{Power saved})(\text{Operating hours})/\eta_{motor}$$

$$= (2.563 \text{ kW})(4200 \text{ h/yr})/0.92$$

$$= \textbf{11,700 kWh/yr}$$

$$\text{Cost savings} = (\text{Energy savings})(\text{Unit cost of energy})$$

$$= (11,700 \text{ kWh/yr})(\$0.078/\text{kWh})$$

$$= \textbf{\$913/yr}$$

***Discussion*** Note that the facility will save 11,700 kWh of electricity worth $913 a year when this air leak is fixed. This is a substantial amount for a single leak whose equivalent diameter is 3 mm.

## 2 Installing High-Efficiency Motors

Practically all compressors are powered by electric motors, and the *electrical energy* a motor draws for a specified power output is *inversely proportional* to its efficiency. Electric motors cannot convert the electrical energy they consume into mechanical energy completely, and the ratio of the mechanical power supplied to the electrical power consumed during operation is called the **motor efficiency**, $\eta_{motor}$. Therefore, the electric power consumed by the motor and the mechanical (shaft) power supplied to the compressor are related to each other by (Fig. 7–75)

$$\dot{W}_{electric} = \dot{W}_{comp}/\eta_{motor} \tag{7–92}$$

For example, assuming no transmission losses, a motor that is 80 percent efficient will draw $1/0.8 = 1.25$ kW of electric power for each kW of shaft power it delivers to the compressor, whereas a motor that is 95 percent efficient will draw only $1/0.95 = 1.05$ kW to deliver 1 kW. Therefore, high-efficiency motors cost less to operate than their standard counterparts, but they also usually cost more to purchase. However, the energy savings usually make up for the price differential during the first few years. This is especially true for large compressors that operate more than one regular shift. The *electric power saved* by replacing the existing standard motor of efficiency $\eta_{standard}$ by a high-efficiency one of efficiency $\eta_{efficient}$ is determined from

$$\dot{W}_{electric,saved} = \dot{W}_{electric,standard} - \dot{W}_{electric,efficient}$$
$$= \dot{W}_{comp}(1/\eta_{standard} - 1/\eta_{efficient}) \qquad \text{(7–93)}$$
$$= (\text{Rated power})(\text{Load factor})(1/\eta_{standard} - 1/\eta_{efficient})$$

where *rated power* is the nominal power of the motor listed on its label (the power the motor delivers at full load) and the *load factor* is the fraction of the rated power at which the motor normally operates. Then, the annual energy savings as a result of replacing a motor by a high-efficiency motor instead of a comparable standard one is

$$\text{Energy savings} = \dot{W}_{electric,saved} \times \text{Annual operating hours} \qquad \text{(7–94)}$$

The efficiencies of motors used to power compressors usually range from about 70 percent to over 96 percent. The portion of electric energy not converted to mechanical energy is converted to heat. The amount of heat generated by the motors may reach high levels, especially at part load, and it may cause overheating if not dissipated effectively. It may also cause the air temperature in the compressor room to rise to undesirable levels. For example, a 90-percent-efficient 100-kW motor generates as much heat as a 10-kW resistance heater in the confined space of the compressor room, and it contributes greatly to the heating of the air in the room. If this heated air is not vented properly, and the air into the compressor is drawn from inside the compressor room, the performance of the compressor will also decline, as explained later.

Important considerations in the selection of a motor for a compressor are the operating profile of the compressor (i.e., the variation of the load with time), and the efficiency of the motor at part-load conditions. The part-load efficiency of a motor is as important as the full-load efficiency if the compressor is expected to operate at part load during a significant portion of the total operating time. A typical motor has a nearly flat efficiency curve between half load and full load, and peak efficiency is usually at about 75% load. Efficiency falls off pretty steeply below half load, and thus operation below 50% load should be avoided as much as possible. For example, the efficiency of a motor may drop from 90 percent at full load to 87 percent at half load and 80 percent at quarter load (Fig. 7–76). The efficiency of another motor of similar specifications, on the other hand, may drop from 91 percent at full load to 75 percent at quarter load. The first motor is obviously better suited for a situation

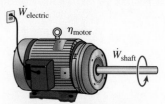

| Motor efficiency $\eta_{motor}$ | Electrical power consumed per kW of mechanical (shaft) power output, $\dot{W}_{electric} = \dot{W}_{shaft}/\eta_{motor}$ |
|---|---|
| 100% | 1.00 kW |
| 90 | 1.11 |
| 80 | 1.25 |
| 70 | 1.43 |
| 60 | 1.67 |
| 50 | 2.00 |
| 40 | 2.50 |
| 30 | 3.33 |
| 20 | 5.00 |
| 10 | 10.00 |

**FIGURE 7–75**
The electrical energy consumed by a motor is inversely proportional to its efficiency.

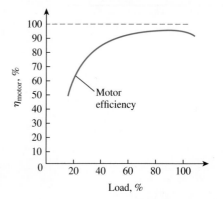

**FIGURE 7–76**
The efficiency of an electric motor decreases at part load.

in which a compressor is expected to operate at quarter load during a significant portion of the time. The efficiency at part-load conditions can be improved greatly by installing variable voltage controllers if it is economical to do so. Also, *oversizing* a motor just to be on the safe side and to have some excess power just in case is a bad practice since this will cause the motor to operate almost always at *part load* and thus at a *lower efficiency*. Besides, oversized motors have a higher initial cost. However, oversized motors waste little energy as long as they operate at loads above 50% of design.

## Using a Smaller Motor at High Capacity

We tend to purchase *larger equipment* than needed for reasons like having a safety margin or anticipated future expansion, and compressors are no exception. The uncertainties in plant operation are partially responsible for opting for a larger compressor, since it is preferred to have an oversized compressor than an undersized one. Sometimes compressors that have several times the required capacity are purchased with the perception that the extra capacity may be needed some day. The result is a compressor that runs intermittently at full load, or one that runs continuously at part load.

A compressor that operates at part load also causes the motor to operate less efficiently since the efficiency of an electric motor decreases as the point of operation shifts down from its rated power, as explained above. The result is a motor that consumes more electricity per unit power delivered, and thus a more expensive operation. The operating costs can be reduced by switching to a smaller motor that runs at rated power and thus at a higher efficiency.

## 3   Using Outside Air for Compressor Intake

We have pointed out earlier that the power consumed by a compressor is proportional to the *specific volume,* which is proportional to the *absolute temperature* of the gas at a given pressure. It is also clear from Eq. 7–89 that the compressor work is directly proportional to the *inlet temperature* of air. Therefore, the lower the inlet temperature of the air, the smaller the compressor work. Then the *power reduction factor,* which is the fraction of compressor power reduced as a result of taking intake air from the outside, becomes

$$f_{\text{reduction}} = \frac{w_{\text{comp,inside}} - w_{\text{comp,outside}}}{w_{\text{comp,inside}}} = \frac{T_{\text{inside}} - T_{\text{outside}}}{T_{\text{inside}}} = 1 - \frac{T_{\text{outside}}}{T_{\text{inside}}} \qquad (7\text{–}95)$$

where $T_{\text{inside}}$ and $T_{\text{outside}}$ are the absolute temperatures (in K) of the ambient air inside and outside the facility, respectively. Thus, reducing the absolute inlet temperature by 5 percent, for example, will reduce the compressor power input by 5 percent. As a rule of thumb, for a specified amount of compressed air, the power consumption of the compressor decreases (or, for a fixed power input, the amount of compressed air increases) by 1 percent for each 3°C drop in the temperature of the inlet air to the compressor.

Compressors are usually located inside the production facilities or in adjacent shelters specifically built outside these facilities. The intake air is normally drawn from inside the building or the shelter. However, in many locations the air temperature in the building is higher than the outside air temperature, because of space heaters in the winter and the heat given up by a large number of mechanical and electrical equipment as well as the furnaces year round. The temperature rise in the shelter is also due to the heat dissipation from the compressor and its motor. The outside air is generally *cooler* and thus *denser* than the air in the compressor room even on hot summer days. Therefore, it is advisable to install an *intake duct* to the compressor inlet so that the air is supplied directly from the outside of the building instead of the inside, as shown in Fig. 7–77. This will reduce the energy consumption of the compressor since it takes less energy to compress a specified amount of cool air than the same amount of warm air. Compressing the warm air in a building in winter also wastes the energy used to heat the air.

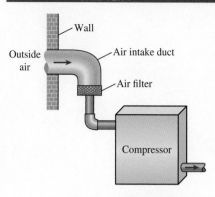

**FIGURE 7–77**
The power consumption of a compressor can be reduced by taking in air from the outside.

## 4  Reducing the Air Pressure Setting

Another source of energy waste in compressed-air systems is compressing the air to a higher pressure than required by the air-driven equipment since it takes more energy to compress air to a higher pressure. In such cases considerable energy savings can be realized by determining the minimum required pressure and then reducing the air pressure control setting on the compressor accordingly. This can be done on both screw-type and reciprocating compressors by simply adjusting the pressure setting to match the needs.

The amount of energy it takes to compress a unit mass of air is determined from Eq. 7–89. We note from that relation that the higher the pressure $P_2$ at the compressor exit, the larger the work required for compression. Reducing the exit pressure of the compressor to $P_{2,\text{reduced}}$ will reduce the power input requirements of the compressor by a factor of

$$f_{\text{reduction}} = \frac{w_{\text{comp,current}} - w_{\text{comp,reduced}}}{w_{\text{comp,current}}} = 1 - \frac{(P_{2,\text{reduced}}/P_1)^{(n-1)/n} - 1}{(P_2/P_1)^{(n-1)/n} - 1} \quad \text{(7–96)}$$

A power reduction (or savings) factor of $f_{\text{reduction}} = 0.08$, for example, indicates that the power consumption of the compressor is reduced by 8 percent as a result of reducing the pressure setting.

Some applications require slightly compressed air. In such cases, the need can be met by a blower instead of a compressor. Considerable energy can be saved in this manner since a blower requires a small fraction of the power needed by a compressor for a specified mass flow rate.

---

### ◼ *EXAMPLE 7–23*    **Reducing the Pressure Setting to Reduce Cost**

◼ The compressed-air requirements of a plant located at 1400-m elevation is
◼ being met by a 75-hp compressor that takes in air at the local atmospheric
◼ pressure of 85.6 kPa and the average temperature of 15°C and compresses
it to 900 kPa gauge (Fig. 7–78). The plant is currently paying $12,000
a year in electricity costs to run the compressor. An investigation of the

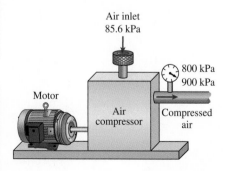

**FIGURE 7–78**
Schematic for Example 7–23.

compressed-air system and the equipment using the compressed air reveals that compressing the air to 800 kPa is sufficient for this plant. Determine how much money will be saved as a result of reducing the pressure of the compressed air.

**SOLUTION** It is observed that the compressor of a facility compresses the air to much higher pressures than needed. The cost savings associated with pressure reduction of the compressor are to be determined.
*Assumptions* **1** Air is an ideal gas. **2** Compression process is isentropic, and thus $n = k = 1.4$.
*Analysis* The fraction of energy saved as a result of reducing the pressure setting of the compressor is

$$f_{\text{reduction}} = 1 - \frac{(P_{2,\text{reduced}}/P_1)^{(n-1)/n} - 1}{(P_2/P_1)^{(n-1)/n} - 1}$$

$$= 1 - \frac{(885.6/85.6)^{(1.4-1)/1.4} - 1}{(985.6/85.6)^{(1.4-1)/1.4} - 1} = 0.060$$

That is, reducing the pressure setting will reduce the energy consumed by the compressor by about 6 percent. Then,

$$\text{Cost savings} = (\text{Current cost})f_{\text{reduction}} = (\$12,000/\text{yr})(0.06) = \mathbf{\$720/yr}$$

Therefore, reducing the pressure setting by 100 kPa will result in annual savings of \$720 in this case.

There are also other ways to reduce the cost of compressed air in industrial facilities. An obvious way is *turning the compressor off* during nonproduction periods such as lunch hours, nights, and even weekends. A considerable amount of power can be wasted during this stand-by mode. This is especially the case for screw-type compressors since they consume up to 85 percent of their rated power in this mode. The reciprocating compressors are not immune from this deficiency, however, since they also must cycle on and off because of the air leaks present in the compressed-air lines. The system can be shut down manually during nonproduction periods to save energy, but installing a timer (with manual override) is preferred to do this automatically since it is human nature to put things off when the benefits are not obvious or immediate.

The compressed air is sometimes cooled considerably below its dew point in *refrigerated dryers* in order to condense and remove a large fraction of the water vapor in the air as well as other noncondensable gases such as oil vapors. The temperature of air rises considerably as it is compressed, sometimes exceeding 250°C at compressor exit when compressed adiabatically to just 700 kPa. Therefore, it is desirable to cool air after compression in order to minimize the amount of power consumed by the refrigeration system, just as it is desirable to let the hot food in a pan cool to the ambient temperature before putting it into the refrigerator. The cooling can be done by either ambient air or water, and the heat picked up by the cooling medium can be used for space heating, feedwater heating, or process-related heating.

Compressors are commonly cooled directly by air or by circulating a liquid such as oil or water through them in order to minimize the power consumption. The heat picked up by the oil or water is usually rejected to the ambient in a liquid-to-air heat exchanger. This *heat rejected* usually amounts to 60 to 90 percent of the power input, and thus it represents a huge amount of energy that can be used for a useful purpose such as *space heating* in winter, *preheating* the air or water in a furnace, or other process-related purposes (Fig. 7–79). For example, assuming 80 percent of the power input is converted to heat, a 150-hp compressor can reject as much heat as a 90-kW electric resistance heater or a 105 kW natural gas heater when operating at full load. Thus, the proper utilization of the waste heat from a compressor can result in significant energy and cost savings.

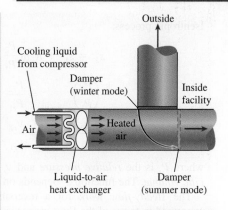

**FIGURE 7–79**
Waste heat from a compressor can be used to heat a building in winter.

## SUMMARY

The second law of thermodynamics leads to the definition of a new property called *entropy,* which is a quantitative measure of microscopic disorder for a system. Any quantity whose cyclic integral is zero is a property, and entropy is defined as

$$dS = \left(\frac{dQ}{T}\right)_{\text{int rev}}$$

For the special case of an internally reversible, isothermal process, it gives

$$\Delta S = \frac{Q}{T_0}$$

The inequality part of the Clausius inequality combined with the definition of entropy yields an inequality known as the *increase of entropy principle,* expressed as

$$S_{\text{gen}} \geq 0$$

where $S_{\text{gen}}$ is the *entropy generated* during the process. Entropy change is caused by heat transfer, mass flow, and irreversibilities. Heat transfer to a system increases the entropy, and heat transfer from a system decreases it. The effect of irreversibilities is always to increase the entropy.

The *entropy-change* and *isentropic relations* for a process can be summarized as follows:

**1.** *Pure substances:*

Any process:  $\Delta s = s_2 - s_1$

Isentropic process:  $s_2 = s_1$

**2.** *Incompressible substances:*

Any process:  $s_2 - s_1 = c_{\text{avg}} \ln \dfrac{T_2}{T_1}$

Isentropic process:  $T_2 = T_1$

**3.** *Ideal gases:*
  *a.* Constant specific heats (approximate treatment):
Any process:

$$s_2 - s_1 = c_{v,\text{avg}} \ln \frac{T_2}{T_1} + R \ln \frac{v_2}{v_1}$$

$$s_2 - s_1 = c_{p,\text{avg}} \ln \frac{T_2}{T_1} - R \ln \frac{P_2}{P_1}$$

Isentropic process:

$$\left(\frac{T_2}{T_1}\right)_{s=\text{const.}} = \left(\frac{v_1}{v_2}\right)^{k-1}$$

$$\left(\frac{T_2}{T_1}\right)_{s=\text{const.}} = \left(\frac{P_2}{P_1}\right)^{(k-1)/k}$$

$$\left(\frac{P_2}{P_1}\right)_{s=\text{const.}} = \left(\frac{v_1}{v_2}\right)^{k}$$

  *b.* Variable specific heats (exact treatment):
Any process:

$$s_2 - s_1 = s_2^\circ - s_1^\circ - R \ln \frac{P_2}{P_1}$$

Isentropic process:

$$s_2^\circ = s_1^\circ + R \ln \frac{P_2}{P_1}$$

$$\left(\frac{P_2}{P_1}\right)_{s=\text{const.}} = \frac{P_{r2}}{P_{r1}}$$

$$\left(\frac{v_2}{v_1}\right)_{s=\text{const.}} = \frac{v_{r2}}{v_{r1}}$$

where $P_r$ is the *relative pressure* and $v_r$ is the *relative specific volume*. The function $s^\circ$ depends on temperature only.

The *steady-flow work* for a reversible process can be expressed in terms of the fluid properties as

$$w_{\text{rev}} = -\int_1^2 v\,dP - \Delta\text{ke} - \Delta\text{pe}$$

For incompressible substances ($v$ = constant) it simplifies to

$$w_{\text{rev}} = -v(P_2 - P_1) - \Delta\text{ke} - \Delta\text{pe}$$

The work done during a steady-flow process is proportional to the specific volume. Therefore, $v$ should be kept as small as possible during a compression process to minimize the work input and as large as possible during an expansion process to maximize the work output.

The reversible work inputs to a compressor compressing an ideal gas from $T_1$, $P_1$ to $P_2$ in an isentropic ($Pv^k$ = constant), polytropic ($Pv^n$ = constant), or isothermal ($Pv$ = constant) manner, are determined by integration for each case with the following results:

Isentropic: $\quad w_{\text{comp,in}} = \dfrac{kR(T_2 - T_1)}{k-1} = \dfrac{kRT_1}{k-1}\left[\left(\dfrac{P_2}{P_1}\right)^{(k-1)/k} - 1\right]$

Polytropic: $\quad w_{\text{comp,in}} = \dfrac{nR(T_2 - T_1)}{n-1} = \dfrac{nRT_1}{n-1}\left[\left(\dfrac{P_2}{P_1}\right)^{(n-1)/n} - 1\right]$

Isothermal: $\quad w_{\text{comp,in}} = RT \ln \dfrac{P_2}{P_1}$

The work input to a compressor can be reduced by using multistage compression with intercooling. For maximum savings from the work input, the pressure ratio across each stage of the compressor must be the same.

Most steady-flow devices operate under adiabatic conditions, and the ideal process for these devices is the isentropic process. The parameter that describes how efficiently a device approximates a corresponding isentropic device is called *isentropic* or *adiabatic efficiency*. It is expressed for turbines, compressors, and nozzles as follows:

$$\eta_T = \frac{\text{Actual turbine work}}{\text{Isentropic turbine work}} = \frac{w_a}{w_s} \cong \frac{h_1 - h_{2a}}{h_1 - h_{2s}}$$

$$\eta_C = \frac{\text{Isentropic compressor work}}{\text{Actual compressor work}} = \frac{w_s}{w_a} \cong \frac{h_{2s} - h_1}{h_{2a} - h_1}$$

$$\eta_N = \frac{\text{Actual KE at nozzle exit}}{\text{Isentropic KE at nozzle exit}} = \frac{V_{2a}^2}{V_{2s}^2} \cong \frac{h_1 - h_{2a}}{h_1 - h_{2s}}$$

In the relations above, $h_{2a}$ and $h_{2s}$ are the enthalpy values at the exit state for actual and isentropic *processes*, respectively.

The entropy balance for any system undergoing any process can be expressed in the general form as

$$\underbrace{S_{\text{in}} - S_{\text{out}}}_{\substack{\text{Net entropy transfer} \\ \text{by heat and mass}}} + \underbrace{S_{\text{gen}}}_{\substack{\text{Entropy} \\ \text{generation}}} = \underbrace{\Delta S_{\text{system}}}_{\substack{\text{Change} \\ \text{in entropy}}}$$

or, in the *rate form*, as

$$\underbrace{\dot{S}_{\text{in}} - \dot{S}_{\text{out}}}_{\substack{\text{Rate of net entropy} \\ \text{transfer by heat} \\ \text{and mass}}} + \underbrace{\dot{S}_{\text{gen}}}_{\substack{\text{Rate of Entropy} \\ \text{generation}}} = \underbrace{dS_{\text{system}}/dt}_{\substack{\text{Rate of change} \\ \text{in entropy}}}$$

For a general *steady-flow process* it simplifies to

$$\dot{S}_{\text{gen}} = \sum \dot{m}_e s_e - \sum \dot{m}_i s_i - \sum \frac{\dot{Q}_k}{T_k}$$

## REFERENCES AND SUGGESTED READINGS

**1.** A. Bejan. *Advanced Engineering Thermodynamics.* 3rd ed. New York: Wiley Interscience, 2006.

**2.** A. Bejan. *Entropy Generation through Heat and Fluid Flow.* New York: Wiley Interscience, 1982.

**3.** Y. A. Çengel and H. Kimmel. "Optimization of Expansion in Natural Gas Liquefaction Processes." *LNG Journal,* U.K., May–June, 1998.

**4.** Y. Çerci, Y. A. Çengel, and R. H. Turner, "Reducing the Cost of Compressed Air in Industrial Facilities." *International Mechanical Engineering Congress and Exposition,* San Francisco, California, November 12–17, 1995.

**5.** W. F. E. Feller. *Air Compressors: Their Installation, Operation, and Maintenance.* New York: McGraw-Hill, 1944.

**6.** D. W. Nutter, A. J. Britton, and W. M. Heffington. "Conserve Energy to Cut Operating Costs." *Chemical Engineering,* September 1993, pp. 127–137.

**7.** J. Rifkin. *Entropy.* New York: The Viking Press, 1980.

**8.** M. Kostic, "Revisiting The Second Law of Energy Degradation and Entropy Generation: From Sadi Carnot's Ingenious Reasoning to Holistic Generalization." *AIP Conf. Proc.* 1411, pp. 327–350, 2011; doi: 10.1063/1.3665247.

# PROBLEMS*

## Entropy and the Increase of Entropy Principle

**7–1C** Does the temperature in the Clausius inequality relation have to be absolute temperature? Why?

**7–2C** Does the cyclic integral of heat have to be zero (i.e., does a system have to reject as much heat as it receives to complete a cycle)? Explain.

**7–3C** Is a quantity whose cyclic integral is zero necessarily a property?

**7–4C** To determine the entropy change for an irreversible process between states 1 and 2, should the integral $\int_1^2 \delta Q/T$ be performed along the actual process path or an imaginary reversible path? Explain.

**7–5C** Is an isothermal process necessarily internally reversible? Explain your answer with an example.

**7–6C** How do the values of the integral $\int_1^2 \delta Q/T$ compare for a reversible and irreversible process between the same end states?

**7–7C** The entropy of a hot baked potato decreases as it cools. Is this a violation of the increase of entropy principle? Explain.

**7–8C** When a system is adiabatic, what can be said about the entropy change of the substance in the system?

**7–9C** Work is entropy free, and sometimes the claim is made that work will not change the entropy of a fluid passing through an adiabatic steady-flow system with a single inlet and outlet. Is this a valid claim?

**7–10C** A piston–cylinder device contains helium gas. During a reversible, isothermal process, the entropy of the helium will (*never, sometimes, always*) increase.

**7–11C** A piston–cylinder device contains nitrogen gas. During a reversible, adiabatic process, the entropy of the nitrogen will (*never, sometimes, always*) increase.

**7–12C** A piston–cylinder device contains superheated steam. During an actual adiabatic process, the entropy of the steam will (*never, sometimes, always*) increase.

**7–13C** The entropy of steam will (*increase, decrease, remain the same*) as it flows through an actual adiabatic turbine.

**7–14C** The entropy of the working fluid of the ideal Carnot cycle (*increases, decreases, remains the same*) during the isothermal heat addition process.

**7–15C** The entropy of the working fluid of the ideal Carnot cycle (*increases, decreases, remains the same*) during the isothermal heat rejection process.

**7–16C** During a heat transfer process, the entropy of a system (*always, sometimes, never*) increases.

**7–17C** Steam is accelerated as it flows through an actual adiabatic nozzle. The entropy of the steam at the nozzle exit will be (*greater than, equal to, less than*) the entropy at the nozzle inlet.

**7–18C** What three different mechanisms can cause the entropy of a control volume to change?

**7–19** Air is compressed by a 15-kW compressor from $P_1$ to $P_2$. The air temperature is maintained constant at 25°C during this process as a result of heat transfer to the surrounding medium at 20°C. Determine the rate of entropy change of the air. State the assumptions made in solving this problem. *Answer:* −0.0503 kW/K

**7–20** Heat in the amount of 100 kJ is transferred directly from a hot reservoir at 1200 K to a cold reservoir at 600 K. Calculate the entropy change of the two reservoirs and determine if the increase of entropy principle is satisfied.

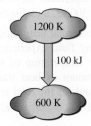

1200 K

100 kJ

600 K

**FIGURE P7–20**

**7–21** In Prob. 7-20, assume that the heat is transferred from the cold reservoir to the hot reservoir contrary to the Clausius statement of the second law. Prove that this violates the increase of entropy principle—as it must according to Clausius.

**7–22** A completely reversible heat pump produces heat at a rate of 300 kW to warm a house maintained at 24°C. The exterior air, which is at 7°C, serves as the source. Calculate the rate of entropy change of the two reservoirs and determine if this heat pump satisfies the second law according to the increase of entropy principle.

---

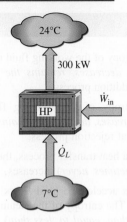

**FIGURE P7–22**

**7–23** During the isothermal heat addition process of a Carnot cycle, 900 kJ of heat is added to the working fluid from a source at 400°C. Determine (a) the entropy change of the working fluid, (b) the entropy change of the source, and (c) the total entropy change for the process.

**7–24** 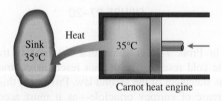 Reconsider Prob. 7-23. Using EES (or other) software, study the effects of the varying heat added to the working fluid and the source temperature on the entropy change of the working fluid, the entropy change of the source, and the total entropy change for the process. Let the source temperature vary from 100 to 1000°C. Plot the entropy changes of the source and of the working fluid against the source temperature for heat transfer amounts of 500 kJ, 900 kJ, and 1300 kJ, and discuss the results.

**7–25** During the isothermal heat rejection process of a Carnot cycle, the working fluid experiences an entropy change of −1.3 kJ/K. If the temperature of the heat sink is 35°C, determine (a) the amount of heat transfer, (b) the entropy change of the sink, and (c) the total entropy change for this process. *Answers:* (a) 400 kJ, (b) 1.3 kJ/K, (c) 0

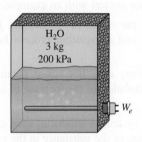

**FIGURE P7–25**

**7–26** Refrigerant-134a enters the coils of the evaporator of a refrigeration system as a saturated liquid–vapor mixture at a pressure of 140 kPa. The refrigerant absorbs 180 kJ of heat from the cooled space, which is maintained at −10°C, and leaves as saturated vapor at the same pressure. Determine (a) the entropy change of the refrigerant, (b) the entropy change of the cooled space, and (c) the total entropy change for this process.

## Entropy Changes of Pure Substances

**7–27C** Is a process that is internally reversible and adiabatic necessarily isentropic? Explain.

**7–28** 1-kg of water at 2 MPa fill a weighted piston-cylinder device whose volume is 0.07 m$^3$. The water is then heated at constant pressure until the temperature reaches 250°C. Determine the resulting change in the water's total entropy. *Answer:* 1.38 kJ/K

**7–29** A well-insulated rigid tank contains 3 kg of a saturated liquid–vapor mixture of water at 200 kPa. Initially, three-quarters of the mass is in the liquid phase. An electric resistance heater placed in the tank is now turned on and kept on until all the liquid in the tank is vaporized. Determine the entropy change of the steam during this process. *Answer:* 11.1 kJ/K

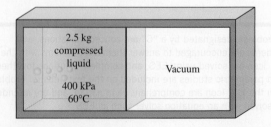

**FIGURE P7–29**

**7–30** The radiator of a steam heating system has a volume of 20 L and is filled with superheated water vapor at 200 kPa and 150°C. At this moment both the inlet and the exit valves to the radiator are closed. After a while the temperature of the steam drops to 40°C as a result of heat transfer to the room air. Determine the entropy change of the steam during this process. *Answer:* −0.132 kJ/K

**7–31**  A rigid tank is divided into two equal parts by a partition. One part of the tank contains 2.5 kg of compressed liquid water at 400 kPa and 60°C while the other part is evacuated. The partition is now removed, and the water expands to fill the entire tank. Determine the entropy change of water during this process, if the final pressure in the tank is 40 kPa. *Answer:* 0.492 kJ/K

**FIGURE P7–31**

**7–32** An insulated piston–cylinder device contains 5 L of saturated liquid water at a constant pressure of 150 kPa. An electric resistance heater inside the cylinder is now turned on, and 2200 kJ of energy is transferred to the steam. Determine the entropy change of the water during this process.
*Answer: 5.72 kJ/K*

**7–33** Water vapor enters a turbine at 6 MPa and 400°C, and leaves the turbine at 100 kPa with the same specific entropy as that at the inlet. Calculate the difference between the specific enthalpy of the water at the turbine inlet and exit.

**7–34** 1-kg of R-134a initially at 600 kPa and 25°C undergoes a process during which the entropy is kept constant until the pressure drops to 100 kPa. Determine the final temperature of the R-134a and the final specific internal energy.

**7–35** Refrigerant-134a is expanded isentropically from 600 kPa and 70°C at the inlet of a steady-flow turbine to 100 kPa at the outlet. The outlet area is 1 m², and the inlet area is 0.5 m². Calculate the inlet and outlet velocities when the mass flow rate is 0.75 kg/s. *Answers: 0.0646 m/s, 0.171 m/s*

**7–36** A piston–cylinder device contains 1.2 kg of saturated water vapor at 200°C. Heat is now transferred to steam, and steam expands reversibly and isothermally to a final pressure of 800 kPa. Determine the heat transferred and the work done during this process.

**7–37**  Reconsider Prob. 7–36. Using EES (or other) software, evaluate and plot the heat transferred to the steam and the work done as a function of final pressure as the pressure varies from the initial value to the final value of 800 kPa.

**7–38** Refrigerant-134a at 320 kPa and 40°C undergoes an isothermal process in a closed system until its quality is 45 percent. On per unit mass basis, determine how much work and heat transfer are required. *Answers: 40.6 kJ/kg, 130 kJ/kg*

R-134a
320 kPa
40°C

**FIGURE P7–38**

**7–39** A rigid tank contains 5 kg of saturated vapor steam at 100°C. The steam is cooled to the ambient temperature of 25°C.

(*a*) Sketch the process with respect to the saturation lines on a *T-v* diagram.

(*b*) Determine the entropy change of the steam, in kJ/K.

(*c*) For the steam and its surroundings, determine the total entropy change associated with this process, in kJ/K.

**7–40** A 0.5-m³ rigid tank contains refrigerant-134a initially at 200 kPa and 40 percent quality. Heat is transferred now to the refrigerant from a source at 35°C until the pressure rises to 400 kPa. Determine (*a*) the entropy change of the refrigerant, (*b*) the entropy change of the heat source, and (*c*) the total entropy change for this process.

**7–41**  Reconsider Prob. 7–40. Using EES (or other) software, investigate the effects of the source temperature and final pressure on the total entropy change for the process. Let the source temperature vary from 30 to 210°C, and the final pressure vary from 250 to 500 kPa. Plot the total entropy change for the process as a function of the source temperature for final pressures of 250 kPa, 400 kPa, and 500 kPa, and discuss the results.

**7–42** Determine the heat transfer, in kJ/kg, for the reversible process 1-3 shown in Fig. P7–42.

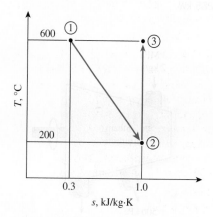

**FIGURE P7–42**

**7–43** Calculate the heat transfer, in kJ/kg, for the reversible process 1-3 shown in Fig. P7–43. *Answer: 71.5 kJ/kg*

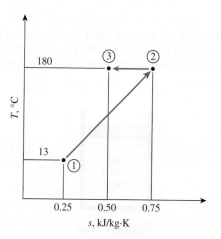

**FIGURE P7–43**

**7–44** Steam enters an adiabatic diffuser at 150 kPa and 120°C with a velocity of 550 m/s. Determine the minimum velocity that the steam can have at the outlet when the outlet pressure is 300 kPa.

**7–45** Steam enters an adiabatic turbine at 6 MPa and 500°C and leaves at a pressure of 0.3 MPa. Determine the maximum amount of work that can be delivered by this turbine.

**7–46**  Reconsider Prob. 7–45. Using EES (or other) software, evaluate and plot the work done by the steam as a function of final pressure as it varies from 6 to 0.3 MPa. Also investigate the effect of varying the turbine inlet temperature from the saturation temperature at 6 MPa to 500°C on the turbine work.

**7–47** An isentropic steam turbine processes 2 kg/s of steam at 3 MPa, which is exhausted at 50 kPa and 100°C. 5 percent of this flow is diverted for feedwater heating at 500 kPa. Determine the power produced by this turbine, in kW.
*Answer: 2285 kW*

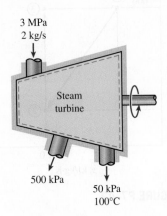

3 MPa
2 kg/s

Steam
turbine

500 kPa

50 kPa
100°C

**FIGURE P7–47**

**7–48** Water at 70 kPa and 100°C is compressed isentropically in a closed system to 4 MPa. Determine the final temperature of the water and the work required, in kJ/kg, for this compression.

H₂O
70 kPa
100°C

**FIGURE P7–48**

**7–49** 0.7-kg of R-134a is expanded isentropically from 800 kPa and 50°C to 140 kPa. Determine the total heat transfer and work production for this expansion.

**7–50** 2-kg of saturated water vapor at 600 kPa are contained in a piston-cylinder device. The water expands adiabatically until the pressure is 100 kPa and is said to produce 700 kJ of work output.

(a) Determine the entropy change of the water, in kJ/kg·K.

(b) Is this process realistic? Using the *T-s* diagram for the process and the concepts of second law, support your answer.

**7–51** Steam enters a steady-flow adiabatic nozzle with a low inlet velocity as a saturated vapor at 6 MPa and expands to 1.2 MPa.

(a) Under the conditions that the exit velocity is to be the maximum possible value, sketch the *T-s* diagram with respect to the saturation lines for this process.

(b) Determine the maximum exit velocity of the steam, in m/s.

*Answer: 764 m/s*

**7–52** A rigid, 20-L steam cooker is arranged with a pressure relief valve set to release vapor and maintain the pressure once the pressure inside the cooker reaches 150 kPa. Initially, this cooker is filled with water at 175 kPa with a quality of 10 percent. Heat is now added until the quality inside the cooker is 40 percent. Determine the minimum entropy change of the thermal energy reservoir supplying this heat.

**7–53** In Prob. 7–52, the water is stirred at the same time that it is being heated. Determine the minimum entropy change of the heat-supplying source if 100 kJ of work is done on the water as it is being heated.

**7–54** A piston–cylinder device contains 5 kg of steam at 100°C with a quality of 50 percent. This steam undergoes two processes as follows:

1-2   Heat is transferred to the steam in a reversible manner while the temperature is held constant until the steam exists as a saturated vapor.

2-3   The steam expands in an adiabatic, reversible process until the pressure is 15 kPa.

(a) Sketch these processes with respect to the saturation lines on a single *T-s* diagram.

(b) Determine the heat transferred to the steam in process 1-2, in kJ.

(c) Determine the work done by the steam in process 2-3, in kJ.

**7–55** An electric windshield defroster is used to remove 0.6-cm of ice from a windshield. The properties of the ice are $T_{sat}$ = 0°C, $u_{if} = h_{if}$ = 335 kJ/kg, and $v$ = 0.001 m³/kg. Determine the electrical energy required per square meter of windshield surface area to melt this ice and remove it as liquid water at

0°C. What is the minimum temperature at which the defroster may be operated? Assume that no heat is transferred from the defroster or ice to the surroundings.

## Entropy Change of Incompressible Substances

**7–56C** Consider two solid blocks, one hot and the other cold, brought into contact in an adiabatic container. After a while, thermal equilibrium is established in the container as a result of heat transfer. The first law requires that the amount of energy lost by the hot solid be equal to the amount of energy gained by the cold one. Does the second law require that the decrease in entropy of the hot solid be equal to the increase in entropy of the cold one?

**7–57** A 50-kg copper block initially at 140°C is dropped into an insulated tank that contains 90 L of water at 10°C. Determine the final equilibrium temperature and the total entropy change for this process.

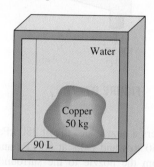

**FIGURE P7–57**

**7–58** 10-grams of computer chips with a specific heat of 0.3 kJ/kg·K are initially at 20°C. These chips are cooled by placement in 5 grams of saturated liquid R-134a at −40°C. Presuming that the pressure remains constant while the chips are being cooled, determine the entropy change of (a) the chips, (b) the R-134a, and (c) the entire system. Is this process possible? Why?

**7–59** A 25-kg iron block initially at 350°C is quenched in an insulated tank that contains 100 kg of water at 18°C. Assuming the water that vaporizes during the process condenses back in the tank, determine the total entropy change during this process.

**7–60** A 30-kg aluminum block initially at 140°C is brought into contact with a 40-kg block of iron at 60°C in an insulated enclosure. Determine the final equilibrium temperature and the total entropy change for this process. *Answers:* 109°C, 0.251 kJ/K

**7–61** [EES] Reconsider Prob. 7–60. Using EES (or other) software, study the effect of the mass of the iron block on the final equilibrium temperature and the total entropy change for the process. Let the mass of the iron vary from 10 to 100 kg. Plot the equilibrium temperature and the

total entropy change as a function of iron mass, and discuss the results.

**7–62** A 30-kg iron block and a 40-kg copper block, both initially at 80°C, are dropped into a large lake at 15°C. Thermal equilibrium is established after a while as a result of heat transfer between the blocks and the lake water. Determine the total entropy change for this process.

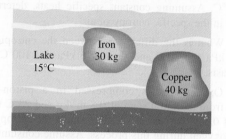

**FIGURE P7–62**

**7–63** An adiabatic pump is to be used to compress saturated liquid water at 10 kPa to a pressure to 15 MPa in a reversible manner. Determine the work input using (a) entropy data from the compressed liquid table, (b) inlet specific volume and pressure values, (c) average specific volume and pressure values. Also, determine the errors involved in parts (b) and (c).

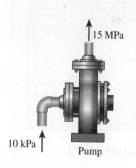

**FIGURE P7–63**

## Entropy Change of Ideal Gases

**7–64C** Some properties of ideal gases such as internal energy and enthalpy vary with temperature only [that is, $u = u(T)$ and $h = h(T)$]. Is this also the case for entropy?

**7–65C** Can the entropy of an ideal gas change during an isothermal process?

**7–66C** An ideal gas undergoes a process between two specified temperatures, first at constant pressure and then at constant volume. For which case will the ideal gas experience a larger entropy change? Explain.

**7–67** Prove that the two relations for entropy change of ideal gases under the constant-specific-heat assumption (Eqs. 7–33 and 7–34) are equivalent.

**7–68** Starting with the second $T\,ds$ relation (Eq. 7–26), obtain Eq. 7–34 for the entropy change of ideal gases under the constant-specific-heat assumption.

**7–69** Which of the two gases—helium or nitrogen—experiences the greatest entropy change as its state is changed from 2000 kPa and 427°C to 200 kPa and 27°C?

**7–70** Air is expanded from 2000 kPa and 500°C to 100 kPa and 50°C. Assuming constant specific heats, determine the change in the specific entropy of air.

**7–71** What is the difference between the entropies of air at 105 kPa and 30°C and air at 275 kPa and 100°C per unit mass basis.

**7–72** Oxygen gas is compressed in a piston–cylinder device from an initial state of 0.8 m³/kg and 25°C to a final state of 0.1 m³/kg and 287°C. Determine the entropy change of the oxygen during this process. Assume constant specific heats.

**7–73** A 1.5-m³ insulated rigid tank contains 2.7 kg of carbon dioxide at 100 kPa. Now paddle-wheel work is done on the system until the pressure in the tank rises to 150 kPa. Determine the entropy change of carbon dioxide during this process. Assume constant specific heats. *Answer: 0.719 kJ/K*

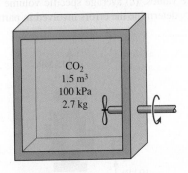

**FIGURE P7–73**

**7–74** An insulated piston–cylinder device initially contains 300 L of air at 120 kPa and 17°C. Air is now heated for 15 min by a 200-W resistance heater placed inside the cylinder. The pressure of air is maintained constant during this process. Determine the entropy change of air, assuming (a) constant specific heats and (b) variable specific heats.

**7–75** A piston–cylinder device contains 0.75 kg of nitrogen gas at 140 kPa and 37°C. The gas is now compressed slowly in a polytropic process during which $PV^{1.3}$ = constant. The process ends when the volume is reduced by one-half. Determine the entropy change of nitrogen during this process. *Answer: −0.0385 kJ/K*

**7–76** Reconsider Prob. 7–75. Using EES (or other) software, investigate the effect of varying the

polytropic exponent from 1 to 1.4 on the entropy change of the nitrogen. Show the processes on a common $P\text{-}v$ diagram.

**7–77** Air is compressed steadily by a 5-kW compressor from 100 kPa and 17°C to 600 kPa and 167°C at a rate of 1.6 kg/min. During this process, some heat transfer takes place between the compressor and the surrounding medium at 17°C. Determine the rate of entropy change of air during this process. *Answer: −0.0025 kW/K*

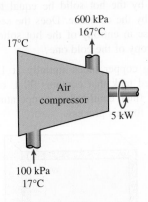

**FIGURE P7–77**

**7–78** Air enters a nozzle steadily at 280 kPa and 77°C with a velocity of 50 m/s and exits at 85 kPa and 320 m/s. The heat losses from the nozzle to the surrounding medium at 20°C are estimated to be 3.2 kJ/kg. Determine (a) the exit temperature and (b) the total entropy change for this process.

**7–79** Reconsider Prob. 7–78. Using EES (or other) software, study the effect of varying the surrounding medium temperature from 10 to 40°C on the exit temperature and the total entropy change for this process, and plot the results.

**7–80** 1-kg of air at 200 kPa and 127°C is contained in a piston-cylinder device. Air is now allowed to expand in a reversible, isothermal process until its pressure is 100 kPa. Determine the amount of heat transferred to the air during this expansion.

**7–81** Nitrogen is compressed isentropically from 100 kPa and 27°C to 1000 kPa in a piston-cylinder device. Determine its final temperature.

**7–82** Air at 3.5 MPa and 500°C is expanded in an adiabatic gas turbine to 0.2 MPa. Calculate the maximum work that this turbine can produce, in kJ/kg.

**7–83** Air is compressed in an isentropic compressor from 100 kPa and 20°C to 1500 kPa. Determine the outlet temperature and the work consumed by this compressor per unit mass of air. *Answers: 627 K, 341 kJ/kg*

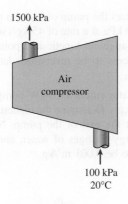

**FIGURE P7–83**

**7–84** An insulated rigid tank is divided into two equal parts by a partition. Initially, one part contains 12 kmol of an ideal gas at 330 kPa and 50°C, and the other side is evacuated. The partition is now removed, and the gas fills the entire tank. Determine the total entropy change during this process. *Answer:* 69.2 kJ/K

**7–85** An insulated rigid tank contains 4 kg of argon gas at 450 kPa and 30°C. A valve is now opened, and argon is allowed to escape until the pressure inside drops to 200 kPa. Assuming the argon remaining inside the tank has undergone a reversible, adiabatic process, determine the final mass in the tank. *Answer:* 2.46 kg

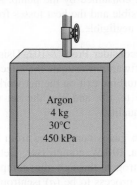

**FIGURE P7–85**

**7–86** Reconsider Prob. 7–85. Using EES (or other) software, investigate the effect of the final pressure on the final mass in the tank as the pressure varies from 450 to 150 kPa, and plot the results.

**7–87** Air enters an adiabatic nozzle at 400 kPa, 277°C, and 60 m/s and exits at 80 kPa. Assuming air to be an ideal gas with variable specific heats and disregarding any irreversibilities, determine the exit velocity of the air.

**7–88** Air at 257°C and 400 kPa is contained in a piston-cylinder device. The air expands adiabatically until the pressure is 100 kPa. Determine the mass of air needed to produce maximum work of 1000 kJ. Assume air has constant specific heats evaluated at 300 K. *Answer:* 8.04 kg

**7–89** Air at 27°C and 100 kPa is contained in a piston-cylinder device. When the air is compressed adiabatically, a minimum work input of 1000 kJ will increase the pressure to 600 kPa. Assuming air has constant specific heats evaluated at 300 K, determine the mass of air in the device.

**7–90** Air is compressed in a piston-cylinder device from 90 kPa and 20°C to 400 kPa in a reversible isothermal process. Determine (*a*) the entropy change of air and (*b*) the work done.

**7–91** Helium gas is compressed from 90 kPa and 30°C to 450 kPa in a reversible, adiabatic process. Determine the final temperature and the work done, assuming the process takes place (*a*) in a piston-cylinder device and (*b*) in a steady-flow compressor.

**7–92** 5-kg of air at 427°C and 600 kPa are contained in a piston-cylinder device. The air expands adiabatically until the pressure is 100 kPa and produces 600 kJ of work output. Assume air has constant specific heats evaluated at 300 K.

(*a*) Determine the entropy change of the air, in kJ/kg·K

(*b*) Since the process is adiabatic, is the process realistic? Using concepts of the second law, support your answer.

**7–93** A container filled with 45 kg of liquid water at 95°C is placed in a 90-m³ room that is initially at 12°C. Thermal equilibrium is established after a while as a result of heat transfer between the water and the air in the room. Using constant specific heats, determine (*a*) the final equilibrium temperature, (*b*) the amount of heat transfer between the water and the air in the room, and (*c*) the entropy generation. Assume the room is well sealed and heavily insulated.

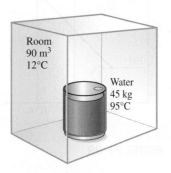

**FIGURE P7–93**

## Reversible Steady-Flow Work

**7–94C** In large compressors, the gas is frequently cooled while being compressed to reduce the power consumed by the compressor. Explain how cooling the gas during a compression process reduces the power consumption.

**7–95C** The turbines in steam power plants operate essentially under adiabatic conditions. A plant engineer suggests to end this practice. She proposes to run cooling water through the outer surface of the casing to cool the steam as it flows through the turbine. This way, she reasons, the entropy of the steam will decrease, the performance of the turbine will improve, and as a result the work output of the turbine will increase. How would you evaluate this proposal?

**7–96C** It is well known that the power consumed by a compressor can be reduced by cooling the gas during compression. Inspired by this, somebody proposes to cool the liquid as it flows through a pump, in order to reduce the power consumption of the pump. Would you support this proposal? Explain.

**7–97** Air is compressed isothermally from 91 kPa and 32°C to 550 kPa in a reversible steady-flow device. Calculate the work required, in kJ/kg, for this compression. *Answer:* 157 kJ/kg

**7–98** Saturated water vapor at 150°C is compressed in a reversible steady-flow device to 1000 kPa while its specific volume remains constant. Determine the work required, in kJ/kg.

**7–99** Calculate the work produced, in kJ/kg, for the reversible steady-flow process 1-3 shown in Fig. P7–99.

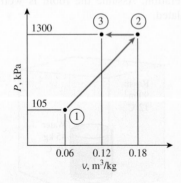

**FIGURE P7–99**

**7–100** Water enters the pump of a steam power plant as saturated liquid at 20 kPa at a rate of 45 kg/s and exits at 6 MPa. Neglecting the changes in kinetic and potential energies and assuming the process to be reversible, determine the power input to the pump.

**7–101** Liquid water enters a 16-kW pump at 100-kPa pressure at a rate of 5 kg/s. Determine the highest pressure the liquid water can have at the exit of the pump. Neglect the kinetic and potential energy changes of water, and take the specific volume of water to be 0.001 m³/kg. *Answer:* 3300 kPa

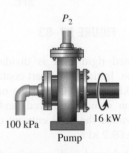

**FIGURE P7–101**

**7–102** Consider a steam power plant that operates between the pressure limits of 5 MPa and 10 kPa. Steam enters the pump as saturated liquid and leaves the turbine as saturated vapor. Determine the ratio of the work delivered by the turbine to the work consumed by the pump. Assume the entire cycle to be reversible and the heat losses from the pump and the turbine to be negligible.

**7–103** [EES] Reconsider Prob. 7–102. Using EES (or other) software, investigate the effect of the quality of the steam at the turbine exit on the net work output. Vary the quality from 0.5 to 1.0, and plot the net work output as a function of this quality.

**7–104** Nitrogen gas is compressed from 80 kPa and 27°C to 480 kPa by a 10-kW compressor. Determine the mass flow rate of nitrogen through the compressor, assuming the compression process to be (*a*) isentropic, (*b*) polytropic with *n* = 1.3, (*c*) isothermal, and (*d*) ideal two-stage polytropic with *n* = 1.3. *Answers:* (*a*) 0.048 kg/s, (*b*) 0.051 kg/s, (*c*) 0.063 kg/s, (*d*) 0.056 kg/s

**7–105** Saturated refrigerant-134a vapor at 100 kPa is compressed reversibly in an adiabatic compressor to 600 kPa. Determine the work input to the compressor. What would your answer be if the refrigerant were first condensed at constant pressure before it was compressed?

## Isentropic Efficiencies of Steady-Flow Devices

**7–106** Describe the ideal process for an (a) adiabatic turbine, (b) adiabatic compressor, and (c) adiabatic nozzle, and define the isentropic efficiency for each device.

**7–107C** Is the isentropic process a suitable model for compressors that are cooled intentionally? Explain.

**7–108C** On a $T$-$s$ diagram, does the actual exit state (state 2) of an adiabatic turbine have to be on the right-hand side of the isentropic exit state (state 2$s$)? Why?

**7–109** Steam enters an adiabatic turbine at 5 MPa, 650°C, and 80 m/s and leaves at 50 kPa, 150°C, and 140 m/s. If the power output of the turbine is 8 MW, determine (a) the mass flow rate of the steam flowing through the turbine and (b) the isentropic efficiency of the turbine. *Answers:* (a) 8.03 kg/s, (b) 82.8 percent

**7–110** Combustion gases enter an adiabatic gas turbine at 827°C and 850 kPa and leave at 425 kPa with a low velocity. Treating the combustion gases as air and assuming an isentropic efficiency of 82 percent, determine the work output of the turbine. *Answer:* 165 kJ/kg

**7–111** Steam at 4 MPa and 350°C is expanded in an adiabatic turbine to 120 kPa. What is the isentropic efficiency of this turbine if the steam is exhausted as a saturated vapor?

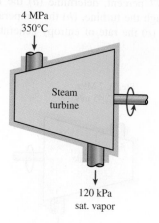

4 MPa
350°C

Steam
turbine

120 kPa
sat. vapor

**FIGURE P7–111**

**7–112** Steam enters an adiabatic turbine at 8 MPa and 500°C with a mass flow rate of 3 kg/s and leaves at 30 kPa. The isentropic efficiency of the turbine is 0.90. Neglecting the kinetic energy change of the steam, determine (a) the temperature at the turbine exit and (b) the power output of the turbine. *Answers:* (a) 69.1°C, (b) 3054 kW

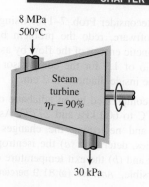

8 MPa
500°C

Steam
turbine
$\eta_T = 90\%$

30 kPa

**FIGURE P7–112**

**7–113** Reconsider Prob. 7–112. Using EES (or other) software, study the effect of varying the turbine isentropic efficiency from 0.75 to 1.0 on both the work done and the exit temperature of the steam, and plot your results.

**7–114** Carbon dioxide enters an adiabatic compressor at 100 kPa and 300 K at a rate of 1.8 kg/s and exits at 600 kPa and 450 K. Neglecting the kinetic energy changes, determine the isentropic efficiency of the compressor.

**7–115** A refrigeration unit compresses saturated R-134a vapor at 10°C to 1000 kPa. How much power is required to compress 0.9 kg/s of R-134a with a compressor efficiency of 85 percent? *Answer:* 19.3 kW

**7–116** Refrigerant-134a enters an adiabatic compressor as saturated vapor at 100 kPa at a rate of 0.7 m³/min and exits at 1-MPa pressure. If the isentropic efficiency of the compressor is 87 percent, determine (a) the temperature of the refrigerant at the exit of the compressor and (b) the power input, in kW. Also, show the process on a $T$-$s$ diagram with respect to saturation lines.

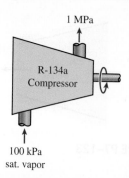

1 MPa

R-134a
Compressor

100 kPa
sat. vapor

**FIGURE P7–116**

**7–117** Reconsider Prob. 7–116. Using EES (or other) software, redo the problem by including the effects of the kinetic energy of the flow by assuming an inlet-to-exit area ratio of 1.5 for the compressor when the compressor exit pipe inside diameter is 2 cm.

**7–118** Air is compressed by an adiabatic compressor from 95 kPa and 27°C to 600 kPa and 277°C. Assuming variable specific heats and neglecting the changes in kinetic and potential energies, determine (a) the isentropic efficiency of the compressor and (b) the exit temperature of air if the process were reversible. *Answers:* (a) 81.9 percent, (b) 506 K

**7–119** Argon gas enters an adiabatic compressor at 98 kPa and 25°C with a velocity of 20 m/s, and it exits at 1400 kPa and 75 m/s. If the isentropic efficiency of the compressor is 87 percent, determine (a) the exit temperature of the argon and (b) the work input to the compressor.

**7–120** Air enters an adiabatic nozzle at 400 kPa and 547°C with low velocity and exits at 240 m/s. If the isentropic efficiency of the nozzle is 90 percent, determine the exit temperature and pressure of the air.

**7–121** Reconsider Prob. 7–120. Using EES (or other) software, study the effect of varying the nozzle isentropic efficiency from 0.8 to 1.0 on both the exit temperature and pressure of the air, and plot the results.

**7–122** The exhaust nozzle of a jet engine expands air at 300 kPa and 180°C adiabatically to 100 kPa. Determine the air velocity at the exit when the inlet velocity is low and the nozzle isentropic efficiency is 96 percent.

**7–123** Hot combustion gases enter the nozzle of a turbojet engine at 260 kPa, 747°C, and 80 m/s, and they exit at a pressure of 85 kPa. Assuming an isentropic efficiency of 92 percent and treating the combustion gases as air, determine (a) the exit velocity and (b) the exit temperature. *Answers:* (a) 728 m/s, (b) 786 K

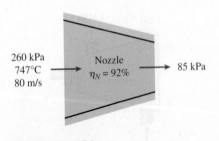

260 kPa
747°C
80 m/s

Nozzle
$\eta_N = 92\%$

85 kPa

**FIGURE P7–123**

### Entropy Balance

**7–124** Refrigerant-134a is expanded adiabatically from 700 kPa and 30°C to a saturated vapor at 60 kPa. Determine the entropy generation for this process, in kJ/kg·K.

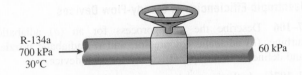

R-134a
700 kPa
30°C

60 kPa

**FIGURE P7–124**

**7–125** Oxygen enters an insulated 12-cm-diameter pipe with a velocity of 70 m/s. At the pipe entrance, the oxygen is at 240 kPa and 20°C; and, at the exit, it is at 200 kPa and 18°C. Calculate the rate at which entropy is generated in the pipe.

**7–126** Nitrogen is compressed by an adiabatic compressor from 100 kPa and 25°C to 600 kPa and 290°C. Calculate the entropy generation for this process, in kJ/kg·K.

**7–127** Air enters a compressor steadily at the ambient conditions of 100 kPa and 22°C and leaves at 800 kPa. Heat is lost from the compressor in the amount of 120 kJ/kg and the air experiences an entropy decrease of 0.40 kJ/kg·K. Using constant specific heats, determine (a) the exit temperature of the air, (b) the work input to the compressor, and (c) entropy generation during this process.

**7–128** Steam enters an adiabatic turbine steadily at 7 MPa, 500°C, and 45 m/s, and leaves at 100 kPa and 75 m/s. If the power output of the turbine is 5 MW and the isentropic efficiency is 77 percent, determine (a) the mass flow rate of steam through the turbine, (b) the temperature at the turbine exit, and (c) the rate of entropy generation during this process.

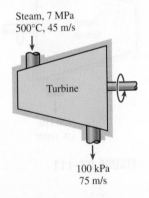

Steam, 7 MPa
500°C, 45 m/s

Turbine

100 kPa
75 m/s

**FIGURE P7–128**

**7–129** In an ice-making plant, water at 0°C is frozen at atmospheric pressure by evaporating saturated R-134a liquid at −16°C. The refrigerant leaves this evaporator as a saturated vapor, and the plant is sized to produce ice at 0°C at a rate of 2500 kg/h. Determine the rate of entropy generation in this plant. *Answer:* 0.0528 kW/K

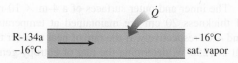

**FIGURE P7–129**

**7–130** Water at 200 kPa and 10°C enters a mixing chamber at a rate of 135 kg/min where it is mixed steadily with steam entering at 200 kPa and 150°C. The mixture leaves the chamber at 200 kPa and 55°C, and heat is lost to the surrounding air at 20°C at a rate of 180 kJ/min. Neglecting the changes in kinetic and potential energies, determine the rate of entropy generation during this process?

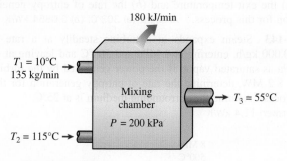

**FIGURE P7–130**

**7–131** A well-insulated heat exchanger is to heat water ($c_p = 4.18$ kJ/kg·°C) from 25 to 60°C at a rate of 0.50 kg/s. The heating is to be accomplished by geothermal water ($c_p = 4.31$ kJ/kg·°C) available at 140°C at a mass flow rate of 0.75 kg/s. Determine (a) the rate of heat transfer and (b) the rate of entropy generation in the heat exchanger.

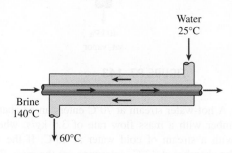

**FIGURE P7–131**

**7–132** An adiabatic heat exchanger is to cool ethylene glycol ($c_p = 2.56$ kJ/kg·°C) flowing at a rate of 2 kg/s from 80 to 40°C by water ($c_p = 4.18$ kJ/kg·°C) that enters at 20°C and leaves at 55°C. Determine (a) the rate of heat transfer and (b) the rate of entropy generation in the heat exchanger.

**7–133** A well-insulated, thin-walled, double-pipe, counter-flow heat exchanger is to be used to cool oil

($c_p = 2.20$ kJ/kg·°C) from 150°C to 40°C at a rate of 2 kg/s by water ($c_p = 4.18$ kJ/kg·°C) that enters at 22°C at a rate of 1.5 kg/s. Determine (a) the rate of heat transfer and (b) the rate of entropy generation in the heat exchanger.

**7–134** In a dairy plant, milk at 4°C is pasteurized continuously at 72°C at a rate of 12 L/s for 24 hours a day and 365 days a year. The milk is heated to the pasteurizing temperature by hot water heated in a natural-gas-fired boiler that has an efficiency of 82 percent. The pasteurized milk is then cooled by cold water at 18°C before it is finally refrigerated back to 4°C. To save energy and money, the plant installs a regenerator that has an effectiveness of 82 percent. If the cost of natural gas is $1.30/therm (1 therm = 105,500 kJ), determine how much energy and money the regenerator will save this company per year and the annual reduction in entropy generation.

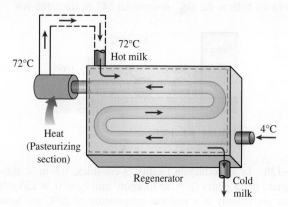

**FIGURE P7–134**

**7–135** An ordinary egg can be approximated as a 5.5-cm-diameter sphere. The egg is initially at a uniform temperature of 8°C and is dropped into boiling water at 97°C. Taking the properties of the egg to be $\rho = 1020$ kg/m³ and $c_p = 3.32$ kJ/kg·°C, determine (a) how much heat is transferred to the egg by the time the average temperature of the egg rises to 70°C and (b) the amount of entropy generation associated with this heat transfer process.

**FIGURE P7–135**

**7–136** Chickens with an average mass of 2.2 kg and average specific heat of 3.54 kJ/kg·°C are to be cooled by chilled water that enters a continuous-flow-type immersion chiller at 0.5°C and leaves at 2.5°C. Chickens are dropped into the chiller at a uniform temperature of 15°C at a rate of 250 chickens per hour and are cooled to an average temperature of 3°C before they are taken out. The chiller gains heat from the surroundings at 25°C at a rate of 150 kJ/h. Determine (a) the rate of heat removal from the chickens, in kW, and (b) the rate of entropy generation during this chilling process.

**7–137** Carbon-steel balls ($\rho$ = 7833 kg/m³ and $c_p$ = 0.465 kJ/kg·°C) 8 mm in diameter are annealed by heating them first to 900°C in a furnace and then allowing them to cool slowly to 100°C in ambient air at 35°C. If 2500 balls are to be annealed per hour, determine (a) the rate of heat transfer from the balls to the air and (b) the rate of entropy generation due to heat loss from the balls to the air. *Answers:* (a) 542 W, (b) 0.986 W/K

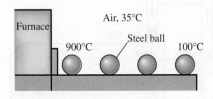

**FIGURE P7–137**

**7–138** In a production facility, 3-cm-thick, 0.6-m × 0.6-m square brass plates ($\rho$ = 8530 kg/m³ and $c_p$ = 0.38 kJ/kg·°C) that are initially at a uniform temperature of 25°C are heated by passing them through an oven at 700°C at a rate of 450 per minute. If the plates remain in the oven until their average temperature rises to 500°C, determine (a) the rate of heat transfer to the plates in the furnace and (b) the rate of entropy generation associated with this heat transfer process.

**7–139** Long cylindrical steel rods ($\rho$ = 7833 kg/m³ and $c_p$ = 0.465 kJ/kg·°C) of 10-cm diameter are heat treated by drawing them at a velocity of 3 m/min through a 7-m-long oven maintained at 900°C. If the rods enter the oven at 30°C and leave at 700°C, determine (a) the rate of heat transfer to the rods in the oven and (b) the rate of entropy generation associated with this heat transfer process.

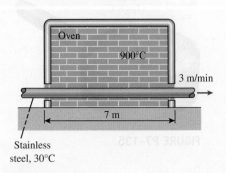

**FIGURE P7–139**

**7–140** The inner and outer surfaces of a 4-m × 10-m brick wall of thickness 20 cm are maintained at temperatures of 16°C and 4°C, respectively. If the rate of heat transfer through the wall is 1250 W, determine the rate of entropy generation within the wall.

**7–141** A frictionless piston–cylinder device contains saturated liquid water at 275-kPa pressure. Now 600 kJ of heat is transferred to water from a source at 537°C, and part of the liquid vaporizes at constant pressure. Determine the total entropy generated during this process, in kJ/K.

**7–142** Steam enters an adiabatic nozzle at 2 MPa and 350°C with a velocity of 55 m/s and exits at 0.8 MPa and 390 m/s. If the nozzle has an inlet area of 7.5 cm², determine (a) the exit temperature and (b) the rate of entropy generation for this process. *Answers:* (a) 303°C, (b) 0.0854 kW/K

**7–143** Steam expands in a turbine steadily at a rate of 40,000 kg/h, entering at 8 MPa and 500°C and leaving at 40 kPa as saturated vapor. If the power generated by the turbine is 8.2 MW, determine the rate of entropy generation for this process. Assume the surrounding medium is at 25°C. *Answer:* 11.4 kW/K

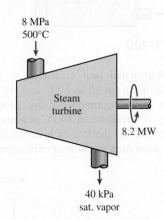

**FIGURE P7–143**

**7–144** A hot-water stream at 70°C enters an adiabatic mixing chamber with a mass flow rate of 3.6 kg/s, where it is mixed with a stream of cold water at 20°C. If the mixture leaves the chamber at 42°C, determine (a) the mass flow rate of the cold water and (b) the rate of entropy generation during this adiabatic mixing process. Assume all the streams are at a pressure of 200 kPa.

**7–145** Liquid water at 200 kPa and 15°C is heated in a chamber by mixing it with superheated steam at 200 kPa and 150°C. Liquid water enters the mixing chamber at a rate of 4.3 kg/s, and the chamber is estimated to lose heat to the surrounding air at 20°C at a rate of 1200 kJ/min. If the mixture leaves the mixing chamber at 200 kPa and 80°C, determine (a) the mass flow rate of the superheated steam and (b) the rate of entropy generation during this mixing process. *Answers:* (a) 0.481 kg/s, (b) 0.746 kW/K

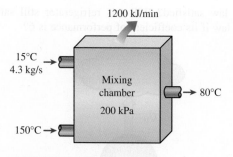

**FIGURE P7–145**

**7–146** A 0.18-m³ rigid tank is filled with saturated liquid water at 120°C. A valve at the bottom of the tank is now opened, and one-half of the total mass is withdrawn from the tank in the liquid form. Heat is transferred to water from a source at 230°C so that the temperature in the tank remains constant. Determine (*a*) the amount of heat transfer and (*b*) the total entropy generation for this process.

**7–147** An iron block of unknown mass at 85°C is dropped into an insulated tank that contains 20 L of water at 20°C. At the same time, a paddle wheel driven by a 200-W motor is activated to stir the water. Thermal equilibrium is established after 10 min with a final temperature of 25°C. Determine (*a*) the mass of the iron block and (*b*) the entropy generated during this process.

## Special Topic: Reducing the Cost of Compressed Air

**7–148** Compressed air is one of the key utilities in manufacturing facilities, and the total installed power of compressed-air systems in the United States is estimated to be about 20 million horsepower. Assuming the compressors to operate at full load during one-third of the time on average and the average motor efficiency to be 90 percent, determine how much energy and money will be saved per year if the energy consumed by compressors is reduced by 5 percent as a result of implementing some conservation measures. Take the unit cost of electricity to be $0.11/kWh.

**7–149** The compressed-air requirements of a plant at sea level are being met by a 90-hp compressor that takes in air at the local atmospheric pressure of 101.3 kPa and the average temperature of 15°C and compresses it to 1100 kPa. An investigation of the compressed-air system and the equipment using the compressed air reveals that compressing the air to 750 kPa is sufficient for this plant. The compressor operates 3500 h/yr at 75 percent of the rated load and is driven by an electric motor that has an efficiency of 94 percent. Taking the price of electricity to be $0.105/kWh, determine the amount of energy and money saved as a result of reducing the pressure of the compressed air.

**7–150** A 150-hp compressor in an industrial facility is housed inside the production area where the average temperature during operating hours is 25°C. The average temperature of outdoors during the same hours is 10°C. The compressor operates 4500 h/yr at 85 percent of rated load and is driven by an electric motor that has an efficiency of 90 percent. Taking the price of electricity to be $0.12/kWh, determine the amount of energy and money saved as a result of drawing outside air to the compressor instead of using the inside air.

**7–151** The compressed-air requirements of a plant are being met by a 100-hp screw compressor that runs at full load during 40 percent of the time and idles the rest of the time during operating hours. The compressor consumes 35 percent of the rated power when idling and 90 percent of the power when compressing air. The annual operating hours of the facility are 3800 h, and the unit cost of electricity is $0.115/kWh.

It is determined that the compressed-air requirements of the facility during 60 percent of the time can be met by a 25-hp reciprocating compressor that consumes 95 percent of the rated power when compressing air and no power when not compressing air. It is estimated that the 25-hp compressor runs 85 percent of the time. The efficiencies of the motors of the large and the small compressors at or near full load are 0.90 and 0.88, respectively. The efficiency of the large motor at 35 percent load is 0.82. Determine the amount of energy and money saved as a result of switching to the 25-hp compressor during 60 percent of the time.

**7–152** The compressed-air requirements of a plant are being met by a 90-hp screw compressor. The facility stops production for one hour every day, including weekends, for lunch break, but the compressor is kept operating. The compressor consumes 35 percent of the rated power when idling, and the unit cost of electricity is $0.11/kWh. Determine the amount of energy and money saved per year as a result of turning the compressor off during lunch break. Take the efficiency of the motor at part load to be 84 percent.

**7–153** The compressed-air requirements of a plant are met by a 150-hp compressor equipped with an intercooler, an aftercooler, and a refrigerated dryer. The plant operates 6300 h/yr, but the compressor is estimated to be compressing air during only one-third of the operating hours, that is, 2100 hours a year. The compressor is either idling or is shut off the rest of the time. Temperature measurements and calculations indicate that 25 percent of the energy input to the compressor is removed from the compressed air as heat in the aftercooler. The COP of the refrigeration unit is 2.5, and the cost of electricity is $0.12/kWh. Determine the amount of the energy and money saved per year as a result of cooling the compressed air before it enters the refrigerated dryer.

**7–154** The 1800-rpm, 150-hp motor of a compressor is burned out and is to be replaced by either a standard motor that has a full-load efficiency of 93.0 percent and costs $9031 or a high-efficiency motor that has an efficiency of 96.2

percent and costs $10,942. The compressor operates 4368 h/ yr at full load, and its operation at part load is negligible. If the cost of electricity is $0.125/kWh, determine the amount of energy and money this facility will save by purchasing the high-efficiency motor instead of the standard motor. Also, determine if the savings from the high-efficiency motor justify the price differential if the expected life of the motor is 10 years. Ignore any possible rebates from the local power company.

**7–155** The space heating of a facility is accomplished by nat-ural gas heaters that are 85 percent efficient. The compressed air needs of the facility are met by a large liquid-cooled com-pressor. The coolant of the compressor is cooled by air in a liquid-to-air heat exchanger whose airflow section is 1.0-m high and 1.0-m wide. During typical operation, the air is heated from 20 to 52°C as it flows through the heat exchanger. The average velocity of air on the inlet side is measured to be 3 m/s. The compressor operates 20 hours a day and 5 days a week throughout the year. Taking the heating season to be 6 months (26 weeks) and the cost of the natural gas to be $1.25/ therm (1 therm = 100,000 Btu = 105,500 kJ), determine how much money will be saved by diverting the compressor waste heat into the facility during the heating season.

**7–156** The compressors of a production facility maintain the compressed-air lines at a (gage) pressure of 700 kPa at 1400-m elevation, where the atmospheric pressure is 85.6 kPa. The average temperature of air is 15°C at the compressor inlet and 25°C in the compressed-air lines. The facility operates 4200 h/yr, and the average price of electric-ity is $0.12/kWh. Taking the compressor efficiency to be 0.8, the motor efficiency to be 0.93, and the discharge coefficient to be 0.65, determine the energy and money saved per year by sealing a leak equivalent to a 3-mm-diameter hole on the compressed-air line.

**7–157** The energy used to compress air in the United States is estimated to exceed one-half quadrillion ($0.5 \times 10^{15}$) kJ per year. It is also estimated that 10 to 40 percent of the compressed air is lost through leaks. Assuming, on average, 20 percent of the compressed air is lost through air leaks and the unit cost of electricity is $0.13/kWh, determine the amount and cost of electricity wasted per year due to air leaks.

## Review Problems

**7–158** A heat engine whose thermal efficiency is 35 percent uses a hot reservoir at 600 K and a cold reservoir at 300 K. Calculate the entropy change of the two reservoirs when 1 kJ of heat is transferred from the hot reservoir to the engine. Does this engine satisfy the increase of entropy principle? If the thermal efficiency of the heat engine is increased to 60 percent, will the increase of entropy principle still be satisfied?

**7–159** A refrigerator with a coefficient of performance of 4 transfers heat from a cold region at −20°C to a hot region at 30°C. Calculate the total entropy change of the regions when 1 kJ of heat is transferred from the cold region. Is the

second law satisfied? Will this refrigerator still satisfy the second law if its coefficient of performance is 6?

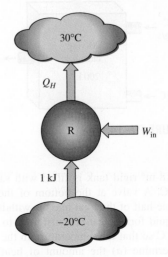

**FIGURE P7–159**

**7–160** It has been suggested that air at 100 kPa and 25°C can be cooled by first compressing it adiabatically in a closed system to 1000 kPa and then expanding it adiabatically back to 100 kPa. Is this possible?

**7–161** 0.5-kg of air at 70 kPa and 20°C is contained in a piston-cylinder device. Next, the air is compressed revers-ibly to 700 kPa while the temperature is maintained constant. Determine the total amount of heat transferred to the air dur-ing this compression.

**7–162** Can saturated water vapor at 200 kPa be condensed to a saturated liquid in an isobaric, closed system process while only exchanging heat with an isothermal energy reser-voir at 90°C? (Hint: Determine the entropy generation.)

**7–163** A horizontal cylinder is separated into two com-partments by an adiabatic, frictionless piston. One side con-tains 0.2 m³ of nitrogen and the other side contains 0.1 kg of helium, both initially at 20°C and 95 kPa. The sides of the cylinder and the helium end are insulated. Now heat is added to the nitrogen side from a reservoir at 500°C until the pressure of the helium rises to 120 kPa. Determine (a) the final temperature of the helium, (b) the final volume of the nitrogen, (c) the heat transferred to the nitrogen, and (d) the entropy generation during this process.

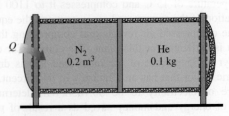

**FIGURE P7–163**

**7–164** A piston–cylinder device contains air that undergoes a reversible thermodynamic cycle. Initially, air is at 400 kPa and 300 K with a volume of 0.3 m³ Air is first expanded isothermally to 150 kPa, then compressed adiabatically to the initial pressure, and finally compressed at the constant pressure to the initial state. Accounting for the variation of specific heats with temperature, determine the work and heat transfer for each process.

**7–165** A piston–cylinder device initially contains 0.43 m³ of helium gas at 175 kPa and 20°C. Helium is now compressed in a polytropic process ($PV^n$ = constant) to 500 kPa and 150°C. Determine (a) the entropy change of helium, (b) the entropy change of the surroundings, and (c) whether this process is reversible, irreversible, or impossible. Assume the surroundings are at 20°C. *Answers:* (a) –0.0339 kJ/K, (b) 0.0409 kJ/K, (c) irreversible

**7–166** A piston–cylinder device contains steam that undergoes a reversible thermodynamic cycle. Initially the steam is at 400 kPa and 350°C with a volume of 0.3 m³. The steam is first expanded isothermally to 150 kPa, then compressed adiabatically to the initial pressure, and finally compressed at the constant pressure to the initial state. Determine the net work and heat transfer for the cycle after you calculate the work and heat interaction for each process.

**7–167** A 0.8-m³ rigid tank contains carbon dioxide ($CO_2$) gas at 250 K and 100 kPa. A 500-W electric resistance heater placed in the tank is now turned on and kept on for 40 min after which the pressure of $CO_2$ is measured to be 175 kPa. Assuming the surroundings to be at 300 K and using constant specific heats, determine (a) the final temperature of $CO_2$, (b) the net amount of heat transfer from the tank, and (c) the entropy generation during this process.

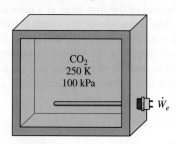

**FIGURE P7–167**

**7–168** Helium gas is throttled steadily from 400 kPa and 60°C. Heat is lost from the helium in the amount of 1.75 kJ/kg to the surroundings at 25°C and 100 kPa. If the entropy of the helium increases by 0.34 kJ/kg·K in the valve, determine (a) the exit pressure and temperature and (b) the entropy generation during this process. *Answers:* (a) 339 kPa, 59.7°C, (b) 0.346 kJ/kg·K

**7–169** Air enters the evaporator section of a window air conditioner at 100 kPa and 27°C with a volume flow rate

of 6 m³/min. The refrigerant-134a at 120 kPa with a quality of 0.3 enters the evaporator at a rate of 2 kg/min and leaves as saturated vapor at the same pressure. Determine the exit temperature of the air and the rate of entropy generation for this process, assuming (a) the outer surfaces of the air conditioner are insulated and (b) heat is transferred to the evaporator of the air conditioner from the surrounding medium at 32°C at a rate of 30 kJ/min.
*Answers:* (a) –15.9°C, 0.00196 kW/K, (b) –11.6°C, 0.00225 kW/K

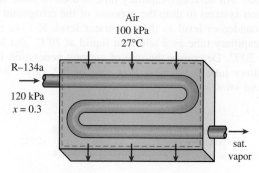

**FIGURE P7–169**

**7–170** Refrigerant-134a enters a compressor as a saturated vapor at 160 kPa at a rate of 0.03 m³/s and leaves at 800 kPa. The power input to the compressor is 10 kW. If the surroundings at 20°C experience an entropy increase of 0.008 kW/K, determine (a) the rate of heat loss from the compressor, (b) the exit temperature of the refrigerant, and (c) the rate of entropy generation.

**7–171** Air at 500 kPa and 400 K enters an adiabatic nozzle at a velocity of 30 m/s and leaves at 300 kPa and 350 K. Using variable specific heats, determine (a) the isentropic efficiency, (b) the exit velocity, and (c) the entropy generation.

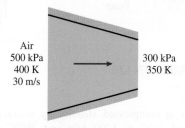

**FIGURE P7–171**

**7–172** 3-kg of helium gas at 100 kPa and 27°C are adiabatically compressed to 900 kPa. If the isentropic compression efficiency is 80 percent, determine the required work input and the final temperature of helium.

**7–173** An inventor claims to have invented an adiabatic steady-flow device with a single inlet-outlet that produces 230 kW when expanding 1 kg/s of air from 1200 kPa and 300°C to 100 kPa. Is this claim valid?

**7–174** You are to expand a gas adiabatically from 3 MPa and 300°C to 80 kPa in a piston-cylinder device. Which of the two choices – air with an isentropic expansion efficiency of 90 percent or neon with an isentropic expansion efficiency of 80 percent – will produce the most work?

**7–175** An adiabatic capillary tube is used in some refrigeration systems to drop the pressure of the refrigerant from the condenser level to the evaporator level. R-134a enters the capillary tube as a saturated liquid at 70°C, and leaves at −20°C. Determine the rate of entropy generation in the capillary tube for a mass flow rate of 0.2 kg/s.  *Answer:* 0.0166 kW/K

R-134a
70°C
sat. liq.  →  Capillary tube  →  −20°C

**FIGURE P7–175**

**7–176** Determine the work input and entropy generation during the compression of steam from 100 kPa to 1 MPa in (*a*) an adiabatic pump and (*b*) an adiabatic compressor if the inlet state is saturated liquid in the pump and saturated vapor in the compressor and the isentropic efficiency is 85 percent for both devices.

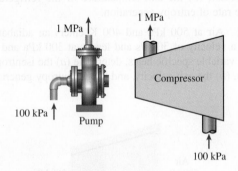

1 MPa

1 MPa

100 kPa

Pump

Compressor

100 kPa

**FIGURE P7–176**

**7–177** Air is compressed steadily by a compressor from 100 kPa and 20°C to 1200 kPa and 300°C at a rate of 0.4 kg/s. The compressor is intentionally cooled by utilizing fins on the surface of the compressor and heat is lost from the compressor at a rate of 15 kW to the surroundings at 20°C. Using constant specific heats at room temperature, determine (*a*) the power input to the compressor, (*b*) the isothermal efficiency, and (*c*) the entropy generation during this process.

**7–178** Air is compressed steadily by a compressor from 100 kPa and 17°C to 700 kPa at a rate of 5 kg/min. Determine the minimum power input required if the process is (*a*) adiabatic and (*b*) isothermal. Assume air to be an ideal gas with variable specific heats, and neglect the changes in kinetic and potential energies.  *Answers:* (*a*) 18.0 kW, (*b*) 13.5 kW

**7–179** Air enters a two-stage compressor at 100 kPa and 27°C and is compressed to 625 kPa. The pressure ratio across each stage is the same, and the air is cooled to the initial temperature between the two stages. Assuming the compression process to be isentropic, determine the power input to the compressor for a mass flow rate of 0.15 kg/s. What would your answer be if only one stage of compression were used?  *Answers:* 27.1 kW, 31.1 kW

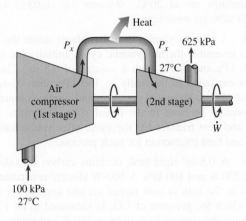

Heat

$P_x$        $P_x$  625 kPa

27°C

Air compressor (1st stage)        (2nd stage)

$\dot{W}$

100 kPa
27°C

**FIGURE P7–179**

**7–180** Steam at 6 MPa and 500°C enters a two-stage adiabatic turbine at a rate of 15 kg/s. 10 percent of the steam is extracted at the end of the first stage at a pressure of 1.2 MPa for other use. The remainder of the steam is further expanded in the second stage and leaves the turbine at 20 kPa. Determine the power output of the turbine, assuming (*a*) the process is reversible and (*b*) the turbine has an isentropic efficiency of 88 percent.  *Answers:* (*a*) 16,290 kW, (*b*) 14,335 kW

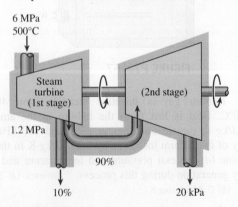

6 MPa
500°C

Steam turbine (1st stage)        (2nd stage)

1.2 MPa

90%

10%        20 kPa

**FIGURE P7–180**

**7–181** Refrigerant-134a at 140 kPa and −10°C is compressed by an adiabatic 1.3-kW compressor to an exit state of 700 kPa and 60°C. Neglecting the changes in kinetic and potential energies, determine (a) the isentropic efficiency of the compressor, (b) the volume flow rate of the refrigerant at the compressor inlet, in L/min, and (c) the maximum volume flow rate at the inlet conditions that this adiabatic 1.3-kW compressor can handle without violating the second law.

**7–182** An adiabatic air compressor is to be powered by a direct-coupled adiabatic steam turbine that is also driving a generator. Steam enters the turbine at 12.5 MPa and 500°C at a rate of 25 kg/s and exits at 10 kPa and a quality of 0.92. Air enters the compressor at 98 kPa and 295 K at a rate of 10 kg/s and exits at 1 MPa and 620 K. Determine (a) the net power delivered to the generator by the turbine and (b) the rate of entropy generation within the turbine and the compressor during this process.

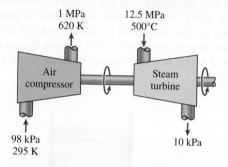

FIGURE P7–182

**7–183** Reconsider Prob. 7–182. Using EES (or other) software, determine the isentropic efficiencies for the compressor and turbine. Then use EES to study how varying the compressor efficiency over the range 0.6 to 0.8 and the turbine efficiency over the range 0.7 to 0.95 affect the net work for the cycle and the entropy generated for the process. Plot the net work as a function of the compressor efficiency for turbine efficiencies of 0.7, 0.8, and 0.9, and discuss your results.

**7–184** Air is expanded in an adiabatic turbine of 85 percent isentropic efficiency from an inlet state of 2200 kPa and 300°C to an outlet pressure of 200 kPa. Calculate the outlet temperature of air and the work produced by this turbine per unit mass of air.

**7–185** Air is expanded in an adiabatic turbine of 90 percent isentropic efficiency from an inlet state of 2800 kPa and 400°C to an outlet pressure of 150 kPa. Calculate the outlet temperature of air, the work produced by this turbine, and the entropy generation. *Answers:* 332 K, 346 kJ/kg, 0.123 kJ/kg·K

**7–186** To control the power output of an isentropic steam turbine, a throttle valve is placed in the steam line supplying the turbine inlet, as shown in the figure. Steam at 6 MPa and 400°C is supplied to the throttle inlet, and the turbine exhaust pressure is set at 70 kPa. Compare the work produced by this steam turbine, in kJ/kg, when the throttle valve is completely open (so that there is no pressure loss) and when it is partially closed so that the pressure at the turbine inlet is 3 MPa.

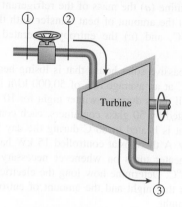

FIGURE P7–186

**7–187** Two rigid tanks are connected by a valve. Tank A is insulated and contains 0.3 m³ of steam at 400 kPa and 60 percent quality. Tank B is uninsulated and contains 2 kg of steam at 200 kPa and 250°C. The valve is now opened, and steam flows from tank A to tank B until the pressure in tank A drops to 200 kPa. During this process 300 kJ of heat is transferred from tank B to the surroundings at 17°C. Assuming the steam remaining inside tank A to have undergone a reversible adiabatic process, determine (a) the final temperature in each tank and (b) the entropy generated during this process. *Answers:* (a) 120.2°C, 116.1°C, (b) 0.498 kJ/K

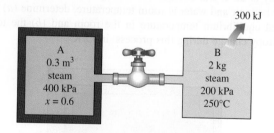

FIGURE P7–187

**7–188** A 1200-W electric resistance heating element whose diameter is 0.5 cm is immersed in 40 kg of water initially at 20°C. Assuming the water container is well-insulated, determine how long it will take for this heater to raise the water temperature to 50°C. Also, determine the entropy generated during this process, in kJ/K.

**7–189** A 0.15-m³ rigid tank initially contains refrigerant-134a at 0.4 MPa and 100 percent quality. The tank is connected by a valve to a supply line that carries refrigerant-134a at 1 MPa and 26°C. The valve is now opened, allowing the refrigerant to enter the tank, and is closed when it is observed that the tank contains only saturated liquid at 0.7 MPa. Determine (a) the mass of the refrigerant that entered the tank, (b) the amount of heat transfer with the surroundings at 20°C, and (c) the entropy generated during this process.

**7–190** A passive solar house that is losing heat to the outdoors at 3°C at an average rate of 50,000 kJ/h is maintained at 22°C at all times during a winter night for 10 h. The house is to be heated by 50 glass containers, each containing 20 L of water that is heated to 80°C during the day by absorbing solar energy. A thermostat controlled 15 kW backup electric resistance heater turns on whenever necessary to keep the house at 22°C. Determine how long the electric heating system was on that night and the amount of entropy generated during the night.

**7–191** A 0.42-m³ steel container that has a mass of 34 kg when empty is filled with liquid water. Initially, both the steel tank and the water are at 50°C. Now heat is transferred, and the entire system cools to the surrounding air temperature of 20°C. Determine the total entropy generated during this process.

**7–192** In order to cool 1-ton of water at 20°C in an insulated tank, a person pours 80 kg of ice at −5°C into the water. Determine (a) the final equilibrium temperature in the tank and (b) the entropy generation during this process. The melting temperature and the heat of fusion of ice at atmospheric pressure are 0°C and 333.7 kJ/kg.

**7–193** One ton of liquid water at 80°C is brought into a well-insulated and well-sealed 4-m × 5-m × 7-m room initially at 22°C and 100 kPa. Assuming constant specific heats for both air and water at room temperature, determine (a) the final equilibrium temperature in the room and (b) the total entropy change during this process, in kJ/K.

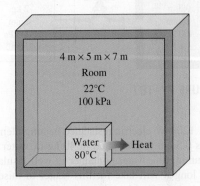

**FIGURE P7–193**

**7–194** A well-insulated 4-m × 4-m × 5-m room initially at 10°C is heated by the radiator of a steam heating system. The radiator has a volume of 15 L and is filled with superheated vapor at 200 kPa and 200°C. At this moment both the inlet and the exit valves to the radiator are closed. A 120-W fan is used to distribute the air in the room. The pressure of the steam is observed to drop to 100 kPa after 30 min as a result of heat transfer to the room. Assuming constant specific heats for air at room temperature, determine (a) the average temperature of air in 30 min, (b) the entropy change of the steam, (c) the entropy change of the air in the room, and (d) the entropy generated during this process, in kJ/K. Assume the air pressure in the room remains constant at 100 kPa at all times.

**7–195** An insulated piston–cylinder device initially contains 0.02 m³ of saturated liquid–vapor mixture of water with a quality of 0.1 at 100°C. Now some ice at −18°C is dropped into the cylinder. If the cylinder contains saturated liquid at 100°C when thermal equilibrium is established, determine (a) the amount of ice added and (b) the entropy generation during this process. The melting temperature and the heat of fusion of ice at atmospheric pressure are 0°C and 333.7 kJ/kg.

Ice −18°C     0.02 m³ 100°C

**FIGURE P7–195**

**7–196** Consider a 50-L evacuated rigid bottle that is surrounded by the atmosphere at 95 kPa and 27°C. A valve at the neck of the bottle is now opened and the atmospheric air is allowed to flow into the bottle. The air trapped in the bottle eventually reaches thermal equilibrium with the atmosphere as a result of heat transfer through the wall of the bottle. The valve remains open during the process so that the trapped air also reaches mechanical equilibrium with the atmosphere. Determine the net heat transfer through the wall of the bottle and the entropy generation during this filling process. *Answers:* 4.75 kJ, 0.0158 kJ/K

**7–197** (a) Water flows through a shower head steadily at a rate of 10 L/min. An electric resistance heater placed in the water pipe heats the water from 16 to 43°C. Taking the density of water to be 1 kg/L, determine the electric power input to the heater, in kW, and the rate of entropy generation during this process, in kW/K.

(*b*) In an effort to conserve energy, it is proposed to pass the drained warm water at a temperature of 39°C through a heat exchanger to preheat the incoming cold water. If the heat exchanger has an effectiveness of 0.50 (that is, it recovers only half of the energy that can possibly be transferred from the drained water to incoming cold water), determine the electric power input required in this case and the reduction in the rate of entropy generation in the resistance heating section.

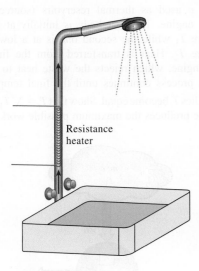

Resistance heater

**FIGURE P7–197**

**7–198** [EES] Using EES (or other) software, determine the work input to a multistage compressor for a given set of inlet and exit pressures for any number of stages. Assume that the pressure ratio across each stage is identical and the compression process is polytropic. List and plot the compressor work against the number of stages for $P_1 = 100$ kPa, $T_1 = 25$°C, $P_2 = 1000$ kPa, and $n = 1.35$ for air. Based on this chart, can you justify using compressors with more than three stages?

**7–199** The inner and outer surfaces of a 2-m × 2-m window glass in winter are 10°C and 3°C, respectively. If the rate of heat loss through the window is 3.2 kJ/s, determine the amount of heat loss, in kJ, through the glass over a period of 5 h. Also, determine the rate of entropy generation during this process within the glass.

**7–200** The inner and outer glasses of a 2-m × 2-m double-pane window are at 18°C and 6°C, respectively. If the glasses are very nearly isothermal and the rate of heat transfer through the window is 110 W, determine the rates of entropy transfer through both sides of the window and the rate of entropy generation within the window, in W/K.

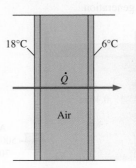

18°C 6°C

$\dot{Q}$

Air

**FIGURE P7–200**

**7–201** A hot-water pipe at 80°C is losing heat to the surrounding air at 5°C at a rate of 2200 W. Determine the rate of entropy generation in the surrounding air, in W/K.

**7–202** Consider the turbocharger of an internal combustion engine. The exhaust gases enter the turbine at 450°C at a rate of 0.02 kg/s and leave at 400°C. Air enters the compressor at 70°C and 95 kPa at a rate of 0.018 kg/s and leaves at 135 kPa. The mechanical efficiency between the turbine and the compressor is 95 percent (5 percent of turbine work is lost during its transmission to the compressor). Using air properties for the exhaust gases, determine (*a*) the air temperature at the compressor exit and (*b*) the isentropic efficiency of the compressor. *Answers:* (*a*) 126°C, (*b*) 0.642

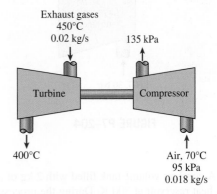

Exhaust gases
450°C
0.02 kg/s                    135 kPa

Turbine                      Compressor

400°C              Air, 70°C
                  95 kPa
                  0.018 kg/s

**FIGURE P7–202**

**7–203** A 0.40-m³ insulated piston–cylinder device initially contains 1.3 kg of air at 30°C. At this state, the piston is free to move. Now air at 500 kPa and 70°C is allowed to enter the cylinder from a supply line until the volume increases by 50 percent. Using constant specific heats at room temperature, determine (*a*) the final temperature, (*b*) the amount of mass that has entered, (*c*) the work done, and (*d*) the entropy generation.

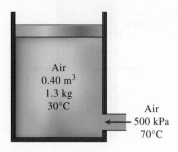

**FIGURE P7–203**

**7–204** When the transportation of natural gas in a pipeline is not feasible for economic reasons, it is first liquefied using nonconventional refrigeration techniques and then transported in super-insulated tanks. In a natural gas liquefaction plant, the liquefied natural gas (LNG) enters a cryogenic turbine at 30 bar and $-160°C$ at a rate of 20 kg/s and leaves at 3 bar. If 115 kW power is produced by the turbine, determine the efficiency of the turbine. Take the density of LNG to be 423.8 kg/m³.   *Answer:* 90.3 percent

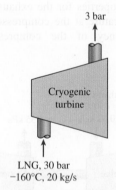

**FIGURE P7–204**

**7–205** A constant volume tank filled with 2 kg of air rejects heat to a heat reservoir at 300 K. During the process the temperature of the air in the tank decreases to the reservoir temperature. Determine the expressions for the entropy changes for the tank and reservoir and the total entropy change or entropy generated of this isolated system. Plot these entropy changes as functions of the initial temperature of the air. Comment on your results. Assume constant specific heats for air at 300 K.

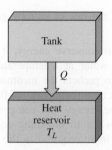

**FIGURE P7–205**

**7–206** Consider two bodies of identical mass $m$ and specific heat $c$ used as thermal reservoirs (source and sink) for a heat engine. The first body is initially at an absolute temperature $T_1$ while the second one is at a lower absolute temperature $T_2$. Heat is transferred from the first body to the heat engine, which rejects the waste heat to the second body. The process continues until the final temperatures of the two bodies $T_f$ become equal. Show that $T_f = \sqrt{T_1 T_2}$ when the heat engine produces the maximum possible work.

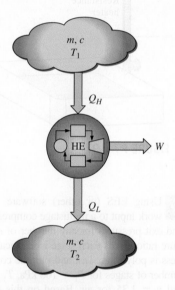

**FIGURE P7–206**

**7–207** A heat engine receives heat from a constant volume tank filled with 2 kg of air. The engine produces work that is stored in a work reservoir and rejects 400 kJ of heat to a heat reservoir at 300 K. During the process the temperature of the air in the tank decreases to 300 K. (*a*) Determine the initial temperature of the air that will maximize the work and the thermal efficiency of the engine. (*b*) Evaluate the total entropy change of this isolated system, the work produced, and the thermal efficiency for the initial air temperature in the tank from part (*a*) and at 100 K above and below the answer to part (a). (c) Plot the thermal efficiency and the entropy generation as functions of the initial temperature of the air. Comment on your answers. Assume constant specific heats for air at 300 K. *Answers:* (a) 759 K, 0.393

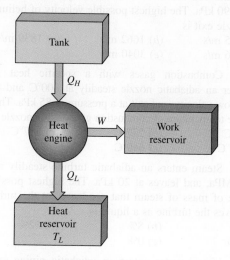

**FIGURE P7–207**

**7–208** For an ideal gas with constant specific heats show that the compressor and turbine isentropic efficiencies may be written as

$$\eta_C = \frac{(P_2/P_1)^{(k-1)/k}}{(T_2/T_1) - 1} \quad \text{and} \quad \eta_T = \frac{(T_4/T_3) - 1}{(P_4/P_3)^{(k-1)/k} - 1}$$

The states 1 and 2 represent the compressor inlet and exit states and the states 3 and 4 represent the turbine inlet and exit states.

**7–209** Starting with the Gibbs equation $dh = T ds + v dP$, obtain the expression for the change in enthalpy of an ideal gas having constant specific heats during the isentropic process $Pv^k = $ constant.

**7–210** An initially empty rigid vessel is filled with a fluid from a source whose properties remain constant. Determine the entropy generation if this is done adiabatically and without any work, and the fluid is an ideal gas. Your answer should be in terms of the vessel's volume, the properties of the gas, the dead state, temperature, the initial and final gas pressure and temperatures, and the pressure and temperature of the gas-supplying source.

**7–211** The temperature of an ideal gas having constant specific heats is given as a function of specific entropy and specific volume as $T(s,v) = Av^{1-k}\exp(s/c_v)$ where $A$ is a constant. For a reversible, constant volume process, find the expression for heat transfer per unit mass as a function of $c_v$ and $T$ using $Q = \int T dS$. Compare this result with that

obtained by applying the first law to a closed system undergoing a constant volume process.

**7–212** An ideal gas undergoes a reversible, steady-flow process in which pressure and volume are related by the polytropic equation $Pv^n = $ constant. Neglecting the changes in kinetic and potential energies of the flow and assuming constant specific heats, (a) obtain the expression for the heat transfer per unit mass flow for the process and (b) evaluate this expression for the special case where $n = k = c_p/c_v$.

**7–213** The polytropic or small stage efficiency of a compressor $\eta_{\infty,C}$ is defined as the ratio of the actual differential work done on the fluid to the isentropic differential work done on the flowing through the compressor $\eta_{\infty,C} = dh_s/dh$. Consider an ideal gas with constant specific heats as the working fluid undergoing a process in a compressor in which the polytropic efficiency is constant. Show that the temperature ratio across the compressor is related to the pressure ratio across the compressor by

$$\frac{T_2}{T_1} = \left(\frac{P_2}{P_1}\right)^{\left(\frac{1}{\eta_{\infty,C}}\right)\left(\frac{R}{c_p}\right)} = \left(\frac{P_2}{P_1}\right)^{\left(\frac{1}{\eta_{\infty,C}}\right)\left(\frac{k-1}{k}\right)}$$

## Fundamentals of Engineering (FE) Exam Problems

**7–214** Steam is compressed from 6 MPa and 300°C to 10 MPa isentropically. The final temperature of the steam is
(a) 290°C  (b) 300°C  (c) 311°C
(d) 371°C  (e) 422°C

**7–215** An apple with an average mass of 0.12 kg and average specific heat of 3.65 kJ/kg·°C is cooled from 25°C to 5°C. The entropy change of the apple is
(a) −0.705 kJ/K  (b) −0.254 kJ/K  (c) −0.0304 kJ/K
(d) 0 kJ/K  (e) 0.348 kJ/K

**7–216** A piston–cylinder device contains 5 kg of saturated water vapor at 3 MPa. Now heat is rejected from the cylinder at constant pressure until the water vapor completely condenses so that the cylinder contains saturated liquid at 3 MPa at the end of the process. The entropy change of the system during this process is
(a) 0 kJ/K  (b) −3.5 kJ/K  (c) −12.5 kJ/K
(d) −17.7 kJ/K  (e) −19.5 kJ/K

**7–217** Steam expands in an adiabatic turbine from 4 MPa and 500°C to 0.1 MPa at a rate of 2 kg/s. If steam leaves the turbine as saturated vapor, the power output of the turbine is
(a) 2058 kW  (b) 1910 kW  (c) 1780 kW
(d) 1674 kW  (e) 1542 kW

**7–218** Argon gas expands in an adiabatic turbine from 3 MPa and 750°C to 0.2 MPa at a rate of 5 kg/s. The maximum power output of the turbine is
(a) 1.06 MW  (b) 1.29 MW  (c) 1.43 MW
(d) 1.76 MW  (e) 2.08 MW

**7–219** A unit mass of a substance undergoes an irreversible process from state 1 to state 2 while gaining heat from the surroundings at temperature $T$ in the amount of $q$. If the entropy of the substance is $s_1$ at state 1, and $s_2$ at state 2, the entropy change of the substance $\Delta s$ during this process is

(a) $\Delta s < s_2 - s_1$      (b) $\Delta s > s_2 - s_1$
(c) $\Delta s = s_2 - s_1$      (d) $\Delta s = s_2 - s_1 + q/T$
(e) $\Delta s > s_2 - s_1 + q/T$

**7–220** A unit mass of an ideal gas at temperature $T$ undergoes a reversible isothermal process from pressure $P_1$ to pressure $P_2$ while losing heat to the surroundings at temperature $T$ in the amount of $q$. If the gas constant of the gas is $R$, the entropy change of the gas $\Delta s$ during this process is

(a) $\Delta s = R \ln(P_2/P_1)$      (b) $\Delta s = R \ln(P_2/P_1) - q/T$
(c) $\Delta s = R \ln(P_1/P_2)$      (d) $\Delta s = R \ln(P_1/P_2) - q/T$
(e) $\Delta s = 0$

**7–221** Helium gas is compressed from 27°C and 3.50 m³/kg to 0.775 m³/kg in a reversible and adiabatic manner. The temperature of helium after compression is

(a) 74°C      (b) 122°C      (c) 547°C
(d) 709°C      (e) 1082°C

**7–222** Heat is lost through a plane wall steadily at a rate of 600 W. If the inner and outer surface temperatures of the wall are 20°C and 5°C, respectively, the rate of entropy generation within the wall is

(a) 0.11 W/K      (b) 4.21 W/K      (c) 2.10 W/K
(d) 42.1 W/K      (e) 90.0 W/K

**7–223** Air is compressed steadily and adiabatically from 17°C and 90 kPa to 200°C and 400 kPa. Assuming constant specific heats for air at room temperature, the isentropic efficiency of the compressor is

(a) 0.76      (b) 0.94      (c) 0.86
(d) 0.84      (e) 1.00

**7–224** Argon gas expands in an adiabatic turbine steadily from 600°C and 800 kPa to 80 kPa at a rate of 2.5 kg/s. For isentropic efficiency of 88 percent, the power produced by the turbine is

(a) 240 kW      (b) 361 kW      (c) 414 kW
(d) 602 kW      (e) 777 kW

**7–225** Water enters a pump steadily at 100 kPa at a rate of 35 L/s and leaves at 800 kPa. The flow velocities at the inlet and the exit are the same, but the pump exit where the discharge pressure is measured is 6.1 m above the inlet section. The minimum power input to the pump is

(a) 34 kW      (b) 22 kW      (c) 27 kW
(d) 52 kW      (e) 44 kW

**7–226** Air is to be compressed steadily and isentropically from 1 atm to 16 atm by a two-stage compressor. To minimize the total compression work, the intermediate pressure between the two stages must be

(a) 3 atm      (b) 4 atm      (c) 8.5 atm
(d) 9 atm      (e) 12 atm

**7–227** Helium gas enters an adiabatic nozzle steadily at 500°C and 600 kPa with a low velocity, and exits at a pressure of 90 kPa. The highest possible velocity of helium gas at the nozzle exit is

(a) 1475 m/s      (b) 1662 m/s      (c) 1839 m/s
(d) 2066 m/s      (e) 3040 m/s

**7–228** Combustion gases with a specific heat ratio of 1.3 enter an adiabatic nozzle steadily at 800°C and 800 kPa with a low velocity, and exit at a pressure of 85 kPa. The lowest possible temperature of combustion gases at the nozzle exit is

(a) 43°C      (b) 237°C      (c) 367°C
(d) 477°C      (e) 640°C

**7–229** Steam enters an adiabatic turbine steadily at 400°C and 5 MPa, and leaves at 20 kPa. The highest possible percentage of mass of steam that condenses at the turbine exit and leaves the turbine as a liquid is

(a) 4%      (b) 8%      (c) 12%
(d) 18%      (e) 0%

**7–230** Liquid water enters an adiabatic piping system at 15°C at a rate of 8 kg/s. If the water temperature rises by 0.2°C during flow due to friction, the rate of entropy generation in the pipe is

(a) 23 W/K      (b) 55 W/K      (c) 68 W/K
(d) 220 W/K      (e) 443 W/K

**7–231** Liquid water is to be compressed by a pump whose isentropic efficiency is 75 percent from 0.2 MPa to 5 MPa at a rate of 0.15 m³/min. The required power input to this pump is

(a) 4.8 kW      (b) 6.4 kW      (c) 9.0 kW
(d) 16.0 kW      (e) 12 kW

**7–232** Steam enters an adiabatic turbine at 8 MPa and 500°C at a rate of 18 kg/s, and exits at 0.2 MPa and 300°C. The rate of entropy generation in the turbine is

(a) 0 kW/K      (b) 7.2 kW/K      (c) 21 kW/K
(d) 15 kW/K      (e) 17 kW/K

**7–233** Helium gas is compressed steadily from 90 kPa and 25°C to 800 kPa at a rate of 2 kg/min by an adiabatic compressor. If the compressor consumes 80 kW of power while operating, the isentropic efficiency of this compressor is

(a) 54.0%      (b) 80.5%      (c) 75.8%
(d) 90.1%      (e) 100%

## Design and Essay Problems

**7–234** Compressors powered by natural gas engines are increasing in popularity. Several major manufacturing facilities have already replaced the electric motors that drive their compressors by gas driven engines in order to reduce their energy bills since the cost of natural gas is much lower than the cost of electricity. Consider a facility that has a 130-kW compressor that runs 4400 h/yr at an average load factor of 0.6. Making reasonable assumptions and using unit costs for natural gas and electricity at your location, determine the potential cost savings per year by switching to gas driven engines.

**7–235** It is well-known that the temperature of a gas rises while it is compressed as a result of the energy input in the form of compression work. At high compression ratios, the air temperature may rise above the autoignition temperature of some hydrocarbons, including some lubricating oil. Therefore, the presence of some lubricating oil vapor in high-pressure air raises the possibility of an explosion, creating a fire hazard. The concentration of the oil within the compressor is usually too low to create a real danger. However, the oil that collects on the inner walls of exhaust piping of the compressor may cause an explosion. Such explosions have largely been eliminated by using the proper lubricating oils, carefully designing the equipment, intercooling between compressor stages, and keeping the system clean.

A compressor is to be designed for an industrial application in Los Angeles. If the compressor exit temperature is not to exceed 250°C for safety consideration, determine the maximum allowable compression ratio that is safe for all possible weather conditions for that area.

**7–236** Identify the major sources of entropy generation in your house and propose ways of reducing them.

**7–237** Obtain the following information about a power plant that is closest to your town: the net power output; the type and amount of fuel; the power consumed by the pumps, fans, and other auxiliary equipment; stack gas losses; temperatures at several locations; and the rate of heat rejection at the condenser. Using these and other relevant data, determine the rate of entropy generation in that power plant.

**7–238** You are designing a closed-system, isentropic-expansion process using an ideal gas that operates between the pressure limits of $P_1$ and $P_2$. The gases under consideration are hydrogen, nitrogen, air, helium, argon, and carbon dioxide. Which of these gases will produce the greatest amount of work? Which will require the least amount of work in a compression process?

# EXERGY

The increased awareness that the world's energy resources are limited has caused many countries to reexamine their energy policies and take drastic measures in eliminating waste. It has also sparked interest in the scientific community to take a closer look at the energy conversion devices and to develop new techniques to better utilize the existing limited resources. The first law of thermodynamics deals with the *quantity* of energy and asserts that energy cannot be created or destroyed. This law merely serves as a necessary tool for the bookkeeping of energy during a process and offers no challenges to the engineer. The second law, however, deals with the *quality* of energy. More specifically, it is concerned with the degradation of energy during a process, the entropy generation, and the lost opportunities to do work; and it offers plenty of room for improvement.

The second law of thermodynamics has proved to be a very powerful tool in the optimization of complex thermodynamic systems. In this chapter, we examine the performance of engineering devices in light of the second law of thermodynamics. We start our discussions with the introduction of *exergy* (also called *availability*), which is the maximum useful work that could be obtained from the system at a given state in a specified environment, and we continue with the *reversible work,* which is the maximum useful work that can be obtained as a system undergoes a process between two specified states. Next we discuss the *irreversibility* (also called the *exergy destruction* or *lost work*), which is the wasted work potential during a process as a result of irreversibilities, and we define a *second-law efficiency.* We then develop the *exergy balance* relation and apply it to closed systems and control volumes.

## OBJECTIVES

The objectives of Chapter 8 are to:

- Examine the performance of engineering devices in light of the second law of thermodynamics.
- Define exergy, which is the maximum useful work that could be obtained from the system at a given state in a specified environment.
- Define reversible work, which is the maximum useful work that can be obtained as a system undergoes a process between two specified states.
- Define the exergy destruction, which is the wasted work potential during a process as a result of irreversibilities.
- Define the second-law efficiency.
- Develop the exergy balance relation.
- Apply exergy balance to closed systems and control volumes.

# 8–1 · EXERGY: WORK POTENTIAL OF ENERGY

When a new energy source, such as a geothermal well, is discovered, the first thing the explorers do is estimate the amount of energy contained in the source. This information alone, however, is of little value in deciding whether to build a power plant on that site. What we really need to know is the *work potential* of the source—that is, the amount of energy we can extract as useful work. The rest of the energy is eventually discarded as waste energy and is not worthy of our consideration. Thus, it would be very desirable to have a property to enable us to determine the useful work potential of a given amount of energy at some specified state. This property is *exergy*, which is also called the *availability* or *available energy*.

The work potential of the energy contained in a system at a specified state is simply the maximum useful work that can be obtained from the system. You will recall that the work done during a process depends on the initial state, the final state, and the process path. That is,

$$\text{Work} = f(\text{initial state, process path, final state})$$

In an exergy analysis, the *initial state* is specified, and thus it is not a variable. The work output is maximized when the process between two specified states is executed in a *reversible manner*, as shown in Chap. 7. Therefore, all the irreversibilities are disregarded in determining the work potential. Finally, the system must be in the *dead state* at the end of the process to maximize the work output.

A system is said to be in the **dead state** when it is in thermodynamic equilibrium with the environment it is in (Fig. 8–1). At the dead state, a system is at the temperature and pressure of its environment (in thermal and mechanical equilibrium); it has no kinetic or potential energy relative to the environment (zero velocity and zero elevation above a reference level); and it does not react with the environment (chemically inert). Also, there are no unbalanced magnetic, electrical, and surface tension effects between the system and its surroundings, if these are relevant to the situation at hand. The properties of a system at the dead state are denoted by subscript zero, for example, $P_0$, $T_0$, $h_0$, $u_0$, and $s_0$. Unless specified otherwise, the dead-state temperature and pressure are taken to be $T_0 = 25°C$ and $P_0 = 1$ atm (101.325 kPa). A system has zero exergy at the dead state.

Distinction should be made between the *surroundings, immediate surroundings,* and the *environment*. By definition, **surroundings** are everything outside the system boundaries. The **immediate surroundings** refer to the portion of the surroundings that is affected by the process, and **environment** refers to the region beyond the immediate surroundings whose properties are not affected by the process at any point. Therefore, any irreversibilities during a process occur within the system and its immediate surroundings, and the environment is free of any irreversibilities. When analyzing the cooling of a hot baked potato in a room at 25°C, for example, the warm air that surrounds the potato is the immediate surroundings, and the remaining part of the room air at 25°C is the environment. Note that the temperature of the immediate surroundings changes from the temperature of the potato at the boundary to the environment temperature of 25°C (Fig. 8–2).

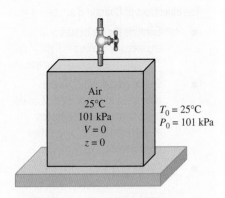

**FIGURE 8–1**

A system that is in equilibrium with its environment is said to be at the dead state.

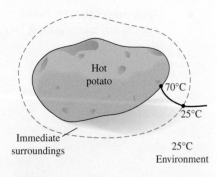

**FIGURE 8–2**

The immediate surroundings of a hot potato are simply the temperature gradient zone of the air next to the potato.

The notion that a system must go to the dead state at the end of the process to maximize the work output can be explained as follows: If the system temperature at the final state is greater than (or less than) the temperature of the environment it is in, we can always produce additional work by running a heat engine between these two temperature levels. If the final pressure is greater than (or less than) the pressure of the environment, we can still obtain work by letting the system expand to the pressure of the environment. If the final velocity of the system is not zero, we can catch that extra kinetic energy by a turbine and convert it to rotating shaft work, and so on. No work can be produced from a system that is initially at the dead state. The atmosphere around us contains a tremendous amount of energy. However, the atmosphere is in the dead state, and the energy it contains has no work potential (Fig. 8–3).

Therefore, we conclude that a *system delivers the maximum possible work as it undergoes a reversible process from the specified initial state to the state of its environment, that is, the dead state*. This represents the *useful work potential* of the system at the specified state and is called **exergy**. It is important to realize that exergy does not represent the amount of work that a work-producing device will actually deliver upon installation. Rather, it represents the *upper limit on the amount of work a device can deliver without violating any thermodynamic laws*. There will always be a difference, large or small, between exergy and the actual work delivered by a device. This difference represents the room engineers have for improvement.

Note that the exergy of a system at a specified state depends on the conditions of the environment (the dead state) as well as the properties of the system. Therefore, exergy is a property of the *system–environment combination* and not of the system alone. Altering the environment is another way of increasing exergy, but it is definitely not an easy alternative.

The term *availability* was made popular in the United States by the M.I.T. School of Engineering in the 1940s. Today, an equivalent term, *exergy,* introduced in Europe in the 1950s, has found global acceptance partly because it is shorter, it rhymes with energy and entropy, and it can be adapted without requiring translation. In this text the preferred term is *exergy.*

**FIGURE 8–3**
The atmosphere contains a tremendous amount of energy, but no exergy.
©*Jeremy Woodhouse/Getty Images RF*

## Exergy (Work Potential) Associated with Kinetic and Potential Energy

Kinetic energy is a form of *mechanical energy,* and thus it can be converted to work entirely. Therefore, the *work potential* or *exergy* of the kinetic energy of a system is equal to the kinetic energy itself regardless of the temperature and pressure of the environment. That is,

*Exergy of kinetic energy:* $\qquad x_{ke} = ke = \dfrac{V^2}{2} \quad$ (kJ/kg) $\qquad$ **(8–1)**

where $V$ is the velocity of the system relative to the environment.

Potential energy is also a form of *mechanical energy,* and thus it can be converted to work entirely. Therefore, the *exergy* of the potential energy of

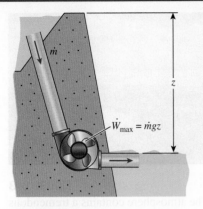

**FIGURE 8–4**
The *work potential* or *exergy* of potential energy is equal to the potential energy itself.

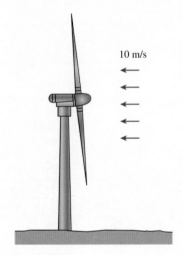

**FIGURE 8–5**
Schematic for Example 8–1.

a system is equal to the potential energy itself regardless of the temperature and pressure of the environment (Fig. 8–4). That is,

*Exergy of potential energy:*   $x_{pe} = pe = gz$   (kJ/kg)   **(8–2)**

where $g$ is the gravitational acceleration and $z$ is the elevation of the system relative to a reference level in the environment.

Therefore, the exergies of kinetic and potential energies are equal to themselves, and they are entirely available for work. However, the internal energy $u$ and enthalpy $h$ of a system are not entirely available for work, as shown later.

---

**EXAMPLE 8–1     Maximum Power Generation by a Wind Turbine**

A wind turbine with a 12-m-diameter rotor, as shown in Fig. 8–5, is to be installed at a location where the wind is blowing steadily at an average velocity of 10 m/s. Determine the maximum power that can be generated by the wind turbine.

**SOLUTION**  A wind turbine is being considered for a specified location. The maximum power that can be generated by the wind turbine is to be determined.
**Assumptions**  Air is at standard conditions of 1 atm and 25°C, and thus its density is 1.18 kg/m³.
**Analysis**  The air flowing with the wind has the same properties as the stagnant atmospheric air except that it possesses a velocity and thus some kinetic energy. This air will reach the dead state when it is brought to a complete stop. Therefore, the exergy of the blowing air is simply the kinetic energy it possesses:

$$ke = \frac{V^2}{2} = \frac{(10 \text{ m/s})^2}{2} \left( \frac{1 \text{ kJ/kg}}{1000 \text{ m}^2/\text{s}^2} \right) = 0.05 \text{ kJ/kg}$$

That is, every unit mass of air flowing at a velocity of 10 m/s has a work potential of 0.05 kJ/kg. In other words, a perfect wind turbine will bring the air to a complete stop and capture that 0.05 kJ/kg of work potential. To determine the maximum power, we need to know the amount of air passing through the rotor of the wind turbine per unit time, that is, the mass flow rate, which is determined to be

$$\dot{m} = \rho AV = \rho \frac{\pi D^2}{4} V = (1.18 \text{ kg/m}^3) \frac{\pi (12 \text{ m})^2}{4} (10 \text{ m/s}) = 1335 \text{ kg/s}$$

Thus,

Maximum power $= \dot{m}(ke) = (1335 \text{ kg/s})(0.05 \text{ kJ/kg}) = \textbf{66.8 kW}$

This is the maximum power available to the wind turbine. Assuming a conversion efficiency of 30 percent, an actual wind turbine will convert 20.0 kW to electricity. Notice that the work potential for this case is equal to the entire kinetic energy of the air.
**Discussion**  It should be noted that although the entire kinetic energy of the wind is available for power production, Betz's law states that the power output of a wind machine is at maximum when the wind is slowed to one-third of its initial velocity. Therefore, for maximum power (and thus minimum cost

per installed power), the highest efficiency of a wind turbine is about 59 percent. In practice, the actual efficiency ranges between 20 and 40 percent and is about 35 percent for many wind turbines.

Wind power is suitable for harvesting when there are steady winds with an average velocity of at least 6 m/s. Recent improvements in wind turbine design have brought the cost of generating wind power to about 5 cents per kWh, which is competitive with electricity generated from other resources.

---

### EXAMPLE 8–2    Exergy Transfer from a Furnace

Consider a large furnace that can transfer heat at a temperature of 1100 K at a steady rate of 3000 kW. Determine the rate of exergy flow associated with this heat transfer. Assume an environment temperature of 25°C.

**SOLUTION**   Heat is being supplied by a large furnace at a specified temperature. The rate of exergy flow is to be determined.

**Analysis**   The furnace in this example can be modeled as a heat reservoir that supplies heat indefinitely at a constant temperature. The exergy of this heat energy is its useful work potential, that is, the maximum possible amount of work that can be extracted from it. This corresponds to the amount of work that a reversible heat engine operating between the furnace and the environment can produce.

The thermal efficiency of this reversible heat engine is

$$\eta_{th,max} = \eta_{th,rev} = 1 - \frac{T_L}{T_H} = 1 - \frac{T_0}{T_H} = 1 - \frac{298 \text{ K}}{1100 \text{ K}} = 0.729 \text{ (or 72.9\%)}$$

That is, a heat engine can convert, at best, 72.9 percent of the heat received from this furnace to work. Thus, the exergy of this furnace is equivalent to the power produced by the reversible heat engine:

$$\dot{W}_{max} = \dot{W}_{rev} = \eta_{th,rev} \dot{Q}_{in} = (0.729)(3000 \text{ kW}) = \textbf{2187 kW}$$

**Discussion**   Notice that 27.1 percent of the heat transferred from the furnace is not available for doing work. The portion of energy that cannot be converted to work is called **unavailable energy** (Fig. 8–6). Unavailable energy is simply the difference between the total energy of a system at a specified state and the exergy of that energy.

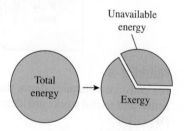

**FIGURE 8–6**
Unavailable energy is the portion of energy that cannot be converted to work by even a reversible heat engine.

## 8–2 · REVERSIBLE WORK AND IRREVERSIBILITY

The property exergy serves as a valuable tool in determining the quality of energy and comparing the work potentials of different energy sources or systems. The evaluation of exergy alone, however, is not sufficient for studying engineering devices operating between two fixed states. This is because when evaluating exergy, the final state is always assumed to be the *dead state*, which is hardly ever the case for actual engineering systems. The isentropic efficiencies discussed in Chap. 7 are also of limited use because

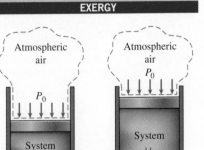

**FIGURE 8–7**
As a closed system expands, some work needs to be done to push the atmospheric air out of the way ($W_{surr}$).

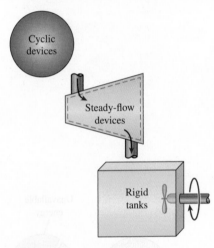

**FIGURE 8–8**
For constant-volume systems, the total actual and useful works are identical ($W_u = W$).

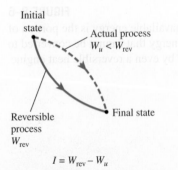

$$I = W_{rev} - W_u$$

**FIGURE 8–9**
The difference between reversible work and actual useful work is the irreversibility.

the exit state of the model (isentropic) process is not the same as the actual exit state and it is limited to adiabatic processes.

In this section, we describe two quantities that are related to the actual initial and final states of processes and serve as valuable tools in the thermodynamic analysis of components or systems. These two quantities are the *reversible work* and *irreversibility* (or *exergy destruction*). But first we examine the **surroundings work**, which is the work done by or against the surroundings during a process.

The work done by work-producing devices is not always entirely in a usable form. For example, when a gas in a piston–cylinder device expands, part of the work done by the gas is used to push the atmospheric air out of the way of the piston (Fig. 8–7). This work, which cannot be recovered and utilized for any useful purpose, is equal to the atmospheric pressure $P_0$ times the volume change of the system,

$$W_{surr} = P_0(V_2 - V_1) \tag{8–3}$$

The difference between the actual work $W$ and the surroundings work $W_{surr}$ is called the **useful work** $W_u$:

$$W_u = W - W_{surr} = W - P_0(V_2 - V_1) \tag{8–4}$$

When a system is expanding and doing work, part of the work done is used to overcome the atmospheric pressure, and thus $W_{surr}$ represents a loss. When a system is compressed, however, the atmospheric pressure helps the compression process, and thus $W_{surr}$ represents a gain.

Note that the work done by or against the atmospheric pressure has significance only for systems whose volume changes during the process (i.e., systems that involve moving boundary work). It has no significance for cyclic devices and systems whose boundaries remain fixed during a process such as rigid tanks and steady-flow devices (turbines, compressors, nozzles, heat exchangers, etc.), as shown in Fig. 8–8.

**Reversible work** $W_{rev}$ is defined as *the maximum amount of useful work that can be produced (or the minimum work that needs to be supplied) as a system undergoes a process between the specified initial and final states.* This is the useful work output (or input) obtained (or expended) when the process between the initial and final states is executed in a totally reversible manner. When the final state is the dead state, the reversible work equals exergy. For processes that require work, reversible work represents the minimum amount of work necessary to carry out that process. For convenience in presentation, the term *work* is used to denote both work and power throughout this chapter.

Any difference between the reversible work $W_{rev}$ and the useful work $W_u$ is due to the irreversibilities present during the process, and this difference is called **irreversibility** $I$. It is expressed as (Fig. 8–9)

$$I = W_{rev,out} - W_{u,out} \quad \text{or} \quad I = W_{u,in} - W_{rev,in} \tag{8–5}$$

The irreversibility is equivalent to the *exergy destroyed,* discussed in Sec. 8–4. For a totally reversible process, the actual and reversible work terms are identical, and thus the irreversibility is zero. This is expected since totally reversible processes generate no entropy. Irreversibility is a *positive quantity* for all actual (irreversible) processes since $W_{rev} \geq W_u$ for work-producing devices and $W_{rev} \leq W_u$ for work-consuming devices.

Irreversibility can be viewed as the *wasted work potential* or the *lost opportunity* to do work. It represents the energy that could have been converted to work but was not. The smaller the irreversibility associated with a process, the greater the work that is produced (or the smaller the work that is consumed). The performance of a system can be improved by minimizing the irreversibility associated with it.

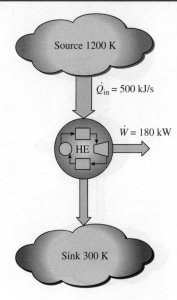

**FIGURE 8–10**
Schematic for Example 8–3.

### ■ *EXAMPLE 8–3*   The Rate of Irreversibility of a Heat Engine

A heat engine receives heat from a source at 1200 K at a rate of 500 kJ/s and rejects the waste heat to a medium at 300 K (Fig. 8–10). The power output of the heat engine is 180 kW. Determine the reversible power and the irreversibility rate for this process.

**SOLUTION**   The operation of a heat engine is considered. The reversible power and the irreversibility rate associated with this operation are to be determined.

*Analysis*   The reversible power for this process is the amount of power that a reversible heat engine, such as a Carnot heat engine, would produce when operating between the same temperature limits, and is determined to be:

$$\dot{W}_{rev,out} = \eta_{th,rev}\,\dot{Q}_{in} = \left(1 - \frac{T_{sink}}{T_{source}}\right)\dot{Q}_{in} = \left(1 - \frac{300\text{ K}}{1200\text{ K}}\right)(500\text{ kW}) = \textbf{375 kW}$$

This is the maximum power that can be produced by a heat engine operating between the specified temperature limits and receiving heat at the specified rate. This would also represent the *available power* if 300 K were the lowest temperature available for heat rejection.

The irreversibility rate is the difference between the reversible power (maximum power that could have been produced) and the useful power output:

$$\dot{I} = \dot{W}_{rev,out} - \dot{W}_{u,out} = 375 - 180 = \textbf{195 kW}$$

*Discussion*   Note that 195 kW of power potential is wasted during this process as a result of irreversibilities. Also, the $500 - 375 = 125$ kW of heat rejected to the sink is not available for converting to work and thus is not part of the irreversibility.

### ■ *EXAMPLE 8–4*   Irreversibility during the Cooling of an Iron Block

A 500-kg iron block shown in Fig. 8–11 is initially at 200°C and is allowed to cool to 27°C by transferring heat to the surrounding air at 27°C. Determine the reversible work and the irreversibility for this process.

**SOLUTION**   A hot iron block is allowed to cool in air. The reversible work and irreversibility associated with this process are to be determined.

*Assumptions*   **1** The kinetic and potential energies are negligible. **2** The process involves no work interactions.

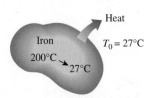

**FIGURE 8–11**
Schematic for Example 8–4.

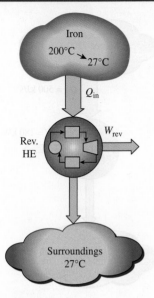

**FIGURE 8–12**

An irreversible heat transfer process can be made reversible by the use of a reversible heat engine.

*Analysis* We take the *iron block* as the system. This is a *closed system* since no mass crosses the system boundary. We note that heat is lost from the system.

It probably came as a surprise to you that we are asking to find the "reversible work" for a process that does not involve any work interactions. Well, even if no attempt is made to produce work during this process, the potential to do work still exists, and the reversible work is a quantitative measure of this potential.

The reversible work in this case is determined by considering a series of imaginary reversible heat engines operating between the source (at a variable temperature $T$) and the sink (at a constant temperature $T_0$), as shown in Fig. 8–12. Summing their work output:

$$\delta W_{rev} = \eta_{th,rev}\, \delta Q_{in} = \left(1 - \frac{T_{sink}}{T_{source}}\right)\delta Q_{in} = \left(1 - \frac{T_0}{T}\right)\delta Q_{in}$$

and

$$W_{rev} = \int\left(1 - \frac{T_0}{T}\right)\delta Q_{in}$$

The source temperature $T$ changes from $T_1 = 200°C = 473$ K to $T_0 = 27°C = 300$ K during this process. A relation for the differential heat transfer from the iron block can be obtained from the differential form of the energy balance applied on the iron block,

$$\underbrace{\delta E_{in} - \delta E_{out}}_{\substack{\text{Net energy transfer} \\ \text{by heat, work, and mass}}} = \underbrace{dE_{system}}_{\substack{\text{Change in internal, kinetic,} \\ \text{potential, etc., energies}}}$$

$$-\delta Q_{out} = dU = mc_{avg}dT$$

Then,

$$\delta Q_{in,heat\ engine} = \delta Q_{out,system} = -mc_{avg}dT$$

since heat transfers from the iron and to the heat engine are equal in magnitude and opposite in direction. Substituting and performing the integration, the reversible work is determined to be

$$W_{rev} = \int_{T_1}^{T_0}\left(1 - \frac{T_0}{T}\right)(-mc_{avg}\,dT) = mc_{avg}(T_1 - T_0) - mc_{avg}T_0 \ln\frac{T_1}{T_0}$$

$$= (500\ kg)(0.45\ kJ/kg{\cdot}K)\left[(473 - 300)\ K - (300\ K)\ln\frac{473\ K}{300\ K}\right]$$

$$= \mathbf{8191\ kJ}$$

where the specific heat value is obtained from Table A–3. The first term in the above equation [$Q = mc_{avg}(T_1 - T_0) = 38,925$ kJ] is the total heat transfer from the iron block to the heat engine. The reversible work for this problem is found to be 8191 kJ, which means that 8191 (21 percent) of the 38,925 kJ of heat transferred from the iron block to the ambient air *could* have been converted to work. If the specified ambient temperature of 27°C is the lowest available environment temperature, the reversible work determined above also represents the exergy, which is the maximum work potential of the sensible energy contained in the iron block.

The irreversibility for this process is determined from its definition,

$$I = W_{rev} - W_u = 8191 - 0 = \textbf{8191 kJ}$$

***Discussion*** Notice that the reversible work and irreversibility (the wasted work potential) are the same for this case since the entire work potential is wasted. The source of irreversibility in this process is the heat transfer through a finite temperature difference.

## ■ *EXAMPLE 8–5*  Heating Potential of a Hot Iron Block

■ The iron block discussed in Example 8–4 is to be used to maintain a
■ house at 27°C when the outdoor temperature is 5°C. Determine the maxi-
■ mum amount of heat that can be supplied to the house as the iron cools
to 27°C.

**SOLUTION** The iron block is now reconsidered for heating a house. The maximum amount of heating this block can provide is to be determined.
***Analysis*** Probably the first thought that comes to mind to make the most use of the energy stored in the iron block is to take it inside and let it cool in the house, as shown in Fig. 8–13, transferring its sensible energy as heat to the indoors air (provided that it meets the approval of the household, of course). The iron block can keep "losing" heat until its temperature drops to the indoor temperature of 27°C, transferring a total of 38,925 kJ of heat. Since we utilized the entire energy of the iron block available for heating without wasting a single kilojoule, it seems like we have a 100-percent-efficient operation, and nothing can beat this, right? Well, not quite.

In Example 8–4 we determined that this process has an irreversibility of 8191 kJ, which implies that things are not as "perfect" as they seem. A "perfect" process is one that involves "zero" irreversibility. The irreversibility in this process is associated with the heat transfer through a finite temperature difference that can be eliminated by running a reversible heat engine between the iron block and the indoor air. This heat engine produces (as determined in Example 8–4) 8191 kJ of work and reject the remaining 38,925 − 8191 = 30,734 kJ of heat to the house. Now we managed to eliminate the irreversibility and ended up with 8191 kJ of work. What can we do with this work? Well, at worst we can convert it to heat by running a paddle wheel, for example, creating an equal amount of irreversibility. Or we can supply this work to a heat pump that transports heat from the outdoors at 5°C to the indoors at 27°C. Such a heat pump, if reversible, has a coefficient of performance of

$$COP_{HP} = \frac{1}{1 - T_L/T_H} = \frac{1}{1 - (278 \text{ K})/(300 \text{ K})} = 13.6$$

That is, this heat pump can supply the house with 13.6 times the energy it consumes as work. In our case, it will consume the 8191 kJ of work and deliver 8191 × 13.6 = 111,398 kJ of heat to the house. Therefore, the hot iron block has the potential to supply

$$(30,734 + 111,398) \text{ kJ} = 142,132 \text{ kJ} \cong \textbf{142 MJ}$$

**FIGURE 8–13**
Schematic for Example 8–5.

of heat to the house. The irreversibility for this process is zero, and this is *the best* we can do under the specified conditions. A similar argument can be given for the electric heating of residential or commercial buildings.

**Discussion** Now try to answer the following question: What would happen if the heat engine were operated between the iron block and the outside air instead of the house until the temperature of the iron block fell to 27°C? Would the amount of heat supplied to the house still be 142 MJ? Here is a hint: The initial and final states in both cases are the same, and the irreversibility for both cases is zero.

## 8–3 · SECOND-LAW EFFICIENCY

In Chap. 6 we defined the *thermal efficiency* and the *coefficient of performance* for devices as a measure of their performance. They are defined on the basis of the first law only, and they are sometimes referred to as the *first-law efficiencies*. The first law efficiency, however, makes no reference to the best possible performance, and thus it may be misleading.

Consider two heat engines, both having a thermal efficiency of 30 percent, as shown in Fig. 8–14. One of the engines (engine A) is supplied with heat from a source at 600 K, and the other one (engine B) from a source at 1000 K. Both engines reject heat to a medium at 300 K. At first glance, both engines seem to convert to work the same fraction of heat that they receive; thus they are performing equally well. When we take a second look at these engines in light of the second law of thermodynamics, however, we see a totally different picture. These engines, at best, can perform as reversible engines, in which case their efficiencies would be

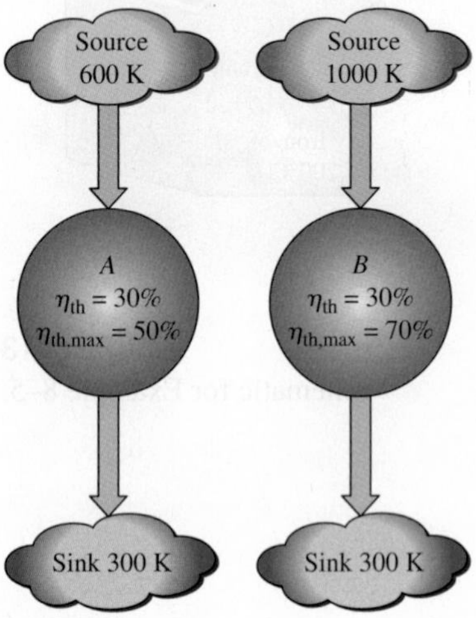

$$\eta_{\text{rev},A} = \left(1 - \frac{T_L}{T_H}\right)_A = 1 - \frac{300 \text{ K}}{600 \text{ K}} = 0.50 \text{ or } 50\%$$

$$\eta_{\text{rev},B} = \left(1 - \frac{T_L}{T_H}\right)_B = 1 - \frac{300 \text{ K}}{1000 \text{ K}} = 0.70 \text{ or } 70\%$$

Now it is becoming apparent that engine B has a greater work potential available to it (70 percent of the heat supplied as compared to 50 percent for engine A), and thus should do a lot better than engine A. Therefore, we can say that engine B is performing poorly relative to engine A even though both have the same thermal efficiency.

It is obvious from this example that the first-law efficiency alone is not a realistic measure of performance of engineering devices. To overcome this deficiency, we define a **second-law efficiency** $\eta_{\text{II}}$ as the ratio of the actual thermal efficiency to the maximum possible (reversible) thermal efficiency under the same conditions (Fig. 8–15):

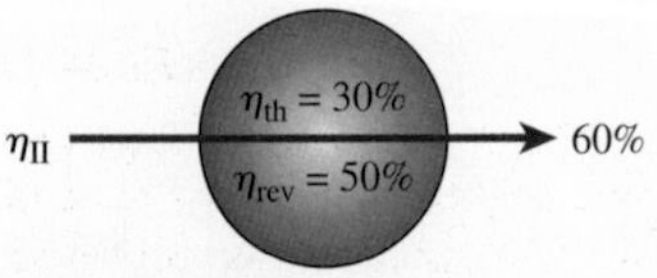

$$\eta_{\text{II}} = \frac{\eta_{\text{th}}}{\eta_{\text{th,rev}}} \quad \text{(heat engines)} \tag{8–6}$$

Based on this definition, the second-law efficiencies of the two heat engines discussed above are

$$\eta_{\text{II},A} = \frac{0.30}{0.50} = 0.60 \quad \text{and} \quad \eta_{\text{II},B} = \frac{0.30}{0.70} = 0.43$$

**FIGURE 8–14**
Two heat engines that have the same thermal efficiency, but different maximum thermal efficiencies.

**FIGURE 8–15**
Second-law efficiency is a measure of the performance of a device relative to its performance under reversible conditions.

That is, engine A is converting 60 percent of the available work potential to useful work. This ratio is only 43 percent for engine B.

The second-law efficiency can also be expressed as the ratio of the useful work output and the maximum possible (reversible) work output:

$$\eta_{II} = \frac{W_u}{W_{rev}} \quad \text{(work-producing devices)} \quad \text{(8–7)}$$

This definition is more general since it can be applied to processes (in turbines, piston–cylinder devices, etc.) as well as to cycles. Note that the second-law efficiency cannot exceed 100 percent (Fig. 8–16).

We can also define a second-law efficiency for work-consuming noncyclic (such as compressors) and cyclic (such as refrigerators) devices as the ratio of the minimum (reversible) work input to the useful work input:

$$\eta_{II} = \frac{W_{rev}}{W_u} \quad \text{(work-consuming devices)} \quad \text{(8–8)}$$

For cyclic devices such as refrigerators and heat pumps, it can also be expressed in terms of the coefficients of performance as

$$\eta_{II} = \frac{COP}{COP_{rev}} \quad \text{(refrigerators and heat pumps)} \quad \text{(8–9)}$$

Again, because of the way we defined the second-law efficiency, its value cannot exceed 100 percent. In the above relations, the reversible work $W_{rev}$ should be determined by using the same initial and final states as in the actual process.

The definitions above for the second-law efficiency do not apply to devices that are not intended to produce or consume work. Therefore, we need a more general definition. However, there is some disagreement on a general definition of the second-law efficiency, and thus a person may encounter different definitions for the same device. The second-law efficiency is intended to serve as a measure of approximation to reversible operation, and thus its value should range from zero in the worst case (complete destruction of exergy) to one in the best case (no destruction of exergy). With this in mind, we define the second-law efficiency of a system during a process as (Fig. 8–17)

$$\eta_{II} = \frac{\text{Exergy recovered}}{\text{Exergy expended}} = 1 - \frac{\text{Exergy destroyed}}{\text{Exergy expended}} \quad \text{(8–10)}$$

Therefore, when determining the second-law efficiency, the first thing we need to do is determine how much exergy or work potential is expended or consumed during a process. In a reversible operation, we should be able to recover entirely the exergy expended during the process, and the irreversibility in this case should be zero. The second-law efficiency is zero when we recover none of the exergy expended by the system. Note that the exergy can be supplied or recovered at various amounts in various forms such as heat, work, kinetic energy, potential energy, internal energy, and enthalpy. Sometimes there are differing (though valid) opinions on what constitutes expended exergy, and this causes differing definitions for second-law efficiency. At all times, however, the exergy recovered and the exergy destroyed (the irreversibility) must add up to the exergy expended. Also, we need to define the system precisely in order to identify correctly any interactions between the system and its surroundings.

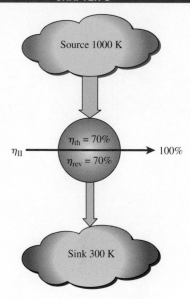

**FIGURE 8–16**
Second-law efficiency of all reversible devices is 100 percent.

**FIGURE 8–17**
The second-law efficiency of naturally occurring processes is zero if none of the work potential is recovered.

For a *heat engine,* the exergy expended is the decrease in the exergy of the heat transferred to the engine, which is the difference between the exergy of the heat supplied and the exergy of the heat rejected. (The exergy of the heat rejected at the temperature of the surroundings is zero.) The net work output is the recovered exergy.

For a *refrigerator* or *heat pump,* the exergy expended is the work input since the work supplied to a cyclic device is entirely consumed. The recovered exergy is the exergy of the heat transferred to the high-temperature medium for a heat pump, and the exergy of the heat transferred from the low-temperature medium for a refrigerator.

For a *heat exchanger* with two unmixed fluid streams, usually the exergy expended is the decrease in the exergy of the higher-temperature fluid stream, and the exergy recovered is the increase in the exergy of the lower-temperature fluid stream. This is discussed further in Sec. 8–8.

In the case of *electric resistance heating,* the exergy expended is the electrical energy the resistance heater consumes from the resource of electric grid. The exergy recovered is the exergy content of the heat supplied to the room, which is the work that can be produced by a Carnot engine receiving this heat. If the heater maintains the heated space at a constant temperature of $T_H$ in an environment at $T_0$, the second-law efficiency for the electric heater becomes

$$\eta_{\text{II,electric heater}} = \frac{\dot{X}_{\text{recovered}}}{\dot{X}_{\text{expended}}} = \frac{\dot{X}_{\text{heat}}}{\dot{W}_e} = \frac{\dot{Q}_e(1 - T_0/T_H)}{\dot{W}_e} = 1 - \frac{T_0}{T_H} \quad (8\text{–}11)$$

since, from the first law considerations, $\dot{Q}_e = \dot{W}_e$. Note that the second-law efficiency of a resistance heater becomes zero when the heater is outdoors (as in a radiant heater) and thus the exergy of the heat supplied to the environment is not recoverable.

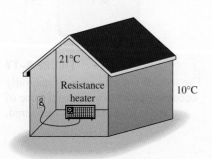

**FIGURE 8–18**
Schematic for Example 8–6.

---

**EXAMPLE 8–6**    **Second-Law Efficiency of Resistance Heaters**

A dealer advertises that he has just received a shipment of electric resistance heaters for residential buildings that have an efficiency of 100 percent (Fig. 8–18). Assuming an indoor temperature of 21°C and outdoor temperature of 10°C, determine the second-law efficiency of these heaters.

**SOLUTION**   Electric resistance heaters are being considered for residential buildings. The second-law efficiency of these heaters is to be determined.
**Analysis**   Obviously the efficiency that the dealer is referring to is the first-law efficiency, meaning that for each unit of electric energy (work) consumed, the heater will supply the house with 1 unit of energy (heat). That is, the advertised heater has a COP of 1.

At the specified conditions, a reversible heat pump would have a coefficient of the performance of

$$\text{COP}_{\text{HP,rev}} = \frac{1}{1 - T_L/T_H} = \frac{1}{1 - (10 + 273 \text{ K})/(21 + 273 \text{ K})} = 26.7$$

That is, it would supply the house with 26.7 units of heat (extracted from the cold outside air) for each unit of electric energy it consumes.

The second-law efficiency of this resistance heater is

$$\eta_{II} = \frac{COP}{COP_{rev}} = \frac{1.0}{26.7} = 0.037 \text{ or } 3.7\%$$

which does not look so impressive. The dealer will not be happy to see this value. Considering the high price of electricity, a consumer will probably be better off with a "less" efficient gas heater.

**Discussion** The second-law efficiency of this electric heater can also be determined directly from Eq. 8–11 to be

$$\eta_{II,electric\ heater} = 1 - \frac{T_0}{T_H} = 1 - \frac{(10 + 273)\,K}{(21 + 273)\,K} = 0.037 \text{ or } 3.7\%$$

Therefore, if we change our mind and decide to convert the heat back to electricity, the best we can do is 3.7 percent. That is, 96.3 percent of the heat can never be converted to electrical energy.

# 8–4 · EXERGY CHANGE OF A SYSTEM

The property *exergy* is the work potential of a system in a specified environment and represents the maximum amount of useful work that can be obtained as the system is brought to equilibrium with the environment. Unlike energy, the value of exergy depends on the state of the environment as well as the state of the system. Therefore, exergy is a combination property. The exergy of a system that is in equilibrium with its environment is zero. The state of the environment is referred to as the "dead state" since the system is practically "dead" (cannot do any work) from a thermodynamic point of view when it reaches that state.

In this section we limit the discussion to **thermo-mechanical exergy**, and thus disregard any mixing and chemical reactions. Therefore, a system at this "restricted dead state" is at the temperature and pressure of the environment and it has no kinetic or potential energies relative to the environment. However, it may have a different chemical composition than the environment. Exergy associated with different chemical compositions and chemical reactions is discussed in later chapters.

Below we develop relations for the exergies and exergy changes for a fixed mass and a flow stream.

## Exergy of a Fixed Mass:
## Nonflow (or Closed System) Exergy

In general, internal energy consists of *sensible, latent, chemical,* and *nuclear* energies. However, in the absence of any chemical or nuclear reactions, the chemical and nuclear energies can be disregarded and the internal energy can be considered to consist of only sensible and latent energies that can be transferred to or from a system as *heat* whenever there is a temperature difference across the system boundary. The second law of thermodynamics states that heat cannot be converted to work entirely, and thus the work potential of internal energy must be less than the internal energy itself. But how much less?

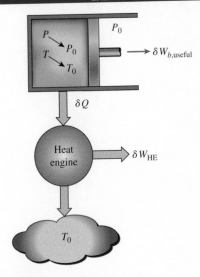

**FIGURE 8–19**
The *exergy* of a specified mass at a specified state is the useful work that can be produced as the mass undergoes a reversible process to the state of the environment.

To answer that question, we need to consider a stationary closed system at a specified state that undergoes a *reversible* process to the state of the environment (that is, the final temperature and pressure of the system should be $T_0$ and $P_0$, respectively). The useful work delivered during this process is the exergy of the system at its initial state (Fig. 8–19).

Consider a piston–cylinder device that contains a fluid of mass $m$ at temperature $T$ and pressure $P$. The system (the mass inside the cylinder) has a volume $V$, internal energy $U$, and entropy $S$. The system is now allowed to undergo a differential change of state during which the volume changes by a differential amount $dV$ and heat is transferred in the differential amount of $\delta Q$. Taking the direction of heat and work transfers to be *from* the system (heat and work outputs), the energy balance for the system during this differential process can be expressed as

$$\underbrace{\delta E_{in} - \delta E_{out}}_{\substack{\text{Net energy transfer} \\ \text{by heat, work, and mass}}} = \underbrace{dE_{system}}_{\substack{\text{Change in internal, kinetic,} \\ \text{potential, etc., energies}}} \qquad (8\text{–}12)$$

$$-\delta Q - \delta W = dU$$

since the only form of energy the system contains is *internal energy,* and the only forms of energy transfer a fixed mass can involve are heat and work. Also, the only form of work a simple compressible system can involve during a reversible process is the boundary work, which is given to be $\delta W = P\, dV$ when the direction of work is taken to be from the system (otherwise it would be $-P\, dV$). The pressure $P$ in the $P\, dV$ expression is the absolute pressure, which is measured from absolute zero. Any useful work delivered by a piston–cylinder device is due to the pressure above the atmospheric level. Therefore,

$$\delta W = P\, dV = (P - P_0)\, dV + P_0\, dV = \delta W_{b,useful} + P_0\, dV \qquad (8\text{–}13)$$

A reversible process cannot involve any heat transfer through a finite temperature difference, and thus any heat transfer between the system at temperature $T$ and its surroundings at $T_0$ must occur through a reversible heat engine. Noting that $dS = \delta Q/T$ for a reversible process, and the thermal efficiency of a reversible heat engine operating between the temperatures of $T$ and $T_0$ is $\eta_{th} = 1 - T_0/T$, the differential work produced by the engine as a result of this heat transfer is

$$\delta W_{HE} = \left(1 - \frac{T_0}{T}\right)\delta Q = \delta Q - \frac{T_0}{T}\delta Q = \delta Q - (-T_0\, dS) \rightarrow$$

$$\delta Q = \delta W_{HE} - T_0\, dS \qquad (8\text{–}14)$$

Substituting the $\delta W$ and $\delta Q$ expressions in Eqs. 8–13 and 8–14 into the energy balance relation (Eq. 8–12) gives, after rearranging,

$$\delta W_{total\ useful} = \delta W_{HE} + \delta W_{b,useful} = -dU - P_0\, dV + T_0\, dS$$

Integrating from the given state (no subscript) to the dead state (0 subscript) we obtain

$$W_{total\ useful} = (U - U_0) + P_0(V - V_0) - T_0(S - S_0)$$

where $W_{total\ useful}$ is the total useful work delivered as the system undergoes a reversible process from the given state to the dead state, which is *exergy* by definition.

A closed system, in general, may possess kinetic and potential energies, and the total energy of a closed system is equal to the sum of its internal, kinetic, and potential energies. Noting that kinetic and potential energies themselves are forms of exergy, the exergy of a closed system of mass $m$ is

$$X = (U - U_0) + P_0(V - V_0) - T_0(S - S_0) + m\frac{V^2}{2} + mgz \qquad \textbf{(8-15)}$$

On a unit mass basis, the **closed system** (or **nonflow**) **exergy** $\phi$ is expressed as

$$\phi = (u - u_0) + P_0(v - v_0) - T_0(s - s_0) + \frac{V^2}{2} + gz \qquad \textbf{(8-16)}$$

$$= (e - e_0) + P_0(v - v_0) - T_0(s - s_0)$$

where $u_0$, $v_0$, and $s_0$ are the properties of the *system* evaluated at the dead state. Note that the exergy of a system is zero at the dead state since $e = e_0$, $v = v_0$, and $s = s_0$ at that state.

The exergy change of a closed system during a process is simply the difference between the final and initial exergies of the system,

$$\Delta X = X_2 - X_1 = m(\phi_2 - \phi_1) = (E_2 - E_1) + P_0(V_2 - V_1) - T_0(S_2 - S_1) \qquad \textbf{(8–17)}$$

$$= (U_2 - U_1) + P_0(V_2 - V_1) - T_0(S_2 - S_1) + m\frac{V_2^2 - V_1^2}{2} + mg(z_2 - z_1)$$

or, on a unit mass basis,

$$\Delta\phi = \phi_2 - \phi_1 = (u_2 - u_1) + P_0(v_2 - v_1) - T_0(s_2 - s_1) + \frac{V_2^2 - V_1^2}{2} + g(z_2 - z_1)$$

$$= (e_2 - e_1) + P_0(v_2 - v_1) - T_0(s_2 - s_1) \qquad \textbf{(8–18)}$$

For *stationary* closed systems, the kinetic and potential energy terms drop out.

When the properties of a system are not uniform, the exergy of the system can be determined by integration from

$$X_{\text{system}} = \int \phi\, \delta m = \int_V \phi\rho\, dV \qquad \textbf{(8–19)}$$

where $V$ is the volume of the system and $\rho$ is density.

Note that exergy is a property, and the value of a property does not change unless the *state* changes. Therefore, the *exergy change* of a system is zero if the state of the system or the environment does not change during the process. For example, the exergy change of steady flow devices such as nozzles, compressors, turbines, pumps, and heat exchangers in a given environment is zero during steady operation.

The exergy of a closed system is either *positive* or *zero*. It is never negative. Even a medium at *low temperature* ($T < T_0$) and/or *low pressure* ($P < P_0$) contains exergy since a cold medium can serve as the heat sink to a heat engine that absorbs heat from the environment at $T_0$, and an evacuated space makes it possible for the atmospheric pressure to move a piston and do useful work (Fig. 8–20).

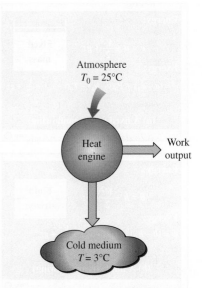

**FIGURE 8–20**

The *exergy* of a cold medium is also a *positive* quantity since work can be produced by transferring heat to it.

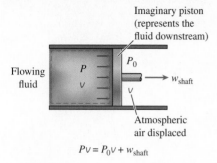

**FIGURE 8–21**

The *exergy* associated with *flow energy* is the useful work that would be delivered by an imaginary piston in the flow section.

# Exergy of a Flow Stream: Flow (or Stream) Exergy

In Chap. 5 it was shown that a flowing fluid has an additional form of energy, called the *flow energy,* which is the energy needed to maintain flow in a pipe or duct, and was expressed as $w_{flow} = Pv$ where $v$ is the specific volume of the fluid, which is equivalent to the *volume change* of a unit mass of the fluid as it is displaced during flow. The flow work is essentially the boundary work done by a fluid on the fluid downstream, and thus the exergy associated with flow work is equivalent to the exergy associated with the boundary work, which is the boundary work in excess of the work done against the atmospheric air at $P_0$ to displace it by a volume $v$ (Fig. 8–21). Noting that the flow work is $Pv$ and the work done against the atmosphere is $P_0v$, the *exergy* associated with flow energy can be expressed as

$$x_{flow} = Pv - P_0v = (P - P_0)v \qquad (8\text{–}20)$$

Therefore, the exergy associated with flow energy is obtained by replacing the pressure $P$ in the flow work relation by the pressure in excess of the atmospheric pressure, $P - P_0$. Then the exergy of a flow stream is determined by simply adding the flow exergy relation above to the exergy relation in Eq. 8–16 for a nonflowing fluid,

$$x_{flowing\ fluid} = x_{nonflowing\ fluid} + x_{flow} \qquad (8\text{–}21)$$

$$= (u - u_0) + P_0(v - v_0) - T_0(s - s_0) + \frac{V^2}{2} + gz + (P - P_0)v$$

$$= (u + Pv) - (u_0 + P_0v_0) - T_0(s - s_0) + \frac{V^2}{2} + gz$$

$$= (h - h_0) - T_0(s - s_0) + \frac{V^2}{2} + gz$$

The final expression is called **flow (or stream) exergy**, and is denoted by $\psi$ (Fig. 8–22).

*Flow exergy:* $\qquad \psi = (h - h_0) - T_0(s - s_0) + \frac{V^2}{2} + gz \qquad (8\text{–}22)$

Then the *exergy change* of a fluid stream as it undergoes a process from state 1 to state 2 becomes

$$\Delta\psi = \psi_2 - \psi_1 = (h_2 - h_1) - T_0(s_2 - s_1) + \frac{V_2^2 - V_1^2}{2} + g(z_2 - z_1) \qquad (8\text{–}23)$$

For fluid streams with negligible kinetic and potential energies, the kinetic and potential energy terms drop out.

Note that the *exergy change* of a closed system or a fluid stream represents the *maximum* amount of useful work that can be done (or the *minimum* amount of useful work that needs to be supplied if it is negative) as the system changes from state 1 to state 2 in a specified environment, and represents the *reversible work* $W_{rev}$. It is independent of the type of process executed, the kind of system used, and the nature of energy interactions with the surroundings. Also note that the exergy of a closed system cannot be negative, but the exergy of a flow stream can at pressures below the environment pressure $P_0$.

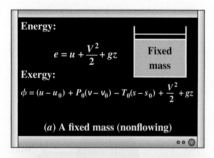

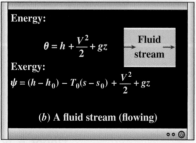

**FIGURE 8–22**

The *energy* and *exergy* contents of (*a*) a fixed mass and (*b*) a fluid stream.

**FIGURE 8-23**
Schematic for Example 8-7.

## EXAMPLE 8–7　Work Potential of Compressed Air in a Tank

A 200-m³ rigid tank contains compressed air at 1 MPa and 300 K. Determine how much work can be obtained from this air if the environment conditions are 100 kPa and 300 K.

**SOLUTION** Compressed air stored in a large tank is considered. The work potential of this air is to be determined.

**Assumptions** 1 Air is an ideal gas. 2 The kinetic and potential energies are negligible.

**Analysis** We take the air in the rigid tank as the system (Fig. 8–23). This is a *closed system* since no mass crosses the system boundary during the process. Here the question is the work potential of a fixed mass, which is the nonflow exergy by definition.

Taking the state of the air in the tank to be state 1 and noting that $T_1 = T_0 = 300$ K, the mass of air in the tank is

$$m_1 = \frac{P_1 V}{R T_1} = \frac{(1000 \text{ kPa})(200 \text{ m}^3)}{(0.287 \text{ kPa·m}^3/\text{kg·K})(300 \text{ K})} = 2323 \text{ kg}$$

The exergy content of the compressed air can be determined from

$$X_1 = m\phi_1$$

$$= m\left[ (u_1 - u_0)^{\nearrow 0} + P_0(v_1 - v_0) - T_0(s_1 - s_0) + \frac{V_1^2}{2}^{\nearrow 0} + gz_1^{\nearrow 0} \right]$$

$$= m[P_0(v_1 - v_0) - T_0(s_1 - s_0)]$$

We note that

$$P_0(v_1 - v_0) = P_0\left( \frac{RT_1}{P_1} - \frac{RT_0}{P_0} \right) = RT_0\left( \frac{P_0}{P_1} - 1 \right) \quad (\text{since } T_1 = T_0)$$

$$T_0(s_1 - s_0) = T_0\left( c_p \ln \frac{T_1}{T_0} - R \ln \frac{P_1}{P_0} \right) = -RT_0 \ln \frac{P_1}{P_0} \quad (\text{since } T_1 = T_0)$$

Therefore,

$$\phi_1 = RT_0\left( \frac{P_0}{P_1} - 1 \right) + RT_0 \ln \frac{P_1}{P_0} = RT_0\left( \ln \frac{P_1}{P_0} + \frac{P_0}{P_1} - 1 \right)$$

$$= (0.287 \text{ kJ/kg·K})(300 \text{ K})\left( \ln \frac{1000 \text{ kPa}}{100 \text{ kPa}} + \frac{100 \text{ kPa}}{1000 \text{ kPa}} - 1 \right)$$

$$= 120.76 \text{ kJ/kg}$$

and

$$X_1 = m_1\phi_1 = (2323 \text{ kg})(120.76 \text{ kJ/kg}) = 280{,}525 \text{ kJ} \cong \mathbf{281 \text{ MJ}}$$

**Discussion** The work potential of the system is 281 MJ, and thus a maximum of 281 MJ of useful work can be obtained from the compressed air stored in the tank in the specified environment.

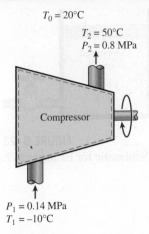

$T_0 = 20°C$

$T_2 = 50°C$
$P_2 = 0.8$ MPa

Compressor

$P_1 = 0.14$ MPa
$T_1 = -10°C$

**FIGURE 8–24**
Schematic for Example 8–8.

*EXAMPLE 8–8*    **Exergy Change During a Compression Process**

Refrigerant-134a is to be compressed from 0.14 MPa and $-10°C$ to 0.8 MPa and 50°C steadily by a compressor. Taking the environment conditions to be 20°C and 95 kPa, determine the exergy change of the refrigerant during this process and the minimum work input that needs to be supplied to the compressor per unit mass of the refrigerant.

**SOLUTION**   Refrigerant-134a is being compressed from a specified inlet state to a specified exit state. The exergy change of the refrigerant and the minimum compression work per unit mass are to be determined.
*Assumptions*   **1** Steady operating conditions exist. **2** The kinetic and potential energies are negligible.
*Analysis*   We take the *compressor* as the system (Fig. 8–24). This is a *control volume* since mass crosses the system boundary during the process. Here the question is the exergy change of a fluid stream, which is the change in the flow exergy $\psi$.
   The properties of the refrigerant at the inlet and the exit states are

*Inlet state:*    $P_1 = 0.14$ MPa $\}$ $h_1 = 246.37$ kJ/kg
                  $T_1 = -10°C$      $s_1 = 0.9724$ kJ/kg·K

*Exit state:*    $P_2 = 0.8$ MPa $\}$ $h_2 = 286.71$ kJ/kg
                 $T_2 = 50°C$      $s_2 = 0.9803$ kJ/kg·K

The exergy change of the refrigerant during this compression process is determined directly from Eq. 8–23 to be

$$\Delta\psi = \psi_2 - \psi_1 = (h_2 - h_1) - T_0(s_2 - s_1) + \frac{V_2^2 - V_1^{2}{}^{\,0}}{2} + g(z_2 - z_1)^{\,0}$$

$$= (h_2 - h_1) - T_0(s_2 - s_1)$$

$$= (286.71 - 246.37)\ \text{kJ/kg} - (293\ \text{K})[(0.9803 - 0.9724)\text{kJ/kg·K}]$$

$$= \textbf{38.0 kJ/kg}$$

Therefore, the exergy of the refrigerant increases during compression by 38.0 kJ/kg.
   The exergy change of a system in a specified environment represents the reversible work in that environment, which is the minimum work input required for work-consuming devices such as compressors. Therefore, the increase in exergy of the refrigerant is equal to the minimum work that needs to be supplied to the compressor:

$$w_{\text{in,min}} = \psi_2 - \psi_1 = \textbf{38.0 kJ/kg}$$

*Discussion*   Note that if the compressed refrigerant at 0.8 MPa and 50°C were to be expanded to 0.14 MPa and $-10°C$ in a turbine in the same environment in a reversible manner, 38.0 kJ/kg of work would be produced.

# 8–5 · EXERGY TRANSFER BY HEAT, WORK, AND MASS

Exergy, like energy, can be transferred to or from a system in three forms: *heat, work,* and *mass flow.* Exergy transfer is recognized at the system boundary as exergy crosses it, and it represents the exergy gained or lost by a system during a process. The only two forms of exergy interactions associated with a fixed mass or closed system are *heat transfer* and *work*.

## Exergy by Heat Transfer, *Q*

Recall from Chap. 6 that the work potential of the energy transferred from a heat source at temperature $T$ is the maximum work that can be obtained from that energy in an environment at temperature $T_0$ and is equivalent to the work produced by a Carnot heat engine operating between the source and the environment. Therefore, the Carnot efficiency $\eta_C = 1 - T_0/T$ represents the fraction of energy of a heat source at temperature $T$ that can be converted to work (Fig. 8–25). For example, only 70 percent of the energy transferred from a heat source at $T = 1000$ K can be converted to work in an environment at $T_0 = 300$ K.

Heat is a form of disorganized energy, and thus only a portion of it can be converted to work, which is a form of organized energy (the second law). We can always produce work from heat at a temperature above the environment temperature by transferring it to a heat engine that rejects the waste heat to the environment. Therefore, heat transfer is always accompanied by exergy transfer. Heat transfer $Q$ at a location at thermodynamic temperature $T$ is always accompanied by *exergy transfer* $X_{\text{heat}}$ in the amount of

*Exergy transfer by heat:*
$$X_{\text{heat}} = \left(1 - \frac{T_0}{T}\right)Q \quad \text{(kJ)} \qquad \text{(8–24)}$$

This relation gives the exergy transfer accompanying heat transfer $Q$ whether $T$ is greater than or less than $T_0$. When $T > T_0$, heat transfer to a system increases the exergy of that system and heat transfer from a system decreases it. But the opposite is true when $T < T_0$. In this case, the heat transfer $Q$ is the heat rejected to the cold medium (the waste heat), and it should not be confused with the heat supplied by the environment at $T_0$. The exergy transferred with heat is zero when $T = T_0$ at the point of transfer.

Perhaps you are wondering what happens when $T < T_0$. That is, what if we have a medium that is at a lower temperature than the environment? In this case it is conceivable that we can run a heat engine between the environment and the "cold" medium, and thus a cold medium offers us an opportunity to produce work. However, this time the environment serves as the heat source and the cold medium as the heat sink. In this case, the relation above gives the negative of the exergy transfer associated with the heat $Q$ transferred to the cold medium. For example, for $T = 100$ K and a heat transfer of $Q = 1$ kJ to the medium, Eq. 8–24 gives $X_{\text{heat}} = (1 - 300/100)(1 \text{ kJ}) = -2$ kJ, which means that the exergy of the cold medium decreases by 2 kJ. It also means that this exergy can be recovered, and the cold medium–environment combination has the potential to produce 2 units of work for each unit of heat rejected to the cold medium at 100 K. That is, a Carnot heat engine operating between $T_0 = 300$ K and $T = 100$ K produces 2 units of work while rejecting 1 unit of heat for each 3 units of heat it receives from the environment.

When $T > T_0$, the exergy and heat transfer are in the same direction. That is, both the exergy and energy content of the medium to which heat is transferred increase. When $T < T_0$ (cold medium), however, the exergy and heat transfer are in opposite directions. That is, the energy of the cold medium increases as a result of heat transfer, but its exergy decreases. The exergy

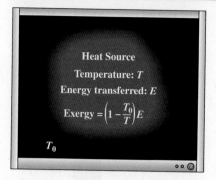

**FIGURE 8–25**
The Carnot efficiency $\eta_C = 1 - T_0/T$ represents the fraction of the energy transferred from a heat source at temperature $T$ that can be converted to work in an environment at temperature $T_0$.

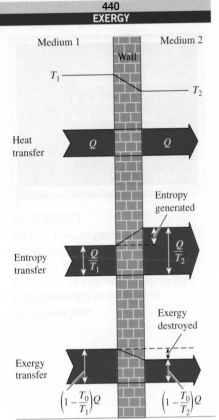

Medium 1     Wall     Medium 2

$T_1$

$T_2$

Heat transfer     $Q$     $Q$

Entropy generated

Entropy transfer     $\dfrac{Q}{T_1}$     $\dfrac{Q}{T_2}$

Exergy destroyed

Exergy transfer     $\left(1 - \dfrac{T_0}{T_1}\right)Q$     $\left(1 - \dfrac{T_0}{T_2}\right)Q$

**FIGURE 8–26**
The transfer and destruction of exergy during a heat transfer process through a finite temperature difference.

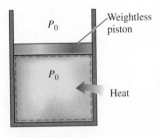

$P_0$     Weightless piston

$P_0$

Heat

**FIGURE 8–27**
There is no useful work transfer associated with boundary work when the pressure of the system is maintained constant at atmospheric pressure.

of the cold medium eventually becomes zero when its temperature reaches $T_0$. Equation 8–24 can also be viewed as the *exergy associated with thermal energy Q* at temperature *T*.

When the temperature $T$ at the location where heat transfer is taking place is not constant, the exergy transfer accompanying heat transfer is determined by integration to be

$$X_{heat} = \int\left(1 - \frac{T_0}{T}\right)\delta Q \tag{8–25}$$

Note that heat transfer through a finite temperature difference is irreversible, and some entropy is generated as a result. The entropy generation is always accompanied by exergy destruction, as illustrated in Fig. 8–26. Also note that *heat transfer Q* at a location at temperature *T* is always accompanied by *entropy transfer* in the amount of $Q/T$ and *exergy transfer* in the amount of $(1 - T_0/T)Q$.

## Exergy Transfer by Work, *W*

Exergy is the useful work potential, and the exergy transfer by work can simply be expressed as

$$\textit{Exergy transfer by work: } X_{work} = \begin{cases} W - W_{surr} & \text{(for boundary work)} \\ W & \text{(for other forms of work)} \end{cases} \tag{8–26}$$

where $W_{surr} = P_0(V_2 - V_1)$, $P_0$ is atmospheric pressure, and $V_1$ and $V_2$ are the initial and final volumes of the system. Therefore, the exergy transfer with work such as shaft work and electrical work is equal to the work $W$ itself. In the case of a system that involves boundary work, such as a piston–cylinder device, the work done to push the atmospheric air out of the way during expansion cannot be transferred, and thus it must be subtracted. Also, during a compression process, part of the work is done by the atmospheric air, and thus we need to supply less useful work from an external source.

To clarify this point further, consider a vertical cylinder fitted with a weightless and frictionless piston (Fig. 8–27). The cylinder is filled with a gas that is maintained at the atmospheric pressure $P_0$ at all times. Heat is now transferred to the system and the gas in the cylinder expands. As a result, the piston rises and boundary work is done. However, this work cannot be used for any useful purpose since it is just enough to push the atmospheric air aside. (If we connect the piston to an external load to extract some useful work, the pressure in the cylinder will have to rise above $P_0$ to beat the resistance offered by the load.) When the gas is cooled, the piston moves down, compressing the gas. Again, no work is needed from an external source to accomplish this compression process. Thus we conclude that the work done by or against the atmosphere is not available for any useful purpose, and should be excluded from available work.

## Exergy Transfer by Mass, *m*

Mass contains *exergy* as well as energy and entropy, and the exergy, energy, and entropy contents of a system are proportional to mass. Also, the rates of exergy, entropy, and energy transport into or out of a system are proportional to the mass flow rate. Mass flow is a mechanism to transport exergy, entropy, and energy into or out of a system. When mass in the amount of $m$ enters or leaves a system, exergy in the amount of $m\psi$, where $\psi = (h - h_0) - T_0(s - s_0) + V^2/2 + gz$, accompanies it. That is,

*Exergy transfer by mass:* $\qquad\qquad X_{\text{mass}} = m\psi$ $\qquad\qquad$ (8–27)

Therefore, the exergy of a system increases by $m\psi$ when mass in the amount of $m$ enters, and decreases by the same amount when the same amount of mass at the same state leaves the system (Fig. 8–28).

Exergy flow associated with a fluid stream when the fluid properties are variable can be determined by integration from

$$\dot{X}_{\text{mass}} = \int_{A_c} \psi\rho V_n \, dA_c \quad \text{and} \quad X_{\text{mass}} = \int_{\Delta t} \psi \, \delta m = \int_{\Delta t} \dot{X}_{\text{mass}} \, dt \quad (8\text{–}28)$$

where $A_c$ is the cross-sectional area of the flow and $V_n$ is the local velocity normal to $dA_c$.

Note that exergy transfer by heat $X_{\text{heat}}$ is zero for adiabatic systems, and the exergy transfer by mass $X_{\text{mass}}$ is zero for systems that involve no mass flow across their boundaries (i.e., closed systems). The total exergy transfer is zero for isolated systems since they involve no heat, work, or mass transfer.

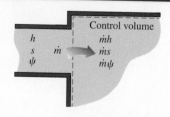

**FIGURE 8–28**
Mass contains energy, entropy, and exergy, and thus mass flow into or out of a system is accompanied by energy, entropy, and exergy transfer.

# 8–6 ▪ THE DECREASE OF EXERGY PRINCIPLE AND EXERGY DESTRUCTION

In Chap. 2 we presented the *conservation of energy principle* and indicated that energy cannot be created or destroyed during a process. In Chap. 7 we established the *increase of entropy principle,* which can be regarded as one of the statements of the second law, and indicated that entropy can be created but cannot be destroyed. That is, entropy generation $S_{\text{gen}}$ must be positive (actual processes) or zero (reversible processes), but it cannot be negative. Now we are about to establish an alternative statement of the second law of thermodynamics, called the *decrease of exergy principle,* which is the counterpart of the increase of entropy principle.

Consider an *isolated system* shown in Fig. 8–29. By definition, no heat, work, or mass can cross the boundaries of an isolated system, and thus there is no energy and entropy transfer. Then the *energy* and *entropy* balances for an isolated system can be expressed as

*Energy balance:* $\qquad E_{\text{in}}^{\,\nearrow 0} - E_{\text{out}}^{\,\nearrow 0} = \Delta E_{\text{system}} \rightarrow 0 = E_2 - E_1$

*Entropy balance:* $\qquad S_{\text{in}}^{\,\nearrow 0} - S_{\text{out}}^{\,\nearrow 0} + S_{\text{gen}} = \Delta S_{\text{system}} \rightarrow S_{\text{gen}} = S_2 - S_1$

Multiplying the second relation by $T_0$ and subtracting it from the first one gives

$$-T_0 S_{\text{gen}} = E_2 - E_1 - T_0(S_2 - S_1) \qquad (8\text{–}29)$$

From Eq. 8–17 we have

$$X_2 - X_1 = (E_2 - E_1) + P_0(V_2 - V_1)^{\nearrow 0} - T_0(S_2 - S_1) \qquad (8\text{–}30)$$

$$= (E_2 - E_1) - T_0(S_2 - S_1)$$

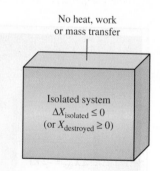

**FIGURE 8–29**
The isolated system considered in the development of the decrease of exergy principle.

since $V_2 = V_1$ for an isolated system (it cannot involve any moving boundary and thus any boundary work). Combining Eqs. 8–29 and 8–30 gives

$$-T_0 S_{gen} = X_2 - X_1 \leq 0 \qquad \text{(8–31)}$$

since $T_0$ is the thermodynamic temperature of the environment and thus a positive quantity, $S_{gen} \geq 0$, and thus $T_0 S_{gen} \geq 0$. Then we conclude that

$$\Delta X_{\text{isolated}} = (X_2 - X_1)_{\text{isolated}} \leq 0 \qquad \text{(8–32)}$$

This equation can be expressed as *the exergy of an isolated system during a process always decreases or, in the limiting case of a reversible process, remains constant.* In other words, it *never* increases and *exergy is destroyed* during an actual process. This is known as the **decrease of exergy principle.** For an isolated system, the decrease in exergy equals exergy destroyed.

## Exergy Destruction

Irreversibilities such as friction, mixing, chemical reactions, heat transfer through a finite temperature difference, unrestrained expansion, nonquasi-equilibrium compression or expansion always *generate entropy,* and anything that generates entropy always *destroys exergy.* The **exergy destroyed** is proportional to the entropy generated, as can be seen from Eq. 8–31, and is expressed as

$$X_{\text{destroyed}} = T_0 S_{gen} \geq 0 \qquad \text{(8–33)}$$

Note that exergy destroyed is a *positive quantity* for any actual process and becomes *zero* for a reversible process. Exergy destroyed represents the lost work potential and is also called the *irreversibility* or *lost work.*

Equations 8–32 and 8–33 for the decrease of exergy and the exergy destruction are applicable to *any kind of system* undergoing *any kind of process* since any system and its surroundings can be enclosed by a sufficiently large arbitrary boundary across which there is no heat, work, and mass transfer, and thus any system and its surroundings constitute an *isolated system.*

No actual process is truly reversible, and thus some exergy is destroyed during a process. Therefore, the exergy of the universe, which can be considered to be an isolated system, is continuously decreasing. The more irreversible a process is, the larger the exergy destruction during that process. No exergy is destroyed during a reversible process ($X_{\text{destroyed,rev}} = 0$).

The decrease of exergy principle does not imply that the exergy of a system cannot increase. The exergy change of a system *can* be positive or negative during a process (Fig. 8–30), but exergy destroyed cannot be negative. The decrease of exergy principle can be summarized as follows:

$$X_{\text{destroyed}} \begin{cases} > 0 & \text{Irreversible process} \\ = 0 & \text{Reversible process} \\ < 0 & \text{Impossible process} \end{cases} \qquad \text{(8–34)}$$

This relation serves as an alternative criterion to determine whether a process is reversible, irreversible, or impossible.

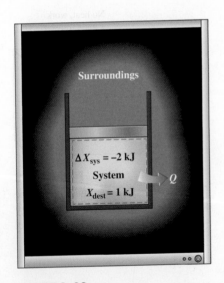

**FIGURE 8–30**
The exergy change of a system can be negative, but the exergy destruction cannot.

# 8–7 · EXERGY BALANCE: CLOSED SYSTEMS

The nature of exergy is opposite to that of entropy in that exergy can be *destroyed*, but it cannot be created. Therefore, the *exergy change* of a system during a process is less than the *exergy transfer* by an amount equal to the *exergy destroyed* during the process within the system boundaries. Then the *decrease of exergy principle* can be expressed as (Fig. 8–31)

$$\begin{pmatrix} \text{Total} \\ \text{exergy} \\ \text{entering} \end{pmatrix} - \begin{pmatrix} \text{Total} \\ \text{exergy} \\ \text{leaving} \end{pmatrix} - \begin{pmatrix} \text{Total} \\ \text{exergy} \\ \text{destroyed} \end{pmatrix} = \begin{pmatrix} \text{Change in the} \\ \text{total exergy} \\ \text{of the system} \end{pmatrix}$$

or

$$X_{\text{in}} - X_{\text{out}} - X_{\text{destroyed}} = \Delta X_{\text{system}} \qquad \text{(8–35)}$$

This relation is referred to as the **exergy balance** and can be stated as *the exergy change of a system during a process is equal to the difference between the net exergy transfer through the system boundary and the exergy destroyed within the system boundaries as a result of irreversibilities.*

We mentioned earlier that exergy can be transferred to or from a system by heat, work, and mass transfer. Then the exergy balance for *any system* undergoing *any process* can be expressed more explicitly as

*General:*
$$\underbrace{X_{\text{in}} - X_{\text{out}}}_{\substack{\text{Net exergy transfer} \\ \text{by heat, work, and mass}}} - \underbrace{X_{\text{destroyed}}}_{\substack{\text{Exergy} \\ \text{destruction}}} = \underbrace{\Delta X_{\text{system}}}_{\substack{\text{Change} \\ \text{in exergy}}} \quad \text{(kJ)} \qquad \text{(8–36)}$$

or, in the **rate form**, as

*General, rate form:*
$$\underbrace{\dot{X}_{\text{in}} - \dot{X}_{\text{out}}}_{\substack{\text{Rate of net exergy transfer} \\ \text{by heat, work, and mass}}} - \underbrace{\dot{X}_{\text{destroyed}}}_{\substack{\text{Rate of exergy} \\ \text{destruction}}} = \underbrace{dX_{\text{system}}/dt}_{\substack{\text{Rate of change} \\ \text{in exergy}}} \quad \text{(kW)} \qquad \text{(8–37)}$$

where the rates of exergy transfer by heat, work, and mass are expressed as $\dot{X}_{\text{heat}} = (1 - T_0/T)\dot{Q}$, $\dot{X}_{\text{work}} = \dot{W}_{\text{useful}}$, and $\dot{X}_{\text{mass}} = \dot{m}\psi$, respectively. The exergy balance can also be expressed per unit mass as

*General, unit-mass basis:* $\quad (x_{\text{in}} - x_{\text{out}}) - x_{\text{destroyed}} = \Delta x_{\text{system}} \quad \text{(kJ/kg)} \qquad \text{(8–38)}$

where all the quantities are expressed per unit mass of the system. Note that for a *reversible process*, the exergy destruction term $X_{\text{destroyed}}$ drops out from all of the relations above. Also, it is usually more convenient to find the entropy generation $S_{\text{gen}}$ first, and then to evaluate the exergy destroyed directly from Eq. 8–33. That is,

$$X_{\text{destroyed}} = T_0 S_{\text{gen}} \quad \text{or} \quad \dot{X}_{\text{destroyed}} = T_0 \dot{S}_{\text{gen}} \qquad \text{(8–39)}$$

When the environment conditions $P_0$ and $T_0$ and the end states of the system are specified, the exergy change of the system $\Delta X_{\text{system}} = X_2 - X_1$ can be determined directly from Eq. 8–17 regardless of how the process is executed. However, the determination of the exergy transfers by heat, work, and mass requires a knowledge of these interactions.

A *closed system* does not involve any mass flow and thus any exergy transfer associated with mass flow. Taking the positive direction of heat transfer to be to the system and the positive direction of work transfer to be

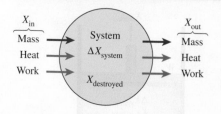

**FIGURE 8–31**
Mechanisms of exergy transfer.

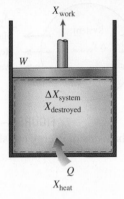

$$X_{\text{heat}} - X_{\text{work}} - X_{\text{destroyed}} = \Delta X_{\text{system}}$$

**FIGURE 8–32**

Exergy balance for a closed system when the direction of heat transfer is taken to be to the system and the direction of work from the system.

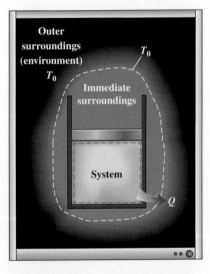

**FIGURE 8–33**

Exergy destroyed outside system boundaries can be accounted for by writing an exergy balance on the extended system that includes the system and its immediate surroundings.

from the system, the exergy balance for a closed system can be expressed more explicitly as (Fig. 8–32)

*Closed system:* $\quad\quad X_{\text{heat}} - X_{\text{work}} - X_{\text{destroyed}} = \Delta X_{\text{system}}$ **(8–40)**

or

*Closed system:* $\sum \left(1 - \dfrac{T_0}{T_k}\right)Q_k - [W - P_0(V_2 - V_1)] - T_0 S_{\text{gen}} = X_2 - X_1$ **(8–41)**

where $Q_k$ is the heat transfer through the boundary at temperature $T_k$ at location $k$. Dividing the previous equation by the time interval $\Delta t$ and taking the limit as $\Delta t \rightarrow 0$ gives the *rate form* of the exergy balance for a closed system,

*Rate form:* $\sum \left(1 - \dfrac{T_0}{T_k}\right)\dot{Q}_k - \left(\dot{W} - P_0\dfrac{dV_{\text{system}}}{dt}\right) - T_0 \dot{S}_{\text{gen}} = \dfrac{dX_{\text{system}}}{dt}$ **(8–42)**

Note that the relations above for a closed system are developed by taking the heat transfer to a system and work done by the system to be positive quantities. Therefore, heat transfer from the system and work done on the system should be taken to be negative quantities when using those relations.

The exergy balance relations presented above can be used to determine the *reversible work* $W_{\text{rev}}$ by setting the exergy destruction term equal to zero. The work $W$ in that case becomes the reversible work. That is, $W = W_{\text{rev}}$ when $X_{\text{destroyed}} = T_0 S_{\text{gen}} = 0$.

Note that $X_{\text{destroyed}}$ represents the exergy destroyed *within the system boundary* only, and not the exergy destruction that may occur outside the system boundary during the process as a result of external irreversibilities. Therefore, a process for which $X_{\text{destroyed}} = 0$ is *internally reversible* but not necessarily *totally* reversible. The *total* exergy destroyed during a process can be determined by applying the exergy balance to an *extended system* that includes the system itself and its immediate surroundings where external irreversibilities might be occurring (Fig. 8–33). Also, the exergy change in this case is equal to the sum of the exergy changes of the system and the *exergy change* of the immediate surroundings. Note that under steady conditions, the state and thus the exergy of the immediate surroundings (the "buffer zone") at any point does not change during the process, and thus the exergy change of the immediate surroundings is zero. When evaluating the exergy transfer between an extended system and the environment, the boundary temperature of the extended system is simply taken to be the environment temperature $T_0$.

For a *reversible process*, the *entropy generation* and thus the *exergy destruction* are *zero*, and the exergy balance relation in this case becomes analogous to the energy balance relation. That is, the exergy change of the system becomes equal to the exergy transfer.

Note that the *energy change* of a system equals the *energy transfer* for *any* process, but the *exergy change* of a system equals the *exergy transfer* only for a *reversible* process. The *quantity* of energy is always preserved during an actual process (the first law), but the *quality* is bound to decrease (the second law). This decrease in quality is always accompanied by an increase in entropy and a decrease in exergy. When 10 kJ of heat is transferred from a hot medium to a cold one, for example, we still have 10 kJ of energy at the end of the process, but at a lower temperature, and thus at a lower quality and at a lower potential to do work.

■
■ **EXAMPLE 8–9**    **General Exergy Balance for Closed Systems**
■
■ Starting with energy and entropy balances, derive the general exergy balance
■ relation for a closed system (Eq. 8–41).

**SOLUTION**  Starting with energy and entropy balance relations, a general relation for exergy balance for a closed system is to be obtained.
**Analysis**  We consider a general closed system (a fixed mass) that is free to exchange heat and work with its surroundings (Fig. 8–34). The system undergoes a process from state 1 to state 2. Taking the positive direction of heat transfer to be *to* the system and the positive direction of work transfer to be *from* the system, the energy and entropy balances for this closed system can be expressed as

*Energy balance:*    $E_{in} - E_{out} = \Delta E_{system} \rightarrow Q - W = E_2 - E_1$

*Entropy balance:*    $S_{in} - S_{out} + S_{gen} = \Delta S_{system} \rightarrow \int_1^2 \left(\frac{\delta Q}{T}\right)_{boundary} + S_{gen} = S_2 - S_1$

Multiplying the second relation by $T_0$ and subtracting it from the first one gives

$$Q - T_0 \int_1^2 \left(\frac{\delta Q}{T}\right)_{boundary} - W - T_0 S_{gen} = E_2 - E_1 - T_0(S_2 - S_1)$$

However, the heat transfer for the process 1-2 can be expressed as $Q = \int_1^2 \delta Q$ and the right side of the above equation is, from Eq. 8–17, $(X_2 - X_1) - P_0(V_2 - V_1)$. Thus,

$$\int_1^2 \delta Q - T_0 \int_1^2 \left(\frac{\delta Q}{T}\right)_{boundary} - W - T_0 S_{gen} = X_2 - X_1 - P_0(V_2 - V_1)$$

Letting $T_b$ denote the boundary temperature and rearranging give

$$\int_1^2 \left(1 - \frac{T_0}{T_b}\right)\delta Q - [W - P_0(V_2 - V_1)] - T_0 S_{gen} = X_2 - X_1 \qquad \text{(8–43)}$$

which is equivalent to Eq. 8–41 for the exergy balance except that the integration is replaced by summation in that equation for convenience. This completes the proof.
**Discussion**  Note that the exergy balance relation above is obtained by adding the energy and entropy balance relations, and thus it is not an independent equation. However, it can be used in place of the entropy balance relation as an alternative second law expression in exergy analysis.

**FIGURE 8–34**
A general closed system considered in Example 8–9.

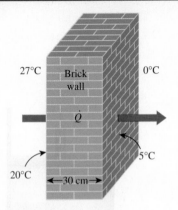

**FIGURE 8–35**
Schematic for Example 8–10.

---

### EXAMPLE 8–10    Exergy Destruction during Heat Conduction

Consider steady heat transfer through a 5-m × 6-m brick wall of a house of thickness 30 cm. On a day when the temperature of the outdoors is 0°C, the house is maintained at 27°C. The temperatures of the inner and outer surfaces of the brick wall are measured to be 20°C and 5°C, respectively, and the rate of heat transfer through the wall is 1035 W. Determine the rate of exergy destruction in the wall, and the rate of total exergy destruction associated with this heat transfer process.

**SOLUTION**  Steady heat transfer through a wall is considered. For specified heat transfer rate, wall surface temperatures, and environment conditions, the rate of exergy destruction within the wall and the rate of total exergy destruction are to be determined.

**Assumptions**  **1** The process is steady, and thus the rate of heat transfer through the wall is constant. **2** The exergy change of the wall is zero during this process since the state and thus the exergy of the wall do not change anywhere in the wall. **3** Heat transfer through the wall is one-dimensional.

**Analysis**  We first take the *wall* as the system (Fig. 8–35). This is a *closed system* since no mass crosses the system boundary during the process. We note that heat and exergy are entering from one side of the wall and leaving from the other side.

Applying the rate form of the exergy balance to the wall gives

$$\underbrace{\dot{X}_{in} - \dot{X}_{out}}_{\substack{\text{Rate of net exergy transfer} \\ \text{by heat, work, and mass}}} - \underbrace{\dot{X}_{destroyed}}_{\substack{\text{Rate of exergy} \\ \text{destruction}}} = \underbrace{dX_{system}/dt}_{\substack{\text{Rate of change} \\ \text{in exergy}}}^{\nearrow^{0 \text{ (steady)}}} = 0$$

$$\dot{Q}\left(1 - \frac{T_0}{T}\right)_{in} - \dot{Q}\left(1 - \frac{T_0}{T}\right)_{out} - \dot{X}_{destroyed} = 0$$

$$(1035 \text{ W})\left(1 - \frac{273 \text{ K}}{293 \text{ K}}\right) - (1035 \text{ W})\left(1 - \frac{273 \text{ K}}{278 \text{ K}}\right) - \dot{X}_{destroyed} = 0$$

Solving, the rate of exergy destruction in the wall is determined to be

$$\dot{X}_{destroyed} = \textbf{52.0 W}$$

Note that exergy transfer with heat at any location is $(1 - T_0/T)Q$ at that location, and the direction of exergy transfer is the same as the direction of heat transfer.

To determine the rate of total exergy destruction during this heat transfer process, we extend the system to include the regions on both sides of the wall that experience a temperature change. Then one side of the system boundary becomes room temperature while the other side, the temperature of the outdoors. The exergy balance for this *extended system* (system + immediate surroundings) is the same as that given above, except the two boundary temperatures are 300 and 273 K instead of 293 and 278 K, respectively. Then the rate of total exergy destruction becomes

$$\dot{X}_{destroyed,total} = (1035 \text{ W})\left(1 - \frac{273 \text{ K}}{300 \text{ K}}\right) - (1035 \text{ W})\left(1 - \frac{273 \text{ K}}{273 \text{ K}}\right) = \textbf{93.2 W}$$

The difference between the two exergy destructions is 41.2 W and represents the exergy destroyed in the air layers on both sides of the wall. The exergy destruction in this case is entirely due to irreversible heat transfer through a finite temperature difference.

**Discussion** This problem was solved in Chap. 7 for entropy generation. We could have determined the exergy destroyed by simply multiplying the entropy generation by the environment temperature of $T_0 = 273$ K.

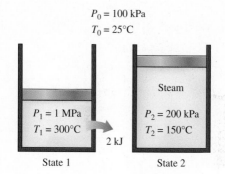

$P_0 = 100$ kPa
$T_0 = 25°C$

$P_1 = 1$ MPa
$T_1 = 300°C$

Steam

$P_2 = 200$ kPa
$T_2 = 150°C$

2 kJ

State 1                State 2

**FIGURE 8–36**
Schematic for Example 8–11.

■ **EXAMPLE 8–11**    **Exergy Destruction During Expansion of Steam**
■
■ A piston–cylinder device contains 0.05 kg of steam at 1 MPa and 300°C.
■ Steam now expands to a final state of 200 kPa and 150°C, doing work. Heat losses from the system to the surroundings are estimated to be 2 kJ during this process. Assuming the surroundings to be at $T_0 = 25°C$ and $P_0 = 100$ kPa, determine (a) the exergy of the steam at the initial and the final states, (b) the exergy change of the steam, (c) the exergy destroyed, and (d) the second-law efficiency for the process.

**SOLUTION** Steam in a piston–cylinder device expands to a specified state. The exergies of steam at the initial and final states, the exergy change, the exergy destroyed, and the second-law efficiency for this process are to be determined.

**Assumptions** The kinetic and potential energies are negligible.

**Analysis** We take the *steam* contained within the piston–cylinder device as the system (Fig. 8–36). This is a *closed system* since no mass crosses the system boundary during the process. We note that boundary work is done by the system and heat is lost from the system during the process.

(a) First we determine the properties of the steam at the initial and final states as well as the state of the surroundings:

State 1:
$$\left. \begin{array}{l} P_1 = 1 \text{ MPa} \\ T_1 = 300°C \end{array} \right\}$$
$u_1 = 2793.7$ kJ/kg
$v_1 = 0.25799$ m³/kg     (Table A–6)
$s_1 = 7.1246$ kJ/kg·K

State 2:
$$\left. \begin{array}{l} P_2 = 200 \text{ kPa} \\ T_2 = 150°C \end{array} \right\}$$
$u_2 = 2577.1$ kJ/kg
$v_2 = 0.95986$ m³/kg     (Table A–6)
$s_2 = 7.2810$ kJ/kg·K

Dead state:
$$\left. \begin{array}{l} P_0 = 100 \text{ kPa} \\ T_0 = 25°C \end{array} \right\}$$
$u_0 \cong u_{f @ 25°C} = 104.83$ kJ/kg
$v_0 \cong v_{f @ 25°C} = 0.00103$ m³/kg     (Table A–4)
$s_0 \cong s_{f @ 25°C} = 0.3672$ kJ/kg·K

The exergies of the system at the initial state $X_1$ and the final state $X_2$ are determined from Eq. 8–15 to be

$$X_1 = m[(u_1 - u_0) - T_0(s_1 - s_0) + P_0(v_1 - v_0)]$$

$$= (0.05 \text{ kg})\{(2793.7 - 104.83) \text{ kJ/kg}$$

$$- (298 \text{ K})[(7.1246 - 0.3672) \text{ kJ/kg·K}]$$

$$+ (100 \text{ kPa})[(0.25799 - 0.00103) \text{ m}^3/\text{kg}]\}(\text{kJ/kPa·m}^3)$$

$$= \textbf{35.0 kJ}$$

and

$$X_2 = m[(u_2 - u_0) - T_0(s_2 - s_0) + P_0(v_2 - v_0)]$$

$$= (0.05 \text{ kg})\{(2577.1 - 104.83) \text{ kJ/kg}$$

$$- (298 \text{ K})[(7.2810 - 0.3672) \text{ kJ/kg·K}]$$

$$+ (100 \text{ kPa})[(0.95986 - 0.00103) \text{ m}^3\text{/kg}]\}(\text{kJ/kPa·m}^3)$$

$$= \textbf{25.4 kJ}$$

That is, steam initially has an exergy content of 35 kJ, which drops to 25.4 kJ at the end of the process. In other words, if the steam were allowed to undergo a reversible process from the initial state to the state of the environment, it would produce 35 kJ of useful work.

(b) The exergy change for a process is simply the difference between the exergy at the initial and final states of the process,

$$\Delta X = X_2 - X_1 = 25.4 - 35.0 = \textbf{−9.6 kJ}$$

That is, if the process between states 1 and 2 were executed in a reversible manner, the system would deliver 9.6 kJ of useful work.

(c) The total exergy destroyed during this process can be determined from the exergy balance applied on the *extended system* (system + immediate surroundings) whose boundary is at the environment temperature of $T_0$ (so that there is no exergy transfer accompanying heat transfer to or from the environment),

$$\underbrace{X_{\text{in}} - X_{\text{out}}}_{\substack{\text{Net exergy transfer} \\ \text{by heat, work, and mass}}} - \underbrace{X_{\text{destroyed}}}_{\substack{\text{Exergy} \\ \text{destruction}}} = \underbrace{\Delta X_{\text{system}}}_{\substack{\text{Change} \\ \text{in exergy}}}$$

$$-X_{\text{work,out}} - X_{\text{heat,out}}{\nearrow}^{0} - X_{\text{destroyed}} = X_2 - X_1$$

$$X_{\text{destroyed}} = X_1 - X_2 - W_{u,\text{out}}$$

where $W_{u,\text{out}}$ is the useful boundary work delivered as the system expands. By writing an energy balance on the system, the total boundary work done during the process is determined to be

$$\underbrace{E_{\text{in}} - E_{\text{out}}}_{\substack{\text{Net energy transfer} \\ \text{by heat, work, and mass}}} = \underbrace{\Delta E_{\text{system}}}_{\substack{\text{Change in internal, kinetic,} \\ \text{potential, etc., energies}}}$$

$$-Q_{\text{out}} - W_{b,\text{out}} = \Delta U$$

$$W_{b,\text{out}} = -Q_{\text{out}} - \Delta U = -Q_{\text{out}} - m(u_2 - u_1)$$

$$= -(2 \text{ kJ}) - (0.05 \text{ kg})(2577.1 - 2793.7) \text{ kJ/kg}$$

$$= 8.8 \text{ kJ}$$

This is the total boundary work done by the system, including the work done against the atmosphere to push the atmospheric air out of the way during the expansion process. The useful work is the difference between the two:

$$W_u = W - W_{\text{surr}} = W_{b,\text{out}} - P_0(V_2 - V_1) = W_{b,\text{out}} - P_0 m(v_2 - v_1)$$

$$= 8.8 \text{ kJ} - (100 \text{ kPa})(0.05 \text{ kg})[(0.9599 - 0.25799) \text{ m}^3\text{/kg}]\left(\frac{1 \text{ kJ}}{1 \text{ kPa·m}^3}\right)$$

$$= 5.3 \text{ kJ}$$

Substituting, the exergy destroyed is determined to be

$$X_{destroyed} = X_1 - X_2 - W_{u,out} = 35.0 - 25.4 - 5.3 = \textbf{4.3 kJ}$$

That is, 4.3 kJ of work potential is wasted during this process. In other words, an additional 4.3 kJ of energy *could have been* converted to work during this process, but was not.

The exergy destroyed could also be determined from

$$X_{destroyed} = T_0 S_{gen} = T_0\left[m(s_2 - s_1) + \frac{Q_{surr}}{T_0}\right]$$

$$= (298 \text{ K})\left\{(0.05 \text{ kg})[(7.2810 - 7.1246) \text{ kJ/kg·K}] + \frac{2 \text{ kJ}}{298 \text{ K}}\right\}$$

$$= 4.3 \text{ kJ}$$

which is the same result obtained before.

(*d*) Noting that the decrease in the exergy of the steam is the exergy expended and the useful work output is the exergy recovered, the second-law efficiency for this process can be determined from

$$\eta_{II} = \frac{\text{Exergy recovered}}{\text{Exergy expended}} = \frac{W_u}{X_1 - X_2} = \frac{5.3}{35.0 - 25.4} = \textbf{0.552 or 55.2\%}$$

That is, 44.8 percent of the work potential of the steam is wasted during this process.

---

## EXAMPLE 8–12    Exergy Destroyed During Stirring of a Gas

An insulated rigid tank contains 0.9 kg of air at 150 kPa and 20°C. A paddle wheel inside the tank is now rotated by an external power source until the temperature in the tank rises to 55°C (Fig. 8–37). If the surrounding air is at $T_0 = 20°C$, determine (*a*) the exergy destroyed and (*b*) the reversible work for this process.

**SOLUTION** The air in an adiabatic rigid tank is heated by stirring it by a paddle wheel. The exergy destroyed and the reversible work for this process are to be determined.

**Assumptions** 1 Air at about atmospheric conditions can be treated as an ideal gas with constant specific heats at room temperature. 2 The kinetic and potential energies are negligible. 3 The volume of a rigid tank is constant, and thus there is no boundary work. 4 The tank is well insulated and thus there is no heat transfer.

**Analysis** We take the *air* contained within the tank as the system. This is a *closed system* since no mass crosses the system boundary during the process. We note that shaft work is done on the system.

(*a*) The exergy destroyed during a process can be determined from an exergy balance, or directly from $X_{destroyed} = T_0 S_{gen}$. We will use the second approach since it is usually easier. But first we determine the entropy generated from an entropy balance,

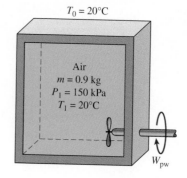

**FIGURE 8–37**
Schematic for Example 8–12.

$$\underbrace{S_{in} - S_{out}}_{\substack{\text{Net entropy transfer} \\ \text{by heat and mass}}} + \underbrace{S_{gen}}_{\substack{\text{Entropy} \\ \text{generation}}} = \underbrace{\Delta S_{system}}_{\substack{\text{Change} \\ \text{in entropy}}}$$

$$0 + S_{gen} = \Delta S_{system} = m\left( c_v \ln\frac{T_2}{T_1} + R \ln\frac{V_2}{V_1}^{\,0} \right)$$

$$S_{gen} = mc_v \ln\frac{T_2}{T_1}$$

Taking $c_v = 0.718$ kJ/kg·°C and substituting, the exergy destroyed becomes

$$X_{destroyed} = T_0 S_{gen} = T_0 mc_v \ln\frac{T_2}{T_1}$$

$$= (293 \text{ K})(0.9 \text{ kg})(0.718 \text{ kJ/kg·°C})\ln\frac{328 \text{ K}}{293 \text{ K}}$$

$$= \mathbf{21.4 \text{ kJ}}$$

(b) The reversible work, which represents the minimum work input $W_{rev,in}$ in this case, can be determined from the exergy balance by setting the exergy destruction equal to zero,

$$\underbrace{X_{in} - X_{out}}_{\substack{\text{Net exergy transfer} \\ \text{by heat, work, and mass}}} - \underbrace{X_{destroyed}}_{\substack{\text{Exergy} \\ \text{destruction}}}^{\,0 \text{ (reversible)}} = \underbrace{\Delta X_{system}}_{\substack{\text{Change} \\ \text{in exergy}}}$$

$$W_{rev,in} = X_2 - X_1$$

$$= (E_2 - E_1) + P_0(V_2 - V_1)^{\,0} - T_0(S_2 - S_1)$$

$$= (U_2 - U_1) - T_0(S_2 - S_1)$$

since $\Delta KE = \Delta PE = 0$ and $V_2 = V_1$. Noting that $T_0(S_2 - S_1) = T_0 \Delta S_{system} = 21.4$ kJ, the reversible work becomes

$$W_{rev,in} = mc_v(T_2 - T_1) - T_0(S_2 - S_1)$$

$$= (0.9 \text{ kg})(0.718 \text{ kJ/kg·°C})(55 - 20)\text{°C} - 21.4 \text{ kJ}$$

$$= (22.6 - 21.4) \text{ Btu}$$

$$= \mathbf{1.2 \text{ kJ}}$$

Therefore, a work input of just 1.2 kJ would be sufficient to accomplish this process (raise the temperature of air in the tank from 20 to 55°C) if all the irreversibilities were eliminated.

**Discussion** The solution is complete at this point. However, to gain some physical insight, we will set the stage for a discussion. First, let us determine the actual work (the paddle-wheel work $W_{pw}$) done during this process. Applying the energy balance on the system,

$$\underbrace{E_{in} - E_{out}}_{\substack{\text{Net energy transfer} \\ \text{by heat, work, and mass}}} = \underbrace{\Delta E_{system}}_{\substack{\text{Change in internal, kinetic,} \\ \text{potential, etc., energies}}}$$

$$W_{pw,in} = \Delta U = 22.6 \text{ kJ} \qquad \text{[from part (b)]}$$

since the system is adiabatic ($Q = 0$) and involves no moving boundaries ($W_b = 0$).

To put the information into perspective, 22.6 kJ of work is consumed during the process, 21.4 kJ of exergy is destroyed, and the reversible work input for the process is 1.2 kJ. What does all this mean? It simply means that we could have created the same effect on the closed system (raising its temperature to 55°C at constant volume) by consuming 1.2 kJ of work only instead of 22.6 kJ, and thus saving 21.4 kJ of work from going to waste. This would have been accomplished by a reversible heat pump.

To prove what we have just said, consider a Carnot heat pump that absorbs heat from the surroundings at $T_0 = 293$ K and transfers it to the air in the rigid tank until the air temperature $T$ rises from 293 to 328 K, as shown in Fig. 8–38. The system involves no direct work interactions in this case, and the heat supplied to the system can be expressed in differential form as

$$\delta Q_H = dU = mc_v \, dT$$

The coefficient of performance of a reversible heat pump is given by

$$COP_{HP} = \frac{\delta Q_H}{\delta W_{net,in}} = \frac{1}{1 - T_0/T}$$

Thus,

$$\delta W_{net,in} = \frac{\delta Q_H}{COP_{HP}} = \left(1 - \frac{T_0}{T}\right) mc_v \, dT$$

Integrating, we get

$$W_{net,in} = \int_1^2 \left(1 - \frac{T_0}{T}\right) mc_v \, dT$$

$$= mc_{v,avg}(T_2 - T_1) - T_0 mc_{v,avg} \ln \frac{T_2}{T_1}$$

$$= (22.6 - 21.4) \text{ kJ} = 1.2 \text{ kJ}$$

The first term on the right-hand side of the final expression above is recognized as $\Delta U$ and the second term as the exergy destroyed, whose values were determined earlier. By substituting those values, the total work input to the heat pump is determined to be 1.2 kJ, proving our claim. Notice that the system is still supplied with 22.6 kJ of energy; all we did in the latter case is replace the 21.4 kJ of valuable work by an equal amount of "useless" energy captured from the surroundings.

*Discussion* It is also worth mentioning that the exergy of the system as a result of 22.6 kJ of paddle-wheel work done on it has increased by 1.2 kJ only, that is, by the amount of the reversible work. In other words, if the system were returned to its initial state, it would produce, at most, 1.2 kJ of work.

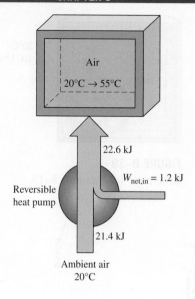

**FIGURE 8–38**

The same effect on the system can be accomplished by a reversible heat pump that consumes only 1.2 kJ of work.

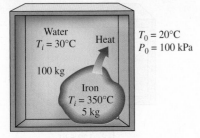

**FIGURE 8–39**
Schematic for Example 8–13.

*EXAMPLE 8–13*    **Dropping a Hot Iron Block into Water**

A 5-kg iron block initially at 350°C is quenched in an insulated tank that contains 100 kg of water at 30°C (Fig. 8–39). Assuming the water that vaporizes during the process condenses back in the tank and the surroundings are at 20°C and 100 kPa, determine (*a*) the final equilibrium temperature, (*b*) the exergy of the combined system at the initial and the final states, and (*c*) the wasted work potential during this process.

**SOLUTION**    A hot iron block is quenched in an insulated tank by water. The final equilibrium temperature, the initial and final exergies, and the wasted work potential are to be determined.

*Assumptions*    **1** Both water and the iron block are incompressible substances. **2** Constant specific heats at room temperature can be used for both the water and the iron. **3** The system is stationary and thus the kinetic and potential energy changes are zero, $\Delta KE = \Delta PE = 0$. **4** There are no electrical, shaft, or other forms of work involved. **5** The system is well-insulated and thus there is no heat transfer.

*Analysis*    We take the entire contents of the tank, *water + iron block,* as the *system.* This is a *closed system* since no mass crosses the system boundary during the process. We note that the volume of a rigid tank is constant, and thus there is no boundary work.

(*a*) Noting that no energy enters or leaves the system during the process, the application of the energy balance gives

$$\underbrace{E_{\text{in}} - E_{\text{out}}}_{\substack{\text{Net energy transfer} \\ \text{by heat, work, and mass}}} = \underbrace{\Delta E_{\text{system}}}_{\substack{\text{Change in internal, kinetic,} \\ \text{potential, etc., energies}}}$$

$$0 = \Delta U$$

$$0 = (\Delta U)_{\text{iron}} + (\Delta U)_{\text{water}}$$

$$0 = [mc(T_f - T_i)]_{\text{iron}} + [mc(T_f - T_i)]_{\text{water}}$$

By using the specific-heat values for water and iron at room temperature (from Table A–3), the final equilibrium temperature $T_f$ becomes

$$0 = (5 \text{ kg})(0.45 \text{ kJ/kg·°C})(T_f - 350°C)$$
$$+ (100 \text{ kg})(4.18 \text{ kJ/kg·°C})(T_f - 30°C)$$

which yields

$$T_f = \textbf{31.7°C}$$

(*b*) Exergy *X* is an extensive property, and the exergy of a composite system at a specified state is the sum of the exergies of the components of that system at that state. It is determined from Eq. 8–15, which for an incompressible substance reduces to

$$X = (U - U_0) - T_0(S - S_0) + P_0(V \overset{\nearrow^0}{-} V_0)$$

$$= mc(T - T_0) - T_0 mc \ln \frac{T}{T_0} + 0$$

$$= mc\left(T - T_0 - T_0 \ln \frac{T}{T_0}\right)$$

where $T$ is the temperature at the specified state and $T_0$ is the temperature of the surroundings. At the initial state,

$$X_{1,\text{iron}} = (5 \text{ kg})(0.45 \text{ kJ/kg·K})\left[(623 - 293) \text{ K} - (293 \text{ K}) \ln \frac{623 \text{ K}}{293 \text{ K}}\right]$$

$$= 245.2 \text{ kJ}$$

$$X_{1,\text{water}} = (100 \text{ kg})(4.18 \text{ kJ/kg·K})\left[(303 - 293) \text{ K} - (293 \text{ K}) \ln \frac{303 \text{ K}}{293 \text{ K}}\right]$$

$$= 69.8 \text{ kJ}$$

$$X_{1,\text{total}} = X_{1,\text{iron}} + X_{1,\text{water}} = (245.2 + 69.8)kJ = \mathbf{315 \text{ kJ}}$$

Similarly, the total exergy at the final state is

$$X_{2,\text{iron}} = 0.5 \text{ kJ}$$

$$X_{2,\text{water}} = 95.1 \text{ kJ}$$

$$X_{2,\text{total}} = X_{2,\text{iron}} + X_{2,\text{water}} = 0.5 + 95.1 = \mathbf{95.6 \text{ kJ}}$$

That is, the exergy of the combined system (water + iron) decreased from 315 to 95.6 kJ as a result of this irreversible heat transfer process.

(c) The wasted work potential is equivalent to the exergy destroyed, which can be determined from $X_{\text{destroyed}} = T_0 S_{\text{gen}}$ or by performing an exergy balance on the system. The second approach is more convenient in this case since the initial and final exergies of the system are already evaluated.

$$\underbrace{X_{\text{in}} - X_{\text{out}}}_{\substack{\text{Net exergy transfer} \\ \text{by heat, work, and mass}}} - \underbrace{X_{\text{destroyed}}}_{\substack{\text{Exergy} \\ \text{destruction}}} = \underbrace{\Delta X_{\text{system}}}_{\substack{\text{Change} \\ \text{in exergy}}}$$

$$0 - X_{\text{destroyed}} = X_2 - X_1$$

$$X_{\text{destroyed}} = X_1 - X_2 = 315 - 95.6 = \mathbf{219.4 \text{ kJ}}$$

**Discussion** Note that 219.4 kJ of work could have been produced as the iron was cooled from 350 to 31.7°C and water was heated from 30 to 31.7°C, but was not.

### EXAMPLE 8–14    Work Potential of Heat Transfer Between Two Tanks

Two constant-volume tanks, each filled with 30 kg of air, have temperatures of 900 K and 300 K (Fig. 8–40). A heat engine placed between the two tanks extracts heat from the high temperature tank, produces work, and rejects heat to the low temperature tank. Determine the maximum work that can be produced by the heat engine and the final temperatures of the tanks. Assume constant specific heats at room temperature.

**SOLUTION** A heat engine operates between two tanks filled with air at different temperatures. The maximum work that can be produced and the final temperature of the tanks are to be determined.
**Assumptions** Air is an ideal gas with constant specific heats at room temperature.

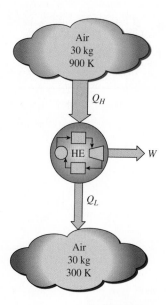

**FIGURE 8–40**
Schematic for Example 8-14

**Properties**  The gas constant of air is 0.287 kPa·m³/kg·K (Table A-1). The constant volume specific heat of air at room temperature is $c_v = 0.718$ kJ/kg·K (Table A-2a)

**Analysis**  For maximum work production, the process must be reversible and thus the entropy generation must be zero. We take the two tanks (the heat source and heat sink) and the heat engine as the system. Noting that the system involves no heat and mass transfer and that the entropy change for cyclic devices is zero, the entropy balance can be expressed as

$$\underbrace{S_{in} - S_{out}}_{\substack{\text{Net entropy transfer} \\ \text{by heat and mass}}} + \underbrace{S_{gen}^{\nearrow 0}}_{\substack{\text{Entropy} \\ \text{generation}}} = \underbrace{\Delta S_{system}}_{\substack{\text{Change} \\ \text{in entropy}}}$$

$$0 + S_{gen}^{\nearrow 0} = \Delta S_{tank,source} + \Delta S_{tank,sink} + \Delta S_{heat\,engine}^{\nearrow 0}$$

$$\Delta S_{tank,source} + \Delta S_{tank,sink} = 0$$

$$\left(mc_v \ln\frac{T_2}{T_1} + mR \ln\frac{V_2}{V_1}^{\nearrow 0}\right)_{source} + \left(mc_v \ln\frac{T_2}{T_1} + mR \ln\frac{V_2}{V_1}^{\nearrow 0}\right)_{sink} = 0$$

$$\ln\frac{T_2}{T_{1,A}}\frac{T_2}{T_{1,B}} = 0 \rightarrow T_2^2 = T_{1,A}T_{1,B}$$

where $T_{1,A}$ and $T_{1,B}$ are the initial temperatures of the source and the sink, respectively, and $T_2$ is the common final temperature. Therefore, the final temperature of the tanks for maximum power production is

$$T_2 = \sqrt{T_{1,A}T_{1,B}} = \sqrt{(900\,\text{K})(300\,\text{K})} = \mathbf{519.6\,K}$$

The energy balance $E_{in} - E_{out} = \Delta E_{system}$ for the source and sink can be expressed as follows:

*Source:*

$$-Q_{source,out} = \Delta U = mc_v(T_2 - T_{1A}) \rightarrow Q_{source,out} = mc_v(T_{1,A} - T_2)$$

$$Q_{source,out} = mc_v(T_{1,A} - T_2) = (30\text{ kg})(0.718\text{ kJ/kg·K})(900 - 519.6)\text{K} = 8193\text{ kJ}$$

*Sink:*

$$Q_{sink,in} = mc_v(T_2 - T_{1,B}) = (30\text{ kg})(0.718\text{ kJ/kg·K})(519.6 - 300)\text{ K} = 4731\text{ kJ}$$

Then the work produced in this case becomes

$$W_{max,out} = Q_H - Q_L = Q_{source,out} - Q_{sink,in} = 8193 - 4731 = \mathbf{3463\,kJ}$$

**Discussion**  Note that 3463 kJ of the 8193 kJ heat transferred from the source can be converted to work, and is the best that can be done. This corresponds to a first law efficiency of 3463/8193 = 0.423 or 42.3 percent, but to a second-law efficiency of 100 percent since the process involves no entropy generation and thus no exergy destruction.

# 8–8 · EXERGY BALANCE: CONTROL VOLUMES

The exergy balance relations for control volumes differ from those for closed systems in that they involve one more mechanism of exergy transfer: *mass flow across the boundaries.* As mentioned earlier, mass possesses exergy as well as energy and entropy, and the amounts of these three extensive properties are proportional to the amount of mass (Fig. 8–41). Again taking the positive direction of heat transfer to be to the system and the positive direction of work transfer to be from the system, the general exergy balance relations (Eqs. 8–36 and 8–37) can be expressed for a control volume more explicitly as

$$X_{heat} - X_{work} + X_{mass,in} - X_{mass,out} - X_{destroyed} = (X_2 - X_1)_{CV} \qquad \text{(8–44)}$$

or

$$\sum\left(1 - \frac{T_0}{T_k}\right)Q_k - [W - P_0(V_2 - V_1)] + \sum_{in} m\psi - \sum_{out} m\psi - X_{destroyed} = (X_2 - X_1)_{CV}$$

$$\text{(8–45)}$$

It can also be expressed in the **rate form** as

$$\sum\left(1 - \frac{T_0}{T_k}\right)\dot{Q}_k - \left(\dot{W} - P_0\frac{dV_{CV}}{dt}\right) + \sum_{in} \dot{m}\psi - \sum_{out} \dot{m}\psi - \dot{X}_{destroyed} = \frac{dX_{CV}}{dt}$$

$$\text{(8–46)}$$

The exergy balance relation above can be stated as *the rate of exergy change within the control volume during a process is equal to the rate of net exergy transfer through the control volume boundary by heat, work, and mass flow minus the rate of exergy destruction within the boundaries of the control volume.*

When the initial and final states of the control volume are specified, the exergy change of the control volume is $X_2 - X_1 = m_2\phi_2 - m_1\phi_1$.

## Exergy Balance for Steady-Flow Systems

Most control volumes encountered in practice such as turbines, compressors, nozzles, diffusers, heat exchangers, pipes, and ducts operate steadily, and thus they experience no changes in their mass, energy, entropy, and exergy contents as well as their volumes. Therefore, $dV_{CV}/dt = 0$ and $dX_{CV}/dt = 0$ for such systems, and the amount of exergy entering a steady-flow system in all forms (heat, work, mass transfer) must be equal to the amount of exergy leaving plus the exergy destroyed. Then the rate form of the general exergy balance (Eq. 8–46) reduces for a **steady-flow process** to (Fig. 8–42)

$$\textit{Steady-flow:} \qquad \sum\left(1 - \frac{T_0}{T_k}\right)\dot{Q}_k - \dot{W} + \sum_{in} \dot{m}\psi - \sum_{out} \dot{m}\psi - \dot{X}_{destroyed} = 0 \quad \text{(8–47)}$$

For a *single-stream* (one-inlet, one-exit) steady-flow device, the relation above further reduces to

$$\textit{Single-stream:} \quad \sum\left(1 - \frac{T_0}{T_k}\right)\dot{Q}_k - \dot{W} + \dot{m}(\psi_1 - \psi_2) - \dot{X}_{destroyed} = 0 \qquad \text{(8–48)}$$

where the subscripts 1 and 2 represent inlet and exit states, $\dot{m}$ is the mass flow rate, and the change in the flow exergy is given by Eq. 8–23 as

$$\psi_1 - \psi_2 = (h_1 - h_2) - T_0(s_1 - s_2) + \frac{V_1^2 - V_2^2}{2} + g(z_1 - z_2)$$

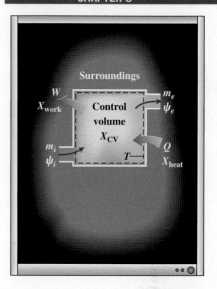

**FIGURE 8–41**
Exergy is transferred into or out of a control volume by mass as well as heat and work transfer.

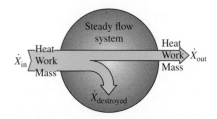

**FIGURE 8–42**
The exergy transfer to a steady-flow system is equal to the exergy transfer from it plus the exergy destruction within the system.

Dividing Eq. 8–48 by $\dot{m}$ gives the exergy balance on a *unit-mass basis* as

$$\sum \left(1 - \frac{T_0}{T_k}\right)q_k - w + (\psi_1 - \psi_2) - x_{\text{destroyed}} = 0 \quad \text{(kJ/kg)} \qquad \textbf{(8–49)}$$

where $q = \dot{Q}/\dot{m}$ and $w = \dot{W}/\dot{m}$ are the heat transfer and work done per unit mass of the working fluid, respectively.

For the case of an adiabatic single-stream device with no work interactions, the exergy balance relation further simplifies to $\dot{X}_{\text{destroyed}} = \dot{m}(\psi_1 - \psi_2)$, which indicates that the specific exergy of the fluid must decrease as it flows through a work-free adiabatic device or remain the same ($\psi_2 = \psi_1$) in the limiting case of a reversible process regardless of the changes in other properties of the fluid.

## Reversible Work

The exergy balance relations presented above can be used to determine the reversible work $W_{\text{rev}}$ by setting the exergy destroyed equal to zero. The work $W$ in that case becomes the reversible work. That is,

*General:* $\qquad\qquad W = W_{\text{rev}} \qquad$ when $X_{\text{destroyed}} = 0 \qquad\qquad\qquad$ **(8–50)**

For example, the reversible power for a single-stream steady-flow device is, from Eq. 8–48,

*Single-stream:* $\quad \dot{W}_{\text{rev}} = \dot{m}(\psi_1 - \psi_2) + \sum\left(1 - \frac{T_0}{T_k}\right)\dot{Q}_k \quad \text{(kW)} \qquad$ **(8–51)**

which reduces for an adiabatic device to

*Adiabatic, single-stream:* $\qquad\qquad \dot{W}_{\text{rev}} = \dot{m}(\psi_1 - \psi_2) \qquad\qquad\qquad$ **(8–52)**

Note that the exergy destroyed is zero only for a reversible process, and reversible work represents the maximum work output for work-producing devices such as turbines and the minimum work input for work-consuming devices such as compressors.

## Second-Law Efficiency of Steady-Flow Devices

The *second-law efficiency* of various steady-flow devices can be determined from its general definition, $\eta_{\text{II}} = $ (Exergy recovered)/(Exergy expended). When the changes in kinetic and potential energies are negligible, the second-law efficiency of an *adiabatic turbine* can be determined from

$$\eta_{\text{II,turb}} = \frac{w_{\text{out}}}{\psi_1 - \psi_2} = \frac{h_1 - h_2}{\psi_1 - \psi_2} = \frac{w_{\text{out}}}{w_{\text{rev,out}}} \quad \text{or} \quad \eta_{\text{II,turb}} = 1 - \frac{T_0 s_{\text{gen}}}{\psi_1 - \psi_2} \qquad \textbf{(8–53)}$$

where $s_{\text{gen}} = s_2 - s_1$. For an *adiabatic compressor* with negligible kinetic and potential energies, the second-law efficiency becomes

$$\eta_{\text{II,comp}} = \frac{\psi_2 - \psi_1}{w_{\text{in}}} = \frac{\psi_2 - \psi_1}{h_2 - h_1} = \frac{w_{\text{in,rev}}}{w_{\text{in}}} \quad \text{or} \quad \eta_{\text{II,comp}} = 1 - \frac{T_0 s_{\text{gen}}}{h_2 - h_1} \qquad \textbf{(8–54)}$$

where again $s_{\text{gen}} = s_2 - s_1$. Note that in the case of turbine, the exergy resource utilized is steam, and the expended exergy is simply the decrease in the exergy of the steam. The recovered exergy is the turbine shaft work. In the case of compressor, the exergy resource is mechanical work, and the expended exergy is the work consumed by the compressor. The recovered exergy in this case is the increase in the exergy of the compressed fluid.

For an adiabatic *heat exchanger* with two unmixed fluid streams (Fig. 8–43), the exergy expended is the decrease in the exergy of the hot stream, and the exergy recovered is the increase in the exergy of the cold stream, provided that the cold stream is not at a lower temperature than the surroundings. Then the second-law efficiency of the heat exchanger becomes

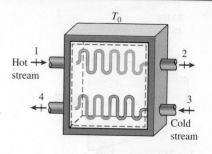

**FIGURE 8–43**
A heat exchanger with two unmixed fluid streams.

$$\eta_{\text{II,HX}} = \frac{\dot{m}_{\text{cold}}(\psi_4 - \psi_3)}{\dot{m}_{\text{hot}}(\psi_1 - \psi_2)} \quad \text{or} \quad \eta_{\text{II,HX}} = 1 - \frac{T_0 \dot{S}_{\text{gen}}}{\dot{m}_{\text{hot}}(\psi_1 - \psi_2)} \qquad (8\text{–}55)$$

where $\dot{S}_{\text{gen}} = \dot{m}_{\text{hot}}(s_2 - s_1) + \dot{m}_{\text{cold}}(s_4 - s_3)$. Perhaps you are wondering what happens if the heat exchanger is not adiabatic; that is, it is losing some heat to its surroundings at $T_0$. If the temperature of the boundary (the outer surface of the heat exchanger) $T_b$ is equal $T_0$, the definition above still holds (except the entropy generation term needs to be modified if the second definition is used). However, if $T_b > T_0$, then the exergy of the lost heat at the boundary should be included in the recovered exergy:

$$\eta_{\text{II,HX}} = \frac{\dot{m}_{\text{cold}}(\psi_4 - \psi_3) + \dot{Q}_{\text{loss}}(1 - T_0/T_b)}{\dot{m}_{\text{hot}}(\psi_1 - \psi_2)} = 1 - \frac{T_0 \dot{S}_{\text{gen}}}{\dot{m}_{\text{hot}}(\psi_1 - \psi_2)} \qquad (8\text{–}56)$$

where $T_b$ is the temperature of the system boundary through which the lost heat crosses at a rate of $\dot{Q}_{\text{loss}}$. Also, $\dot{S}_{\text{gen}} = \dot{m}_{\text{hot}}(s_2 - s_1) + \dot{m}_{\text{cold}}(s_4 - s_3) + \dot{Q}_{\text{loss}}/T_b$ in this case.

Although no attempt is made in practice to utilize this exergy associated with lost heat and it is allowed to be destroyed, the heat exchanger should not be held responsible for this destruction, which occurs outside its boundaries. If we are interested in the exergy destroyed during the process, not just within the boundaries of the device, then it makes sense to consider an *extended system* that includes the immediate surroundings of the device such that the boundaries of the new enlarged system are at $T_0$. The second-law efficiency of the extended system reflects the effects of the irreversibilities that occur within and just outside the device.

An interesting situation arises when the temperature of the cold stream remains below the temperature of the surroundings at all times. In that case the exergy of the cold stream actually decreases instead of increasing. In such cases it is better to define the second-law efficiency as the ratio of the sum of the exergies of the outgoing streams to the sum of the exergies of the incoming streams.

For an adiabatic *mixing chamber* where a hot steam 1 is mixed with a cold stream 2, forming a mixture 3, the exergy resource is the hot fluid. Then the exergy expended is the exergy decrease of the hot fluid, and the exergy recovered is the exegy increase of the cold fluid. Noting that state 3 is the common state of the mixture, the second-law efficiency can be expressed as

$$\eta_{\text{II,mix}} = \frac{\dot{m}_{\text{cold}}(\psi_3 - \psi_2)}{\dot{m}_{\text{hot}}(\psi_1 - \psi_3)} \quad \text{or} \quad \eta_{\text{II,mix}} = 1 - \frac{T_0 \dot{S}_{\text{gen}}}{\dot{m}_{\text{hot}}(\psi_1 - \psi_3)} \qquad (8\text{–}57)$$

where $\dot{S}_{\text{gen}} = (\dot{m}_{\text{hot}} + \dot{m}_{\text{cold}})s_3 - \dot{m}_{\text{hot}}s_1 - \dot{m}_{\text{cold}}s_2$.

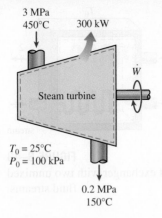

3 MPa
450°C

300 kW

$\dot{W}$

Steam turbine

$T_0 = 25°C$
$P_0 = 100$ kPa

0.2 MPa
150°C

**FIGURE 8–44**
Schematic for Example 8–15.

### EXAMPLE 8–15    Second-Law Analysis of a Steam Turbine

Steam enters a turbine steadily at 3 MPa and 450°C at a rate of 8 kg/s and exits at 0.2 MPa and 150°C, (Fig. 8–44). The steam is losing heat to the surrounding air at 100 kPa and 25°C at a rate of 300 kW, and the kinetic and potential energy changes are negligible. Determine (a) the actual power output, (b) the maximum possible power output, (c) the second-law efficiency, (d) the exergy destroyed, and (e) the exergy of the steam at the inlet conditions.

**SOLUTION**  A steam turbine operating steadily between specified inlet and exit states is considered. The actual and maximum power outputs, the second-law efficiency, the exergy destroyed, and the inlet exergy are to be determined.
**Assumptions**  **1** This is a steady-flow process since there is no change with time at any point and thus $\Delta m_{CV} = 0$, $\Delta E_{CV} = 0$, and $\Delta X_{CV} = 0$. **2** The kinetic and potential energies are negligible.
**Analysis**  We take the *turbine* as the system. This is a *control volume* since mass crosses the system boundary during the process. We note that there is only one inlet and one exit and thus $\dot{m}_1 = \dot{m}_2 = m$. Also, heat is lost to the surrounding air and work is done by the system.
   The properties of the steam at the inlet and exit states and the state of the environment are

$$\text{Inlet state:} \quad \left.\begin{array}{l} P_1 = 3\,\text{MPa} \\ T_1 = 450°C \end{array}\right\} \quad \begin{array}{l} h_1 = 3344.9 \text{ kJ/kg} \\ s_1 = 7.0856 \text{ kJ/kg·K} \end{array} \quad \text{(Table A–6)}$$

$$\text{Exit state:} \quad \left.\begin{array}{l} P_2 = 0.2 \text{ MPa} \\ T_2 = 150°C \end{array}\right\} \quad \begin{array}{l} h_2 = 2769.1 \text{ kJ/kg} \\ s_2 = 7.2810 \text{ kJ/kg·K} \end{array} \quad \text{(Table A–6)}$$

$$\text{Dead state:} \quad \left.\begin{array}{l} P_0 = 100 \text{ kPa} \\ T_0 = 25°C \end{array}\right\} \quad \begin{array}{l} h_0 \cong h_{f\,@\,25°C} = 104.83 \text{ kJ/kg} \\ s_0 \cong s_{f\,@\,25°C} = 0.3672 \text{ kJ/kg·K} \end{array} \quad \text{(Table A–4)}$$

(a) The actual power output of the turbine is determined from the rate form of the energy balance,

$$\underbrace{\dot{E}_{in} - \dot{E}_{out}}_{\substack{\text{Rate of net energy transfer} \\ \text{by heat, work, and mass}}} = \underbrace{dE_{system}/dt}_{\substack{\text{Rate of change in internal, kinetic,} \\ \text{potential, etc., energies}}}^{\nearrow 0 \text{(steady)}} = 0$$

$$\dot{E}_{in} = \dot{E}_{out}$$

$$\dot{m}h_1 = \dot{W}_{out} + \dot{Q}_{out} + \dot{m}h_2 \quad (\text{since ke} \cong \text{pe} \cong 0)$$

$$\dot{W}_{out} = \dot{m}(h_1 - h_2) - \dot{Q}_{out}$$

$$= (8 \text{ kg/s})[(3344.9 - 2769.1) \text{ kJ/kg}] - 300 \text{ kW}$$

$$= \textbf{4306 kW}$$

(b) The maximum power output (reversible power) is determined from the rate form of the exergy balance applied on the *extended system* (system + immediate surroundings), whose boundary is at the environment temperature of $T_0$, and by setting the exergy destruction term equal to zero,

$$\underbrace{\dot{X}_{in} - \dot{X}_{out}}_{\substack{\text{Rate of net exergy transfer} \\ \text{by heat, work, and mass}}} - \underbrace{\dot{X}_{destroyed}}_{\substack{\text{Rate of exergy} \\ \text{destruction}}}^{\nearrow 0 \text{(reversible)}} = \underbrace{dX_{system}/dt}_{\substack{\text{Rate of change} \\ \text{in exergy}}}^{\nearrow 0 \text{(steady)}} = 0$$

$$\dot{X}_{in} = \dot{X}_{out}$$

$$\dot{m}\psi_1 = \dot{W}_{rev,out} + \overset{0}{\cancel{\dot{X}_{heat}}} + \dot{m}\psi_2$$

$$\dot{W}_{rev,out} = \dot{m}(\psi_1 - \psi_2)$$

$$= \dot{m}[(h_1 - h_2) - T_0(s_1 - s_2) - \overset{0}{\cancel{\Delta ke}} - \overset{0}{\cancel{\Delta pe}}]$$

Note that exergy transfer with heat is zero when the temperature at the point of transfer is the environment temperature $T_0$. Substituting,

$$\dot{W}_{rev,out} = (8 \text{ kg/s})[(3344.9 - 2769.1) \text{ kJ/kg}$$
$$- (298 \text{ K})(7.0856 - 7.2810)\text{kJ/kg·K}]$$
$$= \textbf{5072 kW}$$

(c) The second-law efficiency of a turbine is the ratio of the actual work delivered to the reversible work,

$$\eta_{II} = \frac{\dot{W}_{out}}{\dot{W}_{rev,out}} = \frac{4306 \text{ kW}}{5072 \text{ kW}} = \textbf{0.849 or 84.9\%}$$

That is, 15.1 percent of the work potential is wasted during this process.

(d) The difference between the reversible work and the actual useful work is the exergy destroyed, which is determined to be

$$\dot{X}_{destroyed} = \dot{W}_{rev,out} - \dot{W}_{out} = 5072 - 4306 = \textbf{776 kW}$$

That is, the potential to produce useful work is wasted at a rate of 776 kW during this process. The exergy destroyed could also be determined by first calculating the rate of entropy generation $\dot{S}_{gen}$ during the process.

(e) The exergy (maximum work potential) of the steam at the inlet conditions is simply the stream exergy, and is determined from

$$\psi_1 = (h_1 - h_0) - T_0(s_1 - s_0) + \frac{\overset{0}{\cancel{V_1^2}}}{2} + \overset{0}{\cancel{gz_1}}$$

$$= (h_1 - h_0) - T_0(s_1 - s_0)$$

$$= (3344.9 - 104.83)\text{kJ/kg} - (298 \text{ K})(7.0856 - 0.3672) \text{ kJ/kg·K}$$

$$= \textbf{1238 kJ/kg}$$

That is, not counting the kinetic and potential energies, every kilogram of the steam entering the turbine has a work potential of 1238 kJ. This corresponds to a power potential of (8 kg/s)(1238 kJ/kg) = 9904 kW. Obviously, the turbine is converting 4306/9904 = 43.5 percent of the available work potential of the steam to work.

### EXAMPLE 8–16    Exergy Destroyed During Mixing of Fluid Streams

Water at 200 kPa and 10°C enters a mixing chamber at a rate of 150 kg/min, where it is mixed steadily with steam entering at 200 kPa and 150°C. The mixture leaves the chamber at 200 kPa and 70°C, and heat is being lost to the surrounding air at $T_0 = 20°C$ at a rate of 190 kJ/min (Fig. 8–45).

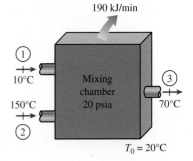

**FIGURE 8–45**
Schematic for Example 8–16.

Neglecting the changes in kinetic and potential energies, determine the reversible power and the rate of exergy destruction for this process.

**SOLUTION**  Liquid water and steam are mixed in a chamber that is losing heat at a specified rate. The reversible power and the rate of exergy destruction are to be determined.

*Assumptions*  **1** This is a steady-flow process since there is no change with time at any point and thus $\Delta m_{CV} = 0$, $\Delta E_{CV} = 0$, and $\Delta S_{CV} = 0$. **2** There are no work interactions involved. **3** The kinetic and potential energies are negligible, $ke \cong pe \cong 0$.

*Analysis*  We take the *mixing chamber* as the system (Fig. 8–45). This is a *control volume* since mass crosses the system boundary during the process. We note that there are two inlets and one exit.

Under the stated assumptions and observations, the mass and energy balances for this steady-flow system can be expressed in the rate form as follows:

*Mass balance:*  $\quad \dot{m}_{in} - \dot{m}_{out} = dm_{system}/dt \nearrow^{0(steady)} = 0 \rightarrow \dot{m}_1 + \dot{m}_2 = \dot{m}_3$

*Energy balance:*  $\quad \underbrace{\dot{E}_{in} - \dot{E}_{out}}_{\substack{\text{Rate of net energy transfer} \\ \text{by heat, work, and mass}}} = \underbrace{dE_{system}/dt \nearrow^{0(steady)}}_{\substack{\text{Rate of change in internal, kinetic,} \\ \text{potential, etc., energies}}} = 0$

$$\dot{E}_{in} = \dot{E}_{out}$$

$$\dot{m}_1 h_1 + \dot{m}_2 h_2 = \dot{m}_3 h_3 + \dot{Q}_{out} \text{ (since } \dot{W} = 0, ke \cong pe \cong 0)$$

Combining the mass and energy balances gives

$$\dot{Q}_{out} = \dot{m}_1 h_1 + \dot{m}_2 h_2 - (\dot{m}_1 + \dot{m}_2)h_3$$

The desired properties at the specified states are determined from the steam tables to be

*State 1:*  $\quad \left. \begin{array}{l} P_1 = 200 \text{ kPa} \\ T_1 = 10°C \end{array} \right\} \quad \begin{array}{l} h_1 = h_{f @ 10°C} = 42.022 \text{ kJ/kg} \\ s_1 = s_{f @ 10°C} = 0.1511 \text{ kJ/kg·K} \end{array}$

*State 2:*  $\quad \left. \begin{array}{l} P_2 = 200 \text{ kPa} \\ T_2 = 150°C \end{array} \right\} \quad \begin{array}{l} h_2 = 2769.1 \text{ kJ/kg} \\ s_2 = 7.2810 \text{ kJ/kg·K} \end{array}$

*State 3:*  $\quad \left. \begin{array}{l} P_3 = 200 \text{ kPa} \\ T_3 = 70°C \end{array} \right\} \quad \begin{array}{l} h_3 = h_{f@70°C} = 293.07 \text{ kJ/kg} \\ s_3 = s_{f@70°C} = 0.9551 \text{ kJ/kg·K} \end{array}$

Substituting,

190 kJ/min = [150 × 42.022 + $\dot{m}_2$ × 2769.1 − (150 + $\dot{m}_2$) × 293.07] kJ/min which gives

$$\dot{m}_2 = 15.29 \text{ kg/min}$$

The maximum power output (reversible power) is determined from the rate form of the exergy balance applied on the *extended system* (system + immediate surroundings), whose boundary is at the environment temperature of $T_0$, and by setting the exergy destruction term equal to zero,

$$\underbrace{\dot{X}_{in} - \dot{X}_{out}}_{\substack{\text{Rate of net exergy transfer} \\ \text{by heat, work, and mass}}} - \underbrace{\dot{X}_{destroyed} \nearrow^{0(reversible)}}_{\substack{\text{Rate of exergy} \\ \text{destruction}}} = \underbrace{dX_{system}/dt \nearrow^{0(steady)}}_{\substack{\text{Rate of change} \\ \text{in exergy}}} = 0$$

$$\dot{X}_{in} = \dot{X}_{out}$$

$$\dot{m}_1 \psi_1 + \dot{m}_2 \psi_2 = \dot{W}_{rev,out} + \dot{X}_{heat}^{\,\,0} + \dot{m}_3 \psi_3$$

$$\dot{W}_{rev,out} = \dot{m}_1 \psi_1 + \dot{m}_2 \psi_2 - \dot{m}_3 \psi_3$$

Note that exergy transfer by heat is zero when the temperature at the point of transfer is the environment temperature $T_0$, and the kinetic and potential energies are negligible. Therefore,

$$
\begin{aligned}
\dot{W}_{rev,out} &= \dot{m}_1(h_1 - T_0 s_1) + \dot{m}_2(h_2 - T_0 s_2) - \dot{m}_3(h_3 - T_0 s_3) \\
&= (150 \text{ kg/min})[42.022 \text{ kJ/kg} - (293 \text{ K})(0.1511 \text{ kJ/kg·K})] \\
&\quad + (15.29 \text{ kg/min})[2769.1 \text{ kJ/kg} - (293 \text{ K})(7.2810 \text{ kJ/kg·K})] \\
&\quad - (165.29 \text{ kg/min})[293.07 \text{ kJ/kg} - (293 \text{ K})(0.9551 \text{ kJ/kg·K})] \\
&= 7197 \text{ kJ/min}
\end{aligned}
$$

That is, we could have produced work at a rate of 7197 kJ/min if we ran a heat engine between the hot and the cold fluid streams instead of allowing them to mix directly.

The exergy destroyed is determined from

$$\dot{X}_{destroyed} = \dot{W}_{rev,out} - \dot{W}_u^{\,\,0} = T_0 \dot{S}_{gen}$$

Thus,

$$\dot{X}_{destroyed} = \dot{W}_{rev,out} = 7197 \text{ kJ/min}$$

since there is no actual work produced during the process.

**Discussion** The entropy generation rate for this process can be shown to be $\dot{S}_{gen}$ = 24.53 kJ/min·K. Thus the exergy destroyed could also be determined from the second part of the above equation:

$$\dot{X}_{destroyed} = T_0 \dot{S}_{gen} = (293 \text{ K})(24.53 \text{ kJ/min·K}) = 7187 \text{ kJ/min}$$

The slight difference between the two results is due to roundoff error.

■ **EXAMPLE 8–17**     **Charging a Compressed Air Storage System**

A 200-m³ rigid tank initially contains atmospheric air at 100 kPa and 300 K and is to be used as a storage vessel for compressed air at 1 MPa and 300 K (Fig. 8–46). Compressed air is to be supplied by a compressor that takes in atmospheric air at $P_0$ = 100 kPa and $T_0$ = 300 K. Determine the minimum work requirement for this process.

**SOLUTION** Air is to be compressed and stored at high pressure in a large tank. The minimum work required is to be determined.
**Assumptions** 1 Air is an ideal gas. 2 The kinetic and potential energies are negligible. 3 The properties of air at the inlet remain constant during the entire charging process.

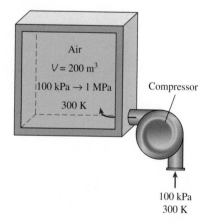

**FIGURE 8–46**
Schematic for Example 8–17.

*Analysis* We take the *rigid tank combined with the compressor* as the system. This is a *control volume* since mass crosses the system boundary during the process. We note that this is an unsteady-flow process since the mass content of the system changes as the tank is charged. Also, there is only one inlet and no exit.

The minimum work required for a process is the *reversible work,* which can be determined from the exergy balance applied on the *extended system* (system + immediate surroundings) whose boundary is at the environment temperature of $T_0$ (so that there is no exergy transfer accompanying heat transfer to or from the environment) and by setting the exergy destruction term equal to zero,

$$\underbrace{X_{\text{in}} - X_{\text{out}}}_{\substack{\text{Net exergy transfer} \\ \text{by heat, work, and mass}}} - \underbrace{X_{\text{destroyed}}^{\nearrow 0\,\text{(reversible)}}}_{\substack{\text{Exergy} \\ \text{destruction}}} = \underbrace{\Delta X_{\text{system}}}_{\substack{\text{Change} \\ \text{in exergy}}}$$

$$X_{\text{in}} - X_{\text{out}} = X_2 - X_1$$

$$W_{\text{rev,in}} + m_1\psi_1^{\nearrow 0} = m_2\phi_2 - m_1\phi_1^{\nearrow 0}$$

$$W_{\text{rev,in}} = m_2\phi_2$$

Note that $\phi_1 = \psi_1 = 0$ since the initial air in the tank and the air entering are at the state of the environment, and the exergy of a substance at the state of the environment is zero. The final mass of air and the exergy of the pressurized air in the tank at the end of the process are

$$m_2 = \frac{P_2 V}{RT_2} = \frac{(1000\ \text{kPa})\,(200\ \text{m}^3)}{(0.287\ \text{kPa·m}^3/\text{kg·K})\,(300\text{K})} = 2323\ \text{kg}$$

$$\phi_2 = (u_2 - u_0)^{\nearrow 0\,\text{(since } T_2 = T_0)} + P_0(v_2 - v_0) - T_0(s_2 - s_0) + \frac{V_2^2}{2}^{\nearrow 0} + gz_2^{\nearrow 0}$$

$$= P_0(v_2 - v_0) - T_0(s_2 - s_0)$$

We note that

$$P_0(v_2 - v_0) = P_0\left(\frac{RT_2}{P_2} - \frac{RT_0}{P_0}\right) = RT_0\left(\frac{P_0}{P_2} - 1\right) \qquad (\text{since } T_2 = T_0)$$

$$T_0(s_2 - s_0) = T_0\left(c_p \ln\frac{T_2}{T_0}^{\nearrow 0} - R\ln\frac{P_2}{P_0}\right) = -RT_0\ln\frac{P_2}{P_0} \qquad (\text{since } T_2 = T_0)$$

Therefore,

$$\phi_2 = RT_0\left(\frac{P_0}{P_2} - 1\right) + RT_0\ln\frac{P_2}{P_0} = RT_0\left(\ln\frac{P_2}{P_0} + \frac{P_0}{P_2} - 1\right)$$

$$= (0.287\ \text{kJ/kg·K})(300\ \text{K})\left(\ln\frac{1000\ \text{kPa}}{100\ \text{kPa}} + \frac{100\ \text{kPa}}{1000\ \text{kPa}} - 1\right)$$

$$= 120.76\ \text{kJ/kg}$$

and

$$W_{\text{rev,in}} = m_2\phi_2 = (2323\ \text{kg})(120.76\ \text{kJ/kg}) \cong \mathbf{281\ MJ}$$

*Discussion* Note that a minimum of 281 MJ of work input is required to fill the tank with compressed air at 300 K and 1 MPa. In reality, the required work input will be greater by an amount equal to the exergy destruction during the process. Compare this to the result of Example 8–7. What can you conclude?

**TOPIC OF SPECIAL INTEREST\***     Second-Law Aspects of Daily Life

Thermodynamics is a fundamental natural science that deals with various aspects of energy, and even nontechnical people have a basic understanding of energy and the first law of thermodynamics since there is hardly any aspect of life that does not involve the transfer or transformation of energy in different forms. All the *dieters,* for example, base their lifestyle on the conservation of energy principle. Although the first-law aspects of thermodynamics are readily understood and easily accepted by most people, there is not a public awareness about the second law of thermodynamics, and the second-law aspects are not fully appreciated even by people with technical backgrounds. This causes some students to view the second law as something that is of theoretical interest rather than an important and practical engineering tool. As a result, students show little interest in a detailed study of the second law of thermodynamics. This is unfortunate because the students end up with a one-sided view of thermodynamics and miss the balanced, complete picture.

Many *ordinary events* that go unnoticed can serve as excellent vehicles to convey important concepts of thermodynamics. Below we attempt to demonstrate the relevance of the second-law concepts such as exergy, reversible work, irreversibility, and the second-law efficiency to various aspects of daily life using examples with which even nontechnical people can identify. Hopefully, this will enhance our understanding and appreciation of the second law of thermodynamics and encourage us to use it more often in technical and even nontechnical areas. The critical reader is reminded that the concepts presented below are *soft* and *difficult to quantize,* and that they are offered here to stimulate interest in the study of the second law of thermodynamics and to enhance our understanding and appreciation of it.

The second-law concepts are implicitly used in various aspects of daily life. Many successful people seem to make extensive use of them without even realizing it. There is growing awareness that quality plays as important a role as quantity in even ordinary daily activities. The following appeared in an article in the *Reno Gazette-Journal* on March 3, 1991:

> Dr. Held considers himself a survivor of the tick-tock conspiracy. About four years ago, right around his 40th birthday, he was putting in 21-hour days—working late, working out, taking care of his three children and getting involved in sports. He got about four or five hours of sleep a night. . . . "Now I'm in bed by 9:30 and I'm up by 6," he says. "I get twice as much done as I used to. I don't have to do things twice or read things three times before I understand them."

*This section can be skipped without a loss in continuity.

This statement has a strong relevance to the second-law discussions. It indicates that the problem is not how much time we have (the first law), but, rather, how effectively we use it (the second law). For a person to get more *done in less time* is no different than for a car to go *more miles on less fuel*.

In thermodynamics, *reversible work* for a process is defined as the maximum useful work output (or minimum work input) for that process. It is the useful work that a system would deliver (or consume) during a process between two specified states if that process is executed in a reversible (perfect) manner. The difference between the reversible work and the actual useful work is due to imperfections and is called *irreversibility* (the wasted work potential). For the special case of the final state being the dead state or the state of the surroundings, the reversible work becomes a maximum and is called the *exergy* of the system at the initial state. The irreversibility for a reversible or perfect process is zero.

The *exergy* of a person in daily life can be viewed as the best job that person can do under the most favorable conditions. The *reversible work* in daily life, on the other hand, can be viewed as the best job a person can do under some specified conditions. Then the difference between the reversible work and the actual work done under those conditions can be viewed as the *irreversibility* or the *exergy destroyed*. In engineering systems, we try to identify the major sources of irreversibilities and minimize them in order to maximize performance. In daily life, a person should do just that to maximize his or her performance.

The exergy of a person at a given time and place can be viewed as the maximum amount of work he or she can do at that time and place. Exergy is certainly difficult to quantify because of the interdependence of physical and intellectual capabilities of a person. The ability to perform physical and intellectual tasks simultaneously complicates things even further. *Schooling* and *training* obviously increase the exergy of a person. *Aging* decreases the physical exergy. Unlike most mechanical things, the exergy of human beings is a function of time, and the physical and/or intellectual exergy of a person goes to waste if it is not utilized at the time. A barrel of oil loses nothing from its exergy if left unattended for 40 years. However, a person will lose much of his or her entire exergy during that time period if he or she just sits back.

A hard-working farmer, for example, may make full use of his *physical exergy* but very little use of his *intellectual exergy*. That farmer, for example, could learn a foreign language or a science by listening to some educational CDs at the same time he is doing his physical work. This is also true for people who spend considerable time in the car commuting to work. It is hoped that some day we will be able to do exergy analysis for people and their activities. Such an analysis will point out the way for people to minimize their exergy destruction, and get more done in less time. Computers can perform several tasks at once. Why shouldn't human beings be able to do the same?

*Children* are born with different levels of *exergies* (talents) in different areas. Giving aptitude tests to children at an early age is simply an attempt to uncover the extent of their "hidden" exergies, or talents. The children are then directed to areas in which they have the greatest exergy. As adults, they are more likely to perform at high levels without stretching the limits if they are naturally fit to be in that area.

We can view the level of *alertness* of a person as his or her *exergy* for intellectual affairs. When a person is well-rested, the degree of alertness, and thus intellectual exergy, is at a maximum and this exergy decreases with time as the person gets tired, as illustrated in Fig. 8–47. Different tasks in daily life require different levels of intellectual exergy, and the difference between available and required alertness can be viewed as the *wasted alertness* or *exergy destruction*. To minimize exergy destruction, there should be a close match between available alertness and required alertness.

Consider a well-rested student who is planning to spend her next 4 h studying and watching a 2-h-long movie. From the *first-law* point of view, it makes no difference in what order these tasks are performed. But from the *second-law* point of view, it makes a lot of difference. Of these two tasks, studying requires more intellectual alertness than watching a movie does, and thus it makes thermodynamic sense to study first when the alertness is high and to watch the movie later when the alertness is lower, as shown in the figure. A student who does it backwards wastes a lot of alertness while watching the movie, as illustrated in Fig. 8–47, and she has to keep going back and forth while studying because of insufficient alertness, thus getting less done in the same time period.

In thermodynamics, *the first-law efficiency* (or thermal efficiency) of a heat engine is defined as the ratio of net work output to total heat input. That is, it is the fraction of the heat supplied that is converted to net work. In general, the first-law efficiency can be viewed as the ratio of the desired output to the required input. The first-law efficiency makes no reference to the *best possible performance,* and thus the first-law efficiency alone is not a realistic measure of performance. To overcome this deficiency, we defined the second-law efficiency, which is a measure of actual performance relative to the best possible performance under the same conditions. For heat engines, the second-law efficiency is defined as the ratio of the actual thermal efficiency to the maximum possible (reversible) thermal efficiency under the same conditions.

In daily life, the *first-law efficiency* or *performance* of a person can be viewed as the accomplishment of that person relative to the effort he or she puts in. The *second-law efficiency* of a person, on the other hand, can be viewed as the performance of that person relative to the best possible performance under the circumstances.

*Happiness* is closely related to the *second-law efficiency*. Small children are probably the happiest human beings because there is so little they can do, but they do it so well, considering their limited capabilities. That is, children have very high second-law efficiencies in their daily lives. The term "full life" also refers to second-law efficiency. A person is considered to have a full life, and thus a very high second-law efficiency, if he or she has utilized all of his or her abilities to the limit during a lifetime.

Even a person with some disabilities has to put in considerably more effort to accomplish what a physically fit person accomplishes. Yet, despite

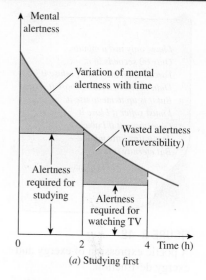

(a) Studying first

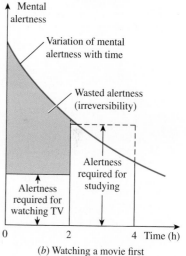

(b) Watching a movie first

**FIGURE 8–47**

The irreversibility associated with a student studying and watching a movie on television, each for two hours.

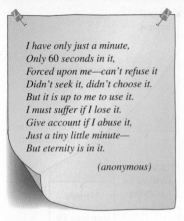

*I have only just a minute,
Only 60 seconds in it,
Forced upon me—can't refuse it
Didn't seek it, didn't choose it.
But it is up to me to use it.
I must suffer if I lose it.
Give account if I abuse it,
Just a tiny little minute—
But eternity is in it.*

*(anonymous)*

**FIGURE 8–48**
A poetic expression of exergy and exergy destruction.

accomplishing less with more effort, the person with disabilities who gives an impressive performance often gets more praise. Thus we can say that this person with disabilities had a low first-law efficiency (accomplishing little with a lot of effort) but a very high second-law efficiency (accomplishing as much as possible under the circumstances).

In daily life, exergy can also be viewed as the *opportunities that we have* and the exergy destruction as the *opportunities wasted.* Time is the biggest asset that we have, and the time wasted is the wasted opportunity to do something useful (Fig. 8–48).

The second law of thermodynamics also has interesting philosophical ramifications. Mass and energy are conserved quantities and are associated with the first law of thermodynamics, while entropy and exergy are non-conserved quantities and are associated with the second law. The universe we perceive through our five senses consists of conserved quantities, and thus we tend to view the non-conserved quantities as being non-real and even out of this universe. The widely accepted big bang theory about the origin of the universe gave rise to the notion that this is an all-material universe, and everything is made of matter (more correctly, mass-energy) only. As conserved quantities, mass and energy fit into the description of truly physical quantities, but entropy and exergy do not since entropy can be created and exergy can be destroyed. Thus entropy and exergy are not truly physical quanities although they are closely related to the physical quantities of mass and energy. Therefore, the second law deals with quantities that are of a different kind of existence—a universe in which things come into existence out of nothing and go out of existence into nothing—and opens up a universe that is beyond the conserved all-material universe we know of.

A similar argument can be given for the laws of nature that rule over matter. There is no question that both the first and the second laws of thermodynamics exist, and these and other laws like Newton's laws of motion govern the physical universe behind the scenes. As Alfred Montapert puts it, *"Nature's laws are the invisible government of the earth."* Albert Einstein expresses this phenomenon as *"A spirit is manifest in the laws of the universe."* Yet these laws that constitute the core of sciences cannot be sensed by our five senses and do not have a material existence, and thus they are not subject to the limitations of time and space. As such, the laws that seem to have infused all matter like a spirit rule everywhere, but they are not anywhere. It appears that quantities like entropy and exergy that come into existence out of nothing and go out of existence into nothing together with the laws of nature like the first and the second laws that govern the big-bang universe with an invisible powerful hand, are pointing the way for a broadened definition of existence that is more in line with perceived and observed phenomena.

The arguments presented here are exploratory in nature, and they are hoped to initiate some interesting discussions and research that may lead into better understanding of performance in various aspects of daily life. The second law may eventually be used to determine quantitatively the most effective way to improve the quality of life and performance in daily life, as it is presently used to improve the performance of engineering systems.

## SUMMARY

The energy content of the universe is constant, just as its mass content is. Yet at times of crisis we are bombarded with speeches and articles on how to "conserve" energy. As engineers, we know that energy is already conserved. What is not conserved is *exergy,* which is the useful work potential of the energy. Once the exergy is wasted, it can never be recovered. When we use energy (to heat our homes for example), we are not destroying any energy; we are merely converting it to a less useful form, a form of less exergy.

The useful work potential of a system at the specified state is called *exergy.* Exergy is a property and is associated with the state of the system and the environment. A system that is in equilibrium with its surroundings has zero exergy and is said to be at the *dead state.* The exergy of heat supplied by thermal energy reservoirs is equivalent to the work output of a Carnot heat engine operating between the reservoir and the environment.

*Reversible work* $W_{rev}$ is defined as the maximum amount of useful work that can be produced (or the minimum work that needs to be supplied) as a system undergoes a process between the specified initial and final states. This is the useful work output (or input) obtained when the process between the initial and final states is executed in a totally reversible manner. The difference between the reversible work $W_{rev}$ and the useful work $W_u$ is due to the irreversibilities present during the process and is called the *irreversibility I.* It is equivalent to the *exergy destroyed* and is expressed as

$$I = X_{destroyed} = T_0 S_{gen} = W_{rev,out} - W_{u,out} = W_{u,in} - W_{rev,in}$$

where $S_{gen}$ is the entropy generated during the process. For a totally reversible process, the useful and reversible work terms are identical and thus exergy destruction is zero. Exergy destroyed represents the lost work potential and is also called the *wasted work* or *lost work.*

The *second-law efficiency* is a measure of the performance of a device relative to the performance under reversible conditions for the same end states and is given by

$$\eta_{II} = \frac{\eta_{th}}{\eta_{th,rev}} = \frac{W_u}{W_{rev}}$$

for heat engines and other work-producing devices and

$$\eta_{II} = \frac{COP}{COP_{rev}} = \frac{W_{rev}}{W_u}$$

for refrigerators, heat pumps, and other work-consuming devices. In general, the second-law efficiency is expressed as

$$\eta_{II} = \frac{\text{Exergy recovered}}{\text{Exergy expended}} = 1 - \frac{\text{Exergy destroyed}}{\text{exergy expended}}$$

The exergies of a fixed mass (nonflow exergy) and of a flow stream are expressed as

*Nonflow exergy:*

$$\psi = (u - u_0) + P_0(v - v_0) - T_0(s - s_0) + \frac{V^2}{2} + gz$$
$$= (e - e_0) + P_0(v - v_0) - T_0(s - s_0)$$

*Flow exergy:*
$$\psi = (h - h_0) - T_0(s - s_0) + \frac{V^2}{2} + gz$$

Then the *exergy change* of a fixed mass or fluid stream as it undergoes a process from state 1 to state 2 is given by

$$\Delta X = X_2 - X_1 = m(\phi_2 - \phi_1)$$
$$= (E_2 - E_1) + P_0(V_2 - V_1) - T_0(S_2 - S_1)$$
$$= (U_2 - U_1) + P_0(V_2 - V_1) - T_0(S_2 - S_1)$$
$$+ m\frac{V_2^2 - V_1^2}{2} + mg(z_2 - z_1)$$

$$\Delta\psi = \psi_2 - \psi_1 = (h_2 - h_1) - T_0(s_2 - s_1)$$
$$+ \frac{V_2^2 - V_1^2}{2} + g(z_2 - z_1)$$

Exergy can be transferred by heat, work, and mass flow, and exergy transfer accompanied by heat, work, and mass transfer are given by

*Exergy transfer by heat:*
$$X_{heat} = \left(1 - \frac{T_0}{T}\right)Q$$

*Exergy transfer by work:*
$$X_{work} = \begin{cases} W - W_{surr} & \text{(for boundary work)} \\ W & \text{(for other forms of work)} \end{cases}$$

*Exergy transfer by mass:*
$$X_{mass} = m\psi$$

The exergy of an isolated system during a process always decreases or, in the limiting case of a reversible process, remains constant. This is known as the *decrease of exergy principle* and is expressed as

$$\Delta X_{isolated} = (X_2 - X_1)_{isolated} \leq 0$$

Exergy balance for *any system* undergoing *any process* can be expressed as

*General:*
$$\underbrace{X_{in} - X_{out}}_{\substack{\text{Net exergy transfer} \\ \text{by heat, work, and mass}}} - \underbrace{X_{destroyed}}_{\substack{\text{Exergy} \\ \text{destruction}}} = \underbrace{\Delta X_{system}}_{\substack{\text{Change} \\ \text{in exergy}}}$$

| General, rate form: | $\underbrace{\dot{X}_{in} - \dot{X}_{out}}_{\substack{\text{Rate of net exergy transfer} \\ \text{by heat, work, and mass}}} - \underbrace{\dot{X}_{destroyed}}_{\substack{\text{Rate of exergy} \\ \text{destruction}}} = \underbrace{dX_{system}/dt}_{\substack{\text{Rate of change} \\ \text{in exergy}}}$ |

General, unit-mass basis:

$$(x_{in} - x_{out}) - x_{destroyed} = \Delta x_{system}$$

where

$$\dot{X}_{heat} = (1 - T_0/T)\dot{Q}$$

$$\dot{X}_{work} = \dot{W}_{useful}$$

$$\dot{X}_{mass} = \dot{m}\psi$$

For a *reversible process,* the exergy destruction term $X_{destroyed}$ drops out. Taking the positive direction of heat transfer to be

to the system and the positive direction of work transfer to be from the system, the general exergy balance relations can be expressed more explicitly as

$$\sum\left(1 - \frac{T_0}{T_k}\right)Q_k - [W - P_0(V_2 - V_1)]$$
$$+ \sum_{in} m\psi - \sum_{out} m\psi - X_{destroyed} = X_2 - X_1$$

$$\sum\left(1 - \frac{T_0}{T_k}\right)\dot{Q}_k - \left(\dot{W} - P_0\frac{dV_{CV}}{dt}\right)$$
$$+ \sum_{in} \dot{m}\psi - \sum_{out} \dot{m}\psi - \dot{X}_{destroyed} = \frac{dX_{CV}}{dt}$$

# REFERENCES AND SUGGESTED READINGS

**1.** J. E. Ahern. *The Exergy Method of Energy Systems Analysis.* New York: John Wiley & Sons, 1980.

**2.** A. Bejan. *Advanced Engineering Thermodynamics.* 3rd ed. New York: Wiley Interscience, 2006.

**3.** A. Bejan. *Entropy Generation through Heat and Fluid Flow.* New York: John Wiley & Sons, 1982.

**4.** Y. A. Çengel. "A Unified and Intuitive Approach to Teaching Thermodynamics." ASME International Congress and Exposition, Atlanta, Georgia, November 17–22, 1996.

# PROBLEMS*

## Exergy, Irreversibility, Reversible Work, and Second-Law Efficiency

**8–1C** What final state will maximize the work output of a device?

**8–2C** Is the exergy of a system different in different environments?

**8–3C** How does useful work differ from actual work? For what kind of systems are these two identical?

**8–4C** Consider a process that involves no irreversibilities. Will the actual useful work for that process be equal to the reversible work?

**8–5C** Consider two geothermal wells whose energy contents are estimated to be the same. Will the exergies of these wells necessarily be the same? Explain.

**8–6C** Consider two systems that are at the same pressure as the environment. The first system is at the same temperature as the environment, whereas the second system is at a lower temperature than the environment. How would you compare the exergies of these two systems?

**8–7C** What is the second-law efficiency? How does it differ from the first-law efficiency?

**8–8C** Does a power plant that has a higher thermal efficiency necessarily have a higher second-law efficiency than one with a lower thermal efficiency? Explain.

**8–9C** Does a refrigerator that has a higher COP necessarily have a higher second-law efficiency than one with a lower COP? Explain.

**8–10C** Can a process for which the reversible work is zero be reversible? Can it be irreversible? Explain.

**8–11C** Consider a process during which no entropy is generated ($S_{gen} = 0$). Does the exergy destruction for this process have to be zero?

**8–12** The electric power needs of a community are to be met by windmills with 40-m-diameter rotors. The windmills

* Problems designated by a "C" are concept questions, and students are encouraged to answer them all. Problems with the icon are solved using EES, and complete solutions together with parametric studies are included on the text website. Problems with the icon are comprehensive in nature, and are intended to be solved with an equation solver such as EES.

are to be located where the wind is blowing steadily at an average velocity of 6 m/s. Determine the minimum number of windmills that need to be installed if the required power output is 1500 kW.

**8–13** One method of meeting the extra electric power demand at peak periods is to pump some water from a large body of water (such as a lake) to a water reservoir at a higher elevation at times of low demand and to generate electricity at times of high demand by letting this water run down and rotate a turbine (i.e., convert the electric energy to potential energy and then back to electric energy). For an energy storage capacity of $5 \times 10^6$ kWh, determine the minimum amount of water that needs to be stored at an average elevation (relative to the ground level) of 75 m. *Answer:* $2.45 \times 10^{10}$ kg

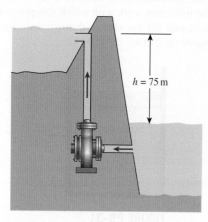

$h = 75$ m

**FIGURE P8–13**

**8–14** How much of the 100 kJ of thermal energy at 650 K can be converted to useful work? Assume the environment to be at 25°C.

**8–15** A heat engine that receives heat from a furnace at 1200°C and rejects waste heat to a river at 20°C has a thermal efficiency of 40 percent. Determine the second-law efficiency of this power plant.

**8–16** Consider a thermal energy reservoir at 1500 K that can supply heat at a rate of 150,000 kJ/h. Determine the exergy of this supplied energy, assuming an environmental temperature of 25°C.

**8–17** A heat engine receives heat from a source at 1100 K at a rate of 400 kJ/s, and it rejects the waste heat to a medium at 320 K. The measured power output of the heat engine is 120 kW, and the environment temperature is 25°C. Determine (*a*) the reversible power, (*b*) the rate of irreversibility, and (*c*) the second-law efficiency of this heat engine. *Answers:* (*a*) 284 kW, (*b*) 164 kW, (*c*) 42.3 percent

**8–18** Reconsider Prob. 8–17. Using EES (or other) software, study the effect of reducing the temperature at which the waste heat is rejected on the reversible power, the rate of irreversibility, and the second-law efficiency as the rejection temperature is varied from 500 to 298 K, and plot the results.

**8–19** A heat engine that rejects waste heat to a sink at 280 K has a thermal efficiency of 25 percent and a second-law efficiency of 50 percent. Determine the temperature of the source that supplies heat to this engine. *Answer:* 560 K

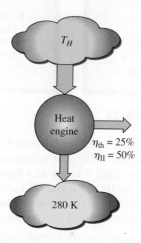

$T_H$

Heat engine

$\eta_{th} = 25\%$
$\eta_{II} = 50\%$

280 K

**FIGURE P8–19**

**8–20** A house that is losing heat at a rate of 50,000 kJ/h when the outside temperature drops to 4°C is to be heated by electric resistance heaters. If the house is to be maintained at 25°C at all times, determine the reversible work input for this process and the irreversibility. *Answers:* 0.978 kW, 12.91 kW

**8–21** A freezer is maintained at –7°C by removing heat from it at a rate of 80 kJ/min. The power input to the freezer is 0.50 kW, and the surrounding air is at 25°C. Determine (*a*) the reversible power, (*b*) the irreversibility, and (*c*) the second-law efficiency of this freezer *Answers:* (*a*) 0.16 kW, (*b*) 0.34 kW, (*c*) 32.0 percent

**8–22** Show that the power produced by a wind turbine is proportional to the cube of the wind velocity and to the square of the blade span diameter.

## Exergy Analysis of Closed Systems

**8–23C** Can a system have a higher second-law efficiency than the first-law efficiency during a process? Give examples.

**8–24** A mass of 8 kg of helium undergoes a process from an initial state of 3 m³/kg and 15°C to a final state of 0.5 m³/kg and 80°C. Assuming the surroundings to be at 25°C and 100 kPa, determine the increase in the useful work potential of the helium during this process.

**8–25** Air is expanded in an adiabatic closed system from 1250 kPa and 60°C to 140 kPa with an isentropic expansion efficiency of 95 percent. What is the second-law efficiency of this expansion? Take $T_0 = 25°C$ and $P_0 = 100$ kPa.

**8–26** Which has the capability to produce the most work in a closed system – 1 kg of steam at 800 kPa and 180°C or 1 kg of R–134a at 800 kPa and 180°C? Take $T_0 = 25°C$ and $P_0 = 100$ kPa. *Answers:* 623 kJ, 5.0 kJ

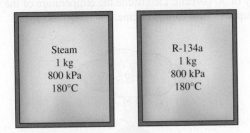

**FIGURE P8–26**

**8–27** A piston–cylinder device contains 8 kg of refrigerant-134a at 0.7 MPa and 60°C. The refrigerant is now cooled at constant pressure until it exists as a liquid at 20°C. If the surroundings are at 100 kPa and 20°C, determine (a) the exergy of the refrigerant at the initial and the final states and (b) the exergy destroyed during this process.

**8–28** The radiator of a steam heating system has a volume of 20 L and is filled with superheated water vapor at 200 kPa and 200°C. At this moment both the inlet and the exit valves to the radiator are closed. After a while it is observed that the temperature of the steam drops to 80°C as a result of heat transfer to the room air, which is at 21°C. Assuming the surroundings to be at 0°C, determine (a) the amount of heat transfer to the room and (b) the maximum amount of heat that can be supplied to the room if this heat from the radiator is supplied to a heat engine that is driving a heat pump. Assume the heat engine operates between the radiator and the surroundings. *Answers:* (a) 30.3 kJ, (b) 116 kJ

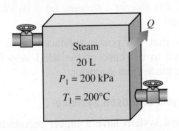

**FIGURE P8–28**

**8–29** Reconsider Prob. 8–28. Using EES (or other) software, investigate the effect of the final steam temperature in the radiator on the amount of actual heat transfer and the maximum amount of heat that can be transferred. Vary the final steam temperature from 80 to 21°C and plot the actual and maximum heat transferred to the room as functions of final steam temperature.

**8–30** A well-insulated rigid tank contains 3 kg of saturated liquid–vapor mixture of water at 250 kPa. Initially, three-quarters of the mass is in the liquid phase. An electric resistance heater placed in the tank is turned on and kept on until all the liquid in the tank is vaporized. Assuming the surroundings to be at 25°C and 100 kPa, determine (a) the exergy destruction and (b) the second-law efficiency for this process.

**8–31** An insulated piston–cylinder device contains 8 L of saturated liquid water at a constant pressure of 120 kPa. An electric resistance heater inside the cylinder is turned on, and electrical work is done on the water in the amount of 1400 kJ. Assuming the surroundings to be at 25°C and 100 kPa, determine (a) the minimum work with which this process could be accomplished and (b) the exergy destroyed during this process. *Answers:* (a) 278 kJ, (b) 1104 kJ

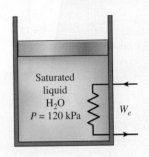

**FIGURE P8–31**

**8–32** Reconsider Prob. 8–31. Using EES (or other) software, investigate the effect of the amount of electrical work supplied to the device on the minimum work and the exergy destroyed as the electrical work is varied from 0 to 2000 kJ, and plot your results.

**8–33** An insulated piston–cylinder device contains 0.03 m³ of saturated refrigerant-134a vapor at 0.6 MPa pressure. The refrigerant is now allowed to expand in a reversible manner until the pressure drops to 0.16 MPa. Determine the change in the exergy of the refrigerant during this process and the reversible work. Assume the surroundings to be at 25°C and 100 kPa.

**8–34** A piston–cylinder device initially contains 2 L of air at 100 kPa and 25°C. Air is now compressed to a final state of 600 kPa and 150°C. The useful work input is 1.2 kJ. Assuming the surroundings are at 100 kPa and 25°C, determine (a) the exergy of the air at the initial and the final states, (b) the minimum work that must be supplied to accomplish this compression process, and (c) the second-law efficiency of this process. *Answers:* (a) 0, 0.171 kJ, (b) 0.171 kJ, (c) 14.3 percent

**FIGURE P8–34**

**8–35** A 0.8-m³ insulated rigid tank contains 1.54 kg of carbon dioxide at 100 kPa. Now paddle-wheel work is done on the system until the pressure in the tank rises to 135 kPa. Determine (a) the actual paddle-wheel work done during this process and (b) the minimum paddle-wheel work with which this process (between the same end states) could be accomplished. Take $T_0 = 298$ K. *Answers:* (a) 101 kJ, (b) 7.18 kJ

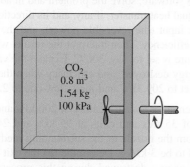

**FIGURE P8–35**

**8–36** An insulated piston–cylinder device initially contains 20 L of air at 140 kPa and 27°C. Air is now heated for 10 min by a 100-W resistance heater placed inside the cylinder. The pressure of air is maintained constant during this process, and the surroundings are at 27°C and 100 kPa. Determine the exergy destroyed during this process. *Answer:* 19.9 kJ

**8–37** An insulated rigid tank is divided into two equal parts by a partition. Initially, one part contains 3 kg of argon gas at 300 kPa and 70°C, and the other side is evacuated. The partition is now removed, and the gas fills the entire tank. Assuming the surroundings to be at 25°C, determine the exergy destroyed during this process. *Answer:* 129 kJ

**8–38** A 30-kg copper block initially at 105°C is dropped into an insulated tank that contains 0.035 m³ of water at 17°C. Determine (a) the final equilibrium temperature and (b) the work potential wasted during this process. Assume the surroundings to be at 17°C.

**8–39** An iron block of unknown mass at 85°C is dropped into an insulated tank that contains 100 L of water at 20°C. At the same time, a paddle wheel driven by a 200-W motor is activated to stir the water. It is observed that thermal equilibrium is established after 20 min with a final temperature of 24°C. Assuming the surroundings to be at 20°C, determine (a) the mass of the iron block and (b) the exergy destroyed during this process. *Answers:* (a) 52.0 kg, (b) 375 kJ

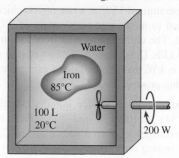

**FIGURE P8–39**

**8–40** A 0.35-m³ rigid tank contains refrigerant-134a at 200 kPa and 55 percent quality. Heat is transferred now to the refrigerant from a source at 50°C until the pressure rises to 360 kPa. Assuming the surroundings to be at 25°C, determine (a) the amount of heat transfer between the source and the refrigerant and (b) the exergy destroyed during this process.

**8–41** Stainless steel ball bearings ($\rho = 8085$ kg/m³ and $c_p = 0.480$ kJ/kg·°C) having a diameter of 1.2 cm are to be quenched in water at a rate of 1400 per minute. The balls leave the oven at a uniform temperature of 900°C and are exposed to air at 30°C for a while before they are dropped into the water. If the temperature of the balls drops to 850°C prior to quenching, determine (a) the rate of heat transfer from the balls to the air and (b) the rate of exergy destruction due to heat loss from the balls to the air.

**8–42** An ordinary egg can be approximated as a 5.5-cm-diameter sphere. The egg is initially at a uniform temperature of 8°C and is dropped into boiling water at 97°C. Taking the properties of egg to be $\rho = 1020$ kg/m³ and $c_p = 3.32$ kJ/kg·°C, determine how much heat is transferred to the egg by the time the average temperature of the egg rises to 70°C and the amount of exergy destruction associated with this heat transfer process. Take $T_0 = 25$°C.

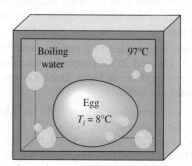

**FIGURE P8–42**

**8–43** Chickens with an average mass of 1.6 kg and average specific heat of 3.54 kJ/kg·°C are to be cooled by chilled water that enters a continuous-flow-type immersion chiller at 0.5°C and leaves at 2.5°C. Chickens are dropped into the chiller at a uniform temperature of 15°C at a rate of 700 chickens per hour and are cooled to an average temperature of 3°C before they are taken out. The chiller gains heat from the surroundings at a rate of 400 kJ/h. Determine (a) the rate of heat removal from the chicken, in kW, and (b) the rate of exergy destruction during this chilling process. Take $T_0 = 25°C$.

**8–44** A piston–cylinder device initially contains 1.4 kg of refrigerant-134a at 100 kPa and 20°C. Heat is now transferred to the refrigerant from a source at 150°C, and the piston which is resting on a set of stops, starts moving when the pressure inside reaches 120 kPa. Heat transfer continues until the temperature reaches 80°C. Assuming the surroundings to be at 25°C and 100 kPa, determine (a) the work done, (b) the heat transfer, (c) the exergy destroyed, and (d) the second-law efficiency of this process. *Answers:* (a) 0.497 kJ, (b) 67.9 kJ, (c) 14.8 kJ, (d) 26.2 percent

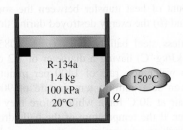

R-134a
1.4 kg
100 kPa
20°C

$Q$

150°C

**FIGURE P8–44**

## Exergy Analysis of Control Volumes

**8–45** Refrigerant-134a at 1 MPa and 100°C is throttled to a pressure of 0.8 MPa. Determine the reversible work and exergy destroyed during this throttling process. Assume the surroundings to be at 30°C.

**8–46** Reconsider Prob. 8–45. Using EES (or other) software, investigate the effect of exit pressure on the reversible work and exergy destruction. Vary the throttle exit pressure from 1 to 0.1 MPa and plot the reversible work and exergy destroyed as functions of the exit pressure. Discuss the results.

**8–47** Helium is expanded in a turbine from 1500 kPa and 300°C to 100 kPa and 25°C. Determine the maximum work this turbine can produce, in kJ/kg. Does the maximum work require an adiabatic turbine?

**8–48** Air is compressed steadily by an 8-kW compressor from 100 kPa and 17°C to 600 kPa and 167°C at a rate of 2.1 kg/min. Neglecting the changes in kinetic and potential energies, determine (a) the increase in the exergy of the air and (b) the rate of exergy destroyed during this process. Assume the surroundings to be at 17°C.

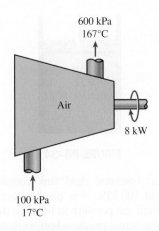

600 kPa
167°C

Air

8 kW

100 kPa
17°C

**FIGURE P8–48**

**8–49** Reconsider Prob. 8–48. Using EES (or other) software, solve the problem and in addition determine the actual heat transfer, if any, and its direction, the minimum power input (the reversible power), and the compressor second-law efficiency. Then interpret the results when the outlet temperature is set to, say, 300°C. Explain the values of heat transfer, exergy destroyed, and efficiency when the outlet temperature is set to 209.31°C and mass flow rate to 2.466 kg/min.

**8–50** Air enters a nozzle steadily at 200 kPa and 65°C with a velocity of 35 m/s and exits at 95 kPa and 240 m/s. The heat loss from the nozzle to the surrounding medium at 17°C is estimated to be 3 kJ/kg. Determine (a) the exit temperature and (b) the exergy destroyed during this process. *Answers:* (a) 34.0°C, (b) 36.9 kJ/kg

**8–51** Reconsider Prob. 8–50. Using EES (or other) software, study the effect of varying the nozzle exit velocity from 100 to 300 m/s on both the exit temperature and exergy destroyed, and plot the results.

**8–52** Steam enters a diffuser at 10 kPa and 60°C with a velocity of 375 m/s and exits as saturated vapor at 50°C and 70 m/s. The exit area of the diffuser is 3 m². Determine (a) the mass flow rate of the steam and (b) the wasted work potential during this process. Assume the surroundings to be at 25°C.

**8–53** Air is compressed steadily by a compressor from 100 kPa and 17°C to 700 kPa and 247°C at a rate of 10 kg/min. Assuming the surroundings to be at 17°C, determine the minimum power input to the compressor. Assume air to be an ideal gas with variable specific heats, and neglect the changes in kinetic and potential energies.

**8–54** Argon gas enters an adiabatic compressor at 120 kPa and 30°C with a velocity of 20 m/s and exits at 1.2 MPa, 530°C, and 80 m/s. The inlet area of the compressor is 130 cm². Assuming the surroundings to be at 25°C, determine the reversible power input and exergy destroyed. *Answers:* 126 kW, 4.12 kW

**8–55** Steam enters an adiabatic turbine at 6 MPa, 600°C, and 80 m/s and leaves at 50 kPa, 100°C, and 140 m/s. If the power output of the turbine is 5 MW, determine (a) the reversible power output and (b) the second-law efficiency of the turbine. Assume the surroundings to be at 25°C. *Answers:* (a) 5.81 MW, (b) 86.1 percent

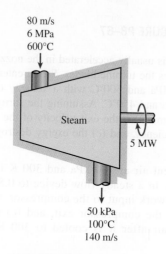

80 m/s
6 MPa
600°C

Steam

5 MW

50 kPa
100°C
140 m/s

**FIGURE P8–55**

**8–56** Steam is throttled from 7 MPa and 500°C to a pressure of 1 MPa. Determine the decrease in exergy of the steam during this process. Assume the surroundings to be at 25°C. *Answer:* 261 kJ/kg

**8–57** Carbon dioxide enters a compressor at 100 kPa and 300 K at a rate of 0.2 kg/s and exits at 600 kPa and 450 K. Determine the power input to the compressor if the process involved no irreversibilities. Assume the surroundings to be at 25°C. *Answer:* 25.5 kW

**8–58** Combustion gases enter a gas turbine at 900°C, 800 kPa, and 100 m/s and leave at 650°C, 400 kPa, and 220 m/s. Taking $c_p = 1.15$ kJ/kg·°C and $k = 1.3$ for the combustion gases, determine (a) the exergy of the combustion gases at the turbine inlet and (b) the work output of the turbine under reversible conditions. Assume the surroundings to be at 25°C and 100 kPa. Can this turbine be adiabatic?

**8–59** Refrigerant-134a enters an adiabatic compressor at −30°C as a saturated vapor at a rate of 0.45 m³/min and leaves at 900 kPa and 55°C. Determine (a) the power input to the compressor, (b) the isentropic efficiency of the compressor, and (c) the rate of exergy destruction and the second-law efficiency of the compressor. Take $T_0 = 27$°C. *Answers:* (a) 1.92 kW, (b) 85.3 percent, (c) 0.261 kW, 86.4 percent

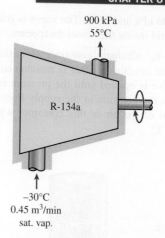

900 kPa
55°C

R-134a

−30°C
0.45 m³/min
sat. vap.

**FIGURE P8–59**

**8–60** Refrigerant-134a is condensed in a refrigeration system by rejecting heat to ambient air at 25°C. R-134a enters the condenser at 700 kPa and 50°C at a rate of 0.05 kg/s and leaves at the same pressure as a saturated liquid. Determine (a) the rate of heat rejected in the condenser, (b) the COP of this refrigeration cycle if the cooling load at these conditions is 6 kW, and (c) the rate of exergy destruction in the condenser.

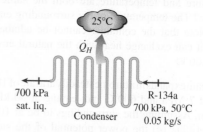

25°C

$\dot{Q}_H$

700 kPa
sat. liq.

R-134a
700 kPa, 50°C
0.05 kg/s

Condenser

**FIGURE P8–60**

**8–61** Air enters the evaporator section of a window air conditioner at 100 kPa and 27°C with a volume flow rate of 6 m³/min. Refrigerant-134a at 120 kPa with a quality of 0.3 enters the evaporator at a rate of 2 kg/min and leaves as saturated vapor at the same pressure. Determine the exit temperature of the air and the exergy destruction for this process, assuming (a) the outer surfaces of the air conditioner are insulated and (b) heat is transferred to the evaporator of the air conditioner from the surrounding medium at 32°C at a rate of 30 kJ/min.

**8–62** How much exergy is lost in a rigid vessel filled with 1 kg of liquid R-134a, whose temperature remains constant at 24°C, as R-134a vapor is released from the vessel? This vessel may exchange heat with the surrounding atmosphere, which is at 100 kPa and 24°C. The vapor is released until the last of the liquid inside the vessel disappears.

**8–63** A 1.1-m³ adiabatic container is initially evacuated. The supply line contains air that is maintained at 1 MPa and 30°C. The valve is opened until the pressure in the container is the same as the pressure in the supply line. Determine the work potential of the air in this container when it is filled. Take $T_0 = 27°C$.

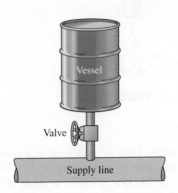

**FIGURE P8–63**

**8–64** What is the work potential of the air in the filled container of Prob. 8-63 if it is filled in such a way that the final pressure and temperature are both the same as in the supply line? The temperature of the surrounding environment is 27°C. Note that the container cannot be adiabatic in this case, and it can exchange heat with the natural environment. *Answer:* 1100 kJ

**8–65** Steam expands in a turbine steadily at a rate of 18,000 kg/h, entering at 7 MPa and 600°C and leaving at 50 kPa as saturated vapor. Assuming the surroundings to be at 100 kPa and 25°C, determine (a) the power potential of the steam at the inlet conditions and (b) the power output of the turbine if there were no irreversibilities present. *Answers:* (a) 7710 kW, (b) 5775 kW

**8–66** Air enters a compressor at ambient conditions of 100 kPa and 17°C with a low velocity and exits at 1 MPa, 327°C, and 105 m/s. The compressor is cooled by the ambient air at 17°C at a rate of 1500 kJ/min. The power input to the compressor is 300 kW. Determine (a) the mass flow rate of air and (b) the portion of the power input that is used just to overcome the irreversibilities.

**8–67** Hot combustion gases enter the nozzle of a turbojet engine at 230 kPa, 627°C, and 60 m/s and exit at 70 kPa and 450°C. Assuming the nozzle to be adiabatic and the surroundings to be at 20°C, determine (a) the exit velocity and (b) the decrease in the exergy of the gases. Take $k = 1.3$ and $c_p = 1.15$ kJ/kg·°C for the combustion gases.

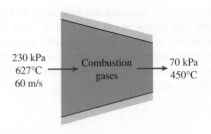

**FIGURE P8–67**

**8–68** Steam is usually accelerated in the nozzle of a turbine before it strikes the turbine blades. Steam enters an adiabatic nozzle at 7 MPa and 500°C with a velocity of 70 m/s and exits at 5 MPa and 450°C. Assuming the surroundings to be at 25°C, determine (a) the exit velocity of the steam, (b) the isentropic efficiency, and (c) the exergy destroyed within the nozzle.

**8–69** Ambient air at 100 kPa and 300 K is compressed isentropically in a steady-flow device to 0.8 MPa. Determine (a) the work input to the compressor, (b) the exergy of the air at the compressor exit, and (c) the exergy of compressed air after it is cooled to 300 K at 0.8 MPa pressure.

**8–70** A 0.6-m³ rigid tank is filled with saturated liquid water at 170°C. A valve at the bottom of the tank is now opened, and one-half of the total mass is withdrawn from the tank in liquid form. Heat is transferred to water from a source of 210°C so that the temperature in the tank remains constant. Determine (a) the amount of heat transfer and (b) the reversible work and exergy destruction for this process. Assume the surroundings to be at 25°C and 100 kPa. *Answers:* (a) 2545 kJ, (b) 141.2 kJ, 141.2 kJ

**8–71** A 0.1-m³ rigid tank contains saturated refrigerant-134a at 800 kPa. Initially, 30 percent of the volume is occupied by liquid and the rest by vapor. A valve at the bottom of the tank is opened, and liquid is withdrawn from the tank. Heat is transferred to the refrigerant from a source at 60°C so that the pressure inside the tank remains constant. The valve is closed when no liquid is left in the tank and vapor starts to come out. Assuming the surroundings to be at 25°C, determine (a) the final mass in the tank and (b) the reversible work associated with this process. *Answers:* (a) 3.90 kg, (b) 16.9 kJ

**8–72** A vertical piston–cylinder device initially contains 0.12 m³ of helium at 20°C. The mass of the piston is such that it maintains a constant pressure of 200 kPa inside. A valve is now opened, and helium is allowed to escape until the volume inside the cylinder is decreased by one-half. Heat transfer takes place between the helium and its surroundings at 20°C and 95 kPa so that the temperature of helium in the cylinder remains constant. Determine (a) the maximum work potential of the helium at the initial state and (b) the exergy destroyed during this process.

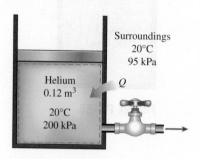

**FIGURE P8–72**

**8–73** An insulated vertical piston–cylinder device initially contains 15 kg of water, 13 kg of which is in the vapor phase. The mass of the piston is such that it maintains a constant pressure of 300 kPa inside the cylinder. Now steam at 2 MPa and 400°C is allowed to enter the cylinder from a supply line until all the liquid in the cylinder is vaporized. Assuming the surroundings to be at 25°C and 100 kPa, determine (a) the amount of steam that has entered and (b) the exergy destroyed during this process. *Answers: (a) 8.27 kg, (b) 2832 kJ*

**8–74** Consider a family of four, with each person taking a 6-minute shower every morning. The average flow rate through the shower head is 10 L/min. City water at 15°C is heated to 55°C in an electric water heater and tempered to 42°C by cold water at the T-elbow of the shower before being routed to the shower head. Determine the amount of exergy destroyed by this family per year as a result of taking daily showers. Take $T_0 = 25°C$.

**8–75** Cold water ($c_p = 4.18$ kJ/kg·°C) leading to a shower enters a well-insulated, thin-walled, double-pipe, counter-flow heat exchanger at 15°C at a rate of 0.25 kg/s and is heated to 45°C by hot water ($c_p = 4.19$ kJ/kg·°C) that enters at 100°C at a rate of 3 kg/s. Determine (a) the rate of heat transfer and (b) the rate of exergy destruction in the heat exchanger. Take $T_0 = 25°C$.

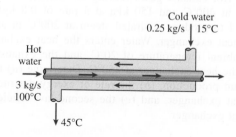

**FIGURE P8–75**

**8–76** Outdoor air ($c_p = 1.005$ kJ/kg·°C) is to be preheated by hot exhaust gases in a cross-flow heat exchanger before it enters the furnace. Air enters the heat exchanger at 101 kPa and 30°C at a rate of 0.5 m³/s. The combustion gases ($c_p = 1.10$ kJ/kg·°C) enter at 350°C at a rate of 0.85 kg/s and leave at 260°C. Determine the rate of heat transfer to the air and the rate of exergy destruction in the heat exchanger.

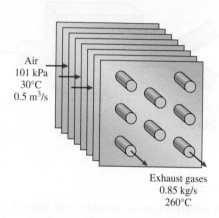

**FIGURE P8–76**

**8–77** Steam is to be condensed on the shell side of a heat exchanger at 50°C. Cooling water enters the tubes at 15°C at a rate of 55 kg/s and leaves at 23°C. Assuming the heat exchanger to be well insulated, determine (a) the rate of heat transfer in the heat exchanger and (b) the rate of exergy destruction in the heat exchanger. Take $T_0 = 25°C$.

**8–78** Air enters a compressor at ambient conditions of 100 kPa and 20°C at a rate of 4.5 m³/s with a low velocity, and exits at 900 kPa, 60°C, and 80 m/s. The compressor is cooled by cooling water that experiences a temperature rise of 10°C. The isothermal efficiency of the compressor is 70 percent. Determine (a) the actual and reversible power inputs, (b) the second-law efficiency, and (c) the mass flow rate of the cooling water.

**8–79** A hot-water stream at 70°C enters an adiabatic mixing chamber with a mass flow rate of 2 kg/s, where it is mixed with a stream of cold water at 20°C. If the mixture leaves the chamber at 45°C, determine (a) the mass flow rate of the cold water and (b) the exergy destroyed during this adiabatic mixing process. Assume all the streams are at a pressure of 350 kPa and the surroundings are at 25°C. *Answers: (a) 2.0 kg/s, (b) 15.5 kW*

**8–80** Liquid water at 20°C is heated in a chamber by mixing it with saturated steam. Liquid water enters the chamber at the steam pressure at a rate of 4.6 kg/s and the saturated steam enters at a rate of 0.19 kg/s. The mixture leaves the mixing chamber as a liquid at 45°C. If the surroundings are at 20°C, determine (a) the temperature of saturated steam entering the chamber, (b) the exergy destruction during this mixing process, and (c) the second-law efficiency of the mixing chamber. *Answers:* (a) 129.2°C, (b) 105 kW, (c) 15.8 percent

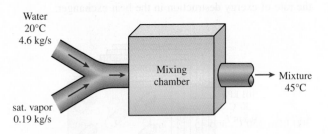

Water
20°C
4.6 kg/s

Mixing chamber

Mixture
45°C

sat. vapor
0.19 kg/s

**FIGURE P8–80**

## Review Problems

**8–81**  A refrigerator has a second-law efficiency of 28 percent, and heat is removed from the refrigerated space at a rate of 800 kJ/min. If the space is maintained at –4°C while the surrounding air temperature is 32°C, determine the power input to the refrigerator.

**8–82**  Refrigerant-134a is expanded adiabatically in an expansion valve from 700 kPa and 25°C to 160 kPa. For environment conditions of 100 kPa and 20°C, determine (a) the work potential of R-134a at the inlet, (b) the exergy destruction during the process, and (c) the second-law efficiency.

**8–83**  Steam enters an adiabatic nozzle at 3.5 MPa and 300°C with a low velocity and leaves at 1.6 MPa and 250°C at a rate of 0.4 kg/s. If the ambient state is 100 kPa and 18°C, determine (a) the exit velocity, (b) the rate of exergy destruction, and (c) the second-law efficiency.

**8–84**  Steam is condensed in a closed system at a constant pressure of 75 kPa from a saturated vapor to a saturated liquid by rejecting heat to a thermal energy reservoir at 37°C. Determine the second-law efficiency of this process. Take $T_0 = 25°C$ and $P_0 = 100$ kPa.

**8–85**  Refrigerant-134a is converted from a saturated liquid to a saturated vapor in a closed system using a reversible constant pressure process by transferring heat from a heat reservoir at 6°C. From second-law point of view, is it more effective to do this phase change at 100 kPa or 180 kPa? Take $T_0 = 25°C$ and $P_0 = 100$ kPa.

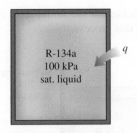

R-134a
100 kPa
sat. liquid

q

**FIGURE P8–85**

**8–86**  An adiabatic heat exchanger is to cool ethylene glycol ($c_p = 2.56$ kJ/kg·°C) flowing at a rate of 2 kg/s from 80 to 40°C by water ($c_p = 4.18$ kJ/kg·°C) that enters at 20°C and leaves at 55°C. Determine (a) the rate of heat transfer and (b) the rate of exergy destruction in the heat exchanger.

**8–87**  A well-insulated, thin-walled, counter-flow heat exchanger is to be used to cool oil ($c_p = 2.20$ kJ/kg·°C) from 150 to 40°C at a rate of 2 kg/s by water ($c_p = 4.18$ kJ/kg·°C) that enters at 22°C at a rate of 1.5 kg/s. The diameter of the tube is 2.5 cm, and its length is 6 m. Determine (a) the rate of heat transfer and (b) the rate of exergy destruction in the heat exchanger.

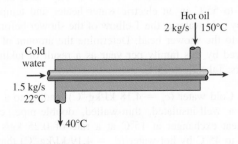

Hot oil
2 kg/s  150°C

Cold water
1.5 kg/s
22°C

40°C

**FIGURE P8–87**

**8–88**  Hot exhaust gases leaving an internal combustion engine at 400°C and 150 kPa at a rate of 0.8 kg/s is to be used to produce saturated steam at 200°C in an insulated heat exchanger. Water enters the heat exchanger at the ambient temperature of 20°C, and the exhaust gases leave the heat exchanger at 350°C. Determine (a) the rate of steam production, (b) the rate of exergy destruction in the heat exchanger, and (c) the second-law efficiency of the heat exchanger.

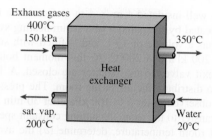

Exhaust gases
400°C
150 kPa

350°C

Heat
exchanger

sat. vap.
200°C

Water
20°C

**FIGURE P8–88**

**8–89** A crater lake has a base area of 20,000 m², and the water it contains is 12 m deep. The ground surrounding the crater is nearly flat and is 140 m below the base of the lake. Determine the maximum amount of electrical work, in kWh, that can be generated by feeding this water to a hydroelectric power plant. *Answer: 95,500 kWh*

**8–90** The inner and outer surfaces of a 5-m × 6-m brick wall of thickness 30 cm are maintained at temperatures of 20°C and 5°C, respectively, and the rate of heat transfer through the wall is 900 W. Determine the rate of exergy destruction associated with this process. Take $T_0 = 0°C$.

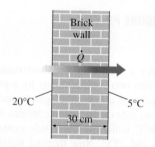

Brick
wall

$\dot{Q}$

20°C

5°C

30 cm

**FIGURE P8–90**

**8–91** A 1000-W iron is left on the ironing board with its base exposed to the air at 20°C. If the temperature of the base of the iron is 150°C, determine the rate of exergy destruction for this process due to heat transfer, in steady operation.

**8–92** A 30-cm-long, 1500-W electric resistance heating element whose diameter is 1.2 cm is immersed in 70 kg of water initially at 20°C. Assuming the water container is well-insulated, determine how long it will take for this heater to raise the water temperature to 80°C. Also, determine the minimum work input required and exergy destruction for this process, in kJ. Take $T_0 = 20°C$.

---

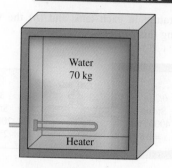

Water
70 kg

Heater

**FIGURE P8–92**

**8–93** An adiabatic steam nozzle has steam entering at 300 kPa, 150°C, and 45 m/s, and leaving as a saturated vapor at 150 kPa. Calculate the actual and maximum outlet velocity. Take $T_0 = 25°C$. *Answers: 372 m/s, 473 m/s*

**8–94** To control an isentropic steam turbine, a throttle valve is placed in the steam line leading to the turbine inlet. Steam at 6 MPa and 600°C is supplied to the throttle inlet, and the turbine exhaust pressure is set at 40 kPa. What is the effect on the stream exergy at the turbine inlet when the throttle valve is partially closed such that the pressure at the turbine inlet is 2 MPa. Compare the second-law efficiency of this system when the valve is partially open to when it is fully open. Take $T_0 = 25°C$.

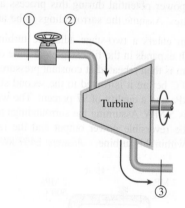

① ②

Turbine

③

**FIGURE P8–94**

**8–95** Two rigid tanks are connected by a valve. Tank *A* is insulated and contains 0.2 m³ of steam at 400 kPa and 80 percent quality. Tank *B* is uninsulated and contains 3 kg of steam at 200 kPa and 250°C. The valve is now opened, and steam flows from tank *A* to tank *B* until the pressure in tank *A* drops to 300 kPa. During this process 900 kJ of heat is transferred from tank *B* to the surroundings at 0°C. Assuming the steam remaining inside tank *A* to have undergone a reversible adiabatic process, determine (*a*) the final temperature in each tank and (b) the work potential wasted during this process.

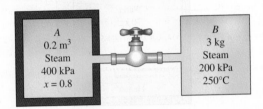

**FIGURE P8–95**

**8–96** A piston–cylinder device initially contains 0.22 m³ of helium gas at 280 kPa and 20°C. Helium is now compressed in a polytropic process ($Pv^n$ = constant) to 980 kPa and 160°C. Assuming the surroundings to be at 100 kPa and 20°C, determine (a) the actual useful work consumed and (b) the minimum useful work input needed for this process. *Answers:* (a) 52.3 kJ, (b) 48.5 kJ

**8–97** Steam at 7 MPa and 400°C enters a two-stage adiabatic turbine at a rate of 15 kg/s. Ten percent of the steam is extracted at the end of the first stage at a pressure of 1.8 MPa for other use. The remainder of the steam is further expanded in the second stage and leaves the turbine at 10 kPa. If the turbine has an isentropic efficiency of 88 percent, determine the wasted power potential during this process as a result of irreversibilities. Assume the surroundings to be at 25°C.

**8–98** Steam enters a two-stage adiabatic turbine at 8 MPa and 500°C. It expands in the first stage to a state of 2 MPa and 350°C. Steam is then reheated at constant pressure to a temperature of 500°C before it is routed to the second stage, where it exits at 30 kPa and a quality of 97 percent. The work output of the turbine is 5 MW. Assuming the surroundings to be at 25°C, determine the reversible power output and the rate of exergy destruction within this turbine. *Answers:* 5457 kW, 457 kW

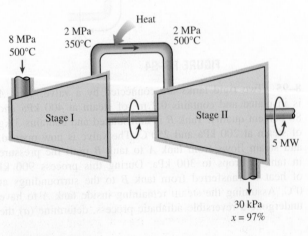

**FIGURE P8–98**

**8–99** A well-insulated 4-m × 4-m × 5-m room initially at 10°C is heated by the radiator of a steam heating system. The radiator has a volume of 15 L and is filled with superheated vapor at 200 kPa and 200°C. At this moment both the inlet and the exit valves to the radiator are closed. A 150-W fan is used to distribute the air in the room. The pressure of the steam is observed to drop to 100 kPa after 30 min as a result of heat transfer to the room. Assuming constant specific heats for air at room temperature, determine (a) the average temperature of room air in 24 min, (b) the entropy change of the steam, (c) the entropy change of the air in the room, and (d) the exergy destruction for this process, in kJ. Assume the air pressure in the room remains constant at 100 kPa at all times, and take $T_0 = 10°C$.

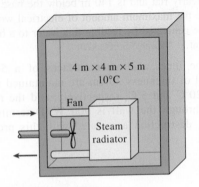

**FIGURE P8–99**

**8–100** Consider a well-insulated horizontal rigid cylinder that is divided into two compartments by a piston that is free to move but does not allow either gas to leak into the other side. Initially, one side of the piston contains 1 m³ of $N_2$ gas at 500 kPa and 80°C while the other side contains 1 m³ of He gas at 500 kPa and 25°C. Now thermal equilibrium is established in the cylinder as a result of heat transfer through the piston. Using constant specific heats at room temperature, determine (a) the final equilibrium temperature in the cylinder and (b) the wasted work potential during this process. What would your answer be if the piston were not free to move? Take $T_0 = 25°C$.

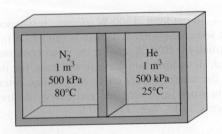

**FIGURE P8–100**

**8–101** Repeat Prob. 8–100 by assuming the piston is made of 5 kg of copper initially at the average temperature of the two gases on both sides.

**8–102** Argon gas enters an adiabatic turbine at 700°C and 1400 kPa at a rate of 20 kg/min and exhausts at 150 kPa. If the power output of the turbine is 80 kW, determine (a) the isentropic efficiency and (b) the second-law efficiency of the turbine. Assume the surroundings to be at 25°C.

**8–103** In large steam power plants, the feedwater is frequently heated in closed feedwater heaters, which are basically heat exchangers, by steam extracted from the turbine at some stage. Steam enters the feedwater heater at 1.6 MPa and 250°C and leaves as saturated liquid at the same pressure. Feedwater enters the heater at 4 MPa and 30°C and leaves 10°C below the exit temperature of the steam. Neglecting any heat losses from the outer surfaces of the heater, determine (a) the ratio of the mass flow rates of the extracted steam and the feedwater heater and (b) the reversible work for this process per unit mass of the feedwater. Assume the surroundings to be at 25°C. *Answers:* (a) 0.333, (b) 110 kJ/kg

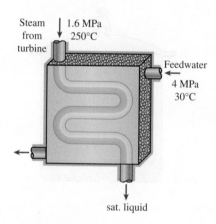

Steam from turbine | 1.6 MPa 250°C

Feedwater
4 MPa
30°C

sat. liquid

**FIGURE P8–103**

**8–104** Reconsider Prob. 8–103. Using EES (or other) software, investigate the effect of the state of the steam at the inlet of the feedwater heater on the ratio of mass flow rates and the reversible power. Vary the extracted steam pressure between 200 and 2000 kPa. Plot both the ratio of the mass flow rates of the extracted steam and the feedwater heater and the reversible work for this process per unit mass of feedwater as functions of the extraction pressure.

**8–105** In order to cool 1 ton of water at 20°C in an insulated tank, a person pours 80 kg of ice at −5°C into the water. Determine (a) the final equilibrium temperature in the tank and (b) the exergy destroyed during this process. The melting temperature and the heat of fusion of ice at atmospheric pressure are 0°C and 333.7 kJ/kg, respectively. Take $T_0 = 20°C$.

**8–106** One method of passive solar heating is to stack gallons of liquid water inside the buildings and expose them to the sun. The solar energy stored in the water during the day is released at night to the room air, providing some heating. Consider a house that is maintained at 22°C and whose heating is assisted by a 350-L water storage system. If the water is heated to 45°C during the day, determine the amount of heating this water will provide to the house at night. Assuming an outside temperature of 5°C, determine the exergy destruction associated with this process. *Answers:* 33,550 kJ, 1172 kJ

**8–107** A passive solar house that is losing heat to the outdoors at 5°C at an average rate of 50,000 kJ/h is maintained at 22°C at all times during a winter night for 10 h. The house is to be heated by 50 glass containers, each containing 20 L of water that is heated to 80°C during the day by absorbing solar energy. A thermostat-controlled 15-kW back-up electric resistance heater turns on whenever necessary to keep the house at 22°C. Determine (a) how long the electric heating system was on that night, (b) the exergy destruction, and (c) the minimum work input required for that night, in kJ.

**8–108** Consider a 20-L evacuated rigid bottle that is surrounded by the atmosphere at 100 kPa and 25°C. A valve at the neck of the bottle is now opened and the atmospheric air is allowed to flow into the bottle. The air trapped in the bottle eventually reaches thermal equilibrium with the atmosphere as a result of heat transfer through the wall of the bottle. The valve remains open during the process so that the trapped air also reaches mechanical equilibrium with the atmosphere. Determine the net heat transfer through the wall of the bottle and the exergy destroyed during this filling process.

20 L
Evacuated

100 kPa
25°C

**FIGURE P8–108**

**8–109** A frictionless piston-cylinder device, shown in Fig. P8-109, initially contains 0.01 m³ of argon gas at 400 K and 350 kPa. Heat is now transferred to the argon from a furnace at 1200 K, and the argon expands isothermally until its volume is doubled. No heat transfer takes place between the argon and the surrounding atmospheric air, which is at 300 K and 100 kPa. Determine (a) the useful work output, (b) the exergy destroyed, and (c) the maximum work that can be produced during this process.

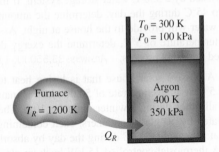

**FIGURE P8–109**

**8–110** Two constant-pressure devices, each filled with 30 kg of air, have temperatures of 900 K and 300 K. A heat engine placed between the two devices extracts heat from the high-temperature device, produces work, and rejects heat to the low-temperature device. Determine the maximum work that can be produced by the heat engine and the final temperatures of the devices. Assume constant specific heats at room temperature.

**8–111** A constant-volume tank contains 30 kg of nitrogen at 900 K, and a constant-pressure device contains 15 kg of argon at 300 K. A heat engine placed between the tank and device extracts heat from the high-temperature tank, produces work, and rejects heat to the low-temperature device. Determine the maximum work that can be produced by the heat engine and the final temperatures of the nitrogen and argon. Assume constant specific heats at room temperature.

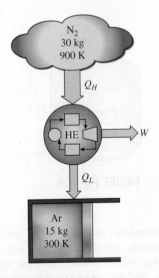

**FIGURE P8–111**

**8–112** A 100-L well-insulated rigid tank is initially filled with nitrogen at 1000 kPa and 20°C. Now a valve is opened and one-half of nitrogen's mass is allowed to escape. Determine the change in the exergy content of the tank.

**8–113** A 4-L pressure cooker has an operating pressure of 175 kPa. Initially, one-half of the volume is filled with liquid water and the other half by water vapor. The cooker is now placed on top of a 750-W electrical heating unit that is kept on for 20 min. Assuming the surroundings to be at 25°C and 100 kPa, determine (a) the amount of water that remained in the cooker and (b) the exergy destruction associated with the entire process, including the conversion of electric energy to heat energy. *Answers:* (a) 1.507 kg, (b) 689 kJ

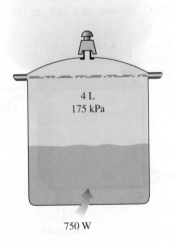

**FIGURE P8–113**

**8–114** What would your answer to Prob. 8–113 be if heat were supplied to the pressure cooker from a heat source at 180°C instead of the electrical heating unit?

**8–115** Steam is to be condensed in the condenser of a steam power plant at a temperature of 50°C with cooling water from a nearby lake that enters the tubes of the condenser at 12°C at a rate of 240 kg/s and leaves at 20°C. Assuming the condenser to be perfectly insulated, determine (a) the rate of condensation of the steam and (b) the rate of exergy destruction in the condenser. *Answers:* (a) 3.37 kg/s, (b) 837 kW

**8–116** The compressed-air storage tank shown in Fig. P8–116 has a volume of 500,000 m³, and it initially contains air at 100 kPa and 20°C. The isentropic compressor proceeds to compress air that enters the compressor at 100 kPa and 20°C until the tank is filled at 600 kPa and 20°C. All heat exchanges are with the surrounding air at 20°C. Calculate the change in the work potential of the air stored in the tank. How does this compare to the work required to compress the air as the tank was being filled?

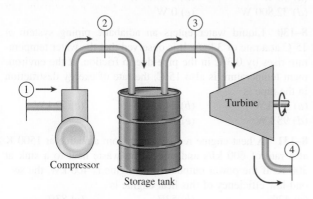

**FIGURE P8–116**

**8–117** The air stored in the tank of Prob. 8–116 is now released through the isentropic turbine until the tank contents are at 100 kPa and 20°C. The pressure is always 100 kPa at the turbine outlet, and all heat exchanges are with the surrounding air, which is at 20°C. How does the total work produced by the turbine compare to the change in the work potential of the air in the storage tank?

**8–118** A constant-volume tank has a temperature of 600 K and a constant-pressure device has a temperature of 280 K. Both the tank and device are filled with 40 kg of air. A heat engine placed between the tank and device receives heat from the high-temperature tank, produces work, and rejects heat to the low-temperature device. Determine the maximum work that can be produced by the heat engine and the final temperatures of the tank and device. Assume constant specific heats at room temperature.

**8–119** In a dairy plant, milk at 4°C is pasteurized continuously at 72°C at a rate of 12 L/s for 24 h/day and 365 days/yr. The milk is heated to the pasteurizing temperature by hot water heated in a natural gas-fired boiler having an efficiency of 82 percent. The pasteurized milk is then cooled by cold water at 18°C before it is finally refrigerated back to 4°C. To save energy and money, the plant installs a regenerator that has an effectiveness of 82 percent. If the cost of natural gas is $1.30/therm (1 therm = 105,500 kJ), determine how much

energy and money the regenerator will save this company per year and the annual reduction in exergy destruction.

**8–120** Combustion gases enter a gas turbine at 627°C and 1.2 MPa at a rate of 2.5 kg/s and leave at 527°C and 500 kPa. It is estimated that heat is lost from the turbine at a rate of 20 kW. Using air properties for the combustion gases and assuming the surroundings to be at 25°C and 100 kPa, determine (a) the actual and reversible power outputs of the turbine, (b) the exergy destroyed within the turbine, and (c) the second-law efficiency of the turbine.

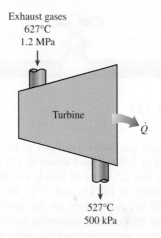

Exhaust gases
627°C
1.2 MPa

Turbine

$\dot{Q}$

527°C
500 kPa

**FIGURE P8–120**

**8–121** Refrigerant-134a enters an adiabatic compressor at 120 kPa superheated by 2.3°C, and leaves at 0.7 MPa. If the compressor has a second-law efficiency of 85 percent, determine (a) the actual work input, (b) the isentropic efficiency, and (c) the exergy destruction. Take the environment temperature to be 25°C. *Answers:* (a) 43.9 kJ/kg, (b) 0.842, (c) 6.58 kJ/kg

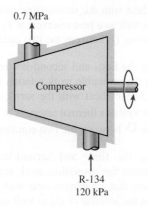

0.7 MPa

Compressor

R-134
120 kPa

**FIGURE P8–121**

**8–122** Water enters a pump at 100 kPa and 30°C at a rate of 1.35 kg/s, and leaves at 4 MPa. If the pump has an isentropic efficiency of 70 percent, determine (a) the actual power input, (b) the rate of frictional heating, (c) the exergy destruction, and (d) the second-law efficiency for an environment temperature of 20°C.

**8–123** Argon gas expands from 3.5 MPa and 100°C to 500 kPa in an adiabatic expansion valve. For environment conditions of 100 kPa and 25°C, determine (a) the exergy of argon at the inlet, (b) the exergy destruction during the process, and (c) the second-law efficiency.

Argon
3.5 MPa
100°C                                    500 kPa

**FIGURE P8–123**

**8–124** Nitrogen gas enters a diffuser at 100 kPa and 110°C with a velocity of 205 m/s, and leaves at 110 kPa and 45 m/s. It is estimated that 2.5 kJ/kg of heat is lost from the diffuser to the surroundings at 100 kPa and 27°C. The exit area of the diffuser is 0.04 m². Accounting for the variation of the specific heats with temperature, determine (a) the exit temperature, (b) the rate of exergy destruction, and (c) the second-law efficiency of the diffuser. *Answers:* (a) 127°C, (b) 12.4 kW, (c) 76.1 percent

**8–125** Obtain a relation for the second-law efficiency of a heat engine that receives heat $Q_H$ from a source at temperature $T_H$ and rejects heat $Q_L$ to a sink at $T_L$, which is higher than $T_0$ (the temperature of the surroundings), while producing work in the amount of $W$.

**8–126** Writing the first- and second-law relations and simplifying, obtain the reversible work relation for a closed system that exchanges heat with the surrounding medium at $T_0$ in the amount of $Q_0$ as well as a heat reservoir at $T_R$ in the amount of $Q_R$. (*Hint:* Eliminate $Q_0$ between the two equations.)

**8–127** Writing the first- and second-law relations and simplifying, obtain the reversible work relation for a steady-flow system that exchanges heat with the surrounding medium at $T_0$ a rate of $\dot{Q}_0$ as well as a thermal reservoir at $T_R$ at a rate of $\dot{Q}_R$. (*Hint:* Eliminate $\dot{Q}_0$ between the two equations.)

**8–128** Writing the first- and second-law relations and simplifying, obtain the reversible work relation for a uniform-flow system that exchanges heat with the surrounding medium at $T_0$ in the amount of $Q_0$ as well as a heat reservoir

at $T_R$ in the amount of $Q_R$. (*Hint:* Eliminate $Q_0$ between the two equations.)

## Fundamentals of Engineering (FE) Exam Problems

**8–129** Heat is lost through a plane wall steadily at a rate of 800 W. If the inner and outer surface temperatures of the wall are 20°C and 5°C, respectively, and the environment temperature is 0°C, the rate of exergy destruction within the wall is
(a) 40 W          (b) 17,500 W          (c) 765 W
(d) 32,800 W      (e) 0 W

**8–130** Liquid water enters an adiabatic piping system at 15°C at a rate of 3 kg/s. It is observed that the water temperature rises by 0.3°C in the pipe due to friction. If the environment temperature is also 15°C, the rate of exergy destruction in the pipe is
(a) 3.8 kW          (b) 24 kW          (c) 72 kW
(d) 98 kW          (e) 124 kW

**8–131** A heat engine receives heat from a source at 1500 K at a rate of 600 kJ/s and rejects the waste heat to a sink at 300 K. If the power output of the engine is 400 kW, the second-law efficiency of this heat engine is
(a) 42%          (b) 53%          (c) 83%
(d) 67%          (e) 80%

**8–132** A water reservoir contains 100 tons of water at an average elevation of 60 m. The maximum amount of electric power that can be generated from this water is
(a) 8 kWh          (b) 16 kWh          (c) 1630 kWh
(d) 16,300 kWh     (e) 58,800 kWh

**8–133** A house is maintained at 21°C in winter by electric resistance heaters. If the outdoor temperature is 9°C, the second-law efficiency of the resistance heaters is
(a) 0%          (b) 4.1%          (c) 5.7%
(d) 25%          (e) 100%

**8–134** A 12-kg solid whose specific heat is 2.8 kJ/kg·°C is at a uniform temperature of −10°C. For an environment temperature of 20°C, the exergy content of this solid is
(a) Less than zero          (b) 0 kJ          (c) 4.6 kJ
(d) 55 kJ                    (e) 1008 kJ

**8–135** Keeping the limitations imposed by the second law of thermodynamics in mind, choose the *wrong* statement below:
(a) A heat engine cannot have a thermal efficiency of 100%.
(b) For all reversible processes, the second-law efficiency is 100%.
(c) The second-law efficiency of a heat engine cannot be greater than its thermal efficiency.
(d) The second-law efficiency of a process is 100% if no entropy is generated during that process.
(e) The coefficient of performance of a refrigerator can be greater than 1.

**8–136** A furnace can supply heat steadily at a 1300 K at a rate of 500 kJ/s. The maximum amount of power that can be produced by using the heat supplied by this furnace in an environment at 300 K is

(a) 115 kW          (b) 192 kW          (c) 385 kW
(d) 500 kW          (e) 650 kW

**8–137** Air is throttled from 50°C and 800 kPa to a pressure of 200 kPa at a rate of 0.5 kg/s in an environment at 25°C. The change in kinetic energy is negligible, and no heat transfer occurs during the process. The power potential wasted during this process is

(a) 0               (b) 0.20 kW         (c) 47 kW
(d) 59 kW           (e) 119 kW

**8–138** Steam enters a turbine steadily at 4 MPa and 400°C and exits at 0.2 MPa and 150°C in an environment at 25°C. The decrease in the exergy of the steam as it flows through the turbine is

(a) 58 kJ/kg        (b) 445 kJ/kg       (c) 458 kJ/kg
(d) 518 kJ/kg       (e) 597 kJ/kg

## Design and Essay Problems

**8–139** Obtain the following information about a power plant that is closest to your town: the net power output; the type and amount of fuel used; the power consumed by the pumps, fans, and other auxiliary equipment; stack gas losses; temperatures at several locations; and the rate of heat rejection at the condenser. Using these and other relevant data, determine the rate of irreversibility in that power plant.

**8–140** Human beings are probably the most capable creatures, and they have a high level of physical, intellectual, emotional, and spiritual potentials or exergies. Unfortunately people make little use of their exergies, letting most of their exergies go to waste. Draw four exergy versus time charts, and plot your physical, intellectual, emotional, and spiritual exergies on each of these charts for a 24-h period using your best judgment based on your experience. On these four charts, plot your respective exergies that you have utilized during the last 24 h. Compare the two plots on each chart and determine if you are living a "full" life or if you are wasting your life away. Can you think of any ways to reduce the mismatch between your exergies and your utilization of them?

**8–141** The domestic hot-water systems involve a high level of irreversibility and thus they have low second-law efficiencies. The water in these systems is heated from about 15°C to about 60°C, and most of the hot water is mixed with cold water to reduce its temperature to 45°C or even lower before it is used for any useful purpose such as taking a shower or washing clothes at a warm setting. The water is discarded at about the same temperature at which it was used and replaced by fresh cold water at 15°C. Redesign a typical residential hot-water system such that the irreversibility is greatly reduced. Draw a sketch of your proposed design.

**8–142** Consider natural gas, electric resistance, and heat pump heating systems. For a specified heating load, which one of these systems will do the job with the least irreversibility? Explain.

**8–143** The temperature of the air in a building can be maintained at a desirable level during winter by using different methods of heating. Compare heating this air in a heat exchanger unit with condensing steam to heating it with an electric-resistance heater. Perform a second-law analysis to determine the heating method that generates the least entropy and thus causes the least exergy destruction.

# GAS POWER CYCLES

■ ■ ■ ■ ■ ■ ■

**OBJECTIVES**

The objectives of Chapter 9 are to:

■    Evaluate the performance of gas power cycles for which the working fluid remains a gas throughout the entire cycle.

■    Develop simplifying assumptions applicable to gas power cycles.

■    Review the operation of reciprocating engines.

■    Analyze both closed and open gas power cycles.

■    Solve problems based on the Otto, Diesel, Stirling, and Ericsson cycles.

■    Solve problems based on the Brayton cycle; the Brayton cycle with regeneration; and the Brayton cycle with intercooling, reheating, and regeneration.

■    Analyze jet-propulsion cycles.

■    Perform second-law analysis of gas power cycles.

**T**wo important areas of application for thermodynamics are power generation and refrigeration. Both are usually accomplished by systems that operate on a thermodynamic cycle. Thermodynamic cycles can be divided into two general categories: *power cycles,* which are discussed in this chapter and Chap. 10, and *refrigeration cycles,* which are discussed in Chap. 11. The devices or systems used to produce a net power output are often called *engines,* and the thermodynamic cycles they operate on are called *power cycles.* The devices or systems used to produce a refrigeration effect are called *refrigerators, air conditioners,* or *heat pumps,* and the cycles they operate on are called *refrigeration cycles.*

Thermodynamic cycles can also be categorized as *gas cycles* and *vapor cycles,* depending on the *phase* of the working fluid. In gas cycles, the working fluid remains in the gaseous phase throughout the entire cycle, whereas in vapor cycles the working fluid exists in the vapor phase during one part of the cycle and in the liquid phase during another part.

Thermodynamic cycles can be categorized yet another way: *closed* and *open cycles.* In closed cycles, the working fluid is returned to the initial state at the end of the cycle and is recirculated. In open cycles, the working fluid is renewed at the end of each cycle instead of being recirculated. In automobile engines, the combustion gases are exhausted and replaced by fresh air–fuel mixture at the end of each cycle. The engine operates on a mechanical cycle, but the working fluid does not go through a complete thermodynamic cycle.

Heat engines are categorized as *internal combustion* and *external combustion engines,* depending on how the heat is supplied to the working fluid. In external combustion engines (such as steam power plants), heat is supplied to the working fluid from an external source such as a furnace, a geothermal well, a nuclear reactor, or even the sun. In internal combustion engines (such as automobile engines), this is done by burning the fuel within the system boundaries. In this chapter, various gas power cycles are analyzed under some simplifying assumptions.

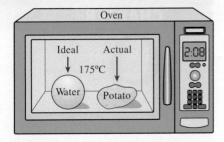

**FIGURE 9–1**

Modeling is a powerful engineering tool that provides great insight and simplicity at the expense of some loss in accuracy.

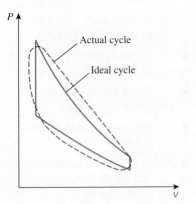

**FIGURE 9–2**

The analysis of many complex processes can be reduced to a manageable level by utilizing some idealizations.

# 9–1 · BASIC CONSIDERATIONS IN THE ANALYSIS OF POWER CYCLES

Most power-producing devices operate on cycles, and the study of power cycles is an exciting and important part of thermodynamics. The cycles encountered in actual devices are difficult to analyze because of the presence of complicating effects, such as friction, and the absence of sufficient time for establishment of the equilibrium conditions during the cycle. To make an analytical study of a cycle feasible, we have to keep the complexities at a manageable level and utilize some idealizations (Fig. 9–1). When the actual cycle is stripped of all the internal irreversibilities and complexities, we end up with a cycle that resembles the actual cycle closely but is made up totally of internally reversible processes. Such a cycle is called an **ideal cycle** (Fig. 9–2).

A simple idealized model enables engineers to study the effects of the major parameters that dominate the cycle without getting bogged down in the details. The cycles discussed in this chapter are somewhat idealized, but they still retain the general characteristics of the actual cycles they represent. The conclusions reached from the analysis of ideal cycles are also applicable to actual cycles. The thermal efficiency of the Otto cycle, the ideal cycle for spark-ignition automobile engines, for example, increases with the compression ratio. This is also the case for actual automobile engines. The numerical values obtained from the analysis of an ideal cycle, however, are not necessarily representative of the actual cycles, and care should be exercised in their interpretation. The simplified analysis presented in this chapter for various power cycles of practical interest may also serve as the starting point for a more in-depth study.

Heat engines are designed for the purpose of converting thermal energy to work, and their performance is expressed in terms of the **thermal efficiency** $\eta_{th}$, which is the ratio of the net work produced by the engine to the total heat input:

$$\eta_{th} = \frac{W_{net}}{Q_{in}} \quad \text{or} \quad \eta_{th} = \frac{w_{net}}{q_{in}} \tag{9–1}$$

Recall that heat engines that operate on a totally reversible cycle, such as the Carnot cycle, have the highest thermal efficiency of all heat engines operating between the same temperature levels. That is, nobody can develop a cycle more efficient than the *Carnot cycle*. Then the following question arises naturally: If the Carnot cycle is the best possible cycle, why do we not use it as the model cycle for all the heat engines instead of bothering with several so-called *ideal* cycles? The answer to this question is hardware-related. Most cycles encountered in practice differ significantly from the Carnot cycle, which makes it unsuitable as a realistic model. Each ideal cycle discussed in this chapter is related to a specific work-producing device and is an *idealized* version of the actual cycle.

The ideal cycles are internally reversible, but, unlike the Carnot cycle, they are not necessarily externally reversible. That is, they may involve irreversibilities external to the system such as heat transfer through a finite temperature difference. Therefore, the thermal efficiency of an ideal cycle, in

**FIGURE 9–3**
An automotive engine with the combustion chamber exposed.
*©Idealink Photography/Alamy RF*

general, is less than that of a totally reversible cycle operating between the same temperature limits. However, it is still considerably higher than the thermal efficiency of an actual cycle because of the idealizations utilized (Fig. 9–3).

The idealizations and simplifications commonly employed in the analysis of power cycles can be summarized as follows:

1. The cycle does not involve any *friction*. Therefore, the working fluid does not experience any pressure drop as it flows in pipes or devices such as heat exchangers.
2. All expansion and compression processes take place in a *quasi-equilibrium* manner.
3. The pipes connecting the various components of a system are well insulated, and *heat transfer* through them is negligible.

Neglecting the changes in *kinetic* and *potential energies* of the working fluid is another commonly utilized simplification in the analysis of power cycles. This is a reasonable assumption since in devices that involve shaft work, such as turbines, compressors, and pumps, the kinetic and potential energy terms are usually very small relative to the other terms in the energy equation. Fluid velocities encountered in devices such as condensers, boilers, and mixing chambers are typically low, and the fluid streams experience little change in their velocities, again making kinetic energy changes negligible. The only devices where the changes in kinetic energy are significant are the nozzles and diffusers, which are specifically designed to create large changes in velocity.

In the preceding chapters, *property diagrams* such as the $P$-$v$ and $T$-$s$ diagrams have served as valuable aids in the analysis of thermodynamic processes. On both the $P$-$v$ and $T$-$s$ diagrams, the area enclosed by the process curves of a cycle represents the net work produced during the cycle (Fig. 9–4), which is also equivalent to the net heat transfer for that cycle. The $T$-$s$ diagram is particularly useful as a visual aid in the analysis of ideal power cycles. An ideal power cycle does not involve any internal irreversibilities, and so the only effect that can change the entropy of the working fluid during a process is heat transfer.

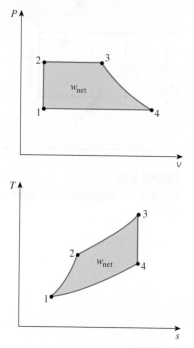

**FIGURE 9–4**
On both $P$-$v$ and $T$-$s$ diagrams, the area enclosed by the process curve represents the net work of the cycle.

On a *T-s* diagram, a *heat-addition* process proceeds in the direction of increasing entropy, a *heat-rejection* process proceeds in the direction of decreasing entropy, and an *isentropic* (internally reversible, adiabatic) process proceeds at constant entropy. The area under the process curve on a *T-s* diagram represents the heat transfer for that process. The area under the heat addition process on a *T-s* diagram is a geometric measure of the total heat supplied during the cycle $q_{in}$, and the area under the heat rejection process is a measure of the total heat rejected $q_{out}$. The difference between these two (the area enclosed by the cyclic curve) is the net heat transfer, which is also the net work produced during the cycle. Therefore, on a *T-s* diagram, the ratio of the area enclosed by the cyclic curve to the area under the heat-addition process curve represents the thermal efficiency of the cycle. *Any modification that increases the ratio of these two areas will also increase the thermal efficiency of the cycle.*

Although the working fluid in an ideal power cycle operates on a closed loop, the type of individual processes that comprises the cycle depends on the individual devices used to execute the cycle. In the Rankine cycle, which is the ideal cycle for steam power plants, the working fluid flows through a series of steady-flow devices such as the turbine and condenser, whereas in the Otto cycle, which is the ideal cycle for the spark-ignition automobile engine, the working fluid is alternately expanded and compressed in a piston–cylinder device. Therefore, equations pertaining to steady-flow systems should be used in the analysis of the Rankine cycle, and equations pertaining to closed systems should be used in the analysis of the Otto cycle.

# 9–2 · THE CARNOT CYCLE AND ITS VALUE IN ENGINEERING

The Carnot cycle is composed of four totally reversible processes: isothermal heat addition, isentropic expansion, isothermal heat rejection, and isentropic compression. The *P-v* and *T-s* diagrams of a Carnot cycle are replotted in Fig. 9–5. The Carnot cycle can be executed in a closed system (a piston–cylinder device) or a steady-flow system (utilizing two turbines and two compressors, as shown in Fig. 9–6), and either a gas or a vapor can be utilized as the working fluid. The Carnot cycle is the most efficient cycle that can be executed between a heat source at temperature $T_H$ and a sink at temperature $T_L$, and its thermal efficiency is expressed as

$$\eta_{th,Carnot} = 1 - \frac{T_L}{T_H} \tag{9-2}$$

Reversible isothermal heat transfer is very difficult to achieve in reality because it would require very large heat exchangers and it would take a very long time (a power cycle in a typical engine is completed in a fraction of a second). Therefore, it is not practical to build an engine that would operate on a cycle that closely approximates the Carnot cycle.

The real value of the Carnot cycle comes from its being a standard against which the actual or the ideal cycles can be compared. The thermal efficiency of the Carnot cycle is a function of the sink and source

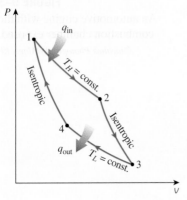

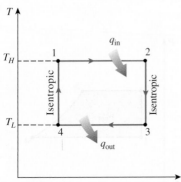

**FIGURE 9–5**

*P-v* and *T-s* diagrams of a Carnot cycle.

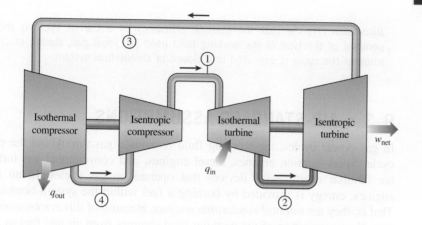

**FIGURE 9–6**
A steady-flow Carnot engine.

temperatures only, and the thermal efficiency relation for the Carnot cycle (Eq. 9–2) conveys an important message that is equally applicable to both ideal and actual cycles: *Thermal efficiency increases with an increase in the average temperature at which heat is supplied to the system or with a decrease in the average temperature at which heat is rejected from the system.*

The source and sink temperatures that can be used in practice are not without limits, however. The highest temperature in the cycle is limited by the maximum temperature that the components of the heat engine, such as the piston or the turbine blades, can withstand. The lowest temperature is limited by the temperature of the cooling medium utilized in the cycle such as a lake, a river, or the atmospheric air.

■ **EXAMPLE 9–1**    **Derivation of the Efficiency of the Carnot Cycle**

Show that the thermal efficiency of a Carnot cycle operating between the temperature limits of $T_H$ and $T_L$ is solely a function of these two temperatures and is given by Eq. 9–2.

**SOLUTION**   It is to be shown that the efficiency of a Carnot cycle depends on the source and sink temperatures alone.
**Analysis**   The $T$-$s$ diagram of a Carnot cycle is redrawn in Fig. 9–7. All four processes that comprise the Carnot cycle are reversible, and thus the area under each process curve represents the heat transfer for that process. Heat is transferred to the system during process 1-2 and rejected during process 3-4. Therefore, the amount of heat input and heat output for the cycle can be expressed as

$$q_{in} = T_H(s_2 - s_1) \quad \text{and} \quad q_{out} = T_L(s_3 - s_4) = T_L(s_2 - s_1)$$

since processes 2-3 and 4-1 are isentropic, and thus $s_2 = s_3$ and $s_4 = s_1$. Substituting these into Eq. 9–1, we see that the thermal efficiency of a Carnot cycle is

$$\eta_{th} = \frac{w_{net}}{q_{in}} = 1 - \frac{q_{out}}{q_{in}} = 1 - \frac{T_L(s_2 - s_1)}{T_H(s_2 - s_1)} = 1 - \frac{T_L}{T_H}$$

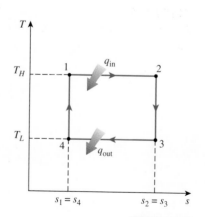

**FIGURE 9–7**
$T$-$s$ diagram for Example 9–1.

## 9–3 · AIR-STANDARD ASSUMPTIONS

In gas power cycles, the working fluid remains a gas throughout the entire cycle. Spark-ignition engines, diesel engines, and conventional gas turbines are familiar examples of devices that operate on gas cycles. In all these engines, energy is provided by burning a fuel within the system boundaries. That is, they are *internal combustion engines*. Because of this combustion process, the composition of the working fluid changes from air and fuel to combustion products during the course of the cycle. However, considering that air is predominantly nitrogen that undergoes hardly any chemical reactions in the combustion chamber, the working fluid closely resembles air at all times.

Even though internal combustion engines operate on a mechanical cycle (the piston returns to its starting position at the end of each revolution), the working fluid does not undergo a complete thermodynamic cycle. It is thrown out of the engine at some point in the cycle (as exhaust gases) instead of being returned to the initial state. Working on an open cycle is the characteristic of all internal combustion engines.

The actual gas power cycles are rather complex. To reduce the analysis to a manageable level, we utilize the following approximations, commonly known as the **air-standard assumptions**:

1. The working fluid is air, which continuously circulates in a closed loop and always behaves as an ideal gas.
2. All the processes that make up the cycle are internally reversible.
3. The combustion process is replaced by a heat-addition process from an external source (Fig. 9–8).
4. The exhaust process is replaced by a heat-rejection process that restores the working fluid to its initial state.

Another assumption that is often utilized to simplify the analysis even more is that air has constant specific heats whose values are determined at *room temperature* (25°C). When this assumption is utilized, the air-standard assumptions are called the **cold-air-standard assumptions**. A cycle for which the air-standard assumptions are applicable is frequently referred to as an **air-standard cycle**.

The air-standard assumptions previously stated provide considerable simplification in the analysis without significantly deviating from the actual cycles. This simplified model enables us to study qualitatively the influence of major parameters on the performance of the actual engines.

## 9–4 · AN OVERVIEW OF RECIPROCATING ENGINES

Despite its simplicity, the reciprocating engine (basically a piston–cylinder device) is one of the rare inventions that has proved to be very versatile and to have a wide range of applications. It is the powerhouse of the vast majority

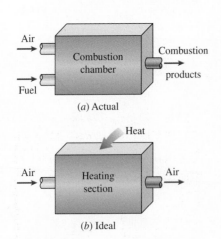

**FIGURE 9–8**

The combustion process is replaced by a heat-addition process in ideal cycles.

of automobiles, trucks, light aircraft, ships, and electric power generators, as well as many other devices.

The basic components of a reciprocating engine are shown in Fig. 9–9. The piston reciprocates in the cylinder between two fixed positions called the **top dead center** (TDC)—the position of the piston when it forms the smallest volume in the cylinder—and the **bottom dead center** (BDC)—the position of the piston when it forms the largest volume in the cylinder. The distance between the TDC and the BDC is the largest distance that the piston can travel in one direction, and it is called the **stroke** of the engine. The diameter of the piston is called the **bore**. The air or air–fuel mixture is drawn into the cylinder through the **intake valve**, and the combustion products are expelled from the cylinder through the **exhaust valve**.

The minimum volume formed in the cylinder when the piston is at TDC is called the **clearance volume** (Fig. 9–10). The volume displaced by the piston as it moves between TDC and BDC is called the **displacement volume**. The ratio of the maximum volume formed in the cylinder to the minimum (clearance) volume is called the **compression ratio** $r$ of the engine:

$$r = \frac{V_{max}}{V_{min}} = \frac{V_{BDC}}{V_{TDC}} \qquad (9\text{–}3)$$

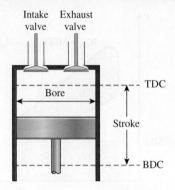

**FIGURE 9–9**
Nomenclature for reciprocating engines.

Notice that the compression ratio is a *volume ratio* and should not be confused with the pressure ratio.

Another term frequently used in conjunction with reciprocating engines is the **mean effective pressure** (MEP). It is a fictitious pressure that, if it acted on the piston during the entire power stroke, would produce the same amount of net work as that produced during the actual cycle (Fig. 9–11). That is,

$$W_{net} = \text{MEP} \times \text{Piston area} \times \text{Stroke} = \text{MEP} \times \text{Displacement volume}$$

or

$$\text{MEP} = \frac{W_{net}}{V_{max} - V_{min}} = \frac{w_{net}}{V_{max} - V_{min}} \quad (\text{kPa}) \qquad (9\text{–}4)$$

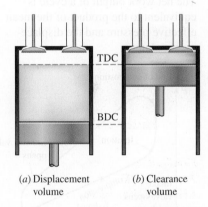

(a) Displacement volume   (b) Clearance volume

**FIGURE 9–10**
Displacement and clearance volumes of a reciprocating engine.

The mean effective pressure can be used as a parameter to compare the performances of reciprocating engines of equal size. The engine with a larger value of MEP delivers more net work per cycle and thus performs better.

Reciprocating engines are classified as **spark-ignition (SI) engines** or **compression-ignition (CI) engines**, depending on how the combustion process in the cylinder is initiated. In SI engines, the combustion of the air–fuel mixture is initiated by a spark plug. In CI engines, the air–fuel mixture is self-ignited as a result of compressing the mixture above its self-ignition temperature. In the next two sections, we discuss the *Otto* and *Diesel cycles,* which are the ideal cycles for the SI and CI reciprocating engines, respectively.

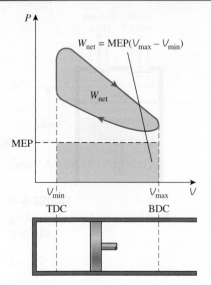

**FIGURE 9–11**

The net work output of a cycle is equivalent to the product of the mean effective pressure and the displacement volume.

# 9–5 · OTTO CYCLE: THE IDEAL CYCLE FOR SPARK-IGNITION ENGINES

The Otto cycle is the ideal cycle for spark-ignition reciprocating engines. It is named after Nikolaus A. Otto, who built a successful four-stroke engine in 1876 in Germany using the cycle proposed by Frenchman Beau de Rochas in 1862. In most spark-ignition engines, the piston executes four complete strokes (two mechanical cycles) within the cylinder, and the crankshaft completes two revolutions for each thermodynamic cycle. These engines are called **four-stroke** internal combustion engines. A schematic of each stroke as well as a *P-v* diagram for an actual four-stroke spark-ignition engine is given in Fig. 9–12a.

Initially, both the intake and the exhaust valves are closed, and the piston is at its lowest position (BDC). During the *compression stroke,* the piston moves upward, compressing the air–fuel mixture. Shortly before the piston reaches its highest position (TDC), the spark plug fires and the mixture ignites, increasing the pressure and temperature of the system. The high-pressure gases force the piston down, which in turn forces the crankshaft to rotate, producing a useful work output during the *expansion* or *power stroke.* Towards the end of expansion stroke, the exhaust valve opens and the combustion gases that are above the atmospheric pressure rush out of the cylinder through the open exhaust valve. This process is called **exhaust blowdown**, and most combustion gases

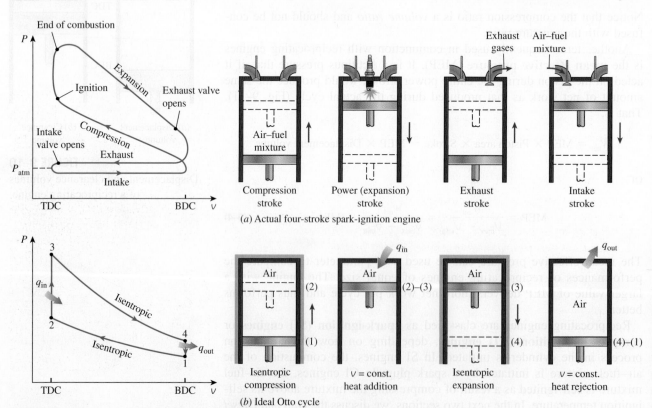

(a) Actual four-stroke spark-ignition engine

(b) Ideal Otto cycle

**FIGURE 9–12**

Actual and ideal cycles in spark-ignition engines and their *P-v* diagrams.

leave the cylinder by the time the piston reaches BDC. The cylinder is still filled by the exhaust gases at a lower pressure at BDC. Now the piston moves upward one more time, purging the exhaust gases through the exhaust valve (the *exhaust stroke*), and down a second time, drawing in fresh air–fuel mixture through the intake valve (the *intake stroke*). Notice that the pressure in the cylinder is slightly above the atmospheric value during the exhaust stroke and slightly below during the intake stroke.

In **two-stroke engines**, all four functions described above are executed in just two strokes: the power stroke and the compression stroke. In these engines, the crankcase is sealed, and the outward motion of the piston is used to slightly pressurize the air–fuel mixture in the crankcase, as shown in Fig. 9–13. Also, the intake and exhaust valves are replaced by openings in the lower portion of the cylinder wall. During the latter part of the power stroke, the piston uncovers first the exhaust port, allowing the exhaust gases to be partially expelled, and then the intake port, allowing the fresh air–fuel mixture to rush in and drive most of the remaining exhaust gases out of the cylinder. This mixture is then compressed as the piston moves upward during the compression stroke and is subsequently ignited by a spark plug.

The two-stroke engines are generally less efficient than their four-stroke counterparts because of the incomplete expulsion of the exhaust gases and the partial expulsion of the fresh air–fuel mixture with the exhaust gases. However, they are relatively simple and inexpensive, and they have high power-to-weight and power-to-volume ratios, which make them suitable for applications requiring small size and weight such as for motorcycles, chain saws, and lawn mowers (Fig. 9–14).

Advances in several technologies—such as direct fuel injection, stratified charge combustion, and electronic controls—brought about a renewed interest in two-stroke engines that can offer high performance and fuel economy while satisfying the stringent emission requirements. For a given weight and displacement, a well-designed two-stroke engine can provide significantly more power than its four-stroke counterpart because two-stroke engines produce power on every engine revolution instead of every other one. In the new two-stroke engines, the highly atomized fuel spray that is injected into the combustion chamber toward the end of the compression stroke burns much more completely. The fuel is sprayed after the exhaust valve is closed, which prevents unburned fuel from being ejected into the atmosphere. With stratified combustion, the flame that is initiated by igniting a small amount of the rich fuel–air mixture near the spark plug propagates through the combustion chamber filled with a much leaner mixture, and this results in much cleaner combustion. Also, the advances in electronics have made it possible to ensure the optimum operation under varying engine load and speed conditions. Major car companies have research programs underway on two-stroke engines which are expected to make a comeback in the future.

The thermodynamic analysis of the actual four-stroke or two-stroke cycles described is not a simple task. However, the analysis can be simplified significantly if the air-standard assumptions are utilized. The resulting cycle, which closely resembles the actual operating conditions, is the ideal **Otto cycle**. It consists of four internally reversible processes:

1-2  Isentropic compression

2-3  Constant-volume heat addition

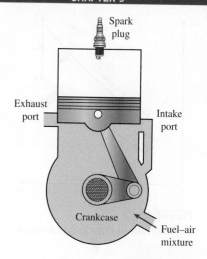

**FIGURE 9–13**
Schematic of a two-stroke reciprocating engine.

**FIGURE 9–14**
Two-stroke engines are commonly used in motorcycles and lawn mowers.
*©John A. Rizzo/Getty Images RF*

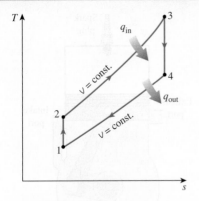

**FIGURE 9–15**
*T-s* diagram of the ideal Otto cycle.

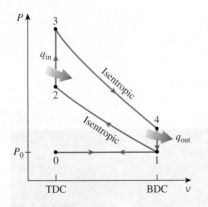

**FIGURE 9–16**
*P-v* diagram of the ideal Otto cycle that includes intake and exhaust strokes.

3-4  Isentropic expansion

4-1  Constant-volume heat rejection

The execution of the Otto cycle in a piston–cylinder device together with a *P-v* diagram is illustrated in Fig. 9–12*b*. The *T-s* diagram of the Otto cycle is given in Fig. 9–15.

The ideal Otto cycle shown in Fig. 9–12*b* has one shortcoming. This ideal cycle consists of two strokes equivalent to one mechanical cycle or one crankshaft rotation. The actual engine operation shown in Fig. 9–12*a*, on the other hand, involves four strokes equivalent to two mechanical cycles or two crankshaft rotations. This can be corrected by including intake and exhaust strokes in the ideal Otto cycle, as shown in Fig. 9–16. In this modified cycle, air-fuel mixture (approximated as air due to air-standard assumptions) enters the cylinder through the open intake valve at atmospheric pressure $P_0$ during process 0-1 as the piston moves from TDC to BDC. The intake valve is closed at state 1 and air is compressed isentropically to state 2. Heat is transferred at constant volume (process 2-3); it is expanded isentropically to state 4; and heat is rejected at constant volume (process 4-1). Exhaust gases (again approximated as air) are expelled through the open exhaust valve (process 1-0) as the pressure remains constant at $P_0$.

The modified Otto cycle shown in Fig. 9–16 is executed in an open system during the intake and exhaust processes and in a closed system during the remaining four processes. We should point out that the constant-volume heat addition process (2-3) in the ideal Otto cycle replaces the combustion process of the actual engine operation while the constant-volume heat rejection process (4-1) replaces the exhaust blowdown.

The work interactions during the constant-pressure intake (0-1) and constant-pressure exhaust (1-0) processes can be expressed as

$$W_{\text{out},0-1} = P_0(v_1 - v_0)$$

$$W_{\text{in},1-0} = P_0(v_1 - v_0)$$

These two processes cancel each other as the work output during the intake is equal to work input during the exhaust. Then, the cycle reduces to the one in Fig. 9–12*b*. Therefore, inclusion of the intake and exhaust processes has no effect on the net work output from the cycle. However, when calculating power output from the cycle during an ideal Otto cycle analysis, we must consider the fact that the ideal Otto cycle has four strokes just like actual four-stroke spark-ignition engine. This is illustrated in the last part of Example 9–2.

The Otto cycle is executed in a closed system, and disregarding the changes in kinetic and potential energies, the energy balance for any of the processes is expressed, on a unit-mass basis, as

$$(q_{\text{in}} - q_{\text{out}}) + (w_{\text{in}} - w_{\text{out}}) = \Delta u \quad \text{(kJ/kg)} \tag{9–5}$$

No work is involved during the two heat transfer processes since both take place at constant volume. Therefore, heat transfer to and from the working fluid can be expressed as

$$q_{\text{in}} = u_3 - u_2 = c_v(T_3 - T_2) \tag{9–6a}$$

and

$$q_{\text{out}} = u_4 - u_1 = c_v(T_4 - T_1) \tag{9–6b}$$

Then the thermal efficiency of the ideal Otto cycle under the cold air standard assumptions becomes

$$\eta_{\text{th,Otto}} = \frac{w_{\text{net}}}{q_{\text{in}}} = 1 - \frac{q_{\text{out}}}{q_{\text{in}}} = 1 - \frac{T_4 - T_1}{T_3 - T_2} = 1 - \frac{T_1(T_4/T_1 - 1)}{T_2(T_3/T_2 - 1)}$$

Processes 1-2 and 3-4 are isentropic, and $v_2 = v_3$ and $v_4 = v_1$. Thus,

$$\frac{T_1}{T_2} = \left(\frac{v_2}{v_1}\right)^{k-1} = \left(\frac{v_3}{v_4}\right)^{k-1} = \frac{T_4}{T_3} \tag{9–7}$$

Substituting these equations into the thermal efficiency relation and simplifying give

$$\eta_{\text{th,Otto}} = 1 - \frac{1}{r^{k-1}} \tag{9–8}$$

where

$$r = \frac{V_{\text{max}}}{V_{\text{min}}} = \frac{V_1}{V_2} = \frac{v_1}{v_2} \tag{9–9}$$

is the **compression ratio** and $k$ is the specific heat ratio $c_p/c_v$.

Equation 9–8 shows that under the cold-air-standard assumptions, the thermal efficiency of an ideal Otto cycle depends on the compression ratio of the engine and the specific heat ratio of the working fluid. The thermal efficiency of the ideal Otto cycle increases with both the compression ratio and the specific heat ratio. This is also true for actual spark-ignition internal combustion engines. A plot of thermal efficiency versus the compression ratio is given in Fig. 9–17 for $k = 1.4$, which is the specific heat ratio value of air at room temperature. For a given compression ratio, the thermal efficiency of an actual spark-ignition engine is less than that of an ideal Otto cycle because of the irreversibilities, such as friction, and other factors such as incomplete combustion.

We can observe from Fig. 9–17 that the thermal efficiency curve is rather steep at low compression ratios but flattens out starting with a compression ratio value of about 8. Therefore, the increase in thermal efficiency with the compression ratio is not as pronounced at high compression ratios. Also, when high compression ratios are used, the temperature of the air–fuel mixture rises above the autoignition temperature of the fuel (the temperature at which the fuel ignites without the help of a spark) during the combustion process, causing an early and rapid burn of the fuel at some point or points ahead of the flame front, followed by almost instantaneous inflammation of the end gas. This premature ignition of the fuel, called **autoignition**, produces an audible noise, which is called **engine knock**. Autoignition in spark-ignition engines cannot be tolerated because it hurts performance and can cause engine damage. The requirement that autoignition not be allowed places an upper limit on the compression ratios that can be used in spark-ignition internal combustion engines.

Improvement of the thermal efficiency of gasoline engines by utilizing higher compression ratios (up to about 12) without facing the autoignition

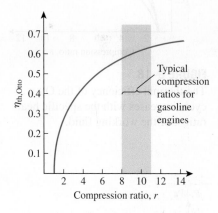

**FIGURE 9–17**

Thermal efficiency of the ideal Otto cycle as a function of compression ratio ($k = 1.4$).

problem has been made possible by using gasoline blends that have good antiknock characteristics, such as gasoline mixed with tetraethyl lead. Tetraethyl lead had been added to gasoline since the 1920s because it is an inexpensive method of raising the *octane rating,* which is a measure of the engine knock resistance of a fuel. Leaded gasoline, however, has a very undesirable side effect: it forms compounds during the combustion process that are hazardous to health and pollute the environment. In an effort to combat air pollution, the government adopted a policy in the mid-1970s that resulted in the eventual phase-out of leaded gasoline. Unable to use lead, the refiners developed other techniques to improve the antiknock characteristics of gasoline. Most cars made since 1975 have been designed to use unleaded gasoline, and the compression ratios had to be lowered to avoid engine knock. The ready availability of high octane fuels made it possible to raise the compression ratios again in recent years. Also, owing to the improvements in other areas (reduction in overall automobile weight, improved aerodynamic design, etc.), today's cars have better fuel economy. This is an example of how engineering decisions involve compromises, and efficiency is only one of the considerations in final design.

The second parameter affecting the thermal efficiency of an ideal Otto cycle is the specific heat ratio $k$. For a given compression ratio, an ideal Otto cycle using a monatomic gas (such as argon or helium, $k = 1.667$) as the working fluid will have the highest thermal efficiency. The specific heat ratio $k$, and thus the thermal efficiency of the ideal Otto cycle, decreases as the molecules of the working fluid get larger (Fig. 9–18). At room temperature it is 1.4 for air, 1.3 for carbon dioxide, and 1.2 for ethane. The working fluid in actual engines contains larger molecules such as carbon dioxide, and the specific heat ratio decreases with temperature, which is one of the reasons that the actual cycles have lower thermal efficiencies than the ideal Otto cycle. The thermal efficiencies of actual spark-ignition engines range from about 25 to 30 percent.

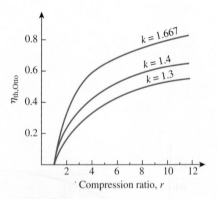

**FIGURE 9–18**
The thermal efficiency of the Otto cycle increases with the specific heat ratio $k$ of the working fluid.

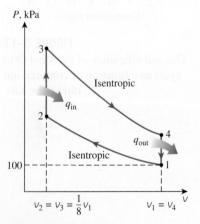

**FIGURE 9–19**
$P$-$v$ diagram for the Otto cycle discussed in Example 9–2.

### EXAMPLE 9–2   The Ideal Otto Cycle

An ideal Otto cycle has a compression ratio of 8. At the beginning of the compression process, air is at 100 kPa and 17°C, and 800 kJ/kg of heat is transferred to air during the constant-volume heat-addition process. Accounting for the variation of specific heats of air with temperature, determine (a) the maximum temperature and pressure that occur during the cycle, (b) the net work output, (c) the thermal efficiency, and (d) the mean effective pressure for the cycle.

(e) Also, determine the power output from the cycle, in kW, for an engine speed of 4000 rpm (rev/min). Assume that this cycle is operated on an engine that has four cylinders with a total displacement volume of 1.6 L.

**SOLUTION**   An ideal Otto cycle is considered. The maximum temperature and pressure, the net work output, the thermal efficiency, and the mean effective pressure are to be determined.

**Assumptions**   1 The air-standard assumptions are applicable. 2 Kinetic and potential energy changes are negligible. 3 The variation of specific heats with temperature is to be accounted for.

**Analysis**   The $P$-$v$ diagram of the ideal Otto cycle described is shown in Fig. 9–19. We note that the air contained in the cylinder forms a closed system.

(a) The maximum temperature and pressure in an Otto cycle occur at the end of the constant-volume heat-addition process (state 3). But first we need to determine the temperature and pressure of air at the end of the isentropic compression process (state 2), using data from Table A–17:

$$T_1 = 290 \text{ K} \rightarrow u_1 = 206.91 \text{ kJ/kg}$$
$$v_{r1} = 676.1$$

Process 1-2 (isentropic compression of an ideal gas):

$$\frac{v_{r2}}{v_{r1}} = \frac{v_2}{v_1} = \frac{1}{r} \rightarrow v_{r2} = \frac{v_{r1}}{r} = \frac{676.1}{8} = 84.51 \rightarrow T_2 = 652.4 \text{ K}$$
$$u_2 = 475.11 \text{ kJ/kg}$$

$$\frac{P_2 v_2}{T_2} = \frac{P_1 v_1}{T_1} \rightarrow P_2 = P_1 \left(\frac{T_2}{T_1}\right)\left(\frac{v_1}{v_2}\right)$$
$$= (100 \text{ kPa})\left(\frac{652.4 \text{ K}}{290 \text{ K}}\right)(8) = 1799.7 \text{ kPa}$$

Process 2-3 (constant-volume heat addition):

$$q_{in} = u_3 - u_2$$
$$800 \text{ kJ/kg} = u_3 - 475.11 \text{ kJ/kg}$$
$$u_3 = 1275.11 \text{ kJ/kg} \rightarrow T_3 = \mathbf{1575.1 \text{ K}}$$
$$v_{r3} = 6.108$$

$$\frac{P_3 v_3}{T_3} = \frac{P_2 v_2}{T_2} \rightarrow P_3 = P_2 \left(\frac{T_3}{T_2}\right)\left(\frac{v_2}{v_3}\right)$$
$$= (1.7997 \text{ MPa})\left(\frac{1575.1 \text{ K}}{652.4 \text{ K}}\right)(1) = \mathbf{4.345 \text{ MPa}}$$

(b) The net work output for the cycle is determined either by finding the boundary ($P \, dV$) work involved in each process by integration and adding them or by finding the net heat transfer that is equivalent to the net work done during the cycle. We take the latter approach. However, first we need to find the internal energy of the air at state 4:
  Process 3-4 (isentropic expansion of an ideal gas):

$$\frac{v_{r4}}{v_{r3}} = \frac{v_4}{v_3} = r \rightarrow v_{r4} = r v_{r3} = (8)(6.108) = 48.864 \rightarrow T_4 = 795.6 \text{ K}$$
$$u_4 = 588.74 \text{ kJ/kg}$$

Process 4-1 (constant-volume heat rejection):

$$-q_{out} = u_1 - u_4 \rightarrow q_{out} = u_4 - u_1$$
$$q_{out} = 588.74 - 206.91 = 381.83 \text{ kJ/kg}$$

Thus,

$$w_{net} = q_{net} = q_{in} - q_{out} = 800 - 381.83 = \mathbf{418.17 \text{ kJ/kg}}$$

(*c*) The thermal efficiency of the cycle is determined from its definition:

$$\eta_{th} = \frac{w_{net}}{q_{in}} = \frac{418.17 \text{ kJ/kg}}{800 \text{ kJ/kg}} = 0.523 \text{ or } \textbf{52.3\%}$$

Under the cold-air-standard assumptions (constant specific heat values at room temperature), the thermal efficiency would be (Eq. 9–8)

$$\eta_{th,Otto} = 1 - \frac{1}{r^{k-1}} = 1 - r^{1-k} = 1 - (8)^{1-1.4} = 0.565 \text{ or } 56.5\%$$

which is considerably different from the value obtained above. Therefore, care should be exercised in utilizing the cold-air-standard assumptions.

(*d*) The mean effective pressure is determined from its definition, Eq. 9–4:

$$\text{MEP} = \frac{w_{net}}{v_1 - v_2} = \frac{w_{net}}{v_1 - v_1/r} = \frac{w_{net}}{v_1(1 - 1/r)}$$

where

$$v_1 = \frac{RT_1}{P_1} = \frac{(0.287 \text{ kPa} \cdot \text{m}^3/\text{kg} \cdot \text{K})(290 \text{ K})}{100 \text{ kPa}} = 0.8323 \text{ m}^3/\text{kg}$$

Thus,

$$\text{MEP} = \frac{418.17 \text{ kJ/kg}}{(0.8323 \text{ m}^3/\text{kg})(1 - \frac{1}{8})}\left(\frac{1 \text{ kPa} \cdot \text{m}^3}{1 \text{ kJ}}\right) = \textbf{574 kPa}$$

(*e*) The total air mass taken by all four cylinders when they are charged is

$$m = \frac{V_d}{v_1} = \frac{0.0016 \text{ m}^3}{0.8323 \text{ m}^3/\text{kg}} = 0.001922 \text{ kg}$$

The net work produced by the cycle is

$$W_{net} = m w_{net} = (0.001922 \text{ kg})(418.17 \text{ kJ/kg}) = 0.8037 \text{ kJ}$$

That is, the net work produced per thermodynamic cycle is 0.8037 kJ/cycle. Noting that there are two revolutions per thermodynamic cycle ($n_{rev} = 2$ rev/cycle) in a four-stroke engine (or in the ideal Otto cycle including intake and exhaust strokes), the power produced by the engine is determined from

$$\dot{W}_{net} = \frac{W_{net}\dot{n}}{n_{rev}} = \frac{(0.8037 \text{ kJ/cycle})(4000 \text{ rev/min})}{2 \text{ rev/cycle}}\left(\frac{1 \text{ min}}{60 \text{ s}}\right) = \textbf{26.8 kW}$$

**Discussion** If we analyzed a two-stroke engine operating on an ideal Otto cycle with the same values, the power output would be calculated as

$$\dot{W}_{net} = \frac{W_{net}\dot{n}}{n_{rev}} = \frac{(0.8037 \text{ kJ/cycle})(4000 \text{ rev/min})}{1 \text{ rev/cycle}}\left(\frac{1 \text{ min}}{60 \text{ s}}\right) = 53.6 \text{ kW}$$

Note that there is one revolution in one thermodynamic cycle in two-stroke engines.

## 9–6 · DIESEL CYCLE: THE IDEAL CYCLE FOR COMPRESSION-IGNITION ENGINES

The Diesel cycle is the ideal cycle for CI reciprocating engines. The CI engine, first proposed by Rudolph Diesel in the 1890s, is very similar to the SI engine discussed in the last section, differing mainly in the method of initiating combustion. In spark-ignition engines (also known as *gasoline engines*), the air–fuel mixture is compressed to a temperature that is below the autoignition temperature of the fuel, and the combustion process is initiated by firing a spark plug. In CI engines (also known as *diesel engines*), the air is compressed to a temperature that is above the autoignition temperature of the fuel, and combustion starts on contact as the fuel is injected into this hot air. Therefore, the spark plug is replaced by a fuel injector in diesel engines (Fig. 9–20).

In gasoline engines, a mixture of air and fuel is compressed during the compression stroke, and the compression ratios are limited by the onset of autoignition or engine knock. In diesel engines, only air is compressed during the compression stroke, eliminating the possibility of autoignition. Therefore, diesel engines can be designed to operate at much higher compression ratios, typically between 12 and 24. Not having to deal with the problem of autoignition has another benefit: many of the stringent requirements placed on the gasoline can now be removed, and fuels that are less refined (thus less expensive) can be used in diesel engines.

The fuel injection process in diesel engines starts when the piston approaches TDC and continues during the first part of the power stroke. Therefore, the combustion process in these engines takes place over a longer interval. Because of this longer duration, the combustion process in the ideal Diesel cycle is approximated as a constant-pressure heat-addition process. In fact, this is the only process where the Otto and the Diesel cycles differ. The remaining three processes are the same for both ideal cycles. That is, process 1-2 is isentropic compression, 2-3 is constant-pressure heat addition, 3-4 is isentropic expansion, and 4-1 is constant-volume heat rejection. The similarity between the two cycles is also apparent from the *P-v* and *T-s* diagrams of the Diesel cycle, shown in Fig. 9–21.

Noting that the Diesel cycle is executed in a piston–cylinder device, which forms a closed system, the amount of heat transferred to the working fluid at constant pressure and rejected from it at constant volume can be expressed as

$$q_{in} - w_{b,out} = u_3 - u_2 \rightarrow q_{in} = P_2(v_3 - v_2) + (u_3 - u_2)$$
$$= h_3 - h_2 = c_p(T_3 - T_2) \quad \text{(9–10a)}$$

and

$$-q_{out} = u_1 - u_4 \rightarrow q_{out} = u_4 - u_1 = c_v(T_4 - T_1) \quad \text{(9–10b)}$$

Then the thermal efficiency of the ideal Diesel cycle under the cold-air-standard assumptions becomes

$$\eta_{th,Diesel} = \frac{w_{net}}{q_{in}} = 1 - \frac{q_{out}}{q_{in}} = 1 - \frac{T_4 - T_1}{k(T_3 - T_2)} = 1 - \frac{T_1(T_4/T_1 - 1)}{kT_2(T_3/T_2 - 1)}$$

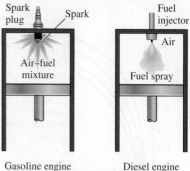

**FIGURE 9–20**

In diesel engines, the spark plug is replaced by a fuel injector, and only air is compressed during the compression process.

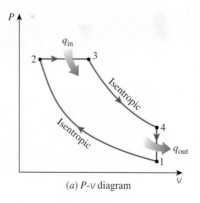

(a) P-v diagram

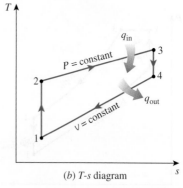

(b) T-s diagram

**FIGURE 9–21**

T-s and P-v diagrams for the ideal Diesel cycle.

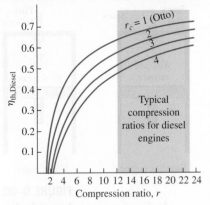

**FIGURE 9–22**
Thermal efficiency of the ideal Diesel cycle as a function of compression and cutoff ratios ($k = 1.4$).

We now define a new quantity, the **cutoff ratio** $r_c$, as the ratio of the cylinder volumes after and before the combustion process:

$$r_c = \frac{V_3}{V_2} = \frac{v_3}{v_2} \quad (9\text{–}11)$$

Utilizing this definition and the isentropic ideal-gas relations for processes 1-2 and 3-4, we see that the thermal efficiency relation reduces to

$$\eta_{th,Diesel} = 1 - \frac{1}{r^{k-1}}\left[\frac{r_c^k - 1}{k(r_c - 1)}\right] \quad (9\text{–}12)$$

where $r$ is the compression ratio defined by Eq. 9–9. Looking at Eq. 9–12 carefully, one would notice that under the cold-air-standard assumptions, the efficiency of a Diesel cycle differs from the efficiency of an Otto cycle by the quantity in the brackets. This quantity is always greater than 1. Therefore,

$$\eta_{th,Otto} > \eta_{th,Diesel} \quad (9\text{–}13)$$

when both cycles operate on the same compression ratio. Also, as the cutoff ratio decreases, the efficiency of the Diesel cycle increases (Fig. 9–22). For the limiting case of $r_c = 1$, the quantity in the brackets becomes unity (can you prove it?), and the efficiencies of the Otto and Diesel cycles become identical. Remember, though, that diesel engines operate at much higher compression ratios and thus are usually more efficient than the spark-ignition (gasoline) engines. The diesel engines also burn the fuel more completely since they usually operate at lower revolutions per minute and the air–fuel mass ratio is much higher than spark-ignition engines. Thermal efficiencies of large diesel engines range from about 35 to 40 percent.

The higher efficiency and lower fuel costs of diesel engines make them attractive in applications requiring relatively large amounts of power, such as in locomotive engines, emergency power generation units, large ships, and heavy trucks. As an example of how large a diesel engine can be, a 12-cylinder diesel engine built in 1964 by the Fiat Corporation of Italy had a normal power output of 25,200 hp (18.8 MW) at 122 rpm, a cylinder bore of 90 cm, and a stroke of 91 cm.

In modern high-speed compression ignition engines, fuel is injected into the combustion chamber much sooner compared to the early diesel engines. Fuel starts to ignite late in the compression stroke, and consequently part of the combustion occurs almost at constant volume. Fuel injection continues until the piston reaches the top dead center, and combustion of the fuel keeps the pressure high well into the expansion stroke. Thus, the entire combustion process can better be modeled as the combination of constant-volume and constant-pressure processes. The ideal cycle based on this concept is called the **dual cycle** and $P$-$v$ diagram for it is given in Fig. 9–23. The relative amounts of heat transferred during each process can be adjusted to approximate the actual cycle more closely. Note that both the Otto and the Diesel cycles can be obtained as special cases of the dual cycle. Dual cycle is a more realistic model than diesel cycle for representing modern, high-speed compression ignition engines.

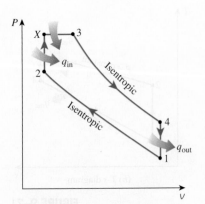

**FIGURE 9–23**
$P$-$v$ diagram of an ideal dual cycle.

■ **EXAMPLE 9–3**    **The Ideal Diesel Cycle**

An ideal Diesel cycle with air as the working fluid has a compression ratio of 18 and a cutoff ratio of 2. At the beginning of the compression process, the working fluid is at 100 kPa, 27°C, and 1917 cm³. Utilizing the cold-air-standard assumptions, determine (a) the temperature and pressure of air at the end of each process, (b) the net work output and the thermal efficiency, and (c) the mean effective pressure.

**SOLUTION**   An ideal Diesel cycle is considered. The temperature and pressure at the end of each process, the net work output, the thermal efficiency, and the mean effective pressure are to be determined.

**Assumptions**   **1** The cold-air-standard assumptions are applicable and thus air can be assumed to have constant specific heats at room temperature. **2** Kinetic and potential energy changes are negligible.

**Properties**   The gas constant of air is $R = 0.287$ kPa·m³/kg·K and its other properties at room temperature are $c_p = 1.005$ kJ/kg·K, $c_v = 0.718$ kJ/kg·K, and $k = 1.4$ (Table A–2a).

**Analysis**   The P-V diagram of the ideal Diesel cycle described is shown in Fig. 9–24. We note that the air contained in the cylinder forms a closed system.

(a) The temperature and pressure values at the end of each process can be determined by utilizing the ideal-gas isentropic relations for processes 1-2 and 3-4. But first we determine the volumes at the end of each process from the definitions of the compression ratio and the cutoff ratio:

$$V_2 = \frac{V_1}{r} = \frac{1917 \text{ cm}^3}{18} = 106.5 \text{ cm}^3$$

$$V_3 = r_c V_2 = (2)(106.5 \text{ cm}^3) = 213 \text{ cm}^3$$

$$V_4 = V_1 = 1917 \text{ cm}^3$$

Process 1-2 (isentropic compression of an ideal gas, constant specific heats):

$$T_2 = T_1 \left(\frac{V_1}{V_2}\right)^{k-1} = (300 \text{ K})(18)^{1.4-1} = \textbf{953 K}$$

$$P_2 = P_1 \left(\frac{V_1}{V_2}\right)^{k} = (100 \text{ kPa})(18)^{1.4} = \textbf{5720 kPa}$$

Process 2-3 (constant-pressure heat addition to an ideal gas):

$$P_3 = P_2 = \textbf{5720 kPa}$$

$$\frac{P_2 V_2}{T_2} = \frac{P_3 V_3}{T_3} \rightarrow T_3 = T_2 \left(\frac{V_3}{V_2}\right) = (953 \text{ K})(2) = \textbf{1906 K}$$

Process 3-4 (isentropic expansion of an ideal gas, constant specific heats):

$$T_4 = T_3 \left(\frac{V_3}{V_4}\right)^{k-1} = (1906 \text{ K}) \left(\frac{213 \text{ cm}^3}{1917 \text{ cm}^3}\right)^{1.4-1} = \textbf{791 K}$$

$$P_4 = P_3 \left(\frac{V_3}{V_4}\right)^{k} = (5720 \text{ kPa}) \left(\frac{213 \text{ cm}^3}{1917 \text{ cm}^3}\right)^{1.4} = \textbf{264 kPa}$$

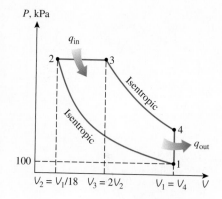

**FIGURE 9–24**
P-V diagram for the ideal Diesel cycle discussed in Example 9–3.

(b) The net work for a cycle is equivalent to the net heat transfer. But first we find the mass of air:

$$m = \frac{P_1 V_1}{RT_1} = \frac{(100 \text{ kPa})(1917 \times 10^{-6} \text{ m}^3)}{(0.287 \text{ kPa·m}^3/\text{kg·K})(300 \text{ K})} = 0.00223 \text{ kg}$$

Process 2-3 is a constant-pressure heat-addition process, for which the boundary work and $\Delta u$ terms can be combined into $\Delta h$. Thus,

$$\begin{aligned} Q_{in} &= m(h_3 - h_2) = mc_p(T_3 - T_2) \\ &= (0.00223 \text{ kg})(1.005 \text{ kJ/kg·K})[(1906 - 953) \text{ K}] \\ &= 2.136 \text{ kJ} \end{aligned}$$

Process 4-1 is a constant-volume heat-rejection process (it involves no work interactions), and the amount of heat rejected is

$$\begin{aligned} Q_{out} &= m(u_4 - u_1) = mc_v(T_4 - T_1) \\ &= (0.00223 \text{ kg})(0.718 \text{ kJ/kg·K})[(791 - 300) \text{ K}] \\ &= 0.786 \text{ kJ} \end{aligned}$$

Thus,

$$W_{net} = Q_{in} - Q_{out} = 2.136 - 0.186 = \mathbf{1.35 \text{ kJ}}$$

Then the thermal efficiency becomes

$$\eta_{th} = \frac{W_{net}}{Q_{in}} = \frac{1.35 \text{ kJ}}{2.136 \text{ kJ}} = \mathbf{0.632 \text{ or } 63.2\%}$$

The thermal efficiency of this Diesel cycle under the cold-air-standard assumptions could also be determined from Eq. 9–12.

(c) The mean effective pressure is determined from its definition, Eq. 9–4:

$$\text{MEP} = \frac{W_{net}}{V_{max} - V_{min}} = \frac{W_{net}}{V_1 - V_2} = \frac{1.35 \text{ kJ}}{(1917 - 106.5) \times 10^{-6} \text{ m}^3} \left( \frac{1 \text{ kPa·m}^3}{1 \text{ kJ}} \right)$$

$$= \mathbf{746 \text{ kPa}}$$

**Discussion** Note that a constant pressure of 746 kPa during the power stroke would produce the same net work output as the entire Diesel cycle.

# 9–7 · STIRLING AND ERICSSON CYCLES

The ideal Otto and Diesel cycles discussed in the preceding sections are composed entirely of internally reversible processes and thus are internally reversible cycles. These cycles are not totally reversible, however, since they involve heat transfer through a finite temperature difference during the non-isothermal heat-addition and heat-rejection processes, which are irreversible. Therefore, the thermal efficiency of an Otto or Diesel engine will be less than that of a Carnot engine operating between the same temperature limits.

Consider a heat engine operating between a heat source at $T_H$ and a heat sink at $T_L$. For the heat-engine cycle to be totally reversible, the temperature difference between the working fluid and the heat source (or sink) should never exceed a differential amount $dT$ during any heat-transfer process. That is, both the heat-addition and heat-rejection processes during the cycle must take place isothermally, one at a temperature of $T_H$ and the other at a temperature of $T_L$. This is precisely what happens in a Carnot cycle.

There are two other cycles that involve an isothermal heat-addition process at $T_H$ and an isothermal heat-rejection process at $T_L$: the *Stirling cycle* and the *Ericsson cycle*. They differ from the Carnot cycle in that the two isentropic processes are replaced by two constant-volume regeneration processes in the Stirling cycle and by two constant-pressure regeneration processes in the Ericsson cycle. Both cycles utilize **regeneration**, a process during which heat is transferred to a thermal energy storage device (called a *regenerator*) during one part of the cycle and is transferred back to the working fluid during another part of the cycle (Fig. 9–25).

Figure 9–26(*b*) shows the *T-s* and *P-v* diagrams of the **Stirling cycle**, which is made up of four totally reversible processes:

1-2   $T = constant$ expansion (heat addition from the external source)
2-3   $v = constant$ regeneration (internal heat transfer from the working fluid to the regenerator)
3-4   $T = constant$ compression (heat rejection to the external sink)
4-1   $v = constant$ regeneration (internal heat transfer from the regenerator back to the working fluid)

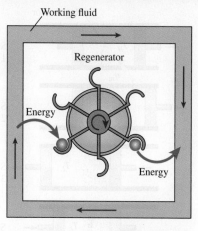

**FIGURE 9–25**
A regenerator is a device that borrows energy from the working fluid during one part of the cycle and pays it back (without interest) during another part.

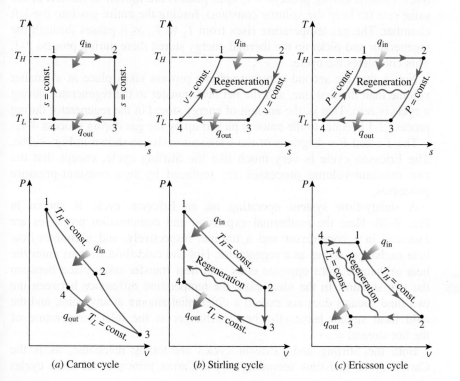

(*a*) Carnot cycle    (*b*) Stirling cycle    (*c*) Ericsson cycle

**FIGURE 9–26**
*T-s* and *P-v* diagrams of Carnot, Stirling, and Ericsson cycles.

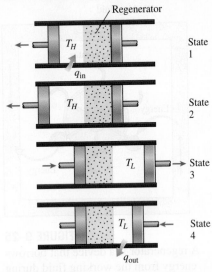

**FIGURE 9–27**
The execution of the Stirling cycle.

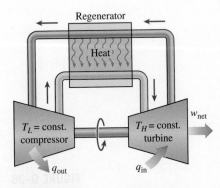

**FIGURE 9–28**
A steady-flow Ericsson engine.

The execution of the Stirling cycle requires rather innovative hardware. The actual Stirling engines, including the original one patented by Robert Stirling, are heavy and complicated. To spare the reader the complexities, the execution of the Stirling cycle in a closed system is explained with the help of the hypothetical engine shown in Fig. 9–27.

This system consists of a cylinder with two pistons on each side and a regenerator in the middle. The regenerator can be a wire or a ceramic mesh or any kind of porous plug with a high thermal mass (mass times specific heat). It is used for the temporary storage of thermal energy. The mass of the working fluid contained within the regenerator at any instant is considered negligible.

Initially, the left chamber houses the entire working fluid (a gas), which is at a high temperature and pressure. During process 1-2, heat is transferred to the gas at $T_H$ from a source at $T_H$. As the gas expands isothermally, the left piston moves outward, doing work, and the gas pressure drops. During process 2-3, both pistons are moved to the right at the same rate (to keep the volume constant) until the entire gas is forced into the right chamber. As the gas passes through the regenerator, heat is transferred to the regenerator and the gas temperature drops from $T_H$ to $T_L$. For this heat transfer process to be reversible, the temperature difference between the gas and the regenerator should not exceed a differential amount $dT$ at any point. Thus, the temperature of the regenerator will be $T_H$ at the left end and $T_L$ at the right end of the regenerator when state 3 is reached. During process 3-4, the right piston is moved inward, compressing the gas. Heat is transferred from the gas to a sink at temperature $T_L$ so that the gas temperature remains constant at $T_L$ while the pressure rises. Finally, during process 4-1, both pistons are moved to the left at the same rate (to keep the volume constant), forcing the entire gas into the left chamber. The gas temperature rises from $T_L$ to $T_H$ as it passes through the regenerator and picks up the thermal energy stored there during process 2-3. This completes the cycle.

Notice that the second constant-volume process takes place at a smaller volume than the first one, and the net heat transfer to the regenerator during a cycle is zero. That is, the amount of energy stored in the regenerator during process 2-3 is equal to the amount picked up by the gas during process 4-1.

The $T$-$s$ and $P$-$v$ diagrams of the **Ericsson cycle** are shown in Fig. 9–26c. The Ericsson cycle is very much like the Stirling cycle, except that the two constant-volume processes are replaced by two constant-pressure processes.

A steady-flow system operating on an Ericsson cycle is shown in Fig. 9–28. Here the isothermal expansion and compression processes are executed in a compressor and a turbine, respectively, and a counter-flow heat exchanger serves as a regenerator. Hot and cold fluid streams enter the heat exchanger from opposite ends, and heat transfer takes place between the two streams. In the ideal case, the temperature difference between the two fluid streams does not exceed a differential amount at any point, and the cold fluid stream leaves the heat exchanger at the inlet temperature of the hot stream.

Both the Stirling and Ericsson cycles are totally reversible, as is the Carnot cycle, and thus according to the Carnot principle, all three cycles

must have the same thermal efficiency when operating between the same temperature limits:

$$\eta_{th,Stirling} = \eta_{th,Ericsson} = \eta_{th,Carnot} = 1 - \frac{T_L}{T_H} \tag{9-14}$$

This is proved for the Carnot cycle in Example 9–1 and can be proved in a similar manner for both the Stirling and Ericsson cycles.

---

■ **EXAMPLE 9–4**    **Thermal Efficiency of the Ericsson Cycle**

Using an ideal gas as the working fluid, show that the thermal efficiency of an Ericsson cycle is identical to the efficiency of a Carnot cycle operating between the same temperature limits.

**SOLUTION**   It is to be shown that the thermal efficiencies of Carnot and Ericsson cycles are identical.

*Analysis*  Heat is transferred to the working fluid isothermally from an external source at temperature $T_H$ during process 1-2, and it is rejected again isothermally to an external sink at temperature $T_L$ during process 3-4. For a reversible isothermal process, heat transfer is related to the entropy change by

$$q = T \Delta s$$

The entropy change of an ideal gas during an isothermal process is

$$\Delta s = c_p \ln \frac{T_e}{T_i}^{\nearrow 0} - R \ln \frac{P_e}{P_i} = -R \ln \frac{P_e}{P_i}$$

The heat input and heat output can be expressed as

$$q_{in} = T_H(s_2 - s_1) = T_H\left(-R \ln \frac{P_2}{P_1}\right) = RT_H \ln \frac{P_1}{P_2}$$

and

$$q_{out} = T_L(s_4 - s_3) = -T_L\left(-R \ln \frac{P_4}{P_3}\right) = RT_L \ln \frac{P_4}{P_3}$$

Then the thermal efficiency of the Ericsson cycle becomes

$$\eta_{th,Ericsson} = 1 - \frac{q_{out}}{q_{in}} = 1 - \frac{RT_L \ln(P_4/P_3)}{RT_H \ln(P_1/P_2)} = 1 - \frac{T_L}{T_H}$$

since $P_1 = P_4$ and $P_3 = P_2$. Notice that this result is independent of whether the cycle is executed in a closed or steady-flow system.

---

Stirling and Ericsson cycles are difficult to achieve in practice because they involve heat transfer through a differential temperature difference in all components including the regenerator. This would require providing infinitely large surface areas for heat transfer or allowing an infinitely long time for the process. Neither is practical. In reality, all heat transfer processes take

place through a finite temperature difference, the regenerator does not have an efficiency of 100 percent, and the pressure losses in the regenerator are considerable. Because of these limitations, both Stirling and Ericsson cycles have long been of only theoretical interest. However, there is renewed interest in engines that operate on these cycles because of their potential for higher efficiency and better emission control. The Ford Motor Company, General Motors Corporation, and the Phillips Research Laboratories of the Netherlands have successfully developed Stirling engines suitable for trucks, buses, and even automobiles. More research and development are needed before these engines can compete with the gasoline or diesel engines.

Both the Stirling and the Ericsson engines are *external combustion* engines. That is, the fuel in these engines is burned outside the cylinder, as opposed to gasoline or diesel engines, where the fuel is burned inside the cylinder.

External combustion offers several advantages. First, a variety of fuels can be used as a source of thermal energy. Second, there is more time for combustion, and thus the combustion process is more complete, which means less air pollution and more energy extraction from the fuel. Third, these engines operate on closed cycles, and thus a working fluid that has the most desirable characteristics (stable, chemically inert, high thermal conductivity) can be utilized as the working fluid. Hydrogen and helium are two gases commonly employed in these engines.

Despite the physical limitations and impracticalities associated with them, both the Stirling and Ericsson cycles give a strong message to design engineers: *Regeneration can increase efficiency.* It is no coincidence that modern gas-turbine and steam power plants make extensive use of regeneration. In fact, the Brayton cycle with intercooling, reheating, and regeneration, which is utilized in large gas-turbine power plants and discussed later in this chapter, closely resembles the Ericsson cycle.

# 9–8 · BRAYTON CYCLE: THE IDEAL CYCLE FOR GAS-TURBINE ENGINES

The Brayton cycle was first proposed by George Brayton for use in the reciprocating oil-burning engine that he developed around 1870. Today, it is used for gas turbines only where both the compression and expansion processes take place in rotating machinery. Gas turbines usually operate on an *open cycle,* as shown in Fig. 9–29. Fresh air at ambient conditions is drawn into the compressor, where its temperature and pressure are raised. The high-pressure air proceeds into the combustion chamber, where the fuel is burned at constant pressure. The resulting high-temperature gases then enter the turbine, where they expand to the atmospheric pressure while producing power. The exhaust gases leaving the turbine are thrown out (not recirculated), causing the cycle to be classified as an open cycle.

The open gas-turbine cycle described above can be modeled as a *closed cycle,* as shown in Fig. 9–30, by utilizing the air-standard assumptions. Here the compression and expansion processes remain the same, but the combustion process is replaced by a constant-pressure heat-addition process from an external source, and the exhaust process is replaced by a constant-pressure

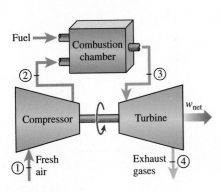

**FIGURE 9–29**
An open-cycle gas-turbine engine.

heat-rejection process to the ambient air. The ideal cycle that the working fluid undergoes in this closed loop is the **Brayton cycle**, which is made up of four internally reversible processes:

1-2    Isentropic compression (in a compressor)

2-3    Constant-pressure heat addition

3-4    Isentropic expansion (in a turbine)

4-1    Constant-pressure heat rejection

The *T-s* and *P-v* diagrams of an ideal Brayton cycle are shown in Fig. 9–31. Notice that all four processes of the Brayton cycle are executed in steady-flow devices; thus, they should be analyzed as steady-flow processes. When the changes in kinetic and potential energies are neglected, the energy balance for a steady-flow process can be expressed, on a unit–mass basis, as

$$(q_{in} - q_{out}) + (w_{in} - w_{out}) = h_{exit} - h_{inlet} \tag{9–15}$$

Therefore, heat transfers to and from the working fluid are

$$q_{in} = h_3 - h_2 = c_p(T_3 - T_2) \tag{9–16a}$$

and

$$q_{out} = h_4 - h_1 = c_p(T_4 - T_1) \tag{9–16b}$$

Then the thermal efficiency of the ideal Brayton cycle under the cold-air-standard assumptions becomes

$$\eta_{th,Brayton} = \frac{w_{net}}{q_{in}} = 1 - \frac{q_{out}}{q_{in}} = 1 - \frac{c_p(T_4 - T_1)}{c_p(T_3 - T_2)} = 1 - \frac{T_1(T_4/T_1 - 1)}{T_2(T_3/T_2 - 1)}$$

Processes 1-2 and 3-4 are isentropic, and $P_2 = P_3$ and $P_4 = P_1$. Thus,

$$\frac{T_2}{T_1} = \left(\frac{P_2}{P_1}\right)^{(k-1)/k} = \left(\frac{P_3}{P_4}\right)^{(k-1)/k} = \frac{T_3}{T_4}$$

Substituting these equations into the thermal efficiency relation and simplifying give

$$\eta_{th,Brayton} = 1 - \frac{1}{r_p^{(k-1)/k}} \tag{9–17}$$

where

$$r_p = \frac{P_2}{P_1} \tag{9–18}$$

is the **pressure ratio** and *k* is the specific heat ratio. Equation 9–17 shows that under the cold-air-standard assumptions, the thermal efficiency of an ideal Brayton cycle depends on the pressure ratio of the gas turbine and the specific heat ratio of the working fluid. The thermal efficiency increases with both of these parameters, which is also the case for actual gas turbines. A plot of thermal efficiency versus the pressure ratio is given in Fig. 9–32 for $k = 1.4$, which is the specific-heat-ratio value of air at room temperature.

The highest temperature in the cycle occurs at the end of the combustion process (state 3), and it is limited by the maximum temperature that the

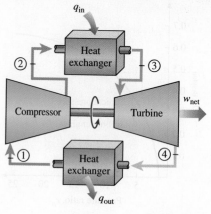

**FIGURE 9–30**

A closed-cycle gas-turbine engine.

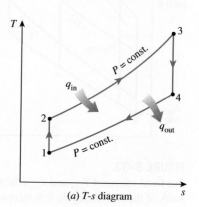

*(a) T-s diagram*

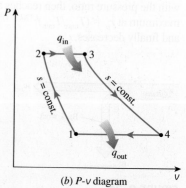

*(b) P-v diagram*

**FIGURE 9–31**

*T-s* and *P-v* diagrams for the ideal Brayton cycle.

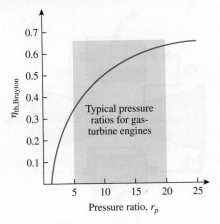

**FIGURE 9–32**
Thermal efficiency of the ideal Brayton cycle as a function of the pressure ratio.

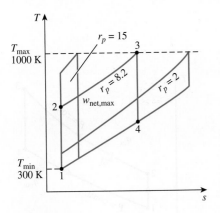

**FIGURE 9–33**
For fixed values of $T_{min}$ and $T_{max}$, the net work of the Brayton cycle first increases with the pressure ratio, then reaches a maximum at $r_p = (T_{max}/T_{min})^{k/[2(k-1)]}$, and finally decreases.

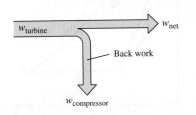

**FIGURE 9–34**
The fraction of the turbine work used to drive the compressor is called the back work ratio.

turbine blades can withstand. This also limits the pressure ratios that can be used in the cycle. For a fixed turbine inlet temperature $T_3$, the net work output per cycle increases with the pressure ratio, reaches a maximum, and then starts to decrease, as shown in Fig. 9–33. Therefore, there should be a compromise between the pressure ratio (thus the thermal efficiency) and the net work output. With less work output per cycle, a larger mass flow rate (thus a larger system) is needed to maintain the same power output, which may not be economical. In most common designs, the pressure ratio of gas turbines ranges from about 11 to 16.

The air in gas turbines performs two important functions: It supplies the necessary oxidant for the combustion of the fuel, and it serves as a coolant to keep the temperature of various components within safe limits. The second function is accomplished by drawing in more air than is needed for the complete combustion of the fuel. In gas turbines, an air–fuel mass ratio of 50 or above is not uncommon. Therefore, in a cycle analysis, treating the combustion gases as air does not cause any appreciable error. Also, the mass flow rate through the turbine is greater than that through the compressor, the difference being equal to the mass flow rate of the fuel. Thus, assuming a constant mass flow rate throughout the cycle yields conservative results for open-loop gas-turbine engines.

The two major application areas of gas-turbine engines are *aircraft propulsion* and *electric power generation*. When it is used for aircraft propulsion, the gas turbine produces just enough power to drive the compressor and a small generator to power the auxiliary equipment. The high-velocity exhaust gases are responsible for producing the necessary thrust to propel the aircraft. Gas turbines are also used as stationary power plants to generate electricity as stand-alone units or in conjunction with steam power plants on the high-temperature side. In these plants, the exhaust gases of the gas turbine serve as the heat source for the steam. The gas-turbine cycle can also be executed as a closed cycle for use in nuclear power plants. This time the working fluid is not limited to air, and a gas with more desirable characteristics (such as helium) can be used.

The majority of the Western world's naval fleets already use gas-turbine engines for propulsion and electric power generation. The General Electric LM2500 gas turbines used to power ships have a simple-cycle thermal efficiency of 37 percent. The General Electric WR-21 gas turbines equipped with intercooling and regeneration have a thermal efficiency of 43 percent and produce 21.6 MW. The regeneration also reduces the exhaust temperature from 600°C to 350°C. Air is compressed to 3 atm before it enters the intercooler. Compared to steam-turbine and diesel-propulsion systems, the gas turbine offers greater power for a given size and weight, high reliability, long life, and more convenient operation. The engine start-up time has been reduced from 4 h required for a typical steam-propulsion system to less than 2 min for a gas turbine. Many modern marine propulsion systems use gas turbines together with diesel engines because of the high fuel consumption of simple-cycle gas-turbine engines. In combined diesel and gas-turbine systems, diesel is used to provide for efficient low-power and cruise operation, and gas turbine is used when high speeds are needed.

In gas-turbine power plants, the ratio of the compressor work to the turbine work, called the **back work ratio**, is very high (Fig. 9–34). Usually more

than one-half of the turbine work output is used to drive the compressor. The situation is even worse when the isentropic efficiencies of the compressor and the turbine are low. This is quite in contrast to steam power plants, where the back work ratio is only a few percent. This is not surprising, however, since a liquid is compressed in steam power plants instead of a gas, and the steady-flow work is proportional to the specific volume of the working fluid.

A power plant with a high back work ratio requires a larger turbine to provide the additional power requirements of the compressor. Therefore, the turbines used in gas-turbine power plants are larger than those used in steam power plants of the same net power output.

## Development of Gas Turbines

The gas turbine has experienced phenomenal progress and growth since its first successful development in the 1930s. The early gas turbines built in the 1940s and even 1950s had simple-cycle efficiencies of about 17 percent because of the low compressor and turbine efficiencies and low turbine inlet temperatures due to metallurgical limitations of those times. Therefore, gas turbines found only limited use despite their versatility and their ability to burn a variety of fuels. The efforts to improve the cycle efficiency concentrated in three areas:

**1. Increasing the turbine inlet (or firing) temperatures** This has been the primary approach taken to improve gas-turbine efficiency. The turbine inlet temperatures have increased steadily from about 540°C in the 1940s to 1425°C and even higher today. These increases were made possible by the development of new materials and the innovative cooling techniques for the critical components such as coating the turbine blades with ceramic layers and cooling the blades with the discharge air from the compressor. Maintaining high turbine inlet temperatures with an air-cooling technique requires the combustion temperature to be higher to compensate for the cooling effect of the cooling air. However, higher combustion temperatures increase the amount of nitrogen oxides ($NO_x$), which are responsible for the formation of ozone at ground level and smog. Using steam as the coolant allowed an increase in the turbine inlet temperatures by 110°C without an increase in the combustion temperature. Steam is also a much more effective heat transfer medium than air.

**2. Increasing the efficiencies of turbomachinery components** The performance of early turbines suffered greatly from the inefficiencies of turbines and compressors. However, the advent of computers and advanced techniques for computer-aided design made it possible to design these components aerodynamically with minimal losses. The increased efficiencies of the turbines and compressors resulted in a significant increase in the cycle efficiency.

**3. Adding modifications to the basic cycle** The simple-cycle efficiencies of early gas turbines were practically doubled by incorporating intercooling, regeneration (or recuperation), and reheating, discussed in the next two sections. These improvements, of course, come at the expense of increased initial and operation costs, and they cannot be justified unless the decrease in fuel costs offsets the increase in other costs. The relatively low fuel prices, the general desire in the industry to minimize installation costs, and the tremendous increase in the simple-cycle efficiency to about 40 percent left little desire for opting for these modifications.

The first gas turbine for an electric utility was installed in 1949 in Oklahoma as part of a combined-cycle power plant. It was built by General Electric and produced 3.5 MW of power. Gas turbines installed until the mid-1970s suffered from low efficiency and poor reliability. In the past, the base-load electric power generation was dominated by large coal and nuclear power plants. However, there has been an historic shift toward natural gas–fired gas turbines because of their higher efficiencies, lower capital costs, shorter installation times, and better emission characteristics, and the abundance of natural gas supplies, and more and more electric utilities are using gas turbines for base-load power production as well as for peaking. The construction costs for gas-turbine power plants are roughly half that of comparable conventional fossil-fuel steam power plants, which were the primary base-load power plants until the early 1980s. More than half of all power plants to be installed in the foreseeable future are forecast to be gas-turbine or combined gas–steam turbine types.

A gas turbine manufactured by General Electric in the early 1990s had a pressure ratio of 13.5 and generated 135.7 MW of net power at a thermal efficiency of 33 percent in simple-cycle operation. A more recent gas turbine manufactured by General Electric uses a turbine inlet temperature of 1425°C and produces up to 282 MW while achieving a thermal efficiency of 39.5 percent in the simple-cycle mode. A 1.3-ton small-scale gas turbine labeled OP-16, built by the Dutch firm Opra Optimal Radial Turbine, can run on gas or liquid fuel and can replace a 16-ton diesel engine. It has a pressure ratio of 6.5 and produces up to 2 MW of power. Its efficiency is 26 percent in the simple-cycle operation, which rises to 37 percent when equipped with a regenerator.

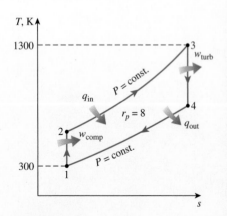

**FIGURE 9–35**

$T$-$s$ diagram for the Brayton cycle discussed in Example 9–5.

---

### EXAMPLE 9–5    The Simple Ideal Brayton Cycle

A gas-turbine power plant operating on an ideal Brayton cycle has a pressure ratio of 8. The gas temperature is 300 K at the compressor inlet and 1300 K at the turbine inlet. Utilizing the air-standard assumptions, determine (a) the gas temperature at the exits of the compressor and the turbine, (b) the back work ratio, and (c) the thermal efficiency.

**SOLUTION**    A power plant operating on the ideal Brayton cycle is considered. The compressor and turbine exit temperatures, back work ratio, and the thermal efficiency are to be determined.

**Assumptions**    **1** Steady operating conditions exist. **2** The air-standard assumptions are applicable. **3** Kinetic and potential energy changes are negligible. **4** The variation of specific heats with temperature is to be considered.

**Analysis**    The $T$-$s$ diagram of the ideal Brayton cycle described is shown in Fig. 9–35. We note that the components involved in the Brayton cycle are steady-flow devices.

(a) The air temperatures at the compressor and turbine exits are determined from isentropic relations:

Process 1–2 (isentropic compression of an ideal gas):

$$T_1 = 300 \text{ K} \rightarrow h_1 = 300.19 \text{ kJ/kg}$$

$$P_{r1} = 1.386$$

$$P_{r2} = \frac{P_2}{P_1} P_{r1} = (8)(1.386) = 11.09 \rightarrow T_2 = \mathbf{540 \text{ K}} \quad \text{(at compressor exit)}$$

$$h_2 = 544.35 \text{ kJ/kg}$$

Process 3–4 (isentropic expansion of an ideal gas):

$$T_3 = 1300 \text{ K} \rightarrow h_3 = 1395.97 \text{ kJ/kg}$$

$$P_{r3} = 330.9$$

$$P_{r4} = \frac{P_4}{P_3} P_{r3} = \left(\frac{1}{8}\right)(330.9) = 41.36 \rightarrow T_4 = \mathbf{770 \text{ K}} \quad \text{(at turbine exit)}$$

$$h_4 = 789.37 \text{ kJ/kg}$$

(b) To find the back work ratio, we need to find the work input to the compressor and the work output of the turbine:

$$w_{\text{comp,in}} = h_2 - h_1 = 544.35 - 300.19 = 244.16 \text{ kJ/kg}$$

$$w_{\text{turb,out}} = h_3 - h_4 = 1395.97 - 789.37 = 606.60 \text{ kJ/kg}$$

Thus,

$$r_{\text{bw}} = \frac{w_{\text{comp,in}}}{w_{\text{turb,out}}} = \frac{244.16 \text{ kJ/kg}}{606.60 \text{ kJ/kg}} = \mathbf{0.403}$$

That is, 40.3 percent of the turbine work output is used just to drive the compressor.

(c) The thermal efficiency of the cycle is the ratio of the net power output to the total heat input:

$$q_{\text{in}} = h_3 - h_2 = 1395.97 - 544.35 = 851.62 \text{ kJ/kg}$$

$$w_{\text{net}} = w_{\text{out}} - w_{\text{in}} = 606.60 - 244.16 = 362.4 \text{ kJ/kg}$$

Thus,

$$\eta_{\text{th}} = \frac{w_{\text{net}}}{q_{\text{in}}} = \frac{362.4 \text{ kJ/kg}}{851.62 \text{ kJ/kg}} = \mathbf{0.426 \text{ or } 42.6\%}$$

The thermal efficiency could also be determined from

$$\eta_{\text{th}} = 1 - \frac{q_{\text{out}}}{q_{\text{in}}}$$

where

$$q_{\text{out}} = h_4 - h_1 = 789.37 - 300.19 = 489.2 \text{ kJ/kg}$$

**Discussion** Under the cold-air-standard assumptions (constant specific heat values at room temperature), the thermal efficiency would be, from Eq. 9–17,

$$\eta_{th,Brayton} = 1 - \frac{1}{r_p^{(k-1)/k}} = 1 - \frac{1}{8^{(1.4-1)/1.4}} = 0.448 \text{ or } 44.8\%$$

which is sufficiently close to the value obtained by accounting for the variation of specific heats with temperature.

## Deviation of Actual Gas-Turbine Cycles from Idealized Ones

The actual gas-turbine cycle differs from the ideal Brayton cycle on several accounts. For one thing, some pressure drop during the heat-addition and heat-rejection processes is inevitable. More importantly, the actual work input to the compressor is more, and the actual work output from the turbine is less because of irreversibilities. The deviation of actual compressor and turbine behavior from the idealized isentropic behavior can be accurately accounted for by utilizing the isentropic efficiencies of the turbine and compressor as

$$\eta_C = \frac{w_s}{w_a} \cong \frac{h_{2s} - h_1}{h_{2a} - h_1} \tag{9-19}$$

and

$$\eta_T = \frac{w_a}{w_s} \cong \frac{h_3 - h_{4a}}{h_3 - h_{4s}} \tag{9-20}$$

where states 2a and 4a are the actual exit states of the compressor and the turbine, respectively, and 2s and 4s are the corresponding states for the isentropic case, as illustrated in Fig. 9–36. The effect of the turbine and compressor efficiencies on the thermal efficiency of the gas-turbine engines is illustrated below with an example.

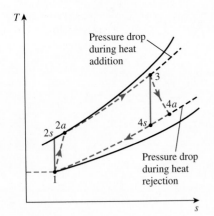

**FIGURE 9–36**

The deviation of an actual gas-turbine cycle from the ideal Brayton cycle as a result of irreversibilities.

---

**EXAMPLE 9–6** **An Actual Gas-Turbine Cycle**

Assuming a compressor efficiency of 80 percent and a turbine efficiency of 85 percent, determine (a) the back work ratio, (b) the thermal efficiency, and (c) the turbine exit temperature of the gas-turbine cycle discussed in Example 9–5.

**SOLUTION** The Brayton cycle discussed in Example 9–5 is reconsidered. For specified turbine and compressor efficiencies, the back work ratio, the thermal efficiency, and the turbine exit temperature are to be determined.

**Analysis** (a) The *T-s* diagram of the cycle is shown in Fig. 9–37. The actual compressor work and turbine work are determined by using the definitions of compressor and turbine efficiencies, Eqs. 9–19 and 9–20:

*Compressor:*
$$w_{comp,in} = \frac{w_s}{\eta_C} = \frac{244.16 \text{ kJ/kg}}{0.80} = 305.20 \text{ kJ/kg}$$

*Turbine:*
$$w_{turb,out} = \eta_T w_s = (0.85)(606.60 \text{ kJ/kg}) = 515.61 \text{ kJ/kg}$$

Thus,

$$r_{bw} = \frac{w_{comp,in}}{w_{turb,out}} = \frac{305.20 \text{ kJ/kg}}{515.61 \text{ kJ/kg}} = 0.592$$

That is, the compressor is now consuming 59.2 percent of the work produced by the turbine (up from 40.3 percent). This increase is due to the irreversibilities that occur within the compressor and the turbine.

(b) In this case, air leaves the compressor at a higher temperature and enthalpy, which are determined to be

$$w_{comp,in} = h_{2a} - h_1 \rightarrow h_{2a} = h_1 + w_{comp,in}$$

$$= 300.19 + 305.20$$

$$= 605.39 \text{ kJ/kg} \quad (\text{and } T_{2a} = 598 \text{ K})$$

Thus,

$$q_{in} = h_3 - h_{2a} = 1395.97 - 605.39 = 790.58 \text{ kJ/kg}$$

$$w_{net} = w_{out} - w_{in} = 515.61 - 305.20 = 210.41 \text{ kJ/kg}$$

and

$$\eta_{th} = \frac{w_{net}}{q_{in}} = \frac{210.41 \text{ kJ/kg}}{790.58 \text{ kJ/kg}} = 0.266 \text{ or } 26.6\%$$

That is, the irreversibilities occurring within the turbine and compressor caused the thermal efficiency of the gas turbine cycle to drop from 42.6 to 26.6 percent. This example shows how sensitive the performance of a gas-turbine power plant is to the efficiencies of the compressor and the turbine. In fact, gas-turbine efficiencies did not reach competitive values until significant improvements were made in the design of gas turbines and compressors.

(c) The air temperature at the turbine exit is determined from an energy balance on the turbine:

$$w_{turb,out} = h_3 - h_{4a} \rightarrow h_{4a} = h_3 - w_{turb,out}$$

$$= 1395.97 - 515.61$$

$$= 880.36 \text{ kJ/kg}$$

Then, from Table A–17,

$$T_{4a} = 853 \text{ K}$$

**Discussion** The temperature at turbine exit is considerably higher than that at the compressor exit ($T_{2a}$ = 598 K), which suggests the use of regeneration to reduce fuel cost.

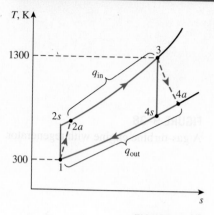

**FIGURE 9–37**
*T-s* diagram of the gas-turbine cycle discussed in Example 9–6.

## 9–9 · THE BRAYTON CYCLE WITH REGENERATION

In gas-turbine engines, the temperature of the exhaust gas leaving the turbine is often considerably higher than the temperature of the air leaving the compressor. Therefore, the high-pressure air leaving the compressor can be heated by transferring heat to it from the hot exhaust gases in a counter-flow

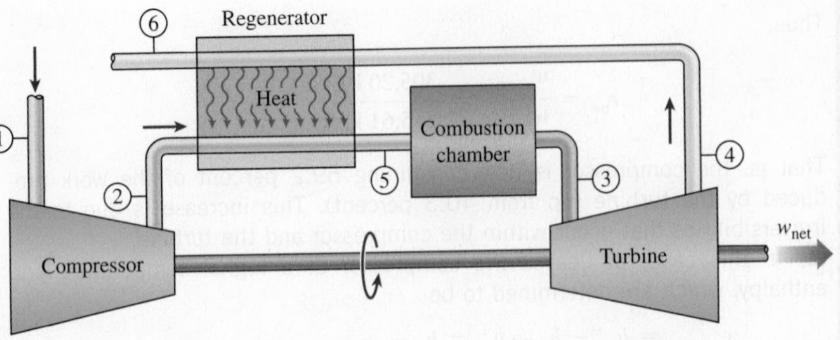

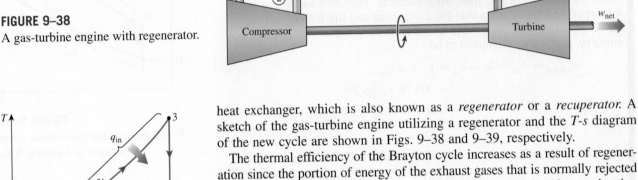

**FIGURE 9–38**
A gas-turbine engine with regenerator.

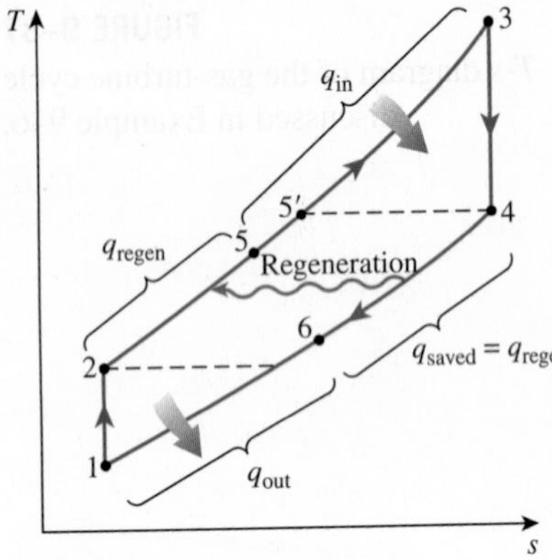

**FIGURE 9–39**
*T-s* diagram of a Brayton cycle with regeneration.

heat exchanger, which is also known as a *regenerator* or a *recuperator*. A sketch of the gas-turbine engine utilizing a regenerator and the *T-s* diagram of the new cycle are shown in Figs. 9–38 and 9–39, respectively.

The thermal efficiency of the Brayton cycle increases as a result of regeneration since the portion of energy of the exhaust gases that is normally rejected to the surroundings is now used to preheat the air entering the combustion chamber. This, in turn, decreases the heat input (thus fuel) requirements for the same net work output. Note, however, that the use of a regenerator is recommended only when the turbine exhaust temperature is higher than the compressor exit temperature. Otherwise, heat will flow in the reverse direction (*to* the exhaust gases), decreasing the efficiency. This situation is encountered in gas-turbine engines operating at very high pressure ratios.

The highest temperature occurring within the regenerator is $T_4$, the temperature of the exhaust gases leaving the turbine and entering the regenerator. Under no conditions can the air be preheated in the regenerator to a temperature above this value. Air normally leaves the regenerator at a lower temperature, $T_5$. In the limiting (ideal) case, the air exits the regenerator at the inlet temperature of the exhaust gases $T_4$. Assuming the regenerator to be well insulated and any changes in kinetic and potential energies to be negligible, the actual and maximum heat transfers from the exhaust gases to the air can be expressed as

$$q_{\text{regen,act}} = h_5 - h_2 \tag{9–21}$$

and

$$q_{\text{regen,max}} = h_{5'} - h_2 = h_4 - h_2 \tag{9–22}$$

The extent to which a regenerator approaches an ideal regenerator is called the **effectiveness** $\epsilon$ and is defined as

$$\epsilon = \frac{q_{\text{regen,act}}}{q_{\text{regen,max}}} = \frac{h_5 - h_2}{h_4 - h_2} \tag{9–23}$$

When the cold-air-standard assumptions are utilized, it reduces to

$$\epsilon \cong \frac{T_5 - T_2}{T_4 - T_2} \tag{9–24}$$

A regenerator with a higher effectiveness obviously saves a greater amount of fuel since it preheats the air to a higher temperature prior to combustion.

However, achieving a higher effectiveness requires the use of a larger regenerator, which carries a higher price tag and causes a larger pressure drop. Therefore, the use of a regenerator with a very high effectiveness cannot be justified economically unless the savings from the fuel costs exceed the additional expenses involved. The effectiveness of most regenerators used in practice is below 0.85.

Under the cold-air-standard assumptions, the thermal efficiency of an ideal Brayton cycle with regeneration is

$$\eta_{th,regen} = 1 - \left(\frac{T_1}{T_3}\right)(r_p)^{(k-1)/k} \tag{9-25}$$

Therefore, the thermal efficiency of an ideal Brayton cycle with regeneration depends on the ratio of the minimum to maximum temperatures as well as the pressure ratio. The thermal efficiency is plotted in Fig. 9–40 for various pressure ratios and minimum-to-maximum temperature ratios. This figure shows that regeneration is most effective at lower pressure ratios and low minimum-to-maximum temperature ratios.

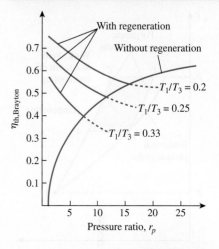

**FIGURE 9–40**

Thermal efficiency of the ideal Brayton cycle with and without regeneration.

---

■ **EXAMPLE 9–7**    **Actual Gas-Turbine Cycle with Regeneration**

■ Determine the thermal efficiency of the gas-turbine described in Example 9–6
■ if a regenerator having an effectiveness of 80 percent is installed.

**SOLUTION**    The gas-turbine discussed in Example 9–6 is equipped with a regenerator. For a specified effectiveness, the thermal efficiency is to be determined.

**Analysis**    The *T-s* diagram of the cycle is shown in Fig. 9–41. We first determine the enthalpy of the air at the exit of the regenerator, using the definition of effectiveness:

$$\epsilon = \frac{h_5 - h_{2a}}{h_{4a} - h_{2a}}$$

$$0.80 = \frac{(h_5 - 605.39)\ \text{kJ/kg}}{(880.36 - 605.39)\ \text{kJ/kg}} \rightarrow h_5 = 825.37\ \text{kJ/kg}$$

Thus,

$$q_{in} = h_3 - h_5 = (1395.97 - 825.37)\ \text{kJ/kg} = 570.60\ \text{kJ/kg}$$

This represents a savings of 220.0 kJ/kg from the heat input requirements. The addition of a regenerator (assumed to be frictionless) does not affect the net work output. Thus,

$$\eta_{th} = \frac{w_{net}}{q_{in}} = \frac{210.41\ \text{kJ/kg}}{570.60\ \text{kJ/kg}} = \textbf{0.369 or 36.9\%}$$

**Discussion**    Note that the thermal efficiency of the gas turbine has gone up from 26.6 to 36.9 percent as a result of installing a regenerator that helps to recuperate some of the thermal energy of the exhaust gases.

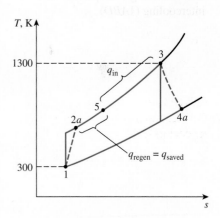

**FIGURE 9–41**

*T-s* diagram of the regenerative Brayton cycle described in Example 9–7.

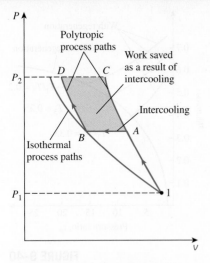

**FIGURE 9–42**

Comparison of work inputs to a single-stage compressor (1AC) and a two-stage compressor with intercooling (1ABD).

# 9–10 · THE BRAYTON CYCLE WITH INTERCOOLING, REHEATING, AND REGENERATION

The net work of a gas-turbine cycle is the difference between the turbine work output and the compressor work input, and it can be increased by either decreasing the compressor work or increasing the turbine work, or both. It was shown in Chap. 7 that the work required to compress a gas between two specified pressures can be decreased by carrying out the compression process in stages and cooling the gas in between (Fig. 9–42)—that is, using *multistage compression with intercooling.* As the number of stages is increased, the compression process becomes nearly isothermal at the compressor inlet temperature, and the compression work decreases.

Likewise, the work output of a turbine operating between two pressure levels can be increased by expanding the gas in stages and reheating it in between—that is, utilizing *multistage expansion with reheating.* This is accomplished without raising the maximum temperature in the cycle. As the number of stages is increased, the expansion process becomes nearly isothermal. The foregoing argument is based on a simple principle: *The steady-flow compression or expansion work is proportional to the specific volume of the fluid. Therefore, the specific volume of the working fluid should be as low as possible during a compression process and as high as possible during an expansion process.* This is precisely what intercooling and reheating accomplish.

Combustion in gas turbines typically occurs at four times the amount of air needed for complete combustion to avoid excessive temperatures. Therefore, the exhaust gases are rich in oxygen, and reheating can be accomplished by simply spraying additional fuel into the exhaust gases between two expansion states.

The working fluid leaves the compressor at a lower temperature, and the turbine at a higher temperature, when intercooling and reheating are utilized. This makes regeneration more attractive since a greater potential for regeneration exists. Also, the gases leaving the compressor can be heated to a higher temperature before they enter the combustion chamber because of the higher temperature of the turbine exhaust.

A schematic of the physical arrangement and the *T-s* diagram of an ideal two-stage gas-turbine cycle with intercooling, reheating, and regeneration are shown in Figs. 9–43 and 9–44. The gas enters the first stage of the compressor at state 1, is compressed isentropically to an intermediate pressure $P_2$, is cooled at constant pressure to state 3 ($T_3 = T_1$), and is compressed in the second stage isentropically to the final pressure $P_4$. At state 4 the gas enters the regenerator, where it is heated to $T_5$ at constant pressure. In an ideal regenerator, the gas leaves the regenerator at the temperature of the turbine exhaust, that is, $T_5 = T_9$. The primary heat addition (or combustion) process takes place between states 5 and 6. The gas enters the first stage of the turbine at state 6 and expands isentropically to state 7, where it enters the reheater. It is reheated at constant pressure to state 8 ($T_8 = T_6$), where it enters the second stage of the turbine. The gas exits the turbine at state 9 and enters the regenerator, where it is cooled to state 10 at constant pressure. The cycle is completed by cooling the gas to the initial state (or purging the exhaust gases).

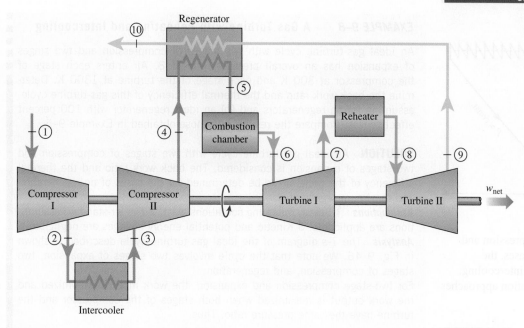

**FIGURE 9–43**

A gas-turbine engine with two-stage compression with intercooling, two-stage expansion with reheating, and regeneration.

It was shown in Chap. 7 that the work input to a two-stage compressor is minimized when equal pressure ratios are maintained across each stage. It can be shown that this procedure also maximizes the turbine work output. Thus, for best performance we have

$$\frac{P_2}{P_1} = \frac{P_4}{P_3} \quad \text{and} \quad \frac{P_6}{P_7} = \frac{P_8}{P_9} \tag{9–26}$$

In the analysis of the actual gas-turbine cycles, the irreversibilities that are present within the compressor, the turbine, and the regenerator as well as the pressure drops in the heat exchangers should be taken into consideration.

The back work ratio of a gas-turbine cycle improves as a result of intercooling and reheating. However, this does not mean that the thermal efficiency also improves. The fact is, intercooling and reheating always decreases the thermal efficiency unless they are accompanied by regeneration. This is because intercooling decreases the average temperature at which heat is added, and reheating increases the average temperature at which heat is rejected. This is also apparent from Fig. 9–44. Therefore, in gas-turbine power plants, intercooling and reheating are always used in conjunction with regeneration.

If the number of compression and expansion stages is increased, the ideal gas-turbine cycle with intercooling, reheating, and regeneration approaches the Ericsson cycle, as illustrated in Fig. 9–45, and the thermal efficiency approaches the theoretical limit (the Carnot efficiency). However, the contribution of each additional stage to the thermal efficiency is less and less, and the use of more than two or three stages cannot be justified economically.

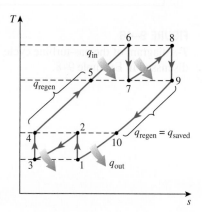

**FIGURE 9–44**

*T-s* diagram of an ideal gas-turbine cycle with intercooling, reheating, and regeneration.

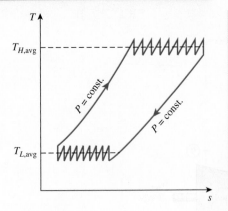

**FIGURE 9–45**

As the number of compression and expansion stages increases, the gas-turbine cycle with intercooling, reheating, and regeneration approaches the Ericsson cycle.

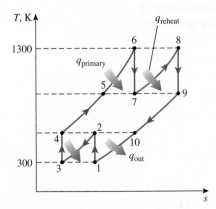

**FIGURE 9–46**

$T$-$s$ diagram of the gas-turbine cycle discussed in Example 9–8.

## EXAMPLE 9–8    A Gas Turbine with Reheating and Intercooling

An ideal gas-turbine cycle with two stages of compression and two stages of expansion has an overall pressure ratio of 8. Air enters each stage of the compressor at 300 K and each stage of the turbine at 1300 K. Determine the back work ratio and the thermal efficiency of this gas-turbine cycle, assuming (*a*) no regenerators and (*b*) an ideal regenerator with 100 percent effectiveness. Compare the results with those obtained in Example 9–5.

**SOLUTION**    An ideal gas-turbine cycle with two stages of compression and two stages of expansion is considered. The back work ratio and the thermal efficiency of the cycle are to be determined for the cases of no regeneration and maximum regeneration.

*Assumptions*    **1** Steady operating conditions exist. **2** The air-standard assumptions are applicable. **3** Kinetic and potential energy changes are negligible.

*Analysis*    The *T-s* diagram of the ideal gas-turbine cycle described is shown in Fig. 9–46. We note that the cycle involves two stages of expansion, two stages of compression, and regeneration.

For two-stage compression and expansion, the work input is minimized and the work output is maximized when both stages of the compressor and the turbine have the same pressure ratio. Thus,

$$\frac{P_2}{P_1} = \frac{P_4}{P_3} = \sqrt{8} = 2.83 \quad \text{and} \quad \frac{P_6}{P_7} = \frac{P_8}{P_9} = \sqrt{8} = 2.83$$

Air enters each stage of the compressor at the same temperature, and each stage has the same isentropic efficiency (100 percent in this case). Therefore, the temperature (and enthalpy) of the air at the exit of each compression stage will be the same. A similar argument can be given for the turbine. Thus,

At inlets:    $T_1 = T_3, \ h_1 = h_3$ and $T_6 = T_8, \ h_6 = h_8$

At exits:    $T_2 = T_4, \ h_2 = h_4$ and $T_7 = T_9, \ h_7 = h_9$

Under these conditions, the work input to each stage of the compressor will be the same, and so will the work output from each stage of the turbine.

(*a*) In the absence of any regeneration, the back work ratio and the thermal efficiency are determined by using data from Table A–17 as follows:

$$T_1 = 300 \text{ K} \rightarrow h_1 = 300.19 \text{ kJ/kg}$$

$$P_{r1} = 1.386$$

$$P_{r2} = \frac{P_2}{P_1}P_{r1} = \sqrt{8}(1.386) = 3.92 \rightarrow T_2 = 403.3 \text{ K}$$

$$h_2 = 404.31 \text{ kJ/kg}$$

$$T_6 = 1300 \text{ K} \rightarrow h_6 = 1395.97 \text{ kJ/kg}$$

$$P_{r6} = 330.9$$

$$P_{r7} = \frac{P_7}{P_6}P_{r6} = \frac{1}{\sqrt{8}}(330.9) = 117.0 \rightarrow T_7 = 1006.4 \text{ K}$$

$$h_7 = 1053.33 \text{ kJ/kg}$$

Then

$$w_{comp,in} = 2(w_{comp,in,I}) = 2(h_2 - h_1) = 2(404.31 - 300.19) = 208.24 \text{ kJ/kg}$$

$$w_{turb,out} = 2(w_{turb,out,I}) = 2(h_6 - h_7) = 2(1395.97 - 1053.33) = 685.28 \text{ kJ/kg}$$

$$w_{net} = w_{turb,out} - w_{comp,in} = 685.28 - 208.24 = 477.04 \text{ kJ/kg}$$

$$q_{in} = q_{primary} + q_{reheat} = (h_6 - h_4) + (h_8 - h_7)$$

$$= (1395.97 - 404.31) + (1395.97 - 1053.33) = 1334.30 \text{ kJ/kg}$$

Thus,

$$r_{bw} = \frac{w_{comp,in}}{w_{turb,out}} = \frac{208.24 \text{ kJ/kg}}{685.28 \text{ kJ/kg}} = \mathbf{0.304}$$

and

$$\eta_{th} = \frac{w_{net}}{q_{in}} = \frac{477.04 \text{ kJ/kg}}{1334.30 \text{ kJ/kg}} = \mathbf{0.358 \text{ or } 35.8\%}$$

A comparison of these results with those obtained in Example 9–5 (single-stage compression and expansion) reveals that multistage compression with intercooling and multistage expansion with reheating improve the back work ratio (it drops from 0.403 to 0.304) but hurt the thermal efficiency (it drops from 42.6 to 35.8 percent). Therefore, intercooling and reheating are not recommended in gas-turbine power plants unless they are accompanied by regeneration.

(b) The addition of an ideal regenerator (no pressure drops, 100 percent effectiveness) does not affect the compressor work and the turbine work. Therefore, the net work output and the back work ratio of an ideal gas-turbine cycle are identical whether there is a regenerator or not. A regenerator, however, reduces the heat input requirements by preheating the air leaving the compressor, using the hot exhaust gases. In an ideal regenerator, the compressed air is heated to the turbine exit temperature $T_9$ before it enters the combustion chamber. Thus, under the air-standard assumptions, $h_5 = h_7 = h_9$.

The heat input and the thermal efficiency in this case are

$$q_{in} = q_{primary} + q_{reheat} = (h_6 - h_5) + (h_8 - h_7)$$

$$= (1395.97 - 1053.33) + (1395.97 - 1053.33) = 685.28 \text{ kJ/kg}$$

and

$$\eta_{th} = \frac{w_{net}}{q_{in}} = \frac{477.04 \text{ kJ/kg}}{685.28 \text{ kJ/kg}} = \mathbf{0.696 \text{ or } 69.6\%}$$

*Discussion* Note that the thermal efficiency almost doubles as a result of regeneration compared to the no-regeneration case. The overall effect of two-stage compression and expansion with intercooling, reheating, and regeneration on the thermal efficiency is an increase of 63 percent. As the number of compression and expansion stages is increased, the cycle will approach the Ericsson cycle, and the thermal efficiency will approach

$$\eta_{th,Ericsson} = \eta_{th,Carnot} = 1 - \frac{T_L}{T_H} = 1 - \frac{300 \text{ K}}{1300 \text{ K}} = 0.769$$

Adding a second stage increases the thermal efficiency from 42.6 to 69.6 percent, an increase of 27 percentage points. This is a significant increase in efficiency, and usually it is well worth the extra cost associated with the second stage. Adding more stages, however (no matter how many), can increase the efficiency an additional 7.3 percentage points at most, and usually cannot be justified economically.

**FIGURE 9–47**

In jet engines, the high-temperature and high-pressure gases leaving the turbine are accelerated in a nozzle to provide thrust.

*Photo by Yunus Çengel*

# 9–11 · IDEAL JET-PROPULSION CYCLES

Gas-turbine engines are widely used to power aircraft because they are light and compact and have a high power-to-weight ratio. Aircraft gas turbines operate on an open cycle called a **jet-propulsion cycle**. The ideal jet-propulsion cycle differs from the simple ideal Brayton cycle in that the gases are not expanded to the ambient pressure in the turbine. Instead, they are expanded to a pressure such that the power produced by the turbine is just sufficient to drive the compressor and the auxiliary equipment, such as a small generator and hydraulic pumps. That is, the net work output of a jet-propulsion cycle is zero. The gases that exit the turbine at a relatively high pressure are subsequently accelerated in a nozzle to provide the thrust to propel the aircraft (Fig. 9–47). Also, aircraft gas turbines operate at higher pressure ratios (typically between 10 and 25), and the fluid passes through a diffuser first, where it is decelerated and its pressure is increased before it enters the compressor.

Aircraft are propelled by accelerating a fluid in the opposite direction to motion. This is accomplished by either slightly accelerating a large mass of fluid (*propeller-driven engine*) or greatly accelerating a small mass of fluid (*jet* or *turbojet engine*) or both (*turboprop engine*).

A schematic of a turbojet engine and the *T-s* diagram of the ideal turbojet cycle are shown in Fig. 9–48. The pressure of air rises slightly as it is decelerated in the diffuser. Air is compressed by the compressor. It is mixed with fuel in the combustion chamber, where the mixture is burned at constant pressure.

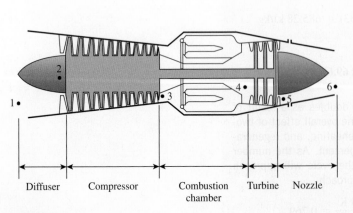

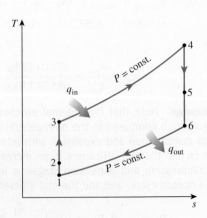

**FIGURE 9–48**

Basic components of a turbojet engine and the *T-s* diagram for the ideal turbojet cycle.

The high-pressure and high-temperature combustion gases partially expand in the turbine, producing enough power to drive the compressor and other equipment. Finally, the gases expand in a nozzle to the ambient pressure and leave the engine at a high velocity.

In the ideal case, the turbine work is assumed to equal the compressor work. Also, the processes in the diffuser, the compressor, the turbine, and the nozzle are assumed to be isentropic. In the analysis of actual cycles, however, the irreversibilities associated with these devices should be considered. The effect of the irreversibilities is to reduce the thrust that can be obtained from a turbojet engine.

The thrust developed in a turbojet engine is the unbalanced force that is caused by the difference in the momentum of the low-velocity air entering the engine and the high-velocity exhaust gases leaving the engine, and it is determined from Newton's second law. The pressures at the inlet and the exit of a turbojet engine are identical (the ambient pressure); thus, the net thrust developed by the engine is

$$F = (\dot{m}V)_{\text{exit}} - (\dot{m}V)_{\text{inlet}} = \dot{m}(V_{\text{exit}} - V_{\text{inlet}}) \quad (\text{N}) \qquad (9\text{–}27)$$

where $V_{\text{exit}}$ is the exit velocity of the exhaust gases and $V_{\text{inlet}}$ is the inlet velocity of the air, both relative to the aircraft. Thus, for an aircraft cruising in still air, $V_{\text{inlet}}$ is the aircraft velocity. In reality, the mass flow rates of the gases at the engine exit and the inlet are different, the difference being equal to the combustion rate of the fuel. However, the air–fuel mass ratio used in jet-propulsion engines is usually very high, making this difference very small. Thus, $\dot{m}$ in Eq. 9–27 is taken as the mass flow rate of air through the engine. For an aircraft cruising at a constant speed, the thrust is used to overcome air drag, and the net force acting on the body of the aircraft is zero. Commercial airplanes save fuel by flying at higher altitudes during long trips since air at higher altitudes is thinner and exerts a smaller drag force on aircraft.

The power developed from the thrust of the engine is called the **propulsive power** $\dot{W}_P$, which is the *propulsive force* (*thrust*) times the *distance* this force acts on the aircraft per unit time, that is, the thrust times the aircraft velocity (Fig. 9–49):

$$\dot{W}_P = FV_{\text{aircraft}} = \dot{m}(V_{\text{exit}} - V_{\text{inlet}})V_{\text{aircraft}} \quad (\text{kW}) \qquad (9\text{–}28)$$

The net work developed by a turbojet engine is zero. Thus, we cannot define the efficiency of a turbojet engine in the same way as stationary gas-turbine engines. Instead, we should use the general definition of efficiency, which is the ratio of the desired output to the required input. The desired output in a turbojet engine is the *power produced* to propel the aircraft $\dot{W}_P$, and the required input is the *heating value of the fuel* $\dot{Q}_{\text{in}}$. The ratio of these two quantities is called the **propulsive efficiency** and is given by

$$\eta_P = \frac{\text{Propulsive power}}{\text{Energy input rate}} = \frac{\dot{W}_P}{\dot{Q}_{\text{in}}} \qquad (9\text{–}29)$$

Propulsive efficiency is a measure of how efficiently the thermal energy released during the combustion process is converted to propulsive energy. The remaining part of the energy released shows up as the kinetic energy of the exhaust gases relative to a fixed point on the ground and as an increase in the enthalpy of the gases leaving the engine.

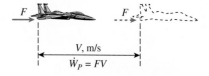

**FIGURE 9–49**

Propulsive power is the thrust acting on the aircraft through a distance per unit time.

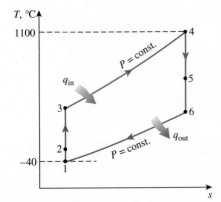

**FIGURE 9–50**

*T-s* diagram for the turbojet cycle
described in Example 9–9.

*EXAMPLE 9–9* **The Ideal Jet-Propulsion Cycle**

A turbojet aircraft flies with a velocity of 260 m/s at an altitude where the
air is at 35 kPa and −40°C. The compressor has a pressure ratio of 10, and
the temperature of the gases at the turbine inlet is 1100°C. Air enters the
compressor at a rate of 45 kg/s. Utilizing the cold-air-standard assumptions,
determine (a) the temperature and pressure of the gases at the turbine exit,
(b) the velocity of the gases at the nozzle exit, and (c) the propulsive efficiency of the cycle.

**SOLUTION** The operating conditions of a turbojet aircraft are specified. The
temperature and pressure at the turbine exit, the velocity of gases at the
nozzle exit, and the propulsive efficiency are to be determined.
*Assumptions* **1** Steady operating conditions exist. **2** The cold-air-standard
assumptions are applicable and thus air can be assumed to have constant
specific heats at room temperature ($c_p$ = 1.005 kJ/kg·°C and $k$ = 1.4).
**3** Kinetic and potential energies are negligible, except at the diffuser inlet
and the nozzle exit. **4** The turbine work output is equal to the compressor
work input.
*Analysis* The *T-s* diagram of the ideal jet propulsion cycle described is
shown in Fig. 9–50. We note that the components involved in the jet-propulsion
cycle are steady-flow devices.
(a) Before we can determine the temperature and pressure at the turbine
exit, we need to find the temperatures and pressures at other states:
*Process 1–2* (isentropic compression of an ideal gas in a diffuser): For
convenience, we can assume that the aircraft is stationary and the air is
moving toward the aircraft at a velocity of $V_1$ = 260 m/s. Ideally, the air
exits the diffuser with a negligible velocity ($V_2 \cong 0$):

$$h_2 + \frac{V_2^{2\,\nearrow 0}}{2} = h_1 + \frac{V_1^2}{2}$$

$$0 = c_p(T_2 - T_1) - \frac{V_1^2}{2}$$

$$T_2 = T_1 + \frac{V_1^2}{2c_p}$$

$$= 233 \text{ K} + \frac{(260 \text{ m/s})^2}{2(1.005 \text{ kJ/kg·K})}\left(\frac{1 \text{ kJ/kg}}{1000 \text{ m}^2/\text{s}^2}\right)$$

$$= 267 \text{ K}$$

$$P_2 = P_1\left(\frac{T_2}{T_1}\right)^{k/(k-1)} = (35 \text{ kPa})\left(\frac{267 \text{ K}}{233 \text{ K}}\right)^{1.4/(1.4-1)} = 56.4 \text{ kPa}$$

*Process 2–3* (isentropic compression of an ideal gas in a compressor):

$$P_3 = (r_p)(P_2) = (10)(56.4 \text{ kPa}) = 564 \text{ kPa } (= P_4)$$

$$T_3 = T_2\left(\frac{P_3}{P_2}\right)^{(k-1)/k} = (267 \text{ K})(10)^{(1.4-1)/1.4} = 515 \text{ K}$$

*Process 4–5* (isentropic expansion of an ideal gas in a turbine): Neglecting the kinetic energy changes across the compressor and the turbine and assuming the turbine work to be equal to the compressor work, we find the temperature and pressure at the turbine exit to be

$$w_{comp,in} = w_{turb,out}$$

$$h_3 - h_2 = h_4 - h_5$$

$$c_p(T_3 - T_2) = c_p(T_4 - T_5)$$

$$T_5 = T_4 - T_3 + T_2 = 1373 - 515 + 267 = \textbf{1125 K}$$

$$P_5 = P_4\left(\frac{T_5}{T_4}\right)^{k/(k-1)} = (564\ \text{kPa})\left(\frac{1125\ \text{K}}{1373\ \text{K}}\right)^{1.4/(1.4-1)} = \textbf{281 kPa}$$

(*b*) To find the air velocity at the nozzle exit, we need to first determine the nozzle exit temperature and then apply the steady-flow energy equation. *Process 5-6* (isentropic expansion of an ideal gas in a nozzle):

$$T_6 = T_5\left(\frac{P_6}{P_5}\right)^{(k-1)/k} = (1125\ \text{K})\left(\frac{35\ \text{kPa}}{281\ \text{kPa}}\right)^{(1.4-1)/1.4} = 620\ \text{K}$$

$$h_6 + \frac{V_6^2}{2} = h_5 + \overset{0}{\cancel{\frac{V_5^2}{2}}}$$

$$0 = c_p(T_6 - T_5) + \frac{V_6^2}{2}$$

$$V_6 = \sqrt{2c_p(T_5 - T_6)}$$

$$= \sqrt{2(1.005\ \text{kJ/kg·K})\ [(1125 - 620)\ \text{K]}\left(\frac{1000\ \text{m}^2/\text{s}^2}{1\ \text{kJ/kg}}\right)}$$

$$= \textbf{1007 m/s}$$

(*c*) The propulsive efficiency of a turbojet engine is the ratio of the propulsive power developed $\dot{W}_P$ to the total heat transfer rate to the working fluid:

$$\dot{W}_P = \dot{m}(V_{exit} - V_{inlet})V_{aircraft}$$

$$= (45\ \text{kg/s})[(1007 - 260)\ \text{m/s}](260\ \text{m/s})\left(\frac{1\ \text{kJ/kg}}{1000\ \text{m}^2/\text{s}^2}\right)$$

$$= 8740\ \text{kW}$$

$$\dot{Q}_{in} = \dot{m}(h_4 - h_3) = \dot{m}c_p(T_4 - T_3)$$

$$= (45\ \text{kg/s})(1.005\ \text{kJ/kg·K})[(1373 - 515)\ \text{K}]$$

$$= 38,803\ \text{kW}$$

$$\eta_P = \frac{\dot{W}_P}{\dot{Q}_{in}} = \frac{8740\ \text{kW}}{38,803\ \text{kW}} = \textbf{0.225 or 22.5\%}$$

That is, 22.5 percent of the energy input is used to propel the aircraft and to overcome the drag force exerted by the atmospheric air.

*Discussion* For those who are wondering what happened to the rest of the energy, here is a brief account:

$$\dot{KE}_{out} = \dot{m}\frac{V_g^2}{2} = (45 \text{ kg/s}) \left\{ \frac{[(1007 - 260)\text{m/s}]^2}{2} \right\} \left( \frac{1 \text{ kJ/kg}}{1000 \text{ m}^2/\text{s}^2} \right)$$

$$= 12{,}555 \text{ kW} \quad (32.4\%)$$

$$\dot{Q}_{out} = \dot{m}(h_6 - h_1) = \dot{m}c_p(T_6 - T_1)$$

$$= (45 \text{ kg/s})(1.005 \text{ kJ/kg·K})[(620 - 233) \text{ K}]$$

$$= 17{,}502 \text{ kW} \quad (45.1\%)$$

Thus, 32.4 percent of the energy shows up as excess kinetic energy (kinetic energy of the gases relative to a fixed point on the ground). Notice that for the highest propulsion efficiency, the velocity of the exhaust gases relative to the ground $V_g$ should be zero. That is, the exhaust gases should leave the nozzle at the velocity of the aircraft. The remaining 45.1 percent of the energy shows up as an increase in enthalpy of the gases leaving the engine. These last two forms of energy eventually become part of the internal energy of the atmospheric air (Fig. 9–51).

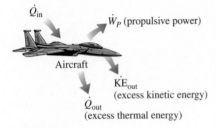

**FIGURE 9–51**
Energy supplied to an aircraft (from the burning of a fuel) manifests itself in various forms.

## Modifications to Turbojet Engines

The first airplanes built were all propeller-driven, with propellers powered by engines essentially identical to automobile engines. The major break-through in commercial aviation occurred with the introduction of the turbo-jet engine in 1952. Both propeller-driven engines and jet-propulsion-driven engines have their own strengths and limitations, and several attempts have been made to combine the desirable characteristics of both in one engine. Two such modifications are the *propjet engine* and the *turbofan engine*.

The most widely used engine in aircraft propulsion is the **turbofan** (or *fanjet*) engine wherein a large fan driven by the turbine forces a consider-able amount of air through a duct (cowl) surrounding the engine, as shown in Figs. 9–52 and 9–53. The fan exhaust leaves the duct at a higher velocity, enhancing the total thrust of the engine significantly. A turbofan engine is based on the principle that for the same power, a large volume of slower-moving air

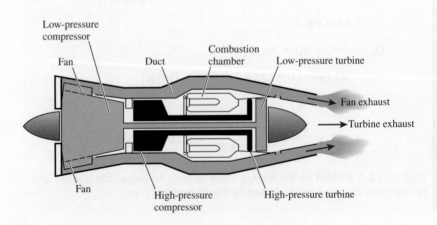

**FIGURE 9–52**
A turbofan engine.

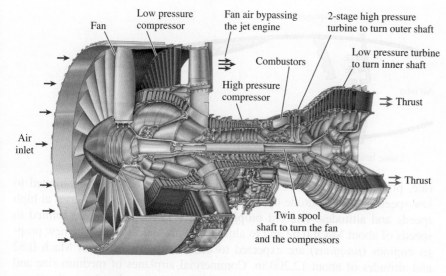

Fan

Low pressure compressor

Fan air bypassing the jet engine

2-stage high pressure turbine to turn outer shaft

Combustors

Low pressure turbine to turn inner shaft

High pressure compressor

Air inlet

Thrust

Thrust

Twin spool shaft to turn the fan and the compressors

**FIGURE 9–53**
A modern jet engine used to power Boeing 777 aircraft. This is a Pratt & Whitney PW4084 turbofan capable of producing 375 kN of thrust. It is 4.87 m long, has a 2.84 m diameter fan, and it weighs 6800 kg.

*Reproduced by permission of United Technologies Corporation, Pratt & Whitney.*

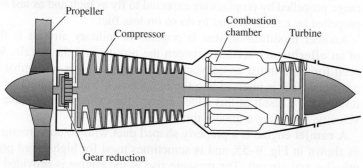

Propeller

Compressor

Combustion chamber

Turbine

Gear reduction

**FIGURE 9–54**
A turboprop engine.

produces more thrust than a small volume of fast-moving air. The first commercial turbofan engine was successfully tested in 1955.

The turbofan engine on an airplane can be distinguished from the less efficient turbojet engine by its fat cowling covering the large fan. All the thrust of a turbojet engine is due to the exhaust gases leaving the engine at about twice the speed of sound. In a turbofan engine, the high-speed exhaust gases are mixed with the lower-speed air, which results in a considerable reduction in noise.

New cooling techniques have resulted in considerable increases in efficiencies by allowing gas temperatures at the burner exit to reach over 1500°C, which is more than 100°C above the melting point of the turbine blade materials. Turbofan engines deserve most of the credit for the success of jumbo jets that weigh almost 400,000 kg and are capable of carrying over 400 passengers for up to a distance of 10,000 km at speeds over 950 km/h with less fuel per passenger kilometer.

The ratio of the mass flow rate of air bypassing the combustion chamber to that of air flowing through it is called the *bypass ratio*. The first commercial high-bypass-ratio engines had a bypass ratio of 5. Increasing the bypass ratio of a turbofan engine increases thrust. Thus, it makes sense to remove the cowl from the fan. The result is a **propjet** engine, as shown in Fig. 9–54. Turbofan and propjet engines differ primarily in their bypass ratios: 5 or 6 for turbofans and as high as 100 for propjets. As a general

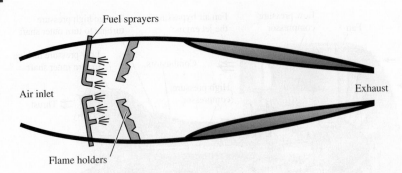

**FIGURE 9–55**
A ramjet engine.

rule, propellers are more efficient than jet engines, but they are limited to low-speed and low-altitude operation since their efficiency decreases at high speeds and altitudes. The old propjet engines (*turboprops*) were limited to speeds of about Mach 0.62 and to altitudes of around 9100 m. The new propjet engines (*propfans*) are expected to achieve speeds of about Mach 0.82 and altitudes of about 12,200 m. Commercial airplanes of medium size and range propelled by propfans are expected to fly as high and as fast as the planes propelled by turbofans, and to do so on less fuel.

Another modification that is popular in military aircraft is the addition of an **afterburner** section between the turbine and the nozzle. Whenever a need for extra thrust arises, such as for short takeoffs or combat conditions, additional fuel is injected into the oxygen-rich combustion gases leaving the turbine. As a result of this added energy, the exhaust gases leave at a higher velocity, providing a greater thrust.

A **ramjet** engine is a properly shaped duct with no compressor or turbine, as shown in Fig. 9–55, and is sometimes used for high-speed propulsion of missiles and aircraft. The pressure rise in the engine is provided by the ram effect of the incoming high-speed air being rammed against a barrier. Therefore, a ramjet engine needs to be brought to a sufficiently high speed by an external source before it can be fired.

The ramjet performs best in aircraft flying above Mach 2 or 3 (two or three times the speed of sound). In a ramjet, the air is slowed down to about Mach 0.2, fuel is added to the air and burned at this low velocity, and the combustion gases are expended and accelerated in a nozzle.

A **scramjet** engine is essentially a ramjet in which air flows through at supersonic speeds (above the speed of sound). Ramjets that convert to scramjet configurations at speeds above Mach 6 are successfully tested at speeds of about Mach 8.

Finally, a **rocket** is a device where a solid or liquid fuel and an oxidizer react in the combustion chamber. The high-pressure combustion gases are then expanded in a nozzle. The gases leave the rocket at very high velocities, producing the thrust to propel the rocket.

## 9–12 · SECOND-LAW ANALYSIS OF GAS POWER CYCLES

The ideal Carnot, Ericsson, and Stirling cycles are *totally reversible*; thus they do not involve any irreversibilities. The ideal Otto, Diesel, and Brayton cycles, however, are only *internally reversible,* and they may involve irreversibilities external to the system. A second-law analysis of these cycles reveals where the largest irreversibilities occur and where to start improvements.

Relations for *exergy* and *exergy destruction* for both closed and steady-flow systems are developed in Chap. 8. The exergy destruction for a closed system can be expressed as

$$X_{\text{dest}} = T_0 S_{\text{gen}} = T_0(\Delta S_{\text{sys}} - S_{\text{in}} + S_{\text{out}})$$

$$= T_0 \left[ (S_2 - S_1)_{\text{sys}} - \frac{Q_{\text{in}}}{T_{b,\text{in}}} + \frac{Q_{\text{out}}}{T_{b,\text{out}}} \right] \quad \text{(kJ)} \qquad \textbf{(9–30)}$$

where $T_{b,\text{in}}$ and $T_{b,\text{out}}$ are the temperatures of the system boundary where heat is transferred into and out of the system, respectively. A similar relation for steady-flow systems can be expressed, in rate form, as

$$\dot{X}_{\text{dest}} = T_0 \dot{S}_{\text{gen}} = T_0(\dot{S}_{\text{out}} - \dot{S}_{\text{in}}) = T_0 \left( \sum_{\text{out}} \dot{m}s - \sum_{\text{in}} \dot{m}s - \frac{\dot{Q}_{\text{in}}}{T_{b,\text{in}}} + \frac{\dot{Q}_{\text{out}}}{T_{b,\text{out}}} \right) \quad \text{(kW)}$$

$$\textbf{(9–31)}$$

or, on a unit–mass basis for a one-inlet, one-exit steady-flow device, as

$$x_{\text{dest}} = T_0 s_{\text{gen}} = T_0 \left( s_e - s_i - \frac{q_{\text{in}}}{T_{b,\text{in}}} + \frac{q_{\text{out}}}{T_{b,\text{out}}} \right) \quad \text{(kJ/kg)} \qquad \textbf{(9–32)}$$

where subscripts $i$ and $e$ denote the inlet and exit states, respectively.

The exergy destruction of a *cycle* is the sum of the exergy destructions of the processes that compose that cycle. The exergy destruction of a cycle can also be determined without tracing the individual processes by considering the entire cycle as a single process and using one of the relations above. Entropy is a property, and its value depends on the state only. For a cycle, reversible or actual, the initial and the final states are identical; thus $s_e = s_i$. Therefore, the exergy destruction of a cycle depends on the magnitude of the heat transfer with the high- and low-temperature reservoirs involved and on their temperatures. It can be expressed on a unit–mass basis as

$$x_{\text{dest}} = T_0 \left( \sum \frac{q_{\text{out}}}{T_{b,\text{out}}} - \sum \frac{q_{\text{in}}}{T_{b,\text{in}}} \right) \quad \text{(kJ/kg)} \qquad \textbf{(9–33)}$$

For a cycle that involves heat transfer only with a source at $T_H$ and a sink at $T_L$, the exergy destruction becomes

$$x_{\text{dest}} = T_0 \left( \frac{q_{\text{out}}}{T_L} - \frac{q_{\text{in}}}{T_H} \right) \quad \text{(kJ/kg)} \qquad \textbf{(9–34)}$$

The exergies of a closed system $\phi$ and a fluid stream $\psi$ at any state can be determined from

$$\phi = (u - u_0) - T_0(s - s_0) + P_0(v - v_0) + \frac{V^2}{2} + gz \quad \text{(kJ/kg)} \qquad \textbf{(9–35)}$$

and

$$\psi = (h - h_0) - T_0(s - s_0) + \frac{V^2}{2} + gz \quad \text{(kJ/kg)} \qquad \textbf{(9–36)}$$

where subscript "0" denotes the state of the surroundings.

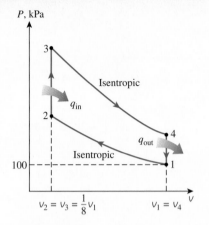

**FIGURE 9–56**
Schematic for Example 9–10.

### EXAMPLE 9–10

Consider an engine operating on the ideal Otto cycle with a compression ratio of 8 (Fig. 9–56). At the beginning of the compression process, air is at 100 kPa and 17°C. During the constant-volume heat-addition process, 800 kJ/kg of heat is transferred to air from a source at 1700 K and waste heat is rejected to the surroundings at 300 K. Accounting for the variation of specific heats of air with temperature, determine (a) the exergy destruction associated with each of the four processes and the cycle and (b) the second-law efficiency of this cycle.

**SOLUTION** An engine operating on the ideal Otto cycle is considered. For specific source and sink temperatures, the exergy destruction associated with this cycle and the second-law efficiency are to be determined.
**Assumptions** 1 Steady operating conditions exist. 2 Kinetic and potential energy changes are negligible.
**Analysis** We take the engine bordering the heat source at temperature $T_H$ and the environment at temperature $T_0$ as the system. This cycle was analyzed in Example 9-2, and various quantities were given or determined to be

$$r = 8 \qquad P_2 = 1.7997 \text{ MPa}$$
$$T_0 = 290 \text{ K} \qquad P_3 = 4.345 \text{ MPa}$$
$$T_1 = 290 \text{ K} \qquad q_{in} = 800 \text{ kJ/kg}$$
$$T_2 = 652.4 \text{ K} \qquad q_{out} = 381.83 \text{ kJ/kg}$$
$$T_3 = 1575.1 \text{ K} \qquad w_{net} = 418.17 \text{ kJ/kg}$$

Processes 1–2 and 3–4 are isentropic ($s_1 = s_2$, $s_3 = s_4$) and therefore do not involve any internal or external irreversibilities; that is, $X_{dest,12} = 0$ and $X_{dest,34} = 0$.

Processes 2–3 and 4–1 are constant-volume heat-addition and heat-rejection processes, respectively, and are internally reversible. However, the heat transfer between the working fluid and the source or the sink takes place through a finite temperature difference, rendering both processes irreversible. The exergy destruction associated with each process is determined from Eq. 9–32. However, first we need to determine the entropy change of air during these processes:

$$s_3 - s_2 = s_3^\circ - s_2^\circ - R \ln \frac{P_3}{P_2}$$

$$= (3.5045 - 2.4975) \text{ kJ/kg·K} - (0.287 \text{ kJ/kg·K}) \ln \frac{4.345 \text{ MPa}}{1.7997 \text{ MPa}}$$

$$= 0.7540 \text{ kJ/kg·K}$$

Also,

$$q_{in} = 800 \text{ kJ/kg} \quad \text{and} \quad T_{source} = 1700 \text{ K}$$

Thus

$$x_{dest,23} = T_0 \left[ (s_3 - s_2)_{sys} - \frac{q_{in}}{T_{source}} \right]$$

$$= (290 \text{ K}) \left[ 0.7540 \text{ kJ/kg·K} - \frac{800 \text{ kJ/kg}}{1700 \text{ K}} \right]$$

$$= 82.2 \text{ kJ/kg}$$

For process 4-1, $s_1 - s_4 = s_2 - s_3 = -0.7540$ kJ/kg·K, $q_{R,41} = q_{out} = 381.83$ kJ/kg, and $T_{sink} = 290$ K. Thus,

$$x_{dest,41} = T_0 \left[ (s_1 - s_4)_{sys} + \frac{q_{out}}{T_{sink}} \right]$$

$$= (290\,K) \left[ -0.7540\,kJ/kg\cdot K + \frac{381.83\,kJ/kg}{290\,K} \right]$$

$$= 163.2\ kJ/kg$$

Therefore, the irreversibility of the cycle is

$$x_{dest,cycle} = x_{dest,12} + x_{dest,23} + x_{dest,34} + x_{dest,41}$$

$$= 0 + 82.2\ kJ/kg + 0 + 163.2\,kJ/kg$$

$$= 245.4\ kJ/kg$$

The exergy destruction of the cycle could also be determined from Eq. 9-34. Notice that the largest exergy destruction in the cycle occurs during the heat-rejection process. Therefore, any attempt to reduce the exergy destruction should start with this process.

(b) The second-law efficiency is defined as

$$\eta_{II} = \frac{\text{Exergy recovered}}{\text{Exergy expended}} = \frac{x_{recovered}}{x_{expended}} = 1 - \frac{x_{destroyed}}{x_{expended}}$$

Here the expended energy is the energy content of the heat supplied to the air in the engine (which is its work potential) and the energy recovered is the net work output:

$$x_{expended} = x_{heat,\,in} = \left(1 - \frac{T_0}{T_H}\right) q_{in}$$

$$= \left(1 - \frac{290\,K}{1700\,K}\right)(800\ kJ/kg) = 663.5\ kJ/kg$$

$$x_{recovered} = w_{net,\,out} = 418.17\ kJ/kg$$

Substituting, the second-law efficiency of this power plant is determined to be

$$\eta_{II} = \frac{x_{recovered}}{x_{expended}} = \frac{418.17\ kJ/kg}{663.5\ kJ/kg} = 0.630 \text{ or } 63.0\%$$

**Discussion** The second-law efficiency can also be determined using the exergy destruction data,

$$\eta_{II} = 1 - \frac{x_{destroyed}}{x_{expended}} = 1 - \frac{245.4\ kJ/kg}{663.5\ kJ/kg} = 0.630 \text{ or } 63.0\%$$

Note that the exergy destruction associated with heat transfer involving both the heat source and the environment are accounted for in the results.

**TOPIC OF SPECIAL INTEREST\***     Saving Fuel and Money by Driving Sensibly

**FIGURE 9–57**

The average car in the United States is driven about 21,700 km a year, uses about 2200 liters of gasoline, worth $2200 at $1.00/L.

7.8 L/100 km

$1800/yr

11.8 L/100 km

$2700/yr

**FIGURE 9–58**

Under average driving conditions, the owner of a 7.8 L/100 km vehicle spends $900 less each year on gasoline than the owner of a 11.8 L/100 km vehicle (assuming $1.00/L and 21,700 km/yr).

Two-thirds of the oil used in the United States is used for transportation. Half of this oil is consumed by passenger cars and light trucks that are used to commute to and from work (38 percent), run a family business (35 percent), and for recreational, social, and religious activities (27 percent). The overall fuel efficiency of the vehicles has increased considerably over the years due to improvements primarily in aerodynamics, materials, and electronic controls. However, the average fuel consumption of new vehicles has not changed much from about 12 L/100 km because of the increasing consumer trend toward purchasing larger and less fuel-efficient cars, trucks, and sport utility vehicles. Motorists also continue to drive more each year: 21,680 km in 2010 compared to 16,540 km in 1990. Also, the annual gasoline use per vehicle in the United States has increased to 2200 liters in 2010 (worth $2200 at $1.00/L) from 1915 liters in 1990 (Fig. 9–57).

Saving fuel is not limited to good driving habits. It also involves purchasing the right car, using it responsibly, and maintaining it properly. A  car does not burn any fuel when it is not running, and thus a sure way to save fuel is not to drive the car at all—but this is not the reason we buy a car. We can reduce driving and thus fuel consumption by considering viable alternatives such as *living close to work and shopping areas, working at home, working longer hours in fewer days, joining a car pool or starting one, using public transportation, combining errands into a single trip and planning ahead, avoiding rush hours and roads with heavy traffic and many traffic lights,* and simply *walking* or *bicycling* instead of driving to nearby places, with the added benefit of good health and physical fitness. Driving only when necessary is the best way to save fuel, money, and the environment too.

Driving efficiently starts before buying a car, just like raising good children starts before getting married. The buying decision made now will affect the fuel consumption for many years. Under average driving conditions, the owner of a 7.8 L/100 km vehicle will spend $900 less each year on fuel than the owner of a 11.8 L/100 km vehicle (assuming a fuel cost of $1.00/L and 21,700 km of driving per year). If the vehicle is owned for 5 years, the 7.8 L/100 km vehicle will save $4500 during this period (Fig. 9–58). The fuel consumption of a car depends on many factors such as *the type of the vehicle, the weight, the transmission type, the size and efficiency of the engine,* and *the accessories* and *the options installed.* The most fuel-efficient cars are aerodynamically designed compact cars with a small engine, manual transmission, low frontal area (the height times the width of the car), and bare essentials.

At highway speeds, most fuel is used to overcome aerodynamic drag or air resistance to motion, which is the force needed to move the vehicle through the air. This resistance force is proportional to the drag coefficient and the frontal area.

---

\*This section can be skipped without a loss in continuity. Information in this section is based largely on the publications of the U.S. Department of Energy, Environmental Protection Agency, and the American Automotive Association.

Therefore, for a given frontal area, a sleek-looking aerodynamically designed vehicle with contoured lines that coincide with the streamlines of air flow has a smaller drag coefficient and thus better fuel economy than a boxlike vehicle with sharp corners (Fig. 9–59). For the same overall shape, a compact car has a smaller frontal area and thus better fuel economy compared to a large car.

Moving around the *extra weight* requires more fuel, and thus it hurts fuel economy. Therefore, the lighter the vehicle, the more fuel-efficient it is. Also as a general rule, the larger the engine is, the greater its rate of fuel consumption is. So you can expect a car with a 1.8 L engine to be more fuel efficient than one with a 3.0 L engine. For a given engine size, *diesel engines* operate on much higher compression ratios than the gasoline engines, and thus they are inherently more fuel-efficient. *Manual transmissions* are usually more efficient than the automatic ones, but this is not always the case. A car with automatic transmission generally uses 10 percent more fuel than a car with manual transmission because of the losses associated with the hydraulic connection between the engine and the transmission, and the added weight. Transmissions with an *overdrive gear* (found in four-speed automatic transmissions and five-speed manual transmissions) save fuel and reduce noise and engine wear during highway driving by decreasing the engine rpm while maintaining the same vehicle speed.

*Front wheel drive* offers better traction (because of the engine weight on top of the front wheels), reduced vehicle weight and thus better fuel economy, with an added benefit of increased space in the passenger compartment. Four-wheel drive mechanisms provide better traction and braking thus safer driving on slippery roads and loose gravel by transmitting torque to all four wheels. However, the added safety comes with increased weight, noise, and cost, and decreased fuel economy. *Radial tires* usually reduce the fuel consumption by 5 to 10 percent by reducing the rolling resistance, but their pressure should be checked regularly since they can look normal and still be underinflated. *Cruise control* saves fuel during long trips on open roads by maintaining steady speed. *Tinted windows* and light interior and exterior colors reduce solar heat gain, and thus the need for air-conditioning.

## BEFORE DRIVING

Certain things done before driving can make a significant difference on the fuel cost of the vehicle while driving. Below we discuss some measures such as using the right kind of fuel, minimizing idling, removing extra weight, and keeping the tires properly inflated.

## Use Fuel with the Minimum Octane Number Recommended by the Vehicle Manufacturer

Many motorists buy higher-priced premium fuel, thinking that it is better for the engine. Most of today's cars are designed to operate on regular unleaded fuel. If the owner's manual does not call for premium fuel, using anything other than regular gas is simply a waste of money. Octane number is not a measure of the "power" or "quality" of the fuel, it is simply a measure of fuel's resistance to engine knock caused by premature ignition. Despite

**FIGURE 9–59**
Aerodynamically designed vehicles have a smaller drag coefficient and thus better fuel economy than boxlike vehicles with sharp corners.

**FIGURE 9–60**

Despite the implications of flashy names, a fuel with a higher octane number is not a better fuel; it is simply more expensive.

©S. Meltzer/PhotoLink/Getty Images RF

the implications of flashy names like "premium," "super," or "power plus," a fuel with a higher octane number is not a better fuel; it is simply more expensive because of the extra processing involved to raise the octane number (Fig. 9–60). Older cars may need to go up one grade level from the recommended new car octane number if they start knocking.

## Do Not Overfill the Gas Tank

Topping off the gas tank may cause the fuel to backflow during pumping. In hot weather, an overfilled tank may also cause the fuel to overflow due to thermal expansion. This wastes fuel, pollutes the environment, and may damage the car's paint. Also, fuel tank caps that do not close tightly allow some gasoline to be lost by evaporation. Buying fuel in cool weather such as early in the mornings minimizes evaporative losses. Each liter of spilled or evaporated fuel emits as much hydrocarbon to the air as 2000 km of driving.

## Park in the Garage

The engine of a car parked in a garage overnight is warmer the next morning. This reduces the problems associated with the warming-up period such as starting, excessive fuel consumption, and environmental pollution. In hot weather, a garage blocks the direct sunlight and reduces the need for air conditioning.

## Start the Car Properly and Avoid Extended Idling

With today's cars, it is not necessary to prime the engine first by pumping the accelerator pedal repeatedly before starting. This only wastes fuel. Warming up the engine isn't necessary either. Keep in mind that an idling engine wastes fuel and pollutes the environment. Don't race a cold engine to warm it up. An engine warms up faster on the road under a light load, and the catalytic converter begins to function sooner. Start driving as soon as the engine is started, but avoid rapid acceleration and highway driving before the engine and thus the oil fully warms up to prevent engine wear.

In cold weather, the warm-up period is much longer, the fuel consumption during warm-up is much higher, and the exhaust emissions are much larger. At −20°C, for example, a car needs to be driven at least 5 km to warm up fully. A gasoline engine uses up to 50 percent more fuel during warm-up than it does after it is warmed up. Exhaust emissions from a cold engine during warm-up are much higher since the catalytic converters do not function properly before reaching their normal operating temperature of about 390°C.

## Don't Carry Unnecessary Weight in or on the Vehicle

Remove any snow or ice from the vehicle, and avoid carrying unneeded items, especially heavy ones (such as snow chains, old tires, books) in the passenger compartment, trunk, or the cargo area of the vehicle (Fig. 9–61). This wastes fuel since it requires extra fuel to carry around the extra weight. An extra 50 kg decreases fuel economy of a car by about 1–2 percent.

Some people find it convenient to use a roof rack or carrier for additional cargo space. However, if you must carry some extra items, place them inside the vehicle rather than on roof racks to reduce drag. Any snow that accumulates on a vehicle

**FIGURE 9–61**

A loaded roof rack can increase fuel consumption by up to 5 percent in highway driving.

and distorts its shape must be removed for the same reason. A loaded roof rack can increase fuel consumption by up to 5 percent in highway driving. Even the most streamlined empty rack increases aerodynamic drag and thus fuel consumption. Therefore, the roof rack should be removed when it is no longer needed.

## Keep Tires Inflated to the Recommended Maximum Pressure

Keeping the tires inflated properly is one of the easiest and most important things one can do to improve fuel economy. If a range is recommended by the manufacturer, the higher pressure should be used to maximize fuel efficiency. Tire pressure should be checked when the tire is cold since tire pressure changes with temperature (it increases by 7 kPa (1 psi) for every 6°C rise in temperature due to a rise in ambient temperature or just road friction). Underinflated tires run hot and jeopardize safety, cause the tires to wear prematurely, affect the vehicle's handling adversely, and hurt the fuel economy by increasing the rolling resistance. Overinflated tires cause unpleasant bumpy rides, and cause the tires to wear unevenly. Tires lose about 7 kPa (1 psi) pressure per month due to air loss caused by the tire hitting holes, bumps, and curbs. Therefore, the tire pressure should be checked at least once a month. Just one tire underinflated by 14 kPa (2 psi) results in a 1 percent increase in fuel consumption (Fig. 9–62). Underinflated tires often cause fuel consumption of vehicles to increase by 5 or 6 percent.

It is also important to keep the wheels aligned. Driving a vehicle with the front wheels out of alignment increases rolling resistance and thus fuel consumption while causing handling problems and uneven tire wear. Therefore, the wheels should be aligned properly whenever necessary.

**FIGURE 9–62**
Underinflated tires often cause fuel consumption of vehicles to increase by 5 or 6 percent.
*©McGraw-Hill Education/Lars A. Niki RF*

## WHILE DRIVING

The driving habits can make a significant difference in the amount of fuel used. Driving sensibly and practicing some fuel-efficient driving techniques such as those discussed below can improve fuel economy easily by more than 10 percent.

### Avoid Quick Starts and Sudden Stops

Despite the attention they may get, the abrupt, aggressive "jackrabbit" starts waste fuel, wear the tires, jeopardize safety, and are harder on vehicle components and connectors. The squealing stops wear the brake pads prematurely, and may cause the driver to lose control of the vehicle. Easy starts and stops save fuel, reduce wear and tear, reduce pollution, and are safer and more courteous to other drivers.

### Drive at Moderate Speeds

Avoiding high speeds on open roads results in safer driving and better fuel economy. In highway driving, over 50 percent of the power produced by the engine is used to overcome aerodynamic drag (i.e., to push air out of the way). Aerodynamic drag and thus fuel consumption increase rapidly at speeds above 88 km/h, as shown in Fig. 9–63. On average, a car uses about

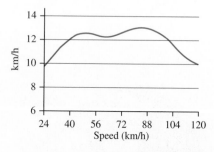

**FIGURE 9–63**
Aerodynamic drag increases and thus fuel economy decreases rapidly at speeds above 88 km/h.
*Source: EPA and U.S. Dept. of Energy.*

10 percent more fuel at 100 km/h and 20 percent more fuel at 110 km/h than it does at 90 km/h.

The discussion above should not lead one to conclude that the lower the speed, the better the fuel economy—because it is not. The number of kilometers that can be driven per liter of fuel drops sharply at speeds below 50 km/h, as shown in the chart. Besides, speeds slower than the flow of traffic can create a traffic hazard. Therefore, a car should be driven at moderate speeds for safety and best fuel economy.

## Maintain a Constant Speed

The fuel consumption remains at a minimum during steady driving at a moderate speed. Keep in mind that every time the accelerator is hard pressed, more fuel is pumped into the engine. The vehicle should be accelerated gradually and smoothly since extra fuel is squirted into the engine during quick acceleration. Using cruise control on highway trips can help maintain a constant speed and reduce fuel consumption. Steady driving is also safer, easier on the nerves, and better for the heart.

## Anticipate Traffic Ahead and Avoid Tailgating

A driver can reduce fuel consumption by up to 10 percent by anticipating traffic conditions ahead and adjusting the speed accordingly, and avoiding tailgating and thus unnecessary braking and acceleration (Fig. 9–64). Accelerations and decelerations waste fuel. Braking and abrupt stops can be minimized, for example, by not following too closely, and slowing down gradually by releasing the gas pedal when approaching a red light, a stop sign, or slow traffic. This relaxed driving style is safer, saves fuel and money, reduces pollution, reduces wear on the tires and brakes, and is appreciated by other drivers. Allowing sufficient time to reach the destination makes it easier to resist the urge to tailgate.

## Avoid Sudden Acceleration and Sudden Braking (Except in Emergencies)

Accelerate gradually and smoothly when passing other vehicles or merging with faster traffic. Pumping or hard pressing the accelerator pedal while driving causes the engine to switch to a "fuel enrichment mode" of operation that wastes fuel. In city driving, nearly half of the engine power is used for acceleration. When accelerating with stick-shifts, the RPM of the engine should be kept to a minimum. Braking wastes the mechanical energy produced by the engine and wears the brake pads.

## Avoid Resting Feet on the Clutch or Brake Pedal while Driving

Resting the left foot on the brake pedal increases the temperature of the brake components, and thus reduces their effectiveness and service life while wasting fuel. Similarly, resting the left foot on the clutch pedal lessens the pressure on the clutch pads, causing them to slip and wear prematurely, wasting fuel.

**FIGURE 9–64**
Fuel consumption can be decreased by up to 10 percent by anticipating traffic conditions ahead and adjusting accordingly.
©PhotoDisc/Getty Images RF

## Use Highest Gear (Overdrive) During Highway Driving

Overdrive improves fuel economy during highway driving by decreasing the vehicle's engine speed (or RPM). The lower engine speed reduces fuel consumption per unit time as well as engine wear. Therefore, overdrive (the fifth gear in cars with overdrive manual transmission) should be used as soon as the vehicle's speed is high enough.

## Turn the Engine Off Rather Than Letting It Idle

Unnecessary idling during lengthy waits (such as waiting for someone or for service at a drive-up window, being stuck in traffic, etc.) wastes fuel, pollutes the air, and causes engine wear (more wear than driving) (Fig. 9–65). Therefore, the engine should be turned off rather than letting it idle. Idling for more than a minute consumes much more fuel than restarting the engine. Fuel consumption in the lines of drive-up windows and the pollution emitted can be avoided altogether by simply parking the car and going inside.

**FIGURE 9–65**
Unnecessary idling during lengthy waits wastes fuel, costs money, and pollutes the air.

## Use the Air Conditioner Sparingly

Air conditioning consumes considerable power and thus increases fuel consumption by 3 to 4 percent during highway driving, and by as much as 10 percent during city driving (Fig. 9–66). The best alternative to air conditioning is to supply fresh outdoor air to the car through the vents by turning on the flow-through ventilation system (usually by running the air conditioner in the "economy" mode) while keeping the windows and the sunroof closed. This measure is adequate to achieve comfort in pleasant weather, and it saves the most fuel since the compressor of the air conditioner is off. In warmer weather, however, ventilation cannot provide adequate cooling effect. In that case we can attempt to achieve comfort by rolling down the windows or opening the sunroof. This is certainly a viable alternative for city driving, but not so on highways since the aerodynamic drag caused by wide-open windows or sunroof at highway speeds consumes more fuel than does the air conditioner. Therefore, at highway speeds, the windows or the sunroof should be closed and the air conditioner should be turned on instead to save fuel. This is especially the case for the newer, aerodynamically designed cars.

Most air conditioners have a "maximum" or "recirculation" setting that reduces the amount of hot outside air that must be cooled, and thus the fuel consumption for air-conditioning. A passive measure to reduce the need for air conditioning is to park the vehicle in the shade, and to leave the windows slightly open to allow for air circulation.

**FIGURE 9–66**
Air conditioning increases fuel consumption by 3 to 4 percent during highway driving, and by as much as 10 percent during city driving.

## AFTER DRIVING

You cannot be an efficient person and accomplish much unless you take good care of yourself (eating right, maintaining physical fitness, having checkups, etc.), and the cars are no exception. Regular maintenance improves performance, increases gas mileage, reduces pollution, lowers repair costs, and extends engine life. A little time and money saved now may cost a lot later in increased fuel, repair, and replacement costs.

**FIGURE 9–67**
Proper maintenance maximizes fuel
efficiency and extends engine life.

Proper maintenance such as *checking the levels of fluids (engine oil, coolant, transmission, brake, power steering, windshield washer, etc.), the tightness of all belts, and formation of cracks or frays on hoses, belts, and wires, keeping tires properly inflated, lubricating the moving components, and replacing clogged air, fuel, or oil filters* maximizes fuel efficiency (Fig. 9–67). Clogged air filters increase fuel consumption (by up to 10 percent) and pollution by restricting airflow to the engine, and thus they should be replaced. The car should be tuned up regularly unless it has electronic controls and a fuel-injection system. High temperatures (which may be due to a malfunction of the cooling fan) should be avoided as they may cause the break down of the engine oil and thus excessive wear of the engine, and low temperatures (which may be due to a malfunction of the thermostat) may extend the engine's warm-up period and may prevent the engine from reaching the optimum operating conditions. Both effects reduce fuel economy.

Clean oil extends engine life by reducing engine wear caused by friction, removes acids, sludge, and other harmful substances from the engine, improves performance, reduces fuel consumption, and decreases air pollution. Oil also helps to cool the engine, provides a seal between the cylinder walls and the pistons, and prevents the engine from rusting. Therefore, oil and oil filter should be changed as recommended by the vehicle manufacturer. Fuel-efficient oils (indicated by "Energy Efficient API" label) contain certain additives that reduce friction and increase a vehicle's fuel economy by 3 percent or more.

In summary, a person can save fuel, money, and the environment by *purchasing an energy-efficient vehicle, minimizing the amount of driving, being fuel-conscious while driving, and maintaining the car properly.* These measures have the added benefits of enhanced safety, reduced maintenance costs, and extended vehicle life.

## SUMMARY

A cycle during which a net amount of work is produced is called a *power cycle*, and a power cycle during which the working fluid remains a gas throughout is called a *gas power cycle*. The most efficient cycle operating between a heat source at temperature $T_H$ and a sink at temperature $T_L$ is the Carnot cycle, and its thermal efficiency is given by

$$\eta_{\text{th,Carnot}} = 1 - \frac{T_L}{T_H}$$

The actual gas cycles are rather complex. The approximations used to simplify the analysis are known as the *air-standard assumptions*. Under these assumptions, all the processes are assumed to be internally reversible; the working fluid is assumed to be air, which behaves as an ideal gas; and the combustion and exhaust processes are replaced by heat-addition and heat-rejection processes, respectively. The air-standard assumptions are called *cold-air-standard assumptions* if air is also assumed to have constant specific heats at room temperature.

In reciprocating engines, the *compression ratio r* and the *mean effective pressure* MEP are defined as

$$r = \frac{V_{\text{max}}}{V_{\text{min}}} = \frac{V_{\text{BDC}}}{V_{\text{TDC}}}$$

$$\text{MEP} = \frac{w_{\text{net}}}{V_{\text{max}} - V_{\text{min}}}$$

The *Otto cycle* is the ideal cycle for the spark-ignition reciprocating engines, and it consists of four internally reversible processes: isentropic compression, constant-volume

heat addition, isentropic expansion, and constant-volume heat rejection. Under cold-air-standard assumptions, the thermal efficiency of the ideal Otto cycle is

$$\eta_{th,Otto} = 1 - \frac{1}{r^{k-1}}$$

where $r$ is the compression ratio and $k$ is the specific heat ratio $c_p/c_v$.

The *Diesel cycle* is the ideal cycle for the compression-ignition reciprocating engines. It is very similar to the Otto cycle, except that the constant-volume heat-addition process is replaced by a constant-pressure heat-addition process. Its thermal efficiency under cold-air-standard assumptions is

$$\eta_{th,Diesel} = 1 - \frac{1}{r^{k-1}}\left[\frac{r_c^k - 1}{k(r_c - 1)}\right]$$

where $r_c$ is the *cutoff ratio,* defined as the ratio of the cylinder volumes after and before the combustion process.

*Stirling* and *Ericsson cycles* are two totally reversible cycles that involve an isothermal heat-addition process at $T_H$ and an isothermal heat-rejection process at $T_L$. They differ from the Carnot cycle in that the two isentropic processes are replaced by two constant-volume regeneration processes in the Stirling cycle and by two constant-pressure regeneration processes in the Ericsson cycle. Both cycles utilize *regeneration,* a process during which heat is transferred to a thermal energy storage device (called a *regenerator*) during one part of the cycle that is then transferred back to the working fluid during another part of the cycle.

The ideal cycle for modern gas-turbine engines is the *Brayton cycle,* which is made up of four internally reversible processes: isentropic compression, constant-pressure heat addition, isentropic expansion, and constant-pressure heat rejection. Under cold-air-standard assumptions, its thermal efficiency is

$$\eta_{th,Brayton} = 1 - \frac{1}{r_p^{(k-1)/k}}$$

where $r_p = P_{max}/P_{min}$ is the pressure ratio and $k$ is the specific heat ratio. The thermal efficiency of the simple Brayton cycle increases with the pressure ratio.

The deviation of the actual compressor and the turbine from the idealized isentropic ones can be accurately accounted for by utilizing their isentropic efficiencies, defined as

$$\eta_C = \frac{w_s}{w_a} \cong \frac{h_{2s} - h_1}{h_{2a} - h_1}$$

and

$$\eta_T = \frac{w_a}{w_s} \cong \frac{h_3 - h_{4a}}{h_3 - h_{4s}}$$

where states 1 and 3 are the inlet states, 2a and 4a are the actual exit states, and 2s and 4s are the isentropic exit states.

In gas-turbine engines, the temperature of the exhaust gas leaving the turbine is often considerably higher than the temperature of the air leaving the compressor. Therefore, the high-pressure air leaving the compressor can be heated by transferring heat to it from the hot exhaust gases in a counter-flow heat exchanger, which is also known as a *regenerator.* The extent to which a regenerator approaches an ideal regenerator is called the *effectiveness* $\epsilon$ and is defined as

$$\epsilon = \frac{q_{regen,act}}{q_{regen,max}}$$

Under cold-air-standard assumptions, the thermal efficiency of an ideal Brayton cycle with regeneration becomes

$$\eta_{th,regen} = 1 - \left(\frac{T_1}{T_3}\right)(r_p)^{(k-1)/k}$$

where $T_1$ and $T_3$ are the minimum and maximum temperatures, respectively, in the cycle.

The thermal efficiency of the Brayton cycle can also be increased by utilizing *multistage compression with intercooling, regeneration, and multistage expansion with reheating.* The work input to the compressor is minimized when equal pressure ratios are maintained across each stage. This procedure also maximizes the turbine work output.

Gas-turbine engines are widely used to power aircraft because they are light and compact and have a high power-to-weight ratio. The ideal *jet-propulsion cycle* differs from the simple ideal Brayton cycle in that the gases are partially expanded in the turbine. The gases that exit the turbine at a relatively high pressure are subsequently accelerated in a nozzle to provide the thrust needed to propel the aircraft.

The *net thrust* developed by the engine is

$$F = \dot{m}(V_{exit} - V_{inlet})$$

where $\dot{m}$ is the mass flow rate of gases, $V_{exit}$ is the exit velocity of the exhaust gases, and $V_{inlet}$ is the inlet velocity of the air, both relative to the aircraft.

The power developed from the thrust of the engine is called the *propulsive power* $\dot{W}_P$, and it is given by

$$\dot{W}_P = \dot{m}(V_{exit} - V_{inlet})V_{aircraft}$$

*Propulsive efficiency* is a measure of how efficiently the energy released during the combustion process is converted to propulsive energy, and it is defined as

$$\eta_P = \frac{\text{Propulsive power}}{\text{Energy input rate}} = \frac{\dot{W}_P}{\dot{Q}_{in}}$$

For an ideal cycle that involves heat transfer only with a source at $T_H$ and a sink at $T_L$, the exergy destruction is

$$x_{dest} = T_0\left(\frac{q_{out}}{T_L} - \frac{q_{in}}{T_H}\right)$$

## REFERENCES AND SUGGESTED READINGS

1. V. D. Chase. "Propfans: A New Twist for the Propeller." *Mechanical Engineering,* November 1986, pp. 47–50.

2. C. R. Ferguson and A. T. Kirkpatrick, *Internal Combustion Engines: Applied Thermosciences,* 2nd ed., New York: Wiley, 2000.

3. R. A. Harmon. "The Keys to Cogeneration and Combined Cycles." *Mechanical Engineering,* February 1988, pp. 64–73.

4. J. Heywood, *Internal Combustion Engine Fundamentals,* New York: McGraw-Hill, 1988.

5. L. C. Lichty. *Combustion Engine Processes.* New York: McGraw-Hill, 1967.

6. H. McIntosh. "Jumbo Jet." *10 Outstanding Achievements 1964–1989.* Washington, D.C.: National Academy of Engineering, 1989, pp. 30–33.

7. W. Pulkrabek, *Engineering Fundamentals of the Internal Combustion Engine,* 2nd ed., Upper Saddle River, NJ: Prentice-Hall, 2004.

8. W. Siuru. "Two-stroke Engines: Cleaner and Meaner." *Mechanical Engineering.* June 1990, pp. 66–69.

9. C. F. Taylor. *The Internal Combustion Engine in Theory and Practice.* Cambridge, MA: M.I.T. Press, 1968.

## PROBLEMS*

### Actual and Ideal Cycles, Carnot Cycle, Air-Standard Assumptions, Reciprocating Engines

**9–1C**  What are the air-standard assumptions?

**9–2C**  What is the difference between air-standard assumptions and the cold-air-standard assumptions?

**9–3C**  How does the thermal efficiency of an ideal cycle, in general, compare to that of a Carnot cycle operating between the same temperature limits?

**9–4C**  What does the area enclosed by the cycle represent on a $P$-$v$ diagram? How about on a $T$-$s$ diagram?

**9–5C**  Define the compression ratio for reciprocating engines.

**9–6C**  How is the mean effective pressure for reciprocating engines defined?

**9–7C**  Can the mean effective pressure of an automobile engine in operation be less than the atmospheric pressure?

**9–8C**  As a car gets older, will its compression ratio change? How about the mean effective pressure?

**9–9C**  What is the difference between spark-ignition and compression-ignition engines?

**9–10C**  Define the following terms related to reciprocating engines: stroke, bore, top dead center, and clearance volume.

*Problems designated by a "C" are concept questions, and students are encouraged to answer them all. Problems with the 🖸 icon are solved using EES, and complete solutions together with parametric studies are included on the text website. Problems with the 📀 icon are comprehensive in nature, and are intended to be solved with an equation solver such as EES.

**9–11**  An air-standard cycle is executed within a closed piston-cylinder system and consists of three processes as follows:

1-2  $V = constant$ heat addition from 100 kPa and 27°C to 700 kPa

2-3  Isothermal expansion until $V_3 = 7V_2$

3-1  $P = constant$ heat rejection to the initial state

Assume air has constant properties with $c_v = 0.718$ kJ/kg·K, $c_p = 1.005$ kJ/kg·K, $R = 0.287$ kJ/kg·K, and $k = 1.4$.

(*a*)  Sketch the $P$-$v$ and $T$-$s$ diagrams for the cycle.

(*b*)  Determine the ratio of the compression work to the expansion work (the back work ratio).

(*c*)  Determine the cycle thermal efficiency.

*Answers:* (*b*) 0.440, (*c*) 26.6 percent

**9–12**  An air-standard cycle with variable specific heats is executed in a closed system and is composed of the following four processes:

1-2  Isentropic compression from 100 kPa and 22°C to 600 kPa

2-3  $v = constant$ heat addition to 1500 K

3-4  Isentropic expansion to 100 kPa

4-1  $P = constant$ heat rejection to initial state

(*a*)  Show the cycle on $P$-$v$ and $T$-$s$ diagrams.

(*b*)  Calculate the net work output per unit mass.

(*c*)  Determine the thermal efficiency.

**9–13**  🖸  Reconsider Prob. 9–12. Using EES (or other) software, study the effect of varying the temperature after the constant-volume heat addition from 1500 K to 2500 K. Plot the net work output and thermal efficiency as a function of the maximum temperature of the cycle. Plot the $T$-$s$ and $P$-$v$ diagrams for the cycle when the maximum temperature of the cycle is 1500 K.

**9–14** An air-standard cycle is executed in a closed system with 0.5 kg of air and consists of the following three processes:

1-2   Isentropic compression from 100 kPa and 27°C to 1 MPa

2-3   $P = constant$ heat addition in the amount of 416 kJ

3-1   $P = c_1v + c_2$ heat rejection to initial state ($c_1$ and $c_2$ are constants)

(a)   Show the cycle on $P$-$v$ and $T$-$s$ diagrams.
(b)   Calculate the heat rejected.
(c)   Determine the thermal efficiency.

Assume constant specific heats at room temperature.
*Answers:* (b) 272 kJ, (c) 34.7 percent

**9–15** An air-standard cycle with variable specific heats is executed in a closed system and is composed of the following four processes:

1-2   $v = constant$ heat addition from 100 kPa and 27°C in the amount of 700 kJ/kg

2-3   $P = constant$ heat addition to 1800 K

3-4   Isentropic expansion to 100 kPa

4-1   $P = constant$ heat rejection to initial state

(a)   Show the cycle on $P$-$v$ and $T$-$s$ diagrams.
(b)   Calculate the total heat input per unit mass.
(c)   Determine the thermal efficiency.

*Answers:* (b) 1451 kJ/kg, (c) 24.3 percent

**9–16** Repeat Prob. 9–15 using constant specific heats at room temperature.

**9–17** An air-standard Carnot cycle is executed in a closed system between the temperature limits of 350 and 1200 K. The pressures before and after the isothermal compression are 150 and 300 kPa, respectively. If the net work output per cycle is 0.5 kJ, determine (a) the maximum pressure in the cycle, (b) the heat transfer to air, and (c) the mass of air. Assume variable specific heats for air. *Answers:* (a) 30.0 MPa, (b) 0.706 kJ, (c) 0.00296 kg

**9–18** Repeat Problem 9–17 using helium as the working fluid.

**9–19** Consider a Carnot cycle executed in a closed system with 0.6 kg of air. The temperature limits of the cycle are 300 and 1100 K, and the minimum and maximum pressures that occur during the cycle are 20 and 3000 kPa. Assuming constant specific heats, determine the net work output per cycle.

**9–20** Consider a Carnot cycle executed in a closed system with air as the working fluid. The maximum pressure in the cycle is 1300 kPa while the maximum temperature is 950 K.

If the entropy increase during the isothermal heat rejection process is 0.25 kJ/kg·K and the net work output is 100 kJ/kg, determine (a) the minimum pressure in the cycle, (b) the heat rejection from the cycle, and (c) the thermal efficiency of the cycle. (d) If an actual heat engine cycle operates between the same temperature limits and produces 5200 kW of power for an air flow rate of 95 kg/s, determine the second law efficiency of this cycle.

**9–21** An ideal gas is contained in a piston-cylinder device and undergoes a power cycle as follows:

1-2   isentropic compression from an initial temperature $T_1 = 20°C$ with a compression ratio $r = 5$

2-3   constant pressure heat addition

3-1   constant volume heat rejection

The gas has constant specific heats with $c_v = 0.7$ kJ/kg·K and $R = 0.3$ kJ/kg·K.

(a)   Sketch the $P$-$v$ and $T$-$s$ diagrams for the cycle.
(b)   Determine the heat and work interactions for each process, in kJ/kg.
(c)   Determine the cycle thermal efficiency.
(d)   Obtain the expression for the cycle thermal efficiency as a function of the compression ratio $r$ and ratio of specific heats $k$.

## Otto Cycle

**9–22C** What four processes make up the ideal Otto cycle?

**9–23C** Are the processes that make up the Otto cycle analyzed as closed-system or steady-flow processes? Why?

**9–24C** How do the efficiencies of the ideal Otto cycle and the Carnot cycle compare for the same temperature limits? Explain.

**9–25C** How does the thermal efficiency of an ideal Otto cycle change with the compression ratio of the engine and the specific heat ratio of the working fluid?

**9–26C** How is the rpm (revolutions per minute) of an actual four-stroke gasoline engine related to the number of thermodynamic cycles? What would your answer be for a two-stroke engine?

**9–27C** Why are high compression ratios not used in spark-ignition engines?

**9–28C** An ideal Otto cycle with a specified compression ratio is executed using (a) air, (b) argon, and (c) ethane as the working fluid. For which case will the thermal efficiency be the highest? Why?

**9–29C** What is the difference between fuel-injected gasoline engines and diesel engines?

**9–30** An ideal Otto cycle has a compression ratio of 10.5, takes in air at 90 kPa and 40°C, and is repeated 2500 times per minute. Using constant specific heats at room temperature, determine the thermal efficiency of this cycle and the rate of heat input if the cycle is to produce 90 kW of power.

**9–31** Repeat Prob. 9–30 for a compression ratio of 8.5.

**9–32** An ideal Otto cycle has a compression ratio of 8. At the beginning of the compression process, air is at 95 kPa and 27°C, and 750 kJ/kg of heat is transferred to air during the constant-volume heat-addition process. Taking into account the variation of specific heats with temperature, determine (a) the pressure and temperature at the end of the heat-addition process, (b) the net work output, (c) the thermal efficiency, and (d) the mean effective pressure for the cycle. *Answers:* (a) 3898 kPa, 1539 K, (b) 392.4 kJ/kg, (c) 52.3 percent, (d) 495 kPa

**9–33** [EES] Reconsider Problem 9–32. Using EES (or other) software, study the effect of varying the compression ratio from 5 to 10. Plot the net work output and thermal efficiency as a function of the compression ratio. Plot the *T-s* and *P-v* diagrams for the cycle when the compression ratio is 8.

**9–34** Repeat Problem 9–32 using constant specific heats at room temperature.

**9–35** A six-cylinder, four-stroke, spark-ignition engine operating on the ideal Otto cycle takes in air at 95 kPa and 40°C, and is limited to a maximum cycle temperature of 1300°C. Each cylinder has a bore of 8.9 cm, and each piston has a stroke of 9.9 cm. The minimum enclosed volume is 9.8 percent of the maximum enclosed volume. How much power will this engine produce when operated at 2500 rpm? Use constant specific heats at room temperature.

**9–36** An ideal Otto cycle with air as the working fluid has a compression ratio of 8. The minimum and maximum temperatures in the cycle are 300 and 1340 K. Accounting for the variation of specific heats with temperature, determine (a) the amount of heat transferred to the air during the heat-addition process, (b) the thermal efficiency, and (c) the thermal efficiency of a Carnot cycle operating between the same temperature limits.

**9–37** Repeat Prob. 9–36 using argon as the working fluid.

**9–38** When we double the compression ratio of an ideal Otto cycle, what happens to the maximum gas temperature and pressure when the state of the air at the beginning of the compression and the amount of heat addition remain the same? Use constant specific heats at room temperature.

**9–39** In a spark-ignition engine, some cooling occurs as the gas is expanded. This may be modeled by using a polytropic process in lieu of the isentropic process. Determine if the polytropic exponent used in this model will be greater than or less than the isentropic exponent.

## Diesel Cycle

**9–40C** How does a diesel engine differ from a gasoline engine?

**9–41C** How does the ideal Diesel cycle differ from the ideal Otto cycle?

**9–42C** For a specified compression ratio, is a diesel or gasoline engine more efficient?

**9–43C** Do diesel or gasoline engines operate at higher compression ratios? Why?

**9–44** An air-standard Diesel cycle has a compression ratio of 16 and a cutoff ratio of 2. At the beginning of the compression process, air is at 95 kPa and 27°C. Accounting for the variation of specific heats with temperature, determine (a) the temperature after the heat-addition process, (b) the thermal efficiency, and (c) the mean effective pressure. *Answers:* (a) 1725 K, (b) 56.3 percent, (c) 675.9 kPa

**9–45** Repeat Problem 9–44 using constant specific heats at room temperature.

**9–46** An ideal Diesel cycle has a compression ratio of 17 and a cutoff ratio of 1.3. Determine the maximum temperature of the air and the rate of heat addition to this cycle when it produces 140 kW of power and the state of the air at the beginning of the compression is 90 kPa and 57°C. Use constant specific heats at room temperature.

**9–47** An air-standard dual cycle has a compression ratio of 14 and a cutoff ratio of 1.2. The pressure ratio during the constant-volume heat addition process is 1.5. Determine the thermal efficiency, amount of heat added, the maximum gas pressure and temperature when this cycle is operated at 80 kPa and 20°C at the beginning of the compression. Use constant specific heats at room temperature.

**9–48** Repeat Prob. 9–47 when the state of the air at the beginning of the compression is 80 kPa and −20°C.

**9–49** An air-standard Diesel cycle has a compression ratio of 18.2. Air is at 47°C and 100 kPa at the beginning of the compression process and at 1800 K at the end of the heat-addition process. Accounting for the variation of specific heats with temperature, determine (a) the cutoff ratio, (b) the heat rejection per unit mass, and (c) the thermal efficiency.

**9–50** Repeat Prob. 9–49 using constant specific heats at room temperature.

**9–51** An ideal diesel engine has a compression ratio of 20 and uses air as the working fluid. The state of air at the beginning of the compression process is 95 kPa and 20°C. If the maximum temperature in the cycle is not to exceed 2200 K, determine (a) the thermal efficiency and (b) the mean effective pressure. Assume constant specific heats for air at room temperature. *Answers:* (a) 63.5 percent, (b) 933 kPa

**9–52** Repeat Prob. 9–51, but replace the isentropic expansion process by polytropic expansion process with the polytropic exponent n = 1.35. Use variable specific heats.

**9–53** Reconsider Prob. 9–52. Using EES (or other) software, study the effect of varying the compression ratio from 14 to 24. Plot the net work output, mean effective pressure, and thermal efficiency as a function of the compression ratio. Plot the T-s and P-v diagrams for the cycle when the compression ratio is 20.

**9–54** A four-cylinder two-stroke 2.4-L diesel engine that operates on an ideal Diesel cycle has a compression ratio of 22 and a cutoff ratio of 1.8. Air is at 70°C and 97 kPa at the beginning of the compression process. Using the cold-air-standard assumptions, determine how much power the engine will deliver at 3500 rpm.

**9–55** Repeat Prob. 9–54 using nitrogen as the working fluid.

**9–56** An ideal dual cycle has a compression ratio of 15 and a cutoff ratio of 1.4. The pressure ratio during constant-volume heat addition process is 1.1. The state of the air at the beginning of the compression is $P_1 = 98$ kPa and $T_1 = 24$°C. Calculate the cycle's net specific work, specific heat addition, and thermal efficiency. Use constant specific heats at room temperature.

**9–57** The compression ratio of an ideal dual cycle is 14. Air is at 100 kPa and 300 K at the beginning of the compression process and at 2200 K at the end of the heat-addition process. Heat transfer to air takes place partly at constant volume and partly at constant pressure, and it amounts to 1520.4 kJ/kg. Assuming variable specific heats for air, determine (a) the fraction of heat transferred at constant volume and (b) the thermal efficiency of the cycle.

**9–58** Reconsider Problem 9–57. Using EES (or other) software, study the effect of varying the compression ratio from 10 to 18. For the compression ratio equal to 14, plot the T-s and P-v diagrams for the cycle.

**9–59** Repeat Problem 9–57 using constant specific heats at room temperature. Is the constant specific heat assumption reasonable in this case?

**9–60** Develop an expression for cutoff ratio $r_c$ which expresses it in terms of $q_{in}/(c_p T_1 r^{k-1})$ for an air-standard Diesel cycle.

**9–61** An air-standard cycle, called the dual cycle, with constant specific heats is executed in a closed piston-cylinder system and is composed of the following five processes:

1-2 Isentropic compression with a compression ratio, $r = V_1/V_2$

2-3 Constant volume heat addition with a pressure ratio, $r_p = P_3/P_2$

3-4 Constant pressure heat addition with a volume ratio, $r_c = V_4/V_3$

4-5 Isentropic expansion while work is done until $V_5 = V_1$

5-1 Constant volume heat rejection to the initial state

(a) Sketch the P-v and T-s diagrams for this cycle.

(b) Obtain an expression for the cycle thermal efficiency as a function of $k$, $r$, $r_c$, and $r_p$.

(c) Evaluate the limit of the efficiency as $r_p$ approaches unity and compare your answer with the expression for the Diesel cycle efficiency.

(d) Evaluate the limit of the efficiency as $r_c$ approaches unity and compare your answer with the expression for the Otto cycle efficiency.

## Stirling and Ericsson Cycles

**9–62C** What cycle is composed of two isothermal and two constant-volume processes?

**9–63C** How does the ideal Ericsson cycle differ from the Carnot cycle?

**9–64C** Consider the ideal Otto, Stirling, and Carnot cycles operating between the same temperature limits. How would you compare the thermal efficiencies of these three cycles?

**9–65C** Consider the ideal Diesel, Ericsson, and Carnot cycles operating between the same temperature limits. How would you compare the thermal efficiencies of these three cycles?

**9–66** An ideal Ericsson engine using helium as the working fluid operates between temperature limits of 305 and 1665 K and pressure limits of 175 and 1400 kPa. Assuming a mass flow rate of 6 kg/s, determine (a) the thermal efficiency of the cycle, (b) the heat transfer rate in the regenerator, and (c) the power delivered.

**9–67** An ideal Stirling engine using helium as the working fluid operates between temperature limits of 300 and 2000 K and pressure limits of 150 kPa and 3 MPa. Assuming the mass of the helium used in the cycle is 0.12 kg, determine (a) the thermal efficiency of the cycle, (b) the amount of heat transfer in the regenerator, and (c) the work output per cycle.

**9–68** Consider an ideal Ericsson cycle with air as the working fluid executed in a steady-flow system. Air is at 27°C and 120 kPa at the beginning of the isothermal compression process, during which 150 kJ/kg of heat is rejected. Heat transfer to air occurs at 1200 K. Determine (a) the maximum pressure in the cycle, (b) the net work output per unit mass of air, and (c) the thermal efficiency of the cycle. *Answers:* (a) 685 kPa, (b) 450 kJ/kg, (c) 75.0 percent

**9–69** An air-standard Stirling cycle operates with a maximum pressure of 3600 kPa and a minimum pressure of 50 kPa. The maximum volume is 12 times the minimum volume, and the low-temperature reservoir is at 20°C. Allowing a 5°C temperature difference between the external reservoirs and the air when appropriate, calculate the specific heat added to the cycle and its net specific work.

**9–70** How much heat is stored (and recovered) in the regenerator of Prob. 9–69. Use constant specific heats at room temperature.

## Ideal and Actual Gas-Turbine (Brayton) Cycles

**9–71C** For fixed maximum and minimum temperatures, what is the effect of the pressure ratio on (a) the thermal efficiency and (b) the net work output of a simple ideal Brayton cycle?

**9–72C** What is the back work ratio? What are typical back work ratio values for gas-turbine engines?

**9–73C** Why are the back work ratios relatively high in gas-turbine engines?

**9–74C** How do the inefficiencies of the turbine and the compressor affect (a) the back work ratio and (b) the thermal efficiency of a gas-turbine engine?

**9–75** A simple ideal Brayton cycle with air as the working fluid has a pressure ratio of 10. The air enters the compressor at 290 K and the turbine at 1100 K. Accounting for the variation of specific heats with temperature, determine (a) the air temperature at the compressor exit, (b) the back work ratio, and (c) the thermal efficiency.

**9–76** A gas-turbine power plant operates on the simple Brayton cycle with air as the working fluid and delivers 32 MW of power. The minimum and maximum temperatures in the cycle are 310 and 900 K, and the pressure of air at the compressor exit is 8 times the value at the compressor inlet. Assuming an isentropic efficiency of 80 percent for the compressor and 86 percent for the turbine, determine the mass flow rate of air through the cycle. Account for the variation of specific heats with temperature.

**9–77** Repeat Problem 9–76 using constant specific heats at room temperature.

**9–78** A simple Brayton cycle using air as the working fluid has a pressure ratio of 10. The minimum and maximum temperatures in the cycle are 295 and 1240 K. Assuming an isentropic efficiency of 83 percent for the compressor and 87 percent for the turbine, determine (a) the air temperature at the turbine exit, (b) the net work output, and (c) the thermal efficiency.

**9–79** Reconsider Prob. 9–78. Using EES (or other) software, allow the mass flow rate, pressure ratio, turbine inlet temperature, and the isentropic efficiencies of the turbine and compressor to vary. Assume the compressor inlet pressure is 100 kPa. Develop a general solution for the problem by taking advantage of the diagram window method for supplying data to EES software.

**9–80** Repeat Prob. 9–78 using constant specific heats at room temperature.

**9–81** Consider a simple Brayton cycle using air as the working fluid; has a pressure ratio of 12; has a maximum cycle temperature of 600°C; and operates the compressor inlet at 100 kPa and 15°C. Which will have the greatest impact on the back-work ratio: a compressor isentropic efficiency of 80 percent or a turbine isentropic efficiency of 80 percent? Use constant specific heats at room temperature.

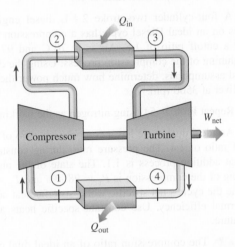

**FIGURE P9–81**

**9–82** Air is used as the working fluid in a simple ideal Brayton cycle that has a pressure ratio of 12, a compressor inlet temperature of 300 K, and a turbine inlet temperature of 1000 K. Determine the required mass flow rate of air for a net power output of 70 MW, assuming both the compressor and the turbine have an isentropic efficiency of (a) 100 percent and (b) 85 percent. Assume constant specific heats at room temperature. *Answers:* (a) 352 kg/s, (b) 1037 kg/s

**9–83** An aircraft engine operates on a simple ideal Brayton cycle with a pressure ratio of 10. Heat is added to the cycle at a rate of 500 kW; air passes through the engine at a rate of 1 kg/s; and the air at the beginning of the compression is at 70 kPa and 0°C. Determine the power produced by this engine and its thermal efficiency. Use constant specific heats at room temperature.

**9–84** Repeat Prob. 9–83 for a pressure ratio of 15.

**9–85** A gas-turbine power plant operates on the simple Brayton cycle between the pressure limits of 100 and 1600 kPa. The working fluid is air, which enters the compressor at 40°C at a rate of 850 m³/min and leaves the turbine at 650°C. Using variable specific heats for air and assuming a compressor isentropic efficiency of 85 percent and a turbine isentropic efficiency of 88 percent, determine (a) the net power output, (b) the back work ratio, and (c) the thermal efficiency. *Answers:* (a) 6081 kW, (b) 0.536, (c) 37.4 percent

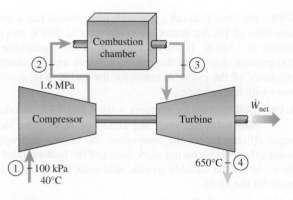

1.6 MPa

650°C – ④

① – 100 kPa
40°C

**FIGURE P9–85**

(c) If the net power output is 200 MW, determine mass flow rate of the air into the compressor, in kg/s.

*Answers:* (b) 1279 K, 457 kPa, (c) 442 kg/s

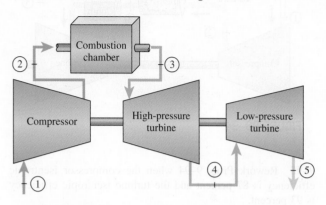

**FIGURE P9–89**

**9–86** A gas-turbine power plant operates on a simple Brayton cycle with air as the working fluid. The air enters the turbine at 800 kPa and 1100 K and leaves at 100 kPa and 670 K. Heat is rejected to the surroundings at a rate of 6700 kW, and air flows through the cycle at a rate of 18 kg/s. Assuming the turbine to be isentropic and the compressor to have an isentropic efficiency of 80 percent, determine the net power output of the plant. Account for the variation of specific heats with temperature. *Answer:* 2979 kW

**9–87** For what compressor efficiency will the gas-turbine power plant in Problem 9–86 produce zero net work?

**9–88** A gas-turbine power plant operates on the simple Brayton cycle between the pressure limits of 100 and 800 kPa. Air enters the compressor at 30°C and leaves at 330°C at a mass flow rate of 200 kg/s. The maximum cycle temperature is 1400 K. During operation of the cycle, the net power output is measured experimentally to be 60 MW. Assume constant properties for air at 300 K with $c_v = 0.718$ kJ/kg·K, $c_p = 1.005$ kJ/kg·K, $R = 0.287$ kJ/kg·K, $k = 1.4$.

(a) Sketch the T-s diagram for the cycle.

(b) Determine the isentropic efficiency of the turbine for these operating conditions.

(c) Determine the cycle thermal efficiency.

**9–89** A gas-turbine power plant operates on a modified Brayton cycle shown in the figure with an overall pressure ratio of 8. Air enters the compressor at 0°C and 100 kPa. The maximum cycle temperature is 1500 K. The compressor and the turbines are isentropic. The high pressure turbine develops just enough power to run the compressor. Assume constant properties for air at 300 K with $c_v = 0.718$ kJ/kg·K, $c_p = 1.005$ kJ/kg·K, $R = 0.287$ kJ/kg·K, $k = 1.4$.

(a) Sketch the T-s diagram for the cycle. Label the data states.

(b) Determine the temperature and pressure at state 4, the exit of the high pressure turbine.

**Brayton Cycle with Regeneration**

**9–90C** How does regeneration affect the efficiency of a Brayton cycle, and how does it accomplish it?

**9–91C** Somebody claims that at very high pressure ratios, the use of regeneration actually decreases the thermal efficiency of a gas-turbine engine. Is there any truth in this claim? Explain.

**9–92C** In an ideal regenerator, is the air leaving the compressor heated to the temperature at (a) turbine inlet, (b) turbine exit, (c) slightly above turbine exit?

**9–93C** In 1903, Aegidius Elling of Norway designed and built an 11-hp gas turbine that used steam injection between the combustion chamber and the turbine to cool the combustion gases to a safe temperature for the materials available at the time. Currently there are several gas-turbine power plants that use steam injection to augment power and improve thermal efficiency. For example, the thermal efficiency of the General Electric LM5000 gas turbine is reported to increase from 35.8 percent in simple-cycle operation to 43 percent when steam injection is used. Explain why steam injection increases the power output and the efficiency of gas turbines. Also, explain how you would obtain the steam.

**9–94** A gas turbine for an automobile is designed with a regenerator. Air enters the compressor of this engine at 100 kPa and 30°C. The compressor pressure ratio is 10; the maximum cycle temperature is 800°C; and the cold air stream leaves the regenerator 10°C cooler than the hot air stream at the inlet of the regenerator. Assuming both the compressor and the turbine to be isentropic, determine the rates of heat addition and rejection for this cycle when it produces 115 kW. Use constant specific heats at room temperature. *Answers:* 258 kW, 143 kW

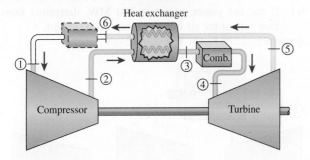

**FIGURE P9–94**

**9–95** Rework Prob. 9–94 when the compressor isentropic efficiency is 87 percent and the turbine isentropic efficiency is 93 percent.

**9–96** A gas turbine engine operates on the ideal Brayton cycle with regeneration, as shown in Fig. P9–99. Now the regenerator is rearranged so that the air streams of states 2 and 5 enter at one end of the regenerator and streams 3 and 6 exit at the other end (i.e., parallel flow arrangement of a heat exchanger). Consider such a system when air enters the compressor at 100 kPa and 20°C; the compressor pressure ratio is 7; the maximum cycle temperature is 727°C; and the difference between the hot and cold air stream temperatures is 6°C at the end of the regenerator where the cold stream leaves the regenerator. Is the cycle arrangement shown in the figure more or less efficient than this arrangement? Assume both the compressor and the turbine are isentropic, and use constant specific heats at room temperature.

**9–97** The idea of using gas turbines to power automobiles was conceived in the 1930s, and considerable research was done in the 1940s and 1950s to develop automotive gas turbines by major automobile manufacturers such as the Chrysler and Ford corporations in the United States and Rover in the United Kingdom. The world's first gas-turbine-powered automobile, the 200-hp Rover Jet 1, was built in 1950 in the United Kingdom. This was followed by the production of the Plymouth Sport Coupe by Chrysler in 1954 under the leadership of G. J. Huebner. Several hundred gas-turbine-powered Plymouth cars were built in the early 1960s for demonstration purposes and were loaned to a select group of people to gather field experience. The users had no complaints other than slow acceleration. But the cars were never mass-produced because of the high production (especially material) costs and the failure to satisfy the provisions of the 1966 Clean Air Act.

A gas-turbine-powered Plymouth car built in 1960 had a turbine inlet temperature of 927°C, a pressure ratio of 4, and a regenerator effectiveness of 0.9. Using isentropic efficiencies of 80 percent for both the compressor and the turbine, determine the thermal efficiency of this car. Also, determine the mass flow rate of air for a net power output of 97 kW. Assume the ambient air to be at 300 K and 100 kPa.

**9–98** An ideal Brayton cycle with regeneration has a pressure ratio of 10. Air enters the compressor at 300 K and the turbine at 1200 K. If the effectiveness of the regenerator is 100 percent, determine the net work output and the thermal efficiency of the cycle. Account for the variation of specific heats with temperature.

**9–99** Reconsider Problem 9–98. Using EES (or other) software, study the effects of varying the isentropic efficiencies for the compressor and turbine and regenerator effectiveness on net work done and the heat supplied to the cycle for the variable specific heat case. Plot the *T-s* diagram for the cycle.

**9–100** Repeat Problem 9–98 using constant specific heats at room temperature.

**9–101** A Brayton cycle with regeneration using air as the working fluid has a pressure ratio of 7. The minimum and maximum temperatures in the cycle are 310 and 1150 K. Assuming an isentropic efficiency of 75 percent for the compressor and 82 percent for the turbine and an effectiveness of 65 percent for the regenerator, determine (a) the air temperature at the turbine exit, (b) the net work output, and (c) the thermal efficiency. *Answers:* (a) 783 K, (b) 108 kJ/kg, (c) 22.5 percent

**9–102** A stationary gas-turbine power plant operates on an ideal regenerative Brayton cycle ($\epsilon = 100$ percent) with air as the working fluid. Air enters the compressor at 95 kPa and 290 K and the turbine at 880 kPa and 1100 K. Heat is transferred to air from an external source at a rate of 30,000 kJ/s. Determine the power delivered by this plant (a) assuming constant specific heats for air at room temperature and (b) accounting for the variation of specific heats with temperature.

**9–103** Air enters the compressor of a regenerative gas-turbine engine at 310 K and 100 kPa, where it is compressed to 900 kPa and 650 K. The regenerator has an effectiveness of 80 percent, and the air enters the turbine at 1400 K. For a turbine efficiency of 90 percent, determine (a) the amount of heat transfer in the regenerator and (b) the thermal efficiency. Assume variable specific heats for air. *Answers:* (a) 193 kJ/kg, (b) 40.0 percent

**9–104** Repeat Prob. 9–103 using constant specific heats at room temperature.

**9–105** Repeat Prob. 9–103 for a regenerator effectiveness of 70 percent.

**9–106** Develop an expression for the thermal efficiency of an ideal Brayton cycle with an ideal regenerator of effectiveness 100 percent. Use constant specific heats at room temperature.

### Brayton Cycle with Intercooling, Reheating, and Regeneration

**9–107C** For a specified pressure ratio, why does multistage compression with intercooling decrease the compressor work, and multistage expansion with reheating increase the turbine work?

**9–108C** The single-stage compression process of an ideal Brayton cycle without regeneration is replaced by a multistage compression process with intercooling between the same pressure limits. As a result of this modification,

(a) Does the compressor work increase, decrease, or remain the same?

(b) Does the back work ratio increase, decrease, or remain the same?

(c) Does the thermal efficiency increase, decrease, or remain the same?

**9–109C** The single-stage expansion process of an ideal Brayton cycle without regeneration is replaced by a multistage expansion process with reheating between the same pressure limits. As a result of this modification,

(a) Does the turbine work increase, decrease, or remain the same?

(b) Does the back work ratio increase, decrease, or remain the same?

(c) Does the thermal efficiency increase, decrease, or remain the same?

**9–110C** A simple ideal Brayton cycle without regeneration is modified to incorporate multistage compression with intercooling and multistage expansion with reheating, without changing the pressure or temperature limits of the cycle. As a result of these two modifications,

(a) Does the net work output increase, decrease, or remain the same?

(b) Does the back work ratio increase, decrease, or remain the same?

(c) Does the thermal efficiency increase, decrease, or remain the same?

(d) Does the heat rejected increase, decrease, or remain the same?

**9–111C** A simple ideal Brayton cycle is modified to incorporate multistage compression with intercooling, multistage expansion with reheating, and regeneration without changing the pressure limits of the cycle. As a result of these modifications,

(a) Does the net work output increase, decrease, or remain the same?

(b) Does the back work ratio increase, decrease, or remain the same?

(c) Does the thermal efficiency increase, decrease, or remain the same?

(d) Does the heat rejected increase, decrease, or remain the same?

**9–112C** In an ideal gas-turbine cycle with intercooling, reheating, and regeneration, as the number of compression and expansion stages is increased, the cycle thermal efficiency approaches (a) 100 percent, (b) the Otto cycle efficiency, or (c) the Carnot cycle efficiency.

**9–113** Consider a regenerative gas-turbine power plant with two stages of compression and two stages of expansion. The overall pressure ratio of the cycle is 9. The air enters each stage of the compressor at 300 K and each stage of the turbine at 1200 K. Accounting for the variation of specific heats with temperature, determine the minimum mass flow rate of air needed to develop a net power output of 110 MW. *Answer:* 250 kg/s

**9–114** Repeat Problem 9–113 using argon as the working fluid.

**9–115** Consider an ideal gas-turbine cycle with two stages of compression and two stages of expansion. The pressure ratio across each stage of the compressor and turbine is 3. The air enters each stage of the compressor at 300 K and each stage of the turbine at 1200 K. Determine the back work ratio and the thermal efficiency of the cycle, assuming (a) no regenerator is used and (b) a regenerator with 75 percent effectiveness is used. Use variable specific heats.

**9–116** Repeat Problem 9–115, assuming an efficiency of 86 percent for each compressor stage and an efficiency of 90 percent for each turbine stage.

**9–117** Air enters a gas turbine with two stages of compression and two stages of expansion at 100 kPa and 17°C. This system uses a regenerator as well as reheating and intercooling. The pressure ratio across each compressor is 4; 300 kJ/kg of heat are added to the air in each combustion chamber; and the regenerator operates perfectly while increasing the temperature of the cold air by 20°C. Determine this system's thermal efficiency. Assume isentropic operations for all compressor and the turbine stages and use constant specific heats at room temperature.

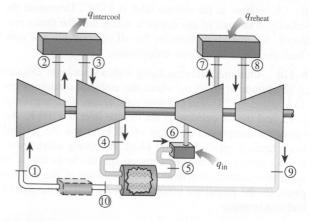

**FIGURE P9–117**

**9–118** Repeat Prob. 9–117 for the case of three stages of compression with intercooling and three stages with expansion with reheating. *Answer:* 40.1 percent

**9–119** How much would the thermal efficiency of the cycle in Prob. 9–118 change if the temperature of the cold-air stream leaving the regenerator is 80°C lower than the temperature of the hot-air stream entering the regenerator?

## Jet-Propulsion Cycles

**9–120C** What is propulsive power? How is it related to thrust?

**9–121C** What is propulsive efficiency? How is it determined?

**9–122C** Is the effect of turbine and compressor irreversibilities of a turbojet engine to reduce (a) the net work, (b) the thrust, or (c) the fuel consumption rate?

**9–123** A turbojet is flying with a velocity of 275 m/s at an altitude of 6100 m, where the ambient conditions are 50 kPa and −12°C. The pressure ratio across the compressor is 13, and the temperature at the turbine inlet is 1330 K. Assuming ideal operation for all components and constant specific heats for air at room temperature, determine (a) the pressure at the turbine exit, (b) the velocity of the exhaust gases, and (c) the propulsive efficiency.

**9–124** A turbofan engine operating on an aircraft flying at 200 m/s at an altitude where the air is at 50 kPa and −20°C, is to produce 50,000 N of thrust. The inlet diameter of this engine is 2.5 m; the compressor pressure ratio is 12; and the mass flow rate ratio is 8. Determine the air temperature at the fan outlet needed to produce this thrust. Assume ideal operation for all components and constant specific heats at room temperature. *Answer:* 233 K

**9–125** A pure jet engine propels an aircraft at 240 m/s through air at 45 kPa and −13°C. The inlet diameter of this engine is 1.6 m, the compressor pressure ratio is 13, and the temperature at the turbine inlet is 557°C. Determine the velocity at the exit of this engine's nozzle and the thrust produced. Assume ideal operation for all components and constant specific heats at room temperature.

**9–126** A turbojet aircraft is flying with a velocity of 280 m/s at an altitude of 9150 m, where the ambient conditions are 32 kPa and −32°C. The pressure ratio across the compressor is 12, and the temperature at the turbine inlet is 1100 K. Air enters the compressor at a rate of 50 kg/s, and the jet fuel has a heating value of 42,700 kJ/kg. Assuming ideal operation for all components and constant specific heats for air at room temperature, determine (a) the velocity of the exhaust gases, (b) the propulsive power developed, and (c) the rate of fuel consumption.

**9–127** Repeat Prob. 9–126 using a compressor efficiency of 80 percent and a turbine efficiency of 85 percent.

**9–128** Consider an aircraft powered by a turbojet engine that has a pressure ratio of 9. The aircraft is stationary on the ground, held in position by its brakes. The ambient air is at 7°C and 95 kPa and enters the engine at a rate of 20 kg/s. The jet fuel has a heating value of 42,700 kJ/kg, and it is burned completely at a rate of 0.5 kg/s. Neglecting the effect of the diffuser and disregarding the slight increase in mass at the engine exit as well as the inefficiencies of engine components, determine the force that must be applied on the brakes to hold the plane stationary. *Answer:* 19,370 N

**9–129** Reconsider Prob. 9–128. In the problem statement, replace the inlet mass flow rate by an inlet volume flow rate of 18.1 m³/s. Using EES (or other) software, investigate the effect of compressor inlet temperature in the range of −20 to 30°C on the force that must be applied to the brakes to hold the plane stationary. Plot this force as a function of compressor inlet temperature.

**9–130** Air at 7°C enters a turbojet engine at a rate of 16 kg/s and at a velocity of 300 m/s (relative to the engine). Air is heated in the combustion chamber at a rate 15,000 kJ/s and it leaves the engine at 427°C. Determine the thrust produced by this turbojet engine. (*Hint:* Choose the entire engine as your control volume.)

## Second-Law Analysis of Gas Power Cycles

**9–131** Determine the total exergy destruction associated with the Otto cycle described in Problem 9–32, assuming a source temperature of 2000 K and a sink temperature of 300 K. Also, determine the energy at the end of the power stroke. *Answers:* 245.1 kJ/kg, 145.2 kJ/kg

**9–132** Determine the total exergy destruction associated with the Diesel cycle described in Problem 9–44, assuming a source temperature of 2000 K and a sink temperature of 300 K. Also, determine the energy at the end of the isentropic compression process. *Answers:* 292.7 kJ/kg, 348.6 kJ/kg

**9–133** Calculate the exergy destruction for each process of Stirling cycle of Prob. 9–69, in kJ/kg.

**9–134** Calculate the exergy destruction associated with each of the processes of the Brayton cycle described in Prob. 9–78, assuming a source temperature of 1600 K and a sink temperature of 295 K.

**9–135** Repeat Prob. 9–81 using exergy analysis.

**9–136** Determine the total exergy destruction associated with the Brayton cycle described in Prob. 9–101, assuming a source temperature of 1500 K and a sink temperature of 290 K. Also, determine the energy of the exhaust gases at the exit of the regenerator.

**9–137** Reconsider Prob. 9–136. Using EES (or other) software, investigate the effect of varying the cycle pressure ratio from 6 to 14 on the total exergy destruction for the cycle and the energy of the exhaust gas leaving the regenerator. Plot these results as functions of pressure ratio. Discuss the results.

**9–138** Determine the exergy destruction associated with each of the processes of the Brayton cycle described in Prob. 9–103, assuming a source temperature of 1260 K and a sink temperature of 300 K. Also, determine the exergy of the exhaust gases at the exit of the regenerator. Take $P_{exhaust} = P_0 = 100$ kPa.

**9–139** Calculate the lost work potential for each process of Prob. 9–119. The temperature of the hot reservoir is the same as the maximum cycle temperature and the temperature of the cold reservoir is the same as the minimum cycle temperature.

**9–140** A gas-turbine power plant operates on the regenerative Brayton cycle between the pressure limits of 100 and 700 kPa. Air enters the compressor at 30°C at a rate of 12.6 kg/s and leaves at 260°C. It is then heated in a regenerator to 400°C by the hot combustion gases leaving the turbine. A diesel fuel with a heating value of 42,000 kJ/kg is burned in the combustion chamber with a combustion efficiency of 97 percent. The combustion gases leave the combustion chamber at 871°C and enter the turbine whose isentropic efficiency is 85 percent. Treating combustion gases as air and using constant specific heats at 500°C, determine (a) the isentropic efficiency of the compressor, (b) the effectiveness of the regenerator, (c) the air–fuel ratio in the combustion chamber, (d) the net power output and the back work ratio, (e) the thermal efficiency, and (f) the second-law efficiency of the plant. Also determine (g) the second-law efficiencies of the compressor, the turbine, and the regenerator, and (h) the rate of the energy flow with the combustion gases at the regenerator exit. *Answers:* (a) 0.881, (b) 0.632, (c) 78.1, (d) 2267 kW, 0.583, (e) 0.345, (f) 0.469, (g) 0.929, 0.932, 0.890, (h) 1351 kW

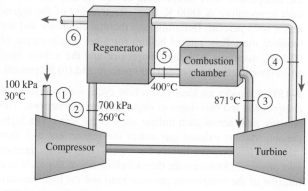

**FIGURE P9–140**

**9–141** A four-cylinder, four-stroke, 1.8-liter modern, high-speed compression-ignition engine operates on the ideal dual cycle with a compression ratio of 16. The air is at 95 kPa and 70°C at the beginning of the compression process and the engine speed is 2200 rpm. Equal amounts of fuel are burned

at constant volume and at constant pressure. The maximum allowable pressure in the cycle is 7.5 MPa due to material strength limitations. Using constant specific heats at 1000 K, determine (a) the maximum temperature in the cycle, (b) the net work output and the thermal efficiency, (c) the mean effective pressure, and (d) the net power output. Also, determine (e) the second-law efficiency of the cycle and the rate of energy output with the exhaust gases when they are purged. *Answers:* (a) 2308 K, (b) 836 kJ/kg, 59.5 percent, (c) 860 kPa, (d) 28.4 kW, (e) 68.3 percent, 10.3 kW

**Review Problems**

**9–142** A Carnot cycle is executed in a closed system and uses 0.0025 kg of air as the working fluid. The cycle efficiency is 60 percent, and the lowest temperature in the cycle is 300 K. The pressure at the beginning of the isentropic expansion is 700 kPa, and at the end of the isentropic compression it is 1 MPa. Determine the net work output per cycle.

**9–143** An air-standard cycle with variable coefficients is executed in a closed system and is composed of the following four processes:

1-2   $V$ = constant heat addition from 100 kPa and 27°C to 300 kPa
2-3   $P$ = constant heat addition to 1027°C
3-4   Isentropic expansion to 100 kPa
4-1   $P$ = constant heat rejection to initial state

(a) Show the cycle on $P$-$V$ and $T$-$s$ diagrams.
(b) Calculate the net work output per unit mass.
(c) Determine the thermal efficiency.

**9–144** Repeat Problem 9–143 using constant specific heats at room temperature.

**9–145** An Otto cycle with a compression ratio of 10.5 begins its compression at 90 kPa and 35°C. The maximum cycle temperature is 1000°C. Utilizing air-standard assumptions, determine the thermal efficiency of this cycle using (a) constant specific heats at room temperature and (b) variable specific heats. *Answers:* (a) 61.0 percent, (b) 57.7 percent

**9–146** A Diesel cycle has a compression ratio of 20 and begins its compression at 91 kPa and 7°C. The maximum cycle temperature is 987°C. Utilizing air-standard assumptions, determine the thermal efficiency of this cycle using (a) constant specific heats at room temperature and (b) variable specific heats.

**9–147** A Brayton cycle with a pressure ratio of 12 operates with air entering the compressor at 90 kPa and −3°C, and the turbine at 527°C. Calculate the net specific work produced by this cycle treating the air as an ideal gas with (a) constant specific heats at room temperature and (b) variable specific heats.

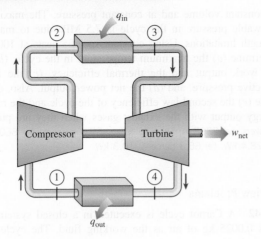

$q_{in}$

Compressor

Turbine

$w_{net}$

$q_{out}$

**FIGURE P9–147**

**9–148** A four-stroke turbocharged V-16 diesel engine built by GE Transportation Systems to power fast trains produces 4400 hp at 1500 rpm. Determine the amount of work produced per cylinder per (a) mechanical cycle and (b) thermodynamic cycle.

**9–149** Consider a simple ideal Brayton cycle operating between the temperature limits of 300 and 1500 K. Using constant specific heats at room temperature, determine the pressure ratio for which the compressor and the turbine exit temperatures of air are equal.

**9–150** A four-cylinder, four-stroke spark-ignition engine operates on the ideal Otto cycle with a compression ratio of 11 and a total displacement volume of 1.8 liter. The air is at 90 kPa and 50°C at the beginning of the compression process. The heat input is 1.5 kJ per cycle per cylinder. Accounting for the variation of specific heats of air with temperature, determine (a) the maximum temperature and pressure that occur during the cycle, (b) the net work per cycle per cylinder and the thermal efficiency of the cycle, (c) the mean effective pressure, and (d) the power output for an engine speed of 3000 rpm.

**9–151** A four-cylinder spark-ignition engine has a compression ratio of 10.5, and each cylinder has a maximum volume of 0.4 L. At the beginning of the compression process, the air is at 98 kPa and 37°C, and the maximum temperature in the cycle is 2100 K. Assuming the engine to operate on the ideal Otto cycle, determine (a) the amount of heat supplied per cylinder, (b) the thermal efficiency, and (c) the number of revolutions per minute required for a net power output of 45 kW. Assume variable specific heats for air.

**9–152** Reconsider Prob. 9–151. Using EES (or other) software, study the effect of varying the compression ratio from 5 to 11 on the net work done and the efficiency of the cycle. Plot the $P$-$v$ and $T$-$s$ diagrams for the cycle, and discuss the results.

**9–153** A typical hydrocarbon fuel produces 43,000 kJ/kg of heat when used in a spark-ignition engine. Determine the compression ratio required for an ideal Otto cycle to use 0.039 grams of fuel to produce 1 kJ of work. Use constant specific heats at room temperature. *Answer: 9.66*

**9–154** An ideal dual cycle has a compression ratio of 14 and uses air as the working fluid. At the beginning of the compression process, air is at 100 kPa and 50°C, and occupies a volume of 1600 cm$^3$. During the heat-addition process, 0.6 kJ of heat is transferred to air at constant volume and 1.1 kJ at constant pressure. Using constant specific heats evaluated at room temperature, determine the thermal efficiency of the cycle.

**9–155** Consider an ideal Stirling cycle using air as the working fluid. Air is at 400 K and 200 kPa at the beginning of the isothermal compression process, and heat is supplied to air from a source at 1800 K in the amount of 750 kJ/kg. Determine (a) the maximum pressure in the cycle and (b) the net work output per unit mass of air. *Answers: (a) 3844 kPa, (b) 583 kJ/kg*

**9–156** Consider a simple ideal Brayton cycle with air as the working fluid. The pressure ratio of the cycle is 6, and the minimum and maximum temperatures are 300 and 1300 K, respectively. Now the pressure ratio is doubled without changing the minimum and maximum temperatures in the cycle. Determine the change in (a) the net work output per unit mass and (b) the thermal efficiency of the cycle as a result of this modification. Assume variable specific heats for air. *Answers: (a) 41.5 kJ/kg, (b) 10.6 percent*

**9–157** Repeat Prob. 9–156 using constant specific heats at room temperature.

**9–158** Helium is used as the working fluid in a Brayton cycle with regeneration. The pressure ratio of the cycle is 8, the compressor inlet temperature is 300 K, and the turbine inlet temperature is 1800 K. The effectiveness of the regenerator is 75 percent. Determine the thermal efficiency and the required mass flow rate of helium for a net power output of 60 MW, assuming both the compressor and the turbine have an isentropic efficiency of (a) 100 percent and (b) 80 percent.

**9–159** Consider an ideal gas-turbine cycle with one stage of compression and two stages of expansion and regeneration. The pressure ratio across each turbine stage is the same. The high-pressure turbine exhaust gas enters the regenerator and then enters the low-pressure turbine for expansion to the compressor inlet pressure. Determine the thermal efficiency of this cycle as a function of the compressor pressure ratio and the high-pressure turbine to compressor inlet temperature ratio. Compare your result with the efficiency of the standard regenerative cycle.

**9–160** A gas-turbine plant operates on the regenerative Brayton cycle with two stages of reheating and two-stages of intercooling between the pressure limits of 100 and 1200 kPa. The working fluid is air. The air enters the first and the second stages of the compressor at 300 K and 350 K, respectively, and the first and the second stages of the turbine at 1400 K and 1300 K, respectively. Assuming both the compressor and the turbine have an isentropic efficiency of 80 percent and the regenerator has an effectiveness of 75 percent and using variable specific heats, determine (a) the back work ratio and the net work output, (b) the thermal efficiency, and (c) the second-law efficiency of the cycle. Also determine (d) the exergies at the exits of the combustion chamber (state 6) and the regenerator (state 10) (See Fig. 9–43 in the text). *Answers:* (a) 0.523, 317 kJ/kg, (b) 0.553, (c) 0.704, (d) 931 kJ/kg, 129 kJ/kg

**9–161** Compare the thermal efficiency of a two-stage gas turbine with regeneration, reheating and intercooling to that of a three-stage gas turbine with the same equipment when (a) all components operate ideally, (b) air enters the first compressor at 100 kPa and 20°C, (c) the total pressure ratio across all stages of compression is 16, and (d) the maximum cycle temperature is 800°C.

**9–162** Electricity and process heat requirements of a manufacturing facility are to be met by a cogeneration plant consisting of a gas turbine and a heat exchanger for steam production. The plant operates on the simple Brayton cycle between the pressure limits of 100 and 1000 kPa with air as the working fluid. Air enters the compressor at 20°C. Combustion gases leave the turbine and enter the heat exchanger at 450°C, and leave the heat exchanger of 325°C, while the liquid water enters the heat exchanger at 15°C and leaves at 200°C as a saturated vapor. The net power produced by the gas-turbine cycle is 1500 kW. Assuming a compressor isentropic efficiency of 86 percent and a turbine isentropic efficiency of 88 percent and using variable specific heats, determine (a) the mass flow rate of air, (b) the back work ratio and the thermal efficiency, and (c) the rate at which steam is produced in the heat exchanger. Also determine (d) the utilization efficiency of the cogeneration plant, defined as the ratio of the total energy utilized to the energy supplied to the plant.

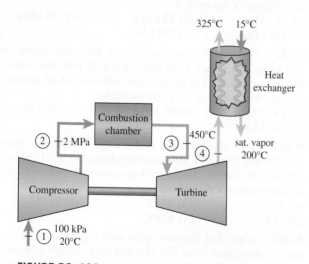

**FIGURE P9–162**

**9–163** A turbojet aircraft flies with a velocity of 1100 km/h at an altitude where the air temperature and pressure are −35°C and 40 kPa. Air leaves the diffuser at 50 kPa with a velocity of 15 m/s, and combustion gases enter the turbine at 450 kPa and 950°C. The turbine produces 800 kW of power, all of which is used to drive the compressor. Assuming an isentropic efficiency of 83 percent for the compressor, turbine, and nozzle, and using variable specific heats, determine (a) the pressure of combustion gases at the turbine exit, (b) the mass flow rate of air through the compressor, (c) the velocity of the gases at the nozzle exit, and (d) the propulsive power and the propulsive efficiency for this engine. *Answers:* (a) 137 kPa, (b) 2.66 kg/s, (c) 696 m/s, (d) 317 kW, 0.166

**9–164** An air standard cycle with constant specific heats is executed in a closed piston-cylinder system and is composed of the following three processes:

1-2   Constant volume heat addition
2-3   Isentropic expansion with an expansion ratio $r = V_3/V_2$
3-1   Constant pressure heat rejection

(a)   Sketch the $P$-$v$ and $T$-$s$ diagrams for this cycle
(b)   Obtain an expression for the back work ratio as a function of $k$ and $r$
(c)   Obtain an expression for the cycle thermal efficiency as a function of $k$ and $r$
(d)   Determine the value of the back work ratio and efficiency as $r$ goes to unity

What do your results imply about the net work done by the cycle?

**9–165** Consider the ideal regenerative Brayton cycle. Determine the pressure ratio that maximizes the thermal efficiency of the cycle and compare this value with the pressure ratio that maximizes the cycle net work. For the same maximum-to-minimum temperature ratios, explain why the pressure ratio for maximum efficiency is less than the pressure ratio for maximum work.

**9–166** [EES] Using EES (or other) software, study the effect of variable specific heats on the thermal efficiency of the ideal Otto cycle using air as the working fluid. At the beginning of the compression process, air is at 100 kPa and 300 K. Determine the percentage of error involved in using constant specific heat values at room temperature for the following combinations of compression ratios and maximum cycle temperatures: $r = 6, 8, 10, 12,$ and $T_{max} = 1000, 1500, 2000, 2500$ K.

**9–167** [EES] Using EES (or other) software, determine the effects of pressure ratio, maximum cycle temperature, and compressor and turbine efficiencies on the net work output per unit mass and the thermal efficiency of a simple Brayton cycle with air as the working fluid. Air is at 100 kPa and 300 K at the compressor inlet. Also, assume constant specific heats for air at room temperature. Determine the net work output and the thermal efficiency for all combinations of the following parameters, and draw conclusions from the results.

Pressure ratio:                 5, 8, 14
Maximum cycle temperature:     800, 1200, 1600 K

Compressor isentropic efficiency:    80, 100 percent
Turbine isentropic efficiency:    80, 100 percent

**9–168** 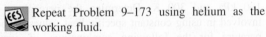 Repeat Problem 9–167 by considering the variation of specific heats of air with temperature.

**9–169** Repeat Problem 9–167 using helium as the working fluid.

**9–170** Using EES (or other) software, determine the effects of pressure ratio, maximum cycle temperature, regenerator effectiveness, and compressor and turbine efficiencies on the net work output per unit mass and on the thermal efficiency of a regenerative Brayton cycle with air as the working fluid. Air is at 100 kPa and 300 K at the compressor inlet. Also, assume constant specific heats for air at room temperature. Determine the net work output and the thermal efficiency for all combinations of the following parameters.

Pressure ratio:    6, 10
Maximum cycle temperature:    1500, 2000 K
Compressor isentropic efficiency:    80, 100 percent
Turbine isentropic efficiency:    80, 100 percent
Regenerator effectiveness:    70, 90 percent

**9–171** Repeat Problem 9–170 by considering the variation of specific heats of air with temperature.

**9–172** Repeat Problem 9–170 using helium as the working fluid.

**9–173** Using EES (or other) software, determine the effect of the number of compression and expansion stages on the thermal efficiency of an ideal regenerative Brayton cycle with multistage compression and expansion. Assume that the overall pressure ratio of the cycle is 18, and the air enters each stage of the compressor at 300 K and each stage of the turbine at 1200 K. Using constant specific heats for air at room temperature, determine the thermal efficiency of the cycle by varying the number of stages from 1 to 22 in increments of 3. Plot the thermal efficiency versus the number of stages. Compare your results to the efficiency of an Ericsson cycle operating between the same temperature limits.

**9–174** Repeat Problem 9–173 using helium as the working fluid.

## Fundamentals of Engineering (FE) Exam Problems

**9–175** An Otto cycle with air as the working fluid has a compression ratio of 10.4. Under cold-air-standard conditions, the thermal efficiency of this cycle is
(a) 10 percent    (b) 39 percent    (c) 61 percent
(d) 79 percent    (e) 82 percent

**9–176** For specified limits for the maximum and minimum temperatures, the ideal cycle with the lowest thermal efficiency is
(a) Carnot    (b) Stirling    (c) Ericsson
(d) Otto    (e) All are the same

**9–177** A Carnot cycle operates between the temperature limits of 300 and 2000 K, and produces 600 kW of net power. The rate of entropy change of the working fluid during the heat addition process is
(a) 0    (b) 0.300 kW/K    (c) 0.353 kW/K
(d) 0.261 kW/K    (e) 2.0 kW/K

**9–178** Air in an ideal Diesel cycle is compressed from 2 to 0.13 L, and then it expands during the constant pressure heat addition process to 0.30 L. Under cold air standard conditions, the thermal efficiency of this cycle is
(a) 41 percent    (b) 59 percent    (c) 66 percent
(d) 70 percent    (e) 78 percent

**9–179** Helium gas in an ideal Otto cycle is compressed from 20°C and 2.5 to 0.25 L, and its temperature increases by an additional 700°C during the heat addition process. The temperature of helium before the expansion process is
(a) 1790°C    (b) 2060°C    (c) 1240°C
(d) 620°C    (e) 820°C

**9–180** In an ideal Otto cycle, air is compressed from 1.20 kg/m³ and 2.2 to 0.26 L, and the net work output of the cycle is 440 kJ/kg. The mean effective pressure (MEP) for this cycle is
(a) 612 kPa    (b) 599 kPa    (c) 528 kPa
(d) 416 kPa    (e) 367 kPa

**9–181** In an ideal Brayton cycle, air is compressed from 95 kPa and 25°C to 1100 kPa. Under cold-air-standard conditions, the thermal efficiency of this cycle is
(a) 45 percent    (b) 50 percent    (c) 62 percent
(d) 73 percent    (e) 86 percent

**9–182** Consider an ideal Brayton cycle executed between the pressure limits of 1200 and 100 kPa and temperature limits of 20 and 1000°C with argon as the working fluid. The net work output of the cycle is
(a) 68 kJ/kg    (b) 93 kJ/kg    (c) 158 kJ/kg
(d) 186 kJ/kg    (e) 310 kJ/kg

**9–183** An ideal Brayton cycle has a net work output of 150 kJ/kg and a back work ratio of 0.4. If both the turbine and the compressor had an isentropic efficiency of 85 percent, the net work output of the cycle would be
(a) 74 kJ/kg    (b) 95 kJ/kg    (c) 109 kJ/kg
(d) 128 kJ/kg    (e) 177 kJ/kg

**9–184** In an ideal Brayton cycle, air is compressed from 100 kPa and 25°C to 1 MPa, and then heated to 927°C before entering the turbine. Under cold-air-standard conditions, the air temperature at the turbine exit is
(a) 349°C    (b) 426°C    (c) 622°C
(d) 733°C    (e) 825°C

**9–185** In an ideal Brayton cycle with regeneration, argon gas is compressed from 100 kPa and 25°C to 400 kPa, and then heated to 1200°C before entering the turbine. The highest temperature that argon can be heated in the regenerator is

(a) 246°C      (b) 846°C      (c) 689°C
(d) 368°C      (e) 573°C

**9–186** In an ideal Brayton cycle with regeneration, air is compressed from 80 kPa and 10°C to 400 kPa and 175°C, is heated to 450°C in the regenerator, and then further heated to 1000°C before entering the turbine. Under cold-air-standard conditions, the effectiveness of the regenerator is
(a) 33 percent      (b) 44 percent      (c) 62 percent
(d) 77 percent      (e) 89 percent

**9–187** Consider a gas turbine that has a pressure ratio of 6 and operates on the Brayton cycle with regeneration between the temperature limits of 20 and 900°C. If the specific heat ratio of the working fluid is 1.3, the highest thermal efficiency this gas turbine can have is
(a) 38 percent      (b) 46 percent      (c) 62 percent
(d) 58 percent      (e) 97 percent

**9–188** An ideal gas turbine cycle with many stages of compression and expansion and a regenerator of 100 percent effectiveness has an overall pressure ratio of 10. Air enters every stage of compressor at 290 K, and every stage of turbine at 1200 K. The thermal efficiency of this gas-turbine cycle is
(a) 36 percent      (b) 40 percent      (c) 52 percent
(d) 64 percent      (e) 76 percent

**9–189** Air enters a turbojet engine at 320 m/s at a rate of 30 kg/s, and exits at 650 m/s relative to the aircraft. The thrust developed by the engine is
(a) 5 kN      (b) 10 kN      (c) 15 kN
(d) 20 kN      (e) 26 kN

## Design and Essay Problems

**9–190** The weight of a diesel engine is directly proportional to the compression ratio ($W = kr$) because extra metal must be used to strengthen the engine for the higher pressures. Examine the net specific work produced by a diesel engine per unit of weight as the pressure ratio is varied and the specific heat input remains fixed. Do this for several heat inputs and proportionality constants $k$. Are there any optimal combinations of $k$ and specific heat inputs.

**9–191** In response to concerns about the environment, some major car manufacturers are currently marketing electric cars. Write an essay on the advantages and disadvantages of electric cars, and discuss when it is advisable to purchase an electric car instead of a traditional internal combustion car.

**9–192** Intense research is underway to develop adiabatic engines that require no cooling of the engine block. Such engines are based on ceramic materials because of the ability of such materials to withstand high temperatures. Write an essay on the current status of adiabatic engine development. Also determine the highest possible efficiencies with these engines, and compare them to the highest possible efficiencies of current engines.

**9–193** Write an essay on the most recent developments on the two-stroke engines, and find out when we might be seeing cars powered by two-stroke engines in the market. Why do the major car manufacturers have a renewed interest in two-stroke engines?

**9–194** Exhaust gases from the turbine of a simple Brayton cycle are quite hot and may be used for other thermal purposes. One proposed use is generating saturated steam at 110°C from water at 30°C in a boiler. This steam will be distributed to several buildings on a college campus for space heating. A Brayton cycle with a pressure ratio of 6 is to be used for this purpose. Plot the power produced, the flow rate of produced steam, and the maximum cycle temperature as functions of the rate at which heat is added to the cycle. The temperature at the turbine inlet is not to exceed 2000°C.

**9–195** A gas turbine operates with a regenerator and two stages of reheating and intercooling. This system is designed so that when air enters the compressor at 100 kPa and 15°C, the pressure ratio for each stage of compression is 3; the air temperature when entering a turbine is 500°C; and the regenerator operates perfectly. At full load, this engine produces 800 kW. For this engine to service a partial load, the heat addition in both combustion chambers is reduced. Develop an optimal schedule of heat addition to the combustion chambers for partial loads ranging from 400 to 800 kW.

**9–196** Since its introduction in 1903 by Aegidius Elling of Norway, steam injection between the combustion chamber and the turbine is used even in some modern gas turbines currently in operation to cool the combustion gases to a metallurgical-safe temperature while increasing the mass flow rate through the turbine. Currently, there are several gas-turbine power plants that use steam injection to augment power and improve thermal efficiency.

Consider a gas-turbine power plant whose pressure ratio is 8. The isentropic efficiencies of the compressor and the turbine are 80 percent, and there is a regenerator with an effectiveness of 70 percent. When the mass flow rate of air through the compressor is 40 kg/s, the turbine inlet temperature becomes 1700 K. But the turbine inlet temperature is limited to 1500 K, and thus steam injection into the combustion gases is being considered. However, to avoid the complexities associated with steam injection, it is proposed to use excess air (that is, to take in much more air than needed for complete combustion) to lower the combustion and thus turbine inlet temperature while increasing the mass flow rate and thus power output of the turbine. Evaluate this proposal, and compare the thermodynamic performance of "high air flow" to that of a "steam-injection" gas-turbine power plant under the following design conditions: the ambient air is at 100 kPa and 25°C, adequate water supply is available at 20°C, and the amount of fuel supplied to the combustion chamber remains constant.

# VAPOR AND COMBINED POWER CYCLES

I n Chap. 9 we discussed gas power cycles for which the working fluid remains a gas throughout the entire cycle. In this chapter, we consider *vapor power cycles* in which the working fluid is alternatively vaporized and condensed. We also consider power generation coupled with process heating called *cogeneration*.

The continued quest for higher thermal efficiencies has resulted in some innovative modifications to the basic vapor power cycle. Among these, we discuss the *reheat* and *regenerative cycles,* as well as combined gas–vapor power cycles.

Steam is the most common working fluid used in vapor power cycles because of its many desirable characteristics, such as low cost, availability, and high enthalpy of vaporization. Therefore, this chapter is mostly devoted to the discussion of steam power plants. Steam power plants are commonly referred to as *coal plants, nuclear plants,* or *natural gas plants,* depending on the type of fuel used to supply heat to the steam. However, the steam goes through the same basic cycle in all of them. Therefore, all can be analyzed in the same manner.

■ ■ ■ ■ ■ ■ ■
## OBJECTIVES
The objectives of Chapter 10 are to:

- Analyze vapor power cycles in which the working fluid is alternately vaporized and condensed.

- Perform second-law analysis of vapor power cycles.

- Analyze power generation coupled with process heating called cogeneration.

- Investigate ways to modify the basic Rankine vapor power cycle to increase the cycle thermal efficiency.

- Analyze the reheat and regenerative vapor power cycles.

- Analyze power cycles that consist of two separate cycles known as combined cycles.

# 10–1 · THE CARNOT VAPOR CYCLE

We have mentioned repeatedly that the Carnot cycle is the most efficient cycle operating between two specified temperature limits. Thus it is natural to look at the Carnot cycle first as a prospective ideal cycle for vapor power plants. If we could, we would certainly adopt it as the ideal cycle. As explained below, however, the Carnot cycle is not a suitable model for power cycles. Throughout the discussions, we assume *steam* to be the working fluid since it is the working fluid predominantly used in vapor power cycles.

Consider a steady-flow *Carnot cycle* executed within the saturation dome of a pure substance, as shown in Fig. 10–1a. The fluid is heated reversibly and isothermally in a boiler (process 1-2), expanded isentropically in a turbine (process 2-3), condensed reversibly and isothermally in a condenser (process 3-4), and compressed isentropically by a compressor to the initial state (process 4-1).

Several impracticalities are associated with this cycle:

**1.** Isothermal heat transfer to or from a two-phase system is not difficult to achieve in practice since maintaining a constant pressure in the device automatically fixes the temperature at the saturation value. Therefore, processes 1-2 and 3-4 can be approached closely in actual boilers and condensers. Limiting the heat transfer processes to two-phase systems, however, severely limits the maximum temperature that can be used in the cycle (it has to remain under the critical-point value, which is 374°C for water). Limiting the maximum temperature in the cycle also limits the thermal efficiency. Any attempt to raise the maximum temperature in the cycle involves heat transfer to the working fluid in a single phase, which is not easy to accomplish isothermally.

**2.** The isentropic expansion process (process 2-3) can be approximated closely by a well-designed turbine. However, the quality of the steam decreases during this process, as shown on the *T-s* diagram in Fig. 10–1a. Thus the turbine has to handle steam with low quality, that is, steam with a high moisture content. The impingement of liquid droplets on the turbine blades causes erosion and is a major source of wear. Thus steam with qualities less than about 90 percent cannot be tolerated in the operation of power plants. This problem could be eliminated by using a working fluid with a very steep saturated vapor line.

**3.** The isentropic compression process (process 4-1) involves the compression of a liquid–vapor mixture to a saturated liquid. There are two difficulties associated with this process. First, it is not easy to control the condensation process so precisely as to end up with the desired quality at state 4. Second, it is not practical to design a compressor that handles two phases.

Some of these problems could be eliminated by executing the Carnot cycle in a different way, as shown in Fig. 10–1b. This cycle, however, presents other problems such as isentropic compression to extremely high pressures and isothermal heat transfer at variable pressures. Thus we conclude that the Carnot cycle cannot be approximated in actual devices and is not a realistic model for vapor power cycles.

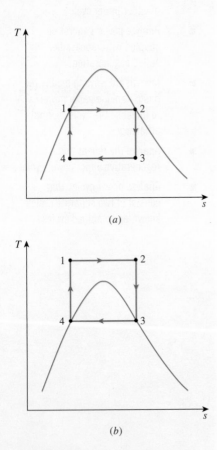

(a)

(b)

**FIGURE 10–1**

*T-s* diagram of two Carnot vapor cycles.

# 10–2 · RANKINE CYCLE: THE IDEAL CYCLE FOR VAPOR POWER CYCLES

Many of the impracticalities associated with the Carnot cycle can be eliminated by superheating the steam in the boiler and condensing it completely in the condenser, as shown schematically on a *T-s* diagram in Fig. 10–2. The cycle that results is the **Rankine cycle**, which is the ideal cycle for vapor power plants. The ideal Rankine cycle does not involve any internal irreversibilities and consists of the following four processes:

1-2   Isentropic compression in a pump

2-3   Constant pressure heat addition in a boiler

3-4   Isentropic expansion in a turbine

4-1   Constant pressure heat rejection in a condenser

Water enters the *pump* at state 1 as saturated liquid and is compressed isentropically to the operating pressure of the boiler. The water temperature increases somewhat during this isentropic compression process due to a slight decrease in the specific volume of water. The vertical distance between states 1 and 2 on the *T-s* diagram is greatly exaggerated for clarity. (If water were truly incompressible, would there be a temperature change at all during this process?)

Water enters the *boiler* as a compressed liquid at state 2 and leaves as a superheated vapor at state 3. The boiler is basically a large heat exchanger where the heat originating from combustion gases, nuclear reactors, or other sources is transferred to the water essentially at constant pressure. The boiler, together with the section where the steam is superheated (the superheater), is often called the *steam generator*.

The superheated vapor at state 3 enters the *turbine,* where it expands isentropically and produces work by rotating the shaft connected to an electric generator. The pressure and the temperature of steam drop during this process to the values at state 4, where steam enters the *condenser.* At this state, steam is usually a saturated liquid–vapor mixture with a high quality. Steam is condensed at constant pressure in the condenser, which is basically a large heat exchanger, by rejecting heat to a cooling medium such as a lake, a river, or the atmosphere. Steam leaves the condenser as saturated liquid and enters the pump, completing the cycle. In areas where water is precious, the power plants are cooled by air instead of water. This method of cooling, which is also used in car engines, is called *dry cooling*. Several power plants in the world, including some in the United States, use dry cooling to conserve water.

Remembering that the area under the process curve on a *T-s* diagram represents the heat transfer for internally reversible processes, we see that the area under process curve 2-3 represents the heat transferred to the water in the boiler and the area under the process curve 4-1 represents the heat rejected in the condenser. The difference between these two (the area enclosed by the cycle curve) is the net work produced during the cycle.

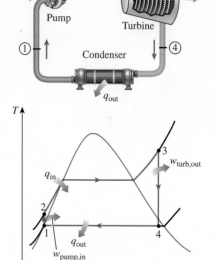

**FIGURE 10–2**
The simple ideal Rankine cycle.

## Energy Analysis of the Ideal Rankine Cycle

All four components associated with the Rankine cycle (the pump, boiler, turbine, and condenser) are steady-flow devices, and thus all four processes that make up the Rankine cycle can be analyzed as steady-flow processes. The kinetic and potential energy changes of the steam are usually small relative to the work and heat transfer terms and are therefore usually neglected. Then the *steady-flow energy equation* per unit mass of steam reduces to

$$(q_{\text{in}} - q_{\text{out}}) + (w_{\text{in}} - w_{\text{out}}) = h_e - h_i \quad \text{(kJ/kg)} \tag{10-1}$$

The boiler and the condenser do not involve any work, and the pump and the turbine are assumed to be isentropic. Then the conservation of energy relation for each device can be expressed as follows:

*Pump* ($q = 0$): $$w_{\text{pump,in}} = h_2 - h_1 \tag{10-2}$$

or,

$$w_{\text{pump,in}} = \nu(P_2 - P_1) \tag{10-3}$$

where

$$h_1 = h_{f @ P_1} \quad \text{and} \quad \nu \cong \nu_1 = \nu_{f @ P_1} \tag{10-4}$$

*Boiler* ($w = 0$): $$q_{\text{in}} = h_3 - h_2 \tag{10-5}$$

*Turbine* ($q = 0$): $$w_{\text{turb,out}} = h_3 - h_4 \tag{10-6}$$

*Condenser* ($w = 0$): $$q_{\text{out}} = h_4 - h_1 \tag{10-7}$$

The *thermal efficiency* of the Rankine cycle is determined from

$$\eta_{\text{th}} = \frac{w_{\text{net}}}{q_{\text{in}}} = 1 - \frac{q_{\text{out}}}{q_{\text{in}}} \tag{10-8}$$

where

$$w_{\text{net}} = q_{\text{in}} - q_{\text{out}} = w_{\text{turb,out}} - w_{\text{pump,in}}$$

The conversion efficiency of power plants in the United States is often expressed in terms of **heat rate**, which is the amount of heat supplied, in Btu's, to generate 1 kWh of electricity. The smaller the heat rate, the greater the efficiency. Considering that 1 kWh = 3412 Btu and disregarding the losses associated with the conversion of shaft power to electric power, the relation between the heat rate and the thermal efficiency can be expressed as

$$\eta_{\text{th}} = \frac{3412 \ (\text{Btu/kWh})}{\text{Heat rate (Btu/kWh)}} \tag{10-9}$$

For example, a heat rate of 11,363 Btu/kWh is equivalent to 30 percent efficiency.

The thermal efficiency can also be interpreted as the ratio of the area enclosed by the cycle on a *T-s* diagram to the area under the heat-addition process. The use of these relations is illustrated in the following example.

## EXAMPLE 10–1 The Simple Ideal Rankine Cycle

Consider a steam power plant operating on the simple ideal Rankine cycle. Steam enters the turbine at 3 MPa and 350°C and is condensed in the condenser at a pressure of 75 kPa. Determine the thermal efficiency of this cycle.

**SOLUTION** A steam power plant operating on the simple ideal Rankine cycle is considered. The thermal efficiency of the cycle is to be determined.
**Assumptions** 1 Steady operating conditions exist. 2 Kinetic and potential energy changes are negligible.
**Analysis** The schematic of the power plant and the *T-s* diagram of the cycle are shown in Fig. 10–3. We note that the power plant operates on the ideal Rankine cycle. Therefore, the pump and the turbine are isentropic, there are no pressure drops in the boiler and condenser, and steam leaves the condenser and enters the pump as saturated liquid at the condenser pressure.

First we determine the enthalpies at various points in the cycle, using data from steam tables (Tables A–4, A–5, and A–6):

State 1: $\left. \begin{array}{l} P_1 = 75 \text{ kPa} \\ \text{Sat. liquid} \end{array} \right\} \begin{array}{l} h_1 = h_{f\text{ @ 75 kPa}} = 384.44 \text{ kJ/kg} \\ v_1 = v_{f\text{ @ 75 kPa}} = 0.001037 \text{ m}^3\text{/kg} \end{array}$

State 2: $P_2 = 3 \text{ MPa}$

$s_2 = s_1$

$w_{\text{pump,in}} = v_1(P_2 - P_1) = (0.001037 \text{ m}^3\text{/kg})[(3000 - 75) \text{ kPa}]\left( \dfrac{1 \text{ kJ}}{1 \text{ kPa·m}^3} \right)$

$= 3.03 \text{ kJ/kg}$

$h_2 = h_1 + w_{\text{pump,in}} = (384.44 + 3.03) \text{ kJ/kg} = 387.47 \text{ kJ/kg}$

State 3: $\left. \begin{array}{l} P_3 = 3 \text{ MPa} \\ T_3 = 350°C \end{array} \right\} \begin{array}{l} h_3 = 3116.1 \text{ kJ/kg} \\ s_3 = 6.7450 \text{ kJ/kg·K} \end{array}$

State 4: $P_4 = 75 \text{ kPa}$ (sat. mixture)

$s_4 = s_3$

$x_4 = \dfrac{s_4 - s_f}{s_{fg}} = \dfrac{6.7450 - 1.2132}{6.2426} = 0.8861$

$h_4 = h_f + x_4 h_{fg} = 384.44 + 0.8861(2278.0) = 2403.0 \text{ kJ/kg}$

Thus,

$q_{\text{in}} = h_3 - h_2 = (3116.1 - 387.47) \text{ kJ/kg} = 2728.6 \text{ kJ/kg}$

$q_{\text{out}} = h_4 - h_1 = (2403.0 - 384.44) \text{ kJ/kg} = 2018.6 \text{ kJ/kg}$

and

$\eta_{\text{th}} = 1 - \dfrac{q_{\text{out}}}{q_{\text{in}}} = 1 - \dfrac{2018.6 \text{ kJ/kg}}{2728.6 \text{ kJ/kg}} = \mathbf{0.260 \text{ or } 26.0\%}$

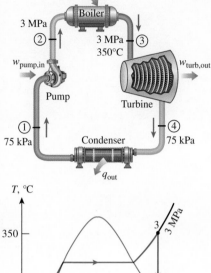

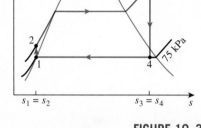

**FIGURE 10–3**
Schematic and *T-s* diagram for Example 10–1.

The thermal efficiency could also be determined from

$$w_{\text{turb,out}} = h_3 - h_4 = (3116.1 - 2403.0) \text{ kJ/kg} = 713.1 \text{ kJ/kg}$$

$$w_{\text{net}} = w_{\text{turb,out}} - w_{\text{pump,in}} = (713.1 - 3.03) \text{ kJ/kg} = 710.1 \text{ kJ/kg}$$

or

$$w_{\text{net}} = q_{\text{in}} - q_{\text{out}} = (2728.6 - 2018.6) \text{ kJ/kg} = 710.0 \text{ kJ/kg}$$

and

$$\eta_{\text{th}} = \frac{w_{\text{net}}}{q_{\text{in}}} = \frac{710.0 \text{ kJ/kg}}{2728.6 \text{ kJ/kg}} = \mathbf{0.260 \text{ or } 26.0\%}$$

That is, this power plant converts 26 percent of the heat it receives in the boiler to net work. An actual power plant operating between the same temperature and pressure limits will have a lower efficiency because of the irreversibilities such as friction.

***Discussion*** Notice that the back work ratio ($r_{\text{bw}} = w_{\text{in}}/w_{\text{out}}$) of this power plant is 0.004, and thus only 0.4 percent of the turbine work output is required to operate the pump. Having such low back work ratios is characteristic of vapor power cycles. This is in contrast to the gas power cycles, which typically involve very high back work ratios (about 40 to 80 percent).

It is also interesting to note the thermal efficiency of a Carnot cycle operating between the same temperature limits

$$\eta_{\text{th,Carnot}} = 1 - \frac{T_{\text{min}}}{T_{\text{max}}} = 1 - \frac{(91.76 + 273) \text{ K}}{(350 + 273) \text{ K}} = 0.415$$

The difference between the two efficiencies is due to the large external irreversibility in the Rankine cycle caused by the large temperature difference between steam and combustion gases in the furnace.

# 10–3 · DEVIATION OF ACTUAL VAPOR POWER CYCLES FROM IDEALIZED ONES

The actual vapor power cycle differs from the ideal Rankine cycle, as illustrated in Fig. 10–4a, as a result of irreversibilities in various components. Fluid friction and heat loss to the surroundings are the two common sources of irreversibilities.

*Fluid friction* causes pressure drops in the boiler, the condenser, and the piping between various components. As a result, steam leaves the boiler at a somewhat lower pressure. Also, the pressure at the turbine inlet is somewhat lower than that at the boiler exit due to the pressure drop in the connecting pipes. The pressure drop in the condenser is usually very small. To compensate for these pressure drops, the water must be pumped to a sufficiently higher pressure than the ideal cycle calls for. This requires a larger pump and larger work input to the pump.

The other major source of irreversibility is the *heat loss* from the steam to the surroundings as the steam flows through various components. To maintain the same level of net work output, more heat needs to be transferred to the steam in the boiler to compensate for these undesired heat losses. As a result, cycle efficiency decreases.

Of particular importance are the irreversibilities occurring within the pump and the turbine. A pump requires a greater work input, and a turbine produces a smaller work output as a result of irreversibilities. Under ideal conditions, the flow through these devices is isentropic. The deviation of actual pumps and turbines from the isentropic ones can be accounted for by utilizing *isentropic efficiencies*, defined as

$$\eta_P = \frac{w_s}{w_a} = \frac{h_{2s} - h_1}{h_{2a} - h_1} \qquad (10\text{–}10)$$

and

$$\eta_T = \frac{w_a}{w_s} = \frac{h_3 - h_{4a}}{h_3 - h_{4s}} \qquad (10\text{–}11)$$

where states 2a and 4a are the actual exit states of the pump and the turbine, respectively, and 2s and 4s are the corresponding states for the isentropic case (Fig. 10-4b).

Other factors also need to be considered in the analysis of actual vapor power cycles. In actual condensers, for example, the liquid is usually sub-cooled to prevent the onset of *cavitation*, the rapid vaporization and con-densation of the fluid at the low-pressure side of the pump impeller, which may damage it. Additional losses occur at the bearings between the moving parts as a result of friction. Steam that leaks out during the cycle and air that leaks into the condenser represent two other sources of loss. Finally, the power consumed by the auxiliary equipment such as fans that supply air to the furnace should also be considered in evaluating the overall performance of power plants.

The effect of irreversibilities on the thermal efficiency of a steam power cycle is illustrated below with an example.

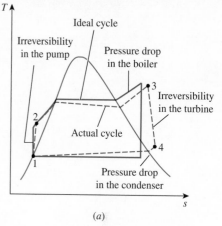

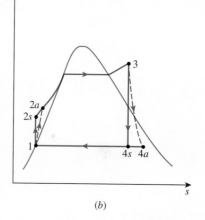

**FIGURE 10–4**
(*a*) Deviation of actual vapor power cycle from the ideal Rankine cycle. (*b*) The effect of pump and turbine irreversibilities on the ideal Rankine cycle.

---

### EXAMPLE 10–2    An Actual Steam Power Cycle

A steam power plant operates on the cycle shown in Fig. 10–5. If the isen-tropic efficiency of the turbine is 87 percent and the isentropic efficiency of the pump is 85 percent, determine (*a*) the thermal efficiency of the cycle and (*b*) the net power output of the plant for a mass flow rate of 15 kg/s.

**SOLUTION**  A steam power cycle with specified turbine and pump efficien-cies is considered. The thermal efficiency and the net power output are to be determined.
**Assumptions**  **1** Steady operating conditions exist. **2** Kinetic and potential energy changes are negligible.
**Analysis**  The schematic of the power plant and the *T-s* diagram of the cycle are shown in Fig. 10–5. The temperatures and pressures of steam at various points are also indicated on the figure. We note that the power plant involves steady-flow components and operates on the Rankine cycle, but the imper-fections at various components are accounted for.
(*a*) The thermal efficiency of a cycle is the ratio of the net work output to the heat input, and it is determined as follows:

*Pump work input:*

$$w_{pump,in} = \frac{w_{s,pump,in}}{\eta_P} = \frac{v_1(P_2 - P_1)}{\eta_P}$$

$$= \frac{(0.001009 \text{ m}^3/\text{kg})[(16{,}000 - 9) \text{ kPa}]}{0.85}\left(\frac{1 \text{ kJ}}{1 \text{ kPa·m}^3}\right)$$

$$= 19.0 \text{ kJ/kg}$$

*Turbine work output:*

$$w_{turb,out} = \eta_T w_{s,turb,out}$$

$$= \eta_T(h_5 - h_{6s}) = 0.87(3583.1 - 2115.3) \text{ kJ/kg}$$

$$= 1277.0 \text{ kJ/kg}$$

*Boiler heat input:* $\quad q_{in} = h_4 - h_3 = (3647.6 - 160.1) \text{ kJ/kg} = 3487.5 \text{ kJ/kg}$

Thus,

$$w_{net} = w_{turb,out} - w_{pump,in} = (1277.0 - 19.0) \text{ kJ/kg} = 1258.0 \text{ kJ/kg}$$

$$\eta_{th} = \frac{w_{net}}{q_{in}} = \frac{1258.0 \text{ kJ/kg}}{3487.5 \text{ kJ/kg}} = \textbf{0.361 or 36.1\%}$$

(*b*) The power produced by this power plant is

$$\dot{W}_{net} = \dot{m}\, w_{net} = (15 \text{ kg/s})(1258.0 \text{ kJ/kg}) = \textbf{18.9 MW}$$

**Discussion** Without the irreversibilities, the thermal efficiency of this cycle would be 43.0 percent (see Example 10–3c).

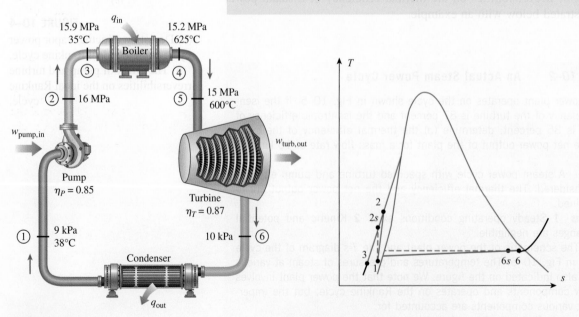

**FIGURE 10–5**
Schematic and *T-s* diagram for Example 10–2.

# 10–4 · HOW CAN WE INCREASE THE EFFICIENCY OF THE RANKINE CYCLE?

Steam power plants are responsible for the production of most electric power in the world, and even small increases in thermal efficiency can mean large savings from the fuel requirements. Therefore, every effort is made to improve the efficiency of the cycle on which steam power plants operate.

The basic idea behind all the modifications to increase the thermal efficiency of a power cycle is the same: *Increase the average temperature at which heat is transferred to the working fluid in the boiler, or decrease the average temperature at which heat is rejected from the working fluid in the condenser.* That is, the average fluid temperature should be as high as possible during heat addition and as low as possible during heat rejection. Next we discuss three ways of accomplishing this for the simple ideal Rankine cycle.

## Lowering the Condenser Pressure (*Lowers* $T_{low,avg}$)

Steam exists as a saturated mixture in the condenser at the saturation temperature corresponding to the pressure inside the condenser. Therefore, lowering the operating pressure of the condenser automatically lowers the temperature of the steam, and thus the temperature at which heat is rejected.

The effect of lowering the condenser pressure on the Rankine cycle efficiency is illustrated on a *T-s* diagram in Fig. 10–6. For comparison purposes, the turbine inlet state is maintained the same. The colored area on this diagram represents the increase in net work output as a result of lowering the condenser pressure from $P_4$ to $P_4'$. The heat input requirements also increase (represented by the area under curve 2′-2), but this increase is very small. Thus the overall effect of lowering the condenser pressure is an increase in the thermal efficiency of the cycle.

To take advantage of the increased efficiencies at low pressures, the condensers of steam power plants usually operate well below the atmospheric pressure. This does not present a major problem since the vapor power cycles operate in a closed loop. However, there is a lower limit on the condenser pressure that can be used. It cannot be lower than the saturation pressure corresponding to the temperature of the cooling medium. Consider, for example, a condenser that is to be cooled by a nearby river at 15°C. Allowing a temperature difference of 10°C for effective heat transfer, the steam temperature in the condenser must be above 25°C; thus the condenser pressure must be above 3.2 kPa, which is the saturation pressure at 25°C.

Lowering the condenser pressure is not without any side effects, however. For one thing, it creates the possibility of air leakage into the condenser. More importantly, it increases the moisture content of the steam at the final stages of the turbine, as can be seen from Fig. 10–6. The presence of large quantities of moisture is highly undesirable in turbines because it decreases the turbine efficiency and erodes the turbine blades. Fortunately, this problem can be corrected, as discussed next.

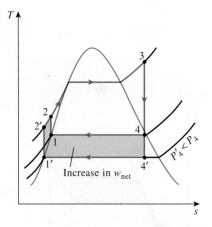

**FIGURE 10–6**

The effect of lowering the condenser pressure on the ideal Rankine cycle.

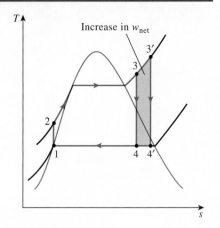

**FIGURE 10–7**

The effect of superheating the steam to higher temperatures on the ideal Rankine cycle.

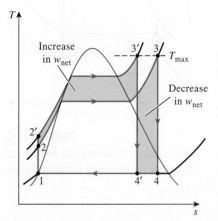

**FIGURE 10–8**

The effect of increasing the boiler pressure on the ideal Rankine cycle.

## Superheating the Steam to High Temperatures (*Increases* $T_{high,avg}$)

The average temperature at which heat is transferred to steam can be increased without increasing the boiler pressure by superheating the steam to high temperatures. The effect of superheating on the performance of vapor power cycles is illustrated on a $T$-$s$ diagram in Fig. 10–7. The colored area on this diagram represents the increase in the net work. The total area under the process curve 3-3′ represents the increase in the heat input. Thus both the net work and heat input increase as a result of superheating the steam to a higher temperature. The overall effect is an increase in thermal efficiency, however, since the average temperature at which heat is added increases.

Superheating the steam to higher temperatures has another very desirable effect: It decreases the moisture content of the steam at the turbine exit, as can be seen from the $T$-$s$ diagram (the quality at state 4′ is higher than that at state 4).

The temperature to which steam can be superheated is limited, however, by metallurgical considerations. Presently the highest steam temperature allowed at the turbine inlet is about 620°C. Any increase in this value depends on improving the present materials or finding new ones that can withstand higher temperatures. Ceramics are very promising in this regard.

## Increasing the Boiler Pressure (*Increases* $T_{high,avg}$)

Another way of increasing the average temperature during the heat-addition process is to increase the operating pressure of the boiler, which automatically raises the temperature at which boiling takes place. This, in turn, raises the average temperature at which heat is transferred to the steam and thus raises the thermal efficiency of the cycle.

The effect of increasing the boiler pressure on the performance of vapor power cycles is illustrated on a $T$-$s$ diagram in Fig. 10–8. Notice that for a fixed turbine inlet temperature, the cycle shifts to the left and the moisture content of steam at the turbine exit increases. This undesirable side effect can be corrected, however, by reheating the steam, as discussed in the next section.

Operating pressures of boilers have gradually increased over the years from about 2.7 MPa in 1922 to over 30 MPa today, generating enough steam to produce a net power output of 1000 MW or more in a large power plant. Today many modern steam power plants operate at supercritical pressures ($P > 22.06$ MPa) and have thermal efficiencies of about 40 percent for fossil-fuel plants and 34 percent for nuclear plants. There are over 150 supercritical-pressure steam power plants in operation in the United States. The lower efficiencies of nuclear power plants are due to the lower maximum temperatures used in those plants for safety reasons. The $T$-$s$ diagram of a supercritical Rankine cycle is shown in Fig. 10–9.

The effects of lowering the condenser pressure, superheating to a higher temperature, and increasing the boiler pressure on the thermal efficiency of the Rankine cycle are illustrated below with an example.

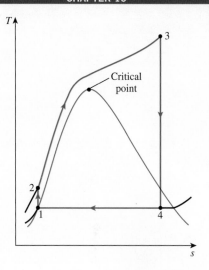

FIGURE 10–9
A supercritical Rankine cycle.

■ **EXAMPLE 10–3**   **Effect of Boiler Pressure and Temperature on Efficiency**

■ Consider a steam power plant operating on the ideal Rankine cycle. Steam enters the turbine at 3 MPa and 350°C and is condensed in the condenser at a pressure of 10 kPa. Determine (*a*) the thermal efficiency of this power plant, (*b*) the thermal efficiency if steam is superheated to 600°C instead of 350°C, and (*c*) the thermal efficiency if the boiler pressure is raised to 15 MPa while the turbine inlet temperature is maintained at 600°C.

**SOLUTION**   A steam power plant operating on the ideal Rankine cycle is considered. The effects of superheating the steam to a higher temperature and raising the boiler pressure on thermal efficiency are to be investigated.
*Analysis*   The *T-s* diagrams of the cycle for all three cases are given in Fig. 10–10.

(*a*) This is the steam power plant discussed in Example 10–1, except that the condenser pressure is lowered to 10 kPa. The thermal efficiency is determined in a similar manner:

*State 1:*   $\left.\begin{array}{l} P_1 = 10 \text{ kPa} \\ \text{Sat. liquid} \end{array}\right\}$   $\begin{array}{l} h_1 = h_{f\,@\,10\,\text{kPa}} = 191.81 \text{ kJ/kg} \\ v_1 = v_{f\,@\,10\,\text{kPa}} = 0.00101 \text{ m}^3\text{/kg} \end{array}$

*State 2:*   $P_2 = 3 \text{ MPa}$
$s_2 = s_1$

$$w_{\text{pump,in}} = v_1(P_2 - P_1) = (0.00101 \text{ m}^3\text{/kg})[(3000 - 10) \text{ kPa}]\left(\frac{1 \text{ kJ}}{1 \text{ kPa·m}^3}\right)$$

$$= 3.02 \text{ kJ/kg}$$

$$h_2 = h_1 + w_{\text{pump,in}} = (191.81 + 3.02) \text{ kJ/kg} = 194.83 \text{ kJ/kg}$$

*State 3:*   $\left.\begin{array}{l} P_3 = 3 \text{ MPa} \\ T_3 = 350°C \end{array}\right\}$   $\begin{array}{l} h_3 = 3116.1 \text{ kJ/kg} \\ s_3 = 6.7450 \text{ kJ/kg·K} \end{array}$

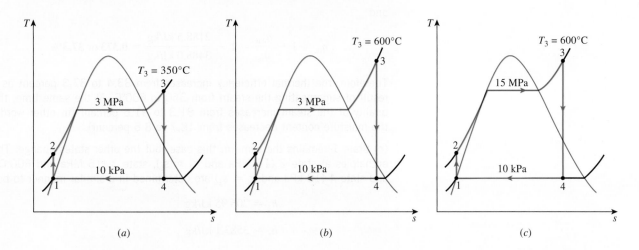

(*a*)          (*b*)          (*c*)

**FIGURE 10–10**
*T-s* diagrams of the three cycles discussed in Example 10–3.

*State 4:* $\quad P_4 = 10 \text{ kPa}$ (sat. mixture)

$$s_4 = s_3$$

$$x_4 = \frac{s_4 - s_f}{s_{fg}} = \frac{6.7450 - 0.6492}{7.4996} = 0.8128$$

Thus,

$$h_4 = h_f + x_4 h_{fg} = 191.81 + 0.8128(2392.1) = 2136.1 \text{ kJ/kg}$$

$$q_{in} = h_3 - h_2 = (3116.1 - 194.83) \text{ kJ/kg} = 2921.3 \text{ kJ/kg}$$

$$q_{out} = h_4 - h_1 = (2136.1 - 191.81) \text{ kJ/kg} = 1944.3 \text{ kJ/kg}$$

and

$$\eta_{th} = 1 - \frac{q_{out}}{q_{in}} = 1 - \frac{1944.3 \text{ kJ/kg}}{2921.3 \text{ kJ/kg}} = \mathbf{0.334} \text{ or } \mathbf{33.4\%}$$

Therefore, the thermal efficiency increases from 26.0 to 33.4 percent as a result of lowering the condenser pressure from 75 to 10 kPa. At the same time, however, the quality of the steam decreases from 88.6 to 81.3 percent (in other words, the moisture content increases from 11.4 to 18.7 percent).

(*b*) States 1 and 2 remain the same in this case, and the enthalpies at state 3 (3 MPa and 600°C) and state 4 (10 kPa and $s_4 = s_3$) are determined to be

$$h_3 = 3682.8 \text{ kJ/kg}$$

$$h_4 = 2380.3 \text{ kJ/kg} \quad (x_4 = 0.915)$$

Thus,

$$q_{in} = h_3 - h_2 = 3682.8 - 194.83 = 3488.0 \text{ kJ/kg}$$

$$q_{out} = h_4 - h_1 = 2380.3 - 191.81 = 2188.5 \text{ kJ/kg}$$

and

$$\eta_{th} = 1 - \frac{q_{out}}{q_{in}} = 1 - \frac{2188.5 \text{ kJ/kg}}{3488.0 \text{ kJ/kg}} = \mathbf{0.373} \text{ or } \mathbf{37.3\%}$$

Therefore, the thermal efficiency increases from 33.4 to 37.3 percent as a result of superheating the steam from 350 to 600°C. At the same time, the quality of the steam increases from 81.3 to 91.5 percent (in other words, the moisture content decreases from 18.7 to 8.5 percent).

(*c*) State 1 remains the same in this case, but the other states change. The enthalpies at state 2 (15 MPa and $s_2 = s_1$), state 3 (15 MPa and 600°C), and state 4 (10 kPa and $s_4 = s_3$) are determined in a similar manner to be

$$h_2 = 206.95 \text{ kJ/kg}$$

$$h_3 = 3583.1 \text{ kJ/kg}$$

$$h_4 = 2115.3 \text{ kJ/kg} \quad (x_4 = 0.804)$$

Thus,

$$q_{in} = h_3 - h_2 = 3583.1 - 206.95 = 3376.2 \text{ kJ/kg}$$

$$q_{out} = h_4 - h_1 = 2115.3 - 191.81 = 1923.5 \text{ kJ/kg}$$

and

$$\eta_{th} = 1 - \frac{q_{out}}{q_{in}} = 1 - \frac{1923.5 \text{ kJ/kg}}{3376.2 \text{ kJ/kg}} = 0.430 \text{ or } 43.0\%$$

**Discussion**   The thermal efficiency increases from 37.3 to 43.0 percent as a result of raising the boiler pressure from 3 to 15 MPa while maintaining the turbine inlet temperature at 600°C. At the same time, however, the quality of the steam decreases from 91.5 to 80.4 percent (in other words, the moisture content increases from 8.5 to 19.6 percent).

## 10–5 · THE IDEAL REHEAT RANKINE CYCLE

We noted in the last section that increasing the boiler pressure increases the thermal efficiency of the Rankine cycle, but it also increases the moisture content of the steam to unacceptable levels. Then it is natural to ask the following question:

*How can we take advantage of the increased efficiencies at higher boiler pressures without facing the problem of excessive moisture at the final stages of the turbine?*

Two possibilities come to mind:

**1.** Superheat the steam to very high temperatures before it enters the turbine. This would be the desirable solution since the average temperature at which heat is added would also increase, thus increasing the cycle efficiency. This is not a viable solution, however, since it requires raising the steam temperature to metallurgically unsafe levels.

**2.** Expand the steam in the turbine in two stages, and reheat it in between. In other words, modify the simple ideal Rankine cycle with a **reheat** process. Reheating is a practical solution to the excessive moisture problem in turbines, and it is commonly used in modern steam power plants.

The *T-s* diagram of the ideal reheat Rankine cycle and the schematic of the power plant operating on this cycle are shown in Fig. 10–11. The ideal reheat Rankine cycle differs from the simple ideal Rankine cycle in that the expansion process takes place in two stages. In the first stage (the high-pressure turbine), steam is expanded isentropically to an intermediate pressure and sent back to the boiler where it is reheated at constant pressure, usually to the inlet temperature of the first turbine stage. Steam then expands isentropically in the second stage (low-pressure turbine) to the condenser pressure. Thus the total heat input and the total turbine work output for a reheat cycle become

$$q_{in} = q_{primary} + q_{reheat} = (h_3 - h_2) + (h_5 - h_4) \qquad \text{(10–12)}$$

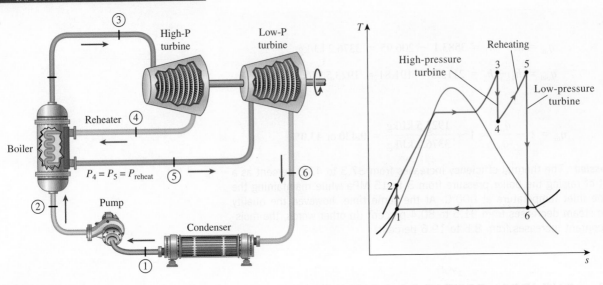

**FIGURE 10–11**
The ideal reheat Rankine cycle.

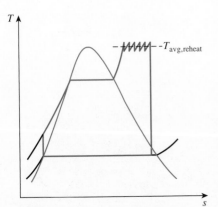

**FIGURE 10–12**
The average temperature at which
heat is transferred during reheating
increases as the number of reheat
stages is increased.

and

$$w_{\text{turb,out}} = w_{\text{turb,I}} + w_{\text{turb,II}} = (h_3 - h_4) + (h_5 - h_6) \quad (10\text{–}13)$$

The incorporation of the single reheat in a modern power plant improves the cycle efficiency by 4 to 5 percent by increasing the average temperature at which heat is transferred to the steam.

The average temperature during the reheat process can be increased by increasing the number of expansion and reheat stages. As the number of stages is increased, the expansion and reheat processes approach an isothermal process at the maximum temperature, as shown in Fig. 10–12. The use of more than two reheat stages, however, is not practical. The theoretical improvement in efficiency from the second reheat is about half of that which results from a single reheat. If the turbine inlet pressure is not high enough, double reheat would result in superheated exhaust. This is undesirable as it would cause the average temperature for heat rejection to increase and thus the cycle efficiency to decrease. Therefore, double reheat is used only on supercritical-pressure ($P > 22.06$ MPa) power plants. A third reheat stage would increase the cycle efficiency by about half of the improvement attained by the second reheat. This gain is too small to justify the added cost and complexity.

The reheat cycle was introduced in the mid-1920s, but it was abandoned in the 1930s because of the operational difficulties. The steady increase in boiler pressures over the years made it necessary to reintroduce single reheat in the late 1940s and double reheat in the early 1950s.

The reheat temperatures are very close or equal to the turbine inlet temperature. The optimum reheat pressure is about one-fourth of the maximum cycle pressure. For example, the optimum reheat pressure for a cycle with a boiler pressure of 12 MPa is about 3 MPa.

Remember that the sole purpose of the reheat cycle is to reduce the moisture content of the steam at the final stages of the expansion process. If we had materials that could withstand sufficiently high temperatures, there would be no need for the reheat cycle.

■ **EXAMPLE 10–4**    **The Ideal Reheat Rankine Cycle**

Consider a steam power plant operating on the ideal reheat Rankine cycle. Steam enters the high-pressure turbine at 15 MPa and 600°C and is condensed in the condenser at a pressure of 10 kPa. If the moisture content of the steam at the exit of the low-pressure turbine is not to exceed 10.4 percent, determine (a) the pressure at which the steam should be reheated and (b) the thermal efficiency of the cycle. Assume the steam is reheated to the inlet temperature of the high-pressure turbine.

**SOLUTION**   A steam power plant operating on the ideal reheat Rankine cycle is considered. For a specified moisture content at the turbine exit, the reheat pressure and the thermal efficiency are to be determined.
**Assumptions** **1** Steady operating conditions exist. **2** Kinetic and potential energy changes are negligible.
**Analysis** The schematic of the power plant and the $T$-$s$ diagram of the cycle are shown in Fig. 10–13. We note that the power plant operates on the ideal reheat Rankine cycle. Therefore, the pump and the turbines are isentropic, there are no pressure drops in the boiler and condenser, and steam leaves the condenser and enters the pump as saturated liquid at the condenser pressure.

(a) The reheat pressure is determined from the requirement that the entropies at states 5 and 6 be the same:

*State 6:*    $P_6 = 10$ kPa
$$x_6 = 0.896 \quad \text{(sat. mixture)}$$
$$s_6 = s_f + x_6 s_{fg} = 0.6492 + 0.896(7.4996) = 7.3688 \text{ kJ/kg·K}$$

Also,

$$h_6 = h_f + x_6 h_{fg} = 191.81 + 0.896(2392.1) = 2335.1 \text{ kJ/kg}$$

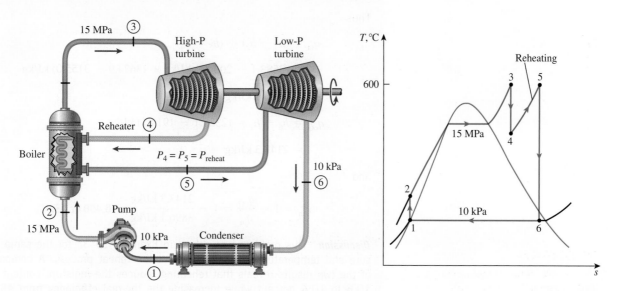

**FIGURE 10–13**
Schematic and $T$-$s$ diagram for Example 10–4.

Thus,

*State 5:* $\left.\begin{array}{l} T_5 = 600°C \\ s_5 = s_6 \end{array}\right\}$ $\begin{array}{l} P_5 = \textbf{4.0 MPa} \\ h_5 = 3674.9 \text{ kJ/kg} \end{array}$

Therefore, steam should be reheated at a pressure of 4 MPa or lower to prevent a moisture content above 10.4 percent.

(*b*) To determine the thermal efficiency, we need to know the enthalpies at all other states:

*State 1:* $\left.\begin{array}{l} P_1 = 10 \text{ kPa} \\ \text{Sat. liquid} \end{array}\right\}$ $\begin{array}{l} h_1 = h_{f @ 10 \text{ kPa}} = 191.81 \text{ kJ/kg} \\ v_1 = v_{f @ 10 \text{ kPa}} = 0.00101 \text{ m}^3/\text{kg} \end{array}$

*State 2:* $\quad P_2 = 15 \text{ MPa}$

$\quad\quad\quad s_2 = s_1$

$w_{\text{pump,in}} = v_1(P_2 - P_1) = (0.00101 \text{ m}^3/\text{kg})$

$$\times [(15,000 - 10)\text{kPa}]\left(\frac{1 \text{ kJ}}{1 \text{ kPa·m}^3}\right)$$

$$= 15.14 \text{ kJ/kg}$$

$$h_2 = h_1 + w_{\text{pump,in}} = (191.81 + 15.14) \text{ kJ/kg} = 206.95 \text{ kJ/kg}$$

*State 3:* $\left.\begin{array}{l} P_3 = 15 \text{ MPa} \\ T_3 = 600°C \end{array}\right\}$ $\begin{array}{l} h_3 = 3583.1 \text{ kJ/kg} \\ s_3 = 6.6796 \text{ kJ/kg·K} \end{array}$

*State 4:* $\left.\begin{array}{l} P_4 = 4 \text{ MPa} \\ s_4 = s_3 \end{array}\right\}$ $\begin{array}{l} h_4 = 3155.0 \text{ kJ/kg} \\ (T_4 = 375.5°C) \end{array}$

Thus

$$q_{\text{in}} = (h_3 - h_2) + (h_5 - h_4)$$

$$= (3583.1 - 206.95) \text{ kJ/kg} + (3674.9 - 3155.0) \text{ kJ/kg}$$

$$= 3896.1 \text{ kJ/kg}$$

$$q_{\text{out}} = h_6 - h_1 = (2335.1 - 191.81) \text{ kJ/kg}$$

$$= 2143.3 \text{ kJ/kg}$$

and

$$\eta_{\text{th}} = 1 - \frac{q_{\text{out}}}{q_{\text{in}}} = 1 - \frac{2143.3 \text{ kJ/kg}}{3896.1 \text{ kJ/kg}} = \textbf{0.450 or 45.0\%}$$

***Discussion*** This problem was solved in Example 10–3*c* for the same pressure and temperature limits but without the reheat process. A comparison of the two results reveals that reheating reduces the moisture content from 19.6 to 10.4 percent while increasing the thermal efficiency from 43.0 to 45.0 percent.

# 10–6 ▪ THE IDEAL REGENERATIVE RANKINE CYCLE

A careful examination of the *T-s* diagram of the Rankine cycle redrawn in Fig. 10–14 reveals that heat is transferred to the working fluid during process 2-2′ at a relatively low temperature. This lowers the average heat-addition temperature and thus the cycle efficiency.

To remedy this shortcoming, we look for ways to raise the temperature of the liquid leaving the pump (called the *feedwater*) before it enters the boiler. One such possibility is to transfer heat to the feedwater from the expanding steam in a counterflow heat exchanger built into the turbine, that is, to use **regeneration**. This solution is also impractical because it is difficult to design such a heat exchanger and because it would increase the moisture content of the steam at the final stages of the turbine.

A practical regeneration process in steam power plants is accomplished by extracting, or "bleeding," steam from the turbine at various points. This steam, which could have produced more work by expanding further in the turbine, is used to heat the feedwater instead. The device where the feedwater is heated by regeneration is called a **regenerator**, or a **feedwater heater (FWH)**.

Regeneration not only improves cycle efficiency, but also provides a convenient means of deaerating the feedwater (removing the air that leaks in at the condenser) to prevent corrosion in the boiler. It also helps control the large volume flow rate of the steam at the final stages of the turbine (due to the large specific volumes at low pressures). Therefore, regeneration has been used in all modern steam power plants since its introduction in the early 1920s.

A feedwater heater is basically a heat exchanger where heat is transferred from the steam to the feedwater either by mixing the two fluid streams (open feedwater heaters) or without mixing them (closed feedwater heaters). Regeneration with both types of feedwater heaters is discussed below.

## Open Feedwater Heaters

An **open** (or **direct-contact**) **feedwater heater** is basically a *mixing chamber*, where the steam extracted from the turbine mixes with the feedwater exiting the pump. Ideally, the mixture leaves the heater as a saturated liquid at the heater pressure. The schematic of a steam power plant with one open feedwater heater (also called *single-stage regenerative cycle*) and the *T-s* diagram of the cycle are shown in Fig. 10–15.

In an ideal regenerative Rankine cycle, steam enters the turbine at the boiler pressure (state 5) and expands isentropically to an intermediate pressure (state 6). Some steam is extracted at this state and routed to the feedwater heater, while the remaining steam continues to expand isentropically to the condenser pressure (state 7). This steam leaves the condenser as a saturated liquid at the condenser pressure (state 1). The condensed water, which is also called the *feedwater,* then enters an isentropic pump, where it is compressed to the feedwater heater pressure (state 2) and is routed to the feedwater heater, where it mixes with the steam extracted from the turbine. The fraction of the steam extracted is such that the mixture leaves the heater as a saturated liquid at the heater pressure (state 3). A second pump raises the pressure of the water to the boiler pressure (state 4). The cycle is completed by heating the water in the boiler to the turbine inlet state (state 5).

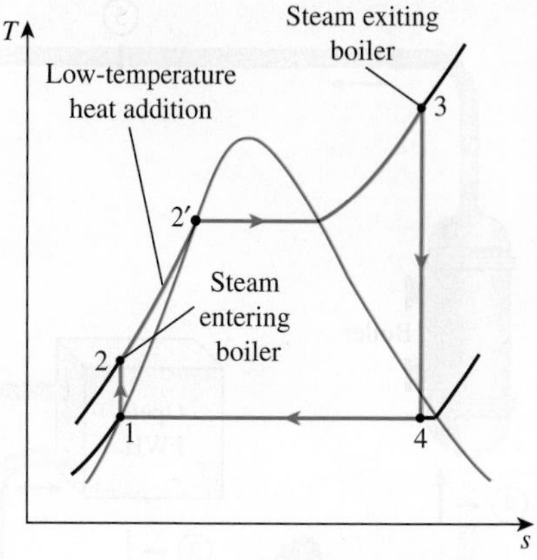

**FIGURE 10–14**

The first part of the heat-addition process in the boiler takes place at relatively low temperatures.

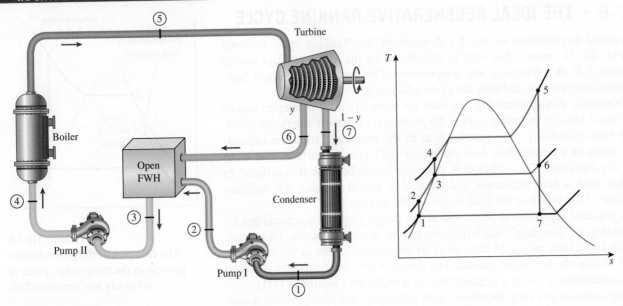

**FIGURE 10–15**
The ideal regenerative Rankine cycle with an open feedwater heater.

In the analysis of steam power plants, it is more convenient to work with quantities expressed per unit mass of the steam flowing through the boiler. For each 1 kg of steam leaving the boiler, $y$ kg expands partially in the turbine and is extracted at state 6. The remaining $(1 - y)$ kg expands completely to the condenser pressure. Therefore, the mass flow rates are different in different components. If the mass flow rate through the boiler is $\dot{m}$, for example, it is $(1 - y)\dot{m}$ through the condenser. This aspect of the regenerative Rankine cycle should be considered in the analysis of the cycle as well as in the interpretation of the areas on the $T$-$s$ diagram. In light of Fig. 10–15, the heat and work interactions of a regenerative Rankine cycle with one feedwater heater can be expressed per unit mass of steam flowing through the boiler as follows:

$$q_{in} = h_5 - h_4 \tag{10-14}$$

$$q_{out} = (1 - y)(h_7 - h_1) \tag{10-15}$$

$$w_{turb,out} = (h_5 - h_6) + (1 - y)(h_6 - h_7) \tag{10-16}$$

$$w_{pump,in} = (1 - y)w_{pump\,I,in} + w_{pump\,II,in} \tag{10-17}$$

where

$$y = \dot{m}_6/\dot{m}_5 \quad \text{(fraction of steam extracted)}$$

$$w_{pump\,I,in} = \upsilon_1(P_2 - P_1)$$

$$w_{pump\,II,in} = \upsilon_3(P_4 - P_3)$$

The thermal efficiency of the Rankine cycle increases as a result of regeneration. This is because regeneration raises the average temperature at which heat is transferred to the steam in the boiler by raising the temperature of the water before it enters the boiler. The cycle efficiency increases further as the number of feedwater heaters is increased. Many large plants in operation today use as many as eight feedwater heaters. The optimum number of feedwater heaters is determined from economical considerations. The use of an additional feedwater heater cannot be justified unless it saves more from the fuel costs than its own cost.

## Closed Feedwater Heaters

Another type of feedwater heater frequently used in steam power plants is the **closed feedwater heater**, in which heat is transferred from the extracted steam to the feedwater without any mixing taking place. The two streams now can be at different pressures, since they do not mix. The schematic of a steam power plant with one closed feedwater heater and the *T-s* diagram of the cycle are shown in Fig. 10–16. In an ideal closed feedwater heater, the feedwater is heated to the exit temperature of the extracted steam, which ideally leaves the heater as a saturated liquid at the extraction pressure. In actual power plants, the feedwater leaves the heater below the exit temperature of the extracted steam because a temperature difference of at least a few degrees is required for any effective heat transfer to take place.

The condensed steam is then either pumped to the feedwater line or routed to another heater or to the condenser through a device called a **trap**. A trap allows the liquid to be throttled to a lower pressure region but *traps* the vapor. The enthalpy of steam remains constant during this throttling process.

The open and closed feedwater heaters can be compared as follows. Open feedwater heaters are simple and inexpensive and have good heat

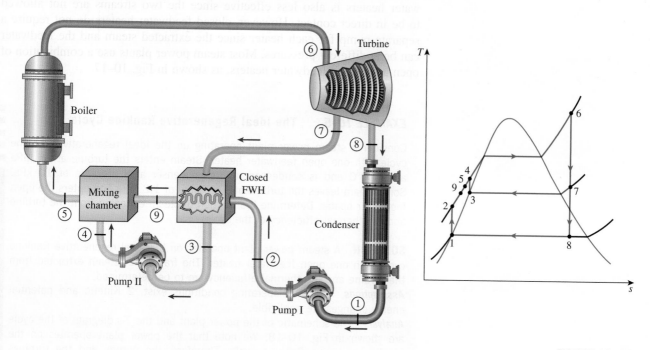

**FIGURE 10–16**

The ideal regenerative Rankine cycle with a closed feedwater heater.

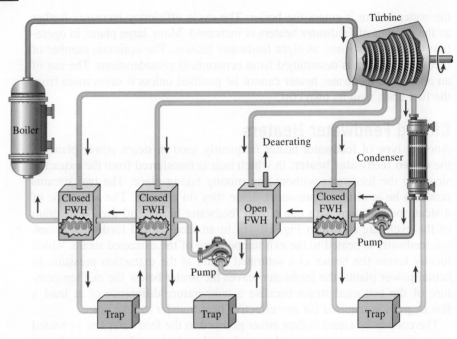

**FIGURE 10–17**
A steam power plant with one open and three closed feedwater heaters.

transfer characteristics. They also bring the feedwater to the saturation state. For each heater, however, a pump is required to handle the feedwater. The closed feedwater heaters are more complex because of the internal tubing network, and thus they are more expensive. Heat transfer in closed feedwater heaters is also less effective since the two streams are not allowed to be in direct contact. However, closed feedwater heaters do not require a separate pump for each heater since the extracted steam and the feedwater can be at different pressures. Most steam power plants use a combination of open and closed feedwater heaters, as shown in Fig. 10–17.

---

*EXAMPLE 10–5*  **The Ideal Regenerative Rankine Cycle**

Consider a steam power plant operating on the ideal regenerative Rankine cycle with one open feedwater heater. Steam enters the turbine at 15 MPa and 600°C and is condensed in the condenser at a pressure of 10 kPa. Some steam leaves the turbine at a pressure of 1.2 MPa and enters the open feedwater heater. Determine the fraction of steam extracted from the turbine and the thermal efficiency of the cycle.

**SOLUTION**  A steam power plant operates on the ideal regenerative Rankine cycle with one open feedwater heater. The fraction of steam extracted from the turbine and the thermal efficiency are to be determined.
*Assumptions*  **1** Steady operating conditions exist. **2** Kinetic and potential energy changes are negligible.
*Analysis*  The schematic of the power plant and the *T-s* diagram of the cycle are shown in Fig. 10–18. We note that the power plant operates on the ideal regenerative Rankine cycle. Therefore, the pumps and the turbines

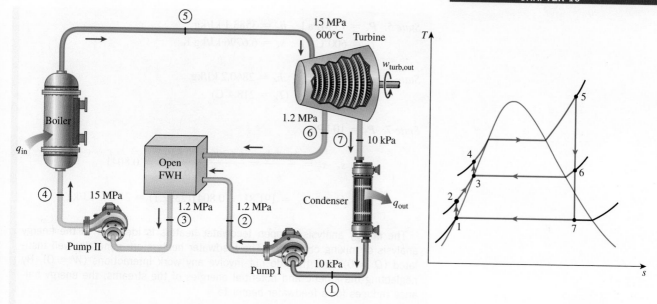

**FIGURE 10–18**
Schematic and $T$-$s$ diagram for
Example 10–5.

are isentropic; there are no pressure drops in the boiler, condenser, and
feedwater heater; and steam leaves the condenser and the feedwater heater
as saturated liquid. First, we determine the enthalpies at various states:

*State 1:* $\quad P_1 = 10 \text{ kPa}$ $\Big\}$ $\quad h_1 = h_{f \text{ @ } 10 \text{ kPa}} = 191.81 \text{ kJ/kg}$
$\qquad\qquad$ Sat. liquid $\qquad v_1 = v_{f \text{ @ } 10 \text{ kPa}} = 0.00101 \text{ m}^3/\text{kg}$

*State 2:* $\quad P_2 = 1.2 \text{ MPa}$

$\qquad\qquad s_2 = s_1$

$\qquad w_{\text{pump I,in}} = v_1(P_2 - P_1) = (0.00101 \text{ m}^3/\text{kg})[(1200 - 10) \text{ kPa}]\left(\dfrac{1 \text{ kJ}}{1 \text{ kPa}\cdot\text{m}^3}\right)$

$\qquad\qquad\quad = 1.20 \text{ kJ/kg}$

$\qquad\quad h_2 = h_1 + w_{\text{pump I,in}} = (191.81 + 1.20) \text{ kJ/kg} = 193.01 \text{ kJ/kg}$

*State 3:* $\quad P_3 = 1.2 \text{ MPa}$ $\Big\}$ $\quad v_3 = v_{f \text{ @ } 1.2 \text{ MPa}} = 0.001138 \text{ m}^3/\text{kg}$
$\qquad\qquad$ Sat. liquid $\qquad h_3 = h_{f \text{ @ } 1.2 \text{ MPa}} = 798.33 \text{ kJ/kg}$

*State 4:* $\quad P_4 = 15 \text{ MPa}$

$\qquad\qquad s_4 = s_3$

$\qquad w_{\text{pump II,in}} = v_3(P_4 - P_3)$

$\qquad\qquad\quad = (0.001138 \text{ m}^3/\text{kg})[(15{,}000 - 1200) \text{ kPa}]\left(\dfrac{1 \text{ kJ}}{1 \text{ kPa}\cdot\text{m}^3}\right)$

$\qquad\qquad\quad = 15.70 \text{ kJ/kg}$

$\qquad\quad h_4 = h_3 + w_{\text{pump II,in}} = (798.33 + 15.70) \text{ kJ/kg} = 814.03 \text{ kJ/kg}$

*State 5:* $\left. \begin{array}{l} P_5 = 15 \text{ MPa} \\ T_5 = 600°\text{C} \end{array} \right\}$ $\begin{array}{l} h_5 = 3583.1 \text{ kJ/kg} \\ s_5 = 6.6796 \text{ kJ/kg·K} \end{array}$

*State 6:* $\left. \begin{array}{l} P_6 = 1.2 \text{ MPa} \\ s_6 = s_5 \end{array} \right\}$ $\begin{array}{l} h_6 = 2860.2 \text{ kJ/kg} \\ (T_6 = 218.4°\text{C}) \end{array}$

*State 7:* $P_7 = 10 \text{ kPa}$

$$s_7 = s_5 \quad x_7 = \frac{s_7 - s_f}{s_{fg}} = \frac{6.6796 - 0.6492}{7.4996} = 0.8041$$

$$h_7 = h_f + x_7 h_{fg} = 191.81 + 0.8041(2392.1) = 2115.3 \text{ kJ/kg}$$

The energy analysis of open feedwater heaters is identical to the energy analysis of mixing chambers. The feedwater heaters are generally well insulated ($\dot{Q} = 0$), and they do not involve any work interactions ($\dot{W} = 0$). By neglecting the kinetic and potential energies of the streams, the energy balance reduces for a feedwater heater to

$$\dot{E}_{\text{in}} = \dot{E}_{\text{out}} \longrightarrow \sum_{\text{in}} \dot{m} h = \sum_{\text{out}} \dot{m} h$$

or

$$y h_6 + (1 - y) h_2 = 1(h_3)$$

where $y$ is the fraction of steam extracted from the turbine ($= \dot{m}_6 / \dot{m}_5$). Solving for $y$ and substituting the enthalpy values, we find

$$y = \frac{h_3 - h_2}{h_6 - h_2} = \frac{798.33 - 193.01}{2860.2 - 193.01} = \textbf{0.2270}$$

Thus,

$$q_{\text{in}} = h_5 - h_4 = (3583.1 - 814.03) \text{ kJ/kg} = 2769.1 \text{ kJ/kg}$$

$$q_{\text{out}} = (1 - y)(h_7 - h_1) = (1 - 0.2270)(2115.3 - 191.81) \text{ kJ/kg}$$

$$= 1486.9 \text{ kJ/kg}$$

and

$$\eta_{\text{th}} = 1 - \frac{q_{\text{out}}}{q_{\text{in}}} = 1 - \frac{1486.9 \text{ kJ/kg}}{2769.1 \text{ kJ/kg}} = \textbf{0.463 or 46.3\%}$$

***Discussion*** This problem was worked out in Example 10–3c for the same pressure and temperature limits but without the regeneration process. A comparison of the two results reveals that the thermal efficiency of the cycle has increased from 43.0 to 46.3 percent as a result of regeneration. The net work output decreased by 171 kJ/kg, but the heat input decreased by 607 kJ/kg, which results in a net increase in the thermal efficiency.

### EXAMPLE 10–6    The Ideal Reheat-Regenerative Rankine Cycle

Consider a steam power plant that operates on an ideal reheat–regenerative Rankine cycle with one open feedwater heater, one closed feedwater heater, and one reheater. Steam enters the turbine at 15 MPa and 600°C and is condensed in the condenser at a pressure of 10 kPa. Some steam is extracted from the turbine at 4 MPa for the closed feedwater heater, and the remaining steam is reheated at the same pressure to 600°C. The extracted steam is completely condensed in the heater and is pumped to 15 MPa before it mixes with the feedwater at the same pressure. Steam for the open feedwater heater is extracted from the low-pressure turbine at a pressure of 0.5 MPa. Determine the fractions of steam extracted from the turbine as well as the thermal efficiency of the cycle.

**SOLUTION**   A steam power plant operates on the ideal reheat–regenerative Rankine cycle with one open feedwater heater, one closed feedwater heater, and one reheater. The fractions of steam extracted from the turbine and the thermal efficiency are to be determined.

*Assumptions*   **1** Steady operating conditions exist. **2** Kinetic and potential energy changes are negligible. **3** In both open and closed feedwater heaters, feedwater is heated to the saturation temperature at the feedwater heater pressure. (Note that this is a conservative assumption since extracted steam enters the closed feedwater heater at 376°C and the saturation temperature at the closed feedwater pressure of 4 MPa is 250°C).

*Analysis*   The schematic of the power plant and the *T-s* diagram of the cycle are shown in Fig. 10–19. The power plant operates on the ideal reheat–regenerative Rankine cycle and thus the pumps and the turbines are isentropic; there are no pressure drops in the boiler, reheater, condenser,

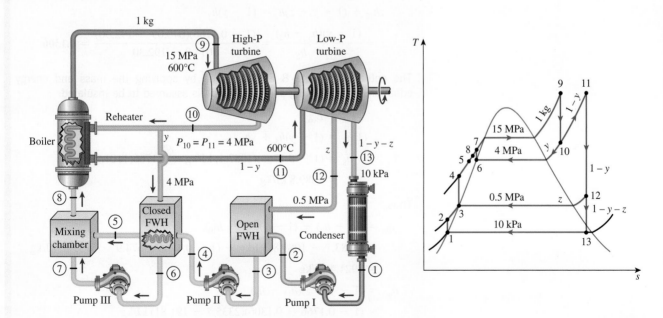

**FIGURE 10–19**
Schematic and *T-s* diagram for Example 10–6.

and feedwater heaters; and steam leaves the condenser and the feedwater heaters as saturated liquid.

The enthalpies at the various states and the pump work per unit mass of fluid flowing through them are

$$h_1 = 191.81 \text{ kJ/kg} \qquad\qquad h_9 = 3155.0 \text{ kJ/kg}$$
$$h_2 = 192.30 \text{ kJ/kg} \qquad\qquad h_{10} = 3155.0 \text{ kJ/kg}$$
$$h_3 = 640.09 \text{ kJ/kg} \qquad\qquad h_{11} = 3674.9 \text{ kJ/kg}$$
$$h_4 = 643.92 \text{ kJ/kg} \qquad\qquad h_{12} = 3014.8 \text{ kJ/kg}$$
$$h_5 = 1087.4 \text{ kJ/kg} \qquad\qquad h_{13} = 2335.7 \text{ kJ/kg}$$
$$h_6 = 1087.4 \text{ kJ/kg} \qquad\qquad w_{\text{pump I,in}} = 0.49 \text{ kJ/kg}$$
$$h_7 = 1101.2 \text{ kJ/kg} \qquad\qquad w_{\text{pump II,in}} = 3.83 \text{ kJ/kg}$$
$$h_8 = 1089.8 \text{ kJ/kg} \qquad\qquad w_{\text{pump III,in}} = 13.77 \text{ kJ/kg}$$

The fractions of steam extracted are determined from the mass and energy balances of the feedwater heaters:

*Closed feedwater heater:*

$$\dot{E}_{\text{in}} = \dot{E}_{\text{out}}$$
$$yh_{10} + (1 - y)h_4 = (1 - y)h_5 + yh_6$$
$$y = \frac{h_5 - h_4}{(h_{10} - h_6) + (h_5 - h_4)} = \frac{1087.4 - 643.92}{(3155.0 - 1087.4) + (1087.4 - 643.92)} = \mathbf{0.1766}$$

*Open feedwater heater:*

$$\dot{E}_{\text{in}} = \dot{E}_{\text{out}}$$
$$zh_{12} + (1 - y - z)h_2 = (1 - y)h_3$$
$$z = \frac{(1 - y)(h_3 - h_2)}{h_{12} - h_2} = \frac{(1 - 0.1766)(640.09 - 192.30)}{3014.8 - 192.30} = \mathbf{0.1306}$$

The enthalpy at state 8 is determined by applying the mass and energy equations to the mixing chamber, which is assumed to be insulated:

$$\dot{E}_{\text{in}} = \dot{E}_{\text{out}}$$
$$(1)h_8 = (1 - y)h_5 + yh_7$$
$$h_8 = (1 - 0.1766)(1087.4) \text{ kJ/kg} + 0.1766(1101.2) \text{ kJ/kg}$$
$$= 1089.8 \text{ kJ/kg}$$

Thus,

$$q_{\text{in}} = (h_9 - h_8) + (1 - y)(h_{11} - h_{10})$$
$$= (3583.1 - 1089.8) \text{ kJ/kg} + (1 - 0.1766)(3674.9 - 3155.0) \text{ kJ/kg}$$
$$= 2921.4 \text{ kJ/kg}$$

$$q_{\text{out}} = (1 - y - z)(h_{13} - h_1)$$
$$= (1 - 0.1766 - 0.1306)(2335.7 - 191.81) \text{ kJ/kg}$$
$$= 1485.3 \text{ kJ/kg}$$

and

$$\eta_{th} = 1 - \frac{q_{out}}{q_{in}} = 1 - \frac{1485.3 \text{ kJ/kg}}{2921.4 \text{ kJ/kg}} = 0.492 \text{ or } 49.2\%$$

**Discussion** This problem was worked out in Example 10–4 for the same pressure and temperature limits with reheat but without the regeneration process. A comparison of the two results reveals that the thermal efficiency of the cycle has increased from 45.0 to 49.2 percent as a result of regeneration.

The thermal efficiency of this cycle could also be determined from

$$\eta_{th} = \frac{w_{net}}{q_{in}} = \frac{w_{turb,out} - w_{pump,in}}{q_{in}}$$

where

$$w_{turb,out} = (h_9 - h_{10}) + (1 - y)(h_{11} - h_{12}) + (1 - y - z)(h_{12} - h_{13})$$

$$w_{pump,in} = (1 - y - z)w_{pump\,I,in} + (1 - y)w_{pump\,II,in} + (y)w_{pump\,III,in}$$

Also, if we assume that the feedwater leaves the closed FWH as a saturated liquid at 15 MPa (and thus at $T_5 = 342°C$ and $h_5 = 1610.3 \text{ kJ/kg}$), it can be shown that the thermal efficiency would be 50.6 percent.

# 10–7 · SECOND-LAW ANALYSIS OF VAPOR POWER CYCLES

The ideal Carnot cycle is a *totally reversible cycle,* and thus it does not involve any irreversibilities. The ideal Rankine cycles (simple, reheat, or regenerative), however, are only *internally reversible,* and they may involve irreversibilities external to the system, such as heat transfer through a finite temperature difference. A second-law analysis of these cycles reveals where the largest irreversibilities occur and what their magnitudes are.

Relations for exergy and exergy destruction for steady-flow systems are developed in Chap. 8. The exergy destruction for a steady-flow system can be expressed, in the rate form, as

$$\dot{X}_{dest} = T_0 \dot{S}_{gen} = T_0(\dot{S}_{out} - \dot{S}_{in}) = T_0 \left( \sum_{out} \dot{m}s + \frac{\dot{Q}_{out}}{T_{b,out}} - \sum_{in} \dot{m}s - \frac{\dot{Q}_{in}}{T_{b,in}} \right) \quad (kW)$$

**(10–18)**

or on a unit mass basis for a one-inlet, one-exit, steady-flow device as

$$x_{dest} = T_0 s_{gen} = T_0 \left( s_e - s_i + \frac{q_{out}}{T_{b,out}} - \frac{q_{in}}{T_{b,in}} \right) \quad (kJ/kg)$$

**(10–19)**

where $T_{b,in}$ and $T_{b,out}$ are the temperatures of the system boundary where heat is transferred into and out of the system, respectively.

The exergy destruction associated with a *cycle* depends on the magnitude of the heat transfer with the high- and low-temperature reservoirs involved, and their temperatures. It can be expressed on a unit mass basis as

$$x_{dest} = T_0 \left( \sum \frac{q_{out}}{T_{b,out}} - \sum \frac{q_{in}}{T_{b,in}} \right) \quad (kJ/kg)$$

**(10–20)**

For a cycle that involves heat transfer only with a source at $T_H$ and a sink at $T_L$, the exergy destruction become

$$x_{dest} = T_0\left(\frac{q_{out}}{T_L} - \frac{q_{in}}{T_H}\right) \quad (kJ/kg) \qquad (10\text{--}21)$$

The exergy of a fluid stream $\psi$ at any state can be determined from

$$\psi = (h - h_0) - T_0(s - s_0) + \frac{V^2}{2} + gz \quad (kJ/kg) \qquad (10\text{--}22)$$

where the subscript "0" denotes the state of the surroundings.

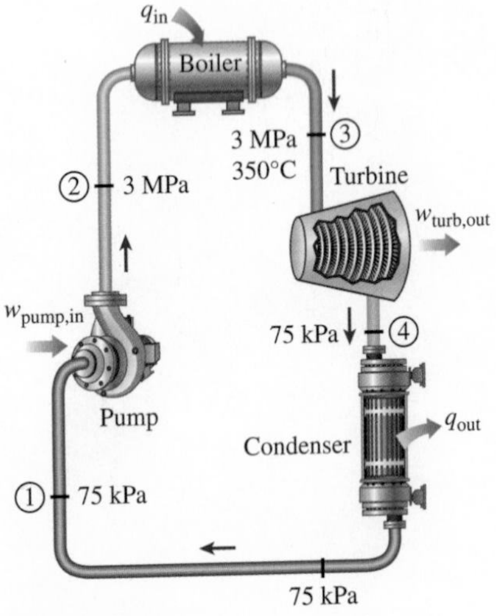

**FIGURE 10–20**
Schematic for Example 10–7.

### EXAMPLE 10–7 Second-Law Analysis of an Ideal Rankine Cycle

Consider a steam power plant operating on the simple ideal Rankine cycle (Fig. 10–20). Steam enters the turbine at 3 MPa and 350°C and is condensed in the condenser at a pressure of 75 kPa. Heat is supplied to the steam in a furnace maintained at 800 K, and waste heat is rejected to the surroundings at 300 K. Determine (a) the exergy destruction associated with each of the four processes and the whole cycle and (b) the second-law efficiency of this cycle.

**SOLUTION** A steam power plant operating on the simple ideal Rankine cycle is considered. For specified source and sink temperatures, the exergy destruction associated with this cycle and the second-law efficiency are to be determined.

**Assumptions** 1 Steady operating conditions exist. 2 Kinetic and potential energy changes are negligible.

**Analysis** We take the power plant bordering the furnace at temperature $T_H$ and the environment at temperature $T_0$ as the control volume. This cycle was analyzed in Example 10–1, and various quantities were determined to be $q_{in} = 2729$ kJ/kg, $w_{pump,in} = 3.0$ kJ/kg, $w_{turb,out} = 713$ kJ/kg, $q_{out} = 2019$ kJ/kg, and $\eta_{th} = 26.0$ percent.

(a) Processes 1-2 and 3-4 are isentropic ($s_1 = s_2$, $s_3 = s_4$) and therefore do not involve any internal or external irreversibilities, that is,

$$x_{dest,12} = 0 \quad \text{and} \quad x_{dest,34} = 0$$

Processes 2-3 and 4-1 are constant-pressure heat-addition and heat-rejection processes, respectively, and they are internally reversible. But the heat transfer between the working fluid and the source or the sink takes place through a finite temperature difference, rendering both processes irreversible. The irreversibility associated with each process is determined from Eq. 10–19. The entropy of the steam at each state is determined from the steam tables:

$$s_2 = s_1 = s_{f\ @\ 75\ kPa} = 1.2132\ kJ/kg{\cdot}K$$

$$s_4 = s_3 = 6.7450\ kJ/kg{\cdot}K \quad \text{(at 3 MPa, 350°C)}$$

Thus,

$$x_{dest,23} = T_0\left(s_3 - s_2 - \frac{q_{in,23}}{T_{source}}\right)$$

$$= (300\ K)\left[(6.7450 - 1.2132)\ kJ/kg{\cdot}K - \frac{2729\ kJ/kg}{800\ K}\right]$$

$$= 636\ kJ/kg$$

$$x_{\text{dest},41} = T_0\left(s_1 - s_4 + \frac{q_{\text{out},41}}{T_{\text{sink}}}\right)$$

$$= (300 \text{ K})\left[(1.2132 - 6.7450) \text{ kJ/kg·K} + \frac{2019 \text{ kJ/kg}}{300 \text{ K}}\right]$$

$$= 360 \text{ kJ/kg}$$

Therefore, the irreversibility of the cycle is

$$x_{\text{dest,cycle}} = x_{\text{dest},12} + x_{\text{dest},23} + x_{\text{dest},34} + x_{\text{dest},41}$$

$$= 0 + 636 \text{ kJ/kg} + 0 + 360 \text{ kJ/kg}$$

$$= 996 \text{ kJ/kg}$$

The total exergy destroyed during the cycle could also be determined from Eq. 10–21. Notice that the largest exergy destruction in the cycle occurs during the heat-addition process. Therefore, any attempt to reduce the exergy destruction should start with this process. Raising the turbine inlet temperature of the steam, for example, would reduce the temperature difference and thus the exergy destruction.

(b) The second-law efficiency is defined as

$$\eta_{\text{II}} = \frac{\text{Exergy recovered}}{\text{Exergy expended}} = \frac{x_{\text{recovered}}}{x_{\text{expended}}} = 1 - \frac{x_{\text{destroyed}}}{x_{\text{expended}}}$$

Here the expended exergy is the exergy content of the heat supplied to steam in boiler (which is its work potential) and the pump input, and the exergy recovered is the work output of the turbine:

$$x_{\text{heat,in}} = \left(1 - \frac{T_0}{T_H}\right)q_{\text{in}} = \left(1 - \frac{300 \text{ K}}{800 \text{ K}}\right)(2729 \text{ kJ/kg}) = 1706 \text{ kJ/kg}$$

$$x_{\text{expended}} = x_{\text{heat,in}} + x_{\text{pump,in}} = 1706 + 3.0 = 1709 \text{ kJ/kg}$$

$$x_{\text{recovered}} = w_{\text{turbine,out}} = 713 \text{ kJ/kg}$$

Substituting, the second-law efficiency of this power plant is determined to be

$$\eta_{\text{II}} = \frac{x_{\text{recovered}}}{x_{\text{expended}}} = \frac{713 \text{ kJ/kg}}{1709 \text{ kJ/kg}} = 0.417 \quad \text{or} \quad 41.7\%$$

**Discussion** The second-law efficiency can also be determined using the exergy destruction data,

$$\eta_{\text{II}} = 1 - \frac{x_{\text{destroyed}}}{x_{\text{expended}}} = 1 - \frac{996 \text{ kJ/kg}}{1709 \text{ kJ/kg}} = 0.417 \quad \text{or} \quad 41.7\%$$

Also, the system considered contains both the furnace and the condenser, and thus the exergy destruction associated with heat transfer involving both the furnace and the condenser are accounted for.

# 10–8 · COGENERATION

In all the cycles discussed so far, the sole purpose was to convert a portion of the heat transferred to the working fluid to work, which is the most valuable form of energy. The remaining portion of the heat is rejected to

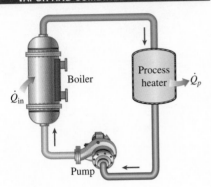

**FIGURE 10–21**
A simple process-heating plant.

rivers, lakes, oceans, or the atmosphere as waste heat, because its quality (or grade) is too low to be of any practical use. Wasting a large amount of heat is a price we have to pay to produce work, because electrical or mechanical work is the only form of energy on which many engineering devices (such as a fan) can operate.

Many systems or devices, however, require energy input in the form of heat, called *process heat*. Some industries that rely heavily on process heat are chemical, pulp and paper, oil production and refining, steel making, food processing, and textile industries. Process heat in these industries is usually supplied by steam at 5 to 7 atm and 150 to 200°C. Energy is usually transferred to the steam by burning coal, oil, natural gas, or another fuel in a furnace.

Now let us examine the operation of a process-heating plant closely. Disregarding any heat losses in the piping, all the heat transferred to the steam in the boiler is used in the process-heating units, as shown in Fig. 10–21. Therefore, process heating seems like a perfect operation with practically no waste of energy. From the second-law point of view, however, things do not look so perfect. The temperature in furnaces is typically very high (around 1400°C), and thus the energy in the furnace is of very high quality. This high-quality energy is transferred to water to produce steam at about 200°C or below (a highly irreversible process). Associated with this irreversibility is, of course, a loss in exergy or work potential. It is simply not wise to use high-quality energy to accomplish a task that could be accomplished with low-quality energy.

Industries that use large amounts of process heat also consume a large amount of electric power. Therefore, it makes economical as well as engineering sense to use the already-existing work potential to produce power instead of letting it go to waste. The result is a plant that produces electricity while meeting the process-heat requirements of certain industrial processes. Such a plant is called a *cogeneration plant*. In general, **cogeneration** is *the production of more than one useful form of energy (such as process heat and electric power) from the same energy source*.

Either a steam-turbine (Rankine) cycle or a gas-turbine (Brayton) cycle or even a combined cycle (discussed later) can be used as the power cycle in a cogeneration plant. The schematic of an ideal steam-turbine cogeneration plant is shown in Fig. 10–22. Let us say this plant is to supply process heat $\dot{Q}_p$ at 500 kPa at a rate of 100 kW. To meet this demand, steam is expanded in the turbine to a pressure of 500 kPa, producing power at a rate of, say, 20 kW. The flow rate of the steam can be adjusted such that steam leaves the process-heating section as a saturated liquid at 500 kPa. Steam is then pumped to the boiler pressure and is heated in the boiler to state 3. The pump work is usually very small and can be neglected. Disregarding any heat losses, the rate of heat input in the boiler is determined from an energy balance to be 120 kW.

Probably the most striking feature of the ideal steam-turbine cogeneration plant shown in Fig. 10–22 is the absence of a condenser. Thus no heat is rejected from this plant as waste heat. In other words, all the energy transferred to the steam in the boiler is utilized as either process heat or electric power. Thus it is appropriate to define a **utilization factor** $\epsilon_u$ for a cogeneration plant as

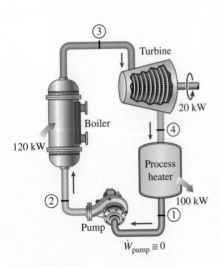

**FIGURE 10–22**
An ideal cogeneration plant.

$$\epsilon_u = \frac{\text{Net power output} + \text{Process heat delivered}}{\text{Total heat input}} = \frac{\dot{W}_{net} + \dot{Q}_p}{\dot{Q}_{in}} \quad (10\text{–}23)$$

or

$$\epsilon_u = 1 - \frac{\dot{Q}_{out}}{\dot{Q}_{in}} \qquad (10\text{-}24)$$

where $\dot{Q}_{out}$ represents the heat rejected in the condenser. Strictly speaking, $\dot{Q}_{out}$ also includes all the undesirable heat losses from the piping and other components, but they are usually small and thus neglected. It also includes combustion inefficiencies such as incomplete combustion and stack losses when the utilization factor is defined on the basis of the heating value of the fuel. The utilization factor of the ideal steam-turbine cogeneration plant is obviously 100 percent. Actual cogeneration plants have utilization factors as high as 80 percent. Some recent cogeneration plants have even higher utilization factors.

Notice that without the turbine, we would need to supply heat to the steam in the boiler at a rate of only 100 kW instead of at 120 kW. The additional 20 kW of heat supplied is converted to work. Therefore, a cogeneration power plant is equivalent to a process-heating plant combined with a power plant that has a thermal efficiency of 100 percent.

The ideal steam-turbine cogeneration plant described above is not practical because it cannot adjust to the variations in power and process-heat loads. The schematic of a more practical (but more complex) cogeneration plant is shown in Fig. 10–23. Under normal operation, some steam is extracted from the turbine at some predetermined intermediate pressure $P_6$. The rest of the steam expands to the condenser pressure $P_7$ and is then cooled at constant pressure. The heat rejected from the condenser represents the waste heat for the cycle.

At times of high demand for process heat, all the steam is routed to the process-heating units and none to the condenser ($\dot{m}_7 = 0$). The waste heat is zero in this mode. If this is not sufficient, some steam leaving the boiler is throttled by an expansion or pressure-reducing valve (PRV) to the extraction pressure $P_6$ and is directed to the process-heating unit. Maximum process heating is realized when all the steam leaving the boiler passes through the expansion valve ($\dot{m}_5 = \dot{m}_4$). No power is produced in this mode. When there is no demand for process heat, all the steam passes through the turbine and the condenser ($\dot{m}_5 = \dot{m}_6 = 0$), and the cogeneration plant operates as an ordinary steam power plant. The rates of heat input, heat rejected, and process heat supply as well as the power produced for this cogeneration plant can be expressed as follows:

$$\dot{Q}_{in} = \dot{m}_3(h_4 - h_3) \qquad (10\text{-}25)$$

$$\dot{Q}_{out} = \dot{m}_7(h_7 - h_1) \qquad (10\text{-}26)$$

$$\dot{Q}_p = \dot{m}_5 h_5 + \dot{m}_6 h_6 - \dot{m}_8 h_8 \qquad (10\text{-}27)$$

$$\dot{W}_{turb} = (\dot{m}_4 - \dot{m}_5)(h_4 - h_6) + \dot{m}_7(h_6 - h_7) \qquad (10\text{-}28)$$

Under optimum conditions, a cogeneration plant simulates the ideal cogeneration plant discussed earlier. That is, all the steam expands in the turbine to the extraction pressure and continues to the process-heating unit. No steam passes through the expansion valve or the condenser; thus, no waste heat is rejected ($\dot{m}_4 = \dot{m}_6$ and $\dot{m}_5 = \dot{m}_7 = 0$). This condition may be difficult to achieve in practice because of the constant variations in the

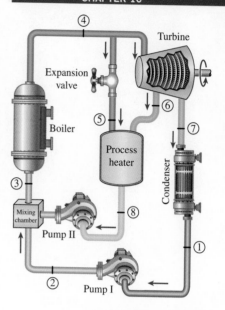

**FIGURE 10–23**
A cogeneration plant with adjustable loads.

process-heat and power loads. But the plant should be designed so that the optimum operating conditions are approximated most of the time.

The use of cogeneration dates to the beginning of this century when power plants were integrated to a community to provide district heating, that is, space, hot water, and process heating for residential and commercial buildings. The district heating systems lost their popularity in the 1940s owing to low fuel prices. However, the rapid rise in fuel prices in the 1970s brought about renewed interest in district heating.

Cogeneration plants have proved to be economically very attractive. Consequently, more and more such plants have been installed in recent years, and more are being installed.

---

### EXAMPLE 10–8    An Ideal Cogeneration Plant

Consider the cogeneration plant shown in Fig. 10–24. Steam enters the turbine at 7 MPa and 500°C. Some steam is extracted from the turbine at 500 kPa for process heating. The remaining steam continues to expand to 5 kPa. Steam is then condensed at constant pressure and pumped to the boiler pressure of 7 MPa. At times of high demand for process heat, some steam leaving the boiler is throttled to 500 kPa and is routed to the process heater. The extraction fractions are adjusted so that steam leaves the process heater as a saturated liquid at 500 kPa. It is subsequently pumped to 7 MPa. The mass flow rate of steam through the boiler is 15 kg/s. Disregarding any pressure drops and heat losses in the piping and assuming the turbine and the pump to be isentropic, determine (a) the maximum rate at which process heat can be supplied, (b) the power produced and the utilization factor when no process heat is supplied, and (c) the rate of process heat supply when 10 percent of the steam is extracted before it enters the turbine and 70 percent of the steam is extracted from the turbine at 500 kPa for process heating.

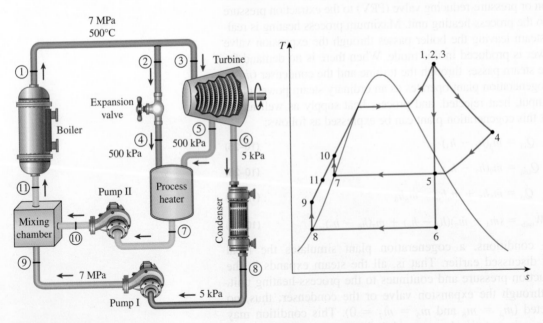

**FIGURE 10–24**

Schematic and *T-s* diagram for Example 10–8.

**SOLUTION** A cogeneration plant is considered. The maximum rate of process heat supply, the power produced and the utilization factor when no process heat is supplied, and the rate of process heat supply when steam is extracted from the steam line and turbine at specified ratios are to be determined.

**Assumptions** 1 Steady operating conditions exist. 2 Pressure drops and heat losses in piping are negligible. 3 Kinetic and potential energy changes are negligible.

**Analysis** The schematic of the cogeneration plant and the *T-s* diagram of the cycle are shown in Fig. 10–24. The power plant operates on an ideal cycle and thus the pumps and the turbines are isentropic; there are no pressure drops in the boiler, process heater, and condenser; and steam leaves the condenser and the process heater as saturated liquid.

The work inputs to the pumps and the enthalpies at various states are as follows:

$$w_{\text{pump I,in}} = v_8(P_9 - P_8) = (0.001005 \text{ m}^3/\text{kg})[(7000 - 5)\text{kPa}]\left(\frac{1 \text{ kJ}}{1 \text{ kPa·m}^3}\right)$$
$$= 7.03 \text{ kJ/kg}$$

$$w_{\text{pump II,in}} = v_7(P_{10} - P_7) = (0.001093 \text{ m}^3/\text{kg})[(7000 - 500) \text{ kPa}]\left(\frac{1 \text{ kJ}}{1 \text{ kPa·m}^3}\right)$$
$$= 7.10 \text{ kJ/kg}$$

$$h_1 = h_2 = h_3 = h_4 = 3411.4 \text{ kJ/kg}$$

$$h_5 = 2739.3 \text{ kJ/kg}$$

$$h_6 = 2073.0 \text{ kJ/kg}$$

$$h_7 = h_{f \text{ @ } 500 \text{ kPa}} = 640.09 \text{ kJ/kg}$$

$$h_8 = h_{f \text{ @ } 5 \text{ kPa}} = 137.75 \text{ kJ/kg}$$

$$h_9 = h_8 + w_{\text{pump I,in}} = (137.75 + 7.03) \text{ kJ/kg} = 144.78 \text{ kJ/kg}$$

$$h_{10} = h_7 + w_{\text{pump II,in}} = (640.09 + 7.10) \text{ kJ/kg} = 647.19 \text{ kJ/kg}$$

(*a*) The maximum rate of process heat is achieved when all the steam leaving the boiler is throttled and sent to the process heater and none is sent to the turbine (that is, $\dot{m}_4 = \dot{m}_7 = \dot{m}_1 = 15$ kg/s and $\dot{m}_3 = \dot{m}_5 = \dot{m}_6 = 0$). Thus,

$$\dot{Q}_{p,\text{max}} = \dot{m}_1(h_4 - h_7) = (15 \text{ kg/s})[(3411.4 - 640.09) \text{ kJ/kg}] = \textbf{41,570 kW}$$

The utilization factor is 100 percent in this case since no heat is rejected in the condenser, heat losses from the piping and other components are assumed to be negligible, and combustion losses are not considered.

(*b*) When no process heat is supplied, all the steam leaving the boiler passes through the turbine and expands to the condenser pressure of 5 kPa (that is, $\dot{m}_3 = \dot{m}_6 = \dot{m}_1 = 15$ kg/s and $\dot{m}_2 = \dot{m}_5 = 0$). Maximum power is produced in this mode, which is determined to be

$$\dot{W}_{\text{turb,out}} = \dot{m}(h_3 - h_6) = (15 \text{ kg/s})[(3411.4 - 2073.0) \text{ kJ/kg}] = 20,076 \text{ kW}$$

$$\dot{W}_{\text{pump,in}} = (15 \text{ kg/s})(7.03 \text{ kJ/kg}) = 105 \text{ kW}$$

$$\dot{W}_{\text{net,out}} = \dot{W}_{\text{turb,out}} - \dot{W}_{\text{pump,in}} = (20,076 - 105) \text{ kW} = 19,971 \text{ kW} \cong \textbf{20.0 MW}$$

$$\dot{Q}_{\text{in}} = \dot{m}_1(h_1 - h_{11}) = (15 \text{ kg/s})[(3411.4 - 144.78) \text{ kJ/kg}] = 48,999 \text{ kW}$$

Thus,

$$\epsilon_u = \frac{\dot{W}_{\text{net}} + \dot{Q}_p}{\dot{Q}_{\text{in}}} = \frac{(19{,}971 + 0)\ \text{kW}}{48{,}999\ \text{kW}} = 0.408 \text{ or } 40.8\%$$

That is, 40.8 percent of the energy is utilized for a useful purpose. Notice that the utilization factor is equivalent to the thermal efficiency in this case.

(c) Neglecting any kinetic and potential energy changes, an energy balance on the process heater yields

$$\dot{E}_{\text{in}} = \dot{E}_{\text{out}}$$

$$\dot{m}_4 h_4 + \dot{m}_5 h_5 = \dot{Q}_{p,\text{out}} + \dot{m}_7 h_7$$

or

$$\dot{Q}_{p,\text{out}} = \dot{m}_4 h_4 + \dot{m}_5 h_5 - \dot{m}_7 h_7$$

where

$$\dot{m}_4 = (0.1)(15\ \text{kg/s}) = 1.5\ \text{kg/s}$$
$$\dot{m}_5 = (0.7)(15\ \text{kg/s}) = 10.5\ \text{kg/s}$$
$$\dot{m}_7 = \dot{m}_4 + \dot{m}_5 = 1.5 + 10.5 = 12\ \text{kg/s}$$

Thus

$$\dot{Q}_{p,\text{out}} = (1.5\ \text{kg/s})(3411.4\ \text{kJ/kg}) + (10.5\ \text{kg/s})(2739.3\ \text{kJ/kg})$$
$$- (12\ \text{kg/s})(640.09\ \text{kJ/kg})$$
$$= 26.2\ \text{MW}$$

*Discussion* Note that 26.2 MW of the heat transferred will be utilized in the process heater. We could also show that 11.0 MW of power is produced in this case, and the rate of heat input in the boiler is 43.0 MW. Thus the utilization factor is 86.5 percent.

# 10–9 · COMBINED GAS–VAPOR POWER CYCLES

The continued quest for higher thermal efficiencies has resulted in rather innovative modifications to conventional power plants. The *binary vapor cycle* discussed later is one such modification. A more popular modification involves a gas power cycle topping a vapor power cycle, which is called the **combined gas–vapor cycle**, or just the **combined cycle**. The combined cycle of greatest interest is the gas-turbine (Brayton) cycle topping a steam-turbine (Rankine) cycle, which has a higher thermal efficiency than either of the cycles executed individually.

Gas-turbine cycles typically operate at considerably higher temperatures than steam cycles. The maximum fluid temperature at the turbine inlet is about 620°C for modern steam power plants, but over 1425°C for gas-turbine power plants. It is over 1500°C at the burner exit of turbojet engines. The use of higher temperatures in gas turbines is made possible by developments in cooling the turbine blades and coating the blades with high-temperature-resistant materials such as ceramics. Because of the higher average temperature at which heat is supplied, gas-turbine cycles have a greater potential for higher thermal efficiencies. However, the gas-turbine cycles have one inherent disadvantage: The gas leaves the gas turbine at very high temperatures (usually above 500°C), which erases any potential gains in the thermal efficiency. The situation can be improved somewhat by using regeneration, but the improvement is limited.

It makes engineering sense to take advantage of the very desirable characteristics of the gas-turbine cycle at high temperatures *and* to use the high-temperature exhaust gases as the energy source for the bottoming cycle such as a steam power cycle. The result is a combined gas–steam cycle, as shown in Fig. 10–25. In this cycle, energy is recovered from the exhaust gases by transferring it to the steam in a heat exchanger that serves as the boiler. In general, more than one gas turbine is needed to supply sufficient heat to the steam. Also, the steam cycle may involve regeneration as well as reheating. Energy for the reheating process can be supplied by burning some additional fuel in the oxygen-rich exhaust gases.

Developments in gas-turbine technology have made the combined gas–steam cycle economically very attractive. The combined cycle increases the efficiency without increasing the initial cost greatly. Consequently, many new power plants operate on combined cycles, and many more existing steam- or

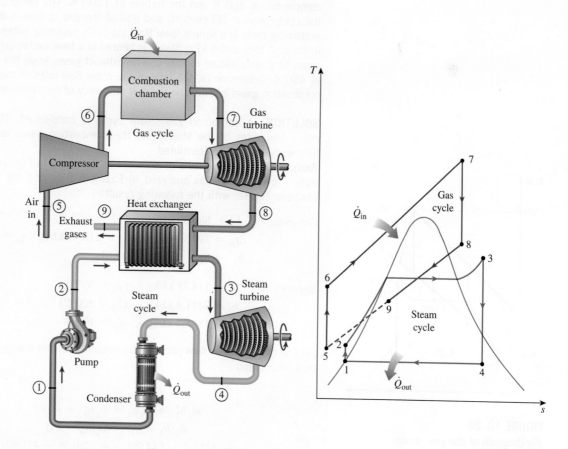

**FIGURE 10–25**
Combined gas–steam power plant.

gas-turbine plants are being converted to combined-cycle power plants. Thermal efficiencies well over 40 percent are reported as a result of conversion.

A 1090-MW Tohoku combined plant that was put in commercial operation in 1985 in Niigata, Japan, is reported to operate at a thermal efficiency of 44 percent. This plant has two 191-MW steam turbines and six 118-MW gas turbines. Hot combustion gases enter the gas turbines at 1154°C, and steam enters the steam turbines at 500°C. Steam is cooled in the condenser by cooling water at an average temperature of 15°C. The compressors have a pressure ratio of 14, and the mass flow rate of air through the compressors is 443 kg/s.

A 1350-MW combined-cycle power plant built in Ambarli, Turkey, in 1988 by Siemens of Germany is the first commercially operating thermal plant in the world to attain an efficiency level as high as 52.5 percent at design operating conditions. This plant has six 150-MW gas turbines and three 173-MW steam turbines. Some recent combined-cycle power plants have achieved efficiencies above 60 percent.

### EXAMPLE 10–9    A Combined Gas-Steam Power Cycle

Consider the combined gas–steam power cycle shown in Fig. 10–26. The topping cycle is a gas-turbine cycle that has a pressure ratio of 8. Air enters the compressor at 300 K and the turbine at 1300 K. The isentropic efficiency of the compressor is 80 percent, and that of the gas turbine is 85 percent. The bottoming cycle is a simple ideal Rankine cycle operating between the pressure limits of 7 MPa and 5 kPa. Steam is heated in a heat exchanger by the exhaust gases to a temperature of 500°C. The exhaust gases leave the heat exchanger at 450 K. Determine (a) the ratio of the mass flow rates of the steam and the combustion gases and (b) the thermal efficiency of the combined cycle.

**SOLUTION**  A combined gas–steam cycle is considered. The ratio of the mass flow rates of the steam and the combustion gases and the thermal efficiency are to be determined.

**Analysis**  The T-s diagrams of both cycles are given in Fig. 10–26. The gas-turbine cycle alone was analyzed in Example 9–6, and the steam cycle in Example 10–8b, with the following results:

*Gas cycle:*  $h_4' = 880.36$ kJ/kg    $(T_4' = 853$ K$)$

$q_{in} = 790.58$ kJ/kg  $w_{net} = 210.41$ kJ/kg  $\eta_{th} = 26.6\%$

$h_5' = h_{@\,450\,K} = 451.80$ kJ/kg

*Steam cycle:*  $h_2 = 144.78$ kJ/kg    $(T_2 = 33°C)$

$h_3 = 3411.4$ kJ/kg    $(T_3 = 500°C)$

$w_{net} = 1331.4$ kJ/kg    $\eta_{th} = 40.8\%$

(a) The ratio of mass flow rates is determined from an energy balance on the heat exchanger:

$$\dot{E}_{in} = \dot{E}_{out}$$

$$\dot{m}_g h_5' + \dot{m}_s h_3 = \dot{m}_g h_4' + \dot{m}_s h_2$$

$$\dot{m}_s(h_3 - h_2) = \dot{m}_g(h_4' - h_5')$$

$$\dot{m}_s(3411.4 - 144.78) = \dot{m}_g(880.36 - 451.80)$$

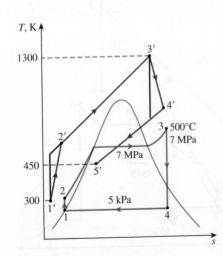

**FIGURE 10–26**

T-s diagram of the gas–steam combined cycle described in Example 10–9.

Thus,

$$\frac{\dot{m}_s}{\dot{m}_g} = y = 0.131$$

That is, 1 kg of exhaust gases can heat only 0.131 kg of steam from 33 to 500°C as they are cooled from 853 to 450 K. Then the total net work output per kilogram of combustion gases becomes

$$w_{net} = w_{net,gas} + y w_{net,steam}$$

$$= (210.41 \text{ kJ/kg gas}) + (0.131 \text{ kg steam/kg gas})(1331.4 \text{ kJ/kg steam})$$

$$= 384.8 \text{ kJ/kg gas}$$

Therefore, for each kg of combustion gases produced, the combined plant will deliver 384.8 kJ of work. The net power output of the plant is determined by multiplying this value by the mass flow rate of the working fluid in the gas-turbine cycle.

(b) The thermal efficiency of the combined cycle is determined from

$$\eta_{th} = \frac{w_{net}}{q_{in}} = \frac{384.8 \text{ kJ/kg gas}}{790.6 \text{ kJ/kg gas}} = 0.487 \text{ or } 48.7\%$$

**Discussion** Note that this combined cycle converts to useful work 48.7 percent of the energy supplied to the gas in the combustion chamber. This value is considerably higher than the thermal efficiency of the gas-turbine cycle (26.6 percent) or the steam-turbine cycle (40.8 percent) operating alone.

---

## TOPIC OF SPECIAL INTEREST*    Binary Vapor Cycles

With the exception of a few specialized applications, the working fluid predominantly used in vapor power cycles is water. Water is the *best* working fluid presently available, but it is far from being the *ideal* one. The binary cycle is an attempt to overcome some of the shortcomings of water and to approach the *ideal* working fluid by using two fluids. Before we discuss the binary cycle, let us list the characteristics of a working fluid most suitable for vapor power cycles:

**1.** A high critical temperature and a safe maximum pressure. A critical temperature above the metallurgically allowed maximum temperature (about 620°C) makes it possible to transfer a considerable portion of the heat isothermally at the maximum temperature as the fluid changes phase. This makes the cycle approach the Carnot cycle. Very high pressures at the maximum temperature are undesirable because they create material-strength problems.

---

*This section can be skipped without a loss in continuity.

**2.** Low triple-point temperature. A triple-point temperature below the temperature of the cooling medium prevents any solidification problems.

**3.** A condenser pressure that is not too low. Condensers usually operate below atmospheric pressure. Pressures well below the atmospheric pressure create air-leakage problems. Therefore, a substance whose saturation pressure at the ambient temperature is too low is not a good candidate.

**4.** A high enthalpy of vaporization ($h_{fg}$) so that heat transfer to the working fluid is nearly isothermal and large mass flow rates are not needed.

**5.** A saturation dome that resembles an inverted U. This eliminates the formation of excessive moisture in the turbine and the need for reheating.

**6.** Good heat transfer characteristics (high thermal conductivity).

**7.** Other properties such as being inert, inexpensive, readily available, and nontoxic.

Not surprisingly, no fluid possesses all these characteristics. Water comes the closest, although it does not fare well with respect to characteristics 1, 3, and 5. We can cope with its subatmospheric condenser pressure by careful sealing, and with the inverted V-shaped saturation dome by reheating, but there is not much we can do about item 1. Water has a low critical temperature (374°C, well below the metallurgical limit) and very high saturation pressures at high temperatures (16.5 MPa at 350°C).

Well, we cannot change the way water behaves during the high-temperature part of the cycle, but we certainly can replace it with a more suitable fluid. The result is a power cycle that is actually a combination of two cycles, one in the high-temperature region and the other in the low-temperature region. Such a cycle is called a **binary vapor cycle**. In binary vapor cycles, the condenser of the high-temperature cycle (also called the *topping cycle*) serves as the boiler of the low-temperature cycle (also called the *bottoming cycle*). That is, the heat output of the high-temperature cycle is used as the heat input to the low-temperature one.

Some working fluids found suitable for the high-temperature cycle are mercury, sodium, potassium, and sodium–potassium mixtures. The schematic and *T-s* diagram for a mercury–water binary vapor cycle are shown in Fig. 10–27. The critical temperature of mercury is 898°C (well above the current metallurgical limit), and its critical pressure is only about 18 MPa. This makes mercury a very suitable working fluid for the topping cycle. Mercury is not suitable as the sole working fluid for the entire cycle, however, since at a condenser temperature of 32°C its saturation pressure is 0.07 Pa. A power plant cannot operate at this vacuum because of air-leakage problems. At an acceptable condenser pressure of 7 kPa, the saturation temperature of mercury is 237°C, which is too high as the minimum temperature in the cycle. Therefore, the use of mercury as a working fluid is limited to the high-temperature cycles. Other disadvantages of mercury are its toxicity and high cost. The mass flow rate of mercury in binary vapor cycles is several times that of water because of its low enthalpy of vaporization.

It is evident from the *T-s* diagram in Fig. 10–27 that the binary vapor cycle approximates the Carnot cycle more closely than the steam cycle

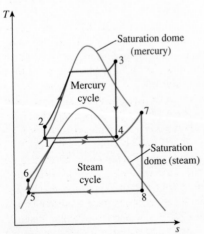

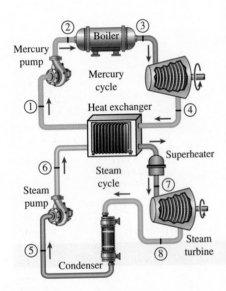

**FIGURE 10–27**
Mercury–water binary vapor cycle.

for the same temperature limits. Therefore, the thermal efficiency of a power plant can be increased by switching to binary cycles. The use of mercury–water binary cycles in the United States dates back to 1928. Several such plants have been built since then in the New England area, where fuel costs are typically higher. A small (40-MW) mercury–steam power plant that was in service in New Hampshire in 1950 had a higher thermal efficiency than most of the large modern power plants in use at that time.

Studies show that thermal efficiencies of 50 percent or higher are possible with binary vapor cycles. However, binary vapor cycles are not economically attractive because of their high initial cost and the competition offered by the combined gas–steam power plants.

## SUMMARY

The *Carnot cycle* is not a suitable model for vapor power cycles because it cannot be approximated in practice. The model cycle for vapor power cycles is the *Rankine cycle,* which is composed of four internally reversible processes: constant-pressure heat addition in a boiler, isentropic expansion in a turbine, constant-pressure heat rejection in a condenser, and isentropic compression in a pump. Steam leaves the condenser as a saturated liquid at the condenser pressure.

The thermal efficiency of the Rankine cycle can be increased by increasing the average temperature at which heat is transferred to the working fluid and/or by decreasing the average temperature at which heat is rejected to the cooling medium. The average temperature during heat rejection can be decreased by lowering the turbine exit pressure. Consequently, the condenser pressure of most vapor power plants is well below the atmospheric pressure. The average temperature during heat addition can be increased by raising the boiler pressure or by superheating the fluid to high temperatures. There is a limit to the degree of superheating, however, since the fluid temperature is not allowed to exceed a metallurgically safe value.

Superheating has the added advantage of decreasing the moisture content of the steam at the turbine exit. Lowering the exhaust pressure or raising the boiler pressure, however, increases the moisture content. To take advantage of the improved efficiencies at higher boiler pressures and lower condenser pressures, steam is usually *reheated* after expanding partially in the high-pressure turbine. This is done by extracting the steam after partial expansion in the high-pressure turbine, sending it back to the boiler where it is reheated at constant pressure, and returning it to the low-pressure turbine for complete expansion to the condenser pressure. The average temperature during the reheat process, and thus the thermal efficiency of the cycle, can be increased by increasing the number of expansion and reheat stages. As the number of stages is increased, the expansion and reheat processes approach an isothermal process at maximum temperature. Reheating also decreases the moisture content at the turbine exit.

Another way of increasing the thermal efficiency of the Rankine cycle is *regeneration.* During a regeneration process, liquid water (feedwater) leaving the pump is heated by steam bled off the turbine at some intermediate pressure in devices called *feedwater heaters.* The two streams are mixed in open feedwater heaters, and the mixture leaves as a saturated liquid at the heater pressure. In closed feedwater heaters, heat is transferred from the steam to the feedwater without mixing.

The production of more than one useful form of energy (such as process heat and electric power) from the same energy source is called *cogeneration.* Cogeneration plants produce electric power while meeting the process heat requirements of certain industrial processes. This way, more of the energy transferred to the fluid in the boiler is utilized for a useful purpose. The fraction of energy that is used for either process heat or power generation is called the *utilization factor* of the cogeneration plant.

The overall thermal efficiency of a power plant can be increased by using a *combined cycle.* The most common combined cycle is the gas–steam combined cycle where a gas-turbine cycle operates at the high-temperature range and a steam-turbine cycle at the low-temperature range. Steam is heated by the high-temperature exhaust gases leaving the gas turbine. Combined cycles have a higher thermal efficiency than the steam- or gas-turbine cycles operating alone.

## REFERENCES AND SUGGESTED READINGS

1. R. L. Bannister and G. J. Silvestri. "The Evolution of Central Station Steam Turbines." *Mechanical Engineering,* February 1989, pp. 70–78.

2. R. L. Bannister, G. J. Silvestri, A. Hizume, and T. Fujikawa. "High Temperature Supercritical Steam Turbines." *Mechanical Engineering,* February 1987, pp. 60–65.

3. M. M. El-Wakil. *Powerplant Technology.* New York: McGraw-Hill, 1984.

4. K. W. Li and A. P. Priddy. *Power Plant System Design.* New York: John Wiley & Sons, 1985.

5. H. Sorensen. *Energy Conversion Systems.* New York: John Wiley & Sons, 1983.

6. *Steam, Its Generation and Use.* 39th ed. New York: Babcock and Wilcox Co., 1978.

7. *Turbomachinery* 28, no. 2 (March/April 1987). Norwalk, CT: Business Journals, Inc.

8. J. Weisman and R. Eckart. *Modern Power Plant Engineering.* Englewood Cliffs, NJ: Prentice-Hall, 1985.

## PROBLEMS*

### Carnot Vapor Cycle

**10–1C** Why is the Carnot cycle not a realistic model for steam power plants?

**10–2** A steady-flow Carnot cycle uses water as the working fluid. Water changes from saturated liquid to saturated vapor as heat is transferred to it from a source at 250°C. Heat rejection takes place at a pressure of 20 kPa. Show the cycle on a $T$-$s$ diagram relative to the saturation lines, and determine (*a*) the thermal efficiency, (*b*) the amount of heat rejected, and (*c*) the net work output.

**10–3** Repeat Prob. 10–2 for a heat rejection pressure of 10 kPa.

**10–4** Consider a steady-flow Carnot cycle with water as the working fluid. The maximum and minimum temperatures in the cycle are 350 and 60°C. The quality of water is 0.891 at the beginning of the heat-rejection process and 0.1 at the end. Show the cycle on a $T$-$s$ diagram relative to the saturation lines, and determine (*a*) the thermal efficiency, (*b*) the

pressure at the turbine inlet, and (*c*) the net work output. *Answers:* (*a*) 0.465, (*b*) 1.40 MPa, (*c*) 1623 kJ/kg

### The Simple Rankine Cycle

**10–5C** Consider a simple ideal Rankine cycle with fixed turbine inlet conditions. What is the effect of lowering the condenser pressure on

| | |
|---|---|
| Pump work input: | (*a*) increases, (*b*) decreases, (*c*) remains the same |
| Turbine work output: | (*a*) increases, (*b*) decreases, (*c*) remains the same |
| Heat supplied: | (*a*) increases, (*b*) decreases, (*c*) remains the same |
| Heat rejected: | (*a*) increases, (*b*) decreases, (*c*) remains the same |
| Cycle efficiency: | (*a*) increases, (*b*) decreases, (*c*) remains the same |
| Moisture content at turbine exit: | (*a*) increases, (*b*) decreases, (*c*) remains the same |

**10–6C** Consider a simple ideal Rankine cycle with fixed turbine inlet temperature and condenser pressure. What is the effect of increasing the boiler pressure on

*Problems designated by a "C" are concept questions, and students are encouraged to answer them all. Problems with the 🖭 icon are solved using EES, and complete solutions together with parametric studies are included on the text website. Problems with the 🖭 icon are comprehensive in nature, and are intended to be solved with an equation solver such as EES.

| Pump work input: | (a) increases, | (b) decreases, (c) remains the same |
| Turbine work output: | (a) increases, | (b) decreases, (c) remains the same |
| Heat supplied: | (a) increases, | (b) decreases, (c) remains the same |
| Heat rejected: | (a) increases, | (b) decreases, (c) remains the same |
| Cycle efficiency: | (a) increases, | (b) decreases, (c) remains the same |
| Moisture content at turbine exit: | (a) increases, | (b) decreases, (c) remains the same |

**10–7C**  Consider a simple ideal Rankine cycle with fixed boiler and condenser pressures. What is the effect of super-heating the steam to a higher temperature on

| Pump work input: | (a) increases, | (b) decreases, (c) remains the same |
| Turbine work output: | (a) increases, | (b) decreases, (c) remains the same |
| Heat supplied: | (a) increases, | (b) decreases, (c) remains the same |
| Heat rejected: | (a) increases, | (b) decreases, (c) remains the same |
| Cycle efficiency: | (a) increases, | (b) decreases, (c) remains the same |
| Moisture content at turbine exit: | (a) increases, | (b) decreases, (c) remains the same |

**10–8C**  How do actual vapor power cycles differ from idealized ones?

**10–9C**  The entropy of steam increases in actual steam turbines as a result of irreversibilities. In an effort to control entropy increase, it is proposed to cool the steam in the turbine by running cooling water around the turbine casing. It is argued that this will reduce the entropy and the enthalpy of the steam at the turbine exit and thus increase the work output. How would you evaluate this proposal?

**10–10C**  Is it possible to maintain a pressure of 10 kPa in a condenser that is being cooled by river water entering at 20°C?

**10–11**  A steam power plant operates on a simple ideal Rankine cycle between the pressure limits of 3 MPa and 50 kPa. The temperature of the steam at the turbine inlet is 300°C, and the mass flow rate of steam through the cycle is 35 kg/s. Show the cycle on a T-s diagram with respect to

saturation lines, and determine (a) the thermal efficiency of the cycle and (b) the net power output of the power plant.

**10–12**  Refrigerant-134a is used as the working fluid in a simple ideal Rankine cycle which operates the boiler at 2000 kPa and the condenser at 24°C. The mixture at the exit of the turbine has a quality of 93 percent. Determine the turbine inlet temperature, the cycle thermal efficiency, and the back-work ratio of this cycle.

**10–13**  A simple ideal Rankine cycle which uses water as the working fluid operates its condenser at 40°C and its boiler at 300°C. Calculate the work produced by the turbine, the heat supplied in the boiler, and the thermal efficiency of this cycle when the steam enters the turbine without any superheating.

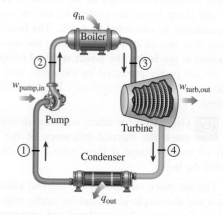

**FIGURE P10–13**

**10–14**  A simple ideal Rankine cycle with water as the working fluid operates between the pressure limits of 17.5 MPa in the boiler and 30 kPa in the condenser. What is the minimum temperature required at the turbine inlet such that the quality of the steam leaving the turbine is not below 80 percent. When operated at this temperature, what is the thermal efficiency of this cycle?

**10–15**  Consider a 210-MW steam power plant that operates on a simple ideal Rankine cycle. Steam enters the turbine at 10 MPa and 500°C and is cooled in the condenser at a pressure of 10 kPa. Show the cycle on a T-s diagram with respect to saturation lines, and determine (a) the quality of the steam at the turbine exit, (b) the thermal efficiency of the cycle, and (c) the mass flow rate of the steam. *Answers:* (a) 0.793, (b) 40.2 percent, (c) 165 kg/s

**10–16**  Repeat Prob. 10–15 assuming an isentropic efficiency of 85 percent for both the turbine and the pump.  *Answers:* (*a*) 0.874, (*b*) 34.1 percent, (*c*) 194 kg/s

**10–17**  A steam power plant operates on a simple ideal Rankine cycle between the pressure limits of 9000 and 15 kPa. The mass flow rate of steam through the cycle is 35 kg/s. The moisture content of the steam at the turbine exit is not to exceed 10 percent. Show the cycle on a *T-s* diagram with respect to saturation lines, and determine (*a*) the minimum turbine inlet temperature, (*b*) the rate of heat input in the boiler, and (*c*) the thermal efficiency of the cycle.

**10–18**  Repeat Prob. 10–17 assuming an isentropic efficiency of 85 percent for both the turbine and the pump.

**10–19**  A simple Rankine cycle uses water as the working fluid. The boiler operates at 6000 kPa and the condenser at 50 kPa. At the entrance to the turbine, the temperature is 450°C. The isentropic efficiency of the turbine is 94 percent, pressure and pump losses are negligible, and the water leaving the condenser is subcooled by 6.3°C. The boiler is sized for a mass flow rate of 20 kg/s. Determine the rate at which heat is added in the boiler, the power required to operate the pumps, the net power produced by the cycle, and the thermal efficiency.  *Answers:* 59,660 kW, 122 kW, 18,050 kW, 30.3 percent.

**10–20**  [EES]  Using EES (or other) software, determine how much the thermal efficiency of the cycle in Prob. 10–19 would change if there were a 50 kPa pressure drop across the boiler.

**10–21**  The net work output and the thermal efficiency for the Carnot and the simple ideal Rankine cycles with steam as the working fluid are to be calculated and compared. Steam enters the turbine in both cases at 5 MPa as a saturated vapor, and the condenser pressure is 50 kPa. In the Rankine cycle, the condenser exit state is saturated liquid and in the Carnot cycle, the boiler inlet state is saturated liquid. Draw the *T-s* diagrams for both cycles.

**10–22**  A binary geothermal power plant uses geothermal water at 160°C as the heat source. The cycle operates on the simple Rankine cycle with isobutane as the working fluid. Heat is transferred to the cycle by a heat exchanger in which geothermal liquid water enters at 160°C at a rate of 555.9 kg/s and leaves at 90°C. Isobutane enters the turbine at 3.25 MPa and 147°C at a rate of 305.6 kg/s, and leaves at 79.5°C and 410 kPa. Isobutane is condensed in an air-cooled condenser and pumped to the heat exchanger pressure. Assuming the pump to have an isentropic efficiency of 90 percent, determine (*a*) the isentropic efficiency of the turbine, (*b*) the net power output of the plant, and (*c*) the thermal efficiency of the cycle. The properties of isobutane may be obtained from EES.

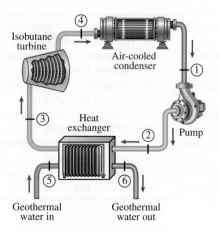

**FIGURE P10–22**

**10–23**  Consider a coal-fired steam power plant that produces 175 MW of electric power. The power plant operates on a simple ideal Rankine cycle with turbine inlet conditions of 7 MPa and 550°C and a condenser pressure of 15 kPa. The coal has a heating value (energy released when the fuel is burned) of 29,300 kJ/kg. Assuming that 85 percent of this energy is transferred to the steam in the boiler and that the electric generator has an efficiency of 96 percent, determine (*a*) the overall plant efficiency (the ratio of net electric power output to the energy input as fuel) and (*b*) the required rate of coal supply.  *Answers:* (*a*) 31.5 percent, (*b*) 68.3 t/h

## The Reheat Rankine Cycle

**10–24C**  Show the ideal Rankine cycle with three stages of reheating on a *T-s* diagram. Assume the turbine inlet temperature is the same for all stages. How does the cycle efficiency vary with the number of reheat stages?

**10–25C**  How do the following quantities change when a simple ideal Rankine cycle is modified with reheating? Assume the mass flow rate is maintained the same.

| Pump work input: | (*a*) increases, | (*b*) decreases, |
| --- | --- | --- |
| | (*c*) remains the same | |
| Turbine work output: | (*a*) increases, | (*b*) decreases, |
| | (*c*) remains the same | |
| Heat supplied: | (*a*) increases, | (*b*) decreases, |
| | (*c*) remains the same | |
| Heat rejected: | (*a*) increases, | (*b*) decreases, |
| | (*c*) remains the same | |
| Moisture content at turbine exit: | (*a*) increases, | (*b*) decreases, |
| | (*c*) remains the same | |

**10–26C** Consider a simple ideal Rankine cycle and an ideal Rankine cycle with three reheat stages. Both cycles operate between the same pressure limits. The maximum temperature is 700°C in the simple cycle and 450°C in the reheat cycle. Which cycle do you think will have a higher thermal efficiency?

**10–27** An ideal reheat Rankine cycle with water as the working fluid operates the boiler at 15,000 kPa, the reheater at 2000 kPa, and the condenser at 100 kPa. The temperature is 450°C at the entrance of the high-pressure and low-pressure turbines. The mass flow rate through the cycle is 1.74 kg/s. Determine the power used by pumps, the power produced by the cycle, the rate of heat transfer in the reheater, and the thermal efficiency of this system.

**10–28** A steam power plant operates on the ideal reheat Rankine cycle. Steam enters the high-pressure turbine at 6 MPa and 400°C and leaves at 2 MPa. Steam is then reheated at constant pressure to 400°C before it expands to 20 kPa in the low-pressure turbine. Determine the turbine work output, in kJ/kg, and the thermal efficiency of the cycle. Also, show the cycle on a *T-s* diagram with respect to saturation lines.

**10–29** Reconsider Prob. 10–28. Using EES (or other) software, solve this problem by the diagram window data entry feature of EES. Include the effects of the turbine and pump efficiencies and also show the effects of reheat on the steam quality at the low-pressure turbine exit. Plot the cycle on a *T-s* diagram with respect to the saturation lines. Discuss the results of your parametric studies.

**10–30** Steam enters the high-pressure turbine of a steam power plant that operates on the ideal reheat Rankine cycle at 6 MPa and 500°C and leaves as saturated vapor. Steam is then reheated to 400°C before it expands to a pressure of 10 kPa. Heat is transferred to the steam in the boiler at a rate of $6 \times 10^4$ kW. Steam is cooled in the condenser by the cooling water from a nearby river, which enters the condenser at 7°C. Show the cycle on a *T-s* diagram with respect to saturation lines, and determine (a) the pressure at which reheating takes place, (b) the net power output and thermal efficiency, and (c) the minimum mass flow rate of the cooling water required.

**10–31** Consider a steam power plant that operates on the ideal reheat Rankine cycle. The plant maintains the boiler at 5000 kPa, the reheat section at 1200 kPa, and the condenser at 20 kPa. The mixture quality at the exit of both turbines is 96 percent. Determine the temperature at the inlet of each turbine and the cycle's thermal efficiency. *Answers:* 327°C, 481°C, 35.0 percent

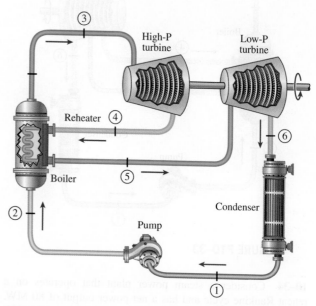

**FIGURE P10–31**

**10–32** A steam power plant operates on an ideal reheat Rankine cycle between the pressure limits of 15 MPa and 10 kPa. The mass flow rate of steam through the cycle is 12 kg/s. Steam enters both stages of the turbine at 500°C. If the moisture content of the steam at the exit of the low-pressure turbine is not to exceed 10 percent, determine (a) the pressure at which reheating takes place, (b) the total rate of heat input in the boiler, and (c) the thermal efficiency of the cycle. Also, show the cycle on a *T-s* diagram with respect to saturation lines.

**10–33** A steam power plant operates on the reheat Rankine cycle. Steam enters the high-pressure turbine at 12.5 MPa and 550°C at a rate of 7.7 kg/s and leaves at 2 MPa. Steam is then reheated at constant pressure to 450°C before it expands in the low-pressure turbine. The isentropic efficiencies of the turbine and the pump are 85 percent and 90 percent, respectively. Steam leaves the condenser as a saturated liquid. If the moisture content of the steam at the exit of the turbine is not to exceed 5 percent, determine (a) the condenser pressure, (b) the net power output, and (c) the thermal efficiency. *Answers:* (a) 9.73 kPa, (b) 10.2 MW, (c) 36.9 percent

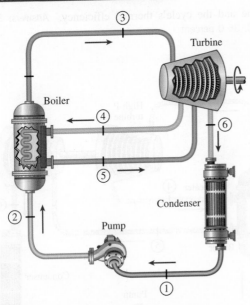

**FIGURE P10–33**

**10–34** Consider a steam power plant that operates on a reheat Rankine cycle and has a net power output of 80 MW. Steam enters the high-pressure turbine at 10 MPa and 500°C and the low-pressure turbine at 1 MPa and 500°C. Steam leaves the condenser as a saturated liquid at a pressure of 10 kPa. The isentropic efficiency of the turbine is 80 percent, and that of the pump is 95 percent. Show the cycle on a $T$-$s$ diagram with respect to saturation lines, and determine (a) the quality (or temperature, if superheated) of the steam at the turbine exit, (b) the thermal efficiency of the cycle, and (c) the mass flow rate of the steam. *Answers:* (a) 88.1°C, (b) 34.1 percent, (c) 62.7 kg/s

**10–35** Repeat Prob. 10–34 assuming both the pump and the turbine are isentropic. *Answers:* (a) 0.949, (b) 41.3 percent, (c) 50.0 kg/s

## Regenerative Rankine Cycle

**10–36C** During a regeneration process, some steam is extracted from the turbine and is used to heat the liquid water leaving the pump. This does not seem like a smart thing to do since the extracted steam could produce some more work in the turbine. How do you justify this action?

**10–37C** Consider a simple ideal Rankine cycle and an ideal regenerative Rankine cycle with one open feedwater heater. The two cycles are very much alike, except the feedwater in the regenerative cycle is heated by extracting some steam just before it enters the turbine. How would you compare the efficiencies of these two cycles?

**10–38C** How do open feedwater heaters differ from closed feedwater heaters?

**10–39C** How do the following quantities change when the simple ideal Rankine cycle is modified with regeneration? Assume the mass flow rate through the boiler is the same.

| | | |
|---|---|---|
| Turbine work output: | (a) increases, | (b) decreases, |
| | (c) remains the same | |
| Heat supplied: | (a) increases, | (b) decreases, |
| | (c) remains the same | |
| Heat rejected: | (a) increases, | (b) decreases, |
| | (c) remains the same | |
| Moisture content at turbine exit: | (a) increases, | (b) decreases, |
| | (c) remains the same | |

**10–40** Turbine bleed steam enters an open feedwater heater of a regenerative Rankine cycle at 200 kPa and 150°C while the cold feedwater enters at 40°C. Determine the ratio of the bleed steam mass flow rate to the inlet feedwater mass flow rate required to heat the feedwater to 110°C.

**10–41** The closed feedwater heater of a regenerative Rankine cycle is to heat 7000 kPa feedwater from 260°C to a saturated liquid. The turbine supplies bleed steam at 6000 kPa and 325°C to this unit. This steam is condensed to a saturated liquid before entering the pump. Calculate the amount of bleed steam required to heat 1 kg of feedwater in this unit. *Answer:* 0.0779 kg/s

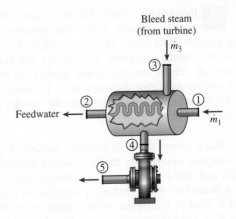

**FIGURE P10–41**

**10–42** A steam power plant operates on an ideal regenerative Rankine cycle. Steam enters the turbine at 6 MPa and 450°C and is condensed in the condenser at 20 kPa. Steam is extracted from the turbine at 0.4 MPa to heat the feedwater in an open feedwater heater. Water leaves the feedwater heater as a saturated liquid. Show the cycle on a $T$-$s$ diagram, and determine (a) the net work output per kilogram of steam flowing through the boiler and (b) the thermal efficiency of the cycle. *Answers:* (a) 1017 kJ/kg, (b) 37.8 percent

**10–43** Repeat Prob. 10–42 by replacing the open feedwater heater with a closed feedwater heater. Assume that the feedwater leaves the heater at the condensation temperature of the extracted steam and that the extracted steam leaves the heater as a saturated liquid and is pumped to the line carrying the feedwater.

**10–44** A steam power plant operates on an ideal regenerative Rankine cycle with two open feedwater heaters. Steam enters the turbine at 8 MPa and 550°C and exhausts to the condenser at 10 kPa. Steam is extracted from the turbine at 0.6 and 0.2 MPa. Water leaves both feedwater heaters as a saturated liquid. The mass flow rate of steam through the boiler is 16 kg/s. Show the cycle on a *T-s* diagram, and determine (*a*) the net power output of the power plant and (*b*) the thermal efficiency of the cycle.   *Answers:* (*a*) 19.8 MW, (*b*) 43.5 percent

**10–45** Consider a steam power plant that operates on the ideal regenerative Rankine cycle with a closed feedwater heater as shown in the figure. The plant maintains the turbine inlet at 3000 kPa and 350°C; and operates the condenser at 20 kPa. Steam is extracted at 1000 kPa to serve the closed feedwater heater, which discharges into the condenser after being throttled to condenser pressure. Calculate the work produced by the turbine, the work consumed by the pump, and the heat supply in the boiler for this cycle per unit of boiler flow rate.   *Answers:* 741 kJ/kg, 3.0 kJ/kg, 2353 kJ/kg

**10–48** Determine the thermal efficiency of the regenerative Rankine cycle of Prob. 10–45 when the isentropic efficiency of the turbine before and after steam extraction point is 90 percent and the condenser condensate is subcooled by 10°C.

**10–49** [EES] Reconsider Prob. 10–45. Using EES (or other) software, determine how much additional heat must be supplied to the boiler when the turbine isentropic efficiency before and after the extraction point is 90 percent and there is a 10 kPa pressure drop across the boiler?

**10–50** Consider an ideal steam regenerative Rankine cycle with two feedwater heaters, one closed and one open. Steam enters the turbine at 10 MPa and 600°C and exhausts to the condenser at 10 kPa. Steam is extracted from the turbine at 1.2 MPa for the closed feedwater heater and at 0.6 MPa for the open one. The feedwater is heated to the condensation temperature of the extracted steam in the closed feedwater heater. The extracted steam leaves the closed feedwater heater as a saturated liquid, which is subsequently throttled to the open feedwater heater. Show the cycle on a *T-s* diagram with respect to saturation lines, and determine (*a*) the mass flow rate of steam through the boiler for a net power output of 400 MW and (*b*) the thermal efficiency of the cycle.

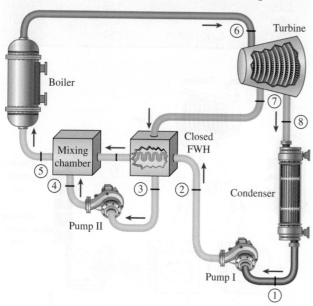

**FIGURE P10–45**

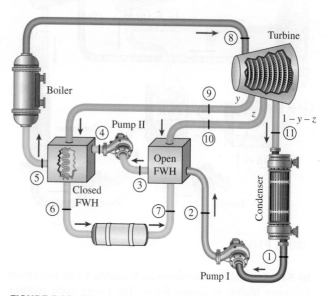

**FIGURE P10–50**

**10–46** [EES] Reconsider Prob. 10–45. Using EES (or other) software, determine the optimum bleed pressure for the closed feedwater heater that maximizes the thermal efficiency of the cycle.   *Answer:* 220 kPa

**10–47** Determine the thermal efficiency of the regenerative Rankine cycle of Prob. 10–45 when the isentropic efficiency of the turbine is 90 percent before and after steam extraction point.

**10–51** [EES] Reconsider Prob. 10–50. Using EES (or other) software, investigate the effects of turbine and pump efficiencies as they are varied from 70 percent to 100 percent on the mass flow rate and thermal efficiency. Plot the mass flow rate and the thermal efficiency as a function of turbine efficiency for pump efficiencies of 70, 85, and 100 percent, and discuss the results. Also plot the *T-s* diagram for turbine and pump efficiencies of 85 percent.

**10–52** A steam power plant operates on an ideal reheat–regenerative Rankine cycle and has a net power output of 80 MW. Steam enters the high-pressure turbine at 10 MPa and 550°C and leaves at 0.8 MPa. Some steam is extracted at this pressure to heat the feedwater in an open feedwater heater. The rest of the steam is reheated to 500°C and is expanded in the low-pressure turbine to the condenser pressure of 10 kPa. Show the cycle on a *T-s* diagram with respect to saturation lines, and determine (*a*) the mass flow rate of steam through the boiler and (*b*) the thermal efficiency of the cycle.   *Answers:* (*a*) 54.5 kg/s, (*b*) 44.4 percent

**10–53** Repeat Prob. 10–52, but replace the open feedwater heater with a closed feedwater heater. Assume that the feedwater leaves the heater at the condensation temperature of the extracted steam and that the extracted steam leaves the heater as a saturated liquid and is pumped to the line carrying the feedwater.

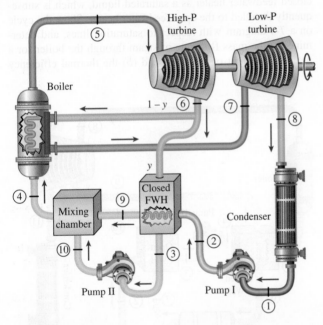

**FIGURE P10–53**

**10–54** An ideal Rankine steam cycle modified with two closed feedwater heaters is shown below. The power cycle receives 75 kg/s of steam at the high pressure inlet to the turbine. The feedwater heater exit states for the boiler feedwater and the condensed steam are the normally assumed ideal states. The fraction of mass entering the high pressure turbine at state 5 that is extracted for the feedwater heater operating at 1400 kPa is $y = 0.1446$. Use the data provided in the tables given below to

(*a*) Sketch the *T-s* diagram for the ideal cycle.
(*b*) Determine the fraction of mass, *z*, that is extracted for the closed feedwater heater operating at the 245 kPa extraction pressure.

(*c*) Determine the required cooling water flow rate, in kg/s, to keep the cooling water temperature rise in the condenser to 10°C. Assume $c_p = 4.18$ kJ/kg·K for cooling water.
(*d*) Determine the net power output and the thermal efficiency of the plant.

| Process states and selected data | | | | |
|---|---|---|---|---|
| State | P, kPa | T, °C | h, kJ/kg | s, kJ/kg·K |
| 1 | 20 | | | |
| 2 | 5000 | | | |
| 3 | 5000 | | | |
| 4 | 5000 | | | |
| 5 | 5000 | 700 | 3900 | 7.512 |
| 6 | 1400 | | 3406 | 7.512 |
| 7 | 245 | | 2918 | 7.512 |
| 8 | 20 | | 2477 | 7.512 |

| Saturation data | | | |
|---|---|---|---|
| P, kPa | $v_f$, m³/kg | $h_f$, kJ/kg | $s_g$, kJ/kg·K |
| 20 | 0.00102 | 251 | 7.907 |
| 245 | | 533 | 7.060 |
| 1400 | | 830 | 6.468 |
| 5000 | 0.00129 | 1154 | 5.973 |

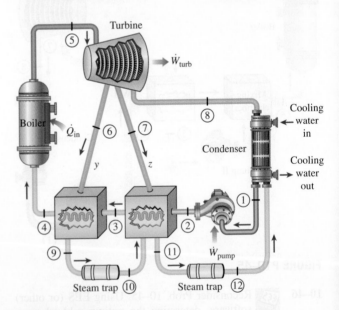

**FIGURE P10–54**

## Second-Law Analysis of Vapor Power Cycles

**10–55** Determine the exergy destruction associated with each of the processes of the Rankine cycle described in

Prob. 10–11, assuming a source temperature of 1500 K and a sink temperature of 290 K.

**10–56** Determine the exergy destruction associated with each of the processes of the Rankine cycle described in Prob. 10–15, assuming a source temperature of 1500 K and a sink temperature of 290 K. *Answers:* 0, 1112 kJ/kg, 0, 172.3 kJ/kg

**10–57** Determine the exergy destruction associated with each of the processes of the reheat Rankine cycle described in Prob. 10–28. Assume a source temperature of 1500 K and a sink temperature of 295 K.

**10–58** [EES] Reconsider Prob. 10–57. Using EES (or other) software, solve this problem by the diagram window data entry feature of EES. Include the effects of the turbine and pump efficiencies to evaluate the irreversibilities associated with each of the processes. Plot the cycle on a *T-s* diagram with respect to the saturation lines. Discuss the results of your parametric studies.

**10–59** Determine the exergy destruction associated with the heat addition process and the expansion process in Prob. 10–34. Assume a source temperature of 1600 K and a sink temperature of 285 K. Also, determine the exergy of the steam at the boiler exit. Take $P_0 = 100$ kPa. *Answers:* 1289 kJ/kg, 247.9 kJ/kg, 1495 kJ/kg

**10–60** Determine the exergy destruction associated with the regenerative cycle described in Prob. 10–42. Assume a source temperature of 1500 K and a sink temperature of 290 K. *Answer:* 1155 kJ/kg

**10–61** Determine the exergy destruction associated with the reheating and regeneration processes described in Prob. 10–52. Assume a source temperature of 1800 K and a sink temperature of 290 K.

**10–62** The schematic of a single-flash geothermal power plant with state numbers is given in Fig. P10–62. Geothermal resource exists as saturated liquid at 230°C. The geothermal liquid is withdrawn from the production well at a rate of 230 kg/s and is flashed to a pressure of 500 kPa by an essentially isenthalpic flashing process where the resulting vapor is separated from the liquid in a separator and is directed to the turbine. The steam leaves the turbine at 10 kPa with a moisture content of 5 percent and enters the condenser where it is condensed; it is routed to a reinjection well along with the liquid coming off the separator. Determine (*a*) the power output of the turbine and the thermal efficiency of the plant, (*b*) the exergy of the geothermal liquid at the exit of the flash chamber, and the exergy destructions and the second-law efficiencies for (*c*) the turbine and (*d*) the entire plant. *Answers:* (*a*) 10.8 MW, 0.053, (*b*) 17.3 MW, (*c*) 10.9 MW, 0.500, (*d*) 39.0 MW, 0.218

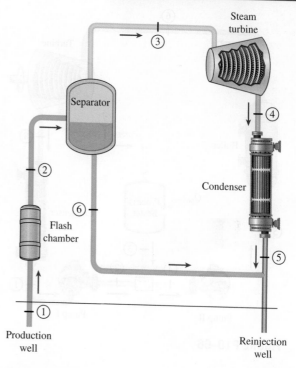

**FIGURE P10–62**

### Cogeneration

**10–63C** How is the utilization factor $\epsilon_u$ for cogeneration plants defined? Could $\epsilon_u$ be unity for a cogeneration plant that does not produce any power?

**10–64C** Consider a cogeneration plant for which the utilization factor is 1. Is the irreversibility associated with this cycle necessarily zero? Explain.

**10–65C** Consider a cogeneration plant for which the utilization factor is 0.5. Can the exergy destruction associated with this plant be zero? If yes, under what conditions?

**10–66** Steam enters the turbine of a cogeneration plant at 4 MPa and 500°C. One-fourth of the steam is extracted from the turbine at 1200-kPa pressure for process heating. The remaining steam continues to expand to 10 kPa. The extracted steam is then condensed and mixed with feedwater at constant pressure and the mixture is pumped to the boiler pressure of 7 MPa. The mass flow rate of steam through the boiler is 55 kg/s. Disregarding any pressure drops and heat losses in the piping, and assuming the turbine and the pump to be isentropic, determine the net power produced and the utilization factor of the plant.

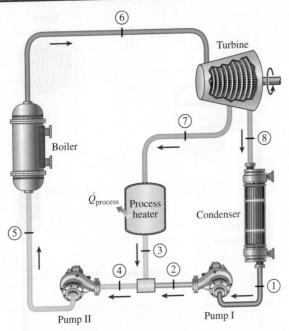

**FIGURE P10–66**

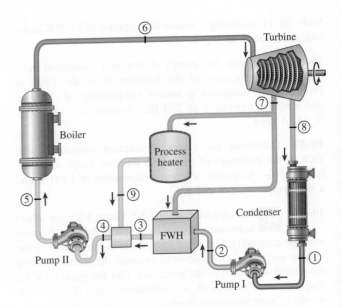

**FIGURE P10–68**

**10–67** Steam is generated in the boiler of a cogeneration plant at 10 MPa and 450°C at a steady rate of 5 kg/s. In normal operation, steam expands in a turbine to a pressure of 0.5 MPa and is then routed to the process heater, where it supplies the process heat. Steam leaves the process heater as a saturated liquid and is pumped to the boiler pressure. In this mode, no steam passes through the condenser, which operates at 20 kPa.

(a) Determine the power produced and the rate at which process heat is supplied in this mode.

(b) Determine the power produced and the rate of process heat supplied if only 60 percent of the steam is routed to the process heater and the remainder is expanded to the condenser pressure.

**10–68** Consider a cogeneration power plant modified with regeneration. Steam enters the turbine at 9 MPa and 400°C and expands to a pressure of 1.6 MPa. At this pressure, 35 percent of the steam is extracted from the turbine, and the remainder expands to 10 kPa. Part of the extracted steam is used to heat the feedwater in an open feedwater heater. The rest of the extracted steam is used for process heating and leaves the process heater as a saturated liquid at 1.6 MPa. It is subsequently mixed with the feedwater leaving the feedwater heater, and the mixture is pumped to the boiler pressure. Assuming the turbines and the pumps to be isentropic, show the cycle on a *T-s* diagram with respect to saturation lines, and determine the mass flow rate of steam through the boiler for a net power output of 25 MW. *Answer:* 29.1 kg/s

**10–69** [EES] Reconsider Prob. 10–68. Using EES (or other) software, investigate the effect of the extraction pressure for removing steam from the turbine to be used for the process heater and open feedwater heater on the required mass flow rate. Plot the mass flow rate through the boiler as a function of the extraction pressure, and discuss the results.

**10–70** Steam is generated in the boiler of a cogeneration plant at 4 MPa and 400°C at a rate of 15 kg/s. The plant is to produce power while meeting the process steam requirements for a certain industrial application. One-third of the steam leaving the boiler is throttled to a pressure of 0.8 MPa and is routed to the process heater. The rest of the steam is expanded in an isentropic turbine to a pressure of 0.8 MPa and is also routed to the process heater. Steam leaves the process heater at 115°C. Neglecting the pump work, determine (a) the net power produced, (b) the rate of process heat supply, and (c) the utilization factor of this plant.

### Combined Gas–Vapor Power Cycles

**10–71C** In combined gas–steam cycles, what is the energy source for the steam?

**10–72C** Why is the combined gas–steam cycle more efficient than either of the cycles operated alone?

**10–73** The gas-turbine portion of a combined gas–steam power plant has a pressure ratio of 16. Air enters the compressor at 300 K at a rate of 14 kg/s and is heated to 1500 K in the combustion chamber. The combustion gases leaving the gas turbine are used to heat the steam to 400°C at 10 MPa in a heat exchanger. The combustion gases leave the heat exchanger at 420 K. The steam leaving the turbine is condensed at 15 kPa. Assuming

all the compression and expansion processes to be isentropic, determine (a) the mass flow rate of the steam, (b) the net power output, and (c) the thermal efficiency of the combined cycle. For air, assume constant specific heats at room temperature.
*Answers:* (a) 1.275 kg/s, (b) 7819 kW, (c) 66.4 percent

**10–74** Consider a combined gas–steam power plant that has a net power output of 450 MW. The pressure ratio of the gas-turbine cycle is 14. Air enters the compressor at 300 K and the turbine at 1400 K. The combustion gases leaving the gas turbine are used to heat the steam at 8 MPa to 400°C in a heat exchanger. The combustion gases leave the heat exchanger at 460 K. An open feedwater heater incorporated with the steam cycle operates at a pressure of 0.6 MPa. The condenser pressure is 20 kPa. Assuming all the compression and expansion processes to be isentropic, determine (a) the mass flow rate ration of air to steam, (b) the required rate of heat input in the combustion chamber, and (c) thermal efficiency of the combined cycle.

**10–75** Reconsider Prob. 10–74. Using EES (or other) software, study the effects of the gas cycle pressure ratio as it is varied from 10 to 20 on the ratio of gas flow rate to steam flow rate and cycle thermal efficiency. Plot your results as functions of gas cycle pressure ratio, and discuss the results.

**10–76** Repeat Prob. 10–74 assuming isentropic efficiencies of 100 percent for the pump, 82 percent for the compressor, and 86 percent for the gas and steam turbines.

**10–77** Reconsider Prob. 10–76. Using EES (or other) software, study the effects of the gas cycle pressure ratio as it is varied from 10 to 20 on the ratio of gas flow rate to steam flow rate and cycle thermal efficiency. Plot your results as functions of gas cycle pressure ratio, and discuss the results.

**10–78** Consider a combined gas–steam power plant that has a net power output of 280 MW. The pressure ratio of the gas-turbine cycle is 11. Air enters the compressor at 300 K and the turbine at 1100 K. The combustion gases leaving the gas turbine are used to heat the steam at 5 MPa to 350°C in a heat exchanger. The combustion gases leave the heat exchanger at 420 K. An open feedwater heater incorporated with the steam cycle operates at a pressure of 0.8 MPa. The condenser pressure is 10 kPa. Assuming isentropic efficiencies of 100 percent for the pump, 82 percent for the compressor, and 86 percent for the gas and steam turbines, determine (a) the mass flow rate ratio of air to steam, (b) the required rate of heat input in the combustion chamber, and (c) the thermal efficiency of the combined cycle.

**10–79** Reconsider Prob. 10–78. Using EES (or other) software, study the effects of the gas cycle pressure ratio as it is varied from 10 to 20 on the ratio of gas flow rate to steam flow rate and cycle thermal efficiency. Plot your results as functions of gas cycle pressure ratio, and discuss the results.

**10–80** Consider a combined gas–steam power cycle. The topping cycle is a simple Brayton cycle that has a pressure ratio of 7. Air enters the compressor at 15°C at a rate of 40 kg/s and the gas turbine at 950°C. The bottoming cycle is a reheat Rankine cycle between the pressure limits of 6 MPa and 10 kPa. Steam is heated in a heat exchanger at a rate of 4.6 kg/s by the exhaust gases leaving the gas turbine, and the exhaust gases leave the heat exchanger at 200°C. Steam leaves the high-pressure turbine at 1.0 MPa and is reheated to 400°C in the heat exchanger before it expands in the low-pressure turbine. Assuming 80 percent isentropic efficiency for all pumps and turbines, determine (a) the moisture content at the exit of the low-pressure turbine, (b) the steam temperature at the inlet of the high-pressure turbine, (c) the net power output and the thermal efficiency of the combined plant.

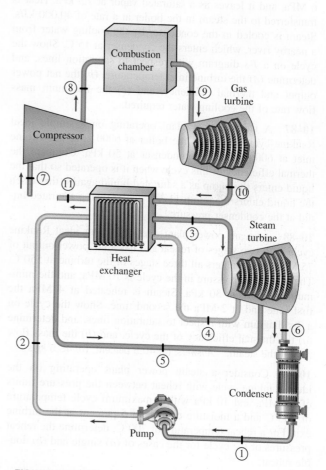

**FIGURE P10–80**

## Special Topic: Binary Vapor Cycles

**10–81C** Why is steam not an ideal working fluid for vapor power cycles?

**10–82C** What is a binary power cycle? What is its purpose?

**10–83C** What is the difference between the binary vapor power cycle and the combined gas–steam power cycle?

**10–84C** Why is mercury a suitable working fluid for the topping portion of a binary vapor cycle but not for the bottoming cycle?

**10–85** By writing an energy balance on the heat exchanger of a binary vapor power cycle, obtain a relation for the ratio of mass flow rates of two fluids in terms of their enthalpies.

## Review Problems

**10–86** Steam enters the turbine of a steam power plant that operates on a simple ideal Rankine cycle at a pressure of 6 MPa, and it leaves as a saturated vapor at 7.5 kPa. Heat is transferred to the steam in the boiler at a rate of 40,000 kJ/s. Steam is cooled in the condenser by the cooling water from a nearby river, which enters the condenser at 15°C. Show the cycle on a $T$-$s$ diagram with respect to saturation lines, and determine ($a$) the turbine inlet temperature, ($b$) the net power output and thermal efficiency, and ($c$) the minimum mass flow rate of the cooling water required.

**10–87** A steam power plant operating on a simple ideal Rankine cycle maintains the boiler at 6000 kPa, the turbine inlet at 600°C, and the condenser at 50 kPa. Compare the thermal efficiency of this cycle when it is operated so that the liquid enters the pump as a saturated liquid against that when the liquid enters the pump 11.3°C cooler than a saturated liquid at the condenser pressure.

**10–88** A steam power plant operates on an ideal Rankine cycle with two stages of reheat and has a net power output of 75 MW. Steam enters all three stages of the turbine at 550°C. The maximum pressure in the cycle is 10 MPa, and the minimum pressure is 30 kPa. Steam is reheated at 4 MPa the first time and at 2 MPa the second time. Show the cycle on a $T$-$s$ diagram with respect to saturation lines, and determine ($a$) the thermal efficiency of the cycle, and ($b$) the mass flow rate of the steam. *Answers: (a) 40.5 percent, (b) 48.5 kg/s*

**10–89** Consider a steam power plant operating on the ideal Rankine cycle with reheat between the pressure limits of 30 MPa and 10 kPa with a maximum cycle temperature of 700°C and a moisture content of 5 percent at the turbine exit. For a reheat temperature of 700°C, determine the reheat pressures of the cycle for the cases of ($a$) single and ($b$) double reheat.

**10–90** Consider a steam power plant that operates on a regenerative Rankine cycle and has a net power output of 150 MW. Steam enters the turbine at 10 MPa and 500°C and the condenser at 10 kPa. The isentropic efficiency of the turbine is 80 percent, and that of the pumps is 95 percent. Steam is extracted from the turbine at 0.5 MPa to heat the feedwater in an open feedwater heater. Water leaves the feedwater heater as a saturated liquid. Show the cycle on a $T$-$s$ diagram, and determine ($a$) the mass flow rate of steam through the

boiler, and ($b$) the thermal efficiency of the cycle. Also, determine the exergy destruction associated with the regeneration process. Assume a source temperature of 1300 K and a sink temperature of 303 K.

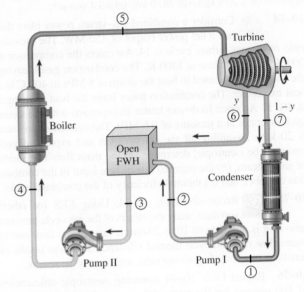

**FIGURE P10–90**

**10–91** Repeat Prob. 10–90 assuming both the pump and the turbine are isentropic.

**10–92** Consider an ideal reheat–regenerative Rankine cycle with one open feedwater heater. The boiler pressure is 10 MPa, the condenser pressure is 15 kPa, the reheater pressure is 1 MPa, and the feedwater pressure is 0.6 MPa. Steam enters both the high- and low-pressure turbines at 500°C. Show the cycle on a $T$-$s$ diagram with respect to saturation lines, and determine ($a$) the fraction of steam extracted for regeneration and ($b$) the thermal efficiency of the cycle. *Answers: (a) 0.144, (b) 42.1 percent*

**10–93** Repeat Prob. 10–92 assuming an isentropic efficiency of 84 percent for the turbines and 100 percent for the pumps.

**10–94** Steam is to be supplied from a boiler to a high-pressure turbine whose isentropic efficiency is 85 percent at conditions to be determined. The steam is to leave the high-pressure turbine as a saturated vapor at 1.4 MPa, and the turbine is to produce 5.5 MW of power. Steam at the turbine exit is extracted at a rate of 1000 kg/min and routed to a process heater while the rest of the steam is supplied to a low-pressure turbine whose isentropic efficiency is 80 percent. The low-pressure turbine allows the steam to expand to 10 kPa pressure and produces 1.5 MW of power. Determine the temperature, pressure, and the flow rate of steam at the inlet of the high-pressure turbine.

**10–95** A textile plant requires 4 kg/s of saturated steam at 2 MPa, which is extracted from the turbine of a cogeneration plant. Steam enters the turbine at 8 MPa and 500°C at a rate of 11 kg/s and leaves at 20 kPa. The extracted steam leaves the process heater as a saturated liquid and mixes with the feedwater at constant pressure. The mixture is pumped to the boiler pressure. Assuming an isentropic efficiency of 88 percent for both the turbine and the pumps, determine (a) the rate of process heat supply, (b) the net power output, and (c) the utilization factor of the plant. *Answers:* (a) 8.56 MW, (b) 8.60 MW, (c) 53.8 percent

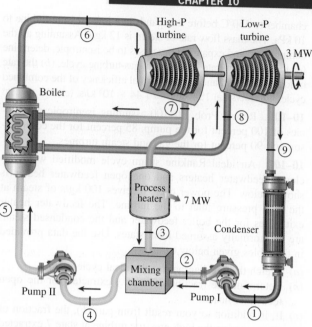

**FIGURE P10–96**

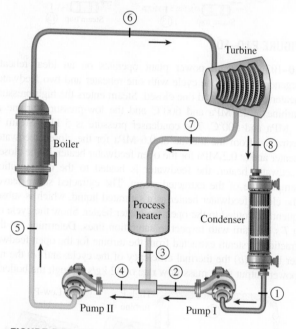

**FIGURE P10–95**

**10–96** Consider a cogeneration power plant that is modified with reheat and that produces 3 MW of power and supplies 7 MW of process heat. Steam enters the high-pressure turbine at 8 MPa and 500°C and expands to a pressure of 1 MPa. At this pressure, part of the steam is extracted from the turbine and routed to the process heater, while the remainder is reheated to 500°C and expanded in the low-pressure turbine to the condenser pressure of 15 kPa. The condensate from the condenser is pumped to 1 MPa and is mixed with the extracted steam, which leaves the process heater as a compressed liquid at 120°C. The mixture is then pumped to the boiler pressure. Assuming the turbine to be isentropic, show the cycle on a *T-s* diagram with respect to saturation lines, and disregarding pump work, determine (a) the rate of heat input in the boiler and (b) the fraction of steam extracted for process heating.

**10–97** Atmospheric air enters the air compressor of a simple combined gas-steam power system at 100 kPa and 27°C. The air compressor's compression ratio is 10; the gas cycle's maximum temperature is 1147°C; and the air compressor and turbine have an isentropic efficiency of 90 percent. The gas leaves the heat exchanger 30°C hotter than the saturation temperature of the steam in the heat exchanger. The steam pressure in the heat exchanger is 6 MPa, and the steam leaves the heat exchanger at 300°C. The steam-condenser pressure is 30 kPa and the isentropic efficiency of the steam turbine is 95 percent. Determine the overall thermal efficiency of this combined cycle. For air, use constant specific heats at room temperature. *Answer:* 46.3 percent

**10–98** It has been suggested that the steam passing through the condenser of the combined cycle in Prob. 10–97 be routed to buildings during the winter to heat them. When this is done, the pressure in the heating system where the steam is now condensed will have to be increased to 60 kPa. How does this change the overall thermal efficiency of the combined cycle?

**10–99** During winter, the system of Prob. 10–98 must supply 585 kW of heat to the buildings. What is the mass flow rate of air through the air compressor and the system's total electrical power production in winter? *Answers:* 3.61 kg/s, 1339 kW

**10–100** The gas-turbine cycle of a combined gas–steam power plant has a pressure ratio of 12. Air enters the compressor at 310 K and the turbine at 1400 K. The combustion gases leaving the gas turbine are used to heat the steam at 12.5 MPa to 500°C in a heat exchanger. The combustion gases leave the heat exchanger at 247°C. Steam expands in a high-pressure turbine to a pressure of 2.5 MPa and is reheated in the combustion chamber to 550°C before it expands in a low-pressure turbine to 10 kPa. The mass flow rate of steam is 12 kg/s. Assuming all the compression and expansion processes to be isentropic, determine (a) the mass flow rate of air in the gas-turbine cycle, (b) the rate of total heat input, and (c) the thermal efficiency of the combined cycle. *Answers:* (a) 154 kg/s, (b) $1.44 \times 10^5$ kJ/s, (c) 59.1 percent

**10–101** Repeat Prob. 10–100 assuming isentropic efficiencies of 100 percent for the pump, 85 percent for the compressor, and 90 percent for the gas and steam turbines.

**10–102** An ideal Rankine steam cycle modified with two closed feedwater heaters and one open feedwater heater is shown below. The power cycle receives 100 kg/s of steam at the high pressure inlet to the turbine. The feedwater heater exit states for the boiler feedwater and the condensed steam are the normally assumed ideal states. Use the data provided in the tables given below to

(a) Sketch the T-s diagram for the ideal cycle.
(b) Determine the fraction of mass y extracted for the open feedwater heater.
(c) If, in addition to your result from part (b), the fraction of mass entering the high pressure turbine at state 7 extracted for the closed feedwater heater operating at 140 kPa is $z = 0.0655$, and at 1910 kPa the extraction fraction is $w = 0.0830$, determine the cooling water temperature rise in the condenser, in °C, when the cooling water flow rate is 4200 kg/s. Assume $c_p = 4.18$ kJ/kg·K for cooling water.
(d) Determine the rate of heat rejected in the condenser and the thermal efficiency of the plant.

**Process states and selected data**

| State | P, kPa | T, °C | h, kJ/kg | s, kJ/kg·K |
|-------|--------|-------|----------|------------|
| 1 | 20 | | | |
| 2 | 620 | | | |
| 3 | 620 | | | |
| 4 | 620 | | | |
| 5 | 5000 | | | |
| 6 | 5000 | | | |
| 7 | 5000 | 700 | 3900 | 7.514 |
| 8 | 1910 | | 3515 | 7.514 |
| 9 | 620 | | 3154 | 7.514 |
| 10 | 140 | | 2799 | 7.514 |
| 11 | 20 | | 2478 | 7.514 |

**Saturation data**

| P, kPa | $T_{sat}$, °C | $v_f$, m³/kg | $h_f$, kJ/kg | $s_g$, kJ/kg·K |
|--------|---------------|--------------|--------------|----------------|
| 20 | 60.1 | 0.00102 | 251 | 7.907 |
| 140 | 109.3 | 0.00105 | 458 | 7.246 |
| 620 | 160.1 | 0.00110 | 676 | 6.748 |
| 1910 | 210.1 | 0.00117 | 898 | 6.356 |
| 5000 | 263.9 | 0.00129 | 1154 | 5.973 |

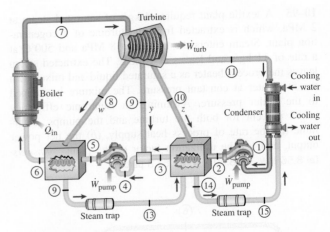

**FIGURE P10–102**

**10–103** A steam power plant operates on an ideal reheat–regenerative Rankine cycle with one reheater and two feedwater heaters, one open and one closed. Steam enters the high-pressure turbine at 15 MPa and 600°C and the low-pressure turbine at 1 MPa and 500°C. The condenser pressure is 5 kPa. Steam is extracted from the turbine at 0.6 MPa for the closed feedwater heater and at 0.2 MPa for the open feedwater heater. In the closed feedwater heater, the feedwater is heated to the condensation temperature of the extracted steam. The extracted steam leaves the closed feedwater heater as a saturated liquid, which is subsequently throttled to the open feedwater heater. Show the cycle on a T-s diagram with respect to saturation lines. Determine (a) the fraction of steam extracted from the turbine for the open feedwater heater, (b) the thermal efficiency of the cycle, and (c) the net power output for a mass flow rate of 42 kg/s through the boiler.

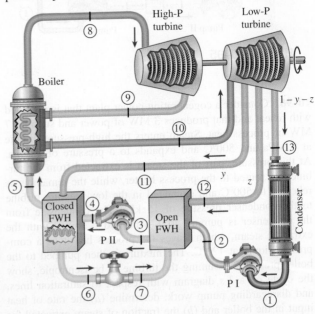

**FIGURE P10–103**

**10–104** [EES] Using EES (or other) software, investigate the effect of the boiler pressure on the performance of a simple ideal Rankine cycle. Steam enters the turbine at 500°C and exits at 10 kPa. The boiler pressure is varied from 0.5 to 20 MPa. Determine the thermal efficiency of the cycle and plot it against the boiler pressure, and discuss the results.

**10–105** [EES] Using EES (or other) software, investigate the effect of the condenser pressure on the performance of a simple ideal Rankine cycle. Turbine inlet conditions of steam are maintained constant at 10 MPa and 550°C while the condenser pressure is varied from 5 to 100 kPa. Determine the thermal efficiency of the cycle and plot it against the condenser pressure, and discuss the results.

**10–106** [EES] Using EES (or other) software, investigate the effect of reheat pressure on the performance of an ideal Rankine cycle. The maximum and minimum pressures in the cycle are 15 MPa and 10 kPa, respectively, and steam enters both stages of the turbine at 500°C. The reheat pressure is varied from 12.5 to 0.5 MPa. Determine the thermal efficiency of the cycle and plot it against the reheat pressure, and discuss the results.

**10–107** [EES] Using EES (or other) software, investigate the effect of extraction pressure on the performance of an ideal regenerative Rankine cycle with one open feedwater heater. Steam enters the turbine at 15 MPa and 600°C and the condenser at 10 kPa. Determine the thermal efficiency of the cycle, and plot it against extraction pressures of 12.5, 10, 7, 5, 2, 1, 0.5, 0.1, and 0.05 MPa, and discuss the results.

**10–108** Show that the thermal efficiency of a combined gas–steam power plant $\eta_{cc}$ can be expressed as

$$\eta_{cc} = \eta_g + \eta_s - \eta_g \eta_s$$

where $\eta_g = W_g/Q_{in}$ and $\eta_s = W_s/Q_{g,out}$ are the thermal efficiencies of the gas and steam cycles, respectively. Using this relation, determine the thermal efficiency of a combined power cycle that consists of a topping gas-turbine cycle with an efficiency of 40 percent and a bottoming steam-turbine cycle with an efficiency of 30 percent.

**10–109** It can be shown that the thermal efficiency of a combined gas–steam power plant $\eta_{cc}$ can be expressed in terms of the thermal efficiencies of the gas- and the steam-turbine cycles as

$$\eta_{cc} = \eta_g + \eta_s - \eta_g \eta_s$$

Prove that the value of $\eta_{cc}$ is greater than either of $\eta_g$ or $\eta_s$. That is, the combined cycle is more efficient than either of the gas-turbine or steam-turbine cycles alone.

**10–110** Starting with Eq. 10–20, show that the exergy destruction associated with a simple ideal Rankine cycle can be expressed as $x_{dest} = q_{in}(\eta_{th,Carnot} - \eta_{th})$, where $\eta_{th}$ is efficiency of the Rankine cycle and $\eta_{th,Carnot}$ is the efficiency of the Carnot cycle operating between the same temperature limits.

**10–111** A solar collector system delivers heat to a power plant. It is well known that the thermal collection efficiency $\eta_{sc}$ of a solar collector diminishes with increasing solar collection output temperature $T_H$, or $\eta_{sc} = A - BT_H$ where $A$ and $B$ are known constants. The thermal efficiency of the power plant $\eta_{th}$ is a fixed fraction of the Carnot thermal efficiency, such that $\eta_{th} = F(1 - T_L/T_H)$ where $F$ is a known constant assumed here independent of temperatures and $T_L$ is the condenser temperature, also constant for this problem. Here, the solar collection temperature $T_H$ is also taken to be the source temperature for the power plant.

(a) At what temperature $T_H$ should the solar collector be operated to obtain the maximum overall system efficiency?

(b) Develop an expression for the maximum overall system efficiency.

## Fundamentals of Engineering (FE) Exam Problems

**10–112** Consider a simple ideal Rankine cycle. If the condenser pressure is lowered while keeping turbine inlet state the same,

(a) the turbine work output will decrease.
(b) the amount of heat rejected will decrease.
(c) the cycle efficiency will decrease.
(d) the moisture content at turbine exit will decrease.
(e) the pump work input will decrease.

**10–113** Consider a simple ideal Rankine cycle with fixed boiler and condenser pressures. If the steam is superheated to a higher temperature,

(a) the turbine work output will decrease.
(b) the amount of heat rejected will decrease.
(c) the cycle efficiency will decrease.
(d) the moisture content at turbine exit will decrease.
(e) the amount of heat input will decrease.

**10–114** Consider a simple ideal Rankine cycle with fixed boiler and condenser pressures. If the cycle is modified with reheating,

(a) the turbine work output will decrease.
(b) the amount of heat rejected will decrease.
(c) the pump work input will decrease.
(d) the moisture content at turbine exit will decrease.
(e) the amount of heat input will decrease.

**10–115** Consider a simple ideal Rankine cycle with fixed boiler and condenser pressures. If the cycle is modified with regeneration that involves one open feedwater heater (select the correct statement per unit mass of steam flowing through the boiler),

(a) the turbine work output will decrease.

(b) the amount of heat rejected will increase.

(c) the cycle thermal efficiency will decrease.

(d) the quality of steam at turbine exit will decrease.

(e) the amount of heat input will increase.

**10–116** Consider a steady-flow Carnot cycle with water as the working fluid executed under the saturation dome between the pressure limits of 3 MPa and 10 kPa. Water changes from saturated liquid to saturated vapor during the heat addition process. The net work output of this cycle is

(a) 666 kJ/kg    (b) 888 kJ/kg    (c) 1040 kJ/kg
(d) 1130 kJ/kg    (e) 1440 kJ/kg

**10–117** A simple ideal Rankine cycle operates between the pressure limits of 10 kPa and 3 MPa, with a turbine inlet temperature of 600°C. Disregarding the pump work, the cycle efficiency is

(a) 24 percent    (b) 37 percent    (c) 52 percent
(d) 63 percent    (e) 71 percent

**10–118** A simple ideal Rankine cycle operates between the pressure limits of 10 kPa and 5 MPa, with a turbine inlet temperature of 600°C. The mass fraction of steam that condenses at the turbine exit is

(a) 6 percent    (b) 9 percent    (c) 12 percent
(d) 15 percent    (e) 18 percent

**10–119** A steam power plant operates on the simple ideal Rankine cycle between the pressure limits of 10 kPa and 5 MPa, with a turbine inlet temperature of 600°C. The rate of heat transfer in the boiler is 300 kJ/s. Disregarding the pump work, the power output of this plant is

(a) 93 kW    (b) 118 kW    (c) 190 kW
(d) 216 kW    (e) 300 kW

**10–120** Consider a combined gas-steam power plant. Water for the steam cycle is heated in a well-insulated heat exchanger by the exhaust gases that enter at 800 K at a rate of 60 kg/s and leave at 400 K. Water enters the heat exchanger at 200°C and 8 MPa and leaves at 350°C and 8 MPa. If the exhaust gases are treated as air with constant specific heats at room temperature, the mass flow rate of water through the heat exchanger becomes

(a) 11 kg/s    (b) 24 kg/s    (c) 46 kg/s
(d) 53 kg/s    (e) 60 kg/s

**10–121** An ideal reheat Rankine cycle operates between the pressure limits of 10 kPa and 8 MPa, with reheat occurring at 4 MPa. The temperature of steam at the inlets of both turbines is 500°C, and the enthalpy of steam is 3185 kJ/kg at the exit of the high-pressure turbine, and 2247 kJ/kg at the exit of the low-pressure turbine. Disregarding the pump work, the cycle efficiency is

(a) 29 percent    (b) 32 percent    (c) 36 percent
(d) 41 percent    (e) 49 percent

**10–122** Pressurized feedwater in a steam power plant is to be heated in an ideal open feedwater heater that operates at a pressure of 2 MPa with steam extracted from the turbine. If the enthalpy of feedwater is 252 kJ/kg and the enthalpy of extracted steam is 2810 kJ/kg, the mass fraction of steam extracted from the turbine is

(a) 10 percent    (b) 14 percent    (c) 26 percent
(d) 36 percent    (e) 50 percent

**10–123** Consider a steam power plant that operates on the regenerative Rankine cycle with one open feedwater heater. The enthalpy of the steam is 3374 kJ/kg at the turbine inlet, 2797 kJ/kg at the location of bleeding, and 2346 kJ/kg at the turbine exit. The net power output of the plant is 120 MW, and the fraction of steam bled off the turbine for regeneration is 0.172. If the pump work is negligible, the mass flow rate of steam at the turbine inlet is

(a) 117 kg/s    (b) 126 kg/s    (c) 219 kg/s
(d) 268 kg/s    (e) 679 kg/s

**10–124** Consider a cogeneration power plant modified with regeneration. Steam enters the turbine at 6 MPa and 450°C at a rate of 20 kg/s and expands to a pressure of 0.4 MPa. At this pressure, 60 percent of the steam is extracted from the turbine, and the remainder expands to a pressure of 10 kPa. Part of the extracted steam is used to heat feedwater in an open feedwater heater. The rest of the extracted steam is used for process heating and leaves the process heater as a saturated liquid at 0.4 MPa. It is subsequently mixed with the feedwater leaving the feedwater heater, and the mixture is pumped to the boiler pressure. The steam in the condenser is cooled and condensed by the cooling water from a nearby river, which enters the adiabatic condenser at a rate of 463 kg/s.

1. The total power output of the turbine is

(a) 17.0 MW    (b) 8.4 MW    (c) 12.2 MW
(d) 20.0 MW    (e) 3.4 MW

2. The temperature rise of the cooling water from the river in the condenser is

(a) 8.0°C    (b) 5.2°C    (c) 9.6°C
(d) 12.9°C    (e) 16.2°C

3. The mass flow rate of steam through the process heater is

(a) 1.6 kg/s    (b) 3.8 kg/s    (c) 5.2 kg/s
(d) 7.6 kg/s    (e) 10.4 kg/s

4. The rate of heat supply from the process heater per unit mass of steam passing through it is

(a) 246 kJ/kg    (b) 893 kJ/kg    (c) 1344 kJ/kg
(d) 1891 kJ/kg    (e) 2060 kJ/kg

5. The rate of heat transfer to the steam in the boiler is

(a) 26.0 MJ/s    (b) 53.8 MJ/s    (c) 39.5 MJ/s
(d) 62.8 MJ/s    (e) 125.4 MJ/s

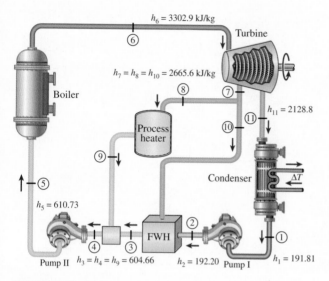

$h_6 = 3302.9 \text{ kJ/kg}$

Turbine

Boiler

$h_7 = h_8 = h_{10} = 2665.6 \text{ kJ/kg}$

Process heater

$h_{11} = 2128.8$

Condenser

$\Delta T$

$h_5 = 610.73$

Pump II  $h_3 = h_4 = h_9 = 604.66$  FWH  $h_2 = 192.20$  Pump I  $h_1 = 191.81$

**FIGURE P10–124**

## Design and Essay Problems

**10–125**  Design a steam power cycle that can achieve a cycle thermal efficiency of at least 40 percent under the conditions that all turbines have isentropic efficiencies of 85 percent and all pumps have isentropic efficiencies of 60 percent. Prepare an engineering report describing your design. Your design report must include, but is not limited to, the following:

(a) Discussion of various cycles attempted to meet the goal as well as the positive and negative aspects of your design.

(b) System figures and *T-s* diagrams with labeled states and temperature, pressure, enthalpy, and entropy information for your design.

(c) Sample calculations.

**10–126**  A natural gas–fired furnace in a textile plant is used to provide steam at 130°C. At times of high demand, the furnace supplies heat to the steam at a rate of 30 MJ/s. The plant also uses up to 6 MW of electrical power purchased from the local power company. The plant management is considering converting the existing process plant into a cogeneration plant to meet both their process-heat and power requirements. Your job is to come up with some designs. Designs based on a gas turbine or a steam turbine are to be considered. First decide whether a system based on a gas turbine or a steam turbine will best serve the purpose, considering the cost and the complexity. Then propose your design for the cogeneration plant complete with pressures and temperatures and the mass flow rates. Show that the proposed design meets the power and process-heat requirements of the plant.

**10–127**  Design the condenser of a steam power plant that has a thermal efficiency of 40 percent and generates 10 MW of net electric power. Steam enters the condenser as saturated vapor at 10 kPa, and it is to be condensed outside horizontal tubes through which cooling water from a nearby river flows. The temperature rise of the cooling water is limited to 8°C, and the velocity of the cooling water in the pipes is limited to 6 m/s to keep the pressure drop at an acceptable level. From prior experience, the average heat flux based on the outer surface of the tubes can be taken to be 12,000 W/m². Specify the pipe diameter, total pipe length, and the arrangement of the pipes to minimize the condenser volume.

**10–128**  Several geothermal power plants are in operation in the United States and more are being built since the heat source of a geothermal plant is hot geothermal water, which is "free energy." An 8-MW geothermal power plant is being considered at a location where geothermal water at 160°C is available. Geothermal water is to serve as the heat source for a closed Rankine power cycle with refrigerant-134a as the working fluid. Specify suitable temperatures and pressures for the cycle, and determine the thermal efficiency of the cycle. Justify your selections.

**10–129**  A photographic equipment manufacturer uses a flow of 29,200 kg/h of steam in its manufacturing process. Presently the spent steam at 27 kPa and 107°C is exhausted to the atmosphere. Do the preliminary design of a system to use the energy in the waste steam economically. If electricity is produced, it can be generated about 8000 h/yr and its value is $0.08/kWh. If the energy is used for space heating, the value is also $0.08/kWh, but it can only be used about 3000 h/yr (only during the "heating season"). If the steam is condensed and the liquid $H_2O$ is recycled through the process, its value is $0.18/100 L. Make all assumptions as realistic as possible. Sketch the system you propose. Make a separate list of required components and their specifications (capacity, efficiency, etc.). The final result will be the calculated annual dollar value of the energy use plan (actually a *saving* because it will replace electricity or heat and/or water that would otherwise have to be purchased).

**10–130**  Stack gases exhausting from electrical power plants are at approximately 150°C. Design a basic Rankine cycle that uses water, refrigerant-134a, or ammonia as the working fluid and that produces the maximum amount of work from this energy source while rejecting heat to the ambient air at 40°C. You are to use a turbine whose efficiency is 92 percent and whose exit quality cannot be less than 85 percent.

**10–131**  Contact your power company and obtain information on the thermodynamic aspects of their most recently built power plant. If it is a conventional power plant, find out why it is preferred over a highly efficient combined power plant.

**10–132**  Steam boilers have long been used to provide process heat as well as to generate power. Write an essay on the history of steam boilers and the evolution of modern supercritical steam power plants. What was the role of the American Society of Mechanical Engineers in this development?

# REFERIGERATION CYCLES

A major application area of thermodynamics is *refrigeration,* which is the transfer of heat from a lower temperature region to a higher temperature one. Devices that produce refrigeration are called *refrigerators,* and the cycles on which they operate are called *refrigeration cycles.* The most frequently used refrigeration cycle is the *vapor-compression refrigeration cycle* in which the refrigerant is vaporized and condensed alternately and is compressed in the vapor phase. Another well-known refrigeration cycle, and is the *gas refrigeration cycle* in which the refrigerant remains in the gaseous phase throughout. Other refrigeration cycles discussed in this chapter are *cascade refrigeration,* where more than one refrigeration cycle is used: and *absorption refrigeration,* where the refrigerant is dissolved in a liquid before it is compressed.

## OBJECTIVES

The objectives of Chapter 11 are to:

- Introduce the concepts of refrigerators and heat pumps and the measure of their performance.

- Analyze the ideal vapor-compression refrigeration cycle.

- Analyze the actual vapor-compression refrigeration cycle.

- Perform second-law analysis of vapor-compression refrigeration cycle.

- Review the factors involved in selecting the right refrigerant for an application.

- Discuss the operation of refrigeration and heat pump systems.

- Evaluate the performance of innovative vapor-compression refrigeration systems.

- Analyze gas refrigeration systems.

- Introduce the concepts of absorption-refrigeration systems.

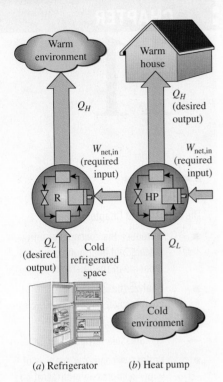

*(a)* Refrigerator    *(b)* Heat pump

**FIGURE 11–1**

The objective of a refrigerator is to remove heat ($Q_L$) from the cold medium; the objective of a heat pump is to supply heat ($Q_H$) to a warm medium.

# 11–1 · REFRIGERATORS AND HEAT PUMPS

We all know from experience that heat flows in the direction of decreasing temperature, that is, from high-temperature regions to low-temperature ones. This heat-transfer process occurs in nature without requiring any devices. The reverse process, however, cannot occur by itself. The transfer of heat from a low-temperature region to a high-temperature one requires special devices called **refrigerators**.

Refrigerators are cyclic devices, and the working fluids used in the refrigeration cycles are called **refrigerants**. A refrigerator is shown schematically in Fig. 11–1*a*. Here $Q_L$ is the magnitude of the heat removed from the refrigerated space at temperature $T_L$, $Q_H$ is the magnitude of the heat rejected to the warm space at temperature $T_H$, and $W_{net,in}$ is the net work input to the refrigerator. As discussed in Chap. 6, $Q_L$ and $Q_H$ represent magnitudes and thus are positive quantities.

Another device that transfers heat from a low-temperature medium to a high-temperature one is the **heat pump**. Refrigerators and heat pumps are essentially the same devices; they differ in their objectives only. The objective of a refrigerator is to maintain the refrigerated space at a low temperature by removing heat from it. Discharging this heat to a higher-temperature medium is merely a necessary part of the operation, not the purpose. The objective of a heat pump, however, is to maintain a heated space at a high temperature. This is accomplished by absorbing heat from a low-temperature source, such as well water or cold outside air in winter, and supplying this heat to a warmer medium such as a house (Fig. 11–1*b*).

The performance of refrigerators and heat pumps is expressed in terms of the **coefficient of performance** (COP), defined as

$$\text{COP}_R = \frac{\text{Desired output}}{\text{Required input}} = \frac{\text{Cooling effect}}{\text{Work input}} = \frac{Q_L}{W_{net,in}} \qquad (11\text{–}1)$$

$$\text{COP}_{HP} = \frac{\text{Desired output}}{\text{Required input}} = \frac{\text{Heating effect}}{\text{Work input}} = \frac{Q_H}{W_{net,in}} \qquad (11\text{–}2)$$

These relations can also be expressed in the rate form by replacing the quantities $Q_L$, $Q_H$, and $W_{net,in}$ by $\dot{Q}_L$, $\dot{Q}_H$, and $\dot{W}_{net,in}$, respectively. Notice that both $\text{COP}_R$ and $\text{COP}_{HP}$ can be greater than 1. A comparison of Eqs. 11–1 and 11–2 reveals that

$$\text{COP}_{HP} = \text{COP}_R + 1 \qquad (11\text{–}3)$$

for fixed values of $Q_L$ and $Q_H$. This relation implies that $\text{COP}_{HP} > 1$ since $\text{COP}_R$ is a positive quantity. That is, a heat pump functions, at worst, as a resistance heater, supplying as much energy to the house as it consumes. In reality, however, part of $Q_H$ is lost to the outside air through piping and other devices, and $\text{COP}_{HP}$ may drop below unity when the outside air temperature is too low. When this happens, the system normally switches to the fuel (natural gas, propane, oil, etc.) or resistance-heating mode.

The *cooling capacity* of a refrigeration system—that is, the rate of heat removal from the refrigerated space—is often expressed in terms of **tons of refrigeration**. The capacity of a refrigeration system that can freeze 1 ton

(2000 lbm) of liquid water at 0°C (32°F) into ice at 0°C in 24 h is said to be 1 ton. One ton of refrigeration is equivalent to 211 kJ/min or 200 Btu/min. The cooling load of a typical 200-m² residence is in the 3-ton (10-kW) range.

# 11–2 · THE REVERSED CARNOT CYCLE

Recall from Chap. 6 that the Carnot cycle is a totally reversible cycle that consists of two reversible isothermal and two isentropic processes. It has the maximum thermal efficiency for given temperature limits, and it serves as a standard against which actual power cycles can be compared.

Since it is a reversible cycle, all four processes that comprise the Carnot cycle can be reversed. Reversing the cycle does also reverse the directions of any heat and work interactions. The result is a cycle that operates in the counterclockwise direction on a $T$-$s$ diagram, which is called the **reversed Carnot cycle**. A refrigerator or heat pump that operates on the reversed Carnot cycle is called a **Carnot refrigerator** or a **Carnot heat pump**.

Consider a reversed Carnot cycle executed within the saturation dome of a refrigerant, as shown in Fig. 11–2. The refrigerant absorbs heat isothermally from a low-temperature source at $T_L$ in the amount of $Q_L$ (process 1-2), is compressed isentropically to state 3 (temperature rises to $T_H$), rejects heat isothermally to a high-temperature sink at $T_H$ in the amount of $Q_H$ (process 3-4), and expands isentropically to state 1 (temperature drops to $T_L$). The refrigerant changes from a saturated vapor state to a saturated liquid state in the condenser during process 3-4.

The coefficients of performance of Carnot refrigerators and heat pumps are expressed in terms of temperatures as

$$\text{COP}_{R,\text{Carnot}} = \frac{1}{T_H/T_L - 1} \tag{11–4}$$

and

$$\text{COP}_{HP,\text{Carnot}} = \frac{1}{1 - T_L/T_H} \tag{11–5}$$

Notice that both COPs increase as the difference between the two temperatures decreases, that is, as $T_L$ rises or $T_H$ falls.

The reversed Carnot cycle is the *most efficient* refrigeration cycle operating between two specified temperature levels. Therefore, it is natural to look at it first as a prospective ideal cycle for refrigerators and heat pumps. If we could, we certainly would adapt it as the ideal cycle. As explained below, however, the reversed Carnot cycle is not a suitable model for refrigeration cycles.

The two isothermal heat transfer processes are not difficult to achieve in practice since maintaining a constant pressure automatically fixes the temperature of a two-phase mixture at the saturation value. Therefore, processes 1-2 and 3-4 can be approached closely in actual evaporators and condensers. However, processes 2-3 and 4-1 cannot be approximated closely in practice. This is because process 2-3 involves the compression of a liquid–vapor mixture, which requires a compressor that will handle two phases, and process 4-1 involves the expansion of high-moisture-content refrigerant in a turbine.

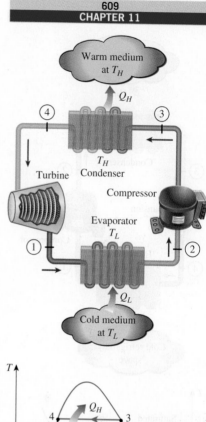

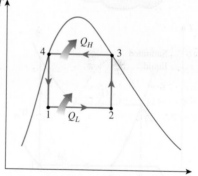

**FIGURE 11–2**

Schematic of a Carnot refrigerator and $T$-$s$ diagram of the reversed Carnot cycle.

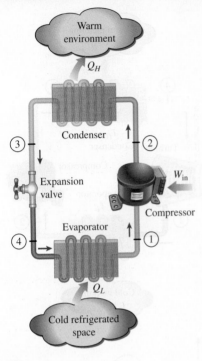

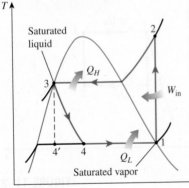

**FIGURE 11–3**

Schematic and $T$-$s$ diagram for the ideal vapor-compression refrigeration cycle.

It seems as if these problems could be eliminated by executing the reversed Carnot cycle outside the saturation region. But in this case we have difficulty in maintaining isothermal conditions during the heat-absorption and heat-rejection processes. Therefore, we conclude that the reversed Carnot cycle cannot be approximated in actual devices and is not a realistic model for refrigeration cycles. However, the reversed Carnot cycle can serve as a standard against which actual refrigeration cycles are compared.

## 11–3 ▪ THE IDEAL VAPOR-COMPRESSION REFRIGERATION CYCLE

Many of the impracticalities associated with the reversed Carnot cycle can be eliminated by vaporizing the refrigerant completely before it is compressed and by replacing the turbine with a throttling device, such as an expansion valve or capillary tube. The cycle that results is called the **ideal vapor-compression refrigeration cycle**, and it is shown schematically and on a $T$-$s$ diagram in Fig. 11–3. The vapor-compression refrigeration cycle is the most widely used cycle for refrigerators, air-conditioning systems, and heat pumps. It consists of four processes:

| 1-2 | Isentropic compression in a compressor |
| 2-3 | Constant-pressure heat rejection in a condenser |
| 3-4 | Throttling in an expansion device |
| 4-1 | Constant-pressure heat absorption in an evaporator |

In an ideal vapor-compression refrigeration cycle, the refrigerant enters the compressor at state 1 as saturated vapor and is compressed isentropically to the condenser pressure. The temperature of the refrigerant increases during this isentropic compression process to well above the temperature of the surrounding medium. The refrigerant then enters the condenser as superheated vapor at state 2 and leaves as saturated liquid at state 3 as a result of heat rejection to the surroundings. The temperature of the refrigerant at this state is still above the temperature of the surroundings.

The saturated liquid refrigerant at state 3 is throttled to the evaporator pressure by passing it through an expansion valve or capillary tube. The temperature of the refrigerant drops below the temperature of the refrigerated space during this process. The refrigerant enters the evaporator at state 4 as a low-quality saturated mixture, and it completely evaporates by absorbing heat from the refrigerated space. The refrigerant leaves the evaporator as saturated vapor and reenters the compressor, completing the cycle.

In a household refrigerator, the tubes in the freezer compartment where heat is absorbed by the refrigerant serves as the evaporator. The coils behind the refrigerator, where heat is dissipated to the kitchen air, serve as the condenser (Fig. 11–4).

Remember that the area under the process curve on a $T$-$s$ diagram represents the heat transfer for internally reversible processes. The area under the process curve 4-1 represents the heat absorbed by the refrigerant in the evaporator, and the area under the process curve 2-3 represents the heat rejected in the condenser. A rule of thumb is that the *COP improves by 2 to 4 percent for each °C the evaporating temperature is raised or the condensing temperature is lowered.*

Another diagram frequently used in the analysis of vapor-compression refrigeration cycles is the *P-h* diagram, as shown in Fig. 11–5. On this diagram, three of the four processes appear as straight lines, and the heat transfer in the condenser and the evaporator is proportional to the lengths of the corresponding process curves.

Notice that unlike the ideal cycles discussed before, the ideal vapor-compression refrigeration cycle is not an internally reversible cycle since it involves an irreversible (throttling) process. This process is maintained in the cycle to make it a more realistic model for the actual vapor-compression refrigeration cycle. If the throttling device were replaced by an isentropic turbine, the refrigerant would enter the evaporator at state 4' instead of state 4. As a result, the refrigeration capacity would increase (by the area under process curve 4'-4 in Fig. 11–3) and the net work input would decrease (by the amount of work output of the turbine). Replacing the expansion valve by a turbine is not practical, however, since the added benefits cannot justify the added cost and complexity.

All four components associated with the vapor-compression refrigeration cycle are steady-flow devices, and thus all four processes that make up the cycle can be analyzed as steady-flow processes. The kinetic and potential energy changes of the refrigerant are usually small relative to the work and heat transfer terms, and therefore they can be neglected. Then the steady-flow energy equation on a unit–mass basis reduces to

$$(q_{in} - q_{out}) + (w_{in} - w_{out}) = h_e - h_i \tag{11–6}$$

The condenser and the evaporator do not involve any work, and the compressor can be approximated as adiabatic. Then the COPs of refrigerators and heat pumps operating on the vapor-compression refrigeration cycle can be expressed as

$$COP_R = \frac{q_L}{w_{net,in}} = \frac{h_1 - h_4}{h_2 - h_1} \tag{11–7}$$

and

$$COP_{HP} = \frac{q_H}{w_{net,in}} = \frac{h_2 - h_3}{h_2 - h_1} \tag{11–8}$$

where $h_1 = h_{g\,@\,P_1}$ and $h_3 = h_{f\,@\,P_3}$ for the ideal case.

Vapor-compression refrigeration dates back to 1834 when the Englishman Jacob Perkins received a patent for a closed-cycle ice machine using ether or other volatile fluids as refrigerants. A working model of this machine was built, but it was never produced commercially. In 1850, Alexander Twining began to design and build vapor-compression ice machines using ethyl ether, which is a commercially used refrigerant in vapor-compression systems. Initially, vapor-compression refrigeration systems were large and were mainly used for ice making, brewing, and cold storage. They lacked automatic controls and were steam-engine driven. In the 1890s, electric motor-driven smaller machines equipped with automatic controls started to replace the older units, and refrigeration systems began to appear in butcher shops and households. By 1930, the continued improvements made it possible to have vapor-compression refrigeration systems that were relatively efficient, reliable, small, and inexpensive.

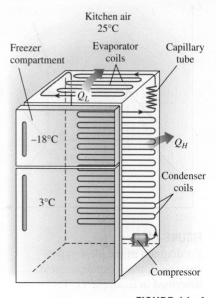

**FIGURE 11–4**
An ordinary household refrigerator.

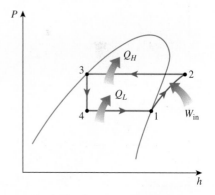

**FIGURE 11–5**
The *P-h* diagram of an ideal vapor-compression refrigeration cycle.

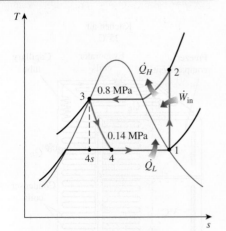

**FIGURE 11–6**
*T-s* diagram of the ideal vapor-compression refrigeration cycle described in Example 11–1.

**EXAMPLE 11–1**  **The Ideal Vapor-Compression Refrigeration Cycle**

A refrigerator uses refrigerant-134a as the working fluid and operates on an ideal vapor-compression refrigeration cycle between 0.14 and 0.8 MPa. If the mass flow rate of the refrigerant is 0.05 kg/s, determine (a) the rate of heat removal from the refrigerated space and the power input to the compressor, (b) the rate of heat rejection to the environment, and (c) the COP of the refrigerator.

**SOLUTION** A refrigerator operates on an ideal vapor-compression refrigeration cycle between two specified pressure limits. The rate of refrigeration, the power input, the rate of heat rejection, and the COP are to be determined.

**Assumptions** 1 Steady operating conditions exist. 2 Kinetic and potential energy changes are negligible.

**Analysis** The *T-s* diagram of the refrigeration cycle is shown in Fig. 11–6. We note that this is an ideal vapor-compression refrigeration cycle, and thus the compressor is isentropic and the refrigerant leaves the condenser as a saturated liquid and enters the compressor as saturated vapor. From the refrigerant-134a tables, the enthalpies of the refrigerant at all four states are determined as follows:

$$P_1 = 0.14 \text{ MPa} \longrightarrow h_1 = h_{g\,@\,0.14\,\text{MPa}} = 239.19 \text{ kJ/kg}$$
$$s_1 = s_{g\,@\,0.14\,\text{MPa}} = 0.94467 \text{ kJ/kg·K}$$

$$\left. \begin{array}{l} P_2 = 0.8 \text{ MPa} \\ s_2 = s_1 \end{array} \right\} h_2 = 275.40 \text{ kJ/kg}$$

$$P_3 = 0.8 \text{ MPa} \longrightarrow h_3 = h_{f\,@\,0.8\,\text{MPa}} = 95.48 \text{ kJ/kg}$$
$$h_4 \cong h_3 \text{ (throttling)} \longrightarrow h_4 = 95.48 \text{ kJ/kg}$$

(a) The rate of heat removal from the refrigerated space and the power input to the compressor are determined from their definitions:

$$\dot{Q}_L = \dot{m}(h_1 - h_4) = (0.05 \text{ kg/s})[(239.19 - 95.48) \text{ kJ/kg}] = \textbf{7.19 kW}$$

and

$$\dot{W}_{in} = \dot{m}(h_2 - h_1) = (0.05 \text{ kg/s})[(275.40 - 239.19) \text{ kJ/kg}] = \textbf{1.81 kW}$$

(b) The rate of heat rejection from the refrigerant to the environment is

$$\dot{Q}_H = \dot{m}(h_2 - h_3) = (0.05 \text{ kg/s})[(275.40 - 95.48) \text{ kJ/kg}] = \textbf{9.00 kW}$$

It could also be determined from

$$\dot{Q}_H = \dot{Q}_L + \dot{W}_{in} = 7.19 + 1.81 = 9.00 \text{ kW}$$

(c) The coefficient of performance of the refrigerator is

$$\text{COP}_R = \frac{\dot{Q}_L}{\dot{W}_{in}} = \frac{7.19 \text{ kW}}{1.81 \text{ kW}} = \textbf{3.97}$$

That is, this refrigerator removes about 4 units of thermal energy from the refrigerated space for each unit of electric energy it consumes.

**Discussion** It would be interesting to see what happens if the throttling valve were replaced by an isentropic turbine. The enthalpy at state 4s (the turbine exit with $P_{4s} = 0.14$ MPa, and $s_{4s} = s_3 = 0.35408$ kJ/kg·K) is 88.95 kJ/kg, and the turbine would produce 0.33 kW of power. This would decrease the power input to the refrigerator from 1.81 to 1.48 kW and increase the rate of heat removal from the refrigerated space from 7.19 to 7.51 kW. As a result, the COP of the refrigerator would increase from 3.97 to 5.07, an increase of 28 percent.

# 11–4 · ACTUAL VAPOR-COMPRESSION REFRIGERATION CYCLE

An actual vapor-compression refrigeration cycle differs from the ideal one in several ways, owing mostly to the irreversibilities that occur in various components. Two common sources of irreversibilities are fluid friction (causes pressure drops) and heat transfer to or from the surroundings. The *T-s* diagram of an actual vapor-compression refrigeration cycle is shown in Fig. 11–7.

In the ideal cycle, the refrigerant leaves the evaporator and enters the compressor as *saturated vapor*. In practice, however, it may not be possible to control the state of the refrigerant so precisely. Instead, it is easier to design the system so that the refrigerant is slightly superheated at the compressor inlet. This slight overdesign ensures that the refrigerant is completely vaporized when it enters the compressor. Also, the line connecting the evaporator to the compressor is usually very long; thus the pressure drop caused by fluid friction and heat transfer from the surroundings to the refrigerant can be very significant. The result of superheating, heat gain in the connecting line, and pressure drops in the evaporator and the connecting line is an increase in the specific volume, thus an increase in the power input requirements to the compressor since steady-flow work is proportional to the specific volume.

The *compression process* in the ideal cycle is internally reversible and adiabatic, and thus isentropic. The actual compression process, however, involves frictional effects, which increase the entropy, and heat transfer, which may increase or decrease the entropy, depending on the direction. Therefore, the entropy of the refrigerant may increase (process 1-2) or decrease (process 1-2′) during an actual compression process, depending on which effects dominate. The compression process 1-2′ may be even more desirable than the isentropic compression process since the specific volume of the refrigerant and thus the work input requirement are smaller in this case. Therefore, the refrigerant should be cooled during the compression process whenever it is practical and economical to do so.

In the ideal case, the refrigerant is assumed to leave the condenser as *saturated liquid* at the compressor exit pressure. In reality, however, it is unavoidable to have some pressure drop in the condenser as well as in the lines connecting the condenser to the compressor and to the throttling valve. Also, it is not easy to execute the condensation process with such precision

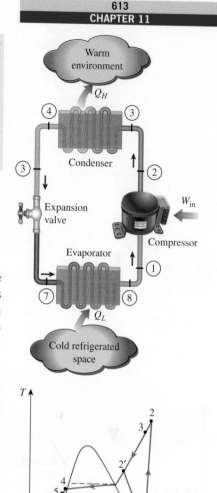

**FIGURE 11–7**
Schematic and *T-s* diagram for the actual vapor-compression refrigeration cycle.

that the refrigerant is a saturated liquid at the end, and it is undesirable to route the refrigerant to the throttling valve before the refrigerant is completely condensed. Therefore, the refrigerant is subcooled somewhat before it enters the throttling valve. We do not mind this at all, however, since the refrigerant in this case enters the evaporator with a lower enthalpy and thus can absorb more heat from the refrigerated space. The throttling valve and the evaporator are usually located very close to each other, so the pressure drop in the connecting line is small.

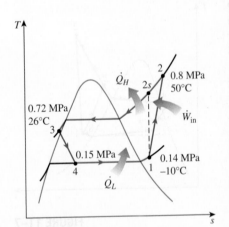

**FIGURE 11–8**
$T$-$s$ diagram for Example 11–2.

**EXAMPLE 11–2**  **The Actual Vapor-Compression Refrigeration Cycle**

Refrigerant-134a enters the compressor of a refrigerator as superheated vapor at 0.14 MPa and −10°C at a rate of 0.05 kg/s and leaves at 0.8 MPa and 50°C. The refrigerant is cooled in the condenser to 26°C and 0.72 MPa and is throttled to 0.15 MPa. Disregarding any heat transfer and pressure drops in the connecting lines between the components, determine (a) the rate of heat removal from the refrigerated space and the power input to the compressor, (b) the isentropic efficiency of the compressor, and (c) the coefficient of performance of the refrigerator.

**SOLUTION** A refrigerator operating on a vapor-compression cycle is considered. The rate of refrigeration, the power input, the compressor efficiency, and the COP are to be determined.
**Assumptions** 1 Steady operating conditions exist. 2 Kinetic and potential energy changes are negligible.
**Analysis** The $T$-$s$ diagram of the refrigeration cycle is shown in Fig. 11–8. We note that the refrigerant leaves the condenser as a compressed liquid and enters the compressor as superheated vapor. The enthalpies of the refrigerant at various states are determined from the refrigerant tables to be

$$\left. \begin{array}{l} P_1 = 0.14 \text{ MPa} \\ T_1 = -10°C \end{array} \right\} \quad h_1 = 246.37 \text{ kJ/kg}$$

$$\left. \begin{array}{l} P_2 = 0.8 \text{ MPa} \\ T_2 = 50°C \end{array} \right\} \quad h_2 = 286.71 \text{ kJ/kg}$$

$$\left. \begin{array}{l} P_3 = 0.72 \text{ MPa} \\ T_3 = 26°C \end{array} \right\} \quad h_3 \cong h_{f\,@\,26°C} = 87.83 \text{ kJ/kg}$$

$$h_4 \cong h_3 \text{ (throttling)} \quad \longrightarrow \quad h_4 = 87.83 \text{ kJ/kg}$$

(a) The rate of heat removal from the refrigerated space and the power input to the compressor are determined from their definitions:

$$\dot{Q}_L = \dot{m}(h_1 - h_4) = (0.05 \text{ kg/s})[(246.37 - 87.83) \text{ kJ/kg}] = \textbf{7.93 kW}$$

and

$$\dot{W}_{in} = \dot{m}(h_2 - h_1) = (0.05 \text{ kg/s})[(286.71 - 246.37) \text{ kJ/kg}] = \textbf{2.02 kW}$$

(b) The isentropic efficiency of the compressor is determined from

$$\eta_C \cong \frac{h_{2s} - h_1}{h_2 - h_1}$$

where the enthalpy at state 2s ($P_{2s}$ = 0.8 MPa and $s_{2s} = s_1$ = 0.9724 kJ/kg·K) is 284.20 kJ/kg. Thus,

$$\eta_C = \frac{284.20 - 246.37}{286.71 - 246.37} = 0.938 \text{ or } 93.8\%$$

(c) The coefficient of performance of the refrigerator is

$$COP_R = \frac{\dot{Q}_L}{\dot{W}_{in}} = \frac{7.93 \text{ kW}}{2.02 \text{ kW}} = 3.93$$

**Discussion** This problem is identical to the one worked out in Example 11–1, except that the refrigerant is slightly superheated at the compressor inlet and subcooled at the condenser exit. Also, the compressor is not isentropic. As a result, the heat removal rate from the refrigerated space increases (by 10.3 percent), but the power input to the compressor increases even more (by 11.6 percent). Consequently, the COP of the refrigerator decreases from 3.97 to 3.93.

# 11–5 · SECOND-LAW ANALYSIS OF VAPOR-COMPRESSION REFRIGERATION CYCLE[1]

Consider the vapor-compression refrigeration cycle operating between a low-temperature medium at $T_L$ and a high-temperature medium at $T_H$ as shown in Fig. 11–9. The maximum COP of a refrigeration cycle operating between temperature limits of $T_L$ and $T_H$ was given in Eq. 11–4 as

$$COP_{R,max} = COP_{R,rev} = COP_{R,Carnot} = \frac{T_L}{T_H - T_L} = \frac{1}{T_H/T_L - 1} \qquad (11\text{–}9)$$

Actual refrigeration cycles are not as efficient as ideal ones like the Carnot cycle because of the irreversibilities involved. But the conclusion we can draw from Eq. 11–9 that the COP is inversely proportional to the temperature difference $T_H - T_L$ is equally valid for actual refrigeration cycles.

The goal of a second-law or exergy analysis of a refrigeration system is to determine the components that can benefit the most by improvements. This is done identifying the locations of greatest exergy destruction and the components with the lowest exergy or second-law efficiency. Exergy destruction in a component can be determined directly from an exergy balance or indirectly by first calculating the entropy generation and then using the relation

$$\dot{X}_{dest} = T_0 \dot{S}_{gen} \qquad (11\text{–}10)$$

where $T_0$ is the environment (the dead state) temperature. For a refrigerator, $T_0$ is usually the temperature of the high-temperature medium $T_H$ (for a heat pump it is $T_L$). Exergy destructions and the second-law efficiencies for

[1]This section is contributed by Professor Mehmet Kanoglu of the University of Gaziantep.

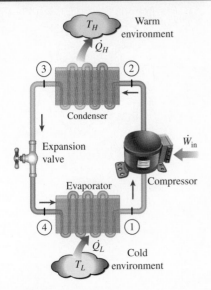

**FIGURE 11–9**
The vapor-compression refrigeration cycle considered in the second-law analysis.

major components of a refrigeration system operating on the cycle shown in Fig. 11–9 may be written as follows:

*Compressor:*

$$\dot{X}_{\text{dest},1-2} = T_0 \dot{S}_{\text{gen},1-2} = \dot{m} T_0(s_2 - s_1) \tag{11-11}$$

$$\eta_{\text{II,Comp}} = \frac{\dot{X}_{\text{recovered}}}{\dot{X}_{\text{expended}}} = \frac{\dot{W}_{\text{rev}}}{\dot{W}_{\text{act,in}}} = \frac{\dot{m}[h_2 - h_1 - T_0(s_2 - s_1)]}{\dot{m}(h_2 - h_1)} = \frac{\psi_2 - \psi_1}{h_2 - h_1}$$

$$= 1 - \frac{\dot{X}_{\text{dest},1-2}}{\dot{W}_{\text{act,in}}} \tag{11-12}$$

*Condenser:*

$$\dot{X}_{\text{dest},2-3} = T_0 \dot{S}_{\text{gen},2-3} = T_0 \left[ \dot{m}(s_3 - s_2) + \frac{\dot{Q}_H}{T_H} \right] \tag{11-13}$$

$$\eta_{\text{II,Cond}} = \frac{\dot{X}_{\text{recovered}}}{\dot{X}_{\text{expended}}} = \frac{\dot{X}_{Q_H}}{\dot{X}_2 - \dot{X}_3} = \frac{\dot{Q}_H(1 - T_0/T_H)}{\dot{X}_2 - \dot{X}_3}$$

$$= \frac{\dot{Q}_H(1 - T_0/T_H)}{\dot{m}[h_2 - h_3 - T_0(s_2 - s_3)]} = 1 - \frac{\dot{X}_{\text{dest},2-3}}{\dot{X}_2 - \dot{X}_3} \tag{11-14}$$

Note that when $T_H = T_0$, which is usually the case for refrigerators, $\eta_{\text{II,Cond}} = 0$ since there is no *recoverable* exergy in this case.

*Expansion valve:*

$$\dot{X}_{\text{dest},3-4} = T_0 \dot{S}_{\text{gen},3-4} = \dot{m} T_0(s_4 - s_3) \tag{11-15}$$

$$\eta_{\text{II,ExpValve}} = \frac{\dot{X}_{\text{recovered}}}{\dot{X}_{\text{expended}}} = \frac{0}{\dot{X}_3 - \dot{X}_4} = 0$$

or $\qquad \eta_{\text{II,ExpValve}} = 1 - \frac{\dot{X}_{\text{dest},3-4}}{\dot{X}_{\text{expended}}} = 1 - \frac{\dot{X}_3 - \dot{X}_4}{\dot{X}_3 - \dot{X}_4} = 0 \tag{11-16}$

*Evaporator:*

$$\dot{X}_{\text{dest},4-1} = T_0 \dot{S}_{\text{gen},4-1} = T_0 \left[ \dot{m}(s_1 - s_4) - \frac{\dot{Q}_L}{T_L} \right] \tag{11-17}$$

$$\eta_{\text{II,Evap}} = \frac{\dot{X}_{\text{recovered}}}{\dot{X}_{\text{expended}}} = \frac{\dot{X}_{Q_L}}{\dot{X}_4 - \dot{X}_1} = \frac{\dot{Q}_L(T_0 - T_L)/T_L}{\dot{X}_4 - \dot{X}_1}$$

$$= \frac{\dot{Q}_L(T_0 - T_L)/T_L}{\dot{m}[h_4 - h_1 - T_0(s_4 - s_1)]} = 1 - \frac{\dot{X}_{\text{dest},4-1}}{\dot{X}_4 - \dot{X}_1} \tag{11-18}$$

Here $\dot{X}_{Q_L}$ represents the positive of the exergy rate associated with the withdrawal of heat from the low-temperature medium at $T_L$ at a rate $\dot{Q}_L$. Note that the directions of heat and exergy transfer become opposite when $T_L < T_0$ (that is, the exergy of the low-temperature medium increases as it loses heat). Also, $\dot{X}_{Q_L}$ is equivalent to the power that can be produced by a Carnot heat engine receiving heat from the environment at $T_0$ and rejecting heat to the low temperature medium at $T_L$ at a rate of $\dot{Q}_L$, which can be shown to be

$$\dot{X}_{\dot{Q}_L} = \dot{Q}_L \frac{T_0 - T_L}{T_L} \tag{11–19}$$

From the definition of reversibility, this is equivalent to the minimum or reversible power input required to remove heat at a rate of $\dot{Q}_L$ and reject it to the environment at $T_0$. That is, $\dot{W}_{\text{rev,in}} = \dot{W}_{\text{min,in}} = \dot{X}_{\dot{Q}_L}$.

Note that when $T_L = T_0$, which is often the case for heat pumps, $\eta_{\text{II,Evap}} = 0$ since there is no recoverable exergy in this case.

The total exergy destruction associated with the cycle is the sum of the exergy destructions:

$$\dot{X}_{\text{dest,total}} = \dot{X}_{\text{dest},1-2} + \dot{X}_{\text{dest},2-3} + \dot{X}_{\text{dest},3-4} + \dot{X}_{\text{dest},4-1} \tag{11–20}$$

It can be shown that the total exergy destruction associated with a refrigeration cycle can also be obtained by taking the difference between the exergy supplied (power input) and the exergy recovered (the exergy of the heat withdrawn from the low-temperature medium):

$$\dot{X}_{\text{dest,total}} = \dot{W}_{\text{in}} - \dot{X}_{\dot{Q}_L} \tag{11–21}$$

The second-law or exergy efficiency of the cycle can then be expressed as

$$\eta_{\text{II,cycle}} = \frac{\dot{X}_{\dot{Q}_L}}{\dot{W}_{\text{in}}} = \frac{\dot{W}_{\text{min,in}}}{\dot{W}_{\text{in}}} = 1 - \frac{\dot{X}_{\text{dest,total}}}{\dot{W}_{\text{in}}} \tag{11–22}$$

Substituting $\dot{W}_{\text{in}} = \dfrac{\dot{Q}_L}{\text{COP}_R}$ and $\dot{X}_{\dot{Q}_L} = \dot{Q}_L \dfrac{T_0 - T_L}{T_L}$ into Eq. 11-22 gives

$$\eta_{\text{II,cycle}} = \frac{\dot{X}_{\dot{Q}_L}}{\dot{W}_{\text{in}}} = \frac{\dot{Q}_L(T_0 - T_L)/T_L}{\dot{Q}_L/\text{COP}_R} = \frac{\text{COP}_R}{T_L/(T_H - T_L)} = \frac{\text{COP}_R}{\text{COP}_{R,\text{rev}}} \tag{11–23}$$

since $T_0 = T_H$ for a refrigeration cycle. Thus, the second-law efficiency is also equal to the ratio of actual and maximum COPs for the cycle. This second-law efficiency definition accounts for all irreversibilities associated within the refrigerator, including the heat transfers with the refrigerated space and the environment.

---

**■ EXAMPLE 11–3    Exergy Analysis of Vapor-Compression Refrigeration Cycle**

■ A vapor-compression refrigeration cycle with refrigerant-134a as the working fluid is used to maintain a space at −13°C by rejecting heat to ambient air at 27°C. R-134a enters the compressor at 100 kPa superheated by 6.4°C at a rate of 0.05 kg/s. The isentropic efficiency of the compressor is 85 percent. The refrigerant leaves the condenser at 39.4°C as a saturated liquid. Determine (a) the rate of cooling provided and the COP of the system, (b) the exergy destruction in each basic component, (c) the minimum power input and the second-law efficiency of the cycle, and (d) the rate of total exergy destruction.

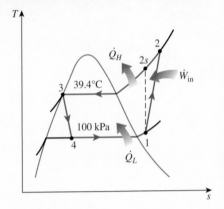

**FIGURE 11–10**
Temperature-entropy diagram of the vapor-compression refrigeration cycle considered in Example 11–3.

**SOLUTION** A vapor-compression refrigeration cycle is considered. The cooling rate, the COP, the exergy destructions, the minimum power input, the second-law efficiency, and the total exergy destruction are to be determined.

**Assumptions** **1** Steady operating conditions exist. **2** Kinetic and potential energy changes are negligible.

**Analysis** (a) The T-s diagram of the cycle is given in Fig. 11–10. The properties of R-134a are (Tables A–11 through A–13)

$$P_1 = 100 \text{ kPa} \\ T_1 = T_{\text{sat@100 kPa}} + \Delta T_{\text{superheat}} \\ = -26.4 + 6.4 = -20°C \Bigg\} \begin{array}{l} h_1 = 239.52 \text{ kJ/kg} \\ s_1 = 0.9721 \text{ kJ/kg·K} \end{array}$$

$$P_3 = P_{\text{sat@39.4°C}} = 1000 \text{ kPa}$$

$$\left. \begin{array}{l} P_2 = P_3 = 1000 \text{ kPa} \\ s_{2s} = s_1 = 0.9721 \text{ kJ/kg·K} \end{array} \right\} h_{2s} = 289.14 \text{ kJ/kg}$$

$$\left. \begin{array}{l} P_3 = 1000 \text{ kPa} \\ x_3 = 0 \end{array} \right\} \begin{array}{l} h_3 = 107.34 \text{ kJ/kg} \\ s_3 = 0.39196 \end{array}$$

$$h_4 = h_3 = 107.34 \text{ kJ/kg}$$

$$\left. \begin{array}{l} P_4 = 100 \text{ kPa} \\ h_4 = 107.34 \text{ kJ/kg} \end{array} \right\} s_4 = 0.4368 \text{ kJ/kg·K}$$

From the definition of isentropic efficiency,

$$\eta_C = \frac{h_{2s} - h_1}{h_2 - h_1}$$

$$0.85 = \frac{289.14 - 239.52}{h_2 - 239.52} \longrightarrow h_2 = 297.90 \text{ kJ/kg}$$

$$\left. \begin{array}{l} P_2 = 1000 \text{ kPa} \\ h_2 = 297.90 \text{ kJ/kg} \end{array} \right\} s_2 = 0.9984 \text{ kJ/kg·K}$$

The refrigeration load, the rate of heat rejected, and the power input are

$$\dot{Q}_L = \dot{m}(h_1 - h_4) = (0.05 \text{ kg/s})(239.52 - 107.34)\text{kJ/kg} = \textbf{6.609 kW}$$
$$\dot{Q}_H = \dot{m}(h_2 - h_3) = (0.05 \text{ kg/s})(297.90 - 107.34)\text{kJ/kg} = 9.528 \text{ kW}$$
$$\dot{W}_{\text{in}} = \dot{m}(h_2 - h_1) = (0.05 \text{ kg/s})(297.90 - 239.52)\text{kJ/kg} = 2.919 \text{ kW}$$

Then the COP of the refrigeration cycle becomes

$$\text{COP}_R = \frac{\dot{Q}_L}{\dot{W}_{\text{in}}} = \frac{6.609 \text{ kW}}{2.919 \text{ kW}} = \textbf{2.264}$$

(b) Noting that the dead-state temperature is $T_0 = T_H = 27 + 273 = 300$ K, the exergy destruction in each component of the cycle is determined as follows:

*Compressor:*

$$\dot{X}_{\text{dest,1-2}} = T_0 \dot{S}_{\text{gen1-2}} = T_0 \dot{m}(s_2 - s_1)$$

$$= (300 \text{ K})(0.05 \text{ kg/s})(0.9984 - 0.9721)\text{kJ/kg·K}$$

$$= \textbf{0.3945 kW}$$

*Condenser:*

$$\dot{X}_{\text{dest},2-3} = T_0 \dot{S}_{\text{gen},2-3} = T_0\left[\dot{m}(s_3 - s_2) + \frac{\dot{Q}_H}{T_H}\right]$$

$$= (300 \text{ K})\left[(0.05 \text{ kg/s})(0.39196 - 0.9984) \text{ kJ/kg·K} + \frac{9.528 \text{ kW}}{300 \text{ K}}\right]$$

$$= \mathbf{0.4314 \text{ kW}}$$

*Expansion valve:*

$$\dot{X}_{\text{dest},3-4} = T_0 \dot{S}_{\text{gen},3-4} = T_0\dot{m}(s_4 - s_3)$$

$$= (300 \text{ K})(0.05 \text{ kg/s})(0.4368 - 0.39196)\text{kJ/kg·K}$$

$$= \mathbf{0.6726 \text{ kW}}$$

*Evaporator:*

$$\dot{X}_{\text{dest},4-1} = T_0 \dot{S}_{\text{gen},4-1} = T_0\left[\dot{m}(s_1 - s_4) - \frac{\dot{Q}_L}{T_L}\right]$$

$$= (300 \text{ K})\left[(0.05 \text{ kg/s})(0.9721 - 0.4368)\text{kJ/kg·K} - \frac{6.609 \text{ kW}}{260 \text{ K}}\right]$$

$$= \mathbf{0.4037 \text{ kW}}$$

(*c*) Exergy flow associated with heat transferred from the low-temperature medium is

$$\dot{X}_{\dot{Q}_L} = \dot{Q}_L\frac{T_0 - T_L}{T_L} = (6.609 \text{ kW})\frac{300 \text{ K} - 260 \text{ K}}{260 \text{ K}} = 1.017 \text{ kW}$$

This is also the minimum or reversible power input for the cycle:

$$\dot{W}_{\text{min,in}} = \dot{X}_{\dot{Q}_L} = \mathbf{1.017 \text{ kW}}$$

The second-law efficiency of the cycle is

$$\eta_{\text{II}} = \frac{\dot{X}_{\dot{Q}_L}}{\dot{W}_{\text{in}}} = \frac{1.017 \text{ kW}}{2.919 \text{ kW}} = 0.348 = \mathbf{34.8\%}$$

This efficiency may also be determined from $\eta_{\text{II}} = \text{COP}_R/\text{COP}_{R,\text{rev}}$ where

$$\text{COP}_{R,\text{rev}} = \frac{T_L}{T_H - T_L} = \frac{(-13 + 273) \text{ K}}{[27 - (-13)]\text{K}} = 6.500$$

Substituting,

$$\eta_{\text{II}} = \frac{\text{COP}_R}{\text{COP}_{R,\text{rev}}} = \frac{2.264}{6.500} = 0.348 = 34.8\%$$

The results are identical, as expected.

(*d*) The total exergy destruction is the difference between the exergy expended (power input) and the exergy recovered (the exergy of the heat transferred from the low-temperature medium):

$$\dot{X}_{\text{dest,total}} = \dot{W}_{\text{in}} - \dot{X}_{\dot{Q}_L} = 2.919 \text{ kW} - 1.017 \text{ kW} = \mathbf{1.902 \text{ kW}}$$

The total exergy destruction can also be determined by adding exergy destruction in each component:

$$\dot{X}_{dest,total} = \dot{X}_{dest,1-2} + \dot{X}_{dest,2-3} + \dot{X}_{dest,3-4} + \dot{X}_{dest,4-1}$$
$$= 0.3945 + 0.4314 + 0.6726 + 0.4037$$
$$= 1.902 \text{ kW}$$

The two results are again identical, as expected.

**Discussion**  The exergy input to the cycle is equal to the actual work input, which is 2.92 kW. The same cooling load could have been accomplished by only 34.8 percent of this power (1.02 kW) if a reversible system were used. The difference between the two is the exergy destructed in the cycle (1.90 kW). The expansion valve appears to be the most irreversible component, which accounts for 35.4 percent of the irreversibilities in the cycle. Replacing the expansion valve by a turbine would decrease the irreversibilites while decreasing the net power input. However, this may or may not be practical in an actual system. It can be shown that increasing the evaporating temperature and decreasing the condensing temperature would also decrease the exergy destruction in these components.

## 11–6 · SELECTING THE RIGHT REFRIGERANT

When designing a refrigeration system, there are several refrigerants from which to choose, such as chlorofluorocarbons (CFCs), ammonia, hydrocarbons (propane, ethane, ethylene, etc.), carbon dioxide, air (in the air-conditioning of aircraft), and even water (in applications above the freezing point). The right choice of refrigerant depends on the situation at hand. Of these, refrigerants such as R-11, R-12, R-22, R-134a, and R-502 account for over 90 percent of the market in the United States.

*Ethyl ether* was the first commercially used refrigerant in vapor-compression systems in 1850, followed by ammonia, carbon dioxide, methyl chloride, sulphur dioxide, butane, ethane, propane, isobutane, gasoline, and chlorofluorocarbons, among others.

The industrial and heavy-commercial sectors were very satisfied with *ammonia,* and still are, although ammonia is toxic. The advantages of ammonia over other refrigerants are its low cost, higher COPs (and thus lower energy cost), more favorable thermodynamic and transport properties and thus higher heat transfer coefficients (requires smaller and lower-cost heat exchangers), greater detectability in the event of a leak, and no effect on the ozone layer. The major drawback of ammonia is its toxicity, which makes it unsuitable for domestic use. Ammonia is predominantly used in food refrigeration facilities such as the cooling of fresh fruits, vegetables, meat, and fish; refrigeration of beverages and dairy products such as beer, wine, milk, and cheese; freezing of ice cream and other foods; ice production; and low-temperature refrigeration in the pharmaceutical and other process industries.

It is remarkable that the early refrigerants used in the light-commercial and household sectors such as sulfur dioxide, ethyl chloride, and methyl chloride were highly toxic. The widespread publicity of a few instances of leaks that resulted in serious illnesses and death in the 1920s caused a public cry to ban or limit the use of these refrigerants, creating a need for the development of a safe refrigerant for household use. At the request of Frigidaire Corporation, General Motors' research laboratory developed R-21, the first member of the CFC family of refrigerants, within three days in 1928. Of several CFCs developed, the research team settled on R-12 as the refrigerant most suitable for commercial use and gave the CFC family the trade name "Freon." Commercial production of R-11 and R-12 was started in 1931 by a company jointly formed by General Motors and E. I. du Pont de Nemours and Co., Inc. The versatility and low cost of CFCs made them the refrigerants of choice. CFCs were also widely used in aerosols, foam insulations, and the electronic industry as solvents to clean computer chips.

R-11 is used primarily in large-capacity water chillers serving air-conditioning systems in buildings. R-12 is used in domestic refrigerators and freezers, as well as automotive air conditioners. R-22 is used in window air conditioners, heat pumps, air conditioners of commercial buildings, and large industrial refrigeration systems, and offers strong competition to ammonia. R-502 (a blend of R-115 and R-22) is the dominant refrigerant used in commercial refrigeration systems such as those in supermarkets because it allows low temperatures at evaporators while operating at single-stage compression.

The ozone crisis has caused a major stir in the refrigeration and air-conditioning industry and has triggered a critical look at the refrigerants in use. It was realized in the mid-1970s that CFCs allow more ultraviolet radiation into the earth's atmosphere by destroying the protective ozone layer and thus contributing to the greenhouse effect that causes global warming. As a result, the use of some CFCs is banned by international treaties. Fully halogenated CFCs (such as R-11, R-12, and R-115) do the most damage to the ozone layer. The nonfully halogenated refrigerants such as R-22 have about 5 percent of the ozone-depleting capability of R-12. Refrigerants that are friendly to the ozone layer that protects the earth from harmful ultraviolet rays have been developed. The once popular refrigerant R-12 has largely been replaced by the recently developed chlorine-free R-134a.

Two important parameters that need to be considered in the selection of a refrigerant are the temperatures of the two media (the refrigerated space and the environment) with which the refrigerant exchanges heat.

To have heat transfer at a reasonable rate, a temperature difference of 5 to 10°C should be maintained between the refrigerant and the medium with which it is exchanging heat. If a refrigerated space is to be maintained at −10°C, for example, the temperature of the refrigerant should remain at about −20°C while it absorbs heat in the evaporator. The lowest pressure in a refrigeration cycle occurs in the evaporator, and this pressure should be above atmospheric pressure to prevent any air leakage into the refrigeration system. Therefore, a refrigerant should have a saturation pressure of 1 atm or higher at −20°C in this particular case. Ammonia and R-134a are two such substances.

The temperature (and thus the pressure) of the refrigerant on the condenser side depends on the medium to which heat is rejected. Lower temperatures in the condenser (thus higher COPs) can be maintained if the refrigerant is cooled by liquid water instead of air. The use of water cooling cannot be justified economically, however, except in large industrial refrigeration systems. The temperature of the refrigerant in the condenser cannot fall below the temperature of the cooling medium (about 20°C for a household refrigerator), and the saturation pressure of the refrigerant at this temperature should be well below its critical pressure if the heat rejection process is to be approximately isothermal. If no single refrigerant can meet the temperature requirements, then two or more refrigeration cycles with different refrigerants can be used in series. Such a refrigeration system is called a *cascade system* and is discussed later in this chapter.

Other desirable characteristics of a refrigerant include being nontoxic, noncorrosive, nonflammable, and chemically stable; having a high enthalpy of vaporization (minimizes the mass flow rate); and, of course, being available at low cost.

In the case of heat pumps, the minimum temperature (and pressure) for the refrigerant may be considerably higher since heat is usually extracted from media that are well above the temperatures encountered in refrigeration systems.

## 11–7 · HEAT PUMP SYSTEMS

Heat pumps are generally more expensive to purchase and install than other heating systems, but they save money in the long run in some areas because they lower the heating bills. Despite their relatively higher initial costs, the popularity of heat pumps is increasing. About one-third of all single-family homes built in the United States in recent years are heated by heat pumps.

The most common energy source for heat pumps is atmospheric air (air-to-air systems), although water and soil are also used. The major problem with air-source systems is *frosting,* which occurs in humid climates when the temperature falls below 2 to 5°C. The frost accumulation on the evaporator coils is highly undesirable since it seriously disrupts heat transfer. The coils can be defrosted, however, by reversing the heat pump cycle (running it as an air conditioner). This results in a reduction in the efficiency of the system. Water-source systems usually use well water from depths of up to 80 m in the temperature range of 5 to 18°C, and they do not have a frosting problem. They typically have higher COPs but are more complex and require easy access to a large body of water such as underground water. Ground-source systems are also rather involved since they require long tubing placed deep in the ground where the soil temperature is relatively constant. The COP of heat pumps usually ranges between 1.5 and 4, depending on the particular system used and the temperature of the source. A new class of recently developed heat pumps that use variable-speed electric motor drives are at least twice as energy efficient as their predecessors.

Both the capacity and the efficiency of a heat pump fall significantly at low temperatures. Therefore, most air-source heat pumps require a supplementary heating system such as electric resistance heaters or an oil or gas furnace. Since water and soil temperatures do not fluctuate much, supplementary heating may not be required for water-source or ground-source

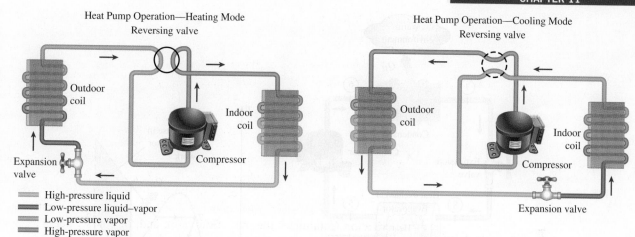

Heat Pump Operation—Heating Mode
Reversing valve

Outdoor coil

Indoor coil

Compressor

Expansion valve

▬ High-pressure liquid
▬ Low-pressure liquid–vapor
▬ Low-pressure vapor
▬ High-pressure vapor

Heat Pump Operation—Cooling Mode
Reversing valve

Outdoor coil

Indoor coil

Compressor

Expansion valve

**FIGURE 11–11**
A heat pump can be used to heat a house in winter and to cool it in summer.

systems. However, the heat pump system must be large enough to meet the maximum heating load.

Heat pumps and air conditioners have the same mechanical components. Therefore, it is not economical to have two separate systems to meet the heating and cooling requirements of a building. One system can be used as a heat pump in winter and an air conditioner in summer. This is accomplished by adding a reversing valve to the cycle, as shown in Fig. 11–11. As a result of this modification, the condenser of the heat pump (located indoors) functions as the evaporator of the air conditioner in summer. Also, the evaporator of the heat pump (located outdoors) serves as the condenser of the air conditioner. This feature increases the competitiveness of the heat pump. Such dual-purpose units are commonly used in motels.

Heat pumps are most competitive in areas that have a large cooling load during the cooling season and a relatively small heating load during the heating season, such as in the southern parts of the United States. In these areas, the heat pump can meet the entire cooling and heating needs of residential or commercial buildings. The heat pump is least competitive in areas where the heating load is very large and the cooling load is small, such as in the northern parts of the United States.

# 11–8 · INNOVATIVE VAPOR-COMPRESSION REFRIGERATION SYSTEMS

The simple vapor-compression refrigeration cycle discussed above is the most widely used refrigeration cycle, and it is adequate for most refrigeration applications. The ordinary vapor-compression refrigeration systems are simple, inexpensive, reliable, and practically maintenance-free (when was the last time you serviced your household refrigerator?). However, for large industrial applications *efficiency,* not simplicity, is the major concern. Also, for some applications the simple vapor-compression refrigeration cycle is inadequate and needs to be modified. We now discuss a few such modifications and refinements.

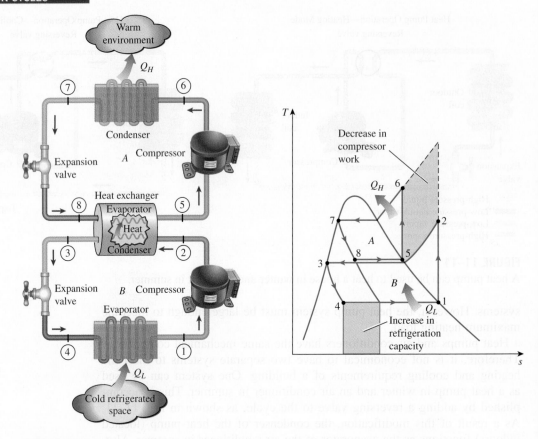

**FIGURE 11–12**
A two-stage cascade refrigeration system with the same refrigerant in both stages.

## Cascade Refrigeration Systems

Some industrial applications require moderately low temperatures, and the temperature range they involve may be too large for a single vapor-compression refrigeration cycle to be practical. A large temperature range also means a large pressure range in the cycle and a poor performance for a reciprocating compressor. One way of dealing with such situations is to perform the refrigeration process in stages, that is, to have two or more refrigeration cycles that operate in series. Such refrigeration cycles are called **cascade refrigeration cycles**.

A two-stage cascade refrigeration cycle is shown in Fig. 11–12. The two cycles are connected through the heat exchanger in the middle, which serves as the evaporator for the topping cycle (cycle A) and the condenser for the bottoming cycle (cycle B). Assuming the heat exchanger is well insulated and the kinetic and potential energies are negligible, the heat transfer from the fluid in the bottoming cycle should be equal to the heat transfer to the fluid in the topping cycle. Thus, the ratio of mass flow rates through each cycle should be

$$\dot{m}_A(h_5 - h_8) = \dot{m}_B(h_2 - h_3) \longrightarrow \frac{\dot{m}_A}{\dot{m}_B} = \frac{h_2 - h_3}{h_5 - h_8} \qquad \textbf{(11–24)}$$

Also,

$$COP_{R,cascade} = \frac{\dot{Q}_L}{\dot{W}_{net,in}} = \frac{\dot{m}_B(h_1 - h_4)}{\dot{m}_A(h_6 - h_5) + \dot{m}_B(h_2 - h_1)} \quad (11\text{–}25)$$

In the cascade system shown in the figure, the refrigerants in both cycles are assumed to be the same. This is not necessary, however, since there is no mixing taking place in the heat exchanger. Therefore, refrigerants with more desirable characteristics can be used in each cycle. In this case, there would be a separate saturation dome for each fluid, and the T-s diagram for one of the cycles would be different. Also, in actual cascade refrigeration systems, the two cycles would overlap somewhat since a temperature difference between the two fluids is needed for any heat transfer to take place.

It is evident from the T-s diagram in Fig. 11–12 that the compressor work decreases and the amount of heat absorbed from the refrigerated space increases as a result of cascading. Therefore, cascading improves the COP of a refrigeration system. Some refrigeration systems use three or four stages of cascading.

■ **EXAMPLE 11–4**    **A Two-Stage Cascade Refrigeration Cycle**

Consider a two-stage cascade refrigeration system operating between the pressure limits of 0.8 and 0.14 MPa. Each stage operates on an ideal vapor-compression refrigeration cycle with refrigerant-134a as the working fluid. Heat rejection from the lower cycle to the upper cycle takes place in an adiabatic counterflow heat exchanger where both streams enter at about 0.32 MPa. (In practice, the working fluid of the lower cycle is at a higher pressure and temperature in the heat exchanger for effective heat transfer.) If the mass flow rate of the refrigerant through the upper cycle is 0.05 kg/s, determine (a) the mass flow rate of the refrigerant through the lower cycle, (b) the rate of heat removal from the refrigerated space and the power input to the compressor, and (c) the coefficient of performance of this cascade refrigerator.

**SOLUTION** A cascade refrigeration system operating between the specified pressure limits is considered. The mass flow rate of the refrigerant through the lower cycle, the rate of refrigeration, the power input, and the COP are to be determined.
**Assumptions** 1 Steady operating conditions exist. 2 Kinetic and potential energy changes are negligible. 3 The heat exchanger is adiabatic.
**Properties** The enthalpies of the refrigerant at all eight states are determined from the refrigerant tables and are indicated on the T-s diagram.
**Analysis** The T-s diagram of the refrigeration cycle is shown in Fig. 11–13. The topping cycle is labeled cycle A and the bottoming one, cycle B. For both cycles, the refrigerant leaves the condenser as a saturated liquid and enters the compressor as saturated vapor.

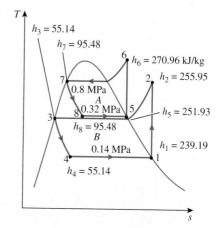

**FIGURE 11–13**
T-s diagram of the cascade refrigeration cycle described in Example 11–4.

(*a*) The mass flow rate of the refrigerant through the lower cycle is determined from the steady-flow energy balance on the adiabatic heat exchanger,

$$\dot{E}_{out} = \dot{E}_{in} \longrightarrow \dot{m}_A h_5 + \dot{m}_B h_3 = \dot{m}_A h_8 + \dot{m}_B h_2$$

$$\dot{m}_A(h_5 - h_8) = \dot{m}_B(h_2 - h_3)$$

$$(0.05 \text{ kg/s})[(251.93 - 95.48) \text{ kJ/kg}] = \dot{m}_B[(255.95 - 55.14) \text{ kJ/kg}]$$

$$\dot{m}_B = \mathbf{0.0390 \ kg/s}$$

(*b*) The rate of heat removal by a cascade cycle is the rate of heat absorption in the evaporator of the lowest stage. The power input to a cascade cycle is the sum of the power inputs to all of the compressors:

$$\dot{Q}_L = \dot{m}_B(h_1 - h_4) = (0.0390 \text{ kg/s})[(239.19 - 55.14) \text{ kJ/kg}] = \mathbf{7.18 \ kW}$$

$$\dot{W}_{in} = \dot{W}_{comp\,I,in} + \dot{W}_{comp\,II,in} = \dot{m}_A(h_6 - h_5) + \dot{m}_B(h_2 - h_1)$$

$$= (0.05 \text{ kg/s})[(270.96 - 251.93) \text{ kJ/kg}]$$

$$+ (0.039 \text{ kg/s})[(255.95 - 239.19) \text{ kJ/kg}]$$

$$= \mathbf{1.61 \ kW}$$

(*c*) The COP of a refrigeration system is the ratio of the refrigeration rate to the net power input:

$$\text{COP}_R = \frac{\dot{Q}_L}{\dot{W}_{net,in}} = \frac{7.18 \text{ kW}}{1.61 \text{ kW}} = \mathbf{4.46}$$

**Discussion** This problem was worked out in Example 11–1 for a single-stage refrigeration system. Notice that the COP of the refrigeration system increases from 3.97 to 4.46 as a result of cascading. The COP of the system can be increased even more by increasing the number of cascade stages.

## Multistage Compression Refrigeration Systems

When the fluid used throughout the cascade refrigeration system is the same, the heat exchanger between the stages can be replaced by a mixing chamber (called a *flash chamber*) since it has better heat transfer characteristics. Such systems are called **multistage compression refrigeration systems**. A two-stage compression refrigeration system is shown in Fig. 11–14.

In this system, the liquid refrigerant expands in the first expansion valve to the flash chamber pressure, which is the same as the compressor interstage pressure. Part of the liquid vaporizes during this process. This saturated vapor (state 3) is mixed with the superheated vapor from the low-pressure compressor (state 2), and the mixture enters the high-pressure compressor at state 9. This is, in essence, a regeneration process. The saturated liquid (state 7) expands through the second expansion valve into the evaporator, where it picks up heat from the refrigerated space.

The compression process in this system resembles a two-stage compression with intercooling, and the compressor work decreases. Care should be exercised in the interpretations of the areas on the *T-s* diagram in this case since the mass flow rates are different in different parts of the cycle.

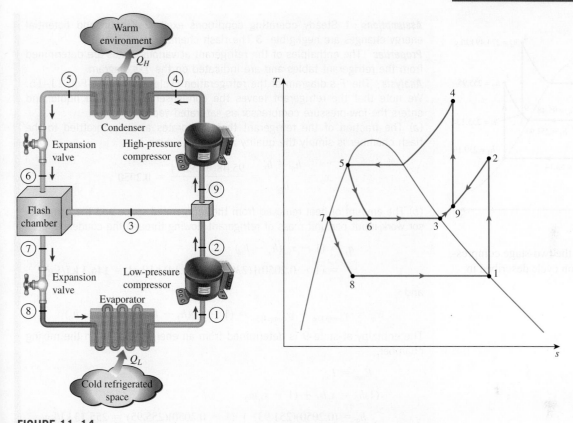

**FIGURE 11–14**
A two-stage compression refrigeration system with a flash chamber.

---

**EXAMPLE 11–5**  **A Two-Stage Refrigeration Cycle with a Flash Chamber**

Consider a two-stage compression refrigeration system operating between the pressure limits of 0.8 and 0.14 MPa. The working fluid is refrigerant-134a. The refrigerant leaves the condenser as a saturated liquid and is throttled to a flash chamber operating at 0.32 MPa. Part of the refrigerant evaporates during this flashing process, and this vapor is mixed with the refrigerant leaving the low-pressure compressor. The mixture is then compressed to the condenser pressure by the high-pressure compressor. The liquid in the flash chamber is throttled to the evaporator pressure and cools the refrigerated space as it vaporizes in the evaporator. Assuming the refrigerant leaves the evaporator as a saturated vapor and both compressors are isentropic, determine (a) the fraction of the refrigerant that evaporates as it is throttled to the flash chamber, (b) the amount of heat removed from the refrigerated space and the compressor work per unit mass of refrigerant flowing through the condenser, and (c) the coefficient of performance.

**SOLUTION** A two-stage compression refrigeration system operating between specified pressure limits is considered. The fraction of the refrigerant that evaporates in the flash chamber, the refrigeration and work input per unit mass, and the COP are to be determined.

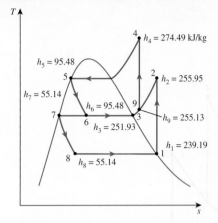

**FIGURE 11–15**
*T-s* diagram of the two-stage compression refrigeration cycle described in Example 11–5.

**Assumptions** **1** Steady operating conditions exist. **2** Kinetic and potential energy changes are negligible. **3** The flash chamber is adiabatic.

**Properties** The enthalpies of the refrigerant at various states are determined from the refrigerant tables and are indicated on the *T-s* diagram.

**Analysis** The *T-s* diagram of the refrigeration cycle is shown in Fig. 11–15. We note that the refrigerant leaves the condenser as saturated liquid and enters the low-pressure compressor as saturated vapor.

(*a*) The fraction of the refrigerant that evaporates as it is throttled to the flash chamber is simply the quality at state 6, which is

$$x_6 = \frac{h_6 - h_f}{h_{fg}} = \frac{95.48 - 55.14}{196.78} = \textbf{0.2050}$$

(*b*) The amount of heat removed from the refrigerated space and the compressor work input per unit mass of refrigerant flowing through the condenser are

$$q_L = (1 - x_6)(h_1 - h_8)$$
$$= (1 - 0.2050)[(239.19 - 55.14) \text{ kJ/kg}] = \textbf{146.3 kJ/kg}$$

and

$$w_{in} = w_{comp\,I,in} + w_{comp\,II,in} = (1 - x_6)(h_2 - h_1) + (1)(h_4 - h_9)$$

The enthalpy at state 9 is determined from an energy balance on the mixing chamber,

$$\dot{E}_{out} = \dot{E}_{in}$$
$$(1)h_9 = x_6 h_3 + (1 - x_6)h_2$$
$$h_9 = (0.2050)(251.93) + (1 - 0.2050)(255.95) = 255.13 \text{ kJ/kg}$$

Also, $s_9 = 0.9417$ kJ/kg·K. Thus the enthalpy at state 4 (0.8 MPa, $s_4 = s_9$) is $h_4 = 274.49$ kJ/kg. Substituting,

$$w_{in} = (1 - 0.2050)[(255.95 - 239.19) \text{ kJ/kg}] + (274.49 - 255.13) \text{ kJ/kg}$$
$$= \textbf{32.68 kJ/kg}$$

(*c*) The coefficient of performance is

$$COP_R = \frac{q_L}{w_{in}} = \frac{146.3 \text{ kJ/kg}}{32.68 \text{ kJ/kg}} = \textbf{4.48}$$

**Discussion** This problem was worked out in Example 11–1 for a single-stage refrigeration system (COP = 3.97) and in Example 11–4 for a two-stage cascade refrigeration system (COP = 4.46). Notice that the COP of the refrigeration system increased considerably relative to the single-stage compression but did not change much relative to the two-stage cascade compression.

# Multipurpose Refrigeration Systems with a Single Compressor

Some applications require refrigeration at more than one temperature. This could be accomplished by using a separate throttling valve and a separate compressor for each evaporator operating at different temperatures. However, such

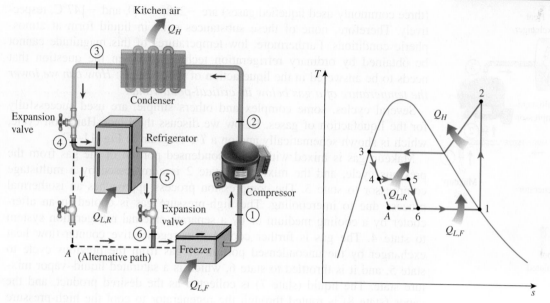

**FIGURE 11–16**

Schematic and $T$-$s$ diagram for a refrigerator–freezer unit with one compressor.

a system is bulky and probably uneconomical. A more practical and economical approach would be to route all the exit streams from the evaporators to a single compressor and let it handle the compression process for the entire system.

Consider, for example, an ordinary refrigerator–freezer unit. A simplified schematic of the unit and the $T$-$s$ diagram of the cycle are shown in Fig. 11–16. Most refrigerated goods have a high water content, and the refrigerated space must be maintained above the ice point to prevent freezing. The freezer compartment, however, is maintained at about $-18°C$. Therefore, the refrigerant should enter the freezer at about $-25°C$ to have heat transfer at a reasonable rate in the freezer. If a single expansion valve and evaporator were used, the refrigerant would have to circulate in both compartments at about $-25°C$, which would cause ice formation in the neighborhood of the evaporator coils and dehydration of the produce. This problem can be eliminated by throttling the refrigerant to a higher pressure (hence temperature) for use in the refrigerated space and then throttling it to the minimum pressure for use in the freezer. The entire refrigerant leaving the freezer compartment is subsequently compressed by a single compressor to the condenser pressure.

## Liquefaction of Gases

The liquefaction of gases has always been an important area of refrigeration since many important scientific and engineering processes at cryogenic temperatures (temperatures below about $-100°C$) depend on liquefied gases. Some examples of such processes are the separation of oxygen and nitrogen from air, preparation of liquid propellants for rockets, the study of material properties at low temperatures, and the study of some exciting phenomena such as superconductivity.

At temperatures above the critical-point value, a substance exists in the gas phase only. The critical temperatures of helium, hydrogen, and nitrogen

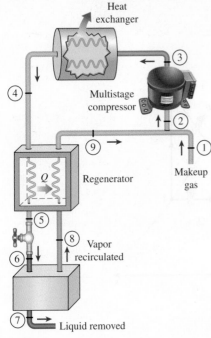

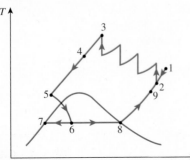

**FIGURE 11–17**

Linde-Hampson system for liquefying gases.

(three commonly used liquefied gases) are $-268$, $-240$, and $-147°C$, respectively. Therefore, none of these substances exist in liquid form at atmospheric conditions. Furthermore, low temperatures of this magnitude cannot be obtained by ordinary refrigeration techniques. Then the question that needs to be answered in the liquefaction of gases is this: *How can we lower the temperature of a gas below its critical-point value?*

Several cycles, some complex and others simple, are used successfully for the liquefaction of gases. Below we discuss the Linde-Hampson cycle, which is shown schematically and on a *T-s* diagram in Fig. 11–17.

Makeup gas is mixed with the uncondensed portion of the gas from the previous cycle, and the mixture at state 2 is compressed by a multistage compressor to state 3. The compression process approaches an isothermal process due to intercooling. The high-pressure gas is cooled in an aftercooler by a cooling medium or by a separate external refrigeration system to state 4. The gas is further cooled in a regenerative counter-flow heat exchanger by the uncondensed portion of gas from the previous cycle to state 5, and it is throttled to state 6, which is a saturated liquid–vapor mixture state. The liquid (state 7) is collected as the desired product, and the vapor (state 8) is routed through the regenerator to cool the high-pressure gas approaching the throttling valve. Finally, the gas is mixed with fresh makeup gas, and the cycle is repeated.

This and other refrigeration cycles used for the liquefaction of gases can also be used for the solidification of gases.

# 11–9 ■ GAS REFRIGERATION CYCLES

As explained in Sec. 11–2, the Carnot cycle (the standard of comparison for power cycles) and the reversed Carnot cycle (the standard of comparison for refrigeration cycles) are identical, except that the reversed Carnot cycle operates in the reverse direction. This suggests that the power cycles discussed in earlier chapters can be used as refrigeration cycles by simply reversing them. In fact, the vapor-compression refrigeration cycle is essentially a modified Rankine cycle operating in reverse. Another example is the reversed Stirling cycle, which is the cycle on which Stirling refrigerators operate. In this section, we discuss the *reversed Brayton cycle,* better known as the **gas refrigeration cycle**.

Consider the gas refrigeration cycle shown in Fig. 11–18. The surroundings are at $T_0$, and the refrigerated space is to be maintained at $T_L$. The gas is compressed during process 1-2. The high-pressure, high-temperature gas at state 2 is then cooled at constant pressure to $T_0$ by rejecting heat to the surroundings. This is followed by an expansion process in a turbine, during which the gas temperature drops to $T_4$. (Can we achieve the cooling effect by using a throttling valve instead of a turbine?) Finally, the cool gas absorbs heat from the refrigerated space until its temperature rises to $T_1$.

All the processes described are internally reversible, and the cycle executed is the *ideal* gas refrigeration cycle. In actual gas refrigeration cycles, the compression and expansion processes deviate from the isentropic ones, and $T_3$ is higher than $T_0$ unless the heat exchanger is infinitely large.

On a *T-s* diagram, the area under process curve 4-1 represents the heat removed from the refrigerated space, and the enclosed area 1-2-3-4-1 represents the net work input. The ratio of these areas is the COP for the cycle, which may be expressed as

$$\text{COP}_R = \frac{q_L}{w_{net,in}} = \frac{q_L}{w_{comp,in} - w_{turb,out}} \quad (11\text{–}26)$$

where

$$q_L = h_1 - h_4$$
$$w_{turb,out} = h_3 - h_4$$
$$w_{comp,in} = h_2 - h_1$$

The gas refrigeration cycle deviates from the reversed Carnot cycle because the heat transfer processes are not isothermal. In fact, the gas temperature varies considerably during heat transfer processes. Consequently, the gas refrigeration cycles have lower COPs relative to the vapor-compression refrigeration cycles or the reversed Carnot cycle. This is also evident from the $T$-$s$ diagram in Fig. 11–19. The reversed Carnot cycle consumes a fraction of the net work (rectangular area $1A3B$) but produces a greater amount of refrigeration (triangular area under $B1$).

Despite their relatively low COPs, the gas refrigeration cycles have two desirable characteristics: They involve simple, lighter components, which make them suitable for aircraft cooling, and they can incorporate regeneration, which makes them suitable for liquefaction of gases and cryogenic applications. An open-cycle aircraft cooling system is shown in Fig. 11–20. Atmospheric air is compressed by a compressor, cooled by the surrounding air, and expanded in a turbine. The cool air leaving the turbine is then directly routed to the cabin.

The regenerative gas cycle is shown in Fig. 11–21. Regenerative cooling is achieved by inserting a counter-flow heat exchanger into the cycle. Without regeneration, the lowest turbine inlet temperature is $T_0$, the temperature of the surroundings or any other cooling medium. With regeneration, the high-pressure gas is further cooled to $T_4$ before expanding in the turbine. Lowering the turbine inlet temperature automatically lowers the turbine exit temperature, which is the minimum temperature in the cycle. Extremely low temperatures can be achieved by repeating this process.

---

### EXAMPLE 11–6    The Simple Ideal Gas Refrigeration Cycle

An ideal gas refrigeration cycle using air as the working medium is to maintain a refrigerated space at –18°C while rejecting heat to the surrounding medium at 27°C. The pressure ratio of the compressor is 4. Determine (a) the maximum and minimum temperatures in the cycle, (b) the coefficient of performance, and (c) the rate of refrigeration for a mass flow rate of 0.05 kg/s.

**SOLUTION**    An ideal gas refrigeration cycle using air as the working fluid is considered. The maximum and minimum temperatures, the COP, and the rate of refrigeration are to be determined.

*Assumptions*  **1** Steady operating conditions exist. **2** Air is an ideal gas with variable specific heats. **3** Kinetic and potential energy changes are negligible.

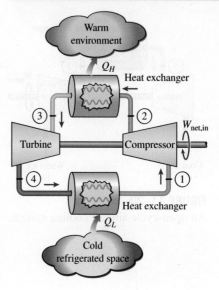

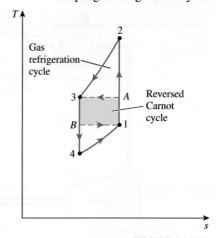

**FIGURE 11–18**
Simple gas refrigeration cycle.

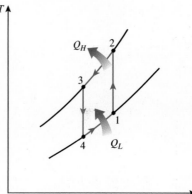

**FIGURE 11–19**
A reserved Carnot cycle produces more refrigeration (area under $B1$) with less work input (area $1A3B$).

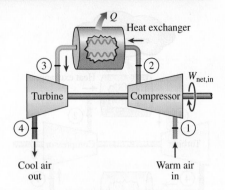

**FIGURE 11–20**
An open-cycle aircraft cooling system.

*Analysis* The *T-s* diagram of the gas refrigeration cycle is shown in Fig. 11–22. We note that this is an ideal gas-compression refrigeration cycle, and thus, both the compressor and the turbine are isentropic, and the air is cooled to the environment temperature before it enters the turbine.

(a) The maximum and minimum temperatures in the cycle are determined from the isentropic relations of ideal gases for the compression and expansion processes. From Table A–17,

$$T_1 = 255 \text{ K} \longrightarrow h_1 = 255.07 \text{ kJ/kg and } P_{r1} = 0.7867$$

$$P_{r2} = \frac{P_2}{P_1} P_{r1} = (4)(0.7867) = 3.147 \longrightarrow \begin{array}{l} h_2 = 379.74 \text{ kJ/kg} \\ T_2 = \textbf{379 K (or 106°C)} \end{array}$$

$$T_3 = 300 \text{ K} \longrightarrow h_3 = 300.19 \text{ kJ/kg} \qquad \text{and} \quad P_{r3} = 1.3860$$

$$P_{r4} = \frac{P_4}{P_3} P_{r3} = (0.25)(1.386) = 0.3465 \longrightarrow \begin{array}{l} h_4 = 201.60 \text{ kJ/kg} \\ T_4 = \textbf{202 K (or −71°C)} \end{array}$$

Therefore, the highest and the lowest temperatures in the cycle are 106 and −71°C, respectively.

(b) The COP of this ideal gas refrigeration cycle is

$$\text{COP}_R = \frac{q_L}{w_{net,in}} = \frac{q_L}{w_{comp,in} - W_{turb,out}}$$

where

$$q_L = h_1 - h_4 = 255.07 - 201.60 = 53.47 \text{ kJ/kg}$$

$$w_{turb,out} = h_3 - h_4 = 300.19 - 201.60 = 98.59 \text{ kJ/kg}$$

$$w_{comp,in} = h_2 - h_1 = 379.74 - 255.07 = 124.67 \text{ kJ/kg}$$

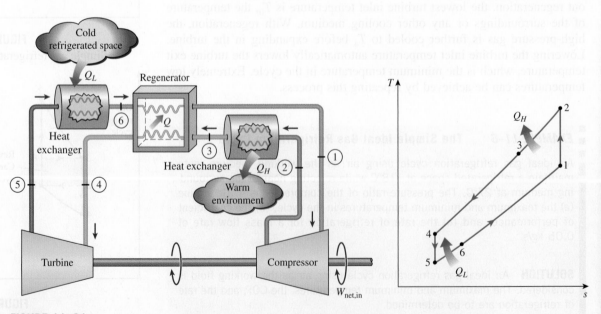

**FIGURE 11–21**
Gas refrigeration cycle with regeneration.

Thus,

$$COP_R = \frac{53.47}{124.67 - 98.59} = 2.05$$

(c) The rate of refrigeration is

$$\dot{Q}_{refrig} = \dot{m}q_L = (0.05 \text{ kg/s})(53.47 \text{ kJ/kg}) = 2.67 \text{ kW}$$

**Discussion** It is worth noting that an ideal vapor-compression cycle working under similar conditions would have a COP greater than 3.

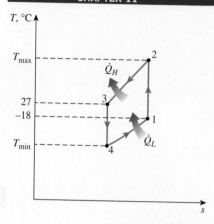

**FIGURE 11–22**

T-s diagram of the ideal-gas refrigeration cycle described in Example 11–6.

# 11–10 · ABSORPTION REFRIGERATION SYSTEMS

Another form of refrigeration that becomes economically attractive when there is a source of inexpensive thermal energy at a temperature of 100 to 200°C is **absorption refrigeration**. Some examples of inexpensive thermal energy sources include geothermal energy, solar energy, and waste heat from cogeneration or process steam plants, and even natural gas when it is available at a relatively low price.

As the name implies, absorption refrigeration systems involve the absorption of a *refrigerant* by a *transport medium*. The most widely used absorption refrigeration system is the ammonia–water system, where ammonia ($NH_3$) serves as the refrigerant and water ($H_2O$) as the transport medium. Other absorption refrigeration systems include water–lithium bromide and water–lithium chloride systems, where water serves as the refrigerant. The latter two systems are limited to applications such as air-conditioning where the minimum temperature is above the freezing point of water.

To understand the basic principles involved in absorption refrigeration, we examine the $NH_3-H_2O$ system shown in Fig. 11–23. The ammonia–water refrigeration machine was patented by the Frenchman Ferdinand Carre in 1859. Within a few years, the machines based on this principle were being built in the United States primarily to make ice and store food. You will immediately notice from the figure that this system looks very much like the vapor-compression system, except that the compressor has been replaced by a complex absorption mechanism consisting of an absorber, a pump, a generator, a regenerator, a valve, and a rectifier. Once the pressure of $NH_3$ is raised by the components in the box (this is the only thing they are set up to do), it is cooled and condensed in the condenser by rejecting heat to the surroundings, is throttled to the evaporator pressure, and absorbs heat from the refrigerated space as it flows through the evaporator. So, there is nothing new there. Here is what happens in the box:

Ammonia vapor leaves the evaporator and enters the absorber, where it dissolves and reacts with water to form $NH_3 \cdot H_2O$. This is an exothermic reaction; thus heat is released during this process. The amount of $NH_3$ that can be dissolved in $H_2O$ is inversely proportional to the temperature. Therefore, it is necessary to cool the absorber to maintain its temperature as low as possible, hence to maximize the amount of $NH_3$ dissolved in water.

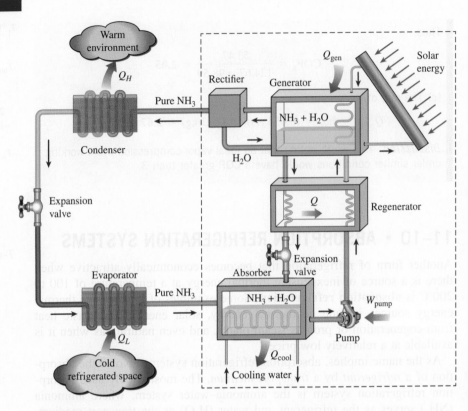

**FIGURE 11–23**
Ammonia absorption refrigeration cycle.

The liquid $NH_3 + H_2O$ solution, which is rich in $NH_3$, is then pumped to the generator. Heat is transferred to the solution from a source to vaporize some of the solution. The vapor, which is rich in $NH_3$, passes through a rectifier, which separates the water and returns it to the generator. The high-pressure pure $NH_3$ vapor then continues its journey through the rest of the cycle. The hot $NH_3 + H_2O$ solution, which is weak in $NH_3$, then passes through a regenerator, where it transfers some heat to the rich solution leaving the pump, and is throttled to the absorber pressure.

Compared with vapor-compression systems, absorption refrigeration systems have one major advantage: A liquid is compressed instead of a vapor. The steady-flow work is proportional to the specific volume, and thus the work input for absorption refrigeration systems is very small (on the order of one percent of the heat supplied to the generator) and often neglected in the cycle analysis. The operation of these systems is based on heat transfer from an external source. Therefore, absorption refrigeration systems are often classified as *heat-driven systems*.

The absorption refrigeration systems are much more expensive than the vapor-compression refrigeration systems. They are more complex and occupy more space, they are much less efficient thus requiring much larger cooling towers to reject the waste heat, and they are more difficult to service since they are less common. Therefore, absorption refrigeration systems should be considered only when the unit cost of thermal energy is low and is projected to remain low relative to electricity. Absorption refrigeration systems are primarily used in large commercial and industrial installations.

The COP of absorption refrigeration systems is defined as

$$COP_{absorption} = \frac{\text{Desired output}}{\text{Required input}} = \frac{Q_L}{Q_{gen} + W_{pump}} \cong \frac{Q_L}{Q_{gen}} \quad (11\text{-}27)$$

The maximum COP of an absorption refrigeration system is determined by assuming that the entire cycle is totally reversible (i.e., the cycle involves no irreversibilities and any heat transfer is through a differential temperature difference). The refrigeration system would be reversible if the heat from the source ($Q_{gen}$) were transferred to a Carnot heat engine, and the work output of this heat engine ($W = \eta_{th,rev}Q_{gen}$) is supplied to a Carnot refrigerator to remove heat from the refrigerated space. Note that $Q_L = W \times COP_{R,rev} = \eta_{th,rev}Q_{gen}COP_{R,rev}$. Then the overall COP of an absorption refrigeration system under reversible conditions becomes (Fig. 11–24)

$$COP_{rev,absorption} = \frac{Q_L}{Q_{gen}} = \eta_{th,rev}COP_{R,rev} = \left(1 - \frac{T_0}{T_s}\right)\left(\frac{T_L}{T_0 - T_L}\right) \quad (11\text{-}28)$$

where $T_L$, $T_0$, and $T_s$ are the thermodynamic temperatures of the refrigerated space, the environment, and the heat source, respectively. Any absorption refrigeration system that receives heat from a source at $T_s$ and removes heat from the refrigerated space at $T_L$ while operating in an environment at $T_0$ has a lower COP than the one determined from Eq. 11–28. For example, when the source is at 120°C, the refrigerated space is at $-10$°C, and the environment is at 25°C, the maximum COP that an absorption refrigeration system can have is 1.8. The COP of actual absorption refrigeration systems is usually less than 1.

Air-conditioning systems based on absorption refrigeration, called *absorption chillers,* perform best when the heat source can supply heat at a high temperature with little temperature drop. The absorption chillers are typically rated at an input temperature of 116°C. The chillers perform at lower temperatures, but their cooling capacity decreases sharply with decreasing source temperature, about 12.5 percent for each 6°C drop in the source temperature. For example, the capacity goes down to 50 percent when the supply water temperature drops to 93°C. In that case, one needs to double the size (and thus the cost) of the chiller to achieve the same cooling. The COP of the chiller is affected less by the decline of the source temperature. The COP drops by 2.5 percent for each 6°C drop in the source temperature. The nominal COP of single-stage absorption chillers at 116°C is 0.65 to 0.70. Therefore, for each ton of refrigeration, a heat input of (12,660 kJ/h)/0.65 = 19,480 kJ/h is required. At 88°C, the COP drops by 12.5 percent and thus the heat input increases by 12.5 percent for the same cooling effect. Therefore, the economic aspects must be evaluated carefully before any absorption refrigeration system is considered, especially when the source temperature is below 93°C.

Another absorption refrigeration system that is quite popular with campers is a propane-fired system invented by two Swedish undergraduate students. In this system, the pump is replaced by a third fluid (hydrogen), which makes it a truly portable unit.

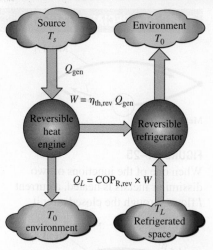

$$W = \eta_{th,\,rev}\,Q_{gen} = \left(1 - \frac{T_0}{T_s}\right)Q_{gen}$$

$$Q_L = COP_{R,rev}W = \left(\frac{T_L}{T_0 - T_L}\right)W$$

$$COP_{rev,absorption} = \frac{Q_L}{Q_{gen}} = \left(1 - \frac{T_0}{T_s}\right)\left(\frac{T_L}{T_0 - T_L}\right)$$

**FIGURE 11–24**

Determining the maximum COP of an absorption refrigeration system.

**TOPIC OF SPECIAL INTEREST***      Thermoelectric Power Generation and Refrigeration Systems

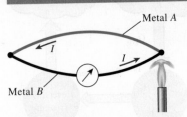

**FIGURE 11–25**
When one of the junctions of two dissimilar metals is heated, a current $I$ flows through the closed circuit.

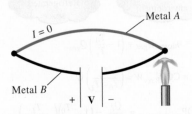

**FIGURE 11–26**
When a thermoelectric circuit is broken, a potential difference is generated.

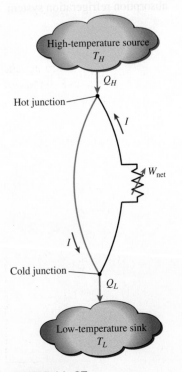

**FIGURE 11–27**
Schematic of a simple thermoelectric power generator.

All the refrigeration systems discussed above involve many moving parts and bulky, complex components. Then this question comes to mind: Is it really necessary for a refrigeration system to be so complex? Can we not achieve the same effect in a more direct way? The answer to this question is *yes*. It is possible to use electric energy more directly to produce cooling without involving any refrigerants and moving parts. Below we discuss one such system, called *thermoelectric refrigerator*.

Consider two wires made from different metals joined at both ends (junctions), forming a closed circuit. Ordinarily, nothing will happen. However, when one of the ends is heated, something interesting happens: A current flows continuously in the circuit, as shown in Fig. 11–25. This is called the **Seebeck effect**, in honor of Thomas Seebeck, who made this discovery in 1821. The circuit that incorporates both thermal and electrical effects is called a **thermoelectric circuit**, and a device that operates on this circuit is called a **thermoelectric device**.

The Seebeck effect has two major applications: temperature measurement and power generation. When the thermoelectric circuit is broken, as shown in Fig. 11–26, the current ceases to flow, and we can measure the driving force (the electromotive force) or the voltage generated in the circuit by a voltmeter. The voltage generated is a function of the temperature difference and the materials of the two wires used. Therefore, temperature can be measured by simply measuring voltages. The two wires used to measure the temperature in this manner form a *thermocouple*, which is the most versatile and most widely used temperature measurement device. A common T-type thermocouple, for example, consists of copper and constantan wires, and it produces about 40 μV per °C difference.

The Seebeck effect also forms the basis for thermoelectric power generation. The schematic diagram of a **thermoelectric generator** is shown in Fig. 11–27. Heat is transferred from a high-temperature source to the hot junction in the amount of $Q_H$, and it is rejected to a low-temperature sink from the cold junction in the amount of $Q_L$. The difference between these two quantities is the net electrical work produced, that is, $W_e = Q_H - Q_L$. It is evident from Fig. 11–27 that the thermoelectric power cycle closely resembles an ordinary heat engine cycle, with electrons serving as the working fluid. Therefore, the thermal efficiency of a thermoelectric generator operating between the temperature limits of $T_H$ and $T_L$ is limited by the efficiency of a Carnot cycle operating between the same temperature limits. Thus, in the absence of any irreversibilities (such as $I^2R$ heating, where $R$ is the total electrical resistance of the wires), the thermoelectric generator will have the Carnot efficiency.

The major drawback of thermoelectric generators is their low efficiency. The future success of these devices depends on finding materials with more desirable characteristics. For example, the voltage output of thermoelectric devices has been increased several times by switching from metal pairs to semiconductors. A practical thermoelectric generator using *n*-type (heavily

---

*This section can be skipped without a loss in continuity

doped to create excess electrons) and *p*-type (heavily doped to create a deficiency of electrons) materials connected in series is shown in Fig. 11–28. Despite their low efficiencies, thermoelectric generators have definite weight and reliability advantages and are presently used in rural areas and in space applications. For example, silicon–germanium-based thermoelectric generators have been powering *Voyager* spacecraft since 1980 and are expected to continue generating power for many more years.

If Seebeck had been fluent in thermodynamics, he would probably have tried reversing the direction of flow of electrons in the thermoelectric circuit (by externally applying a potential difference in the reverse direction) to create a refrigeration effect. But this honor belongs to Jean Charles Athanase Peltier, who discovered this phenomenon in 1834. He noticed during his experiments that when a small current was passed through the junction of two dissimilar wires, the junction was cooled, as shown in Fig. 11–29. This is called the **Peltier effect**, and it forms the basis for **thermoelectric refrigeration**. A practical thermoelectric refrigeration circuit using semiconductor materials is shown in Fig. 11–30. Heat is absorbed from the refrigerated space in the amount of $Q_L$ and rejected to the warmer environment in the amount of $Q_H$. The difference between these two quantities is the net electrical work that needs to be supplied; that is, $W_e = Q_H - Q_L$. Thermoelectric refrigerators presently cannot compete with vapor-compression refrigeration systems because of their low coefficient of performance. They are available in the market, however, and are preferred in some applications because of their small size, simplicity, quietness, and reliability.

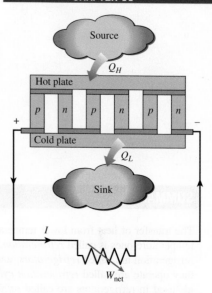

**FIGURE 11–28**
A thermoelectric power generator.

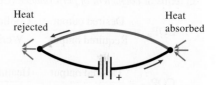

**FIGURE 11–29**
When a current is passed through the junction of two dissimilar materials, the junction is cooled.

### EXAMPLE 11–7     Cooling of a Canned Drink by a Thermoelectric Refrigerator

A thermoelectric refrigerator that resembles a small ice chest is powered by a car battery and has a COP of 0.1. If the refrigerator cools a 0.350-L canned drink from 20 to 4°C in 30 min, determine the average electric power consumed by the thermoelectric refrigerator.

**SOLUTION** A thermoelectric refrigerator with a specified COP is used to cool canned drinks. The power consumption of the refrigerator is to be determined.
**Assumptions** Heat transfer through the walls of the refrigerator is negligible during operation.
**Properties** The properties of canned drinks are the same as those of water at room temperature, $\rho = 1$ kg/L and $c = 4.18$ kJ/kg·°C (Table A–3).
**Analysis** The cooling rate of the refrigerator is simply the rate of decrease of the energy of the canned drinks,

$$m = \rho V = (1 \text{ kg/L})(0.350 \text{ L}) = 0.350 \text{ kg}$$

$$Q_{\text{cooling}} = mc\Delta T = (0.350 \text{ kg})(4.18 \text{ kJ/kg·°C})(20 - 4)°C = 23.4 \text{ kJ}$$

$$\dot{Q}_{\text{cooling}} = \frac{Q_{\text{cooling}}}{\Delta t} = \frac{23.4 \text{ kJ}}{30 \times 60 \text{ s}} = 0.0130 \text{ kW} = 13 \text{ W}$$

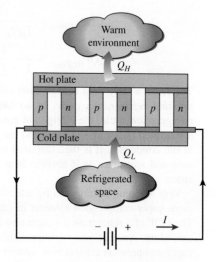

**FIGURE 11–30**
A thermoelectric refrigerator.

Then the average power consumed by the refrigerator becomes

$$\dot{W}_{in} = \frac{\dot{Q}_{cooling}}{COP_R} = \frac{13 \ W}{0.10} = 130 \ W$$

**Discussion** In reality, the power consumption will be larger because of the heat gain through the walls of the refrigerator.

## SUMMARY

The transfer of heat from lower temperature regions to higher temperature ones is called *refrigeration*. Devices that produce refrigeration are called *refrigerators,* and the cycles on which they operate are called *refrigeration cycles*. The working fluids used in refrigerators are called *refrigerants*. Refrigerators used for the purpose of heating a space by transferring heat from a cooler medium are called *heat pumps*.

The performance of refrigerators and heat pumps is expressed in terms of *coefficient of performance* (COP), defined as

$$COP_R = \frac{\text{Desired output}}{\text{Required output}} = \frac{\text{Cooling effect}}{\text{Work input}} = \frac{Q_L}{W_{net,in}}$$

$$COP_{HP} = \frac{\text{Desired output}}{\text{Required input}} = \frac{\text{Heating effect}}{\text{Work input}} = \frac{Q_H}{W_{net,in}}$$

The standard of comparison for refrigeration cycles is the *reversed Carnot cycle*. A refrigerator or heat pump that operates on the reversed Carnot cycle is called a *Carnot refrigerator* or a *Carnot heat pump,* and their COPs are

$$COP_{R,Carnot} = \frac{1}{T_H/T_L - 1}$$

$$COP_{HP,Carnot} = \frac{1}{1 - T_L/T_H}$$

The most widely used refrigeration cycle is the *vapor-compression refrigeration cycle*. In an ideal vapor-compression refrigeration cycle, the refrigerant enters the compressor as a saturated vapor and is cooled to the saturated liquid state in the condenser. It is then throttled to the evaporator pressure and vaporizes as it absorbs heat from the refrigerated space.

Very low temperatures can be achieved by operating two or more vapor-compression systems in series, called *cascading*. The COP of a refrigeration system also increases as a result of cascading. Another way of improving the performance of a vapor-compression refrigeration system is by using *multistage compression with regenerative cooling*. A refrigerator with a

single compressor can provide refrigeration at several temperatures by throttling the refrigerant in stages. The vapor-compression refrigeration cycle can also be used to liquefy gases after some modifications.

The power cycles can be used as refrigeration cycles by simply reversing them. Of these, the *reversed Brayton cycle,* which is also known as the *gas refrigeration cycle,* is used to cool aircraft and to obtain very low (cryogenic) temperatures after it is modified with regeneration. The work output of the turbine can be used to reduce the work input requirements to the compressor. Thus the COP of a gas refrigeration cycle is

$$COP_R = \frac{q_L}{w_{net,in}} = \frac{q_L}{w_{comp,in} - w_{turb,out}}$$

Another form of refrigeration that becomes economically attractive when there is a source of inexpensive thermal energy at a temperature of 100 to 200°C is *absorption refrigeration*, where the refrigerant is absorbed by a transport medium and compressed in liquid form. The most widely used absorption refrigeration system is the ammonia–water system, where ammonia serves as the refrigerant and water as the transport medium. The work input to the pump is usually very small, and the COP of absorption refrigeration systems is defined as

$$COP_{absorption} = \frac{\text{Desired output}}{\text{Required input}} = \frac{Q_L}{Q_{gen} + W_{pump}} \cong \frac{Q_L}{Q_{gen}}$$

The maximum COP an absorption refrigeration system can have is determined by assuming totally reversible conditions, which yields

$$COP_{rev,absorption} = \eta_{th,rev} COP_{R,rev} = \left(1 - \frac{T_0}{T_s}\right)\left(\frac{T_L}{T_0 - T_L}\right)$$

where $T_0$, $T_L$, and $T_s$ are the thermodynamic temperatures of the environment, the refrigerated space, and the heat source, respectively.

# REFERENCES AND SUGGESTED READINGS

1. *ASHRAE, Handbook of Fundamentals.* Atlanta: American Society of Heating, Refrigerating, and Air-Conditioning Engineers, 1985.

2. *Heat Pump Systems—A Technology Review.* OECD Report, Paris, 1982.

3. B. Nagengast. "A Historical Look at CFC Refrigerants." *ASHRAE Journal* 30, no. 11 (November 1988), pp. 37–39.

4. W. F. Stoecker. "Growing Opportunities for Ammonia Refrigeration." *Proceedings of the Meeting of the International Institute of Ammonia Refrigeration*, Austin, Texas, 1989.

5. W. F. Stoecker and J. W. Jones. *Refrigeration and Air Conditioning.* 2nd ed. New York: McGraw-Hill, 1982.

# PROBLEMS*

## The Reversed Carnot Cycle

**11–1C** Why is the reversed Carnot cycle executed within the saturation dome not a realistic model for refrigeration cycles?

**11–2C** Why do we study the reversed Carnot cycle even though it is not a realistic model for refrigeration cycles?

**11–3** A steady-flow Carnot refrigeration cycle uses refrigerant- 134a as the working fluid. The refrigerant changes from saturated vapor to saturated liquid at 60°C in the condenser as it rejects heat. The evaporator pressure is 140 kPa. Show the cycle on a *T-s* diagram relative to saturation lines, and determine (*a*) the coefficient of performance, (*b*) the amount of heat absorbed from the refrigerated space, and (*c*) the net work input. *Answers:* (*a*) 3.23, (*b*) 106 kJ/kg, (*c*) 32.9 kJ/kg

## Ideal and Actual Vapor-Compression Refrigeration Cycles

**11–4C** Why is the throttling valve not replaced by an isentropic turbine in the ideal vapor-compression refrigeration cycle?

**11–5C** It is proposed to use water instead of refrigerant-134a as the working fluid in air-conditioning applications where the minimum temperature never falls below the freezing point. Would you support this proposal? Explain.

**11–6C** In a refrigeration system, would you recommend condensing the refrigerant-134a at a pressure of 0.7 or 1.0 MPa if heat is to be rejected to a cooling medium at 15°C? Why?

**11–7C** Does the area enclosed by the cycle on a *T-s* diagram represent the net work input for the reversed Carnot cycle? How about for the ideal vapor-compression refrigeration cycle?

**11–8C** Consider two vapor-compression refrigeration cycles. The refrigerant enters the throttling valve as a saturated liquid at 30°C in one cycle and as subcooled liquid at 30°C in the other one. The evaporator pressure for both cycles is the same. Which cycle do you think will have a higher COP?

**11–9C** The COP of vapor-compression refrigeration cycles improves when the refrigerant is subcooled before it enters the throttling valve. Can the refrigerant be subcooled indefinitely to maximize this effect, or is there a lower limit? Explain.

**11–10** An ice-making machine operates on the ideal vapor-compression cycle, using refrigerant-134a. The refrigerant enters the compressor as saturated vapor at 140 kPa and leaves the condenser as saturated liquid at 600 kPa. Water enters the ice machine at 13°C and leaves as ice at −4°C. For an ice production rate of 7 kg/h, determine the power input to the ice machine (393 kJ of heat needs to be removed from each 1 kg of water at 13°C to turn it into ice at −4°C).

**11–11** A refrigerator operates on the ideal vapor-compression refrigeration cycle and uses refrigerant-134a as the working fluid. The condenser operates at 1.6 MPa and the evaporator at -6°C. If an adiabatic, reversible expansion device were available and used to expand the liquid leaving the condenser, how much would the COP improve by using this device instead of the throttle device? *Answer:* 9.7 percent

---

\* Problems designated by a "C" are concept questions, and students are encouraged to answer them all. Problems with the 🖱 icon are solved using EES, and complete solutions together with parametric studies are included on the text website. Problems with the 📖 icon are comprehensive in nature, and are intended to be solved with an equation solver such as EES.

**11–12** An ideal vapor-compression refrigeration cycle that uses refrigerant-134a as its working fluid maintains a condenser at 800 kPa and the evaporator at −12°C. Determine this system's COP and the amount of power required to service a 150 kW cooling load. *Answers:* 4.87, 30.8 kW

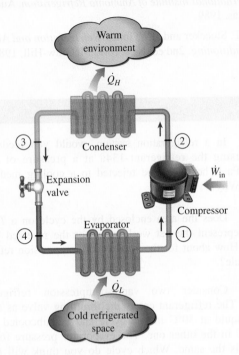

**FIGURE P11–12**

**11–13** Consider a 300 kJ/min refrigeration system that operates on an ideal vapor-compression refrigeration cycle with refrigerant-134a as the working fluid. The refrigerant enters the compressor as saturated vapor at 140 kPa and is compressed to 800 kPa. Show the cycle on a *T-s* diagram with respect to saturation lines, and determine (*a*) the quality of the refrigerant at the end of the throttling process, (*b*) the coefficient of performance, and (*c*) the power input to the compressor.

**11–14** Reconsider Prob. 11–13. Using EES (or other) software, investigate the effect of evaporator pressure on the COP and the power input. Let the evaporator pressure vary from 100 to 400 kPa. Plot the COP and the power input as functions of evaporator pressure, and discuss the results.

**11–15** Repeat Prob. 11–13 assuming an isentropic efficiency of 85 percent for the compressor. Also, determine the rate of exergy destruction associated with the compression process in this case. Take $T_0 = 298$ K.

**11–16** Refrigerant-134a enters the compressor of a refrigerator as superheated vapor at 0.20 MPa and −5°C at a rate of 0.07 kg/s, and it leaves at 1.2 MPa and 70°C. The refrigerant is cooled in the condenser to 44°C and 1.15 MPa, and it is throttled to 0.21 MPa. Disregarding any heat transfer and

pressure drops in the connecting lines between the components, show the cycle on a *T-s* diagram with respect to saturation lines, and determine (*a*) the rate of heat removal from the refrigerated space and the power input to the compressor, (*b*) the isentropic efficiency of the compressor, and (*c*) the COP of the refrigerator. *Answers:* (*a*) 9.42 kW, 3.63 kW, (*b*) 74.1 percent, (*c*) 2.60

**11–17** A commercial refrigerator with refrigerant-134a as the working fluid is used to keep the refrigerated space at −30°C by rejecting its waste heat to cooling water that enters the condenser at 18°C at a rate of 0.25 kg/s and leaves at 26°C. The refrigerant enters the condenser at 1.2 MPa and 65°C and leaves at 42°C. The inlet state of the compressor is 60 kPa and −34°C and the compressor is estimated to gain a net heat of 450 W from the surroundings. Determine (*a*) the quality of the refrigerant at the evaporator inlet, (*b*) the refrigeration load, (*c*) the COP of the refrigerator, and (*d*) the theoretical maximum refrigeration load for the same power input to the compressor.

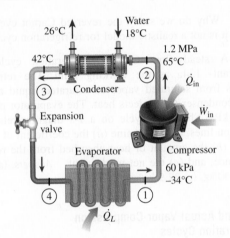

**FIGURE P11–17**

**11–18** Refrigerant-134a enters the compressor of a refrigerator at 100 kPa and −20°C at a rate of 0.5 m³/min and leaves at 0.8 MPa. The isentropic efficiency of the compressor is 78 percent. The refrigerant enters the throttling valve at 0.75 MPa and 26°C and leaves the evaporator as saturated vapor at −26°C. Show the cycle on a *T-s* diagram with respect to saturation lines, and determine (*a*) the power input to the compressor, (*b*) the rate of heat removal from the refrigerated space, and (*c*) the pressure drop and rate of heat gain in the line between the evaporator and the compressor. *Answers:* (*a*) 2.40 kW, (*b*) 6.17 kW, (*c*) 1.73 kPa, 0.203 kW

**11–19** Reconsider Prob. 11–18. Using EES (or other) software, investigate the effects of varying the compressor isentropic efficiency over the range 60 to 100 percent and the compressor inlet volume flow rate from 0.1 to 1.0 m³/min on the power input and the rate of refrigeration. Plot the rate of refrigeration and the power input to the compressor as functions of compressor efficiency for compressor inlet volume flow rates of 0.1, 0.5, and 1.0 m³/min, and discuss the results.

**11–20** A refrigerator uses refrigerant-134a as the working fluid and operates on the ideal vapor-compression refrigeration cycle except for the compression process. The refrigerant enters the evaporator at 120 kPa with a quality of 34 percent and leaves the compressor at 70°C. If the compressor consumes 450 W of power, determine (a) the mass flow rate of the refrigerant, (b) the condenser pressure, and (c) the COP of the refrigerator. *Answers:* (a) 0.00644 kg/s, (b) 800 kPa, (c) 2.03

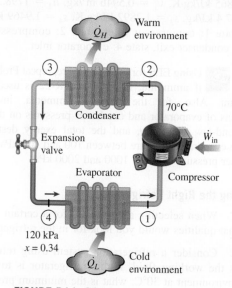

**FIGURE P11–20**

**11–21** The manufacturer of an air conditioner claims a seasonal energy efficiency ratio (SEER) of 16 (Btu/h)/W for one of its units. This unit operates on the normal vapor compression refrigeration cycle and uses refrigerant-22 as the working fluid. This SEER is for the operating conditions when the evaporator saturation temperature is −5°C and the condenser saturation temperature is 45°C. Selected data for refrigerant-22 are provided in the table below.

| $T$, °C | $P_{sat}$, kPa | $h_f$, kJ/kg | $h_g$, kJ/kg | $s_g$, kJ/kg·K |
|------|------|------|------|------|
| −5 | 421.2 | 38.76 | 248.1 | 0.9344 |
| 45 | 1728 | 101 | 261.9 | 0.8682 |

(a) Sketch the hardware and the *T-s* diagram for this air conditioner.
(b) Determine the heat absorbed by the refrigerant in the evaporator per unit mass of refrigerant-22, in kJ/kg.
(c) Determine the work input to the compressor and the heat rejected in the condenser per unit mass of refrigerant-22, in kJ/kg.

**11–22** An actual refrigerator operates on the vapor-compression refrigeration cycle with refrigerant-22 as the working fluid. The refrigerant evaporates at −15°C and condenses at 40°C. The isentropic efficiency of the compressor is 83 percent. The refrigerant is superheated by 5°C at the compressor inlet and subcooled by 5°C at the exit of the condenser. Determine (a) the heat removed from the cooled space and the work input, in kJ/kg and the COP of the cycle. Determine (b) the same parameters if the cycle operated on the ideal vapor-compression refrigeration cycle between the same evaporating and condensing temperatures.

The properties of R-22 in the case of actual operation are: $h_1$ = 402.49 kJ/kg, $h_2$ =454.00 kJ/kg, $h_3$ = 243.19 kJ/kg

The properties of R-22 in the case of ideal operation are: $h_1$ = 399.04 kJ/kg, $h_2$ = 440.71 kJ/kg, $h_3$ = 249.80 kJ/kg

Note: state 1: compressor inlet, state 2: compressor exit, state 3: condenser exit, state 4: evaporator inlet.

## Second-Law Analysis of Vapor-Compression Refrigeration Cycle

**11–23C** How is the second-law efficiency of a refrigerator operating on the vapor-compression refrigeration cycle defined? Provide two alternative definitions and explain each term.

**11–24C** How is the second-law efficiency of a heat pump operating on the vapor-compression refrigeration cycle defined? Provide two alternative definitions and show that one can be derived from the other.

**11–25C** Consider isentropic compressor of a vapor-compression refrigeration cycle. What are the isentropic efficiency and second-law efficiency of this compressor? Justify your answers. Is the second-law efficiency of a compressor necessarily equal to its isentropic efficiency? Explain.

**11–26** A space is kept at −15°C by a vapor-compression refrigeration system in an ambient at 25°C. The space gains heat steadily at a rate of 3500 kJ/h and the rate of heat rejection in the condenser is 5500 kJ/h. Determine the power input, in kW, the COP of the cycle and the second-law efficiency of the system.

**11–27** Bananas are to be cooled from 28°C to 12°C at a rate of 1140 kg/h by a refrigerator that operates on a vapor-compression refrigeration cycle. The power input to the refrigerator is 8.6 kW. Determine (a) the rate of heat absorbed from the bananas, in kJ/h, and the COP, (b) the minimum power input to the refrigerator, and (c) the second-law efficiency and the exergy destruction for the cycle. The specific heat of bananas above freezing is 3.35 kJ/kg·°C. *Answers:* (a) 61,100 kJ/h, 1.97, (b) 0.463 kW, (c) 5.4 percent, 8.14 kW

**11–28** A vapor-compression refrigeration system absorbs heat from a space at 0°C at a rate of 24,000 Btu/h and rejects heat to water in the condenser. The water experiences a temperature rise of 12°C in the condenser. The COP of the system is estimated to be 2.05. Determine (a) the power input to the system, in kW, (b) the mass flow rate of water through the condenser, and (c) the second-law efficiency and the exergy destruction for the refrigerator. Take $T_0 = 20°C$ and $c_{p,water} = 4.18$ kJ/kg·°C.

**11–29** A refrigerator operating on the vapor-compression refrigeration cycle using refrigerant-134a as the refrigerant is considered. The temperature of the cooled space and the ambient air are at −13°C and 27°C, respectively. R-134a enters the compressor at 140 kPa as a saturated vapor and leaves at 1 MPa and 70°C. The refrigerant leaves the condenser as a saturated liquid. The rate of cooling provided by the system is 13 kW. Determine (a) the mass flow rate of R-134a and the COP, (b) the exergy destruction in each component of the cycle and the second-law efficiency of the compressor, and (c) the second-law efficiency of the cycle and the total exergy destruction in the cycle.

**11–30** A room is kept at −5°C by a vapor-compression refrigeration cycle with R-134a as the refrigerant. Heat is rejected to cooling water that enters the condenser at 20°C at a rate of 0.13 kg/s and leaves at 28°C. The refrigerant enters the condenser at 1.2 MPa and 50°C and leave as a saturated liquid. If the compressor consumes 1.9 kW of power, determine (a) the refrigeration load, in Btu/h and the COP, (b) the second-law efficiency of the refrigerator and the total exergy destruction in the cycle, and (c) the exergy destruction in the condenser. Take $T_0 = 20°C$ and $c_{p,water} = 4.18$ kJ/kg·°C. *Answers:* (a) 8350 Btu/h, 1.29, (b) 12.0 percent, 1.67 kW, (c) 0.303 kW

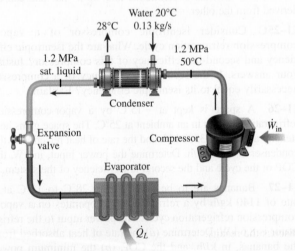

**FIGURE P11–30**

**11–31** A refrigerator operates on the ideal vapor-compression refrigeration cycle with refrigerant-134a as the working fluid. The refrigerant evaporates at −10°C and condenses at 57.9°C. The refrigerant absorbs heat from a space at 5°C and rejects heat to ambient air at 25°C. Determine (a) the cooling load, in kJ/kg, and the COP, (b) the exergy destruction in each component of the cycle and the total exergy destruction in the cycle, and (c) the second-law efficiency of the compressor, evaporator, and the cycle.

**11–32** A refrigeration system operates on the ideal vapor-compression refrigeration cycle with ammonia as the refrigerant. The evaporator and condenser pressures are 200 kPa and 2000 kPa, respectively. The temperatures of the low-temperature and high-temperature mediums are −9°C and 27°C, respectively. If the rate of heat rejected in the condenser is 18.0 kW, determine (a) the volume flow rate of ammonia at the compressor inlet, in L/s, (b) the power input and the COP, and (c) the second-law efficiency of the cycle and the total exergy destruction in the cycle. The properties of ammonia at various states are given as follows: $h_1 = 1439.3$ kJ/kg, $s_1 = 5.8865$ kJ/kg·K, $v_1 = 0.5946$ m³/kg, $h_2 = 1798.3$ kJ/kg, $h_3 = 437.4$ kJ/kg, $s_3 = 1.7892$ kJ/kg·K, $s_4 = 1.9469$ kJ/kg·K. Note: state 1: compressor inlet, state 2: compressor exit, state 3: condenser exit, state 4: evaporator inlet.

**11–33** [EES] Using EES (or other) software, repeat Prob. 11–32 if ammonia, R-134a, and R-22 is used as the refrigerant. Also, for the case of ammonia, investigate the effects of evaporator and condenser pressures on the COP, the second-law efficiency, and the total exergy destruction. Vary the evaporator pressure between 100 and 400 kPa and the condenser pressure between 1000 and 2000 kPa.

## Selecting the Right Refrigerant

**11–34C** When selecting a refrigerant for a certain application, what qualities would you look for in the refrigerant?

**11–35C** Consider a refrigeration system using refrigerant-134a as the working fluid. If this refrigerator is to operate in an environment at 30°C, what is the minimum pressure to which the refrigerant should be compressed? Why?

**11–36C** A refrigerant-134a refrigerator is to maintain the refrigerated space at −10°C. Would you recommend an evaporator pressure of 0.12 or 0.14 MPa for this system? Why?

**11–37** A refrigerator that operates on the ideal vapor-compression cycle with refrigerant-134a is to maintain the refrigerated space at −10°C while rejecting heat to the environment at 25°C. Select reasonable pressures for the evaporator and the condenser, and explain why you chose those values.

**11–38** A heat pump that operates on the ideal vapor-compression cycle with refrigerant-134a is used to heat a house and maintain it at 26°C by using underground water at 14°C as the heat source. Select reasonable pressures for the evaporator and the condenser, and explain why you chose those values.

## Heat Pump Systems

**11–39C** Do you think a heat pump system will be more cost-effective in New York or in Miami? Why?

**11–40C** What is a water-source heat pump? How does the COP of a water-source heat pump system compare to that of an air-source system?

**11–41** A heat pump that operates on the ideal vapor-compression cycle with refrigerant-134a is used to heat water from 15 to 45°C at a rate of 0.12 kg/s. The condenser and evaporator pressures are 1.4 and 0.32 MPa, respectively. Determine the power input to the heat pump.

**11–42** A heat pump with refrigerant-134a as the working fluid is used to keep a space at 25°C by absorbing heat from geothermal water that enters the evaporator at 50°C at a rate of 0.065 kg/s and leaves at 40°C. The refrigerant enters the evaporator at 20°C with a quality of 23 percent and leaves at the inlet pressure as saturated vapor. The refrigerant loses 300 W of heat to the surroundings as it flows through the compressor and the refrigerant leaves the compressor at 1.4 MPa at the same entropy as the inlet. Determine (a) the degrees of subcooling of the refrigerant in the condenser, (b) the mass flow rate of the refrigerant, (c) the heating load and the COP of the heat pump, and (d) the theoretical minimum power input to the compressor for the same heating load. *Answers:* (a) 3.8°C, (b) 0.0194 kg/s, (c) 3.07 kW, 4.68, (d) 0.238 kW

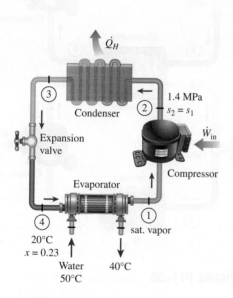

**FIGURE P11–42**

**11–43** Refrigerant-134a enters the condenser of a residential heat pump at 800 kPa and 50°C at a rate of 0.022 kg/s and leaves at 750 kPa subcooled by 3°C. The refrigerant enters the compressor at 200 kPa superheated by 4°C. Determine (a) the isentropic efficiency of the compressor, (b) the rate of heat supplied to the heated room, and (c) the COP of the

heat pump. Also, determine (d) the COP and the rate of heat supplied to the heated room if this heat pump operated on the ideal vapor-compression cycle between the pressure limits of 200 and 800 kPa.

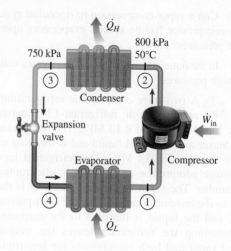

**FIGURE P11–43**

**11–44** A heat pump using refrigerant-134a heats a house by using underground water at 8°C as the heat source. The house is losing heat at a rate of 60,000 kJ/h. The refrigerant enters the compressor at 280 kPa and 0°C, and it leaves at 1 MPa and 60°C. The refrigerant exits the condenser at 30°C. Determine (a) the power input to the heat pump, (b) the rate of heat absorption from the water, and (c) the increase in electric power input if an electric resistance heater is used instead of a heat pump. *Answers:* (a) 3.55 kW, (b) 13.12 kW, (c) 13.12 kW

**11–45** Reconsider Prob. 11–44. Using EES (or other) software, investigate the effect of varying the compressor isentropic efficiency over the range 60 to 100 percent. Plot the power input to the compressor and the electric power saved by using a heat pump rather than electric resistance heating as functions of compressor efficiency, and discuss the results.

### Innovative Refrigeration Systems

**11–46C** How does the COP of a cascade refrigeration system compare to the COP of a simple vapor-compression cycle operating between the same pressure limits?

**11–47C** A certain application requires maintaining the refrigerated space at −32°C. Would you recommend a simple refrigeration cycle with refrigerant-134a or a two-stage cascade refrigeration cycle with a different refrigerant at the bottoming cycle? Why?

**11–48C** Consider a two-stage cascade refrigeration cycle and a two-stage compression refrigeration cycle with a flash chamber. Both cycles operate between the same pressure limits and use the same refrigerant. Which system would you favor? Why?

**11–49C** Can a vapor-compression refrigeration system with a single compressor handle several evaporators operating at different pressures? How?

**11–50C** In the liquefaction process, why are gases compressed to very high pressures?

**11–51** A two-stage compression refrigeration system operates with refrigerant-134a between the pressure limits of 1.4 and 0.10 MPa. The refrigerant leaves the condenser as a saturated liquid and is throttled to a flash chamber operating at 0.4 MPa. The refrigerant leaving the low-pressure compressor at 0.4 MPa is also routed to the flash chamber. The vapor in the flash chamber is then compressed to the condenser pressure by the high-pressure compressor, and the liquid is throttled to the evaporator pressure. Assuming the refrigerant leaves the evaporator as saturated vapor and both compressors are isentropic, determine (a) the fraction of the refrigerant that evaporates as it is throttled to the flash chamber, (b) the rate of heat removed from the refrigerated space for a mass flow rate of 0.25 kg/s through the condenser, and (c) the coefficient of performance.

**11–52** Repeat Prob. 11–51 for a flash chamber pressure of 0.6 MPa.

**11–53** Reconsider Prob. 11–51. Using EES (or other) software, investigate the effect of the various refrigerants for compressor efficiencies of 80, 90, and 100 percent. Compare the performance of the refrigeration system with different refrigerants.

**11–54** Consider a two-stage cascade refrigeration system operating between the pressure limits of 0.8 and 0.14 MPa. Each stage operates on the ideal vapor-compression refrigeration cycle with refrigerant-134a as the working fluid. Heat rejection from the lower cycle to the upper cycle takes place in an adiabatic counterflow heat exchanger where both streams enter at about 0.4 MPa. If the mass flow rate of the refrigerant through the upper cycle is 0.24 kg/s, determine (a) the mass flow rate of the refrigerant through the lower cycle, (b) the rate of heat removal from the refrigerated space and the power input to the compressor, and (c) the coefficient of performance of this cascade refrigerator. *Answers:* (a) 0.195 kg/s, (b) 34.2 kW, 7.63 kW, (c) 4.49

**11–55** Repeat Prob. 11–54 for a heat exchanger pressure of 0.55 MPa.

**11–56** Consider a two-stage cascade refrigeration system operating between the pressure limits of 1.4 MPa and 160 kPa with refrigerant-134a as the working fluid. Heat rejection from the lower cycle to the upper cycle takes place in an adiabatic counterflow heat exchanger where the pressure in the upper and lower cycles are 0.4 and 0.5 MPa, respectively. In both cycles, the refrigerant is a saturated liquid at the condenser exit and a saturated vapor at the compressor inlet, and the isentropic efficiency of the compressor is 80 percent. If the mass flow rate of the refrigerant through the lower cycle is 0.11 kg/s, determine (a) the mass flow rate of the refrigerant through the upper cycle, (b) the rate of heat removal from the refrigerated space, and (c) the COP of this refrigerator. *Answers:* (a) 0.169 kg/s, (b) 18.5 kW, (c) 2.12

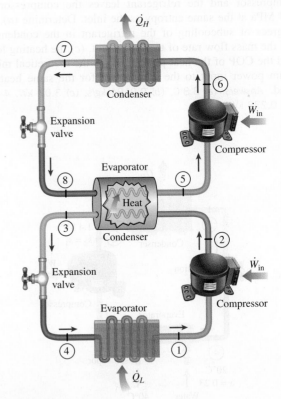

**FIGURE P11–56**

**11–57** Consider a two-stage cascade refrigeration system operating between the pressure limits of 1.2 MPa and 200 kPa with refrigerant-134a as the working fluid. The refrigerant leaves the condenser as a saturated liquid and is throttled to a flash chamber operating at 0.45 MPa. Part of the refrigerant evaporates during this flashing process, and this vapor is mixed with the refrigerant leaving the low-pressure

compressor. The mixture is then compressed to the condenser pressure by the high-pressure compressor. The liquid in the flash chamber is throttled to the evaporator pressure and cools the refrigerated space as it vaporizes in the evaporator. The mass flow rate of the refrigerant through the low-pressure compressor is 0.15 kg/s. Assuming the refrigerant leaves the evaporator as a saturated vapor and the isentropic efficiency is 80 percent for both compressors, determine (a) the mass flow rate of the refrigerant through the high-pressure compressor, (b) the rate of heat removal from the refrigerated space, and (c) the COP of this refrigerator. Also, determine (d) the rate of heat removal and the COP if this refrigerator operated on a single-stage cycle between the same pressure limits with the same compressor efficiency and the same flow rate as in part (a).

the low-temperature evaporator serves a cooling load of 8 kW. Determine the cooling rate of the high-temperature evaporator, the power required by the compressor, and the COP of the system. The refrigerant is saturated liquid at the exit of the condenser and saturated vapor at the exit of each evaporator, and the compressor is isentropic. *Answers:* 6.58 kW, 4.51 kW, 3.24

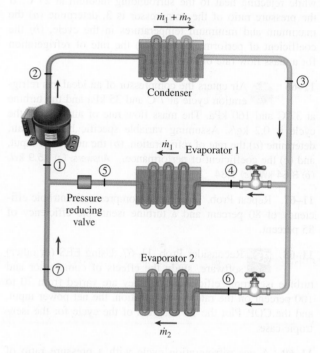

**FIGURE P11–58**

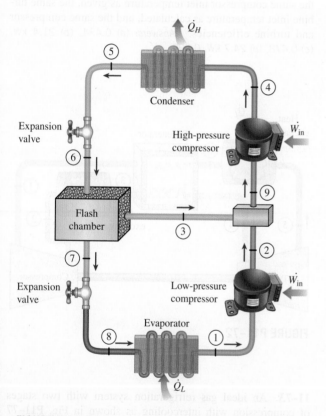

**FIGURE P11–57**

**11–58** A two-evaporator compression refrigeration system as shown in Fig. P11-60 uses refrigerant-134a as the working fluid. The system operates evaporator 1 at 0°C, evaporator 2 at −26.4°C, and the condenser at 800 kPa. The refrigerant is circulated through the compressor at a rate of 0.1 kg/s and

## Gas Refrigeration Cycle

**11–59C** How does the ideal-gas refrigeration cycle differ from the Carnot refrigeration cycle?

**11–60C** Devise a refrigeration cycle that works on the reversed Stirling cycle. Also, determine the COP for this cycle.

**11–61C** How is the ideal-gas refrigeration cycle modified for aircraft cooling?

**11–62C** In gas refrigeration cycles, can we replace the turbine by an expansion valve as we did in vapor-compression refrigeration cycles? Why?

**11–63C** How do we achieve very low temperatures with gas refrigeration cycles?

**11–64** Air enters the compressor of an ideal gas refrigeration cycle at 70°C and 70 kPa and the turbine at 47°C and 210 kPa. The mass flow rate of air through the cycle is 0.2 kg/s Determine (a) the rate of refrigeration, (b) the net power input, and (c) the coefficient of performance.

**11–65** An ideal gas refrigeration cycle using air as the working fluid is to maintain a refrigerated space at −23°C while rejecting heat to the surrounding medium at 27°C. If the pressure ratio of the compressor is 3, determine (a) the maximum and minimum temperatures in the cycle, (b) the coefficient of performance, and (c) the rate of refrigeration for a mass flow rate of 0.08 kg/s.

**11–66** Air enters the compressor of an ideal gas refrigeration cycle at 7°C and 35 kPa and the turbine at 37°C and 160 kPa. The mass flow rate of air through the cycle is 0.2 kg/s. Assuming variable specific heats for air, determine (a) the rate of refrigeration, (b) the net power input, and (c) the coefficient of performance. *Answers:* (a) 15.9 kW, (b) 8.64 kW, (c) 1.84

**11–67** Repeat Prob. 11–66 for a compressor isentropic efficiency of 80 percent and a turbine isentropic efficiency of 85 percent.

**11–68** Reconsider Prob. 11–67. Using EES (or other) software, study the effects of compressor and turbine isentropic efficiencies as they are varied from 70 to 100 percent on the rate of refrigeration, the net power input, and the COP. Plot the *T-s* diagram of the cycle for the isentropic case.

**11–69** A gas refrigeration cycle with a pressure ratio of 4 uses helium as the working fluid. The temperature of the helium is −6°C at the compressor inlet and 50°C at the turbine inlet. Assuming isentropic efficiencies of 85 percent for both the turbine and the compressor, determine (a) the minimum temperature in the cycle, (b) the coefficient of performance, and (c) the mass flow rate of the helium for a refrigeration rate of 25 kW.

**11–70** A gas refrigeration system using air as the working fluid has a pressure ratio of 4. Air enters the compressor at −7°C. The high-pressure air is cooled to 27°C by rejecting heat to the surroundings. It is further cooled to −15°C by regenerative cooling before it enters the turbine. Assuming both the turbine and the compressor to be isentropic and using constant specific heats at room temperature, determine (a) the lowest temperature that can be obtained by this cycle, (b) the coefficient of performance of the cycle, and (c) the mass flow rate of air for a refrigeration rate of 12 kW. *Answers:* (a) −99.4°C, (b) 1.12, (c) 0.237 kg/s

**11–71** Repeat Prob. 11–70 assuming isentropic efficiencies of 75 percent for the compressor and 80 percent for the turbine.

**11–72** A gas refrigeration system using air as the working fluid has a pressure ratio of 5. Air enters the compressor at 0°C. The high-pressure air is cooled to 35°C by rejecting heat to the surroundings. The refrigerant leaves the turbine at −80°C and then it absorbs heat from the refrigerated space before entering the regenerator. The mass flow rate of air is 0.4 kg/s. Assuming isentropic efficiencies of 80 percent for the compressor and 85 percent for the turbine and using constant specific heats at room temperature, determine (a) the effectiveness of the regenerator, (b) the rate of heat removal from the refrigerated space, and (c) the COP of the cycle. Also, determine (d) the refrigeration load and the COP if this system operated on the simple gas refrigeration cycle. Use the same compressor inlet temperature as given, the same turbine inlet temperature as calculated, and the same compressor and turbine efficiencies. *Answers:* (a) 0.434, (b) 21.4 kW, (c) 0.478, (d) 24.7 kW, 0.599

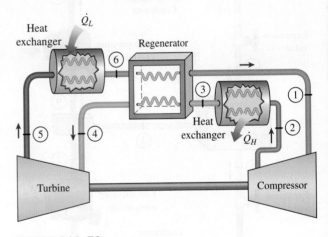

**FIGURE P11–72**

**11–73** An ideal gas refrigeration system with two stages of compression with intercooling as shown in Fig. P11–77 operates with air entering the first compressor at 90 kPa and −24°C. Each compression stage has a pressure ratio of 3 and the two intercoolers can cool the air to 5°C. Calculate the coefficient of performance of this system and the rate at which air must be circulated through this system to service a 45,000 kJ/h cooling load. Use constant specific heats at room temperature. *Answers:* 1.56, 0.124 kg/s

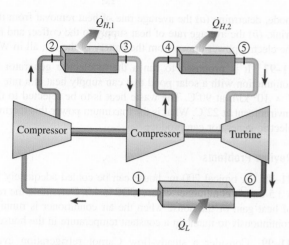

**FIGURE P11–73**

**11–74** How will the answers of Prob. 11–73 change when the isentropic efficiency of each compressor is 85 percent and the isentropic efficiency of the turbine is 95 percent?

## Absorption Refrigeration Systems

**11–75C** What is absorption refrigeration? How does an absorption refrigeration system differ from a vapor-compression refrigeration system?

**11–76C** What are the advantages and disadvantages of absorption refrigeration?

**11–77C** In absorption refrigeration cycles, why is the fluid in the absorber cooled and the fluid in the generator heated?

**11–78C** How is the coefficient of performance of an absorption refrigeration system defined?

**11–79** An absorption refrigeration system that receives heat from a source at 95°C and maintains the refrigerated space at 0°C is claimed to have a COP of 3.1. If the environmental temperature is 19°C, can this claim be valid? Justify your answer.

**11–80** An absorption refrigeration system receives heat from a source at 120°C and maintains the refrigerated space at 0°C. If the temperature of the environment is 25°C, what is the maximum COP this absorption refrigeration system can have?

**11–81** Heat is supplied to an absorption refrigeration system from a geothermal well at 110°C at a rate of $5 \times 10^5$ kJ/h. The environment is at 25°C, and the refrigerated space is maintained at −18°C. Determine the maximum rate at which this system can remove heat from the refrigerated space. *Answer:* $6.58 \times 10^5$ kJ/h

**11–82** A reversible absorption refrigerator consists of a reversible heat engine and a reversible refrigerator. The system removes heat from a cooled space at −15°C at a rate of 70 kW. The refrigerator operates in an environment at 25°C. If the heat is supplied to the cycle by condensing saturated steam at 150°C, determine (a) the rate at which the steam condenses, and (b) the power input to the reversible refrigerator. (c) If the COP of an actual absorption chiller at the same temperature limits has a COP of 0.8, determine the second-law efficiency of this chiller. *Answers:* (a) 0.0174 kg/s, (b) 10.9 kW, (c) 42.0 percent

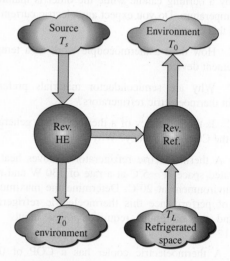

**FIGURE P11–82**

**11–83** An ammonia-water absorption refrigeration cycle is used to keep a space at −4°C when the ambient temperature is 20°C. Pure ammonia enters the condenser at 2100 kPa and 60°C at a rate of 0.02 kg/s. Ammonia leaves the condenser as a saturated liquid and is expanded to 200 kPa. Ammonia leaves the evaporator as a saturated vapor. Heat is supplied to the generator by geothermal liquid water that enters at 115°C at a rate of 0.25 kg/s and leaves at 95°C. Determine (a) the rate of cooling provided by the system, in kW, the COP, and (b) the second-law efficiency of the system. The enthalpies of ammonia at various states of the system are: condenser inlet $h_2 = 1523.2$ kJ/kg, evaporator inlet $h_4 = 446.9$ kJ/kg, evaporator exit $h_1 = 1439.3$ kJ/kg. Also, take the specific heat of geothermal water to be 4.18 kJ/kg·°C.

## Special Topic: Thermoelectric Power Generation and Refrigeration Systems

**11–84C** What is a thermoelectric circuit?

**11–85C** Describe the Seebeck and the Peltier effects.

**11–86C** Consider a circular copper wire formed by connecting the two ends of a copper wire. The connection point is now heated by a burning candle. Do you expect any current to flow through the wire?

**11–87C** An iron and a constantan wire are formed into a closed circuit by connecting the ends. Now both junctions are heated and are maintained at the same temperature. Do you expect any electric current to flow through this circuit?

**11–88C** A copper and a constantan wire are formed into a closed circuit by connecting the ends. Now one junction is heated by a burning candle while the other is maintained at room temperature. Do you expect any electric current to flow through this circuit?

**11–89C** How does a thermocouple work as a temperature measurement device?

**11–90C** Why are semiconductor materials preferable to metals in thermoelectric refrigerators?

**11–91C** Is the efficiency of a thermoelectric generator limited by the Carnot efficiency? Why?

**11–92** A thermoelectric refrigerator removes heat from a refrigerated space at $-5°C$ at a rate of 130 W and rejects it to an environment at 20°C. Determine the maximum coefficient of performance this thermoelectric refrigerator can have and the minimum required power input. *Answers:* 10.72, 12.1 W

**11–93** A thermoelectric cooler has a COP of 0.15 and removes heat from a refrigerated space at a rate of 180 W. Determine the required power input to the thermoelectric cooler, in W.

**11–94** A thermoelectric cooler has a COP of 0.18 and the power input to the cooler is 1.3 kW. Determine the rate of heat removed from the refrigerated space, in kJ/min.

**11–95** A thermoelectric refrigerator is powered by a 12-V car battery that draws 3 A of current when running. The refrigerator resembles a small ice chest and is claimed to cool nine canned drinks, 0.350-L each, from 25 to 3°C in 12 h. Determine the average COP of this refrigerator.

**FIGURE P11–95**

**11–96** Thermoelectric coolers that plug into the cigarette lighter of a car are commonly available. One such cooler is claimed to cool a 350 g drink from 26 to 3°C or to heat a cup of coffee from 24 to 54°C in about 15 min in a well-insulated cup holder. Assuming an average COP of 0.2 in the cooling mode, determine (*a*) the average rate of heat removal from the drink, (*b*) the average rate of heat supply to the coffee, and (*c*) the electric power drawn from the battery of the car, all in W.

**11–97** It is proposed to run a thermoelectric generator in conjunction with a solar pond that can supply heat at a rate of $7 \times 10^6$ kJ/h at 90°C. The waste heat is to be rejected to the environment at 22°C. What is the maximum power this thermoelectric generator can produce?

## Review Problems

**11–98** A typical 200-m$^2$ house can be cooled adequately by a 3.5-ton air conditioner whose COP is 4.0. Determine the rate of heat gain of the house when the air conditioner is running continuously to maintain a constant temperature in the house.

**11–99** Consider a steady-flow Carnot refrigeration cycle that uses refrigerant-134a as the working fluid. The maximum and minimum temperatures in the cycle are 30 and $-20°C$, respectively. The quality of the refrigerant is 0.15 at the beginning of the heat absorption process and 0.80 at the end. Show the cycle on a *T-s* diagram relative to saturation lines, and determine (*a*) the coefficient of performance, (*b*) the condenser and evaporator pressures, and (*c*) the net work input.

**11–100** A heat pump water heater (HPWH) heats water by absorbing heat from the ambient air and transferring it to water. The heat pump has a COP of 3.4 and consumes 6 kW of electricity when running. Determine if this heat pump can be used to meet the cooling needs of a room most of the time for "free" by absorbing heat from the air in the room. The rate of heat gain of a room is usually less than 45,000 kJ/h.

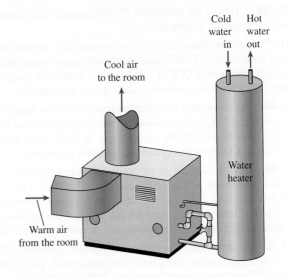

**FIGURE P11–100**

**11–101** A heat pump that operates on the ideal vapor-compression cycle with refrigerant-134a is used to heat a

house. The mass flow rate of the refrigerant is 0.25 kg/s. The condenser and evaporator pressures are 1400 and 320 kPa, respectively. Show the cycle on a *T-s* diagram with respect to saturation lines, and determine (*a*) the rate of heat supply to the house, (*b*) the volume flow rate of the refrigerant at the compressor inlet, and (*c*) the COP of this heat pump.

**11–102** A large refrigeration plant is to be maintained at −15°C, and it requires refrigeration at a rate of 100 kW. The condenser of the plant is to be cooled by liquid water, which experiences a temperature rise of 8°C as it flows over the coils of the condenser. Assuming the plant operates on the ideal vapor-compression cycle using refrigerant-134a between the pressure limits of 120 and 700 kPa, determine (*a*) the mass flow rate of the refrigerant, (*b*) the power input to the compressor, and (*c*) the mass flow rate of the cooling water.

**11–103** ![EES] Reconsider Prob. 11–102. Using EES (or other) software, investigate the effect of evaporator pressure on the COP and the power input. Let the evaporator pressure vary from 120 to 380 kPa. Plot the COP and the power input as functions of evaporator pressure, and discuss the results.

**11–104** Repeat Prob. 11–102 assuming the compressor has an isentropic efficiency of 75 percent. Also, determine the rate of exergy destruction associated with the compression process in this case. Take $T_0 = 25$°C.

**11–105** A refrigeration unit operates on the ideal vapor compression refrigeration cycle and uses refrigerant-22 as the working fluid. The operating conditions for this unit are evaporator saturation temperature of −5°C and the condenser saturation temperature of 45°C. Selected data for refrigerant-22 are provided in the table below.

| *T*, °C | $P_{sat}$, kPa | $h_f$, kJ/kg | $h_g$, kJ/kg | $s_g$, kJ/kg·K |
|---|---|---|---|---|
| −5 | 421.2 | 38.76 | 248.1 | 0.9344 |
| 45 | 1728 | 101 | 261.9 | 0.8682 |

For R-22 at $P = 1728$ kPa and $s = 0.9344$ kJ/kg·K, $T = 68.15$°C and $h = 283.7$ kJ/kg. Also, take $c_{p,\text{air}} = 1.005$ kJ/kg·K.

(*a*) Sketch the hardware and the *T-s* diagram for this heat pump application.

(*b*) Determine the COP for this refrigeration unit.

(*c*) The evaporator of this unit is located inside the air handler of the building. The air flowing through the air handler enters the air handler at 27°C and is limited to a 20°C temperature drop. Determine the ratio of volume flow rate of air entering the air handler ($m^3_{\text{air}}$/min) to mass flow rate of R-22 ($\text{kg}_{\text{R-22}}$/s) through the air handler, in ($m^3_{\text{air}}$/min)/($\text{kg}_{\text{R-22}}$/s). Assume the air pressure is 100 kPa.

**11–106** An air conditioner with refrigerant-134a as the working fluid is used to keep a room at 26°C by rejecting

the waste heat to the outside air at 34°C. The room is gaining heat through the walls and the windows at a rate of 250 kJ/min while the heat generated by the computer, TV, and lights amounts to 900 W. An unknown amount of heat is also generated by the people in the room. The condenser and evaporator pressures are 1200 and 500 kPa, respectively. The refrigerant is saturated liquid at the condenser exit and saturated vapor at the compressor inlet. If the refrigerant enters the compressor at a rate of 100 L/min and the isentropic efficiency of the compressor is 75 percent, determine (*a*) the temperature of the refrigerant at the compressor exit, (*b*) the rate of heat generation by the people in the room, (*c*) the COP of the air conditioner, and (*d*) the minimum volume flow rate of the refrigerant at the compressor inlet for the same compressor inlet and exit conditions.

*Answers:* (*a*) 54.5°C, (*b*) 0.665 kW, (*c*) 5.87, (*d*) 15.7 L/min

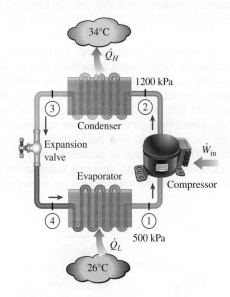

**FIGURE P11–106**

**11–107** An air-conditioner operates on the vapor-compression refrigeration cycle with refrigerant-134a as the refrigerant. The air conditioner is used to keep a space at 21°C while rejecting the waste heat to the ambient air at 37°C. The refrigerant enters the compressor at 180 kPa superheated by 2.7°C at a rate of 0.06 kg/s and leaves the compressor at 1200 kPa and 60°C. R-134a is subcooled by 6.3°C at the exit of the condenser. Determine (*a*) the rate of cooling provided to the space, in Btu/h, and the COP, (*b*) the isentropic efficiency and the exergy efficiency of the compressor, (*c*) the exergy destruction in each component of the cycle and the total exergy destruction in the cycle, and (*d*) the minimum power input and the second-law efficiency of the cycle.

**11–108** Consider a two-stage compression refrigeration system operating between the pressure limits of 1.4

and 0.12 MPa. The working fluid is refrigerant-134a. The refrigerant leaves the condenser as a saturated liquid and is throttled to a flash chamber operating at 0.5 MPa. Part of the refrigerant evaporates during this flashing process, and this vapor is mixed with the refrigerant leaving the low-pressure compressor. The mixture is then compressed to the condenser pressure by the high-pressure compressor. The liquid in the flash chamber is throttled to the evaporator pressure, and it cools the refrigerated space as it vaporizes in the evaporator. Assuming the refrigerant leaves the evaporator as saturated vapor and both compressors are isentropic, determine (a) the fraction of the refrigerant that evaporates as it is throttled to the flash chamber, (b) the amount of heat removed from the refrigerated space and the compressor work per unit mass of refrigerant flowing through the condenser, and (c) the coefficient of performance. *Answers: (a)* 0.290, *(b)* 116 kJ/kg, 42.7 kJ/kg, *(c)* 2.72

**11–109** A two-stage compression refrigeration system with an adiabatic liquid-vapor separation unit as shown in Fig. P11–109 uses refrigerant-134a as the working fluid. The system operates the evaporator at −32°C, the condenser at 1400 kPa, and the separator at 8.9°C. The refrigerant is circulated through the condenser at a rate of 2 kg/s. Determine the rate of cooling and power requirement for this system. The refrigerant is saturated liquid at the inlet of each expansion valve and saturated vapor at the inlet of each compressor, and the compressors are isentropic.

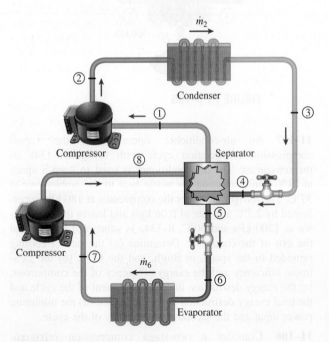

**FIGURE P11–109**

**11–110** Which process of the cycle in Prob. 11–109 has the greatest rate of exergy destruction when the low-temperature reservoir is at −22°C and the high-temperature reservoir is at 20°C? *Answer:* 30.9 kW (condenser)

**11–111** An aircraft on the ground is to be cooled by a gas refrigeration cycle operating with air on an open cycle. Air enters the compressor at 30°C and 100 kPa and is compressed to 250 kPa. Air is cooled to 70°C before it enters the turbine. Assuming both the turbine and the compressor to be isentropic, determine the temperature of the air leaving the turbine and entering the cabin. *Answer:* −9°C

**11–112** Consider a regenerative gas refrigeration cycle using helium as the working fluid. Helium enters the compressor at 100 kPa and −10°C and is compressed to 300 kPa. Helium is then cooled to 20°C by water. It then enters the regenerator where it is cooled further before it enters the turbine. Helium leaves the refrigerated space at −25°C and enters the regenerator. Assuming both the turbine and the compressor to be isentropic, determine (a) the temperature of the helium at the turbine inlet, (b) the coefficient of performance of the cycle, and (c) the net power input required for a mass flow rate of 0.45 kg/s.

**11–113** An absorption refrigeration system is to remove heat from the refrigerated space at 2°C at a rate of 28 kW while operating in an environment at 25°C. Heat is to be supplied from a solar pond at 95°C. What is the minimum rate of heat supply required? *Answer:* 12.3 kW

**11–114** ❯❯ Reconsider Prob. 11–113. Using EES (or other) software, investigate the effect of the source temperature on the minimum rate of heat supply. Let the source temperature vary from 50 to 250°C. Plot the minimum rate of heat supply as a function of source temperature, and discuss the results.

**11–115** A gas refrigeration system using air as the working fluid has a pressure ratio of 5. Air enters the compressor at 0°C. The high-pressure air is cooled to 35°C by rejecting heat to the surroundings. The refrigerant leaves the turbine at −80°C and enters the refrigerated space where it absorbs heat before entering the regenerator. The mass flow rate of air is 0.4 kg/s. Assuming isentropic efficiencies of 80 percent for the compressor and 85 percent for the turbine and using variable specific heats, determine (a) the effectiveness of the regenerator, (b) the rate of heat removal from the refrigerated space, and (c) the COP of the cycle. Also, determine (d) the refrigeration load and the COP if this system operated on the simple gas refrigeration cycle. Use the same compressor inlet temperature as given, the same turbine inlet temperature as calculated, and the same compressor and turbine efficiencies.

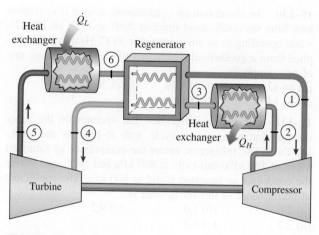

**FIGURE P11–115**

**11–116** The refrigeration system of Fig. P11–116 is another variation of the basic vapor-compression refrigeration system which attempts to reduce the compression work. In this system, a heat exchanger is used to superheat the vapor entering the compressor while subcooling the liquid exiting from the condenser. Consider a system of this type that uses refrigerant-134a as its refrigerant and operates the evaporator at −10.09°C, and the condenser at 900 kPa. Determine the system COP when the heat exchanger provides 5.51°C of subcooling at the throttle valve entrance. Assume the refrigerant leaves the evaporator as a saturated vapor and the compressor is isentropic. *Answer: 4.60*

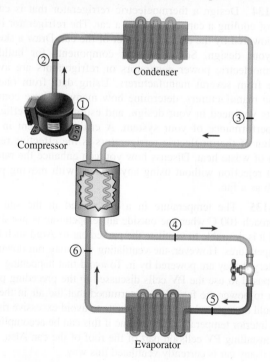

**FIGURE P11–116**

**11–117** Repeat Prob. 11–116 if the heat exchanger provides 9.51°C of subcooling.

**11–118** An ideal gas refrigeration system with three stages of compression with intercooling operates with air entering the first compressor at 50 kPa and −30°C. Each compressor in this system has a pressure ratio of 7, and the air temperature at the outlet of all intercoolers is 15°C. Calculate the COP of this system. Use constant specific heats at room temperature.

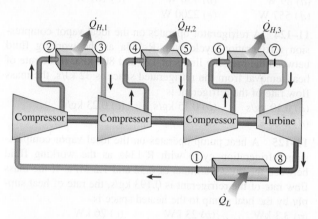

**FIGURE P11–118**

**11–119** [EES] Using EES (or other) software, investigate the effect of the evaporator pressure on the COP of an ideal vapor-compression refrigeration cycle with R-134a as the working fluid. Assume the condenser pressure is kept constant at 1.4 MPa while the evaporator pressure is varied from 100 kPa to 500 kPa. Plot the COP of the refrigeration cycle against the evaporator pressure, and discuss the results.

**11–120** [EES] Using EES (or other) software, investigate the effect of the condenser pressure on the COP of an ideal vapor-compression refrigeration cycle with R-134a as the working fluid. Assume the evaporator pressure is kept constant at 150 kPa while the condenser pressure is varied from 400 to 1400 kPa. Plot the COP of the refrigeration cycle against the condenser pressure, and discuss the results.

**11–121** Derive a relation for the COP of the two-stage refrigeration system with a flash chamber as shown in Fig. 11–14 in terms of the enthalpies and the quality at state 6. Consider a unit mass in the condenser.

### Fundamentals of Engineering (FE) Exam Problems

**11–122** Consider a heat pump that operates on the reversed Carnot cycle with R-134a as the working fluid executed under the saturation dome between the pressure limits of 140 and 800 kPa. R-134a changes from saturated vapor to saturated liquid during the heat rejection process. The net work input for this cycle is

(a) 28 kJ/kg  (b) 34 kJ/kg  (c) 49 kJ/kg
(d) 144 kJ/kg  (e) 275 kJ/kg

**11–123** A refrigerator removes heat from a refrigerated space at 0°C at a rate of 2.2 kJ/s and rejects it to an environment at 20°C. The minimum required power input is

(a) 89 W  (b) 150 W  (c) 161 W
(d) 557 W  (e) 2200 W

**11–124** A refrigerator operates on the ideal vapor compression refrigeration cycle with R-134a as the working fluid between the pressure limits of 120 and 800 kPa. If the rate of heat removal from the refrigerated space is 32 kJ/s, the mass flow rate of the refrigerant is

(a) 0.19 kg/s  (b) 0.15 kg/s  (c) 0.23 kg/s
(d) 0.28 kg/s  (e) 0.81 kg/s

**11–125** A heat pump operates on the ideal vapor compression refrigeration cycle with R-134a as the working fluid between the pressure limits of 0.32 and 1.2 MPa. If the mass flow rate of the refrigerant is 0.193 kg/s, the rate of heat supply by the heat pump to the heated space is

(a) 3.3 kW  (b) 23 kW  (c) 26 kW
(d) 31 kW  (e) 45 kW

**11–126** An ideal vapor compression refrigeration cycle with R-134a as the working fluid operates between the pressure limits of 120 kPa and 700 kPa. The mass fraction of the refrigerant that is in the liquid phase at the inlet of the evaporator is

(a) 0.69  (b) 0.63  (c) 0.58
(d) 0.43  (e) 0.35

**11–127** Consider a heat pump that operates on the ideal vapor compression refrigeration cycle with R-134a as the working fluid between the pressure limits of 0.32 and 1.2 MPa. The coefficient of performance of this heat pump is

(a) 0.17  (b) 1.2  (c) 3.1
(d) 4.9  (e) 5.9

**11–128** An ideal gas refrigeration cycle using air as the working fluid operates between the pressure limits of 80 and 280 kPa. Air is cooled to 35°C before entering the turbine. The lowest temperature of this cycle is

(a) −58°C  (b) −26°C  (c) 5°C
(d) 11°C  (e) 24°C

**11–129** Consider an ideal gas refrigeration cycle using helium as the working fluid. Helium enters the compressor at 100 kPa and 17°C and compressed to 400 kPa. Helium is then cooled to 20°C before it enters the turbine. For a mass flow rate of 0.2 kg/s, the net power input required is

(a) 28.3 kW  (b) 40.5 kW  (c) 64.7 kW
(d) 93.7 kW  (e) 113 kW

**11–130** An absorption air-conditioning system is to remove heat from the conditioned space at 20°C at a rate of 150 kJ/s while operating in an environment at 35°C. Heat is to be supplied from a geothermal source at 140°C. The minimum rate of heat supply is

(a) 86 kJ/s  (b) 21 kJ/s  (c) 30 kJ/s
(d) 61 kJ/s  (e) 150 kJ/s

**11–131** Consider a refrigerator that operates on the vapor compression refrigeration cycle with R-134a as the working fluid. The refrigerant enters the compressor as saturated vapor at 160 kPa, and exits at 800 kPa and 50°C, and leaves the condenser as saturated liquid at 800 kPa. The coefficient of performance of this refrigerator is

(a) 2.6  (b) 1.0  (c) 4.2
(d) 3.2  (e) 4.4

## Design and Essay Problems

**11–132** Write an essay on air-, water-, and soil-based heat pumps. Discuss the advantages and the disadvantages of each system. For each system identify the conditions under which that system is preferable over the other two. In what situations would you not recommend a heat pump heating system?

**11–133** Design a vapor-compression refrigeration system that will maintain the refrigerated space at −15°C while operating in an environment at 20°C using refrigerant-134a as the working fluid.

**11–134** Design a thermoelectric refrigerator that is capable of cooling a canned drink in a car. The refrigerator is to be powered by the cigarette lighter of the car. Draw a sketch of your design. Semiconductor components for building thermoelectric power generators or refrigerators are available from several manufacturers. Using data from one of these manufacturers, determine how many of these components you need in your design, and estimate the coefficient of performance of your system. A critical problem in the design of thermoelectric refrigerators is the effective rejection of waste heat. Discuss how you can enhance the rate of heat rejection without using any devices with moving parts such as a fan.

**11–135** The temperature in a car parked in the sun can approach 100°C when the outside air temperature is just 25°C, and it is desirable to ventilate the parked car to avoid such high temperatures. However, the ventilating fans may run down the battery if they are powered by it. To avoid that happening, it is proposed to use the PV cells discussed in the preceding problem to power the fans. It is determined that the air in the car should be replaced once every minute to avoid excessive rise in the interior temperature. Determine if this can be accomplished by installing PV cells on part of the roof of the car. Also, find out if any car is currently ventilated this way.

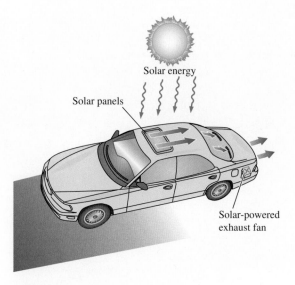

**FIGURE P11–135**

**11–136** It is proposed to use a solar-powered thermoelectric system installed on the roof to cool residential buildings. The system consists of a thermoelectric refrigerator that is powered by a thermoelectric power generator whose top surface is a solar collector. Discuss the feasibility and the cost of such a system, and determine if the proposed system installed on one side of the roof can meet a significant portion of the cooling requirements of a typical house in your area.

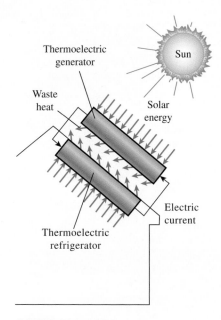

**FIGURE P11–136**

**11–137** A refrigerator using R-12 as the working fluid keeps the refrigerated space at −15°C in an environment at 30°C. You are asked to redesign this refrigerator by replacing R-12 with the ozone-friendly R-134a. What changes in the pressure levels would you suggest in the new system? How do you think the COP of the new system will compare to the COP of the old system?

**11–138** A company owns a refrigeration system whose refrigeration capacity is 200 tons (1 ton of refrigeration = 211 kJ/min), and you are to design a forced-air cooling system for fruits whose diameters do not exceed 7 cm under the following conditions: The fruits are to be cooled from 28°C to an average temperature of 8°C. The air temperature is to remain above −2°C and below 10°C at all times, and the velocity of air approaching the fruits must remain under 2 m/s. The cooling section can be as wide as 3.5 m and as high as 2 m.

Assuming reasonable values for the average fruit density, specific heat, and porosity (the fraction of air volume in a box), recommend reasonable values for (*a*) the air velocity approaching the cooling section, (*b*) the product-cooling capacity of the system, in kg·fruit/h, and (*c*) the volume flow rate of air.

**11–139** In the 1800s, before the development of modern air-conditioning, it was proposed to cool air for buildings with the following procedure using a large piston–cylinder device ["John Gorrie: Pioneer of Cooling and Ice Making," *ASHRAE Journal* 33, no. 1 (Jan. 1991)]:

1. Pull in a charge of outdoor air.
2. Compress it to a high pressure.
3. Cool the charge of air using outdoor air.
4. Expand it back to atmospheric pressure.
5. Discharge the charge of air into the space to be cooled.

Suppose the goal is to cool a room 6 m × 10 m × 2.5 m. Outdoor air is at 30°C, and it has been determined that 10 air changes per hour supplied to the room at 10°C could provide adequate cooling. Do a preliminary design of the system and do calculations to see if it would be feasible. (You may make optimistic assumptions for the analysis.)

(*a*) Sketch the system showing how you will drive it and how step 3 will be accomplished.

(*b*) Determine what pressure will be required (step 2).

(*c*) Estimate (guess) how long step 3 will take and what size will be needed for the piston–cylinder to provide the required air changes and temperature.

(*d*) Determine the work required in step 2 for one cycle and per hour.

(*e*) Discuss any problems you see with the concept of your design. (Include discussion of any changes that may be required to offset optimistic assumptions.)

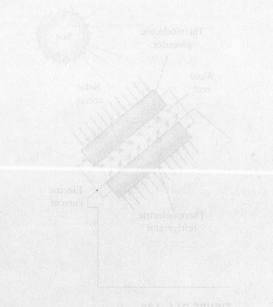

FIGURE P11–135

FIGURE P11–139

# THERMODYNAMIC PROPERTY RELATIONS

In the preceding chapters we made extensive use of the property tables. We tend to take the property tables for granted, but thermodynamic laws and principles are of little use to engineers without them. In this chapter, we focus our attention on how the property tables are prepared and how some unknown properties can be determined from limited available data.

It will come as no surprise that some properties such as temperature, pressure, volume, and mass can be measured directly. Other properties such as density and specific volume can be determined from these using some simple relations. However, properties such as internal energy, enthalpy, and entropy are not so easy to determine because they cannot be measured directly or related to easily measurable properties through some simple relations. Therefore, it is essential that we develop some fundamental relations between commonly encountered thermodynamic properties and express the properties that cannot be measured directly in terms of easily measurable properties.

By the nature of the material, this chapter makes extensive use of partial derivatives. Therefore, we start by reviewing them. Then we develop the Maxwell relations, which form the basis for many thermodynamic relations. Next we discuss the Clapeyron equation, which enables us to determine the enthalpy of vaporization from $P$, $v$, and $T$ measurements alone, and we develop general relations for $c_v$, $c_p$, $du$, $dh$, and $ds$ that are valid for all pure substances under all conditions. Then we discuss the Joule-Thomson coefficient, which is a measure of the temperature change with pressure during a throttling process. Finally, we develop a method of evaluating the $\Delta h$, $\Delta u$, and $\Delta s$ of real gases through the use of generalized enthalpy and entropy departure charts.

## OBJECTIVES

The objectives of Chapter 12 are to:

■ Develop fundamental relations between commonly encountered thermodynamic properties and express the properties that cannot be measured directly in terms of easily measurable properties.

■ Develop the Maxwell relations, which form the basis for many thermodynamic relations.

■ Develop the Clapeyron equation and determine the enthalpy of vaporization from $P$, $v$, and $T$ measurements alone.

■ Develop general relations for $c_v$, $c_p$, $du$, $dh$, and $ds$ that are valid for all pure substances.

■ Discuss the Joule-Thomson coefficient.

■ Develop a method of evaluating the $\Delta h$, $\Delta u$, and $\Delta s$ of real gases through the use of generalized enthalpy and entropy departure charts.

# 12–1 · A LITTLE MATH—PARTIAL DERIVATIVES AND ASSOCIATED RELATIONS

Many of the expressions developed in this chapter are based on the state postulate, which expresses that the state of a simple, compressible substance is completely specified by any two independent, intensive properties. All other properties at that state can be expressed in terms of those two properties. Mathematically speaking,

$$z = z(x, y)$$

where $x$ and $y$ are the two independent properties that fix the state and $z$ represents any other property. Most basic thermodynamic relations involve differentials. Therefore, we start by reviewing the derivatives and various relations among derivatives to the extent necessary in this chapter.

Consider a function $f$ that depends on a single variable $x$, that is, $f = f(x)$. Figure 12–1 shows such a function that starts out flat but gets rather steep as $x$ increases. The steepness of the curve is a measure of the degree of dependence of $f$ on $x$. In our case, the function $f$ depends on $x$ more strongly at larger $x$ values. The steepness of a curve at a point is measured by the slope of a line tangent to the curve at that point, and it is equivalent to the **derivative** of the function at that point defined as

$$\frac{df}{dx} = \lim_{\Delta x \to 0} \frac{\Delta f}{\Delta x} = \lim_{\Delta x \to 0} \frac{f(x + \Delta x) - f(x)}{\Delta x} \tag{12–1}$$

Therefore, *the derivative of a function f(x) with respect to x represents the rate of change of f with x.*

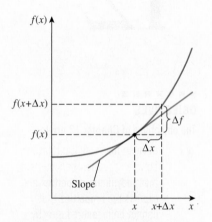

**FIGURE 12–1**
The derivative of a function at a specified point represents the slope of the function at that point.

---

**EXAMPLE 12–1**    **Approximating Differential Quantities by Differences**

The $c_p$ of ideal gases depends on temperature only, and it is expressed as $c_p(T) = dh(T)/dT$. Determine the $c_p$ of air at 300 K, using the enthalpy data from Table A–17, and compare it to the value listed in Table A–2b.

**SOLUTION**  The $c_p$ value of air at a specified temperature is to be determined using enthalpy data.
**Analysis**  The $c_p$ value of air at 300 K is listed in Table A–2b to be 1.005 kJ/kg·K. This value could also be determined by differentiating the function $h(T)$ with respect to $T$ and evaluating the result at $T = 300$ K. However, the function $h(T)$ is not available. But, we can still determine the $c_p$ value approximately by replacing the differentials in the $c_p(T)$ relation by differences in the neighborhood of the specified point (Fig. 12–2):

$$c_p(300 \text{ K}) = \left[\frac{dh(T)}{dT}\right]_{T=300\text{ K}} \cong \left[\frac{\Delta h(T)}{\Delta T}\right]_{T \cong 300\text{ K}} = \frac{h(305 \text{ K}) - h(295 \text{ K})}{(305 - 295) \text{ K}}$$

$$= \frac{(305.22 - 295.17) \text{ kJ/kg}}{(305 - 295) \text{ K}} = \mathbf{1.005 \text{ kJ/kg·K}}$$

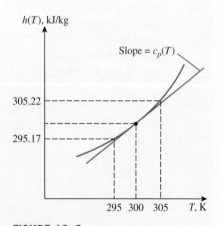

**FIGURE 12–2**
Schematic for Example 12–1.

**Discussion** Note that the calculated $c_p$ value is identical to the listed value. Therefore, differential quantities can be viewed as differences. They can even be replaced by differences, whenever necessary, to obtain approximate results. The widely used finite difference numerical method is based on this simple principle.

## Partial Differentials

Now consider a function that depends on two (or more) variables, such as $z = z(x, y)$. This time the value of $z$ depends on both $x$ and $y$. It is sometimes desirable to examine the dependence of $z$ on only one of the variables. This is done by allowing one variable to change while holding the others constant and observing the change in the function. The variation of $z(x, y)$ with $x$ when $y$ is held constant is called the **partial derivative** of $z$ with respect to $x$, and it is expressed as

$$\left(\frac{\partial z}{\partial x}\right)_y = \lim_{\Delta x \to 0} \left(\frac{\Delta z}{\Delta x}\right)_y = \lim_{\Delta x \to 0} \frac{z(x + \Delta x, y) - z(x, y)}{\Delta x} \qquad (12\text{–}2)$$

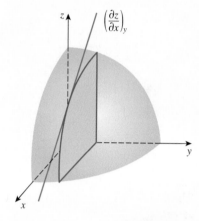

**FIGURE 12–3**
Geometric representation of partial derivative $(\partial z / \partial x)_y$.

This is illustrated in Fig. 12–3. The symbol $\partial$ represents differential changes, just like the symbol $d$. They differ in that the symbol $d$ represents the *total* differential change of a function and reflects the influence of all variables, whereas $\partial$ represents the *partial* differential change due to the variation of a single variable.

Note that the changes indicated by $d$ and $\partial$ are identical for independent variables, but not for dependent variables. For example, $(\partial x)_y = dx$ but $(\partial z)_y \neq dz$. [In our case, $dz = (\partial z)_x + (\partial z)_y$.] Also note that the value of the partial derivative $(\partial z / \partial x)_y$, in general, is different at different $y$ values.

To obtain a relation for the total differential change in $z(x, y)$ for simultaneous changes in $x$ and $y$, consider a small portion of the surface $z(x, y)$ shown in Fig. 12–4. When the independent variables $x$ and $y$ change by $\Delta x$ and $\Delta y$, respectively, the dependent variable $z$ changes by $\Delta z$, which can be expressed as

$$\Delta z = z(x + \Delta x, y + \Delta y) - z(x, y)$$

Adding and subtracting $z(x, y + \Delta y)$, we get

$$\Delta z = z(x + \Delta x, y + \Delta y) - z(x, y + \Delta y) + z(x, y + \Delta y) - z(x, y)$$

or

$$\Delta z = \frac{z(x + \Delta x, y + \Delta y) - z(x, y + \Delta y)}{\Delta x} \Delta x + \frac{z(x, y + \Delta y) - z(x, y)}{\Delta y} \Delta y$$

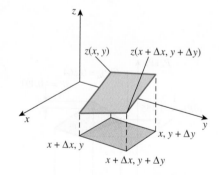

**FIGURE 12–4**
Geometric representation of total derivative $dz$ for a function $z(x, y)$.

Taking the limits as $\Delta x \to 0$ and $\Delta y \to 0$ and using the definitions of partial derivatives, we obtain

$$dz = \left(\frac{\partial z}{\partial x}\right)_y dx + \left(\frac{\partial z}{\partial y}\right)_x dy \qquad (12\text{–}3)$$

Equation 12–3 is the fundamental relation for the **total differential** of a dependent variable in terms of its partial derivatives with respect to the independent variables. This relation can easily be extended to include more independent variables.

---

**EXAMPLE 12–2**   **Total Differential versus Partial Differential**

Consider air at 300 K and 0.86 m³/kg. The state of air changes to 302 K and 0.87 m³/kg as a result of some disturbance. Using Eq. 12–3, estimate the change in the pressure of air.

**SOLUTION**  The temperature and specific volume of air changes slightly during a process. The resulting change in pressure is to be determined.
**Assumption**  Air is an ideal gas.
**Analysis**  Strictly speaking, Eq. 12–3 is valid for differential changes in variables. However, it can also be used with reasonable accuracy if these changes are small. The changes in $T$ and $v$, respectively, can be expressed as

$$dT \cong \Delta T = (302 - 300) \text{ K} = 2 \text{ K}$$

and

$$dv \cong \Delta v = (0.87 - 0.86) \text{ m}^3/\text{kg} = 0.01 \text{ m}^3/\text{kg}$$

An ideal gas obeys the relation $Pv = RT$. Solving for $P$ yields

$$P = \frac{RT}{v}$$

Note that $R$ is a constant and $P = P(T, v)$. Applying Eq. 12–3 and using average values for $T$ and $v$,

$$dP = \left(\frac{\partial P}{\partial T}\right)_v dT + \left(\frac{\partial P}{\partial v}\right)_T dv = \frac{R \, dT}{v} - \frac{RT \, dv}{v^2}$$

$$= (0.287 \text{ kPa·m}^3/\text{kg·K})\left[\frac{2 \text{ K}}{0.865 \text{ m}^3/\text{kg}} - \frac{(301 \text{ K})(0.01 \text{ m}^3/\text{kg})}{(0.865 \text{ m}^3/\text{kg})^2}\right]$$

$$= 0.664 \text{ kPa} - 1.155 \text{ kPa}$$

$$= -0.491 \text{ kPa}$$

Therefore, the pressure will decrease by 0.491 kPa as a result of this disturbance. Notice that if the temperature had remained constant ($dT = 0$), the pressure would decrease by 1.155 kPa as a result of the 0.01 m³/kg increase in specific volume. However, if the specific volume had remained constant ($dv = 0$), the pressure would increase by 0.664 kPa as a result of the 2-K rise in temperature (Fig. 12–5). That is,

$$\left(\frac{\partial P}{\partial T}\right)_v dT = (\partial P)_v = 0.664 \text{ kPa}$$

$$\left(\frac{\partial P}{\partial v}\right)_T dv = (\partial P)_T = -1.155 \text{ kPa}$$

and

$$dP = (\partial P)_v + (\partial P)_T = 0.664 - 1.155 = -0.491 \text{ kPa}$$

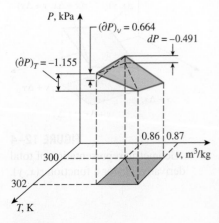

**FIGURE 12–5**
Geometric representation of the disturbance discussed in Example 12–2.

*Discussion* Of course, we could have solved this problem easily (and exactly) by evaluating the pressure from the ideal-gas relation $P = RT/v$ at the final state (302 K and 0.87 m³/kg) and the initial state (300 K and 0.86 m³/kg) and taking their difference. This yields −0.491 kPa, which is exactly the value obtained above. Thus the small finite quantities (2 K, 0.01 m³/kg) can be approximated as differential quantities with reasonable accuracy.

## Partial Differential Relations

Now let us rewrite Eq. 12–3 as

$$dz = M\,dx + N\,dy \tag{12-4}$$

where

$$M = \left(\frac{\partial z}{\partial x}\right)_y \quad \text{and} \quad N = \left(\frac{\partial z}{\partial y}\right)_x$$

Taking the partial derivative of $M$ with respect to $y$ and of $N$ with respect to $x$ yields

$$\left(\frac{\partial M}{\partial y}\right)_x = \frac{\partial^2 z}{\partial x\,\partial y} \quad \text{and} \quad \left(\frac{\partial N}{\partial x}\right)_y = \frac{\partial^2 z}{\partial y\,\partial x}$$

The order of differentiation is immaterial for properties since they are continuous point functions and have exact differentials. Therefore, the two relations above are identical:

$$\left(\frac{\partial M}{\partial y}\right)_x = \left(\frac{\partial N}{\partial x}\right)_y \tag{12-5}$$

This is an important relation for partial derivatives, and it is used in calculus to test whether a differential $dz$ is exact or inexact. In thermodynamics, this relation forms the basis for the development of the Maxwell relations discussed in the next section.

Finally, we develop two important relations for partial derivatives—the reciprocity and the cyclic relations. The function $z = z(x, y)$ can also be expressed as $x = x(y, z)$ if $y$ and $z$ are taken to be the independent variables. Then the total differential of $x$ becomes, from Eq. 12–3,

$$dx = \left(\frac{\partial x}{\partial y}\right)_z dy + \left(\frac{\partial x}{\partial z}\right)_y dz \tag{12-6}$$

Eliminating $dx$ by combining Eqs. 12–3 and 12–6, we have

$$dz = \left[\left(\frac{\partial z}{\partial x}\right)_y \left(\frac{\partial x}{\partial y}\right)_z + \left(\frac{\partial z}{\partial y}\right)_x\right] dy + \left(\frac{\partial x}{\partial z}\right)_y \left(\frac{\partial z}{\partial x}\right)_y dz$$

Rearranging,

$$\left[\left(\frac{\partial z}{\partial x}\right)_y \left(\frac{\partial x}{\partial y}\right)_z + \left(\frac{\partial z}{\partial y}\right)_x\right] dy = \left[1 - \left(\frac{\partial x}{\partial z}\right)_y \left(\frac{\partial z}{\partial x}\right)_y\right] dz \tag{12-7}$$

The variables $y$ and $z$ are independent of each other and thus can be varied independently. For example, $y$ can be held constant ($dy = 0$), and $z$ can be

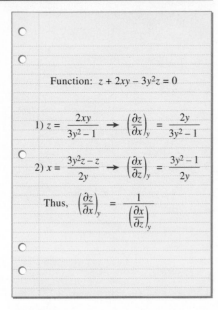

**FIGURE 12–6**
Demonstration of the reciprocity
relation for the function
$z + 2xy - 3y^2z = 0$.

varied over a range of values ($dz \neq 0$). Therefore, for this equation to be valid at all times, the terms in the brackets must equal zero, regardless of the values of $y$ and $z$. Setting the terms in each bracket equal to zero gives

$$\left(\frac{\partial x}{\partial z}\right)_y\left(\frac{\partial z}{\partial x}\right)_y = 1 \rightarrow \left(\frac{\partial x}{\partial z}\right)_y = \frac{1}{(\partial z/\partial x)_y} \qquad (12\text{–}8)$$

$$\left(\frac{\partial z}{\partial x}\right)_y\left(\frac{\partial x}{\partial y}\right)_z = -\left(\frac{\partial x}{\partial y}\right)_x \rightarrow \left(\frac{\partial x}{\partial y}\right)_z\left(\frac{\partial y}{\partial z}\right)_x\left(\frac{\partial z}{\partial x}\right)_y = -1 \qquad (12\text{–}9)$$

The first relation is called the **reciprocity relation,** and it shows that the inverse of a partial derivative is equal to its reciprocal (Fig. 12–6). The second relation is called the **cyclic relation,** and it is frequently used in thermodynamics.

---

***EXAMPLE 12–3***    **Verification of Cyclic and Reciprocity Relations**

Using the ideal-gas equation of state, verify (*a*) the cyclic relation, and (*b*) the reciprocity relation at constant *P*.

**SOLUTION**   The cyclic and reciprocity relations are to be verified for an ideal gas.
***Analysis***   The ideal-gas equation of state $Pv = RT$ involves the three variables $P$, $v$, and $T$. Any two of these can be taken as the independent variables, with the remaining one being the dependent variable.
(*a*) Replacing *x*, *y*, and *z* in Eq. 12–9 by *P*, *v*, and *T*, respectively, we can express the cyclic relation for an ideal gas as

$$\left(\frac{\partial P}{\partial v}\right)_T\left(\frac{\partial v}{\partial T}\right)_P\left(\frac{\partial T}{\partial P}\right)_v = -1$$

where

$$P = P(v, T) = \frac{RT}{v} \rightarrow \left(\frac{\partial P}{\partial v}\right)_T = -\frac{RT}{v^2}$$

$$v = v(P, T) = \frac{RT}{P} \rightarrow \left(\frac{\partial v}{\partial T}\right)_P = \frac{R}{P}$$

$$T = T(P, v) = \frac{Pv}{R} \rightarrow \left(\frac{\partial T}{\partial P}\right)_v = \frac{v}{R}$$

Substituting yields

$$\left(-\frac{RT}{v^2}\right)\left(\frac{R}{P}\right)\left(\frac{v}{R}\right) = -\frac{RT}{Pv} = -1$$

which is the desired result.
(*b*) The reciprocity rule for an ideal gas at $P$ = constant can be expressed as

$$\left(\frac{\partial v}{\partial T}\right)_P = \frac{1}{(\partial T/\partial v)_P}$$

Performing the differentiations and substituting, we have

$$\frac{R}{P} = \frac{1}{P/R} \rightarrow \frac{R}{P} = \frac{R}{P}$$

Thus the proof is complete.

# 12–2 ▪ THE MAXWELL RELATIONS

The equations that relate the partial derivatives of properties $P$, $v$, $T$, and $s$ of a simple compressible system to each other are called the *Maxwell relations*. They are obtained from the four Gibbs equations by exploiting the exactness of the differentials of thermodynamic properties.

Two of the Gibbs relations were derived in Chap. 7 and expressed as

$$du = T\,ds - P\,dv \tag{12–10}$$

$$dh = T\,ds + v\,dP \tag{12–11}$$

The other two Gibbs relations are based on two new combination properties—the **Helmholtz function** $a$ and the **Gibbs function** $g$, defined as

$$a = u - Ts \tag{12–12}$$

$$g = h - Ts \tag{12–13}$$

Differentiating, we get

$$da = du - T\,ds - s\,dT$$

$$dg = dh - T\,ds - s\,dT$$

Simplifying the above relations by using Eqs. 12–10 and 12–11, we obtain the other two Gibbs relations for simple compressible systems:

$$da = -s\,dT - P\,dv \tag{12–14}$$

$$dg = -s\,dT + v\,dP \tag{12–15}$$

A careful examination of the four Gibbs relations reveals that they are of the form

$$dz = M\,dx + N\,dy \tag{12–4}$$

with

$$\left(\frac{\partial M}{\partial y}\right)_x = \left(\frac{\partial N}{\partial x}\right)_y \tag{12–5}$$

since $u$, $h$, $a$, and $g$ are properties and thus have exact differentials. Applying Eq. 12–5 to each of them, we obtain

$$\left(\frac{\partial T}{\partial v}\right)_s = -\left(\frac{\partial P}{\partial s}\right)_v \tag{12–16}$$

$$\left(\frac{\partial T}{\partial P}\right)_s = \left(\frac{\partial v}{\partial s}\right)_P \tag{12–17}$$

$$\left(\frac{\partial s}{\partial v}\right)_T = \left(\frac{\partial P}{\partial T}\right)_v \tag{12–18}$$

$$\left(\frac{\partial s}{\partial P}\right)_T = -\left(\frac{\partial v}{\partial T}\right)_P \tag{12–19}$$

These are called the **Maxwell relations** (Fig. 12–7). They are extremely valuable in thermodynamics because they provide a means of determining the change in entropy, which cannot be measured directly, by simply measuring the changes in properties $P$, $v$, and $T$. Note that the Maxwell relations given

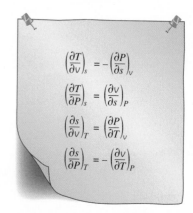

$$\left(\frac{\partial T}{\partial v}\right)_s = -\left(\frac{\partial P}{\partial s}\right)_v$$

$$\left(\frac{\partial T}{\partial P}\right)_s = \left(\frac{\partial v}{\partial s}\right)_P$$

$$\left(\frac{\partial s}{\partial v}\right)_T = \left(\frac{\partial P}{\partial T}\right)_v$$

$$\left(\frac{\partial s}{\partial P}\right)_T = -\left(\frac{\partial v}{\partial T}\right)_P$$

**FIGURE 12–7**
Maxwell relations are extremely valuable in thermodynamic analysis.

above are limited to simple compressible systems. However, other similar relations can be written just as easily for nonsimple systems such as those involving electrical, magnetic, and other effects.

---

**EXAMPLE 12–4**     **Verification of the Maxwell Relations**

Verify the validity of the last Maxwell relation (Eq. 12–19) for steam at 250°C and 300 kPa.

**SOLUTION**   The validity of the last Maxwell relation is to be verified for steam at a specified state.

**Analysis**   The last Maxwell relation states that for a simple compressible substance, the change in entropy with pressure at constant temperature is equal to the negative of the change in specific volume with temperature at constant pressure.

If we had explicit analytical relations for the entropy and specific volume of steam in terms of other properties, we could easily verify this by performing the indicated derivations. However, all we have for steam are tables of properties listed at certain intervals. Therefore, the only course we can take to solve this problem is to replace the differential quantities in Eq. 12–19 with corresponding finite quantities, using property values from the tables (Table A–6 in this case) at or about the specified state.

$$\left(\frac{\partial s}{\partial P}\right)_T \stackrel{?}{=} -\left(\frac{\partial v}{\partial T}\right)_P$$

$$\left(\frac{\Delta s}{\Delta P}\right)_{T = 250°C} \stackrel{?}{\cong} -\left(\frac{\Delta v}{\Delta T}\right)_{P = 300 \text{ kPa}}$$

$$\left[\frac{s_{400 \text{ kPa}} - s_{200 \text{ kPa}}}{(400 - 200) \text{ kPa}}\right]_{T = 250°C} \stackrel{?}{\cong} -\left[\frac{v_{300°C} - v_{200°C}}{(300 - 200)°C}\right]_{P = 300 \text{ kPa}}$$

$$\frac{(7.3804 - 7.7100) \text{ kJ/kg·K}}{(400 - 200) \text{ kPa}} \stackrel{?}{\cong} -\frac{(0.87535 - 0.71643) \text{ m}^3/\text{kg}}{(300 - 200)°C}$$

$$-0.00165 \text{ m}^3/\text{kg·K} \cong -0.00159 \text{ m}^3/\text{kg·K}$$

since kJ = kPa·m$^3$ and K ≡ °C for temperature differences. The two values are within 4 percent of each other. This difference is due to replacing the differential quantities by relatively large finite quantities. Based on the close agreement between the two values, the steam seems to satisfy Eq. 12–19 at the specified state.

**Discussion**   This example shows that the entropy change of a simple compressible system during an isothermal process can be determined from a knowledge of the easily measurable properties $P$, $v$, and $T$ alone.

---

# 12–3 ▪ THE CLAPEYRON EQUATION

The Maxwell relations have far-reaching implications in thermodynamics and are frequently used to derive useful thermodynamic relations. The Clapeyron equation is one such relation, and it enables us to determine the enthalpy change associated with a phase change (such as the enthalpy of vaporization $h_{fg}$) from a knowledge of $P$, $v$, and $T$ data alone.

Consider the third Maxwell relation, Eq. 12–18:

$$\left(\frac{\partial P}{\partial T}\right)_v = \left(\frac{\partial s}{\partial v}\right)_T$$

During a phase-change process, the pressure is the saturation pressure, which depends on the temperature only and is independent of the specific volume. That is, $P_{sat} = f(T_{sat})$. Therefore, the partial derivative $(\partial P/\partial T)_v$ can be expressed as a total derivative $(dP/dT)_{sat}$, which is the slope of the saturation curve on a $P$-$T$ diagram at a specified saturation state (Fig. 12–8). This slope is independent of the specific volume, and thus it can be treated as a constant during the integration of Eq. 12–18 between two saturation states at the same temperature. For an isothermal liquid–vapor phase-change process, for example, the integration yields

$$s_g - s_f = \left(\frac{dP}{dT}\right)_{sat} (v_g - v_f) \tag{12–20}$$

or

$$\left(\frac{dP}{dT}\right)_{sat} = \frac{s_{fg}}{v_{fg}} \tag{12–21}$$

During this process the pressure also remains constant. Therefore, from Eq. 12–11,

$$dh = T\,ds + v\,dP \nearrow^{0} \rightarrow \int_f^g dh = \int_f^g T\,ds \rightarrow h_{fg} = Ts_{fg}$$

Substituting this result into Eq. 12–21, we obtain

$$\left(\frac{dP}{dT}\right)_{sat} = \frac{h_{fg}}{Tv_{fg}} \tag{12–22}$$

which is called the **Clapeyron equation** after the French engineer and physicist E. Clapeyron (1799–1864). This is an important thermodynamic relation since it enables us to determine the enthalpy of vaporization $h_{fg}$ at a given temperature by simply measuring the slope of the saturation curve on a $P$-$T$ diagram and the specific volume of saturated liquid and saturated vapor at the given temperature.

The Clapeyron equation is applicable to any phase-change process that occurs at constant temperature and pressure. It can be expressed in a general form as

$$\left(\frac{dP}{dT}\right)_{sat} = \frac{h_{12}}{Tv_{12}} \tag{12–23}$$

where the subscripts 1 and 2 indicate the two phases.

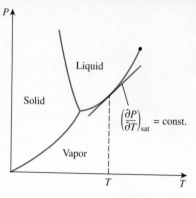

**FIGURE 12–8**
The slope of the saturation curve on a $P$-$T$ diagram is constant at a constant $T$ or $P$.

■ **EXAMPLE 12–5**    **Evaluating the $h_{fg}$ of a Substance from the $P$-$v$-$T$ Data**

■ Using the Clapeyron equation, estimate the value of the enthalpy of vaporization of refrigerant-134a at 20°C, and compare it with the tabulated value.

**SOLUTION** The $h_{fg}$ of refrigerant-134a is to be determined using the Clapeyron equation.

**Analysis** From Eq. 12–22,

$$h_{fg} = T v_{fg} \left( \frac{dP}{dT} \right)_{sat}$$

where, from Table A–11,

$$v_{fg} = (v_g - v_f)_{@ 20°C} = 0.036012 - 0.0008160 = 0.035196 \text{ m}^3/\text{kg}$$

$$\left( \frac{dP}{dT} \right)_{sat,20°C} \cong \left( \frac{\Delta P}{\Delta T} \right)_{sat,20°C} = \frac{P_{sat @ 24°C} - P_{sat @ 16°C}}{24°C - 16°C}$$

$$= \frac{646.18 - 504.58 \text{ kPa}}{8°C} = 17.70 \text{ kPa/K}$$

since $\Delta T(°C) \equiv \Delta T(K)$. Substituting, we get

$$h_{fg} = (293.15 \text{ K})(0.035196 \text{ m}^3/\text{kg})(17.70 \text{ kPa/K}) \left( \frac{1 \text{ kJ}}{1 \text{ kPa·m}^3} \right)$$

$$= 182.62 \text{ kJ/kg}$$

The tabulated value of $h_{fg}$ at 20°C is 182.33 kJ/kg. The small difference between the two values is due to the approximation used in determining the slope of the saturation curve at 20°C.

The Clapeyron equation can be simplified for liquid–vapor and solid–vapor phase changes by utilizing some approximations. At low pressures $v_g \gg v_f$, and thus $v_{fg} \cong v_g$. By treating the vapor as an ideal gas, we have $v_g = RT/P$. Substituting these approximations into Eq. 12–22, we find

$$\left( \frac{dP}{dT} \right)_{sat} = \frac{P h_{fg}}{RT^2}$$

or

$$\left( \frac{dP}{P} \right)_{sat} = \frac{h_{fg}}{R} \left( \frac{dT}{T^2} \right)_{sat}$$

For small temperature intervals $h_{fg}$ can be treated as a constant at some average value. Then integrating this equation between two saturation states yields

$$\ln \left( \frac{P_2}{P_1} \right)_{sat} \cong \frac{h_{fg}}{R} \left( \frac{1}{T_1} - \frac{1}{T_2} \right)_{sat} \tag{12–24}$$

This equation is called the **Clapeyron–Clausius equation**, and it can be used to determine the variation of saturation pressure with temperature. It can also be used in the solid–vapor region by replacing $h_{fg}$ by $h_{ig}$ (the enthalpy of sublimation) of the substance.

**■ EXAMPLE 12–6**   **Extrapolating Tabular Data with the Clapeyron Equation**

Estimate the saturation pressure of refrigerant-134a at −45°C, using the data available in the refrigerant tables.

**SOLUTION**  The saturation pressure of refrigerant-134a is to be determined using other tabulated data.

**Analysis**  Table A–11 lists saturation data at temperatures −40°C and above. Therefore, we should either resort to other sources or use extrapolation to obtain saturation data at lower temperatures. Equation 12–24 provides an intelligent way to extrapolate:

$$\ln\left(\frac{P_2}{P_1}\right)_{sat} \cong \frac{h_{fg}}{R}\left(\frac{1}{T_1} - \frac{1}{T_2}\right)_{sat}$$

In our case $T_1 = -40°C$ and $T_2 = -45°C$. For refrigerant-134a, $R = 0.08149$ kJ/kg·K. Also from Table A–11 at −40°C, we read $h_{fg} = 225.86$ kJ/kg and $P_1 = P_{sat\ @\ -40°C} = 51.25$ kPa. Substituting these values into Eq. 12–24 gives

$$\ln\left(\frac{P_2}{51.25\ \text{kPa}}\right) \cong \frac{225.86\ \text{kJ/kg}}{0.08149\ \text{kJ/kg·K}}\left(\frac{1}{223\ \text{K}} - \frac{1}{228\ \text{K}}\right)$$

$$P_2 \cong 39.48\ \text{kPa}$$

Therefore, according to Eq. 12–24, the saturation pressure of refrigerant-134a at −45°C is 39.48 kPa. The actual value, obtained from another source, is 39.15 kPa. Thus the value predicted by Eq. 12–24 is in error by about 1 percent, which is quite acceptable for most purposes. (If we had used linear extrapolation instead, we would have obtained 37.23 kPa, which is in error by 5 percent.)

# 12–4 ■ GENERAL RELATIONS FOR $du$, $dh$, $ds$, $c_v$, AND $c_P$

The state postulate established that the state of a simple compressible system is completely specified by two independent, intensive properties. Therefore, at least theoretically, we should be able to calculate all the properties of a system at any state once two independent, intensive properties are available. This is certainly good news for properties that cannot be measured directly such as internal energy, enthalpy, and entropy. However, the calculation of these properties from measurable ones depends on the availability of simple and accurate relations between the two groups.

In this section we develop general relations for changes in internal energy, enthalpy, and entropy in terms of pressure, specific volume, temperature, and specific heats alone. We also develop some general relations involving specific heats. The relations developed will enable us to determine the *changes* in these properties. The property values at specified states can be determined only after the selection of a reference state, the choice of which is quite arbitrary.

# Internal Energy Changes

We choose the internal energy to be a function of $T$ and $v$; that is, $u = u(T, v)$ and take its total differential (Eq. 12–3):

$$du = \left(\frac{\partial u}{\partial T}\right)_v dT + \left(\frac{\partial u}{\partial v}\right)_T dv$$

Using the definition of $c_v$, we have

$$du = c_v \, dT + \left(\frac{\partial u}{\partial v}\right)_T dv \qquad (12\text{–}25)$$

Now we choose the entropy to be a function of $T$ and $v$; that is, $s = s(T, v)$ and take its total differential,

$$ds = \left(\frac{\partial s}{\partial T}\right)_v dT + \left(\frac{\partial s}{\partial v}\right)_T dv \qquad (12\text{–}26)$$

Substituting this into the $T \, ds$ relation $du = T \, ds - P \, dv$ yields

$$du = T\left(\frac{\partial s}{\partial T}\right)_v dT + \left[T\left(\frac{\partial s}{\partial v}\right)_T - P\right] dv \qquad (12\text{–}27)$$

Equating the coefficients of $dT$ and $dv$ in Eqs. 12–25 and 12–27 gives

$$\left(\frac{\partial s}{\partial T}\right)_v = \frac{c_v}{T}$$

$$\left(\frac{\partial u}{\partial v}\right)_T = T\left(\frac{\partial s}{\partial v}\right)_T - P \qquad (12\text{–}28)$$

Using the third Maxwell relation (Eq. 12–18), we get

$$\left(\frac{\partial u}{\partial v}\right)_T = T\left(\frac{\partial P}{\partial T}\right)_v - P$$

Substituting this into Eq. 12–25, we obtain the desired relation for $du$:

$$du = c_v dT + \left[T\left(\frac{\partial P}{\partial T}\right)_v - P\right] dv \qquad (12\text{–}29)$$

The change in internal energy of a simple compressible system associated with a change of state from $(T_1, v_1)$ to $(T_2, v_2)$ is determined by integration:

$$u_2 - u_1 = \int_{T_1}^{T_2} c_v dT + \int_{v_1}^{v_2}\left[T\left(\frac{\partial P}{\partial T}\right)_v - P\right] dv \qquad (12\text{–}30)$$

# Enthalpy Changes

The general relation for $dh$ is determined in exactly the same manner. This time we choose the enthalpy to be a function of $T$ and $P$, that is, $h = h(T, P)$, and take its total differential,

$$dh = \left(\frac{\partial h}{\partial T}\right)_P dT + \left(\frac{\partial h}{\partial P}\right)_T dP$$

Using the definition of $c_p$, we have

$$dh = c_p \, dT + \left(\frac{\partial h}{\partial P}\right)_T dP \qquad (12\text{–}31)$$

Now we choose the entropy to be a function of $T$ and $P$; that is, we take $s = s(T, P)$ and take its total differential,

$$ds = \left(\frac{\partial s}{\partial T}\right)_P dT + \left(\frac{\partial s}{\partial P}\right)_T dP \qquad (12\text{-}32)$$

Substituting this into the $T\,ds$ relation $dh = T\,ds + v\,dP$ gives

$$dh = T\left(\frac{\partial s}{\partial T}\right)_P dT + \left[v + T\left(\frac{\partial s}{\partial P}\right)_T\right] dP \qquad (12\text{-}33)$$

Equating the coefficients of $dT$ and $dP$ in Eqs. 12–31 and 12–33, we obtain

$$\left(\frac{\partial s}{\partial T}\right)_P = \frac{c_p}{T}$$

$$\left(\frac{\partial h}{\partial P}\right)_T = v + T\left(\frac{\partial s}{\partial P}\right)_T \qquad (12\text{-}34)$$

Using the fourth Maxwell relation (Eq. 12–19), we have

$$\left(\frac{\partial h}{\partial P}\right)_T = v - T\left(\frac{\partial v}{\partial T}\right)_P$$

Substituting this into Eq. 12–31, we obtain the desired relation for $dh$:

$$dh = c_p\,dT + \left[v - T\left(\frac{\partial v}{\partial T}\right)_P\right] dP \qquad (12\text{-}35)$$

The change in enthalpy of a simple compressible system associated with a change of state from $(T_1, P_1)$ to $(T_2, P_2)$ is determined by integration:

$$h_2 - h_1 = \int_{T_1}^{T_2} c_p\,dT + \int_{P_1}^{P_2} \left[v - T\left(\frac{\partial v}{\partial T}\right)_P\right] dP \qquad (12\text{-}36)$$

In reality, one needs only to determine either $u_2 - u_1$ from Eq. 12–30 or $h_2 - h_1$ from Eq. 12–36, depending on which is more suitable to the data at hand. The other can easily be determined by using the definition of enthalpy $h = u + Pv$:

$$h_2 - h_1 = u_2 - u_1 + (P_2 v_2 - P_1 v_1) \qquad (12\text{-}37)$$

## Entropy Changes

Below we develop two general relations for the entropy change of a simple compressible system.

The first relation is obtained by replacing the first partial derivative in the total differential $ds$ (Eq. 12–26) by Eq. 12–28 and the second partial derivative by the third Maxwell relation (Eq. 12–18), yielding

$$ds = \frac{c_v}{T} dT + \left(\frac{\partial P}{\partial T}\right)_v dv \qquad (12\text{-}38)$$

and

$$s_2 - s_1 = \int_{T_1}^{T_2} \frac{c_v}{T} dT + \int_{v_1}^{v_2} \left(\frac{\partial P}{\partial T}\right)_v dv \qquad (12\text{-}39)$$

The second relation is obtained by replacing the first partial derivative in the total differential of $ds$ (Eq. 12–32) by Eq. 12–34, and the second partial derivative by the fourth Maxwell relation (Eq. 12–19), yielding

$$ds = \frac{c_P}{T} dT - \left(\frac{\partial v}{\partial T}\right)_P dP \qquad (12\text{–}40)$$

and

$$s_2 - s_1 = \int_{T_1}^{T_2} \frac{c_p}{T} dT - \int_{P_1}^{P_2} \left(\frac{\partial v}{\partial T}\right)_P dP \qquad (12\text{–}41)$$

Either relation can be used to determine the entropy change. The proper choice depends on the available data.

## Specific Heats $c_v$ and $c_p$

Recall that the specific heats of an ideal gas depend on temperature only. For a general pure substance, however, the specific heats depend on specific volume or pressure as well as the temperature. Below we develop some general relations to relate the specific heats of a substance to pressure, specific volume, and temperature.

At low pressures gases behave as ideal gases, and their specific heats essentially depend on temperature only. These specific heats are called *zero pressure*, or *ideal-gas, specific heats* (denoted $c_{v0}$ and $c_{p0}$), and they are relatively easier to determine. Thus it is desirable to have some general relations that enable us to calculate the specific heats at higher pressures (or lower specific volumes) from a knowledge of $c_{v0}$ or $c_{p0}$ and the $P$-$v$-$T$ behavior of the substance. Such relations are obtained by applying the test of exactness (Eq. 12–5) on Eqs. 12–38 and 12–40, which yields

$$\left(\frac{\partial c_v}{\partial v}\right)_T = T\left(\frac{\partial^2 P}{\partial T^2}\right)_v \qquad (12\text{–}42)$$

and

$$\left(\frac{\partial c_p}{\partial P}\right)_T = -T\left(\frac{\partial^2 v}{\partial T^2}\right)_P \qquad (12\text{–}43)$$

The deviation of $c_p$ from $c_{p0}$ with increasing pressure, for example, is determined by integrating Eq. 12–43 from zero pressure to any pressure $P$ along an isothermal path:

$$(c_p - c_{p0})_T = -T \int_0^P \left(\frac{\partial^2 v}{\partial T^2}\right)_P dP \qquad (12\text{–}44)$$

The integration on the right-hand side requires a knowledge of the $P$-$v$-$T$ behavior of the substance alone. The notation indicates that $v$ should be differentiated twice with respect to $T$ while $P$ is held constant. The resulting expression should be integrated with respect to $P$ while $T$ is held constant.

Another desirable general relation involving specific heats is one that relates the two specific heats $c_p$ and $c_v$. The advantage of such a relation is obvious: We will need to determine only one specific heat (usually $c_p$) and calculate the other one using that relation and the $P$-$v$-$T$ data of the substance. We start the development of such a relation by equating the two $ds$ relations (Eqs. 12–38 and 12–40) and solving for $dT$:

$$dT = \frac{T(\partial P/\partial T)_v}{c_p - c_v}\, dv + \frac{T(\partial v/\partial T)_P}{c_p - c_v}\, dP$$

Choosing $T = T(v, P)$ and differentiating, we get

$$dT = \left(\frac{\partial T}{\partial v}\right)_P dv + \left(\frac{\partial T}{\partial P}\right)_v dP$$

Equating the coefficient of either $dv$ or $dP$ of the above two equations gives the desired result:

$$c_p - c_v = T\left(\frac{\partial v}{\partial T}\right)_P\left(\frac{\partial P}{\partial T}\right)_v \tag{12-45}$$

An alternative form of this relation is obtained by using the cyclic relation:

$$\left(\frac{\partial P}{\partial T}\right)_v\left(\frac{\partial T}{\partial v}\right)_P\left(\frac{\partial v}{\partial P}\right)_T = -1 \rightarrow \left(\frac{\partial P}{\partial T}\right)_v = -\left(\frac{\partial v}{\partial T}\right)_P\left(\frac{\partial P}{\partial v}\right)_T$$

Substituting the result into Eq. 12–45 gives

$$c_p - c_v = -T\left(\frac{\partial v}{\partial T}\right)_P^2\left(\frac{\partial P}{\partial v}\right)_T \tag{12-46}$$

This relation can be expressed in terms of two other thermodynamic properties called the **volume expansivity** $\beta$ and the **isothermal compressibility** $\alpha$, which are defined as (Fig. 12–9)

$$\beta = \frac{1}{v}\left(\frac{\partial v}{\partial T}\right)_P \tag{12-47}$$

and

$$\alpha = -\frac{1}{v}\left(\frac{\partial v}{\partial P}\right)_T \tag{12-48}$$

Substituting these two relations into Eq. 12–46, we obtain a third general relation for $c_p - c_v$:

$$c_p - c_v = \frac{vT\beta^2}{\alpha} \tag{12-49}$$

It is called the **Mayer relation** in honor of the German physician and physicist J. R. Mayer (1814–1878). We can draw several conclusions from this equation:

1. The isothermal compressibility a is a positive quantity for all substances in all phases. The volume expansivity could be negative for some substances (such as liquid water below 4°C), but its square is always positive or zero. The temperature $T$ in this relation is thermodynamic temperature, which is also positive. Therefore we conclude that *the constant-pressure specific heat is always greater than or equal to the constant-volume specific heat:*

$$c_p \geq c_v \tag{12-50}$$

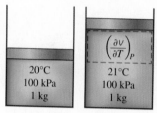

(a) A substance with a large $\beta$

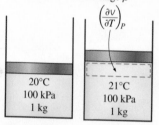

(b) A substance with a small $\beta$

**FIGURE 12–9**

The coefficient of volume expansion is a measure of the change in volume of a substance with temperature at constant pressure.

2. The difference between $c_p$ and $c_v$ approaches zero as the absolute temperature approaches zero.
3. The two specific heats are identical for truly incompressible substances since $v = $ constant. The difference between the two specific heats is very small and is usually disregarded for substances that are *nearly* incompressible, such as liquids and solids.

---

**EXAMPLE 12–7    Internal Energy Change of a van der Waals Gas**

Derive a relation for the internal energy change as a gas that obeys the van der Waals equation of state. Assume that in the range of interest $c_v$ varies according to the relation $c_v = c_1 + c_2 T$, where $c_1$ and $c_2$ are constants.

**SOLUTION**   A relation is to be obtained for the internal energy change of a van der Waals gas.
**Analysis**   The change in internal energy of any simple compressible system in any phase during any process can be determined from Eq. 12–30:

$$u_2 - u_1 = \int_{T_1}^{T_2} c_v \, dT + \int_{v_1}^{v_2} \left[ T \left( \frac{\partial P}{\partial T} \right)_v - P \right] dv$$

The van der Waals equation of state is

$$P = \frac{RT}{v - b} - \frac{a}{v^2}$$

Then

$$\left( \frac{\partial P}{\partial T} \right)_v = \frac{R}{v - b}$$

Thus,

$$T \left( \frac{\partial P}{\partial T} \right)_v - P = \frac{RT}{v - b} - \frac{RT}{v - b} + \frac{a}{v^2} = \frac{a}{v^2}$$

Substituting gives

$$u_2 - u_1 = \int_{T_1}^{T_2} (c_1 + c_2 T) \, dT + \int_{v_1}^{v_2} \frac{a}{v^2} \, dv$$

Integrating yields

$$u_2 - u_1 = c_1(T_2 - T_1) + \frac{c_2}{2}(T_2^2 - T_1^2) + a \left( \frac{1}{v_1} - \frac{1}{v_2} \right)$$

which is the desired relation.

---

**EXAMPLE 12–8    Internal Energy as a Function of Temperature Alone**

Show that the internal energy of (*a*) an ideal gas and (*b*) an incompressible substance is a function of temperature only, $u = u(T)$.

**SOLUTION**   It is to be shown that $u = u(T)$ for ideal gases and incompressible substances.

*Analysis* The differential change in the internal energy of a general simple compressible system is given by Eq. 12–29 as

$$du = c_v\, dT + \left[ T\left(\frac{\partial P}{\partial T}\right)_v - P \right] dv$$

(a) For an ideal gas $Pv = RT$. Then

$$T\left(\frac{\partial P}{\partial T}\right)_v - P = T\left(\frac{R}{v}\right) - P = P - P = 0$$

Thus,

$$du = c_v\, dT$$

To complete the proof, we need to show that $c_v$ is not a function of $v$ either. This is done with the help of Eq. 12–42:

$$\left(\frac{\partial c_v}{\partial v}\right)_T = T\left(\frac{\partial^2 P}{\partial T^2}\right)_v$$

For an ideal gas $P = RT/v$. Then

$$\left(\frac{\partial P}{\partial T}\right)_v = \frac{R}{v} \quad \text{and} \quad \left(\frac{\partial^2 P}{\partial T^2}\right)_v = \left[\frac{\partial (R/v)}{\partial T}\right]_v = 0$$

Thus,

$$\left(\frac{\partial c_v}{\partial v}\right)_T = 0$$

which states that $c_v$ does not change with specific volume. That is, $c_v$ is not a function of specific volume either. Therefore we conclude that the internal energy of an ideal gas is a function of temperature only (Fig. 12–10).

(b) For an incompressible substance, $v = $ constant and thus $dv = 0$. Also from Eq. 12–49, $c_p = c_v = c$ since $\alpha = \beta = 0$ for incompressible substances. Then Eq. 12–29 reduces to

$$du = c\, dT$$

Again we need to show that the specific heat $c$ depends on temperature only and not on pressure or specific volume. This is done with the help of Eq. 12–43:

$$\left(\frac{\partial c_p}{\partial P}\right)_T = -T\left(\frac{\partial^2 v}{\partial T^2}\right)_P = 0$$

since $v = $ constant. Therefore, we conclude that the internal energy of a truly incompressible substance depends on temperature only.

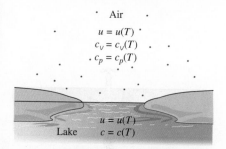

**FIGURE 12–10**
The internal energies and specific heats of ideal gases and incompressible substances depend on temperature only.

■ **EXAMPLE 12–9    The Specific Heat Difference of an Ideal Gas**

■ Show that $c_p - c_v = R$ for an ideal gas.

**SOLUTION** It is to be shown that the specific heat difference for an ideal gas is equal to its gas constant.

*Analysis*  This relation is easily proved by showing that the right-hand side of Eq. 12–46 is equivalent to the gas constant $R$ of the ideal gas:

$$c_p - c_v = -T\left(\frac{\partial v}{\partial T}\right)_P^2 \left(\frac{\partial P}{\partial v}\right)_T$$

$$P = \frac{RT}{v} \rightarrow \left(\frac{\partial P}{\partial v}\right)_T = -\frac{RT}{v^2} = -\frac{P}{v}$$

$$v = \frac{RT}{P} \rightarrow \left(\frac{\partial v}{\partial T}\right)_P^2 = \left(\frac{R}{P}\right)^2$$

Substituting,

$$-T\left(\frac{\partial v}{\partial T}\right)_P^2 \left(\frac{\partial P}{\partial v}\right)_T = -T\left(\frac{R}{P}\right)^2\left(-\frac{P}{v}\right) = R$$

Therefore,

$$c_p - c_v = R$$

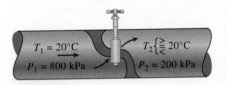

**FIGURE 12–11**

The temperature of a fluid may increase, decrease, or remain constant during a throttling process.

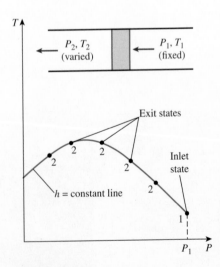

**FIGURE 12–12**

The development of an $h$ = constant line on a $P$-$T$ diagram.

## 12–5 · THE JOULE-THOMSON COEFFICIENT

When a fluid passes through a restriction such as a porous plug, a capillary tube, or an ordinary valve, its pressure decreases. As we have shown in Chap. 5, the enthalpy of the fluid remains approximately constant during such a throttling process. You will remember that a fluid may experience a large drop in its temperature as a result of throttling, which forms the basis of operation for refrigerators and air conditioners. This is not always the case, however. The temperature of the fluid may remain unchanged, or it may even increase during a throttling process (Fig. 12–11).

The temperature behavior of a fluid during a throttling ($h$ = constant) process is described by the **Joule-Thomson coefficient**, defined as

$$\mu = \left(\frac{\partial T}{\partial P}\right)_h \tag{12–51}$$

Thus the Joule-Thomson coefficient is a measure of the change in temperature with pressure during a constant-enthalpy process. Notice that if

$$\mu_{JT} \begin{cases} < 0 & \text{temperature increases} \\ = 0 & \text{temperature remains constant} \\ > 0 & \text{temperature decreases} \end{cases}$$

during a throttling process.

A careful look at its defining equation reveals that the Joule-Thomson coefficient represents the slope of $h$ = constant lines on a $T$-$P$ diagram. Such diagrams can be easily constructed from temperature and pressure measurements alone during throttling processes. A fluid at a fixed temperature and pressure $T_1$ and $P_1$ (thus fixed enthalpy) is forced to flow through a porous plug, and its temperature and pressure downstream ($T_2$ and $P_2$) are measured. The experiment is repeated for different sizes of porous plugs, each giving a different set of $T_2$ and $P_2$. Plotting the temperatures against the pressures gives us an $h$ = constant line on a $T$-$P$ diagram, as shown in Fig. 12–12. Repeating the experiment for different sets of inlet pressure and temperature and plotting the results, we can construct a *T-P* diagram for a substance with several *h* = constant lines, as shown in Fig. 12–13.

Some constant-enthalpy lines on the *T-P* diagram pass through a point of zero slope or zero Joule-Thomson coefficient. The line that passes through these points is called the **inversion line**, and the temperature at a point where a constant-enthalpy line intersects the inversion line is called the **inversion temperature**. The temperature at the intersection of the *P* = 0 line (ordinate) and the upper part of the inversion line is called the **maximum inversion temperature**. Notice that the slopes of the *h* = constant lines are negative ($\mu_{JT} < 0$) at states to the right of the inversion line and positive ($\mu_{JT} > 0$) to the left of the inversion line.

A throttling process proceeds along a constant-enthalpy line in the direction of decreasing pressure, that is, from right to left. Therefore, the temperature of a fluid increases during a throttling process that takes place on the right-hand side of the inversion line. However, the fluid temperature decreases during a throttling process that takes place on the left-hand side of the inversion line. It is clear from this diagram that a cooling effect cannot be achieved by throttling unless the fluid is below its maximum inversion temperature. This presents a problem for substances whose maximum inversion temperature is well below room temperature. For hydrogen, for example, the maximum inversion temperature is −68°C. Thus hydrogen must be cooled below this temperature if any further cooling is to be achieved by throttling.

Next we would like to develop a general relation for the Joule-Thomson coefficient in terms of the specific heats, pressure, specific volume, and temperature. This is easily accomplished by modifying the generalized relation for enthalpy change (Eq. 12–35)

$$dh = c_p\, dT + \left[ v - T\left(\frac{\partial v}{\partial T}\right)_P \right] dP$$

For an *h* = constant process we have *dh* = 0. Then this equation can be rearranged to give

$$-\frac{1}{c_p}\left[ v - T\left(\frac{\partial v}{\partial T}\right)_P \right] = \left(\frac{\partial T}{\partial P}\right)_h = \mu_{JT} \qquad \textbf{(12–52)}$$

which is the desired relation. Thus, the Joule-Thomson coefficient can be determined from a knowledge of the constant-pressure specific heat and the *P-v-T* behavior of the substance. Of course, it is also possible to predict the constant-pressure specific heat of a substance by using the Joule-Thomson coefficient, which is relatively easy to determine, together with the *P-v-T* data for the substance.

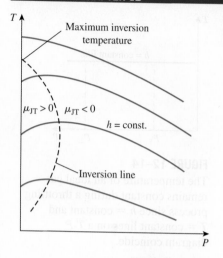

**FIGURE 12–13**
Constant-enthalpy lines of a substance on a *T-P* diagram.

---

■ **EXAMPLE 12–10**　**Joule-Thomson Coefficient of an Ideal Gas**

■ Show that the Joule-Thomson coefficient of an ideal gas is zero.

■ **SOLUTION**　It is to be shown that $\mu_{JT} = 0$ for an ideal gas.
**Analysis**　For an ideal gas $v = RT/P$, and thus

$$\left(\frac{\partial v}{\partial T}\right)_P = \frac{R}{P}$$

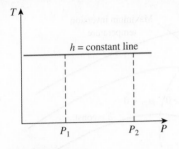

**FIGURE 12–14**

The temperature of an ideal gas remains constant during a throttling process since $h$ = constant and $T$ = constant lines on a $T$-$P$ diagram coincide.

Substituting this into Eq. 12–52 yields

$$\mu_{JT} = \frac{-1}{c_p}\left[v - T\left(\frac{\partial v}{\partial T}\right)_P\right] = \frac{-1}{c_p}\left[v - T\frac{R}{P}\right] = -\frac{1}{c_p}(v - v) = 0$$

**Discussion** This result is not surprising since the enthalpy of an ideal gas is a function of temperature only, $h = h(T)$, which requires that the temperature remain constant when the enthalpy remains constant. Therefore, a throttling process cannot be used to lower the temperature of an ideal gas (Fig. 12–14).

## 12–6 · THE $\Delta h$, $\Delta u$, AND $\Delta s$ OF REAL GASES

We have mentioned many times that gases at low pressures behave as ideal gases and obey the relation $Pv = RT$. The properties of ideal gases are relatively easy to evaluate since the properties $u$, $h$, $c_v$, and $c_p$ depend on temperature only. At high pressures, however, gases deviate considerably from ideal-gas behavior, and it becomes necessary to account for this deviation. In Chap. 3 we accounted for the deviation in properties $P$, $v$, and $T$ by either using more complex equations of state or evaluating the compressibility factor $Z$ from the compressibility charts. Now we extend the analysis to evaluate the changes in the enthalpy, internal energy, and entropy of nonideal (real) gases, using the general relations for $du$, $dh$, and $ds$ developed earlier.

### Enthalpy Changes of Real Gases

The enthalpy of a real gas, in general, depends on the pressure as well as on the temperature. Thus the enthalpy change of a real gas during a process can be evaluated from the general relation for $dh$ (Eq. 12–36)

$$h_2 - h_1 = \int_{T_1}^{T_2} c_p\, dT + \int_{P_1}^{P_2}\left[v - T\left(\frac{\partial v}{\partial T}\right)_P\right] dP$$

where $P_1$, $T_1$ and $P_2$, $T_2$ are the pressures and temperatures of the gas at the initial and the final states, respectively. For an isothermal process $dT = 0$, and the first term vanishes. For a constant-pressure process, $dP = 0$, and the second term vanishes.

Properties are point functions, and thus the change in a property between two specified states is the same no matter which process path is followed. This fact can be exploited to greatly simplify the integration of Eq. 12–36. Consider, for example, the process shown on a $T$-$s$ diagram in Fig. 12–15. The enthalpy change during this process $h_2 - h_1$ can be determined by performing the integrations in Eq. 12–36 along a path that consists of two isothermal ($T_1$ = constant and $T_2$ = constant) lines and one isobaric ($P_0$ = constant) line instead of the actual process path, as shown in Fig. 12–15.

Although this approach increases the number of integrations, it also simplifies them since one property remains constant now during each part of the process. The pressure $P_0$ can be chosen to be very low or zero, so that the gas can be treated as an ideal gas during the $P_0$ = constant process. Using a superscript asterisk (*) to denote an ideal-gas state, we can express the enthalpy change of a real gas during process 1-2 as

$$h_2 - h_1 = (h_2 - h_2^*) + (h_2^* - h_1^*) + (h_1^* - h_1) \tag{12–53}$$

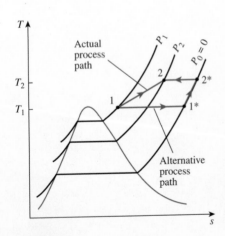

**FIGURE 12–15**

An alternative process path to evaluate the enthalpy changes of real gases.

where, from Eq. 12–36,

$$h_2 - h_2^* = 0 + \int_{P_2^*}^{P_2} \left[ v - T\left(\frac{\partial v}{\partial T}\right)_P \right]_{T=T_2} dP = \int_{P_0}^{P_2} \left[ v - T\left(\frac{\partial v}{\partial T}\right)_P \right]_{T=T_2} dP \quad (12\text{–}54)$$

$$h_2^* - h_1^* = \int_{T_1}^{T_2} c_p \, dT + 0 = \int_{T_1}^{T_2} c_{p0}(T) \, dT \quad (12\text{–}55)$$

$$h_1^* - h_1 = 0 + \int_{P_1}^{P_1^*} \left[ v - T\left(\frac{\partial v}{\partial T}\right)_P \right]_{T=T_1} dP = -\int_{P_0}^{P_1} \left[ v - T\left(\frac{\partial v}{\partial T}\right)_P \right]_{T=T_1} dP \quad (12\text{–}56)$$

The difference between $h$ and $h^*$ is called the **enthalpy departure**, and it represents the variation of the enthalpy of a gas with pressure at a fixed temperature. The calculation of enthalpy departure requires a knowledge of the $P$-$v$-$T$ behavior of the gas. In the absence of such data, we can use the relation $Pv = ZRT$, where $Z$ is the compressibility factor. Substituting $v = ZRT/P$ and simplifying Eq. 12–56, we can write the enthalpy departure at any temperature $T$ and pressure $P$ as

$$(h^* - h)_T = -RT^2 \int_0^P \left(\frac{\partial Z}{\partial T}\right)_P \frac{dP}{P}$$

The above equation can be generalized by expressing it in terms of the reduced coordinates, using $T = T_{cr}T_R$ and $P = P_{cr}P_R$. After some manipulations, the enthalpy departure can be expressed in a nondimensionalized form as

$$Z_h = \frac{(\bar{h}^* - \bar{h})_T}{R_u T_{cr}} = T_R^2 \int_0^{P_R} \left(\frac{\partial Z}{\partial T_R}\right)_{P_R} d(\ln P_R) \quad (12\text{–}57)$$

where $Z_h$ is called the **enthalpy departure factor**. The integral in the above equation can be performed graphically or numerically by employing data from the compressibility charts for various values of $P_R$ and $T_R$. The values of $Z_h$ are presented in graphical form as a function of $P_R$ and $T_R$ in Fig. A–29. This graph is called the **generalized enthalpy departure chart**, and it is used to determine the deviation of the enthalpy of a gas at a given $P$ and $T$ from the enthalpy of an ideal gas at the same $T$. By replacing $h^*$ by $h_{ideal}$ for clarity, Eq. 12–53 for the enthalpy change of a gas during a process 1–2 can be rewritten as

$$\bar{h}_2 - \bar{h}_1 = (\bar{h}_2 - \bar{h}_1)_{ideal} - R_u T_{cr}(Z_{h_2} - Z_{h_1}) \quad (12\text{–}58)$$

or

$$h_2 - h_1 = (h_2 - h_1)_{ideal} - RT_{cr}(Z_{h_2} - Z_{h_1}) \quad (12\text{–}59)$$

where the values of $Z_h$ are determined from the generalized enthalpy departure chart and $(\bar{h}_2 - \bar{h}_1)_{ideal}$ is determined from the ideal-gas tables. Notice that the last terms on the right-hand side are zero for an ideal gas.

## Internal Energy Changes of Real Gases

The internal energy change of a real gas is determined by relating it to the enthalpy change through the definition $\bar{h} = \bar{u} + P\bar{v} = \bar{u} + ZR_u T$:

$$\bar{u}_2 - \bar{u}_1 = (\bar{h}_2 - \bar{h}_1) - R_u(Z_2 T_2 - Z_1 T_1) \quad (12\text{–}60)$$

# Entropy Changes of Real Gases

The entropy change of a real gas is determined by following an approach similar to that used above for the enthalpy change. There is some difference in derivation, however, owing to the dependence of the ideal-gas entropy on pressure as well as the temperature.

The general relation for $ds$ was expressed as (Eq. 12–41)

$$s_2 - s_1 = \int_{T_1}^{T_2} \frac{c_p}{T} \, dT - \int_{P_1}^{P_2} \left( \frac{\partial v}{\partial T} \right)_P dP$$

where $P_1$, $T_1$ and $P_2$, $T_2$ are the pressures and temperatures of the gas at the initial and the final states, respectively. The thought that comes to mind at this point is to perform the integrations in the previous equation first along a $T_1 = $ constant line to zero pressure, then along the $P = 0$ line to $T_2$, and finally along the $T_2 = $ constant line to $P_2$, as we did for the enthalpy. This approach is not suitable for entropy-change calculations, however, since it involves the value of entropy at zero pressure, which is infinity. We can avoid this difficulty by choosing a different (but more complex) path between the two states, as shown in Fig. 12–16. Then the entropy change can be expressed as

$$s_2 - s_1 = (s_2 - s_b^*) + (s_b^* - s_2^*) + (s_2^* - s_1^*) + (s_1^* - s_a^*) + (s_a^* - s_1) \quad \text{(12–61)}$$

States 1 and 1* are identical ($T_1 = T_1^*$ and $P_1 = P_1^*$) and so are states 2 and 2*. The gas is assumed to behave as an ideal gas at the imaginary states 1* and 2* as well as at the states between the two. Therefore, the entropy change during process 1*-2* can be determined from the entropy-change relations for ideal gases. The calculation of entropy change between an actual state and the corresponding imaginary ideal-gas state is more involved, however, and requires the use of generalized entropy departure charts, as explained below.

Consider a gas at a pressure $P$ and temperature $T$. To determine how much different the entropy of this gas would be if it were an ideal gas at the same temperature and pressure, we consider an isothermal process from the actual state $P$, $T$ to zero (or close to zero) pressure and back to the imaginary ideal-gas state $P^*$, $T^*$ (denoted by superscript *), as shown in Fig. 12–16. The entropy change during this isothermal process can be expressed as

$$(s_P - s_P^*)_T = (s_P - s_0^*)_T + (s_0^* - s_P^*)_T$$

$$= -\int_0^P \left( \frac{\partial v}{\partial T} \right)_P dP - \int_P^0 \left( \frac{\partial v^*}{\partial T} \right)_P dP$$

where $v = ZRT/P$ and $v^* = v_{\text{ideal}} = RT/P$. Performing the differentiations and rearranging, we obtain

$$(s_P - s_P^*)_T = \int_0^P \left[ \frac{(1 - Z)R}{P} - \frac{RT}{P} \left( \frac{\partial Z r}{\partial T} \right)_P \right] dP$$

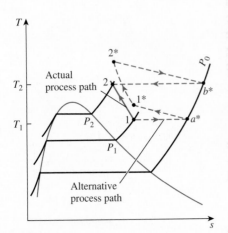

**FIGURE 12–16**

An alternative process path to evaluate the entropy changes of real gases during process 1-2.

By substituting $T = T_{cr}T_R$ and $P = P_{cr}P_R$ and rearranging, the entropy departure can be expressed in a nondimensionalized form as

$$Z_s = \frac{(\bar{s}^* - \bar{s})_{T,P}}{R_u} = \int_0^{P_R} \left[ Z - 1 + T_R \left( \frac{\partial Z}{\partial T_R} \right)_{P_R} \right] d(\ln P_R) \qquad \text{(12–62)}$$

The difference $(\bar{s}^* - \bar{s})_{T,P}$ is called the **entropy departure** and $Z_s$ is called the **entropy departure factor**. The integral in the above equation can be performed by using data from the compressibility charts. The values of $Z_s$ are presented in graphical form as a function of $P_R$ and $T_R$ in Fig. A–30. This graph is called the **generalized entropy departure chart**, and it is used to determine the deviation of the entropy of a gas at a given $P$ and $T$ from the entropy of an ideal gas at the same $P$ and $T$. Replacing $s^*$ by $s_{ideal}$ for clarity, we can rewrite Eq. 12–61 for the entropy change of a gas during a process 1–2 as

$$\bar{s}_2 - \bar{s}_1 = (\bar{s}_2 - \bar{s}_1)_{ideal} - R_u(Z_{s_2} - Z_{s_1}) \qquad \text{(12–63)}$$

or

$$s_2 - s_1 = (s_2 - s_1)_{ideal} - R(Z_{s_2} - Z_{s_1}) \qquad \text{(12–64)}$$

where the values of $Z_s$ are determined from the generalized entropy departure chart and the entropy change $(s_2 - s_1)_{ideal}$ is determined from the ideal-gas relations for entropy change. Notice that the last terms on the right-hand side are zero for an ideal gas.

---

### ■ EXAMPLE 12–11    Thermodynamic Analysis with Non-Ideal Gas Properties

■ Propane is compressed isothermally by a piston–cylinder device from 95°C and 1400 kPa to 5500 kPa (Fig. 12–17). Using the generalized charts, determine the work done and the heat transfer per unit mass of propane.

**SOLUTION** Propane is compressed isothermally by a piston-cylinder device. The work done and the heat transfer are to be determined using the generalized charts.

*Assumptions* **1** The compression process is quasi-equilibrium. **2** Kinetic and potential energy changes are negligible.

*Analysis* The critical temperature and pressure of propane are $T_{cr} = 370$ K and $P_{cr} = 4.26$ MPa (Table A–1), respectively. Propane remains close to its critical temperature, and is compressed to a pressure above its critical value. Therefore, propane is expected to deviate from the ideal-gas behavior, and thus it should be treated as a real gas.

The enthalpy departure and the compressibility factors of propane at the initial and the final states are determined from the generalized charts to be (Figs. A–29 and A–15).

$$\left. \begin{array}{l} T_{R_1} = \dfrac{T_1}{T_{cr}} = \dfrac{368 \text{ K}}{370 \text{ K}} = 0.995 \\[2mm] P_{R_1} = \dfrac{P_1}{P_{cr}} = \dfrac{1400 \text{ kPa}}{4260 \text{ kPa}} = 0.329 \end{array} \right\} \rightarrow Z_{h_1} = 0.37 \text{ and } Z_1 = 0.88$$

**FIGURE 12–17**
Schematic for Example 12–11.

and

$$T_{R_2} = \frac{T_2}{T_{cr}} = \frac{368 \text{ K}}{370 \text{ K}} = 0.995$$

$$P_{R_2} = \frac{P_2}{P_{cr}} = \frac{5500 \text{ kPa}}{4260 \text{ kPa}} = 1.291$$

$$\left. \right\} \rightarrow Z_{h_2} = 4.2 \text{ and } Z_2 = 0.22$$

As an approximation, treating propane as a real gas with $Z_{avg} = (Z_1 + Z_2)/2 = (0.88 + 0.22)/2 = 0.55$ gives

$$P_v = ZRT \cong Z_{avg}RT = C = \text{constant}$$

Then the boundary work becomes

$$w_{b,in} = -\int_1^2 P dv = -\int_1^2 \frac{C}{v} dv = -C \ln \frac{v_2}{v_1} = -Z_{avg}RT \ln \frac{Z_2RT/P_2}{Z_1RT/P_1}$$

$$= -Z_{avg}RT \ln \frac{Z_2 P_1}{Z_1 P_2}$$

$$= -(0.55)(0.1885 \text{ kJ/kg·K})(368 \text{ K}) \ln \frac{(0.22)(1400 \text{ kPa})}{(0.88)(5500 \text{ kPa})}$$

$$= \mathbf{105.1 \text{ kJ/kg}}$$

Also,

$$h_2 - h_1 = RT_{cr}(Z_{h_1} - Z_{h_2}) + (h_2 - h_1)_{ideal}{\nearrow}^0$$

$$= (0.1885 \text{ kJ/kg·K})(370 \text{ K})(0.37 - 4.2) + 0$$

$$= -267.1 \text{ kJ/kg}$$

$$u_2 - u_1 = (h_2 - h_1) - R(Z_2 T_2 - Z_1 T_1)$$

$$= (-267.1 \text{ kJ/kg}) - (0.1885 \text{ kJ/kg·K})$$

$$\times [(0.22)(368 \text{ K}) - (0.88)(368 \text{ K})]$$

$$= -221.3 \text{ kJ/kg}$$

Then the heat transfer during this process is determined from the closed-system energy balance equation for the piston-cylinder device to be

$$E_{in} - E_{out} = \Delta E_{system}$$

$$q_{in} + w_{b,in} = \Delta u = u_2 - u_1$$

$$q_{in} = (u_2 - u_1) - w_{b,in} = -221.3 - 105.1 = -326.4 \text{ kJ/kg}$$

The negative sign indicates heat rejection. Therefore, heat transfer out of the system during this process is

$$q_{out} = \mathbf{326.4 \text{ kJ/kg}}$$

**Discussion** Note that if the ideal-gas assumption were used for propane, the magnitudes of boundary work and heat transfer would have been the same (94.9 kJ/kg). Therefore, the ideal-gas approximation would underestimate boundary work by 10 percent, and the heat transfer by 71 percent.

# SUMMARY

Some thermodynamic properties can be measured directly, but many others cannot. Therefore, it is necessary to develop some relations between these two groups so that the properties that cannot be measured directly can be evaluated. The derivations are based on the fact that properties are point functions, and the state of a simple, compressible system is completely specified by any two independent, intensive properties.

The equations that relate the partial derivatives of properties $P$, $v$, $T$, and $s$ of a simple compressible substance to each other are called the *Maxwell relations*. They are obtained from the *four Gibbs equations*, expressed as

$$du = T\,ds - P\,dv$$
$$dh = T\,ds + v\,dP$$
$$da = -s\,dT - P\,dv$$
$$dg = -s\,dT + v\,dP$$

The *Maxwell relations* are

$$\left(\frac{\partial T}{\partial v}\right)_s = -\left(\frac{\partial P}{\partial s}\right)_v$$

$$\left(\frac{\partial T}{\partial P}\right)_s = \left(\frac{\partial v}{\partial s}\right)_P$$

$$\left(\frac{\partial s}{\partial v}\right)_T = \left(\frac{\partial P}{\partial T}\right)_v$$

$$\left(\frac{\partial s}{\partial P}\right)_T = -\left(\frac{\partial v}{\partial T}\right)_P$$

The *Clapeyron equation* enables us to determine the enthalpy change associated with a phase change from a knowledge of $P$, $v$, and $T$ data alone. It is expressed as

$$\left(\frac{dP}{dT}\right)_{\text{sat}} = \frac{h_{fg}}{Tv_{fg}}$$

For liquid–vapor and solid–vapor phase-change processes at low pressures, it can be approximated as

$$\ln\left(\frac{P_2}{P_1}\right)_{\text{sat}} \cong \frac{h_{fg}}{R}\left(\frac{T_2 - T_1}{T_1T_2}\right)_{\text{sat}}$$

The changes in internal energy, enthalpy, and entropy of a simple compressible substance can be expressed in terms of pressure, specific volume, temperature, and specific heats alone as

$$du = c_v\,dT + \left[T\left(\frac{\partial P}{\partial T}\right)_v - P\right]dv$$

$$dh = c_p\,dT + \left[v - T\left(\frac{\partial v}{\partial T}\right)_P\right]dP$$

$$ds = \frac{c_v}{T}\,dT + \left(\frac{\partial P}{\partial T}\right)_v dv$$

or

$$ds = \frac{c_p}{T}\,dT - \left(\frac{\partial v}{\partial T}\right)_P dP$$

For specific heats, we have the following general relations:

$$\left(\frac{\partial c_v}{\partial v}\right)_T = T\left(\frac{\partial^2 P}{\partial T^2}\right)_v$$

$$\left(\frac{\partial c_p}{\partial P}\right)_T = -T\left(\frac{\partial^2 v}{\partial T^2}\right)_P$$

$$c_{p,T} - c_{p0,T} = -T\int_0^P \left(\frac{\partial^2 v}{\partial T^2}\right)_P dP$$

$$c_p - c_v = -T\left(\frac{\partial v}{\partial T}\right)_P^2\left(\frac{\partial P}{\partial v}\right)_T$$

$$c_p - c_v = \frac{vT\beta^2}{\alpha}$$

where $\beta$ is the *volume expansivity* and $\alpha$ is the *isothermal compressibility*, defined as

$$\beta = \frac{1}{v}\left(\frac{\partial v}{\partial T}\right)_P \quad \text{and} \quad \alpha = -\frac{1}{v}\left(\frac{\partial v}{\partial P}\right)_T$$

The difference $c_p - c_v$ is equal to $R$ for ideal gases and to zero for incompressible substances.

The temperature behavior of a fluid during a throttling ($h = $ constant) process is described by the *Joule-Thomson coefficient*, defined as

$$\mu_{\text{JT}} = \left(\frac{\partial T}{\partial P}\right)_h$$

The Joule-Thomson coefficient is a measure of the change in temperature of a substance with pressure during a constant-enthalpy process, and it can also be expressed as

$$\mu_{\text{JT}} = -\frac{1}{c_p}\left[v - T\left(\frac{\partial v}{\partial T}\right)_P\right]$$

The enthalpy, internal energy, and entropy changes of real gases can be determined accurately by utilizing *generalized enthalpy* or *entropy departure charts* to account for the deviation from the ideal-gas behavior by using the following relations:

$$\bar{h}_2 - \bar{h}_1 = (\bar{h}_2 - \bar{h}_1)_{\text{ideal}} - R_u T_{\text{cr}}(Z_{h_2} - Z_{h_1})$$

$$\bar{u}_2 - \bar{u}_1 = (\bar{h}_2 - \bar{h}_1) - R_u(Z_2 T_2 - Z_1 T_1)$$

$$\bar{s}_2 - \bar{s}_1 = (\bar{s}_2 - \bar{s}_1)_{\text{ideal}} - R_u(Z_{s_2} - Z_{s_1})$$

where the values of $Z_h$ and $Z_s$ are determined from the generalized charts.

## REFERENCES AND SUGGESTED READINGS

**1.** A. Bejan. *Advanced Engineering Thermodynamics.* 3rd ed. New York: Wiley, 2006.

**2.** K. Wark, Jr. *Advanced Thermodynamics for Engineers.* New York: McGraw-Hill, 1995.

## PROBLEMS*

### Partial Derivatives and Associated Relations

**12–1C** What is the difference between partial differentials and ordinary differentials?

**12–2C** Consider a function $z(x, y)$ and its partial derivative $(\partial z/\partial y)_x$. Under what conditions is this partial derivative equal to the total derivative $dz/dy$?

**12–3C** Consider a function $z(x, y)$ and its partial derivative $(\partial z/\partial y)_x$. If this partial derivative is equal to zero for all values of $x$, what does it indicate?

**12–4C** Consider the function $z(x, y)$, its partial derivatives $(\partial z/\partial x)_y$ and $(\partial z/\partial y)_x$, and the total derivative $dz/dx$.

(a) How do the magnitudes $(\partial x)_y$ and $dx$ compare?

(b) How do the magnitudes $(\partial z)_y$ and $dz$ compare?

(c) Is there any relation among $dz$, $(\partial z)_x$, and $(\partial z)_y$?

**12–5** Consider air at 350 K and 0.75 m³/kg. Using Eq. 12–3, determine the change in pressure corresponding to an increase of (a) 1 percent in temperature at constant specific volume, (b) 1 percent in specific volume at constant temperature, and (c) 1 percent in both the temperature and specific volume.

**12–6** Repeat Problem 12–5 for helium.

**12–7** Nitrogen gas at 400 K and 300 kPa behaves as an ideal gas. Estimate the $c_p$ and $c_v$ of the nitrogen at this state, using enthalpy and internal energy data from Table A–18, and compare them to the values listed in Table A–2b.

**12–8** Consider an ideal gas at 300 K and 100 kPa. As a result of some disturbance, the conditions of the gas change to 305 K and 96 kPa. Estimate the change in the specific volume of the gas using (a) Eq. 12–3 and (b) the ideal-gas relation at each state.

**12–9** Using the equation of state $P(v - a) = RT$, verify (a) the cyclic relation and (b) the reciprocity relation at constant $v$.

**12–10** Derive a relation for the slope of the $v =$ constant lines on a $T$-$P$ diagram for a gas that obeys the van der Waals equation of state. *Answer:* $(v - b)/R$

### The Maxwell Relations

**12–11** Verify the validity of the last Maxwell relation (Eq. 12–19) for refrigerant-134a at 50°C and 0.7 MPa.

**12–12** Reconsider Prob. 12–11. Using EES (or other) software, verify the validity of the last Maxwell relation for refrigerant-134a at the specified state.

**12–13** Verify the validity of the last Maxwell relation (Eq. 12–19) for steam at 300°C and 2 MPa.

**12–14** Using the Maxwell relations, determine a relation for $(\partial s/\partial P)_T$ for a gas whose equation of state is $P(v - b) = RT$. *Answer:* $-R/P$

**12–15** Using the Maxwell relations, determine a relation for $(\partial s/\partial v)_T$ for a gas whose equation of state is $(P - a/v^2)(v - b) = RT$.

**12–16** Using the Maxwell relations and the ideal-gas equation of state, determine a relation for $(\partial s/\partial v)_T$ for an ideal gas. *Answer:* $R/v$

**12–17** Prove that $\left(\dfrac{\partial P}{\partial T}\right)_s = \dfrac{k}{k-1}\left(\dfrac{\partial P}{\partial T}\right)_v$.

### The Clapeyron Equation

**12–18C** What is the value of the Clapeyron equation in thermodynamics?

**12–19C** Does the Clapeyron equation involve any approximations, or is it exact?

**12–20** Using the Clapeyron equation, estimate the enthalpy of vaporization of refrigerant-134a at 40°C, and compare it to the tabulated value.

**12–21** Reconsider Prob. 12–20. Using EES (or other) software, plot the enthalpy of vaporization of refrigerant-134a as a function of temperature over the temperature range −20 to 80°C by using the Clapeyron equation and the refrigerant-134a data in EES. Discuss your results.

---

* Problems designated by a "C" are concept questions, and students are encouraged to answer them all. Problems with the 🌐 icon are solved using EES, and complete solutions together with parametric studies are included on the text website. Problems with the 💻 icon are comprehensive in nature, and are intended to be solved with an equation solver such as EES.

**12–22** Using the Clapeyron equation, estimate the enthalpy of vaporization of steam at 300 kPa, and compare it to the tabulated value.

**12–23** Determine the $h_{fg}$ of refrigerant-134a at $-10°C$ on the basis of (a) the Clapeyron equation and (b) the Clapeyron-Clausius equation. Compare your results to the tabulated $h_{fg}$ value.

**12–24** 0.22-kg of a saturated vapor is converted to a saturated liquid by being cooled in a weighted piston-cylinder device maintained at 350 kPa. During the phase conversion, the system volume decreases by 0.04 m³; 250 kJ of heat are removed; and the temperature remains fixed at $-10°C$. Estimate the boiling point temperature of this substance when its pressure is 420 kPa. *Answer:* 266 K

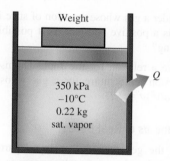

Weight

350 kPa
$-10°C$
0.22 kg
sat. vapor

$Q$

**FIGURE P12–24**

**12–25** Estimate the $s_{fg}$ of the substance in Problem 12–24 at $-10°C$. *Answer:* 4.32 kJ/kg·K

**12–26** A table of properties for methyl chloride lists the saturation pressure as 805 kPa at 38°C. At 38°C, this table also lists $h_{fg}$ = 360 kJ/kg, and $v_{fg}$ = 0.05390 m³/kg. Estimate the saturation pressure $P_{sat}$ of methyl chloride at 33°C and 43°C.

**12–27** Using the Clapeyron-Clausius equation and the triple-point data of water, estimate the sublimation pressure of water at $-30°C$ and compare to the value in Table A–8.

**12–28** Show that $c_{p,g} - c_{p,f} = T\left(\frac{\partial (h_{fg}/T)}{\partial T}\right)_P + v_{fg}\left(\frac{\partial P}{\partial T}\right)_{sat}$.

## General Relations for du, dh, ds, $c_v$, and $c_p$

**12–29C** Can the variation of specific heat $c_p$ with pressure at a given temperature be determined from a knowledge of $P$-$v$-$T$ data alone?

**12–30** Estimate the volume expansivity $\beta$ and the isothermal compressibility $\alpha$ of refrigerant-134a at 200 kPa and 30°C.

**12–31** Estimate the specific heat difference $c_p - c_v$ for liquid water at 15 MPa and 80°C. *Answer:* 0.32 kJ/kg·K

**12–32** Determine the change in the internal energy of air, in kJ/kg, as it undergoes a change of state from 100 kPa and 20°C to 600 kPa and 300°C using the equation of state $P(v - a) = RT$ where $a = 1$ m³/kg, and compare the result to the value obtained by using the ideal gas equation of state.

**12–33** Determine the change in the enthalpy of air, in kJ/kg, as it undergoes a change of state from 100 kPa and 34°C to 800 kPa and 420°C using the equation of state $P(v - a) = RT$ where $a = 0.01$ m³/kg, and compare the result to the value obtained by using the ideal gas equation of state. *Answers:* 404 kJ/kg, 397 kJ/kg

**12–34** Determine the change in the entropy of air, in kJ/kg·K, as it undergoes a change of state from 100 kPa and 20°C to 600 kPa and 300°C using the equation of state $P(v - a) = RT$ where $a = 0.01$ m³/kg, and compare the result to the value obtained by using the ideal gas equation of state.

**12–35** Determine the change in the internal energy of helium, in kJ/kg, as it undergoes a change of state from 100 kPa and 20°C to 600 kPa and 300°C using the equation of state $P(v - a) = RT$ where $a = 0.01$ m³/kg, and compare the result to the value obtained by using the ideal gas equation of state.

**12–36** Determine the change in the enthalpy of helium, in kJ/kg, as it undergoes a change of state from 150 kPa and 20°C to 750 kPa and 380°C using the equation of state $P(v - a) = RT$ where $a = 0.01$ m³/kg, and compare the result to the value obtained by using the ideal gas equation of state.

**12–37** Determine the change in the entropy of helium, in kJ/kg·K, as it undergoes a change of state from 100 kPa and 20°C to 600 kPa and 300°C using the equation of state $P(v - a) = RT$ where $a = 0.01$ m³/kg, and compare the result to the value obtained by using the ideal gas equation of state. *Answers:* −0.239 kJ/kg·K, −0.239 kJ/kg·K

**12–38** Derive expressions for (a) $\Delta u$, (b) $\Delta h$, and (c) $\Delta s$ for a gas whose equation of state is $P(v - a) = RT$ for an isothermal process. *Answers:* (a) 0, (b) $a(P_2 - P_1)$, (c) $-R \ln (P_2/P_1)$

**12–39** Derive expressions for (a) $\Delta u$, (b) $\Delta h$, and (c) $\Delta s$ for a gas that obeys the van der Waals equation of state for an isothermal process.

**12–40** Derive an expression for the specific heat difference $c_p - c_v$ for (a) an ideal gas, (b) a van der Waals gas, and (c) an incompressible substance.

**12–41** Show that $c_p - c_v = T\left(\frac{\partial P}{\partial T}\right)_v\left(\frac{\partial v}{\partial T}\right)_P$.

**12–42** Temperature may alternatively be defined as

$$T = \left(\frac{\partial u}{\partial s}\right)_v$$

Prove that this definition reduces the net entropy change of two constant-volume systems filled with simple compressible substances to zero as the two systems approach thermal equilibrium.

**12–43** Derive a relation for the volume expansivity $\beta$ and the isothermal compressibility $\alpha$ (a) for an ideal gas and (b) for a gas whose equation of state is $P(v - a) = RT$.

**12–44** Derive an expression for the isothermal compressibility of a substance whose equation of state is

$$P = \frac{RT}{v - b} - \frac{a}{v(v + b)T^{1/2}}$$

where $a$ and $b$ are empirical constants.

**12–45** Derive an expression for the volume expansivity of a substance whose equation of state is

$$P = \frac{RT}{v - b} - \frac{a}{v^2 T}$$

where $a$ and $b$ are empirical constants.

**12–46** Show that $\beta = \alpha(\partial P/\partial T)_v$.

**12–47** Demonstrate that $k = \dfrac{c_p}{c_v} = -\dfrac{v\alpha}{(\partial v/\partial P)_s}$.

**12–48** The Helmholtz function of a substance has the form

$$a = -RT \ln \frac{v}{v_0} - cT_0\left(1 - \frac{T}{T_0} + \frac{T}{T_0} \ln \frac{T}{T_0}\right)$$

where $T_0$ and $v_0$ are the temperature and specific volume at a reference state. Show how to obtain $P$, $h$, $s$, $c_v$, and $c_p$ from this expression.

**12–49** Show that the enthalpy of an ideal gas is a function of temperature only and that for an incompressible substance it also depends on pressure.

## The Joule-Thomson Coefficient

**12–50C** What does the Joule-Thomson coefficient represent?

**12–51C** Describe the inversion line and the maximum inversion temperature.

**12–52C** The pressure of a fluid always decreases during an adiabatic throttling process. Is this also the case for the temperature?

**12–53C** Does the Joule-Thomson coefficient of a substance change with temperature at a fixed pressure?

**12–54C** Will the temperature of helium change if it is throttled adiabatically from 300 K and 600 kPa to 150 kPa?

**12–55** Estimate the Joule-Thomson coefficient of nitrogen at (a) 850 kPa and 195 K, and (b) 8500 kPa and 400 K. Use nitrogen properties from EES or other source.

**12–56** Reconsider Prob. 12–55. Using EES (or other) software, plot the Joule-Thomson coefficient for nitrogen over the pressure range 0.7 to 10 MPa at the enthalpy values 230, 400, and 525 kJ/kg. Discuss the results.

**12–57** Steam is throttled slightly from 2 MPa and 500°C. Will the temperature of the steam increase, decrease, or remain the same during this process?

**12–58** Estimate the Joule-Thomson coefficient of steam at (a) 3 MPa and 300°C and (b) 6 MPa and 500°C.

**12–59** Demonstrate that the Joule-Thomson coefficient is given by

$$\mu = \frac{T^2}{c_p}\left[\frac{\partial(v/T)}{\partial T}\right]_P.$$

**12–60** Consider a gas whose equation of state is $P(v - a) = RT$, where $a$ is a positive constant. Is it possible to cool this gas by throttling?

**12–61** Derive a relation for the Joule-Thomson coefficient and the inversion temperature for a gas whose equation of state is $(P + a/v^2)v = RT$.

## The $dh$, $du$, and $ds$ of Real Gases

**12–62C** On the generalized enthalpy departure chart, the normalized enthalpy departure values seem to approach zero as the reduced pressure $P_R$ approaches zero. How do you explain this behavior?

**12–63C** Why is the generalized enthalpy departure chart prepared by using $P_R$ and $T_R$ as the parameters instead of $P$ and $T$?

**12–64** Determine the enthalpy of nitrogen, in kJ/kg, at 175 K and 8 MPa using (a) data from the ideal-gas nitrogen table and (b) the generalized enthalpy departure chart. Compare your results to the actual value of 125.5 kJ/kg. *Answers:* (a) 181.5 kJ/kg, (b) 121.6 kJ/kg

**12–65** Determine the enthalpy change and the entropy change of $CO_2$ per unit mass as it undergoes a change of state from 250 K and 7 MPa to 280 K and 12 MPa, (a) by assuming ideal-gas behavior and (b) by accounting for the deviation from ideal-gas behavior.

**12–66** Saturated water vapor at 260°C is expanded while its pressure is kept constant until its temperature is 550°C. Calculate the change in the specific enthalpy and entropy using (a) the departure charts, and (b) the property tables. *Answers:* (a) 655 kJ/kg, 0.983 kJ/kg·K (b) 757 kJ/kg, 1.154 kJ/kg·K

**12–67** Water vapor at 1000 kPa and 600°C is expanded to 500 kPa and 400°C. Calculate the change in the specific entropy and enthalpy of this water vapor using the departure charts and the property tables.

**12–68** Methane is compressed adiabatically by a steady-flow compressor from 0.8 MPa and −10°C to 6 MPa and 175°C at a rate of 0.33 kg/s. Using the generalized charts, determine the required power input to the compressor. *Answer: 132 kW*

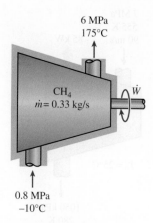

6 MPa
175°C

CH$_4$
$\dot{m}$ = 0.33 kg/s

$\dot{W}$

0.8 MPa
−10°C

**FIGURE P12–68**

**12–69** Carbon dioxide enters an adiabatic nozzle at 8 MPa and 450 K with a low velocity and leaves at 2 MPa and 350 K. Using the generalized enthalpy departure chart, determine the exit velocity of the carbon dioxide. *Answer: 384 m/s*

**12–70** Reconsider Prob. 12–69. Using EES (or other) software, compare the exit velocity to the nozzle assuming ideal-gas behavior, the generalized chart data, and EES data for carbon dioxide.

**12–71** Oxygen is adiabatically and reversibly expanded in a nozzle from 1400 kPa and 317°C to 500 kPa. Determine the velocity at which the oxygen leaves the nozzle, assuming that it enters with negligible velocity, treating the oxygen as an ideal gas with temperature variable specific heats and using the departure charts. *Answers: 525.8 m/s, 526.4 m/s*

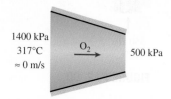

1400 kPa
317°C
≈ 0 m/s

O$_2$

500 kPa

**FIGURE P12–71**

**12–72** Propane is compressed isothermally by a piston–cylinder device from 100°C and 1 MPa to 4 MPa. Using the generalized charts, determine the work done and the heat transfer per unit mass of propane.

**12–73** Reconsider Prob. 12–72. Using EES (or other) software, extend the problem to compare the solutions based on the ideal-gas assumption, generalized chart data, and real fluid data. Also extend the solution to methane.

**12–74** Determine the exergy destruction associated with the process described in Prob. 12–72. Assume $T_0 = 25°C$.

**12–75** A 0.05-m³ well-insulated rigid tank contains oxygen at 175 K and 6 MPa. A paddle wheel placed in the tank is turned on, and the temperature of the oxygen rises to 225 K. Using the generalized charts, determine (a) the final pressure in the tank, and (b) the paddle-wheel work done during this process. *Answers: (a) 9652 kPa, (b) 423 kJ*

## Review Problems

**12–76** Derive relations for (a) $\Delta u$, (b) $\Delta h$, and (c) $\Delta s$ of a gas that obeys the equation of state $(P + a/v^2)v = RT$ for an isothermal process.

**12–77** Starting with the relation $dh = T\,ds + v\,dP$, show that the slope of a constant-pressure line on an $h$-$s$ diagram (a) is constant in the saturation region, and (b) increases with temperature in the superheated region.

**12–78** Show that

$$c_v = -T\left(\frac{\partial v}{\partial T}\right)_s\left(\frac{\partial P}{\partial T}\right)_v \quad \text{and} \quad c_p = T\left(\frac{\partial P}{\partial T}\right)_s\left(\frac{\partial v}{\partial T}\right)_P$$

**12–79** Temperature and pressure may be defined as

$$T = \left(\frac{\partial u}{\partial s}\right)_v \quad \text{and} \quad P = -\left(\frac{\partial u}{\partial v}\right)_s$$

Using these definitions, prove that for a simple compressible substance

$$\left(\frac{\partial s}{\partial v}\right)_u = \frac{P}{T}$$

**12–80** For ideal gases, the development of the constant-pressure specific heat yields

$$\left(\frac{\partial h}{\partial P}\right)_T = 0$$

Prove this by using the definitions of pressure and temperature, $T = (\partial u/\partial s)_v$ and $P = -(\partial u/\partial v)_s$.

**12–81** Starting with $\mu_{JT} = (1/c_p)[T(\partial v/\partial T)_p - v]$ and noting that $Pv = ZRT$, where $Z = Z(P, T)$ is the compressibility factor, show that the position of the Joule-Thomson coefficient inversion curve on the $T$-$P$ plane is given by the equation $(\partial Z/\partial T)_P = 0$.

**12–82** For a homogeneous (single-phase) simple pure substance, the pressure and temperature are independent properties, and any property can be expressed as a function of these two properties. Taking $v = v(P, T)$, show that the change in specific volume can be expressed in terms of the volume expansivity $\beta$ and isothermal compressibility $\alpha$ as

$$\frac{dv}{v} = \beta\,dT = \alpha\,dP$$

Also, assuming constant average values for $\beta$ and $\alpha$, obtain a relation for the ratio of the specific volumes $v_2/v_1$ as a homogeneous system undergoes a process from state 1 to state 2.

**12–83** Repeat Prob. 12–82 for an isobaric process.

**12–84** Consider an infinitesimal reversible adiabatic compression or expansion process. By taking $s = s(P, v)$ and using the Maxwell relations, show that for this process $Pv^k =$ constant, where $k$ is the *isentropic expansion exponent* defined as

$$k = \frac{v}{P}\left(\frac{\partial P}{\partial v}\right)_s$$

Also, show that the isentropic expansion exponent $k$ reduces to the specific heat ratio $c_p/c_v$ for an ideal gas.

**12–85** Estimate the $c_p$ of nitrogen at 300 kPa and 400 K, using (a) the relation in Prob. 12–84, and (b) its definition. Compare your results to the value listed in Table A–2b.

**12–86** Steam is throttled from 2.5 MPa and 400°C to 1.2 MPa. Estimate the temperature change of the steam during this process and the average Joule-Thomson coefficient. *Answers:* −9.9°C, 7.6°C/MPa

**12–87** The volume expansivity $\beta$ values of copper at 300 K and 500 K are $49.2 \times 10^{-6}\ \text{K}^{-1}$ and $54.2 \times 10^{-6}\ \text{K}^{-1}$, respectively, and $\beta$ varies almost linearly in this temperature range. Determine the percent change in the volume of a copper block as it is heated from 300 K to 500 K at atmospheric pressure.

**12–88** An adiabatic 0.2-m³ storage tank that is initially evacuated is connected to a supply line that carries nitrogen at 225 K and 10 MPa. A valve is opened, and nitrogen flows into the tank from the supply line. The valve is closed when the pressure in the tank reaches 10 MPa. Determine the final temperature in the tank (a) treating nitrogen as an ideal gas, and (b) using generalized charts. Compare your results to the actual value of 293 K.

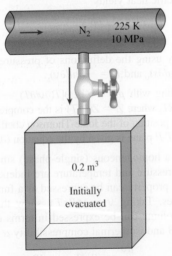

**FIGURE P12–88**

**12–89** Argon gas enters a turbine at 7 MPa and 555 K with a velocity of 90 m/s and leaves at 1050 kPa and 280 K with a velocity of 135 m/s at a rate of 5 kg/s. Heat is being lost to the surroundings at 25°C at a rate of 85 kW. Using the generalized charts, determine (a) the power output of the turbine and

(b) the exergy destruction associated with the process. *Answers:* (a) 611 kW, (b) 137 kW

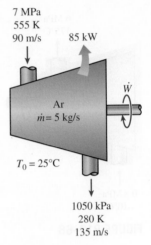

**FIGURE P12–89**

**12–90** Methane is to be adiabatically and reversibly compressed from 350 kPa and 37°C to 3500 kPa. Calculate the specific work required for this compression treating the methane as an ideal gas with variable specific heats and using the departure charts.

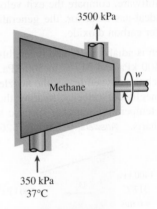

**FIGURE P12–90**

**12–91** 〔EES〕 Refrigerant-134a undergoes an isothermal process at 40°C from 2 to 0.1 MPa in a closed system. Determine the work done by the refrigerant-134a by using the tabular (EES) data and the generalized charts, in kJ/kg.

**12–92** A rigid tank contains 1.2 m³ of argon at −100°C and 1 MPa. Heat is now transferred to argon until the temperature in the tank rises to 0°C. Using the generalized charts, determine (a) the mass of the argon in the tank, (b) the final pressure, and (c) the heat transfer. *Answers:* (a) 35.1 kg, (b) 1531 kPa, (c) 1251 kJ

**12–93** Methane is contained in a piston–cylinder device and is heated at constant pressure of 5 MPa from 100 to 250°C.

Determine the heat transfer, work and entropy change per unit mass of the methane using (*a*) the ideal-gas assumption, (*b*) the generalized charts, and (*c*) real fluid data from EES or other sources.

## Fundamentals of Engineering (FE) Exam Problems

**12–94** A substance whose Joule-Thomson coefficient is negative is throttled to a lower pressure. During this process, (select the correct statement)

(*a*) the temperature of the substance will increase.
(*b*) the temperature of the substance will decrease.
(*c*) the entropy of the substance will remain constant.
(*d*) the entropy of the substance will decrease.
(*e*) the enthalpy of the substance will decrease.

**12–95** Consider the liquid–vapor saturation curve of a pure substance on the *P-T* diagram. The magnitude of the slope of the tangent line to this curve at a temperature *T* (in Kelvin) is

(*a*) proportional to the enthalpy of vaporization $h_{fg}$ at that temperature.
(*b*) proportional to the temperature *T*.
(*c*) proportional to the square of the temperature *T*.
(*d*) proportional to the volume change $v_{fg}$ at that temperature.
(*e*) inversely proportional to the entropy change $s_{fg}$ at that temperature.

**12–96** Based on the generalized charts, the error involved in the enthalpy of $CO_2$ at 300 K and 5 MPa if it is assumed to be an ideal gas is

(*a*) 0%   (*b*) 9%   (*c*) 16%   (*d*) 22%   (*e*) 27%

**12–97** Based on data from the refrigerant-134a tables, the Joule-Thompson coefficient of refrigerant-134a at 0.8 MPa and 100°C is approximately

(*a*) 0   (*b*) −5°C/MPa   (*c*) 11°C/MPa
(*d*) 8°C/MPa   (*e*) 26°C/MPa

**12–98** For a gas whose equation of state is $P(v - b) = RT$, the specified heat difference $c_p - c_v$ is equal to

(*a*) *R*   (*b*) *R* − *b*   (*c*) *R* + *b*   (*d*) 0   (*e*) $R(1 + v/b)$

## Design and Essay Problems

**12–99** Consider the function $z = z(x, y)$. Write an essay on the physical interpretation of the ordinary derivative $dz/dx$ and the partial derivative $(\partial z/\partial x)_y$. Explain how these two derivatives are related to each other and when they become equivalent.

**12–100** There have been several attempts to represent the thermodynamic relations geometrically, the best known of these being Koenig's thermodynamic square shown in the figure. There is a systematic way of obtaining the four Maxwell relations as well as the four relations for *du*, *dh*, *dg*, and *da* from this figure. By comparing these relations to Koenig's diagram, come up with the rules to obtain these eight thermodynamic relations from this diagram.

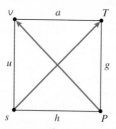

**FIGURE P12–100**

**12–101** Several attempts have been made to express the partial derivatives of the most common thermodynamic properties in a compact and systematic manner in terms of measurable properties. The work of P. W. Bridgman is perhaps the most fruitful of all, and it resulted in the well-known Bridgman's table. The 28 entries in that table are sufficient to express the partial derivatives of the eight common properties *P*, *T*, *v*, *s*, *u*, *h*, *f*, and *g* in terms of the six properties *P*, *v*, *T*, $c_p$, $\beta$, and $\alpha$, which can be measured directly or indirectly with relative ease. Obtain a copy of Bridgman's table and explain, with examples, how it is used.

# GAS MIXTURES

**U**p to this point, we have limited our consideration to thermodynamic systems that involve a single pure substance such as water. Many important thermodynamic applications, however, involve *mixtures* of several pure substances rather than a single pure substance. Therefore, it is important to develop an understanding of mixtures and learn how to handle them.

In this chapter, we deal with nonreacting gas mixtures. A nonreacting gas mixture can be treated as a pure substance since it is usually a homogeneous mixture of different gases. The properties of a gas mixture obviously depend on the properties of the individual gases (called *components* or *constituents*) as well as on the amount of each gas in the mixture. Therefore, it is possible to prepare tables of properties for mixtures. This has been done for common mixtures such as air. It is not practical to prepare property tables for every conceivable mixture composition, however, since the number of possible compositions is endless. Therefore, we need to develop rules for determining mixture properties from a knowledge of mixture composition and the properties of the individual components. We do this first for ideal-gas mixtures and then for real-gas mixtures. The basic principles involved are also applicable to liquid or solid mixtures, called *solutions*.

■ ■ ■ ■ ■ ■ ■
## OBJECTIVES
The objectives of Chapter 13 are to:

- Develop rules for determining nonreacting gas mixture properties from knowledge of mixture composition and the properties of the individual components.

- Define the quantities used to describe the composition of a mixture, such as mass fraction, mole fraction, and volume fraction.

- Apply the rules for determining mixture properties of ideal-gas mixtures and real-gas mixtures.

- Predict the $P$-$v$-$T$ behavior of gas mixtures based on Dalton's law of additive pressures and Amagat's law of additive volumes.

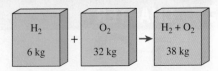

**FIGURE 13–1**
The mass of a mixture is equal to the sum of the masses of its components.

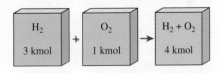

**FIGURE 13–2**
The number of moles of a nonreacting mixture is equal to the sum of the number of moles of its components.

**FIGURE 13–3**
The sum of the mole fractions of a mixture is equal to 1.

## 13–1 · COMPOSITION OF A GAS MIXTURE: MASS AND MOLE FRACTIONS

To determine the properties of a mixture, we need to know the *composition* of the mixture as well as the properties of the individual components. There are two ways to describe the composition of a mixture: either by specifying the number of moles of each component, called **molar analysis**, or by specifying the mass of each component, called **gravimetric analysis**.

Consider a gas mixture composed of $k$ components. The mass of the mixture $m_m$ is the sum of the masses of the individual components, and the mole number of the mixture $N_m$ is the sum of the mole numbers of the individual components[*] (Figs. 13–1 and 13–2). That is,

$$m_m = \sum_{i=1}^{k} m_i \quad \text{and} \quad N_m = \sum_{i=1}^{k} N_i \tag{13–1a, b}$$

The ratio of the mass of a component to the mass of the mixture is called the **mass fraction** mf, and the ratio of the mole number of a component to the mole number of the mixture is called the **mole fraction** $y$:

$$\text{mf}_i = \frac{m_i}{m_m} \quad \text{and} \quad y_i = \frac{N_i}{N_m} \tag{13–2a, b}$$

Dividing Eq. 13–1a by $m_m$ or Eq. 13–1b by $N_m$, we can easily show that the sum of the mass fractions or mole fractions for a mixture is equal to 1 (Fig. 13–3):

$$\sum_{i=1}^{k} \text{mf}_i = 1 \quad \text{and} \quad \sum_{i=1}^{k} y_i = 1$$

The mass of a substance can be expressed in terms of the mole number $N$ and molar mass $M$ of the substance as $m = NM$. Then the **apparent** (or **average**) **molar mass** and the **gas constant** of a mixture can be expressed as

$$M_m = \frac{m_m}{N_m} = \frac{\sum m_i}{N_m} = \frac{\sum N_i M_i}{N_m} = \sum_{i=1}^{k} y_i M_i \quad \text{and} \quad R_m = \frac{R_u}{M_m} \tag{13–3a, b}$$

The molar mass of a mixture can also be expressed as

$$M_m = \frac{m_m}{N_m} = \frac{m_m}{\sum m_i/M_i} = \frac{1}{\sum m_i/(m_m M_i)} = \frac{1}{\displaystyle\sum_{i=1}^{k} \frac{\text{mf}_i}{M_i}} \tag{13–4}$$

Mass and mole fractions of a mixture are related by

$$\text{mf}_i = \frac{m_i}{m_m} = \frac{N_i M_i}{N_m M_m} = y_i \frac{M_i}{M_m} \tag{13–5}$$

---

[*]Throughout this chapter, the subscript $m$ denotes the gas mixture and the subscript $i$ denotes any single component of the mixture.

## EXAMPLE 13–1    Mass and Mole Fractions of a Gas Mixture

Consider a gas mixture that consists of 3 kg of $O_2$, 5 kg of $N_2$, and 12 kg of $CH_4$, as shown in Fig. 13–4. Determine (a) the mass fraction of each component, (b) the mole fraction of each component, and (c) the average molar mass and gas constant of the mixture.

**SOLUTION**   The masses of components of a gas mixture are given. The mass fractions, the mole fractions, the molar mass, and the gas constant of the mixture are to be determined.

**Analysis**   (a) The total mass of the mixture is

$$m_m = m_{O_2} + m_{N_2} + m_{CH_4} = 3 + 5 + 12 = 20 \text{ kg}$$

Then, the mass fraction of each component becomes

$$\text{mf}_{O_2} = \frac{m_{O_2}}{m_m} = \frac{3 \text{ kg}}{20 \text{ kg}} = \textbf{0.15}$$

$$\text{mf}_{N_2} = \frac{m_{N_2}}{m_m} = \frac{5 \text{ kg}}{20 \text{ kg}} = \textbf{0.25}$$

$$\text{mf}_{CH_4} = \frac{m_{CH_4}}{m_m} = \frac{12 \text{ kg}}{20 \text{ kg}} = \textbf{0.60}$$

(b) To find the mole fractions, we need to determine the mole numbers of each component first:

$$N_{O_2} = \frac{m_{O_2}}{M_{O_2}} = \frac{3 \text{ kg}}{32 \text{ kg/kmol}} = 0.094 \text{ kmol}$$

$$N_{N_2} = \frac{m_{N_2}}{M_{N_2}} = \frac{5 \text{ kg}}{28 \text{ kg/kmol}} = 0.179 \text{ kmol}$$

$$N_{CH_4} = \frac{m_{CH_4}}{M_{CH_4}} = \frac{12 \text{ kg}}{16 \text{ kg/kmol}} = 0.750 \text{ kmol}$$

Thus,

$$N_m = N_{O_2} + N_{N_2} + N_{CH_4} = 0.094 + 0.179 + 0.750 = 1.023 \text{ kmol}$$

and

$$y_{O_2} = \frac{N_{O_2}}{N_m} = \frac{0.094 \text{ kmol}}{1.023 \text{ kmol}} = \textbf{0.092}$$

$$y_{N_2} = \frac{N_{N_2}}{N_m} = \frac{0.179 \text{ kmol}}{1.023 \text{ kmol}} = \textbf{0.175}$$

$$y_{CH_4} = \frac{N_{CH_4}}{N_m} = \frac{0.750 \text{ kmol}}{1.023 \text{ kmol}} = \textbf{0.733}$$

(c) The average molar mass and gas constant of the mixture are determined from their definitions,

---

3 kg $O_2$
5 kg $N_2$
12 kg $CH_4$

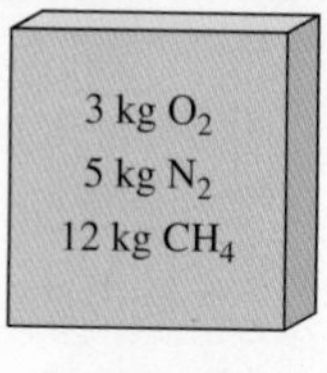

**FIGURE 13–4**
Schematic for Example 13–1.

$$M_m = \frac{m_m}{N_m} = \frac{20 \text{ kg}}{1.023 \text{ kmol}} = \mathbf{19.6 \text{ kg/kmol}}$$

or

$$M_m = \sum y_i M_i = y_{O_2} M_{O_2} + y_{N_2} M_{N_2} + y_{CH_4} M_{CH_4}$$
$$= (0.092)(32) + (0.175)(28) + (0.733)(16)$$
$$= 19.6 \text{ kg/kmol}$$

Also,

$$R_m = \frac{R_u}{M_m} = \frac{8.314 \text{ kJ/kmol·K}}{19.6 \text{ kg/kmol}} = \mathbf{0.424 \text{ kJ/kg·K}}$$

**Discussion** When mass fractions are available, the molar mass and mole fractions could also be determined directly from Eqs. 13–4 and 13–5.

# 13–2 ■ *P-v-T* BEHAVIOR OF GAS MIXTURES: IDEAL AND REAL GASES

An ideal gas is defined as a gas whose molecules are spaced far apart so that the behavior of a molecule is not influenced by the presence of other molecules—a situation encountered at low densities. We also mentioned that real gases approximate this behavior closely when they are at a low pressure or high temperature relative to their critical-point values. The *P-v-T* behavior of an ideal gas is expressed by the simple relation $Pv = RT$, which is called the *ideal-gas equation of state*. The *P-v-T* behavior of real gases is expressed by more complex equations of state or by $Pv = ZRT$, where $Z$ is the compressibility factor.

When two or more ideal gases are mixed, the behavior of a molecule normally is not influenced by the presence of other similar or dissimilar molecules, and therefore a nonreacting mixture of ideal gases also behaves as an ideal gas. Air, for example, is conveniently treated as an ideal gas in the range where nitrogen and oxygen behave as ideal gases. When a gas mixture consists of real (nonideal) gases, however, the prediction of the *P-v-T* behavior of the mixture becomes rather involved.

The prediction of the *P-v-T* behavior of gas mixtures is usually based on two models: *Dalton's law of additive pressures* and *Amagat's law of additive volumes*. Both models are described and discussed below.

*Dalton's law of additive pressures: The pressure of a gas mixture is equal to the sum of the pressures each gas would exert if it existed alone at the mixture temperature and volume (Fig. 13–5).*

*Amagat's law of additive volumes: The volume of a gas mixture is equal to the sum of the volumes each gas would occupy if it existed alone at the mixture temperature and pressure (Fig. 13–6).*

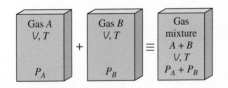

**FIGURE 13–5**
Dalton's law of additive pressures for a mixture of two ideal gases.

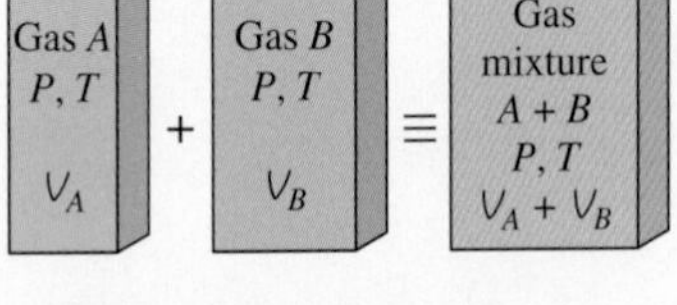

**FIGURE 13–6**
Amagat's law of additive volumes for a mixture of two ideal gases.

Dalton's and Amagat's laws hold exactly for ideal-gas mixtures, but only approximately for real-gas mixtures. This is due to intermolecular forces that may be significant for real gases at high densities. For ideal gases, these two laws are identical and give identical results.

Dalton's and Amagat's laws can be expressed as follows:

*Dalton's law:* $\qquad\qquad P_m = \sum_{i=1}^{k} P_i(T_m, V_m)$    exact for ideal gases,    **(13–6)**

                                                               approximate

*Amagat's law:* $\qquad\qquad V_m = \sum_{i=1}^{k} V_i(T_m, P_m)$    for real gases    **(13–7)**

In these relations, $P_i$ is called the **component pressure** and $V_i$ is called the **component volume** (Fig. 13–7). Note that $V_i$ is the volume a component *would* occupy if it existed alone at $T_m$ and $P_m$, not the actual volume occupied by the component in the mixture. (In a vessel that holds a gas mixture, each component fills the entire volume of the vessel. Therefore, the volume of each component is equal to the volume of the vessel.) Also, the ratio $P_i/P_m$ is called the **pressure fraction** and the ratio $V_i/V_m$ is called the **volume fraction** of component $i$.

## Ideal-Gas Mixtures

For ideal gases, $P_i$ and $V_i$ can be related to $y_i$ by using the ideal-gas relation for both the components and the gas mixture:

$$\frac{P_i(T_m, V_m)}{P_m} = \frac{N_i R_u T_m/V_m}{N_m R_u T_m/V_m} = \frac{N_i}{N_m} = y_i$$

$$\frac{V_i(T_m, P_m)}{V_m} = \frac{N_i R_u T_m/P_m}{N_m R_u T_m/P_m} = \frac{N_i}{N_m} = y_i$$

Therefore,

$$\frac{P_i}{P_m} = \frac{V_i}{V_m} = \frac{N_i}{N_m} = y_i \qquad\qquad \textbf{(13–8)}$$

Equation 13–8 is strictly valid for ideal-gas mixtures since it is derived by assuming ideal-gas behavior for the gas mixture and each of its components. The quantity $y_i P_m$ is called the **partial pressure** (identical to the *component pressure* for ideal gases), and the quantity $y_i V_m$ is called the **partial volume** (identical to the *component volume* for ideal gases). *Note that for an ideal-gas mixture, the mole fraction, the pressure fraction, and the volume fraction of a component are identical.*

The composition of an ideal-gas mixture (such as the exhaust gases leaving a combustion chamber) is frequently determined by a volumetric analysis (called the Orsat Analysis) and Eq. 13–8. A sample gas at a known volume, pressure, and temperature is passed into a vessel containing reagents that absorb one of the gases. The volume of the remaining gas is then measured at the original pressure and temperature. The ratio of the reduction in volume to the original volume (volume fraction) represents the mole fraction of that particular gas.

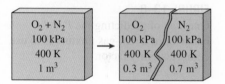

**FIGURE 13–7**

The volume a component would occupy if it existed alone at the mixture $T$ and $P$ is called the *component volume* (for ideal gases, it is equal to the partial volume $y_i V_m$).

**FIGURE 13–8**
One way of predicting the $P$-$v$-$T$ behavior of a real-gas mixture is to use compressibility factor.

# Real-Gas Mixtures

Dalton's law of additive pressures and Amagat's law of additive volumes can also be used for real gases, often with reasonable accuracy. This time, however, the component pressures or component volumes should be evaluated from relations that take into account the deviation of each component from ideal-gas behavior. One way of doing that is to use more exact equations of state (van der Waals, Beattie–Bridgeman, Benedict–Webb–Rubin, etc.) instead of the ideal-gas equation of state. Another way is to use the compressibility factor (Fig. 13–8) as

$$PV = ZNR_uT \tag{13–9}$$

The compressibility factor of the mixture $Z_m$ can be expressed in terms of the compressibility factors of the individual gases $Z_i$ by applying Eq. 13–9 to both sides of Dalton's law or Amagat's law expression and simplifying. We obtain

$$Z_m = \sum_{i=1}^{k} y_i Z_i \tag{13–10}$$

where $Z_i$ is determined either at $T_m$ and $V_m$ (Dalton's law) or at $T_m$ and $P_m$ (Amagat's law) for each individual gas. It may seem that using either law gives the same result, but it does not.

The compressibility-factor approach, in general, gives more accurate results when the $Z_i$'s in Eq. 13–10 are evaluated by using Amagat's law instead of Dalton's law. This is because Amagat's law involves the use of mixture pressure $P_m$, which accounts for the influence of intermolecular forces between the molecules of different gases. Dalton's law disregards the influence of dissimilar molecules in a mixture on each other. As a result, it tends to underpredict the pressure of a gas mixture for a given $V_m$ and $T_m$. Therefore, Dalton's law is more appropriate for gas mixtures at low pressures. Amagat's law is more appropriate at high pressures.

Note that there is a significant difference between using the compressibility factor for a single gas and for a mixture of gases. The compressibility factor predicts the $P$-$v$-$T$ behavior of single gases rather accurately, as discussed in Chapter 3, but not for mixtures of gases. When we use compressibility factors for the components of a gas mixture, we account for the influence of like molecules on each other; the influence of dissimilar molecules remains largely unaccounted for. Consequently, a property value predicted by this approach may be considerably different from the experimentally determined value.

Another approach for predicting the $P$-$v$-$T$ behavior of a gas mixture is to treat the gas mixture as a pseudopure substance (Fig. 13–9). One such method, proposed by W. B. Kay in 1936 and called **Kay's rule**, involves the use of a *pseudocritical pressure* $P'_{cr,m}$ and *pseudocritical temperature* $T'_{cr,m}$ for the mixture, defined in terms of the critical pressures and temperatures of the mixture components as

$$P'_{cr,m} = \sum_{i=1}^{k} y_i P_{cr,i} \quad \text{and} \quad T'_{cr,m} = \sum_{i=1}^{k} y_i T_{cr,i} \tag{13–11a, b}$$

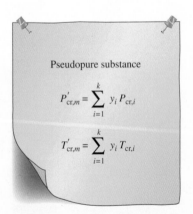

**FIGURE 13–9**
Another way of predicting the $P$-$v$-$T$ behavior of a real-gas mixture is to treat it as a pseudopure substance with critical properties $P'_{cr}$ and $T'_{cr}$.

The compressibility factor of the mixture $Z_m$ is then easily determined by using these pseudocritical properties. The result obtained by using Kay's rule is accurate to within about 10 percent over a wide range of temperatures and pressures, which is acceptable for most engineering purposes.

Another way of treating a gas mixture as a pseudopure substance is to use a more accurate equation of state such as the van der Waals, Beattie–Bridgeman, or Benedict–Webb–Rubin equation for the mixture, and to determine the constant coefficients in terms of the coefficients of the components. In the van der Waals equation, for example, the two constants for the mixture are determined from

$$a_m = \left( \sum_{i=1}^{k} y_i a_i^{1/2} \right)^2 \quad \text{and} \quad b_m = \sum_{i=1}^{k} y_i b_i \qquad \textbf{(13–12a, b)}$$

where expressions for $a_i$ and $b_i$ are given in Chapter 3.

■ **EXAMPLE 13–2**  **$P$-$v$-$T$ Behavior of Nonideal Gas Mixtures**
■
■ A rigid tank contains 2 kmol of $N_2$ and 6 kmol of $CO_2$ gases at 300 K
■ and 15 MPa (Fig. 13–10). Estimate the volume of the tank on the basis of
(a) the ideal-gas equation of state, (b) Kay's rule, (c) compressibility factors and Amagat's law, and (d) compressibility factors and Dalton's law.

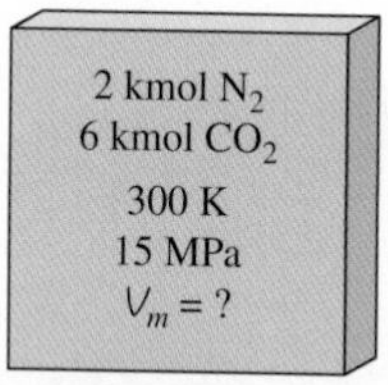

**SOLUTION**  The composition of a mixture in a rigid tank is given. The volume of the tank is to be determined using four different approaches.
*Assumptions*  Stated in each section.
*Analysis*  (a) When the mixture is assumed to behave as an ideal gas, the volume of the mixture is easily determined from the ideal-gas relation for the mixture:

$$V_m = \frac{N_m R_u T_m}{P_m} = \frac{(8 \text{ kmol})(8.314 \text{ kPa·m}^3/\text{kmol·K})(300 \text{ K})}{15,000 \text{ kPa}} = \textbf{1.330 m}^3$$

since

$$N_m = N_{N_2} + N_{CO_2} = 2 + 6 = 8 \text{ kmol}$$

(b) To use Kay's rule, we need to determine the pseudocritical temperature and pseudocritical pressure of the mixture by using the critical-point properties of $N_2$ and $CO_2$ from Table A–1. However, first we need to determine the mole fraction of each component:

$$y_{N_2} = \frac{N_{N_2}}{N_m} = \frac{2 \text{ kmol}}{8 \text{ kmol}} = 0.25 \quad \text{and} \quad y_{CO_2} = \frac{N_{CO_2}}{N_m} = \frac{6 \text{ kmol}}{8 \text{ kmol}} = 0.75$$

$$T'_{cr,m} = \sum y_i T_{cr,i} = y_{N_2} T_{cr,N_2} + y_{CO_2} T_{cr,CO_2}$$

$$= (0.25)(126.2 \text{ K}) + (0.75)(304.2 \text{ K}) = 259.7 \text{ K}$$

$$P'_{cr,m} = \sum y_i P_{cr,i} = y_{N_2} P_{cr,N_2} + y_{CO_2} P_{cr,CO_2}$$

$$= (0.25)(3.39 \text{ MPa}) + (0.75)(7.39 \text{ MPa}) = 6.39 \text{ MPa}$$

2 kmol $N_2$
6 kmol $CO_2$
300 K
15 MPa
$V_m = ?$

**FIGURE 13–10**
Schematic for Example 13–2.

Then,

$$T_R = \frac{T_m}{T'_{cr,m}} = \frac{300 \text{ K}}{259.7 \text{ K}} = 1.16 \\ P_R = \frac{P_m}{P'_{cr,m}} = \frac{15 \text{ MPa}}{6.39 \text{ MPa}} = 2.35 \Bigg\} Z_m = 0.49 \quad \text{(Fig. A–15}b\text{)}$$

Thus,

$$V_m = \frac{Z_m N_m R_u T_m}{P_m} = Z_m V_{ideal} = (0.49)(1.330 \text{ m}^3) = \mathbf{0.652 \text{ m}^3}$$

(c) When Amagat's law is used in conjunction with compressibility factors, $Z_m$ is determined from Eq. 13–10. But first we need to determine the $Z$ of each component on the basis of Amagat's law:

$$\text{N}_2\text{:} \quad T_{R,\text{N}_2} = \frac{T_m}{T_{cr,\text{N}_2}} = \frac{300 \text{ K}}{126.2 \text{ K}} = 2.38 \\ P_{R,\text{N}_2} = \frac{P_m}{P_{cr,\text{N}_2}} = \frac{15 \text{ MPa}}{3.39 \text{ MPa}} = 4.42 \Bigg\} Z_{\text{N}_2} = 1.02 \quad \text{(Fig. A–15}b\text{)}$$

$$\text{CO}_2\text{:} \quad T_{R,\text{CO}_2} = \frac{T_m}{T_{cr,\text{CO}_2}} = \frac{300 \text{ K}}{304.2 \text{ K}} = 0.99 \\ P_{R,\text{CO}_2} = \frac{P_m}{P_{cr,\text{CO}_2}} = \frac{15 \text{ MPa}}{7.39 \text{ MPa}} = 2.03 \Bigg\} Z_{\text{CO}_2} = 0.30 \quad \text{(Fig. A–15}b\text{)}$$

Mixture: $\qquad Z_m = \sum y_i Z_i = y_{\text{N}_2} Z_{\text{N}_2} + y_{\text{CO}_2} Z_{\text{CO}_2}$

$$= (0.25)(1.02) + (0.75)(0.30) = 0.48$$

Thus,

$$V_m = \frac{Z_m N_m R_u T_m}{P_m} = Z_m V_{ideal} = (0.48)(1.330 \text{ m}^3) = \mathbf{0.638 \text{ m}^3}$$

The compressibility factor in this case turned out to be almost the same as the one determined by using Kay's rule.

(d) When Dalton's law is used in conjunction with compressibility factors, $Z_m$ is again determined from Eq. 13–10. However, this time the $Z$ of each component is to be determined at the mixture temperature and volume, which is not known. Therefore, an iterative solution is required. We start the calculations by assuming that the volume of the gas mixture is 1.330 m³, the value determined by assuming ideal-gas behavior.

The $T_R$ values in this case are identical to those obtained in part (c) and remain constant. The pseudoreduced volume is determined from its definition in Chap. 3:

$$v_{R,\text{N}_2} = \frac{\bar{v}_{\text{N}_2}}{R_u T_{cr,\text{N}_2}/P_{cr,\text{N}_2}} = \frac{V_m/N_{\text{N}_2}}{R_u T_{cr,\text{N}_2}/P_{cr,\text{N}_2}}$$

$$= \frac{(1.33 \text{ m}^3)/(2 \text{ kmol})}{(8.314 \text{ kPa}\cdot\text{m}^3/\text{kmol}\cdot\text{K})(126.2 \text{ K})/(3390 \text{ kPa})} = 2.15$$

Similarly,

$$v_{R,CO_2} = \frac{(1.33 \text{ m}^3)/(6 \text{ kmol})}{(8.314 \text{ kPa·m}^3/\text{kmol·K})(304.2 \text{ K})/(7390 \text{ kPa})} = 0.648$$

From Fig. A–15, we read $Z_{N_2} = 0.99$ and $Z_{CO_2} = 0.56$. Thus,

$$Z_m = y_{N_2}Z_{N_2} + y_{CO_2}Z_{CO_2} = (0.25)(0.99) + (0.75)(0.56) = 0.67$$

and

$$V_m = \frac{Z_m N_m R T_m}{P_m} = Z_m V_{ideal} = (0.67)(1.330 \text{ m}^3) = 0.891 \text{ m}^3$$

This is 33 percent lower than the assumed value. Therefore, we should repeat the calculations, using the new value of $V_m$. When the calculations are repeated we obtain 0.738 m³ after the second iteration, 0.678 m³ after the third iteration, and 0.648 m³ after the fourth iteration. This value does not change with more iterations. Therefore,

$$V_m = \textbf{0.648 m}^3$$

**Discussion** Notice that the results obtained in parts (b), (c), and (d) are very close. But they are very different from the ideal-gas values. Therefore, treating a mixture of gases as an ideal gas may yield unacceptable errors at high pressures.

# 13–3 · PROPERTIES OF GAS MIXTURES: IDEAL AND REAL GASES

Consider a gas mixture that consists of 2 kg of $N_2$ and 3 kg of $CO_2$. The total mass (an *extensive property*) of this mixture is 5 kg. How did we do it? Well, we simply added the mass of each component. This example suggests a simple way of evaluating the **extensive properties** of a nonreacting ideal- or real-gas mixture: *Just add the contributions of each component of the mixture* (Fig. 13–11). Then the total internal energy, enthalpy, and entropy of a gas mixture can be expressed, respectively, as

$$U_m = \sum_{i=1}^{k} U_i = \sum_{i=1}^{k} m_i u_i = \sum_{i=1}^{k} N_i \bar{u}_i \quad \text{(kJ)} \tag{13–13}$$

$$H_m = \sum_{i=1}^{k} H_i = \sum_{i=1}^{k} m_i h_i = \sum_{i=1}^{k} N_i \bar{h}_i \quad \text{(kJ)} \tag{13–14}$$

$$S_m = \sum_{i=1}^{k} S_i = \sum_{i=1}^{k} m_i s_i = \sum_{i=1}^{k} N_i \bar{s}_i \quad \text{(kJ/K)} \tag{13–15}$$

By following a similar logic, the changes in internal energy, enthalpy, and entropy of a gas mixture during a process can be expressed, respectively, as

$$\Delta U_m = \sum_{i=1}^{k} \Delta U_i = \sum_{i=1}^{k} m_i \, \Delta u_i = \sum_{i=1}^{k} N_i \, \Delta \bar{u}_i \quad \text{(kJ)} \tag{13–16}$$

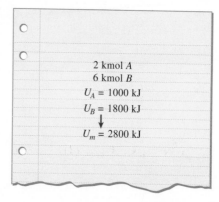

2 kmol A
6 kmol B
$U_A = 1000$ kJ
$U_B = 1800$ kJ
$U_m = 2800$ kJ

**FIGURE 13–11**
The extensive properties of a mixture are determined by simply adding the properties of the components.

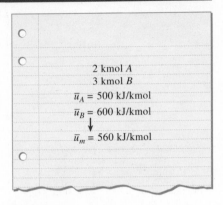

**FIGURE 13–12**
The intensive properties of a mixture are determined by weighted averaging.

$$\Delta H_m = \sum_{i=1}^{k} \Delta H_i = \sum_{i=1}^{k} m_i \, \Delta h_i = \sum_{i=1}^{k} N_i \, \Delta \bar{h}_i \quad \text{(kJ)} \tag{13–17}$$

$$\Delta S_m = \sum_{i=1}^{k} \Delta S_i = \sum_{i=1}^{k} m_i \, \Delta s_i = \sum_{i=1}^{k} N_i \, \Delta \bar{s}_i \quad \text{(kJ/K)} \tag{13–18}$$

Now reconsider the same mixture, and assume that both $N_2$ and $CO_2$ are at 25°C. The temperature (an *intensive* property) of the mixture is, as you would expect, also 25°C. Notice that we did not add the component temperatures to determine the mixture temperature. Instead, we used some kind of averaging scheme, a characteristic approach for determining the **intensive properties** of a mixture. The internal energy, enthalpy, and entropy of a mixture *per unit mass* or *per unit mole* of the mixture can be determined by dividing the equations above by the mass or the mole number of the mixture ($m_m$ or $N_m$). We obtain (Fig. 13–12)

$$u_m = \sum_{i=1}^{k} \text{mf}_i \, u_i \quad \text{(kJ/kg)} \quad \text{and} \quad \bar{u}_m = \sum_{i=1}^{k} y_i \bar{u}_i \ \text{(kJ/kmol)} \tag{13–19}$$

$$h_m = \sum_{i=1}^{k} \text{mf}_i \, h_i \quad \text{(kJ/kg)} \quad \text{and} \quad \bar{h}_m = \sum_{i=1}^{k} y_i \bar{h}_i \ \text{(kJ/kmol)} \tag{13–20}$$

$$s_m = \sum_{i=1}^{k} \text{mf}_i \, s_i \quad \text{(kJ/kg·K)} \quad \text{and} \quad \bar{s}_m = \sum_{i=1}^{k} y_i \bar{s}_i \ \text{(kJ/kmol·K)} \tag{13–21}$$

Similarly, the specific heats of a gas mixture can be expressed as

$$c_{v,m} = \sum_{i=1}^{k} \text{mf}_i \, c_{v,i} \quad \text{(kJ/kg·K)} \quad \text{and} \quad \bar{c}_{v,m} = \sum_{i=1}^{k} y_i \bar{c}_{v,i} \ \text{(kJ/kmol·K)} \tag{13–22}$$

$$c_{p,m} = \sum_{i=1}^{k} \text{mf}_i \, c_{p,i} \quad \text{(kJ/kg·K)} \quad \text{and} \quad \bar{c}_{p,m} = \sum_{i=1}^{k} y_i \bar{c}_{p,i} \ \text{(kJ/kmol·K)} \tag{13–23}$$

Notice that *properties per unit mass involve mass fractions* ($\text{mf}_i$) *and properties per unit mole involve mole fractions* ($y_i$).

The relations given above are exact for ideal-gas mixtures, and approximate for real-gas mixtures. (In fact, they are also applicable to nonreacting liquid and solid solutions especially when they form an "ideal solution.") The only major difficulty associated with these relations is the determination of properties for each individual gas in the mixture. The analysis can be simplified greatly, however, by treating the individual gases as ideal gases, if doing so does not introduce a significant error.

## Ideal-Gas Mixtures

The gases that comprise a mixture are often at a high temperature and low pressure relative to the critical-point values of individual gases. In such cases, the gas mixture and its components can be treated as ideal gases with negligible error. Under the ideal-gas approximation, the properties of a gas are not influenced by the presence of other gases, and each gas component in the mixture behaves as if it exists alone at the mixture temperature $T_m$ and mixture volume $V_m$. This principle is known as the **Gibbs–Dalton law**, which is an extension of Dalton's law of additive pressures. Also, the $h$, $u$, $c_v$, and $c_p$ of an ideal gas depend on temperature only and are independent of the pressure or the volume of the ideal-gas mixture. The partial pressure of a component in an ideal-gas mixture is simply $P_i = y_i P_m$, where $P_m$ is the mixture pressure.

Evaluation of $\Delta u$ or $\Delta h$ of the components of an ideal-gas mixture during a process is relatively easy since it requires only a knowledge of the initial and final temperatures. Care should be exercised, however, in evaluating the $\Delta s$ of the components since the entropy of an ideal gas depends on the pressure or volume of the component as well as on its temperature. The entropy change of individual gases in an ideal-gas mixture during a process can be determined from

$$\Delta s_i = s_{i,2}^\circ - s_{i,1}^\circ - R_i \ln \frac{P_{i,2}}{P_{i,1}} \cong c_{p,i} \ln \frac{T_{i,2}}{T_{i,1}} - R_i \ln \frac{P_{i,2}}{P_{i,1}} \qquad (13\text{-}24)$$

or

$$\Delta \bar{s}_i = \bar{s}_{i,2}^\circ - \bar{s}_{i,1}^\circ - R_u \ln \frac{P_{i,2}}{P_{i,1}} \cong \bar{c}_{p,i} \ln \frac{T_{i,2}}{T_{i,1}} - R_u \ln \frac{P_{i,2}}{P_{i,1}} \qquad (13\text{-}25)$$

where $P_{i,2} = y_{i,2} P_{m,2}$ and $P_{i,1} = y_{i,1} P_{m,1}$. Notice that the partial pressure $P_i$ of each component is used in the evaluation of the entropy change, not the mixture pressure $P_m$ (Fig. 13–13).

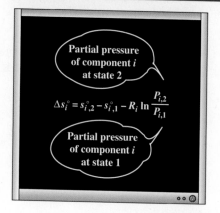

$$\Delta s_i^\circ = s_{i,2}^\circ - s_{i,1}^\circ - R_i \ln \frac{P_{i,2}}{P_{i,1}}$$

**FIGURE 13–13**
Partial pressures (not the mixture pressure) are used in the evaluation of entropy changes of ideal-gas mixtures.

---

**EXAMPLE 13–3    Expansion of an Ideal Gas Mixture in a Turbine**

A mixture of oxygen ($O_2$), carbon dioxide ($CO_2$), and helium (He) gases with mass fractions of 0.0625, 0.625, and 0.3125, respectively, enter an adiabatic turbine at 1000 kPa and 600 K steadily and expand to 100 kPa pressure (Fig. 13–14). The isentropic efficiency of the turbine is 90 percent. For gas components assuming constant specific heats at room temperature, determine (a) the work output her unit mass of mixture and (b) the exergy destruction and the second-law efficiency of the turbine. Take the environment temperature to be $T_0 = 25°C$.

**SOLUTION** The mass fractions of the components of a gas mixture that expands in an adiabatic turbine are given. The work output, the exergy destruction, and the second-law efficiency are to be determined.

**Assumptions** All gases will be modeled as ideal gases with constant specific heats.

**Analysis** (a) The mass fractions of mixture components are given to be $mf_{O_2} = 0.0625$, $mf_{CO_2} = 0.625$, and $mf_{He} = 0.3125$. The specific heats of these gases at room temperature are (Table A–2a):

| | $c_v$, kJ/kg·K | $c_p$, kJ/kg·K |
|---|---|---|
| $O_2$: | 0.658 | 0.918 |
| $CO_2$: | 0.657 | 0.846 |
| He: | 3.1156 | 5.1926 |

Then, the constant-pressure and constant-volume specific heats of the mixture become

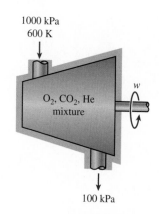

1000 kPa
600 K

$O_2$, $CO_2$, He
mixture

$w$

100 kPa

**FIGURE 13–14**
Schematic for Example 13–3.

$$c_p = mf_{O_2}c_{p,O_2} + mf_{CO_2}c_{p,CO_2} + mf_{He}c_{p,He}$$
$$= 0.0625 \times 0.918 + 0.625 \times 0.846 + 0.3125 \times 5.1926$$
$$= 2.209 \text{ kJ/kg·K}$$

$$c_v = mf_{O_2}c_{v,O_2} + mf_{CO_2}c_{v,CO_2} + mf_{He}c_{v,He}$$
$$= 0.0625 \times 0.658 + 0.625 \times 0.657 + 0.3125 \times 3.1156$$
$$= 1.425 \text{ kJ/kg·K}$$

The apparent gas constant of the mixture and the specific heat ratio are

$$R = c_p - c_v = 2.209 - 1.425 = 0.7836 \text{ kJ/kg·K}$$

$$k = \frac{c_p}{c_v} = \frac{2.209 \text{ kJ/kg·K}}{1.425 \text{ kJ/kg·K}} = 1.550$$

The temperature at the end of the expansion for the isentropic process is

$$T_{2s} = T_1 \left(\frac{P_2}{P_1}\right)^{(k-1)/k} = (600 \text{ K})\left(\frac{100 \text{ kPa}}{1000 \text{ kPa}}\right)^{0.55/1.55} = 265.0 \text{ K}$$

Using the definition of turbine isentropic efficiency, the actual outlet temperature is

$$T_2 = T_1 - \eta_{turb}(T_1 - T_{2s}) = (600 \text{ K}) - (0.90)(600 - 265) \text{ K} = 298.5 \text{ K}$$

Noting that the turbine is adiabatic and thus there is no heat transfer, the actual work output is determined to be

$$w_{out} = h_1 - h_2 = c_p(T_1 - T_2) = (2.209 \text{ kJ/kg·K})(600 - 298.5)$$
$$= \mathbf{666.0 \text{ kJ/kg}}$$

(b) The entropy change of the gas mixture and the exergy destruction in the turbine are

$$s_2 - s_1 = c_p \ln \frac{T_2}{T_1} - R \ln \frac{P_2}{P_1} = (2.209 \text{ kJ/kg·K}) \ln \frac{298.5 \text{ K}}{600 \text{ K}}$$

$$- (0.7836 \text{ kJ/kg·K}) \ln \frac{100 \text{ kPa}}{1000 \text{ kPa}} = 0.2658 \text{ kJ/kg·K}$$

$$x_{dest} = T_0 s_{gen} = T_0(s_2 - s_1) = (298 \text{ K})(0.2658 \text{ kJ/kg·K}) = \mathbf{79.2 \text{ kJ/kg}}$$

The expended exergy is the sum of the work output of turbine (exergy recovered) and the exergy destruction (exergy wasted),

$$x_{expended} = x_{recovered} + x_{dest} = w_{out} + x_{dest} = 666.0 + 79.2 = 745.2 \text{ kJ/kg}$$

The second-law efficiency is the ratio of the recovered to expended exergy,

$$\eta_{II} = \frac{x_{recovered}}{x_{expended}} = \frac{w_{out}}{x_{expended}} = \frac{666.0 \text{ kJ/kg}}{745.2 \text{ kJ/kg}} = \mathbf{0.894} \text{ or } \mathbf{89.4 \text{ percent}}$$

**Discussion** The second-law efficiency is a measure of thermodynamic perfection. A process that generates no entropy and thus destroys no exergy always has a second-law efficiency of 100 percent.

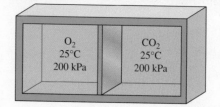

**FIGURE 13–15**
Schematic for Example 13–4.

■ **EXAMPLE 13–4**    **Exergy Destruction during Mixing of Ideal Gases**

An insulated rigid tank is divided into two compartments by a partition, as shown in Fig. 13–15. One compartment contains 3 kmol of $O_2$, and the other compartment contains 5 kmol of $CO_2$. Both gases are initially at 25°C and 200 kPa. Now the partition is removed, and the two gases are allowed to mix. Assuming the surroundings are at 25°C and both gases behave as ideal gases, determine the entropy change and exergy destruction associated with this process.

**SOLUTION**    A rigid tank contains two gases separated by a partition. The entropy change and exergy destroyed after the partition is removed are to be determined.

**Assumptions**    Both gases and their mixture are ideal gases.

**Analysis**    We take the entire contents of the tank (both compartments) as the system. This is a *closed system* since no mass crosses the boundary during the process. We note that the volume of a rigid tank is constant, and there is no energy transfer as heat or work. Also, both gases are initially at the same temperature and pressure.

When two ideal gases initially at the same temperature and pressure are mixed by removing a partition between them, the mixture will also be at the same temperature and pressure. (Can you prove it? Will this be true for nonideal gases?) Therefore, the temperature and pressure in the tank will still be 25°C and 200 kPa, respectively, after the mixing. The entropy change of each component gas can be determined from Eqs. 13–18 and 13–25:

$$\Delta S_m = \sum \Delta S_i = \sum N_i \, \Delta \bar{s}_i = \sum N_i \left( \bar{c}_{p,i} \ln \frac{T_{i,2}}{T_{i,1}}^{\,0} - R_u \ln \frac{P_{i,2}}{P_{i,1}} \right)$$

$$= -R_u \sum N_i \ln \frac{y_{i,2} P_{m,2}}{P_{i,1}} = -R_u \sum N_i \ln y_{i,2}$$

since $P_{m,2} = P_{i,1} = 200$ kPa. It is obvious that the entropy change is independent of the composition of the mixture in this case and depends on only the mole fraction of the gases in the mixture. What is not so obvious is that if the same gas in two different chambers is mixed at constant temperature and pressure, the entropy change is zero.

Substituting the known values, the entropy change becomes

$$N_m = N_{O_2} + N_{CO_2} = (3 + 5) \text{ kmol} = 8 \text{ kmol}$$

$$y_{O_2} = \frac{N_{O_2}}{N_m} = \frac{3 \text{ kmol}}{8 \text{ kmol}} = 0.375$$

$$y_{CO_2} = \frac{N_{CO_2}}{N_m} = \frac{5 \text{ kmol}}{8 \text{ kmol}} = 0.625$$

$$\Delta S_m = -R_u(N_{O_2} \ln y_{O_2} + N_{CO_2} \ln y_{CO_2})$$

$$= -(8.314 \text{ kJ/kmol·K})[(3 \text{ kmol})(\ln 0.375) + (5 \text{ kmol})(\ln 0.625)]$$

$$= \textbf{44.0 kJ/K}$$

The exergy destruction associated with this mixing process is determined from

$$X_{\text{destroyed}} = T_0 S_{\text{gen}} = T_0 \Delta S_{\text{sys}}$$
$$= (298 \text{ K})(44.0 \text{ kJ/K})$$
$$= \textbf{13.1 MJ}$$

*Discussion* This large value of exergy destruction shows that mixing processes are highly irreversible.

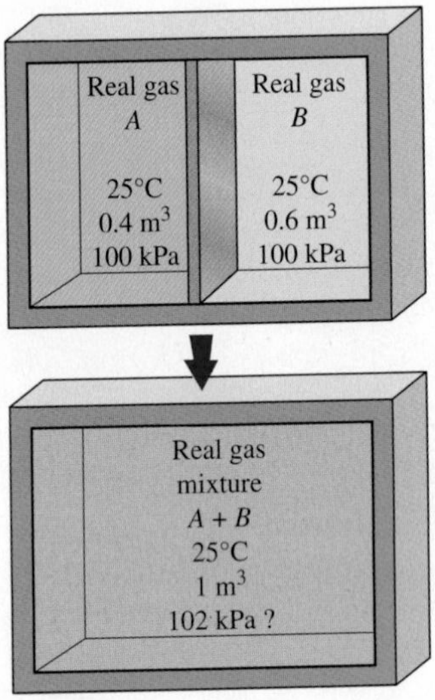

**FIGURE 13–16**
It is difficult to predict the behavior of nonideal-gas mixtures because of the influence of dissimilar molecules on each other.

## Real-Gas Mixtures

When the components of a gas mixture do not behave as ideal gases, the analysis becomes more complex because the properties of real (nonideal) gases such as $u$, $h$, $c_v$, and $c_p$ depend on the pressure (or specific volume) as well as on the temperature. In such cases, the effects of deviation from ideal-gas behavior on the mixture properties should be accounted for.

Consider two nonideal gases contained in two separate compartments of an adiabatic rigid tank at 100 kPa and 25°C. The partition separating the two gases is removed, and the two gases are allowed to mix. What do you think the final pressure in the tank will be? You are probably tempted to say 100 kPa, which would be true for ideal gases. However, this is not true for nonideal gases because of the influence of the molecules of different gases on each other (deviation from Dalton's law, Fig. 13–16).

When real-gas mixtures are involved, it may be necessary to account for the effect of nonideal behavior on the mixture properties such as enthalpy and entropy. One way of doing that is to use compressibility factors in conjunction with generalized equations and charts developed in Chapter 12 for real gases.

Consider the following $T \, ds$ relation for a gas mixture:

$$dh_m = T_m \, ds_m + v_m \, dP_m$$

It can also be expressed as

$$d\left(\sum \text{mf}_i h_i\right) = T_m \, d\left(\sum \text{mf}_i s_i\right) + \left(\sum \text{mf}_i v_i\right) dP_m$$

or

$$\sum \text{mf}_i(dh_i - T_m \, ds_i - v_i \, dP_m) = 0$$

which yields

$$dh_i = T_m \, ds_i + v_i \, dP_m \qquad (13\text{–}26)$$

This is an important result because Eq. 13–26 is the starting equation in the development of the generalized relations and charts for enthalpy and entropy. It suggests that the generalized property relations and charts for real gases developed in Chapter 12 can also be used for the components of real-gas mixtures. But the reduced temperature $T_R$ and reduced pressure $P_R$

for each component should be evaluated by using the mixture temperature $T_m$ and mixture pressure $P_m$. This is because Eq. 13–26 involves the mixture pressure $P_m$, not the component pressure $P_i$.

The approach described above is somewhat analogous to Amagat's law of additive volumes (evaluating mixture properties at the mixture pressure and temperature), which holds exactly for ideal-gas mixtures and approximately for real-gas mixtures. Therefore, the mixture properties determined with this approach are not exact, but they are sufficiently accurate.

What if the mixture volume and temperature are specified instead of the mixture pressure and temperature? Well, there is no need to panic. Just evaluate the mixture pressure, using Dalton's law of additive pressures, and then use this value (which is only approximate) as the mixture pressure.

Another way of evaluating the properties of a real-gas mixture is to treat the mixture as a pseudopure substance having pseudocritical properties, determined in terms of the critical properties of the component gases by using Kay's rule. The approach is quite simple, and the accuracy is usually acceptable.

---

■ **EXAMPLE 13–5**   **Cooling of a Nonideal Gas Mixture**

Air is a mixture of $N_2$, $O_2$, and small amounts of other gases, and it can be approximated as 79 percent $N_2$ and 21 percent $O_2$ on mole basis. During a steady-flow process, air is cooled from 220 to 160 K at a constant pressure of 10 MPa (Fig. 13–17). Determine the heat transfer during this process per kmol of air, using (a) the ideal-gas approximation, (b) Kay's rule, and (c) Amagat's law.

**SOLUTION**   Air at a low temperature and high pressure is cooled at constant pressure. The heat transfer is to be determined using three different approaches.

**Assumptions**   **1** This is a steady-flow process since there is no change with time at any point and thus $\Delta m_{CV} = 0$ and $\Delta E_{CV} = 0$. **2** The kinetic and potential energy changes are negligible.

**Analysis**   We take the *cooling section* as the system. This is a *control volume* since mass crosses the system boundary during the process. We note that heat is transferred out of the system.

The critical properties are $T_{cr} = 126.2$ K and $P_{cr} = 3.39$ MPa for $N_2$ and $T_{cr} = 154.8$ K and $P_{cr} = 5.08$ MPa for $O_2$. Both gases remain above their critical temperatures, but they are also above their critical pressures. Therefore, air will probably deviate from ideal-gas behavior, and thus it should be treated as a real-gas mixture.

The energy balance for this steady-flow system can be expressed on a unit mole basis as

$$e_{in} - e_{out} = \Delta e_{system}^{\,0} = 0 \rightarrow e_{in} = e_{out} \rightarrow \bar{h}_1 = \bar{h}_2 + \bar{q}_{out}$$

$$\bar{q}_{out} = \bar{h}_1 - \bar{h}_2 = y_{N_2}(\bar{h}_1 - \bar{h}_2)_{N_2} + y_{O_2}(\bar{h}_1 - \bar{h}_2)_{O_2}$$

where the enthalpy change for either component can be determined from the generalized enthalpy departure chart (Fig. A–29) and Eq. 12–58:

$$\bar{h}_1 - \bar{h}_2 = \bar{h}_{1,ideal} - \bar{h}_{2,ideal} - R_u T_{cr}(Z_{h_1} - Z_{h_2})$$

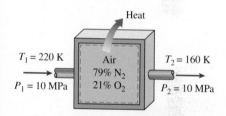

$T_1 = 220$ K

$P_1 = 10$ MPa

Air
79% $N_2$
21% $O_2$

Heat

$T_2 = 160$ K

$P_2 = 10$ MPa

**FIGURE 13–17**
Schematic for Example 13–5.

The first two terms on the right-hand side of this equation represent the ideal-gas enthalpy change of the component. The terms in parentheses represent the deviation from the ideal-gas behavior, and their evaluation requires a knowledge of reduced pressure $P_R$ and reduced temperature $T_R$, which are calculated at the mixture temperature $T_m$ and mixture pressure $P_m$.

(a) If the $N_2$ and $O_2$ mixture is assumed to behave as an ideal gas, the enthalpy of the mixture will depend on temperature only, and the enthalpy values at the initial and the final temperatures can be determined from the ideal-gas tables of $N_2$ and $O_2$ (Tables A–18 and A–19):

$$T_1 = 220 \text{ K} \rightarrow \bar{h}_{1,\text{ideal},N_2} = 6391 \text{ kJ/kmol}$$

$$\bar{h}_{1,\text{ideal},O_2} = 6404 \text{ kJ/kmol}$$

$$T_2 = 160 \text{ K} \rightarrow \bar{h}_{2,\text{ideal},N_2} = 4648 \text{ kJ/kmol}$$

$$\bar{h}_{2,\text{ideal},O_2} = 4657 \text{ kJ/kmol}$$

$$\bar{q}_{\text{out}} = y_{N_2}(\bar{h}_1 - \bar{h}_2)_{N_2} + y_{O_2}(\bar{h}_1 - \bar{h}_2)_{O_2}$$

$$= (0.79)(6391 - 4648) \text{ kJ/kmol} + (0.21)(6404 - 4657) \text{ kJ/kmol}$$

$$= \textbf{1744 kJ/kmol}$$

(b) Kay's rule is based on treating a gas mixture as a pseudopure substance whose critical temperature and pressure are

$$T'_{cr,m} = \sum y_i T_{cr,i} = y_{N_2} T_{cr,N_2} + y_{O_2} T_{cr,O_2}$$

$$= (0.79)(126.2 \text{ K}) + (0.21)(154.8 \text{ K}) = 132.2 \text{ K}$$

and

$$P'_{cr,m} = \sum y_i P_{cr,i} = y_{N_2} P_{cr,N_2} + y_{O_2} P_{cr,O_2}$$

$$= (0.79)(3.39 \text{ MPa}) + (0.21)(5.08 \text{ MPa}) = 3.74 \text{ MPa}$$

Then,

$$\left.\begin{array}{r} T_{R,1} = \dfrac{T_{m,1}}{T_{cr,m}} = \dfrac{220 \text{ K}}{132.2 \text{ K}} = 1.66 \\[4mm] P_R = \dfrac{P_m}{P_{cr,m}} = \dfrac{10 \text{ MPa}}{3.74 \text{ MPa}} = 2.67 \end{array}\right\} Z_{h_1,m} = 1.0$$

$$\left.\begin{array}{r} P_R = \dfrac{P_m}{P_{cr,m}} = \dfrac{10 \text{ MPa}}{3.74 \text{ MPa}} = 2.67 \\[4mm] T_{R,2} = \dfrac{T_{m,2}}{T_{cr,m}} = \dfrac{160 \text{ K}}{132.2 \text{ K}} = 1.21 \end{array}\right\} Z_{h_2,m} = 2.6$$

Also,

$$\bar{h}_{m_1,\text{ideal}} = y_{N_2}\bar{h}_{1,\text{ideal},N_2} + y_{O_2}\bar{h}_{1,\text{ideal},O_2}$$

$$= (0.79)(6391 \text{ kJ/kmol}) + (0.21)(6404 \text{ kJ/kmol})$$

$$= 6394 \text{ kJ/kmol}$$

$$\bar{h}_{m_2,\text{ideal}} = y_{N_2}\bar{h}_{2,\text{ideal},N_2} + y_{O_2}\bar{h}_{2,\text{ideal},O_2}$$

$$= (0.79)(4648 \text{ kJ/kmol}) + (0.21)(4657 \text{ kJ/kmol})$$

$$= 4650 \text{ kJ/kmol}$$

Therefore,

$$\bar{q}_{\text{out}} = (\bar{h}_{m_1,\text{ideal}} - \bar{h}_{m_2,\text{ideal}}) - R_u T_{\text{cr}}(Z_{h_1} - Z_{h_2})_m$$

$$= [(6394 - 4650) \text{ kJ/kmol}] - (8.314 \text{ kJ/kmol·K})(132.2 \text{ K})(1.0 - 2.6)$$

$$= \textbf{3503 kJ/kmol}$$

(c) The reduced temperatures and pressures for both $N_2$ and $O_2$ at the initial and final states and the corresponding enthalpy departure factors are, from Fig. A–29,

$$N_2: \qquad T_{R_1,N_2} = \frac{T_{m,1}}{T_{\text{cr},N_2}} = \frac{220 \text{ K}}{126.2 \text{ K}} = 1.74 \quad \Big\} \quad Z_{h_1,N_2} = 0.9$$

$$P_{R,N_2} = \frac{P_m}{P_{\text{cr},N_2}} = \frac{10 \text{ MPa}}{3.39 \text{ MPa}} = 2.95 \quad \Big\}$$

$$T_{R_2,N_2} = \frac{T_{m,2}}{T_{\text{cr},N_2}} = \frac{160 \text{ K}}{126.2 \text{ K}} = 1.27 \quad \Big\} \quad Z_{h_2,N_2} = 2.4$$

$$O_2: \qquad T_{R_1,O_2} = \frac{T_{m,1}}{T_{\text{cr},O_2}} = \frac{220 \text{ K}}{154.8 \text{ K}} = 1.42 \quad \Big\} \quad Z_{h_1,O_2} = 1.3$$

$$P_{R,O_2} = \frac{P_m}{P_{\text{cr},O_2}} = \frac{10 \text{ MPa}}{5.08 \text{ MPa}} = 1.97 \quad \Big\}$$

$$T_{R_1,O_2} = \frac{T_{m,2}}{T_{\text{cr},O_2}} = \frac{160 \text{ K}}{154.8 \text{ K}} = 1.03 \quad \Big\} \quad Z_{h_2,O_2} = 4.0$$

From Eq. 12–58,

$$(\bar{h}_1 - \bar{h}_2)_{N_2} = (\bar{h}_{1,\text{ideal}} - \bar{h}_{2,\text{ideal}})_{N_2} - R_u T_{\text{cr}}(Z_{h_1} - Z_{h_2})_{N_2}$$

$$= [(6391 - 4648) \text{ kJ/kmol}] - (8.314 \text{ kJ/kmol·K})(126.2 \text{ K})(0.9 - 2.4)$$

$$= 3317 \text{ kJ/kmol}$$

$$(\bar{h}_1 - \bar{h}_2)_{O_2} = (\bar{h}_{1,\text{ideal}} - \bar{h}_{2,\text{ideal}})_{O_2} - R_u T_{\text{cr}}(Z_{h_1} - Z_{h_2})_{O_2}$$

$$= [(6404 - 4657) \text{ kJ/kmol}] - (8.314 \text{ kJ/kmol·K})(154.8 \text{ K})(1.3 - 4.0)$$

$$= 5222 \text{ kJ/kmol}$$

Therefore,

$$\bar{q}_{\text{out}} = y_{N_2}(\bar{h}_1 - \bar{h}_2)_{N_2} + y_{O_2}(\bar{h}_1 - \bar{h}_2)_{O_2}$$

$$= (0.79)(3317 \text{ kJ/kmol}) + (0.21)(5222 \text{ kJ/kmol})$$

$$= \textbf{3717 kJ/kmol}$$

**Discussion** This result is about 6 percent greater than the result obtained in part (b) by using Kay's rule. But it is more than twice the result obtained by assuming the mixture to be an ideal gas.

When two gases or two miscible liquids are brought into contact, they mix and form a homogeneous mixture or solution without requiring any work input. That is, the natural tendency of miscible substances brought into contact is to mix with each other. As such, these are irreversible processes, and thus it is impossible for the reverse process of separation to occur spontaneously. For example, pure nitrogen and oxygen gases readily mix when brought into contact, but a mixture of nitrogen and oxygen (such as air) never separates into pure nitrogen and oxygen when left unattended.

Mixing and separation processes are commonly used in practice. Separation processes require a work (or, more generally, exergy) input, and minimizing this required work input is an important part of the design process of separation plants. The presence of dissimilar molecules in a mixture affect each other, and therefore the influence of composition on the properties must be taken into consideration in any thermodynamic analysis. In this section we analyze the general mixing processes, with particular emphasis on ideal solutions, and determine the entropy generation and exergy destruction. We then consider the reverse process of separation, and determine the minimum (or reversible) work input needed for separation.

The *specific Gibbs function (or Gibbs free energy) g* is defined as the combination property $g = h - Ts$. Using the relation $dh = v \, dP + T \, ds$, the differential change of the Gibbs function of a pure substance is obtained by differentiation to be

$$dg = v \, dP - s \, dT \quad \text{or} \quad dG = V \, dP - S \, dT \quad \text{(pure substance)} \quad \textbf{(13–27)}$$

For a mixture, the total Gibbs function is a function of two independent intensive properties as well as the composition, and thus it can be expressed as $G = G(P, T, N_1, N_2, \ldots, N_i)$. Its differential is

$$dG = \left(\frac{\partial G}{\partial P}\right)_{T,N} dP + \left(\frac{\partial G}{\partial T}\right)_{P,N} dT + \sum_i \left(\frac{\partial G}{\partial N_i}\right)_{P,T,N_j} dN_i \quad \text{(mixture)} \quad \textbf{(13–28)}$$

where the subscript $N_j$ indicates that the mole numbers of all components in the mixture other than component $i$ are to be held constant during differentiation. For a pure substance, the last term drops out since the composition is fixed, and the equation above must reduce to the one for a pure substance. Comparing Eqs. 13–27 and 13–28 gives

$$dG = V \, dP - S \, dT + \sum_i \mu_i \, dN_i \quad \text{or} \quad d\bar{g} = \bar{v} \, dP - \bar{s} \, dT + \sum_i m_i \, dy_i \quad \textbf{(13–29)}$$

where $y_i = N_i/N_m$ is the mole fraction of component $i$ ($N_m$ is the total number of moles of the mixture) and

$$\mu_i = \left(\frac{\partial G}{\partial N_i}\right)_{P,T,N_j} = \bar{g}_i = \bar{h}_i - T\bar{s}_i \quad \text{(for component } i \text{ of a mixture)} \quad \textbf{(13–30)}$$

---

\*This section can be skipped without a loss in continuity.

is the **chemical potential,** which is *the change in the Gibbs function of the mixture in a specified phase when a unit amount of component i in the same phase is added as pressure, temperature, and the amounts of all other components are held constant.* The symbol tilde (as in $\tilde{v}$, $\tilde{h}$, and $\tilde{s}$) is used to denote the **partial molar properties** of the components. Note that the summation term in Eq. 13–29 is zero for a single component system and thus the chemical potential of a pure system in a given phase is equivalent to the molar Gibbs function (Fig. 13–18) since $G = Ng = N\mu$, where

$$\mu = \left(\frac{\partial G}{\partial N}\right)_{P,T} = \bar{g} = \bar{h} - T\bar{s} \quad \text{(pure substance)} \quad \text{(13–31)}$$

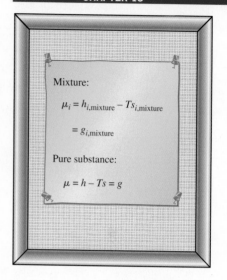

Therefore, the difference between the chemical potential and the Gibbs function is due to the effect of dissimilar molecules in a mixture on each other. It is because of this molecular effect that the volume of the mixture of two miscible liquids may be more or less than the sum of the initial volumes of the individual liquids. Likewise, the total enthalpy of the mixture of two components at the same pressure and temperature, in general, is not equal to the sum of the total enthalpies of the individual components before mixing, the difference being the enthalpy (or heat) of mixing, which is the heat released or absorbed as two or more components are mixed isothermally. For example, the volume of an ethyl alcohol–water mixture is a few percent less than the sum of the volumes of the individual liquids before mixing. Also, when water and flour are mixed to make dough, the temperature of the dough rises noticeably due to the enthalpy of mixing released.

**FIGURE 13–18**
For a pure substance, the chemical potential is equivalent to the Gibbs function.

For reasons explained above, the partial molar properties of the components (denoted by an tilde) should be used in the evaluation of the extensive properties of a mixture instead of the specific properties of the pure components. For example, the total volume, enthalpy, and entropy of a mixture should be determined from, respectively,

$$V = \sum_i N_i \tilde{v}_i \quad H = \sum_i N_i \tilde{h}_i \quad \text{and} \quad S = \sum_i N_i \tilde{s}_i \quad \text{(mixture)} \quad \text{(13–32)}$$

instead of

$$V^* = \sum_i N_i \bar{v}_i \quad H^* = \sum_i N_i \bar{h}_i \quad \text{and} \quad S^* = \sum_i N_i \bar{s}_i \quad \text{(13–33)}$$

Then, the changes in these extensive properties during mixing become

$$\Delta V_{\text{mixing}} = \sum_i N_i(\tilde{v}_i - \bar{v}_i), \, \Delta H_{\text{mixing}} = \sum_i N_i(\tilde{h}_i - \bar{h}_i), \, \Delta S_{\text{mixing}} = \sum_i N_i(\tilde{s}_i - \bar{s}_i)$$

$$\text{(13–34)}$$

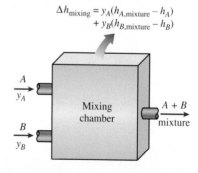

**FIGURE 13–19**
The amount of heat released or absorbed during a mixing process is called the enthalpy (or heat) of mixing, which is zero for ideal solutions.

where $\Delta H_{\text{mixing}}$ is the **enthalpy of mixing** and $\Delta S_{\text{mixing}}$ is the **entropy of mixing** (Fig. 13–19). The enthalpy of mixing is positive for exothermic mixing processes, negative for endothermic mixing processes, and zero for isothermal mixing processes during which no heat is absorbed or released. Note that mixing is an irreversible process, and thus the entropy of mixing must be a positive quantity during an adiabatic process. The specific volume, enthalpy, and entropy of a mixture are determined from

$$\bar{v} = \sum_i y_i \bar{v}_i \quad \bar{h} = \sum_i y_i \bar{h}_i \quad \text{and} \quad \bar{s} = \sum_i y_i \bar{s}_i \tag{13-35}$$

where $y_i$ is the mole fraction of component $i$ in the mixture.

Reconsider Eq. 13–29 for $dG$. Recall that properties are point functions, and they have exact differentials. Therefore, the test of exactness can be applied to the right-hand side of Eq. 13–29 to obtain some important relations. For the differential $dz = M\,dx + N\,dy$ of a function $z(x, y)$, the test of exactness is expressed as $(\partial M/\partial y)_x = (\partial N/\partial x)_y$. When the amount of component $i$ in a mixture is varied at constant pressure or temperature while other components (indicated by $j$) are held constant, Eq. 13–29 simplifies to

$$dG = -S\,dT + \mu_i\,dN_i \quad \text{(for } P = \text{constant and } N_j = \text{constant)} \tag{13-36}$$

$$dG = V\,dP + \mu_i\,dN_i \quad \text{(for } T = \text{constant and } N_j = \text{constant)} \tag{13-37}$$

Applying the test of exactness to both of these relations gives

$$\left(\frac{\partial \mu_i}{\partial T}\right)_{P,N} = -\left(\frac{\partial S}{\partial N_i}\right)_{T,P,N_j} = -\bar{s}_i \quad \text{and} \quad \left(\frac{\partial \mu_i}{\partial P}\right)_{T,N} = \left(\frac{\partial V}{\partial N_i}\right)_{T,P,N_j} = \bar{v}_i \tag{13-38}$$

where the subscript $N$ indicates that the mole numbers of all components (and thus the composition of the mixture) is to remain constant. Taking the chemical potential of a component to be a function of temperature, pressure, and composition and thus $\mu_i = \mu_i\,(P, T, y_1, y_2, \ldots, y_j \ldots)$, its total differential can be expressed as

$$d\mu_i = d\bar{g}_i = \left(\frac{\partial \mu_i}{\partial P}\right)_{T,y} dP + \left(\frac{\partial \mu_i}{\partial T}\right)_{P,y} dT + \sum_i \left(\frac{\partial \mu_i}{\partial y_i}\right)_{P,T,y_j} dy_i \tag{13-39}$$

where the subscript $y$ indicates that the mole fractions of all components (and thus the composition of the mixture) is to remain constant. Substituting Eqs. 13–38 into the above relation gives

$$d\mu_i = \bar{v}_i\,dP - \bar{s}_i\,dT + \sum_i \left(\frac{\partial \mu_i}{\partial y_i}\right)_{P,T,y_j} dy_i \tag{13-40}$$

For a mixture of fixed composition undergoing an isothermal process, it simplifies to

$$d\mu_i = \bar{v}_i\,dP \quad (T = \text{constant}, y_i = \text{constant}) \tag{13-41}$$

## Ideal-Gas Mixtures and Ideal Solutions

When the effect of dissimilar molecules in a mixture on each other is negligible, the mixture is said to be an **ideal mixture** or **ideal solution** and the *chemical potential of a component in such a mixture equals the Gibbs function of the pure component.* Many liquid solutions encountered in practice, especially dilute ones, satisfy this condition very closely and can be considered to be ideal solutions with negligible error. As expected, the ideal solution approximation greatly simplifies the thermodynamic analysis of mixtures. In an ideal solution, a molecule treats the molecules of all components in the mixture the same way—no extra attraction or repulsion for the molecules of other components. This is usually the case for mixtures of similar substances such as those of petroleum products. Very dissimilar substances such as water and oil won't even mix at all to form a solution.

For an ideal-gas mixture at temperature $T$ and total pressure $P$, the partial molar volume of a component $i$ is $\bar{v}_i = v_i = R_u T/P$. Substituting this relation into Eq. 13–41 gives

$$d\mu_i = \frac{R_u T}{P} dP = R_u T d \ln P = R_u T d \ln P_i \; (T = \text{constant}, y_i = \text{constant, ideal gas})$$

(13–42)

since, from Dalton's law of additive pressures, $P_i = y_i P$ for an ideal gas mixture and

$$d \ln P_i = d \ln(y_i P) = d(\ln y_i + \ln P) = d \ln P \quad (y_i = \text{constant}) \qquad \textbf{(13–43)}$$

for constant $y_i$. Integrating Eq. 13–42 at constant temperature from the total mixture pressure $P$ to the component pressure $P_i$ of component $i$ gives

$$\mu_i(T, P_i) = \mu_i(T, P) + R_u T \ln \frac{P_i}{P} = \mu_i(T, P) + R_u T \ln y_i \quad (\text{ideal gas})$$

(13–44)

For $y_i = 1$ (i.e., a pure substance of component $i$ alone), the last term in the above equation drops out and we end up with $\mu_i(T, P_i) = \mu_i(T, P)$, which is the value for the pure substance $i$. Therefore, the term $\mu_i(T, P)$ is simply the chemical potential of the pure substance $i$ when it exists alone at total mixture pressure and temperature, which is equivalent to the Gibbs function since the chemical potential and the Gibbs function are identical for pure substances. The term $\mu_i(T, P)$ is independent of mixture composition and mole fractions, and its value can be determined from the property tables of pure substances. Then, Eq. 13–44 can be rewritten more explicitly as

$$\mu_{i,\text{mixture,ideal}}(T, P_i) = \mu_{i,\text{pure}}(T, P) + R_u T \ln y_i \qquad \textbf{(13–45)}$$

Note that *the chemical potential of a component of an ideal gas mixture depends on the mole fraction of the components as well as the mixture temperature and pressure, and is independent of the identity of the other constituent gases.* This is not surprising since the molecules of an ideal gas behave like they exist alone and are not influenced by the presence of other molecules.

Eq. 13–45 is developed for an ideal-gas mixture, but it is also applicable to mixtures or solutions that behave the same way—that is, mixtures or solutions in which the effects of molecules of different components on each other are negligible. The class of such mixtures is called *ideal solutions* (or *ideal mixtures*), as discussed before. The ideal-gas mixture described is just one category of ideal solutions. Another major category of ideal solutions is the *dilute liquid solutions,* such as the saline water.

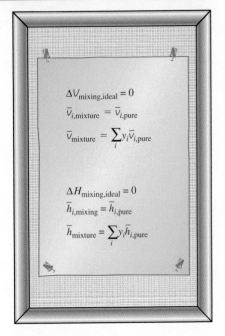

**FIGURE 13–20**
The specific volume and enthalpy of individual components do not change during mixing if they form an ideal solution (this is not the case for entropy).

It can be shown that the enthalpy of mixing and the volume change due to mixing are zero for ideal solutions (*see* Wark, 1995). That is,

$$\Delta V_{\text{mixing,ideal}} = \sum_i N_i(\tilde{v}_i - \bar{v}_i) = 0 \quad \text{and} \quad \Delta H_{\text{mixing,ideal}} = \sum_i N_i(\tilde{h}_i - \bar{h}_i) = 0$$

$$(13\text{–}46)$$

Then it follows that $\tilde{v}_i = \bar{v}_i$ and $\tilde{h}_i = \bar{h}_i$. That is, the partial molar volume and the partial molar enthalpy of a component in a solution equal the specific volume and enthalpy of that component when it existed alone as a pure substance at the mixture temperature and pressure. Therefore, the specific volume and enthalpy of individual components do not change during mixing if they form an ideal solution. Then, the specific volume and enthalpy of an ideal solution can be expressed as (Fig. 13–20)

$$\bar{v}_{\text{mixing,ideal}} = \sum_i y_i \tilde{v}_i = \sum_i y_i \bar{v}_{i,\text{pure}} \quad \text{and} \quad \bar{h}_{\text{mixture,ideal}} = \sum_i y_i \tilde{h}_i = \sum_i y_i \bar{h}_{i,\text{pure}}$$

$$(13\text{–}47)$$

Note that this is not the case for entropy and the properties that involve entropy such as the Gibbs function, even for ideal solutions. To obtain a relation for the entropy of a mixture, we differentiate Eq. 13–45 with respect to temperature at constant pressure and mole fraction,

$$\left(\frac{\partial \mu_{i,\text{mixing}}(T, P_i)}{\partial T}\right)_{P,y} = \left(\frac{\partial \mu_{i,\text{pure}}(T, P)}{\partial T}\right)_{P,y} + R_u \ln y_i \qquad (13\text{–}48)$$

We note from Eq. 13–38 that the two partial derivatives above are simply the negative of the partial molar entropies. Substituting,

$$\bar{s}_{i,\text{mixture,ideal}}(T, P_i) = \bar{s}_{i,\text{pure}}(T, P) - R_u \ln y_1 \qquad \text{(ideal solution)} \qquad (13\text{–}49)$$

Note that $\ln y_i$ is a negative quantity since $y_i < 1$, and thus $-R_u \ln y_i$ is always positive. Therefore, the entropy of a component in a mixture is always greater than the entropy of that component when it exists alone at the mixture temperature and pressure. Then the **entropy of mixing** of an ideal solution is determined by substituting Eq. 13–49 into Eq. 13–34 to be

$$\Delta S_{\text{mixing,ideal}} = \sum_i N_i(\bar{s}_i - \bar{s}_i) = -R_u \sum_i N_i \ln y_i \qquad \text{(ideal solution)} \qquad (13\text{–}50a)$$

or, dividing by the total number of moles of the mixture $N_m$,

$$\Delta \bar{s}_{\text{mixing,ideal}} = \sum_i y_i(\bar{s}_i - \bar{s}_i) = -R_u \sum_i y_i \ln y_i \qquad \text{(per unit mole of mixture)}$$

$$(13\text{–}50b)$$

## Minimum Work of Separation of Mixtures

The entropy balance for a steady-flow system simplifies to $S_{\text{in}} - S_{\text{out}} + S_{\text{gen}} = 0$. Noting that entropy can be transferred by heat and mass only, the entropy generation during an adiabatic mixing process that forms an ideal solution becomes

$$S_{gen} = S_{out} - S_{in} = \Delta S_{mixing} = -R_u \sum_i N_i \ln y_i \quad \text{(ideal solution)} \quad \textbf{(13–51a)}$$

or

$$\bar{s}_{gen} = \bar{s}_{out} - \bar{s}_{in} = \Delta s_{mixing} = -R_u \sum_i y_i \ln y_i \quad \text{(per unit mole of mixture)}$$

$$\textbf{(13–51b)}$$

Also noting that $X_{destroyed} = T_0 S_{gen}$, the exergy destroyed during this (and any other) process is obtained by multiplying the entropy generation by the temperature of the environment $T_0$. It gives

$$X_{destroyed} = T_0 S_{gen} = -R_u T_0 \sum_i N_i \ln y_i \quad \text{(ideal soluton)} \quad \textbf{(13–52a)}$$

or

$$\bar{x}_{destroyed} = T_0 \bar{s}_{gen} = -R_u T_0 \sum_i y_i \ln y_i \quad \text{(per unit mole of mixture)} \quad \textbf{(13–52b)}$$

Exergy destroyed represents the wasted work potential—the work that would be produced if the mixing process occurred reversibly. For a reversible or "thermodynamically perfect" process, the entropy generation and thus the exergy destroyed is zero. Also, for reversible processes, the work output is a maximum (or, the work input is a minimum if the process does not occur naturally and requires input). The difference between the reversible work and the actual useful work is due to irreversibilities and is equal to the exergy destruction. Therefore, $X_{destroyed} = W_{rev} - W_{actual}$. Then, it follows that for a naturally occurring process during which no work is produced, the reversible work is equal to the exergy destruction (Fig. 13–21). Therefore, for the adiabatic mixing process that forms an ideal solution, the reversible work (total and per unit mole of mixture) is, from Eq. 13–52,

$$W_{rev} = -R_u T_0 \sum_i N_i \ln y_i \quad \text{and} \quad \bar{w}_{rev} = -R_u T_0 \sum_i y_i \ln y_i \quad \textbf{(13–53)}$$

A reversible process, by definition, is a process that can be reversed without leaving a net effect on the surroundings. This requires that the direction of all interactions be reversed while their magnitudes remain the same when the process is reversed. Therefore, the work input during a reversible separation process must be equal to the work output during the reverse process of mixing. A violation of this requirement will be a violation of the second law of thermodynamics. The required work input for a reversible separation process is the minimum work input required to accomplish that separation since the work input for reversible processes is always less than the work input of corresponding irreversible processes. Then the minimum work input required for the separation process can be expressed as

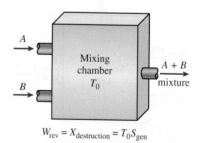

$$W_{rev} = X_{destruction} = T_0 S_{gen}$$

**FIGURE 13–21**
For a naturally occurring process during which no work is produced or consumed, the reversible work is equal to the exergy destruction.

$$W_{\text{min,in}} = -R_u T_0 \sum_i N_i \ln y_i \quad \text{and} \quad \bar{w}_{\text{min,in}} = -R_u T_0 \sum_i y_i \ln y_i \qquad (13\text{-}54)$$

It can also be expressed in the rate form as

$$\dot{W}_{\text{min,in}} = -R_u T_0 \sum_i \dot{N}_i \ln y_i = -\dot{N}_m R_u T_0 \sum_i y_i \ln y_i \qquad (\text{kW}) \qquad (13\text{-}55)$$

where $\dot{W}_{\text{min,in}}$ is the minimum power input required to separate a solution that approaches at a rate of $\dot{N}_m$ kmol/s (or $\dot{m}_m = \dot{N}_m M_m$ kg/s) into its components. The work of separation per unit mass of mixture can be determined from $w_{\text{min,in}} = \bar{w}_{\text{min,in}}/M_m$, where $M_m$ is the apparent molar mass of the mixture.

The minimum work relations above are for complete separation of the components in the mixture. The required work input will be less if the exiting streams are not pure. The reversible work for incomplete separation can be determined by calculating the minimum separation work for the incoming mixture and the minimum separation works for the outgoing mixtures, and then taking their difference.

## Reversible Mixing Processes

The mixing processes that occur naturally are irreversible, and all the work potential is wasted during such processes. For example, when the fresh water from a river mixes with the saline water in an ocean, an opportunity to produce work is lost. If this mixing is done reversibly (through the use of semipermeable membranes, for example) some work can be produced. The maximum amount of work that can be produced during a mixing process is equal to the minimum amount of work input needed for the corresponding separation process (Fig. 13–22). That is,

$$W_{\text{max,out,mixing}} = W_{\text{min,in,separation}} \qquad (13\text{-}56)$$

Therefore, the minimum work input relations given above for separation can also be used to determine the maximum work output for mixing.

The minimum work input relations are independent of any hardware or process. Therefore, the relations developed above are applicable to any separation process regardless of actual hardware, system, or process, and can be used for a wide range of separation processes including the desalination of sea or brackish water.

## Second-Law Efficiency

The second-law efficiency is a measure of how closely a process approximates a corresponding reversible process, and it indicates the range available for potential improvements. Noting that the second-law efficiency ranges from 0 for a totally irreversible process to 100 percent for a totally reversible process, the second-law efficiency for separation and mixing processes can be defined as

$$\eta_{\text{II,separation}} = \frac{\dot{W}_{\text{min,in}}}{\dot{W}_{\text{act,in}}} = \frac{w_{\text{min,in}}}{w_{\text{act,in}}} \quad \text{and} \quad \eta_{\text{II,mixing}} = \frac{\dot{W}_{\text{act,out}}}{\dot{W}_{\text{max,out}}} = \frac{w_{\text{act,out}}}{w_{\text{max,out}}} \qquad (13\text{-}57)$$

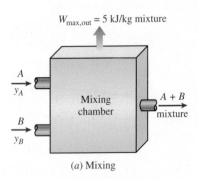

$W_{\text{max,out}} = 5$ kJ/kg mixture

$A$, $y_A$

$B$, $y_B$

Mixing chamber

$A + B$ mixture

(a) Mixing

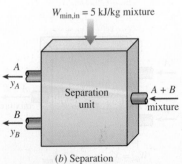

$W_{\text{min,in}} = 5$ kJ/kg mixture

$A$, $y_A$

$B$, $y_B$

Separation unit

$A + B$ mixture

(b) Separation

**FIGURE 13–22**
Under reversible conditions, the work consumed during separation is equal to the work produced during the reverse process of mixing.

where $\dot{W}_{act,in}$ is the actual power input (or exergy consumption) of the separation plant and $\dot{W}_{act,out}$ is the actual power produced during mixing. Note that the second-law efficiency is always less than 1 since the actual separation process requires a greater amount of work input because of irreversibilities. Therefore, the minimum work input and the second-law efficiency provide a basis for comparison of actual separation processes to the "idealized" ones and for assessing the thermodynamic performance of separation plants.

A second-law efficiency for mixing processes can also be defined as the actual work produced during mixing divided by the maximum work potential available. This definition does not have much practical value, however, since no effort is done to produce work during most mixing processes and thus the second-law efficiency is zero.

## Special Case: Separation of a Two Component Mixture

Consider a mixture of two components $A$ and $B$ whose mole fractions are $y_A$ and $y_B$. Noting that $y_B = 1 - y_A$, the minimum work input required to separate 1 kmol of this mixture at temperature $T_0$ completely into pure $A$ and pure $B$ is, from Eq. 13–54,

$$\overline{w}_{min,in} = -R_u T_0 (y_A \ln y_A + y_B \ln y_B) \quad \text{(kJ/kmol mixture)} \qquad \text{(13–58a)}$$

or

$$W_{min,in} = -R_u T_0 (N_A \ln y_A + N_B \ln y_B) \quad \text{(kJ)} \qquad \text{(13–58b)}$$

or, from Eq. 13–55,

$$\dot{W}_{min,in} = -\dot{N}_m R_u T_0 (y_A \ln y_A + y_B \ln y_B)$$
$$= -\dot{m}_m R_m T_0 (y_A \ln y_A + y_B \ln y_B) \quad \text{(kW)} \qquad \text{(13–58c)}$$

Some separation processes involve the extraction of just one of the components from a large amount of mixture so that the composition of the remaining mixture remains practically the same. Consider a mixture of two components $A$ and $B$ whose mole fractions are $y_A$ and $y_B$, respectively. The minimum work required to separate 1 kmol of pure component $A$ from the mixture of $N_m = N_A + N_B$ kmol (with $N_A \gg 1$) is determined by subtracting the minimum work required to separate the remaining mixture $-R_u T_0[(N_A - 1) \ln y_A + N_B \ln y_B]$ from the minimum work required to separate the initial mixture $W_{min,in} = -R_u T_0 (N_A \ln y_A + N_B \ln y_B)$. It gives (Fig. 13–23)

$$\overline{w}_{min,in} = -R_u T_0 \ln y_A = R_u T_0 \ln(1/y_A) \quad \text{(kJ/kmol } A) \qquad \text{(13–59)}$$

The minimum work needed to separate a unit mass (1 kg) of component $A$ is determined from the above relation by replacing $R_u$ by $R_A$ (or by dividing the relation above by the molar mass of component $A$) since $R_A = R_u/M_A$. Eq. 13–59 also gives the maximum amount of work that can be done as one unit of pure component $A$ mixes with a large amount of $A + B$ mixture.

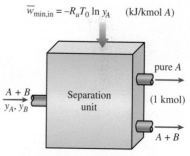

$$\overline{w}_{min,in} = -R_u T_0 \ln y_A \quad \text{(kJ/kmol } A)$$

(a) Separating 1 kmol of $A$ from a large body of mixture

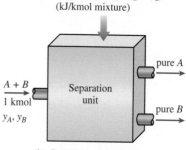

$$\overline{w}_{min,in} = -R_u T_0 (y_A \ln y_A + y_B \ln y_B)$$
$$\text{(kJ/kmol mixture)}$$

(b) Complete separation of 1 kmol mixture into its components $A$ and $B$

**FIGURE 13–23**
The minimum work required to separate a two-component mixture for the two limiting cases.

# An Application: Desalination Processes

The potable water needs of the world is increasing steadily due to population growth, rising living standards, industrialization, and irrigation in agriculture. There are over 10,000 desalination plants in the world, with a total desalted water capacity of over 20 billion liters a day. Saudi Arabia is the largest user of desalination with about 25 percent of the world capacity, and the United States is the second largest user with 10 percent. The major desalination methods are distillation and reverse osmosis. The relations can be used directly for desalination processes, by taking the water (the solvent) to be component $A$ and the dissolved salts (the solute) to be component $B$. Then the minimum work needed to produce 1 kg of pure water from a large reservoir of brackish or seawater at temperature $T_0$ in an environment at $T_0$ is, from Eq. 13–59,

*Desalination*:   $w_{min,in} = -R_w T_0 \ln(1/y_w)$    (kJ/kg pure water)    **(13–60)**

where $R_w = 0.4615$ kJ/kg·K is the gas constant of water and $y_w$ is the mole fraction of water in brackish or seawater. The relation above also gives the maximum amount of work that can be produced as 1 kg of fresh water (from a river, for example) mixes with seawater whose water mole fraction is $y_w$.

The reversible work associated with liquid flow can also be expressed in terms of pressure difference $\Delta P$ and elevation difference $\Delta z$ (potential energy) as $w_{min,in} = \Delta P/\rho = g \, \Delta z$ where $\rho$ is the density of the liquid. Combining these relations with Eq. 13–60 gives

$$\Delta P_{min} = \rho w_{min,in} = \rho R_w T_0 \ln(1/y_w) \quad \text{(kPa)} \qquad \textbf{(13–61)}$$

and

$$\Delta z_{min} = w_{min,in}/g = R_w T_0 \ln(1/y_w)/g \quad \text{(m)} \qquad \textbf{(13–62)}$$

where $\Delta P_{min}$ is the **osmotic pressure**, which represents the pressure difference across a semipermeable membrane that separates fresh water from the saline water under equilibrium conditions, $\rho$ is the density of saline water, and $\Delta z_{min}$ is the **osmotic rise**, which represents the vertical distance the saline water would rise when separated from the fresh water by a membrane that is permeable to water molecules alone (again at equilibrium). For desalination processes, $\Delta P_{min}$ represents the minimum pressure that the saline water must be compressed in order to force the water molecules in saline water through the membrane to the fresh water side during a reverse osmosis desalination process. Alternately, $\Delta z_{min}$ represents the minimum height above the fresh water level that the saline water must be raised to produce the required osmotic pressure difference across the membrane to produce fresh water. The $\Delta z_{min}$ also represents the height that the water with dissolved organic matter inside the roots will rise through a tree when the roots are surrounded by fresh water with the roots acting as semipermeable membranes. The reverse osmosis process with semipermeable membranes is also used in dialysis machines to purify the blood of patients with failed kidneys.

■ **EXAMPLE 13–6**　　**Obtaining Fresh Water from Seawater**

Fresh water is to be obtained from seawater at 15°C with a salinity of 3.48 percent on mass basis (or TDS = 34,800 ppm). Determine (a) the mole fractions of the water and the salts in the seawater, (b) the minimum work input required to separate 1 kg of seawater completely into pure water and pure salts, (c) the minimum work input required to obtain 1 kg of fresh water from the sea, and (d) the minimum gauge pressure that the seawater must be raised if fresh water is to be obtained by reverse osmosis using semipermeable membranes.

**SOLUTION**　Fresh water is to be obtained from seawater. The mole fractions of seawater, the minimum works of separation needed for two limiting cases, and the required pressurization of seawater for reverse osmosis are to be determined.

**Assumptions**　**1** The seawater is an ideal solution since it is dilute. **2** The total dissolved solids in water can be treated as table salt (NaCl). **3** The environment temperature is also 15°C.

**Properties**　The molar masses of water and salt are $M_w = 18.0$ kg/kmol and $M_s = 58.44$ kg/kmol. The gas constant of pure water is $R_w = 0.4615$ kJ/kg·K (Table A–1). The density of seawater is 1028 kg/m³.

**Analysis**　(a) Noting that the mass fractions of salts and water in seawater are $mf_s = 0.0348$ and $mf_w = 1 - mf_s = 0.9652$, the mole fractions are determined from Eqs. 13–4 and 13–5 to be

$$M_m = \frac{1}{\sum \dfrac{mf_i}{M_i}} = \frac{1}{\dfrac{mf_s}{M_s} + \dfrac{mf_w}{M_w}} = \frac{1}{\dfrac{0.0348}{58.44} + \dfrac{0.9652}{18.0}} = 18.44 \text{ kg/kmol}$$

$$y_w = mf_w \frac{M_m}{M_w} = 0.9652 \frac{18.44 \text{ kg/kmol}}{18.0 \text{ kg/kmol}} = \mathbf{0.9888}$$

$$y_s = 1 - y_w = 1 - 0.9888 = \mathbf{0.0112} = 1.12\%$$

(b) The minimum work input required to separate 1 kg of seawater completely into pure water and pure salts is

$$\bar{w}_{\text{min,in}} = -R_u T_0 (y_A \ln y_A + y_B \ln y_B) = -R_u T_0 (y_w \ln y_w + y_s \ln y_s)$$

$$= -(8.314 \text{ kJ/kmol·K})(288.15 \text{ K})(0.9888 \ln 0.9888 + 0.0112 \ln 0.0112)$$

$$= 147.2 \text{ kJ/kmol}$$

$$w_{\text{min,in}} = \frac{\bar{w}_{\text{min,in}}}{M_m} = \frac{147.2 \text{ kJ/kmol}}{18.44 \text{ kg/kmol}} = \mathbf{7.98 \text{ kJ/kg seawater}}$$

Therefore, it takes a minimum of 7.98 kJ of work input to separate 1 kg of seawater into 0.0348 kg of salt and 0.9652 kg (nearly 1 kg) of fresh water.

(c) The minimum work input required to produce 1 kg of fresh water from seawater is

$$w_{\text{min,in}} = R_w T_0 \ln(1/y_w)$$

$$= (0.4615 \text{ kJ/kg·K})(288.15 \text{ K}) \ln(1/0.9888)$$

$$= \mathbf{1.50 \text{ kJ/kg fresh water}}$$

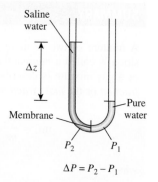

Saline water

$\Delta z$

Pure water

Membrane

$P_2$　　$P_1$

$\Delta P = P_2 - P_1$

**FIGURE 13–24**
The osmotic pressure and the osmotic rise of saline water.

Note that it takes about 5 times more work to separate 1 kg of seawater completely into fresh water and salt than it does to produce 1 kg of fresh water from a large amount of seawater.

(*d*) The osmotic pressure in this case is

$$\Delta P_{min} = \rho_m R_w T_0 \ln(1/y_w)$$

$$= (1028 \text{ kg/m}^3)(0.4615 \text{ kPa·m}^3/\text{kg·K})(288.15 \text{ K})\ln(1/0.9888)$$

$$= \textbf{1540 kPa}$$

which is equal to the minimum gauge pressure to which seawater must be compressed if the fresh water is to be discharged at the local atmospheric pressure. As an alternative to pressurizing, the minimum height above the fresh water level that the seawater must be raised to produce fresh water is (Fig. 13–24)

$$\Delta z_{min} = \frac{w_{min,in}}{g} = \frac{1.50 \text{ kJ/kg}}{9.81 \text{ m/s}^2}\left(\frac{1 \text{ kg·m/s}^2}{1 \text{ N}}\right)\left(\frac{1000 \text{ N·m}}{1 \text{ kJ}}\right) = 153 \text{ m}$$

**Discussion** The minimum separation works determined above also represent the maximum works that can be produced during the reverse process of mixing. Therefore, 7.98 kJ of work can be produced when 0.0348 kg of salt is mixed with 0.9652 kg of water reversibly to produce 1 kg of saline water, and 1.50 kJ of work can be produced as 1 kg of fresh water is mixed with seawater reversibly. Therefore, the power that can be generated as a river with a flow rate of $10^6$ m³/s mixes reversibly with seawater through semipermeable membranes is (Fig. 13–25)

$$\dot{W}_{max,out} = \rho \dot{V} w_{max,out} = (1000 \text{ kg/m}^3)(10^6 \text{ m}^3/\text{s})(1.50 \text{ kJ/kg})\left(\frac{1 \text{ MW}}{10^3 \text{ kJ/s}}\right)$$

$$= 1.5 \times 10^6 \text{ MW}$$

which shows the tremendous amount of power potential wasted as the rivers discharge into the seas.

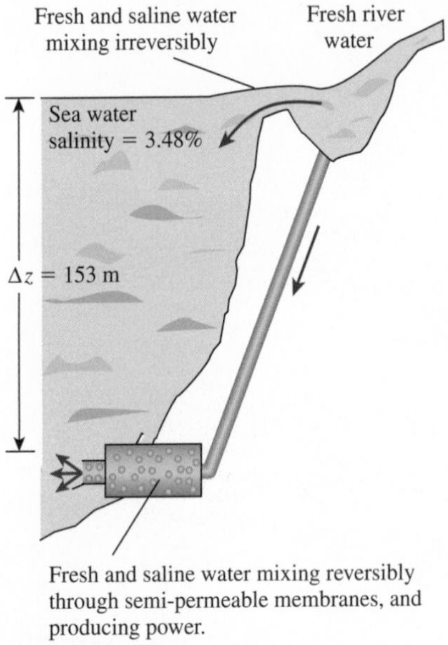

**FIGURE 13–25**
Power can be produced by mixing solutions of different concentrations reversibly.

---

## SUMMARY

A mixture of two or more gases of fixed chemical composition is called a *nonreacting gas mixture*. The composition of a gas mixture is described by specifying either the *mole fraction* or the *mass fraction* of each component, defined as

$$\text{mf}_i = \frac{m_i}{m_m} \quad \text{and} \quad y_i = \frac{N_i}{N_m}$$

where

$$m_m = \sum_{i=1}^{k} m_i \quad \text{and} \quad N_m = \sum_{i=1}^{k} N_i$$

The *apparent* (or average) *molar mass* and *gas constant* of a mixture are expressed as

$$M_m = \frac{m_m}{N_m} = \sum_{i=1}^{k} y_i M_i \quad \text{and} \quad R_m = \frac{R_u}{M_m}$$

Also,

$$\text{mf}_i = y_i \frac{M_i}{M_m} \quad \text{and} \quad M_m = \frac{1}{\sum_{i=1}^{k} \frac{\text{mf}_i}{M_i}}$$

*Dalton's law of additive pressures* states that the pressure of a gas mixture is equal to the sum of the pressures each gas would exert if it existed alone at the mixture temperature and volume. *Amagat's law of additive volumes* states that the volume of a gas mixture is equal to the sum of the volumes each gas would occupy if it existed alone at the mixture temperature and pressure. Dalton's and Amagat's laws hold exactly for ideal-gas mixtures, but only approximately for real-gas mixtures. They can be expressed as

Dalton's law: $\quad P_m = \sum_{i=1}^{k} P_i(T_m, V_m)$

Amagat's law: $\quad V_m = \sum_{i=1}^{k} V_i(T_m, P_m)$

Here $P_i$ is called the *component pressure* and $V_i$ is called the *component volume*. Also, the ratio $P_i/P_m$ is called the *pressure fraction* and the ratio $V_i/V_m$ is called the *volume fraction* of component $i$. For *ideal gases*, $P_i$ and $V_i$ can be related to $y_i$ by

$$\frac{P_i}{P_m} = \frac{V_i}{V_m} = \frac{N_i}{N_m} = y_i$$

The quantity $y_i P_m$ is called the *partial pressure* and the quantity $y_i V_m$ is called the *partial volume*. The $P$-$V$-$T$ behavior of real-gas mixtures can be predicted by using generalized compressibility charts. The compressibility factor of the mixture can be expressed in terms of the compressibility factors of the individual gases as

$$Z_m = \sum_{i=1}^{k} y_i Z_i$$

where $Z_i$ is determined either at $T_m$ and $V_m$ (Dalton's law) or at $T_m$ and $P_m$ (Amagat's law) for each individual gas. The $P$-$V$-$T$ behavior of a gas mixture can also be predicted approximately by *Kay's rule*, which involves treating a gas mixture as a pure substance with pseudocritical properties determined from

$$P'_{cr,m} = \sum_{i=1}^{k} y_i P_{cr,i} \quad \text{and} \quad T'_{cr,m} = \sum_{i=1}^{k} y_i T_{cr,i}$$

The *extensive properties* of a gas mixture, in general, can be determined by summing the contributions of each component of the mixture. The evaluation of *intensive properties* of a gas mixture, however, involves averaging in terms of mass or mole fractions:

$$U_m = \sum_{i=1}^{k} U_i = \sum_{i=1}^{k} m_i u_i = \sum_{i=1}^{k} N_i \bar{u}_i$$

$$H_m = \sum_{i=1}^{k} H_i = \sum_{i=1}^{k} m_i h_i = \sum_{i=1}^{k} N_i \bar{h}_i$$

$$S_m = \sum_{i=1}^{k} S_i = \sum_{i=1}^{k} m_i s_i = \sum_{i=1}^{k} N_i \bar{s}_i$$

and

$$u_m = \sum_{i=1}^{k} \text{mf}_i u_i \quad \text{and} \quad \bar{u}_m = \sum_{i=1}^{k} y_i \bar{u}_i$$

$$h_m = \sum_{i=1}^{k} \text{mf}_i h_i \quad \text{and} \quad \bar{h}_m = \sum_{i=1}^{k} y_i \bar{h}_i$$

$$s_m = \sum_{i=1}^{k} \text{mf}_i s_i \quad \text{and} \quad \bar{s}_m = \sum_{i=1}^{k} y_i \bar{s}_i$$

$$c_{v,m} = \sum_{i=1}^{k} \text{mf}_i c_{v,i} \quad \text{and} \quad \bar{c}_{v,m} = \sum_{i=1}^{k} y_i \bar{c}_{v,i}$$

$$c_{p,m} = \sum_{i=1}^{k} \text{mf}_i c_{p,i} \quad \text{and} \quad \bar{c}_{p,m} = \sum_{i=1}^{k} y_i \bar{c}_{p,i}$$

These relations are exact for ideal-gas mixtures and approximate for real-gas mixtures. The properties or property changes of individual components can be determined by using ideal-gas or real-gas relations developed in earlier chapters.

## REFERENCES AND SUGGESTED READINGS

1. A. Bejan. *Advanced Engineering Thermodynamics.* 3rd ed. New York: Wiley Interscience, 2006.

2. Y. A. Çengel, Y. Cerci, and B. Wood, "Second Law Analysis of Separation Processes of Mixtures," *ASME International Mechanical Engineering Congress and Exposition,* Nashville, Tennessee, 1999.

3. Y. Cerci, Y. A. Çengel, and B. Wood, "The Minimum Separation Work for Desalination Processes," *ASME International Mechanical Engineering Congress and Exposition,* Nashville, Tennessee, 1999.

4. K. Wark, Jr. *Advanced Thermodynamics for Engineers.* New York: McGraw-Hill, 1995.

# PROBLEMS*

## Composition of Gas Mixtures

**13–1C** Consider a mixture of several gases of identical masses. Will all the mass fractions be identical? How about the mole fractions?

**13–2C** The sum of the mole fractions for an ideal-gas mixture is equal to 1. Is this also true for a real-gas mixture?

**13–3C** Somebody claims that the mass and mole fractions for a mixture of $CO_2$ and $N_2O$ gases are identical. Is this true? Why?

**13–4C** Consider a mixture of two gases. Can the apparent molar mass of this mixture be determined by simply taking the arithmetic average of the molar masses of the individual gases? When will this be the case?

**13–5C** What is the *apparent molar mass* for a gas mixture? Does the mass of every molecule in the mixture equal the apparent molar mass?

**13–6** Using the definitions of mass and mole fractions, derive a relation between them.

**13–7** Consider a mixture of two gases A and B. Show that when the mass fractions $mf_A$ and $mf_B$ are known, the mole fractions can be determined from

$$y_A = \frac{M_B}{M_A(1/mf_A - 1) + M_B} \quad \text{and} \quad y_B = 1 - y_A$$

where $M_A$ and $M_B$ are the molar masses of A and B.

**13–8** The composition of moist air is given on a molar basis to be 78 percent $N_2$, 20 percent $O_2$, and 2 percent water vapor. Determine the mass fractions of the constituents of air.

**13–9** A gas mixture has the following composition on a mole basis: 60 percent $N_2$ and 40 percent $CO_2$. Determine the gravimetric analysis of the mixture, its molar mass, and gas constant.

**13–10** Repeat Prob. 13–9 by replacing $N_2$ by $O_2$.

**13–11** A gas mixture consists of 2 kg of $O_2$, 5 kg of $N_2$, and 7 kg of $CO_2$. Determine (a) the mass fraction of each component, (b) the mole fraction of each component, and (c) the average molar mass and gas constant of the mixture.

**13–12** Determine the mole fractions of a gas mixture that consists of 75 percent $CH_4$ and 25 percent $CO_2$ by mass. Also, determine the gas constant of the mixture.

**13–13** A gas mixture consists of 6 kmol of $H_2$ and 2 kmol of $N_2$. Determine the mass of each gas and the apparent gas constant of the mixture. *Answers:* 12 kg, 56 kg, 0.978 kJ/kg·K

## P-v-T Behavior of Gas Mixtures

**13–14C** Is a mixture of ideal gases also an ideal gas? Give an example.

**13–15C** Express Dalton's law of additive pressures. Does this law hold exactly for ideal-gas mixtures? How about nonideal-gas mixtures?

**13–16C** Express Amagat's law of additive volumes. Does this law hold exactly for ideal-gas mixtures? How about nonideal-gas mixtures?

**13–17C** How is the P-v-T behavior of a component in an ideal-gas mixture expressed? How is the P-v-T behavior of a component in a real-gas mixture expressed?

**13–18C** What is the difference between the *component pressure* and the *partial pressure*? When are these two equivalent?

**13–19C** What is the difference between the *component volume* and the *partial volume*? When are these two equivalent?

**13–20C** In a gas mixture, which component will have the higher partial pressure—the one with the higher mole number or the one with the larger molar mass?

**13–21C** Consider a rigid tank that contains a mixture of two ideal gases. A valve is opened and some gas escapes. As a result, the pressure in the tank drops. Will the partial pressure of each component change? How about the pressure fraction of each component?

**13–22C** Consider a rigid tank that contains a mixture of two ideal gases. The gas mixture is heated, and the pressure and temperature in the tank rise. Will the partial pressure of each component change? How about the pressure fraction of each component?

**13–23C** Is this statement correct? *The volume of an ideal-gas mixture is equal to the sum of the volumes of each individual gas in the mixture.* If not, how would you correct it?

**13–24C** Is this statement correct? *The temperature of an ideal-gas mixture is equal to the sum of the temperatures of each individual gas in the mixture.* If not, how would you correct it?

**13–25C** Is this statement correct? *The pressure of an ideal-gas mixture is equal to the sum of the partial pressures of each individual gas in the mixture.* If not, how would you correct it?

*Problems designated by a "C" are concept questions, and students are encouraged to answer them all. Problems with the ⚙ icon are solved using EES, and complete solutions together with parametric studies are included on the text website. Problems with the 🔧 icon are comprehensive in nature, and are intended to be solved with an equation solver such as EES.

**13–26** Atmospheric contaminants are often measured in parts per million (by volume). What would the partial pressure of refrigerant-134a be in atmospheric air at 100 kPa and 20°C to form a 100-ppm contaminant?

**13–27** A mixture of gases consists of 30 percent hydrogen, 40 percent helium, and 30 percent nitrogen by volume. Calculate the mass fractions and apparent molecular weight of this mixture.

**13–28** A gas mixture at 350 K and 300 kPa has the following volumetric analysis: 65 percent $N_2$, 20 percent $O_2$, and 15 percent $CO_2$. Determine the mass fraction and partial pressure of each gas.

**13–29** In an ideal gas mixture the partial pressures of the component gases are as follows: $CO_2$, 20 kPa; $O_2$, 30 kPa; and $N_2$, 50 kPa. Determine the mole fractions and mass fractions of each component. Calculate the apparent molar mass, the apparent gas constant, the constant-volume specific heat, and the specific heat ratio at 300 K for the mixture.

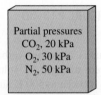

Partial pressures
$CO_2$, 20 kPa
$O_2$, 30 kPa
$N_2$, 50 kPa

**FIGURE P13–29**

**13–30** An engineer has proposed mixing extra oxygen with normal air in internal combustion engines to control some of the exhaust products. If an additional 5 percent (by volume) of oxygen is mixed with standard atmospheric air, how will this change the mixture's molecular weight?

**13–31** A rigid tank that contains 2 kg of $N_2$ at 25°C and 550 kPa is connected to another rigid tank that contains 4 kg of $O_2$ at 25°C and 150 kPa. The valve connecting the two tanks is opened, and the two gases are allowed to mix. If the final mixture temperature is 25°C, determine the volume of each tank and the final mixture pressure. *Answers:* 0.322 m³, 2.07 m³, 204 kPa

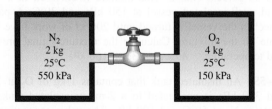

$N_2$
2 kg
25°C
550 kPa

$O_2$
4 kg
25°C
150 kPa

**FIGURE P13–31**

**13–32** A mixture of gases consists of 0.4 kg of oxygen, 0.7 kg of carbon dioxide, and 0.2 kg of helium. This mixture is maintained at 100 kPa and 27°C. Determine the apparent molecular weight of this mixture, the volume it occupies, the partial volume of the oxygen, and the partial pressure of the helium. *Answers:* 16.6 kg/kmol, 1.96 m³, 0.312 m³, 63.8 kPa

**13–33** A mixture of hydrocarbon gases is composed of 60 percent methane, 25 percent propane, and 15 percent butane by weight. Determine the volume occupied by 100 kg of this mixture when its pressure is 3 MPa and its temperature is 37°C.

**13–34** A rigid tank contains 8 kmol of $O_2$ and 10 kmol of $CO_2$ gases at 290 K and 150 kPa. Estimate the volume of the tank. *Answer:* 289 m³

**13–35** Repeat Prob. 13–34 for a temperature of 400 K.

**13–36** A 30 percent (by mass) ethane and 70 percent methane mixture is to be blended in a 100-m³ tank at 130 kPa and 25°C. If the tank is initially evacuated, to what pressure should ethane be added before methane is added?

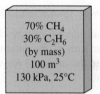

70% $CH_4$
30% $C_2H_6$
(by mass)
100 m³
130 kPa, 25°C

**FIGURE P13–36**

**13–37** A mixture is 35 percent by volume liquid water, whose density is 1000 kg/m³, that is mixed with another fluid, whose density is 800 kg/m³. What is the specific weight, in N/m³, of this mixture at a location where g = .72 m/s²? *Answer:* 8456 N/m³

**13–38** A mixture of air and methane is formed in the inlet manifold of a natural gas-fueled internal combustion engine. The mole fraction of the methane is 15 percent. This engine is operated at 3000 rpm and has a 5-L displacement. Determine the mass flow rate of this mixture in the manifold where the pressure and temperature are 80 kPa and 20°C. *Answer:* 6.65 kg/min

**13–39** Natural gas (95 percent methane and 5 percent ethane by volume) flows through a 90-cm-diameter pipeline with a velocity of 3 m/s. The pressure in the pipeline is 700 kPa, and the temperature is 15°C. Calculate the mass and volumetric flow rates in this pipe.

**13–40** The volumetric analysis of a mixture of gases is 30 percent oxygen, 40 percent nitrogen, 10 percent carbon dioxide, and 20 percent methane. This mixture flows through a 1.6-cm-diameter pipe at 8000 kPa and 15°C with a velocity of 5 m/s. Determine the volumetric and mass flow rates of this mixture (*a*) treating it as an ideal gas mixture, (*b*) using a compressibility factor based on Amagad's law of additive volumes, and (*c*) using Key's psuedocritical pressure and temperature.

**13–41** A rigid tank contains 1 kmol of argon gas at 222 K and 5250 kPa. A valve is now opened, and 3 kmol of $N_2$ gas is allowed to enter the tank at 189 K and 8400 kPa. The final mixture temperature is 200 K. Determine the pressure of the mixture, using (*a*) the ideal-gas equation of state and (*b*) the compressibility chart and Dalton's law.
*Answers:* (*a*) 18.9 MPa, (*b*) 17.0 MPa

## Properties of Gas Mixtures

**13–42C** Is the total internal energy of an ideal-gas mixture equal to the sum of the internal energies of each individual gas in the mixture? Answer the same question for a real-gas mixture.

**13–43C** Is the specific internal energy of a gas mixture equal to the sum of the specific internal energies of each individual gas in the mixture?

**13–44C** Answer Prob. 13–42C and 13–43C for entropy.

**13–45C** Is the total internal energy change of an ideal-gas mixture equal to the sum of the internal energy changes of each individual gas in the mixture? Answer the same question for a real-gas mixture.

**13–46C** When evaluating the entropy change of the components of an ideal-gas mixture, do we have to use the partial pressure of each component or the total pressure of the mixture?

**13–47C** Suppose we want to determine the enthalpy change of a real-gas mixture undergoing a process. The enthalpy change of each individual gas is determined by using the generalized enthalpy chart, and the enthalpy change of the mixture is determined by summing them. Is this an exact approach? Explain.

**13–48** The volumetric analysis of mixture of gases is 30 percent oxygen, 40 percent nitrogen, 10 percent carbon dioxide, and 20 percent methane. This mixture is heated from 20°C to 200°C while flowing through a tube in which the pressure is maintained at 150 kPa. Determine the heat transfer to the mixture per unit mass of the mixture.

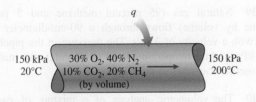

**FIGURE P13–48**

**13–49** A process requires a mixture that is 21 percent oxygen, 78 percent nitrogen, and 1 percent argon by volume. All three gases are supplied from separate tanks to an adiabatic, constant-pressure mixing chamber at 200 kPa but at different temperatures. The oxygen enters at 10°C, the nitrogen at 60°C, and the argon at 200°C. Determine the total entropy change for the mixing process per unit mass of mixture.

**13–50** A mixture of helium and nitrogen with a nitrogen mass fraction of 35 percent is contained in a piston–cylinder device arranged to maintain a fixed pressure of 700 kPa. Determine the work produced, in kJ/kg, as this device is heated from 40°C to 260°C. *Answer:* 320 kJ/kg

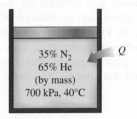

**FIGURE P13–50**

**13–51** A mixture that is 20 percent carbon dioxide, 10 percent oxygen, and 70 percent nitrogen by volume undergoes a process from 300 K and 100 kPa to 500 K and 400 kPa. Determine the makeup of the mixture on a mass basis and the enthalpy change per unit mass of mixture.

**13–52** A 0.9-m³ rigid tank is divided into two equal compartments by a partition. One compartment contains Ne at 20°C and 100 kPa, and the other compartment contains Ar at 50°C and 200 kPa. Now the partition is removed, and the two gases are allowed to mix. Heat is lost to the surrounding air during this process in the amount of 15 kJ. Determine (*a*) the final mixture temperature and (*b*) the final mixture pressure. *Answers:* (*a*) 16.2°C, (*b*) 138.9 kPa

**13–53** Repeat Prob. 13–52 for a heat loss of 8 kJ.

**13–54** The mass fractions of a mixture of gases are 15 percent nitrogen, 5 percent helium, 60 percent methane, and 20 percent ethane. This mixture is enclosed in a 4 m³ rigid, well-insulated vessel at 150 kPa and 30°C. A paddle wheel in the vessel is turned until 200 kJ of work have been done on the mixture. Calculate the mixture's final pressure and temperature. *Answers:* 335 K, 166 kPa

**13–55** An insulated tank that contains 1 kg of $O_2$ at 15°C and 300 kPa is connected to a 2-m³ uninsulated tank that contains $N_2$ at 50°C and 500 kPa. The valve connecting the two tanks is opened, and the two gases form a homogeneous mixture at 25°C. Determine (*a*) the final pressure in the tank, (*b*) the heat transfer, and (*c*) the entropy generated during this process. Assume $T_0 = 25°C$. *Answers:* (*a*) 444.6 kPa, (*b*) 187.2 kJ, (*c*) 0.962 kJ/K

**FIGURE P13–55**

**13–56** Reconsider Prob. 13–55. Using EES (or other) software, compare the results obtained assuming ideal-gas behavior with constant specific heats at the average temperature, and using real-gas data obtained from EES by assuming variable specific heats over the temperature range.

**13–57** A mixture of hydrocarbon gases is composed of 60 percent methane, 25 percent propane, and 15 percent butane by weight. This mixture is compressed from 100 kPa and 20°C to 1000 kPa in a reversible, isothermal, steady-flow compressor. Calculate the work and heat transfer for this compression per unit mass of the mixture.

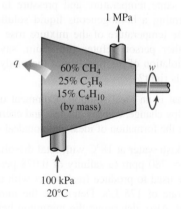

**FIGURE P13–57**

**13–58** An equimolar mixture of helium and argon gases is to be used as the working fluid in a closed-loop gas-turbine cycle. The mixture enters the turbine at 2.5 MPa and 1300 K and expands isentropically to a pressure of 200 kPa. Determine the work output of the turbine per unit mass of the mixture.

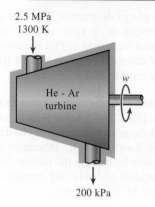

**FIGURE P13–58**

**13–59** A gaseous mixture consists of 75 percent methane and 25 percent ethane by mass. 28,000 m³ of this mixture is trapped in a geological formation as natural gas at 150°C and 14 MPa. This natural gas is pumped 2000 m to the surface. At the surface, the gas pressure is 140 kPa and its temperature is 90°C. Using Kay's rule and the enthalpy-departure charts, calculate the work required to pump this gas. *Answer:* $3.94 \times 10^8$ kJ

**13–60** A mixture of 65 percent $N_2$ and 35 percent $CO_2$ gases (on a mass basis) enters the nozzle of a turbojet engine at 400 kPa and 800 K with a low velocity, and it expands to a pressure of 85 kPa. If the isentropic efficiency of the nozzle is 88 percent, determine (*a*) the exit temperature and (*b*) the exit velocity of the mixture. Assume constant specific heats at room temperature.

**13–61** Reconsider Prob. 13–60. Using EES (or other) software, first solve the stated problem and then, for all other conditions being the same, resolve the problem to determine the composition of the nitrogen and carbon dioxide that is required to have an exit velocity of 670 m/s at the nozzle exit.

**13–62** A piston–cylinder device contains a mixture of 0.8 kg of $H_2$ and 1.2 kg of $N_2$ at 100 kPa and 300 K. Heat is now transferred to the mixture at constant pressure until the volume is doubled. Assuming constant specific heats at the average temperature, determine (*a*) the heat transfer and (*b*) the entropy change of the mixture.

**13–63** Ethane ($C_2H_6$) at 15°C and 300 kPa and methane ($CH_4$) at 60°C and 300 kPa enter an adiabatic mixing chamber. The mass flow rate of ethane is 6 kg/s, which is twice the mass flow rate of methane. Determine (*a*) the mixture temperature and (*b*) the rate of entropy generation during this process, in kW/K.

**13–64** Reconsider Prob. 13–63. Using EES (or other) software, determine the effect of the mass fraction of methane in the mixture on the mixture temperature and the rate of exergy destruction. The total mass flow rate is maintained constant at 9 kg/s, and the mass fraction of methane is varied from 0 to 1. Plot the mixture temperature and the rate of exergy destruction against the mass fraction, and discuss the results. Take $T_0 = 25°C$.

**13–65** A piston–cylinder device contains 6 kg of $H_2$ and 21 kg of $N_2$ at 160 K and 5 MPa. Heat is now transferred to the device, and the mixture expands at constant pressure until the temperature rises to 200 K. Determine the heat transfer during this process by treating the mixture (*a*) as an ideal gas and (*b*) as a nonideal gas and using Amagat's law.
*Answers:* (*a*) 4273 kJ, (*b*) 4745 kJ

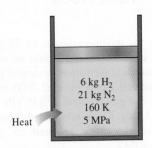

Heat

6 kg $H_2$
21 kg $N_2$
160 K
5 MPa

**FIGURE P13–65**

**13–66** Determine the total entropy change and exergy destruction associated with the process described in Prob. 13–65 by treating the mixture (*a*) as an ideal gas and (*b*) as a nonideal gas and using Amagat's law. Assume constant specific heats at room temperature and take $T_0 = 20°C$.

**13–67** Two mass streams of two different ideal gases are mixed in a steady-flow chamber while receiving energy by heat transfer from the surroundings. The mixing process takes place at constant pressure with no work and negligible changes in kinetic and potential energies. Assume the gases have constant specific heats.

(*a*) Determine the expression for the final temperature of the mixture in terms of the rate of heat transfer to the mixing chamber and the mass flow rates, specific heats, and temperatures of the three mass streams.

(*b*) Obtain an expression for the exit volume flow rate in terms of the rate of heat transfer to the mixing chamber, mixture pressure, universal gas constant, and the specific heats and molar masses of the inlet gases and exit mixture.

(*c*) For the special case of adiabatic mixing, show that the exit volume flow rate is a function of the two inlet volume flow rates and the specific heats and molar masses of the inlets and exit.

(*d*) For the special case of adiabatic mixing of the same ideal gases, show that the exit volume flow rate is a function of the two inlet volume flow rates.

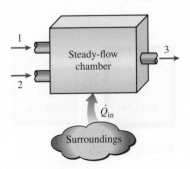

1

Steady-flow chamber

3

2

$\dot{Q}_{in}$

Surroundings

**FIGURE P13–67**

## Special Topic: Chemical Potential and the Separation Work of Mixtures

**13–68C** It is common experience that two gases brought into contact mix by themselves. In the future, could it be possible to invent a process that will enable a mixture to separate into its components by itself without any work (or exergy) input?

**13–69C** A 2-L liquid is mixed with 3 L of another liquid, forming a homogeneous liquid solution at the same temperature and pressure. Can the volume of the solution be more or less than the 5 L? Explain.

**13–70C** A 2-L liquid at 20°C is mixed with 3 L of another liquid at the same temperature and pressure in an adiabatic container, forming a homogeneous liquid solution. Someone claims that the temperature of the mixture rose to 22°C after mixing. Another person refutes the claim, saying that this would be a violation of the first law of thermodynamics. Who do you think is right?

**13–71C** What is an ideal solution? Comment on the volume change, enthalpy change, entropy change, and chemical potential change during the formation of ideal and nonideal solutions.

**13–72** Brackish water at 18°C with total dissolved solid content of TDS = 780 ppm (a salinity of 0.078 percent on mass basis) is to be used to produce fresh water with negligible salt content at a rate of 175 L/s. Determine the minimum power input required. Also, determine the minimum height to which the brackish water must be pumped if fresh water is to be obtained by reverse osmosis using semipermeable membranes.

**13–73** A river is discharging into the ocean at a rate of 150,000 m³/s. Determine the amount of power that can be generated if the river water mixes with the ocean water reversibly. Take the salinity of the ocean to be 2.5 percent on mass basis, and assume both the river and the ocean are at 15°C.

**13–74** [EES] Reconsider Prob. 13–73. Using EES (or other) software, investigate the effect of the salinity of the ocean on the maximum power generated. Let the salinity vary from 0 to 5 percent. Plot the power produced versus the salinity of the ocean, and discuss the results.

**13–75** Fresh water is to be obtained from brackish water at 18°C with a salinity of 0.12 percent on mass basis (or TDS = 1200 ppm). Determine (a) the mole fractions of the water and the salts in the brackish water, (b) the minimum work input required to separate 1 lbm of brackish water completely into pure water and pure salts, and (c) the minimum work input required to obtain 1 kg of fresh water.

**13–76** A desalination plant produces fresh water from seawater at 10°C with a salinity of 3.2 percent on mass basis at a rate of 1.4 $m^3$/s while consuming 8.5 MW of power. The salt content of the fresh water is negligible, and the amount of fresh water produced is a small fraction of the seawater used. Determine the second-law efficiency of this plant.

**13–77** Fresh water is obtained from seawater at a rate of 1.5 $m^3$/s by a desalination plant that consumes 11.5 MW of power and has a second-law efficiency of 20 percent. Determine the power that can be produced if the fresh water produced is mixed with the seawater reversibly.

**13–78** Is it possible for an adiabatic liquid-vapor separator to separate wet steam at 700 kPa and 90 percent quality, so that the pressure of the outlet streams is greater than 700 kPa?

## Review Problems

**13–79** An ideal gas mixture approximation to the makeup of dry air on a percent by volume basis at 100 kPa is as follows: 78 percent $N_2$, 21 percent $O_2$, and 1 percent Ar. Determine the mole fractions, mass fractions, and the partial pressure of each component. Calculate the apparent molar mass, the apparent gas constant, and the constant-pressure specific heat at 300 K for the mixture. Compare your answers with those in Table A-1 and A-2a.

**13–80** The products of combustion of a hydrocarbon fuel and air are composed of 8 kmol $CO_2$, 9 kmol $H_2O$, 4 kmol $O_2$, and 94 kmol $N_2$. If the mixture pressure is 101 kPa, determine the partial pressure of the water vapor in the product gas mixture and the temperature at which the water vapor would begin to condense when the products are cooled a constant pressure.

**13–81** A pipe fitted with a closed valve connects two tanks. One tank contains a 5-kg mixture of 62.5 percent $CO_2$ and 37.5 percent $O_2$ on a mole basis at 30°C and 125 kPa. The second tank contains 10 kg of $N_2$ at 15°C and 200 kPa. The valve in the pipe is opened and the gases are allowed to mix. During the mixing process 100 kJ of heat energy is supplied to the combined tanks. Determine the final pressure and temperature of the mixture and the total volume of the mixture.

**13–82** A piston–cylinder device contains products of combustion from the combustion of a hydrocarbon fuel with air. The combustion process results in a mixture that has the composition on a volume basis as follows: 4.89 percent carbon dioxide, 6.50 percent water vapor, 12.20 percent oxygen, and 76.41 percent nitrogen. This mixture is initially at 1800 K and 1 MPa and expands in an adiabatic, reversible process to 200 kPa. Determine the work done on the piston by the gas, in kJ/kg of mixture. Treat the water vapor as an ideal gas.

**13–83** A mixture of gases consists of 0.1 kg of oxygen, 1 kg of carbon dioxide, and 0.5 kg of helium. This mixture is compressed to 17,500 kPa and 20°C. Determine the mass of this gas contained in a 0.3 $m^3$ tank (a) treating it as an ideal gas mixture, (b) using a compressibility factor based on Dalton's law of additive pressures, (c) using a compressibility factor based on the law of additive volumes, and (d) Kay's psuedocritical pressure and temperature.

**13–84** A gas mixture consists of $O_2$ and $N_2$. The ratio of the mole numbers of $N_2$ to $O_2$ is 3:1. This mixture is heated during a steady-flow process from 180 to 210 K at a constant pressure of 8 MPa. Determine the heat transfer during this process per mole of the mixture, using (a) the ideal-gas approximation and (b) Kay's rule.

**13–85** [EES] Reconsider Prob. 13–84. Using EES (or other) software, investigate the effect of the mole fraction of oxygen in the mixture on heat transfer using real-gas behavior with EES data. Let the mole fraction of oxygen vary from 0 to 1. Plot the heat transfer against the mole fraction, and discuss the results.

**13–86** Determine the total entropy change and exergy destruction associated with the process described in Prob. 13–84, using (a) the ideal-gas approximation and (b) Kay's rule. Assume constant specific heats and $T_0$ = 30°C.

**13–87** A mixture of ideal gases has a specific heat ratio of k = 1.35 and an apparent molecular weight of M = 32 kg/kmol. Determine the work, in kJ/kg, required to compress this mixture isentropically in a closed system from 100 kPa and 15°C to 700 kPa.  *Answer:* 140 kJ/kg

**FIGURE P13–87**

**13–88** A spring-loaded piston–cylinder device contains a mixture of gases whose pressure fractions are 25 percent

Ne, 50 percent $O_2$, and 25 percent $N_2$. The piston diameter and spring are selected for this device such that the volume is 0.1 m³ when the pressure is 200 kPa and 1.0 m³ when the pressure is 1000 kPa. Initially, the gas is added to this device until the pressure is 200 kPa and the temperature is 10°C. The device is now heated until the pressure is 500 kPa. Calculate the total work and heat transfer for this process. *Answers:* 118 kJ, 569 kJ

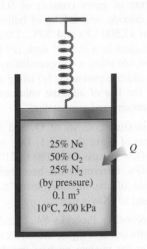

25% Ne
50% $O_2$
25% $N_2$
(by pressure)
0.1 m³
10°C, 200 kPa

$Q$

**FIGURE P13–88**

**13–89** The piston–cylinder device of Prob. 13–88 is filled with a mixture whose mass is 55 percent nitrogen and 45 percent carbon dioxide. Initially, this mixture is at 200 kPa and 45°C. The gas is heated until the volume has doubled. Calculate the total work and heat transfer for this process.

**13–90** Calculate the total work and heat transfer required to triple the initial pressure of the mixture of Prob. 13–89 as it is heated in the spring-loaded piston-cylinder device.

**13–91** A rigid tank contains a mixture of 4 kg of He and 8 kg of $O_2$ at 170 K and 7 MPa. Heat is now transferred to the tank, and the mixture temperature rises to 220 K. Treating the He as an ideal gas and the $O_2$ as a nonideal gas, determine (*a*) the final pressure of the mixture and (*b*) the heat transfer.

**13–92** The mass fractions of a mixture of gases are 15 percent nitrogen, 5 percent helium, 60 percent methane; and 20 percent ethane. This mixture is expanded from 2800 kPa and 260°C to 150 kPa in an adiabatic, steady-flow turbine of 85 percent isentropic efficiency. Calculate the second law efficiency and the exergy destruction during this expansion process. Take $T_0 = 25$°C. *Answers:* 84.2 percent, 91.6 kJ/kg

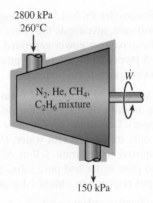

2800 kPa
260°C

$\dot{W}$

$N_2$, He, $CH_4$,
$C_2H_6$ mixture

150 kPa

**FIGURE P13–92**

**13–93** Using EES (or other) software, write a program to determine the mole fractions of the components of a mixture of three gases with known molar masses when the mass fractions are given, and to determine the mass fractions of the components when the mole fractions are given. Run the program for a sample case, and give the results.

**13–94** Using EES (or other) software, write a program to determine the apparent gas constant, constant volume specific heat, and internal energy of a mixture of three ideal gases when the mass fractions and other properties of the constituent gases are given. Run the program for a sample case, and give the results.

**13–95** Using Amagat's law, show that

$$Z_m = \sum_{i=1}^{k} y_i Z_i$$

for a real-gas mixture of $k$ gases, where $Z$ is the compressibility factor.

**Fundamentals of Engineering (FE) Exam Problems**

**13–96** An ideal-gas mixture whose apparent molar mass is 20 kg/kmol consists of $N_2$ and three other gases. If the mole fraction of nitrogen is 0.55, its mass fraction is
(*a*) 0.15          (*b*) 0.23          (*c*) 0.39
(*d*) 0.55          (*e*) 0.77

**13–97** An ideal-gas mixture consists of 2 kmol of $N_2$ and 6 kmol of $CO_2$. The mass fraction of $CO_2$ in the mixture is
(*a*) 0.175          (*b*) 0.250          (*c*) 0.500
(*d*) 0.750          (*e*) 0.875

**13–98** An ideal-gas mixture consists of 2 kmol of $N_2$ and 4 kmol of $CO_2$. The apparent gas constant of the mixture is
(*a*) 0.215 kJ/kg·K          (*b*) 0.225 kJ/kg·K
(*c*) 0.243 kJ/kg·K          (*d*) 0.875 kJ/kg·K
(*e*) 1.24 kJ/kg·K

**13–99** A rigid tank is divided into two compartments by a partition. One compartment contains 3 kmol of $N_2$ at 400 kPa and the other compartment contains 7 kmol of $CO_2$ at 200 kPa. Now the partition is removed, and the two gases form a homogeneous mixture at 250 kPa. The partial pressure of $N_2$ in the mixture is

(a) 75 kPa      (b) 90 kPa      (c) 125 kPa
(d) 175 kPa     (e) 250 kPa

**13–100** An 80-L rigid tank contains an ideal-gas mixture of 5 g of $N_2$ and 5 g of $CO_2$ at a specified pressure and temperature. If $N_2$ were separated from the mixture and stored at mixture temperature and pressure, its volume would be

(a) 32 L      (b) 36 L      (c) 40 L
(d) 49 L      (e) 80 L

**13–101** An ideal-gas mixture consists of 3 kg of Ar and 6 kg of $CO_2$ gases. The mixture is now heated at constant volume from 250 K to 350 K. The amount of heat transfer is

(a) 374 kJ      (b) 436 kJ      (c) 488 kJ
(d) 525 kJ      (e) 664 kJ

**13–102** An ideal-gas mixture consists of 60 percent helium and 40 percent argon gases by mass. The mixture is now expanded isentropically in a turbine from 400°C and 1.2 MPa to a pressure of 200 kPa. The mixture temperature at turbine exit is

(a) 56°C      (b) 195°C      (c) 130°C
(d) 112°C     (e) 400°C

**13–103** One compartment of an insulated rigid tank contains 2 kmol of $CO_2$ at 20°C and 150 kPa while the other compartment contains 5 kmol of $H_2$ gas at 35°C and 300 kPa. Now the partition between the two gases is removed, and the two gases form a homogeneous ideal-gas mixture. The temperature of the mixture is

(a) 25°C      (b) 29°C      (c) 22°C
(d) 32°C      (e) 34°C

**13–104** A piston–cylinder device contains an ideal-gas mixture of 3 kmol of He gas and 7 kmol of Ar gas at 50°C and 400 kPa. Now the gas expands at constant pressure until its volume doubles. The amount of heat transfer to the gas mixture is

(a) 6.2 MJ      (b) 4.2 MJ      (c) 27 MJ
(d) 10 MJ       (e) 67 MJ

**13–105** An ideal-gas mixture of helium and argon gases with identical mass fractions enters a turbine at 1500 K and 1 MPa at a rate of 0.12 kg/s, and expands isentropically to 100 kPa. The power output of the turbine is

(a) 253 kW      (b) 310 kW      (c) 341 kW
(d) 463 kW      (e) 550 kW

## Design and Essay Problems

**13–106** The simple additive rule may not be appropriate for the volume of binary mixtures of gases, Prove this for a pair of gases of your choice at several different temperatures and pressures using Kay's rule and the principle of corresponding states.

**13–107** You have a rigid tank equipped with a pressure gauge. Describe a procedure by which you could use this tank to blend ideal gases in prescribed mole-fraction portions.

**13–108** Prolonged exposure to mercury even at relatively low but toxic concentrations in the air is known to cause permanent mental disorders, insomnia, and pain and numbness in the hands and the feet, among other things. Therefore, the maximum allowable concentration of mercury vapor in the air at work places is regulated by federal agencies. These regulations require that the average level of mercury concentration in the air does not exceed 0.1 mg/m³.

Consider a mercury spill that occurs in an airtight storage room at 20°C in San Francisco during an earthquake. Calculate the highest level of mercury concentration in the air that can occur in the storage room, in mg/m³, and determine if it is within the safe level. The vapor pressure of mercury at 20°C is 0.173 Pa. Propose some guidelines to safeguard against the formation of toxic concentrations of mercury vapor in air in storage rooms and laboratories.

# GAS-VAPOR MIXTURES AND AIR-CONDITIONING

At temperatures below the critical temperature, the gas phase of a substance is frequently referred to as a *vapor*. The term *vapor* implies a gaseous state that is close to the saturation region of the substance, raising the possibility of condensation during a process.

In Chap. 13, we discussed mixtures of gases that are usually above their critical temperatures. Therefore, we were not concerned about any of the gases condensing during a process. Not having to deal with two phases greatly simplified the analysis. When we are dealing with a gas–vapor mixture, however, the vapor may condense out of the mixture during a process, forming a two-phase mixture. This may complicate the analysis considerably. Therefore, a gas–vapor mixture needs to be treated differently from an ordinary gas mixture.

Several gas–vapor mixtures are encountered in engineering. In this chapter, we consider the *air–water vapor mixture,* which is the most commonly encountered gas–vapor mixture in practice. We also discuss *air-conditioning,* which is the primary application area of air–water vapor mixtures.

## OBJECTIVES

The objectives of Chapter 14 are to:

■ Differentiate between dry air and atmospheric air.

■ Define and calculate the specific and relative humidity of atmospheric air.

■ Calculate the dew-point temperature of atmospheric air.

■ Relate the adiabatic saturation temperature and wet-bulb temperatures of atmospheric air.

■ Use the psychrometric chart as a tool to determine the properties of atmospheric air.

■ Apply the principles of the conservation of mass and energy to various air-conditioning processes.

| | Dry air | |
|---|---|---|
| | $T$, °C | $c_p$, kJ/kg·°C |
| | −10 | 1.0038 |
| | 0 | 1.0041 |
| | 10 | 1.0045 |
| | 20 | 1.0049 |
| | 30 | 1.0054 |
| | 40 | 1.0059 |
| | 50 | 1.0065 |

**FIGURE 14–1**

The $c_p$ of air can be assumed to be constant at 1.005 kJ/kg·°C in the temperature range −10 to 50°C with an error under 0.2 percent.

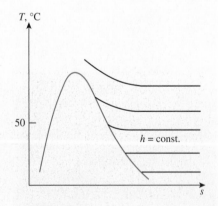

**FIGURE 14–2**

At temperatures below 50°C, the $h$ = constant lines coincide with the $T$ = constant lines in the superheated vapor region of water.

# 14–1 · DRY AND ATMOSPHERIC AIR

Air is a mixture of nitrogen, oxygen, and small amounts of some other gases. Air in the atmosphere normally contains some water vapor (or *moisture*) and is referred to as **atmospheric air**. By contrast, air that contains no water vapor is called **dry air**. It is often convenient to treat air as a mixture of water vapor and dry air since the composition of dry air remains relatively constant, but the amount of water vapor changes as a result of condensation and evaporation from oceans, lakes, rivers, showers, and even the human body. Although the amount of water vapor in the air is small, it plays a major role in human comfort. Therefore, it is an important consideration in air-conditioning applications.

The temperature of air in air-conditioning applications ranges from about −10 to about 50°C. In this range, dry air can be treated as an ideal gas with a constant $c_p$ value of 1.005 kJ/kg·K with negligible error (under 0.2 percent), as illustrated in Fig. 14–1. Taking 0°C as the reference temperature, the enthalpy and enthalpy change of dry air can be determined from

$$h_{\text{dry air}} = c_p T = (1.005 \text{ kJ/kg·°C})T \quad \text{(kJ/kg)} \qquad \text{(14–1a)}$$

and

$$\Delta h_{\text{dry air}} = c_p \Delta T = (1.005 \text{ kJ/kg·°C}) \, \Delta T \quad \text{(kJ/kg)} \qquad \text{(14–1b)}$$

where $T$ is the air temperature in °C and $\Delta T$ is the change in temperature. In air-conditioning processes we are concerned with the *changes* in enthalpy $\Delta h$, which is independent of the reference point selected.

It certainly would be very convenient to also treat the water vapor in the air as an ideal gas and you would probably be willing to sacrifice some accuracy for such convenience. Well, it turns out that we can have the convenience without much sacrifice. At 50°C, the saturation pressure of water is 12.3 kPa. At pressures below this value, water vapor can be treated as an ideal gas with negligible error (under 0.2 percent), even when it is a saturated vapor. Therefore, water vapor in air behaves as if it existed alone and obeys the ideal-gas relation $Pv = RT$. Then the atmospheric air can be treated as an ideal-gas mixture whose pressure is the sum of the partial pressure of dry air* $P_a$ and that of water vapor $P_v$:

$$P = P_a + P_v \quad \text{(kPa)} \qquad \text{(14–2)}$$

The partial pressure of water vapor is usually referred to as the **vapor pressure**. It is the pressure water vapor would exert if it existed alone at the temperature and volume of atmospheric air.

Since water vapor is an ideal gas, the enthalpy of water vapor is a function of temperature only, that is, $h = h(T)$. This can also be observed from the *T-s* diagram of water given in Fig. A–9 and Fig. 14–2 where the constant-enthalpy lines coincide with constant-temperature lines at temperatures below 50°C. Therefore, *the enthalpy of water vapor in air can be taken to be equal to the enthalpy of saturated vapor at the same temperature.* That is,

$$h_v(T, \text{ low } P) \cong h_g(T) \qquad \text{(14–3)}$$

---

* Throughout this chapter, the subscript *a* denotes dry air and the subscript *v* denotes water vapor.

The enthalpy of water vapor at 0°C is 2500.9 kJ/kg. The average $c_p$ value of water vapor in the temperature range −10 to 50°C can be taken to be 1.82 kJ/kg·°C. Then the enthalpy of water vapor can be determined approximately from

$$h_g(T) \cong 2500.9 + 1.82T \quad \text{(kJ/kg)} \quad T \text{ in } °C \tag{14-4}$$

or

$$h_g(T) \cong 1060.9 + 0.435T \quad \text{(Btu/lbm)} \quad T \text{ in } °F \tag{14-5}$$

in the temperature range −10 to 50°C (or 15 to 120°F), with negligible error, as shown in Fig. 14–3.

## 14–2 · SPECIFIC AND RELATIVE HUMIDITY OF AIR

The amount of water vapor in the air can be specified in various ways. Probably the most logical way is to specify directly the mass of water vapor present in a unit mass of dry air. This is called **absolute** or **specific humidity** (also called *humidity ratio*) and is denoted by $\omega$:

$$\omega = \frac{m_v}{m_a} \quad \text{(kg water vapor/kg dry air)} \tag{14-6}$$

The specific humidity can also be expressed as

$$\omega = \frac{m_v}{m_a} = \frac{P_v V/R_v T}{P_a V/R_a T} = \frac{P_v/R_v}{P_a/R_a} = 0.622\frac{P_v}{P_a} \tag{14-7}$$

or

$$\omega = \frac{0.622 P_v}{P - P_v} \quad \text{(kg water vapor/kg dry air)} \tag{14-8}$$

where $P$ is the total pressure.

Consider 1 kg of dry air. By definition, dry air contains no water vapor, and thus its specific humidity is zero. Now let us add some water vapor to this dry air. The specific humidity will increase. As more vapor or moisture is added, the specific humidity will keep increasing until the air can hold no more moisture. At this point, the air is said to be saturated with moisture, and it is called **saturated air**. Any moisture introduced into saturated air will condense. The amount of water vapor in saturated air at a specified temperature and pressure can be determined from Eq. 14–8 by replacing $P_v$ by $P_g$, the saturation pressure of water at that temperature (Fig. 14–4).

The amount of moisture in the air has a definite effect on how comfortable we feel in an environment. However, the comfort level depends more on the amount of moisture the air holds ($m_v$) relative to the maximum amount of moisture the air can hold at the same temperature ($m_g$). The ratio of these two quantities is called the **relative humidity** $\phi$ (Fig. 14–5)

$$\phi = \frac{m_v}{m_g} = \frac{P_v V/R_v T}{P_g V/R_v T} = \frac{P_v}{P_g} \tag{14-9}$$

| | Water vapor | | |
|---|---|---|---|
| | $h_g$, kJ/kg | | Difference, |
| $T$, °C | Table A-4 | Eq. 14-4 | kJ/kg |
| −10 | 2482.1 | 2482.7 | −0.6 |
| 0 | 2500.9 | 2500.9 | 0.0 |
| 10 | 2519.2 | 2519.1 | 0.1 |
| 20 | 2537.4 | 2537.3 | 0.1 |
| 30 | 2555.6 | 2555.5 | 0.1 |
| 40 | 2573.5 | 2573.7 | −0.2 |
| 50 | 2591.3 | 2591.9 | −0.6 |

**FIGURE 14–3**

In the temperature range −10 to 50°C, the $h_g$ of water can be determined from Eq. 14–4 with negligible error.

**FIGURE 14–4**

For saturated air, the vapor pressure is equal to the saturation pressure of water.

**FIGURE 14–5**
Specific humidity is the actual amount of water vapor in 1 kg of dry air, whereas relative humidity is the ratio of the actual amount of moisture in the air at a given temperature to the maximum amount of moisture air can hold at the same temperature.

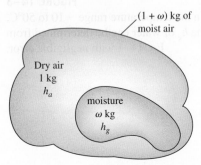

$$h = h_a + \omega h_g, \text{ kJ/kg dry air}$$

**FIGURE 14–6**
The enthalpy of moist (atmospheric) air is expressed per unit mass of dry air, not per unit mass of moist air.

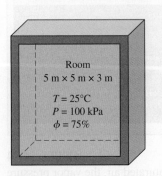

**FIGURE 14–7**
Schematic for Example 14–1.

where

$$P_g = P_{\text{sat @ } T} \tag{14–10}$$

Combining Eqs. 14–8 and 14–9, we can also express the relative humidity as

$$\phi = \frac{\omega P}{(0.622 + \omega)P_g} \quad \text{and} \quad \omega = \frac{0.622\phi P_g}{P - \phi P_g} \tag{14–11a, b}$$

The relative humidity ranges from 0 for dry air to 1 for saturated air. Note that the amount of moisture air can hold depends on its temperature. Therefore, the relative humidity of air changes with temperature even when its specific humidity remains constant.

Atmospheric air is a mixture of dry air and water vapor, and thus the enthalpy of air is expressed in terms of the enthalpies of the dry air and the water vapor. In most practical applications, the amount of dry air in the air–water vapor mixture remains constant, but the amount of water vapor changes. Therefore, the enthalpy of atmospheric air is expressed *per unit mass of dry air* instead of per unit mass of the air–water vapor mixture.

The total enthalpy (an extensive property) of atmospheric air is the sum of the enthalpies of dry air and the water vapor:

$$H = H_a + H_v = m_a h_a + m_v h_v$$

Dividing by $m_a$ gives

$$h = \frac{H}{m_a} = h_a + \frac{m_v}{m_a} h_v = h_a + \omega h_v$$

or

$$h = h_a + \omega h_g \quad \text{(kJ/kg dry air)} \tag{14–12}$$

since $h_v \cong h_g$ (Fig. 14–6).

Also note that the ordinary temperature of atmospheric air is frequently referred to as the **dry-bulb temperature** to differentiate it from other forms of temperatures that shall be discussed.

---

**EXAMPLE 14–1**   **The Amount of Water Vapor in Room Air**

A 5-m × 5-m × 3-m room shown in Fig. 14–7 contains air at 25°C and 100 kPa at a relative humidity of 75 percent. Determine (a) the partial pressure of dry air, (b) the specific humidity, (c) the enthalpy per unit mass of the dry air, and (d) the masses of the dry air and water vapor in the room.

**SOLUTION**  The relative humidity of air in a room is given. The dry air pressure, specific humidity, enthalpy, and the masses of dry air and water vapor in the room are to be determined.
*Assumptions*  The dry air and the water vapor in the room are ideal gases.
*Properties*  The constant-pressure specific heat of air at room temperature is $c_p = 1.005$ kJ/kg·K (Table A–2a). For water at 25°C, we have $T_{\text{sat}} = 3.1698$ kPa and $h_g = 2546.5$ kJ/kg (Table A–4).
*Analysis*  (a) The partial pressure of dry air can be determined from Eq. 14–2:

$$P_a = P - P_v$$

where

$$P_v = \phi P_g = \phi P_{\text{sat @ 25°C}} = (0.75)(3.1698 \text{ kPa}) = 2.38 \text{ kPa}$$

Thus,

$$P_a = (100 - 2.38) \text{ kPa} = \textbf{97.62 kPa}$$

(b) The specific humidity of air is determined from Eq. 14–8:

$$\omega = \frac{0.622 P_v}{P - P_v} = \frac{(0.622)(2.38 \text{ kPa})}{(100 - 2.38) \text{ kPa}} = \textbf{0.0152 kg H}_2\textbf{O/kg dry air}$$

(c) The enthalpy of air per unit mass of dry air is determined from Eq. 14–12:

$$h = h_a + \omega h_v \cong c_p T + \omega h_g$$

$$= (1.005 \text{ kJ/kg·°C})(25°C) + (0.0152)(2546.5 \text{ kJ/kg})$$

$$= \textbf{63.8 kJ/kg dry air}$$

The enthalpy of water vapor (2546.5 kJ/kg) could also be determined from the approximation given by Eq. 14–4:

$$h_{g \text{ @ 25°C}} \cong 2500.9 + 1.82(25) = 2546.4 \text{ kJ/kg}$$

which is almost identical to the value obtained from Table A–4.

(d) Both the dry air and the water vapor fill the entire room completely. Therefore, the volume of each gas is equal to the volume of the room:

$$V_a = V_v = V_{\text{room}} = (5 \text{ m})(5 \text{ m})(3 \text{ m}) = 75 \text{ m}^3$$

The masses of the dry air and the water vapor are determined from the ideal-gas relation applied to each gas separately:

$$m_a = \frac{P_a V_a}{R_a T} = \frac{(97.62 \text{ kPa})(75 \text{ m}^3)}{(0.287 \text{ kPa·m}^3\text{/kg·K})(298 \text{ K})} = \textbf{85.61 kg}$$

$$m_v = \frac{P_v V_v}{R_v T} = \frac{(2.38 \text{ kPa})(75 \text{ m}^3)}{(0.4615 \text{ kPa·m}^3\text{/kg·K})(298 \text{ K})} = \textbf{1.30 kg}$$

The mass of the water vapor in the air could also be determined from Eq. 14–6:

$$m_v = \omega m_a = (0.0152)(85.61 \text{ kg}) = 1.30 \text{ kg}$$

# 14–3 · DEW-POINT TEMPERATURE

If you live in a humid area, you are probably used to waking up most summer mornings and finding the grass wet. You know it did not rain the night before. So what happened? Well, the excess moisture in the air simply condensed on the cool surfaces, forming what we call *dew*. In summer, a considerable amount of water vaporizes during the day. As the temperature falls during the night, so does the "moisture capacity" of air, which is the maximum amount of moisture air can hold. (What happens to the relative humidity during this process?) After a while, the moisture capacity of air equals its moisture content. At this point, air is saturated, and its relative humidity is

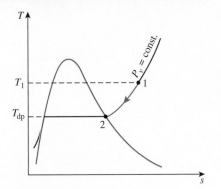

**FIGURE 14–8**
Constant-presssure cooling of moist air and the dew-point temperature on the *T-s* diagram of water.

Moist air

Liquid water droplets (dew)

$T < T_{dp}$

**FIGURE 14–9**
When the temperature of a cold drink is below the dew-point temperature of the surrounding air, it "sweats."

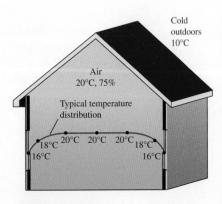

Cold outdoors 10°C

Air 20°C, 75%

Typical temperature distribution

18°C  20°C  20°C  20°C  18°C
16°C                    16°C

**FIGURE 14–10**
Schematic for Example 14–2.

100 percent. Any further drop in temperature results in the condensation of some of the moisture, and this is the beginning of dew formation.

The **dew-point temperature** $T_{dp}$ is defined as *the temperature at which condensation begins when the air is cooled at constant pressure*. In other words, $T_{dp}$ is the saturation temperature of water corresponding to the vapor pressure:

$$T_{dp} = T_{sat @ P_v} \qquad (14\text{–}13)$$

This is also illustrated in Fig. 14–8. As the air cools at constant pressure, the vapor pressure $P_v$ remains constant. Therefore, the vapor in the air (state 1) undergoes a constant-pressure cooling process until it strikes the saturated vapor line (state 2). The temperature at this point is $T_{dp}$, and if the temperature drops any further, some vapor condenses out. As a result, the amount of vapor in the air decreases, which results in a decrease in $P_v$. The air remains saturated during the condensation process and thus follows a path of 100 percent relative humidity (the saturated vapor line). The ordinary temperature and the dew-point temperature of saturated air are identical.

You have probably noticed that when you buy a cold canned drink from a vending machine on a hot and humid day, dew forms on the can. The formation of dew on the can indicates that the temperature of the drink is below the dew-point temperature of the surrounding air (Fig. 14–9).

The dew-point temperature of room air can be determined easily by cooling some water in a metal cup by adding small amounts of ice and stirring. The temperature of the outer surface of the cup when dew starts to form on the surface is the dew-point temperature of the air.

---

**EXAMPLE 14–2     Fogging of the Windows in a House**

In cold weather, condensation frequently occurs on the inner surfaces of the windows due to the lower air temperatures near the window surface. Consider a house, shown in Fig. 14–10, that contains air at 20°C and 75 percent relative humidity. At what window temperature will the moisture in the air start condensing on the inner surfaces of the windows?

**SOLUTION** The interior of a house is maintained at a specified temperature and humidity. The window temperature at which fogging starts is to be determined.

**Properties** The saturation pressure of water at 20°C is $P_{sat} = 2.3392$ kPa (Table A–4).

**Analysis** The temperature distribution in a house, in general, is not uniform. When the outdoor temperature drops in winter, so does the indoor temperature near the walls and the windows. Therefore, the air near the walls and the windows remains at a lower temperature than at the inner parts of a house even though the total pressure and the vapor pressure remain constant throughout the house. As a result, the air near the walls and the windows undergoes a $P_v = constant$ cooling process until the moisture in the air starts condensing. This happens when the air reaches its dew-point temperature $T_{dp}$, which is determined from Eq. 14–13 to be

$$T_{dp} = T_{sat @ P_v}$$

where

$$P_v = \phi P_{g\ @\ 20°C} = (0.75)(2.3392\ \text{kPa}) = 1.754\ \text{kPa}$$

Thus,

$$T_{dp} = T_{sat\ @\ 1.754\ kPa} = \mathbf{15.4\ °C}$$

**Discussion** Note that the inner surface of the window should be maintained above 15.4°C if condensation on the window surfaces is to be avoided.

## 14–4 · ADIABATIC SATURATION AND WET-BULB TEMPERATURES

Relative humidity and specific humidity are frequently used in engineering and atmospheric sciences, and it is desirable to relate them to easily measurable quantities such as temperature and pressure. One way of determining the relative humidity is to determine the dew-point temperature of air, as discussed in the last section. Knowing the dew-point temperature, we can determine the vapor pressure $P_v$ and thus the relative humidity. This approach is simple, but not quite practical.

Another way of determining the absolute or relative humidity is related to an *adiabatic saturation process,* shown schematically and on a *T-s* diagram in Fig. 14–11. The system consists of a long insulated channel that contains a pool of water. A steady stream of unsaturated air that has a specific humidity of $\omega_1$ (unknown) and a temperature of $T_1$ is passed through this channel. As the air flows over the water, some water evaporates and mixes with the airstream. The moisture content of air increases during this process, and its temperature decreases, since part of the latent heat of vaporization of the water that evaporates comes from the air. If the channel is long enough, the airstream exits as saturated air ($\phi = 100$ percent) at temperature $T_2$, which is called the **adiabatic saturation temperature**.

If makeup water is supplied to the channel at the rate of evaporation at temperature $T_2$, the adiabatic saturation process described above can be analyzed as a steady-flow process. The process involves no heat or work interactions, and the kinetic and potential energy changes can be neglected. Then the conservation of mass and conservation of energy relations for this two-inlet, one-exit steady-flow system reduces to the following:

*Mass balance:*

$$\dot{m}_{a_1} = \dot{m}_{a_2} = \dot{m}_a \qquad \text{(The mass flow rate of dry air remains constant)}$$

$$\dot{m}_{w_1} + \dot{m}_f = \dot{m}_{w_2} \qquad \text{(The mass flow rate of vapor in the air increases by an amount equal to the rate of evaporation } \dot{m}_f\text{)}$$

or

$$\dot{m}_a\omega_1 + \dot{m}_f = \dot{m}_a\omega_2$$

Thus,

$$\dot{m}_f = \dot{m}_a(\omega_2 - \omega_1)$$

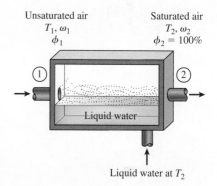

Unsaturated air
$T_1, \omega_1$
$\phi_1$

Saturated air
$T_2, \omega_2$
$\phi_2 = 100\%$

Liquid water

Liquid water at $T_2$

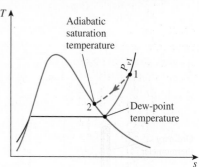

**FIGURE 14–11**
The adiabatic saturation process and its representation on a *T-s* diagram of water.

*Energy balance:*

$$\dot{E}_{in} = \dot{E}_{out} \quad (\text{since } \dot{Q} = 0 \text{ and } \dot{W} = 0)$$

$$\dot{m}_a h_1 + \dot{m}_f h_{f_2} = \dot{m}_a h_2$$

or

$$\dot{m}_a h_1 + \dot{m}_a (\omega_2 - \omega_1) h_{f_2} = \dot{m}_a h_2$$

Dividing by $\dot{m}_a$ gives

$$h_1 + (\omega_2 - \omega_1) h_{f_2} = h_2$$

or

$$(c_p T_1 + \omega_1 h_{g_1}) + (\omega_2 - \omega_1) h_{f_2} = (c_p T_2 + \omega_2 h_{g_2})$$

which yields

$$\omega_1 = \frac{c_p(T_2 - T_1) + \omega_2 h_{fg_2}}{h_{g_1} - h_{f_2}} \tag{14–14}$$

where, from Eq. 14–11*b*,

$$\omega_2 = \frac{0.622 P_{g_2}}{P_2 - P_{g_2}} \tag{14–15}$$

since $\phi_2 = 100$ percent. Thus we conclude that the specific humidity (and relative humidity) of air can be determined from Eqs. 14–14 and 14–15 by measuring the pressure and temperature of air at the inlet and the exit of an adiabatic saturator.

If the air entering the channel is already saturated, then the adiabatic saturation temperature $T_2$ will be identical to the inlet temperature $T_1$, in which case Eq. 14–14 yields $\omega_1 = \omega_2$. In general, the adiabatic saturation temperature is between the inlet and dew-point temperatures.

The adiabatic saturation process discussed above provides a means of determining the absolute or relative humidity of air, but it requires a long channel or a spray mechanism to achieve saturation conditions at the exit. A more practical approach is to use a thermometer whose bulb is covered with a cotton wick saturated with water and to blow air over the wick, as shown in Fig. 14–12. The temperature measured in this manner is called the **wet-bulb temperature** $T_{wb}$, and it is commonly used in air-conditioning applications.

The basic principle involved is similar to that in adiabatic saturation. When unsaturated air passes over the wet wick, some of the water in the wick evaporates. As a result, the temperature of the water drops, creating a temperature difference (which is the driving force for heat transfer) between the air and the water. After a while, the heat loss from the water by evaporation equals the heat gain from the air, and the water temperature stabilizes. The thermometer reading at this point is the wet-bulb temperature. The wet-bulb temperature can also be measured by placing the wet-wicked thermometer in a holder attached to a handle and rotating the holder rapidly, that is, by moving the thermometer instead of the air. A device that works on this principle is called a *sling psychrometer* and is shown in Fig. 14–13. Usually a dry-bulb thermometer is also mounted on the frame of this device so that both the wet- and dry-bulb temperatures can be read simultaneously.

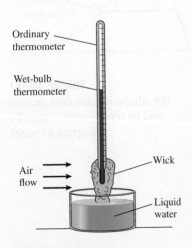

Ordinary thermometer

Wet-bulb thermometer

Air flow

Wick

Liquid water

**FIGURE 14–12**

A simple arrangement to measure the wet-bulb temperature.

Advances in electronics made it possible to measure humidity directly in a fast and reliable way. It appears that sling psychrometers and wet-wicked thermometers are about to become things of the past. Today, hand-held electronic humidity measurement devices based on the capacitance change in a thin polymer film as it absorbs water vapor are capable of sensing and digitally displaying the relative humidity within 1 percent accuracy in a matter of seconds.

In general, the adiabatic saturation temperature and the wet-bulb temperature are not the same. However, for air–water vapor mixtures at atmospheric pressure, the wet-bulb temperature happens to be approximately equal to the adiabatic saturation temperature. Therefore, the wet-bulb temperature $T_{wb}$ can be used in Eq. 14–14 in place of $T_2$ to determine the specific humidity of air.

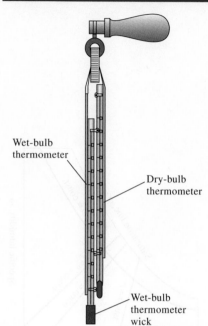

**FIGURE 14–13**
Sling psychrometer.

■ **EXAMPLE 14–3     The Specific and Relative Humidity of Air**

■ The dry- and the wet-bulb temperatures of atmospheric air at 1 atm
■ (101.325 kPa) pressure are measured with a sling psychrometer and
■ determined to be 25 and 15°C, respectively. Determine (*a*) the specific
■ humidity, (*b*) the relative humidity, and (*c*) the enthalpy of the air.

**SOLUTION**   Dry- and wet-bulb temperatures are given. The specific humidity, relative humidity, and enthalpy are to be determined.
**Properties**   The saturation pressure of water is 1.7057 kPa at 15°C, and 3.1698 kPa at 25°C (Table A–4). The constant-pressure specific heat of air at room temperature is $c_p = 1.005$ kJ/kg·K (Table A–2a).
**Analysis**   (*a*) The specific humidity $\omega_1$ is determined from Eq. 14–14,

$$\omega_1 = \frac{c_p(T_2 - T_1) + \omega_2 h_{fg_2}}{h_{g_1} - h_{f_2}}$$

where $T_2$ is the wet-bulb temperature and $\omega_2$ is

$$\omega_2 = \frac{0.622 P_{g_2}}{P_2 - P_{g_2}} = \frac{(0.622)(1.7057 \text{ kPa})}{(101.325 - 1.7057) \text{ kPa}}$$

$$= 0.01065 \text{ kg H}_2\text{O/kg dry air}$$

Thus,

$$\omega_1 = \frac{(1.005 \text{ kJ/kg·°C})[(15 - 25)°C] + (0.01065)(2465.4 \text{ kJ/kg})}{(2546.5 - 62.982) \text{ kJ/kg}}$$

$$= \mathbf{0.00653 \text{ kg H}_2\text{O/kg dry air}}$$

(*b*) The relative humidity $\phi_1$ is determined from Eq. 14–11a to be

$$\phi_1 = \frac{\omega_1 P_2}{(0.622 + \omega_1)P_{g_1}} = \frac{(0.00653)(101.325 \text{ kPa})}{(0.622 + 0.00653)(3.1698 \text{ kPa})} = \mathbf{0.332 \text{ or } 33.2\%}$$

(c) The enthalpy of air per unit mass of dry air is determined from Eq. 14–12:

$$h_1 = h_{a_1} + \omega_1 h_{v_1} \cong c_p T_1 + \omega_1 h_{g_1}$$

$$= (1.005 \text{ kJ/kg·°C})(25°C) + (0.00653)(2546.5 \text{ kJ/kg})$$

$$= \textbf{41.8 kJ/kg dry air}$$

**Discussion** The previous property calculations can be performed easily using EES or other programs with built-in psychrometric functions.

# 14–5 · THE PSYCHROMETRIC CHART

The state of the atmospheric air at a specified pressure is completely specified by two independent intensive properties. The rest of the properties can be calculated easily from the previous relations. The sizing of a typical air-conditioning system involves numerous such calculations, which may eventually get on the nerves of even the most patient engineers. Therefore, there is clear motivation to computerize calculations or to do these calculations once and to present the data in the form of easily readable charts. Such charts are called **psychrometric charts**, and they are used extensively in air-conditioning applications. A psychrometric chart for a pressure of 1 atm (101.325 kPa) is given in Fig. A–31 in SI units. Psychrometric charts at other pressures (for use at considerably higher elevations than sea level) are also available.

The basic features of the psychrometric chart are illustrated in Fig. 14–14. The dry-bulb temperatures are shown on the horizontal axis, and the specific humidity is shown on the vertical axis. (Some charts also show the vapor pressure on the vertical axis since at a fixed total pressure $P$ there is a one-to-one correspondence between the specific humidity $\omega$ and the vapor pressure $P_v$, as can be seen from Eq. 14–8.) On the left end of the chart, there is a curve (called the *saturation line*) instead of a straight line. All the saturated air states are located on this curve. Therefore, it is also the curve of 100 percent relative humidity. Other constant relative-humidity curves have the same general shape.

Lines of constant wet-bulb temperature have a downhill appearance to the right. Lines of constant specific volume (in m³/kg dry air) look similar, except they are steeper. Lines of constant enthalpy (in kJ/kg dry air) lie very nearly parallel to the lines of constant wet-bulb temperature. Therefore, the constant-wet-bulb-temperature lines are used as constant-enthalpy lines in some charts.

For saturated air, the dry-bulb, wet-bulb, and dew-point temperatures are identical (Fig. 14–15). Therefore, the dew-point temperature of atmospheric air at any point on the chart can be determined by drawing a horizontal line (a line of $\omega = $ constant or $P_v = $ constant) from the point to the saturated curve. The temperature value at the intersection point is the dew-point temperature.

The psychrometric chart also serves as a valuable aid in visualizing the air-conditioning processes. An ordinary heating or cooling process, for example, appears as a horizontal line on this chart if no humidification or dehumidification is involved (that is, $\omega = $ constant). Any deviation from a horizontal line indicates that moisture is added or removed from the air during the process.

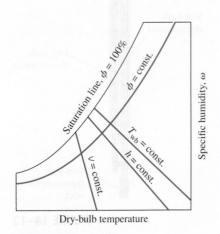

**FIGURE 14–14**
Schematic for a psychrometric chart.

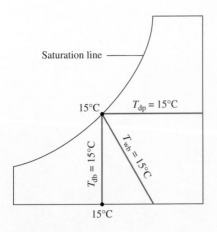

**FIGURE 14–15**
For saturated air, the dry-bulb, wet-bulb, and dew-point temperatures are identical.

■ **EXAMPLE 14–4**    **The Use of the Psychrometric Chart**

Consider a room that contains air at 1 atm, 35°C, and 40 percent relative humidity. Using the psychrometric chart, determine (a) the specific humidity, (b) the enthalpy, (c) the wet-bulb temperature, (d) the dew-point temperature, and (e) the specific volume of the air.

**SOLUTION** The relative humidity of air in a room is given. The specific humidity, enthalpy, wet-bulb temperature, dew-point temperature, and specific volume of the air are to be determined using the psychrometric chart.

**Analysis** At a given total pressure, the state of atmospheric air is completely specified by two independent properties such as the dry-bulb temperature and the relative humidity. Other properties are determined by directly reading their values at the specified state.

(a) The specific humidity is determined by drawing a horizontal line from the specified state to the right until it intersects with the $\omega$ axis, as shown in Fig. 14–16. At the intersection point we read

$$\omega = 0.0142 \text{ kg } H_2O/\text{kg dry air}$$

(b) The enthalpy of air per unit mass of dry air is determined by drawing a line parallel to the $h$ = constant lines from the specific state until it intersects the enthalpy scale, giving

$$h = 71.5 \text{ kJ/kg dry air}$$

(c) The wet-bulb temperature is determined by drawing a line parallel to the $T_{wb}$ = constant lines from the specified state until it intersects the saturation line, giving

$$T_{wb} = 24°C$$

(d) The dew-point temperature is determined by drawing a horizontal line from the specified state to the left until it intersects the saturation line, giving

$$T_{dp} = 19.4°C$$

(e) The specific volume per unit mass of dry air is determined by noting the distances between the specified state and the $v$ = constant lines on both sides of the point. The specific volume is determined by visual interpolation to be

$$v = 0.893 \text{ m}^3/\text{kg dry air}$$

**Discussion** Values read from the psychrometric chart inevitably involve reading errors, and thus are of limited accuracy.

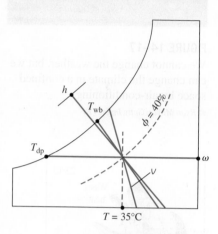

**FIGURE 14–16**
Schematic for Example 14–4.

# 14–6 · HUMAN COMFORT AND AIR-CONDITIONING

Human beings have an inherent weakness—they want to feel comfortable. They want to live in an environment that is neither hot nor cold, neither humid nor dry. However, comfort does not come easily since the desires of the human body and the weather usually are not quite compatible. Achieving comfort requires a constant struggle against the factors that cause discomfort, such as high or low temperatures and high or low humidity. As engineers, it is our duty to help people feel comfortable. (Besides, it keeps us employed.)

**FIGURE 14–17**

We cannot change the weather, but we can change the climate in a confined space by air-conditioning.

© Ryan McVay/Getty Images RF

**FIGURE 14–18**

A body feels comfortable when it can freely dissipate its waste heat, and no more.

It did not take long for people to realize that they could not change the weather in an area. All they can do is change it in a confined space such as a house or a workplace (Fig. 14–17). In the past, this was partially accomplished by fire and simple indoor heating systems. Today, modern air-conditioning systems can heat, cool, humidify, dehumidify, clean, and even deodorize the air–in other words, *condition* the air to peoples' desires. Air-conditioning systems are designed to *satisfy* the needs of the human body; therefore, it is essential that we understand the thermodynamic aspects of the body.

The human body can be viewed as a heat engine whose energy input is food. As with any other heat engine, the human body generates waste heat that must be rejected to the environment if the body is to continue operating. The rate of heat generation depends on the level of the activity. For an average adult male, it is about 87 W when sleeping, 115 W when resting or doing office work, 230 W when bowling, and 440 W when doing heavy physical work. The corresponding numbers for an adult female are about 15 percent less. (This difference is due to the body size, not the body temperature. The deep-body temperature of a healthy person is maintained constant at about 37°C.) A body will feel comfortable in environments in which it can dissipate this waste heat comfortably (Fig. 14–18).

Heat transfer is proportional to the temperature difference. Therefore in cold environments, a body loses more heat than it normally generates, which results in a feeling of discomfort. The body tries to minimize the energy deficit by cutting down the blood circulation near the skin (causing a pale look). This lowers the skin temperature, which is about 34°C for an average person, and thus the heat transfer rate. A low skin temperature causes discomfort. The hands, for example, feel painfully cold when the skin temperature reaches 10°C. We can also reduce the heat loss from the body either by putting barriers (additional clothes, blankets, etc.) in the path of heat or by increasing the rate of heat generation within the body by exercising. For example, the comfort level of a resting person dressed in warm winter clothing in a room at 10°C is roughly equal to the comfort level of an identical person doing moderate work in a room at about −23°C. Or we can just cuddle up and put our hands between our legs to reduce the surface area through which heat flows.

In hot environments, we have the opposite problem—we do not seem to be dissipating enough heat from our bodies, and we feel as if we are going to burst. We dress lightly to make it easier for heat to get away from our bodies, and we reduce the level of activity to minimize the rate of waste heat generation in the body. We also turn on the fan to continuously replace the warmer air layer that forms around our bodies as a result of body heat by the cooler air in other parts of the room. When doing light work or walking slowly, about half of the rejected body heat is dissipated through perspiration as *latent heat* while the other half is dissipated through convection and radiation as *sensible heat.* When resting or doing office work, most of the heat (about 70 percent) is dissipated in the form of sensible heat whereas when doing heavy physical work, most of the heat (about 60 percent) is dissipated in the form of latent heat. The body helps out by perspiring or sweating more. As this sweat evaporates, it absorbs latent heat from the body and cools it. Perspiration is not much help, however, if the relative humidity of the environment is close to 100 percent. Prolonged sweating without any fluid intake causes dehydration and reduced sweating, which may lead to a rise in body temperature and a heat stroke.

Another important factor that affects human comfort is heat transfer by radiation between the body and the surrounding surfaces such as walls and windows. The sun's rays travel through space by radiation. You warm up in front of a fire even if the air between you and the fire is quite cold. Likewise, in a warm room you feel chilly if the ceiling or the wall surfaces are at a considerably lower temperature. This is due to direct heat transfer between your body and the surrounding surfaces by radiation. Radiant heaters are commonly used for heating hard-to-heat places such as car repair shops.

The comfort of the human body depends primarily on three factors: the (dry-bulb) temperature, relative humidity, and air motion. The temperature of the environment is the single most important index of comfort. Most people feel comfortable when the environment temperature is between 22 and 27°C. The relative humidity also has a considerable effect on comfort since it affects the amount of heat a body can dissipate through evaporation. Relative humidity is a measure of air's ability to absorb more moisture. High relative humidity slows down heat rejection by evaporation, and low relative humidity speeds it up. Most people prefer a relative humidity of 40 to 60 percent.

Air motion also plays an important role in human comfort. It removes the warm, moist air that builds up around the body and replaces it with fresh air. Therefore, air motion improves heat rejection by both convection and evaporation. Air motion should be strong enough to remove heat and moisture from the vicinity of the body, but gentle enough to be unnoticed. Most people feel comfortable at an airspeed of about 15 m/min. Very-high-speed air motion causes discomfort instead of comfort. For example, an environment at 10°C with 48 km/h winds feels as cold as an environment at −7°C with 3 km/h winds as a result of the body-chilling effect of the air motion (the *wind-chill factor*). Other factors that affect comfort are air cleanliness, odor, noise, and radiation effect.

# 14–7 · AIR-CONDITIONING PROCESSES

Maintaining a living space or an industrial facility at the desired temperature and humidity requires some processes called air-conditioning processes. These processes include *simple heating* (raising the temperature), *simple cooling* (lowering the temperature), *humidifying* (adding moisture), and *dehumidifying* (removing moisture). Sometimes two or more of these processes are needed to bring the air to a desired temperature and humidity level.

Various air-conditioning processes are illustrated on the psychrometric chart in Fig. 14–19. Notice that simple heating and cooling processes appear as horizontal lines on this chart since the moisture content of the air remains constant ($\omega$ = constant) during these processes. Air is commonly heated and humidified in winter and cooled and dehumidified in summer. Notice how these processes appear on the psychrometric chart.

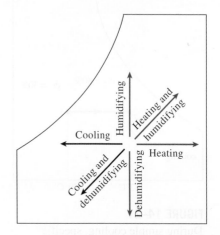

**FIGURE 14–19**
Various air-conditioning processes.

Most air-conditioning processes can be modeled as steady-flow processes, and thus the *mass balance* relation $\dot{m}_{in} = \dot{m}_{out}$ can be expressed for *dry air* and *water* as

*Mass balance for dry air:*   $\sum_{in} \dot{m}_a = \sum_{out} \dot{m}_a$   (kg/s)   (14–16)

*Mass balance for water:*   $\sum_{in} \dot{m}_w = \sum_{out} \dot{m}_w$   or   $\sum_{in} \dot{m}_a \omega = \sum_{out} \dot{m}_a \omega$   (14–17)

Disregarding the kinetic and potential energy changes, the *steady-flow energy balance* relation $\dot{E}_{in} = \dot{E}_{out}$ can be expressed in this case as

$$\dot{Q}_{in} + \dot{W}_{in} + \sum_{in} \dot{m}h = \dot{Q}_{out} + \dot{W}_{out} + \sum_{out} \dot{m}h \quad (14–18)$$

The work term usually consists of the *fan work input,* which is small relative to the other terms in the energy balance relation. Next, we examine some commonly encountered processes in air-conditioning.

## Simple Heating and Cooling ($\omega$ = constant)

Many residential heating systems consist of a stove, a heat pump, or an electric resistance heater. The air in these systems is heated by circulating it through a duct that contains the tubing for the hot gases or the electric resistance wires, as shown in Fig. 14–20. The amount of moisture in the air remains constant during this process since no moisture is added to or removed from the air. That is, the specific humidity of the air remains constant ($\omega$ = constant) during a heating (or cooling) process with no humidification or dehumidification. Such a heating process proceeds in the direction of increasing dry-bulb temperature following a line of constant specific humidity on the psychrometric chart, which appears as a horizontal line.

Notice that the relative humidity of air decreases during a heating process even if the specific humidity $\omega$ remains constant. This is because the relative humidity is the ratio of the moisture content to the moisture capacity of air at the same temperature, and moisture capacity increases with temperature. Therefore, the relative humidity of heated air may be well below comfortable levels, causing dry skin, respiratory difficulties, and an increase in static electricity.

A cooling process at constant specific humidity is similar to the heating process discussed above, except the dry-bulb temperature decreases and the relative humidity increases during such a process, as shown in Fig. 14–21. Cooling can be accomplished by passing the air over some coils through which a refrigerant or chilled water flows.

The conservation of mass equations for a heating or cooling process that involves no humidification or dehumidification reduce to $\dot{m}_{a_1} = \dot{m}_{a_2} = \dot{m}_a$ for dry air and $\omega_1 = \omega_2$ for water. Neglecting any fan work that may be present, the conservation of energy equation in this case reduces to

$$\dot{Q} = \dot{m}_a(h_2 - h_1) \quad \text{or} \quad q = h_2 - h_1$$

where $h_1$ and $h_2$ are enthalpies per unit mass of dry air at the inlet and the exit of the heating or cooling section, respectively.

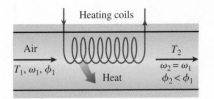

**FIGURE 14–20**

During simple heating, specific humidity remains constant, but relative humidity decreases.

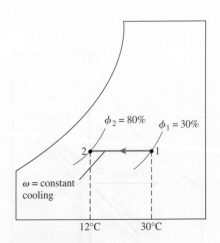

**FIGURE 14–21**

During simple cooling, specific humidity remains constant, but relative humidity increases.

# Heating with Humidification

Problems associated with the low relative humidity resulting from simple heating can be eliminated by humidifying the heated air. This is accomplished by passing the air first through a heating section (process 1-2) and then through a humidifying section (process 2-3), as shown in Fig. 14–22.

The location of state 3 depends on how the humidification is accomplished. If steam is introduced in the humidification section, this will result in humidification with additional heating ($T_3 > T_2$). If humidification is accomplished by spraying water into the airstream instead, part of the latent heat of vaporization comes from the air, which results in the cooling of the heated airstream ($T_3 < T_2$). Air should be heated to a higher temperature in the heating section in this case to make up for the cooling effect during the humidification process.

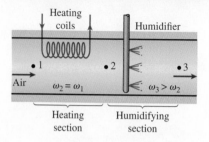

**FIGURE 14–22**
Heating with humidification.

■ **EXAMPLE 14–5** **Heating and Humidification of Air**

An air-conditioning system is to take in outdoor air at 10°C and 30 percent relative humidity at a steady rate of 45 m³/min and to condition it to 25°C and 60 percent relative humidity. The outdoor air is first heated to 22°C in the heating section and then humidified by the injection of hot steam in the humidifying section. Assuming the entire process takes place at a pressure of 100 kPa, determine (a) the rate of heat supply in the heating section and (b) the mass flow rate of the steam required in the humidifying section.

**SOLUTION** Outdoor air is first heated and then humidified by steam injection. The rate of heat transfer and the mass flow rate of steam are to be determined.

**Assumptions** **1** This is a steady-flow process and thus the mass flow rate of dry air remains constant during the entire process. **2** Dry air and water vapor are ideal gases. **3** The kinetic and potential energy changes are negligible.

**Properties** The constant-pressure specific heat of air at room temperature is $c_p = 1.005$ kJ/kg·K, and its gas constant is $R_a = 0.287$ kJ/kg·K (Table A–2a). The saturation pressure of water is 1.2281 kPa at 10°C, and 3.1698 kPa at 25°C. The enthalpy of saturated water vapor is 2519.2 kJ/kg at 10°C, and 2541.0 kJ/kg at 22°C (Table A–4).

**Analysis** We take the system to be the *heating* or the *humidifying section*, as appropriate. The schematic of the system and the psychrometric chart of the process are shown in Fig. 14–23. We note that the amount of water vapor in the air remains constant in the heating section ($\omega_1 = \omega_2$) but increases in the humidifying section ($\omega_3 > \omega_2$).

(a) Applying the mass and energy balances on the heating section gives

*Dry air mass balance:* $\qquad \dot{m}_{a_1} = \dot{m}_{a_2} = \dot{m}_a$

*Water mass balance:* $\qquad \dot{m}_{a_1}\omega_1 = \dot{m}_{a_2}\omega_2 \rightarrow \omega_1 = \omega_2$

*Energy balance:* $\qquad \dot{Q}_{in} + \dot{m}_a h_1 = \dot{m}_a h_2 \rightarrow \dot{Q}_{in} = \dot{m}_a(h_2 - h_1)$

The psychrometric chart offers great convenience in determining the properties of moist air. However, its use is limited to a specified pressure only, which is 1 atm (101.325 kPa) for the one given in the appendix. At pressures other

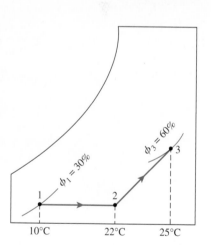

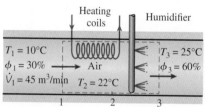

**FIGURE 14–23**
Schematic and psychrometric chart for Example 14–5.

than 1 atm, either other charts for that pressure or the relations developed earlier should be used. In our case, the choice is clear:

$$P_{v_1} = \phi_1 P_{g_1} = \phi P_{\text{sat @ }10°C} = (0.3)(1.2281 \text{ kPa}) = 0.368 \text{ kPa}$$

$$P_{a_1} = P_1 - P_{v_1} = (100 - 0.368) \text{ kPa} = 99.632 \text{ kPa}$$

$$v_1 = \frac{R_a T_1}{P_a} = \frac{(0.287 \text{ kPa·m}^3/\text{kg·K})(283 \text{ K})}{99.632 \text{ kPa}} = 0.815 \text{ m}^3/\text{kg dry air}$$

$$\dot{m}_a = \frac{\dot{V}_1}{v_1} = \frac{45 \text{ m}^3/\text{min}}{0.815 \text{ m}^3/\text{kg}} = 55.2 \text{ kg/min}$$

$$\omega_1 = \frac{0.622 P_{v_1}}{P_1 - P_{v_1}} = \frac{0.622(0.368 \text{ kPa})}{(100 - 0.368) \text{ kPa}} = 0.0023 \text{ kg H}_2\text{O/kg dry air}$$

$$h_1 = c_p T_1 + \omega_1 h_{g_1} = (1.005 \text{ kJ/kg·°C})(10°C) + (0.0023)(2519.2 \text{ kJ/kg})$$

$$= 15.8 \text{ kJ/kg dry air}$$

$$h_2 = c_p T_2 + \omega_2 h_{g_2} = (1.005 \text{ kJ/kg·°C})(22°C) + (0.0023)(2541.0 \text{ kJ/kg})$$

$$= 28.0 \text{ kJ/kg dry air}$$

since $\omega_2 = \omega_1$. Then, the rate of heat transfer to air in the heating section becomes

$$\dot{Q}_{\text{in}} = \dot{m}_a(h_2 - h_1) = (55.2 \text{ kg/min})[(28.0 - 15.8) \text{ kJ/kg}]$$

$$= \mathbf{673 \text{ kJ/min}}$$

(b) The mass balance for water in the humidifying section can be expressed as

$$\dot{m}_{a_2}\omega_2 + \dot{m}_w = \dot{m}_{a_3}\omega_3$$

or

$$\dot{m}_w = \dot{m}_a(\omega_3 - \omega_2)$$

where

$$\omega_3 = \frac{0.622\phi_3 P_{g_3}}{P_3 - \phi_3 P_{g_3}} = \frac{0.622(0.60)(3.1698 \text{ kPa})}{[100 - (0.60)(3.1698)] \text{ kPa}}$$

$$= 0.01206 \text{ kg H}_2\text{O/kg dry air}$$

Thus,

$$\dot{m}_w = (55.2 \text{ kg/min})(0.01206 - 0.0023)$$

$$= \mathbf{0.539 \text{ kg/min}}$$

**Discussion** The result 0.539 kg/min corresponds to a water requirement of close to one ton a day, which is significant.

# Cooling with Dehumidification

The specific humidity of air remains constant during a simple cooling process, but its relative humidity increases. If the relative humidity reaches undesirably high levels, it may be necessary to remove some moisture from the air, that is, to dehumidify it. This requires cooling the air below its dew-point temperature.

The cooling process with dehumidifying is illustrated schematically and on the psychrometric chart in Fig. 14–24 in conjunction with Example 14–6. Hot, moist air enters the cooling section at state 1. As it passes through the cooling coils, its temperature decreases and its relative humidity increases at constant specific humidity. If the cooling section is sufficiently long, air reaches its dew point (state $x$, saturated air). Further cooling of air results in the condensation of part of the moisture in the air. Air remains saturated during the entire condensation process, which follows a line of 100 percent relative humidity until the final state (state 2) is reached. The water vapor that condenses out of the air during this process is removed from the cooling section through a separate channel. The condensate is usually assumed to leave the cooling section at $T_2$.

The cool, saturated air at state 2 is usually routed directly to the room, where it mixes with the room air. In some cases, however, the air at state 2 may be at the right specific humidity but at a very low temperature. In such cases, air is passed through a heating section where its temperature is raised to a more comfortable level before it is routed to the room.

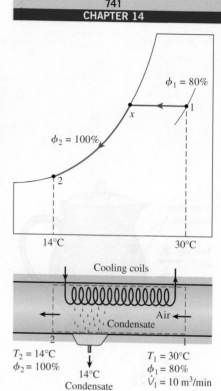

**FIGURE 14–24**
Schematic and psychrometric chart for Example 14–6.

### ■ EXAMPLE 14–6    Cooling and Dehumidification of Air

Air enters a window air conditioner at 1 atm, 30°C, and 80 percent relative humidity at a rate of 10 m³/min, and it leaves as saturated air at 14°C. Part of the moisture in the air that condenses during the process is also removed at 14°C. Determine the rates of heat and moisture removal from the air.

**SOLUTION**   Air is cooled and dehumidified by a window air conditioner. The rates of heat and moisture removal are to be determined.

**Assumptions**   **1** This is a steady-flow process and thus the mass flow rate of dry air remains constant during the entire process. **2** Dry air and the water vapor are ideal gases. **3** The kinetic and potential energy changes are negligible.

**Properties**   The enthalpy of saturated liquid water at 14°C is 58.8 kJ/kg (Table A–4). Also, the inlet and the exit states of the air are completely specified, and the total pressure is 1 atm. Therefore, we can determine the properties of the air at both states from the psychrometric chart to be

$h_1 = 85.4$ kJ/kg dry air        $h_2 = 39.3$ kJ/kg dry air

$\omega_1 = 0.0216$ kg H₂O/kg dry air    and    $\omega_2 = 0.0100$ kg H₂O/kg dry air

$v_1 = 0.889$ m³/kg dry air

**Analysis**   We take the *cooling section* to be the system. The schematic of the system and the psychrometric chart of the process are shown in Fig. 14–24. We note that the amount of water vapor in the air decreases during the process ($\omega_2 < \omega_1$) due to dehumidification. Applying the mass and energy balances on the cooling and dehumidification section gives

Dry air mass balance:    $\dot{m}_{a_1} = \dot{m}_{a_2} = \dot{m}_a$

Water mass balance:    $\dot{m}_{a_1}\omega_1 = \dot{m}_{a_2}\omega_2 + \dot{m}_w \rightarrow \dot{m}_w = \dot{m}_a(\omega_1 - \omega_2)$

Energy balance:    $\sum_{\text{in}} \dot{m}h = \dot{Q}_{\text{out}} + \sum_{\text{out}} \dot{m}h \rightarrow \dot{Q}_{\text{out}} = \dot{m}(h_1 - h_2) - \dot{m}_w h_w$

Then,

$$\dot{m}_a = \frac{\dot{V}_1}{v_1} = \frac{10\ \text{m}^3/\text{min}}{0.889\ \text{m}^3/\text{kg dry air}} = 11.25\ \text{kg/min}$$

$$\dot{m}_w = (11.25\ \text{kg/min})(0.0216 - 0.0100) = \textbf{0.131 kg/min}$$

$$\dot{Q}_{out} = (11.25\ \text{kg/min})[(85.4 - 39.3)\ \text{kJ/kg}] - (0.131\ \text{kg/min})(58.8\ \text{kJ/kg})$$

$$= \textbf{511 kJ/min}$$

Therefore, this air-conditioning unit removes moisture and heat from the air at rates of 0.131 kg/min and 511 kJ/min, respectively.

## Evaporative Cooling

Conventional cooling systems operate on a refrigeration cycle, and they can be used in any part of the world. But they have a high initial and operating cost. In desert (hot and dry) climates, we can avoid the high cost of cooling by using *evaporative coolers,* also known as *swamp coolers.*

Evaporative cooling is based on a simple principle: As water evaporates, the latent heat of vaporization is absorbed from the water body and the surrounding air. As a result, both the water and the air are cooled during the process. This approach has been used for thousands of years to cool water. A porous jug or pitcher filled with water is left in an open, shaded area. A small amount of water leaks out through the porous holes, and the pitcher "sweats." In a dry environment, this water evaporates and cools the remaining water in the pitcher (Fig. 14–25).

You have probably noticed that on a hot, dry day the air feels a lot cooler when the yard is watered. This is because water absorbs heat from the air as it evaporates. An evaporative cooler works on the same principle. The evaporative cooling process is shown schematically and on a psychrometric chart in Fig. 14–26. Hot, dry air at state 1 enters the evaporative cooler, where it is sprayed with liquid water. Part of the water evaporates during this process by absorbing heat from the airstream. As a result, the temperature of the airstream decreases and its humidity increases (state 2). In the limiting case, the air leaves the evaporative cooler saturated at state 2′. This is the lowest temperature that can be achieved by this process.

The evaporative cooling process is essentially identical to the adiabatic saturation process since the heat transfer between the airstream and the surroundings is usually negligible. Therefore, the evaporative cooling process follows a line of constant wet-bulb temperature on the psychrometric chart. (Note that this will not exactly be the case if the liquid water is supplied at a temperature different from the exit temperature of the airstream.) Since the constant-wet-bulb-temperature lines almost coincide with the constant-enthalpy lines, the enthalpy of the airstream can also be assumed to remain constant. That is,

$$T_{wb} \cong \text{constant} \tag{14–19}$$

and

$$h \cong \text{constant} \tag{14–20}$$

during an evaporative cooling process. This is a reasonably accurate approximation, and it is commonly used in air-conditioning calculations.

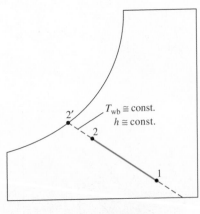

**FIGURE 14–25**
Water in a porous jug left in an open, breezy area cools as a result of evaporative cooling.

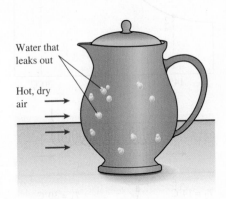

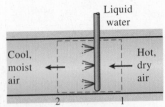

**FIGURE 14–26**
Evaporative cooling.

■ **EXAMPLE 14-7**    **Evaporative Cooling with Soaked Head Cover**

Desert dwellers often wrap their heads with a water-soaked porous cloth (Fig. 14-27). On a desert where the pressure is 1 atm, temperature is 50°C, and relative humidity is 10 percent, what is the temperature of this cloth?

**SOLUTION** Desert dwellers often wrap their heads with a water-soaked porous cloth. The temperature of this cloth on a desert with a specified temperature and relative humidity is to be determined.

***Assumptions*** Air leaves the head covering as saturated.

***Analysis*** Since the cloth behaves as the wick on a wet bulb thermometer, the temperature of the cloth will become the wet-bulb temperature. If we assume the liquid water is supplied at a temperature not much different from the exit temperature of the airstream, the evaporative cooling process follows a line of constant wet-bulb temperature on the psychrometric chart. That is,

$$T_{wb} \approx \text{constant}$$

The wet-bulb temperature at 1 atm, 50°C, and 10 percent relative humidity is determined from the psychrometric chart to be

$$T_2 = T_{wb} = \mathbf{23.8°C}$$

***Discussion*** Note that for saturated air, the dry- and the wet-bulb temperatures are identical. Therefore, the lowest temperature to which air can be cooled is the wet-bulb temperature. Also, note that the temperature of air drops by as much as 26°C in this case by evaporative cooling.

**FIGURE 14-27**
Head wrap discussed in Example 14-7.
© *Glowimages/Getty Images RF*

# Adiabatic Mixing of Airstreams

Many air-conditioning applications require the mixing of two airstreams. This is particularly true for large buildings, most production and process plants, and hospitals, which require that the conditioned air be mixed with a certain fraction of fresh outside air before it is routed into the living space. The mixing is accomplished by simply merging the two airstreams, as shown in Fig. 14-28.

The heat transfer with the surroundings is usually small, and thus the mixing processes can be assumed to be adiabatic. Mixing processes normally involve no work interactions, and the changes in kinetic and potential energies, if any, are negligible. Then, the mass and energy balances for the adiabatic mixing of two airstreams reduce to

*Mass of dry air:*      $\dot{m}_{a_1} + \dot{m}_{a_2} = \dot{m}_{a_3}$         **(14-21)**

*Mass of water vapor:*   $\omega_1 \dot{m}_{a_1} + \omega_2 \dot{m}_{a_2} = \omega_3 \dot{m}_{a_3}$    **(14-22)**

*Energy:*          $\dot{m}_{a_1} h_1 + \dot{m}_{a_2} h_2 = \dot{m}_{a_3} h_3$       **(14-23)**

Eliminating $\dot{m}_{a_3}$ from the relations above, we obtain

$$\frac{\dot{m}_{a_1}}{\dot{m}_{a_2}} = \frac{\omega_2 - \omega_3}{\omega_3 - \omega_1} = \frac{h_2 - h_3}{h_3 - h_1}$$    **(14-24)**

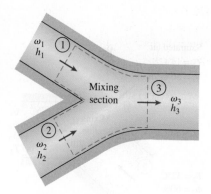

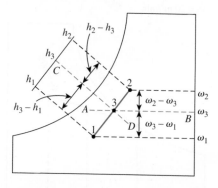

**FIGURE 14-28**
When two airstreams at states 1 and 2 are mixed adiabatically, the state of the mixture lies on the straight line connecting the two states.

This equation has an instructive geometric interpretation on the psychrometric chart. It shows that the ratio of $\omega_2 - \omega_3$ to $\omega_3 - \omega_1$ is equal to the ratio of $\dot{m}_{a_1}$ to $\dot{m}_{a_2}$. The states that satisfy this condition are indicated by the dashed line $AB$. The ratio of $h_2 - h_3$ to $h_3 - h_1$ is also equal to the ratio of $\dot{m}_{a_1}$ to $\dot{m}_{a_2}$, and the states that satisfy this condition are indicated by the dashed line $CD$. The only state that satisfies both conditions is the intersection point of these two dashed lines, which is located on the straight line connecting states 1 and 2. Thus, we conclude that *when two airstreams at two different states (states 1 and 2) are mixed adiabatically, the state of the mixture (state 3) lies on the straight line connecting states 1 and 2 on the psychrometric chart, and the ratio of the distances 2-3 and 3-1 is equal to the ratio of mass flow rates $\dot{m}_{a_1}$ and $\dot{m}_{a_2}$.*

The concave nature of the saturation curve and the conclusion above lead to an interesting possibility. When states 1 and 2 are located close to the saturation curve, the straight line connecting the two states will cross the saturation curve, and state 3 may lie to the left of the saturation curve. In this case, some water will inevitably condense during the mixing process.

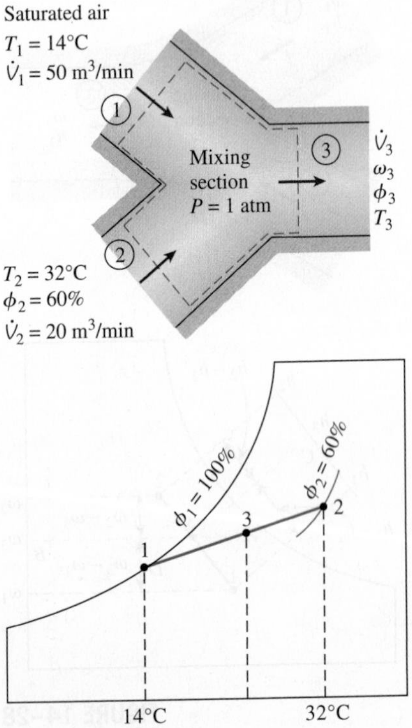

### EXAMPLE 14-8    Mixing of Conditioned Air with Outdoor Air

Saturated air leaving the cooling section of an air-conditioning system at 14°C at a rate of 50 m³/min is mixed adiabatically with the outside air at 32°C and 60 percent relative humidity at a rate of 20 m³/min. Assuming that the mixing process occurs at a pressure of 1 atm, determine the specific humidity, the relative humidity, the dry-bulb temperature, and the volume flow rate of the mixture.

**SOLUTION** Conditioned air is mixed with outside air at specified rates. The specific and relative humidities, dry-bulb temperature, and the flow rate of the mixture are to be determined.

**Assumptions** 1 Steady operating conditions exist. 2 Dry air and water vapor are ideal gases. 3 The kinetic and potential energy changes are negligible. 4 The mixing section is adiabatic.

**Properties** The properties of each inlet stream are determined from the psychrometric chart to be

$$h_1 = 39.4 \text{ kJ/kg dry air}$$
$$\omega_1 = 0.010 \text{ kg H}_2\text{O/kg dry air}$$
$$v_1 = 0.826 \text{ m}^3\text{/kg dry air}$$

and

$$h_2 = 79.0 \text{ kJ/kg dry air}$$
$$\omega_2 = 0.0182 \text{ kg H}_2\text{O/kg dry air}$$
$$v_2 = 0.889 \text{ m}^3\text{/kg dry air}$$

**Analysis** We take the *mixing section* of the streams as the system. The schematic of the system and the psychrometric chart of the process are shown in Fig. 14–29. We note that this is a steady-flow mixing process.

The mass flow rates of dry air in each stream are

$$\dot{m}_{a_1} = \frac{\dot{V}_1}{v_1} = \frac{50 \text{ m}^3\text{/min}}{0.826 \text{ m}^3\text{/kg dry air}} = 60.5 \text{ kg/min}$$

Saturated air
$T_1 = 14°C$
$\dot{V}_1 = 50 \text{ m}^3\text{/min}$

Mixing section
$P = 1$ atm

$\dot{V}_3$
$\omega_3$
$\phi_3$
$T_3$

$T_2 = 32°C$
$\phi_2 = 60\%$
$\dot{V}_2 = 20 \text{ m}^3\text{/min}$

**FIGURE 14–29**
Schematic and psychrometric chart for Example 14–8.

$$\dot{m}_{a_2} = \frac{\dot{V}_2}{v_2} = \frac{20 \text{ m}^3/\text{min}}{0.889 \text{ m}^3/\text{kg dry air}} = 22.5 \text{ kg/min}$$

From the mass balance of dry air,

$$\dot{m}_{a_3} = \dot{m}_{a_1} + \dot{m}_{a_2} = (60.5 + 22.5) \text{ kg/min} = 83 \text{ kg/min}$$

The specific humidity and the enthalpy of the mixture can be determined from Eq. 14–24,

$$\frac{\dot{m}_{a_1}}{\dot{m}_{a_2}} = \frac{\omega_2 - \omega_3}{\omega_3 - \omega_1} = \frac{h_2 - h_3}{h_3 - h_1}$$

$$\frac{60.5}{22.5} = \frac{0.0182 - \omega_3}{\omega_3 - 0.010} = \frac{79.0 - h_3}{h_3 - 39.4}$$

which yield

$$\omega_3 = \textbf{0.0122 kg H}_2\textbf{O/kg dry air}$$

$$h_3 = 50.1 \text{ kJ/kg dry air}$$

These two properties fix the state of the mixture. Other properties of the mixture are determined from the psychrometric chart:

$$T_3 = \textbf{19.0°C}$$

$$\phi_3 = \textbf{89\%}$$

$$v_3 = 0.844 \text{ m}^3/\text{kg dry air}$$

Finally, the volume flow rate of the mixture is determined from

$$\dot{V}_3 = \dot{m}_{a_3} v_3 = (83 \text{ kg/min})(0.844 \text{ m}^3/\text{kg}) = \textbf{70.1 m}^3\textbf{/min}$$

**Discussion**  Notice that the volume flow rate of the mixture is approximately equal to the sum of the volume flow rates of the two incoming streams. This is typical in air-conditioning applications.

# Wet Cooling Towers

Power plants, large air-conditioning systems, and some industries generate large quantities of waste heat that is often rejected to cooling water from nearby lakes or rivers. In some cases, however, the cooling water supply is limited or thermal pollution is a serious concern. In such cases, the waste heat must be rejected to the atmosphere, with cooling water recirculating and serving as a transport medium for heat transfer between the source and the sink (the atmosphere). One way of achieving this is through the use of wet cooling towers.

A **wet cooling tower** is essentially a semienclosed evaporative cooler. An induced-draft counterflow wet cooling tower is shown schematically in Fig. 14–30. Air is drawn into the tower from the bottom and leaves through the top. Warm water from the condenser is pumped to the top of the tower and is sprayed into this airstream. The purpose of spraying is to expose a large surface area of water to the air. As the water droplets fall under the influence of gravity, a small fraction of water (usually a few percent) evaporates and cools the remaining water. The temperature and the moisture content of the air increase during this process. The cooled water collects at the bottom of the

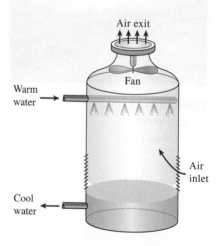

**FIGURE 14–30**
An induced-draft counterflow cooling tower.

tower and is pumped back to the condenser to absorb additional waste heat. Makeup water must be added to the cycle to replace the water lost by evaporation and air draft. To minimize water carried away by the air, drift eliminators are installed in the wet cooling towers above the spray section.

The air circulation in the cooling tower described is provided by a fan, and therefore it is classified as a forced-draft cooling tower. Another popular type of cooling tower is the **natural-draft cooling tower**, which looks like a large chimney and works like an ordinary chimney. The air in the tower has a high water-vapor content, and thus it is lighter than the outside air. Consequently, the light air in the tower rises, and the heavier outside air fills the vacant space, creating an airflow from the bottom of the tower to the top. The flow rate of air is controlled by the conditions of the atmospheric air. Natural-draft cooling towers do not require any external power to induce the air, but they cost a lot more to build than forced-draft cooling towers. The natural-draft cooling towers are hyperbolic in profile, as shown in Fig. 14–31, and some are over 100 m high. The hyperbolic profile is for greater structural strength, not for any thermodynamic reason.

The idea of a cooling tower started with the **spray pond**, where the warm water is sprayed into the air and is cooled by the air as it falls into the pond, as shown in Fig. 14–32. Some spray ponds are still in use today. However, they require 25 to 50 times the area of a cooling tower, water loss due to air drift is high, and they are unprotected against dust and dirt.

We could also dump the waste heat into a still **cooling pond**, which is basically a large artificial lake open to the atmosphere. Heat transfer from the pond surface to the atmosphere is very slow, however, and we would need about 20 times the area of a spray pond in this case to achieve the same cooling.

**FIGURE 14–31**
Two natural draft cooling towers on a roadside.
*Photo by Yunus Çengel*

**FIGURE 14–32**
A spray pond.
*Photo by Yunus Çengel*

---

### EXAMPLE 14–9  Cooling of a Power Plant by a Cooling Tower

Cooling water leaves the condenser of a power plant and enters a wet cooling tower at 35°C at a rate of 100 kg/s. Water is cooled to 22°C in the cooling tower by air that enters the tower at 1 atm, 20°C, and 60 percent relative humidity and leaves saturated at 30°C. Neglecting the power input to the fan, determine (a) the volume flow rate of air into the cooling tower and (b) the mass flow rate of the required makeup water.

**SOLUTION**  Warm cooling water from a power plant is cooled in a wet cooling tower. The flow rates of makeup water and air are to be determined.
**Assumptions**  **1** Steady operating conditions exist and thus the mass flow rate of dry air remains constant during the entire process. **2** Dry air and the water vapor are ideal gases. **3** The kinetic and potential energy changes are negligible. **4** The cooling tower is adiabatic.
**Properties**  The enthalpy of saturated liquid water is 92.28 kJ/kg at 22°C and 146.64 kJ/kg at 35°C (Table A–4). From the psychrometric chart,

$$h_1 = 42.2 \text{ kJ/kg dry air} \qquad h_2 = 100.0 \text{ kJ/kg dry air}$$

$$\omega_1 = 0.0087 \text{ kg H}_2\text{O/kg dry air} \qquad \omega_2 = 0.0273 \text{ kg H}_2\text{O/kg dry air}$$

$$v_1 = 0.842 \text{ m}^3\text{/kg dry air}$$

**Analysis**  We take the entire *cooling tower* to be the system, which is shown schematically in Fig. 14–33. We note that the mass flow rate of liquid water decreases by an amount equal to the amount of water that vaporizes in the tower during the cooling process. The water lost through evaporation must be made up later in the cycle to maintain steady operation.

(a) Applying the mass and energy balances on the cooling tower gives

Dry air mass balance:
$$\dot{m}_{a_1} = \dot{m}_{a_2} = \dot{m}_a$$

Water mass balance:
$$\dot{m}_3 + \dot{m}_{a_1}\omega_1 = \dot{m}_4 + \dot{m}_{a_2}\omega_2$$

or
$$\dot{m}_3 - \dot{m}_4 = \dot{m}_a(\omega_2 - \omega_1) = \dot{m}_{makeup}$$

Energy balance:
$$\sum_{in} \dot{m}h = \sum_{out} \dot{m}h \rightarrow \dot{m}_{a_1}h_1 + \dot{m}_3 h_3 = \dot{m}_{a_2}h_2 + \dot{m}_4 h_4$$

or
$$\dot{m}_3 h_3 = \dot{m}_a(h_2 - h_1) + (\dot{m}_3 - \dot{m}_{makeup})h_4$$

Solving for $\dot{m}_a$ gives

$$\dot{m}_a = \frac{\dot{m}_3(h_3 - h_4)}{(h_2 - h_1) - (\omega_2 - \omega_1)h_4}$$

Substituting,

$$\dot{m}_a = \frac{(100 \text{ kg/s})[(146.64 - 92.28) \text{ kJ/kg}]}{[(100.0 - 42.2) \text{ kJ/kg}] - [(0.0273 - 0.0087)(92.28) \text{ kJ/kg}]} = 96.9 \text{ kg/s}$$

Then the volume flow rate of air into the cooling tower becomes

$$\dot{V}_1 = \dot{m}_a v_1 = (96.9 \text{ kg/s})(0.842 \text{ m}^3/\text{kg}) = \textbf{81.6 m}^3/\textbf{s}$$

(b) The mass flow rate of the required makeup water is determined from

$$\dot{m}_{makeup} = \dot{m}_a(\omega_2 - \omega_1) = (96.9 \text{ kg/s})(0.0273 - 0.0087) = \textbf{1.80 kg/s}$$

**Discussion** Note that over 98 percent of the cooling water is saved and recirculated in this case.

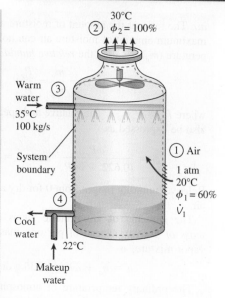

**FIGURE 14–33**
Schematic for Example 14–9.

---

## SUMMARY

In this chapter we discussed the air–water-vapor mixture, which is the most commonly encountered gas–vapor mixture in practice. The air in the atmosphere normally contains some water vapor, and it is referred to as *atmospheric air*. By contrast, air that contains no water vapor is called *dry air*. In the temperature range encountered in air-conditioning applications, both the dry air and the water vapor can be treated as ideal gases. The enthalpy change of dry air during a process can be determined from

$$\Delta h_{dry\ air} = c_p \Delta T = (1.005 \text{ kJ/kg·°C}) \Delta T$$

The atmospheric air can be treated as an ideal-gas mixture whose pressure is the sum of the partial pressure of dry air $P_a$ and that of the water vapor $P_v$,

$$P = P_a + P_v$$

The enthalpy of water vapor in the air can be taken to be equal to the enthalpy of the saturated vapor at the same temperature:

$$h_v(T, \text{low } P) \cong h_g(T) \cong 2500.9 + 1.82T \quad \text{(kJ/kg)} \quad T \text{ in °C}$$
$$\cong 1060.9 + 0.435T \text{ (Btu/lbm)} \quad T \text{ in °F}$$

in the temperature range $-10$ to $50$°C ($15$ to $120$°F).

The mass of water vapor present per unit mass of dry air is called the *specific* or *absolute humidity* $\omega$,

$$\omega = \frac{m_v}{m_a} = \frac{0.622P_v}{P - P_v} \quad \text{(kg H}_2\text{O/kg dry air)}$$

where $P$ is the total pressure of air and $P_v$ is the vapor pressure. There is a limit on the amount of vapor the air can hold at a given temperature. Air that is holding as much moisture as it can at a given temperature is called *saturated* air. The ratio of the amount of moisture air holds ($m_v$) to the maximum amount of moisture air can hold at the same temperature ($m_g$) is called the *relative humidity* $\phi$,

$$\phi = \frac{m_v}{m_g} = \frac{P_v}{P_g}$$

where $P_g = P_{\text{sat @ }T}$. The relative and specific humidities can also be expressed as

$$\phi = \frac{\omega P}{(0.622 + \omega)P_g} \quad \text{and} \quad \omega = \frac{0.622 \phi P_g}{P - \phi P_g}$$

Relative humidity ranges from 0 for dry air to 1 for saturated air.

The enthalpy of atmospheric air is expressed *per unit mass of dry air,* instead of per unit mass of the air–water-vapor mixture, as

$$h = h_a + \omega h_g \quad \text{(kJ/kg dry air)}$$

The ordinary temperature of atmospheric air is referred to as the *dry-bulb temperature* to differentiate it from other forms of temperatures. The temperature at which condensation begins if the air is cooled at constant pressure is called the *dew-point temperature* $T_{\text{dp}}$:

$$T_{\text{dp}} = T_{\text{sat @ }P_v}$$

Relative humidity and specific humidity of air can be determined by measuring the *adiabatic saturation temperature* of air, which is the temperature air attains after flowing over water in a long adiabatic channel until it is saturated,

$$\omega_1 = \frac{c_p(T_2 - T_1) + \omega_2 h_{fg_2}}{h_{g_1} - h_{f_2}}$$

where

$$\omega_2 = \frac{0.622 P_{g_2}}{P_2 - P_{g_2}}$$

and $T_2$ is the adiabatic saturation temperature. A more practical approach in air-conditioning applications is to use a thermometer whose bulb is covered with a cotton wick saturated with water and to blow air over the wick. The temperature measured in this manner is called the *wet-bulb temperature* $T_{\text{wb}}$, and it is used in place of the adiabatic saturation temperature. The properties of atmospheric air at a specified total pressure are presented in the form of easily readable charts, called *psychrometric charts*. The lines of constant enthalpy and the lines of constant wet-bulb temperature are very nearly parallel on these charts.

The needs of the human body and the conditions of the environment are not quite compatible. Therefore, it often becomes necessary to change the conditions of a living space to make it more comfortable. Maintaining a living space or an industrial facility at the desired temperature and humidity may require simple heating (raising the temperature), simple cooling (lowering the temperature), humidifying (adding moisture), or dehumidifying (removing moisture). Sometimes two or more of these processes are needed to bring the air to the desired temperature and humidity level.

Most air-conditioning processes can be modeled as steady-flow processes, and therefore they can be analyzed by applying the steady-flow mass (for both dry air and water) and energy balances,

*Dry air mass:* $\displaystyle\sum_{\text{in}} \dot{m}_a = \sum_{\text{out}} \dot{m}_a$

*Water mass:* $\displaystyle\sum_{\text{in}} \dot{m}_w = \sum_{\text{out}} \dot{m}_w \ \text{or} \ \sum_{\text{in}} \dot{m}_a \omega = \sum_{\text{out}} \dot{m}_a \omega$

*Energy:* $\displaystyle\dot{Q}_{\text{in}} + \dot{W}_{\text{in}} + \sum_{\text{in}} \dot{m} h = \dot{Q}_{\text{out}} + \dot{W}_{\text{out}} + \sum_{\text{out}} \dot{m} h$

The changes in kinetic and potential energies are assumed to be negligible.

During a simple heating or cooling process, the specific humidity remains constant, but the temperature and the relative humidity change. Sometimes air is humidified after it is heated, and some cooling processes include dehumidification. In dry climates, air can be cooled via evaporative cooling by passing it through a section where it is sprayed with water. In locations with limited cooling water supply, large amounts of waste heat can be rejected to the atmosphere with minimum water loss through the use of cooling towers.

## REFERENCES AND SUGGESTED READINGS

**1.** ASHRAE. *1981 Handbook of Fundamentals.* Atlanta, GA: American Society of Heating, Refrigerating, and Air-Conditioning Engineers, 1981.

**2.** S. M. Elonka. "Cooling Towers." *Power,* March 1963.

**3.** W. F. Stoecker and J. W. Jones. *Refrigeration and Air Conditioning.* 2nd ed. New York: McGraw-Hill, 1982.

**4.** L. D. Winiarski and B. A. Tichenor. "Model of Natural Draft Cooling Tower Performance." *Journal of the Sanitary Engineering Division, Proceedings of the American Society of Civil Engineers,* August 1970.

# PROBLEMS*

## Dry and Atmospheric Air: Specific and Relative Humidity

**14–1C**  What is the difference between dry air and atmospheric air?

**14–2C**  What is the difference between the specific humidity and the relative humidity?

**14–3C**  Can the water vapor in air be treated as an ideal gas? Explain.

**14–4C**  Is the relative humidity of saturated air necessarily 100 percent?

**14–5C**  Is it possible to obtain saturated air from unsaturated air without adding any moisture? Explain.

**14–6C**  Moist air is passed through a cooling section where it is cooled and dehumidified. How do (*a*) the specific humidity and (*b*) the relative humidity of air change during this process?

**14–7C**  How will (*a*) the specific humidity and (*b*) the relative humidity of the air contained in a well-sealed room change as it is heated?

**14–8C**  How will (*a*) the specific humidity and (*b*) the relative humidity of the air contained in a well-sealed room change as it is cooled?

**14–9C**  Consider a tank that contains moist air at 3 atm and whose walls are permeable to water vapor. The surrounding air at 1 atm pressure also contains some moisture. Is it possible for the water vapor to flow into the tank from surroundings? Explain.

**14–10C**  Why are the chilled water lines always wrapped with vapor barrier jackets?

**14–11**  A tank contains 15 kg of dry air and 0.17 kg of water vapor at 30°C and 100 kPa total pressure. Determine (*a*) the specific humidity, (*b*) the relative humidity, and (*c*) the volume of the tank.

**14–12**  Repeat Prob. 14–11 for a temperature of 20°C.

**14–13**  A room contains air at 20°C and 98 kPa at a relative humidity of 85 percent. Determine (*a*) the partial pressure of dry air, (*b*) the specific humidity of the air, and (*c*) the enthalpy per unit mass of dry air.

**14–14**  Repeat Prob. 14–13 for a pressure of 85 kPa.

**14–15**  An 8 m³-tank contains saturated air at 30°C, 105 kPa. Determine (*a*) the mass of dry air, (*b*) the specific humidity, and (*c*) the enthalpy of the air per unit mass of the dry air.

**14–16**  Determine the masses of dry air and the water vapor contained in a 90-m³ room at 93 kPa, 26°C, and 50 percent relative humidity.  *Answers:* 95.8 kg, 1.10 kg

**14–17**  Humid air at 100 kPa, 20°C, and 90 percent relative humidity is compressed in a steady-flow, isentropic compressor to 800 kPa. What is the relative humidity of the air at the compressor outlet?

800 kPa

Humid air

100 kPa
20°C, 90%

**FIGURE P14–17**

## Dew-Point, Adiabatic Saturation, and Wet-Bulb Temperatures

**14–18C**  What is the dew-point temperature?

**14–19C**  Andy and Wendy both wear glasses. On a cold winter day, Andy comes from the cold outside and enters the warm house while Wendy leaves the house and goes outside. Whose glasses are more likely to be fogged? Explain.

**14–20C**  In summer, the outer surface of a glass filled with iced water frequently "sweats." How can you explain this sweating?

**14–21C**  In some climates, cleaning the ice off the windshield of a car is a common chore on winter mornings. Explain how ice forms on the windshield during some nights even when there is no rain or snow.

**14–22C**  When are the dry-bulb and dew-point temperatures identical?

**14–23C**  When are the adiabatic saturation and wet-bulb temperatures equivalent for atmospheric air?

---

* Problems designated by a "C" are concept questions, and students are encouraged to answer them all. Problems with the ⊙ icon are solved using EES, and complete solutions together with parametric studies are included on the text website. Problems with the 🔧 icon are comprehensive in nature, and are intended to be solved with an equation solver such as EES.

**14–24** After a long walk in the 12°C outdoors, a person wearing glasses enters a room at 25°C and 55 percent relative humidity. Determine whether the glasses will become fogged.

**14–25** Repeat Prob. 14–24 for a relative humidity of 30 percent.

**14–26** A thirsty woman opens the refrigerator and picks up a cool canned drink at 5°C. Do you think the can will "sweat" as she enjoys the drink in a room at 20°C and 38 percent relative humidity?

**14–27** The dry- and wet-bulb temperatures of atmospheric air at 95 kPa are 25 and 17°C, respectively. Determine (a) the specific humidity, (b) the relative humidity, and (c) the enthalpy of the air, in kJ/kg dry air.

**14–28** The air in a room has a dry-bulb temperature of 26°C and a wet-bulb temperature of 21°C. Assuming a pressure of 100 kPa, determine (a) the specific humidity, (b) the relative humidity, and (c) the dew-point temperature. *Answers:* (a) 0.0138 kg H$_2$O/kg dry air, (b) 64.4 percent, (c) 18.8°C

**14–29** Reconsider Prob. 14–28. Determine the required properties using EES (or other) software. What would the property values be at a pressure of 300 kPa?

**14–30** Atmospheric air at 35°C flows steadily into an adiabatic saturation device and leaves as a saturated mixture at 25°C. Makeup water is supplied to the device at 25°C. Atmospheric pressure is 98 kPa. Determine the relative humidity and specific humidity of the air.

## Psychrometric Chart

**14–31C** How do constant-enthalpy and constant-wet-bulb-temperature lines compare on the psychrometric chart?

**14–32C** At what states on the psychrometric chart are the dry-bulb, wet-bulb, and dew-point temperatures identical?

**14–33C** How is the dew-point temperature at a specified state determined on the psychrometric chart?

**14–34C** Can the enthalpy values determined from a psychrometric chart at sea level be used at higher elevations?

**14–35** A room contains air at 1 atm, 28°C, and 70 percent relative humidity. Using the psychrometric chart, determine (a) the specific humidity, (b) the enthalpy (in kJ/kg dry air), (c) the wet-bulb temperature, (d) the dew-point temperature, and (e) the specific volume of the air (in m$^3$/kg dry air).

**14–36** Reconsider Prob. 14–35. Determine the required properties using EES (or other) software instead of the psychrometric chart. What would the property values be at a location at 1500 m altitude?

**14–37** The air in a room has a pressure of 1 atm, a dry-bulb temperature of 24°C, and a wet-bulb temperature of 17°C. Using the psychrometric chart, determine (a) the specific humidity, (b) the enthalpy, in kJ/kg dry air, (c) the relative humidity, (d) the dew-point temperature, and (e) the specific volume of the air, in m$^3$/kg dry air.

**14–38** Reconsider Prob. 14–37. Determine the required properties using EES (or other) software instead of the psychrometric chart. What would the property values be at a location at 3000 m altitude?

**14–39** Atmospheric air at a pressure of 1 atm and dry-bulb temperature of 28°C has a wet-bulb temperature of 20°C. Using the psychrometric chart, determine (a) the relative humidity, (b) the humidity ratio, (c) the enthalpy, (d) the dew-point temperature, and (e) the water vapor pressure.

Air
1 atm
28°C
$T_{wb}$ = 20°C

**FIGURE P14–39**

**14–40** Determine the adiabatic saturation temperature of the humid air in Prob. 14–39. *Answer:* 20°C

**14–41** Atmospheric air at a pressure of 1 atm and dry-bulb temperature of 30°C has a dew-point temperature of 24°C. Using the psychrometric chart, determine (a) the relative humidity, (b) the humidity ratio, (c) the enthalpy, (d) the wet-bulb temperature, and (e) the water vapor pressure.

**14–42** Determine the adiabatic saturation temperature of the humid air in Prob. 14–41.

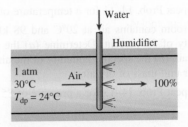

Water

Humidifier

1 atm
30°C          Air                    100%
$T_{dp}$ = 24°C

**FIGURE P14–42**

## Human Comfort and Air-Conditioning

**14–43C** What does a modern air-conditioning system do besides heating or cooling the air?

**14–44C** How does the human body respond to (a) hot weather, (b) cold weather, and (c) hot and humid weather?

**14–45C** What is the radiation effect? How does it affect human comfort?

**14–46C** How does the air motion in the vicinity of the human body affect human comfort?

**14–47C** Consider a tennis match in cold weather where both players and spectators wear the same clothes. Which group of people will feel colder? Why?

**14–48C** Why do you think little babies are more susceptible to cold?

**14–49C** How does humidity affect human comfort?

**14–50C** What are humidification and dehumidification?

**14–51C** What is metabolism? What is the range of metabolic rate for an average man? Why are we interested in the metabolic rate of the occupants of a building when we deal with heating and air-conditioning?

**14–52C** What is sensible heat? How is the sensible heat loss from a human body affected by the (a) skin temperature, (b) environment temperature, and (c) air motion?

**14–53C** What is latent heat? How is the latent heat loss from the human body affected by the (a) skin wettedness and (b) relative humidity of the environment? How is the rate of evaporation from the body related to the rate of latent heat loss?

**14–54** A department store expects to have 225 customers and 20 employees at peak times in summer. Determine the contribution of people to the total cooling load of the store.

**14–55** In a movie theater in winter, 500 people, each generating sensible heat at a rate of 70 W, are watching a movie. The heat losses through the walls, windows, and the roof are estimated to be 140,000 kJ/h. Determine if the theater needs to be heated or cooled.

**14–56** For an infiltration rate of 1.2 air changes per hour (ACH), determine sensible, latent, and total infiltration heat load of a building at sea level, in kW, that is 20 m long, 13 m wide, and 3 m high when the outdoor air is at 32°C and 35 percent relative humidity. The building is maintained at 24°C and 55 percent relative humidity at all times.

**14–57** Repeat Prob. 14–56 for an infiltration rate of 1.8 ACH.

**14–58** An average person produces 0.25 kg of moisture while taking a shower and 0.05 kg while bathing in a tub. Consider a family of four who each shower once a day in a bathroom that is not ventilated. Taking the heat of vaporization of water to be 2450 kJ/kg, determine the contribution of showers to the latent heat load of the air conditioner per day in summer.

**14–59** An average (1.82 kg) chicken has a basal metabolic rate of 5.47 W and an average metabolic rate of 10.2 W (3.78 W sensible and 6.42 W latent) during normal activity. If there are 100 chickens in a breeding room, determine the rate of total heat generation and the rate of moisture production in the room. Take the heat of vaporization of water to be 2430 kJ/kg.

## Simple Heating and Cooling

**14–60C** How do relative and specific humidities change during a simple heating process? Answer the same question for a simple cooling process.

**14–61C** Why does a simple heating or cooling process appear as a horizontal line on the psychrometric chart?

**14–62** Air enters a heating section at 95 kPa, 12°C, and 30 percent relative humidity at a rate of 6 m³/min, and it leaves at 25°C. Determine (a) the rate of heat transfer in the heating section and (b) the relative humidity of the air at the exit. *Answers: (a) 91.1 kJ/min, (b) 13.3 percent*

**14–63** Humid air at 1 atm, 35°C, and 45 percent relative humidity is cooled at constant pressure to the dew-point temperature. Determine the cooling, in kJ/kg dry air, required for this process. *Answer: 14.2 kJ/kg dry air*

**14–64** Humid air at 300 kPa, 10°C, and 90 percent relative humidity is heated in a pipe at constant pressure to 50°C. Calculate the relative humidity at the pipe outlet and the amount of heat, in kJ/kg dry air, required.

**14–65** Air enters a 30-cm-diameter cooling section at 1 atm, 35°C, and 45 percent relative humidity at 18 m/s. Heat is removed from the air at a rate of 750 kJ/min. Determine (a) the exit temperature, (b) the exit relative humidity of the air, and (c) the exit velocity. *Answers: (a) 26.5°C, (b) 73.1 percent, (c) 17.5 m/s*

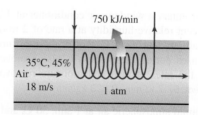

**FIGURE P14–65**

**14–66** Repeat Prob. 14–65 for a heat removal rate of 950 kJ/min.

## Heating with Humidification

**14–67C** Why is heated air sometimes humidified?

**14–68** Air at 1 atm, 15°C, and 60 percent relative humidity is first heated to 20°C in a heating section and then humidified by introducing water vapor. The air leaves the humidifying section at 25°C and 65 percent relative humidity. Determine (a) the amount of steam added to the air, and (b) the amount of heat transfer to the air in the heating section. *Answers:* (a) 0.0065 kg $H_2O$/kg dry air, (b) 5.1 kJ/kg dry air

**14–69** An air-conditioning system operates at a total pressure of 1 atm and consists of a heating section and a humidifier that supplies wet steam (saturated water vapor) at 100°C. Air enters the heating section at 10°C and 70 percent relative humidity at a rate of 35 m³/min, and it leaves the humidifying section at 20°C and 60 percent relative humidity. Determine (a) the temperature and relative humidity of air when it leaves the heating section, (b) the rate of heat transfer in the heating section, and (c) the rate at which water is added to the air in the humidifying section.

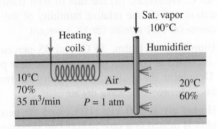

**FIGURE P14–69**

**14–70** Repeat Prob. 14–69 for a total pressure of 95 kPa for the airstream. *Answers:* (a) 19.5°C, 37.7 percent, (b) 391 kJ/min, (c) 0.147 kg/min

## Cooling with Dehumidification

**14–71C** Why is cooled air sometimes reheated in summer before it is discharged to a room?

**14–72** Air enters a window air conditioner at 1 atm, 32°C, and 70 percent relative humidity at a rate of 2 m³/min, and it leaves as saturated air at 15°C. Part of the moisture in the air that condenses during the process is also removed at 15°C. Determine the rates of heat and moisture removal from the air. *Answers:* 97.7 kJ/min, 0.023 kg/min

**14–73** Humid atmospheric air at 1 atm, 30°C, and 90 percent relative humidity is cooled to 10°C while the mixture pressure remains constant. Calculate the amount of water, in kJ/kg dry air, removed from the air and the cooling requirement, in kJ/kg dry air, when the liquid water leaves the system at 15°C.

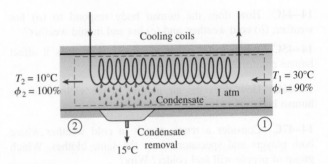

**FIGURE P14–73**

**14–74** Air enters a 40-cm-diameter cooling section at 1 atm, 32°C, and 70 percent relative humidity at 120 m/min. The air is cooled by passing it over a cooling coil through which cold water flows. The water experiences a temperature rise of 6°C. The air leaves the cooling section saturated at 20°C. Determine (a) the rate of heat transfer, (b) the mass flow rate of the water, and (c) the exit velocity of the airstream.

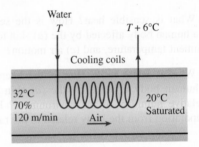

**FIGURE P14–74**

**14–75** Reconsider Prob. 14–74. Using EES (or other) software, develop a general solution of the problem in which the input variables may be supplied and parametric studies performed. For each set of input variables for which the pressure is atmospheric, show the process on the psychrometric chart.

**14–76** Repeat Prob. 14–74 for a total pressure of 95 kPa for air. *Answers:* (a) 466 kJ/min, (b) 18.6 kg/min, (c) 114 m/min

**14–77** Air from a workspace enters an air conditioner unit at 30°C dry bulb and 20°C wet bulb. The air leaves the air conditioner and returns to the space at 20°C dry-bulb and 6.5°C dew-point temperature. If there is any, the condensate leaves the air conditioner at the temperature of the air leaving the cooling coils. The volume flow rate of the air returned to the workspace is 800 m³/min. Atmospheric pressure is 101 kPa. Determine the heat transfer rate from the air, in kW, and the mass flow rate of condensate water, if any, in kg/h.

**14-78** Atmospheric air from the inside of an automobile enters the evaporator section of the air conditioner at 1 atm, 27°C and 50 percent relative humidity. The air returns to the automobile at 10°C and 90 percent relative humidity. The passenger compartment has a volume of 2 m³ and 5 air changes per minute are required to maintain the inside of the automobile at the desired comfort level. Sketch the psychrometric diagram for the atmospheric air flowing through the air conditioning process. Determine the dew point and wet bulb temperatures at the inlet to the evaporator section, in °C. Determine the required heat transfer rate from the atmospheric air to the evaporator fluid, in kW. Determine the rate of condensation of water vapor in the evaporator section, in kg/min.

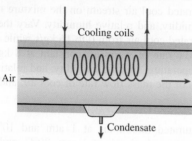

**FIGURE P14–78**

**14-79** Atmospheric air at 1 atm, 32°C, and 95 percent relative humidity is cooled to 24°C and 60 percent relative humidity. A simple ideal vapor-compression refrigeration system using refrigerant-134a as the working fluid is used to provide the cooling required. It operates its evaporator at 4°C and its condenser at a saturation temperature of 39.4°C. The condenser rejects its heat to the atmospheric air. Calculate the exergy destruction, in kJ, in the total system per 1000 m³ of dry air processed.

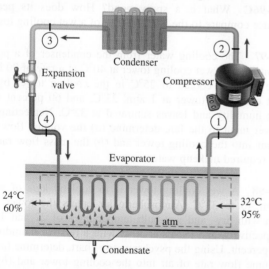

**FIGURE P14–79**

**14-80** Humid air is to be conditioned in a constant pressure process at 1 atm from 39°C dry bulb and 50 percent relative humidity to 17°C dry bulb and 10.8°C wet bulb. The air is first passed over cooling coils to remove all of the moisture necessary to achieve the final moisture content and then is passed over heating coils to achieve the final state.

(a) Sketch the psychometric diagram for the process.

(b) Determine the dew point temperature of the mixture at the inlet of the cooling coils and at the inlet of the heating coils.

(c) What is the net heat transfer for the entire process for this process, in kJ/kg dry air?

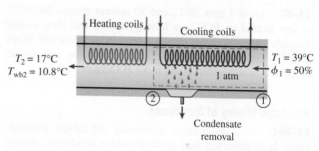

**FIGURE P14–80**

### Evaporative Cooling

**14-81C** What is evaporative cooling? Will it work in humid climates?

**14-82C** During evaporation from a water body to air, under what conditions will the latent heat of vaporization be equal to the heat transfer from the air?

**14-83C** Does an evaporation process have to involve heat transfer? Describe a process that involves both heat and mass transfer.

**14-84** Air enters an evaporative cooler at 95 kPa, 40°C, and 25 percent relative humidity and exits saturated. Determine the exit temperature of air. *Answer:* 23.1°C

**14-85** Air enters an evaporative cooler at 1 atm, 40°C, and 20 percent relative humidity at a rate of 7 m³/min, and it leaves with a relative humidity of 90 percent. Determine (a) the exit temperature of the air and (b) the required rate of water supply to the evaporative cooler.

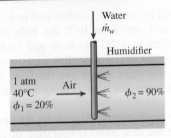

**FIGURE P14–85**

**14–86** Air enters an evaporative cooler at 1 atm, 32°C, and 30 percent relative humidity at a rate of 5 m³/min and leaves at 22°C. Determine (a) the final relative humidity and (b) the amount of water added to air.

**14–87** Air at 1 atm, 20°C, and 50 percent relative humidity is first heated to 35°C in a heating section and then passed through an evaporative cooler where its temperature drops to 25°C. Determine (a) the exit relative humidity and (b) the amount of water added to air, in kg $H_2O$/kg dry air.

## Adiabatic Mixing of Airstreams

**14–88C** Two unsaturated airstreams are mixed adiabatically. It is observed that some moisture condenses during the mixing process. Under what conditions will this be the case?

**14–89C** Consider the adiabatic mixing of two airstreams. Does the state of the mixture on the psychrometric chart have to be on the straight line connecting the two states?

**14–90** Two airstreams are mixed steadily and adiabatically. The first stream enters at 35°C and 30 percent relative humidity at a rate of 15 m³/min, while the second stream enters at 12°C and 90 percent relative humidity at a rate of 25 m³/min. Assuming that the mixing process occurs at a pressure of 1 atm, determine the specific humidity, the relative humidity, the dry-bulb temperature, and the volume flow rate of the mixture. *Answers:* 0.0088 kg $H_2O$/kg dry air, 59.7 percent, 20.2°C, 40.0 m³/min

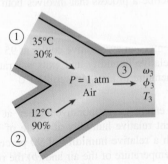

**FIGURE P14–90**

**14–91** Repeat Prob. 14–90 for a total mixing-chamber pressure of 90 kPa.

**14–92** A stream of warm air with a dry-bulb temperature of 36°C and a wet-bulb temperature of 30°C is mixed adiabatically with a stream of saturated cool air at 12°C. The dry air mass flow rates of the warm and cool airstreams are 8 and 10 kg/s, respectively. Assuming a total pressure of 1 atm, determine (a) the temperature, (b) the specific humidity, and (c) the relative humidity of the mixture.

**14–93** Reconsider Prob. 14–92. Using EES (or other) software, determine the effect of the mass flow rate of saturated cool air stream on the mixture temperature, specific humidity, and relative humidity. Vary the mass flow rate of saturated cool air from 0 to 16 kg/s while maintaining the mass flow rate of warm air constant at 8 kg/s. Plot the mixture temperature, specific humidity, and relative humidity as functions of the mass flow rate of cool air, and discuss the results.

**14–94** Saturated humid air at 1 atm and 10°C is to be mixed with atmospheric air at 1 atm, 30°C, and 80 percent relative humidity, to form air at 20°C. Determine the proportions at which these two streams are to be mixed and the relative humidity of the resulting air.

## Wet Cooling Towers

**14–95C** How does a natural-draft wet cooling tower work?

**14–96C** What is a spray pond? How does its performance compare to the performance of a wet cooling tower?

**14–97** The cooling water from the condenser of a power plant enters a wet cooling tower at 40°C at a rate of 90 kg/s. The water is cooled to 25°C in the cooling tower by air that enters the tower at 1 atm, 23°C, and 60 percent relative humidity and leaves saturated at 32°C. Neglecting the power input to the fan, determine (a) the volume flow rate of air into the cooling tower and (b) the mass flow rate of the required makeup water.

**14–98** A wet cooling tower is to cool 60 kg/s of water from 40 to 33°C. Atmospheric air enters the tower at 1 atm with dry- and wet-bulb temperatures of 22 and 16°C, respectively, and leaves at 30°C with a relative humidity of 95 percent. Using the psychrometric chart, determine (a) the volume flow rate of air into the cooling tower and (b) the mass flow rate of the required makeup water. *Answers:* (a) 30.3 m³/s, (b) 0.605 kg/s

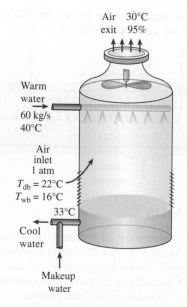

Air   30°C
exit   95%

Warm
water
→
60 kg/s
40°C

Air
inlet
1 atm
$T_{db} = 22°C$
$T_{wb} = 16°C$

33°C
←
Cool
water

Makeup
water

**FIGURE P14–98**

**14–99**  A wet cooling tower is to cool 25 kg/s of cooling water from 40 to 30°C at a location where the atmospheric pressure is 96 kPa. Atmospheric air enters the tower at 20°C and 70 percent relative humidity and leaves saturated at 35°C. Neglecting the power input to the fan, determine (a) the volume flow rate of air into the cooling tower and (b) the mass flow rate of the required makeup water.  *Answers:* (a) 11.2 m³/s, (b) 0.35 kg/s

**14–100**  Water enters a cooling tower at 35°C and at a rate of 1.4 kg/s, and leaves at 25°C. Humid air enters this tower at 1 atm and 17°C with a relative humidity of 30 percent and leaves at 22°C with relative humidity of 80 percent. Determine the mass flow rate of dry air through this tower.

**14–101**  How much work potential, in kJ/kg dry air, is lost in the cooling tower of Prob. 14-100. Take $T_0 = 17°C$.

**Review Problems**

**14–102**  Dry air whose molar analysis is 78.1 percent $N_2$, 20.9 percent $O_2$, and 1 percent Ar flows over a water body until it is saturated. If the pressure and temperature of air remain constant at 1 atm and 25°C during the process, determine (a) the molar analysis of the saturated air and (b) the density of air before and after the process. What do you conclude from your results?

**14–103**  The condensation of the water vapor in compressed-air lines is a major concern in industrial facilities, and the compressed air is often dehumidified to avoid the problems associated with condensation. Consider a compressor that

compresses ambient air from the local atmospheric pressure of 92 kPa to a pressure of 800 kPa (absolute). The compressed air is then cooled to the ambient temperature as it flows through the compressed-air lines. Disregarding any pressure losses, determine if there will be any condensation in the compressed-air lines on a day when the ambient air is at 20°C and 50 percent relative humidity.

**14–104**  The capacity of evaporative coolers is usually expressed in terms of the flow rate of air in ft³/min (or cfm), and a practical way of determining the required size of an evaporative cooler for an 8-ft-high house is to multiply the floor area of the house by 4 (by 3 in dry climates and by 5 in humid climates). For example, the capacity of an evaporative cooler for a 30-ft-long, 40-ft-wide house is 1200 × 4 = 4800 cfm. Develop an equivalent rule of thumb for the selection of an evaporative cooler in SI units for 2.4-m-high houses whose floor areas are given in m².

**14–105**  A cooling tower with a cooling capacity of 30 tons (105 kW) is claimed to evaporate 4000 kg of water per day. Is this a reasonable claim?

**14–106**  The air-conditioning costs of a house can be reduced by up to 10 percent by installing the outdoor unit (the condenser) of the air conditioner at a location shaded by trees and shrubs. If the air-conditioning costs of a house are $500 a year, determine how much the trees will save the home owner in the 20-year life of the system.

**14–107**  The thermostat setting of a house can be lowered by 1°C by wearing a light long-sleeved sweater, or by 2°C by wearing a heavy long-sleeved sweater for the same level of comfort. If each °C reduction in thermostat setting reduces the heating cost of a house by 8 percent at a particular location, determine how much the heating costs of a house can be reduced by wearing heavy sweaters if the annual heating cost of the house is $600.

**14–108**  A typical winter day in Moscow has a temperature of 0°C and a relative humidity of 40 percent. What is the relative humidity inside a dacha that has air that has been heated to 18°C?

**14–109**  The relative humidity inside dacha of Prob. 14-108 is to be brought to 50 percent by evaporating water at 20°C. How much heat, in kJ, is required for this purpose per m³ of air in the dacha?

**14–110**  During a summer day in Phoenix, Arizona, the air is at 1 atm, 43°C, and 15 percent relative humidity. Water at 20°C is evaporated into this air to produce air at 24°C and 80 percent relative humidity. How much water, in kg/kg dry air, is required and how much cooling, in kJ/kg dry air, has been produced?

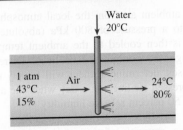

**FIGURE P14–110**

**14–111** If the system of Prob. 14-110 is operated as an adiabatic system and the air produced by this system has a relative humidity of 70 percent, what is the temperature of the air produced? *Answer:* 24.7°C

**14–112** A 1.8-m³ tank contains saturated air at 20°C and 90 kPa. Determine (*a*) the mass of the dry air, (*b*) the specific humidity, and (*c*) the enthalpy of the air per unit mass of the dry air. *Answers:* (*a*) 1.88 kg, (*b*) 0.0166 kg $H_2O$/kg dry air, (*c*) 62.2 kJ/kg dry air

**14–113** [EES] Reconsider Prob. 14–112. Using EES (or other) software, determine the properties of the air at the initial state. Study the effect of heating the air at constant volume until the pressure is 110 kPa. Plot the required heat transfer, in kJ, as a function of pressure.

**14–114** Air at 105 kPa, 16°C, and 70 percent relative humidity flows in an 15-cm diameter duct at a velocity of 10 m/s. Determine (*a*) the dew-point temperature, (*b*) the volume flow rate of air, and (*c*) the mass flow rate of dry air.

**14–115** Air flows steadily through an isentropic nozzle. The air enters the nozzle at 35°C, 200 kPa and 50 percent relative humidity. If no condensation is to occur during the expansion process, determine the pressure, temperature, and velocity of the air at the nozzle exit.

**14–116** Air enters a cooling section at 97 kPa, 35°C, and 30 percent relative humidity at a rate of 6 m³/min, where it is cooled until the moisture in the air starts condensing. Determine (*a*) the temperature of the air at the exit and (*b*) the rate of heat transfer in the cooling section.

**14–117** Outdoor air enters an air-conditioning system at 10°C and 70 percent relative humidity at a steady rate of 26 m³/min, and it leaves at 25°C and 55 percent relative humidity. The outdoor air is first heated to 18°C in the heating section and then humidified by the injection of hot steam in the humidifying section. Assuming the entire process takes place at a pressure of 1 atm, determine (*a*) the rate of heat supply in the heating section and (*b*) the mass flow rate of steam required in the humidifying section.

**14–118** Humid air at 101.3 kPa, 36°C dry bulb and 65 percent relative humidity is cooled at constant pressure to a temperature 10°C below its dew-point temperature. Sketch the psychrometric diagram for the process and determine the heat transfer from the air, in kJ/kg dry air.

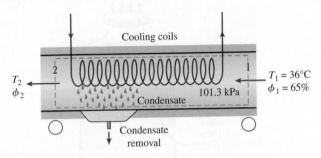

**FIGURE P14–118**

**14–119** Air enters an air-conditioning system that uses refrigerant-134a at 30°C and 70 percent relative humidity at a rate of 4 m³/min. The refrigerant enters the cooling section at 700 kPa with a quality of 20 percent and leaves as saturated vapor. The air is cooled to 20°C at a pressure of 1 atm. Determine (*a*) the rate of dehumidification, (*b*) the rate of heat transfer, and (*c*) the mass flow rate of the refrigerant.

**14–120** Repeat Prob. 14–119 for a total pressure of 90 kPa for air.

**14–121** An air-conditioning system operates at a total pressure of 1 atm and consists of a heating section and an evaporative cooler. Air enters the heating section at 15°C and 55 percent relative humidity at a rate of 30 m³/min, and it leaves the evaporative cooler at 25°C and 45 percent relatively humidity. Determine (*a*) the temperature and relative humidity of the air when it leaves the heating section, (*b*) the rate of heat transfer in the heating section, and (*c*) the rate of water added to air in the evaporative cooler. *Answers:* (*a*) 32.5°C, 19.2 percent, (*b*) 655 kJ/min, (*c*) 0.112 kg/min

**14–122** [EES] Reconsider Prob. 14–121. Using EES (or other) software, study the effect of total pressure in the range 94 to 100 kPa on the results required in the problem. Plot the results as functions of total pressure.

**14–123** Repeat Prob. 14–121 for a total pressure of 96 kPa.

**14–124** Conditioned air at 13°C and 90 percent relative humidity is to be mixed with outside air at 34°C and 40 percent relative humidity at 1 atm. If it is desired that the mixture have a relative humidity of 60 percent, determine (*a*) the ratio of the dry air mass flow rates of the conditioned air to the outside air and (*b*) the temperature of the mixture.

**14–125** [EES] Reconsider Prob. 14–124. Determine the desired quantities using EES (or other) software instead of the psychrometric chart. What would the answers be at a location at an atmospheric pressure of 80 kPa?

**14–126** A natural-draft cooling tower is to remove 70 MW of waste heat from the cooling water that enters the tower at 42°C and leaves at 30°C. Atmospheric air enters the tower at 1 atm with dry- and wet-bulb temperatures of 23 and 16°C, respectively, and leaves saturated at 32°C. Determine (a) the mass flow rate of the cooling water, (b) the volume flow rate of air into the cooling tower, and (c) the mass flow rate of the required makeup water.

**14–127** Reconsider Prob. 14–126. Using EES (or other) software, investigate the effect of air inlet wet-bulb temperature on the required air volume flow rate and the makeup water flow rate when the other input data are the stated values. Plot the results as functions of wet-bulb temperature.

## Fundamentals of Engineering (FE) Exam Problems

**14–128** A room is filled with saturated moist air at 25°C and a total pressure of 100 kPa. If the mass of dry air in the room is 100 kg, the mass of water vapor is
(a) 0.52 kg  (b) 1.97 kg  (c) 2.96 kg
(d) 2.04 kg  (e) 3.17 kg

**14–129** A room contains 65 kg of dry air and 0.6 kg of water vapor at 25°C and 90 kPa total pressure. The relative humidity of air in the room is
(a) 3.5%  (b) 41.5%  (c) 55.2%
(d) 60.9%  (e) 73.0%

**14–130** A 40-m³ room contains air at 30°C and a total pressure of 90 kPa with a relative humidity of 75 percent. The mass of dry air in the room is
(a) 24.7 kg  (b) 29.9 kg  (c) 39.9 kg
(d) 41.4 kg  (e) 52.3 kg

**14–131** A room contains air at 30°C and a total pressure of 96.0 kPa with a relative humidity of 75 percent. The partial pressure of dry air is
(a) 82.0 kPa  (b) 85.8 kPa  (c) 92.8 kPa
(d) 90.6 kPa  (e) 72.0 kPa

**14–132** The air in a house is at 25°C and 65 percent relative humidity. Now the air is cooled at constant pressure. The temperature at which the moisture in the air will start condensing is
(a) 7.4°C  (b) 16.3°C  (c) 18.0°C
(d) 11.3°C  (e) 20.2°C

**14–133** On the psychrometric chart, a cooling and dehumidification process appears as a line that is
(a) horizontal to the left
(b) vertical downward
(c) diagonal upwards to the right (NE direction)
(d) diagonal upwards to the left (NW direction)
(e) diagonal downwards to the left (SW direction)

**14–134** On the psychrometric chart, a heating and humidification process appears as a line that is
(a) horizontal to the right
(b) vertical upward
(c) diagonal upwards to the right (NE direction)
(d) diagonal upwards to the left (NW direction)
(e) diagonal downwards to the right (SE direction)

**14–135** An air stream at a specified temperature and relative humidity undergoes evaporative cooling by spraying water into it at about the same temperature. The lowest temperature the air stream can be cooled to is
(a) the dry bulb temperature at the given state
(b) the wet bulb temperature at the given state
(c) the dew point temperature at the given state
(d) the saturation temperature corresponding to the humidity ratio at the given state
(e) the triple point temperature of water

**14–136** Air is cooled and dehumidified as it flows over the coils of a refrigeration system at 85 kPa from 35°C and a humidity ratio of 0.023 kg/kg dry air to 15°C and a humidity ratio of 0.015 kg/kg dry air. If the mass flow rate of dry air is 0.4 kg/s, the rate of heat removal from the air is
(a) 4 kJ/s  (b) 8 kJ/s  (c) 12 kJ/s
(d) 16 kJ/s  (e) 20 kJ/s

**14–137** Air at a total pressure of 90 kPa, 15°C, and 75 percent relative humidity is heated and humidified to 25°C and 75 percent relative humidity by introducing water vapor. If the mass flow rate of dry air is 4 kg/s, the rate at which steam is added to the air is
(a) 0.032 kg/s  (b) 0.013 kg/s  (c) 0.019 kg/s
(d) 0.0079 kg/s  (e) 0 kg/s

## Design and Essay Problems

**14–138** Write an essay on different humidity measurement devices, including electronic ones, and discuss the advantages and disadvantages of each device.

**14–139** The air-conditioning needs of a large building can be met by a single central system or by several individual window units. Considering that both approaches are commonly used in practice, the right choice depends on the situation on hand. Identify the important factors that need to be considered in decision making, and discuss the conditions under which an air-conditioning system that consists of several window units is preferable over a large single central system, and vice versa.

**14–140** Design an inexpensive evaporative cooling system suitable for use in your house. Show how you would obtain a water spray, how you would provide airflow, and how you would prevent water droplets from drifting into the living space.

**14–141** The daily change in the temperature of the atmosphere tends to be smaller in locations where the relative humidity is high. Demonstrate why this occurs by calculating the change in the temperature of a fixed quantity of air when a fixed quantity of heat is removed from the air. Plot this temperature change as a function of the initial relative humidity and be sure that the air temperature reaches or exceeds the dew-point temperature. Do the same when a fixed amount of heat is added to the air.

**14–142** The condensation and even freezing of moisture in building walls without effective vapor retarders are of real concern in cold climates as they undermine the effectiveness of the insulation. Investigate how the builders in your area are coping with this problem, whether they are using vapor retarders or vapor barriers in the walls, and where they are located in the walls. Prepare a report on your findings, and explain the reasoning for the current practice.

# CHEMICAL REACTIONS

In the preceding chapters we limited our consideration to nonreacting systems—systems whose chemical composition remains unchanged during a process. This was the case even with mixing processes during which a homogeneous mixture is formed from two or more fluids without the occurrence of any chemical reactions. In this chapter, we specifically deal with systems whose chemical composition changes during a process, that is, systems that involve *chemical reactions*.

When dealing with nonreacting systems, we need to consider only the *sensible internal energy* (associated with temperature and pressure changes) and the *latent internal energy* (associated with phase changes). When dealing with reacting systems, however, we also need to consider the *chemical internal energy,* which is the energy associated with the destruction and formation of chemical bonds between the atoms. The energy balance relations developed for nonreacting systems are equally applicable to reacting systems, but the energy terms in the latter case should include the chemical energy of the system.

In this chapter we focus on a particular type of chemical reaction, known as *combustion,* because of its importance in engineering. But the reader should keep in mind, however, that the principles developed are equally applicable to other chemical reactions.

We start this chapter with a general discussion of fuels and combustion. Then we apply the mass and energy balances to reacting systems. In this regard we discuss the adiabatic flame temperature, which is the highest temperature a reacting mixture can attain. Finally, we examine the second-law aspects of chemical reactions.

## OBJECTIVES

The objectives of Chapter 15 are to:

■ Give an overview of fuels and combustion.

■ Apply the conservation of mass to reacting systems to determine balanced reaction equations.

■ Define the parameters used in combustion analysis, such as air–fuel ratio, percent theoretical air, and dew-point temperature.

■ Calculate the enthalpy of reaction, enthalpy of combustion, and the heating values of fuels.

■ Apply energy balances to reacting systems for both steady-flow control volumes and fixed mass systems.

■ Determine the adiabatic flame temperature for reacting mixtures.

■ Evaluate the entropy change of reacting systems.

■ Analyze reacting systems from the second-law perspective.

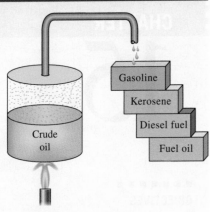

**FIGURE 15–1**

Most liquid hydrocarbon fuels are obtained from crude oil by distillation.

# 15–1 · FUELS AND COMBUSTION

Any material that can be burned to release thermal energy is called a **fuel**. Most familiar fuels consist primarily of hydrogen and carbon. They are called **hydrocarbon fuels** and are denoted by the general formula $C_nH_m$. Hydrocarbon fuels exist in all phases, some examples being coal, gasoline, and natural gas.

The main constituent of coal is carbon. Coal also contains varying amounts of oxygen, hydrogen, nitrogen, sulfur, moisture, and ash. It is difficult to give an exact mass analysis for coal since its composition varies considerably from one geographical area to the next and even within the same geographical location. Most liquid hydrocarbon fuels are a mixture of numerous hydrocarbons and are obtained from crude oil by distillation (Fig. 15–1). The most volatile hydrocarbons vaporize first, forming what we know as gasoline. The less volatile fuels obtained during distillation are kerosene, diesel fuel, and fuel oil. The composition of a particular fuel depends on the source of the crude oil as well as on the refinery.

Although liquid hydrocarbon fuels are mixtures of many different hydrocarbons, they are usually considered to be a single hydrocarbon for convenience. For example, gasoline is treated as **octane**, $C_8H_{18}$, and the diesel fuel as **dodecane**, $C_{12}H_{26}$. Another common liquid hydrocarbon fuel is **methyl alcohol**, $CH_3OH$, which is also called *methanol* and is used in some gasoline blends. The gaseous hydrocarbon fuel natural gas, which is a mixture of methane and smaller amounts of other gases, is often treated as **methane**, $CH_4$, for simplicity.

Natural gas is produced from gas wells or oil wells rich in natural gas. It is composed mainly of methane, but it also contains small amounts of ethane, propane, hydrogen, helium, carbon dioxide, nitrogen, hydrogen sulfate, and water vapor. On vehicles, it is stored either in the gas phase at pressures of 150 to 250 atm as CNG (compressed natural gas), or in the liquid phase at $-162°C$ as LNG (liquefied natural gas). Over a million vehicles in the world, mostly buses, run on natural gas. Liquefied petroleum gas (LPG) is a byproduct of natural gas processing or the crude oil refining. It consists mainly of propane and thus LPG is usually referred to as propane. However, it also contains varying amounts of butane, propylene, and butylenes. Propane is commonly used in fleet vehicles, taxis, school buses, and private cars. Ethanol is obtained from corn, grains, and organic waste. Methonal is produced mostly from natural gas, but it can also be obtained from coal and biomass. Both alcohols are commonly used as additives in oxygenated gasoline and reformulated fuels to reduce air pollution.

Vehicles are a major source of air pollutants such as nitric oxides, carbon monoxide, and hydrocarbons, as well as the greenhouse gas carbon dioxide, and thus there is a growing shift in the transportation industry from the traditional petroleum-based fuels such as gaoline and diesel fuel to the cleaner burning *alternative fuels* friendlier to the environment such as natural gas, alcohols (ethanol and methanol), liquefied petroleum gas (LPG), and hydrogen. The use of electric and hybrid cars is also on the rise. A comparison of some alternative fuels for transportation to gasoline is given in Table 15–1. Note that the energy contents of alternative fuels per unit volume are lower than that of gasoline or diesel fuel, and thus

**TABLE 15–1**

A comparison of some alternative fuels to the traditional petroleum-based fuels used in transportation

| Fuel | Energy content kJ/L | Gasoline equivalence,* L/L-gasoline |
|---|---|---|
| Gasoline | 31,850 | 1 |
| Light diesel | 33,170 | 0.96 |
| Heavy diesel | 35,800 | 0.89 |
| LPG (Liquefied petroleum gas, primarily propane) | 23,410 | 1.36 |
| Ethanol (or ethyl alcohol) | 29,420 | 1.08 |
| Methanol (or methyl alcohol) | 18,210 | 1.75 |
| CNG (Compressed natural gas, primarily methane, at 200 atm) | 8,080 | 3.94 |
| LNG (Liquefied natural gas, primarily methane) | 20,490 | 1.55 |

*Amount of fuel whose energy content is equal to the energy content of 1-L gasoline.

the driving range of a vehicle on a full tank is lower when running on an alternative fuel. Also, when comparing cost, a realistic measure is the cost per unit energy rather than cost per unit volume. For example, methanol at a unit cost of $1.20/L may appear cheaper than gasoline at $1.80/L, but this is not the case since the cost of 10,000 kJ of energy is $0.57 for gasoline and $0.66 for methanol.

A chemical reaction during which a fuel is oxidized and a large quantity of energy is released is called **combustion**. The oxidizer most often used in combustion processes is air, for obvious reasons—it is free and readily available. Pure oxygen $O_2$ is used as an oxidizer only in some specialized applications, such as cutting and welding, where air cannot be used. Therefore, a few words about the composition of air are in order.

On a mole or a volume basis, dry air is composed of 20.9 percent oxygen, 78.1 percent nitrogen, 0.9 percent argon, and small amounts of carbon dioxide, helium, neon, and hydrogen. In the analysis of combustion processes, the argon in the air is treated as nitrogen, and the gases that exist in trace amounts are disregarded. Then dry air can be approximated as 21 percent oxygen and 79 percent nitrogen by mole numbers. Therefore, each mole of oxygen entering a combustion chamber is accompanied by 0.79/0.21 = 3.76 mol of nitrogen (Fig. 15–2). That is,

$$1 \text{ kmol O}_2 + 3.76 \text{ kmol N}_2 = 4.76 \text{ kmol air} \qquad (15\text{–}1)$$

During combustion, nitrogen behaves as an inert gas and does not react with other elements, other than forming a very small amount of nitric oxides. However, even then the presence of nitrogen greatly affects the outcome of a combustion process since nitrogen usually enters a combustion chamber in large quantities at low temperatures and exits at considerably higher temperatures, absorbing a large proportion of the chemical energy released during combustion. Throughout this chapter, nitrogen is assumed to remain perfectly inert. Keep in mind, however, that at very

**FIGURE 15–2**

Each kmol of $O_2$ in air is accompanied by 3.76 kmol of $N_2$.

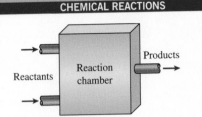

**FIGURE 15–3**
In a steady-flow combustion process, the components that enter the reaction chamber are called reactants and the components that exit are called products.

high temperatures, such as those encountered in internal combustion engines, a small fraction of nitrogen reacts with oxygen, forming hazardous gases such as nitric oxide.

Air that enters a combustion chamber normally contains some water vapor (or moisture), which also deserves consideration. For most combustion processes, the moisture in the air and the $H_2O$ that forms during combustion can also be treated as an inert gas, like nitrogen. At very high temperatures, however, some water vapor dissociates into $H_2$ and $O_2$ as well as into H, O, and OH. When the combustion gases are cooled below the dew-point temperature of the water vapor, some moisture condenses. It is important to be able to predict the dew-point temperature since the water droplets often combine with the sulfur dioxide that may be present in the combustion gases, forming sulfuric acid, which is highly corrosive.

During a combustion process, the components that exist before the reaction are called **reactants** and the components that exist after the reaction are called **products** (Fig. 15–3). Consider, for example, the combustion of 1 kmol of carbon with 1 kmol of pure oxygen, forming carbon dioxide,

$$C + O_2 \rightarrow CO_2 \tag{15–2}$$

Here C and $O_2$ are the reactants since they exist before combustion, and $CO_2$ is the product since it exists after combustion. Note that a reactant does not have to react chemically in the combustion chamber. For example, if carbon is burned with air instead of pure oxygen, both sides of the combustion equation will include $N_2$. That is, the $N_2$ will appear both as a reactant and as a product.

We should also mention that bringing a fuel into intimate contact with oxygen is not sufficient to start a combustion process. (Thank goodness it is not. Otherwise, the whole world would be on fire now.) The fuel must be brought above its **ignition temperature** to start the combustion. The minimum ignition temperatures of various substances in atmospheric air are approximately 260°C for gasoline, 400°C for carbon, 580°C for hydrogen, 610°C for carbon monoxide, and 630°C for methane. Moreover, the proportions of the fuel and air must be in the proper range for combustion to begin. For example, natural gas does not burn in air in concentrations less than 5 percent or greater than about 15 percent.

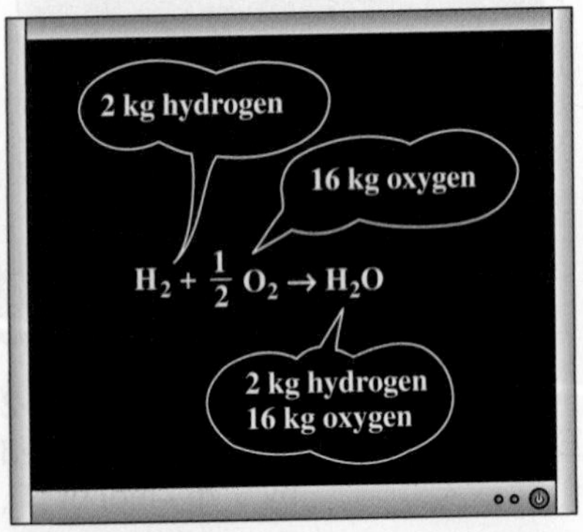

As you may recall from your chemistry courses, chemical equations are balanced on the basis of the **conservation of mass principle** (or the **mass balance**), which can be stated as follows: *The total mass of each element is conserved during a chemical reaction* (Fig. 15–4). That is, the total mass of each element on the right-hand side of the reaction equation (the products) must be equal to the total mass of that element on the left-hand side (the reactants) even though the elements exist in different chemical compounds in the reactants and products. Also, the total number of atoms of each element is conserved during a chemical reaction since the total number of atoms is equal to the total mass of the element divided by its atomic mass.

For example, both sides of Eq. 15–2 contain 12 kg of carbon and 32 kg of oxygen, even though the carbon and the oxygen exist as elements in the reactants and as a compound in the product. Also, the total mass of

**2 kg hydrogen**

**16 kg oxygen**

$$H_2 + \tfrac{1}{2}\, O_2 \rightarrow H_2O$$

**2 kg hydrogen**
**16 kg oxygen**

**FIGURE 15–4**
The mass (and number of atoms) of each element is conserved during a chemical reaction.

reactants is equal to the total mass of products, each being 44 kg. (It is common practice to round the molar masses to the nearest integer if great accuracy is not required.) However, notice that the total mole number of the reactants (2 kmol) is not equal to the total mole number of the products (1 kmol). That is, *the total number of moles is not conserved during a chemical reaction.*

A frequently used quantity in the analysis of combustion processes to quantify the amounts of fuel and air is the **air–fuel ratio** AF. It is usually expressed on a mass basis and is defined as *the ratio of the mass of air to the mass of fuel* for a combustion process (Fig. 15–5). That is,

$$AF = \frac{m_{air}}{m_{fuel}} \tag{15–3}$$

The mass $m$ of a substance is related to the number of moles $N$ through the relation $m = NM$, where $M$ is the molar mass.

The air–fuel ratio can also be expressed on a mole basis as the ratio of the mole numbers of air to the mole numbers of fuel. But we will use the former definition. The reciprocal of air–fuel ratio is called the **fuel–air ratio**.

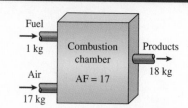

**FIGURE 15–5**
The air–fuel ratio (AF) represents the amount of air used per unit mass of fuel during a combustion process.

### EXAMPLE 15–1  Balancing the Combustion Equation

One kmol of octane ($C_8H_{18}$) is burned with air that contains 20 kmol of $O_2$, as shown in Fig. 15–6. Assuming the products contain only $CO_2$, $H_2O$, $O_2$, and $N_2$, determine the mole number of each gas in the products and the air–fuel ratio for this combustion process.

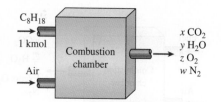

**FIGURE 15–6**
Schematic for Example 15–1.

**SOLUTION** The amount of fuel and the amount of oxygen in the air are given. The amount of the products and the AF are to be determined.
**Assumptions** The combustion products contain $CO_2$, $H_2O$, $O_2$, and $N_2$ only.
**Properties** The molar mass of air is $M_{air}$ = 28.97 kg/kmol ≅ 29.0 kg/kmol (Table A–1).
**Analysis** The chemical equation for this combustion process can be written as

$$C_8H_{18} + 20(O_2 + 3.76N_2) \rightarrow xCO_2 + yH_2O + zO_2 + wN_2$$

where the terms in the parentheses represent the composition of dry air that contains 1 kmol of $O_2$ and $x$, $y$, $z$, and $w$ represent the unknown mole numbers of the gases in the products. These unknowns are determined by applying the mass balance to each of the elements—that is, by requiring that the total mass or mole number of each element in the reactants be equal to that in the products:

| | | |
|---|---|---|
| C: | $8 = x$ | $\rightarrow$  $x = 8$ |
| H: | $18 = 2y$ | $\rightarrow$  $y = 9$ |
| O: | $20 \times 2 = 2x + y + 2z$ | $\rightarrow$  $z = 7.5$ |
| $N_2$: | $(20)(3.76) = w$ | $\rightarrow$  $w = 75.2$ |

Substituting yields

$$C_8H_{18} + 20(O_2 + 3.76N_2) \rightarrow 8CO_2 + 9H_2O + 7.5O_2 + 75.2N_2$$

Note that the coefficient 20 in the balanced equation above represents the number of moles of *oxygen*, not the number of moles of air. The latter is obtained by adding $20 \times 3.76 = 75.2$ moles of nitrogen to the 20 moles of oxygen, giving a total of 95.2 moles of air. The air–fuel ratio (AF) is determined from Eq. 15–3 by taking the ratio of the mass of the air and the mass of the fuel,

$$\begin{aligned}
\text{AF} &= \frac{m_{\text{air}}}{m_{\text{fuel}}} = \frac{(NM)_{\text{air}}}{(NM)_{\text{C}} + (NM)_{\text{H}_2}} \\
&= \frac{(20 \times 4.76 \text{ kmol})(29 \text{ kg/kmol})}{(8 \text{ kmol})(12 \text{ kg/kmol}) + (9 \text{ kmol})(2 \text{ kg/kmol})} \\
&= \textbf{24.2 kg air/kg fuel}
\end{aligned}$$

That is, 24.2 kg of air is used to burn each kilogram of fuel during this combustion process.

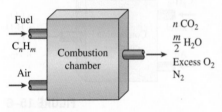

**FIGURE 15–7**
A combustion process is complete if all the combustible components of the fuel are burned to completion.

## 15–2 · THEORETICAL AND ACTUAL COMBUSTION PROCESSES

It is often instructive to study the combustion of a fuel by assuming that the combustion is complete. A combustion process is **complete** if all the carbon in the fuel burns to $CO_2$, all the hydrogen burns to $H_2O$, and all the sulfur (if any) burns to $SO_2$. That is, all the combustible components of a fuel are burned to completion during a complete combustion process (Fig. 15–7). Conversely, the combustion process is **incomplete** if the combustion products contain any unburned fuel or components such as C, $H_2$, CO, or OH.

*Insufficient oxygen* is an obvious reason for incomplete combustion, but it is not the only one. Incomplete combustion occurs even when more oxygen is present in the combustion chamber than is needed for complete combustion. This may be attributed to insufficient mixing in the combustion chamber during the limited time that the fuel and the oxygen are in contact. Another cause of incomplete combustion is *dissociation*, which becomes important at high temperatures.

Oxygen has a much greater tendency to combine with hydrogen than it does with carbon. Therefore, the hydrogen in the fuel normally burns to completion, forming $H_2O$, even when there is less oxygen than needed for complete combustion. Some of the carbon, however, ends up as CO or just as plain C particles (soot) in the products.

The minimum amount of air needed for the complete combustion of a fuel is called the **stoichiometric** or **theoretical air**. Thus, when a fuel is completely burned with theoretical air, no uncombined oxygen is present in the product gases. The theoretical air is also referred to as the *chemically correct amount of air*, or *100 percent theoretical air*. A combustion process with less than the theoretical air is bound to be incomplete. The ideal combustion process during which a fuel is burned completely with theoretical air is called the **stoichiometric** or **theoretical combustion** of that fuel (Fig. 15–8). For example, the theoretical combustion of methane is

$$CH_4 + 2(O_2 + 3.76N_2) \rightarrow CO_2 + 2H_2O + 7.52N_2$$

Notice that the products of the theoretical combustion contain no unburned methane and no C, $H_2$, CO, OH, or free $O_2$.

In actual combustion processes, it is common practice to use more air than the stoichiometric amount to increase the chances of complete combustion or to control the temperature of the combustion chamber. The amount of air in excess of the stoichiometric amount is called **excess air**. The amount of excess air is usually expressed in terms of the stoichiometric air as **percent excess air** or **percent theoretical air**. For example, 50 percent excess air is equivalent to 150 percent theoretical air, and 200 percent excess air is equivalent to 300 percent theoretical air. Of course, the stoichiometric air can be expressed as 0 percent excess air or 100 percent theoretical air. Amounts of air less than the stoichiometric amount are called **deficiency of air** and are often expressed as **percent deficiency of air**. For example, 90 percent theoretical air is equivalent to 10 percent deficiency of air. The amount of air used in combustion processes is also expressed in terms of the **equivalence ratio**, which is the ratio of the actual fuel–air ratio to the stoichiometric fuel–air ratio.

Predicting the composition of the products is relatively easy when the combustion process is assumed to be complete and the exact amounts of the fuel and air used are known. All one needs to do in this case is simply apply the mass balance to each element that appears in the combustion equation, without needing to take any measurements. Things are not so simple, however, when one is dealing with actual combustion processes. For one thing, actual combustion processes are hardly ever complete, even in the presence of excess air. Therefore, it is impossible to predict the composition of the products on the basis of the mass balance alone. Then the only alternative we have is to measure the amount of each component in the products directly.

A commonly used device to analyze the composition of combustion gases is the **Orsat gas analyzer**. In this device, a sample of the combustion gases is collected and cooled to room temperature and pressure, at which point its volume is measured. The sample is then brought into contact with a chemical that absorbs the $CO_2$. The remaining gases are returned to the room temperature and pressure, and the new volume they occupy is measured. The ratio of the reduction in volume to the original volume is the volume fraction of the $CO_2$, which is equivalent to the mole fraction if ideal-gas behavior is assumed (Fig. 15–9). The volume fractions of the other gases are determined by repeating this procedure. In Orsat analysis the gas sample is collected over water and is maintained saturated at all times. Therefore, the vapor pressure of water remains constant during the entire test. For this reason the presence of water vapor in the test chamber is ignored and data are reported on a dry basis. However, the amount of $H_2O$ formed during combustion is easily determined by balancing the combustion equation.

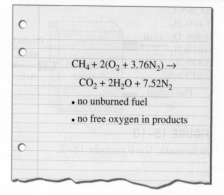

**FIGURE 15–8**
The complete combustion process with no free oxygen in the products is called theoretical combustion.

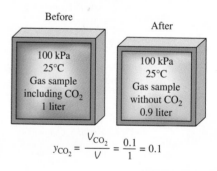

$$y_{CO_2} = \frac{V_{CO_2}}{V} = \frac{0.1}{1} = 0.1$$

**FIGURE 15–9**
Determining the mole fraction of the $CO_2$ in combustion gases by using the Orsat gas analyzer.

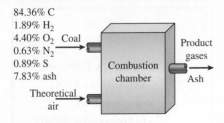

84.36% C
1.89% H₂
4.40% O₂
0.63% N₂
0.89% S
7.83% ash

**FIGURE 15–10**
Schematic for Example 15–2.

**EXAMPLE 15–2**    **Combustion of Coal with Theoretical Air**

Coal from Pennsylvania which has an ultimate analysis (by mass) as 84.36 percent C, 1.89 percent $H_2$, 4.40 percent $O_2$, 0.63 percent $N_2$, 0.89 percent S, and 7.83 percent ash (non-combustibles) is burned with theoretical amount of air (Fig. 15-10). Disregarding the ash content, determine the mole fractions of the products and the apparent molar mass of the product gases. Also determine the air-fuel ratio required for this combustion process.

**SOLUTION**  Coal with known mass analysis is burned with theoretical amount of air. The mole fractions of the product gases, their apparent molar mass, and the air-fuel ratio are to be determined.

**Assumptions**   **1** Combustion is stoichiometric and thus complete. **2** Combustion products contain $CO_2$, $H_2O$, $SO_2$, and $N_2$ only (ash disregarded). **3** Combustion gases are ideal gases.

**Analysis**   The molar masses of C, $H_2$, $O_2$, S, and air are 12, 2, 32, 32, and 29 kg/kmol, respectively (Table A-1). We now consider 100 kg of coal for simplicity. Noting that the mass percentages in this case correspond to the masses of the constituents, the mole numbers of the constituent of the coal are determined to be

$$N_C = \frac{m_C}{M_C} = \frac{84.36 \text{ kg}}{12 \text{ kg/kmol}} = 7.030 \text{ kmol}$$

$$N_{H_2} = \frac{m_{H_2}}{M_{H_2}} = \frac{1.89 \text{ kg}}{2 \text{ kg/kmol}} = 0.9450 \text{ kmol}$$

$$N_{O_2} = \frac{m_{O_2}}{M_{O_2}} = \frac{4.40 \text{ kg}}{32 \text{ kg/kmol}} = 0.1375 \text{ kmol}$$

$$N_{N_2} = \frac{m_{N_2}}{M_{N_2}} = \frac{0.63 \text{ kg}}{28 \text{ kg/kmol}} = 0.0225 \text{ kmol}$$

$$N_S = \frac{m_S}{M_S} = \frac{0.89 \text{ kg}}{32 \text{ kg/kmol}} = 0.0278 \text{ kmol}$$

Ash consists of the non-combustible matter in coal. Therefore, the mass of ash content that enters the combustion chamber is equal to the mass content that leaves. Disregarding this non-reacting component for simplicity, the combustion equation may be written as

$$7.03C + 0.945H_2 + 0.1375O_2 + 0.0225N_2 + 0.0278S + a_{th}(O_2 + 3.76N_2)$$
$$\rightarrow xCO_2 + yH_2O + zSO_2 + wN_2$$

Performing mass balances for the constituents gives

C balance: $x = 7.03$
$H_2$ balance: $y = 0.945$
S balance: $z = 0.0278$
$O_2$ balance: $0.1375 + a_{th} = x + 0.5y + z \rightarrow a_{th} = 7.393$
$N_2$ balance: $w = 0.0225 + 3.76a_{th} = 0.0225 + 3.76 \times 7.393 = 27.82$

Substituting, the balanced combustion equation without the ash becomes

$$7.03C + 0.945H_2 + 0.1375O_2 + 0.0225N_2 + 0.0278S + 7.393(O_2 + 3.76N_2)$$
$$\rightarrow 7.03CO_2 + 0.945H_2O + 0.0278SO_2 + 27.82N_2$$

The mole fractions of the product gases are determined as follows:

$$N_{prod} = 7.03 + 0.945 + 0.0278 + 27.82 = 35.82 \text{ kmol}$$

$$y_{CO_2} = \frac{N_{CO_2}}{N_{prod}} = \frac{7.03 \text{ kmol}}{35.82 \text{ kmol}} = \textbf{0.1963}$$

$$y_{H_2O} = \frac{N_{H_2O}}{N_{prod}} = \frac{0.945 \text{ kmol}}{35.82 \text{ kmol}} = \textbf{0.02638}$$

$$y_{SO_2} = \frac{N_{SO_2}}{N_{prod}} = \frac{0.0278 \text{ kmol}}{35.82 \text{ kmol}} = \textbf{0.000776}$$

$$y_{N_2} = \frac{N_{N_2}}{N_{prod}} = \frac{27.82 \text{ kmol}}{35.82 \text{ kmol}} = \textbf{0.7767}$$

Then, the apparent molar mass of product gases becomes

$$M_{prod} = \frac{m_{prod}}{N_{prod}} = \frac{(7.03 \times 44 + 0.945 \times 18 + 0.0278 \times 64 + 27.82 \times 28)\text{kg}}{35.82 \text{ kmol}}$$

$$= \textbf{30.9 kg/kmol}$$

Finally, the air-fuel mass ratio is determined from its definition to be

$$AF = \frac{m_{air}}{m_{fuel}} = \frac{(7.393 \times 4.76 \text{ kmol})(29 \text{ kg/kmol})}{100 \text{ kg}} = \textbf{10.2 kg air/kg fuel}$$

That is, 10.2 kg of air is supplied for each kg of coal in the furnace.

**Discussion** We could also solve this problem by considering just 1 kg of coal, and still obtain the same results. But we would have to deal with very small fractions in calculations in this case.

---

■ **EXAMPLE 15–3**     **Combustion of a Gaseous Fuel with Moist Air**

A certain natural gas has the following volumetric analysis: 72 percent $CH_4$, 9 percent $H_2$, 14 percent $N_2$, 2 percent $O_2$, and 3 percent $CO_2$. This gas is now burned with the stoichiometric amount of air that enters the combustion chamber at 20°C, 1 atm, and 80 percent relative humidity, as shown in Fig. 15–11. Assuming complete combustion and a total pressure of 1 atm, determine the dew-point temperature of the products.

**SOLUTION** A gaseous fuel is burned with the stoichiometric amount of moist air. The dew point temperature of the products is to be determined.
**Assumptions** **1** The fuel is burned completely and thus all the carbon in the fuel burns to $CO_2$ and all the hydrogen to $H_2O$. **2** The fuel is burned with the stoichiometric amount of air and thus there is no free $O_2$ in the product gases. **3** Combustion gases are ideal gases.
**Properties** The saturation pressure of water at 20°C is 2.3392 kPa (Table A–4).
**Analysis** We note that the moisture in the air does not react with anything; it simply shows up as additional $H_2O$ in the products. Therefore, for simplicity,

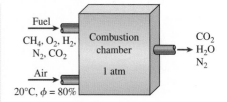

**FIGURE 15–11**
Schematic for Example 15–3.

we balance the combustion equation by using dry air and then add the moisture later to both sides of the equation.
Considering 1 kmol of fuel,

$$\overbrace{(0.72CH_4 + 0.09H_2 + 0.14N_2 + 0.02O_2 + 0.03CO_2)}^{\text{fuel}} + \overbrace{a_{th}(O_2 + 3.76N_2)}^{\text{dry air}} \rightarrow$$
$$xCO_2 + yH_2O + zN_2$$

The unknown coefficients in the above equation are determined from mass balances on various elements,

$$\begin{aligned}
\text{C:} \quad & 0.72 + 0.03 = x \quad \rightarrow \quad x = 0.75 \\
\text{H:} \quad & 0.72 \times 4 + 0.09 \times 2 = 2y \quad \rightarrow \quad y = 1.53 \\
\text{O}_2\text{:} \quad & 0.02 + 0.03 + a_{th} = x + \frac{y}{2} \quad \rightarrow \quad a_{th} = 1.465 \\
\text{N}_2\text{:} \quad & 0.14 + 3.76a_{th} = z \quad \rightarrow \quad z = 5.648
\end{aligned}$$

Next we determine the amount of moisture that accompanies $4.76a_{th} = (4.76)(1.465) = 6.97$ kmol of dry air. The partial pressure of the moisture in the air is

$$P_{v,\text{air}} = \phi_{\text{air}} P_{\text{sat @ 20°C}} = (0.80)(2.3392 \text{ kPa}) = 1.871 \text{ kPa}$$

Assuming ideal-gas behavior, the number of moles of the moisture in the air is

$$N_{v,\text{air}} = \left(\frac{P_{v,\text{air}}}{P_{\text{total}}}\right) N_{\text{total}} = \left(\frac{1.871 \text{ kPa}}{101.325 \text{ kPa}}\right)(6.97 + N_{v,\text{air}})$$

which yields

$$N_{v,\text{air}} = 0.131 \text{ kmol}$$

The balanced combustion equation is obtained by substituting the coefficients determined earlier and adding 0.131 kmol of $H_2O$ to both sides of the equation:

$$\overbrace{(0.72CH_4 + 0.09H_2 + 0.14N_2 + 0.02O_2 + 0.03CO_2)}^{\text{fuel}} + \overbrace{1.465(O_2 + 3.76N_2)}^{\text{dry air}}$$
$$+ \overbrace{0.131H_2O}^{\text{moisture}} \rightarrow 0.75CO_2 + \overbrace{1.661H_2O}^{\text{includes moisture}} + 5.648N_2$$

The dew-point temperature of the products is the temperature at which the water vapor in the products starts to condense as the products are cooled. Again, assuming ideal-gas behavior, the partial pressure of the water vapor in the combustion gases is

$$P_{v,\text{prod}} = \left(\frac{N_{v,\text{prod}}}{N_{\text{prod}}}\right) P_{\text{prod}} = \left(\frac{1.661 \text{ kmol}}{8.059 \text{ kmol}}\right)(101.325 \text{ kPa}) = 20.88 \text{ kPa}$$

Thus,

$$T_{\text{dp}} = T_{\text{sat @ 20.88 kPa}} = \textbf{60.9°C}$$

**Discussion** If the combustion process were achieved with dry air instead of moist air, the products would contain less moisture, and the dew-point temperature in this case would be 59.5°C.

## EXAMPLE 15–4     Reverse Combustion Analysis

Octane ($C_8H_{18}$) is burned with dry air. The volumetric analysis of the products on a dry basis is (Fig. 15–12)

$CO_2$:    10.02 percent
$O_2$:      5.62 percent
CO:       0.88 percent
$N_2$:     83.48 percent

Determine (a) the air–fuel ratio, (b) the percentage of theoretical air used, and (c) the amount of $H_2O$ that condenses as the products are cooled to 25°C at 100 kPa.

**SOLUTION** Combustion products whose composition is given are cooled to 25°C. The AF, the percent theoretical air used, and the fraction of water vapor that condenses are to be determined.
**Assumptions** Combustion gases are ideal gases.
**Properties** The saturation pressure of water at 25°C is 3.1698 kPa (Table A–4).
**Analysis** Note that we know the relative composition of the products, but we do not know how much fuel or air is used during the combustion process. However, they can be determined from mass balances. The $H_2O$ in the combustion gases will start condensing when the temperature drops to the dewpoint temperature.

For ideal gases, the volume fractions are equivalent to the mole fractions. Considering 100 kmol of dry products for convenience, the combustion equation can be written as

$$xC_8H_{18} + a(O_2 + 3.76N_2) \rightarrow 10.02CO_2 + 0.88CO + 5.62O_2 + 83.48N_2 + bH_2O$$

The unknown coefficients x, a, and b are determined from mass balances,

$$
\begin{aligned}
N_2: && 3.76a &= 83.48 & \rightarrow && a &= 22.20 \\
C: && 8x &= 10.02 + 0.88 & \rightarrow && x &= 1.36 \\
H: && 18x &= 2b & \rightarrow && b &= 12.24 \\
O_2: && a &= 10.02 + 0.44 + 5.62 + \frac{b}{2} & \rightarrow && 22.20 &= 22.20
\end{aligned}
$$

The $O_2$ balance is not necessary, but it can be used to check the values obtained from the other mass balances, as we did previously. Substituting, we get

$$1.36C_8H_{18} + 22.2(O_2 + 3.76N_2) \rightarrow$$
$$10.02CO_2 + 0.88CO + 5.62O_2 + 83.48N_2 + 12.24H_2O$$

The combustion equation for 1 kmol of fuel is obtained by dividing the above equation by 1.36,

$$C_8H_{18} + 16.32(O_2 + 3.76N_2) \rightarrow$$
$$7.37CO_2 + 0.65CO + 4.13O_2 + 61.38N_2 + 9H_2O$$

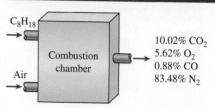

**FIGURE 15–12**
Schematic for Example 15–4.

(a) The air–fuel ratio is determined by taking the ratio of the mass of the air to the mass of the fuel (Eq. 15–3),

$$AF = \frac{m_{air}}{m_{fuel}} = \frac{(16.32 \times 4.76 \text{ kmol})(29 \text{ kg/kmol})}{(8 \text{ kmol})(12 \text{ kg/kmol}) + (9 \text{ kmol})(2 \text{ kg/kmol})}$$

$$= \textbf{19.76 kg air/kg fuel}$$

(b) To find the percentage of theoretical air used, we need to know the theoretical amount of air, which is determined from the theoretical combustion equation of the fuel,

$$C_8H_{18} + a_{th}(O_2 + 3.76N_2) \rightarrow 8CO_2 + 9H_2O + 3.76a_{th}N_2$$

$O_2$:
$$a_{th} = 8 + 4.5 \rightarrow a_{th} = 12.5$$

Then,

$$\text{Percentage of theoretical air} = \frac{m_{air,act}}{m_{air,th}} = \frac{N_{air,act}}{N_{air,th}}$$

$$= \frac{(16.32)(4.76) \text{ kmol}}{(12.50)(4.76) \text{ kmol}}$$

$$= \textbf{131\%}$$

That is, 31 percent excess air was used during this combustion process. Notice that some carbon formed carbon monoxide even though there was considerably more oxygen than needed for complete combustion.

(c) For each kmol of fuel burned, $7.37 + 0.65 + 4.13 + 61.38 + 9 = 82.53$ kmol of products are formed, including 9 kmol of $H_2O$. Assuming that the dew-point temperature of the products is above 25°C, some of the water vapor will condense as the products are cooled to 25°C. If $N_w$ kmol of $H_2O$ condenses, there will be $(9 - N_w)$ kmol of water vapor left in the products. The mole number of the products in the gas phase will also decrease to $82.53 - N_w$ as a result. By treating the product gases (including the remaining water vapor) as ideal gases, $N_w$ is determined by equating the mole fraction of the water vapor to its pressure fraction,

$$\frac{N_v}{N_{prod,gas}} = \frac{P_v}{P_{prod}}$$

$$\frac{9 - N_w}{82.53 - N_w} = \frac{3.1698 \text{ kPa}}{100 \text{ kPa}}$$

$$N_w = \textbf{6.59 kmol}$$

Therefore, the majority of the water vapor in the products (73 percent of it) condenses as the product gases are cooled to 25°C.

# 15–3 · ENTHALPY OF FORMATION AND ENTHALPY OF COMBUSTION

We mentioned in Chap. 2 that the molecules of a system possess energy in various forms such as *sensible* and *latent energy* (associated with a change of state), *chemical energy* (associated with the molecular structure), and *nuclear energy* (associated with the atomic structure), as illustrated in Fig. 15–13. In this text we do not intend to deal with nuclear energy. We also ignored chemical energy until now since the systems considered in previous chapters involved no changes in their chemical structure, and thus no changes in chemical energy. Consequently, all we needed to deal with were the sensible and latent energies.

During a chemical reaction, some chemical bonds that bind the atoms into molecules are broken, and new ones are formed. The chemical energy associated with these bonds, in general, is different for the reactants and the products. Therefore, a process that involves chemical reactions involves changes in chemical energies, which must be accounted for in an energy balance (Fig. 15–14). Assuming the atoms of each reactant remain intact (no nuclear reactions) and disregarding any changes in kinetic and potential energies, the energy change of a system during a chemical reaction is due to a change in state and a change in chemical composition. That is,

$$\Delta E_{\text{sys}} = \Delta E_{\text{state}} + \Delta E_{\text{chem}} \qquad \textbf{(15–4)}$$

Therefore, when the products formed during a chemical reaction exit the reaction chamber at the inlet state of the reactants, we have $\Delta E_{\text{state}} = 0$ and the energy change of the system in this case is due to the changes in its chemical composition only.

In thermodynamics we are concerned with the *changes* in the energy of a system during a process, and not the energy values at the particular states. Therefore, we can choose any state as the reference state and assign a value of zero to the internal energy or enthalpy of a substance at that state. When a process involves no changes in chemical composition, the reference state chosen has no effect on the results. When the process involves chemical reactions, however, the composition of the system at the end of a process is no longer the same as that at the beginning of the process. In this case it becomes necessary to have a common reference state for all substances. The chosen reference state is 25°C and 1 atm, which is known as the **standard reference state**. Property values at the standard reference state are indicated by a superscript (°) (such as $h°$ and $u°$).

When analyzing reacting systems, we must use property values relative to the standard reference state. However, it is not necessary to prepare a new set of property tables for this purpose. We can use the existing tables by subtracting the property values at the standard reference state from the values at the specified state. The ideal-gas enthalpy of $N_2$ at 500 K relative to the standard reference state, for example, is $\bar{h}_{500\text{ K}} - \bar{h}° = 14{,}581 - 8669 = 5912$ kJ/kmol.

Consider the formation of $CO_2$ from its elements, carbon and oxygen, during a steady-flow combustion process (Fig. 15–15). Both the carbon and the oxygen enter the combustion chamber at 25°C and 1 atm. The $CO_2$ formed during this process also leaves the combustion chamber at 25°C and 1 atm. The combustion of carbon is an *exothermic reaction* (a reaction during

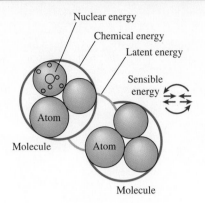

**FIGURE 15–13**
The microscopic form of energy of a substance consists of sensible, latent, chemical, and nuclear energies.

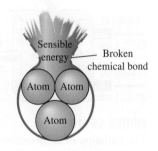

**FIGURE 15–14**
When the existing chemical bonds are destroyed and new ones are formed during a combustion process, usually a large amount of sensible energy is absorbed or released.

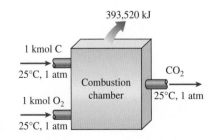

**FIGURE 15–15**
The formation of $CO_2$ during a steady-flow combustion process at 25°C and 1 atm.

which chemical energy is released in the form of heat). Therefore, some heat is transferred from the combustion chamber to the surroundings during this process, which is 393,520 kJ/kmol $CO_2$ formed. (When one is dealing with chemical reactions, it is more convenient to work with quantities per unit mole than per unit time, even for steady-flow processes.)

The process described above involves no work interactions. Therefore, from the steady-flow energy balance relation, the heat transfer during this process must be equal to the difference between the enthalpy of the products and the enthalpy of the reactants. That is,

$$Q = H_{prod} - H_{react} = -393,520 \text{ kJ/kmol} \qquad \textbf{(15–5)}$$

Since both the reactants and the products are at the same state, the enthalpy change during this process is solely due to the changes in the chemical composition of the system. This enthalpy change is different for different reactions, and it is very desirable to have a property to represent the changes in chemical energy during a reaction. This property is the **enthalpy of reaction** $h_R$, which is defined as *the difference between the enthalpy of the products at a specified state and the enthalpy of the reactants at the same state for a complete reaction.*

For combustion processes, the enthalpy of reaction is usually referred to as the **enthalpy of combustion** $h_C$, which represents the amount of heat released during a steady-flow combustion process when 1 kmol (or 1 kg) of fuel is burned completely at a specified temperature and pressure (Fig. 15–16). It is expressed as

$$h_R = h_C = H_{prod} - H_{react} \qquad \textbf{(15–6)}$$

which is −393,520 kJ/kmol for carbon at the standard reference state. The enthalpy of combustion of a particular fuel is different at different temperatures and pressures.

The enthalpy of combustion is obviously a very useful property for analyzing the combustion processes of fuels. However, there are so many different fuels and fuel mixtures that it is not practical to list $h_C$ values for all possible cases. Besides, the enthalpy of combustion is not of much use when the combustion is incomplete. Therefore a more practical approach would be to have a more fundamental property to represent the chemical energy of an element or a compound at some reference state. This property is the **enthalpy of formation** $\bar{h}_f$, which can be viewed as *the enthalpy of a substance at a specified state due to its chemical composition.*

To establish a starting point, we assign the enthalpy of formation of all stable elements (such as $O_2$, $N_2$, $H_2$, and C) a value of zero at the standard reference state of 25°C and 1 atm. That is, $\bar{h}_f = 0$ for all stable elements. (This is no different from assigning the internal energy of saturated liquid water a value of zero at 0.01°C.) Perhaps we should clarify what we mean by *stable*. The stable form of an element is simply the chemically stable form of that element at 25°C and 1 atm. Nitrogen, for example, exists in diatomic form ($N_2$) at 25°C and 1 atm. Therefore, the stable form of nitrogen at the standard reference state is diatomic nitrogen $N_2$, not monatomic nitrogen N. If an element exists in more than one

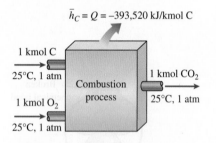

**FIGURE 15–16**
The enthalpy of combustion represents the amount of energy released as a fuel is burned during a steady-flow process at a specified state.

stable form at 25°C and 1 atm, one of the forms should be specified as the stable form. For carbon, for example, the stable form is assumed to be graphite, not diamond.

Now reconsider the formation of $CO_2$ (a compound) from its elements C and $O_2$ at 25°C and 1 atm during a steady-flow process. The enthalpy change during this process was determined to be $-393,520$ kJ/kmol. However, $H_{react} = 0$ since both reactants are elements at the standard reference state, and the products consist of 1 kmol of $CO_2$ at the same state. Therefore, the enthalpy of formation of $CO_2$ at the standard reference state is $-393,520$ kJ/kmol (Fig. 15–17). That is,

$$\bar{h}^{\circ}_{f,CO_2} = -393,520 \text{ kJ/kmol}$$

The negative sign is due to the fact that the enthalpy of 1 kmol of $CO_2$ at 25°C and 1 atm is 393,520 kJ less than the enthalpy of 1 kmol of C and 1 kmol of $O_2$ at the same state. In other words, 393,520 kJ of chemical energy is released (leaving the system as heat) when C and $O_2$ combine to form 1 kmol of $CO_2$. Therefore, a negative enthalpy of formation for a compound indicates that heat is released during the formation of that compound from its stable elements. A positive value indicates heat is absorbed.

You will notice that two $\bar{h}^{\circ}_f$ values are given for $H_2O$ in Table A–26, one for liquid water and the other for water vapor. This is because both phases of $H_2O$ are encountered at 25°C, and the effect of pressure on the enthalpy of formation is small. (Note that under equilibrium conditions, water exists only as a liquid at 25°C *and* 1 atm.) The difference between the two enthalpies of formation is equal to the $h_{fg}$ of water at 25°C, which is 2441.7 kJ/kg or 44,000 kJ/kmol.

Another term commonly used in conjunction with the combustion of fuels is the **heating value** of the fuel, which is defined as the amount of heat released when a fuel is burned completely in a steady-flow process and the products are returned to the state of the reactants. In other words, the heating value of a fuel is equal to the absolute value of the enthalpy of combustion of the fuel. That is,

$$\text{Heating value} = |h_C| \quad \text{(kJ/kg fuel)}$$

The heating value depends on the *phase* of the $H_2O$ in the products. The heating value is called the **higher heating value** (HHV) when the $H_2O$ in the products is in the liquid form, and it is called the **lower heating value** (LHV) when the $H_2O$ in the products is in the vapor form (Fig. 15–18). The two heating values are related by

$$\text{HHV} = \text{LHV} + (mh_{fg})_{H_2O} \quad \text{(kJ/kg fuel)} \tag{15–7}$$

where $m$ is the mass of $H_2O$ in the products per unit mass of fuel and $h_{fg}$ is the enthalpy of vaporization of water at the specified temperature. Higher and lower heating values of common fuels are given in Table A–27.

The heating value or enthalpy of combustion of a fuel can be determined from a knowledge of the enthalpy of formation for the compounds involved. This is illustrated with the following example.

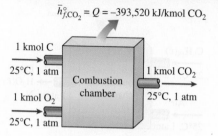

$\bar{h}^{\circ}_{f,CO_2} = Q = -393,520$ kJ/kmol $CO_2$

1 kmol C
25°C, 1 atm

Combustion chamber

1 kmol $CO_2$
25°C, 1 atm

1 kmol $O_2$
25°C, 1 atm

**FIGURE 15–17**
The enthalpy of formation of a compound represents the amount of energy absorbed or released as the component is formed from its stable elements during a steady-flow process at a specified state.

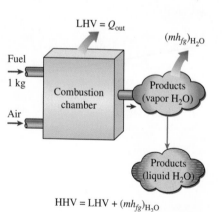

LHV = $Q_{out}$

$(mh_{fg})_{H_2O}$

Fuel
1 kg

Air

Combustion chamber

Products
(vapor $H_2O$)

Products
(liquid $H_2O$)

$\text{HHV} = \text{LHV} + (mh_{fg})_{H_2O}$

**FIGURE 15–18**
The higher heating value of a fuel is equal to the sum of the lower heating value of the fuel and the latent heat of vaporization of the $H_2O$ in the products.

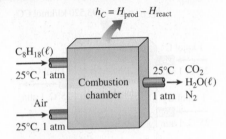

$h_C = H_{prod} - H_{react}$

$C_8H_{18}(\ell)$
25°C, 1 atm

Combustion chamber

25°C | $CO_2$
→ $H_2O(\ell)$
1 atm | $N_2$

Air
25°C, 1 atm

**FIGURE 15–19**
Schematic for Example 15–5.

---

### EXAMPLE 15–5  Evaluation of the Enthalpy of Combustion

Determine the enthalpy of combustion of liquid octane ($C_8H_{18}$) at 25°C and 1 atm, using enthalpy-of-formation data from Table A–26. Assume the water in the products is in the liquid form.

**SOLUTION**  The enthalpy of combustion of a fuel is to be determined using enthalpy of formation data.

**Properties**  The enthalpy of formation at 25°C and 1 atm is −393,520 kJ/kmol for $CO_2$, −285,830 kJ/kmol for $H_2O(\ell)$, and −249,950 kJ/kmol for $C_8H_{18}(\ell)$ (Table A–26).

**Analysis**  The combustion of $C_8H_{18}$ is illustrated in Fig. 15–19. The stoichiometric equation for this reaction is

$$C_8H_{18} + a_{th}(O_2 + 3.76N_2) \rightarrow 8CO_2 + 9H_2O(\ell) + 3.76a_{th}N_2$$

Both the reactants and the products are at the standard reference state of 25°C and 1 atm. Also, $N_2$ and $O_2$ are stable elements, and thus their enthalpy of formation is zero. Then the enthalpy of combustion of $C_8H_{18}$ becomes (Eq. 15–6)

$$\bar{h}_C = H_{prod} - H_{react}$$

$$= \sum N_p \bar{h}_{f,p}^\circ - \sum N_r \bar{h}_{f,r}^\circ = (N\bar{h}_f^\circ)_{CO_2} + (N\bar{h}_f^\circ)_{H_2O} - (N\bar{h}_f^\circ)_{C_8H_{18}}$$

Substituting,

$$\bar{h}_C = (8\text{ kmol})(-393,520\text{ kJ/kmol}) + (9\text{ kmol})(-285,830\text{ kJ/kmol})$$

$$- (1\text{ kmol})(-249,950\text{ kJ/kmol})$$

$$= -5,471,000\text{ kJ/kmol } C_8H_{18} = -47,891\text{ kJ/kg } C_8H_{18}$$

which is practially identical to the listed value of 47,890 kJ/kg in Table A–27. Since the water in the products is assumed to be in the liquid phase, this $h_C$ value corresponds to the HHV of liquid $C_8H_{18}$.

**Discussion**  It can be shown that the result for gaseous octane is −5,512,200 kJ/kmol or −48,255 kJ/kg.

---

When the exact composition of the fuel is known, the *enthalpy of combustion* of that fuel can be determined using enthalpy of formation data as shown above. However, for fuels that exhibit considerable variations in composition depending on the source, such as coal, natural gas, and fuel oil, it is more practical to determine their enthalpy of combustion experimentally by burning them directly in a bomb calorimeter at constant volume or in a steady-flow device.

## 15–4 · FIRST-LAW ANALYSIS OF REACTING SYSTEMS

The energy balance (or the first-law) relations developed in Chaps. 4 and 5 are applicable to both reacting and nonreacting systems. However, chemically reacting systems involve changes in their chemical energy, and thus it is more convenient to rewrite the energy balance relations so that the changes in chemical energies are explicitly expressed. We do this first for steady-flow systems and then for closed systems.

## Steady-Flow Systems

Before writing the energy balance relation, we need to express the enthalpy of a component in a form suitable for use for reacting systems. That is, we need to express the enthalpy such that it is relative to the standard reference state and the chemical energy term appears explicitly. When expressed properly, the enthalpy term should reduce to the enthalpy of formation $\bar{h}_f^\circ$ at the standard reference state. With this in mind, we express the enthalpy of a component on a unit mole basis as (Fig. 15–20)

$$\text{Enthalpy} = \bar{h}_f^\circ + (\bar{h} - \bar{h}^\circ) \quad \text{(kJ/kmol)}$$

where the term in the parentheses represents the sensible enthalpy relative to the standard reference state, which is the difference between $\bar{h}$ (the sensible enthalpy at the specified state) and $\bar{h}^\circ$ (the sensible enthalpy at the standard reference state of 25°C and 1 atm). This definition enables us to use enthalpy values from tables regardless of the reference state used in their construction.

When the changes in kinetic and potential energies are negligible, the steady-flow energy balance relation $\dot{E}_{in} = \dot{E}_{out}$ can be expressed for a *chemically reacting steady-flow system* more explicitly as

$$\underbrace{\dot{Q}_{in} + \dot{W}_{in} + \sum \dot{n}_r (\bar{h}_f^\circ + \bar{h} - \bar{h}^\circ)_r}_{\substack{\text{Rate of net energy transfer in} \\ \text{by heat, work, and mass}}} = \underbrace{\dot{Q}_{out} + \dot{W}_{out} + \sum \dot{n}_p (\bar{h}_f^\circ + \bar{h} - \bar{h}^\circ)_p}_{\substack{\text{Rate of net energy transfer out} \\ \text{by heat, work, and mass}}} \quad \textbf{(15–8)}$$

where $\dot{n}_p$ and $\dot{n}_r$ represent the molal flow rates of the product $p$ and the reactant $r$, respectively.

In combustion analysis, it is more convenient to work with quantities expressed *per mole of fuel*. Such a relation is obtained by dividing each term of the equation above by the molal flow rate of the fuel, yielding

$$\underbrace{Q_{in} + W_{in} + \sum N_r (\bar{h}_f^\circ + \bar{h} - \bar{h}^\circ)_r}_{\substack{\text{Energy transfer in per mole of fuel} \\ \text{by heat, work, and mass}}} = \underbrace{Q_{out} + W_{out} + \sum N_p (\bar{h}_f^\circ + \bar{h} - \bar{h}^\circ)_p}_{\substack{\text{Energy transfer out per mole of fuel} \\ \text{by heat, work, and mass}}} \quad \textbf{(15–9)}$$

where $N_r$ and $N_p$ represent the number of moles of the reactant $r$ and the product $p$, respectively, per mole of fuel. Note that $N_r = 1$ for the fuel, and the other $N_r$ and $N_p$ values can be picked directly from the balanced combustion equation. Taking heat transfer *to* the system and work done *by* the system to be *positive* quantities, the energy balance relation just discussed can be expressed more compactly as

$$Q - W = \sum N_p (\bar{h}_f^\circ + \bar{h} - \bar{h}^\circ)_p - \sum N_r (\bar{h}_f^\circ + \bar{h} - \bar{h}^\circ)_r \quad \textbf{(15–10)}$$

or as

$$Q - W = H_{prod} - H_{react} \quad \text{(kJ/kmol fuel)} \quad \textbf{(15–11)}$$

where

$$H_{prod} = \sum N_p (\bar{h}_f^\circ + \bar{h} - \bar{h}^\circ)_p \quad \text{(kJ/kmol fuel)}$$

$$H_{react} = \sum N_r (\bar{h}_f^\circ + \bar{h} - \bar{h}^\circ)_r \quad \text{(kJ/kmol fuel)}$$

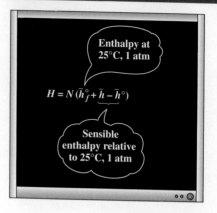

**FIGURE 15–20**
The enthalpy of a chemical component at a specified state is the sum of the enthalpy of the component at 25°C, 1 atm ($\bar{h}_f^\circ$), and the sensible enthalpy of the component relative to 25°C, 1 atm.

If the enthalpy of combustion $\bar{h}_C^\circ$ for a particular reaction is available, the steady-flow energy equation per mole of fuel can be expressed as

$$Q - W = \bar{h}_C^\circ + \sum N_p(\bar{h} - \bar{h}^\circ)_p - \sum N_r(\bar{h} - \bar{h}^\circ)_r \quad \text{(kJ/kmol)} \quad \textbf{(15–12)}$$

The energy balance relations above are sometimes written without the work term since most steady-flow combustion processes do not involve any work interactions.

A combustion chamber normally involves heat output but no heat input. Then the energy balance for a *typical steady-flow combustion process* becomes

$$Q_{out} = \underbrace{\sum N_r(\bar{h}_f^\circ + \bar{h} - \bar{h}^\circ)_r}_{\substack{\text{Energy in by mass} \\ \text{per mole of fuel}}} - \underbrace{\sum N_p(\bar{h}_f^\circ + \bar{h} - \bar{h}^\circ)_p}_{\substack{\text{Energy out by mass} \\ \text{per mole of fuel}}} \quad \textbf{(15–13)}$$

It expresses that the heat output during a combustion process is simply the difference between the energy of the reactants entering and the energy of the products leaving the combustion chamber.

## Closed Systems

The general closed-system energy balance relation $E_{in} - E_{out} = \Delta E_{system}$ can be expressed for a stationary *chemically reacting closed system* as

$$(Q_{in} - Q_{out}) + (W_{in} - W_{out}) = U_{prod} - U_{react} \quad \text{(kJ/kmol fuel)} \quad \textbf{(15–14)}$$

where $U_{prod}$ represents the internal energy of the products and $U_{react}$ represents the internal energy of the reactants. To avoid using another property—*the internal energy of formation* $\bar{u}_f^\circ$—we utilize the definition of enthalpy ($\bar{u} = \bar{h} - P\bar{v}$ or $\bar{u}_f^\circ + \bar{u} - \bar{u}^\circ = \bar{h}_f^\circ + \bar{h} - \bar{h}^\circ - P\bar{v}$) and express the above equation as (Fig. 15–21)

$$Q - W = \sum N_p(\bar{h}_f^\circ + \bar{h} - \bar{h}^\circ - P\bar{v})_p - \sum N_r(\bar{h}_f^\circ + \bar{h} - \bar{h}^\circ - P\bar{v})_r \quad \textbf{(15–15)}$$

where we have taken heat transfer *to* the system and work done *by* the system to be *positive* quantities. The $P\bar{v}$ terms are negligible for solids and liquids, and can be replaced by $R_uT$ for gases that behave as an ideal gas. Also, if desired, the $\bar{h} - P\bar{v}$ terms in Eq. 15–15 can be replaced by $\bar{u}$.

The work term in Eq. 15–15 represents all forms of work, including the boundary work. It was shown in Chap. 4 that $\Delta U + W_b = \Delta H$ for nonreacting closed systems undergoing a quasi-equilibrium $P = \text{constant}$ expansion or compression process. This is also the case for chemically reacting systems.

There are several important considerations in the analysis of reacting systems. For example, we need to know whether the fuel is a solid, a liquid, or a gas since the enthalpy of formation $\bar{h}_f^\circ$ of a fuel depends on the phase of the fuel. We also need to know the state of the fuel when it enters the combustion chamber in order to determine its enthalpy. For entropy calculations it is especially important to know if the fuel and air enter the combustion chamber premixed or separately. When the combustion products are cooled to low temperatures, we need to consider the possibility of condensation of some of the water vapor in the product gases.

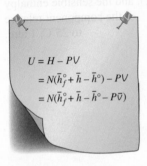

$$U = H - PV$$
$$= N(\bar{h}_f^\circ + \bar{h} - \bar{h}^\circ) - PV$$
$$= N(\bar{h}_f^\circ + \bar{h} - \bar{h}^\circ - P\bar{v})$$

**FIGURE 15–21**
An expression for the internal energy of a chemical component in terms of the enthalpy.

## ■ EXAMPLE 15–6    First-Law Analysis of Steady-Flow Combustion

■ Liquid propane ($C_3H_8$) enters a combustion chamber at 25°C at a rate of
■ 0.05 kg/min where it is mixed and burned with 50 percent excess air that
enters the combustion chamber at 7°C, as shown in Fig. 15–22. An analysis
of the combustion gases reveals that all the hydrogen in the fuel burns to
$H_2O$ but only 90 percent of the carbon burns to $CO_2$, with the remaining
10 percent forming CO. If the exit temperature of the combustion gases
is 1500 K, determine (a) the mass flow rate of air and (b) the rate of heat
transfer from the combustion chamber.

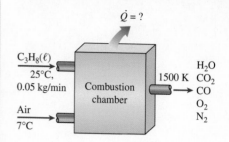

**FIGURE 15–22**
Schematic for Example 15–6.

**SOLUTION**  Liquid propane is burned steadily with excess air. The mass flow
rate of air and the rate of heat transfer are to be determined.
*Assumptions*  **1** Steady operating conditions exist. **2** Air and the combustion
gases are ideal gases. **3** Kinetic and potential energies are negligible.
*Analysis*  We note that all the hydrogen in the fuel burns to $H_2O$ but
10 percent of the carbon burns incompletely and forms CO. Also, the fuel is
burned with excess air and thus there is some free $O_2$ in the product gases.

The theoretical amount of air is determined from the stoichiometric reaction to be

$$C_3H_8(\ell) + a_{th}(O_2 + 3.76N_2) \rightarrow 3CO_2 + 4H_2O + 3.76a_{th}N_2$$

$O_2$ balance:          $a_{th} = 3 + 2 = 5$

Then the balanced equation for the actual combustion process with
50 percent excess air and some CO in the products becomes

$$C_3H_8(\ell) + 7.5(O_2 + 3.76N_2) \rightarrow 2.7CO_2 + 0.3CO + 4H_2O + 2.65O_2 + 28.2N_2$$

(a) The air–fuel ratio for this combustion process is

$$AF = \frac{m_{air}}{m_{fuel}} = \frac{(7.5 \times 4.76 \text{ kmol})(29 \text{ kg/kmol})}{(3 \text{ kmol})(12 \text{ kg/kmol}) + (4 \text{ kmol})(2 \text{ kg/kmol})}$$

$$= 25.53 \text{ kg air/kg fuel}$$

Thus,

$$\dot{m}_{air} = (AF)(\dot{m}_{fuel})$$

$$= (23.53 \text{ kg air/kg fuel})(0.05 \text{ kg fuel/min})$$

$$= \mathbf{1.18 \text{ kg air/min}}$$

(b) The heat transfer for this steady-flow combustion process is determined
from the steady-flow energy balance $E_{out} = E_{in}$ applied on the combustion
chamber per unit mole of the fuel,

$$Q_{out} + \sum N_p(\bar{h}_f^\circ + \bar{h} - \bar{h}^\circ)_p = \sum N_r(\bar{h}_f^\circ + \bar{h} - \bar{h}^\circ)_r$$

or

$$Q_{out} = \sum N_r(\bar{h}_f^\circ + \bar{h} - \bar{h}^\circ)_r - \sum N_p(\bar{h}_f^\circ + \bar{h} - \bar{h}^\circ)_p$$

Assuming the air and the combustion products to be ideal gases, we have $h = h(T)$, and we form the following minitable using data from the property tables:

| Substance | $\bar{h}_f^\circ$ kJ/kmol | $\bar{h}_{280\,K}$ kJ/kmol | $\bar{h}_{298\,K}$ kJ/kmol | $\bar{h}_{1500\,K}$ kJ/kmol |
|---|---|---|---|---|
| $C_3H_8(\ell)$ | −118,910 | — | — | — |
| $O_2$ | 0 | 8150 | 8682 | 49,292 |
| $N_2$ | 0 | 8141 | 8669 | 47,073 |
| $H_2O(g)$ | −241,820 | — | 9904 | 57,999 |
| $CO_2$ | −393,520 | — | 9364 | 71,078 |
| $CO$ | −110,530 | — | 8669 | 47,517 |

The $\bar{h}_f^\circ$ of liquid propane is obtained by subtracting the $\bar{h}_{fg}$ of propane at 25°C from the $\bar{h}_f^\circ$ of gas propane. Substituting gives

$$
\begin{aligned}
Q_{out} = \ & (1 \text{ kmol } C_3H_8)[(-118,910 + \bar{h}_{298} - \bar{h}_{298}) \text{ kJ/kmol } C_3H_8] \\
& + (7.5 \text{ kmol } O_2)[(0 + 8150 - 8682) \text{ kJ/kmol } O_2] \\
& + (28.2 \text{ kmol } N_2)[(0 + 8141 - 8669) \text{ kJ/kmol } N_2] \\
& - (2.7 \text{ kmol } CO_2)[(-393,520 + 71,078 - 9364) \text{ kJ/kmol } CO_2] \\
& - (0.3 \text{ kmol } CO)[(-110,530 + 47,517 - 8669) \text{ kJ/kmol } CO] \\
& - (4 \text{ kmol } H_2O)[(-241,820 + 57,999 - 9904) \text{ kJ/kmol } H_2O] \\
& - (2.65 \text{ kmol } O_2)[(0 + 49,292 - 8682) \text{ kJ/kmol } O_2] \\
& - (28.2 \text{ kmol } N_2)[(0 + 47,073 - 8669) \text{ kJ/kmol } N_2] \\
= \ & 363,880 \text{ kJ/kmol of } C_3H_8
\end{aligned}
$$

Thus 363,880 kJ of heat is transferred from the combustion chamber for each kmol (44 kg) of propane. This corresponds to 363,880/44 = 8270 kJ of heat loss per kilogram of propane. Then the rate of heat transfer for a mass flow rate of 0.05 kg/min for the propane becomes

$$
\dot{Q}_{out} = \dot{m}q_{out} = (0.05 \text{ kg/min})(8270 \text{ kJ/kg}) = 413.5 \text{ kJ/min} = \mathbf{6.89 \ kW}
$$

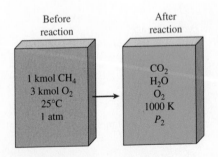

Before reaction

After reaction

1 kmol $CH_4$
3 kmol $O_2$
25°C
1 atm

$CO_2$
$H_2O$
$O_2$
1000 K
$P_2$

**FIGURE 15–23**
Schematic for Example 15–7.

**EXAMPLE 15–7**  **First-Law Analysis of Combustion in a Bomb**

The constant-volume tank shown in Fig. 15–23 contains 1 kmol of methane ($CH_4$) gas and 3 kmol of $O_2$ at 25°C and 1 atm. The contents of the tank are ignited, and the methane gas burns completely. If the final temperature is 1000 K, determine (a) the final pressure in the tank and (b) the heat transfer during this process.

**SOLUTION**  Methane is burned in a rigid tank. The final pressure in the tank and the heat transfer are to be determined.
**Assumptions**  **1** The fuel is burned completely and thus all the carbon in the fuel burns to $CO_2$ and all the hydrogen to $H_2O$. **2** The fuel, the air, and the combustion gases are ideal gases. **3** Kinetic and potential energies are negligible. **4** There are no work interactions involved.

*Analysis* The balanced combustion equation is

$$CH_4(g) + 3O_2 \rightarrow CO_2 + 2H_2O + O_2$$

(*a*) At 1000 K, water exists in the gas phase. Using the ideal-gas relation for both the reactants and the products, the final pressure in the tank is determined to be

$$\left.\begin{array}{l} P_{react}V = N_{react}R_uT_{react} \\ P_{prod}V = N_{prod}R_uT_{prod} \end{array}\right\} \quad P_{prod} = P_{react}\left(\frac{N_{prod}}{N_{react}}\right)\left(\frac{T_{prod}}{T_{react}}\right)$$

Substituting, we get

$$P_{prod} = (1 \text{ atm})\left(\frac{4 \text{ kmol}}{4 \text{ kmol}}\right)\left(\frac{1000 \text{ K}}{298 \text{ K}}\right) = \textbf{3.36 atm}$$

(*b*) Noting that the process involves no work interactions, the heat transfer during this constant-volume combustion process can be determined from the energy balance $E_{in} - E_{out} = \Delta E_{system}$ applied to the tank,

$$-Q_{out} = \sum N_p(\bar{h}_f^\circ + \bar{h} - \bar{h}^\circ - P\bar{v})_p - \sum N_r(\bar{h}_f^\circ + \bar{h} - \bar{h}^\circ - P\bar{v})_r$$

Since both the reactants and the products are assumed to be ideal gases, all the internal energy and enthalpies depend on temperature only, and the $P\bar{v}$ terms in this equation can be replaced by $R_uT$. It yields

$$Q_{out} = \sum N_r(\bar{h}_f^\circ - R_uT)_r - \sum N_p(\bar{h}_f^\circ + \bar{h}_{1000\text{ K}} - \bar{h}_{298\text{ K}} - R_uT)_p$$

since the reactants are at the standard reference temperature of 298 K. From $\bar{h}_f^\circ$ and ideal-gas tables in the Appendix,

| Substance | $\bar{h}_f^\circ$ kJ/kmol | $\bar{h}_{298\text{ K}}$ kJ/kmol | $\bar{h}_{1000\text{ K}}$ kJ/kmol |
|---|---|---|---|
| $CH_4$ | −74,850 | — | — |
| $O_2$ | 0 | 8682 | 31,389 |
| $CO_2$ | −393,520 | 9364 | 42,769 |
| $H_2O(g)$ | −241,820 | 9904 | 35,882 |

Substituting, we have

$$\begin{aligned} Q_{out} =\ & (1 \text{ kmol CH}_4)[(-74,850 - 8.314 \times 298) \text{ kJ/kmol CH}_4] \\ & + (3 \text{ kmol O}_2)[(0 - 8.314 \times 298) \text{ kJ/kmol O}_2] \\ & -(1 \text{ kmol CO}_2)[(-393,520 + 42,769 - 9364 - 8.314 \times 1000) \\ & \quad \text{kJ/kmol CO}_2] \\ & -(2 \text{ kmol H}_2O)[(-241,820 + 35,882 - 9904 - 8.314 \times 1000) \\ & \quad \text{kJ/kmol H}_2O] \\ & -(1 \text{ kmol O}_2)[(0 + 31,389 - 8682 - 8.314 \times 1000) \text{ kJ/kmol O}_2] \\ =\ & \textbf{717,590 kJ/kmol CH}_4 \end{aligned}$$

*Discussion* On a mass basis, the heat transfer from the tank would be 717,590/16 = 44,850 kJ/kg of methane.

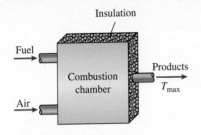

**FIGURE 15–24**

The temperature of a combustion chamber becomes maximum when combustion is complete and no heat is lost to the surroundings ($Q = 0$).

# 15–5 · ADIABATIC FLAME TEMPERATURE

In the absence of any work interactions and any changes in kinetic or potential energies, the chemical energy released during a combustion process either is lost as heat to the surroundings or is used internally to raise the temperature of the combustion products. The smaller the heat loss, the larger the temperature rise. In the limiting case of no heat loss to the surroundings ($Q = 0$), the temperature of the products reaches a maximum, which is called the **adiabatic flame** or **adiabatic combustion temperature** of the reaction (Fig. 15–24).

The adiabatic flame temperature of a steady-flow combustion process is determined from Eq. 15–11 by setting $Q = 0$ and $W = 0$. It yields

$$H_{\text{prod}} = H_{\text{react}} \qquad (15\text{–}16)$$

or

$$\sum N_p(\bar{h}_f^\circ + \bar{h} - \bar{h}^\circ)_p = \sum N_r(\bar{h}_f^\circ + \bar{h} - \bar{h}^\circ)_r \qquad (15\text{–}17)$$

Once the reactants and their states are specified, the enthalpy of the reactants $H_{\text{react}}$ can be easily determined. The calculation of the enthalpy of the products $H_{\text{prod}}$ is not so straightforward, however, because the temperature of the products is not known prior to the calculations. Therefore, the determination of the adiabatic flame temperature requires the use of an iterative technique unless equations for the sensible enthalpy changes of the combustion products are available. A temperature is assumed for the product gases, and the $H_{\text{prod}}$ is determined for this temperature. If it is not equal to $H_{\text{react}}$, calculations are repeated with another temperature. The adiabatic flame temperature is then determined from these two results by interpolation. When the oxidant is air, the product gases mostly consist of $N_2$, and a good first guess for the adiabatic flame temperature is obtained by treating the entire product gases as $N_2$.

In combustion chambers, the highest temperature to which a material can be exposed is limited by metallurgical considerations. Therefore, the adiabatic flame temperature is an important consideration in the design of combustion chambers, gas turbines, and nozzles. The maximum temperatures that occur in these devices are considerably lower than the adiabatic flame temperature, however, since the combustion is usually incomplete, some heat loss takes place, and some combustion gases dissociate at high temperatures (Fig. 15–25). The maximum temperature in a combustion chamber can be controlled by adjusting the amount of excess air, which serves as a coolant.

Note that the adiabatic flame temperature of a fuel is not unique. Its value depends on (1) the state of the reactants, (2) the degree of completion of the reaction, and (3) the amount of air used. For a specified fuel at a specified state burned with air at a specified state, *the adiabatic flame temperature attains its maximum value when complete combustion occurs with the theoretical amount of air.*

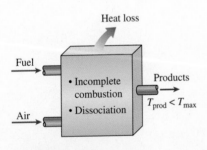

**FIGURE 15–25**

The maximum temperature encountered in a combustion chamber is lower than the theoretical adiabatic flame temperature.

# ■ EXAMPLE 15–8 Adiabatic Flame Temperature in Steady Combustion

■ Liquid octane ($C_8H_{18}$) enters the combustion chamber of a gas turbine steadily at 1 atm and 25°C, and it is burned with air that enters the combustion chamber at the same state, as shown in Fig. 15–26. Determine the adiabatic flame temperature for (a) complete combustion with 100 percent theoretical air, (b) complete combustion with 400 percent theoretical air, and (c) incomplete combustion (some CO in the products) with 90 percent theoretical air.

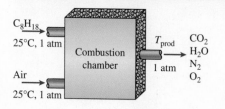

**FIGURE 15–26**
Schematic for Example 15–8.

**SOLUTION** Liquid octane is burned steadily. The adiabatic flame temperature is to be determined for different cases.
**Assumptions** **1** This is a steady-flow combustion process. **2** The combustion chamber is adiabatic. **3** There are no work interactions. **4** Air and the combustion gases are ideal gases. **5** Changes in kinetic and potential energies are negligible.
**Analysis** (a) The balanced equation for the combustion process with the theoretical amount of air is

$$C_8H_{18}(\ell) + 12.5(O_2 + 3.76N_2) \rightarrow 8CO_2 + 9H_2O + 47N_2$$

The adiabatic flame temperature relation $H_{prod} = H_{react}$ in this case reduces to

$$\sum N_p(\bar{h}_f^\circ + \bar{h} - \bar{h}^\circ)_p = \sum N_r \bar{h}_{f,r}^\circ = (N\bar{h}_f^\circ)_{C_8H_{18}}$$

since all the reactants are at the standard reference state and $\bar{h}^\circ = 0$ for $O_2$ and $N_2$. The $\bar{h}_f^\circ$ and $h$ values of various components at 298 K are

| Substance | $\bar{h}_f^\circ$ KJ/kmol | $\bar{h}_{298K}$ KJ/kmol |
|---|---|---|
| $C_8H_{18}(\ell)$ | −249,950 | — |
| $O_2$ | 0 | 8682 |
| $N_2$ | 0 | 8669 |
| $H_2O(g)$ | −241,820 | 9904 |
| $CO_2$ | −393,520 | 9364 |

Substituting, we have

$$(8 \text{ kmol } CO_2)[(-393,520 + \bar{h}_{CO_2} - 9364) \text{ kJ/kmol } CO_2]$$
$$+ (9 \text{ kmol } H_2O)[(-241,820 + \bar{h}_{H_2O} - 9904) \text{ kJ/kmol } H_2O]$$
$$+ (47 \text{ kmol } N_2)[(0 + \bar{h}_{N_2} - 8669) \text{ kJ/kmol } N_2]$$
$$= (1 \text{ kmol } C_8H_{18})(-249,950 \text{ kJ/kmol } C_8H_{18})$$

which yields

$$8\bar{h}_{CO_2} + 9\bar{h}_{H_2O} + 47\bar{h}_{N_2} = 5,646,081 \text{ kJ}$$

It appears that we have one equation with three unknowns. Actually we have only one unknown—the temperature of the products $T_{prod}$—since $h = h(T)$ for ideal gases. Therefore, we have to use an equation solver such as EES or a trial-and-error approach to determine the temperature of the products.

A first guess is obtained by dividing the right-hand side of the equation by the total number of moles, which yields $5,646,081/(8 + 9 + 47) = 88,220$ kJ/kmol. This enthalpy value corresponds to about 2650 K for $N_2$, 2100 K for $H_2O$, and 1800 K for $CO_2$. Noting that the majority of the moles are $N_2$, we see that $T_{prod}$ should be close to 2650 K, but somewhat under it. Therefore, a good first guess is 2400 K. At this temperature,

$$8\bar{h}_{CO_2} + 9\bar{h}_{H_2O} + 47\bar{h}_{N_2} = 8 \times 125,152 + 9 \times 103,508 + 47 \times 79,320$$

$$= 5,660,828 \text{ kJ}$$

This value is higher than 5,646,081 kJ. Therefore, the actual temperature is slightly under 2400 K. Next we choose 2350 K. It yields

$$8 \times 122,091 + 9 \times 100,846 + 47 \times 77,496 = 5,526,654$$

which is lower than 5,646,081 kJ. Therefore, the actual temperature of the products is between 2350 and 2400 K. By interpolation, it is found to be $T_{prod} = $ **2395 K.**

(b) The balanced equation for the complete combustion process with 400 percent theoretical air is

$$C_8H_{18}(\ell) + 50(O_2 + 3.76N_2) \rightarrow 8CO_2 + 9H_2O + 37.5O_2 + 188N_2$$

By following the procedure used in (a), the adiabatic flame temperature in this case is determined to be $T_{prod} = $ **962 K.**

Notice that the temperature of the products decreases significantly as a result of using excess air.

(c) The balanced equation for the incomplete combustion process with 90 percent theoretical air is

$$C_8H_{18}(\ell) + 11.25(O_2 + 3.76N_2) \rightarrow 5.5CO_2 + 2.5CO + 9H_2O + 42.3N_2$$

Following the procedure used in (a), we find the adiabatic flame temperature in this case to be $T_{prod} = $ **2236 K.**

**Discussion** Notice that the adiabatic flame temperature decreases as a result of incomplete combustion or using excess air. Also, *the maximum adiabatic flame temperature is achieved when complete combustion occurs with the theoretical amount of air.*

# 15–6 · ENTROPY CHANGE OF REACTING SYSTEMS

So far we have analyzed combustion processes from the conservation of mass and the conservation of energy points of view. The thermodynamic analysis of a process is not complete, however, without the examination of the second-law aspects. Of particular interest are the exergy and exergy destruction, both of which are related to entropy.

The entropy balance relations developed in Chap. 7 are equally applicable to both reacting and nonreacting systems provided that the entropies of individual constituents are evaluated properly using a common basis.

The **entropy balance** for *any system* (including reacting systems) undergoing *any process* can be expressed as

$$\underbrace{S_{in} - S_{out}}_{\substack{\text{Net entropy transfer} \\ \text{by heat and mass}}} + \underbrace{S_{gen}}_{\substack{\text{Entropy} \\ \text{generation}}} = \underbrace{\Delta S_{system}}_{\substack{\text{Change} \\ \text{in entropy}}} \quad \text{(kJ/K)} \qquad \textbf{(15–18)}$$

Using quantities per unit mole of fuel and taking the positive direction of heat transfer to be *to* the system, the entropy balance relation can be expressed more explicitly for a *closed* or *steady-flow* reacting system as (Fig. 15–27)

$$\sum \frac{Q_k}{T_k} + S_{gen} = S_{prod} - S_{react} \quad \text{(kJ/K)} \qquad \textbf{(15–19)}$$

where $T_k$ is temperature at the boundary where $Q_k$ crosses it. For an *adiabatic process* ($Q = 0$), the entropy transfer term drops out and Eq. 15–19 reduces to

$$S_{gen,adiabatic} = S_{prod} - S_{react} \geq 0 \qquad \textbf{(15–20)}$$

The *total* entropy generated during a process can be determined by applying the entropy balance to an *extended system* that includes the system itself and its immediate surroundings where external irreversibilities might be occurring. When evaluating the entropy transfer between an extended system and the surroundings, the boundary temperature of the extended system is simply taken to be the *environment temperature,* as explained in Chap. 7.

The determination of the entropy change associated with a chemical reaction seems to be straightforward, except for one thing: The entropy relations for the reactants and the products involve the *entropies* of the components, *not entropy changes,* which was the case for nonreacting systems. Thus we are faced with the problem of finding a common base for the entropy of all substances, as we did with enthalpy. The search for such a common base led to the establishment of the **third law of thermodynamics** in the early part of last century. The third law was expressed in Chap. 7 as follows: *The entropy of a pure crystalline substance at absolute zero temperature is zero.*

Therefore, the third law of thermodynamics provides an absolute base for the entropy values for all substances. Entropy values relative to this base are called the **absolute entropy**. The $\bar{s}°$ values listed in Tables A–18 through A–25 for various gases such as $N_2$, $O_2$, $CO$, $CO_2$, $H_2$, $H_2O$, $OH$, and $O$ are the *ideal-gas absolute entropy values* at the specified temperature and *at a pressure of 1 atm.* The absolute entropy values for various fuels are listed in Table A–26 together with the $\bar{h}_f°$ values at the standard reference state of 25°C and 1 atm.

Equation 15–20 is a general relation for the entropy change of a reacting system. It requires the determination of the entropy of each individual component of the reactants and the products, which in general is not very easy to do. The entropy calculations can be simplified somewhat if the gaseous components of the reactants and the products are approximated as ideal gases. However, entropy calculations are never as easy as enthalpy or internal energy calculations, since entropy is a function of both temperature and pressure even for ideal gases.

When evaluating the entropy of a component of an ideal-gas mixture, we should use the temperature and the partial pressure of the component.

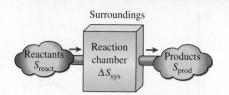

**FIGURE 15–27**
The entropy change associated with a chemical relation.

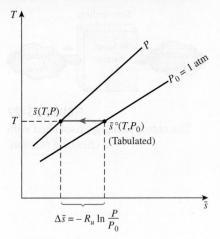

**FIGURE 15–28**

At a specified temperature, the absolute entropy of an ideal gas at pressures other than $P_0 = 1$ atm can be determined by subtracting $R_u \ln (P/P_0)$ from the tabulated value at 1 atm.

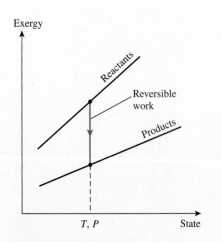

**FIGURE 15–29**

The difference between the exergy of the reactants and of the products during a chemical reaction is the reversible work associated with that reaction.

Note that the temperature of a component is the same as the temperature of the mixture, and the partial pressure of a component is equal to the mixture pressure multiplied by the mole fraction of the component.

Absolute entropy values at pressures other than $P_0 = 1$ atm for any temperature $T$ can be obtained from the ideal-gas entropy change relation written for an imaginary isothermal process between states $(T, P_0)$ and $(T, P)$, as illustrated in Fig. 15–28:

$$\bar{s}(T,P) = \bar{s}°(T,P_0) - R_u \ln \frac{P}{P_0} \qquad (15\text{–}21)$$

For the component $i$ of an ideal-gas mixture, this relation can be written as

$$\bar{s}_i(T,P_i) = \bar{s}_i°(T,P_0) - R_u \ln \frac{y_i P_m}{P_0} \quad (\text{kJ/kmol·K}) \qquad (15\text{–}22)$$

where $P_0 = 1$ atm, $P_i$ is the partial pressure, $y_i$ is the mole fraction of the component, and $P_m$ is the total pressure of the mixture.

If a gas mixture is at a relatively high pressure or low temperature, the deviation from the ideal-gas behavior should be accounted for by incorporating more accurate equations of state or the generalized entropy charts.

## 15–7 ▪ SECOND-LAW ANALYSIS OF REACTING SYSTEMS

Once the total entropy change or the entropy generation is evaluated, the **exergy destroyed** $X_{\text{destroyed}}$ associated with a chemical reaction can be determined from

$$X_{\text{destroyed}} = T_0 S_{\text{gen}} \quad (\text{kJ}) \qquad (15\text{–}23)$$

where $T_0$ is the thermodynamic temperature of the surroundings.

When analyzing reacting systems, we are more concerned with the changes in the exergy of reacting systems than with the values of exergy at various states (Fig. 15–29). Recall from Chap. 8 that the **reversible work** $W_{\text{rev}}$ represents the maximum work that can be done during a process. In the absence of any changes in kinetic and potential energies, the reversible work relation for a steady-flow combustion process that involves heat transfer with only the surroundings at $T_0$ can be obtained by replacing the enthalpy terms by $\bar{h}_f° + \bar{h} - \bar{h}°$, yielding

$$W_{\text{rev}} = \sum N_r (\bar{h}_f° + \bar{h} - \bar{h}° - T_0\bar{s})_r - \sum N_p (\bar{h}_f° + \bar{h} - \bar{h}° - T_0\bar{s})_p \qquad (15\text{–}24)$$

An interesting situation arises when both the reactants and the products are at the temperature of the surroundings $T_0$. In that case, $\bar{h} - T_0\bar{s} = (\bar{h} - T_0\bar{s})_{T_0} = \bar{g}_0$, which is, by definition, the **Gibbs function** of a unit mole of a substance at temperature $T_0$. The $W_{\text{rev}}$ relation in this case can be written as

$$W_{\text{rev}} = \sum N_r \bar{g}_{0,r} - \sum N_p \bar{g}_{0,p} \qquad (15\text{–}25)$$

or

$$W_{\text{rev}} = \sum N_r (\bar{g}_f° + \bar{g}_{T_0} - \bar{g}°)_r - \sum N_p (\bar{g}_f° + \bar{g}_{T_0} - \bar{g}°)_p \qquad (15\text{–}26)$$

where $\bar{g}_f^\circ$ is the Gibbs function of formation ($\bar{g}_f^\circ = 0$ for stable elements like $N_2$ and $O_2$ at the standard reference state of 25°C and 1 atm, just like the enthalpy of formation) and $\bar{g}_{T_0} - \bar{g}^\circ$ represents the value of the sensible Gibbs function of a substance at temperature $T_0$ relative to the standard reference state.

For the very special case of $T_{react} = T_{prod} = T_0 = 25°C$ (i.e., the reactants, the products, and the surroundings are at 25°C) and the partial pressure $P_i = 1$ atm for each component of the reactants and the products, Eq. 15–26 reduces to

$$W_{rev} = \sum N_r \bar{g}_{f,r}^\circ - \sum n_p \bar{g}_{f,p}^\circ \quad \text{(kJ)} \qquad \textbf{(15–27)}$$

We can conclude from the above equation that the $-\bar{g}_f^\circ$ value (the negative of the Gibbs function of formation at 25°C and 1 atm) of a compound represents the *reversible work* associated with the formation of that compound from its stable elements at 25°C and 1 atm in an environment at 25°C and 1 atm (Fig. 15–30). The $\bar{g}_f^\circ$ values of several substances are listed in Table A–26.

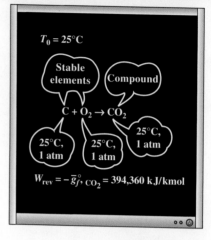

**FIGURE 15–30**
The negative of the Gibbs function of formation of a compound at 25°C, 1 atm represents the reversible work associated with the formation of that compound from its stable elements at 25°C, 1 atm in an environment that is at 25°C, 1 atm.

---

■ **EXAMPLE 15–9** **Reversible Work Associated with a Combustion Process**

■ One lbmol of carbon at 25°C and 1 atm is burned steadily with 1 kmol of oxygen at the same state as shown in Fig. 15–31. The $CO_2$ formed during the process is then brought to 25°C and 1 atm, the conditions of the surroundings. Assuming the combustion is complete, determine the reversible work for this process.

**SOLUTION** Carbon is burned steadily with pure oxygen. The reversible work associated with this process is to be determined.
**Assumptions** **1** Combustion is complete. **2** Steady-flow conditions exist during combustion. **3** Oxygen and the combustion gases are ideal gases. **4** Changes in kinetic and potential energies are negligible.
**Properties** The Gibbs function of formation at 25°C and 1 atm is 0 for C and $O_2$, and −394,360 kJ/kmol for $CO_2$. The enthalpy of formation is 0 for C and $O_2$, and −393,520 kJ/kmol for $CO_2$. The absolute entropy is 5.74 kJ/kmol·K for C, 205.04 kJ/kmol·K for $O_2$, and 213.80 kJ/kmol·K for $CO_2$ (Table A–26).
**Analysis** The combustion equation is

$$C + O_2 \rightarrow CO_2$$

The C, $O_2$, and $CO_2$ are at 25°C and 1 atm, which is the standard reference state and also the state of the surroundings. Therefore, the reversible work in this case is simply the difference between the Gibbs function of formation of the reactants and that of the products (Eq. 15–27):

$$W_{rev} = \sum N_r \bar{g}_{f,r}^\circ - \sum N_p \bar{g}_{f,p}^\circ$$

$$= N_C \bar{g}_{f,C}^{\circ\,0} + N_{O_2} \bar{g}_{f,O_2}^{\circ\,0} - N_{CO_2} \bar{g}_{f,CO_2}^\circ = -N_{CO_2} \bar{g}_{f,CO_2}^\circ$$

$$= (-1 \text{ kmol})(-394{,}360 \text{ kJ/kmol})$$

$$= \textbf{394{,}360 kJ}$$

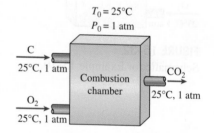

**FIGURE 15–31**
Schematic for Example 15–9.

since the $\bar{g}_f^\circ$ of stable elements at 25°C and 1 atm is zero. Therefore, 394,360 kJ of work could be done as 1 kmol of C is burned with 1 kmol of $O_2$ at 25°C and 1 atm in an environment at the same state. The reversible work in this case represents the exergy of the reactants since the product (the $CO_2$) is at the state of the surroundings.

**Discussion** We could also determine the reversible work without involving the Gibbs function by using Eq. 15–24:

$$W_{rev} = \sum N_r(\bar{h}_f^\circ + \bar{h} - \bar{h}^\circ - T_0\bar{s})_r - \sum N_p(\bar{h}_f^\circ + \bar{h} - \bar{h}^\circ - T_0\bar{s})_p$$
$$= \sum N_r(\bar{h}_f^\circ - T_0\bar{s})_r - \sum N_p(\bar{h}_f^\circ - T_0\bar{s})_p$$
$$= N_C(\bar{h}_f^\circ - T_0\bar{s}^\circ)_C + N_{O_2}(\bar{h}_f^\circ - T_0\bar{s}^\circ)_{O_2} - N_{CO_2}(\bar{h}_f^\circ - T_0\bar{s}^\circ)_{CO_2}$$

Substituting the enthalpy of formation and absolute entropy values, we obtain

$$W_{rev} = (1 \text{ kmol C})[0 - (298 \text{ K})(5.74 \text{ kJ/kmol·K})]$$
$$+ (1 \text{ kmol } O_2)[0 - (298 \text{ K})(205.04 \text{ kJ/kmol·K})]$$
$$- (1 \text{ kmol } CO_2)[-393,520 \text{ kJ/kmol} - (298 \text{ K})(213.80 \text{ kJ/kmol·K})]$$
$$= \mathbf{394,420 \text{ kJ}}$$

which is practically identical to the result obtained before.

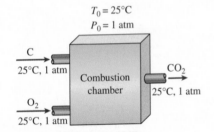

$T_0 = 25°C$
$P_0 = 1$ atm

C
25°C, 1 atm

Combustion chamber

$CO_2$
25°C, 1 atm

$O_2$
25°C, 1 atm

**FIGURE 15–32**
Schematic for Example 15–10.

**EXAMPLE 15–10** **Second-Law Analysis of Adiabatic Combustion**

Methane ($CH_4$) gas enters a steady-flow adiabatic combustion chamber at 25°C and 1 atm. It is burned with 50 percent excess air that also enters at 25°C and 1 atm, as shown in Fig. 15–32. Assuming complete combustion, determine (a) the temperature of the products, (b) the entropy generation, and (c) the reversible work and exergy destruction. Assume that $T_0 = 298$ K and the products leave the combustion chamber at 1 atm pressure.

**SOLUTION** Methane is burned with excess air in a steady-flow combustion chamber. The product temperature, entropy generated, reversible work, and exergy destroyed are to be determined.

**Assumptions** 1 Steady-flow conditions exist during combustion. 2 Air and the combustion gases are ideal gases. 3 Changes in kinetic and potential energies are negligible. 4 The combustion chamber is adiabatic and thus there is no heat transfer. 5 Combustion is complete.

**Analysis** (a) The balanced equation for the complete combustion process with 50 percent excess air is

$$CH_4(g) + 3(O_2 + 3.76N_2) \rightarrow CO_2 + 2H_2O + O_2 + 11.28N_2$$

Under steady-flow conditions, the adiabatic flame temperature is determined from $H_{prod} = H_{react}$, which reduces to

$$\sum N_p(\bar{h}_f^\circ + \bar{h} - \bar{h}^\circ)_p = \sum N_r\bar{h}_{f,r}^\circ = (N\bar{h}_f^\circ)_{CH_4}$$

since all the reactants are at the standard reference state and $\bar{h}_f^\circ = 0$ for $O_2$ and $N_2$. Assuming ideal-gas behavior for air and for the products, the $\bar{h}_f^\circ$ and $h$ values of various components at 298 K can be listed as

| Substance | $\bar{h}_f^\circ$ kJ/kmol | $\bar{h}_{298 K}$ kJ/kmol |
|---|---|---|
| $CH_4(g)$ | −74,850 | — |
| $O_2$ | 0 | 8682 |
| $N_2$ | 0 | 8669 |
| $H_2O(g)$ | −241,820 | 9904 |
| $CO_2$ | −393,520 | 9364 |

Substituting, we have

$$(1 \text{ kmol CO}_2)[(-393,520 + \bar{h}_{CO_2} - 9364) \text{ kJ/kmol CO}_2]$$

$$+ (2 \text{ kmol H}_2\text{O})[(-241,820 + \bar{h}_{H_2O} - 9904) \text{ kJ/kmol H}_2\text{O}]$$

$$+ (11.28 \text{ kmol N}_2)[(0 + \bar{h}_{N_2} - 8669) \text{ kJ/kmol N}_2]$$

$$+ (1 \text{ kmol O}_2)[(0 + \bar{h}_{O_2} - 8682) \text{ kJ/kmol O}_2]$$

$$= (1 \text{ kmol CH}_4)(-74,850 \text{ kJ/kmol CH}_4)$$

which yields

$$\bar{h}_{CO_2} + 2\bar{h}_{H_2O} + \bar{h}_{O_2} + 11.28\bar{h}_{N_2} = 937,950 \text{ kJ}$$

By trial and error, the temperature of the products is found to be

$$T_{prod} = \textbf{1789 K}$$

(b) Noting that combustion is adiabatic, the entropy generation during this process is determined from Eq. 15–20:

$$S_{gen} = S_{prod} - S_{react} = \sum N_p\bar{s}_p - \sum N_r\bar{s}_r$$

The $CH_4$ is at 25°C and 1 atm, and thus its absolute entropy is $\bar{s}_{CH_4} = 186.16$ kJ/kmol·K (Table A–26). The entropy values listed in the ideal-gas tables are for 1 atm pressure. Both the air and the product gases are at a total pressure of 1 atm, but the entropies are to be calculated at the partial pressure of the components, which is equal to $P_i = y_i P_{total}$, where $y_i$ is the mole fraction of component i. From Eq. 15–22:

$$S_i = N_i\bar{s}_i(T, P_i) = N_i[\bar{s}_i^\circ(T, P_0) - R_u \ln y_i P_m]$$

The entropy calculations can be represented in tabular form as follows:

| | $N_i$ | $y_i$ | $\bar{s}_i^\circ(T, 1 \text{ atm})$ | $-R_u \ln y_i P_m$ | $N_i \bar{s}_i^\circ$ |
|---|---|---|---|---|---|
| $CH_4$ | 1 | 1.00 | 186.16 | — | 186.16 |
| $O_2$ | 3 | 0.21 | 205.04 | 12.98 | 654.06 |
| $N_2$ | 11.28 | 0.79 | 191.61 | 1.96 | 2183.47 |
| | | | | $S_{react} =$ | 3023.69 |
| $CO_2$ | 1 | 0.0654 | 302.517 | 22.674 | 325.19 |
| $H_2O$ | 2 | 0.1309 | 258.957 | 16.905 | 551.72 |
| $O_2$ | 1 | 0.0654 | 264.471 | 22.674 | 287.15 |
| $N_2$ | 11.28 | 0.7382 | 247.977 | 2.524 | 2825.65 |
| | | | | $S_{prod} =$ | 3989.71 |

Thus,

$$S_{gen} = S_{prod} - S_{react} = (3989.71 - 3023.69)\text{kJ/kmol·K CH}_4$$

$$= \textbf{966.0 kJ/kmol·K}$$

(c) The exergy destruction or irreversibility associated with this process is determined from Eq. 15–23,

$$X_{destroyed} = T_0 S_{gen} = (298 \text{ K})(966.0 \text{ kJ/kmol·K})$$

$$= \textbf{288 MJ/kmol CH}_4$$

That is, 288 MJ of work potential is wasted during this combustion process for each kmol of methane burned. This example shows that even complete combustion processes are highly irreversible.

This process involves no actual work. Therefore, the reversible work and exergy destroyed are identical:

$$W_{rev} = \textbf{288 MJ/kmol CH}_4$$

That is, 288 MJ of work could be done during this process but is not. Instead, the entire work potential is wasted.

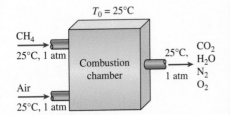

**FIGURE 15–33**
Schematic for Example 15–11.

**EXAMPLE 15–11**    **Second-Law Analysis of Isothermal Combustion**

Methane ($CH_4$) gas enters a steady-flow combustion chamber at 25°C and 1 atm and is burned with 50 percent excess air, which also enters at 25°C and 1 atm, as shown in Fig. 15–33. After combustion, the products are allowed to cool to 25°C. Assuming complete combustion, determine (a) the heat transfer per kmol of $CH_4$, (b) the entropy generation, and (c) the reversible work and exergy destruction. Assume that $T_0 = 298$ K and the products leave the combustion chamber at 1 atm pressure.

**SOLUTION** This is the same combustion process we discussed in Example 15–10, except that the combustion products are brought to the state of the surroundings by transferring heat from them. Thus the combustion equation remains the same:

$$CH_4(g) + 3(O_2 + 3.76N_2) \rightarrow CO_2 + 2H_2O + O_2 + 11.28N_2$$

At 25°C, part of the water will condense. The amount of water vapor that remains in the products is determined from (see Example 15–3)

$$\frac{N_v}{N_{gas}} = \frac{P_v}{P_{total}} = \frac{3.1698 \text{ kPa}}{101.325 \text{ kPa}} = 0.03128$$

and

$$N_v = \left(\frac{P_v}{P_{total}}\right) N_{gas} = (0.03128)(13.28 + N_v) \rightarrow N_v = 0.43 \, \text{kmol}$$

Therefore, 1.57 kmol of the $H_2O$ formed is in the liquid form, which is removed at 25°C and 1 atm. When one is evaluating the partial pressures of the components in the product gases, the only water molecules that need to be considered are those that are in the vapor phase. As before, all the gaseous reactants and products are treated as ideal gases.

(a) Heat transfer during this steady-flow combustion process is determined from the steady-flow energy balance $E_{out} = E_{in}$ on the combustion chamber,

$$Q_{out} + \sum N_p \bar{h}_{f,p}^\circ = \sum N_r \bar{h}_{f,r}^\circ$$

since all the reactants and products are at the standard reference of 25°C and the enthalpy of ideal gases depends on temperature only. Solving for $Q_{out}$ and substituting the $\bar{h}_f^\circ$ values, we have

$$Q_{out} = (1 \text{ kmol CH}_4)(-74{,}850 \text{ kJ/kmol CH}_4)$$

$$-(1 \text{ kmol CO}_2)(-393{,}520 \text{ kJ/kmol CO}_2)$$

$$-[0.43 \text{ kmol H}_2O(g)][-241{,}820 \text{ kJ/kmol H}_2O(g)]$$

$$-[1.57 \text{ kmol H}_2O(\ell)][-285{,}830 \text{ kJ/kmol H}_2O(\ell)]$$

$$= 871{,}400 \text{ kJ/kmol CH}_4$$

(b) The entropy of the reactants was evaluated in Example 15–10 and was determined to be $S_{react} = 3023.69$ kJ/kmol·K $CH_4$. By following a similar approach, the entropy of the products is determined to be

| | $N_i$ | $y_i$ | $\bar{s}_i^\circ(T, 1 \text{ atm})$ | $-R_u \ln y_i P_m$ | $N_i \bar{s}_i$ |
|---|---|---|---|---|---|
| $H_2O(\ell)$ | 1.57 | 1.0000 | 69.92 | — | 109.77 |
| $H_2O$ | 0.43 | 0.0314 | 188.83 | 28.77 | 93.57 |
| $CO_2$ | 1 | 0.0729 | 213.80 | 21.77 | 235.57 |
| $O_2$ | 1 | 0.0729 | 205.04 | 21.77 | 226.81 |
| $N_2$ | 11.28 | 0.8228 | 191.61 | 1.62 | 2179.63 |
| | | | | $S_{prod} =$ | 2845.35 |

Then the total entropy generation during this process is determined from an entropy balance applied on an *extended system* that includes the immediate surroundings of the combustion chamber

$$S_{gen} = S_{prod} - S_{react} + \frac{Q_{out}}{T_{surr}}$$

$$= (2845.35 - 3023.69) \text{ kJ/kmol} + \frac{871{,}400 \text{ kJ/kmol}}{298 \text{ K}}$$

$$= 2746 \text{ kJ/kmol·K CH}_4$$

(c) The exergy destruction and reversible work associated with this process are determined from

$$X_{\text{destroyed}} = T_0 S_{\text{gen}} = (298 \text{ K})(2746 \text{ kJ/kmol·K})$$
$$= \textbf{818 MJ/kmol CH}_4$$

and

$$W_{\text{rev}} = X_{\text{destroyed}} = \textbf{818 MJ/kmol CH}_4$$

since this process involves no actual work. Therefore, 818 MJ of work could be done during this process but is not. Instead, the entire work potential is wasted. The reversible work in this case represents the exergy of the reactants before the reaction starts since the products are in equilibrium with the surroundings, that is, they are at the dead state.

**Discussion** Note that, for simplicity, we calculated the entropy of the product gases before they actually entered the atmosphere and mixed with the atmospheric gases. A more complete analysis would consider the composition of the atmosphere and the mixing of the product gases with the gases in the atmosphere, forming a homogeneous mixture. There is additional entropy generation during this mixing process, and thus additional wasted work potential.

## TOPIC OF SPECIAL INTEREST*       Fuel Cells

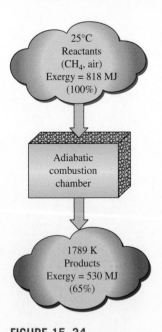

**FIGURE 15–34**
The availability of methane decreases by 35 percent as a result of irreversible combustion process.

Fuels like methane are commonly burned to provide thermal energy at high temperatures for use in heat engines. However, a comparison of the reversible works obtained in the last two examples reveals that the exergy of the reactants (818 MJ/kmol $CH_4$) decreases by 288 MJ/kmol as a result of the irreversible adiabatic combustion process alone. That is, the exergy of the hot combustion gases at the end of the adiabatic combustion process is $818 - 288 = 530$ MJ/kmol $CH_4$. In other words, the work potential of the hot combustion gases is about 65 percent of the work potential of the reactants. It seems that when methane is burned, 35 percent of the work potential is lost before we even start using the thermal energy (Fig. 15–34).

Thus, the second law of thermodynamics suggests that there should be a better way of converting the chemical energy to work. The better way is, of course, the less irreversible way, the best being the reversible case. In chemical reactions, the irreversibility is due to uncontrolled electron exchange between the reacting components. The electron exchange can be controlled by replacing the combustion chamber by electrolytic cells, like car batteries. (This is analogous to replacing unrestrained expansion of a gas in mechanical systems by restrained expansion.) In the electrolytic cells, the electrons are exchanged through conductor wires connected to a load, and the chemical energy is directly converted to electric energy. The energy conversion

*This section can be skipped without a loss in continuity.

devices that work on this principle are called **fuel cells**. Fuel cells are not heat engines, and thus their efficiencies are not limited by the Carnot efficiency. They convert chemical energy to electric energy essentially in an isothermal manner.

A fuel cell functions like a battery, except that it produces its own electricity by combining a fuel with oxygen in a cell electrochemically without combustion, and discards the waste heat. A fuel cell consists of two electrodes separated by an electrolyte such as a solid oxide, phosphoric acid, or molten carbonate. The electric power generated by a single fuel cell is usually too small to be of any practical use. Therefore, fuel cells are usually stacked in practical applications. This modularity gives the fuel cells considerable flexibility in applications: The same design can be used to generate a small amount of power for a remote switching station or a large amount of power to supply electricity to an entire town. Therefore, fuel cells are termed the "microchip of the energy industry."

The operation of a hydrogen–oxygen fuel cell is illustrated in Fig. 15–35. Hydrogen is ionized at the surface of the anode, and hydrogen ions flow through the electrolyte to the cathode. There is a potential difference between the anode and the cathode, and free electrons flow from the anode to the cathode through an external circuit (such as a motor or a generator). Hydrogen ions combine with oxygen and the free electrons at the surface of the cathode, forming water. Therefore, the fuel cell operates like an electrolysis system working in reverse. In steady operation, hydrogen and oxygen continuously enter the fuel cell as reactants, and water leaves as the product. Therefore, the exhaust of the fuel cell is drinkable quality water.

The fuel cell was invented by William Groves in 1839, but it did not receive serious attention until the 1960s, when they were used to produce electricity and water for the Gemini and Apollo spacecraft during their missions to the moon. Today they are used for the same purpose in the space shuttle missions. Despite the irreversible effects such as internal resistance to electron flow, fuel cells have a great potential for much higher conversion efficiencies. Currently fuel cells are available commercially, but they are competitive only in some niche markets because of their higher cost. Fuel cells produce high-quality electric power efficiently and quietly while generating low emissions using a variety of fuels such as hydrogen, natural gas, propane, and biogas. Recently many fuel cells have been installed to generate electricity. For example, a remote police station in Central Park in New York City is powered by a 200-kW phosphoric acid fuel cell that has an efficiency of 40 percent with negligible emissions (it emits 1 ppm $NO_x$ and 5 ppm CO).

Hybrid power systems (HPS) that combine high-temperature fuel cells and gas turbines have the potential for very high efficiency in converting natural gas (or even coal) to electricity. Also, some car manufacturers are planning to introduce cars powered by fuel-cell engines, thus nearly doubling the efficiency from about 30 percent for the gasoline engines to up to 60 percent for fuel cells. Intense research and development programs by major car manufacturers are underway to make fuel cell cars economical and commercially available in the near future.

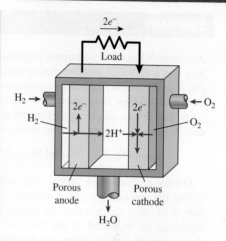

**FIGURE 15–35**
The operation of a hydrogen–oxygen fuel cell.

## SUMMARY

Any material that can be burned to release energy is called a *fuel*, and a chemical reaction during which a fuel is oxidized and a large quantity of energy is released is called *combustion*. The oxidizer most often used in combustion processes is air. The dry air can be approximated as 21 percent oxygen and 79 percent nitrogen by mole numbers. Therefore,

$$1 \text{ kmol } O_2 + 3.76 \text{ kmol } N_2 = 4.76 \text{ kmol air}$$

During a combustion process, the components that exist before the reaction are called *reactants* and the components that exist after the reaction are called *products*. Chemical equations are balanced on the basis of the *conservation of mass principle*, which states that the total mass of each element is conserved during a chemical reaction. The ratio of the mass of air to the mass of fuel during a combustion process is called the *air–fuel ratio* AF:

$$AF = \frac{m_{\text{air}}}{m_{\text{fuel}}}$$

where $m_{\text{air}} = (NM)_{\text{air}}$ and $m_{\text{fuel}} = \Sigma(N_i M_i)_{\text{fuel}}$.

A combustion process is *complete* if all the carbon in the fuel burns to $CO_2$, all the hydrogen burns to $H_2O$, and all the sulfur (if any) burns to $SO_2$. The minimum amount of air needed for the complete combustion of a fuel is called the *stoichiometric* or *theoretical air*. The theoretical air is also referred to as the chemically correct amount of air or 100 percent theoretical air. The ideal combustion process during which a fuel is burned completely with theoretical air is called the *stoichiometric* or *theoretical combustion* of that fuel. The air in excess of the stoichiometric amount is called the *excess air*. The amount of excess air is usually expressed in terms of the stoichiometric air as *percent excess air* or *percent theoretical air*.

During a chemical reaction, some chemical bonds are broken and others are formed. Therefore, a process that involves chemical reactions involves changes in chemical energies. Because of the changed composition, it is necessary to have a *standard reference state* for all substances, which is chosen to be 25°C and 1 atm.

The difference between the enthalpy of the products at a specified state and the enthalpy of the reactants at the same state for a complete reaction is called the *enthalpy of reaction* $h_R$. For combustion processes, the enthalpy of reaction is usually referred to as the *enthalpy of combustion* $h_C$, which represents the amount of heat released during a steady-flow combustion process when 1 kmol (or 1 kg) of fuel is burned completely at a specified temperature and pressure. The enthalpy of a substance at a specified state due to its chemical composition is called the *enthalpy of formation* $\bar{h}_f$. The enthalpy of formation of all stable elements is assigned a value of zero at the standard reference state of

25°C and 1 atm. The *heating value* of a fuel is defined as the amount of heat released when a fuel is burned completely in a steady-flow process and the products are returned to the state of the reactants. The heating value of a fuel is equal to the absolute value of the enthalpy of combustion of the fuel,

$$\text{Heating value} = |h_C| \quad \text{(kJ/kg fuel)}$$

Taking heat transfer *to* the system and work done *by* the system to be positive quantities, the conservation of energy relation for chemically reacting steady-flow systems can be expressed per unit mole of fuel as

$$Q - W = \sum N_p(\bar{h}_f^{\circ} + \bar{h} - \bar{h}^{\circ})_p - \sum N_r(\bar{h}_f^{\circ} + \bar{h} - \bar{h}^{\circ})_r$$

where the superscript ° represents properties at the standard reference state of 25°C and 1 atm. For a closed system, it becomes

$$Q - W = \sum N_p(\bar{h}_f^{\circ} + \bar{h} - \bar{h}^{\circ} - P\bar{v})_p - \sum N_r(\bar{h}_f^{\circ} + \bar{h} - \bar{h}^{\circ} - P\bar{v})_r$$

The $P\bar{v}$ terms are negligible for solids and liquids and can be replaced by $R_u T$ for gases that behave as ideal gases.

In the absence of any heat loss to the surroundings ($Q = 0$), the temperature of the products will reach a maximum, which is called the *adiabatic flame temperature* of the reaction. The adiabatic flame temperature of a steady-flow combustion process is determined from $H_{\text{prod}} = H_{\text{react}}$ or

$$\sum N_p(\bar{h}_f^{\circ} + \bar{h} - \bar{h}^{\circ})_p = \sum N_r(\bar{h}_f^{\circ} + \bar{h} - \bar{h}^{\circ})_r$$

Taking the positive direction of heat transfer to be *to* the system, the entropy balance relation can be expressed for a *closed system* or *steady-flow combustion chamber* as

$$\sum \frac{Q_k}{T_k} + S_{\text{gen}} = S_{\text{prod}} - S_{\text{react}}$$

For an *adiabatic process* it reduces to

$$S_{\text{gen,adiabatic}} = S_{\text{prod}} - S_{\text{react}} \geq 0$$

The *third law of thermodynamics* states that the entropy of a pure crystalline substance at absolute zero temperature is zero. The third law provides a common base for the entropy of all substances, and the entropy values relative to this base are called the *absolute entropy*. The ideal-gas tables list the absolute entropy values over a wide range of temperatures but at a fixed pressure of $P_0 = 1$ atm. Absolute entropy values at other pressures $P$ for any temperature $T$ are determined from

$$\bar{s}(T, P) = \bar{s}^{\circ}(T, P_0) - R_u \ln \frac{P}{P_0}$$

For component $i$ of an ideal-gas mixture, this relation can be written as

$$\bar{s}_i(T, P_i) = \bar{s}_i^\circ(T, P_0) - R_u \ln \frac{y_i P_m}{P_0}$$

where $P_i$ is the partial pressure, $y_i$ is the mole fraction of the component, and $P_m$ is the total pressure of the mixture in atmospheres.

The *exergy destruction* and the *reversible work* associated with a chemical reaction are determined from

$$X_{\text{destroyed}} = W_{\text{rev}} - W_{\text{act}} = T_0 S_{\text{gen}}$$

and

$$W_{\text{rev}} = \sum N_r(\bar{h}_f^\circ + \bar{h} - \bar{h}^\circ - T_0\bar{s})_r - \sum N_p(\bar{h}_f^\circ + \bar{h} - \bar{h}^\circ - T_0\bar{s})_p$$

When both the reactants and the products are at the temperature of the surroundings $T_0$, the reversible work can be expressed in terms of the Gibbs functions as

$$W_{\text{rev}} = \sum N_r(\bar{g}_f^\circ + \bar{g}_{T_0} - \bar{g}^\circ)_r - \sum N_p(\bar{g}_f^\circ + \bar{g}_{T_0} - \bar{g}^\circ)_p$$

## REFERENCES AND SUGGESTED READINGS

**1.** S. W. Angrist. *Direct Energy Conversion.* 4th ed. Boston: Allyn and Bacon, 1982.

**2.** I. Glassman. *Combustion.* New York: Academic Press, 1977.

**3.** R. Strehlow. *Fundamentals of Combustion.* Scranton, PA: International Textbook Co., 1968.

## PROBLEMS*

### Fuels and Combustion

**15–1C**  How does the presence of $N_2$ in air affect the outcome of a combustion process?

**15–2C**  Is the number of atoms of each element conserved during a chemical reaction? How about the total number of moles?

**15–3C**  What is the air–fuel ratio? How is it related to the fuel–air ratio?

**15–4C**  Is the air–fuel ratio expressed on a mole basis identical to the air–fuel ratio expressed on a mass basis?

**15–5C**  What does the dew-point temperature of the product gases represent? How is it determined?

**15–6**  Trace amounts of sulfur (S) in coal are burned in the presence of diatomic oxygen ($O_2$) to form sulfur dioxide ($SO_2$). Determine the minimum mass of oxygen required in the reactants and the mass of sulfur dioxide in the products when 1 kg of sulfur is burned.

### Theoretical and Actual Combustion Processes

**15–7C**  What does 100 percent theoretical air represent?

**15–8C**  Consider a fuel that is burned with (a) 130 percent theoretical air and (b) 70 percent excess air. In which case is the fuel burned with more air?

**15–9C**  Are complete combustion and theoretical combustion identical? If not, how do they differ?

**15–10C**  What are the causes of incomplete combustion?

**15–11C**  Which is more likely to be found in the products of an incomplete combustion of a hydrocarbon fuel, CO or OH? Why?

**15–12**  Methane ($CH_4$) is burned with stoichiometric amount of air during a combustion process. Assuming complete combustion, determine the air–fuel and fuel–air ratios.

**15–13**  Propane fuel ($C_3H_8$) is burned in the presence of air. Assuming that the combustion is theoretical—that is, only nitrogen ($N_2$), water vapor ($H_2O$), and carbon dioxide ($CO_2$) are present in the products—determine (a) the mass fraction of carbon dioxide and (b) the mole and mass fractions of the water vapor in the products.

---

* Problems designated by a "C" are concept questions, and students are encouraged to answer them all. Problems with the 🔘 icon are solved using EES, and complete solutions together with parametric studies are included on the text website. Problems with the 🔘 icon are comprehensive in nature, and are intended to be solved with an equation solver such as an EES.

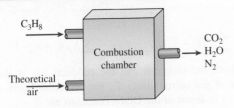

**FIGURE P15–13**

**15–14** n-Butane ($C_4H_{10}$) is burned with stoichiometric amount of oxygen. Determine the mole fraction of carbon dioxide and water in the products. Also, calculate the mole number of carbon dioxide in the products per unit mole of fuel burned.

**15–15** Acetylene ($C_2H_2$) is burned with 25 percent excess oxygen in a cutting torch. Determine the mass fraction of each of the products. Calculate the mass of oxygen used per unit mass of acetylene burned.

**15–16** Propal alcohol ($C_3H_7OH$) is burned with 50 percent excess air. Write the balanced reaction equation for complete combustion and determine the air-to-fuel ratio.
*Answer:* 15.5 kg air/kg fuel

**15–17** n-Octane ($C_8H_{18}$) is burned with 50 percent excess air. Calculate the mass fraction of each product and the mass of water in the products per unit mass of fuel burned. Also, calculate the mass fraction of each reactant.

**15–18** In a combustion chamber, ethane ($C_2H_6$) is burned at a rate of 8 kg/h with air that enters the combustion chamber at a rate of 176 kg/h. Determine the percentage of excess air used during this process. *Answer:* 37 percent

**15–19** One kmol of ethane ($C_2H_6$) is burned with an unknown amount of air during a combustion process. An analysis of the combustion products reveals that the combustion is complete, and there are 3 kmol of free $O_2$ in the products. Determine (*a*) the air–fuel ratio and (*b*) the percentage of theoretical air used during this process.

**15–20** Ethylene ($C_2H_4$) is burned with 175 percent theoretical air during a combustion process. Assuming complete combustion and a total pressure of 100 kPa, determine (*a*) the air–fuel ratio and (*b*) the dew-point temperature of the products. *Answers:* (*a*) 25.9 kg air/kg fuel, (*b*) 40.8°C

**15–21** One kilogram of butane ($C_4H_{10}$) is burned with 25 kg of air that is at 30°C and 90 kPa. Assuming that the combustion is complete and the pressure of the products is 90 kPa, determine (*a*) the percentage of theoretical air used and (*b*) the dew-point temperature of the products.

**15–22** Butane ($C_4H_{10}$) is burned in 200 percent theoretical air. For complete combustion, how many kmol of water must be sprayed into the combustion chamber per kmol of fuel if the products of combustion are to have a dew-point temperature of 50°C when the product pressure is 100 kPa?

**15–23** A fuel mixture of 60 percent by mass methane ($CH_4$) and 40 percent by mass ethanol ($C_2H_6O$), is burned completely with theoretical air. If the total flow rate of the fuel is 10 kg/s, determine the required flow rate of air.
*Answer:* 139 kg/s

**15–24** A certain natural gas has the following volumetric analysis: 65 percent $CH_4$, 8 percent $H_2$, 18 percent $N_2$, 3 percent $O_2$, and 6 percent $CO_2$. This gas is now burned completely with the stoichiometric amount of dry air. What is the air–fuel ratio for this combustion process?

**15–25** Repeat Prob. 15–24 by replacing the dry air by moist air that enters the combustion chamber at 25°C, 1 atm, and 70 percent relative humidity.

**15–26** A gaseous fuel with a volumetric analysis of 45 percent $CH_4$, 35 percent $H_2$, and 20 percent $N_2$ is burned to completion with 130 percent theoretical air. Determine (*a*) the air–fuel ratio and (*b*) the fraction of water vapor that would condense if the product gases were cooled to 25°C at 1 atm. *Answers:* (*a*) 13.9 kg air/kg fuel, (*b*) 84 percent

**15–27** Reconsider Prob. 15–26. Using EES (or other) software, study the effects of varying the percentages of $CH_4$, $H_2$, and $N_2$ making up the fuel and the product gas temperature in the range 5 to 150°C.

**15–28** Methane ($CH_4$) is burned with dry air. The volumetric analysis of the products on a dry basis is 5.20 percent $CO_2$, 0.33 percent $CO$, 11.24 percent $O_2$, and 83.23 percent $N_2$. Determine (*a*) the air–fuel ratio and (*b*) the percentage of theoretical air used. *Answers:* (*a*) 34.5 kg air/kg fuel, (*b*) 200 percent

**15–29** Octane ($C_8H_{18}$) is burned with dry air. The volumetric analysis of the products on a dry basis is 9.21 percent $CO_2$, 0.61 percent $CO$, 7.06 percent $O_2$, and 83.12 percent $N_2$. Determine (*a*) the air-fuel ratio and (*b*) the percentage of theoretical air used.

**15–30** n-Octane ($C_8H_{18}$) is burned with 60 percent excess air with 15 percent of the carbon in the fuel forming carbon monoxide. Calculate the mole fractions of the products and the dew-point temperature of the water vapor in the products when the products are at 1 atm pressure. *Answers:* 0.0678 ($CO_2$), 0.0120 ($CO$), 0.0897 ($H_2O$), 0.0808 ($O_2$), 0.7498 ($N_2$), 44.0°C

**15–31** Methyl alcohol ($CH_3OH$) is burned with 100 percent excess air. During the combustion process, 60 percent of the carbon in the fuel is converted to $CO_2$ and 40 percent is converted to $CO$. Write the balanced reaction equation and determine the air-fuel ratio.

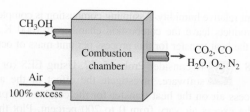

**FIGURE P15–31**

**15–32** Ethyl alcohol ($C_2H_5OH$) is burned with stoichiometric amount of air. The combustion is incomplete with 5 percent (by volume) of the carbon in the fuel forming carbon monoxide and 5 percent of the hydrogen forming OH. Calculate the apparent molecular weight of the products.

**15–33** A coal from Illinois which has an ultimate analysis (by mass) as 67.40 percent C, 5.31 percent $H_2$, 15.11 percent $O_2$, 1.44 percent $N_2$, 2.36 percent S, and 8.38 percent ash (non-combustibles) is burned with 40 percent excess air. Calculate the mass of air required per unit mass of coal burned and the apparent molecular weight of the product gas neglecting the ash constituent. *Answers:* 13.8 kg air/kg fuel, 29.7 kg/kmol

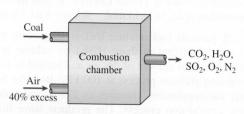

**FIGURE P15–33**

**15–34** A certain coal has the following analysis on a mass basis: 82 percent C, 5 percent $H_2O$, 2 percent $H_2$, 1 percent $O_2$, and 10 percent ash. The coal is burned with 50 percent excess air. Determine the air–fuel ratio. *Answer:* 15.4 kg air/ kg coal

## Enthalpy of Formation and Enthalpy of Combustion

**15–35C** What is enthalpy of formation? How does it differ from the enthalpy of combustion?

**15–36C** What is enthalpy of combustion? How does it differ from the enthalpy of reaction?

**15–37C** What are the higher and the lower heating values of a fuel? How do they differ? How is the heating value of a fuel related to the enthalpy of combustion of that fuel?

**15–38C** The $\bar{h}_f^\circ$ of $N_2$ is listed as zero. Does this mean that $N_2$ contains no chemical energy at the standard reference state?

**15–39C** Which contains more chemical energy, 1 kmol of $H_2$ or 1 kmol of $H_2O$?

**15–40** Determine the enthalpy of combustion of methane ($CH_4$) at 25°C and 1 atm, using the enthalpy of formation data from Table A–26. Assume that the water in the products is in the liquid form. Compare your result to the value listed in Table A–27. *Answer:* −890,330 kJ/kmol

**15–41** [EES] Reconsider Prob. 15–40. Using EES (or other) software, study the effect of temperature on the enthalpy of combustion. Plot the enthalpy of combustion as a function of temperature over the range 25 to 600°C.

**15–42** Repeat Prob. 15–40 for gaseous ethane ($C_2H_6$).

**15–43** Repeat Prob. 15–40 for liquid octane ($C_8H_{18}$).

**15–44** Ethane ($C_2H_6$) is burned at atmospheric pressure with stoichiometric amount of air as the oxidizer. Determine the heat rejected, in kJ/kmol fuel, when the products and reactants are at 25°C, and the water appears in the products as water vapor.

**15–45** What is the minimum pressure of the products of Prob. 15–46 which will assure that the water in the products will be in vapor form?

**15–46** Calculate the HHV and LHV of liquid propane fuel ($C_3H_8$). Compare your results with the values in Table A–27.

**15–47** Calculate the higher and lower heating values of a coal from Illinois which has an ultimate analysis (by mass) as 67.40 percent C, 5.31 percent $H_2$, 15.11 percent $O_2$, 1.44 percent $N_2$, 2.36 percent S, and 8.38 percent ash (non-combustibles). The enthalpy of formation of $SO_2$ is −297,100 kJ/kmol. *Answers:* 32,650 kJ/kg, 31,370 kJ/kg

## First-Law Analysis of Reacting Systems

**15–48C** Consider a complete combustion process during which both the reactants and the products are maintained at the same state. Combustion is achieved with (a) 100 percent theoretical air, (b) 200 percent theoretical air, and (c) the chemically correct amount of pure oxygen. For which case will the amount of heat transfer be the highest? Explain.

**15–49C** Consider a complete combustion process during which the reactants enter the combustion chamber at 20°C and the products leave at 700°C. Combustion is achieved with (a) 100 percent theoretical air, (b) 200 percent theoretical air, and (c) the chemically correct amount of pure oxygen. For which case will the amount of heat transfer be the lowest? Explain.

**15–50C** Derive an energy balance relation for a reacting closed system undergoing a quasi-equilibrium constant pressure expansion or compression process.

**15–51** Acetylene gas ($C_2H_2$) is burned completely with 20 percent excess air during a steady-flow combustion process. The fuel and air enter the combustion chamber at 25°C, and the products leave at 1500 K. Determine (a) the air–fuel ratio and (b) the heat transfer for this process.

**15–52** Liquid propane ($C_3H_8$) enters a combustion chamber at 25°C at a rate of 0.35 kg/min where it is mixed and burned with 150 percent excess air that enters the combustion chamber at 7°C. If the combustion is complete and the exit temperature of the combustion gases is 1000 K, determine (a) the mass flow rate of air and (b) the rate of heat transfer from the combustion chamber. *Answers: (a) 13.7 kg/min, (b) 4890 kJ/min*

**15–53** Propane fuel ($C_3H_8$) is burned with an air-fuel ratio of 25 in an atmospheric pressure heating furnace. Determine the heat transfer per kilogram of fuel burned when the temperature of the products is such that liquid water just begins to form in the products.

**15–54** Hydrogen ($H_2$) is burned completely with the stoichiometric amount of air during a steady-flow combustion process. If both the reactants and the products are maintained at 25°C and 1 atm and the water in the products exists in the liquid form, determine the heat transfer from the combustion chamber during this process. What would your answer be if combustion were achieved with 50 percent excess air?

**15–55** n-Octane gas ($C_8H_{18}$) is burned with 80 percent excess air in a constant pressure burner. The air and fuel enter this burner steadily at standard conditions and the products of combustion leave at 217°C. Calculate the heat transfer, in kJ/ kg fuel, during this combustion.

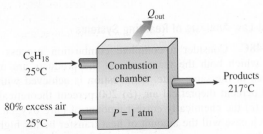

**FIGURE P15–55**

**15–56** A coal from Texas which has an ultimate analysis (by mass) as 39.25 percent C, 6.93 percent $H_2$, 41.11 percent $O_2$, 0.72 percent $N_2$, 0.79 percent S, and 11.20 percent ash (non-combustibles) is burned steadily with 40 percent excess air in a power plant boiler. The coal and air enter this boiler at standard conditions and the products of combustion in the smoke-stack are at 127°C. Calculate the heat transfer, in kJ/kg fuel, in this boiler. Include the effect of the sulfur in the energy analysis by noting that sulfur dioxide has an enthalpy of formation of −297,100 kJ/kmol and an average specific heat at constant pressure of $c_p = 41.7$ kJ/kmol·K.

**15–57** Octane gas ($C_8H_{18}$) at 25°C is burned steadily with 80 percent excess air at 25°C, 1 atm, and 40

percent relative humidity. Assuming combustion is complete and the products leave the combustion chamber at 1000 K, determine the heat transfer for this process per unit mass of octane.

**15–58** Reconsider Prob. 15–57. Using EES (or other) software, investigate the effect of the amount of excess air on the heat transfer for the combustion process. Let the excess air vary from 0 to 200 percent. Plot the heat transfer against excess air, and discuss the results.

**15–59** Diesel fuel ($C_{12}H_{26}$) at 25°C is burned in a steady-flow combustion chamber with 20 percent excess air that also enters at 25°C. The products leave the combustion chamber at 500 K. Assuming combustion is complete, determine the required mass flow rate of the diesel fuel to supply heat at a rate of 2000 kJ/s. *Answer: 49.5 g/s*

**15–60** Liquid ethyl alcohol ($C_2H_5OH(\ell)$) at 25°C is burned in a steady-flow combustion chamber with 40 percent excess air that also enters at 25°C. The products leave the combustion chamber at 600 K. Assuming combustion is complete, determine the required volume flow rate of the liquid ethyl alcohol, to supply heat at a rate of 2000 kJ/s. At 25°C the density of liquid ethyl alcohol is 790 kg/m³, the specific heat at a constant pressure is 114.08 kJ/kmol·K, and the enthalpy of vaporization is 42,340 kJ/kmol. *Answer: 6.81 L/min*

**15–61** A gaseous fuel mixture that is 40 percent propane ($C_3H_8$) and 60 percent methane ($CH_4$) by volume is mixed with the theoretical amount of dry air and burned in a steady-flow, constant pressure process at 100 kPa. Both the fuel and air enter the combustion chamber at 298 K and undergo a complete combustion process. The products leave the combustion chamber at 423 K. Determine
(a) the balanced combustion equation,
(b) the amount of water vapor condensed from the products, and
(c) the required air flow rate, in kg/h, when the combustion process produces a heat transfer output of 140,000 kJ/h.

| | $\bar{h}_f^\circ$, kJ/kmol | M, kg/kmol | $\bar{c}_p$, kJ/kmol·K |
|---|---|---|---|
| $C_3H_8(g)$ | −103,850 | 44 | |
| $CH_4(g)$ | −74,850 | 16 | |
| $CO_2$ | −393,520 | 44 | 41.16 |
| CO | −110,530 | 28 | 29.21 |
| $H_2O(g)$ | −241,820 | 18 | 34.28 |
| $H_2O(l)$ | −285,830 | 18 | 75.24 |
| $O_2$ | | 32 | 30.14 |
| $N_2$ | | 28 | 29.27 |

*Answer: (c) 50.1 kg/h*

**15–62** Gaseous E10 fuel is 10 percent ethanol ($C_2H_6O$) and 90 percent octane ($C_8H_{18}$) on a kmol basis. This fuel is burned with 110 percent theoretical air. During the combustion process, 90 percent of the carbon in the fuel is converted to $CO_2$ and 10 percent is converted to CO. Determine
(a) the balanced combustion equation,

(b) the dew-point temperature of the products, in °C, for a product pressure of 100 kPa,

(c) the heat transfer for the process, in kJ, after 2.5 kg of fuel are burned and the reactants and products are at 25°C with the water in the products remaining a gas, and

(d) the relative humidity of atmospheric air for the case where the atmospheric air is at 25°C and 100 kPa and the products are found to contain 9.57 kmol of water vapor per kmol of fuel burned.

| | $\overline{h}_f^\circ$, kJ/kmol | M, kg/kmol |
|---|---|---|
| $C_3H_{18}(g)$ | −208,450 | 144 |
| $C_2H_6O(g)$ | −235,310 | 46 |
| $CO_2$ | −393,520 | 44 |
| $CO$ | −110,530 | 28 |
| $H_2O(g)$ | −241,820 | 18 |
| $H_2O(l)$ | −285,830 | 18 |
| $O_2$ | | 32 |
| $N_2$ | | 28 |

Answers: (b) 50.5°C, (c) 105.5 MJ, (d) 59.9 percent

**15–63** A constant-volume tank contains a mixture of 120 g of methane ($CH_4$) gas and 600 g of $O_2$ at 25°C and 200 kPa. The contents of the tank are now ignited, and the methane gas burns completely. If the final temperature is 1200 K, determine (a) the final pressure in the tank and (b) the heat transfer during this process.

**15–64** Reconsider Prob. 15–63. Using EES (or other) software, investigate the effect of the final temperature on the final pressure and the heat transfer for the combustion process. Let the final temperature vary from 500 to 1500 K. Plot the final pressure and heat transfer against the final temperature, and discuss the results.

**15–65** One kmol of methane ($CH_4$) undergoes complete combustion with stoichiometric amount of air in a rigid container. Initially, the air and methane are at 100 kPa and 25°C. The products of combustion are at 567°C. How much heat is rejected from the container, in kJ/kmol fuel?

$Q_{out}$

CH$_4$
Theoretical air
100 kPa, 25°C

↓

567°C

**FIGURE P15–65**

**15–66** A closed combustion chamber is designed so that it maintains a constant pressure of 300 kPa during a combustion process. The combustion chamber has an initial volume of 0.5 m³ and contains a stoichiometric mixture of octane ($C_8H_{18}$) gas and air at 25°C. The mixture is now ignited, and the product gases are observed to be at 1000 K at the end of the combustion process. Assuming complete combustion, and treating both the reactants and the products as ideal gases, determine the heat transfer from the combustion chamber during this process. *Answer:* 3610 kJ

**15–67** To supply heated air to a house, a high-efficiency gas furnace burns gaseous propane ($C_3H_8$) with a combustion efficiency of 96 percent. Both the fuel and 140 percent theoretical air are supplied to the combustion chamber at 25°C and 100 kPa, and the combustion is complete. Because this is a high-efficiency furnace, the product gases are cooled to 25°C and 100 kPa before leaving the furnace. To maintain the house at the desired temperature, a heat transfer rate of 31,650 kJ/h is required from the furnace. Determine the volume of water condensed from the product gases per day. *Answer:* 8.7 L/day

## Adiabatic Flame Temperature

**15–68C** A fuel is completely burned first with the stoichiometric amount of air and then with the stoichiometric amount of pure oxygen. For which case will the adiabatic flame temperature be higher?

**15–69C** A fuel at 25°C is burned in a well-insulated steady-flow combustion chamber with air that is also at 25°C. Under what conditions will the adiabatic flame temperature of the combustion process be a maximum?

**15–70** Estimate the adiabatic flame temperature of an acetylene ($C_2H_2$) cutting torch, in °C, which uses a stoichiometric amount of pure oxygen. *Answer:* 8850°C

**15–71** Compare the adiabatic flame temperature of propane fuel ($C_3H_8$) when it is burned with stoichiometric amount of air and when it is burned with 20 percent excess air. The reactants are at 25°C and 1 atm.

**15–72** Acetylene gas ($C_2H_2$) at 25°C is burned during a steady-flow combustion process with 30 percent excess air at 27°C. It is observed that 75,000 kJ of heat is being lost from the combustion chamber to the surroundings per kmol of acetylene. Assuming combustion is complete, determine the exit temperature of the product gases. *Answer:* 2301 K

**15–73** Octane gas ($C_8H_{18}$) at 25°C is burned steadily with 30 percent excess air at 25°C, 1 atm, and 60 percent relative humidity. Assuming combustion is complete and adiabatic, calculate the exit temperature of the product gases.

**15–74** Reconsider Prob. 15–73. Using EES (or other) software, investigate the effect of the relative humidity on the exit temperature of the product gases. Plot the exit temperature of the product gases as a function of relative humidity for $0 < \phi < 100$ percent.

**15–75** A coal from Pennsylvania has an ultimate analysis (by mass) as 84.36 percent C, 1.89 percent $H_2$, 4.40 percent $O_2$, 0.63 percent $N_2$, 0.89 percent S, and 7.83 percent ash (non-combustibles) is burned in an industrial boiler with 100 percent excess air. This combustion is incomplete with 3 percent (by volume) of the carbon in the products forming carbon monoxide. What is the impact of the incomplete combustion on the adiabatic flame temperature, in °C, as compared to when the combustion is complete? Neglect the effect of the sulfur on the energy balance.

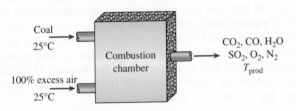

**FIGURE P15–75**

**15–76** An adiabatic constant-volume tank contains a mixture of 1 kmol of hydrogen ($H_2$) gas and the stoichiometric amount of air at 25°C and 1 atm. The contents of the tank are now ignited. Assuming complete combustion, determine the final temperature in the tank.

**15–77** Methane ($CH_4$) is burned with 200 percent excess air in an adiabatic constant volume container. Initially, air and methane are at 1 atm and 25°C. Assuming complete combustion, determine the final pressure and temperature of the combustion products. *Answers: 474 kPa, 1393 K*

**FIGURE P15–77**

### Entropy Change and Second-Law Analysis of Reacting Systems

**15–78C** Express the increase of entropy principle for chemically reacting systems.

**15–79C** How are the absolute entropy values of ideal gases at pressures different from 1 atm determined?

**15–80C** What does the Gibbs function of formation $\bar{g}_f^\circ$ of a compound represent?

**15–81** Liquid octane ($C_8H_{18}$) enters a steady-flow combustion chamber at 25°C and 1 atm at a rate of 0.25 kg/min. It is burned with 50 percent excess air that also enters at 25°C and 1 atm. After combustion, the products are allowed to cool to 25°C. Assuming complete combustion and that all the $H_2O$ in the products is in liquid form, determine (a) the heat transfer rate from the combustion chamber, (b) the entropy generation rate, and (c) the exergy destruction rate. Assume that $T_0 = 298$ K and the products leave the combustion chamber at 1 atm pressure.

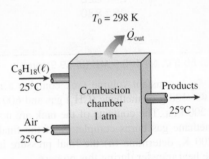

**FIGURE P15–81**

**15–82** Benzene gas ($C_6H_6$) at 1 atm and 25°C is burned during a steady-flow combustion process with 95 percent theoretical air that enters the combustion chamber at 25°C and 1 atm. All the hydrogen in the fuel burns to $H_2O$, but part of the carbon burns to CO. Heat is lost to the surroundings at 25°C, and the products leave the combustion chamber at 1 atm and 850 K. Determine (a) the heat transfer from the combustion chamber and (b) the exergy destruction.

**15–83** Liquid propane ($C_3H_8$) enters a steady-flow combustion chamber at 25°C and 1 atm at a rate of 0.4 kg/min where it is mixed and burned with 150 percent excess air that enters the combustion chamber at 12°C. If the combustion products leave at 1200 K and 1 atm, determine (a) the mass flow rate of air, (b) the rate of heat transfer from the combustion chamber, and (c) the rate of entropy generation during this process. Assume $T_0 = 25°C$. *Answers: (a) 15.7 kg/min, (b) 1732 kJ/min, (c) 34.2 kJ/min·K*

**15–84** Reconsider Prob. 15–83. Using EES (or other) software, study the effect of varying the surroundings temperature from 0 to 38°C on the rate of exergy destruction, and plot it as a function of surroundings temperature.

**15–85** n-Octane ($C_8H_{18}$) is burned in the constant pressure combustor of an aircraft engine with 70 percent excess air. Air enters this combustor at 600 kPa and 327°C, liquid fuel is injected at 25°C, and the products of combustion leave at 600 kPa and 1067°C. Determine the entropy generation and exergy destruction per unit mass of fuel during this combustion process. Take $T_0 = 25°C$.

**15–86** An automobile engine uses methyl alcohol ($CH_3OH$) as fuel with 200 percent excess air. Air enters this engine at 1 atm and 25°C. Liquid fuel at 25°C is mixed with this air before combustion. The exhaust products leave the exhaust system at 1 atm and 77°C. What is the maximum amount of work, in kJ/kg fuel, that can be produced by this engine? Take $T_0 = 25°C$. *Answer:* 22.8 MJ/kg fuel

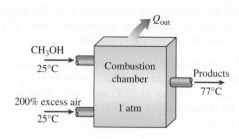

**FIGURE P15–86**

**15–87** A steady-flow combustion chamber is supplied with CO gas at 37°C and 110 kPa at a rate of 0.4 m³/min and air at 25°C and 110 kPa at a rate of 1.5 kg/min. Heat is transferred to a medium at 800 K, and the combustion products leave the combustion chamber at 900 K. Assuming the combustion is complete and $T_0 = 25°C$, determine (a) the rate of heat transfer from the combustion chamber and (b) the rate of exergy destruction. *Answers:* (a) 3567 kJ/min, (b) 1610 kJ/min

**15–88** Acetylene gas ($C_2H_2$) is burned completely with 20 percent excess air during a steady-flow combustion process. The fuel and the air enter the combustion chamber separately at 25°C and 1 atm, and heat is being lost from the combustion chamber to the surroundings at 25°C at a rate of 300,000 kJ/kmol $C_2H_2$. The combustion products leave the combustion chamber at 1 atm pressure. Determine (a) the temperature of the products, (b) the total entropy change per kmol of $C_2H_2$, and (c) the exergy destruction during this process.

## Review Problems

**15–89** A 1-g sample of a certain fuel is burned in a bomb calorimeter that contains 2 kg of water in the presence

of 100 g of air in the reaction chamber. If the water temperature rises by 2.5°C when equilibrium is established, determine the heating value of the fuel, in kJ/kg.

**15–90** A gaseous fuel with 80 percent $CH_4$, 15 percent $N_2$, and 5 percent $O_2$ (on a mole basis) is burned to completion with 120 percent theoretical air that enters the combustion chamber at 30°C, 100 kPa, and 60 percent relative humidity. Determine (a) the air–fuel ratio and (b) the volume flow rate of air required to burn fuel at a rate of 2 kg/min.

**15–91** Hydrogen ($H_2$) is burned with 100 percent excess air that enters the combustion chamber at 32°C, 100 kPa, and 60 percent relative humidity. Assuming complete combustion, determine (a) the air–fuel ratio and (b) the volume flow rate of air required to burn the hydrogen at a rate of 20 kg/h.

**15–92** Propane fuel ($C_3H_8$) is burned with stoichiometric amount of air in a water heater. The products of combustion are at 1 atm pressure and 50°C. What fraction of the water vapor in the products is vapor?

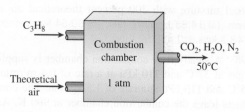

**FIGURE P15–92**

**15–93** A coal from Utah which has an ultimate analysis (by mass) as 61.40 percent C, 5.79 percent $H_2$, 25.31 percent $O_2$, 1.09 percent $N_2$, 1.41 percent S, and 5.00 percent ash (non-combustibles) is burned with 25 percent excess air in an industrial boiler. Assuming complete combustion and that the pressure in the boiler smokestack is 1 atm, calculate the minimum temperature, in °C, of the combustion products before liquid water begins to form in the smokestack. *Answer:* 41.3°C

**15–94** Methane fuel ($CH_4$) is burned with 15 percent excess air in a space-heating furnace. The pressure in the chimney is 1 atm. Presuming complete combustion, determine the temperature of the combustion products at which liquid water will begin to form in the chimney.

**15–95** A mixture of 40 percent by volume methane ($CH_4$), and 60 percent by volume propane ($C_3H_8$), is burned completely with theoretical air and leaves the combustion chamber at 100°C. The products have a pressure of 100 kPa and are cooled at constant pressure to 39°C. Sketch the $T$-$s$ diagram for the water vapor that does not condense, if any. How much of the water formed during the combustion process will be condensed, in kmol $H_2O$/kmol fuel? *Answer:* 1.96

**15–96**  A gaseous fuel mixture of 60 percent propane ($C_3H_8$), and 40 percent butane ($C_4H_{10}$), on a volume basis is burned in air such that the air–fuel ratio is 19 kg air/kg fuel when the combustion process is complete. Determine (*a*) the moles of nitrogen in the air supplied to the combustion process, in kmol/kmol fuel, (*b*) the moles of water formed in the combustion process, in kmol/kmol fuel, and (*c*) the moles of oxygen in the product gases in kmol/kmol fuel. *Answers:* (*a*) 25.7, (*b*) 4.4, (*c*) 1.23

**15–97**  A liquid–gas fuel mixture consists of 90 percent octane ($C_8H_{18}$), and 10 percent alcohol ($C_2H_5OH$), by moles. This fuel is burned with 200 percent theoretical dry air. Write the balanced reaction equation for complete combustion of this fuel mixture. Determine (*a*) the theoretical air–fuel ratio for this reaction, (*b*) the product–fuel ratio for this reaction, (*c*) the air-flow rate for a fuel mixture flow rate of 5 kg/s, and (*d*) the lower heating value of the fuel mixture with 200 percent theoretical air at 25°C. *Answers:* (*a*) 14.83 kg air/kg fuel, (*b*) 30.54 kg product/kg fuel, (*c*) 148.3 kg/s, (*d*) 43,760 kJ/kg fuel

**15–98**  A steady-flow combustion chamber is supplied with CO gas at 37°C and 110 kPa at a rate of 0.4 m³/min and air at 25°C and 110 kPa at a rate of 1.5 kg/min. The combustion products leave the combustion chamber at 900 K. Assuming combustion is complete, determine the rate of heat transfer from the combustion chamber.

**15–99**  A 6-m³ rigid tank initially contains a mixture of 1 kmol of hydrogen ($H_2$) gas and the stoichiometric amount of air at 25°C. The contents of the tank are ignited, and all the hydrogen in the fuel burns to $H_2O$. If the combustion products are cooled to 25°C, determine (*a*) the fraction of the $H_2O$ that condenses and (*b*) the heat transfer from the combustion chamber during this process.

**15–100**  Ethanol gas ($C_2H_6O$) is burned with 110 percent theoretical air. During the combustion process, 90 percent of the carbon in the fuel is converted to $CO_2$ and 10 percent is converted to CO. Determine

(*a*) the theoretical kmols of $O_2$ required for complete combustion of one kmol of ethanol,
(*b*) the balanced combustion equation for the incomplete combustion process, and
(*c*) the rate of heat transfer from the combustion process, in kW, when 3.5 kg/h of fuel are burned when the reactants and products are at 25°C with the water in the products remaining a gas.

| | $\bar{h}^{\circ}_f$, kJ/kmol | M, kg/kmol |
|---|---|---|
| $C_3H_6O(g)$ | −235,310 | 46 |
| $CO_2$ | −393,520 | 44 |
| CO | −110,530 | 28 |
| $H_2O(g)$ | −241,820 | 18 |
| $H_2O(l)$ | −285,830 | 18 |
| $O_2$ | | 32 |
| $N_2$ | | 28 |

**15–101**  Propane gas ($C_3H_8$) enters a steady-flow combustion chamber at 1 atm and 25°C and is burned with air that enters the combustion chamber at the same state. Determine the adiabatic flame temperature for (*a*) complete combustion with 100 percent theoretical air, (*b*) complete combustion with 200 percent theoretical air, and (*c*) incomplete combustion (some CO in the products) with 90 percent theoretical air.

**15–102**  Determine the highest possible temperature that can be obtained when liquid gasoline (assumed $C_8H_{18}$) at 25°C is burned steadily with air at 25°C and 1 atm. What would your answer be if pure oxygen at 25°C were used to burn the fuel instead of air?

**15–103**  Liquid propane ($C_3H_8(\ell)$) enters a combustion chamber at 25°C and 1 atm at a rate of 0.4 kg/min where it is mixed and burned with 150 percent excess air that enters the combustion chamber at 25°C. The heat transfer from the combustion process is 53 kW. Write the balanced combustion equation and determine (*a*) the mass flow rate of air; (*b*) the average molar mass (molecular weight) of the product gases; (*c*) the average specific heat at constant pressure of the product gases; and (*d*) the temperature of the products of combustion. *Answers:* (*a*) 15.63 kg/min, (*b*) 28.63 kg/kmol, (*c*) 36.06 kJ/kmol·K, (*d*) 1282 K

**15–104**  n-Octane ($C_8H_{18}$) is burned with 30 percent excess air, with 10 percent of the carbon forming carbon monoxide. Determine the maximum work that can be produced, in kJ/kg fuel, when the air, fuel products are all at 25°C and 1 atm.

**15–105**  A steam boiler heats liquid water at 200°C to superheated steam at 4 MPa and 400°C. Methane fuel ($CH_4$) is burned at atmospheric pressure with 50 percent excess air. The fuel and air enter the boiler at 25°C and the products of combustion leave at 227°C. Calculate (*a*) the amount of steam generated per unit of fuel mass burned, (*b*) the change in the exergy of the combustion streams, in kJ/fuel, (*c*) the change in the exergy of the steam stream, in kJ/kg steam, and (*d*) the lost work potential, in kJ/kg fuel. Take $T_0 = 25°C$. *Answers:* (*a*) 18.72 kg steam/kg fuel, (*b*) 49,490 kJ/kg fuel, (*c*) 1039 kJ/kg steam, (*d*) 30,040 kJ/kg fuel

**15–106** Repeat Prob. 15–105 using a coal from Utah which has an ultimate analysis (by mass) as 61.40 percent C, 5.79 percent $H_2$, 25.31 percent $O_2$, 1.09 percent $N_2$, 1.41 percent S, and 5.00 percent ash (non-combustibles). Neglect the effect of the sulfur in the energy and entropy balances.

**15–107** Liquid octane ($C_8H_{18}$) enters a steady-flow combustion chamber at 25°C and 8 atm at a rate of 0.8 kg/min. It is burned with 200 percent excess air that is compressed and preheated to 500 K and 8 atm before entering the combustion chamber. After combustion, the products enter an adiabatic turbine at 1300 K and 8 atm and leave at 950 K and 2 atm. Assuming complete combustion and $T_0 = 25°C$, determine (a) the heat transfer rate from the combustion chamber, (b) the power output of the turbine, and (c) the reversible work and exergy destruction for the entire process. *Answers:* (a) 770 kJ/min, (b) 263 kW, (c) 514 kW, 251 kW

**15–108** Develop an expression for the higher heating value of a gaseous alkane $C_nH_{2n+2}$ in terms of $n$.

**15–109** The furnace of a particular power plant can be considered to consist of two chambers: an adiabatic combustion chamber where the fuel is burned completely and adiabatically, and a heat exchanger where heat is transferred to a Carnot heat engine isothermally. The combustion gases in the heat exchanger are well-mixed so that the heat exchanger is at a uniform temperature at all times that is equal to the temperature of the exiting product gases, $T_p$. The work output of the Carnot heat engine can be expressed as

$$W = Q\eta_C = Q\left(1 - \frac{T_0}{T_p}\right)$$

where $Q$ is the magnitude of the heat transfer to the heat engine and $T_0$ is the temperature of the environment. The work output of the Carnot engine will be zero either when $T_p = T_{af}$ (which means the product gases will enter and exit the heat exchanger at the adiabatic flame temperature $T_{af}$, and thus $Q = 0$) or when $T_p = T_0$ (which means the temperature of the product gases in the heat exchanger will be $T_0$, and thus $\eta_C = 0$), and will reach a maximum somewhere in between. Treating the combustion products as ideal gases with constant specific heats and assuming no change in their composition in the heat exchanger, show that the work output of the Carnot heat engine will be maximum when

$$T_p = \sqrt{T_{af}T_0}$$

Also, show that the maximum work output of the Carnot engine in this case becomes

$$W_{max} = CT_{af}\left(1 - \sqrt{\frac{T_0}{T_{af}}}\right)^2$$

where $C$ is a constant whose value depends on the composition of the product gases and their specific heats.

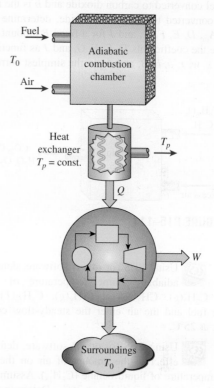

Fuel

$T_0$

Air

Adiabatic combustion chamber

Heat exchanger $T_p$ = const.

$T_p$

$Q$

$W$

Surroundings $T_0$

**FIGURE P15–109**

**15–110** Consider the combustion of hydrocarbon fuel, $C_nH_m$, with excess theoretical air and incomplete combustion according to the chemical reaction as follows:

$$C_nH_m + (1 + B)A_{th}(O_2 + 3.76 N_2) \rightarrow$$
$$D CO_2 + E CO + F H_2O + G O_2 + J N_2$$

where $A_{th}$ is the theoretical $O_2$ required for this fuel and $B$ is the excess amount of air in decimal form. If $a$ is the fraction of carbon in the fuel converted to carbon dioxide and $b$ is the remaining fraction converted to carbon monoxide, determine the coefficients $A_{th}$, $D$, $E$, $F$, $G$, and $J$ for a fixed $B$ amount of excess air. Write the coefficients $D$, $E$, $F$, $G$, and $J$ as functions of $n$, $m$, $a$, $b$, $B$, and $A_{th}$ in the simplest correct forms.

**15–111** Consider the combustion of a mixture of an alcohol, $C_nH_mO_x$, and a hydrocarbon fuel, $C_wH_z$, with excess theoretical air and incomplete combustion according to the chemical reaction as follows:

$$y_1C_nH_mO_x + y_2C_wH_z + (1 + B)A_{th}(O_2 + 3.76 N_2) \rightarrow$$
$$D CO_2 + E CO + F H_2O + G O_2 + J N_2$$

where $y_1$ and $y_2$ are the mole fractions of the fuel mixture, $A_{th}$ is the theoretical $O_2$ required for this fuel, and $B$ is the excess amount of air in decimal form. If $a$ is the fraction of carbon in the fuel converted to carbon dioxide and $b$ is the remaining fraction converted to carbon monoxide, determine the coefficients $A_{th}$, $D$, $E$, $F$, $G$, and $J$ for a fixed $B$ amount of excess air. Write the coefficients $D$, $E$, $F$, $G$, and $J$ as functions of $y_1$, $y_2$, $n$, $m$, $x$, $w$, $z$, $a$, $b$, $B$, and $A_{th}$ in the simplest correct forms.

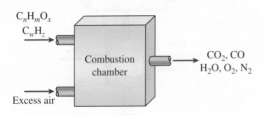

**FIGURE P15–111**

**15–112** Using EES (or other) software, determine the adiabatic flame temperature of the fuels $CH_4(g)$, $C_2H_2(g)$, $CH_3OH(g)$, $C_3H_8(g)$, $C_8H_{18}(\ell)$. Assume both the fuel and the air enter the steady-flow combustion chamber at 25°C.

**15–113** Using EES (or other) software, determine the effect of the amount of air on the adiabatic flame temperature of liquid octane ($C_8H_{18}$). Assume both the air and the octane are initially at 25°C. Determine the adiabatic flame temperature for 75, 90, 100, 120, 150, 200, 300, 500, and 800 percent theoretical air. Assume the hydrogen in the fuel always burns $H_2O$ and the carbon $CO_2$, except when there is a deficiency of air. In the latter case, assume that part of the carbon forms CO. Plot the adiabatic flame temperature against the percent theoretical air, and discuss the results.

**15–114** Using EES (or other) software, determine the fuel among $CH_4(g)$, $C_2H_2(g)$, $C_2H_6(g)$, $C_3H_8(g)$, $C_8H_{18}(\ell)$ that gives the highest temperature when burned completely in an adiabatic constant-volume chamber with the theoretical amount of air. Assume the reactants are at the standard reference state.

**15–115** Using EES (or other) software, write a general program to determine the heat transfer during the complete combustion of a hydrocarbon fuel ($C_nH_m$) at 25°C in a steady-flow combustion chamber when the percent of excess air and the temperatures of air and the products are specified. As a sample case, determine the heat transfer per unit mass of fuel as liquid propane ($C_3H_8$) is burned steadily with 50 percent excess air at 25°C and the combustion products leave the combustion chamber at 1800 K.

**15–116** Using EES (or other) software, determine the rate of heat transfer for the fuels $CH_4(g)$,

$C_2H_2(g)$, $CH_3OH(g)$, $C_3H_8(g)$, and $C_8H_{18}(\ell)$ when they are burned completely in a steady-flow combustion chamber with the theoretical amount of air. Assume the reactants enter the combustion chamber at 298 K and the products leave at 1200 K.

**15–117** Using EES (or other) software, repeat Prob. 15–116 for (a) 50, (b) 100, and (c) 200 percent excess air.

## Fundamentals of Engineering (FE) Exam Problems

**15–118** A fuel is burned steadily in a combustion chamber. The combustion temperature will be the highest except when
(a) the fuel is preheated.
(b) the fuel is burned with a deficiency of air.
(c) the air is dry.
(d) the combustion chamber is well insulated.
(e) the combustion is complete.

**15–119** A fuel is burned with 70 percent theoretical air. This is equivalent to
(a) 30% excess air          (b) 70% excess air
(c) 30% deficiency of air   (d) 70% deficiency of air
(e) stoichiometric amount of air

**15–120** Propane ($C_3H_8$) is burned with 150 percent theoretical air. The air–fuel mass ratio for this combustion process is
(a) 5.3          (b) 10.5          (c) 15.7
(d) 23.4          (e) 39.3

**15–121** One kmol of methane ($CH_4$) is burned with an unknown amount of air during a combustion process. If the combustion is complete and there are 1 kmol of free $O_2$ in the products, the air–fuel mass ratio is
(a) 34.6          (b) 25.7          (c) 17.2
(d) 14.3          (e) 11.9

**15–122** An equimolar mixture of carbon dioxide and water vapor at 1 atm and 60°C enter a dehumidifying section where the entire water vapor is condensed and removed from the mixture, and the carbon dioxide leaves at 1 atm and 60°C. The entropy change of carbon dioxide in the dehumidifying section is
(a) −2.8 kJ/kg·K       (b) −0.13 kJ/kg·K       (c) 0
(d) 0.13 kJ/kg·K       (e) 2.8 kJ/kg·K

**15–123** Methane ($CH_4$) is burned completely with 80 percent excess air during a steady-flow combustion process. If both the reactants and the products are maintained at 25°C and 1 atm and the water in the products exists in the liquid form, the heat transfer from the combustion chamber per unit mass of methane is
(a) 890 MJ/kg          (b) 802 MJ/kg          (c) 75 MJ/kg
(d) 56 MJ/kg          (e) 50 MJ/kg

**15–124** The higher heating value of a hydrocarbon fuel $C_nH_m$ with $m = 8$ is given to be 1560 MJ/kmol of fuel. Then its lower heating value is
(a) 1384 MJ/kmol   (b) 1208 MJ/kmol   (c) 1402 MJ/kmol
(d) 1514 MJ/kmol   (e) 1551 MJ/kmol

**15–125** Acetylene gas ($C_2H_2$) is burned completely during a steady-flow combustion process. The fuel and the air enter the combustion chamber at 25°C, and the products leave at 1500 K. If the enthalpy of the products relative to the standard reference state is $-404$ MJ/kmol of fuel, the heat transfer from the combustion chamber is

(a) 177 MJ/kmol    (b) 227 MJ/kmol    (c) 404 MJ/kmol
(d) 631 MJ/kmol    (e) 751 MJ/kmol

**15–126** Benzene gas ($C_6H_6$) is burned with 95 percent theoretical air during a steady-flow combustion process. The mole fraction of the CO in the products is

(a) 8.3%    (b) 4.7%    (c) 2.1%
(d) 1.9%    (e) 14.3%

**15–127** A fuel is burned during a steady-flow combustion process. Heat is lost to the surroundings at 300 K at a rate of 1120 kW. The entropy of the reactants entering per unit time is 17 kW/K and that of the products is 15 kW/K. The total rate of exergy destruction during this combustion process is

(a) 520 kW    (b) 600 kW    (c) 1120 kW
(d) 340 kW    (e) 739 kW

## Design and Essay Problems

**15–128** Design a combustion process suitable for use in a gas-turbine engine. Discuss possible fuel selections for the several applications of the engine.

**15–129** A promising method of power generation by direct energy conversion is through the use of magnetohydrodynamic (MHD) generators. Write an essay on the current status of MHD generators. Explain their operation principles and how they differ from conventional power plants. Discuss the problems that need to be overcome before MHD generators can become economical.

**15–130** What is oxygenated fuel? How would the heating value of oxygenated fuels compare to those of comparable hydrocarbon fuels on a unit-mass basis? Why is the use of oxygenated fuels mandated in some major cities in winter months?

**15–131** The safe disposal of hazardous waste material is a major environmental concern for industrialized societies and creates challenging problems for engineers. The disposal methods commonly used include landfilling, burying in the ground, recycling, and incineration or burning. Incineration is frequently used as a practical means for the disposal of combustible waste such as organic materials. The EPA regulations require that the waste material be burned almost completely above a specified temperature without polluting the environment. Maintaining the temperature above a certain level, typically about 1100°C, necessitates the use of a fuel when the combustion of the waste material alone is not sufficient to obtain the minimum specified temperature.

A certain industrial process generates a liquid solution of ethanol and water as the waste product at a rate of 10 kg/s. The mass fraction of ethanol in the solution is 0.2. This solution is to be burned using methane ($CH_4$) in a steady-flow combustion chamber. Propose a combustion process that will accomplish this task with a minimal amount of methane. State your assumptions.

**15–132** Constant-volume vessels that contain flammable mixtures of hydrocarbon vapors and air at low pressures are frequently used. Although the ignition of such mixtures is very unlikely as there is no source of ignition in the tank, the Safety and Design Codes require that the tank withstand four times the pressure that may occur should an explosion take place in the tank. For operating gauge pressures under 25 kPa, determine the pressure for which these vessels must be designed in order to meet the requirements of the codes for (a) acetylene $C_2H_2(g)$, (b) propane $C_3H_8(g)$, and (c) n-octane $C_8H_{18}(g)$. Justify any assumptions that you make.

**15–133** An electrical utility uses a Pennsylvania coal which has an ultimate analysis (by mass) as 84.36 percent C, 1.89 percent $H_2$, 4.40 percent $O_2$, 0.63 percent $N_2$, 0.89 percent S, and 7.83 percent ash (non-combustibles) as fuel for its boilers. The utility is changing from the Pennsylvania coal to an Illinois coal which has an ultimate analysis (by mass) as 67.40 percent C, 5.31 percent $H_2$, 15.11 percent $O_2$, 1.44 percent $N_2$, 2.36 percent S, and 8.38 percent ash (non-combustibles) as fuel for its boilers. With the Pennsylvania coal, the boilers used 15 percent excess air. Develop a schedule for the new coal showing the heat released, the smokestack dew-point temperature, adiabatic flame temperature, and carbon dioxide production for various amount of excess air. Use this schedule to determine how to operate with the new coal as closely as possible to the conditions of the old coal. Is there anything else that will have to be changed to use the new coal?

# CHEMICAL AND PHASE EQUILIBRIUM

I n Chapter 15 we analyzed combustion processes under the assumption that combustion is complete when there is sufficient time and oxygen. Often this is not the case, however. A chemical reaction may reach a state of equilibrium before reaching completion even when there is sufficient time and oxygen.

A system is said to be in *equilibrium* if no changes occur within the system when it is isolated from its surroundings. An isolated system is in *mechanical equilibrium* if no changes occur in pressure, in *thermal equilibrium* if no changes occur in temperature, in *phase equilibrium* if no transformations occur from one phase to another, and in *chemical equilibrium* if no changes occur in the chemical composition of the system. The conditions of mechanical and thermal equilibrium are straightforward, but the conditions of chemical and phase equilibrium can be rather involved.

The equilibrium criterion for reacting systems is based on the second law of thermodynamics; more specifically, the increase of entropy principle. For adiabatic systems, chemical equilibrium is established when the entropy of the reacting system reaches a maximum. Most reacting systems encountered in practice are not adiabatic, however. Therefore, we need to develop an equilibrium criterion applicable to any reacting system.

In this chapter, we develop a general criterion for chemical equilibrium and apply it to reacting ideal-gas mixtures. We then extend the analysis to simultaneous reactions. Finally, we discuss phase equilibrium for nonreacting systems.

■ ■ ■ ■ ■ ■ ■
## OBJECTIVES
The objectives of Chapter 16 are to:

■  Develop the equilibrium criterion for reacting systems based on the second law of thermodynamics.

■  Develop a general criterion for chemical equilibrium applicable to any reacting system based on minimizing the Gibbs function for the system.

■  Define and evaluate the chemical equilibrium constant.

■  Apply the general criterion for chemical equilibrium analysis to reacting ideal-gas mixtures.

■  Apply the general criterion for chemical equilibrium analysis to simultaneous reactions.

■  Relate the chemical equilibrium constant to the enthalpy of reaction.

■  Establish the phase equilibrium for nonreacting systems in terms of the specific Gibbs function of the phases of a pure substance.

■  Apply the Gibbs phase rule to determine the number of independent variables associated with a multicomponent, multiphase system.

■  Apply Henry's law and Raoult's law for gases dissolved in liquids.

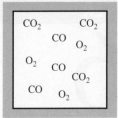

**FIGURE 16–1**
A reaction chamber that contains
a mixture of $CO_2$, CO, and $O_2$ at
a specified temperature and pressure.

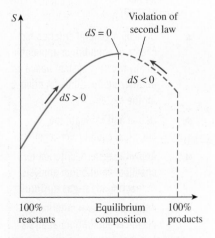

**FIGURE 16–2**
Equilibrium criteria for a chemical
reaction that takes place adiabatically.

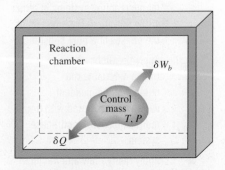

**FIGURE 16–3**
A control mass undergoing a chemical
reaction at a specified temperature and
pressure.

# 16–1 · CRITERION FOR CHEMICAL EQUILIBRIUM

Consider a reaction chamber that contains a mixture of CO, $O_2$, and $CO_2$ at a specified temperature and pressure. Let us try to predict what will happen in this chamber (Fig. 16–1). Probably the first thing that comes to mind is a chemical reaction between CO and $O_2$ to form more $CO_2$:

$$CO + \tfrac{1}{2}O_2 \rightarrow CO_2$$

This reaction is certainly a possibility, but it is not the only possibility. It is also possible that some $CO_2$ in the combustion chamber dissociated into CO and $O_2$. Yet a third possibility would be to have no reactions among the three components at all, that is, for the system to be in chemical equilibrium. It appears that although we know the temperature, pressure, and composition (thus the state) of the system, we are unable to predict whether the system is in chemical equilibrium. In this chapter, we develop the necessary tools to correct this.

Assume that the CO, $O_2$, and $CO_2$ mixture mentioned above is in chemical equilibrium at the specified temperature and pressure. The chemical composition of this mixture does not change unless the temperature or the pressure of the mixture is changed. That is, a reacting mixture, in general, has different equilibrium compositions at different pressures and temperatures. Therefore, when developing a general criterion for chemical equilibrium, we consider a reacting system at a fixed temperature and pressure.

Taking the positive direction of heat transfer to be to the system, the increase of entropy principle for a reacting or nonreacting system was expressed in Chapter 7 as

$$dS_{sys} \geq \frac{\delta Q}{T} \qquad (16\text{–}1)$$

A system and its surroundings form an adiabatic system, and for such systems Eq. 16–1 reduces to $dS_{sys} \geq 0$. That is, a chemical reaction in an adiabatic chamber proceeds in the direction of increasing entropy. When the entropy reaches a maximum, the reaction stops (Fig. 16–2). Therefore, entropy is a very useful property in the analysis of reacting adiabatic systems.

When a reacting system involves heat transfer, the increase of entropy principle relation (Eq. 16–1) becomes impractical to use, however, since it requires a knowledge of heat transfer between the system and its surroundings. A more practical approach would be to develop a relation for the equilibrium criterion in terms of the properties of the reacting system only. Such a relation is developed below.

Consider a reacting (or nonreacting) simple compressible system of fixed mass with only quasi-equilibrium work modes at a specified temperature $T$ and pressure $P$ (Fig. 16–3). Combining the first- and the second-law relations for this system gives

$$\left.\begin{array}{c} \delta Q - P\,dV = dU \\[4pt] dS \geq \dfrac{\delta Q}{T} \end{array}\right\} \quad dU + P\,dV - T\,dS \leq 0 \qquad (16\text{–}2)$$

The differential of the Gibbs function ($G = H - TS$) at constant temperature and pressure is

$$(dG)_{T,P} = dH - T\,dS - S\,dT$$

$$= (dU + P\,dV + V\,dP\nearrow^0) - T\,dS - S\,dT\nearrow^0 \qquad (16\text{-}3)$$

$$= dU + P\,dV - T\,dS$$

From Eqs. 16–2 and 16–3, we have $(dG)_{T,P} \le 0$. Therefore, a chemical reaction at a specified temperature and pressure proceeds in the direction of a decreasing Gibbs function. The reaction stops and chemical equilibrium is established when the Gibbs function attains a minimum value (Fig. 16–4). Therefore, the criterion for chemical equilibrium can be expressed as

$$(dG)_{T,P} = 0 \qquad (16\text{-}4)$$

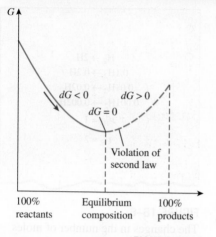

**FIGURE 16–4**
Criteria for chemical equilibrium for a fixed mass at a specified temperature and pressure.

A chemical reaction at a specified temperature and pressure cannot proceed in the direction of the increasing Gibbs function since this will be a violation of the second law of thermodynamics. Notice that if the temperature or the pressure is changed, the reacting system will assume a different equilibrium state, which is the state of the minimum Gibbs function at the new temperature or pressure.

To obtain a relation for chemical equilibrium in terms of the properties of the individual components, consider a mixture of four chemical components $A$, $B$, $C$, and $D$ that exist in equilibrium at a specified temperature and pressure. Let the number of moles of the respective components be $N_A$, $N_B$, $N_C$, and $N_D$. Now consider a reaction that occurs to an infinitesimal extent during which differential amounts of $A$ and $B$ (reactants) are converted to $C$ and $D$ (products) while the temperature and the pressure remain constant (Fig. 16–5):

$$dN_A A + dN_B B \longrightarrow dN_C C + dN_D D$$

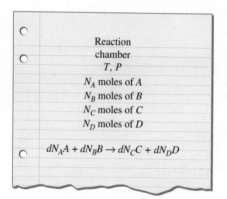

**FIGURE 16–5**
An infinitesimal reaction in a chamber at constant temperature and pressure.

The equilibrium criterion (Eq. 16–4) requires that the change in the Gibbs function of the mixture during this process be equal to zero. That is,

$$(dG)_{T,P} = \sum(dG_i)_{T,P} = \sum(\bar{g}_i\,dN_i)_{T,P} = 0 \qquad (16\text{-}5)$$

or

$$\bar{g}_C\,dN_C + \bar{g}_D\,dN_D + \bar{g}_A\,dN_A + \bar{g}_B\,dN_B = 0 \qquad (16\text{-}6)$$

where the $\bar{g}$'s are the molar Gibbs functions (also called the *chemical potentials*) at the specified temperature and pressure and the $dN$'s are the differential changes in the number of moles of the components.

To find a relation between the $dN$'s, we write the corresponding stoichiometric (theoretical) reaction

$$\nu_A A + \nu_B B \rightleftharpoons \nu_C C + \nu_D D \qquad (16\text{-}7)$$

where the $\nu$'s are the stoichiometric coefficients, which are evaluated easily once the reaction is specified. The stoichiometric reaction plays an important

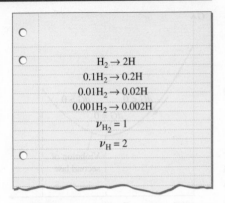

$H_2 \rightarrow 2H$
$0.1H_2 \rightarrow 0.2H$
$0.01H_2 \rightarrow 0.02H$
$0.001H_2 \rightarrow 0.002H$
$\nu_{H_2} = 1$
$\nu_H = 2$

**FIGURE 16–6**

The changes in the number of moles of the components during a chemical reaction are proportional to the stoichiometric coefficients regardless of the extent of the reaction.

role in the determination of the equilibrium composition of the reacting mixtures because the changes in the number of moles of the components are proportional to the stoichiometric coefficients (Fig. 16–6). That is,

$$dN_A = -\varepsilon\nu_A \qquad dN_C = \varepsilon\nu_C$$
$$dN_B = -\varepsilon\nu_B \qquad dN_D = \varepsilon\nu_D \qquad \text{(16–8)}$$

where $\varepsilon$ is the proportionality constant and represents the extent of a reaction. A minus sign is added to the first two terms because the number of moles of the reactants $A$ and $B$ decreases as the reaction progresses.

For example, if the reactants are $C_2H_6$ and $O_2$ and the products are $CO_2$ and $H_2O$, the reaction of 1 $\mu$mol ($10^{-6}$ mol) of $C_2H_6$ results in a 2-$\mu$mol increase in $CO_2$, a 3-$\mu$mol increase in $H_2O$, and a 3.5-$\mu$mol decrease in $O_2$ in accordance with the stoichiometric equation

$$C_2H_6 + 3.5O_2 \rightarrow 2CO_2 + 3H_2O$$

That is, the change in the number of moles of a component is one-millionth ($\varepsilon = 10^{-6}$) of the stoichiometric coefficient of that component in this case.

Substituting the relations in Eq. 16–8 into Eq. 16–6 and canceling $\varepsilon$, we obtain

$$\nu_C\bar{g}_C + \nu_D\bar{g}_D - \nu_A\bar{g}_A - \nu_B\bar{g}_B = 0 \qquad \text{(16–9)}$$

This equation involves the stoichiometric coefficients and the molar Gibbs functions of the reactants and the products, and it is known as the **criterion for chemical equilibrium**. It is valid for any chemical reaction regardless of the phases involved.

Equation 16–9 is developed for a chemical reaction that involves two reactants and two products for simplicity, but it can easily be modified to handle chemical reactions with any number of reactants and products. Next, we analyze the equilibrium criterion for ideal-gas mixtures.

## 16–2 · THE EQUILIBRIUM CONSTANT FOR IDEAL-GAS MIXTURES

Consider a mixture of ideal gases that exists in equilibrium at a specified temperature and pressure. Like entropy, the Gibbs function of an ideal gas depends on both the temperature and the pressure. The Gibbs function values are usually listed versus temperature at a fixed reference pressure $P_0$, which is taken to be 1 atm. The variation of the Gibbs function of an ideal gas with pressure at a fixed temperature is determined by using the definition of the Gibbs function ($\bar{g} = \bar{h} - T\bar{s}$) and the entropy-change relation for isothermal processes [$\Delta\bar{s} = -R_u \ln(P_2/P_1)$]. It yields

$$(\Delta\bar{g})_T = \Delta\bar{h}^{\nearrow 0} - T(\Delta\bar{s})_T = -T(\Delta\bar{s})_T = R_uT \ln\frac{P_2}{P_1}$$

Thus the Gibbs function of component $i$ of an ideal-gas mixture at its partial pressure $P_i$ and mixture temperature $T$ can be expressed as

$$\bar{g}_i(T, P_i) = \bar{g}_i^*(T) + R_uT \ln P_i \qquad \text{(16–10)}$$

where $\bar{g}_i^*(T)$ represents the Gibbs function of component $i$ at 1 atm pressure and temperature $T$, and $P_i$ represents the partial pressure of component $i$ in atmospheres. Substituting the Gibbs function expression for each component into Eq. 16–9, we obtain

$$\nu_C[\bar{g}_C^*(T) + R_u T \ln P_C] + \nu_D[\bar{g}_D^*(T) + R_u T \ln P_D]$$
$$- \nu_A[\bar{g}_A^*(T) + R_u T \ln P_A] - \nu_B[\bar{g}_B^*(T) + R_u T \ln P_B] = 0$$

For convenience, we define the **standard-state Gibbs function change** as

$$\Delta G^*(T) = \nu_C \bar{g}_C^*(T) + \nu_D \bar{g}_D^*(T) - \nu_A \bar{g}_A^*(T) - \nu_B \bar{g}_B^*(T) \qquad \textbf{(16–11)}$$

Substituting, we get

$$\Delta G^*(T) = -R_u T(\nu_C \ln P_C + \nu_D \ln P_D - \nu_A \ln P_A - \nu_B \ln P_B) = -R_u T \ln \frac{P_C^{\nu_C} P_D^{\nu_D}}{P_A^{\nu_A} P_B^{\nu_B}} \qquad \textbf{(16–12)}$$

Now we define the **equilibrium constant $K_P$** for the chemical equilibrium of ideal-gas mixtures as

$$K_P = \frac{P_C^{\nu_C} P_D^{\nu_D}}{P_A^{\nu_A} P_B^{\nu_B}} \qquad \textbf{(16–13)}$$

Substituting into Eq. 16–12 and rearranging, we obtain

$$K_P = e^{-\Delta G^*(T)/R_u T} \qquad \textbf{(16–14)}$$

Therefore, the equilibrium constant $K_P$ of an ideal-gas mixture at a specified temperature can be determined from a knowledge of the standard-state Gibbs function change at the same temperature. The $K_P$ values for several reactions are given in Table A–28.

Once the equilibrium constant is available, it can be used to determine the equilibrium composition of reacting ideal-gas mixtures. This is accomplished by expressing the partial pressures of the components in terms of their mole fractions:

$$P_i = y_i P = \frac{N_i}{N_{\text{total}}} P$$

where $P$ is the total pressure and $N_{\text{total}}$ is the total number of moles present in the reaction chamber, including any *inert gases*. Replacing the partial pressures in Eq. 16–13 by the above relation and rearranging, we obtain (Fig. 16–7)

$$K_P = \frac{N_C^{\nu_C} N_D^{\nu_D}}{N_A^{\nu_A} N_B^{\nu_B}} \left( \frac{P}{N_{\text{total}}} \right)^{\Delta \nu} \qquad \textbf{(16–15)}$$

where

$$\Delta \nu = \nu_C + \nu_D - \nu_A - \nu_B$$

Equation 16–15 is written for a reaction involving two reactants and two products, but it can be extended to reactions involving any number of reactants and products.

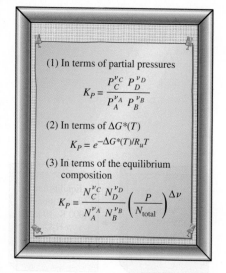

(1) In terms of partial pressures

$$K_P = \frac{P_C^{\nu_C} \, P_D^{\nu_D}}{P_A^{\nu_A} \, P_B^{\nu_B}}$$

(2) In terms of $\Delta G^*(T)$

$$K_P = e^{-\Delta G^*(T)/R_u T}$$

(3) In terms of the equilibrium composition

$$K_P = \frac{N_C^{\nu_C} \, N_D^{\nu_D}}{N_A^{\nu_A} \, N_B^{\nu_B}} \left( \frac{P}{N_{\text{total}}} \right)^{\Delta \nu}$$

**FIGURE 16–7**
Three equivalent $K_P$ relations for reacting ideal-gas mixtures.

*EXAMPLE 16–1*     **Equilibrium Constant of a Dissociation Process**

Using Eq. 16–14 and the Gibbs function data, determine the equilibrium constant $K_P$ for the dissociation process $N_2 \rightarrow 2N$ at 25°C. Compare your result to the $K_P$ value listed in Table A–28.

**SOLUTION**  The equilibrium constant of the reaction $N_2 \rightarrow 2N$ is listed in Table A–28 at different temperatures. It is to be verified using Gibbs function data.

*Assumptions*  **1** The constituents of the mixture are ideal gases. **2** The equilibrium mixture consists of $N_2$ and N only.

*Properties*  The equilibrium constant of this reaction at 298 K is $\ln K_P = -367.5$ (Table A–28). The Gibbs function of formation at 25°C and 1 atm is O for $N_2$ and 455,510 kJ/kmol for N (Table A–26).

*Analysis*  In the absence of $K_P$ tables, $K_P$ can be determined from the Gibbs function data and Eq. 16–14,

$$K_P = e^{-\Delta G^*(T)/R_u T}$$

where, from Eq. 16–11,

$$\Delta G^*(T) = \nu_N \bar{g}_N^*(T) - \nu_{N_2} \bar{g}_{N_2}^*(T)$$
$$= (2)(455{,}510 \text{ kJ/kmol}) - 0$$
$$= 911{,}020 \text{ kJ/kmol}$$

Substituting, we find

$$\ln K_P = -\frac{911{,}020 \text{ kJ/kmol}}{(8.314 \text{ kJ/kmol·K})(298.15 \text{ K})}$$
$$= -367.5$$

or

$$K_P \cong 2 \times 10^{-160}$$

The calculated $K_P$ value is in agreement with the value listed in Table A–28. The $K_P$ value for this reaction is practically zero, indicating that this reaction will not occur at this temperature.

*Discussion*  Note that this reaction involves one product (N) and one reactant ($N_2$), and the stoichiometric coefficients for this reaction are $\nu_N = 2$ and $\nu_{N_2} = 1$. Also, note that the Gibbs function of all stable elements (such as $N_2$) is assigned a value of zero at the standard reference state of 25°C and 1 atm. The Gibbs function values at other temperatures can be calculated from the enthalpy and absolute entropy data by using the definition of the Gibbs function, $\bar{g}^*(T) = \bar{h}(T) - T\bar{s}^*(T)$, where $\bar{h}(T) = \bar{h}_f^o + \bar{h}_T - \bar{h}_{298 K}$.

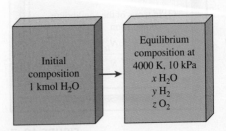

**FIGURE 16–8**
Schematic for Example 16–2.

*EXAMPLE 16–2*     **Producing Hydrogen by Heating Water Vapor to High Temperature**

As an alternative to electrolysis, hydrogen gas can be produced thermally in accordance with the dissociation process $H_2O \rightleftharpoons H_2 + \frac{1}{2}O_2$ by heating water vapor to very high temperatures (Fig. 16–8). Determine the percentage of water vapor that can be separated into hydrogen and oxygen when this reaction occurs at 4000 K and 10 kPa.

**SOLUTION** The reaction $H_2O \rightleftharpoons H_2 + \frac{1}{2}O_2$ is considered at a specified temperature and pressure. The percentage of water vapor that separates into hydrogen and oxygen is to be determined.

**Assumptions** 1 The equilibrium composition consists of $H_2O$, $H_2$, and $O_2$ only, and dissociation into H, OH, and O is negligible. 2 The constituents of the mixture are ideal gases.

**Analysis** This is a dissociation process that is significant at very high temperatures only. For simplicity we consider 1 kmol of $H_2O$. The stoichiometric and actual reactions in this case are as follows:

*Stoichiometric:* $\quad H_2O \rightleftharpoons H_2 + \frac{1}{2}O_2$ (thus $\nu_{H_2O} = 1$, $\nu_{H_2} = 1$, and $\nu_{O_2} = 0.5$)

*Actual:* $\qquad\qquad H_2O \longrightarrow \underbrace{xH_2O}_{\substack{\text{reactants}\\\text{(leftover)}}} + \underbrace{yH_2 + zO_2}_{\text{products}}$

*H balance:* $\qquad\qquad 2 = 2x + 2y \longrightarrow y = 1 - x$

*O balance:* $\qquad\qquad 1 = x + 2z \longrightarrow z = (1 - x)/2$

*Total number of moles:* $\quad N_{\text{total}} = x + y + z = 1.5 - 0.5x$

*Pressure (in atm):* $\quad P = 10\ kPa = 0.09869\ atm$ (since 1 atm $= 101.325\ kPa$)

The equilibrium constant for the reaction $H_2O \rightleftharpoons H_2 + \frac{1}{2}O_2$ at 4000 K is given in Table A-28 to be $\ln K_p = -0.542$ and thus $K_p = 0.5816$.

Assuming ideal gas behavior for all components in equilibrium composition, the equilibrium constant relation in terms of mole numbers can be expressed in this case as

$$K_P = \frac{N_{H_2}^{\nu_{H_2}} N_{O_2}^{\nu_{O_2}}}{N_{H_2O}^{\nu_{H_2O}}} \left(\frac{P}{N_{\text{total}}}\right)^{\nu_{H_2} - \nu_{O_2} - \nu_{H_2O}}$$

Substituting,

$$0.5816 = \frac{(1 - x)[(1 - x)/2]^{1/2}}{x} \left(\frac{0.09869}{1.5 - 0.5x}\right)^{1 + 0.5 - 1}$$

Using an equation solver such as EES or by trial and error, the unknown $x$ is determined to be

$$x = 0.222$$

That is, for each mole of $H_2O$ entering the reaction chamber there is only 0.222 mole of $H_2O$ leaving. Therefore, the fraction of water vapor that dissociated into hydrogen and oxygen when heated to 4000 K is

Fraction of dissociation $= 1 - x = 1 - 0.222 =$ **0.778 or 77.8 percent**

Therefore, hydrogen can be produced at significant rates by heating water vapor to sufficiently high temperatures.

**Discussion** The dissociation of $H_2O$ into atomic H, O, and the compound OH can be significant at high temperatures, and thus the first assumption is very simplistic. This problem can be solved more realistically by considering all possible reactions that are likely to occur simultaneously, as discussed later in this chapter.

A double arrow is used in equilibrium equations as an indication that a chemical reaction does not stop when chemical equilibrium is established; rather, it proceeds in both directions at the same rate. That is, at equilibrium, the reactants are depleted at exactly the same rate as they are replenished from the products by the reverse reaction.

## 16–3 ▪ SOME REMARKS ABOUT THE $K_P$ OF IDEAL-GAS MIXTURES

In the last section we developed three equivalent expressions for the equilibrium constant $K_P$ of reacting ideal-gas mixtures: Eq. 16–13, which expresses $K_P$ in terms of *partial pressures;* Eq. 16–14, which expresses $K_P$ in terms of the *standard-state Gibbs function change $\Delta G^*(T)$*; and Eq. 16–15, which expresses $K_P$ in terms of the *number of moles* of the components. All three relations are equivalent, but sometimes one is more convenient to use than the others. For example, Eq. 16–15 is best suited for determining the equilibrium composition of a reacting ideal-gas mixture at a specified temperature and pressure. On the basis of these relations, we may draw the following conclusions about the equilibrium constant $K_P$ of ideal-gas mixtures:

**1.** *The $K_P$ of a reaction depends on temperature only.* It is independent of the pressure of the equilibrium mixture and is not affected by the presence of inert gases. This is because $K_P$ depends on $\Delta G^*(T)$, which depends on temperature only, and the $\Delta G^*(T)$ of inert gases is zero (see Eq. 16–14). Thus, at a specified temperature the following four reactions have the same $K_P$ value:

$$H_2 + \tfrac{1}{2}O_2 \rightleftharpoons H_2O \qquad \text{at 1 atm}$$

$$H_2 + \tfrac{1}{2}O_2 \rightleftharpoons H_2O \qquad \text{at 5 atm}$$

$$H_2 + \tfrac{1}{2}O_2 + 3N_2 \rightleftharpoons H_2O + 3N_2 \qquad \text{at 3 atm}$$

$$H_2 + 2O_2 + 5N_2 \rightleftharpoons H_2O + 1.5O_2 + 5N_2 \qquad \text{at 2 atm}$$

**2.** *The $K_P$ of the reverse reaction is $1/K_P$.* This is easily seen from Eq. 16–13. For reverse reactions, the products and reactants switch places, and thus the terms in the numerator move to the denominator and vice versa. Consequently, the equilibrium constant of the reverse reaction becomes $1/K_P$. For example, from Table A–28,

$$K_P = 0.1147 \times 10^{11} \quad \text{for} \quad H_2 + \tfrac{1}{2}O_2 \rightleftharpoons H_2O \qquad \text{at 1000 K}$$

$$K_P = 8.718 \times 10^{-11} \quad \text{for} \qquad H_2O \rightleftharpoons H_2 + \tfrac{1}{2}O_2 \text{ at 1000 K}$$

**3.** *The larger the $K_P$, the more complete the reaction.* This is also apparent from Fig. 16–9 and Eq. 16–13. If the equilibrium composition consists largely of product gases, the partial pressures of the products ($P_C$ and $P_D$) are considerably larger than the partial pressures of the reactants ($P_A$ and $P_B$), which results in a large value of $K_P$. In the limiting case of a complete reaction (no leftover reactants in the equilibrium mixture), $K_P$ approaches infinity. Conversely, very small values of $K_P$ indicate that a reaction does not proceed to any appreciable degree. Thus reactions with very small $K_P$ values at a specified temperature can be neglected.

| | $H_2 \rightarrow 2H$ | |
| | $P = 1$ atm | |
| $T$, K | $K_P$ | % mol H |
| 1000 | $5.17 \times 10^{-18}$ | 0.00 |
| 2000 | $2.65 \times 10^{-6}$ | 0.16 |
| 3000 | 0.025 | 14.63 |
| 4000 | 2.545 | 76.80 |
| 5000 | 41.47 | 97.70 |
| 6000 | 267.7 | 99.63 |

**FIGURE 16–9**

The larger the $K_P$, the more complete the reaction.

A reaction with $K_P > 1000$ (or $\ln K_P > 7$) is usually assumed to proceed to completion, and a reaction with $K_P < 0.001$ (or $\ln K_P < -7$) is assumed not to occur at all. For example, $\ln K_P = -6.8$ for the reaction $N_2 \rightleftharpoons 2N$ at 5000 K. Therefore, the dissociation of $N_2$ into monatomic nitrogen (N) can be disregarded at temperatures below 5000 K.

4. *The mixture pressure affects the equilibrium composition* (although it does not affect the equilibrium constant $K_P$). This can be seen from Eq. 16–15, which involves the term $P^{\Delta\nu}$, where $\Delta\nu = \Sigma \nu_P - \Sigma \nu_R$ (the difference between the number of moles of products and the number of moles of reactants in the stoichiometric reaction). At a specified temperature, the $K_P$ value of the reaction, and thus the right-hand side of Eq. 16–15, remains constant. Therefore, the mole numbers of the reactants and the products must change to counteract any changes in the pressure term. The direction of the change depends on the sign of $\Delta\nu$. An increase in pressure at a specified temperature increases the number of moles of the reactants and decreases the number of moles of products if $\Delta\nu$ is positive, have the opposite effect if $\Delta\nu$ is negative, and have no effect if $\Delta\nu$ is zero.

5. *The presence of inert gases affects the equilibrium composition* (although it does not affect the equilibrium constant $K_P$). This can be seen from Eq. 16–15, which involves the term $(1/N_{total})^{\Delta\nu}$, where $N_{total}$ is the total number of moles of the ideal-gas mixture at equilibrium, *including* inert gases. The sign of $\Delta\nu$ determines how the presence of inert gases influences the equilibrium composition (Fig. 16–10). An increase in the number of moles of inert gases at a specified temperature and pressure decreases the number of moles of the reactants and increases the number of moles of products if $\Delta\nu$ is positive, have the opposite effect if $\Delta\nu$ is negative, and have no effect if $\Delta\nu$ is zero.

6. *When the stoichiometric coefficients are doubled, the value of $K_P$ is squared.* Therefore, when one is using $K_P$ values from a table, the stoichiometric coefficients (the $\nu$'s) used in a reaction must be exactly the same ones appearing in the table from which the $K_P$ values are selected. Multiplying all the coefficients of a stoichiometric equation does not affect the mass balance, but it does affect the equilibrium constant calculations since the stoichiometric coefficients appear as exponents of partial pressures in Eq. 16–13. For example,

For $\qquad H_2 + \frac{1}{2}O_2 \rightleftharpoons H_2O \qquad K_{P_1} = \dfrac{P_{H_2O}}{P_{H_2} P_{O_2}^{1/2}}$

But for $\qquad 2H_2 + O_2 \rightleftharpoons 2H_2O \qquad K_{P_2} = \dfrac{P_{H_2O}^2}{P_{H_2}^2 P_{O_2}} = (K_{P_1})^2$

7. *Free electrons in the equilibrium composition can be treated as an ideal gas.* At high temperatures (usually above 2500 K), gas molecules start to dissociate into unattached atoms (such as $H_2 \rightleftharpoons 2H$), and at even higher temperatures atoms start to lose electrons and ionize, for example,

$$H \rightleftharpoons H^+ + e^- \qquad\qquad \text{(16–16)}$$

The dissociation and ionization effects are more pronounced at low pressures. Ionization occurs to an appreciable extent only at very high temperatures, and the mixture of electrons, ions, and neutral atoms can be treated as an ideal gas. Therefore, the equilibrium composition of ionized gas mixtures

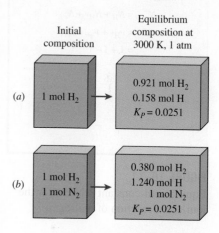

| Initial composition | Equilibrium composition at 3000 K, 1 atm |
|---|---|
| (a) 1 mol $H_2$ | 0.921 mol $H_2$<br>0.158 mol H<br>$K_P = 0.0251$ |
| (b) 1 mol $H_2$<br>1 mol $N_2$ | 0.380 mol $H_2$<br>1.240 mol H<br>1 mol $N_2$<br>$K_P = 0.0251$ |

**FIGURE 16–10**
The presence of inert gases does not affect the equilibrium constant, but it does affect the equilibrium composition.

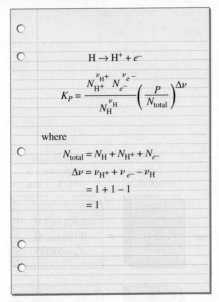

$$H \rightarrow H^+ + e^-$$

$$K_P = \frac{N_{H^+}^{\nu_{H^+}} N_{e^-}^{\nu_{e^-}}}{N_H^{\nu_H}} \left(\frac{P}{N_{total}}\right)^{\Delta\nu}$$

where

$$N_{total} = N_H + N_{H^+} + N_{e^-}$$

$$\Delta\nu = \nu_{H^+} + \nu_{e^-} - \nu_H$$

$$= 1 + 1 - 1$$

$$= 1$$

**FIGURE 16–11**
Equilibrium-constant relation for the ionization reaction of hydrogen.

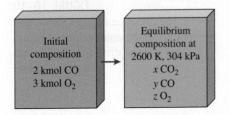

**FIGURE 16–12**
Schematic for Example 16–3.

can be determined from Eq. 16–15 (Fig. 16–11). This treatment may not be adequate in the presence of strong electric fields, however, since the electrons may be at a different temperature than the ions in this case.

**8.** *Equilibrium calculations provide information on the equilibrium composition of a reaction, not on the reaction rate.* Sometimes it may even take years to achieve the indicated equilibrium composition. For example, the equilibrium constant of the reaction $H_2 + \frac{1}{2}O_2 \rightleftharpoons H_2O$ at 298 K is about $10^{40}$, which suggests that a stoichiometric mixture of $H_2$ and $O_2$ at room temperature should react to form $H_2O$, and the reaction should go to completion. However, the rate of this reaction is so slow that it practically does not occur. But when the right catalyst is used, the reaction goes to completion rather quickly to the predicted value.

---

**EXAMPLE 16–3**   **Equilibrium Composition at a Specified Temperature**

A mixture of 2 kmol of CO and 3 kmol of $O_2$ is heated to 2600 K at a pressure of 304 kPa. Determine the equilibrium composition, assuming the mixture consists of $CO_2$, CO, and $O_2$ (Fig. 16–12).

**SOLUTION**   A reactive gas mixture is heated to a high temperature. The equilibrium composition at that temperature is to be determined.
**Assumptions**   **1** The equilibrium composition consists of $CO_2$, CO, and $O_2$. **2** The constituents of the mixture are ideal gases.
**Analysis**   The stoichiometric and actual reactions in this case are as follows:

*Stoichiometric:*   $CO + \frac{1}{2}O_2 \rightleftharpoons CO_2$   (thus $\nu_{CO_2} = 1, \nu_{CO} = 1$, and $\nu_{O_2} = \frac{1}{2}$)

*Actual:*   $2CO + 3O_2 \longrightarrow \underbrace{xCO_2}_{\text{products}} + \underbrace{yCO + zO_2}_{\substack{\text{reactants}\\\text{(leftover)}}}$

*C balance:*   $2 = x + y$   or   $y = 2 - x$

*O balance:*   $8 = 2x + y + 2z$   or   $z = 3 - \dfrac{x}{2}$

*Total number of moles:*   $N_{total} = x + y + z = 5 - \dfrac{x}{2}$

*Pressure:*   $P = 304 \text{ kPa} = 3.0 \text{ atm}$

The closest reaction listed in Table A–28 is $CO_2 \rightleftharpoons CO + \frac{1}{2}O_2$, for which In $K_P = -2.801$ at 2600 K. The reaction we have is the inverse of this, and thus In $K_P = +2.801$, or $K_P = 16.461$ in our case.

Assuming ideal-gas behavior for all components, the equilibrium constant relation (Eq. 16–15) becomes

$$K_P = \frac{N_{CO_2}^{\nu_{CO_2}}}{N_{CO}^{\nu_{CO}} N_{O_2}^{\nu_{O_2}}} \left(\frac{P}{N_{total}}\right)^{\nu_{CO_2} - \nu_{CO} - \nu_{O_2}}$$

Substituting, we get

$$16.461 = \frac{x}{(2 - x)(3 - x/2)^{1/2}} \left(\frac{3}{5 - x/2}\right)^{-1/2}$$

CHAPTER 16

Solving for $x$ yields

$$x = 1.906$$

Then

$$y = 2 - x = 0.094$$

$$z = 3 - \frac{x}{2} = 2.047$$

Therefore, the equilibrium composition of the mixture at 2600 K and 304 kPa is

$$1.906CO_2 + 0.094CO + 2.074O_2$$

**Discussion** In solving this problem, we disregarded the dissociation of $O_2$ into O according to the reaction $O_2 \rightarrow 2O$, which is a real possibility at high temperatures. This is because $\ln K_P = -7.521$ at 2600 K for this reaction, which indicates that the amount of $O_2$ that dissociates into O is negligible. (Besides, we have not learned how to deal with simultaneous reactions yet. We will do so in the next section.)

---

**EXAMPLE 16–4**    **Effect of Inert Gases on Equilibrium Composition**

A mixture of 3 kmol of CO, 2.5 kmol of $O_2$, and 8 kmol of $N_2$ is heated to 2600 K at a pressure of 5 atm. Determine the equilibrium composition of the mixture (Fig. 16–13).

**SOLUTION** A gas mixture is heated to a high temperature. The equilibrium composition at the specified temperature is to be determined.

**Assumptions** **1** The equilibrium composition consists of $CO_2$, CO, $O_2$, and $N_2$. **2** The constituents of the mixture are ideal gases.

**Analysis** This problem is similar to Example 16–3, except that it involves an inert gas $N_2$. At 2600 K, some possible reactions are $O_2 \rightleftharpoons 2O$ ($\ln K_P = -7.521$), $N_2 \rightleftharpoons 2N$ ($\ln K_P = -28.304$), $\frac{1}{2}O_2 + \frac{1}{2}N_2 \rightleftharpoons NO$ ($\ln K_P = -2.671$), and $CO + \frac{1}{2}O_2 \rightleftharpoons CO_2$ ($\ln K_P = 2.801$ or $K_P = 16.461$). Based on these $K_P$ values, we conclude that the $O_2$ and $N_2$ will not dissociate to any appreciable degree, but a small amount will combine to form some oxides of nitrogen. (We disregard the oxides of nitrogen in this example, but they should be considered in a more refined analysis.) We also conclude that most of the CO will combine with $O_2$ to form $CO_2$. Notice that despite the changes in pressure, the number of moles of CO and $O_2$ and the presence of an inert gas, the $K_P$ value of the reaction is the same as that used in Example 16–3.

The stoichiometric and actual reactions in this case are

*Stoichiometric:*    $CO + \frac{1}{2}O_2 \rightleftharpoons CO_2$ (thus $\nu_{CO_2} = 1$, $\nu_{CO} = 1$, and $\nu_{O_2} = \frac{1}{2}$)

*Actual:*    $3CO + 2.5O_2 + 8N_2 \longrightarrow \underbrace{xCO_2}_{\text{products}} + \underbrace{yCO + zO_2}_{\substack{\text{reactants} \\ \text{(leftover)}}} + \underbrace{8N_2}_{\text{inert}}$

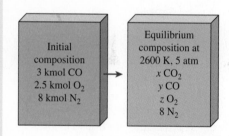

Initial composition
3 kmol CO
2.5 kmol $O_2$
8 kmol $N_2$

Equilibrium composition at 2600 K, 5 atm
$x$ $CO_2$
$y$ CO
$z$ $O_2$
8 $N_2$

**FIGURE 16–13**
Schematic for Example 16–4.

*C balance:* $\qquad$ $3 = x + y \quad$ or $\quad y = 3 - x$

*O balance:* $\qquad$ $8 = 2x + y + 2z \quad$ or $\quad z = 2.5 - \dfrac{x}{2}$

*Total number of moles:* $\qquad$ $N_{\text{total}} = x + y + z + 8 = 13.5 - \dfrac{x}{2}$

Assuming ideal-gas behavior for all components, the equilibrium constant relation (Eq. 16–15) becomes

$$K_P = \frac{N_{CO_2}^{\nu_{CO_2}}}{N_{CO}^{\nu_{CO}} N_{O_2}^{\nu_{O_2}}} \left(\frac{P}{N_{\text{total}}}\right)^{\nu_{CO_2} - \nu_{CO} - \nu_{O_2}}$$

Substituting, we get

$$16.461 = \frac{x}{(3 - x)(2.5 - x/2)^{1/2}} \left(\frac{5}{13.5 - x/2}\right)^{-1/2}$$

Solving for $x$ yields

$$x = 2.754$$

Then

$$y = 3 - x = 0.246$$

$$z = 2.5 - \frac{x}{2} = 1.123$$

Therefore, the equilibrium composition of the mixture at 2600 K and 5 atm is

$$2.754 CO_2 + 0.246 CO + 1.123 O_2 + 8 N_2$$

***Discussion*** Note that the inert gases do not affect the $K_P$ value or the $K_P$ relation for a reaction, but they do affect the equilibrium composition.

# 16–4 · CHEMICAL EQUILIBRIUM FOR SIMULTANEOUS REACTIONS

The reacting mixtures we have considered so far involved only one reaction, and writing a $K_P$ relation for that reaction was sufficient to determine the equilibrium composition of the mixture. However, most practical chemical reactions involve two or more reactions that occur simultaneously, which makes them more difficult to deal with. In such cases, it becomes necessary to apply the equilibrium criterion to all possible reactions that may occur in the reaction chamber. When a chemical species appears in more than one reaction, the application of the equilibrium criterion, together with the mass balance for each chemical species, results in a system of simultaneous equations from which the equilibrium composition can be determined.

We have shown earlier that a reacting system at a specified temperature and pressure achieves chemical equilibrium when its Gibbs function reaches a minimum value, that is, $(dG)_{T,P} = 0$. This is true regardless of the number of reactions that may be occurring. When two or more reactions are involved, this condition is satisfied only when $(dG)_{T,P} = 0$ for each reaction. Assuming ideal-gas behavior, the $K_P$ of each reaction can be determined from Eq. 16–15, with $N_{total}$ being the total number of moles present in the equilibrium mixture.

The determination of the equilibrium composition of a reacting mixture requires that we have as many equations as unknowns, where the unknowns are the number of moles of each chemical species present in the equilibrium mixture. The mass balance of each element involved provides one equation. The rest of the equations must come from the $K_p$ relations written for each reaction. Thus, we conclude that *the number of $K_p$ relations needed to determine the equilibrium composition of a reacting mixture is equal to the number of chemical species minus the number of elements present in equilibrium.* For an equilibrium mixture that consists of $CO_2$, $CO$, $O_2$, and $O$, for example, two $K_p$ relations are needed to determine the equilibrium composition since it involves four chemical species and two elements (Fig. 16–14).

The determination of the equilibrium composition of a reacting mixture in the presence of two simultaneous reactions is here with an example.

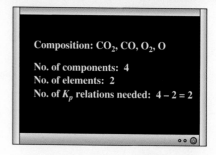

**FIGURE 16–14**

The number of $K_p$ relations needed to determine the equilibrium composition of a reacting mixture is the difference between the number of species and the number of elements.

---

### EXAMPLE 16–5    Equilibrium Composition for Simultaneous Reactions

A mixture of 1 kmol of $H_2O$ and 2 kmol of $O_2$ is heated to 4000 K at a pressure of 1 atm. Determine the equilibrium composition of this mixture, assuming that only $H_2O$, $OH$, $O_2$, and $H_2$ are present (Fig. 16–15).

**SOLUTION**  A gas mixture is heated to a specified temperature at a specified pressure. The equilibrium composition is to be determined.
**Assumptions**  **1** The equilibrium composition consists of $H_2O$, $OH$, $O_2$, and $H_2$. **2** The constituents of the mixture are ideal gases.
**Analysis**  The chemical reaction during this process can be expressed as

$$H_2O + 2O_2 \longrightarrow xH_2O + yH_2 + zO_2 + wOH$$

Mass balances for hydrogen and oxygen yield

*H balance:*          $2 = 2x + 2y + w$          **(1)**

*O balance:*          $5 = x + 2z + w$          **(2)**

The mass balances provide us with only two equations with four unknowns, and thus we need to have two more equations (to be obtained from the $K_p$ relations) to determine the equilibrium composition of the mixture. It appears that part of the $H_2O$ in the products is dissociated into $H_2$ and $OH$ during this process, according to the stoichiometric reactions

$$H_2O \rightleftharpoons H_2 + \tfrac{1}{2}O_2 \quad \text{(reaction 1)}$$
$$H_2O \rightleftharpoons \tfrac{1}{2}H_2 + OH \quad \text{(reaction 2)}$$

The equilibrium constants for these two reactions at 4000 K are determined from Table A–28 to be

$$\ln K_{P_1} = -0.542 \longrightarrow K_{P_1} = 0.5816$$
$$\ln K_{P_2} = -0.044 \longrightarrow K_{P_2} = 0.9570$$

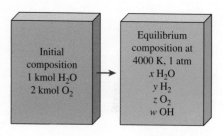

**FIGURE 16–15**
Schematic for Example 16–5.

The $K_P$ relations for these two simultaneous reactions are

$$K_{P_1} = \frac{N_{H_2}^{\nu_{H_2}} N_{O_2}^{\nu_{O_2}}}{N_{H_2O}^{\nu_{H_2O}}} \left(\frac{P}{N_{\text{total}}}\right)^{\nu_{H_2} + \nu_{O_2} - \nu_{H_2O}}$$

$$K_{P_2} = \frac{N_{H_2}^{\nu_{H_2}} N_{OH}^{\nu_{OH}}}{N_{H_2O}^{\nu_{H_2O}}} \left(\frac{P}{N_{\text{total}}}\right)^{\nu_{H_2} + \nu_{OH} - \nu_{H_2O}}$$

where

$$N_{\text{total}} = N_{H_2O} + N_{H_2} + N_{O_2} + N_{OH} = x + y + z + w$$

Substituting yields

$$0.5816 = \frac{(y)(z)^{1/2}}{x} \left(\frac{1}{x + y + z + w}\right)^{1/2} \tag{3}$$

$$0.9570 = \frac{(w)(y)^{1/2}}{x} \left(\frac{1}{x + y + z + w}\right)^{1/2} \tag{4}$$

Solving Eqs. (1), (2), (3), and (4) simultaneously for the four unknowns $x$, $y$, $z$, and $w$ yields

$$x = 0.271 \qquad y = 0.213$$
$$z = 1.849 \qquad w = 1.032$$

Therefore, the equilibrium composition of 1 kmol $H_2O$ and 2 kmol $O_2$ at 1 atm and 4000 K is

$$0.271H_2O + 0.213H_2 + 1.849O_2 + 1.032OH$$

**Discussion** We could also solve this problem by using the $K_P$ relation for the stoichiometric reaction $O_2 \rightleftharpoons 2O$ as one of the two equations.

Solving a system of simultaneous nonlinear equations is extremely tedious and time-consuming if it is done by hand. Thus, it is often necessary to solve these kinds of problems by using an equation solver such as EES.

## 16–5 · VARIATION OF $K_P$ WITH TEMPERATURE

It was shown in Section 16–2 that the equilibrium constant $K_P$ of an ideal gas depends on temperature only, and it is related to the standard-state Gibbs function change $\Delta G^*(T)$ through the relation (Eq. 16–14)

$$\ln K_P = -\frac{\Delta G^*(T)}{R_u T}$$

In this section we develop a relation for the variation of $K_P$ with temperature in terms of other properties.

Substituting $\Delta G^*(T) = \Delta H^*(T) - T \, \Delta S^*(T)$ into the above relation and differentiating with respect to temperature, we get

$$\frac{d(\ln K_p)}{dT} = \frac{\Delta H^*(T)}{R_u T^2} - \frac{d[\Delta H^*(T)]}{R_u T \, dT} + \frac{d[\Delta S^*(T)]}{R_u \, dT}$$

At constant pressure, the second $T\,ds$ relation, $T\,ds = dh - v\,dP$, reduces to $T\,ds = dh$. Also, $T\,d(\Delta S^*) = d(\Delta H^*)$ since $\Delta S^*$ and $\Delta H^*$ consist of entropy and enthalpy terms of the reactants and the products. Therefore, the last two terms in the above relation cancel, and it reduces to

$$\frac{d(\ln K_p)}{dT} = \frac{\Delta H^*(T)}{R_u T^2} = \frac{\bar{h}_R(T)}{R_u T^2} \qquad (16\text{–}17)$$

where $\bar{h}_R(T)$ is the enthalpy of reaction at temperature $T$. Notice that we dropped the superscript * (which indicates a constant pressure of 1 atm) from $\Delta H(T)$, since the enthalpy of an ideal gas depends on temperature only and is independent of pressure. Equation 16–17 is an expression of the variation of $K_P$ with temperature in terms of $\bar{h}_R(T)$, and it is known as the **van't Hoff equation**. To integrate it, we need to know how $\bar{h}_R$ varies with $T$. For small temperature intervals, $\bar{h}_R$ can be treated as a constant and Eq. 16–17 can be integrated to yield

$$\ln \frac{K_{P_2}}{K_{P_1}} \cong \frac{\bar{h}_R}{R_u}\left(\frac{1}{T_1} - \frac{1}{T_2}\right) \qquad (16\text{–}18)$$

This equation has two important implications. First, it provides a means of calculating the $\bar{h}_R$ of a reaction from a knowledge of $K_P$, which is easier to determine. Second, it shows that exothermic reactions ($\bar{h}_R < 0$) such as combustion processes are less complete at higher temperatures since $K_P$ decreases with temperature for such reactions (Fig. 16–16).

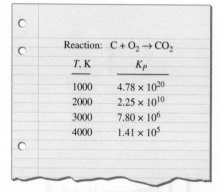

Reaction: $C + O_2 \rightarrow CO_2$

| $T$, K | $K_P$ |
|--------|-------|
| 1000 | $4.78 \times 10^{20}$ |
| 2000 | $2.25 \times 10^{10}$ |
| 3000 | $7.80 \times 10^6$ |
| 4000 | $1.41 \times 10^5$ |

**FIGURE 16–16**
Exothermic reactions are less complete at higher temperatures.

---

**EXAMPLE 16–6    The Enthalpy of Reaction of a Combustion Process**

Estimate the enthalpy of reaction $\bar{h}_R$ for the combustion process of hydrogen $H_2 + 0.5O_2 \rightarrow H_2O$ at 2000 K, using (a) enthalpy data and (b) $K_P$ data.

**SOLUTION** The $\bar{h}_R$ at a specified temperature is to be determined using the enthalpy and $K_p$ data.
**Assumptions** Both the reactants and the products are ideal gases.
**Analysis** (a) The $\bar{h}_R$ of the combustion process of $H_2$ at 2000 K is the amount of energy released as 1 kmol of $H_2$ is burned in a steady-flow combustion chamber at a temperature of 2000 K. It can be determined from Eq. 15–6,

$$\bar{h}_R = \sum N_p(\bar{h}_f^\circ + \bar{h} - \bar{h}^\circ)_p - \sum N_r(\bar{h}_f^\circ + \bar{h} - \bar{h}^\circ)_r$$

$$= N_{H_2O}(\bar{h}_f^\circ + \bar{h}_{2000\,K} - \bar{h}_{298\,K})_{H_2O} - N_{H_2}(\bar{h}_f^\circ + \bar{h}_{2000\,K} - \bar{h}_{298\,K})_{H_2}$$

$$- N_{O_2}(\bar{h}_f^\circ + \bar{h}_{2000\,K} - \bar{h}_{298\,K})_{O_2}$$

Substituting yields

$$\bar{h}_R = (1\ \text{kmol H}_2\text{O})[(-241{,}820 + 82{,}593 - 9904)\ \text{kJ/kmol H}_2\text{O}]$$

$$- (1\ \text{kmol H}_2)[(0 + 61{,}400 - 8468)\ \text{kJ/kmol H}_2]$$

$$- (0.5\ \text{kmol O}_2)[(0 + 67{,}881 - 8682)\ \text{kJ/kmol O}_2]$$

$$= -251{,}663\ \text{kJ/kmol}$$

(*b*) The $\bar{h}_R$ value at 2000 K can be estimated by using $K_P$ values at 1800 and 2200 K (the closest two temperatures to 2000 K for which $K_P$ data are available) from Table A–28. They are $K_{P_1} = 18{,}509$ at $T_1 = 1800$ K and $K_{P_2} = 869.6$ at $T_2 = 2200$ K. By substituting these values into Eq. 16–18, the $\bar{h}_R$ value is determined to be

$$\ln \frac{K_{P_2}}{K_{P_1}} \cong \frac{\bar{h}_R}{R_u} \left( \frac{1}{T_1} - \frac{1}{T_2} \right)$$

$$\ln \frac{869.6}{18{,}509} \cong \frac{\bar{h}_R}{8.314 \text{ kJ/kmol·K}} \left( \frac{1}{1800 \text{ K}} - \frac{1}{2200 \text{ K}} \right)$$

$$\bar{h}_R \cong -251{,}698 \text{ kJ/kmol}$$

**Discussion** Despite the large temperature difference between $T_1$ and $T_2$ (400 K), the two results are almost identical. The agreement between the two results would be even better if a smaller temperature interval were used.

# 16–6 · PHASE EQUILIBRIUM

We showed at the beginning of this chapter that the equilibrium state of a system at a specified temperature and pressure is the state of the minimum Gibbs function, and the equilibrium criterion for a reacting or nonreacting system was expressed as (Eq. 16–4)

$$(dG)_{T,P} = 0$$

In the preceding sections we applied the equilibrium criterion to reacting systems. In this section, we apply it to nonreacting multiphase systems.

We know from experience that a wet T-shirt hanging in an open area eventually dries, a small amount of water left in a glass evaporates, and the aftershave in an open bottle quickly disappears (Fig. 16–17). These examples suggest that there is a driving force between the two phases of a substance that forces the mass to transform from one phase to another. The magnitude of this force depends, among other things, on the relative concentrations of the two phases. A wet T-shirt dries much quicker in dry air than it does in humid air. In fact, it does not dry at all if the relative humidity of the environment is 100 percent. In this case, there is no transformation from the liquid phase to the vapor phase, and the two phases are in **phase equilibrium**. The conditions of phase equilibrium change, however, if the temperature or the pressure is changed. Therefore, we examine phase equilibrium at a specified temperature and pressure.

## Phase Equilibrium for a Single-Component System

The equilibrium criterion for two phases of a pure substance such as water is easily developed by considering a mixture of saturated liquid and saturated vapor in equilibrium at a specified temperature and pressure, such as that shown in Fig. 16–18. The total Gibbs function of this mixture is

$$G = m_f g_f + m_g g_g$$

**FIGURE 16–17**
Wet clothes hung in an open area eventually dry as a result of mass transfer from the liquid phase to the vapor phase.
*© C Squared Studios/Getty Images RF*

**FIGURE 16–18**
A liquid–vapor mixture in equilibrium at a constant temperature and pressure.

where $g_f$ and $g_g$ are the Gibbs functions of the liquid and vapor phases per unit mass, respectively. Now imagine a disturbance during which a differential amount of liquid $dm_f$ evaporates at constant temperature and pressure. The change in the total Gibbs function during this disturbance is

$$(dG)_{T,P} = g_f \, dm_f + g_g \, dm_g$$

since $g_f$ and $g_g$ remain constant at constant temperature and pressure. At equilibrium, $(dG)_{T,P} = 0$. Also from the conservation of mass, $dm_g = -dm_f$. Substituting, we obtain

$$(dG)_{T,P} = (g_f - g_g) \, dm_f$$

which must be equal to zero at equilibrium. It yields

$$g_f = g_g \qquad (16\text{-}19)$$

Therefore, *the two phases of a pure substance are in equilibrium when each phase has the same value of specific Gibbs function.* Also, at the triple point (the state at which all three phases coexist in equilibrium), the specific Gibbs functions of all three phases are equal to each other.

What happens if $g_f > g_g$? Obviously the two phases are not in equilibrium at that moment. The second law requires that $(dG)_{T,P} = (g_f - g_g) \, dm_f \le 0$. Thus, $dm_f$ must be negative, which means that some liquid must vaporize until $g_f = g_g$. Therefore, the Gibbs function difference is the driving force for phase change, just as the temperature difference is the driving force for heat transfer.

---

■ **EXAMPLE 16–7**    **Phase Equilibrium for a Saturated Mixture**

■ Show that a mixture of saturated liquid water and saturated water vapor at
■ 120°C satisfies the criterion for phase equilibrium.

**SOLUTION**  It is to be shown that a saturated mixture satisfies the criterion for phase equilibrium.
**Properties**  The properties of saturated water at 120°C are $h_f = 503.81$ kJ/kg, $s_f = 1.5279$ kJ/kg·K, $h_g = 2706.0$ kJ/kg, and $s_g = 7.1292$ kJ/kg·K (Table A–4).
**Analysis**  Using the definition of Gibbs function together with the enthalpy and entropy data, we have

$$g_f = h_f - Ts_f = 503.81 \text{ kJ/kg} - (393.15 \text{ K})(1.5279 \text{ kJ/kg·K})$$
$$= -96.9 \text{ kJ/kg}$$

and

$$g_g = h_g - Ts_g = 2706.0 \text{ kJ/kg} - (393.15 \text{ K})(7.1292 \text{ kJ/kg·K})$$
$$= -96.8 \text{ kJ/kg}$$

**Discussion**  The two results are in close agreement. They would match exactly if more accurate property data were used. Therefore, the criterion for phase equilibrium is satisfied.

**FIGURE 16–19**
According to the Gibbs phase rule, a single-component, two-phase system can have only one independent variable.

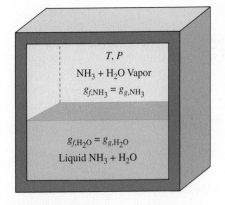

**FIGURE 16–20**
A multicomponent multiphase system is in phase equilibrium when the specific Gibbs function of each component is the same in all phases.

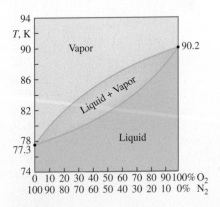

**FIGURE 16–21**
Equilibrium diagram for the two-phase mixture of oxygen and nitrogen at 0.1 MPa.

## The Phase Rule

Notice that a single-component two-phase system may exist in equilibrium at different temperatures (or pressures). However, once the temperature is fixed, the system is locked into an equilibrium state and all intensive properties of each phase (except their relative amounts) are fixed. Therefore, a single-component two-phase system has one independent property, which may be taken to be the temperature or the pressure.

In general, the number of independent variables associated with a multicomponent, multiphase system is given by the **Gibbs phase rule**, expressed as

$$IV = C - PH + 2 \qquad (16\text{–}20)$$

where IV = the number of independent variables, $C$ = the number of components, and PH = the number of phases present in equilibrium. For the single-component ($C = 1$) two-phase (PH = 2) system discussed above, for example, one independent intensive property needs to be specified (IV = 1, Fig. 16–19). At the triple point, however, PH = 3 and thus IV = 0. That is, none of the properties of a pure substance at the triple point can be varied. Also, based on this rule, a pure substance that exists in a single phase (PH = 1) has two independent variables. In other words, two independent intensive properties need to be specified to fix the equilibrium state of a pure substance in a single phase.

## Phase Equilibrium for a Multicomponent System

Many multiphase systems encountered in practice involve two or more components. A multicomponent multiphase system at a specified temperature and pressure is in phase equilibrium when there is no driving force between the different phases of each component. Thus, for phase equilibrium, the specific Gibbs function of each component must be the same in all phases (Fig. 16–20). That is,

$$g_{f,1} = g_{g,1} = g_{s,1} \quad \text{for component 1}$$
$$g_{f,2} = g_{g,2} = g_{s,2} \quad \text{for component 2}$$
$$\cdots\cdots\cdots\cdots\cdots$$
$$g_{f,N} = g_{g,N} = g_{s,N} \quad \text{for component } N$$

We could also derive these relations by using mathematical vigor instead of physical arguments.

Some components may exist in more than one solid phase at the specified temperature and pressure. In this case, the specific Gibbs function of each solid phase of a component must also be the same for phase equilibrium.

In this section we examine the phase equilibrium of two-component systems that involve two phases (liquid and vapor) in equilibrium. For such systems, $C = 2$, PH = 2, and thus IV = 2. That is, a two-component, two-phase system has two independent variables, and such a system will not be in equilibrium unless two independent intensive properties are fixed.

In general, the two phases of a two-component system do not have the same composition in each phase. That is, the mole fraction of a component is different in different phases. This is illustrated in Fig. 16–21 for the two-phase mixture of oxygen and nitrogen at a pressure of 0.1 MPa. On this diagram, the vapor line represents the equilibrium composition of the vapor phase at various temperatures, and the liquid line does the same for the liquid phase. At 84 K, for example, the mole fractions are 30 percent nitrogen and 70 percent oxygen in the liquid phase and 66 percent nitrogen and 34 percent oxygen in the vapor phase. Notice that

$$y_{f,N_2} + y_{f,O_2} = 0.30 + 0.70 = 1 \tag{16-21a}$$

$$y_{g,N_2} + y_{g,O_2} = 0.66 + 0.34 = 1 \tag{16-21b}$$

Therefore, once the temperature and pressure (two independent variables) of a two-component, two-phase mixture are specified, the equilibrium composition of each phase can be determined from the phase diagram, which is based on experimental measurements.

It is interesting to note that temperature is a *continuous* function, but mole fraction (which is a dimensionless concentration), in general, is not. The water and air temperatures at the free surface of a lake, for example, are always the same. The mole fractions of air on the two sides of a water–air interface, however, are obviously very different (in fact, the mole fraction of air in water is close to zero). Likewise, the mole fractions of water on the two sides of a water–air interface are also different even when air is saturated (Fig. 16–22). Therefore, when specifying mole fractions in two-phase mixtures, we need to clearly specify the intended phase.

In most practical applications, the two phases of a mixture are not in phase equilibrium since the establishment of phase equilibrium requires the diffusion of species from higher concentration regions to lower concentration regions, which may take a long time. However, phase equilibrium always exists at the interface of two phases of a species. In the case of air–water interface, the mole fraction of water vapor in the air is easily determined from saturation data, as shown in Example 16–8.

The situation is similar at *solid–liquid* interfaces. Again, at a given temperature, only a certain amount of solid can be dissolved in a liquid, and the solubility of the solid in the liquid is determined from the requirement that thermodynamic equilibrium exists between the solid and the solution at the interface. The **solubility** represents *the maximum amount of solid that can be dissolved in a liquid at a specified temperature* and is widely available in chemistry handbooks. In Table 16–1 we present sample solubility data for sodium chloride (NaCl) and calcium bicarbonate [Ca(HO$_3$)$_2$] at various temperatures. For example, the solubility of salt (NaCl) in water at 310 K is 36.5 kg per 100 kg of water. Therefore, the mass fraction of salt in the saturated brine is simply

$$mf_{salt,liquid\,side} = \frac{m_{salt}}{m} = \frac{36.5\,kg}{(100+36.5)\,kg} = 0.267 \text{ (or 26.7 percent)}$$

whereas the mass fraction of salt in the pure solid salt is mf = 1.0.

Many processes involve the absorption of a gas into a liquid. Most gases are weakly soluble in liquids (such as air in water), and for such dilute solutions the mole fractions of a species $i$ in the gas and liquid phases at the interface are observed to be proportional to each other. That is,

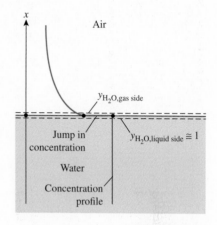

**FIGURE 16–22**
Unlike temperature, the mole fraction of species on the two sides of a liquid–gas (or solid–gas or solid–liquid) interface are usually not the same.

**TABLE 16–1**

Solubility of two inorganic compounds in water at various temperatures, in kg (in 100 kg of water)

(from *Handbook of Chemistry*, McGraw-Hill, 1961)

| | Solute | |
|---|---|---|
| Temperature, K | Salt NaCl | Calcium bicarbonate Ca(HCO$_3$)$_2$ |
| 273.15 | 35.7 | 16.15 |
| 280 | 35.8 | 16.30 |
| 290 | 35.9 | 16.53 |
| 300 | 36.2 | 16.75 |
| 310 | 36.5 | 16.98 |
| 320 | 36.9 | 17.20 |
| 330 | 37.2 | 17.43 |
| 340 | 37.6 | 17.65 |
| 350 | 38.2 | 17.88 |
| 360 | 38.8 | 18.10 |
| 370 | 39.5 | 18.33 |
| 373.15 | 39.8 | 18.40 |

**TABLE 16–2**

Henry's constant $H$ (in bars) for selected gases in water at low to moderate pressures (for gas $i$, $H = P_{i,\text{gas side}}/y_{i,\text{water side}}$) (from Mills, 1995, Table A.21, p. 874)

| Solute | 290 K | 300 K | 310 K | 320 K | 330 K | 340 K |
|---|---|---|---|---|---|---|
| $H_2S$ | 440 | 560 | 700 | 830 | 980 | 1140 |
| $CO_2$ | 1,280 | 1,710 | 2,170 | 2,720 | 3,220 | — |
| $O_2$ | 38,000 | 45,000 | 52,000 | 57,000 | 61,000 | 65,000 |
| $H_2$ | 67,000 | 72,000 | 75,000 | 76,000 | 77,000 | 76,000 |
| CO | 51,000 | 60,000 | 67,000 | 74,000 | 80,000 | 84,000 |
| Air | 62,000 | 74,000 | 84,000 | 92,000 | 99,000 | 104,000 |
| $N_2$ | 76,000 | 89,000 | 101,000 | 110,000 | 118,000 | 124,000 |

$y_{i,\text{gas side}} \propto y_{i,\text{liquid side}}$ or $P_{i,\text{gas side}} \propto P y_{i,\text{liquid side}}$ since $y_i = P_i/P$ for ideal-gas mixtures. This is known as the **Henry's law** and is expressed as

$$y_{i,\text{liquid side}} = \frac{P_{i,\text{gas side}}}{H} \qquad (16\text{–}22)$$

where $H$ is the **Henry's constant**, which is the product of the total pressure of the gas mixture and the proportionality constant. For a given species, it is a function of temperature only and is practically independent of pressure for pressures under about 5 atm. Values of the Henry's constant for a number of aqueous solutions are given in Table 16–2 for various temperatures. From this table and the equation above we make the following observations:

1. The concentration of a gas dissolved in a liquid is inversely proportional to Henry's constant. Therefore, the larger the Henry's constant, the smaller the concentration of dissolved gases in the liquid.
2. The Henry's constant increases (and thus the fraction of a dissolved gas in the liquid decreases) with increasing temperature. Therefore, the dissolved gases in a liquid can be driven off by heating the liquid (Fig. 16–23).
3. The concentration of a gas dissolved in a liquid is proportional to the partial pressure of the gas. Therefore, the amount of gas dissolved in a liquid can be increased by increasing the pressure of the gas. This can be used to advantage in the carbonation of soft drinks with $CO_2$ gas.

Strictly speaking, the result obtained from Eq. 16–22 for the mole fraction of dissolved gas is valid for the liquid layer just beneath the interface, but not necessarily the entire liquid. The latter will be the case only when thermodynamic phase equilibrium is established throughout the entire liquid body.

We mentioned earlier that the use of Henry's law is limited to dilute gas–liquid solutions, that is, liquids with a small amount of gas dissolved in them. Then, the question that arises naturally is, what do we do when the gas is highly soluble in the liquid (or solid), such as ammonia in water? In this case, the linear relationship of Henry's law does not apply, and the mole fraction of a gas dissolved in the liquid (or solid) is usually expressed as a function of the partial pressure of the gas in the gas phase and the temperature.

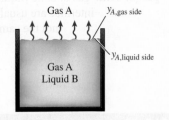

$y_{A,\text{gas side}} \propto y_{A,\text{liquid side}}$

or

$\dfrac{P_{A,\text{gas side}}}{P} \propto y_{A,\text{liquid side}}$

or

$P_{A,\text{gas side}} = H y_{A,\text{liquid side}}$

**FIGURE 16–23**

Dissolved gases in a liquid can be driven off by heating the liquid.

An approximate relation in this case for the *mole fractions* of a species on the *liquid* and *gas sides* of the interface is given by **Raoult's law** as

$$P_{i,\text{gas side}} = y_{i,\text{gas side}} P_{\text{total}} = y_{i,\text{liquid side}} P_{i,\text{sat}}(T) \qquad (16\text{-}23)$$

where $P_{i,\text{sat}}(T)$ is the *saturation pressure* of the species $i$ at the interface temperature and $P_{\text{total}}$ is the *total pressure* on the gas phase side. Tabular data are available in chemical handbooks for common solutions such as the ammonia–water solution that is widely used in absorption-refrigeration systems.

Gases may also dissolve in *solids*, but the diffusion process in this case can be very complicated. The dissolution of a gas may be independent of the structure of the solid, or it may depend strongly on its porosity. Some dissolution processes (such as the dissolution of hydrogen in titanium, similar to the dissolution of $CO_2$ in water) are *reversible,* and thus maintaining the gas content in the solid requires constant contact of the solid with a reservoir of that gas. Some other dissolution processes are *irreversible.* For example, oxygen gas dissolving in titanium forms $TiO_2$ on the surface, and the process does not reverse itself.

The molar density of the gas species $i$ in the solid at the interface $\bar{\rho}_{i,\text{solid side}}$ is proportional to the *partial pressure* of the species $i$ in the gas $P_{i,\text{gas side}}$ on the gas side of the interface and is expressed as

$$\bar{\rho}_{i,\text{solid side}} = \mathcal{S} \times P_{i,\text{gas side}} \qquad (\text{kmol/m}^3) \qquad (16\text{-}24)$$

where $\mathcal{S}$ is the solubility. Expressing the pressure in bars and noting that the unit of molar concentration is kmol of species $i$ per $\text{m}^3$, the unit of solubility is $\text{kmol/m}^3 \cdot \text{bar}$. Solubility data for selected gas–solid combinations are given in Table 16–3. The product of *solubility* of a gas and the *diffusion coefficient* of the gas in a solid is referred to as the *permeability,* which is a measure of the ability of the gas to penetrate a solid. Permeability is inversely proportional to thickness and has the unit $\text{kmol/s} \cdot \text{m} \cdot \text{bar}$.

Finally, if a process involves the *sublimation* of a pure solid such as ice or the *evaporation* of a pure liquid such as water in a different medium such as air, the mole (or mass) fraction of the substance in the liquid or solid phase is simply taken to be 1.0, and the partial pressure and thus the mole fraction of the substance in the gas phase can readily be determined from the saturation data of the substance at the specified temperature. Also, the assumption of thermodynamic equilibrium at the interface is very reasonable for pure solids, pure liquids, and solutions except when chemical reactions are occurring at the interface.

**TABLE 16–3**

Solubility of selected gases and solids (from Barrer, 1941)

(for gas $i$, $\mathcal{S} = \bar{\rho}_{i,\text{solid side}}/P_{i,\text{gas side}}$)

| Gas | Solid | $T$, K | $\mathcal{S}$ kmol/m³·bar |
|-----|-------|--------|-----------|
| $O_2$ | Rubber | 298 | 0.00312 |
| $N_2$ | Rubber | 298 | 0.00156 |
| $CO_2$ | Rubber | 298 | 0.04015 |
| He | $SiO_2$ | 298 | 0.00045 |
| $H_2$ | Ni | 358 | 0.00901 |

■ *EXAMPLE 16–8*    **Mole Fraction of Water Vapor Just over a Lake**

Determine the mole fraction of the water vapor at the surface of a lake whose temperature is 15°C, and compare it to the mole fraction of water in the lake (Fig. 16–24). Take the atmospheric pressure at lake level to be 92 kPa.

**SOLUTION**  The mole fraction of water vapor at the surface of a lake is to be determined and to be compared to the mole fraction of water in the lake. *Assumptions*  **1** Both the air and water vapor are ideal gases. **2** The amount of air dissolved in water is negligible.

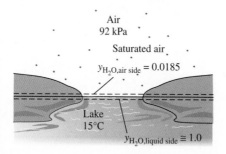

**FIGURE 16–24**
Schematic for Example 16–8.

**Properties** The saturation pressure of water at 15°C is 1.7057 kPa (Table A–4).
**Analysis** There exists phase equilibrium at the free surface of the lake, and thus the air at the lake surface is always saturated at the interface temperature.

The air at the water surface is saturated. Therefore, the partial pressure of water vapor in the air at the lake surface will simply be the saturation pressure of water at 15°C,

$$P_v = P_{\text{sat @ 15°C}} = 1.7057 \text{ kPa}$$

The mole fraction of water vapor in the air at the surface of the lake is determined from Eq. 16–22 to be

$$y_v = \frac{P_v}{P} = \frac{1.7057 \text{ kPa}}{92 \text{ kPa}} = 0.0185 \text{ or } 1.85 \text{ percent}$$

Water contains some dissolved air, but the amount is negligible. Therefore, we can assume the entire lake to be liquid water. Then, its mole fraction becomes

$$y_{\text{water,liquid side}} \cong 1.0 \text{ or } 100 \text{ percent}$$

**Discussion** Note that the concentration of water on a molar basis is 100 percent just beneath the air–water interface and less than 2 percent just above it even though the air is assumed to be saturated (so this is the highest value at 15°C). Therefore, large discontinuities can occur in the concentrations of a species across phase boundaries.

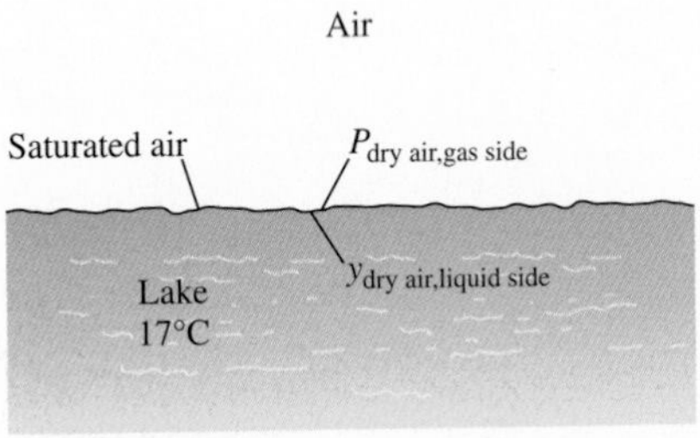

Air

Saturated air $\qquad P_{\text{dry air,gas side}}$

Lake $\qquad y_{\text{dry air,liquid side}}$
17°C

**FIGURE 16–25**
Schematic for Example 16–9.

---

**EXAMPLE 16–9**     **The Amount of Dissolved Air in Water**

Determine the mole fraction of air at the surface of a lake whose temperature is 17°C (Fig. 16–25). Take the atmospheric pressure at lake level to be 92 kPa.

**SOLUTION** The mole fraction of air in lake water is to be determined.
**Assumptions** Both the air and vapor are ideal gases.
**Properties** The saturation pressure of water at 17°C is 1.96 kPa (Table A–4). The Henry's constant for air dissolved in water at 290 K is $H = 62{,}000$ bar (Table 16–2).
**Analysis** This example is similar to the previous example. Again the air at the water surface is saturated, and thus the partial pressure of water vapor in the air at the lake surface is the saturation pressure of water at 17°C,

$$P_v = P_{\text{sat @ 17°C}} = 1.96 \text{ kPa}$$

The partial pressure of dry air is

$$P_{\text{dry air}} = P - P_v = 92 - 1.96 = 90.04 \text{ kPa} = 0.9004 \text{ bar}$$

Note that we could have ignored the vapor pressure since the amount of vapor in air is so small with little loss in accuracy (an error of about 2 percent). The mole fraction of air in the water is, from Henry's law,

$$y_{\text{dry air,liquid side}} = \frac{P_{\text{dry air,gas side}}}{H} = \frac{0.9004 \text{ bar}}{62{,}000 \text{ bar}} = 1.45 \times 10^{-5}$$

*Discussion* This value is very small, as expected. Therefore, the concentration of air in water just below the air–water interface is 1.45 moles per 100,000 moles. But obviously this is enough oxygen for fish and other creatures in the lake. Note that the amount of air dissolved in water will decrease with increasing depth unless phase equilibrium exists throughout the entire lake.

### EXAMPLE 16–10    Diffusion of Hydrogen Gas into a Nickel Plate

Consider a nickel plate that is placed into a tank filled with hydrogen gas at 358 K and 300 kPa. Determine the molar and mass density of hydrogen in the nickel plate when phase equilibrium is established (Fig. 16–26).

**SOLUTION** A nickel plate is exposed to hydrogen gas. The density of hydrogen in the plate is to be determined.
**Properties** The molar mass of hydrogen $H_2$ is $M = 2$ kg/kmol, and the solubility of hydrogen in nickel at the specified temperature is given in Table 16–3 to be 0.00901 kmol/m³·bar.
**Analysis** Noting that 300 kPa = 3 bar, the molar density of hydrogen in the nickel plate is determined from Eq. 16–24 to be

$$\bar{\rho}_{H_2,\text{solid side}} = \mathscr{S} \times P_{H_2,\text{gas side}}$$

$$= (0.00901 \text{ kmol/m}^3 \cdot \text{bar})(3 \text{ bar}) = \mathbf{0.027 \text{ kmol/m}^3}$$

It corresponds to a mass density of

$$\rho_{H_2,\text{solid side}} = \bar{\rho}_{H_2,\text{solid side}} M_{H_2}$$

$$= (0.027 \text{ kmol/m}^3)(2 \text{ kg/kmol}) = \mathbf{0.054 \text{ kg/m}^3}$$

That is, there will be 0.027 kmol (or 0.054 kg) of $H_2$ gas in each m³ volume of nickel plate when phase equilibrium is established.

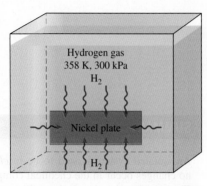

**FIGURE 16–26**
Schematic for Example 16–10.

### EXAMPLE 16–11    Composition of Different Phases of a Mixture

In absorption refrigeration systems, a two-phase equilibrium mixture of liquid ammonia ($NH_3$) and water ($H_2O$) is frequently used. Consider one such mixture at 40°C, shown in Fig. 16–27. If the composition of the liquid phase is 70 percent $NH_3$ and 30 percent $H_2O$ by mole numbers, determine the composition of the vapor phase of this mixture.

**SOLUTION** A two-phase mixture of ammonia and water at a specified temperature is considered. The composition of the liquid phase is given, and the composition of the vapor phase is to be determined.
**Assumptions** The mixture is ideal and thus Raoult's law is applicable.
**Properties** The saturation pressures of $H_2O$ and $NH_3$ at 40°C are $P_{H_2O,\text{sat}} = 7.3851$ kPa and $P_{NH_3,\text{sat}} = 1554.33$ kPa.
**Analysis** The vapor pressures are determined from

$$P_{H_2O,\text{gas side}} = y_{H_2O,\text{liquid side}} P_{H_2O,\text{sat}}(T) = 0.30(7.3851 \text{ kPa}) = 2.22 \text{ kPa}$$

$$P_{NH_3,\text{gas side}} = y_{NH_3,\text{liquid side}} P_{NH_3,\text{sat}}(T) = 0.70(1554.33 \text{ kPa}) = 1088.03 \text{ kPa}$$

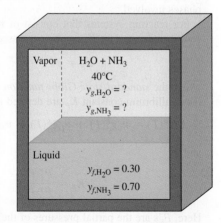

**FIGURE 16–27**
Schematic for Example 16–11.

The total pressure of the mixture is

$$P_{total} = P_{H_2O} + P_{NH_3} = 2.22 + 1088.03 = 1090.25 \text{ kPa}$$

Then, the mole fractions in the gas phase are

$$y_{H_2O, \text{gas side}} = \frac{P_{H_2O, \text{gas side}}}{P_{total}} = \frac{2.22 \text{ kPa}}{1090.25 \text{ kPa}} = 0.0020$$

$$y_{NH_3, \text{gas side}} = \frac{P_{NH_3, \text{gas side}}}{P_{total}} = \frac{1088.03 \text{ kPa}}{1090.25 \text{ kPa}} = 0.9980$$

**Discussion** Note that the gas phase consists almost entirely of ammonia, making this mixture very suitable for absorption refrigeration.

# SUMMARY

An isolated system is said to be in *chemical equilibrium* if no changes occur in the chemical composition of the system. The criterion for chemical equilibrium is based on the second law of thermodynamics, and for a system at a specified temperature and pressure it can be expressed as

$$(dG)_{T,P} = 0$$

For the reaction

$$\nu_A A + \nu_B B \rightleftharpoons \nu_C C + \nu_D D$$

where the $\nu$'s are the stoichiometric coefficients, the *equilibrium criterion* can be expressed in terms of the Gibbs functions as

$$\nu_C \bar{g}_C + \nu_D \bar{g}_D - \nu_A \bar{g}_A - \nu_B \bar{g}_B = 0$$

which is valid for any chemical reaction regardless of the phases involved.

For reacting systems that consist of ideal gases only, the equilibrium constant $K_P$ can be expressed as

$$K_P = e^{-\Delta G^*(T)/R_u T}$$

where the *standard-state Gibbs function change* $\Delta G^*(T)$ and the equilibrium constant $K_P$ are defined as

$$\Delta G^*(T) = \nu_C \bar{g}_C^*(T) + \nu_D \bar{g}_D^*(T) - \nu_A \bar{g}_A^*(T) - \nu_B \bar{g}_B^*(T)$$

and

$$K_P = \frac{P_C^{\nu_C} P_D^{\nu_D}}{P_A^{\nu_A} P_B^{\nu_B}}$$

Here, $P_i$'s are the partial pressures of the components in atm. The $K_P$ of ideal-gas mixtures can also be expressed in terms of the mole numbers of the components as

$$K_P = \frac{N_C^{\nu_C} N_D^{\nu_D}}{N_A^{\nu_A} N_B^{\nu_B}} \left( \frac{P}{N_{total}} \right)^{\Delta \nu}$$

where $\Delta \nu = \nu_C + \nu_D - \nu_A - \nu_B$, $P$ is the total pressure in atm, and $N_{total}$ is the total number of moles present in the reaction chamber, including any inert gases. The equation above is written for a reaction involving two reactants and two products, but it can be extended to reactions involving any number of reactants and products.

The equilibrium constant $K_P$ of ideal-gas mixtures depends on temperature only. It is independent of the pressure of the equilibrium mixture, and it is not affected by the presence of inert gases. The larger the $K_P$, the more complete the reaction. Very small values of $K_P$ indicate that a reaction does not proceed to any appreciable degree. A reaction with $K_P > 1000$ is usually assumed to proceed to completion, and a reaction with $K_P < 0.001$ is assumed not to occur at all. The mixture pressure affects the equilibrium composition, although it does not affect the equilibrium constant $K_P$.

The variation of $K_P$ with temperature is expressed in terms of other thermochemical properties through the *van't Hoff equation*

$$\frac{d(\ln K_P)}{dT} = \frac{\bar{h}_R(T)}{R_u T^2}$$

where $\bar{h}_R(T)$ is the enthalpy of reaction at temperature $T$. For small temperature intervals, it can be integrated to yield

$$\ln \frac{K_{P_2}}{K_{P_1}} \cong \frac{\bar{h}_R}{R_u} \left( \frac{1}{T_1} - \frac{1}{T_2} \right)$$

This equation shows that combustion processes are less complete at higher temperatures since $K_P$ decreases with temperature for exothermic reactions.

Two phases are said to be in *phase equilibrium* when there is no transformation from one phase to the other. Two phases of a pure substance are in equilibrium when each phase has the same value of specific Gibbs function. That is,

$$g_f = g_g$$

In general, the number of independent variables associated with a multicomponent, multiphase system is given by the *Gibbs phase rule*, expressed as

$$IV = C - PH + 2$$

where $IV$ = the number of independent variables, $C$ = the number of components, and $PH$ = the number of phases present in equilibrium.

A multicomponent, multiphase system at a specified temperature and pressure is in phase equilibrium when the specific Gibbs function of each component is the same in all phases.

For a gas $i$ that is weakly soluble in a liquid (such as air in water), the mole fraction of the gas in the liquid

$y_{i,\text{liquid side}}$ is related to the partial pressure of the gas $P_{i,\text{gas side}}$ by Henry's law

$$y_{i,\text{liquid side}} = \frac{P_{i,\text{gas side}}}{H}$$

where $H$ is Henry's constant. When a gas is highly soluble in a liquid (such as ammonia in water), the mole fractions of the species of a two-phase mixture in the liquid and gas phases are given approximately by Raoult's law, expressed as

$$P_{i,\text{gas side}} = y_{i,\text{gas side}} P_{\text{total}} = y_{i,\text{liquid side}} P_{i,\text{sat}}(T)$$

where $P_{\text{total}}$ is the total pressure of the mixture, $P_{i,\text{sat}}(T)$ is the saturation pressure of species $i$ at the mixture temperature, and $y_{i,\text{liquid side}}$ and $y_{i,\text{gas side}}$ are the mole fractions of species $i$ in the liquid and vapor phases, respectively.

## REFERENCES AND SUGGESTED READINGS

1. R. M. Barrer. *Diffusion in and through Solids.* New York: Macmillan, 1941.

2. I. Glassman. *Combustion.* New York: Academic Press, 1977.

3. A. M. Kanury. *Introduction to Combustion Phenomena.* New York: Gordon and Breach, 1975.

4. A. F. Mills. *Basic Heat and Mass Transfer.* Burr Ridge, IL: Richard D. Irwin, 1995.

5. J. M. Smith and H. C. Van Ness. *Introduction to Chemical Engineering Thermodynamics.* 3rd ed. New York: John Wiley & Sons, 1986.

## PROBLEMS*

### $K_P$ and the Equilibrium Composition of Ideal Gases

**16–1C** Write three different $K_P$ relations for reacting ideal-gas mixtures, and state when each relation should be used.

**16–2C** A reaction chamber contains a mixture of $CO_2$, CO, and $O_2$ in equilibrium at a specified temperature and pressure. How will (a) increasing the temperature at constant pressure and (b) increasing the pressure at constant temperature affect the number of moles of $CO_2$?

**16–3C** A reaction chamber contains a mixture of $N_2$ and N in equilibrium at a specified temperature and pressure. How

will (a) increasing the temperature at constant pressure and (b) increasing the pressure at constant temperature affect the number of moles of $N_2$?

**16–4C** A reaction chamber contains a mixture of $CO_2$, CO, and $O_2$ in equilibrium at a specified temperature and pressure. Now some $N_2$ is added to the mixture while the mixture temperature and pressure are kept constant. Will this affect the number of moles of $O_2$? How?

**16–5C** Which element is more likely to dissociate into its monatomic form at 3000 K, $H_2$ or $N_2$? Why?

**16–6C** The equilibrium constant for the $H_2 + \frac{1}{2}O_2 \rightleftharpoons H_2O$ reaction at 1 atm and 1200 K is $K_P$. Use this information to determine the equilibrium constant for the following reactions:

(a) at 1 atm      $H_2 + \frac{1}{2}O_2 \rightleftharpoons H_2O$

(b) at 7 atm      $H_2 + \frac{1}{2}O_2 \rightleftharpoons H_2O$

(c) at 1 atm      $3H_2O \rightleftharpoons 3H_2 + \frac{3}{2}O_2$

(d) at 12 atm      $3H_2O \rightleftharpoons 3H_2 + \frac{3}{2}O_2$

*Problems designated by a "C" are concept questions, and students are encouraged to answer them all. Problems with the ⚙ icon are solved using EES, and complete solutions together with parametric studies are included on the text website. Problems with the 🖥 icon are comprehensive in nature, and are intended to be solved with an equation solver such as EES.

**16–7C** The equilibrium constant of the dissociation reaction $H_2 \rightarrow 2H$ at 3000 K and 1 atm is $K_{P_1}$. Express the equilibrium constants of the following reactions at 3000 K in terms of $K_{P_1}$:

(a) $\qquad H_2 \rightleftharpoons 2H \qquad$ at 2 atm

(b) $\qquad 2H \rightleftharpoons H_2 \qquad$ at 1 atm

(c) $\qquad 2H_2 \rightleftharpoons 4H \qquad$ at 1 atm

(d) $H_2 + 2N_2 \rightleftharpoons 2H + 2N_2 \qquad$ at 2 atm

(e) $\qquad 6H \rightleftharpoons 3H_2 \qquad$ at 4 atm

**16–8C** The equilibrium constant of the reaction $CO + \frac{1}{2}O_2 \rightarrow CO_2$ at 1000 K and 1 atm is $K_{P_1}$. Express the equilibrium constant of the following reactions at 1000 K in terms of $K_{P_1}$:

(a) $\qquad CO + \frac{1}{2}O_2 \rightleftharpoons CO_2 \qquad$ at 3 atm

(b) $\qquad CO_2 \rightleftharpoons CO + \frac{1}{2}O_2 \qquad$ at 1 atm

(c) $\qquad CO + O_2 \rightleftharpoons CO_2 + \frac{1}{2}O_2 \qquad$ at 1 atm

(d) $CO + 2O_2 + 5N_2 \rightleftharpoons CO_2 + 1.5O_2 + 5N_2 \qquad$ at 4 atm

(e) $\qquad 2CO + O_2 \rightleftharpoons 2CO_2 \qquad$ at 1 atm

**16–9C** Consider a mixture of $CO_2$, CO, and $O_2$ in equilibrium at a specified temperature and pressure. Now the pressure is doubled.

(a) Will the equilibrium constant $K_P$ change?

(b) Will the number of moles of $CO_2$, CO, and $O_2$ change? How?

**16–10** A mixture of ideal gases is made up of 30 percent $N_2$, 30 percent $O_2$, and 40 percent $H_2O$ by mole fraction. Determine the Gibbs function of the $N_2$ when the mixture pressure is 5 atm, and its temperature is 600 K.

30% $N_2$
30% $O_2$
40% $H_2O$

5 atm
600 K

**FIGURE P16–10**

**16–11** At what temperature will nitrogen be 0.2 percent dissociated at (a) 1 kPa and (b) 10 kPa? *Answers:* (a) 3628 K, (b) 3909 K

**16–12** Determine the temperature at which 5 percent of diatomic oxygen ($O_2$) dissociates into monatomic oxygen (O) at a pressure of 3 atm. *Answer:* 3133 K

**16–13** Repeat Prob. 16–12 for a pressure of 6 atm.

**16–14** Using the Gibbs function data, determine the equilibrium constant KP for the reaction $H_2O \rightleftharpoons \frac{1}{2}H_2 + OH$ at 25°C. Compare your result with the KP value listed in Table A–28.

**16–15** Use the Gibbs function to determine the equilibrium constant of the $H_2O \rightleftharpoons H_2 + \frac{1}{2}O_2$ reaction at (a) 1000 K and (b) 2000 K. How do these compare to the equilibrium constants of Table A–28?

**16–16** Carbon dioxide is commonly produced through the reaction $C + O_2 \rightleftharpoons CO_2$. Determine the yield of carbon dioxide (mole fraction) when this is done in a reactor maintained at 1 atm and 3800 K. The natural logarithm of the equilibrium constant for the reaction $C + O_2 \rightleftharpoons CO_2$ at 3800 K is $-0.461$. *Answer: 0.122*

**16–17** A gaseous mixture of 30 percent (by mole fraction) methane and 70 percent carbon dioxide is heated at 1 atm pressure to 1200 K. What is the equilibrium composition (by mole fraction) of the resulting mixture? The natural logarithm of the equilibrium constant for the reaction $C + 2H_2 \rightleftharpoons CH_4$ at 1200 K is 4.147. *Answers: 0.000415 ($CH_4$), 0.187 (C), 0.375 ($H_2$), 0.438 ($CO_2$)*

**16–18** Determine the composition of the products of the disassociation reaction $CO_2 \rightleftharpoons CO + O$ when the products are at 1 atm and 2500 K. *Note:* First evaluate the $K_P$ of this reaction using the $K_P$ values of the reactions $CO_2 \rightleftharpoons CO + \frac{1}{2}O_2$ and $0.5O_2 \rightleftharpoons O$.

**16–19** The reaction $N_2 + O_2 \rightleftharpoons 2NO$ occurs in internal combustion engines. Determine the equilibrium mole fraction of NO when the pressure is 101 kPa and the temperature is 1800 K.

**16–20** Determine the equilibrium constant $K_P$ for the process $CO + \frac{1}{2}O_2 = CO_2$ at (a) 298 K and (b) 2000 K. Compare your results with the values for $K_P$ listed in Table A–28.

**16–21** Study the effect of varying the percent excess air during the steady-flow combustion of hydrogen at a pressure of 1 atm. At what temperature will 97 percent of $H_2$ burn into $H_2O$? Assume the equilibrium mixture consists of $H_2O$, $H_2$, $O_2$, and $N_2$.

**16–22** Determine the equilibrium constant $K_P$ for the reaction $CH_4 + 2O_2 \rightleftharpoons CO_2 + 2H_2O$ at 25°C. *Answer: $1.96 \times 10^{140}$*

**16–23** Using the Gibbs function data, determine the equilibrium constant $K_P$ for the dissociation process $CO_2 \rightleftharpoons CO + \frac{1}{2}O_2$ at (a) 298 K and (b) 1800 K. Compare your results with the $K_P$ values listed in Table A–28.

**16–24** Carbon monoxide is burned with 100 percent excess air during a steady-flow process at a pressure of 1 atm. At what temperature will 97 percent of CO burn to $CO_2$? Assume the equilibrium mixture consists of $CO_2$, CO, $O_2$, and $N_2$. *Answer: 2276 K*

**16–25** Reconsider Prob. 16–24. Using EES (or other) software, study the effect of varying the percent excess air during the steady-flow process from 0 to 200 percent on the temperature at which 97 percent of CO burns into $CO_2$. Plot the temperature against the percent excess air, and discuss the results.

**16–26** Air (79 percent $N_2$ and 21 percent $O_2$) is heated to 2000 K at a constant pressure of 2 atm. Assuming the equilibrium mixture consists of $N_2$, $O_2$, and NO, determine the equilibrium composition at this state. Is it realistic to assume that no monatomic oxygen or nitrogen will be present in the equilibrium mixture? Will the equilibrium composition change if the pressure is doubled at constant temperature?

**16–27** Hydrogen ($H_2$) is heated to 3800 K at a constant pressure of 5 atm. Determine the percentage of $H_2$ that will dissociate into H during this process. *Answer 24.0 percent*

**16–28** Carbon dioxide ($CO_2$) is heated to 2400 K at a constant pressure of 3 atm. Determine the percentage of $CO_2$ that will dissociate into CO and $O_2$ during this process.

**16–29** A mixture of 1 mol of CO and 3 mol of $O_2$ is heated to 2200 K at a pressure of 2 atm. Determine the equilibrium composition, assuming the mixture consists of $CO_2$, CO, and $O_2$. *Answers:* $0.995CO_2$, $0.005CO$, $2.5025O_2$

**16–30** A mixture of 3 mol of $N_2$, 1 mol of $O_2$, and 0.1 mol of Ar is heated to 2400 K at a constant pressure of 10 atm. Assuming the equilibrium mixture consists of $N_2$, $O_2$, Ar, and NO, determine the equilibrium composition. *Answers:* $0.0823NO$, $2.9589N_2$, $0.9589O_2$, $0.1Ar$

**16–31** Determine the mole fraction of sodium that ionizes according to the reaction $Na \rightleftharpoons Na^+ + e^-$ at 2000 K and 0.8 atm ($K_P = 0.668$ for this reaction). *Answer: 67.5 percent*

**16–32** A steady-flow combustion chamber is supplied with CO gas at 310 K and 110 kPa at a rate of 0.36 $m^3/min$ and with oxygen ($O_2$) at 298 K and 110 kPa at a rate of 0.3 kg/min. The combustion products leave the combustion chamber at 2000 K and 110 kPa. If the combustion gases consist of $CO_2$, CO, and $O_2$, determine (a) the equilibrium composition of the product gases and (b) the rate of heat transfer from the combustion chamber.

**16–33** Liquid propane ($C_3H_8$) enters a combustion chamber at 25°C at a rate of 1.2 kg/min where it is mixed and burned with 150 percent excess air that enters the combustion chamber at 12°C. If the combustion gases consist of $CO_2$, $H_2O$, CO, $O_2$, and $N_2$ that exit at 1200 K and 2 atm, determine (a) the equilibrium composition of the product gases and (b) the rate of heat transfer from the combustion chamber. Is it realistic to disregard the presence of NO in the product gases? *Answers:* (a) $3CO_2$, $7.5O_2$, $4H_2O$, $47N_2$, (b) 5066 kJ/min

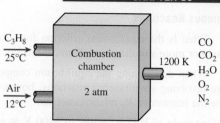

**FIGURE P16–33**

**16–34** Reconsider Prob. 16–33. Using EES (or other) software, investigate if it is realistic to disregard the presence of NO in the product gases?

**16–35** Oxygen ($O_2$) is heated during a steady-flow process at 1 atm from 298 to 3000 K at a rate of 0.5 kg/min. Determine the rate of heat supply needed during this process, assuming (a) some $O_2$ dissociates into O and (b) no dissociation takes place.

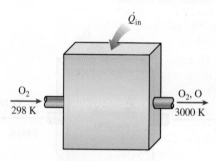

**FIGURE P16–35**

**16–36** Estimate $K_P$ for the following equilibrium reaction at 2500 K:

$$CO + H_2O = CO_2 + H_2$$

At 2000 K it is known that the enthalpy of reaction is $-26{,}176$ kJ/kmol and $K_P$ is 0.2209. Compare your result with the value obtained from the definition of the equilibrium constant.

**16–37** A constant-volume tank contains a mixture of 1 kmol $H_2$ and 1 kmol $O_2$ at 25°C and 1 atm. The contents are ignited. Determine the final temperature and pressure in the tank when the combustion gases are $H_2O$, $H_2$, and $O_2$.

**16–38** Show that as long as the extent of the reaction, $\alpha$, for the disassociation reaction $X_2 \rightleftharpoons 2X$ is smaller than one, $\alpha$ is given by

$$\alpha = \sqrt{\frac{K_P}{4 + K_P}}$$

## Simultaneous Reactions

**16–39C** What is the equilibrium criterion for systems that involve two or more simultaneous chemical reactions?

**16–40C** When determining the equilibrium composition of a mixture involving simultaneous reactions, how would you determine the number of $K_p$ relations needed?

**16–41** One mole of $H_2O$ is heated to 3400 K at a pressure of 1 atm. Determine the equilibrium composition, assuming that only $H_2O$, OH, $O_2$, and $H_2$ are present. *Answers:* $0.574H_2O$, $0.308H_2$, $0.095O_2$, $0.236OH$

**16–42** A mixture of 2 mol of $CO_2$ and 1 mol of $O_2$ is heated to 3200 K at a pressure of 2 atm. Determine the equilibrium composition of the mixture, assuming that only $CO_2$, CO, $O_2$, and O are present.

**16–43** Air (21 percent $O_2$, 79 percent $N_2$) is heated to 3000 K at a pressure of 2 atm. Determine the equilibrium composition, assuming that only $O_2$, $N_2$, O, and NO are present. Is it realistic to assume that no N will be present in the final equilibrium mixture?

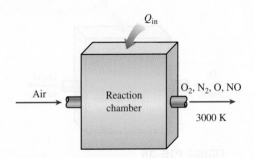

**FIGURE P16–43**

**16–44** Air (21 percent $O_2$, 79 percent $N_2$) is heated to 3000 K at a pressure of 1 atm. Determine the equilibrium composition, assuming that only $O_2$, $N_2$, O, and NO are present. Is it realistic to assume that no N will be present in the final equilibrium mixture?

**16–45** Water vapor ($H_2O$) is heated during a steady-flow process at 1 atm from 298 to 3000 K at a rate of 0.2 kg/min. Determine the rate of heat supply needed during this process, assuming (a) some $H_2O$ dissociates into $H_2$, $O_2$, and OH and (b) no dissociation takes place. *Answers:* (a) 2056 kJ/min, (b) 1404 kJ/min

**16–46** Reconsider Prob. 16–45. Using EES (or other) software, study the effect of the pressure on the rate of heat supplied for the two cases. Let the pressure vary from 1 to 10 atm. For each of the two cases, plot the rate of heat supplied as a function of pressure.

**16–47** Ethyl alcohol ($C_2H_5OH(g)$) at 25°C is burned in a steady-flow adiabatic combustion chamber with 40 percent excess air that also enters at 25°C. Determine the adiabatic flame temperature of the products at 1 atm assuming the significant equilibrium reactions are $CO_2 = CO + \frac{1}{2}O_2$ and $\frac{1}{2}N_2 + \frac{1}{2}O_2 = NO$. Plot the adiabatic flame temperature and kmoles of $CO_2$, CO, and NO at equilibrium for values of percent excess air between 10 and 100 percent.

## Variations of $K_p$ with Temperature

**16–48C** What is the importance of the van't Hoff equation?

**16–49C** Will a fuel burn more completely at 2000 or 2500 K?

**16–50** Estimate the enthalpy of reaction $\bar{h}_R$ for the dissociation process $O_2 \rightleftharpoons 2O$ at 3100 K, using (a) enthalpy data and (b) $K_p$ data. *Answers:* (a) 513,614 kJ/kmol, (b) 512,808 kJ/kmol

**16–51** Estimate the enthalpy of reaction $\bar{h}_R$ for the combustion process of carbon monoxide at 1800 K, using (a) enthalpy data and (b) $K_p$ data.

**16–52** Using the enthalpy of reaction $\bar{h}_R$ data and the $K_p$ value at 2400 K, estimate the $K_p$ value of the combustion process $H_2 + \frac{1}{2}O_2 \rightleftharpoons H_2O$ at 2600 K. *Answer:* 104.1

**16–53** Estimate the enthalpy of reaction $\bar{h}_R$ for the dissociation process $CO_2 \rightleftharpoons CO + \frac{1}{2}O_2$ at 2200 K, using (a) enthalpy data and (b) $K_p$ data.

**16–54** Estimate the enthalpy of reaction for the equilibrium reaction $CH_4 + 2O_2 \rightleftharpoons CO_2 + 2H_2O$ at 2500 K, using (a) enthalpy data and (b) $K_p$ data. Obtain enthalpy and entropy properties from EES.

## Phase Equilibrium

**16–55C** Consider a tank that contains a saturated liquid–vapor mixture of water in equilibrium. Some vapor is now allowed to escape the tank at constant temperature and pressure. Will this disturb the phase equilibrium and cause some of the liquid to evaporate?

**16–56C** Consider a two-phase mixture of ammonia and water in equilibrium. Can this mixture exist in two phases at the same temperature but at a different pressure?

**16–57C** Using the solubility data of a solid in a specified liquid, explain how you would determine the mole fraction of the solid in the liquid at the interface at a specified temperature.

**16–58C** Using solubility data of a gas in a solid, explain how you would determine the molar concentration of the gas in the solid at the solid–gas interface at a specified temperature.

**16–59C** Using the Henry's constant data for a gas dissolved in a liquid, explain how you would determine the mole fraction of the gas dissolved in the liquid at the interface at a specified temperature.

**16–60** Air at 20°C and 700 kPa is blown through a porous media which is saturated with liquid water at 20°C. Determine the maximum partial pressure of the water evaporated into the air as it emerges from the porous media.

**16–61** Water is sprayed into air at 27°C and 100 kPa, and the falling water droplets are collected in a container on the floor. Determine the mass and mole fractions of air dissolved in the water.

**16–62** Show that a saturated liquid–vapor mixture of refrigerant-134a at −10°C satisfies the criterion for phase equilibrium.

**16–63** Show that a mixture of saturated liquid water and saturated water vapor at 300 kPa satisfies the criterion for phase equilibrium.

**16–64** A liquid-vapor mixture of refrigerant-134a is at 280 kPa with a quality of 70 percent. Determine the value of the Gibbs function, in kJ/kg, when the two phases are in equilibrium.

R-134a
280 kPa
x = 0.7

**FIGURE P16–64**

**16–65** Calculate the value of the Gibbs function for saturated steam at 150°C as a saturated liquid, saturated vapor, and a mixture of liquid and vapor with a quality of 60 percent. Demonstrate that phase equilibrium exists.

**16–66** An ammonia-water mixture is at 10°C. Determine the pressure of the ammonia vapor when the mole fraction of the ammonia in the liquid is (a) 20 percent and (b) 80 percent. The saturation pressure of ammonia at 10°C is 615.3 kPa.

**16–67** Using the liquid–vapor equilibrium diagram of an oxygen–nitrogen mixture, determine the composition of each phase at 84 K and 100 kPa.

**16–68** Using the liquid–vapor equilibrium diagram of an oxygen–nitrogen mixture at 100 kPa, determine the temperature at which the composition of the vapor phase is 79 percent $N_2$ and 21 percent $O_2$. *Answer:* 82 K

**16–69** An oxygen-nitrogen mixture consists of 30 kg of oxygen and 40 kg of nitrogen. This mixture is cooled to 84 K at 0.1 MPa pressure. Determine the mass of the oxygen in the liquid and gaseous phase. *Answers:* 8.28 kg, 21.7 kg

**16–70** What is the total mass of the liquid phase of Prob. 16–69. *Answer:* 11.4 kg

**16–71** A wall made of natural rubber separates $O_2$ and $N_2$ gases at 25°C and 500 kPa. Determine the molar concentrations of $O_2$ and $N_2$ in the wall.

**16–72** Consider a rubber plate that is in contact with nitrogen gas at 298 K and 250 kPa. Determine the molar and mass density of nitrogen in the rubber at the interface.

**16–73** In absorption refrigeration systems, a two-phase equilibrium mixture of liquid ammonia ($NH_3$) and water ($H_2O$) is frequently used. Consider a liquid–vapor mixture of ammonia and water in equilibrium at 30°C. If the composition of the liquid phase is 60 percent $NH_3$ and 40 percent $H_2O$ by mole numbers, determine the composition of the vapor phase of this mixture. Saturation pressure of $NH_3$ at 30°C is 1167.4 kPa.

**16–74** An ammonia-water absorption refrigeration unit operates its absorber at 0°C and its generator at 46°C. The vapor mixture in the generator and absorber is to have an ammonia mole fraction of 96 percent. Assuming ideal behavior, determine the operating pressure in the (a) generator and (b) absorber. Also determine the mole fraction of the ammonia in the (c) strong liquid mixture being pumped from the absorber and the (d) weak liquid solution being drained from the generator. The saturation pressure of ammonia at 0°C is 430.6 kPa, and at 46°C it is 1830.2 kPa. *Answers:* (a) 223 kPa, (b) 14.8 kPa, (c) 0.033, (d) 0.117

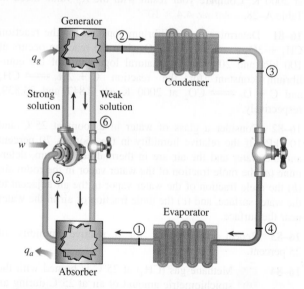

**FIGURE P16–74**

**16–75** Rework Prob. 16–74 when the temperature in the absorber is increased to 6°C and the temperature in the generator is reduced to 40°C. The saturation pressure of ammonia at 6°C is 534.8 kPa, and at 40°C it is 1556.7 kPa.

**16–76** Foam products such as shaving cream are made by liquid mixtures whose ingredients are primarily water and a refrigerant such as refrigerant-134a. Consider a liquid mixture of water and refrigerant-134a with a water mass fraction of 90 percent that is at 20°C. What is the mole fraction of the water and refrigerant-134a vapor in the gas which fills the bubbles that form the foam? *Answers:* 0.173 ($H_2O$), 0.827 (R-134a)

**16–77** Consider a glass of water in a room at 27°C and 97 kPa. If the relative humidity in the room is 100 percent and the water and the air are in thermal and phase equilibrium, determine (a) the mole fraction of the water vapor in the air and (b) the mole fraction of air in the water.

**16–78** Consider a carbonated drink in a bottle at 27°C and 115 kPa. Assuming the gas space above the liquid consists of a saturated mixture of $CO_2$ and water vapor and treating the drink as water, determine (a) the mole fraction of the water vapor in the $CO_2$ gas and (b) the mass of dissolved $CO_2$ in a 300-ml drink.

## Review Problems

**16–79** Determine the mole fraction of argon that ionizes according to the reaction $Ar \rightleftharpoons Ar^+ + e^-$ at 10,000 K and 0.35 atm ($K_P = 0.00042$ for this reaction).

**16–80** Using the Gibbs function data, determine the equilibrium constant $K_P$ for the dissociation process $O_2 \rightleftharpoons 2O$ at 2000 K. Compare your result with the $K_P$ value listed in Table A–28. *Answer:* $4.4 \times 10^{-7}$

**16–81** Determine the equilibrium constant for the reaction $CH_4 + 2O_2 \rightleftharpoons CO_2 + 2H_2O$ when the reaction occurs at 100 kPa and 2000 K. The natural logarithms of the equilibrium constant for the reaction $C + 2H_2 \rightleftharpoons CH_4$ and $C + O_2 \rightleftharpoons CO_2$ at 2000 K are 7.847 and 23.839, respectively.

**16–82** Consider a glass of water in a room at 25°C and 100 kPa. If the relative humidity in the room is 70 percent and the water and the air are in thermal equilibrium, determine (a) the mole fraction of the water vapor in the room air, (b) the mole fraction of the water vapor in the air adjacent to the water surface, and (c) the mole fraction of air in the water near the surface.

**16–83** Repeat Prob. 16–82 for a relative humidity of 25 percent.

**16–84** Methane gas ($CH_4$) at 25°C is burned with the stoichiometric amount of air at 25°C during an adiabatic steady-flow combustion process at 1 atm. Assuming the product gases consist of $CO_2$, $H_2O$, CO, $N_2$, and $O_2$, determine (a) the equilibrium composition of the product gases and (b) the exit temperature.

**16–85** Reconsider Prob. 16–84. Using EES (or other) software, study the effect of excess air on the equilibrium composition and the exit temperature by varying the percent excess air from 0 to 200 percent. Plot the exit temperature against the percent excess air, and discuss the results.

**16–86** Consider the reaction $CH_4 + 2O_2 \rightleftharpoons CO_2 + 2H_2O$ when the reaction occurs at 450 kPa and 3000 K. Determine the equilibrium partial pressure of the carbon dioxide. The natural logarithms of the equilibrium constant for the reactions $C + 2H_2 \rightleftharpoons CH_4$ and $C + O_2 \rightleftharpoons CO_2$ at 3000 K are 9.685 and 15.869, respectively. *Answer:* 148 kPa

**16–87** 10 kmol of methane gas are heated from 1 atm and 298 K to 1 atm and 1000 K. Calculate the total amount of heat transfer required when (a) disassociation is neglected and (b) when disassociation is considered. The natural logarithm of the equilibrium constant for the reaction $C + 2H_2 \rightleftharpoons CH_4$ at 1000 K is 2.328. For the solution of part (a) use empirical coefficients of Table A–2c. For the solution of part (b) use constant specific heats and take the constant-volume specific heats of methane, hydrogen and carbon at 1000 K to be 63.3, 21.7, and 0.711 kJ/kmol·K, respectively. The constant-volume specific heat of methane at 298 K is 27.8 kJ/kmol·K.

**16–88** Solid carbon at 25°C is burned with a stoichiometric amount of air which is at 1 atm pressure and 25°C. Determine the number of moles of $CO_2$ formed per kmol of carbon when only $CO_2$, CO, $O_2$, and $N_2$ are present in the products and the products are at 1 atm and 727°C.

**16–89** Determine the amount of heat released per kilogram of carbon by the combustion of the Prob. 16–88. *Answer:* 23,285 kJ/kg carbon

**16–90** Methane gas is burned with 30 percent excess air. This fuel enters a steady flow combustor at 101 kPa and 25°C, and is mixed with the air. The products of combustion leave this reactor at 101 kPa and 1600 K. Determine the equilibrium composition of the products of combustion, and the amount of heat released by this combustion, in kJ/kmol methane.

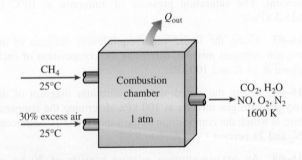

**FIGURE P16–90**

**16–91** Gaseous octane is burned with 40 percent excess air in an automobile engine. During combustion, the pressure is 4 MPa and the temperature reaches 2000 K. Determine the equilibrium composition of the products of combustion.

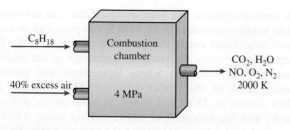

**FIGURE P16–91**

**16–92** Propane gas is burned steadily at 1 atm pressure with a 10 percent excess oxygen supplied by atmospheric air. The reactants enter a steady flow combustor at 25°C. Determine the final temperature of the products if the combustion is done without any heat transfer, and the equilibrium composition. *Answers:* 3 ($CO_2$), 4 ($H_2O$), 0.470 ($O_2$), 0.0611 (NO), 20.7 ($N_2$), 2034°C

**16–93** A constant-volume tank contains a mixture of 1 mol of $H_2$ and 0.5 mol of $O_2$ at 25°C and 1 atm. The contents of the tank are ignited, and the final temperature and pressure in the tank are 2800 K and 5 atm, respectively. If the combustion gases consist of $H_2O$, $H_2$, and $O_2$, determine (*a*) the equilibrium composition of the product gases and (*b*) the amount of heat transfer from the combustion chamber. Is it realistic to assume that no OH will be present in the equilibrium mixture? *Answers:* (*a*) 0.944$H_2O$, 0.056$H_2$, 0.028$O_2$, (*b*) 132,574 J/mol $H_2$

**16–94** A mixture of 2 mol of $H_2O$ and 3 mol of $O_2$ is heated to 3600 K at a pressure of 8 atm. Determine the equilibrium composition of the mixture, assuming that only $H_2O$, OH, $O_2$, and $H_2$ are present.

**16–95** A mixture of 3 mol of $CO_2$ and 3 mol of $O_2$ is heated to 3400 K at a pressure of 2 atm. Determine the equilibrium composition of the mixture, assuming that only $CO_2$, CO, $O_2$, and O are present. *Answers:* 1.313$CO_2$, 1.687CO, 3.187$O_2$, 1.314O

**16–96** Reconsider Prob. 16–95. Using EES (or other) software, study the effect of pressure on the equilibrium composition by varying pressure from 1 atm to 10 atm. Plot the amount of CO present at equilibrium as a function of pressure.

**16–97** Estimate the enthalpy of reaction $\bar{h}_R$ for the combustion process of hydrogen at 2400 K, using (*a*) enthalpy data and (*b*) $K_P$ data. *Answers:* (*a*) −252,377 kJ/kmol, (*b*) −252,047 kJ/kmol

**16–98** Reconsider Prob. 16–97. Using EES (or other) software, investigate the effect of temperature on the enthalpy of reaction using both methods by varying the temperature from 2000 to 3000 K.

**16–99** Using the enthalpy of reaction $\bar{h}_R$ data and the $K_P$ value at 2200 K, estimate the $K_P$ value of the dissociation process $O_2 \rightleftharpoons 2O$ at 2400 K.

**16–100** A carbonated drink is fully charged with $CO_2$ gas at 17°C and 600 kPa such that the entire bulk of the drink is in thermodynamic equilibrium with the $CO_2$–water vapor mixture. Now consider a 2-L soda bottle. If the $CO_2$ gas in that bottle were to be released and stored in a container at 20°C and 100 kPa, determine the volume of the container.

**16–101** Tabulate the natural log of the equilibrium constant as a function of temperature between 298 to 3000 K for the equilibrium reaction CO + $H_2O$ = $CO_2$ + $H_2$. Compare your results to those obtained by combining the ln $K_P$ values for the two equilibrium reactions $CO_2$ = CO + $\frac{1}{2}O_2$ and $H_2O$ = $H_2$ + $\frac{1}{2}O_2$ given in Table A–28.

**16–102** Ethyl alcohol ($C_2H_5OH(g)$) at 25°C is burned in a steady-flow adiabatic combustion chamber with 90 percent excess air that also enters at 25°C. Determine the adiabatic flame temperature of the products at 1 atm assuming the only significant equilibrium reaction is $CO_2$ = CO + $\frac{1}{2}O_2$. Plot the adiabatic flame temperature as the percent excess air varies from 10 to 100 percent.

**16–103** Show that when the three phases of a pure substance are in equilibrium, the specific Gibbs function of each phase is the same.

**16–104** Show that when the two phases of a two-component system are in equilibrium, the specific Gibbs function of each phase of each component is the same.

## Fundamentals of Engineering (FE) Exam Problems

**16–105** If the equilibrium constant for the reaction $H_2$ + $\frac{1}{2}O_2 \rightleftharpoons H_2O$ is $K$, the equilibrium constant for the reaction $2H_2O \rightleftharpoons 2H_2 + O_2$ at the same temperature is

(*a*) 1/$K$        (*b*) 1/(2$K$)        (*c*) 2$K$

(*d*) $K^2$        (*e*) 1/$K^2$

**16–106** If the equilibrium constant for the reaction CO + $\frac{1}{2}O_2 \rightleftharpoons CO_2$ is $K$, the equilibrium constant for the reaction $CO_2$ + $3N_2 \rightleftharpoons CO + \frac{1}{2}O_2 + 3N_2$ at the same temperature is

(*a*) 1/$K$        (*b*) 1/($K$ + 3)        (*c*) 4$K$

(*d*) $K$        (*e*) 1/$K^2$

**16–107** The equilibrium constant for the reaction $H_2$ + $\frac{1}{2}O_2 \rightleftharpoons H_2O$ at 1 atm and 1500°C is given to be $K$. Of the reactions given below, all at 1500°C, the reaction that has a different equilibrium constant is

(*a*) $H_2$ + $\frac{1}{2}O_2 \rightleftharpoons H_2O$        at 5 atm

(*b*) $2H_2$ + $O_2 \rightleftharpoons 2H_2O$        at 1 atm

(*c*) $H_2$ + $O_2 \rightleftharpoons H_2O + \frac{1}{2}O_2$        at 2 atm

(*d*) $H_2$ + $\frac{1}{2}O_2$ + $3N_2 \rightleftharpoons H_2O + 3N_2$        at 5 atm

(*e*) $H_2$ + $\frac{1}{2}O_2$ + $3N_2 \rightleftharpoons H_2O + 3N_2$        at 1 atm

**16–108** Of the reactions given below, the reaction whose equilibrium composition at a specified temperature is not affected by pressure is

(a) $H_2 + \frac{1}{2}O_2 \rightleftharpoons H_2O$
(b) $CO + \frac{1}{2}O_2 \rightleftharpoons CO_2$
(c) $N_2 + O_2 \rightleftharpoons 2NO$
(d) $N_2 \rightleftharpoons 2N$
(e) all of the above

**16–109** Of the reactions given below, the reaction whose number of moles of products increases by the addition of inert gases into the reaction chamber at constant pressure and temperature is
(a) $H_2 + \frac{1}{2}O_2 \rightleftharpoons H_2O$
(b) $CO + \frac{1}{2}O_2 \rightleftharpoons CO_2$
(c) $N_2 + O_2 \rightleftharpoons 2NO$
(d) $N_2 \rightleftharpoons 2N$
(e) all of the above

**16–110** Moist air is heated to a very high temperature. If the equilibrium composition consists of $H_2O$, $O_2$, $N_2$, OH, $H_2$, and NO, the number of equilibrium constant relations needed to determine the equilibrium composition of the mixture is
(a) 1        (b) 2        (c) 3
(d) 4        (e) 5

**16–111** Propane $C_3H_8$ is burned with air, and the combustion products consist of $CO_2$, CO, $H_2O$, $O_2$, $N_2$, OH, $H_2$, and NO. The number of equilibrium constant relations needed to determine the equilibrium composition of the mixture is
(a) 1        (b) 2        (c) 3
(d) 4        (e) 5

**16–112** Consider a gas mixture that consists of three components. The number of independent variables that need to be specified to fix the state of the mixture is
(a) 1        (b) 2        (c) 3
(d) 4        (e) 5

**16–113** The value of Henry's constant for $CO_2$ gas dissolved in water at 290 K is 12.8 MPa. Consider water exposed to atmospheric air at 100 kPa that contains 3 percent $CO_2$ by volume. Under phase equilibrium conditions, the mole fraction of $CO_2$ gas dissolved in water at 290 K is
(a) $2.3 \times 10^{-4}$    (b) $3.0 \times 10^{-4}$    (c) $0.80 \times 10^{-4}$
(d) $2.2 \times 10^{-4}$    (e) $5.6 \times 10^{-4}$

**16–114** The solubility of nitrogen gas in rubber at 25°C is 0.00156 kmol/m³·bar. When phase equilibrium is established, the density of nitrogen in a rubber piece placed in a nitrogen gas chamber at 300 kPa is
(a) 0.005 kg/m³        (b) 0.018 kg/m³        (c) 0.047 kg/m³
(d) 0.13 kg/m³         (e) 0.28 kg/m³

## Design and Essay Problems

**16–115** An engineer suggested that high-temperature disassociation of water be used to produce a hydrogen fuel.

A reactor-separator has been designed that can accommodate temperatures as high as 4000 K and pressures as much as 5 atm. Water enters this reactor-separator at 25°C. The separator separates the various constituents in the mixture into individual streams whose temperature and pressure match those of the reactor-separator. These streams are then cooled to 25°C and stored in atmospheric pressure tanks with the exception of any remaining water, which is returned to the reactor to repeat the process again. Hydrogen gas from these tanks is later burned with a stoichiometric amount of air to provide heat for an electrical power plant. The parameter that characterizes this system is the ratio of the heat released by burning the hydrogen to the amount of heat used to generate the hydrogen gas. Select the operating pressure and temperature for the reactor-separator that maximizes this ratio. Can this ratio ever be bigger than unity?

**16–116** An article that appeared in the *Reno Gazette-Journal* on May 18, 1992, quotes an inventor as saying that he has turned water into motor vehicle fuel in a breakthrough that would increase engine efficiency, save gasoline, and reduce smog. There is also a picture of a car that the inventor has modified to run on half water and half gasoline. The inventor claims that sparks from catalytic poles in the converted engine break down the water into oxygen and hydrogen, which is burned with the gasoline. He adds that hydrogen has a higher energy density than carbon and the high-energy density enables one to get more power. The inventor states that the fuel efficiency of his car increased from 20 mpg (miles per gallon) to more than 50 mpg of gasoline as a result of conversion and notes that the conversion has sharply reduced emissions of hydrocarbons, carbon monoxide, and other exhaust pollutants.

Evaluate the claims made by the inventor, and write a report that is to be submitted to a group of investors who are considering financing this invention.

**16–117** One means of producing liquid oxygen from atmospheric air is to take advantage of the phase-equilibrium properties of oxygen-nitrogen mixtures. This system is illustrated in Fig. P16–117. In this cascaded-reactors system, dry atmospheric air is cooled in the first reactor until liquid is formed. According to the phase-equilibrium properties, this liquid will be richer in oxygen than in the vapor phase. The vapor in the first reactor is discarded while the oxygen enriched liquid leaves the first reactor and is heated in a heat exchanger until it is again a vapor. The vapor mixture enters the second reactor where it is again cooled until a liquid that is further enriched in oxygen is formed. The vapor from the second reactor is routed back to the first reactor while the liquid is routed to another heat exchanger and another reactor to repeat the process once again. The liquid formed in the third reactor is very rich in oxygen. If all three reactors are operated at 1 atm pressure, select the three temperatures that produce the greatest amount of 99 percent pure oxygen.

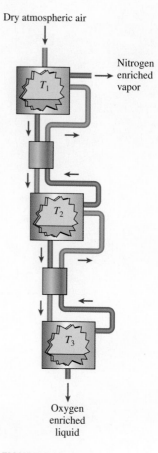

Dry atmospheric air

Nitrogen enriched vapor

$T_1$

$T_2$

$T_3$

Oxygen enriched liquid

**FIGURE P16–117**

**16–118** Automobiles are major emitters of air pollutants such as $NO_x$, CO, and hydrocarbons HC. Find out the legal limits of these pollutants in your area, and estimate the total amount of each pollutant, in kg, that would be produced in your town if all the cars were emitting pollutants at the legal limit. State your assumptions.

# COMPRESSIBLE FLOW

For the most part, we have limited our consideration so far to flows for which density variations and thus compressibility effects are negligible. In this chapter we lift this limitation and consider flows that involve significant changes in density. Such flows are called *compressible flows,* and they are frequently encountered in devices that involve the flow of gases at very high velocities. Compressible flow combines fluid dynamics and thermodynamics in that both are necessary to the development of the required theoretical background. In this chapter, we develop the general relations associated with one-dimensional compressible flows for an ideal gas with constant specific heats.

We start this chapter by introducing the concepts of *stagnation state, speed of sound,* and *Mach number* for compressible flows. The relationships between the static and stagnation fluid properties are developed for isentropic flows of ideal gases, and they are expressed as functions of specific-heat ratios and the Mach number. The effects of area changes for one-dimensional isentropic subsonic and supersonic flows are discussed. These effects are illustrated by considering the isentropic flow through *converging* and *converging–diverging nozzles.* The concept of *shock waves* and the variation of flow properties across normal and oblique shocks are discussed. Finally, we consider the effects of heat transfer on compressible flows and examine steam nozzles.

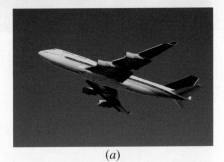

(a)

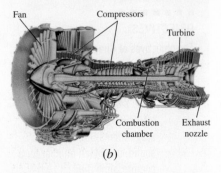

Fan    Compressors
Turbine
Combustion    Exhaust
chamber    nozzle

(b)

**FIGURE 17–1**
Aircraft and jet engines involve high speeds, and thus the kinetic energy term should always be considered when analyzing them.

*(a) © Royalty-Free/Corbis; (b) Reproduced by permission of United Technologies Corporation, Pratt & Whitney.*

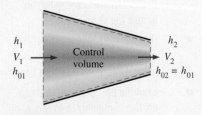

$h_1$
$V_1$
$h_{01}$

Control
volume

$h_2$
$V_2$
$h_{02} = h_{01}$

**FIGURE 17–2**
Steady flow of a fluid through an adiabatic duct.

# 17–1 · STAGNATION PROPERTIES

When analyzing control volumes, we find it very convenient to combine the *internal energy* and the *flow energy* of a fluid into a single term, *enthalpy*, defined per unit mass as $h = u + P/\rho$. Whenever the kinetic and potential energies of the fluid are negligible, as is often the case, the enthalpy represents the *total energy* of a fluid. For high-speed flows, such as those encountered in jet engines (Fig. 17–1), the potential energy of the fluid is still negligible, but the kinetic energy is not. In such cases, it is convenient to combine the enthalpy and the kinetic energy of the fluid into a single term called **stagnation** (or **total**) **enthalpy** $h_0$, defined per unit mass as

$$h_0 = h + \frac{V^2}{2} \quad \text{(kJ/kg)} \tag{17–1}$$

When the potential energy of the fluid is negligible, the stagnation enthalpy represents the *total energy of a flowing fluid stream* per unit mass. Thus it simplifies the thermodynamic analysis of high-speed flows.

Throughout this chapter the ordinary enthalpy $h$ is referred to as the **static enthalpy**, whenever necessary, to distinguish it from the stagnation enthalpy. Notice that the stagnation enthalpy is a combination property of a fluid, just like the static enthalpy, and these two enthalpies are identical when the kinetic energy of the fluid is negligible.

Consider the steady flow of a fluid through a duct such as a nozzle, diffuser, or some other flow passage where the flow takes place adiabatically and with no shaft or electrical work, as shown in Fig. 17–2. Assuming the fluid experiences little or no change in its elevation and its potential energy, the energy balance relation ($\dot{E}_{in} = \dot{E}_{out}$) for this single-stream steady-flow device reduces to

$$h_1 + \frac{V_1^2}{2} = h_2 + \frac{V_2^2}{2} \tag{17–2}$$

or

$$h_{01} = h_{02} \tag{17–3}$$

That is, in the absence of any heat and work interactions and any changes in potential energy, the stagnation enthalpy of a fluid remains constant during a steady-flow process. Flows through nozzles and diffusers usually satisfy these conditions, and any increase in fluid velocity in these devices creates an equivalent decrease in the static enthalpy of the fluid.

If the fluid were brought to a complete stop, then the velocity at state 2 would be zero and Eq. 17–2 would become

$$h_1 + \frac{V_1^2}{2} = h_2 = h_{02}$$

Thus the *stagnation enthalpy* represents the *enthalpy of a fluid when it is brought to rest adiabatically*.

During a stagnation process, the kinetic energy of a fluid is converted to enthalpy (internal energy + flow energy), which results in an increase in the fluid temperature and pressure. The properties of a fluid at the stagnation state are called **stagnation properties** (stagnation temperature, stagnation pressure, stagnation density, etc.). The stagnation state and the stagnation properties are indicated by the subscript 0.

The stagnation state is called the **isentropic stagnation state** when the stagnation process is reversible as well as adiabatic (i.e., isentropic). The entropy of a fluid remains constant during an isentropic stagnation process. The actual (irreversible) and isentropic stagnation processes are shown on an *h-s* diagram in Fig. 17–3. Notice that the stagnation enthalpy of the fluid (and the stagnation temperature if the fluid is an ideal gas) is the same for both cases. However, the actual stagnation pressure is lower than the isentropic stagnation pressure since entropy increases during the actual stagnation process as a result of fluid friction. Many stagnation processes are approximated to be isentropic, and isentropic stagnation properties are simply referred to as stagnation properties.

When the fluid is approximated as an *ideal gas* with constant specific heats, its enthalpy can be replaced by $c_p T$ and Eq. 17–1 is expressed as

$$c_p T_0 = c_p T + \frac{V^2}{2}$$

or

$$T_0 = T + \frac{V^2}{2c_p} \tag{17-4}$$

Here, $T_0$ is called the **stagnation** (or **total**) **temperature**, and it represents *the temperature an ideal gas attains when it is brought to rest adiabatically*. The term $V^2/2c_p$ corresponds to the temperature rise during such a process and is called the **dynamic temperature**. For example, the dynamic temperature of air flowing at 100 m/s is $(100 \text{ m/s})^2/(2 \times 1.005 \text{ kJ/kg·K}) = 5.0 \text{ K}$. Therefore, when air at 300 K and 100 m/s is brought to rest adiabatically (at the tip of a temperature probe, for example), its temperature rises to the stagnation value of 305 K (Fig. 17–4). Note that for low-speed flows, the stagnation and static (or ordinary) temperatures are practically the same. But for high-speed flows, the temperature measured by a stationary probe placed in the fluid (the stagnation temperature) may be significantly higher than the static temperature of the fluid.

The pressure a fluid attains when brought to rest isentropically is called the **stagnation pressure** $P_0$. For ideal gases with constant specific heats, $P_0$ is related to the static pressure of the fluid by

$$\frac{P_0}{P} = \left(\frac{T_0}{T}\right)^{k/(k-1)} \tag{17-5}$$

By noting that $\rho = 1/v$ and using the isentropic relation $Pv^k = P_0 v_0^k$, the ratio of the stagnation density to static density is expressed as

$$\frac{\rho_0}{\rho} = \left(\frac{T_0}{T}\right)^{1/(k-1)} \tag{17-6}$$

When stagnation enthalpies are used, there is no need to refer explicitly to kinetic energy. Then the energy balance $\dot{E}_{in} = \dot{E}_{out}$ for a single-stream, steady-flow device can be expressed as

$$q_{in} + w_{in} + (h_{01} + gz_1) = q_{out} + w_{out} + (h_{02} + gz_2) \tag{17-7}$$

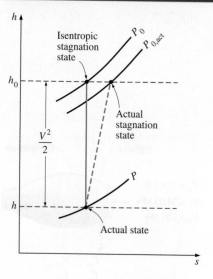

**FIGURE 17–3**
The actual state, actual stagnation state, and isentropic stagnation state of a fluid on an *h-s* diagram.

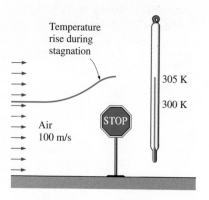

**FIGURE 17–4**
The temperature of an ideal gas flowing at a velocity *V* rises by $V^2/2c_p$ when it is brought to a complete stop.

where $h_{01}$ and $h_{02}$ are the stagnation enthalpies at states 1 and 2, respectively. When the fluid is an ideal gas with constant specific heats, Eq. 17–7 becomes

$$(q_{\text{in}} - q_{\text{out}}) + (w_{\text{in}} - w_{\text{out}}) = c_p(T_{02} - T_{01}) + g(z_2 - z_1) \qquad \text{(17–8)}$$

where $T_{01}$ and $T_{02}$ are the stagnation temperatures.

Notice that kinetic energy terms do not explicitly appear in Eqs. 17–7 and 17–8, but the stagnation enthalpy terms account for their contribution.

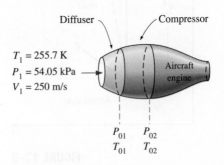

$T_1 = 255.7$ K
$P_1 = 54.05$ kPa
$V_1 = 250$ m/s

$P_{01} \quad P_{02}$
$T_{01} \quad T_{02}$

**FIGURE 17–5**
Schematic for Example 17–1.

**EXAMPLE 17–1** **Compression of High-Speed Air in an Aircraft**

An aircraft is flying at a cruising speed of 250 m/s at an altitude of 5000 m where the atmospheric pressure is 54.05 kPa and the ambient air temperature is 255.7 K. The ambient air is first decelerated in a diffuser before it enters the compressor (Fig. 17–5). Approximating both the diffuser and the compressor to be isentropic, determine (a) the stagnation pressure at the compressor inlet and (b) the required compressor work per unit mass if the stagnation pressure ratio of the compressor is 8.

**SOLUTION** High-speed air enters the diffuser and the compressor of an aircraft. The stagnation pressure of the air and the compressor work input are to be determined.

**Assumptions** 1 Both the diffuser and the compressor are isentropic. 2 Air is an ideal gas with constant specific heats at room temperature.

**Properties** The constant-pressure specific heat $c_p$ and the specific heat ratio $k$ of air at room temperature are

$$c_p = 1.005 \text{ kJ/kg·K} \quad \text{and} \quad k = 1.4$$

**Analysis** (a) Under isentropic conditions, the stagnation pressure at the compressor inlet (diffuser exit) can be determined from Eq. 17–5. However, first we need to find the stagnation temperature $T_{01}$ at the compressor inlet. Under the stated assumptions, $T_{01}$ is determined from Eq. 17–4 to be

$$T_{01} = T_1 + \frac{V_1^2}{2c_p} = 255.7 \text{ K} + \frac{(250 \text{ m/s})^2}{(2)(1.005 \text{ kJ/kg·K})}\left(\frac{1 \text{ kJ/kg}}{1000 \text{ m}^2/\text{s}^2}\right)$$

$$= 286.8 \text{ K}$$

Then from Eq. 17–5,

$$P_{01} = P_1\left(\frac{T_{01}}{T_1}\right)^{k/(k-1)} = (54.05 \text{ kPa})\left(\frac{286.8 \text{ K}}{255.7 \text{ K}}\right)^{1.4/(1.4-1)}$$

$$= 80.77 \text{ kPa}$$

That is, the temperature of air would increase by 31.1°C and the pressure by 26.72 kPa as air is decelerated from 250 m/s to zero velocity. These increases in the temperature and pressure of air are due to the conversion of the kinetic energy into enthalpy.

(b) To determine the compressor work, we need to know the stagnation temperature of air at the compressor exit $T_{02}$. The stagnation pressure ratio across the compressor $P_{02}/P_{01}$ is specified to be 8. Since the compression process is approximated as isentropic, $T_{02}$ can be determined from the ideal-gas isentropic relation (Eq. 17–5):

$$T_{02} = T_{01}\left(\frac{P_{02}}{P_{01}}\right)^{(k-1)/k} = (286.8 \text{ K})(8)^{(1.4-1)/1.4} = 519.5 \text{ K}$$

Disregarding potential energy changes and heat transfer, the compressor work per unit mass of air is determined from Eq. 17–8:

$$w_{in} = c_p(T_{02} - T_{01})$$

$$= (1.005 \text{ kJ/kg·K})(519.5 \text{ K} - 286.8 \text{ K})$$

$$= \textbf{233.9 kJ/kg}$$

Thus the work supplied to the compressor is 233.9 kJ/kg.

**Discussion** Notice that using stagnation properties automatically accounts for any changes in the kinetic energy of a fluid stream.

## 17–2 · SPEED OF SOUND AND MACH NUMBER

An important parameter in the study of compressible flow is the **speed of sound** (or the **sonic speed**), defined as the speed at which an infinitesimally small pressure wave travels through a medium. The pressure wave may be caused by a small disturbance, which creates a slight rise in local pressure.

To obtain a relation for the speed of sound in a medium, consider a duct that is filled with a fluid at rest, as shown in Fig. 17–6. A piston fitted in the duct is now moved to the right with a constant incremental velocity $dV$, creating a sonic wave. The wave front moves to the right through the fluid at the speed of sound $c$ and separates the moving fluid adjacent to the piston from the fluid still at rest. The fluid to the left of the wave front experiences an incremental change in its thermodynamic properties, while the fluid on the right of the wave front maintains its original thermodynamic properties, as shown in Fig. 17–6.

To simplify the analysis, consider a control volume that encloses the wave front and moves with it, as shown in Fig. 17–7. To an observer traveling with the wave front, the fluid to the right appears to be moving toward the wave front with a speed of $c$ and the fluid to the left to be moving away from the wave front with a speed of $c - dV$. Of course, the observer sees the control volume that encloses the wave front (and herself or himself) as stationary, and the observer is witnessing a steady-flow process. The mass balance for this single-stream, steady-flow process is expressed as

$$\dot{m}_{right} = \dot{m}_{left}$$

or

$$\rho Ac = (\rho + d\rho)A(c - dV)$$

By canceling the cross-sectional (or flow) area $A$ and neglecting the higher-order terms, this equation reduces to

$$c \, d\rho - \rho \, dV = 0$$

No heat or work crosses the boundaries of the control volume during this steady-flow process, and the potential energy change can be neglected. Then the steady-flow energy balance $e_{in} = e_{out}$ becomes

$$h + \frac{c^2}{2} = h + dh + \frac{(c - dV)^2}{2}$$

which yields

$$dh - c \, dV = 0$$

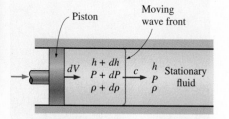

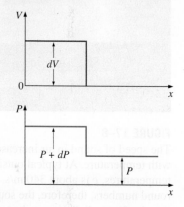

**FIGURE 17–6**

Propagation of a small pressure wave along a duct.

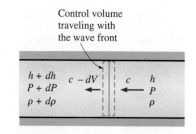

**FIGURE 17–7**

Control volume moving with the small pressure wave along a duct.

**FIGURE 17–8**

The speed of sound in air increases with temperature. At typical outside temperatures, $c$ is about 340 m/s. In round numbers, therefore, the sound of thunder from a lightning strike travels about 1 km in 3 seconds. If you see the lightning and then hear the thunder less than 3 seconds later, you know that the lightning is close, and it is time to go indoors!

© Bear Dancer Studios/Mark Dierker RF

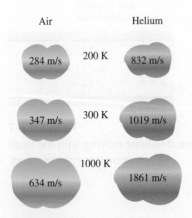

**FIGURE 17–9**

The speed of sound changes with temperature and varies with the fluid.

where we have neglected the second-order term $dV^2$. The amplitude of the ordinary sonic wave is very small and does not cause any appreciable change in the pressure and temperature of the fluid. Therefore, the propagation of a sonic wave is not only adiabatic but also very nearly isentropic. Then the thermodynamic relation $T\,ds = dh - dP/\rho$ reduces to

$$T\,ds^{\,0} = dh - \frac{dP}{\rho}$$

or

$$dh = \frac{dP}{\rho}$$

Combining the above equations yields the desired expression for the speed of sound as

$$c^2 = \frac{dP}{d\rho} \quad \text{at } s = \text{constant}$$

or

$$c^2 = \left(\frac{\partial P}{\partial \rho}\right)_s \tag{17-9}$$

It is left as an exercise for the reader to show, by using thermodynamic property relations, that Eq. 17–9 can also be written as

$$c^2 = k\left(\frac{\partial P}{\partial \rho}\right)_T \tag{17-10}$$

where $k = c_p/c_v$ is the specific heat ratio of the fluid. Note that the speed of sound in a fluid is a function of the thermodynamic properties of that fluid Fig. 17–8.

When the fluid is an ideal gas ($P = \rho RT$), the differentiation in Eq. 17–10 can be performed to yield

$$c^2 = k\left(\frac{\partial P}{\partial \rho}\right)_T = k\left[\frac{\partial(\rho RT)}{\partial \rho}\right]_T = kRT$$

or

$$c = \sqrt{kRT} \tag{17-11}$$

Noting that the gas constant $R$ has a fixed value for a specified ideal gas and the specific heat ratio $k$ of an ideal gas is, at most, a function of temperature, we see that the speed of sound in a specified ideal gas is a function of temperature alone (Fig. 17–9).

A second important parameter in the analysis of compressible fluid flow is the **Mach number** Ma, named after the Austrian physicist Ernst Mach (1838–1916). It is the ratio of the actual speed of the fluid (or an object in still fluid) to the speed of sound in the same fluid at the same state:

$$\text{Ma} = \frac{V}{c} \tag{17-12}$$

Note that the Mach number depends on the speed of sound, which depends on the state of the fluid. Therefore, the Mach number of an aircraft cruising at constant velocity in still air may be different at different locations (Fig. 17–10).

Fluid flow regimes are often described in terms of the flow Mach number. The flow is called **sonic** when Ma = 1, **subsonic** when Ma < 1, **supersonic** when Ma > 1, **hypersonic** when Ma ≫ 1, and **transonic** when Ma ≅ 1.

**FIGURE 17–10**
The Mach number can be different at different temperatures even if the flight speed is the same.

### EXAMPLE 17–2　Mach Number of Air Entering a Diffuser

Air enters a diffuser shown in Fig. 17–11 with a speed of 200 m/s. Determine (*a*) the speed of sound and (*b*) the Mach number at the diffuser inlet when the air temperature is 30°C.

**SOLUTION**　Air enters a diffuser at high speed. The speed of sound and the Mach number are to be determined at the diffuser inlet.
**Assumption**　Air at the specified conditions behaves as an ideal gas.
**Properties**　The gas constant of air is $R$ = 0.287 kJ/kg·K, and its specific heat ratio at 30°C is 1.4.
**Analysis**　We note that the speed of sound in a gas varies with temperature, which is given to be 30°C.

(*a*) The speed of sound in air at 30°C is determined from Eq. 17–11 to be

$$c = \sqrt{kRT} = \sqrt{(1.4)(0.287 \text{ kJ/kg·K})(303 \text{ K})\left(\frac{1000 \text{ m}^2/\text{s}^2}{1 \text{ kJ/kg}}\right)} = \textbf{349 m/s}$$

(*b*) Then the Mach number becomes

$$\text{Ma} = \frac{V}{c} = \frac{200 \text{ m/s}}{349 \text{ m/s}} = \textbf{0.573}$$

**Discussion**　The flow at the diffuser inlet is subsonic since Ma < 1.

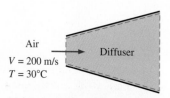

**FIGURE 17–11**
Schematic for Example 17–2.

## 17–3 · ONE-DIMENSIONAL ISENTROPIC FLOW

During fluid flow through many devices such as nozzles, diffusers, and turbine blade passages, flow quantities vary primarily in the flow direction only, and the flow can be approximated as one-dimensional isentropic flow with good accuracy. Therefore, it merits special consideration. Before presenting a formal discussion of one-dimensional isentropic flow, we illustrate some important aspects of it with an example.

### EXAMPLE 17–3　Gas Flow through a Converging–Diverging Duct

Carbon dioxide flows steadily through a varying cross-sectional area duct such as a nozzle shown in Fig. 17–12 at a mass flow rate of 3.00 kg/s. The carbon dioxide enters the duct at a pressure of 1400 kPa and 200°C with a low velocity, and it expands in the nozzle to an exit pressure of 200 kPa. The duct is designed so that the flow can be approximated as isentropic. Determine the density, velocity, flow area, and Mach number at each location along the duct that corresponds to an overall pressure drop of 200 kPa.

**SOLUTION**　Carbon dioxide enters a varying cross-sectional area duct at specified conditions. The flow properties are to be determined along the duct.

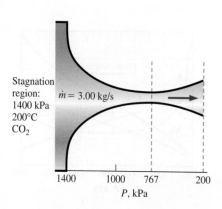

**FIGURE 17–12**
Schematic for Example 17–3.

**Assumptions** **1** Carbon dioxide is an ideal gas with constant specific heats at room temperature. **2** Flow through the duct is steady, one-dimensional, and isentropic.

**Properties** For simplicity we use $c_p = 0.846$ kJ/kg·K and $k = 1.289$ throughout the calculations, which are the constant-pressure specific heat and specific heat ratio values of carbon dioxide at room temperature. The gas constant of carbon dioxide is $R = 0.1889$ kJ/kg·K.

**Analysis** We note that the inlet temperature is nearly equal to the stagnation temperature since the inlet velocity is small. The flow is isentropic, and thus the stagnation temperature and pressure throughout the duct remain constant. Therefore,

$$T_0 \cong T_1 = 200°C = 473 \text{ K}$$

and

$$P_0 \cong P_1 = 1400 \text{ kPa}$$

To illustrate the solution procedure, we calculate the desired properties at the location where the pressure is 1200 kPa, the first location that corresponds to a pressure drop of 200 kPa.

From Eq. 17–5,

$$T = T_0\left(\frac{P}{P_0}\right)^{(k-1)/k} = (473 \text{ K})\left(\frac{1200 \text{ kPa}}{1400 \text{ kPa}}\right)^{(1.289-1)/1.289} = 457 \text{ K}$$

From Eq. 17–4,

$$V = \sqrt{2c_p(T_0 - T)}$$

$$= \sqrt{2(0.846 \text{ kJ/kg·K})(473 \text{ K} - 457 \text{ K})\left(\frac{1000 \text{ m}^2/\text{s}^2}{1 \text{ kJ/kg}}\right)}$$

$$= 164.5 \text{ m/s} \cong \mathbf{164 \text{ m/s}}$$

From the ideal-gas relation,

$$\rho = \frac{P}{RT} = \frac{1200 \text{ kPa}}{(0.1889 \text{ kPa·m}^3/\text{kg·K})(457 \text{ K})} = \mathbf{13.9 \text{ kg/m}^3}$$

From the mass flow rate relation,

$$A = \frac{\dot{m}}{\rho V} = \frac{3.00 \text{ kg/s}}{(13.9 \text{ kg/m}^3)(164.5 \text{ m/s})} = 13.1 \times 10^{-4} \text{ m}^2 = \mathbf{13.1 \text{ cm}^2}$$

From Eqs. 17–11 and 17–12,

$$c = \sqrt{kRT} = \sqrt{(1.289)(0.1889 \text{ kJ/kg·K})(457 \text{ K})\left(\frac{1000 \text{ m}^2/\text{s}^2}{1 \text{ kJ/kg}}\right)} = 333.6 \text{ m/s}$$

$$\text{Ma} = \frac{V}{c} = \frac{164.5 \text{ m/s}}{333.6 \text{ m/s}} = \mathbf{0.493}$$

The results for the other pressure steps are summarized in Table 17–1 and are plotted in Fig. 17–13.

**Discussion** Note that as the pressure decreases, the temperature and speed of sound decrease while the fluid velocity and Mach number increase in the flow direction. The density decreases slowly at first and rapidly later as the fluid velocity increases.

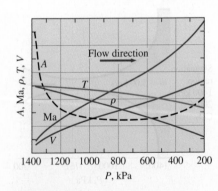

**FIGURE 17–13**
Variation of normalized fluid properties and cross-sectional area along a duct as the pressure drops from 1400 to 200 kPa.

### TABLE 17-1

Variation of fluid properties in flow direction in the duct described in Example 17-3 for $\dot{m} = 3$ kg/s = constant

| P, kPa | T, K | V, m/s | $\rho$, kg/m³ | c, m/s | A, cm² | Ma |
|--------|------|--------|---------------|--------|--------|-------|
| 1400 | 473 | 0 | 15.7 | 339.4 | ∞ | 0 |
| 1200 | 457 | 164.5 | 13.9 | 333.6 | 13.1 | 0.493 |
| 1000 | 439 | 240.7 | 12.1 | 326.9 | 10.3 | 0.736 |
| 800 | 417 | 306.6 | 10.1 | 318.8 | 9.64 | 0.962 |
| 767* | 413 | 317.2 | 9.82 | 317.2 | 9.63 | 1.000 |
| 600 | 391 | 371.4 | 8.12 | 308.7 | 10.0 | 1.203 |
| 400 | 357 | 441.9 | 5.93 | 295.0 | 11.5 | 1.498 |
| 200 | 306 | 530.9 | 3.46 | 272.9 | 16.3 | 1.946 |

* 767 kPa is the critical pressure where the local Mach number is unity.

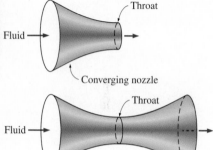

**FIGURE 17-14**
The cross section of a nozzle at the smallest flow area is called the *throat*.

We note from Example 17-3 that the flow area decreases with decreasing pressure down to a critical-pressure value where the Mach number is unity, and then it begins to increase with further reductions in pressure. The Mach number is unity at the location of smallest flow area, called the **throat** (Fig. 17-14). Note that the velocity of the fluid keeps increasing after passing the throat although the flow area increases rapidly in that region. This increase in velocity past the throat is due to the rapid decrease in the fluid density. The flow area of the duct considered in this example first decreases and then increases. Such ducts are called **converging–diverging nozzles**. These nozzles are used to accelerate gases to supersonic speeds and should not be confused with *Venturi nozzles*, which are used strictly for incompressible flow. The first use of such a nozzle occurred in 1893 in a steam turbine designed by a Swedish engineer, Carl G. B. de Laval (1845–1913), and therefore converging–diverging nozzles are often called *Laval nozzles*.

## Variation of Fluid Velocity with Flow Area

It is clear from Example 17-3 that the couplings among the velocity, density, and flow areas for isentropic duct flow are rather complex. In the remainder of this section we investigate these couplings more thoroughly, and we develop relations for the variation of static-to-stagnation property ratios with the Mach number for pressure, temperature, and density.

We begin our investigation by seeking relationships among the pressure, temperature, density, velocity, flow area, and Mach number for one-dimensional isentropic flow. Consider the mass balance for a steady-flow process:

$$\dot{m} = \rho AV = \text{constant}$$

Differentiating and dividing the resultant equation by the mass flow rate, we obtain

$$\frac{d\rho}{\rho} + \frac{dA}{A} + \frac{dV}{V} = 0 \tag{17-13}$$

Neglecting the potential energy, the energy balance for an isentropic flow with no work interactions is expressed in differential form as (Fig. 17-15)

$$\frac{dP}{\rho} + V\,dV = 0 \tag{17-14}$$

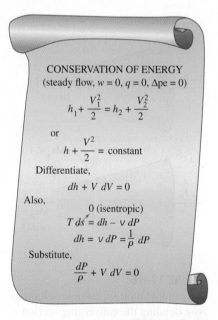

**FIGURE 17-15**
Derivation of the differential form of the energy equation for steady isentropic flow.

CONSERVATION OF ENERGY
(steady flow, $w = 0$, $q = 0$, $\Delta pe = 0$)

$$h_1 + \frac{V_1^2}{2} = h_2 + \frac{V_2^2}{2}$$

or

$$h + \frac{V^2}{2} = \text{constant}$$

Differentiate,

$$dh + V\,dV = 0$$

Also,

$$T\,ds = dh - v\,dP \quad \overset{\text{0 (isentropic)}}{}$$

$$dh = v\,dP = \frac{1}{\rho}\,dP$$

Substitute,

$$\frac{dP}{\rho} + V\,dV = 0$$

This relation is also the differential form of Bernoulli's equation when changes in potential energy are negligible, which is a form of Newton's second law of motion for steady-flow control volumes. Combining Eqs. 17–13 and 17–14 gives

$$\frac{dA}{A} = \frac{dP}{\rho}\left(\frac{1}{V^2} - \frac{d\rho}{dP}\right) \tag{17-15}$$

Rearranging Eq. 17–9 as $(\partial\rho/\partial P)_s = 1/c^2$ and substituting into Eq. 17–15 yield

$$\frac{dA}{A} = \frac{dP}{\rho V^2}(1 - Ma^2) \tag{17-16}$$

This is an important relation for isentropic flow in ducts since it describes the variation of pressure with flow area. We note that $A$, $\rho$, and $V$ are positive quantities. For *subsonic* flow (Ma < 1), the term $1 - Ma^2$ is positive; and thus $dA$ and $dP$ must have the same sign. That is, the pressure of the fluid must increase as the flow area of the duct increases and must decrease as the flow area of the duct decreases. Thus, at subsonic velocities, the pressure decreases in converging ducts (subsonic nozzles) and increases in diverging ducts (subsonic diffusers).

In *supersonic* flow (Ma > 1), the term $1 - Ma^2$ is negative, and thus $dA$ and $dP$ must have opposite signs. That is, the pressure of the fluid must increase as the flow area of the duct decreases and must decrease as the flow area of the duct increases. Thus, at supersonic velocities, the pressure decreases in diverging ducts (supersonic nozzles) and increases in converging ducts (supersonic diffusers).

Another important relation for the isentropic flow of a fluid is obtained by substituting $\rho V = -dP/dV$ from Eq. 17–14 into Eq. 17–16:

$$\frac{dA}{A} = -\frac{dV}{V}(1 - Ma^2) \tag{17-17}$$

This equation governs the shape of a nozzle or a diffuser in subsonic or supersonic isentropic flow. Noting that $A$ and $V$ are positive quantities, we conclude the following:

For subsonic flow (Ma < 1),     $\dfrac{dA}{dV} < 0$

For supersonic flow (Ma > 1),    $\dfrac{dA}{dV} > 0$

For sonic flow (Ma = 1),     $\dfrac{dA}{dV} = 0$

Thus the proper shape of a nozzle depends on the highest velocity desired relative to the sonic velocity. To accelerate a fluid, we must use a converging nozzle at subsonic velocities and a diverging nozzle at supersonic velocities. The velocities encountered in most familiar applications are well below the sonic velocity, and thus it is natural that we visualize a nozzle as a converging duct. However, the highest velocity we can achieve by a converging nozzle is the sonic velocity, which occurs at the exit of the nozzle. If we extend the converging nozzle by further decreasing the flow area, in hopes of accelerating the fluid to supersonic velocities, as shown in Fig. 17–16,

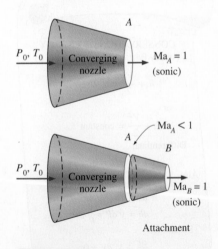

**FIGURE 17–16**

We cannot attain supersonic velocities by extending the converging section of a converging nozzle. Doing so will only move the sonic cross section farther downstream and decrease the mass flow rate.

we are up for disappointment. Now the sonic velocity will occur at the exit of the converging extension, instead of the exit of the original nozzle, and the mass flow rate through the nozzle will decrease because of the reduced exit area.

Based on Eq. 17–16, which is an expression of the conservation of mass and energy principles, we must add a diverging section to a converging nozzle to accelerate a fluid to supersonic velocities. The result is a converging– diverging nozzle. The fluid first passes through a subsonic (converging) section, where the Mach number increases as the flow area of the nozzle decreases, and then reaches the value of unity at the nozzle throat. The fluid continues to accelerate as it passes through a supersonic (diverging) section. Noting that $\dot{m} = \rho AV$ for steady flow, we see that the large decrease in density makes acceleration in the diverging section possible. An example of this type of flow is the flow of hot combustion gases through a nozzle in a gas turbine.

The opposite process occurs in the engine inlet of a supersonic aircraft. The fluid is decelerated by passing it first through a supersonic diffuser, which has a flow area that decreases in the flow direction. Ideally, the flow reaches a Mach number of unity at the diffuser throat. The fluid is further decelerated in a subsonic diffuser, which has a flow area that increases in the flow direction, as shown in Fig. 17–17.

## Property Relations for Isentropic Flow of Ideal Gases

Next we develop relations between the static properties and stagnation properties of an ideal gas in terms of the specific heat ratio $k$ and the Mach number Ma. We assume the flow is isentropic and the gas has constant specific heats.

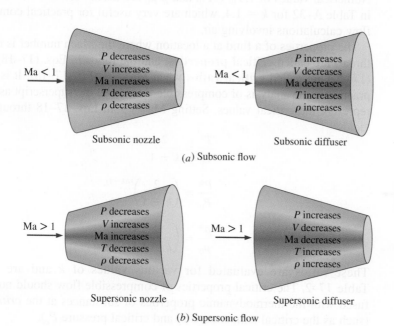

(a) Subsonic flow

(b) Supersonic flow

**FIGURE 17–17**

Variation of flow properties in subsonic and supersonic nozzles and diffusers.

The temperature $T$ of an ideal gas anywhere in the flow is related to the stagnation temperature $T_0$ through Eq. 17–4:

$$T_0 = T + \frac{V^2}{2c_p}$$

or

$$\frac{T_0}{T} = 1 + \frac{V^2}{2c_pT}$$

Noting that $c_p = kR/(k - 1)$, $c^2 = kRT$, and $\mathrm{Ma} = V/c$, we see that

$$\frac{V^2}{2c_pT} = \frac{V^2}{2[kR/(k - 1)]T} = \left(\frac{k - 1}{2}\right)\frac{V^2}{c^2} = \left(\frac{k - 1}{2}\right)\mathrm{Ma}^2$$

Substitution yields

$$\frac{T_0}{T} = 1 + \left(\frac{k - 1}{2}\right)\mathrm{Ma}^2 \tag{17–18}$$

which is the desired relation between $T_0$ and $T$.

The ratio of the stagnation to static pressure is obtained by substituting Eq. 17–18 into Eq. 17–5:

$$\frac{P_0}{P} = \left[1 + \left(\frac{k - 1}{2}\right)\mathrm{Ma}^2\right]^{k/(k - 1)} \tag{17–19}$$

The ratio of the stagnation to static density is obtained by substituting Eq. 17–18 into Eq. 17–6:

$$\frac{\rho_0}{\rho} = \left[1 + \left(\frac{k - 1}{2}\right)\mathrm{Ma}^2\right]^{1/(k - 1)} \tag{17–20}$$

Numerical values of $T/T_0$, $P/P_0$, and $\rho/\rho_0$ are listed versus the Mach number in Table A–32 for $k = 1.4$, which are very useful for practical compressible flow calculations involving air.

The properties of a fluid at a location where the Mach number is unity (the throat) are called **critical properties**, and the ratios in Eqs. (17–18) through (17–20) are called **critical ratios** when $\mathrm{Ma} = 1$ (Fig. 17–18). It is standard practice in the analysis of compressible flow to let the superscript asterisk (*) represent the critical values. Setting $\mathrm{Ma} = 1$ in Eqs. 17–18 through 17–20 yields

$$\frac{T^*}{T_0} = \frac{2}{k + 1} \tag{17–21}$$

$$\frac{P^*}{P_0} = \left(\frac{2}{k + 1}\right)^{k/(k - 1)} \tag{17–22}$$

$$\frac{\rho^*}{\rho_0} = \left(\frac{2}{k + 1}\right)^{1/(k - 1)} \tag{17–23}$$

These ratios are evaluated for various values of $k$ and are listed in Table 17–2. The critical properties of compressible flow should not be confused with the thermodynamic properties of substances at the *critical point* (such as the critical temperature $T_c$ and critical pressure $P_c$).

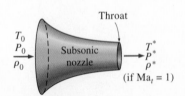

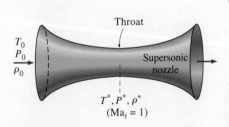

**FIGURE 17–18**

When $\mathrm{Ma}_t = 1$, the properties at the nozzle throat are the critical properties.

**TABLE 17–2**

The critical-pressure, critical-temperature, and critical-density ratios for isentropic flow of some ideal gases

| | Superheated steam, $k = 1.3$ | Hot products of combustion, $k = 1.33$ | Air, $k = 1.4$ | Monatomic gases, $k = 1.667$ |
|---|---|---|---|---|
| $\dfrac{P^*}{P_0}$ | 0.5457 | 0.5404 | 0.5283 | 0.4871 |
| $\dfrac{T^*}{T_0}$ | 0.8696 | 0.8584 | 0.8333 | 0.7499 |
| $\dfrac{\rho^*}{\rho_0}$ | 0.6276 | 0.6295 | 0.6340 | 0.6495 |

---

**EXAMPLE 17–4     Critical Temperature and Pressure in Gas Flow**

Calculate the critical pressure and temperature of carbon dioxide for the flow conditions described in Example 17–3 (Fig. 17–19).

**SOLUTION**  For the flow discussed in Example 17–3, the critical pressure and temperature are to be calculated.

**Assumptions**  1 The flow is steady, adiabatic, and one-dimensional. 2 Carbon dioxide is an ideal gas with constant specific heats.

**Properties**  The specific heat ratio of carbon dioxide at room temperature is $k = 1.289$.

**Analysis**  The ratios of critical to stagnation temperature and pressure are determined to be

$$\frac{T^*}{T_0} = \frac{2}{k+1} = \frac{2}{1.289 + 1} = 0.8737$$

$$\frac{P^*}{P_0} = \left(\frac{2}{k+1}\right)^{k/(k-1)} = \left(\frac{2}{1.289 + 1}\right)^{1.289/(1.289-1)} = 0.5477$$

Noting that the stagnation temperature and pressure are, from Example 17–3, $T_0 = 473$ K and $P_0 = 1400$ kPa, we see that the critical temperature and pressure in this case are

$$T^* = 0.8737 T_0 = (0.8737)(473 \text{ K}) = \textbf{413 K}$$

$$P^* = 0.5477 P_0 = (0.5477)(1400 \text{ kPa}) = \textbf{767 kPa}$$

**Discussion**  Note that these values agree with those listed in the 5th row of Table 17–1, as expected. Also, property values other than these at the throat would indicate that the flow is not critical, and the Mach number is not unity.

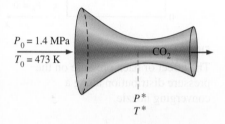

$P_0 = 1.4$ MPa
$T_0 = 473$ K

$P^*$
$T^*$

CO$_2$

**FIGURE 17–19**
Schematic for Example 17–4.

# 17–4 · ISENTROPIC FLOW THROUGH NOZZLES

Converging or converging–diverging nozzles are found in many engineering applications including steam and gas turbines, aircraft and spacecraft propulsion systems, and even industrial blasting nozzles and torch nozzles. In this section we consider the effects of **back pressure** (i.e., the pressure applied

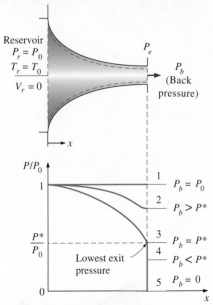

**FIGURE 17–20**
The effect of back pressure on the pressure distribution along a converging nozzle.

at the nozzle discharge region) on the exit velocity, the mass flow rate, and the pressure distribution along the nozzle.

## Converging Nozzles

Consider the subsonic flow through a converging nozzle as shown in Fig. 17–20. The nozzle inlet is attached to a reservoir at pressure $P_r$ and temperature $T_r$. The reservoir is sufficiently large so that the nozzle inlet velocity is negligible. Since the fluid velocity in the reservoir is zero and the flow through the nozzle is approximated as isentropic, the stagnation pressure and stagnation temperature of the fluid at any cross section through the nozzle are equal to the reservoir pressure and temperature, respectively.

Now we begin to reduce the back pressure and observe the resulting effects on the pressure distribution along the length of the nozzle, as shown in Fig. 17–20. If the back pressure $P_b$ is equal to $P_1$, which is equal to $P_r$, there is no flow and the pressure distribution is uniform along the nozzle. When the back pressure is reduced to $P_2$, the exit plane pressure $P_e$ also drops to $P_2$. This causes the pressure along the nozzle to decrease in the flow direction.

When the back pressure is reduced to $P_3$ ($= P^*$, which is the pressure required to increase the fluid velocity to the speed of sound at the exit plane or throat), the mass flow reaches a maximum value and the flow is said to be **choked**. Further reduction of the back pressure to level $P_4$ or below does not result in additional changes in the pressure distribution, or anything else along the nozzle length.

Under steady-flow conditions, the mass flow rate through the nozzle is constant and is expressed as

$$\dot{m} = \rho A V = \left(\frac{P}{RT}\right) A (\mathrm{Ma}\sqrt{kRT}) = PA\,\mathrm{Ma}\sqrt{\frac{k}{RT}}$$

Solving for $T$ from Eq. 17–18 and for $P$ from Eq. 17–19 and substituting,

$$\dot{m} = \frac{A\,\mathrm{Ma}\,P_0\sqrt{k/(RT_0)}}{[1 + (k-1)\mathrm{Ma}^2/2]^{(k+1)/[2(k-1)]}} \qquad (17\text{–}24)$$

Thus the mass flow rate of a particular fluid through a nozzle is a function of the stagnation properties of the fluid, the flow area, and the Mach number. Equation 17–24 is valid at any cross section, and thus $\dot{m}$ can be evaluated at any location along the length of the nozzle.

For a specified flow area $A$ and stagnation properties $T_0$ and $P_0$, the maximum mass flow rate can be determined by differentiating Eq. 17–24 with respect to Ma and setting the result equal to zero. It yields Ma = 1. Since the only location in a nozzle where the Mach number can be unity is the location of minimum flow area (the throat), the mass flow rate through a nozzle is a maximum when Ma = 1 at the throat. Denoting this area by $A^*$, we obtain an expression for the maximum mass flow rate by substituting Ma = 1 in Eq. 17–24:

$$\dot{m}_{\max} = A^* P_0 \sqrt{\frac{k}{RT_0}\left(\frac{2}{k+1}\right)^{(k+1)/[2(k-1)]}} \qquad (17\text{–}25)$$

Thus, for a particular ideal gas, the maximum mass flow rate through a nozzle with a given throat area is fixed by the stagnation pressure and temperature of the inlet flow. The flow rate can be controlled by changing the stagnation pressure or temperature, and thus a converging nozzle can be used as a flowmeter. The flow rate can also be controlled, of course, by varying the throat area. This principle is very important for chemical processes, medical devices, flowmeters, and anywhere the mass flux of a gas must be known and controlled.

A plot of $\dot{m}$ versus $P_b/P_0$ for a converging nozzle is shown in Fig. 17–21. Notice that the mass flow rate increases with decreasing $P_b/P_0$, reaches a maximum at $P_b = P*$, and remains constant for $P_b/P_0$ values less than this critical ratio. Also illustrated on this figure is the effect of back pressure on the nozzle exit pressure $P_e$. We observe that

$$P_e = \begin{cases} P_b & \text{for } P_b \geq P* \\ P* & \text{for } P_b < P* \end{cases}$$

To summarize, for all back pressures lower than the critical pressure $P*$, the pressure at the exit plane of the converging nozzle $P_e$ is equal to $P*$, the Mach number at the exit plane is unity, and the mass flow rate is the maximum (or choked) flow rate. Because the velocity of the flow is sonic at the throat for the maximum flow rate, a back pressure lower than the critical pressure cannot be sensed in the nozzle upstream flow and does not affect the flow rate.

The effects of the stagnation temperature $T_0$ and stagnation pressure $P_0$ on the mass flow rate through a converging nozzle are illustrated in Fig. 17–22 where the mass flow rate is plotted against the static-to-stagnation pressure ratio at the throat $P_t/P_0$. An increase in $P_0$ (or a decrease of $T_0$) will increase the mass flow rate through the converging nozzle; a decrease in $P_0$ (or an increase in $T_0$) will decrease it. We could also conclude this by carefully observing Eqs. 17–24 and 17–25.

A relation for the variation of flow area $A$ through the nozzle relative to throat area $A*$ can be obtained by combining Eqs. 17–24 and 17–25 for the same mass flow rate and stagnation properties of a particular fluid. This yields

$$\frac{A}{A*} = \frac{1}{\text{Ma}}\left[\left(\frac{2}{k+1}\right)\left(1 + \frac{k-1}{2}\text{Ma}^2\right)\right]^{(k+1)/[2(k-1)]} \tag{17–26}$$

Table A–32 gives values of $A/A*$ as a function of the Mach number for air ($k = 1.4$). There is one value of $A/A*$ for each value of the Mach number, but there are two possible values of the Mach number for each value of $A/A*$—one for subsonic flow and another for supersonic flow.

Another parameter sometimes used in the analysis of one-dimensional isentropic flow of ideal gases is Ma*, which is the ratio of the local velocity to the speed of sound at the throat:

$$\text{Ma}* = \frac{V}{c*} \tag{17–27}$$

Equation 17–27 can also be expressed as

$$\text{Ma}* = \frac{V}{c}\frac{c}{c*} = \frac{\text{Ma}\,c}{c*} = \frac{\text{Ma}\sqrt{kRT}}{\sqrt{kRT*}} = \text{Ma}\sqrt{\frac{T}{T*}}$$

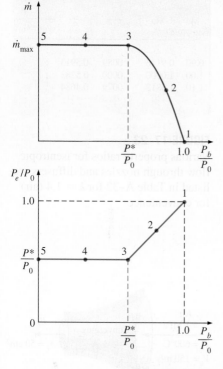

**FIGURE 17–21**
The effect of back pressure $P_b$ on the mass flow rate $\dot{m}$ and the exit pressure $P_e$ of a converging nozzle.

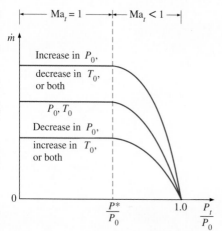

**FIGURE 17–22**
The variation of the mass flow rate through a nozzle with inlet stagnation properties.

| Ma | Ma* | $\dfrac{A}{A^*}$ | $\dfrac{P}{P_0}$ | $\dfrac{\rho}{\rho_0}$ | $\dfrac{T}{T_0}$ |
|------|--------|--------|--------|---|---|
| ⋮ | ⋮ | ⋮ | ⋮ | ⋮ | ⋮ |
| 0.90 | 0.9146 | 1.0089 | 0.5913 | | |
| 1.00 | 1.0000 | 1.0000 | 0.5283 | | |
| 1.10 | 1.0812 | 1.0079 | 0.4684 | | |
| ⋮ | ⋮ | ⋮ | ⋮ | ⋮ | ⋮ |

**FIGURE 17–23**
Various property ratios for isentropic flow through nozzles and diffusers are listed in Table A–32 for $k = 1.4$ (air) for convenience.

Air
$P_i = 1$ MPa
$T_i = 600°C$    Converging    $P_b$
$V_i = 150$ m/s    nozzle    $A_t = 50$ cm²

**FIGURE 17–24**
Schematic for Example 17–5.

where Ma is the local Mach number, $T$ is the local temperature, and $T^*$ is the critical temperature. Solving for $T$ from Eq. 17–18 and for $T^*$ from Eq. 17–21 and substituting, we get

$$\text{Ma}^* = \text{Ma}\sqrt{\frac{k+1}{2+(k-1)\text{Ma}^2}} \qquad (17\text{–}28)$$

Values of Ma* are also listed in Table A–32 versus the Mach number for $k = 1.4$ (Fig. 17–23). Note that the parameter Ma* differs from the Mach number Ma in that Ma* is the local velocity nondimensionalized with respect to the sonic velocity at the *throat*, whereas Ma is the local velocity nondimensionalized with respect to the *local* sonic velocity. (Recall that the sonic velocity in a nozzle varies with temperature and thus with location.)

---

**EXAMPLE 17–5    Effect of Back Pressure on Mass Flow Rate**

Air at 1 MPa and 600°C enters a converging nozzle, shown in Fig. 17–24, with a velocity of 150 m/s. Determine the mass flow rate through the nozzle for a nozzle throat area of 50 cm² when the back pressure is (a) 0.7 MPa and (b) 0.4 MPa.

**SOLUTION**    Air enters a converging nozzle. The mass flow rate of air through the nozzle is to be determined for different back pressures.
**Assumptions**    1 Air is an ideal gas with constant specific heats at room temperature. 2 Flow through the nozzle is steady, one-dimensional, and isentropic.
**Properties**    The constant pressure specific heat and the specific heat ratio of air are $c_p = 1.005$ kJ/kg·K and $k = 1.4$.
**Analysis**    We use the subscripts $i$ and $t$ to represent the properties at the nozzle inlet and the throat, respectively. The stagnation temperature and pressure at the nozzle inlet are determined from Eqs. 17–4 and 17–5:

$$T_{0i} = T_i + \frac{V_i^2}{2c_p} = 873\text{ K} + \frac{(150\text{ m/s})^2}{2(1.005\text{ kJ/kg·K})}\left(\frac{1\text{ kJ/kg}}{1000\text{ m}^2/\text{s}^2}\right) = 884\text{ K}$$

$$P_{0i} = P_i\left(\frac{T_{0i}}{T_i}\right)^{k/(k-1)} = (1\text{ MPa})\left(\frac{884\text{ K}}{873\text{ K}}\right)^{1.4/(1.4-1)} = 1.045\text{ MPa}$$

These stagnation temperature and pressure values remain constant throughout the nozzle since the flow is assumed to be isentropic. That is,

$$T_0 = T_{0i} = 884\text{ K} \quad \text{and} \quad P_0 = P_{0i} = 1.045\text{ MPa}$$

The critical-pressure ratio is determined from Eq. 17–22 to be $P^*/P_0 = 0.5283$.

(a) The back pressure ratio for this case is

$$\frac{P_b}{P_0} = \frac{0.7\text{ MPa}}{1.045\text{ MPa}} = 0.670$$

which is greater than the critical-pressure ratio, 0.5283. Thus the exit plane pressure (or throat pressure $P_t$) is equal to the back pressure in this case. That is, $P_t = P_b = 0.7$ MPa, and $P_t/P_0 = 0.670$. Therefore, the flow is not choked. From Table A–32 at $P_t/P_0 = 0.670$, we read $\text{Ma}_t = 0.778$ and $T_t/T_0 = 0.892$.

The mass flow rate through the nozzle can be calculated from Eq. 17–24. But it can also be determined in a step-by-step manner as follows:

$$T_t = 0.892T_0 = 0.892(884 \text{ K}) = 788.5 \text{ K}$$

$$\rho_t = \frac{P_t}{RT_t} = \frac{700 \text{ kPa}}{(0.287 \text{ kPa·m}^3/\text{kg·K})(788.5 \text{ K})} = 3.093 \text{ kg/m}^3$$

$$V_t = \text{Ma}_t c_t = \text{Ma}_t \sqrt{kRT_t}$$

$$= (0.778)\sqrt{(1.4)(0.287 \text{ kJ/kg·K})(788.5 \text{ K})\left(\frac{1000 \text{ m}^2/\text{s}^2}{1 \text{ kJ/kg}}\right)}$$

$$= 437.9 \text{ m/s}$$

Thus,

$$\dot{m} = \rho_t A_t V_t = (3.093 \text{ kg/m}^3)(50 \times 10^{-4} \text{ m}^2)(437.9 \text{ m/s}) = \mathbf{6.77 \text{ kg/s}}$$

(b) The back pressure ratio for this case is

$$\frac{P_b}{P_0} = \frac{0.4 \text{ MPa}}{1.045 \text{ MPa}} = 0.383$$

which is less than the critical-pressure ratio, 0.5283. Therefore, sonic conditions exist at the exit plane (throat) of the nozzle, and Ma = 1. The flow is choked in this case, and the mass flow rate through the nozzle is calculated from Eq. 17–25:

$$\dot{m} = A^* P_0 \sqrt{\frac{k}{RT_0}\left(\frac{2}{k+1}\right)^{(k+1)/[2(k-1)]}}$$

$$= (50 \times 10^{-4} \text{ m}^2)(1045 \text{ kPa})\sqrt{\frac{1.4}{(0.287 \text{ kJ/kg·K})(884 \text{ K})}\left(\frac{2}{1.4+1}\right)^{2.4/0.8}}$$

$$= \mathbf{7.10 \text{ kg/s}}$$

since $\text{kPa·m}^2\sqrt{\text{kJ/kg}} = \sqrt{1000} \text{ kg/s}$.

**Discussion** This is the maximum mass flow rate through the nozzle for the specified inlet conditions and nozzle throat area.

■ **EXAMPLE 17–6** Air Loss from a Flat Tire

■ Air in an automobile tire is maintained at a pressure of 220 kPa (gage) in an
■ environment where the atmospheric pressure is 94 kPa. The air in the tire is
at the ambient temperature of 25°C. A 4-mm-diameter leak develops in the tire as a result of an accident (Fig. 17–25). Approximating the flow as isentropic determine the initial mass flow rate of air through the leak.

**SOLUTION** A leak develops in an automobile tire as a result of an accident. The initial mass flow rate of air through the leak is to be determined.
**Assumptions** 1 Air is an ideal gas with constant specific heats. 2 Flow of air through the hole is isentropic.

Air

$T = 25°C$
$P_g = 220 \text{ kPa}$

**FIGURE 17–25**
Schematic for Example 17–6.

**Properties** The specific gas constant of air is $R = 0.287$ kPa·m³/kg·K. The specific heat ratio of air at room temperature is $k = 1.4$.

**Analysis** The absolute pressure in the tire is

$$P = P_{gage} + P_{atm} = 220 + 94 = 314 \text{ kPa}$$

The critical pressure is (from Table 17–2)

$$P^* = 0.5283P_o = (0.5283)(314 \text{ kPa}) = 166 \text{ kPa} > 94 \text{ kPa}$$

Therefore, the flow is choked, and the velocity at the exit of the hole is the sonic speed. Then the flow properties at the exit become

$$\rho_0 = \frac{P_0}{RT_0} = \frac{314 \text{ kPa}}{(0.287 \text{ kPa·m}^3/\text{kg·K})(298 \text{ K})} = 3.671 \text{ kg/m}^3$$

$$\rho^* = \rho\left(\frac{2}{k+1}\right)^{1/(k-1)} = (3.671 \text{ kg/m}^3)\left(\frac{2}{1.4+1}\right)^{1/(1.4-1)} = 2.327 \text{ kg/m}^3$$

$$T^* = \frac{2}{k+1}T_0 = \frac{2}{1.4+1}(298 \text{ K}) = 248.3 \text{ K}$$

$$V = c = \sqrt{kRT^*} = \sqrt{(1.4)(0.287 \text{ kJ/kg·K})\left(\frac{1000 \text{ m}^2/\text{s}^2}{1 \text{ kJ/kg}}\right)(248.3 \text{ K})}$$

$$= 315.9 \text{ m/s}$$

Then the initial mass flow rate through the hole is

$$\dot{m} = \rho AV = (2.327 \text{ kg/m}^3)[\pi(0.004 \text{ m})^2/4](315.9 \text{ m/s}) = 0.00924 \text{ kg/s}$$

$$= \mathbf{0.554 \text{ kg/min}}$$

**Discussion** The mass flow rate decreases with time as the pressure inside the tire drops.

## Converging–Diverging Nozzles

When we think of nozzles, we ordinarily think of flow passages whose cross-sectional area decreases in the flow direction. However, the highest velocity to which a fluid can be accelerated in a converging nozzle is limited to the sonic velocity (Ma = 1), which occurs at the exit plane (throat) of the nozzle. Accelerating a fluid to supersonic velocities (Ma > 1) can be accomplished only by attaching a diverging flow section to the subsonic nozzle at the throat. The resulting combined flow section is a converging–diverging nozzle, which is standard equipment in supersonic aircraft and rocket propulsion (Fig. 17–26).

Forcing a fluid through a converging–diverging nozzle is no guarantee that the fluid will be accelerated to a supersonic velocity. In fact, the fluid may find itself decelerating in the diverging section instead of accelerating if the back pressure is not in the right range. The state of the nozzle flow is determined by the overall pressure ratio $P_b/P_0$. Therefore, for given inlet conditions, the flow through a converging–diverging nozzle is governed by the back pressure $P_b$, as will be explained.

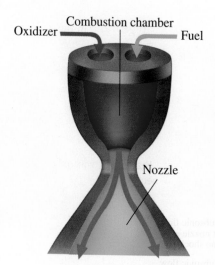

Oxidizer — Combustion chamber — Fuel

Nozzle

**FIGURE 17–26**
Converging–diverging nozzles are commonly used in rocket engines to provide high thrust.
(*Right*) *NASA*

Consider the converging–diverging nozzle shown in Fig. 17–27. A fluid enters the nozzle with a low velocity at stagnation pressure $P_0$. When $P_b = P_0$ (case $A$), there is no flow through the nozzle. This is expected since the flow in a nozzle is driven by the pressure difference between the nozzle inlet and the exit. Now let us examine what happens as the back pressure is lowered.

1. When $P_0 > P_b > P_C$, the flow remains subsonic throughout the nozzle, and the mass flow is less than that for choked flow. The fluid velocity increases in the first (converging) section and reaches a maximum at the throat (but Ma < 1). However, most of the gain in velocity is lost in the second (diverging) section of the nozzle, which acts as a diffuser. The pressure decreases in the converging section, reaches a minimum at the throat, and increases at the expense of velocity in the diverging section.
2. When $P_b = P_C$, the throat pressure becomes $P^*$ and the fluid achieves sonic velocity at the throat. But the diverging section of the nozzle still acts as a diffuser, slowing the fluid to subsonic velocities. The mass flow rate that was increasing with decreasing $P_b$ also reaches its maximum value. Recall that $P^*$ is the lowest pressure that can be obtained at the throat, and the sonic velocity is the highest velocity that can be achieved with a converging nozzle. Thus, lowering $P_b$ further has no influence on the fluid flow in the converging part of the nozzle or the mass flow rate through the nozzle. However, it does influence the character of the flow in the diverging section.
3. When $P_C > P_b > P_E$, the fluid that achieved a sonic velocity at the throat continues accelerating to supersonic velocities in the diverging section as the pressure decreases. This acceleration comes to a sudden stop, however, as a **normal shock** develops at a section between the throat and the exit plane, which causes a sudden drop in velocity to subsonic levels and a sudden increase in pressure. The fluid then continues to decelerate further

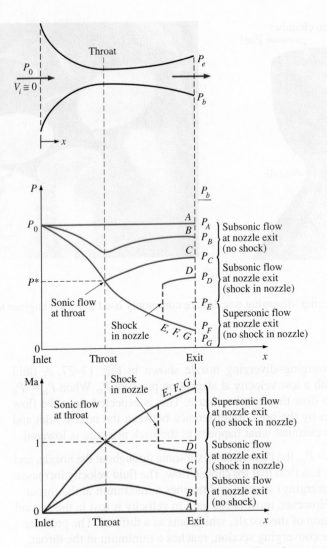

**FIGURE 17–27**
The effects of back pressure on the flow through a converging–diverging nozzle.

in the remaining part of the converging–diverging nozzle. Flow through the shock is highly irreversible, and thus it cannot be approximated as isentropic. The normal shock moves downstream away from the throat as $P_b$ is decreased, and it approaches the nozzle exit plane as $P_b$ approaches $P_E$.

When $P_b = P_E$, the normal shock forms at the exit plane of the nozzle. The flow is supersonic through the entire diverging section in this case, and it can be approximated as isentropic. However, the fluid velocity drops to subsonic levels just before leaving the nozzle as it crosses the normal shock. Normal shock waves are discussed in Section 17–4.

4. When $P_E > P_b > 0$, the flow in the diverging section is supersonic, and the fluid expands to $P_F$ at the nozzle exit with no normal shock forming within the nozzle. Thus, the flow through the nozzle can be approximated as isentropic. When $P_b = P_F$, no shocks occur within or outside the nozzle. When $P_b < P_F$, irreversible mixing and expansion waves occur downstream of the exit plane of the nozzle. When $P_b > P_F$, however, the pressure of the fluid increases from $P_F$ to $P_b$ irreversibly in the wake of the nozzle exit, creating what are called *oblique shocks*.

# EXAMPLE 17–7    Airflow through a Converging–Diverging Nozzle

Air enters a converging–diverging nozzle, shown in Fig. 17–28, at 1.0 MPa and 800 K with negligible velocity. The flow is steady, one-dimensional, and isentropic with $k = 1.4$. For an exit Mach number of Ma = 2 and a throat area of 20 cm², determine (a) the throat conditions, (b) the exit plane conditions, including the exit area, and (c) the mass flow rate through the nozzle.

**SOLUTION**   Air flows through a converging–diverging nozzle. The throat and the exit conditions and the mass flow rate are to be determined.

**Assumptions**   **1** Air is an ideal gas with constant specific heats at room temperature. **2** Flow through the nozzle is steady, one-dimensional, and isentropic.

**Properties**   The specific heat ratio of air is given to be $k = 1.4$. The gas constant of air is 0.287 kJ/kg·K.

**Analysis**   The exit Mach number is given to be 2. Therefore, the flow must be sonic at the throat and supersonic in the diverging section of the nozzle. Since the inlet velocity is negligible, the stagnation pressure and stagnation temperature are the same as the inlet temperature and pressure, $P_0 = 1.0$ MPa and $T_0 = 800$ K. Assuming ideal-gas behavior, the stagnation density is

$$\rho_0 = \frac{P_0}{RT_0} = \frac{1000 \text{ kPa}}{(0.287 \text{ kPa·m}^3/\text{kg·K})(800 \text{ K})} = 4.355 \text{ kg/m}^3$$

(a) At the throat of the nozzle Ma = 1, and from Table A–32 we read

$$\frac{P^*}{P_0} = 0.5283 \qquad \frac{T^*}{T_0} = 0.8333 \qquad \frac{\rho^*}{\rho_0} = 0.6339$$

Thus,

$$P^* = 0.5283P_0 = (0.5283)(1.0 \text{ MPa}) = \textbf{0.5283 MPa}$$

$$T^* = 0.8333T_0 = (0.8333)(800 \text{ K}) = \textbf{666.6 K}$$

$$\rho^* = 0.6339\rho_0 = (0.6339)(4.355 \text{ kg/m}^3) = \textbf{2.761 kg/m}^3$$

Also,

$$V^* = c^* = \sqrt{kRT^*} = \sqrt{(1.4)(0.287 \text{ kJ/kg·K})(666.6 \text{ K})\left(\frac{1000 \text{ m}^2/\text{s}^2}{1 \text{ kJ/kg}}\right)}$$

$$= \textbf{517.5 m/s}$$

(b) Since the flow is isentropic, the properties at the exit plane can also be calculated by using data from Table A–32. For Ma = 2 we read

$$\frac{P_e}{P_0} = 0.1278 \quad \frac{T_e}{T_0} = 0.5556 \quad \frac{\rho_e}{\rho_0} = 0.2300 \quad \text{Ma}_e^* = 1.6330 \quad \frac{A_e}{A^*} = 1.6875$$

Thus,

$$P_e = 0.1278P_0 = (0.1278)(1.0 \text{ MPa}) = \textbf{0.1278 MPa}$$

$$T_e = 0.5556T_0 = (0.5556)(800 \text{ K}) = \textbf{444.5 K}$$

$$\rho_e = 0.2300\rho_0 = (0.2300)(4.355 \text{ kg/m}^3) = \textbf{1.002 kg/m}^3$$

$$A_e = 1.6875A^* = (1.6875)(20 \text{ cm}^2) = \textbf{33.75 cm}^2$$

$T_0 = 800$ K
$P_0 = 1.0$ MPa
$V_i \cong 0$
$\text{Ma}_e = 2$
$A_t = 20 \text{ cm}^2$

**FIGURE 17–28**
Schematic for Example 17–7.

and

$$V_e = \text{Ma}_e^* c^* = (1.6330)(517.5 \text{ m/s}) = \textbf{845.1 m/s}$$

The nozzle exit velocity could also be determined from $V_e = \text{Ma}_e c_e$, where $c_e$ is the speed of sound at the exit conditions:

$$V_e = \text{Ma}_e c_e = \text{Ma}_e \sqrt{kRT_e} = 2\sqrt{(1.4)(0.287 \text{ kJ/kg·K})(444.5 \text{ K})\left(\frac{1000 \text{ m}^2/\text{s}^2}{1 \text{ kJ/kg}}\right)}$$

$$= 845.2 \text{ m/s}$$

(c) Since the flow is steady, the mass flow rate of the fluid is the same at all sections of the nozzle. Thus it may be calculated by using properties at any cross section of the nozzle. Using the properties at the throat, we find that the mass flow rate is

$$\dot{m} = \rho^* A^* V^* = (2.761 \text{ kg/m}^3)(20 \times 10^{-4} \text{ m}^2)(517.5 \text{ m/s}) = \textbf{2.86 kg/s}$$

**Discussion** Note that this is the highest possible mass flow rate that can flow through this nozzle for the specified inlet conditions.

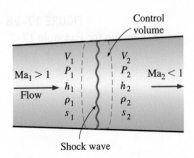

**FIGURE 17–29**
Control volume for flow across a normal shock wave.

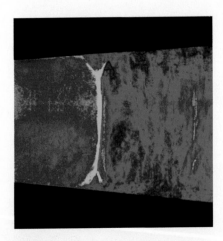

**FIGURE 17–30**
Schlieren image of a normal shock in a Laval nozzle. The Mach number in the nozzle just upstream (to the left) of the shock wave is about 1.3. Boundary layers distort the shape of the normal shock near the walls and lead to flow separation beneath the shock.

*G.S. Settles, Gas Dynamics Lab, Penn State University. Used with permission*

## 17–5 · SHOCK WAVES AND EXPANSION WAVES

We have been that sound waves are caused by infinitesimally small pressure disturbances, and they travel through a medium at the speed of sound. We have also seen in the present chapter that for some back pressure values, abrupt changes in fluid properties occur in a very thin section of a converging–diverging nozzle under supersonic flow conditions, creating a **shock wave**. It is of interest to study the conditions under which shock waves develop and how they affect the flow.

### Normal Shocks

First we consider shock waves that occur in a plane normal to the direction of flow, called **normal shock waves**. The flow process through the shock wave is highly irreversible and *cannot* be approximated as being isentropic.

Next we follow the footsteps of Pierre Laplace (1749–1827), G. F. Bernhard Riemann (1826–1866), William Rankine (1820–1872), Pierre Henry Hugoniot (1851–1887), Lord Rayleigh (1842–1919), and G. I. Taylor (1886– 1975) and develop relationships for the flow properties before and after the shock. We do this by applying the conservation of mass, momentum, and energy relations as well as some property relations to a stationary control volume that contains the shock, as shown in Fig. 17–29. The normal shock waves are extremely thin, so the entrance and exit flow areas for the control volume are approximately equal (Fig 17–30).

We assume steady flow with no heat and work interactions and no potential energy changes. Denoting the properties upstream of the shock by the subscript 1 and those downstream of the shock by 2, we have the following:

*Conservation of mass:* $\qquad \rho_1 A V_1 = \rho_2 A V_2$ **(17–29)**

or

$$\rho_1 V_1 = \rho_2 V_2$$

*Conservation of energy:*
$$h_1 + \frac{V_1^2}{2} = h_2 + \frac{V_2^2}{2} \qquad \textbf{(17–30)}$$

or

$$h_{01} = h_{02} \qquad \textbf{(17–31)}$$

*Linear momentum equation:* Rearranging Eq. 17–14 and integrating yield

$$A(P_1 - P_2) = \dot{m}(V_2 - V_1) \qquad \textbf{(17–32)}$$

*Increase of entropy:*
$$s_2 - s_1 \geq 0 \qquad \textbf{(17–33)}$$

We can combine the conservation of mass and energy relations into a single equation and plot it on an *h-s* diagram, using property relations. The resultant curve is called the **Fanno line,** and it is the locus of states that have the same value of stagnation enthalpy and mass flux (mass flow per unit flow area). Likewise, combining the conservation of mass and momentum equations into a single equation and plotting it on the *h-s* diagram yield a curve called the **Rayleigh line.** Both these lines are shown on the *h-s* diagram in Fig. 17–31. As proved later in Example 17–8, the points of maximum entropy on these lines (points *a* and *b*) correspond to Ma = 1. The state on the upper part of each curve is subsonic and on the lower part supersonic.

The Fanno and Rayleigh lines intersect at two points (points 1 and 2), which represent the two states at which all three conservation equations are satisfied. One of these (state 1) corresponds to the state before the shock, and the other (state 2) corresponds to the state after the shock. Note that the flow is supersonic before the shock and subsonic afterward. Therefore the flow must change from supersonic to subsonic if a shock is to occur. The larger the Mach number before the shock, the stronger the shock will be. In the limiting case of Ma = 1, the shock wave simply becomes a sound wave. Notice from Fig. 17–31 that entropy increases, $s_2 > s_1$. This is expected since the flow through the shock is adiabatic but irreversible.

The conservation of energy principle (Eq. 17–31) requires that the stagnation enthalpy remain constant across the shock; $h_{01} = h_{02}$. For ideal gases $h = h(T)$, and thus

$$T_{01} = T_{02} \qquad \textbf{(17–34)}$$

That is, the stagnation temperature of an ideal gas also remains constant across the shock. Note, however, that the stagnation pressure decreases across the shock because of the irreversibilities, while the ordinary (static) temperature rises drastically because of the conversion of kinetic energy into enthalpy due to a large drop in fluid velocity (see Fig. 17–32).

We now develop relations between various properties before and after the shock for an ideal gas with constant specific heats. A relation for the ratio of the static temperatures $T_2/T_1$ is obtained by applying Eq. 17–18 twice:

$$\frac{T_{01}}{T_1} = 1 + \left(\frac{k-1}{2}\right)Ma_1^2 \quad \text{and} \quad \frac{T_{02}}{T_2} = 1 + \left(\frac{k-1}{2}\right)Ma_2^2$$

Dividing the first equation by the second one and noting that $T_{01} = T_{02}$, we have

$$\frac{T_2}{T_1} = \frac{1 + Ma_1^2(k-1)/2}{1 + Ma_2^2(k-1)/2} \qquad \textbf{(17–35)}$$

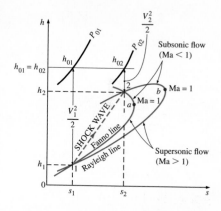

**FIGURE 17–31**

The *h-s* diagram for flow across a normal shock.

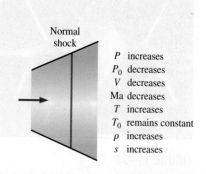

**FIGURE 17–32**

Variation of flow properties across a normal shock in an ideal gas.

From the ideal-gas equation of state,

$$\rho_1 = \frac{P_1}{RT_1} \quad \text{and} \quad \rho_2 = \frac{P_2}{RT_2}$$

Substituting these into the conservation of mass relation $\rho_1 V_1 = \rho_2 V_2$ and noting that $\text{Ma} = V/c$ and $c = \sqrt{kRT}$, we have

$$\frac{T_2}{T_1} = \frac{P_2 V_2}{P_1 V_1} = \frac{P_2 \text{Ma}_2 c_2}{P_1 \text{Ma}_1 c_1} = \frac{P_2 \text{Ma}_2 \sqrt{T_2}}{P_1 \text{Ma}_1 \sqrt{T_1}} = \left(\frac{P_2}{P_1}\right)^2 \left(\frac{\text{Ma}_2}{\text{Ma}_1}\right)^2 \quad (17\text{–}36)$$

Combining Eqs. 17–35 and 12–36 gives the pressure ratio across the shock:

Fanno line:
$$\frac{P_2}{P_1} = \frac{\text{Ma}_1 \sqrt{1 + \text{Ma}_1^2(k-1)/2}}{\text{Ma}_2 \sqrt{1 + \text{Ma}_2^2(k-1)/2}} \quad (17\text{–}37)$$

Equation 17–37 is a combination of the conservation of mass and energy equations; thus, it is also the equation of the Fanno line for an ideal gas with constant specific heats. A similar relation for the Rayleigh line is obtained by combining the conservation of mass and momentum equations. From Eq. 17–32,

$$P_1 - P_2 = \frac{\dot{m}}{A}(V_2 - V_1) = \rho_2 V_2^2 - \rho_1 V_1^2$$

However,

$$\rho V^2 = \left(\frac{P}{RT}\right)(\text{Ma} \, c)^2 = \left(\frac{P}{RT}\right)(\text{Ma}\sqrt{kRT})^2 = Pk\text{Ma}^2$$

Thus,

$$P_1(1 + k\text{Ma}_1^2) = P_2(1 + k\text{Ma}_2^2)$$

or

Rayleigh line:
$$\frac{P_2}{P_1} = \frac{1 + k\text{Ma}_1^2}{1 + k\text{Ma}_2^2} \quad (17\text{–}38)$$

Combining Eqs. 17–37 and 17–38 yields

$$\text{Ma}_2^2 = \frac{\text{Ma}_1^2 + 2/(k-1)}{2\text{Ma}_1^2 k/(k-1) - 1} \quad (17\text{–}39)$$

This represents the intersections of the Fanno and Rayleigh lines and relates the Mach number upstream of the shock to that downstream of the shock.

The occurrence of shock waves is not limited to supersonic nozzles only. This phenomenon is also observed at the engine inlet of supersonic aircraft, where the air passes through a shock and decelerates to subsonic velocities before entering the diffuser of the engine (Fig. 17–33). Explosions also produce powerful expanding spherical normal shocks, which can be very destructive (Fig. 17–34).

Various flow property ratios across the shock are listed in Table A–33 for an ideal gas with $k = 1.4$. Inspection of this table reveals that $\text{Ma}_2$ (the Mach

**FIGURE 17–33**

The air inlet of a supersonic fighter jet is designed such that a shock wave at the inlet decelerates the air to subsonic velocities, increasing the pressure and temperature of the air before it enters the engine.

*© StockTrek/Getty RF*

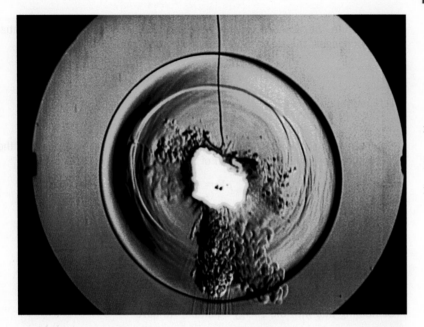

**FIGURE 17–34**
Schlieren image of the blast wave (expanding spherical normal shock) produced by the explosion of a firecracker. The shock expanded radially outward in all directions at a supersonic speed that decreased with radius from the center of the explosion. A microphone sensed the sudden change in pressure of the passing shock wave and triggered the microsecond flashlamp that exposed the photograph.

*G.S. Settles, Gas Dynamics Lab, Penn State University. Used with permission*

number after the shock) is always less than 1 and that the larger the supersonic Mach number before the shock, the smaller the subsonic Mach number after the shock. Also, we see that the static pressure, temperature, and density all increase after the shock while the stagnation pressure decreases.

The entropy change across the shock is obtained by applying the entropy-change equation for an ideal gas across the shock:

$$s_2 - s_1 = c_P \ln \frac{T_2}{T_1} - R \ln \frac{P_2}{P_1} \qquad \textbf{(17–40)}$$

which can be expressed in terms of $k$, $R$, and $Ma_1$ by using the relations developed earlier in this section. A plot of nondimensional entropy change across the normal shock $(s_2 - s_1)/R$ versus $Ma_1$ is shown in Fig. 17–35. Since the flow across the shock is adiabatic and irreversible, the second law of thermodynamics requires that the entropy increase across the shock wave. Thus, a shock wave cannot exist for values of $Ma_1$ less than unity where the entropy change would be negative. For adiabatic flows, shock waves can exist only for supersonic flows, $Ma_1 > 1$.

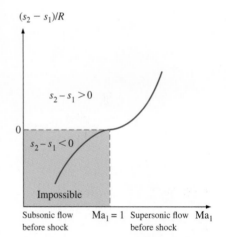

**FIGURE 17–35**
Entropy change across a normal shock.

---

■ **EXAMPLE 17–8**   **The Point of Maximum Entropy**
■                        **on the Fanno Line**
■
■ Show that the point of maximum entropy on the Fanno line (point *a* of
■ Fig. 17–31) for the adiabatic steady flow of a fluid in a duct corresponds to
   the sonic velocity, Ma = 1.

   **SOLUTION**   It is to be shown that the point of maximum entropy on the
   Fanno line for steady adiabatic flow corresponds to sonic velocity.
   *Assumption*   The flow is steady, adiabatic, and one-dimensional.

*Analysis* In the absence of any heat and work interactions and potential energy changes, the steady-flow energy equation reduces to

$$h + \frac{V^2}{2} = \text{constant}$$

Differentiating yields

$$dh + V\,dV = 0$$

For a very thin shock with negligible change of duct area across the shock, the steady-flow continuity (conservation of mass) equation is expressed as

$$\rho V = \text{constant}$$

Differentiating, we have

$$\rho\,dV + V\,d\rho = 0$$

Solving for $dV$ gives

$$dV = -V\frac{d\rho}{\rho}$$

Combining this with the energy equation, we have

$$dh - V^2\frac{d\rho}{\rho} = 0$$

which is the equation for the Fanno line in differential form. At point $a$ (the point of maximum entropy) $ds = 0$. Then from the second $T\,ds$ relation ($T\,ds = dh - v\,dP$) we have $dh = v\,dP = dP/\rho$. Substituting yields

$$\frac{dP}{\rho} - V^2\frac{d\rho}{\rho} = 0 \quad \text{at } s = \text{constant}$$

Solving for $V$, we have

$$V = \left(\frac{\partial P}{\partial \rho}\right)_s^{1/2}$$

which is the relation for the speed of sound, Eq. 17–9. Thus $V = c$ and the proof is complete.

### EXAMPLE 17–9    Shock Wave in a Converging–Diverging Nozzle

If the air flowing through the converging–diverging nozzle of Example 17–7 experiences a normal shock wave at the nozzle exit plane (Fig. 17–36), determine the following after the shock: (*a*) the stagnation pressure, static pressure, static temperature, and static density; (*b*) the entropy change across the shock; (*c*) the exit velocity; and (*d*) the mass flow rate through the nozzle. Approximate the flow as steady, one-dimensional, and isentropic with $k = 1.4$ from the nozzle inlet to the shock location.

**SOLUTION** Air flowing through a converging–diverging nozzle experiences a normal shock at the exit. The effect of the shock wave on various properties is to be determined.

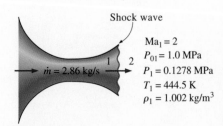

Shock wave

$\text{Ma}_1 = 2$
$P_{01} = 1.0 \text{ MPa}$
$P_1 = 0.1278 \text{ MPa}$
$T_1 = 444.5 \text{ K}$
$\rho_1 = 1.002 \text{ kg/m}^3$

$\dot{m} = 2.86 \text{ kg/s}$

**FIGURE 17–36**
Schematic for Example 17–9.

*Assumptions* **1** Air is an ideal gas with constant specific heats at room temperature. **2** Flow through the nozzle is steady, one-dimensional, and isentropic before the shock occurs. **3** The shock wave occurs at the exit plane.

*Properties* The constant-pressure specific heat and the specific heat ratio of air are $c_p = 1.005$ kJ/kg·K and $k = 1.4$. The gas constant of air is 0.287 kJ/kg·K.

*Analysis* (a) The fluid properties at the exit of the nozzle just before the shock (denoted by subscript 1) are those evaluated in Example 17–7 at the nozzle exit to be

$$P_{01} = 1.0 \text{ MPa} \quad P_1 = 0.1278 \text{ MPa} \quad T_1 = 444.5 \text{ K} \quad \rho_1 = 1.002 \text{ kg/m}^3$$

The fluid properties after the shock (denoted by subscript 2) are related to those before the shock through the functions listed in Table A–33. For $Ma_1 = 2.0$, we read

$$Ma_2 = 0.5774 \quad \frac{P_{02}}{P_{01}} = 0.7209 \quad \frac{P_2}{P_1} = 4.5000 \quad \frac{T_2}{T_1} = 1.6875 \quad \frac{\rho_2}{\rho_1} = 2.6667$$

Then the stagnation pressure $P_{02}$, static pressure $P_2$, static temperature $T_2$, and static density $\rho_2$ after the shock are

$$P_{02} = 0.7209P_{01} = (0.7209)(1.0 \text{ MPa}) = \textbf{0.721 MPa}$$

$$P_2 = 4.5000P_1 = (4.5000)(0.1278 \text{ MPa}) = \textbf{0.575 MPa}$$

$$T_2 = 1.6875T_1 = (1.6875)(444.5 \text{ K}) = \textbf{750 K}$$

$$\rho_2 = 2.6667\rho_1 = (2.6667)(1.002 \text{ kg/m}^3) = \textbf{2.67 kg/m}^3$$

(*b*) The entropy change across the shock is

$$s_2 - s_1 = c_p \ln \frac{T_2}{T_1} - R \ln \frac{P_2}{P_1}$$

$$= (1.005 \text{ kJ/kg·K}) \ln (1.6875) - (0.287 \text{ kJ/kg·K}) \ln (4.5000)$$

$$= \textbf{0.0942 kJ/kg·K}$$

Thus, the entropy of the air increases as it passes through a normal shock, which is highly irreversible.

(*c*) The air velocity after the shock is determined from $V_2 = Ma_2c_2$, where $c_2$ is the speed of sound at the exit conditions after the shock:

$$V_2 = Ma_2c_2 = Ma_2\sqrt{kRT_2}$$

$$= (0.5774)\sqrt{(1.4)(0.287 \text{ kJ/kg·K})(750.1 \text{ K})\left(\frac{1000 \text{ m}^2/\text{s}^2}{1 \text{ kJ/kg}}\right)}$$

$$= \textbf{317 m/s}$$

(*d*) The mass flow rate through a converging–diverging nozzle with sonic conditions at the throat is not affected by the presence of shock waves in the nozzle. Therefore, the mass flow rate in this case is the same as that determined in Example 17–7:

$$\dot{m} = \textbf{2.86 kg/s}$$

*Discussion* This result can easily be verified by using property values at the nozzle exit after the shock at all Mach numbers significantly greater than unity.

**FIGURE 17–37**

When a lion tamer cracks his whip, a weak spherical shock wave forms near the tip and spreads out radially; the pressure inside the expanding shock wave is higher than ambient air pressure, and this is what causes the crack when the shock wave reaches the lion's ear.

*© Joshua Ets-Hokin/Getty RF*

Example 17–9 illustrates that the stagnation pressure and velocity decrease while the static pressure, temperature, density, and entropy increase across the shock (Fig. 17–37). The rise in the temperature of the fluid downstream of a shock wave is of major concern to the aerospace engineer because it creates heat transfer problems on the leading edges of wings and nose cones of space reentry vehicles and the recently proposed hypersonic space planes. Overheating, in fact, led to the tragic loss of the space shuttle *Columbia* in February of 2003 as it was reentering earth's atmosphere.

## Oblique Shocks

Not all shock waves are normal shocks (perpendicular to the flow direction). For example, when the space shuttle travels at supersonic speeds through the atmosphere, it produces a complicated shock pattern consisting of inclined shock waves called **oblique shocks** (Fig. 17–38). As you can see, some portions of an oblique shock are curved, while other portions are straight.

First, we consider straight oblique shocks, like that produced when a uniform supersonic flow ($Ma_1 > 1$) impinges on a slender, two-dimensional wedge of half-angle $\delta$ (Fig. 17–39). Since information about the wedge cannot travel upstream in a supersonic flow, the fluid "knows" nothing about the wedge until it hits the nose. At that point, since the fluid cannot flow *through* the wedge, it turns suddenly through an angle called the **turning angle** or **deflection angle** $\theta$. The result is a straight oblique shock wave, aligned at **shock angle** or **wave angle** $\beta$, measured relative to the oncoming flow (Fig. 17–40). To conserve mass, $\beta$ must obviously be greater than $\delta$. Since the Reynolds number of supersonic flows is typically large, the boundary layer growing along the wedge is very thin, and we ignore its effects. The flow therefore turns by the same angle as the wedge; namely, deflection angle $\theta$ is equal to wedge half-angle $\delta$. If we take into account the displacement thickness effect of the boundary layer, the deflection angle $\theta$ of the oblique shock turns out to be slightly greater than wedge half-angle $\delta$.

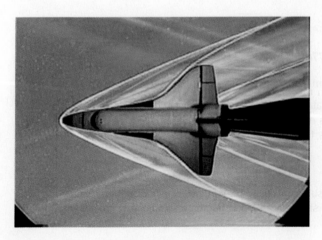

**FIGURE 17–38**

Schlieren image of a small model of the space shuttle orbiter being tested at Mach 3 in the supersonic wind tunnel of the Penn State Gas Dynamics Lab. Several *oblique shocks* are seen in the air surrounding the spacecraft.

*G.S. Settles, Gas Dynamics Lab, Penn State University. Used with permission*

Like normal shocks, the Mach number decreases across an oblique shock, and oblique shocks are possible only if the upstream flow is supersonic. However, unlike normal shocks, in which the downstream Mach number is always subsonic, $Ma_2$ downstream of an oblique shock can be subsonic, sonic, or supersonic, depending on the upstream Mach number $Ma_1$ and the turning angle.

We analyze a straight oblique shock in Fig. 17–40 by decomposing the velocity vectors upstream and downstream of the shock into normal and tangential components, and considering a small control volume around the shock. Upstream of the shock, all fluid properties (velocity, density, pressure, etc.) along the lower left face of the control volume are identical to those along the upper right face. The same is true downstream of the shock. Therefore, the mass flow rates entering and leaving those two faces cancel each other out, and conservation of mass reduces to

$$\rho_1 V_{1,n} A = \rho_2 V_{2,n} A \quad \rightarrow \quad \rho_1 V_{1,n} = \rho_2 V_{2,n} \quad \text{(17–41)}$$

where $A$ is the area of the control surface that is parallel to the shock. Since $A$ is the same on either side of the shock, it has dropped out of Eq. 17–41.

As you might expect, the tangential component of velocity (parallel to the oblique shock) does not change across the shock, i.e., $V_{1,t} = V_{2,t}$. This is easily proven by applying the tangential momentum equation to the control volume.

When we apply conservation of momentum in the direction *normal* to the oblique shock, the only forces are pressure forces, and we get

$$P_1 A - P_2 A = \rho V_{2,n} A V_{2,n} - \rho V_{1,n} A V_{1,n} \quad \rightarrow \quad P_1 - P_2 = \rho_2 V_{2,n}^2 - \rho_1 V_{1,n}^2 \quad \text{(17–42)}$$

Finally, since there is no work done by the control volume and no heat transfer into or out of the control volume, stagnation enthalpy does *not* change across an oblique shock, and conservation of energy yields

$$h_{01} = h_{02} = h_0 \quad \rightarrow \quad h_1 + \frac{1}{2} V_{1,n}^2 + \frac{1}{2} V_{1,t}^2 = h_2 + \frac{1}{2} V_{2,n}^2 + \frac{1}{2} V_{2,t}^2$$

But since $V_{1,t} = V_{2,t}$, this equation reduces to

$$h_1 + \frac{1}{2} V_{1,n}^2 = h_2 + \frac{1}{2} V_{2,n}^2 \quad \text{(17–43)}$$

Careful comparison reveals that the equations for conservation of mass, momentum, and energy (Eqs. 17–41 through 17–43) across an oblique shock are identical to those across a normal shock, except that they are written in terms of the *normal* velocity component only. Therefore, the normal shock relations derived previously apply to oblique shocks as well, but must be written in terms of Mach numbers $Ma_{1,n}$ and $Ma_{2,n}$ normal to the oblique shock. This is most easily visualized by rotating the velocity vectors in Fig. 17–40 by angle $\pi/2 - \beta$, so that the oblique shock appears to be vertical (Fig. 17–41). Trigonometry yields

$$Ma_{1,n} = Ma_1 \sin \beta \quad \text{and} \quad Ma_{2,n} = Ma_2 \sin(\beta - \theta) \quad \text{(17–44)}$$

where $Ma_{1,n} = V_{1,n}/c_1$ and $Ma_{2,n} = V_{2,n}/c_2$. From the point of view shown in Fig. 17–41, we see what looks like a normal shock, but with some superposed tangential flow "coming along for the ride." Thus,

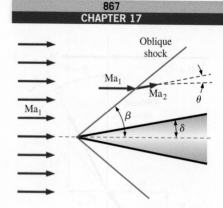

**FIGURE 17–39**
An oblique shock of *shock angle β* formed by a slender, two-dimensional wedge of half-angle δ. The flow is turned by *deflection angle θ* downstream of the shock, and the Mach number decreases.

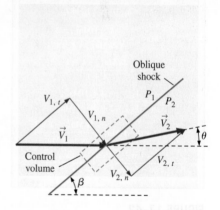

**FIGURE 17–40**
Velocity vectors through an oblique shock of shock angle β and deflection angle θ.

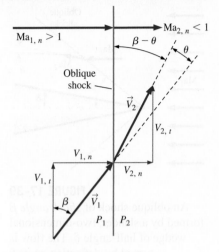

$\mathrm{Ma}_{1, n} > 1$     $\mathrm{Ma}_{2, n} < 1$

**FIGURE 17–41**
The same velocity vectors of Fig. 17–40, but rotated by angle $\pi/2 - \beta$, so that the oblique shock is vertical. Normal Mach numbers $\mathrm{Ma}_{1, n}$ and $\mathrm{Ma}_{2, n}$ are also defined.

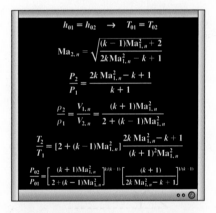

**FIGURE 17–42**
Relationships across an oblique shock for an ideal gas in terms of the normal component of upstream Mach number $\mathrm{Ma}_{1, n}$.

All the equations, shock tables, etc., for normal shocks apply to oblique shocks as well, provided that we use only the **normal** components of the Mach number.

In fact, you may think of normal shocks as special oblique shocks in which shock angle $\beta = \pi/2$, or $90°$. We recognize immediately that an oblique shock can exist only if $\mathrm{Ma}_{1, n} > 1$ and $\mathrm{Ma}_{2, n} < 1$. The normal shock equations appropriate for oblique shocks in an ideal gas are summarized in Fig. 17–42 in terms of $\mathrm{Ma}_{1, n}$.

For known shock angle $\beta$ and known upstream Mach number $\mathrm{Ma}_1$, we use the first part of Eq. 17–44 to calculate $\mathrm{Ma}_{1, n}$, and then use the normal shock tables (or their corresponding equations) to obtain $\mathrm{Ma}_{2, n}$. If we also knew the deflection angle $\theta$, we could calculate $\mathrm{Ma}_2$ from the second part of Eq. 17–44. But, in a typical application, we know either $\beta$ or $\theta$, but not both. Fortunately, a bit more algebra provides us with a relationship between $\theta$, $\beta$, and $\mathrm{Ma}_1$. We begin by noting that $\tan \beta = V_{1, n}/V_{1, t}$ and $\tan(\beta - \theta) = V_{2, n}/V_{2, t}$ (Fig. 17–41). But since $V_{1, t} = V_{2, t}$, we combine these two expressions to yield

$$\frac{V_{2, n}}{V_{1, n}} = \frac{\tan(\beta - \theta)}{\tan \beta} = \frac{2 + (k - 1)\mathrm{Ma}_{1, n}^2}{(k + 1)\mathrm{Ma}_{1, n}^2} = \frac{2 + (k - 1)\mathrm{Ma}_1^2 \sin^2 \beta}{(k + 1)\mathrm{Ma}_1^2 \sin^2 \beta} \quad (17\text{–}45)$$

where we have also used Eq. 17–44 and the fourth equation of Fig. 17–42. We apply trigonometric identities for $\cos 2\beta$ and $\tan(\beta - \theta)$, namely,

$$\cos 2\beta = \cos^2 \beta - \sin^2 \beta \quad \text{and} \quad \tan(\beta - \theta) = \frac{\tan \beta - \tan \theta}{1 + \tan \beta \tan \theta}$$

After some algebra, Eq. 17–45 reduces to

The $\theta$-$\beta$-Ma relationship:     $$\tan \theta = \frac{2 \cot \beta (\mathrm{Ma}_1^2 \sin^2 \beta - 1)}{\mathrm{Ma}_1^2 (k + \cos 2\beta) + 2} \quad (17\text{–}46)$$

Equation 17–46 provides deflection angle $\theta$ as a unique function of shock angle $\beta$, specific heat ratio $k$, and upstream Mach number $\mathrm{Ma}_1$. For air ($k = 1.4$), we plot $\theta$ versus $\beta$ for several values of $\mathrm{Ma}_1$ in Fig. 17–43. We note that this plot is often presented with the axes reversed ($\beta$ versus $\theta$) in compressible flow textbooks, since, physically, shock angle $\beta$ is determined by deflection angle $\theta$.

Much can be learned by studying Fig. 17–43, and we list some observations here:

- Figure 17–43 displays the full range of possible shock waves at a given free-stream Mach number, from the weakest to the strongest. For any value of Mach number $\mathrm{Ma}_1$ greater than 1, the possible values of $\theta$ range from $\theta = 0°$ at some value of $\beta$ between 0 and $90°$, to a maximum value $\theta = \theta_{max}$ at an intermediate value of $\beta$, and then back to $\theta = 0°$ at $\beta = 90°$. Straight oblique shocks for $\theta$ or $\beta$ outside of this range *cannot* and *do not* exist. At $\mathrm{Ma}_1 = 1.5$, for example, straight oblique shocks cannot exist in air with shock angle $\beta$ less than about $42°$, nor with deflection angle $\theta$ greater than about $12°$. If the wedge half-angle is greater than $\theta_{max}$, the shock becomes curved and detaches from the nose of the wedge, forming what is called a **detached oblique shock** or a **bow wave** (Fig. 17–44). The shock angle $\beta$ of the detached shock is $90°$ at the nose, but $\beta$ decreases

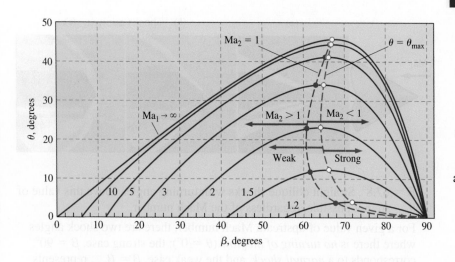

**FIGURE 17–43**

The dependence of straight oblique shock deflection angle $\theta$ on shock angle $\beta$ for several values of upstream Mach number $Ma_1$. Calculations are for an ideal gas with $k = 1.4$. The dashed red line connects points of maximum deflection angle ($\theta = \theta_{max}$). *Weak oblique shocks* are to the left of this line, while *strong oblique shocks* are to the right of this line. The dashed green line connects points where the downstream Mach number is *sonic* ($Ma_2 = 1$). *Supersonic downstream flow* ($Ma_2 > 1$) is to the left of this line, while *subsonic downstream flow* ($Ma_2 < 1$) is to the right of this line.

as the shock curves downstream. Detached shocks are much more complicated than simple straight oblique shocks to analyze. In fact, no simple solutions exist, and prediction of detached shocks requires computational methods.

- Similar oblique shock behavior is observed in *axisymmetric flow* over cones, as in Fig. 17–45, although the $\theta$-$\beta$-Ma relationship for axisymmetric flows differs from that of Eq. 17–46.

- When supersonic flow impinges on a blunt (or bluff) body—a body *without* a sharply pointed nose, the wedge half-angle $\delta$ at the nose is 90°, and an attached oblique shock cannot exist, regardless of Mach number. In fact, a detached oblique shock occurs in front of *all* such blunt-nosed bodies, whether two-dimensional, axisymmetric, or fully three-dimensional. For example, a detached oblique shock is seen in front of the space shuttle model in Fig. 17–38 and in front of a sphere in Fig. 17–46.

- While $\theta$ is a unique function of $Ma_1$ and $\beta$ for a given value of $k$, there are *two* possible values of $\beta$ for $\theta < \theta_{max}$. The dashed red line in Fig. 17–43 passes through the locus of $\theta_{max}$ values, dividing the shocks into **weak oblique shocks** (the smaller value of $\beta$) and **strong oblique shocks** (the larger value of $\beta$). At a given value of $\theta$, the weak shock is more common and is "preferred" by the flow unless the downstream pressure conditions are high enough for the formation of a strong shock.

- For a given upstream Mach number $Ma_1$, there is a unique value of $\theta$ for which the downstream Mach number $Ma_2$ is exactly 1. The dashed green line in Fig. 17–43 passes through the locus of values where $Ma_2 = 1$. To the left of this line, $Ma_2 > 1$, and to the right of this line, $Ma_2 < 1$. Downstream sonic conditions occur on the weak shock side of the plot, with $\theta$ very close to $\theta_{max}$. Thus, the flow downstream of a strong oblique shock is *always subsonic* ($Ma_2 < 1$). The flow downstream of a weak oblique shock remains *supersonic*, except for a narrow range of $\theta$ just below $\theta_{max}$, where it is subsonic, although it is still called a weak oblique shock.

- As the upstream Mach number approaches infinity, straight oblique shocks become possible for any $\beta$ between 0 and 90°, but the maximum possible turning angle for $k = 1.4$ (air) is $\theta_{max} \cong 45.6°$, which occurs at

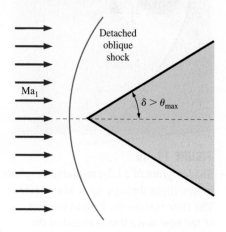

**FIGURE 17–44**

A *detached oblique shock* occurs upstream of a two-dimensional wedge of half-angle $\delta$ when $\delta$ is greater than the maximum possible deflection angle $\theta$. A shock of this kind is called a *bow wave* because of its resemblance to the water wave that forms at the bow of a ship.

**FIGURE 17–45**

Still frames from schlieren video-graphy illustrating the detachment of an oblique shock from a cone with increasing cone half-angle $\delta$ in air at Mach 3. At (a) $\delta = 20°$ and (b) $\delta = 40°$, the oblique shock remains attached, but by (c) $\delta = 60°$, the oblique shock has detached, forming a bow wave.

*G.S. Settles, Gas Dynamics Lab, Penn State University. Used with permission*

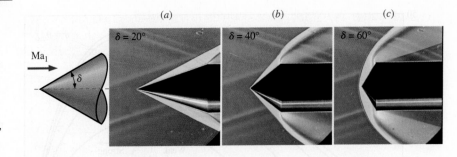

$\beta = 67.8°$. Straight oblique shocks with turning angles above this value of $\theta_{max}$ are not possible, regardless of the Mach number.

- For a given value of upstream Mach number, there are two shock angles where there is *no turning of the flow* ($\theta = 0°$): the strong case, $\beta = 90°$, corresponds to a *normal shock*, and the weak case, $\beta = \beta_{min}$, represents the weakest possible oblique shock at that Mach number, which is called a **Mach wave**. Mach waves are caused, for example, by very small non-uniformities on the walls of a supersonic wind tunnel (several can be seen in Figs. 17–38 and 17–45). Mach waves have no effect on the flow, since the shock is vanishingly weak. In fact, in the limit, Mach waves are *isentropic*. The shock angle for Mach waves is a unique function of the Mach number and is given the symbol $\mu$, not to be confused with the coefficient of viscosity. Angle $\mu$ is called the **Mach angle** and is found by setting $\theta$ equal to zero in Eq. 17–46, solving for $\beta = \mu$, and taking the smaller root. We get

*Mach angle:* $\qquad\qquad \mu = \sin^{-1}(1/Ma_1)$ $\qquad$ **(17–47)**

Since the specific heat ratio appears only in the denominator of Eq. 12–46, $\mu$ is independent of $k$. Thus, we can estimate the Mach number of any supersonic flow simply by measuring the Mach angle and applying Eq. 17–47.

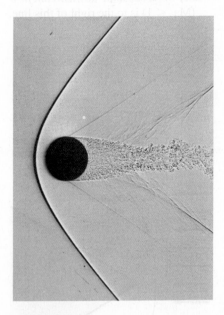

**FIGURE 17–46**

Shadowgram of a 1.3-cm-diameter sphere in free flight through air at Ma = 1.53. The flow is subsonic behind the part of the bow wave that is ahead of the sphere and over its surface back to about 45°. At about 90° the laminar boundary layer separates through an oblique shock wave and quickly becomes turbulent. The fluctuating wake generates a system of weak disturbances that merge into the second "recompression" shock wave.

*A. C. Charters, Army Ballistic Research Laboratory*

## Prandtl–Meyer Expansion Waves

We now address situations where supersonic flow is turned in the *opposite* direction, such as in the upper portion of a two-dimensional wedge at an angle of attack greater than its half-angle $\delta$ (Fig. 17–47). We refer to this type of flow as an **expanding flow**, whereas a flow that produces an oblique shock may be called a **compressing flow**. As previously, the flow changes direction to conserve mass. However, unlike a compressing flow, an expanding flow does *not* result in a shock wave. Rather, a continuous expanding region called an **expansion fan** appears, composed of an infinite number of Mach waves called **Prandtl–Meyer expansion waves**. In other words, the flow does not turn suddenly, as through a shock, but *gradually*—each successive Mach wave turns the flow by an infinitesimal amount. Since each individual expansion wave is nearly isentropic, the flow across the entire expansion fan is also nearly isentropic. The Mach number downstream of the expansion *increases* ($Ma_2 > Ma_1$), while pressure, density, and temperature *decrease*, just as they do in the supersonic (expanding) portion of a converging–diverging nozzle.

Prandtl–Meyer expansion waves are inclined at the local Mach angle $\mu$, as sketched in Fig. 17–47. The Mach angle of the first expansion wave is easily determined as $\mu_1 = \sin^{-1}(1/Ma_1)$. Similarly, $\mu_2 = \sin^{-1}(1/Ma_2)$, where we must be careful to measure the angle relative to the *new* direction of flow downstream of the expansion, namely, parallel to the upper wall of the wedge in Fig. 17–47 if we neglect the influence of the boundary layer along the wall. But how do we determine $Ma_2$? It turns out that the turning angle $\theta$ across the expansion fan can be calculated by integration, making use of the isentropic flow relationships. For an ideal gas, the result is (Anderson, 2003),

*Turning angle across an expansion fan:*     $\theta = \nu(Ma_2) - \nu(Ma_1)$     (17–48)

where $\nu(Ma)$ is an angle called the **Prandtl–Meyer function** (not to be confused with the kinematic viscosity),

$$\nu(Ma) = \sqrt{\frac{k+1}{k-1}} \tan^{-1}\left(\sqrt{\frac{k-1}{k+1}(Ma^2-1)}\right) - \tan^{-1}\left(\sqrt{Ma^2-1}\right) \quad (17\text{–}49)$$

Note that $\nu(Ma)$ is an angle, and can be calculated in either degrees or radians. Physically, $\nu(Ma)$ is the angle through which the flow must expand, starting with $\nu = 0$ at $Ma = 1$, in order to reach a supersonic Mach number, $Ma > 1$.

To find $Ma_2$ for known values of $Ma_1$, $k$, and $\theta$, we calculate $\nu(Ma_1)$ from Eq. 17–49, $\nu(Ma_2)$ from Eq. 17–48, and then $Ma_2$ from Eq. 17–49, noting that the last step involves solving an implicit equation for $Ma_2$. Since there is no heat transfer or work, and the flow can be approximated as isentropic through the expansion, $T_0$ and $P_0$ remain constant, and we use the isentropic flow relations derived previously to calculate other flow properties downstream of the expansion, such as $T_2$, $\rho_2$, and $P_2$.

Prandtl–Meyer expansion fans also occur in axisymmetric supersonic flows, as in the corners and trailing edges of a cone-cylinder (Fig. 17–48).

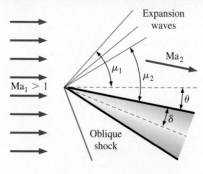

**FIGURE 17–47**
An expansion fan in the upper portion of the flow formed by a two-dimensional wedge at an angle of attack in a supersonic flow. The flow is turned by angle $\theta$, and the Mach number increases across the expansion fan. Mach angles upstream and downstream of the expansion fan are indicated. Only three expansion waves are shown for simplicity, but in fact, there are an infinite number of them. (An oblique shock is also present in the bottom portion of this flow.)

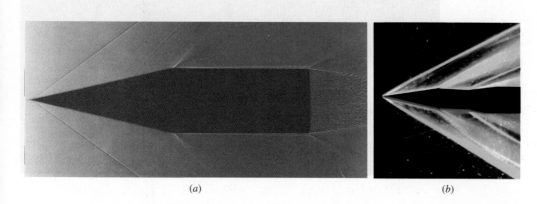

(a)                              (b)

**FIGURE 17–48**
(a) A cone-cylinder of 12.5° half-angle in a Mach number 1.84 flow. The boundary layer becomes turbulent shortly downstream of the nose, generating Mach waves that are visible in this shadowgraph. Expansion waves are seen at the corners and at the trailing edge of the cone. (b) A similar pattern for Mach 3 flow over an 11° 2-D wedge.

*(a) A. C. Charters, Army Ballistic Research Laboratory (b) G.S. Settles, Gas Dynamics Lab, Penn State University. Used with permission*

Some very complex and, to some of us, beautiful interactions involving both shock waves and expansion waves occur in the supersonic jet produced by an "overexpanded" nozzle, as in Fig. 17–49. When such patterns are visible in the exhaust of a jet engine, pilots refer to it as a "tiger tail." Analysis of such flows is beyond the scope of the present text; interested readers are referred to compressible flow textbooks such as Thompson (1972), Leipmann and Roshko (2001), and Anderson (2003).

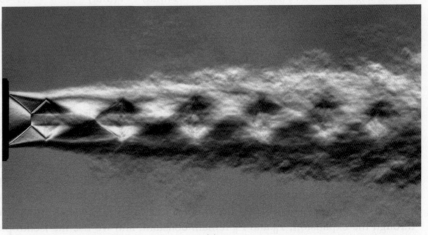

(a)

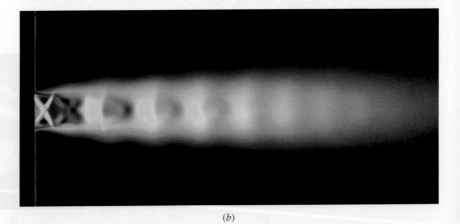

(b)

**FIGURE 17–49**

The complex interactions between shock waves and expansion waves in an "overexpanded" supersonic jet. (a) The flow is visualized by a schlieren-like differential interferogram. (b) Color shlieren image.

(a) Photo by H. Oertel sen. Reproduced by courtesy of the French-German Research Institute of Saint-Louis, ISL. Used with permission. (b) G.S. Settles, Gas Dynamics Lab, Penn State University. Used with permiss

**EXAMPLE 17–10    Estimation of the Mach Number
from Mach Lines**

Estimate the Mach number of the free-stream flow upstream of the space
shuttle in Fig. 17–38 from the figure alone. Compare with the known value
of Mach number provided in the figure caption.

**SOLUTION**  We are to estimate the Mach number from a figure and compare
it to the known value.
**Analysis**  Using a protractor, we measure the angle of the Mach lines
in the free-stream flow: $\mu \cong 19°$. The Mach number is obtained from
Eq. 17–47,

$$\mu = \sin^{-1}\left(\frac{1}{Ma_1}\right) \rightarrow Ma_1 = \frac{1}{\sin 19°} \rightarrow Ma_1 = 3.07$$

Our estimated Mach number agrees with the experimental value of $3.0 \pm 0.1$.
**Discussion**  The result is independent of the fluid properties.

**EXAMPLE 17–11    Oblique Shock Calculations**

Supersonic air at $Ma_1 = 2.0$ and 75.0 kPa impinges on a two-dimensional
wedge of half-angle $\delta = 10°$ (Fig. 17–50). Calculate the two possible oblique
shock angles, $\beta_{weak}$ and $\beta_{strong}$, that could be formed by this wedge. For each
case, calculate the pressure and Mach number downstream of the oblique
shock, compare, and discuss.

**SOLUTION**  We are to calculate the shock angle, Mach number, and pressure
downstream of the weak and strong oblique shock formed by a two-dimensional
wedge.
**Assumptions**  1 The flow is steady. 2 The boundary layer on the wedge is
very thin.
**Properties**  The fluid is air with $k = 1.4$.
**Analysis**  Because of assumption 2, we approximate the oblique shock
deflection angle as equal to the wedge half-angle, i.e., $\theta \cong \delta = 10°$. With
$Ma_1 = 2.0$ and $\theta = 10°$, we solve Eq. 17–46 for the two possible values of
oblique shock angle $\beta$: $\beta_{weak} = \mathbf{39.3°}$ and $\beta_{strong} = \mathbf{83.7°}$. From these values,
we use the first part of Eq. 17–44 to calculate upstream normal Mach
number $Ma_{1,n}$,

*Weak shock*:     $Ma_{1,n} = Ma_1 \sin \beta \rightarrow Ma_{1,n} = 2.0 \sin 39.3° = 1.267$

and

*Strong shock*:     $Ma_{1,n} = Ma_1 \sin \beta \rightarrow Ma_{1,n} = 2.0 \sin 83.7° = 1.988$

We substitute these values of $Ma_{1,n}$ into the second equation of Fig. 17–42
to calculate the downstream normal Mach number $Ma_{2,n}$. For the weak
shock, $Ma_{2,n} = 0.8032$, and for the strong shock, $Ma_{2,n} = 0.5794$. We also
calculate the downstream pressure for each case, using the third equation of
Fig. 17–42, which gives

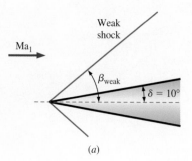

(a)

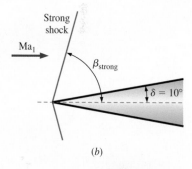

(b)

**FIGURE 17–50**
Two possible oblique shock angles,
(a) $\beta_{weak}$ and (b) $\beta_{strong}$, formed by a
two-dimensional wedge of half-angle
$\delta = 10°$.

*Weak shock:*

$$\frac{P_2}{P_1} = \frac{2k\,\mathrm{Ma}_{1,n}^2 - k + 1}{k + 1} \rightarrow P_2 = (75.0\ \text{kPa})\frac{2(1.4)(1.267)^2 - 1.4 + 1}{1.4 + 1} = \textbf{128 kPa}$$

and

*Strong shock:*

$$\frac{P_2}{P_1} = \frac{2k\,\mathrm{Ma}_{1,n}^2 - k + 1}{k + 1} \rightarrow P_2 = (75.0\ \text{kPa})\frac{2(1.4)(1.988)^2 - 1.4 + 1}{1.4 + 1} = \textbf{333 kPa}$$

Finally, we use the second part of Eq. 17–44 to calculate the downstream Mach number,

*Weak shock:* $\qquad \mathrm{Ma}_2 = \dfrac{\mathrm{Ma}_{2,n}}{\sin(\beta - \theta)} = \dfrac{0.8032}{\sin(39.3° - 10°)} = \textbf{1.64}$

and

*Strong shock:* $\qquad \mathrm{Ma}_2 = \dfrac{\mathrm{Ma}_{2,n}}{\sin(\beta - \theta)} = \dfrac{0.5794}{\sin(83.7° - 10°)} = \textbf{0.604}$

*The changes in Mach number and pressure across the strong shock are much greater than the changes across the weak shock,* as expected.

**Discussion** Since Eq. 12–46 is implicit in $\beta$, we solve it by an iterative approach or with an equation solver such as EES. For both the weak and strong oblique shock cases, $\mathrm{Ma}_{1,n}$ is supersonic and $\mathrm{Ma}_{2,n}$ is subsonic. However, $\mathrm{Ma}_2$ is *supersonic* across the weak oblique shock, but *subsonic* across the strong oblique shock. We could also use the normal shock tables in place of the equations, but with loss of precision.

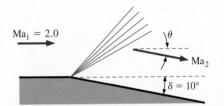

**FIGURE 17–51**
An expansion fan caused by the sudden expansion of a wall with $\delta = 10°$.

**EXAMPLE 17–12    Prandtl–Meyer Expansion Wave Calculations**

Supersonic air at $\mathrm{Ma}_1 = 2.0$ and 230 kPa flows parallel to a flat wall that suddenly expands by $\delta = 10°$ (Fig. 17–51). Ignoring any effects caused by the boundary layer along the wall, calculate downstream Mach number $\mathrm{Ma}_2$ and pressure $P_2$.

**SOLUTION** We are to calculate the Mach number and pressure downstream of a sudden expansion along a wall.
**Assumptions** 1 The flow is steady. 2 The boundary layer on the wall is very thin.
**Properties** The fluid is air with $k = 1.4$.
**Analysis** Because of assumption 2, we approximate the total deflection angle as equal to the wall expansion angle, i.e., $\theta \cong \delta = 10°$. With $\mathrm{Ma}_1 = 2.0$, we solve Eq. 17–49 for the upstream Prandtl–Meyer function,

$$\nu(\mathrm{Ma}) = \sqrt{\frac{k+1}{k-1}}\, \tan^{-1}\left(\sqrt{\frac{k-1}{k+1}(\mathrm{Ma}^2 - 1)}\right) - \tan^{-1}\left(\sqrt{\mathrm{Ma}^2 - 1}\right)$$

$$= \sqrt{\frac{1.4+1}{1.4-1}}\, \tan^{-1}\left(\sqrt{\frac{1.4-1}{1.4+1}(2.0^2 - 1)}\right) - \tan^{-1}\left(\sqrt{2.0^2 - 1}\right) = 26.38°$$

Next, we use Eq. 17–48 to calculate the downstream Prandtl–Meyer function,

$$\theta = \nu(Ma_2) - \nu(Ma_1) \rightarrow \nu(Ma_2) = \theta + \nu(Ma_1) = 10° + 26.38° = 36.38°$$

$Ma_2$ is found by solving Eq. 17–49, which is implicit—an equation solver is helpful. We get $Ma_2 = 2.38$. There are also compressible flow calculators on the Internet that solve these implicit equations, along with both normal and oblique shock equations; e.g., see www.aoe.vt.edu/~devenpor/aoe3114/calc.html.

We use the isentropic relations to calculate the downstream pressure,

$$P_2 = \frac{P_2/P_0}{P_1/P_0} P_1 = \frac{\left[1 + \left(\dfrac{k-1}{2}\right)Ma_2^2\right]^{-k/(k-1)}}{\left[1 + \left(\dfrac{k-1}{2}\right)Ma_1^2\right]^{-k/(k-1)}} (230\ kPa) = 126\ kPa$$

Since this is an expansion, Mach number increases and pressure decreases, as expected.

**Discussion** We could also solve for downstream temperature, density, etc., using the appropriate isentropic relations.

# 17–6 · DUCT FLOW WITH HEAT TRANSFER AND NEGLIGIBLE FRICTION (RAYLEIGH FLOW)

So far we have limited our consideration mostly to *isentropic flow*, also *called reversible adiabatic flow* since it involves no heat transfer and no irreversibilities such as friction. Many compressible flow problems encountered in practice involve chemical reactions such as combustion, nuclear reactions, evaporation, and condensation as well as heat gain or heat loss through the duct wall. Such problems are difficult to analyze exactly since they may involve significant changes in chemical composition during flow, and the conversion of latent, chemical, and nuclear energies to thermal energy (Fig. 17–52).

The essential features of such complex flows can still be captured by a simple analysis by modeling the generation or absorption of thermal energy as heat transfer through the duct wall at the same rate and disregarding any changes in chemical composition. This simplified problem is still too complicated for an elementary treatment of the topic since the flow may involve friction, variations in duct area, and multidimensional effects. In this section, we limit our consideration to one-dimensional flow in a duct of constant cross-sectional area with negligible frictional effects.

Consider steady one-dimensional flow of an ideal gas with constant specific heats through a constant-area duct with heat transfer, but with negligible friction. Such flows are referred to as **Rayleigh flows** after Lord Rayleigh (1842–1919). The conservation of mass, momentum, and

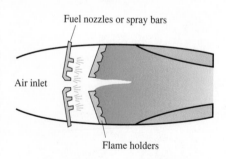

**FIGURE 17–52**
Many practical compressible flow problems involve combustion, which may be modeled as heat gain through the duct wall.

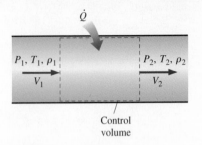

**FIGURE 17–53**
Control volume for flow in a
constant-area duct with heat
transfer and negligible friction.

energy equations for the control volume shown in Fig. 17–53 are written as follows:

**Mass equation**   Noting that the duct cross-sectional area $A$ is constant, the relation $\dot{m}_1 = \dot{m}_2$ or $\rho_1 A_1 V_1 = \rho_2 A_2 V_2$ reduces to

$$\rho_1 V_1 = \rho_2 V_2 \tag{17–50}$$

**x-Momentum equation**   Noting that the frictional effects are negligible and thus there are no shear forces, and assuming there are no external and body forces, the momentum equation $\sum \vec{F} = \sum_{\text{out}} \beta \dot{m} \vec{V} - \sum_{\text{in}} \beta \dot{m} \vec{V}$ in the flow (or $x$-) direction becomes a balance between static pressure forces and momentum transfer. Noting that the flows are high speed and turbulent and we are ignoring friction, the momentum flux correction factor is approximately 1 ($\beta \cong 1$) and thus can be neglected. Then,

$$P_1 A_1 - P_2 A_2 = \dot{m} V_2 - \dot{m} V_1 \rightarrow P_1 - P_2 = (\rho_2 V_2)V_2 - (\rho_1 V_1)V_1$$

or

$$P_1 + \rho_1 V_1^2 = P_2 + \rho_2 V_2^2 \tag{17–51}$$

**Energy equation**   The control volume involves no shear, shaft, or other forms of work, and the potential energy change is negligible. If the rate of heat transfer is $\dot{Q}$ and the heat transfer per unit mass of fluid is $q = \dot{Q}/\dot{m}$, the steady-flow energy balance $\dot{E}_{\text{in}} = \dot{E}_{\text{out}}$ becomes

$$\dot{Q} + \dot{m}\left(h_1 + \frac{V_1^2}{2}\right) = \dot{m}\left(h_2 + \frac{V_2^2}{2}\right) \rightarrow q + h_1 + \frac{V_1^2}{2} = h_2 + \frac{V_2^2}{2} \tag{17–52}$$

For an ideal gas with constant specific heats, $\Delta h = c_p \Delta T$, and thus

$$q = c_p(T_2 - T_1) + \frac{V_2^2 - V_1^2}{2} \tag{17–53}$$

or

$$q = h_{02} - h_{01} = c_p(T_{02} - T_{01}) \tag{17–54}$$

Therefore, the stagnation enthalpy $h_0$ and stagnation temperature $T_0$ change during Rayleigh flow (both increase when heat is transferred to the fluid and thus $q$ is positive, and both decrease when heat is transferred from the fluid and thus $q$ is negative).

**Entropy change**   In the absence of any irreversibilities such as friction, the entropy of a system changes by heat transfer only: it increases with heat gain, and decreases with heat loss. Entropy is a property and thus a state function, and the entropy change of an ideal gas with constant specific heats during a change of state from 1 to 2 is given by

$$s_2 - s_1 = c_p \ln \frac{T_2}{T_1} - R \ln \frac{P_2}{P_1} \tag{17–55}$$

The entropy of a fluid may increase or decrease during Rayleigh flow, depending on the direction of heat transfer.

**Equation of state** Noting that $P = \rho RT$, the properties $P$, $\rho$, and $T$ of an ideal gas at states 1 and 2 are related to each other by

$$\frac{P_1}{\rho_1 T_1} = \frac{P_2}{\rho_2 T_2} \tag{17–56}$$

Consider a gas with known properties $R$, $k$, and $c_p$. For a specified inlet state 1, the inlet properties $P_1$, $T_1$, $\rho_1$, $V_1$, and $s_1$ are known. The five exit properties $P_2$, $T_2$, $\rho_2$, $V_2$, and $s_2$ can be determined from Equations 17–50, 17–51, 17–53, 17–55, and 17–56 for any specified value of heat transfer $q$. When the velocity and temperature are known, the Mach number can be determined from $\text{Ma} = V/c = V/\sqrt{kRT}$.

Obviously there is an infinite number of possible downstream states 2 corresponding to a given upstream state 1. A practical way of determining these downstream states is to assume various values of $T_2$, and calculate all other properties as well as the heat transfer $q$ for each assumed $T_2$ from Eqs. 17–50 through 17–56. Plotting the results on a $T$-$s$ diagram gives a curve passing through the specified inlet state, as shown in Fig. 17–54. The plot of Rayleigh flow on a $T$-$s$ diagram is called the **Rayleigh line**, and several important observations can be made from this plot and the results of the calculations:

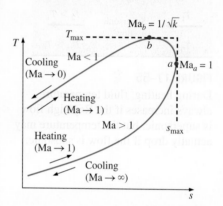

**FIGURE 17–54**
$T$-$s$ diagram for flow in a constant-area duct with heat transfer and negligible friction (Rayleigh flow).

1. All the states that satisfy the conservation of mass, momentum, and energy equations as well as the property relations are on the Rayleigh line. Therefore, for a given initial state, the fluid cannot exist at any downstream state outside the Rayleigh line on a $T$-$s$ diagram. In fact, the Rayleigh line is the locus of all physically attainable downstream states corresponding to an initial state.

2. Entropy increases with heat gain, and thus we proceed to the right on the Rayleigh line as heat is transferred to the fluid. The Mach number is $\text{Ma} = 1$ at point $a$, which is the point of maximum entropy (see Example 17–13 for proof). The states on the upper arm of the Rayleigh line above point $a$ are subsonic, and the states on the lower arm below point $a$ are supersonic. Therefore, a process proceeds to the right on the Rayleigh line with heat addition and to the left with heat rejection regardless of the initial value of the Mach number.

3. Heating increases the Mach number for subsonic flow, but decreases it for supersonic flow. The flow Mach number approaches $\text{Ma} = 1$ in both cases (from 0 in subsonic flow and from $\infty$ in supersonic flow) during heating.

4. It is clear from the energy balance $q = c_p(T_{02} - T_{01})$ that heating increases the stagnation temperature $T_0$ for both subsonic and supersonic flows, and cooling decreases it. (The maximum value of $T_0$ occurs at $\text{Ma} = 1$.) This is also the case for the static temperature $T$ except for the narrow Mach number range of $1/\sqrt{k} < \text{Ma} < 1$ in subsonic flow (see Example 17–13). Both temperature and the Mach

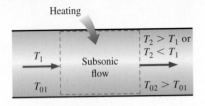

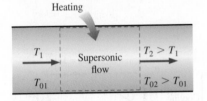

**FIGURE 17–55**

During heating, fluid temperature always increases if the Rayleigh flow is supersonic, but the temperature may actually drop if the flow is subsonic.

number increase with heating in subsonic flow, but $T$ reaches a maximum $T_{max}$ at $Ma = 1/\sqrt{k}$ (which is 0.845 for air), and then decreases. It may seem peculiar that the temperature of a fluid drops as heat is transferred to it. But this is no more peculiar than the fluid velocity increasing in the diverging section of a converging–diverging nozzle. The cooling effect in this region is due to the large increase in the fluid velocity and the accompanying drop in temperature in accordance with the relation $T_0 = T + V^2/2c_p$. Note also that heat rejection in the region $1/\sqrt{k} < Ma < 1$ causes the fluid temperature to increase (Fig. 17–55).

5. The momentum equation $P + KV =$ constant, where $K = \rho V =$ constant (from the continuity equation), reveals that velocity and static pressure have opposite trends. Therefore, static pressure decreases with heat gain in subsonic flow (since velocity and the Mach number increase), but increases with heat gain in supersonic flow (since velocity and the Mach number decrease).

6. The continuity equation $\rho V =$ constant indicates that density and velocity are inversely proportional. Therefore, density decreases with heat transfer to the fluid in subsonic flow (since velocity and the Mach number increase), but increases with heat gain in supersonic flow (since velocity and the Mach number decrease).

7. On the left half of Fig. 17–54, the lower arm of the Rayleigh line is steeper than the upper arm (in terms of $s$ as a function of $T$), which indicates that the entropy change corresponding to a specified temperature change (and thus a given amount of heat transfer) is larger in supersonic flow.

The effects of heating and cooling on the properties of Rayleigh flow are listed in Table 17–3. Note that heating or cooling has opposite effects on most properties. Also, the stagnation pressure decreases during heating and increases during cooling regardless of whether the flow is subsonic or supersonic.

**TABLE 17–3**

The effects of heating and cooling on the properties of Rayleigh flow

| Property | Heating | | Cooling | |
| --- | --- | --- | --- | --- |
| | Subsonic | Supersonic | Subsonic | Supersonic |
| Velocity, $V$ | Increase | Decrease | Decrease | Increase |
| Mach number, Ma | Increase | Decrease | Decrease | Increase |
| Stagnation temperature, $T_0$ | Increase | Increase | Decrease | Decrease |
| Temperature, $T$ | Increase for $Ma < 1/k^{1/2}$ Decrease for $Ma > 1/k^{1/2}$ | Increase | Decrease for $Ma < 1/k^{1/2}$ Increase for $Ma > 1/k^{1/2}$ | Decrease |
| Density, $\rho$ | Decrease | Increase | Increase | Decrease |
| Stagnation pressure, $P_0$ | Decrease | Decrease | Increase | Increase |
| Pressure, $P$ | Decrease | Increase | Increase | Decrease |
| Entropy, $s$ | Increase | Increase | Decrease | Decrease |

## EXAMPLE 17–13 Extrema of Rayleigh Line

Consider the *T-s* diagram of Rayleigh flow, as shown in Fig. 17–56. Using the differential forms of the conservation equations and property relations, show that the Mach number is $Ma_a = 1$ at the point of maximum entropy (point *a*), and $Ma_b = 1\sqrt{k}$ at the point of maximum temperature (point *b*).

**SOLUTION** It is to be shown that $Ma_a = 1$ at the point of maximum entropy and $Ma_b = 1\sqrt{k}$ at the point of maximum temperature on the Rayleigh line.
*Assumptions* The assumptions associated with Rayleigh flow (i.e., steady one-dimensional flow of an ideal gas with constant properties through a constant cross-sectional area duct with negligible frictional effects) are valid.
*Analysis* The differential forms of the continuity ($\rho V$ = constant), momentum [rearranged as $P + (\rho V)V$ = constant], ideal gas ($P = \rho RT$), and enthalpy change ($\Delta h = c_p \Delta T$) equations are expressed as

$$\rho V = \text{constant} \rightarrow \rho\, dV + V\, d\rho = 0 \rightarrow \frac{d\rho}{\rho} = -\frac{dV}{V} \tag{1}$$

$$P + (\rho V)V = \text{constant} \rightarrow dP + (\rho V)\, dV = 0 \rightarrow \frac{dP}{dV} = -\rho V \tag{2}$$

$$P = \rho RT \rightarrow dP = \rho R\, dT + RT\, d\rho \rightarrow \frac{dP}{P} = \frac{dT}{T} + \frac{d\rho}{\rho} \tag{3}$$

The differential form of the entropy change relation (Eq. 17–40) of an ideal gas with constant specific heats is

$$ds = c_p \frac{dT}{T} - R \frac{dP}{P} \tag{4}$$

Substituting Eq. 3 into Eq. 4 gives

$$ds = c_p \frac{dT}{T} - R\left(\frac{dT}{T} + \frac{d\rho}{\rho}\right) = (c_p - R)\frac{dT}{T} - R\frac{d\rho}{\rho} = \frac{R}{k-1}\frac{dT}{T} - R\frac{d\rho}{\rho} \tag{5}$$

since

$$c_p - R = c_v \rightarrow kc_v - R = c_v \rightarrow c_v = R/(k-1)$$

Dividing both sides of Eq. 5 by $dT$ and combining with Eq. 1,

$$\frac{ds}{dT} = \frac{R}{T(k-1)} + \frac{R}{V}\frac{dV}{dT} \tag{6}$$

Dividing Eq. 3 by $dV$ and combining it with Eqs. 1 and 2 give, after rearranging,

$$\frac{dT}{dV} = \frac{T}{V} - \frac{V}{R} \tag{7}$$

Substituting Eq. 7 into Eq. 6 and rearranging,

$$\frac{ds}{dT} = \frac{R}{T(k-1)} + \frac{R}{T - V^2/R} = \frac{R(kRT - V^2)}{T(k-1)(RT - V^2)} \tag{8}$$

Setting $ds/dT = 0$ and solving the resulting equation $R(kRT - V^2) = 0$ for $V$ give the velocity at point *a* to be

$$V_a = \sqrt{kRT_a} \quad \text{and} \quad Ma_a = \frac{V_a}{c_a} = \frac{\sqrt{kRT_a}}{\sqrt{kRT_a}} = 1 \tag{9}$$

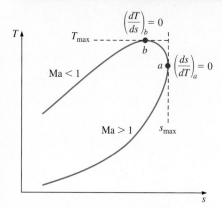

**FIGURE 17–56**
The *T-s* diagram of Rayleigh flow considered in Example 17–13.

Therefore, sonic conditions exist at point a, and thus the Mach number is 1.

Setting $dT/ds = (ds/dT)^{-1} = 0$ and solving the resulting equation $T(k-1) \times (RT - V^2) = 0$ for velocity at point b give

$$V_b = \sqrt{RT_b} \quad \text{and} \quad \text{Ma}_b = \frac{V_b}{c_b} = \frac{\sqrt{RT_b}}{\sqrt{kRT_b}} = \frac{1}{\sqrt{k}} \tag{10}$$

Therefore, the Mach number at point b is $\text{Ma}_b = 1\sqrt{k}$. For air, $k = 1.4$ and thus $\text{Ma}_b = 0.845$.

**Discussion** Note that in Rayleigh flow, sonic conditions are reached as the entropy reaches its maximum value, and maximum temperature occurs during subsonic flow.

---

**EXAMPLE 17–14**   **Effect of Heat Transfer on Flow Velocity**

Starting with the differential form of the energy equation, show that the flow velocity increases with heat addition in subsonic Rayleigh flow, but decreases in supersonic Rayleigh flow.

**SOLUTION** It is to be shown that flow velocity increases with heat addition in subsonic Rayleigh flow and that the opposite occurs in supersonic flow.
**Assumptions** **1** The assumptions associated with Rayleigh flow are valid.
**2** There are no work interactions and potential energy changes are negligible.
**Analysis** Consider heat transfer to the fluid in the differential amount of $\delta q$. The differential forms of the energy equations are expressed as

$$\delta q = dh_0 = d\left(h + \frac{V^2}{2}\right) = c_p\, dT + V\, dV \tag{1}$$

Dividing by $c_p T$ and factoring out $dV/V$ give

$$\frac{\delta q}{c_p T} = \frac{dT}{T} + \frac{V\, dV}{c_p T} = \frac{dV}{V}\left(\frac{V}{dV}\frac{dT}{T} + \frac{(k-1)V^2}{kRT}\right) \tag{2}$$

where we also used $c_p = kR/(k-1)$. Noting that $\text{Ma}^2 = V^2/c^2 = V^2/kRT$ and using Eq. 7 for $dT/dV$ from Example 17–13 give

$$\frac{\delta q}{c_p T} = \frac{dV}{V}\left(\frac{V}{T}\left(\frac{T}{V} - \frac{V}{R}\right) + (k-1)\text{Ma}^2\right) = \frac{dV}{V}\left(1 - \frac{V^2}{TR} + k\,\text{Ma}^2 - \text{Ma}^2\right) \tag{3}$$

Canceling the two middle terms in Eq. 3 since $V^2/TR = k\,\text{Ma}^2$ and rearranging give the desired relation,

$$\frac{dV}{V} = \frac{\delta q}{c_p T} \frac{1}{(1 - \text{Ma}^2)} \tag{4}$$

In subsonic flow, $1 - \text{Ma}^2 > 0$ and thus heat transfer and velocity change have the same sign. As a result, heating the fluid ($\delta q > 0$) increases the flow velocity while cooling decreases it. In supersonic flow, however, $1 - \text{Ma}^2 < 0$ and heat transfer and velocity change have opposite signs. **As a result, heating the fluid ($\delta q > 0$) decreases the flow velocity while cooling increases it** (Fig. 17–57).
**Discussion** Note that heating the fluid has the opposite effect on flow velocity in subsonic and supersonic Rayleigh flows.

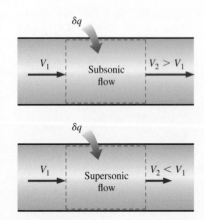

**FIGURE 17–57**
Heating increases the flow velocity in subsonic flow, but decreases it in supersonic flow.

# Property Relations for Rayleigh Flow

It is often desirable to express the variations in properties in terms of the Mach number Ma. Noting that $Ma = V/c = V/\sqrt{kRT}$ and thus $V = Ma\sqrt{kRT}$,

$$\rho V^2 = \rho kRT Ma^2 = kP Ma^2 \tag{17-57}$$

since $P = \rho RT$. Substituting into the momentum equation (Eq. 17–51) gives $P_1 + kP_1 Ma_1^2 = P_2 + kP_2 Ma_2^2$, which can be rearranged as

$$\frac{P_2}{P_1} = \frac{1 + kMa_1^2}{1 + kMa_2^2} \tag{17-58}$$

Again utilizing $V = Ma\sqrt{kRT}$, the continuity equation $\rho_1 V_1 = \rho_2 V_2$ is expressed as

$$\frac{\rho_1}{\rho_2} = \frac{V_2}{V_1} = \frac{Ma_2\sqrt{kRT_2}}{Ma_1\sqrt{kRT_1}} = \frac{Ma_2\sqrt{T_2}}{Ma_1\sqrt{T_1}} \tag{17-59}$$

Then the ideal-gas relation (Eq. 17–56) becomes

$$\frac{T_2}{T_1} = \frac{P_2}{P_1}\frac{\rho_1}{\rho_2} = \left(\frac{1 + kMa_1^2}{1 + kMa_2^2}\right)\left(\frac{Ma_2\sqrt{T_2}}{Ma_1\sqrt{T_1}}\right) \tag{17-60}$$

Solving Eq. 17–60 for the temperature ratio $T_2/T_1$ gives

$$\frac{T_2}{T_1} = \left(\frac{Ma_2(1 + kMa_1^2)}{Ma_1(1 + kMa_2^2)}\right)^2 \tag{17-61}$$

Substituting this relation into Eq. 17–59 gives the density or velocity ratio as

$$\frac{\rho_2}{\rho_1} = \frac{V_1}{V_2} = \frac{Ma_1^2(1 + kMa_2^2)}{Ma_2^2(1 + kMa_1^2)} \tag{17-62}$$

Flow properties at sonic conditions are usually easy to determine, and thus the critical state corresponding to Ma = 1 serves as a convenient reference point in compressible flow. Taking state 2 to be the sonic state ($Ma_2 = 1$, and superscript * is used) and state 1 to be any state (no subscript), the property relations in Eqs. 17–58, 17–61, and 17–62 reduce to (Fig. 17–58)

$$\frac{P}{P^*} = \frac{1 + k}{1 + kMa^2} \quad \frac{T}{T^*} = \left(\frac{Ma(1 + k)}{1 + kMa^2}\right)^2 \quad \text{and} \quad \frac{V}{V^*} = \frac{\rho^*}{\rho} = \frac{(1 + k)Ma^2}{1 + kMa^2} \tag{17-63}$$

Similar relations can be obtained for dimensionless stagnation temperature and stagnation pressure as follows:

$$\frac{T_0}{T_0^*} = \frac{T_0}{T}\frac{T}{T^*}\frac{T^*}{T_0^*} = \left(1 + \frac{k-1}{2}Ma^2\right)\left(\frac{Ma(1+k)}{1 + kMa^2}\right)^2\left(1 + \frac{k-1}{2}\right)^{-1} \tag{17-64}$$

which simplifies to

$$\frac{T_0}{T_0^*} = \frac{(k+1)Ma^2[2 + (k-1)Ma^2]}{(1 + kMa^2)^2} \tag{17-65}$$

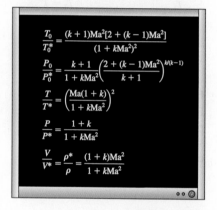

$$\frac{T_0}{T_0^*} = \frac{(k+1)Ma^2[2 + (k-1)Ma^2]}{(1 + kMa^2)^2}$$

$$\frac{P_0}{P_0^*} = \frac{k+1}{1 + kMa^2}\left(\frac{2 + (k-1)Ma^2}{k+1}\right)^{k/(k-1)}$$

$$\frac{T}{T^*} = \left(\frac{Ma(1+k)}{1 + kMa^2}\right)^2$$

$$\frac{P}{P^*} = \frac{1 + k}{1 + kMa^2}$$

$$\frac{V}{V^*} = \frac{\rho^*}{\rho} = \frac{(1+k)Ma^2}{1 + kMa^2}$$

**FIGURE 17–58**

Summary of relations for Rayleigh flow.

Also,

$$\frac{P_0}{P_0^*} = \frac{P_0}{P} \frac{P}{P^*} \frac{P^*}{P_0^*} = \left(1 + \frac{k-1}{2} \mathrm{Ma}^2\right)^{k/(k-1)} \left(\frac{1+k}{1+k\mathrm{Ma}^2}\right) \left(1 + \frac{k-1}{2}\right)^{-k/(k-1)} \tag{17-66}$$

which simplifies to

$$\frac{P_0}{P_0^*} = \frac{k+1}{1+k\mathrm{Ma}^2} \left(\frac{2 + (k-1)\mathrm{Ma}^2}{k+1}\right)^{k/(k-1)} \tag{17-67}$$

The five relations in Eqs. 17–63, 17–65, and 17–67 enable us to calculate the dimensionless pressure, temperature, density, velocity, stagnation temperature, and stagnation pressure for Rayleigh flow of an ideal gas with a specified $k$ for any given Mach number. Representative results are given in tabular and graphical form in Table A–34 for $k = 1.4$.

## Choked Rayleigh Flow

It is clear from the earlier discussions that subsonic Rayleigh flow in a duct may accelerate to sonic velocity (Ma = 1) with heating. What happens if we continue to heat the fluid? Does the fluid continue to accelerate to supersonic velocities? An examination of the Rayleigh line indicates that the fluid at the critical state of Ma = 1 cannot be accelerated to supersonic velocities by heating. Therefore, the flow is *choked*. This is analogous to not being able to accelerate a fluid to supersonic velocities in a converging nozzle by simply extending the converging flow section. If we keep heating the fluid, we will simply move the critical state further downstream and reduce the flow rate since fluid density at the critical state will now be lower. Therefore, for a given inlet state, the corresponding critical state fixes the maximum possible heat transfer for steady flow (Fig. 17–59). That is,

$$q_{max} = h_0^* - h_{01} = c_p(T_0^* - T_{01}) \tag{17-68}$$

Further heat transfer causes choking and thus the inlet state to change (e.g., inlet velocity will decrease), and the flow no longer follows the same Rayleigh line. Cooling the subsonic Rayleigh flow reduces the velocity, and the Mach number approaches zero as the temperature approaches absolute zero. Note that the stagnation temperature $T_0$ is maximum at the critical state of Ma = 1.

In supersonic Rayleigh flow, heating decreases the flow velocity. Further heating simply increases the temperature and moves the critical state farther downstream, resulting in a reduction in the mass flow rate of the fluid. It may seem like supersonic Rayleigh flow can be cooled indefinitely, but it turns out that there is a limit. Taking the limit of Eq. 17–65 as the Mach number approaches infinity gives

$$\lim_{\mathrm{Ma}\to\infty} \frac{T_0}{T_0^*} = 1 - \frac{1}{k^2} \tag{17-69}$$

which yields $T_0/T_0^* = 0.49$ for $k = 1.4$. Therefore, if the critical stagnation temperature is 1000 K, air cannot be cooled below 490 K in Rayleigh flow. Physically this means that the flow velocity reaches infinity by the time the temperature reaches 490 K—a physical impossibility. When supersonic flow cannot be sustained, the flow undergoes a normal shock wave and becomes subsonic.

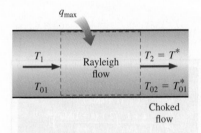

$q_{max}$

$T_1$ ⟶ Rayleigh flow $T_2 = T^*$ ⟶

$T_{01}$ $T_{02} = T_{01}^*$

Choked flow

**FIGURE 17–59**

For a given inlet state, the maximum possible heat transfer occurs when sonic conditions are reached at the exit state.

## EXAMPLE 17–15  Rayleigh Flow in a Tubular Combustor

A combustion chamber consists of tubular combustors of 15-cm diameter. Compressed air enters the tubes at 550 K, 480 kPa, and 80 m/s (Fig. 17–60). Fuel with a heating value of 42,000 kJ/kg is injected into the air and is burned with an air–fuel mass ratio of 40. Approximating combustion as a heat transfer process to air, determine the temperature, pressure, velocity, and Mach number at the exit of the combustion chamber.

**SOLUTION** Fuel is burned in a tubular combustion chamber with compressed air. The exit temperature, pressure, velocity, and Mach number are to be determined.

**Assumptions** **1** The assumptions associated with Rayleigh flow (i.e., steady one-dimensional flow of an ideal gas with constant properties through a constant cross-sectional area duct with negligible frictional effects) are valid. **2** Combustion is complete, and it is treated as a heat addition process, with no change in the chemical composition of the flow. **3** The increase in mass flow rate due to fuel injection is disregarded.

**Properties** We take the properties of air to be $k = 1.4$, $c_p = 1.005$ kJ/kg·K, and $R = 0.287$ kJ/kg·K.

**Analysis** The inlet density and mass flow rate of air are

$$\rho_1 = \frac{P_1}{RT_1} = \frac{480 \text{ kPa}}{(0.287 \text{ kJ/kg·K})(550 \text{ K})} = 3.041 \text{ kg/m}^3$$

$$\dot{m}_{\text{air}} = \rho_1 A_1 V_1 = (3.041 \text{ kg/m}^3) [\pi(0.15 \text{ m})^2/4](80 \text{ m/s}) = 4.299 \text{ kg/s}$$

The mass flow rate of fuel and the rate of heat transfer are

$$\dot{m}_{\text{fuel}} = \frac{\dot{m}_{\text{air}}}{\text{AF}} = \frac{4.299 \text{ kg/s}}{40} = 0.1075 \text{ kg/s}$$

$$\dot{Q} = \dot{m}_{\text{fuel}} \text{ HV} = (0.1075 \text{ kg/s})(42,000 \text{ kJ/kg}) = 4514 \text{ kW}$$

$$q = \frac{\dot{Q}}{\dot{m}_{\text{air}}} = \frac{4514 \text{ kJ/s}}{4.299 \text{ kg/s}} = 1050 \text{ kJ/kg}$$

The stagnation temperature and Mach number at the inlet are

$$T_{01} = T_1 + \frac{V_1^2}{2c_p} = 550 \text{ K} + \frac{(80 \text{ m/s})^2}{2(1.005 \text{ kJ/kg·K})}\left(\frac{1 \text{ kJ/kg}}{1000 \text{ m}^2/\text{s}^2}\right) = 553.2 \text{ K}$$

$$c_1 = \sqrt{kRT_1} = \sqrt{(1.4)(0.287 \text{ kJ/kg·K})(550 \text{ K})\left(\frac{1000 \text{ m}^2/\text{s}^2}{1 \text{ kJ/kg}}\right)} = 470.1 \text{ m/s}$$

$$\text{Ma}_1 = \frac{V_1}{c_1} = \frac{80 \text{ m/s}}{470.1 \text{ m/s}} = 0.1702$$

The exit stagnation temperature is, from the energy equation $q = c_p(T_{02} - T_{01})$,

$$T_{02} = T_{01} + \frac{q}{c_p} = 553.2 \text{ K} + \frac{1050 \text{ kJ/kg}}{1.005 \text{ kJ/kg·K}} = 1598 \text{ K}$$

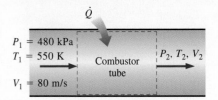

$P_1 = 480$ kPa
$T_1 = 550$ K
$V_1 = 80$ m/s

Combustor tube

$P_2, T_2, V_2$

$\dot{Q}$

**FIGURE 17–60**
Schematic of the combustor tube analyzed in Example 17–15.

The maximum value of stagnation temperature $T_0^*$ occurs at Ma = 1, and its value can be determined from Table A–34 or from Eq. 17–65. At $Ma_1 = 0.1702$ we read $T_0/T_0^* = 0.1291$. Therefore,

$$T_0^* = \frac{T_{01}}{0.1291} = \frac{553.2 \text{ K}}{0.1291} = 4284 \text{ K}$$

The stagnation temperature ratio at the exit state and the Mach number corresponding to it are, from Table A–34,

$$\frac{T_{02}}{T_0^*} = \frac{1598 \text{ K}}{4284 \text{ K}} = 0.3730 \rightarrow Ma_2 = 0.3142 \cong \mathbf{0.314}$$

The Rayleigh flow functions corresponding to the inlet and exit Mach numbers are (Table A–34):

$$Ma_1 = 0.1702: \quad \frac{T_1}{T^*} = 0.1541 \quad \frac{P_1}{P^*} = 2.3065 \quad \frac{V_1}{V^*} = 0.0668$$

$$Ma_2 = 0.3142: \quad \frac{T_2}{T^*} = 0.4389 \quad \frac{P_2}{P^*} = 2.1086 \quad \frac{V_2}{V^*} = 0.2082$$

Then the exit temperature, pressure, and velocity are determined to be

$$\frac{T_2}{T_1} = \frac{T_2/T^*}{T_1/T^*} = \frac{0.4389}{0.1541} = 2.848 \rightarrow T_2 = 2.848 T_1 = 2.848(550 \text{ K}) = \mathbf{1570 \text{ K}}$$

$$\frac{P_2}{P_1} = \frac{P_2/P^*}{P_1/P^*} = \frac{2.1086}{2.3065} = 0.9142 \rightarrow P_2 = 0.9142 P_1 = 0.9142(480 \text{ kPa}) = \mathbf{439 \text{ kPa}}$$

$$\frac{V_2}{V_1} = \frac{V_2/V^*}{V_1/V^*} = \frac{0.2082}{0.0668} = 3.117 \rightarrow V_2 = 3.117 V_1 = 3.117(80 \text{ m/s}) = \mathbf{249 \text{ m/s}}$$

**Discussion** Note that the temperature and velocity increase and pressure decreases during this subsonic Rayleigh flow with heating, as expected. This problem can also be solved using appropriate relations instead of tabulated values, which can likewise be coded for convenient computer solutions.

## 17–7 ▪ STEAM NOZZLES

We have seen in Chapter 3 that water vapor at moderate or high pressures deviates considerably from ideal-gas behavior, and thus most of the relations developed in this chapter are not applicable to the flow of steam through the nozzles or blade passages encountered in steam turbines. Given that the steam properties such as enthalpy are functions of pressure as well as temperature and that no simple property relations exist, an accurate analysis of steam flow through the nozzles is no easy matter. Often it becomes necessary to use steam tables, an $h$-$s$ diagram, or a computer program for the properties of steam.

A further complication in the expansion of steam through nozzles occurs as the steam expands into the saturation region, as shown in Fig. 17–61. As the steam expands in the nozzle, its pressure and temperature drop, and ordinarily one would expect the steam to start condensing when it strikes

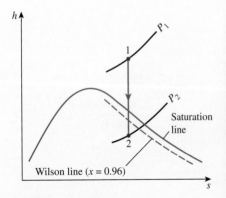

**FIGURE 17–61**
The $h$-$s$ diagram for the isentropic expansion of steam in a nozzle.

the saturation line. However, this is not always the case. Owing to the high speeds, the residence time of the steam in the nozzle is small, and there may not be sufficient time for the necessary heat transfer and the formation of liquid droplets. Consequently, the condensation of the steam may be delayed for a little while. This phenomenon is known as **supersaturation**, and the steam that exists in the wet region without containing any liquid is called **supersaturated steam**. Supersaturation states are nonequilibrium (or metastable) states.

During the expansion process, the steam reaches a temperature lower than that normally required for the condensation process to begin. Once the temperature drops a sufficient amount below the saturation temperature corresponding to the local pressure, groups of steam moisture droplets of sufficient size are formed, and condensation occurs rapidly. The locus of points where condensation takes place regardless of the initial temperature and pressure at the nozzle entrance is called the **Wilson line**. The Wilson line lies between the 4 and 5 percent moisture curves in the saturation region on the *h-s* diagram for steam, and it is often approximated by the 4 percent moisture line. Therefore, steam flowing through a high-velocity nozzle is assumed to begin condensation when the 4 percent moisture line is crossed.

The critical-pressure ratio $P^*/P_0$ for steam depends on the nozzle inlet state as well as on whether the steam is superheated or saturated at the nozzle inlet. However, the ideal-gas relation for the critical-pressure ratio, Eq. 17–22, gives reasonably good results over a wide range of inlet states. As indicated in Table 17–2, the specific heat ratio of superheated steam is approximated as $k = 1.3$. Then the critical-pressure ratio becomes

$$\frac{P^*}{P_0} = \left(\frac{2}{k+1}\right)^{k/(k-1)} = 0.546$$

When steam enters the nozzle as a saturated vapor instead of superheated vapor (a common occurrence in the lower stages of a steam turbine), the critical-pressure ratio is taken to be 0.576, which corresponds to a specific heat ratio of $k = 1.14$.

---

■ *EXAMPLE 17–16*    **Steam Flow through a**
                     **Converging–Diverging Nozzle**

■ Steam enters a converging–diverging nozzle at 2 MPa and 400°C with a negligible velocity and a mass flow rate of 2.5 kg/s, and it exits at a pressure of 300 kPa. The flow is isentropic between the nozzle entrance and throat, and the overall nozzle efficiency is 93 percent. Determine (*a*) the throat and exit areas and (*b*) the Mach number at the throat and the nozzle exit.

**SOLUTION**    Steam enters a converging–diverging nozzle with a low velocity. The throat and exit areas and the Mach number are to be determined.
*Assumptions*  **1** Flow through the nozzle is one-dimensional. **2** The flow is isentropic between the inlet and the throat, and is adiabatic and irreversible between the throat and the exit. **3** The inlet velocity is negligible.
*Analysis*  We denote the entrance, throat, and exit states by 1, *t*, and 2, respectively, as shown in Fig. 17–62.

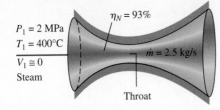

$P_1 = 2$ MPa
$T_1 = 400°C$
$V_1 \cong 0$
Steam

$\eta_N = 93\%$

$\dot{m} = 2.5$ kg/s

Throat

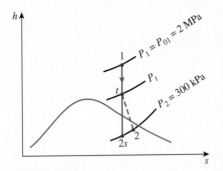

**FIGURE 17–62**
Schematic and *h-s* diagram for Example 17–16.

(*a*) Since the inlet velocity is negligible, the inlet stagnation and static states are identical. The ratio of the exit-to-inlet stagnation pressure is

$$\frac{P_2}{P_{01}} = \frac{300 \text{ kPa}}{2000 \text{ kPa}} = 0.15$$

It is much smaller than the critical-pressure ratio, which is taken to be $P^*/P_{01} = 0.546$ since the steam is superheated at the nozzle inlet. Therefore, the flow surely is supersonic at the exit. Then the velocity at the throat is the sonic velocity, and the throat pressure is

$$P_t = 0.546P_{01} = (0.546)(2 \text{ MPa}) = 1.09 \text{ MPa}$$

At the inlet,

$$\left. \begin{array}{l} P_1 = P_{01} = 2 \text{ MPa} \\ T_1 = T_{01} = 400°C \end{array} \right\} \begin{array}{l} h_1 = h_{01} = 3248.4 \text{ kJ/kg} \\ s_1 = s_t = s_{2s} = 7.1292 \text{ kJ/kg·K} \end{array}$$

Also, at the throat,

$$\left. \begin{array}{l} P_t = 1.09 \text{ MPa} \\ s_t = 7.1292 \text{ kJ/kg·K} \end{array} \right\} \begin{array}{l} h_t = 3076.8 \text{ kJ/kg} \\ v_t = 0.24196 \text{ m}^3\text{/kg} \end{array}$$

Then the throat velocity is determined from Eq. 17–3 to be

$$V_t = \sqrt{2(h_{01} - h_t)} = \sqrt{[2(3248.4 - 3076.8) \text{ kJ/kg}]\left(\frac{1000 \text{ m}^2\text{/s}^2}{1 \text{ kJ/kg}}\right)} = 585.8 \text{ m/s}$$

The flow area at the throat is determined from the mass flow rate relation:

$$A_t = \frac{\dot{m}v_t}{V_t} = \frac{(2.5 \text{ kg/s})(0.2420 \text{ m}^3\text{/kg})}{585.8 \text{ m/s}} = 10.33 \times 10^{-4} \text{ m}^2 = \textbf{10.33 cm}^2$$

At state 2*s*,

$$\left. \begin{array}{l} P_{2s} = P_2 = 300 \text{ kPa} \\ s_{2s} = s_1 = 7.1292 \text{ kJ/kg·K} \end{array} \right\} h_{2s} = 2783.6 \text{ kJ/kg}$$

The enthalpy of the steam at the actual exit state is (see Chap. 7)

$$\eta_N = \frac{h_{01} - h_2}{h_{01} - h_{2s}}$$

$$0.93 = \frac{3248.4 - h_2}{3248.4 - 2783.6} \longrightarrow h_2 = 2816.1 \text{ kJ/kg}$$

Therefore,

$$\left. \begin{array}{l} P_2 = 300 \text{ kPa} \\ h_2 = 2816.1 \text{ kJ/kg} \end{array} \right\} \begin{array}{l} v_2 = 0.67723 \text{ m}^3\text{/kg} \\ s_2 = 7.2019 \text{ kJ/kg·K} \end{array}$$

Then the exit velocity and the exit area become

$$V_2 = \sqrt{2(h_{01} - h_2)} = \sqrt{[2(3248.4 - 2816.1) \text{ kJ/kg}]\left(\frac{1000 \text{ m}^2\text{/s}^2}{1 \text{ kJ/kg}}\right)} = 929.8 \text{ m/s}$$

$$A_2 = \frac{\dot{m}v_2}{V_2} = \frac{(2.5 \text{ kg/s})(0.67723 \text{ m}^3\text{/kg})}{929.8 \text{ m/s}} = 18.21 \times 10^{-4} \text{ m}^2 = \textbf{18.21 cm}^2$$

(b) The velocity of sound and the Mach numbers at the throat and the exit of the nozzle are determined by replacing differential quantities with differences,

$$c = \left(\frac{\partial P}{\partial \rho}\right)_s^{1/2} \cong \left[\frac{\Delta P}{\Delta(1/v)}\right]_s^{1/2}$$

The velocity of sound at the throat is determined by evaluating the specific volume at $s_t = 7.1292$ kJ/kg·K and at pressures of 1.115 and 1.065 MPa ($P_t \pm 25$ kPa):

$$c = \sqrt{\frac{(1115 - 1065)\ \text{kPa}}{(1/0.23776 - 1/0.24633)\ \text{kg/m}^3}\left(\frac{1000\ \text{m}^2/\text{s}^2}{1\ \text{kPa·m}^3/\text{kg}}\right)} = 584.6\ \text{m/s}$$

The Mach number at the throat is determined from Eq. 17–12 to be

$$\text{Ma} = \frac{V}{c} = \frac{585.8\ \text{m/s}}{584.6\ \text{m/s}} = \mathbf{1.002}$$

Thus, the flow at the throat is sonic, as expected. The slight deviation of the Mach number from unity is due to replacing the derivatives by differences.

The velocity of sound and the Mach number at the nozzle exit are determined by evaluating the specific volume at $s_2 = 7.2019$ kJ/kg·K and at pressures of 325 and 275 kPa ($P_2 \pm 25$ kPa):

$$c = \sqrt{\frac{(325 - 275)\ \text{kPa}}{(1/0.63596 - 1/0.72245)\ \text{kg/m}^3}\left(\frac{1000\ \text{m}^2/\text{s}^2}{1\ \text{kPa·m}^3/\text{kg}}\right)} = 515.4\ \text{m/s}$$

and

$$\text{Ma} = \frac{V}{c} = \frac{929.8\ \text{m/s}}{515.4\ \text{m/s}} = \mathbf{1.804}$$

Thus the flow of steam at the nozzle exit is supersonic.

## SUMMARY

In this chapter the effects of compressibility on gas flow are examined. When dealing with compressible flow, it is convenient to combine the enthalpy and the kinetic energy of the fluid into a single term called *stagnation* (or *total*) *enthalpy* $h_0$, defined as

$$h_0 = h + \frac{V^2}{2}$$

The properties of a fluid at the stagnation state are called *stagnation properties* and are indicated by the subscript zero. The *stagnation temperature* of an ideal gas with constant specific heats is

$$T_0 = T + \frac{V^2}{2c_p}$$

which represents the temperature an ideal gas would attain if it is brought to rest adiabatically. The stagnation properties of an ideal gas are related to the static properties of the fluid by

$$\frac{P_0}{P} = \left(\frac{T_0}{T}\right)^{k/(k-1)} \quad \text{and} \quad \frac{\rho_0}{\rho} = \left(\frac{T_0}{T}\right)^{1/(k-1)}$$

The speed at which an infinitesimally small pressure wave travels through a medium is the *speed of sound*. For an ideal gas it is expressed as

$$c = \sqrt{\left(\frac{\partial P}{\partial \rho}\right)_s} = \sqrt{kRT}$$

The *Mach number* is the ratio of the actual velocity of the fluid to the speed of sound at the same state:

$$\text{Ma} = \frac{V}{c}$$

The flow is called *sonic* when Ma $= 1$, *subsonic* when Ma $< 1$, *supersonic* when Ma $> 1$, *hypersonic* when Ma $\gg 1$, and *transonic* when Ma $\cong 1$.

Nozzles whose flow area decreases in the flow direction are called *converging nozzles*. Nozzles whose flow area first decreases and then increases are called *converging–diverging nozzles*. The location of the smallest flow area of a nozzle is called the *throat*. The highest velocity to which a fluid can be accelerated in a converging nozzle is the sonic velocity. Accelerating a fluid to supersonic velocities is possible only in converging–diverging nozzles. In all supersonic converging–diverging nozzles, the flow velocity at the throat is the speed of sound.

The ratios of the stagnation to static properties for ideal gases with constant specific heats can be expressed in terms of the Mach number as

$$\frac{T_0}{T} = 1 + \left(\frac{k - 1}{2}\right)\text{Ma}^2$$

$$\frac{P_0}{P} = \left[1 + \left(\frac{k - 1}{2}\right)\text{Ma}^2\right]^{k/(k-1)}$$

and
$$\frac{\rho_0}{\rho} = \left[1 + \left(\frac{k - 1}{2}\right)\text{Ma}^2\right]^{1/(k-1)}$$

When Ma = 1, the resulting static-to-stagnation property ratios for the temperature, pressure, and density are called *critical ratios* and are denoted by the superscript asterisk:

$$\frac{T^*}{T_0} = \frac{2}{k + 1} \quad \frac{P^*}{P_0} = \left(\frac{2}{k + 1}\right)^{k/(k-1)}$$

and
$$\frac{\rho^*}{\rho_0} = \left(\frac{2}{k + 1}\right)^{1/(k-1)}$$

The pressure outside the exit plane of a nozzle is called the *back pressure*. For all back pressures lower than $P^*$, the pressure at the exit plane of the converging nozzle is equal to $P^*$, the Mach number at the exit plane is unity, and the mass flow rate is the maximum (or choked) flow rate.

In some range of back pressure, the fluid that achieved a sonic velocity at the throat of a converging–diverging nozzle and is accelerating to supersonic velocities in the diverging section experiences a *normal shock*, which causes a sudden rise in pressure and temperature and a sudden drop in velocity to subsonic levels. Flow through the shock is highly irreversible, and thus it cannot be approximated as isentropic. The properties of an ideal gas with constant specific heats before (subscript 1) and after (subscript 2) a shock are related by

$$T_{01} = T_{02} \quad \text{Ma}_2 = \sqrt{\frac{(k - 1)\text{Ma}_1^2 + 2}{2k\text{Ma}_1^2 - k + 1}}$$

$$\frac{T_2}{T_1} = \frac{2 + \text{Ma}_1^2(k - 1)}{2 + \text{Ma}_2^2(k - 1)}$$

and
$$\frac{P_2}{P_1} = \frac{1 + k\text{Ma}_1^2}{1 + k\text{Ma}_2^2} = \frac{2k\text{Ma}_1^2 - k + 1}{k + 1}$$

These equations also hold across an oblique shock, provided that the component of the Mach number *normal* to the oblique shock is used in place of the Mach number.

Steady one-dimensional flow of an ideal gas with constant specific heats through a constant-area duct with heat transfer and negligible friction is referred to as *Rayleigh flow*. The property relations and curves for Rayleigh flow are given in Table A–34. Heat transfer during Rayleigh flow can be determined from

$$q = c_p(T_{02} - T_{01}) = c_p(T_2 - T_1) + \frac{V_2^2 - V_1^2}{2}$$

## REFERENCES AND SUGGESTED READINGS

**1.** J. D. Anderson. *Modern Compressible Flow with Historical Perspective.* 3rd ed. New York: McGraw-Hill, 2003.

**2.** Y. A. Çengel and J. M. Cimbala. *Fluid Mechanics: Fundamentals and Applications.* 3rd ed. New York: McGraw-Hill, 2014.

**3.** H. Cohen, G. F. C. Rogers, and H. I. H. Saravanamuttoo. *Gas Turbine Theory.* 3rd ed. New York: Wiley, 1987.

**4.** W. J. Devenport. Compressible Aerodynamic Calculator, http://www.aoe.vt.edu/~devenpor/aoe3114/calc.html.

**5.** H. Liepmann and A. Roshko. *Elements of Gas Dynamics.* Dover Publications, Mineola, NY, 2001.

**6.** C. E. Mackey, responsible NACA officer and curator. *Equations, Tables, and Charts for Compressible Flow.* NACA Report 1135, http://naca.larc.nasa.gov/reports/1953/naca-report-1135/.

**7.** A. H. Shapiro. *The Dynamics and Thermodynamics of Compressible Fluid Flow.* vol. 1. New York: Ronald Press Company, 1953.

8. P. A. Thompson. *Compressible-Fluid Dynamics.* New York: McGraw-Hill, 1972.

9. United Technologies Corporation. *The Aircraft Gas Turbine and its Operation.* 1982.

10. M. Van Dyke, *An Album of Fluid Motion*, Stanford, CA: The Parabolic Press, 1982.

## PROBLEMS*

### Stagnation Properties

**17–1C** A high-speed aircraft is cruising in still air. How does the temperature of air at the nose of the aircraft differ from the temperature of air at some distance from the aircraft?

**17–2C** What is dynamic temperature?

**17–3C** In air-conditioning applications, the temperature of air is measured by inserting a probe into the flow stream. Thus, the probe actually measures the stagnation temperature. Does this cause any significant error?

**17–4** Air flows through a device such that the stagnation pressure is 0.6 MPa, the stagnation temperature is 400°C, and the velocity is 570 m/s. Determine the static pressure and temperature of the air at this state. *Answers:* 519 K, 0.231 MPa

**17–5** Air at 320 K is flowing in a duct at a velocity of (*a*) 1, (*b*) 10, (*c*) 100, and (*d*) 1000 m/s. Determine the temperature that a stationary probe inserted into the duct will read for each case.

**17–6** Calculate the stagnation temperature and pressure for the following substances flowing through a duct: (*a*) helium at 0.25 MPa, 50°C, and 240 m/s; (*b*) nitrogen at 0.15 MPa, 50°C, and 300 m/s; and (*c*) steam at 0.1 MPa, 350°C, and 480 m/s.

**17–7** Determine the stagnation temperature and stagnation pressure of air that is flowing at 36 kPa, 238 K, and 325 m/s. *Answers:* 291 K, 72.4 kPa

**17–8** Air enters a compressor with a stagnation pressure of 100 kPa and a stagnation temperature of 35°C, and it is compressed to a stagnation pressure of 900 kPa. Assuming the compression process to be isentropic, determine the power

input to the compressor for a mass flow rate of 0.04 kg/s. *Answer:* 10.8 kW

**17–9** Products of combustion enter a gas turbine with a stagnation pressure of 0.75 MPa and a stagnation temperature of 690°C, and they expand to a stagnation pressure of 100 kPa. Taking $k = 1.33$ and $R = 0.287$ kJ/kg·K for the products of combustion, and assuming the expansion process to be isentropic, determine the power output of the turbine per unit mass flow.

### Speed of Sound and Mach Number

**17–10C** What is sound? How is it generated? How does it travel? Can sound waves travel in a vacuum?

**17–11C** In which medium does a sound wave travel faster: in cool air or in warm air?

**17–12C** In which medium will sound travel fastest for a given temperature: air, helium, or argon?

**17–13C** In which medium does a sound wave travel faster: in air at 20°C and 1 atm or in air at 20°C and 5 atm?

**17–14C** Does the Mach number of a gas flowing at a constant velocity remain constant? Explain.

**17–15C** Is it realistic to approximate that the propagation of sound waves is an isentropic process? Explain.

**17–16C** Is the sonic velocity in a specified medium a fixed quantity, or does it change as the properties of the medium change? Explain.

**17–17** The Airbus A-340 passenger plane has a maximum takeoff weight of about 260,000 kg, a length of 64 m, a wing span of 60 m, a maximum cruising speed of 945 km/h, a seating capacity of 271 passengers, a maximum cruising altitude of 14,000 m, and a maximum range of 12,000 km. The air temperature at the cruising altitude is about −60°C. Determine the Mach number of this plane for the stated limiting conditions.

**17–18** Carbon dioxide enters an adiabatic nozzle at 1200 K with a velocity of 50 m/s and leaves at 400 K. Assuming constant specific heats at room temperature, determine the Mach number (*a*) at the inlet and (*b*) at the exit of the nozzle. Assess the accuracy of the constant specific heat approximation. *Answers:* (*a*) 0.0925, (*b*) 3.73

---

* Problems designated by a "C" are concept questions, and students are encouraged to answer them all. Problems with the 🖱 icon are solved using EES, and complete solutions together with parametric studies are included on the text website. Problems with the 🖱 icon are comprehensive in nature and are intended to be solved with an equation solver such as EES.

**17–19** Nitrogen enters a steady-flow heat exchanger at 150 kPa, 10°C, and 100 m/s, and it receives heat in the amount of 120 kJ/kg as it flows through it. Nitrogen leaves the heat exchanger at 100 kPa with a velocity of 200 m/s. Determine the Mach number of the nitrogen at the inlet and the exit of the heat exchanger.

**17–20** Assuming ideal gas behavior, determine the speed of sound in refrigerant-134a at 0.9 MPa and 60°C.

**17–21** Determine the speed of sound in air at (a) 300 K and (b) 800 K. Also determine the Mach number of an aircraft moving in air at a velocity of 330 m/s for both cases.

**17–22** Steam flows through a device with a pressure of 800 kPa, a temperature of 400°C, and a velocity of 275 m/s. Determine the Mach number of the steam at this state by assuming ideal-gas behavior with $k = 1.3$. *Answer:* 0.433

**17–23** [EES] Reconsider Prob. 17–22. Using EES (or other) software, compare the Mach number of steam flow over the temperature range 200 to 400°C. Plot the Mach number as a function of temperature.

**17–24** Air expands isentropically from 2.2 MPa and 77°C to 0.4 MPa. Calculate the ratio of the initial to the final speed of sound. *Answer:* 1.28

**17–25** Repeat Prob. 17–24 for helium gas.

**17–26** The isentropic process for an ideal gas is expressed as $Pv^k = $ constant. Using this process equation and the definition of the speed of sound (Eq. 17–9), obtain the expression for the speed of sound for an ideal gas (Eq. 17–11).

## One-Dimensional Isentropic Flow

**17–27C** Is it possible to accelerate a gas to a supersonic velocity in a converging nozzle? Explain.

**17–28C** A gas initially at a subsonic velocity enters an adiabatic diverging duct. Discuss how this affects (a) the velocity, (b) the temperature, (c) the pressure, and (d) the density of the fluid.

**17–29C** A gas at a specified stagnation temperature and pressure is accelerated to Ma = 2 in a converging–diverging nozzle and to Ma = 3 in another nozzle. What can you say about the pressures at the throats of these two nozzles?

**17–30C** A gas initially at a supersonic velocity enters an adiabatic converging duct. Discuss how this affects (a) the velocity, (b) the temperature, (c) the pressure, and (d) the density of the fluid.

**17–31C** A gas initially at a supersonic velocity enters an adiabatic diverging duct. Discuss how this affects (a) the

velocity, (b) the temperature, (c) the pressure, and (d) the density of the fluid.

**17–32C** Consider a converging nozzle with sonic speed at the exit plane. Now the nozzle exit area is reduced while the nozzle inlet conditions are maintained constant. What will happen to (a) the exit velocity and (b) the mass flow rate through the nozzle?

**17–33C** A gas initially at a subsonic velocity enters an adiabatic converging duct. Discuss how this affects (a) the velocity, (b) the temperature, (c) the pressure, and (d) the density of the fluid.

**17–34** Helium enters a converging–diverging nozzle at 0.7 MPa, 800 K, and 100 m/s. What are the lowest temperature and pressure that can be obtained at the throat of the nozzle?

**17–35** Consider a large commercial airplane cruising at a speed of 1050 km/h in air at an altitude of 10 km where the standard air temperature is −50°C. Determine if the speed of this airplane is subsonic or supersonic.

**17–36** Calculate the critical temperature, pressure, and density of (a) air at 200 kPa, 100°C, and 250 m/s, and (b) helium at 200 kPa, 40°C, and 300 m/s.

**17–37** Air enters a converging–diverging nozzle at a pressure of 1200 kPa with negligible velocity. What is the lowest pressure that can be obtained at the throat of the nozzle? *Answer:* 634 kPa

**17–38** In March 2004, NASA successfully launched an experimental supersonic-combustion ramjet engine (called a *scramjet*) that reached a record-setting Mach number of 7. Taking the air temperature to be −20°C, determine the speed of this engine. *Answer:* 8040 km/h

**17–39** Air at 200 kPa, 100°C, and Mach number Ma = 0.8 flows through a duct. Calculate the velocity and the stagnation pressure, temperature, and density of the air.

**17–40** [EES] Reconsider Prob. 17–39. Using EES (or other) software, study the effect of Mach numbers in the range 0.1 to 2 on the velocity, stagnation pressure, temperature, and density of air. Plot each parameter as a function of the Mach number.

**17–41** An aircraft is designed to cruise at Mach number Ma = 1.1 at 12,000 m where the atmospheric temperature is 236.15 K. Determine the stagnation temperature on the leading edge of the wing.

**17–42** Quiescent carbon dioxide at 1200 kPa and 600 K is accelerated isentropically to a Mach number of 0.6. Determine the temperature and pressure of the carbon dioxide after acceleration. *Answers:* 570 K, 957 kPa

## Isentropic Flow through Nozzles

**17–43C** Is it possible to accelerate a fluid to supersonic velocities with a velocity other than the sonic velocity at the throat? Explain

**17–44C** What would happen if we tried to further accelerate a supersonic fluid with a diverging diffuser?

**17–45C** How does the parameter Ma* differ from the Mach number Ma?

**17–46C** Consider subsonic flow in a converging nozzle with specified conditions at the nozzle inlet and critical pressure at the nozzle exit. What is the effect of dropping the back pressure well below the critical pressure on (a) the exit velocity, (b) the exit pressure, and (c) the mass flow rate through the nozzle?

**17–47C** Consider a converging nozzle and a converging–diverging nozzle having the same throat areas. For the same inlet conditions, how would you compare the mass flow rates through these two nozzles?

**17–48C** Consider gas flow through a converging nozzle with specified inlet conditions. We know that the highest velocity the fluid can have at the nozzle exit is the sonic velocity, at which point the mass flow rate through the nozzle is a maximum. If it were possible to achieve hypersonic velocities at the nozzle exit, how would it affect the mass flow rate through the nozzle?

**17–49C** Consider subsonic flow in a converging nozzle with fixed inlet conditions. What is the effect of dropping the back pressure to the critical pressure on (a) the exit velocity, (b) the exit pressure, and (c) the mass flow rate through the nozzle?

**17–50C** Consider the isentropic flow of a fluid through a converging–diverging nozzle with a subsonic velocity at the throat. How does the diverging section affect (a) the velocity, (b) the pressure, and (c) the mass flow rate of the fluid?

**17–51C** What would happen if we attempted to decelerate a supersonic fluid with a diverging diffuser?

**17–52** Nitrogen enters a converging–diverging nozzle at 700 kPa and 400 K with a negligible velocity. Determine the critical velocity, pressure, temperature, and density in the nozzle.

**17–53** For an ideal gas obtain an expression for the ratio of the speed of sound where Ma = 1 to the speed of sound based on the stagnation temperature, $c*/c_0$.

**17–54** Air enters a converging–diverging nozzle at 1.2 MPa with a negligible velocity. Approximating the flow as isentropic, determine the back pressure that would result in an exit Mach number of 1.8. *Answer:* 209 kPa

**17–55** An ideal gas flows through a passage that first converges and then diverges during an adiabatic, reversible, steady-flow process. For subsonic flow at the inlet, sketch the variation of pressure, velocity, and Mach number along the length of the nozzle when the Mach number at the minimum flow area is equal to unity.

**17–56** Repeat Prob. 17–55 for supersonic flow at the inlet.

**17–57** Explain why the maximum flow rate per unit area for a given ideal gas depends only on $P_0/\sqrt{T_0}$. For an ideal gas with $k = 1.4$ and $R = 0.287$ kJ/kg·K, find the constant $a$ such that $\dot{m}/A* = aP_0/\sqrt{T_0}$.

**17–58** An ideal gas with $k = 1.4$ is flowing through a nozzle such that the Mach number is 1.8 where the flow area is 36 cm². Approximating the flow as isentropic, determine the flow area at the location where the Mach number is 0.9.

**17–59** Repeat Prob. 17–58 for an ideal gas with $k = 1.33$.

**17–60** Air enters a converging–diverging nozzle of a supersonic wind tunnel at 1 MPa and 37°C with a low velocity. The flow area of the test section is equal to the exit area of the nozzle, which is 0.5 m². Calculate the pressure, temperature, velocity, and mass flow rate in the test section for a Mach number Ma = 2. Explain why the air must be very dry for this application. *Answers:* 128 kPa, 172 K, 526 m/s, 680 kg/s

**17–61** Air enters a nozzle at 0.5 MPa, 420 K, and a velocity of 110 m/s. Approximating the flow as isentropic, determine the pressure and temperature of air at a location where the air velocity equals the speed of sound. What is the ratio of the area at this location to the entrance area? *Answers:* 355 K, 278 kPa, 0.428

**17–62** Repeat Prob. 17–61 assuming the entrance velocity is negligible.

**17–63** [EES] Air at 900 kPa and 400 K enters a converging nozzle with a negligible velocity. The throat area of the nozzle is 10 cm². Approximating the flow as isentropic, calculate and plot the exit pressure, the exit velocity, and the mass flow rate versus the back pressure $P_b$ for $0.9 \geq P_b \geq 0.1$ MPa.

**17–64** [EES] Reconsider Prob. 17–63. Using EES (or other) software, solve the problem for the inlet conditions of 0.8 MPa and 1200 K.

## Shock Waves and Expansion Waves

**17–65C** Are the isentropic relations of ideal gases applicable for flows across (a) normal shock waves, (b) oblique shock waves, and (c) Prandtl–Meyer expansion waves?

**17–66C** What do the states on the Fanno line and the Rayleigh line represent? What do the intersection points of these two curves represent?

**17–67C** It is claimed that an oblique shock can be analyzed like a normal shock provided that the normal component of velocity (normal to the shock surface) is used in the analysis. Do you agree with this claim?

**17–68C** How does the normal shock affect (*a*) the fluid velocity, (*b*) the static temperature, (*c*) the stagnation temperature, (*d*) the static pressure, and (*e*) the stagnation pressure?

**17–69C** How do oblique shocks occur? How do oblique shocks differ from normal shocks?

**17–70C** For an oblique shock to occur, does the upstream flow have to be supersonic? Does the flow downstream of an oblique shock have to be subsonic?

**17–71C** Can the Mach number of a fluid be greater than 1 after a normal shock wave? Explain.

**17–72C** Consider supersonic airflow approaching the nose of a two-dimensional wedge and experiencing an oblique shock. Under what conditions does an oblique shock detach from the nose of the wedge and form a bow wave? What is the numerical value of the shock angle of the detached shock at the nose?

**17–73C** Consider supersonic flow impinging on the rounded nose of an aircraft. Is the oblique shock that forms in front of the nose an attached or a detached shock? Explain.

**17–74C** Can a shock wave develop in the converging section of a converging–diverging nozzle? Explain.

**17–75** Air enters a normal shock at 26 kPa, 230 K, and 815 m/s. Calculate the stagnation pressure and Mach number upstream of the shock, as well as pressure, temperature, velocity, Mach number, and stagnation pressure downstream of the shock.

**17–76** Calculate the entropy change of air across the normal shock wave in Problem 17–75.  *Answer:* 0.242 kJ/kg·K

**17–77** For an ideal gas flowing through a normal shock, develop a relation for $V_2/V_1$ in terms of $k$, $Ma_1$, and $Ma_2$.

**17–78** Air enters a converging–diverging nozzle with low velocity at 2.0 MPa and 100°C. If the exit area of the nozzle is 3.5 times the throat area, what must the back pressure be to produce a normal shock at the exit plane of the nozzle? *Answer:* 0.661 MPa

**17–79** What must the back pressure be in Prob. 17–78 for a normal shock to occur at a location where the cross-sectional area is twice the throat area?

**17–80** Air enters a converging–diverging nozzle of a supersonic wind tunnel at 1 MPa and 300 K with a low velocity. If a normal shock wave occurs at the exit plane of the nozzle at Ma = 2.4, determine the pressure, temperature, Mach number, velocity, and stagnation pressure after the shock wave.  *Answers:* 448 kPa, 284 K, 0.523, 177 m/s, 540 kPa

**17–81** Using EES (or other) software, calculate and plot the entropy change of air across the normal shock for upstream Mach numbers between 0.5 and 1.5 in increments of 0.1. Explain why normal shock waves can occur only for upstream Mach numbers greater than Ma = 1.

**17–82** Consider supersonic airflow approaching the nose of a two-dimensional wedge at a Mach number of 5. Using Fig. 17–43, determine the minimum shock angle and the maximum deflection angle a straight oblique shock can have.

**17–83** Air flowing at 32 kPa, 240 K, and $Ma_1 = 3.6$ is forced to undergo an expansion turn of 15°. Determine the Mach number, pressure, and temperature of air after the expansion.  *Answers:* 4.81, 6.65 kPa, 153 K

**17–84** Consider the supersonic flow of air at upstream conditions of 70 kPa and 260 K and a Mach number of 2.4 over a two-dimensional wedge of half-angle 10°. If the axis of the wedge is tilted 25° with respect to the upstream air flow, determine the downstream Mach number, pressure, and temperature above the wedge.  *Answers:* 3.105, 23.8 kPa, 191 K

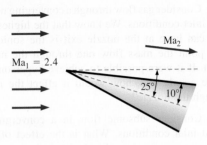

**FIGURE P17–84**

**17–85** Reconsider Prob. 17–84. Determine the downstream Mach number, pressure, and temperature below the wedge for a strong oblique shock for an upstream Mach number of 5.

**17–86** Air at 85 kPa, −1°C, and a Mach number of 2.0 is forced to turn upward by a ramp that makes an 8° angle off the flow direction. As a result, a weak oblique shock forms. Determine the wave angle, Mach number, pressure, and temperature after the shock.

**17–87** Air flowing at 40 kPa, 265 K, and $Ma_1 = 2.0$ is forced to undergo a compression turn of 15°. Determine the Mach number, pressure, and temperature of air after the compression.

**17–88** Air flowing at 60 kPa, 240 K, and a Mach number of 3.4 impinges on a two-dimensional wedge of half-angle 8°. Determine the two possible oblique shock angles, $\beta_{weak}$ and $\beta_{strong}$, that could be formed by this wedge. For each case, calculate the pressure, temperature, and Mach number downstream of the oblique shock.

**17–89** Air flowing steadily in a nozzle experiences a normal shock at a Mach number of Ma = 2.6. If the pressure and temperature of air are 58 kPa and 270 K, respectively, upstream of the shock, calculate the pressure, temperature, velocity, Mach number, and stagnation pressure downstream of the shock. Compare these results to those for helium undergoing a normal shock under the same conditions.

**17–90** Calculate the entropy changes of air and helium across the normal shock wave in Prob. 17–89.

### Duct Flow with Heat Transfer and Negligible Friction (Rayleigh Flow)

**17–91C** What is the effect of heating the fluid on the flow velocity in subsonic Rayleigh flow? Answer the same questions for supersonic Rayleigh flow.

**17–92C** On a *T-s* diagram of Rayleigh flow, what do the points on the Rayleigh line represent?

**17–93C** What is the effect of heat gain and heat loss on the entropy of the fluid during Rayleigh flow?

**17–94C** Consider subsonic Rayleigh flow of air with a Mach number of 0.92. Heat is now transferred to the fluid and the Mach number increases to 0.95. Does the temperature $T$ of the fluid increase, decrease, or remain constant during this process? How about the stagnation temperature $T_0$?

**17–95C** What is the characteristic aspect of Rayleigh flow? What are the main assumptions associated with Rayleigh flow?

**17–96C** Consider subsonic Rayleigh flow that is accelerated to sonic velocity (Ma = 1) at the duct exit by heating. If the fluid continues to be heated, will the flow at duct exit be supersonic, subsonic, or remain sonic?

**17–97** Argon gas enters a constant cross-sectional area duct at $Ma_1 = 0.2$, $P_1 = 320$ kPa, and $T_1 = 400$ K at a rate of 1.2 kg/s. Disregarding frictional losses, determine the highest rate of heat transfer to the argon without reducing the mass flow rate.

**17–98** Air is heated as it flows subsonically through a duct. When the amount of heat transfer reaches 67 kJ/kg, the flow is observed to be choked, and the velocity and the static pressure are measured to be 680 m/s and 270 kPa. Disregarding frictional losses, determine the velocity, static temperature, and static pressure at the duct inlet.

**17–99** Compressed air from the compressor of a gas turbine enters the combustion chamber at $T_1 = 700$ K, $P_1 = 600$ kPa, and $Ma_1 = 0.2$ at a rate of 0.3 kg/s. Via combustion, heat is transferred to the air at a rate of 150 kJ/s as it flows through the duct with negligible friction. Determine the Mach number at the duct exit, and the drop in stagnation pressure $P_{01} - P_{02}$ during this process. *Answers:* 0.271, 12.7 kPa

**17–100** Repeat Prob. 17–99 for a heat transfer rate of 300 kJ/s.

**17–101** Air flows with negligible friction through a 10-cm-diameter duct at a rate of 2.3 kg/s. The temperature and pressure at the inlet are $T_1 = 450$ K and $P_1 = 200$ kPa, and the Mach number at the exit is $Ma_2 = 1$. Determine the rate of heat transfer and the pressure drop for this section of the duct.

**17–102** Air enters an approximately frictionless duct with $V_1 = 70$ m/s, $T_1 = 600$ K, and $P_1 = 350$ kPa. Letting the exit temperature $T_2$ vary from 600 to 5000 K, evaluate the entropy change at intervals of 200 K, and plot the Rayleigh line on a *T-s* diagram.

**17–103** Air is heated as it flows through a 15 cm × 15 cm square duct with negligible friction. At the inlet, air is at $T_1 = 400$ K, $P_1 = 550$ kPa, and $V_1 = 80$ m/s. Determine the rate at which heat must be transferred to the air to choke the flow at the duct exit, and the entropy change of air during this process.

**17–104** Air enters a rectangular duct at $T_1 = 300$ K, $P_1 = 420$ kPa, and $Ma_1 = 2$. Heat is transferred to the air in the amount of 55 kJ/kg as it flows through the duct. Disregarding frictional losses, determine the temperature and Mach number at the duct exit. *Answers:* 386 K, 1.64

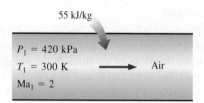

**FIGURE P17–104**

**17–105** Repeat Prob. 17–104 assuming air is cooled in the amount of 55 kJ/kg.

**17–106** Consider a 16-cm-diameter tubular combustion chamber. Air enters the tube at 450 K, 380 kPa, and 55 m/s. Fuel with a heating value of 39,000 kJ/kg is burned by spraying it into the air. If the exit Mach number is 0.8, determine the rate at which the fuel is burned and the exit temperature. Assume complete combustion and disregard the increase in the mass flow rate due to the fuel mass.

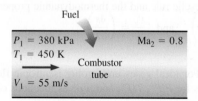

**FIGURE P17–106**

**17–107** Consider supersonic flow of air through a 7-cm-diameter duct with negligible friction. Air enters the duct at $Ma_1 = 1.8$, $P_{01} = 140$ kPa, and $T_{01} = 600$ K, and it is decelerated by heating. Determine the highest temperature that air can be heated by heat addition while the mass flow rate remains constant.

## Steam Nozzles

**17–108C** What is supersaturation? Under what conditions does it occur?

**17–109** Steam enters a converging nozzle at 5.0 MPa and 400°C with a negligible velocity, and it exits at 3.0 MPa. For a nozzle exit area of 60 cm², determine the exit velocity, mass flow rate, and exit Mach number if the nozzle (a) is isentropic and (b) has an efficiency of 94 percent. *Answers:* (a) 529 m/s, 36.9 kg/s, 0.935, (b) 512 m/s, 35.5 kg/s, 0.903

**17–110** Steam enters a converging–diverging nozzle at 1 MPa and 500°C with a negligible velocity at a mass flow rate of 2.5 kg/s, and it exits at a pressure of 200 kPa. Assuming the flow through the nozzle to be isentropic, determine the exit area and the exit Mach number. *Answers:* 31.5 cm², 1.738

**17–111** Repeat Prob. 17–110 for a nozzle efficiency of 85 percent.

## Review Problems

**17–112** The thrust developed by the engine of a Boeing 777 is about 380 kN. Assuming choked flow in the nozzles, determine the mass flow rate of air through the nozzle. Take the ambient conditions to be 220 K and 40 kPa.

**17–113** A stationary temperature probe inserted into a duct where air is flowing at 190 m/s reads 85°C. What is the actual temperature of the air? *Answer:* 67.0°C

**17–114** Nitrogen enters a steady-flow heat exchanger at 150 kPa, 10°C, and 100 m/s, and it receives heat in the amount of 150 kJ/kg as it flows through it. The nitrogen leaves the heat exchanger at 100 kPa with a velocity of 200 m/s. Determine the stagnation pressure and temperature of the nitrogen at the inlet and exit states.

**17–115** Plot the mass flow parameter $\dot{m}\sqrt{RT_0}/(AP_0)$ versus the Mach number for $k = 1.2$, 1.4, and 1.6 in the range of $0 \leq Ma \leq 1$.

**17–116** Obtain Eq. 17–10 by starting with Eq. 17–9 and using the cyclic rule and the thermodynamic property relations $\frac{c_p}{T} = \left(\frac{\partial s}{\partial T}\right)_P$ and $\frac{c_v}{T} = \left(\frac{\partial s}{\partial T}\right)_v$.

**17–117** For ideal gases undergoing isentropic flows, obtain expressions for $P/P^*$, $T/T^*$, and $\rho/\rho^*$ as functions of $k$ and Ma.

**17–118** Using Eqs. 17–4, 17–13, and 17–14, verify that for the steady flow of ideal gases $dT_0/T = dA/A + (1 - Ma^2)\, dV/V$.

Explain the effect of heating and area changes on the velocity of an ideal gas in steady flow for (a) subsonic flow and (b) supersonic flow.

**17–119** A subsonic airplane is flying at a 5000-m altitude where the atmospheric conditions are 54 kPa and 256 K. A Pitot static probe measures the difference between the static and stagnation pressures to be 16 kPa. Calculate the speed of the airplane and the flight Mach number. *Answers:* 199 m/s, 0.620

**17–120** Derive an expression for the speed of sound based on van der Waals' equation of state $P = RT(v - b) - a/v^2$. Using this relation, determine the speed of sound in carbon dioxide at 80°C and 320 kPa, and compare your result to that obtained by assuming ideal-gas behavior. The van der Waals constants for carbon dioxide are $a = 364.3$ kPa·m⁶/kmol² and $b = 0.0427$ m³/kmol.

**17–121** Helium enters a nozzle at 0.6 MPa, 560 K, and a velocity of 120 m/s. Assuming isentropic flow, determine the pressure and temperature of helium at a location where the velocity equals the speed of sound. What is the ratio of the area at this location to the entrance area?

**17–122** Repeat Problem 17–121 assuming the entrance velocity is negligible.

**17–123** [EES] Air at 0.9 MPa and 400 K enters a converging nozzle with a velocity of 180 m/s. The throat area is 10 cm². Assuming isentropic flow, calculate and plot the mass flow rate through the nozzle, the exit velocity, the exit Mach number, and the exit pressure–stagnation pressure ratio versus the back pressure–stagnation pressure ratio for a back pressure range of $0.9 \geq P_b \geq 0.1$ MPa.

**17–124** Nitrogen enters a duct with varying flow area at 400 K, 100 kPa, and a Mach number of 0.3. Assuming a steady, isentropic flow, determine the temperature, pressure, and Mach number at a location where the flow area has been reduced by 20 percent.

**17–125** Repeat Prob. 17–124 for an inlet Mach number of 0.5.

**17–126** Nitrogen enters a converging–diverging nozzle at 620 kPa and 310 K with a negligible velocity, and it experiences a normal shock at a location where the Mach number is Ma = 3.0. Calculate the pressure, temperature, velocity, Mach number, and stagnation pressure downstream of the shock. Compare these results to those of air undergoing a normal shock at the same conditions.

**17–127** An aircraft flies with a Mach number $Ma_1 = 0.9$ at an altitude of 7000 m where the pressure is 41.1 kPa and the temperature is 242.7 K. The diffuser at the engine inlet has an exit Mach number of $Ma_2 = 0.3$. For a mass flow rate of 38 kg/s, determine the static pressure rise across the diffuser and the exit area.

**17–128** Consider an equimolar mixture of oxygen and nitrogen. Determine the critical temperature, pressure, and density for stagnation temperature and pressure of 550 K and 350 kPa.

**17–129** Using the EES software and the relations in Table A–32, calculate the one-dimensional compressible flow functions for an ideal gas with $k = 1.667$, and present your results by duplicating Table A–32.

**17–130** Using the EES software and the relations in Table A–33, calculate the one-dimensional normal shock functions for an ideal gas with $k = 1.667$, and present your results by duplicating Table A–33.

**17–131** Helium expands in a nozzle from 1 MPa, 500 K, and negligible velocity to 0.1 MPa. Calculate the throat and exit areas for a mass flow rate of 0.46 kg/s, assuming the nozzle is isentropic. Why must this nozzle be converging–diverging? *Answers:* 6.46 cm², 10.8 cm²

**17–132** Using EES (or other) software and the relations given in Table A–33, generate the one-dimensional normal shock functions by varying the upstream Mach number from 1 to 10 in increments of 0.5 for air with $k = 1.4$.

**17–133** Repeat Prob. 17–132 for methane with $k = 1.3$.

**17–134** Air is heated as it flows subsonically through a 10 cm × 10 cm square duct. The properties of air at the inlet are maintained at $Ma_1 = 0.6$, $P_1 = 350$ kPa, and $T_1 = 420$ K at all times. Disregarding frictional losses, determine the highest rate of heat transfer to the air in the duct without affecting the inlet conditions. *Answer:* 716 kW

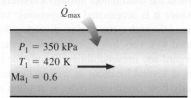

$\dot{Q}_{max}$

$P_1 = 350$ kPa
$T_1 = 420$ K
$Ma_1 = 0.6$

**FIGURE P17–134**

**17–135** Repeat Prob. 17–134 for helium.

**17–136** Air is accelerated as it is heated in a duct with negligible friction. Air enters at $V_1 = 100$ m/s, $T_1 = 400$ K, and $P_1 = 35$ kPa and the exits at a Mach number of $Ma_2 = 0.8$. Determine the heat transfer to the air, in kJ/kg. Also determine the maximum amount of heat transfer without reducing the mass flow rate of air.

**17–137** Air at sonic conditions and at static temperature and pressure of 340 K and 250 kPa, respectively, is to be accelerated to a Mach number of 1.6 by cooling it as it flows through a channel with constant cross-sectional area. Disregarding frictional effects, determine the required heat transfer from the air, in kJ/kg. *Answer:* 47.5 kJ/kg

**17–138** Air is cooled as it flows through a 20-cm-diameter duct. The inlet conditions are $Ma_1 = 1.2$, $T_{01} = 350$ K, and $P_{01} = 240$ kPa and the exit Mach number is $Ma_2 = 2.0$. Disregarding frictional effects, determine the rate of cooling of air.

**17–139** Saturated steam enters a converging–diverging nozzle at 1.75 MPa, 10 percent moisture, and negligible velocity, and it exits at 1.2 MPa. For a nozzle exit area of 25 cm², determine the throat area, exit velocity, mass flow rate, and exit Mach number if the nozzle (a) is isentropic and (b) has an efficiency of 92 percent.

**17–140** Using EES (or other) software, determine the shape of a converging–diverging nozzle for air for a mass flow rate of 3 kg/s and inlet stagnation conditions of 1400 kPa and 200°C. Approximate the flow as isentropic. Repeat the calculations for 50-kPa increments of pressure drop to an exit pressure of 100 kPa. Plot the nozzle to scale. Also, calculate and plot the Mach number along the nozzle.

**17–141** Steam at 6.0 MPa and 700 K enters a converging nozzle with a negligible velocity. The nozzle throat area is 8 cm². Approximating the flow as isentropic, plot the exit pressure, the exit velocity, and the mass flow rate through the nozzle versus the back pressure $P_b$ for $6.0 \geq P_b \geq 3.0$ MPa. Treat the steam as an ideal gas with $k = 1.3$, $c_p = 1.872$ kJ/kg·K, and $R = 0.462$ kJ/kg·K.

**17–142** Find the expression for the ratio of the stagnation pressure after a shock wave to the static pressure before the shock wave as a function of $k$ and the Mach number upstream of the shock wave $Ma_1$.

**17–143** Using EES (or other) software and the relations given in Table A–32, calculate the one-dimensional isentropic compressible-flow functions by varying the upstream Mach number from 1 to 10 in increments of 0.5 for air with $k = 1.4$.

**17–144** Repeat Prob. 17–143 for methane with $k = 1.3$.

### Fundamentals of Engineering (FE) Exam Problems

**17–145** An aircraft is cruising in still air at 5°C at a velocity of 400 m/s. The air temperature at the nose of the aircraft where stagnation occurs is

(a) 5°C      (b) 25°C      (c) 55°C
(d) 80°C      (e) 85°C

**17–146** Air is flowing in a wind tunnel at 25°C, 80 kPa, and 250 m/s. The stagnation pressure at the location of a probe inserted into the flow section is

(a) 87 kPa      (b) 96 kPa      (c) 113 kPa
(d) 119 kPa      (e) 125 kPa

**17–147** An aircraft is reported to be cruising in still air at $-20°C$ and 40 kPa at a Mach number of 0.86. The velocity of the aircraft is

(*a*) 91 m/s      (*b*) 220 m/s      (*c*) 186 m/s
(*d*) 274 m/s      (*e*) 378 m/s

**17–148** Air is flowing in a wind tunnel at 12°C and 66 kPa at a velocity of 230 m/s. The Mach number of the flow is

(*a*) 0.54 m/s      (*b*) 0.87 m/s      (*c*) 3.3 m/s
(*d*) 0.36 m/s      (*e*) 0.68 m/s

**17–149** Consider a converging nozzle with a low velocity at the inlet and sonic velocity at the exit plane. Now the nozzle exit diameter is reduced by half while the nozzle inlet temperature and pressure are maintained the same. The nozzle exit velocity will

(*a*) remain the same    (*b*) double      (*c*) quadruple
(*d*) go down by half    (*e*) go down by one-fourth

**17–150** Air is approaching a converging–diverging nozzle with a low velocity at 12°C and 200 kPa, and it leaves the nozzle at a supersonic velocity. The velocity of air at the throat of the nozzle is

(*a*) 338 m/s      (*b*) 309 m/s      (*c*) 280 m/s
(*d*) 256 m/s      (*e*) 95 m/s

**17–151** Argon gas is approaching a converging–diverging nozzle with a low velocity at 20°C and 120 kPa, and it leaves the nozzle at a supersonic velocity. If the cross-sectional area of the throat is 0.015 m², the mass flow rate of argon through the nozzle is

(*a*) 0.41 kg/s      (*b*) 3.4 kg/s      (*c*) 5.3 kg/s
(*d*) 17 kg/s      (*e*) 22 kg/s

**17–152** Carbon dioxide enters a converging–diverging nozzle at 60 m/s, 310°C, and 300 kPa, and it leaves the nozzle at a supersonic velocity. The velocity of carbon dioxide at the throat of the nozzle is

(*a*) 125 m/s      (*b*) 225 m/s      (*c*) 312 m/s
(*d*) 353 m/s      (*e*) 377 m/s

**17–153** Consider gas flow through a converging–diverging nozzle. Of the five following statements, select the one that is incorrect:

(*a*) The fluid velocity at the throat can never exceed the speed of sound.
(*b*) If the fluid velocity at the throat is below the speed of sound, the diversion section will act like a diffuser.
(*c*) If the fluid enters the diverging section with a Mach number greater than one, the flow at the nozzle exit will be supersonic.
(*d*) There will be no flow through the nozzle if the back pressure equals the stagnation pressure.
(*e*) The fluid velocity decreases, the entropy increases, and stagnation enthalpy remains constant during flow through a normal shock.

**17–154** Combustion gases with $k = 1.33$ enter a converging nozzle at stagnation temperature and pressure of 350°C and 400 kPa, and are discharged into the atmospheric air at 20°C and 100 kPa. The lowest pressure that will occur within the nozzle is

(*a*) 13 kPa      (*b*) 100 kPa      (*c*) 216 kPa
(*d*) 290 kPa      (*e*) 315 kPa

## Design and Essay Problems

**17–155** Find out if there is a supersonic wind tunnel on your campus. If there is, obtain the dimensions of the wind tunnel and the temperatures and pressures as well as the Mach number at several locations during operation. For what typical experiments is the wind tunnel used?

**17–156** Assuming you have a thermometer and a device to measure the speed of sound in a gas, explain how you can determine the mole fraction of helium in a mixture of helium gas and air.

**17–157** Design a 1-m-long cylindrical wind tunnel whose diameter is 25 cm operating at a Mach number of 1.8. Atmospheric air enters the wind tunnel through a converging–diverging nozzle where it is accelerated to supersonic velocities. Air leaves the tunnel through a converging–diverging diffuser where it is decelerated to a very low velocity before entering the fan section. Disregard any irreversibilities. Specify the temperatures and pressures at several locations as well as the mass flow rate of air at steady-flow conditions. Why is it often necessary to dehumidify the air before it enters the wind tunnel?

**FIGURE P17–157**

# PROPERTY TABLES AND CHARTS

## TABLE A–1

Molar mass, gas constant, and critical–point properties

| Substance | Formula | Molar mass, $M$ kg/kmol | Gas constant, $R$ kJ/kg·K* | Critical-point properties | | |
|---|---|---|---|---|---|---|
| | | | | Temperature, K | Pressure, MPa | Volume, $m^3$/kmol |
| Air | — | 28.97 | 0.2870 | 132.5 | 3.77 | 0.0883 |
| Ammonia | $NH_3$ | 17.03 | 0.4882 | 405.5 | 11.28 | 0.0724 |
| Argon | Ar | 39.948 | 0.2081 | 151 | 4.86 | 0.0749 |
| Benzene | $C_6H_6$ | 78.115 | 0.1064 | 562 | 4.92 | 0.2603 |
| Bromine | $Br_2$ | 159.808 | 0.0520 | 584 | 10.34 | 0.1355 |
| $n$–Butane | $C_4H_{10}$ | 58.124 | 0.1430 | 425.2 | 3.80 | 0.2547 |
| Carbon dioxide | $CO_2$ | 44.01 | 0.1889 | 304.2 | 7.39 | 0.0943 |
| Carbon monoxide | CO | 28.011 | 0.2968 | 133 | 3.50 | 0.0930 |
| Carbon tetrachloride | $CCl_4$ | 153.82 | 0.05405 | 556.4 | 4.56 | 0.2759 |
| Chlorine | $Cl_2$ | 70.906 | 0.1173 | 417 | 7.71 | 0.1242 |
| Chloroform | $CHCl_3$ | 119.38 | 0.06964 | 536.6 | 5.47 | 0.2403 |
| Dichlorodifluoromethane (R–12) | $CCl_2F_2$ | 120.91 | 0.06876 | 384.7 | 4.01 | 0.2179 |
| Dichlorofluoromethane (R–21) | $CHCl_2F$ | 102.92 | 0.08078 | 451.7 | 5.17 | 0.1973 |
| Ethane | $C_2H_6$ | 30.070 | 0.2765 | 305.5 | 4.48 | 0.1480 |
| Ethyl alcohol | $C_2H_5OH$ | 46.07 | 0.1805 | 516 | 6.38 | 0.1673 |
| Ethylene | $C_2H_4$ | 28.054 | 0.2964 | 282.4 | 5.12 | 0.1242 |
| Helium | He | 4.003 | 2.0769 | 5.3 | 0.23 | 0.0578 |
| $n$–Hexane | $C_6H_{14}$ | 86.179 | 0.09647 | 507.9 | 3.03 | 0.3677 |
| Hydrogen (normal) | $H_2$ | 2.016 | 4.1240 | 33.3 | 1.30 | 0.0649 |
| Krypton | Kr | 83.80 | 0.09921 | 209.4 | 5.50 | 0.0924 |
| Methane | $CH_4$ | 16.043 | 0.5182 | 191.1 | 4.64 | 0.0993 |
| Methyl alcohol | $CH_3OH$ | 32.042 | 0.2595 | 513.2 | 7.95 | 0.1180 |
| Methyl chloride | $CH_3Cl$ | 50.488 | 0.1647 | 416.3 | 6.68 | 0.1430 |
| Neon | Ne | 20.183 | 0.4119 | 44.5 | 2.73 | 0.0417 |
| Nitrogen | $N_2$ | 28.013 | 0.2968 | 126.2 | 3.39 | 0.0899 |
| Nitrous oxide | $N_2O$ | 44.013 | 0.1889 | 309.7 | 7.27 | 0.0961 |
| Oxygen | $O_2$ | 31.999 | 0.2598 | 154.8 | 5.08 | 0.0780 |
| Propane | $C_3H_8$ | 44.097 | 0.1885 | 370 | 4.26 | 0.1998 |
| Propylene | $C_3H_6$ | 42.081 | 0.1976 | 365 | 4.62 | 0.1810 |
| Sulfur dioxide | $SO_2$ | 64.063 | 0.1298 | 430.7 | 7.88 | 0.1217 |
| Tetrafluoroethane (R–134a) | $CF_3CH_2F$ | 102.03 | 0.08149 | 374.2 | 4.059 | 0.1993 |
| Trichlorofluoromethane (R–11) | $CCl_3F$ | 137.37 | 0.06052 | 471.2 | 4.38 | 0.2478 |
| Water | $H_2O$ | 18.015 | 0.4615 | 647.1 | 22.06 | 0.0560 |
| Xenon | Xe | 131.30 | 0.06332 | 289.8 | 5.88 | 0.1186 |

*The unit kJ/kg·K is equivalent to kPa·$m^3$/kg·K. The gas constant is calculated from $R = R_u/M$, where $R_u = 8.31447$ kJ/kmol·K and $M$ is the molar mass.

Source of Data: K. A. Kobe and R. E. Lynn, Jr., Chemical Review 52 (1953), pp. 117–236; and ASHRAE, Handbook of Fundamentals (Atlanta, GA: American Society of Heating, Refrigerating and Air–Conditioning Engineers, Inc., 1993), pp. 16.4 and 36.1.

## TABLE A–2

Ideal–gas specific heats of various common gases

(a) At 300 K

| Gas | Formula | Gas constant, $R$ kJ/kg·K | $c_p$ kJ/kg·K | $c_v$ kJ/kg·K | $k$ |
|---|---|---|---|---|---|
| Air | — | 0.2870 | 1.005 | 0.718 | 1.400 |
| Argon | Ar | 0.2081 | 0.5203 | 0.3122 | 1.667 |
| Butane | $C_4H_{10}$ | 0.1433 | 1.7164 | 1.5734 | 1.091 |
| Carbon dioxide | $CO_2$ | 0.1889 | 0.846 | 0.657 | 1.289 |
| Carbon monoxide | CO | 0.2968 | 1.040 | 0.744 | 1.400 |
| Ethane | $C_2H_6$ | 0.2765 | 1.7662 | 1.4897 | 1.186 |
| Ethylene | $C_2H_4$ | 0.2964 | 1.5482 | 1.2518 | 1.237 |
| Helium | He | 2.0769 | 5.1926 | 3.1156 | 1.667 |
| Hydrogen | $H_2$ | 4.1240 | 14.307 | 10.183 | 1.405 |
| Methane | $CH_4$ | 0.5182 | 2.2537 | 1.7354 | 1.299 |
| Neon | Ne | 0.4119 | 1.0299 | 0.6179 | 1.667 |
| Nitrogen | $N_2$ | 0.2968 | 1.039 | 0.743 | 1.400 |
| Octane | $C_8H_{18}$ | 0.0729 | 1.7113 | 1.6385 | 1.044 |
| Oxygen | $O_2$ | 0.2598 | 0.918 | 0.658 | 1.395 |
| Propane | $C_3H_8$ | 0.1885 | 1.6794 | 1.4909 | 1.126 |
| Steam | $H_2O$ | 0.4615 | 1.8723 | 1.4108 | 1.327 |

*Note:* The unit kJ/kg·K is equivalent to kJ/kg·°C.

*Source of Data:* B. G. Kyle, *Chemical and Process Thermodynamics, 3rd ed.* (Upper Saddle River, NJ: Prentice Hall, 2000).

## TABLE A–2

Ideal–gas specific heats of various common gases (*Continued*)

(*b*) At various temperatures

| Temperature, K | $c_p$ kJ/kg·K | $c_v$ kJ/kg·K | $k$ | $c_p$ kJ/kg·K | $c_v$ kJ/kg·K | $k$ | $c_p$ kJ/kg·K | $c_v$ kJ/kg·K | $k$ |
|---|---|---|---|---|---|---|---|---|---|
| | | Air | | | Carbon dioxide, $CO_2$ | | | Carbon monoxide, CO | |
| 250 | 1.003 | 0.716 | 1.401 | 0.791 | 0.602 | 1.314 | 1.039 | 0.743 | 1.400 |
| 300 | 1.005 | 0.718 | 1.400 | 0.846 | 0.657 | 1.288 | 1.040 | 0.744 | 1.399 |
| 350 | 1.008 | 0.721 | 1.398 | 0.895 | 0.706 | 1.268 | 1.043 | 0.746 | 1.398 |
| 400 | 1.013 | 0.726 | 1.395 | 0.939 | 0.750 | 1.252 | 1.047 | 0.751 | 1.395 |
| 450 | 1.020 | 0.733 | 1.391 | 0.978 | 0.790 | 1.239 | 1.054 | 0.757 | 1.392 |
| 500 | 1.029 | 0.742 | 1.387 | 1.014 | 0.825 | 1.229 | 1.063 | 0.767 | 1.387 |
| 550 | 1.040 | 0.753 | 1.381 | 1.046 | 0.857 | 1.220 | 1.075 | 0.778 | 1.382 |
| 600 | 1.051 | 0.764 | 1.376 | 1.075 | 0.886 | 1.213 | 1.087 | 0.790 | 1.376 |
| 650 | 1.063 | 0.776 | 1.370 | 1.102 | 0.913 | 1.207 | 1.100 | 0.803 | 1.370 |
| 700 | 1.075 | 0.788 | 1.364 | 1.126 | 0.937 | 1.202 | 1.113 | 0.816 | 1.364 |
| 750 | 1.087 | 0.800 | 1.359 | 1.148 | 0.959 | 1.197 | 1.126 | 0.829 | 1.358 |
| 800 | 1.099 | 0.812 | 1.354 | 1.169 | 0.980 | 1.193 | 1.139 | 0.842 | 1.353 |
| 900 | 1.121 | 0.834 | 1.344 | 1.204 | 1.015 | 1.186 | 1.163 | 0.866 | 1.343 |
| 1000 | 1.142 | 0.855 | 1.336 | 1.234 | 1.045 | 1.181 | 1.185 | 0.888 | 1.335 |
| | | Hydrogen, $H_2$ | | | Nitrogen, $N_2$ | | | Oxygen, $O_2$ | |
| 250 | 14.051 | 9.927 | 1.416 | 1.039 | 0.742 | 1.400 | 0.913 | 0.653 | 1.398 |
| 300 | 14.307 | 10.183 | 1.405 | 1.039 | 0.743 | 1.400 | 0.918 | 0.658 | 1.395 |
| 350 | 14.427 | 10.302 | 1.400 | 1.041 | 0.744 | 1.399 | 0.928 | 0.668 | 1.389 |
| 400 | 14.476 | 10.352 | 1.398 | 1.044 | 0.747 | 1.397 | 0.941 | 0.681 | 1.382 |
| 450 | 14.501 | 10.377 | 1.398 | 1.049 | 0.752 | 1.395 | 0.956 | 0.696 | 1.373 |
| 500 | 14.513 | 10.389 | 1.397 | 1.056 | 0.759 | 1.391 | 0.972 | 0.712 | 1.365 |
| 550 | 14.530 | 10.405 | 1.396 | 1.065 | 0.768 | 1.387 | 0.988 | 0.728 | 1.358 |
| 600 | 14.546 | 10.422 | 1.396 | 1.075 | 0.778 | 1.382 | 1.003 | 0.743 | 1.350 |
| 650 | 14.571 | 10.447 | 1.395 | 1.086 | 0.789 | 1.376 | 1.017 | 0.758 | 1.343 |
| 700 | 14.604 | 10.480 | 1.394 | 1.098 | 0.801 | 1.371 | 1.031 | 0.771 | 1.337 |
| 750 | 14.645 | 10.521 | 1.392 | 1.110 | 0.813 | 1.365 | 1.043 | 0.783 | 1.332 |
| 800 | 14.695 | 10.570 | 1.390 | 1.121 | 0.825 | 1.360 | 1.054 | 0.794 | 1.327 |
| 900 | 14.822 | 10.698 | 1.385 | 1.145 | 0.849 | 1.349 | 1.074 | 0.814 | 1.319 |
| 1000 | 14.983 | 10.859 | 1.380 | 1.167 | 0.870 | 1.341 | 1.090 | 0.830 | 1.313 |

*Source of Data:* Kenneth Wark, *Thermodynamics,* 4th ed. (New York: McGraw–Hill, 1983), p. 783, Table A–4M. Originally published in *Tables of Thermal Properties of Gases,* NBS Circular 564, 1955.

## TABLE A–2

Ideal–gas specific heats of various common gases (*Concluded*)

### (c) As a function of temperature

$$\bar{c}_p = a + bT + cT^2 + dT^3$$

(*T* in K, $c_p$ in kJ/kmol·K)

| Substance | Formula | a | b | c | d | Temperature range, K | % error Max. | % error Avg. |
|---|---|---|---|---|---|---|---|---|
| Nitrogen | $N_2$ | 28.90 | $-0.1571 \times 10^{-2}$ | $0.8081 \times 10^{-5}$ | $-2.873 \times 10^{-9}$ | 273–1800 | 0.59 | 0.34 |
| Oxygen | $O_2$ | 25.48 | $1.520 \times 10^{-2}$ | $-0.7155 \times 10^{-5}$ | $1.312 \times 10^{-9}$ | 273–1800 | 1.19 | 0.28 |
| Air | — | 28.11 | $0.1967 \times 10^{-2}$ | $0.4802 \times 10^{-5}$ | $-1.966 \times 10^{-9}$ | 273–1800 | 0.72 | 0.33 |
| Hydrogen | $H_2$ | 29.11 | $-0.1916 \times 10^{-2}$ | $0.4003 \times 10^{-5}$ | $-0.8704 \times 10^{-9}$ | 273–1800 | 1.01 | 0.26 |
| Carbon monoxide | CO | 28.16 | $0.1675 \times 10^{-2}$ | $0.5372 \times 10^{-5}$ | $-2.222 \times 10^{-9}$ | 273–1800 | 0.89 | 0.37 |
| Carbon dioxide | $CO_2$ | 22.26 | $5.981 \times 10^{-2}$ | $-3.501 \times 10^{-5}$ | $7.469 \times 10^{-9}$ | 273–1800 | 0.67 | 0.22 |
| Water vapor | $H_2O$ | 32.24 | $0.1923 \times 10^{-2}$ | $1.055 \times 10^{-5}$ | $-3.595 \times 10^{-9}$ | 273–1800 | 0.53 | 0.24 |
| Nitric oxide | NO | 29.34 | $-0.09395 \times 10^{-2}$ | $0.9747 \times 10^{-5}$ | $-4.187 \times 10^{-9}$ | 273–1500 | 0.97 | 0.36 |
| Nitrous oxide | $N_2O$ | 24.11 | $5.8632 \times 10^{-2}$ | $-3.562 \times 10^{-5}$ | $10.58 \times 10^{-9}$ | 273–1500 | 0.59 | 0.26 |
| Nitrogen dioxide | $NO_2$ | 22.9 | $5.715 \times 10^{-2}$ | $-3.52 \times 10^{-5}$ | $7.87 \times 10^{-9}$ | 273–1500 | 0.46 | 0.18 |
| Ammonia | $NH_3$ | 27.568 | $2.5630 \times 10^{-2}$ | $0.99072 \times 10^{-5}$ | $-6.6909 \times 10^{-9}$ | 273–1500 | 0.91 | 0.36 |
| Sulfur | $S_2$ | 27.21 | $2.218 \times 10^{-2}$ | $-1.628 \times 10^{-5}$ | $3.986 \times 10^{-9}$ | 273–1800 | 0.99 | 0.38 |
| Sulfur dioxide | $SO_2$ | 25.78 | $5.795 \times 10^{-2}$ | $-3.812 \times 10^{-5}$ | $8.612 \times 10^{-9}$ | 273–1800 | 0.45 | 0.24 |
| Sulfur trioxide | $SO_3$ | 16.40 | $14.58 \times 10^{-2}$ | $-11.20 \times 10^{-5}$ | $32.42 \times 10^{-9}$ | 273–1300 | 0.29 | 0.13 |
| Acetylene | $C_2H_2$ | 21.8 | $9.2143 \times 10^{-2}$ | $-6.527 \times 10^{-5}$ | $18.21 \times 10^{-9}$ | 273–1500 | 1.46 | 0.59 |
| Benzene | $C_6H_6$ | −36.22 | $48.475 \times 10^{-2}$ | $-31.57 \times 10^{-5}$ | $77.62 \times 10^{-9}$ | 273–1500 | 0.34 | 0.20 |
| Methanol | $CH_4O$ | 19.0 | $9.152 \times 10^{-2}$ | $-1.22 \times 10^{-5}$ | $-8.039 \times 10^{-9}$ | 273–1000 | 0.18 | 0.08 |
| Ethanol | $C_2H_6O$ | 19.9 | $20.96 \times 10^{-2}$ | $-10.38 \times 10^{-5}$ | $20.05 \times 10^{-9}$ | 273–1500 | 0.40 | 0.22 |
| Hydrogen chloride | HCl | 30.33 | $-0.7620 \times 10^{-2}$ | $1.327 \times 10^{-5}$ | $-4.338 \times 10^{-9}$ | 273–1500 | 0.22 | 0.08 |
| Methane | $CH_4$ | 19.89 | $5.024 \times 10^{-2}$ | $1.269 \times 10^{-5}$ | $-11.01 \times 10^{-9}$ | 273–1500 | 1.33 | 0.57 |
| Ethane | $C_2H_6$ | 6.900 | $17.27 \times 10^{-2}$ | $-6.406 \times 10^{-5}$ | $7.285 \times 10^{-9}$ | 273–1500 | 0.83 | 0.28 |
| Propane | $C_3H_8$ | −4.04 | $30.48 \times 10^{-2}$ | $-15.72 \times 10^{-5}$ | $31.74 \times 10^{-9}$ | 273–1500 | 0.40 | 0.12 |
| n–Butane | $C_4H_{10}$ | 3.96 | $37.15 \times 10^{-2}$ | $-18.34 \times 10^{-5}$ | $35.00 \times 10^{-9}$ | 273–1500 | 0.54 | 0.24 |
| i–Butane | $C_4H_{10}$ | −7.913 | $41.60 \times 10^{-2}$ | $-23.01 \times 10^{-5}$ | $49.91 \times 10^{-9}$ | 273–1500 | 0.25 | 0.13 |
| n–Pentane | $C_5H_{12}$ | 6.774 | $45.43 \times 10^{-2}$ | $-22.46 \times 10^{-5}$ | $42.29 \times 10^{-9}$ | 273–1500 | 0.56 | 0.21 |
| n–Hexane | $C_6H_{14}$ | 6.938 | $55.22 \times 10^{-2}$ | $-28.65 \times 10^{-5}$ | $57.69 \times 10^{-9}$ | 273–1500 | 0.72 | 0.20 |
| Ethylene | $C_2H_4$ | 3.95 | $15.64 \times 10^{-2}$ | $-8.344 \times 10^{-5}$ | $17.67 \times 10^{-9}$ | 273–1500 | 0.54 | 0.13 |
| Propylene | $C_3H_6$ | 3.15 | $23.83 \times 10^{-2}$ | $-12.18 \times 10^{-5}$ | $24.62 \times 10^{-9}$ | 273–1500 | 0.73 | 0.17 |

*Source of Data*: B. G. Kyle, *Chemical and Process Thermodynamics* (Englewood Cliffs, NJ: Prentice–Hall, 1984).

## TABLE A-3

Properties of common liquids, solids, and foods

### (a) Liquids

| Substance | Boiling data at 1 atm | | Freezing data | | Liquid properties | | |
|---|---|---|---|---|---|---|---|
| | Normal boiling point, °C | Latent heat of vaporization $h_{fg}$, kJ/kg | Freezing point, °C | Latent heat of fusion $h_{if}$, kJ/kg | Temperature, °C | Density $\rho$, kg/m³ | Specific heat $c_p$, kJ/kg·K |
| Ammonia | −33.3 | 1357 | −77.7 | 322.4 | −33.3 | 682 | 4.43 |
| | | | | | −20 | 665 | 4.52 |
| | | | | | 0 | 639 | 4.60 |
| | | | | | 25 | 602 | 4.80 |
| Argon | −185.9 | 161.6 | −189.3 | 28 | −185.6 | 1394 | 1.14 |
| Benzene | 80.2 | 394 | 5.5 | 126 | 20 | 879 | 1.72 |
| Brine (20% sodium chloride by mass) | 103.9 | — | −17.4 | — | 20 | 1150 | 3.11 |
| n−Butane | −0.5 | 385.2 | −138.5 | 80.3 | −0.5 | 601 | 2.31 |
| Carbon dioxide | −78.4* | 230.5 (at 0°C) | −56.6 | | 0 | 298 | 0.59 |
| Ethanol | 78.2 | 838.3 | −114.2 | 109 | 25 | 783 | 2.46 |
| Ethyl alcohol | 78.6 | 855 | −156 | 108 | 20 | 789 | 2.84 |
| Ethylene glycol | 198.1 | 800.1 | −10.8 | 181.1 | 20 | 1109 | 2.84 |
| Glycerine | 179.9 | 974 | 18.9 | 200.6 | 20 | 1261 | 2.32 |
| Helium | −268.9 | 22.8 | — | — | −268.9 | 146.2 | 22.8 |
| Hydrogen | −252.8 | 445.7 | −259.2 | 59.5 | −252.8 | 70.7 | 10.0 |
| Isobutane | −11.7 | 367.1 | −160 | 105.7 | −11.7 | 593.8 | 2.28 |
| Kerosene | 204–293 | 251 | −24.9 | — | 20 | 820 | 2.00 |
| Mercury | 356.7 | 294.7 | −38.9 | 11.4 | 25 | 13,560 | 0.139 |
| Methane | −161.5 | 510.4 | −182.2 | 58.4 | −161.5 | 423 | 3.49 |
| | | | | | −100 | 301 | 5.79 |
| Methanol | 64.5 | 1100 | −97.7 | 99.2 | 25 | 787 | 2.55 |
| Nitrogen | −195.8 | 198.6 | −210 | 25.3 | −195.8 | 809 | 2.06 |
| | | | | | −160 | 596 | 2.97 |
| Octane | 124.8 | 306.3 | −57.5 | 180.7 | 20 | 703 | 2.10 |
| Oil (light) | | | | | 25 | 910 | 1.80 |
| Oxygen | −183 | 212.7 | −218.8 | 13.7 | −183 | 1141 | 1.71 |
| Petroleum | — | 230–384 | | | 20 | 640 | 2.0 |
| Propane | −42.1 | 427.8 | −187.7 | 80.0 | −42.1 | 581 | 2.25 |
| | | | | | 0 | 529 | 2.53 |
| | | | | | 50 | 449 | 3.13 |
| Refrigerant−134a | −26.1 | 217.0 | −96.6 | — | −50 | 1443 | 1.23 |
| | | | | | −26.1 | 1374 | 1.27 |
| | | | | | 0 | 1295 | 1.34 |
| | | | | | 25 | 1207 | 1.43 |
| Water | 100 | 2257 | 0.0 | 333.7 | 0 | 1000 | 4.22 |
| | | | | | 25 | 997 | 4.18 |
| | | | | | 50 | 988 | 4.18 |
| | | | | | 75 | 975 | 4.19 |
| | | | | | 100 | 958 | 4.22 |

* Sublimation temperature. (At pressures below the triple−point pressure of 518 kPa, carbon dioxide exists as a solid or gas. Also, the freezing−point temperature of carbon dioxide is the triple−point temperature of −56.5°C.)

## TABLE A–3

Properties of common liquids, solids, and foods (*Concluded*)

### (*b*) Solids (values are for room temperature unless indicated otherwise)

| Substance | Density, $\rho$ kg/m$^3$ | Specific heat, $c_p$ kJ/kg·K | Substance | Density, $\rho$ kg/m$^3$ | Specific heat, $c_p$ kJ/kg·K |
|---|---|---|---|---|---|
| **Metals** | | | **Nonmetals** | | |
| Aluminum | | | Asphalt | 2110 | 0.920 |
| 200 K | | 0.797 | Brick, common | 1922 | 0.79 |
| 250 K | | 0.859 | Brick, fireclay (500°C) | 2300 | 0.960 |
| 300 K | 2,700 | 0.902 | Concrete | 2300 | 0.653 |
| 350 K | | 0.929 | Clay | 1000 | 0.920 |
| 400 K | | 0.949 | Diamond | 2420 | 0.616 |
| 450 K | | 0.973 | Glass, window | 2700 | 0.800 |
| 500 K | | 0.997 | Glass, pyrex | 2230 | 0.840 |
| Bronze (76% Cu, 2% Zn, 2% Al) | 8,280 | 0.400 | Graphite | 2500 | 0.711 |
| | | | Granite | 2700 | 1.017 |
| Brass, yellow (65% Cu, 35% Zn) | 8,310 | 0.400 | Gypsum or plaster board | 800 | 1.09 |
| Copper | | | Ice | | |
| −173°C | | 0.254 | 200 K | | 1.56 |
| −100°C | | 0.342 | 220 K | | 1.71 |
| −50°C | | 0.367 | 240 K | | 1.86 |
| 0°C | | 0.381 | 260 K | | 2.01 |
| 27°C | 8,900 | 0.386 | 273 K | 921 | 2.11 |
| 100°C | | 0.393 | Limestone | 1650 | 0.909 |
| 200°C | | 0.403 | Marble | 2600 | 0.880 |
| Iron | 7,840 | 0.45 | Plywood (Douglas Fir) | 545 | 1.21 |
| Lead | 11,310 | 0.128 | Rubber (soft) | 1100 | 1.840 |
| Magnesium | 1,730 | 1.000 | Rubber (hard) | 1150 | 2.009 |
| Nickel | 8,890 | 0.440 | Sand | 1520 | 0.800 |
| Silver | 10,470 | 0.235 | Stone | 1500 | 0.800 |
| Steel, mild | 7,830 | 0.500 | Woods, hard (maple, oak, etc.) | 721 | 1.26 |
| Tungsten | 19,400 | 0.130 | Woods, soft (fir, pine, etc.) | 513 | 1.38 |

### (*c*) Foods

| Food | Water content, % (mass) | Freezing point, °C | Specific heat, kJ/kg·K Above freezing | Specific heat, kJ/kg·K Below freezing | Latent heat of fusion, kJ/kg | Food | Water content, % (mass) | Freezing point, °C | Specific heat, kJ/kg·K Above freezing | Specific heat, kJ/kg·K Below freezing | Latent heat of fusion, kJ/kg |
|---|---|---|---|---|---|---|---|---|---|---|---|
| Apples | 84 | −1.1 | 3.65 | 1.90 | 281 | Lettuce | 95 | −0.2 | 4.02 | 2.04 | 317 |
| Bananas | 75 | −0.8 | 3.35 | 1.78 | 251 | Milk, whole | 88 | −0.6 | 3.79 | 1.95 | 294 |
| Beef round | 67 | — | 3.08 | 1.68 | 224 | Oranges | 87 | −0.8 | 3.75 | 1.94 | 291 |
| Broccoli | 90 | −0.6 | 3.86 | 1.97 | 301 | Potatoes | 78 | −0.6 | 3.45 | 1.82 | 261 |
| Butter | 16 | — | — | 1.04 | 53 | Salmon fish | 64 | −2.2 | 2.98 | 1.65 | 214 |
| Cheese, swiss | 39 | −10.0 | 2.15 | 1.33 | 130 | Shrimp | 83 | −2.2 | 3.62 | 1.89 | 277 |
| Cherries | 80 | −1.8 | 3.52 | 1.85 | 267 | Spinach | 93 | −0.3 | 3.96 | 2.01 | 311 |
| Chicken | 74 | −2.8 | 3.32 | 1.77 | 247 | Strawberries | 90 | −0.8 | 3.86 | 1.97 | 301 |
| Corn, sweet | 74 | −0.6 | 3.32 | 1.77 | 247 | Tomatoes, ripe | 94 | −0.5 | 3.99 | 2.02 | 314 |
| Eggs, whole | 74 | −0.6 | 3.32 | 1.77 | 247 | Turkey | 64 | — | 2.98 | 1.65 | 214 |
| Ice cream | 63 | −5.6 | 2.95 | 1.63 | 210 | Watermelon | 93 | −0.4 | 3.96 | 2.01 | 311 |

*Source of Data:* Values are obtained from various handbooks and other sources or are calculated. Water content and freezing–point data of foods are from *ASHRAE, Handbook of Fundamentals,* SI version (Atlanta, GA: American Society of Heating, Refrigerating and Air–Conditioning Engineers, Inc., 1993), Chapter 30, Table 1. Freezing point is the temperature at which freezing starts for fruits and vegetables, and the average freezing temperature for other foods.

## TABLE A–4

Saturated water—Temperature table

| Temp., $T\,°C$ | Sat. press., $P_{sat}$ kPa | Specific volume, m³/kg | | Internal energy, kJ/kg | | | Enthalpy, kJ/kg | | | Entropy, kJ/kg·K | | |
|---|---|---|---|---|---|---|---|---|---|---|---|---|
| | | Sat. liquid, $v_f$ | Sat. vapor, $v_g$ | Sat. liquid, $u_f$ | Evap., $u_{fg}$ | Sat. vapor, $u_g$ | Sat. liquid, $h_f$ | Evap., $h_{fg}$ | Sat. vapor, $h_g$ | Sat. liquid, $s_f$ | Evap., $s_{fg}$ | Sat. vapor, $s_g$ |
| 0.01 | 0.6117 | 0.001000 | 206.00 | 0.000 | 2374.9 | 2374.9 | 0.001 | 2500.9 | 2500.9 | 0.0000 | 9.1556 | 9.1556 |
| 5 | 0.8725 | 0.001000 | 147.03 | 21.019 | 2360.8 | 2381.8 | 21.020 | 2489.1 | 2510.1 | 0.0763 | 8.9487 | 9.0249 |
| 10 | 1.2281 | 0.001000 | 106.32 | 42.020 | 2346.6 | 2388.7 | 42.022 | 2477.2 | 2519.2 | 0.1511 | 8.7488 | 8.8999 |
| 15 | 1.7057 | 0.001001 | 77.885 | 62.980 | 2332.5 | 2395.5 | 62.982 | 2465.4 | 2528.3 | 0.2245 | 8.5559 | 8.7803 |
| 20 | 2.3392 | 0.001002 | 57.762 | 83.913 | 2318.4 | 2402.3 | 83.915 | 2453.5 | 2537.4 | 0.2965 | 8.3696 | 8.6661 |
| 25 | 3.1698 | 0.001003 | 43.340 | 104.83 | 2304.3 | 2409.1 | 104.83 | 2441.7 | 2546.5 | 0.3672 | 8.1895 | 8.5567 |
| 30 | 4.2469 | 0.001004 | 32.879 | 125.73 | 2290.2 | 2415.9 | 125.74 | 2429.8 | 2555.6 | 0.4368 | 8.0152 | 8.4520 |
| 35 | 5.6291 | 0.001006 | 25.205 | 146.63 | 2276.0 | 2422.7 | 146.64 | 2417.9 | 2564.6 | 0.5051 | 7.8466 | 8.3517 |
| 40 | 7.3851 | 0.001008 | 19.515 | 167.53 | 2261.9 | 2429.4 | 167.53 | 2406.0 | 2573.5 | 0.5724 | 7.6832 | 8.2556 |
| 45 | 9.5953 | 0.001010 | 15.251 | 188.43 | 2247.7 | 2436.1 | 188.44 | 2394.0 | 2582.4 | 0.6386 | 7.5247 | 8.1633 |
| 50 | 12.352 | 0.001012 | 12.026 | 209.33 | 2233.4 | 2442.7 | 209.34 | 2382.0 | 2591.3 | 0.7038 | 7.3710 | 8.0748 |
| 55 | 15.763 | 0.001015 | 9.5639 | 230.24 | 2219.1 | 2449.3 | 230.26 | 2369.8 | 2600.1 | 0.7680 | 7.2218 | 7.9898 |
| 60 | 19.947 | 0.001017 | 7.6670 | 251.16 | 2204.7 | 2455.9 | 251.18 | 2357.7 | 2608.8 | 0.8313 | 7.0769 | 7.9082 |
| 65 | 25.043 | 0.001020 | 6.1935 | 272.09 | 2190.3 | 2462.4 | 272.12 | 2345.4 | 2617.5 | 0.8937 | 6.9360 | 7.8296 |
| 70 | 31.202 | 0.001023 | 5.0396 | 293.04 | 2175.8 | 2468.9 | 293.07 | 2333.0 | 2626.1 | 0.9551 | 6.7989 | 7.7540 |
| 75 | 38.597 | 0.001026 | 4.1291 | 313.99 | 2161.3 | 2475.3 | 314.03 | 2320.6 | 2634.6 | 1.0158 | 6.6655 | 7.6812 |
| 80 | 47.416 | 0.001029 | 3.4053 | 334.97 | 2146.6 | 2481.6 | 335.02 | 2308.0 | 2643.0 | 1.0756 | 6.5355 | 7.6111 |
| 85 | 57.868 | 0.001032 | 2.8261 | 355.96 | 2131.9 | 2487.8 | 356.02 | 2295.3 | 2651.4 | 1.1346 | 6.4089 | 7.5435 |
| 90 | 70.183 | 0.001036 | 2.3593 | 376.97 | 2117.0 | 2494.0 | 377.04 | 2282.5 | 2659.6 | 1.1929 | 6.2853 | 7.4782 |
| 95 | 84.609 | 0.001040 | 1.9808 | 398.00 | 2102.0 | 2500.1 | 398.09 | 2269.6 | 2667.6 | 1.2504 | 6.1647 | 7.4151 |
| 100 | 101.42 | 0.001043 | 1.6720 | 419.06 | 2087.0 | 2506.0 | 419.17 | 2256.4 | 2675.6 | 1.3072 | 6.0470 | 7.3542 |
| 105 | 120.90 | 0.001047 | 1.4186 | 440.15 | 2071.8 | 2511.9 | 440.28 | 2243.1 | 2683.4 | 1.3634 | 5.9319 | 7.2952 |
| 110 | 143.38 | 0.001052 | 1.2094 | 461.27 | 2056.4 | 2517.7 | 461.42 | 2229.7 | 2691.1 | 1.4188 | 5.8193 | 7.2382 |
| 115 | 169.18 | 0.001056 | 1.0360 | 482.42 | 2040.9 | 2523.3 | 482.59 | 2216.0 | 2698.6 | 1.4737 | 5.7092 | 7.1829 |
| 120 | 198.67 | 0.001060 | 0.89133 | 503.60 | 2025.3 | 2528.9 | 503.81 | 2202.1 | 2706.0 | 1.5279 | 5.6013 | 7.1292 |
| 125 | 232.23 | 0.001065 | 0.77012 | 524.83 | 2009.5 | 2534.3 | 525.07 | 2188.1 | 2713.1 | 1.5816 | 5.4956 | 7.0771 |
| 130 | 270.28 | 0.001070 | 0.66808 | 546.10 | 1993.4 | 2539.5 | 546.38 | 2173.7 | 2720.1 | 1.6346 | 5.3919 | 7.0265 |
| 135 | 313.22 | 0.001075 | 0.58179 | 567.41 | 1977.3 | 2544.7 | 567.75 | 2159.1 | 2726.9 | 1.6872 | 5.2901 | 6.9773 |
| 140 | 361.53 | 0.001080 | 0.50850 | 588.77 | 1960.9 | 2549.6 | 589.16 | 2144.3 | 2733.5 | 1.7392 | 5.1901 | 6.9294 |
| 145 | 415.68 | 0.001085 | 0.44600 | 610.19 | 1944.2 | 2554.4 | 610.64 | 2129.2 | 2739.8 | 1.7908 | 5.0919 | 6.8827 |
| 150 | 476.16 | 0.001091 | 0.39248 | 631.66 | 1927.4 | 2559.1 | 632.18 | 2113.8 | 2745.9 | 1.8418 | 4.9953 | 6.8371 |
| 155 | 543.49 | 0.001096 | 0.34648 | 653.19 | 1910.3 | 2563.5 | 653.79 | 2098.0 | 2751.8 | 1.8924 | 4.9002 | 6.7927 |
| 160 | 618.23 | 0.001102 | 0.30680 | 674.79 | 1893.0 | 2567.8 | 675.47 | 2082.0 | 2757.5 | 1.9426 | 4.8066 | 6.7492 |
| 165 | 700.93 | 0.001108 | 0.27244 | 696.46 | 1875.4 | 2571.9 | 697.24 | 2065.6 | 2762.8 | 1.9923 | 4.7143 | 6.7067 |
| 170 | 792.18 | 0.001114 | 0.24260 | 718.20 | 1857.5 | 2575.7 | 719.08 | 2048.8 | 2767.9 | 2.0417 | 4.6233 | 6.6650 |
| 175 | 892.60 | 0.001121 | 0.21659 | 740.02 | 1839.4 | 2579.4 | 741.02 | 2031.7 | 2772.7 | 2.0906 | 4.5335 | 6.6242 |
| 180 | 1002.8 | 0.001127 | 0.19384 | 761.92 | 1820.9 | 2582.8 | 763.05 | 2014.2 | 2777.2 | 2.1392 | 4.4448 | 6.5841 |
| 185 | 1123.5 | 0.001134 | 0.17390 | 783.91 | 1802.1 | 2586.0 | 785.19 | 1996.2 | 2781.4 | 2.1875 | 4.3572 | 6.5447 |
| 190 | 1255.2 | 0.001141 | 0.15636 | 806.00 | 1783.0 | 2589.0 | 807.43 | 1977.9 | 2785.3 | 2.2355 | 4.2705 | 6.5059 |
| 195 | 1398.8 | 0.001149 | 0.14089 | 828.18 | 1763.6 | 2591.7 | 829.78 | 1959.0 | 2788.8 | 2.2831 | 4.1847 | 6.4678 |
| 200 | 1554.9 | 0.001157 | 0.12721 | 850.46 | 1743.7 | 2594.2 | 852.26 | 1939.8 | 2792.0 | 2.3305 | 4.0997 | 6.4302 |

## TABLE A–4

Saturated water—Temperature table (Concluded)

| Temp., $T\,°C$ | Sat. press., $P_{sat}$ kPa | Specific volume, m³/kg Sat. liquid, $v_f$ | Sat. vapor, $v_g$ | Internal energy, kJ/kg Sat. liquid, $u_f$ | Evap., $u_{fg}$ | Sat. vapor, $u_g$ | Enthalpy, kJ/kg Sat. liquid, $h_f$ | Evap., $h_{fg}$ | Sat. vapor, $h_g$ | Entropy, kJ/kg·K Sat. liquid, $s_f$ | Evap., $s_{fg}$ | Sat. vapor, $s_g$ |
|---|---|---|---|---|---|---|---|---|---|---|---|---|
| 205 | 1724.3 | 0.001164 | 0.11508 | 872.86 | 1723.5 | 2596.4 | 874.87 | 1920.0 | 2794.8 | 2.3776 | 4.0154 | 6.3930 |
| 210 | 1907.7 | 0.001173 | 0.10429 | 895.38 | 1702.9 | 2598.3 | 897.61 | 1899.7 | 2797.3 | 2.4245 | 3.9318 | 6.3563 |
| 215 | 2105.9 | 0.001181 | 0.094680 | 918.02 | 1681.9 | 2599.9 | 920.50 | 1878.8 | 2799.3 | 2.4712 | 3.8489 | 6.3200 |
| 220 | 2319.6 | 0.001190 | 0.086094 | 940.79 | 1660.5 | 2601.3 | 943.55 | 1857.4 | 2801.0 | 2.5176 | 3.7664 | 6.2840 |
| 225 | 2549.7 | 0.001199 | 0.078405 | 963.70 | 1638.6 | 2602.3 | 966.76 | 1835.4 | 2802.2 | 2.5639 | 3.6844 | 6.2483 |
| 230 | 2797.1 | 0.001209 | 0.071505 | 986.76 | 1616.1 | 2602.9 | 990.14 | 1812.8 | 2802.9 | 2.6100 | 3.6028 | 6.2128 |
| 235 | 3062.6 | 0.001219 | 0.065300 | 1010.0 | 1593.2 | 2603.2 | 1013.7 | 1789.5 | 2803.2 | 2.6560 | 3.5216 | 6.1775 |
| 240 | 3347.0 | 0.001229 | 0.059707 | 1033.4 | 1569.8 | 2603.1 | 1037.5 | 1765.5 | 2803.0 | 2.7018 | 3.4405 | 6.1424 |
| 245 | 3651.2 | 0.001240 | 0.054656 | 1056.9 | 1545.7 | 2602.7 | 1061.5 | 1740.8 | 2802.2 | 2.7476 | 3.3596 | 6.1072 |
| 250 | 3976.2 | 0.001252 | 0.050085 | 1080.7 | 1521.1 | 2601.8 | 1085.7 | 1715.3 | 2801.0 | 2.7933 | 3.2788 | 6.0721 |
| 255 | 4322.9 | 0.001263 | 0.045941 | 1104.7 | 1495.8 | 2600.5 | 1110.1 | 1689.0 | 2799.1 | 2.8390 | 3.1979 | 6.0369 |
| 260 | 4692.3 | 0.001276 | 0.042175 | 1128.8 | 1469.9 | 2598.7 | 1134.8 | 1661.8 | 2796.6 | 2.8847 | 3.1169 | 6.0017 |
| 265 | 5085.3 | 0.001289 | 0.038748 | 1153.3 | 1443.2 | 2596.5 | 1159.8 | 1633.7 | 2793.5 | 2.9304 | 3.0358 | 5.9662 |
| 270 | 5503.0 | 0.001303 | 0.035622 | 1177.9 | 1415.7 | 2593.7 | 1185.1 | 1604.6 | 2789.7 | 2.9762 | 2.9542 | 5.9305 |
| 275 | 5946.4 | 0.001317 | 0.032767 | 1202.9 | 1387.4 | 2590.3 | 1210.7 | 1574.5 | 2785.2 | 3.0221 | 2.8723 | 5.8944 |
| 280 | 6416.6 | 0.001333 | 0.030153 | 1228.2 | 1358.2 | 2586.4 | 1236.7 | 1543.2 | 2779.9 | 3.0681 | 2.7898 | 5.8579 |
| 285 | 6914.6 | 0.001349 | 0.027756 | 1253.7 | 1328.1 | 2581.8 | 1263.1 | 1510.7 | 2773.7 | 3.1144 | 2.7066 | 5.8210 |
| 290 | 7441.8 | 0.001366 | 0.025554 | 1279.7 | 1296.9 | 2576.5 | 1289.8 | 1476.9 | 2766.7 | 3.1608 | 2.6225 | 5.7834 |
| 295 | 7999.0 | 0.001384 | 0.023528 | 1306.0 | 1264.5 | 2570.5 | 1317.1 | 1441.6 | 2758.7 | 3.2076 | 2.5374 | 5.7450 |
| 300 | 8587.9 | 0.001404 | 0.021659 | 1332.7 | 1230.9 | 2563.6 | 1344.8 | 1404.8 | 2749.6 | 3.2548 | 2.4511 | 5.7059 |
| 305 | 9209.4 | 0.001425 | 0.019932 | 1360.0 | 1195.9 | 2555.8 | 1373.1 | 1366.3 | 2739.4 | 3.3024 | 2.3633 | 5.6657 |
| 310 | 9865.0 | 0.001447 | 0.018333 | 1387.7 | 1159.3 | 2547.1 | 1402.0 | 1325.9 | 2727.9 | 3.3506 | 2.2737 | 5.6243 |
| 315 | 10,556 | 0.001472 | 0.016849 | 1416.1 | 1121.1 | 2537.2 | 1431.6 | 1283.4 | 2715.0 | 3.3994 | 2.1821 | 5.5816 |
| 320 | 11,284 | 0.001499 | 0.015470 | 1445.1 | 1080.9 | 2526.0 | 1462.0 | 1238.5 | 2700.6 | 3.4491 | 2.0881 | 5.5372 |
| 325 | 12,051 | 0.001528 | 0.014183 | 1475.0 | 1038.5 | 2513.4 | 1493.4 | 1191.0 | 2684.3 | 3.4998 | 1.9911 | 5.4908 |
| 330 | 12,858 | 0.001560 | 0.012979 | 1505.7 | 993.5 | 2499.2 | 1525.8 | 1140.3 | 2666.0 | 3.5516 | 1.8906 | 5.4422 |
| 335 | 13,707 | 0.001597 | 0.011848 | 1537.5 | 945.5 | 2483.0 | 1559.4 | 1086.0 | 2645.4 | 3.6050 | 1.7857 | 5.3907 |
| 340 | 14,601 | 0.001638 | 0.010783 | 1570.7 | 893.8 | 2464.5 | 1594.6 | 1027.4 | 2622.0 | 3.6602 | 1.6756 | 5.3358 |
| 345 | 15,541 | 0.001685 | 0.009772 | 1605.5 | 837.7 | 2443.2 | 1631.7 | 963.4 | 2595.1 | 3.7179 | 1.5585 | 5.2765 |
| 350 | 16,529 | 0.001741 | 0.008806 | 1642.4 | 775.9 | 2418.3 | 1671.2 | 892.7 | 2563.9 | 3.7788 | 1.4326 | 5.2114 |
| 355 | 17,570 | 0.001808 | 0.007872 | 1682.2 | 706.4 | 2388.6 | 1714.0 | 812.9 | 2526.9 | 3.8442 | 1.2942 | 5.1384 |
| 360 | 18,666 | 0.001895 | 0.006950 | 1726.2 | 625.7 | 2351.9 | 1761.5 | 720.1 | 2481.6 | 3.9165 | 1.1373 | 5.0537 |
| 365 | 19,822 | 0.002015 | 0.006009 | 1777.2 | 526.4 | 2303.6 | 1817.2 | 605.5 | 2422.7 | 4.0004 | 0.9489 | 4.9493 |
| 370 | 21,044 | 0.002217 | 0.004953 | 1844.5 | 385.6 | 2230.1 | 1891.2 | 443.1 | 2334.3 | 4.1119 | 0.6890 | 4.8009 |
| 373.95 | 22,064 | 0.003106 | 0.003106 | 2015.7 | 0 | 2015.7 | 2084.3 | 0 | 2084.3 | 4.4070 | 0 | 4.4070 |

*Source of Data:* Tables A–4 through A–8 are generated using the Engineering Equation Solver (EES) software developed by S. A. Klein and F. L. Alvarado. The routine used in calculations is the highly accurate Steam_IAPWS, which incorporates the 1995 Formulation for the Thermodynamic Properties of Ordinary Water Substance for General and Scientific Use, issued by The International Association for the Properties of Water and Steam (IAPWS). This formulation replaces the 1984 formulation of Haar, Gallagher, and Kell (NBS/NRC Steam Tables, Hemisphere Publishing Co., 1984), which is also available in EES as the routine STEAM. The new formulation is based on the correlations of Saul and Wagner (J. Phys. Chem. Ref. Data, 16, 893, 1987) with modifications to adjust to the International Temperature Scale of 1990. The modifications are described by Wagner and Pruss (J. Phys. Chem. Ref. Data, 22, 783, 1993). The properties of ice are based on Hyland and Wexler, "Formulations for the Thermodynamic Properties of the Saturated Phases of $H_2O$ from 173.15 K to 473.15 K," *ASHRAE Trans.*, Part 2A, Paper 2793, 1983.

## TABLE A–5

Saturated water—Pressure table

| Press., P kPa | Sat. temp., $T_{sat}$ °C | Specific volume, m³/kg | | Internal energy, kJ/kg | | | Enthalpy, kJ/kg | | | Entropy, kJ/kg·K | | |
|---|---|---|---|---|---|---|---|---|---|---|---|---|
| | | Sat. liquid, $v_f$ | Sat. vapor, $v_g$ | Sat. liquid, $u_f$ | Evap., $u_{fg}$ | Sat. vapor, $u_g$ | Sat. liquid, $h_f$ | Evap., $h_{fg}$ | Sat. vapor, $h_g$ | Sat. liquid, $s_f$ | Evap., $s_{fg}$ | Sat. vapor, $s_g$ |
| 1.0 | 6.97 | 0.001000 | 129.19 | 29.302 | 2355.2 | 2384.5 | 29.303 | 2484.4 | 2513.7 | 0.1059 | 8.8690 | 8.9749 |
| 1.5 | 13.02 | 0.001001 | 87.964 | 54.686 | 2338.1 | 2392.8 | 54.688 | 2470.1 | 2524.7 | 0.1956 | 8.6314 | 8.8270 |
| 2.0 | 17.50 | 0.001001 | 66.990 | 73.431 | 2325.5 | 2398.9 | 73.433 | 2459.5 | 2532.9 | 0.2606 | 8.4621 | 8.7227 |
| 2.5 | 21.08 | 0.001002 | 54.242 | 88.422 | 2315.4 | 2403.8 | 88.424 | 2451.0 | 2539.4 | 0.3118 | 8.3302 | 8.6421 |
| 3.0 | 24.08 | 0.001003 | 45.654 | 100.98 | 2306.9 | 2407.9 | 100.98 | 2443.9 | 2544.8 | 0.3543 | 8.2222 | 8.5765 |
| 4.0 | 28.96 | 0.001004 | 34.791 | 121.39 | 2293.1 | 2414.5 | 121.39 | 2432.3 | 2553.7 | 0.4224 | 8.0510 | 8.4734 |
| 5.0 | 32.87 | 0.001005 | 28.185 | 137.75 | 2282.1 | 2419.8 | 137.75 | 2423.0 | 2560.7 | 0.4762 | 7.9176 | 8.3938 |
| 7.5 | 40.29 | 0.001008 | 19.233 | 168.74 | 2261.1 | 2429.8 | 168.75 | 2405.3 | 2574.0 | 0.5763 | 7.6738 | 8.2501 |
| 10 | 45.81 | 0.001010 | 14.670 | 191.79 | 2245.4 | 2437.2 | 191.81 | 2392.1 | 2583.9 | 0.6492 | 7.4996 | 8.1488 |
| 15 | 53.97 | 0.001014 | 10.020 | 225.93 | 2222.1 | 2448.0 | 225.94 | 2372.3 | 2598.3 | 0.7549 | 7.2522 | 8.0071 |
| 20 | 60.06 | 0.001017 | 7.6481 | 251.40 | 2204.6 | 2456.0 | 251.42 | 2357.5 | 2608.9 | 0.8320 | 7.0752 | 7.9073 |
| 25 | 64.96 | 0.001020 | 6.2034 | 271.93 | 2190.4 | 2462.4 | 271.96 | 2345.5 | 2617.5 | 0.8932 | 6.9370 | 7.8302 |
| 30 | 69.09 | 0.001022 | 5.2287 | 289.24 | 2178.5 | 2467.7 | 289.27 | 2335.3 | 2624.6 | 0.9441 | 6.8234 | 7.7675 |
| 40 | 75.86 | 0.001026 | 3.9933 | 317.58 | 2158.8 | 2476.3 | 317.62 | 2318.4 | 2636.1 | 1.0261 | 6.6430 | 7.6691 |
| 50 | 81.32 | 0.001030 | 3.2403 | 340.49 | 2142.7 | 2483.2 | 340.54 | 2304.7 | 2645.2 | 1.0912 | 6.5019 | 7.5931 |
| 75 | 91.76 | 0.001037 | 2.2172 | 384.36 | 2111.8 | 2496.1 | 384.44 | 2278.0 | 2662.4 | 1.2132 | 6.2426 | 7.4558 |
| 100 | 99.61 | 0.001043 | 1.6941 | 417.40 | 2088.2 | 2505.6 | 417.51 | 2257.5 | 2675.0 | 1.3028 | 6.0562 | 7.3589 |
| 101.325 | 99.97 | 0.001043 | 1.6734 | 418.95 | 2087.0 | 2506.0 | 419.06 | 2256.5 | 2675.6 | 1.3069 | 6.0476 | 7.3545 |
| 125 | 105.97 | 0.001048 | 1.3750 | 444.23 | 2068.8 | 2513.0 | 444.36 | 2240.6 | 2684.9 | 1.3741 | 5.9100 | 7.2841 |
| 150 | 111.35 | 0.001053 | 1.1594 | 466.97 | 2052.3 | 2519.2 | 467.13 | 2226.0 | 2693.1 | 1.4337 | 5.7894 | 7.2231 |
| 175 | 116.04 | 0.001057 | 1.0037 | 486.82 | 2037.7 | 2524.5 | 487.01 | 2213.1 | 2700.2 | 1.4850 | 5.6865 | 7.1716 |
| 200 | 120.21 | 0.001061 | 0.88578 | 504.50 | 2024.6 | 2529.1 | 504.71 | 2201.6 | 2706.3 | 1.5302 | 5.5968 | 7.1270 |
| 225 | 123.97 | 0.001064 | 0.79329 | 520.47 | 2012.7 | 2533.2 | 520.71 | 2191.0 | 2711.7 | 1.5706 | 5.5171 | 7.0877 |
| 250 | 127.41 | 0.001067 | 0.71873 | 535.08 | 2001.8 | 2536.8 | 535.35 | 2181.2 | 2716.5 | 1.6072 | 5.4453 | 7.0525 |
| 275 | 130.58 | 0.001070 | 0.65732 | 548.57 | 1991.6 | 2540.1 | 548.86 | 2172.0 | 2720.9 | 1.6408 | 5.3800 | 7.0207 |
| 300 | 133.52 | 0.001073 | 0.60582 | 561.11 | 1982.1 | 2543.2 | 561.43 | 2163.5 | 2724.9 | 1.6717 | 5.3200 | 6.9917 |
| 325 | 136.27 | 0.001076 | 0.56199 | 572.84 | 1973.1 | 2545.9 | 573.19 | 2155.4 | 2728.6 | 1.7005 | 5.2645 | 6.9650 |
| 350 | 138.86 | 0.001079 | 0.52422 | 583.89 | 1964.6 | 2548.5 | 584.26 | 2147.7 | 2732.0 | 1.7274 | 5.2128 | 6.9402 |
| 375 | 141.30 | 0.001081 | 0.49133 | 594.32 | 1956.6 | 2550.9 | 594.73 | 2140.4 | 2735.1 | 1.7526 | 5.1645 | 6.9171 |
| 400 | 143.61 | 0.001084 | 0.46242 | 604.22 | 1948.9 | 2553.1 | 604.66 | 2133.4 | 2738.1 | 1.7765 | 5.1191 | 6.8955 |
| 450 | 147.90 | 0.001088 | 0.41392 | 622.65 | 1934.5 | 2557.1 | 623.14 | 2120.3 | 2743.4 | 1.8205 | 5.0356 | 6.8561 |
| 500 | 151.83 | 0.001093 | 0.37483 | 639.54 | 1921.2 | 2560.7 | 640.09 | 2108.0 | 2748.1 | 1.8604 | 4.9603 | 6.8207 |
| 550 | 155.46 | 0.001097 | 0.34261 | 655.16 | 1908.8 | 2563.9 | 655.77 | 2096.6 | 2752.4 | 1.8970 | 4.8916 | 6.7886 |
| 600 | 158.83 | 0.001101 | 0.31560 | 669.72 | 1897.1 | 2566.8 | 670.38 | 2085.8 | 2756.2 | 1.9308 | 4.8285 | 6.7593 |
| 650 | 161.98 | 0.001104 | 0.29260 | 683.37 | 1886.1 | 2569.4 | 684.08 | 2075.5 | 2759.6 | 1.9623 | 4.7699 | 6.7322 |
| 700 | 164.95 | 0.001108 | 0.27278 | 696.23 | 1875.6 | 2571.8 | 697.00 | 2065.8 | 2762.8 | 1.9918 | 4.7153 | 6.7071 |
| 750 | 167.75 | 0.001111 | 0.25552 | 708.40 | 1865.6 | 2574.0 | 709.24 | 2056.4 | 2765.7 | 2.0195 | 4.6642 | 6.6837 |

## TABLE A-5

Saturated water—Pressure table (*Concluded*)

| Press., $P$ kPa | Sat. temp., $T_{sat}$ °C | Specific volume, m³/kg | | Internal energy, kJ/kg | | | Enthalpy, kJ/kg | | | Entropy, kJ/kg·K | | |
|---|---|---|---|---|---|---|---|---|---|---|---|---|
| | | Sat. liquid, $v_f$ | Sat. vapor, $v_g$ | Sat. liquid, $u_f$ | Evap., $u_{fg}$ | Sat. vapor, $u_g$ | Sat. liquid, $h_f$ | Evap., $h_{fg}$ | Sat. vapor, $h_g$ | Sat. liquid, $s_f$ | Evap., $s_{fg}$ | Sat. vapor, $s_g$ |
| 800 | 170.41 | 0.001115 | 0.24035 | 719.97 | 1856.1 | 2576.0 | 720.87 | 2047.5 | 2768.3 | 2.0457 | 4.6160 | 6.6616 |
| 850 | 172.94 | 0.001118 | 0.22690 | 731.00 | 1846.9 | 2577.9 | 731.95 | 2038.8 | 2770.8 | 2.0705 | 4.5705 | 6.6409 |
| 900 | 175.35 | 0.001121 | 0.21489 | 741.55 | 1838.1 | 2579.6 | 742.56 | 2030.5 | 2773.0 | 2.0941 | 4.5273 | 6.6213 |
| 950 | 177.66 | 0.001124 | 0.20411 | 751.67 | 1829.6 | 2581.3 | 752.74 | 2022.4 | 2775.2 | 2.1166 | 4.4862 | 6.6027 |
| 1000 | 179.88 | 0.001127 | 0.19436 | 761.39 | 1821.4 | 2582.8 | 762.51 | 2014.6 | 2777.1 | 2.1381 | 4.4470 | 6.5850 |
| 1100 | 184.06 | 0.001133 | 0.17745 | 779.78 | 1805.7 | 2585.5 | 781.03 | 1999.6 | 2780.7 | 2.1785 | 4.3735 | 6.5520 |
| 1200 | 187.96 | 0.001138 | 0.16326 | 796.96 | 1790.9 | 2587.8 | 798.33 | 1985.4 | 2783.8 | 2.2159 | 4.3058 | 6.5217 |
| 1300 | 191.60 | 0.001144 | 0.15119 | 813.10 | 1776.8 | 2589.9 | 814.59 | 1971.9 | 2786.5 | 2.2508 | 4.2428 | 6.4936 |
| 1400 | 195.04 | 0.001149 | 0.14078 | 828.35 | 1763.4 | 2591.8 | 829.96 | 1958.9 | 2788.9 | 2.2835 | 4.1840 | 6.4675 |
| 1500 | 198.29 | 0.001154 | 0.13171 | 842.82 | 1750.6 | 2593.4 | 844.55 | 1946.4 | 2791.0 | 2.3143 | 4.1287 | 6.4430 |
| 1750 | 205.72 | 0.001166 | 0.11344 | 876.12 | 1720.6 | 2596.7 | 878.16 | 1917.1 | 2795.2 | 2.3844 | 4.0033 | 6.3877 |
| 2000 | 212.38 | 0.001177 | 0.099587 | 906.12 | 1693.0 | 2599.1 | 908.47 | 1889.8 | 2798.3 | 2.4467 | 3.8923 | 6.3390 |
| 2250 | 218.41 | 0.001187 | 0.088717 | 933.54 | 1667.3 | 2600.9 | 936.21 | 1864.3 | 2800.5 | 2.5029 | 3.7926 | 6.2954 |
| 2500 | 223.95 | 0.001197 | 0.079952 | 958.87 | 1643.2 | 2602.1 | 961.87 | 1840.1 | 2801.9 | 2.5542 | 3.7016 | 6.2558 |
| 3000 | 233.85 | 0.001217 | 0.066667 | 1004.6 | 1598.5 | 2603.2 | 1008.3 | 1794.9 | 2803.2 | 2.6454 | 3.5402 | 6.1856 |
| 3500 | 242.56 | 0.001235 | 0.057061 | 1045.4 | 1557.6 | 2603.0 | 1049.7 | 1753.0 | 2802.7 | 2.7253 | 3.3991 | 6.1244 |
| 4000 | 250.35 | 0.001252 | 0.049779 | 1082.4 | 1519.3 | 2601.7 | 1087.4 | 1713.5 | 2800.8 | 2.7966 | 3.2731 | 6.0696 |
| 5000 | 263.94 | 0.001286 | 0.039448 | 1148.1 | 1448.9 | 2597.0 | 1154.5 | 1639.7 | 2794.2 | 2.9207 | 3.0530 | 5.9737 |
| 6000 | 275.59 | 0.001319 | 0.032449 | 1205.8 | 1384.1 | 2589.9 | 1213.8 | 1570.9 | 2784.6 | 3.0275 | 2.8627 | 5.8902 |
| 7000 | 285.83 | 0.001352 | 0.027378 | 1258.0 | 1323.0 | 2581.0 | 1267.5 | 1505.2 | 2772.6 | 3.1220 | 2.6927 | 5.8148 |
| 8000 | 295.01 | 0.001384 | 0.023525 | 1306.0 | 1264.5 | 2570.5 | 1317.1 | 1441.6 | 2758.7 | 3.2077 | 2.5373 | 5.7450 |
| 9000 | 303.35 | 0.001418 | 0.020489 | 1350.9 | 1207.6 | 2558.5 | 1363.7 | 1379.3 | 2742.9 | 3.2866 | 2.3925 | 5.6791 |
| 10,000 | 311.00 | 0.001452 | 0.018028 | 1393.3 | 1151.8 | 2545.2 | 1407.8 | 1317.6 | 2725.5 | 3.3603 | 2.2556 | 5.6159 |
| 11,000 | 318.08 | 0.001488 | 0.015988 | 1433.9 | 1096.6 | 2530.4 | 1450.2 | 1256.1 | 2706.3 | 3.4299 | 2.1245 | 5.5544 |
| 12,000 | 324.68 | 0.001526 | 0.014264 | 1473.0 | 1041.3 | 2514.3 | 1491.3 | 1194.1 | 2685.4 | 3.4964 | 1.9975 | 5.4939 |
| 13,000 | 330.85 | 0.001566 | 0.012781 | 1511.0 | 985.5 | 2496.6 | 1531.4 | 1131.3 | 2662.7 | 3.5606 | 1.8730 | 5.4336 |
| 14,000 | 336.67 | 0.001610 | 0.011487 | 1548.4 | 928.7 | 2477.1 | 1571.0 | 1067.0 | 2637.9 | 3.6232 | 1.7497 | 5.3728 |
| 15,000 | 342.16 | 0.001657 | 0.010341 | 1585.5 | 870.3 | 2455.7 | 1610.3 | 1000.5 | 2610.8 | 3.6848 | 1.6261 | 5.3108 |
| 16,000 | 347.36 | 0.001710 | 0.009312 | 1622.6 | 809.4 | 2432.0 | 1649.9 | 931.1 | 2581.0 | 3.7461 | 1.5005 | 5.2466 |
| 17,000 | 352.29 | 0.001770 | 0.008374 | 1660.2 | 745.1 | 2405.4 | 1690.3 | 857.4 | 2547.7 | 3.8082 | 1.3709 | 5.1791 |
| 18,000 | 356.99 | 0.001840 | 0.007504 | 1699.1 | 675.9 | 2375.0 | 1732.2 | 777.8 | 2510.0 | 3.8720 | 1.2343 | 5.1064 |
| 19,000 | 361.47 | 0.001926 | 0.006677 | 1740.3 | 598.9 | 2339.2 | 1776.8 | 689.2 | 2466.0 | 3.9396 | 1.0860 | 5.0256 |
| 20,000 | 365.75 | 0.002038 | 0.005862 | 1785.8 | 509.0 | 2294.8 | 1826.6 | 585.5 | 2412.1 | 4.0146 | 0.9164 | 4.9310 |
| 21,000 | 369.83 | 0.002207 | 0.004994 | 1841.6 | 391.9 | 2233.5 | 1888.0 | 450.4 | 2338.4 | 4.1071 | 0.7005 | 4.8076 |
| 22,000 | 373.71 | 0.002703 | 0.003644 | 1951.7 | 140.8 | 2092.4 | 2011.1 | 161.5 | 2172.6 | 4.2942 | 0.2496 | 4.5439 |
| 22,064 | 373.95 | 0.003106 | 0.003106 | 2015.7 | 0 | 2015.7 | 2084.3 | 0 | 2084.3 | 4.4070 | 0 | 4.4070 |

## TABLE A–6

Superheated water

| T | v | u | h | s | v | u | h | s | v | u | h | s |
|---|---|---|---|---|---|---|---|---|---|---|---|---|
| °C | m³/kg | kJ/kg | kJ/kg | kJ/kg·K | m³/kg | kJ/kg | kJ/kg | kJ/kg·K | m³/kg | kJ/kg | kJ/kg | kJ/kg·K |
| | P = 0.01 MPa (45.81°C)* | | | | P = 0.05 MPa (81.32°C) | | | | P = 0.10 MPa (99.61°C) | | | |
| Sat.† | 14.670 | 2437.2 | 2583.9 | 8.1488 | 3.2403 | 2483.2 | 2645.2 | 7.5931 | 1.6941 | 2505.6 | 2675.0 | 7.3589 |
| 50 | 14.867 | 2443.3 | 2592.0 | 8.1741 | | | | | | | | |
| 100 | 17.196 | 2515.5 | 2687.5 | 8.4489 | 3.4187 | 2511.5 | 2682.4 | 7.6953 | 1.6959 | 2506.2 | 2675.8 | 7.3611 |
| 150 | 19.513 | 2587.9 | 2783.0 | 8.6893 | 3.8897 | 2585.7 | 2780.2 | 7.9413 | 1.9367 | 2582.9 | 2776.6 | 7.6148 |
| 200 | 21.826 | 2661.4 | 2879.6 | 8.9049 | 4.3562 | 2660.0 | 2877.8 | 8.1592 | 2.1724 | 2658.2 | 2875.5 | 7.8356 |
| 250 | 24.136 | 2736.1 | 2977.5 | 9.1015 | 4.8206 | 2735.1 | 2976.2 | 8.3568 | 2.4062 | 2733.9 | 2974.5 | 8.0346 |
| 300 | 26.446 | 2812.3 | 3076.7 | 9.2827 | 5.2841 | 2811.6 | 3075.8 | 8.5387 | 2.6389 | 2810.7 | 3074.5 | 8.2172 |
| 400 | 31.063 | 2969.3 | 3280.0 | 9.6094 | 6.2094 | 2968.9 | 3279.3 | 8.8659 | 3.1027 | 2968.3 | 3278.6 | 8.5452 |
| 500 | 35.680 | 3132.9 | 3489.7 | 9.8998 | 7.1338 | 3132.6 | 3489.3 | 9.1566 | 3.5655 | 3132.2 | 3488.7 | 8.8362 |
| 600 | 40.296 | 3303.3 | 3706.3 | 10.1631 | 8.0577 | 3303.1 | 3706.0 | 9.4201 | 4.0279 | 3302.8 | 3705.6 | 9.0999 |
| 700 | 44.911 | 3480.8 | 3929.9 | 10.4056 | 9.9813 | 3480.6 | 3929.7 | 9.6626 | 4.4900 | 3480.4 | 3929.4 | 9.3424 |
| 800 | 49.527 | 3665.4 | 4160.6 | 10.6312 | 9.9047 | 3665.2 | 4160.4 | 9.8883 | 4.9519 | 3665.0 | 4160.2 | 9.5682 |
| 900 | 54.143 | 3856.9 | 4398.3 | 10.8429 | 10.8280 | 3856.8 | 4398.2 | 10.1000 | 5.4137 | 3856.7 | 4398.0 | 9.7800 |
| 1000 | 58.758 | 4055.3 | 4642.8 | 11.0429 | 11.7513 | 4055.2 | 4642.7 | 10.3000 | 5.8755 | 4055.0 | 4642.6 | 9.9800 |
| 1100 | 63.373 | 4260.0 | 4893.8 | 11.2326 | 12.6745 | 4259.9 | 4893.7 | 10.4897 | 6.3372 | 4259.8 | 4893.6 | 10.1698 |
| 1200 | 67.989 | 4470.9 | 5150.8 | 11.4132 | 13.5977 | 4470.8 | 5150.7 | 10.6704 | 6.7988 | 4470.7 | 5150.6 | 10.3504 |
| 1300 | 72.604 | 4687.4 | 5413.4 | 11.5857 | 14.5209 | 4687.3 | 5413.3 | 10.8429 | 7.2605 | 4687.2 | 5413.3 | 10.5229 |
| | P = 0.20 MPa (120.21°C) | | | | P = 0.30 MPa (133.52°C) | | | | P = 0.40 MPa (143.61°C) | | | |
| Sat. | 0.88578 | 2529.1 | 2706.3 | 7.1270 | 0.60582 | 2543.2 | 2724.9 | 6.9917 | 0.46242 | 2553.1 | 2738.1 | 6.8955 |
| 150 | 0.95986 | 2577.1 | 2769.1 | 7.2810 | 0.63402 | 2571.0 | 2761.2 | 7.0792 | 0.47088 | 2564.4 | 2752.8 | 6.9306 |
| 200 | 1.08049 | 2654.6 | 2870.7 | 7.5081 | 0.71643 | 2651.0 | 2865.9 | 7.3132 | 0.53434 | 2647.2 | 2860.9 | 7.1723 |
| 250 | 1.19890 | 2731.4 | 2971.2 | 7.7100 | 0.79645 | 2728.9 | 2967.9 | 7.5180 | 0.59520 | 2726.4 | 2964.5 | 7.3804 |
| 300 | 1.31623 | 2808.8 | 3072.1 | 7.8941 | 0.87535 | 2807.0 | 3069.6 | 7.7037 | 0.65489 | 2805.1 | 3067.1 | 7.5677 |
| 400 | 1.54934 | 2967.2 | 3277.0 | 8.2236 | 1.03155 | 2966.0 | 3275.5 | 8.0347 | 0.77265 | 2964.9 | 3273.9 | 7.9003 |
| 500 | 1.78142 | 3131.4 | 3487.7 | 8.5153 | 1.18672 | 3130.6 | 3486.6 | 8.3271 | 0.88936 | 3129.8 | 3485.5 | 8.1933 |
| 600 | 2.01302 | 3302.2 | 3704.8 | 8.7793 | 1.34139 | 3301.6 | 3704.0 | 8.5915 | 1.00558 | 3301.0 | 3703.3 | 8.4580 |
| 700 | 2.24434 | 3479.9 | 3928.8 | 9.0221 | 1.49580 | 3479.5 | 3928.2 | 8.8345 | 1.12152 | 3479.0 | 3927.6 | 8.7012 |
| 800 | 2.47550 | 3664.7 | 4159.8 | 9.2479 | 1.65004 | 3664.3 | 4159.3 | 9.0605 | 1.23730 | 3663.9 | 4158.9 | 8.9274 |
| 900 | 2.70656 | 3856.3 | 4397.7 | 9.4598 | 1.80417 | 3856.0 | 4397.3 | 9.2725 | 1.35298 | 3855.7 | 4396.9 | 9.1394 |
| 1000 | 2.93755 | 4054.8 | 4642.3 | 9.6599 | 1.95824 | 4054.5 | 4642.0 | 9.4726 | 1.46859 | 4054.3 | 4641.7 | 9.3396 |
| 1100 | 3.16848 | 4259.6 | 4893.3 | 9.8497 | 2.11226 | 4259.4 | 4893.1 | 9.6624 | 1.58414 | 4259.2 | 4892.9 | 9.5295 |
| 1200 | 3.39938 | 4470.5 | 5150.4 | 10.0304 | 2.26624 | 4470.3 | 5150.2 | 9.8431 | 1.69966 | 4470.2 | 5150.0 | 9.7102 |
| 1300 | 3.63026 | 4687.1 | 5413.1 | 10.2029 | 2.42019 | 4686.9 | 5413.0 | 10.0157 | 1.81516 | 4686.7 | 5412.8 | 9.8828 |
| | P = 0.50 MPa (151.83°C) | | | | P = 0.60 MPa (158.83°C) | | | | P = 0.80 MPa (170.41°C) | | | |
| Sat. | 0.37483 | 2560.7 | 2748.1 | 6.8207 | 0.31560 | 2566.8 | 2756.2 | 6.7593 | 0.24035 | 2576.0 | 2768.3 | 6.6616 |
| 200 | 0.42503 | 2643.3 | 2855.8 | 7.0610 | 0.35212 | 2639.4 | 2850.6 | 6.9683 | 0.26088 | 2631.1 | 2839.8 | 6.8177 |
| 250 | 0.47443 | 2723.8 | 2961.0 | 7.2725 | 0.39390 | 2721.2 | 2957.6 | 7.1833 | 0.29321 | 2715.9 | 2950.4 | 7.0402 |
| 300 | 0.52261 | 2803.3 | 3064.6 | 7.4614 | 0.43442 | 2801.4 | 3062.0 | 7.3740 | 0.32416 | 2797.5 | 3056.9 | 7.2345 |
| 350 | 0.57015 | 2883.0 | 3168.1 | 7.6346 | 0.47428 | 2881.6 | 3166.1 | 7.5481 | 0.35442 | 2878.6 | 3162.2 | 7.4107 |
| 400 | 0.61731 | 2963.7 | 3272.4 | 7.7956 | 0.51374 | 2962.5 | 3270.8 | 7.7097 | 0.38429 | 2960.2 | 3267.7 | 7.5735 |
| 500 | 0.71095 | 3129.0 | 3484.5 | 8.0893 | 0.59200 | 3128.2 | 3483.4 | 8.0041 | 0.44332 | 3126.6 | 3481.3 | 7.8692 |
| 600 | 0.80409 | 3300.4 | 3702.5 | 8.3544 | 0.66976 | 3299.8 | 3701.7 | 8.2695 | 0.50186 | 3298.7 | 3700.1 | 8.1354 |
| 700 | 0.89696 | 3478.6 | 3927.0 | 8.5978 | 0.74725 | 3478.1 | 3926.4 | 8.5132 | 0.56011 | 3477.2 | 3925.3 | 8.3794 |
| 800 | 0.98966 | 3663.6 | 4158.4 | 8.8240 | 0.82457 | 3663.2 | 4157.9 | 8.7395 | 0.61820 | 3662.5 | 4157.0 | 8.6061 |
| 900 | 1.08227 | 3855.4 | 4396.6 | 9.0362 | 0.90179 | 3855.1 | 4396.2 | 8.9518 | 0.67619 | 3854.5 | 4395.5 | 8.8185 |
| 1000 | 1.17480 | 4054.0 | 4641.4 | 9.2364 | 0.97893 | 4053.8 | 4641.1 | 9.1521 | 0.73411 | 4053.3 | 4640.5 | 9.0189 |
| 1100 | 1.26728 | 4259.0 | 4892.6 | 9.4263 | 1.05603 | 4258.8 | 4892.4 | 9.3420 | 0.79197 | 4258.3 | 4891.9 | 9.2090 |
| 1200 | 1.35972 | 4470.0 | 5149.8 | 9.6071 | 1.13309 | 4469.8 | 5149.6 | 9.5229 | 0.84980 | 4469.4 | 5149.3 | 9.3898 |
| 1300 | 1.45214 | 4686.6 | 5412.6 | 9.7797 | 1.21012 | 4686.4 | 5412.5 | 9.6955 | 0.90761 | 4686.1 | 5412.2 | 9.5625 |

*The temperature in parentheses is the saturation temperature at the specified pressure.
† Properties of saturated vapor at the specified pressure.

## TABLE A–6

Superheated water (*Concluded*)

| T °C | v m³/kg | u kJ/kg | h kJ/kg | s kJ/kg·K | v m³/kg | u kJ/kg | h kJ/kg | s kJ/kg·K | v m³/kg | u kJ/kg | h kJ/kg | s kJ/kg·K |
|---|---|---|---|---|---|---|---|---|---|---|---|---|
| | *P* = 1.00 MPa (179.88°C) | | | | *P* = 1.20 MPa (187.96°C) | | | | *P* = 1.40 MPa (195.04°C) | | | |
| Sat. | 0.19437 | 2582.8 | 2777.1 | 6.5850 | 0.16326 | 2587.8 | 2783.8 | 6.5217 | 0.14078 | 2591.8 | 2788.9 | 6.4675 |
| 200 | 0.20602 | 2622.3 | 2828.3 | 6.6956 | 0.16934 | 2612.9 | 2816.1 | 6.5909 | 0.14303 | 2602.7 | 2803.0 | 6.4975 |
| 250 | 0.23275 | 2710.4 | 2943.1 | 6.9265 | 0.19241 | 2704.7 | 2935.6 | 6.8313 | 0.16356 | 2698.9 | 2927.9 | 6.7488 |
| 300 | 0.25799 | 2793.7 | 3051.6 | 7.1246 | 0.21386 | 2789.7 | 3046.3 | 7.0335 | 0.18233 | 2785.7 | 3040.9 | 6.9553 |
| 350 | 0.28250 | 2875.7 | 3158.2 | 7.3029 | 0.23455 | 2872.7 | 3154.2 | 7.2139 | 0.20029 | 2869.7 | 3150.1 | 7.1379 |
| 400 | 0.30661 | 2957.9 | 3264.5 | 7.4670 | 0.25482 | 2955.5 | 3261.3 | 7.3793 | 0.21782 | 2953.1 | 3258.1 | 7.3046 |
| 500 | 0.35411 | 3125.0 | 3479.1 | 7.7642 | 0.29464 | 3123.4 | 3477.0 | 7.6779 | 0.25216 | 3121.8 | 3474.8 | 7.6047 |
| 600 | 0.40111 | 3297.5 | 3698.6 | 8.0311 | 0.33395 | 3296.3 | 3697.0 | 7.9456 | 0.28597 | 3295.1 | 3695.5 | 7.8730 |
| 700 | 0.44783 | 3476.3 | 3924.1 | 8.2755 | 0.37297 | 3475.3 | 3922.9 | 8.1904 | 0.31951 | 3474.4 | 3921.7 | 8.1183 |
| 800 | 0.49438 | 3661.7 | 4156.1 | 8.5024 | 0.41184 | 3661.0 | 4155.2 | 8.4176 | 0.35288 | 3660.3 | 4154.3 | 8.3458 |
| 900 | 0.54083 | 3853.9 | 4394.8 | 8.7150 | 0.45059 | 3853.3 | 4394.0 | 8.6303 | 0.38614 | 3852.7 | 4393.3 | 8.5587 |
| 1000 | 0.58721 | 4052.7 | 4640.0 | 8.9155 | 0.48928 | 4052.2 | 4639.4 | 8.8310 | 0.41933 | 4051.7 | 4638.8 | 8.7595 |
| 1100 | 0.63354 | 4257.5 | 4891.4 | 9.1057 | 0.52792 | 4257.5 | 4891.0 | 9.0212 | 0.45247 | 4257.0 | 4890.5 | 8.9497 |
| 1200 | 0.67983 | 4469.0 | 5148.9 | 9.2866 | 0.56652 | 4468.7 | 5148.5 | 9.2022 | 0.48558 | 4468.3 | 5148.1 | 9.1308 |
| 1300 | 0.72610 | 4685.8 | 5411.9 | 9.4593 | 0.60509 | 4685.5 | 5411.6 | 9.3750 | 0.51866 | 4685.1 | 5411.3 | 9.3036 |
| | *P* = 1.60 MPa (201.37°C) | | | | *P* = 1.80 MPa (207.11°C) | | | | *P* = 2.00 MPa (212.38°C) | | | |
| Sat. | 0.12374 | 2594.8 | 2792.8 | 6.4200 | 0.11037 | 2597.3 | 2795.9 | 6.3775 | 0.09959 | 2599.1 | 2798.3 | 6.3390 |
| 225 | 0.13293 | 2645.1 | 2857.8 | 6.5537 | 0.11678 | 2637.0 | 2847.2 | 6.4825 | 0.10381 | 2628.5 | 2836.1 | 6.4160 |
| 250 | 0.14190 | 2692.9 | 2919.9 | 6.6753 | 0.12502 | 2686.7 | 2911.7 | 6.6088 | 0.11150 | 2680.3 | 2903.3 | 6.5475 |
| 300 | 0.15866 | 2781.6 | 3035.4 | 6.8864 | 0.14025 | 2777.4 | 3029.9 | 6.8246 | 0.12551 | 2773.2 | 3024.2 | 6.7684 |
| 350 | 0.17459 | 2866.6 | 3146.0 | 7.0713 | 0.15460 | 2863.6 | 3141.9 | 7.0120 | 0.13860 | 2860.5 | 3137.7 | 6.9583 |
| 400 | 0.19007 | 2950.8 | 3254.9 | 7.2394 | 0.16849 | 2948.3 | 3251.6 | 7.1814 | 0.15122 | 2945.9 | 3248.4 | 7.1292 |
| 500 | 0.22029 | 3120.1 | 3472.6 | 7.5410 | 0.19551 | 3118.5 | 3470.4 | 7.4845 | 0.17568 | 3116.9 | 3468.3 | 7.4337 |
| 600 | 0.24999 | 3293.9 | 3693.9 | 7.8101 | 0.22200 | 3292.7 | 3692.3 | 7.7543 | 0.19962 | 3291.5 | 3690.7 | 7.7043 |
| 700 | 0.27941 | 3473.5 | 3920.5 | 8.0558 | 0.24822 | 3472.6 | 3919.4 | 8.0005 | 0.22326 | 3471.7 | 3918.2 | 7.9509 |
| 800 | 0.30865 | 3659.5 | 4153.4 | 8.2834 | 0.27426 | 3658.8 | 4152.4 | 8.2284 | 0.24674 | 3658.0 | 4151.5 | 8.1791 |
| 900 | 0.33780 | 3852.1 | 4392.6 | 8.4965 | 0.30020 | 3851.5 | 4391.9 | 8.4417 | 0.27012 | 3850.9 | 4391.1 | 8.3925 |
| 1000 | 0.36687 | 4051.2 | 4638.2 | 8.6974 | 0.32606 | 4050.7 | 4637.6 | 8.6427 | 0.29342 | 4050.2 | 4637.1 | 8.5936 |
| 1100 | 0.39589 | 4256.6 | 4890.0 | 8.8878 | 0.35188 | 4256.2 | 4889.6 | 8.8331 | 0.31667 | 4255.7 | 4889.1 | 8.7842 |
| 1200 | 0.42488 | 4467.9 | 5147.7 | 9.0689 | 0.37766 | 4467.6 | 5147.3 | 9.0143 | 0.33989 | 4467.2 | 5147.0 | 8.9654 |
| 1300 | 0.45383 | 4684.8 | 5410.9 | 9.2418 | 0.40341 | 4684.5 | 5410.6 | 9.1872 | 0.36308 | 4684.2 | 5410.3 | 9.1384 |
| | *P* = 2.50 MPa (223.95°C) | | | | *P* = 3.00 MPa (233.85°C) | | | | *P* = 3.50 MPa (242.56°C) | | | |
| Sat. | 0.07995 | 2602.1 | 2801.9 | 6.2558 | 0.06667 | 2603.2 | 2803.2 | 6.1856 | 0.05706 | 2603.0 | 2802.7 | 6.1244 |
| 225 | 0.08026 | 2604.8 | 2805.5 | 6.2629 | | | | | | | | |
| 250 | 0.08705 | 2663.3 | 2880.9 | 6.4107 | 0.07063 | 2644.7 | 2856.5 | 6.2893 | 0.05876 | 2624.0 | 2829.7 | 6.1764 |
| 300 | 0.09894 | 2762.2 | 3009.6 | 6.6459 | 0.08118 | 2750.8 | 2994.3 | 6.5412 | 0.06845 | 2738.8 | 2978.4 | 6.4484 |
| 350 | 0.10979 | 2852.5 | 3127.0 | 6.8424 | 0.09056 | 2844.4 | 3116.1 | 6.7450 | 0.07680 | 2836.0 | 3104.9 | 6.6601 |
| 400 | 0.12012 | 2939.8 | 3240.1 | 7.0170 | 0.09938 | 2933.6 | 3231.7 | 6.9235 | 0.08456 | 2927.2 | 3223.2 | 6.8428 |
| 450 | 0.13015 | 3026.2 | 3351.6 | 7.1768 | 0.10789 | 3021.2 | 3344.9 | 7.0856 | 0.09198 | 3016.1 | 3338.1 | 7.0074 |
| 500 | 0.13999 | 3112.8 | 3462.8 | 7.3254 | 0.11620 | 3108.6 | 3457.2 | 7.2359 | 0.09919 | 3104.5 | 3451.7 | 7.1593 |
| 600 | 0.15931 | 3288.5 | 3686.8 | 7.5979 | 0.13245 | 3285.5 | 3682.8 | 7.5103 | 0.11325 | 3282.5 | 3678.9 | 7.4357 |
| 700 | 0.17835 | 3469.3 | 3915.2 | 7.8455 | 0.14841 | 3467.0 | 3912.2 | 7.7590 | 0.12702 | 3464.7 | 3909.3 | 7.6855 |
| 800 | 0.19722 | 3656.2 | 4149.2 | 8.0744 | 0.16420 | 3654.3 | 4146.9 | 7.9885 | 0.14061 | 3652.5 | 4144.6 | 7.9156 |
| 900 | 0.21597 | 3849.4 | 4389.3 | 8.2882 | 0.17988 | 3847.9 | 4387.5 | 8.2028 | 0.15410 | 3846.4 | 4385.7 | 8.1304 |
| 1000 | 0.23466 | 4049.0 | 4635.6 | 8.4897 | 0.19549 | 4047.7 | 4634.2 | 8.4045 | 0.16751 | 4046.4 | 4632.7 | 8.3324 |
| 1100 | 0.25330 | 4254.7 | 4887.9 | 8.6804 | 0.21105 | 4253.6 | 4886.7 | 8.5955 | 0.18087 | 4252.5 | 4885.6 | 8.5236 |
| 1200 | 0.27190 | 4466.3 | 5146.0 | 8.8618 | 0.22658 | 4465.3 | 5145.1 | 8.7771 | 0.19420 | 4464.4 | 5144.1 | 8.7053 |
| 1300 | 0.29048 | 4683.4 | 5409.5 | 9.0349 | 0.24207 | 4682.6 | 5408.8 | 8.9502 | 0.20750 | 4681.8 | 5408.0 | 8.8786 |

## TABLE A–6

Superheated water (*Continued*)

| T °C | v m³/kg | u kJ/kg | h kJ/kg | s kJ/kg·K | v m³/kg | u kJ/kg | h kJ/kg | s kJ/kg·K | v m³/kg | u kJ/kg | h kJ/kg | s kJ/kg·K |
|---|---|---|---|---|---|---|---|---|---|---|---|---|
| | P = 4.0 MPa (250.35°C) | | | | P = 4.5 MPa (257.44°C) | | | | P = 5.0 MPa (263.94°C) | | | |
| Sat. | 0.04978 | 2601.7 | 2800.8 | 6.0696 | 0.04406 | 2599.7 | 2798.0 | 6.0198 | 0.03945 | 2597.0 | 2794.2 | 5.9737 |
| 275 | 0.05461 | 2668.9 | 2887.3 | 6.2312 | 0.04733 | 2651.4 | 2864.4 | 6.1429 | 0.04144 | 2632.3 | 2839.5 | 6.0571 |
| 300 | 0.05887 | 2726.2 | 2961.7 | 6.3639 | 0.05138 | 2713.0 | 2944.2 | 6.2854 | 0.04535 | 2699.0 | 2925.7 | 6.2111 |
| 350 | 0.06647 | 2827.4 | 3093.3 | 6.5843 | 0.05842 | 2818.6 | 3081.5 | 6.5153 | 0.05197 | 2809.5 | 3069.3 | 6.4516 |
| 400 | 0.07343 | 2920.8 | 3214.5 | 6.7714 | 0.06477 | 2914.2 | 3205.7 | 6.7071 | 0.05784 | 2907.5 | 3196.7 | 6.6483 |
| 450 | 0.08004 | 3011.0 | 3331.2 | 6.9386 | 0.07076 | 3005.8 | 3324.2 | 6.8770 | 0.06332 | 3000.6 | 3317.2 | 6.8210 |
| 500 | 0.08644 | 3100.3 | 3446.0 | 7.0922 | 0.07652 | 3096.0 | 3440.4 | 7.0323 | 0.06858 | 3091.8 | 3434.7 | 6.9781 |
| 600 | 0.09886 | 3279.4 | 3674.9 | 7.3706 | 0.08766 | 3276.4 | 3670.9 | 7.3127 | 0.07870 | 3273.3 | 3666.9 | 7.2605 |
| 700 | 0.11098 | 3462.4 | 3906.3 | 7.6214 | 0.09850 | 3460.0 | 3903.3 | 7.5647 | 0.08852 | 3457.7 | 3900.3 | 7.5136 |
| 800 | 0.12292 | 3650.6 | 4142.3 | 7.8523 | 0.10916 | 3648.8 | 4140.0 | 7.7962 | 0.09816 | 3646.9 | 4137.7 | 7.7458 |
| 900 | 0.13476 | 3844.8 | 4383.9 | 8.0675 | 0.11972 | 3843.3 | 4382.1 | 8.0118 | 0.10769 | 3841.8 | 4380.2 | 7.9619 |
| 1000 | 0.14653 | 4045.1 | 4631.2 | 8.2698 | 0.13020 | 4043.9 | 4629.8 | 8.2144 | 0.11715 | 4042.6 | 4628.3 | 8.1648 |
| 1100 | 0.15824 | 4251.4 | 4884.4 | 8.4612 | 0.14064 | 4250.4 | 4883.2 | 8.4060 | 0.12655 | 4249.3 | 4882.1 | 8.3566 |
| 1200 | 0.16992 | 4463.5 | 5143.2 | 8.6430 | 0.15103 | 4462.6 | 5142.2 | 8.5880 | 0.13592 | 4461.6 | 5141.3 | 8.5388 |
| 1300 | 0.18157 | 4680.9 | 5407.2 | 8.8164 | 0.16140 | 4680.1 | 5406.5 | 8.7616 | 0.14527 | 4679.3 | 5405.7 | 8.7124 |
| | P = 6.0 MPa (275.59°C) | | | | P = 7.0 MPa (285.83°C) | | | | P = 8.0 MPa (295.01°C) | | | |
| Sat. | 0.03245 | 2589.9 | 2784.6 | 5.8902 | 0.027378 | 2581.0 | 2772.6 | 5.8148 | 0.023525 | 2570.5 | 2758.7 | 5.7450 |
| 300 | 0.03619 | 2668.4 | 2885.6 | 6.0703 | 0.029492 | 2633.5 | 2839.9 | 5.9337 | 0.024279 | 2592.3 | 2786.5 | 5.7937 |
| 350 | 0.04225 | 2790.4 | 3043.9 | 6.3357 | 0.035262 | 2770.1 | 3016.9 | 6.2305 | 0.029975 | 2748.3 | 2988.1 | 6.1321 |
| 400 | 0.04742 | 2893.7 | 3178.3 | 6.5432 | 0.039958 | 2879.5 | 3159.2 | 6.4502 | 0.034344 | 2864.6 | 3139.4 | 6.3658 |
| 450 | 0.05217 | 2989.9 | 3302.9 | 6.7219 | 0.044187 | 2979.0 | 3288.3 | 6.6353 | 0.038194 | 2967.8 | 3273.3 | 6.5579 |
| 500 | 0.05667 | 3083.1 | 3423.1 | 6.8826 | 0.048157 | 3074.3 | 3411.4 | 6.8000 | 0.041767 | 3065.4 | 3399.5 | 6.7266 |
| 550 | 0.06102 | 3175.2 | 3541.3 | 7.0308 | 0.051966 | 3167.9 | 3531.6 | 6.9507 | 0.045172 | 3160.5 | 3521.8 | 6.8800 |
| 600 | 0.06527 | 3267.2 | 3658.8 | 7.1693 | 0.055665 | 3261.0 | 3650.6 | 7.0910 | 0.048463 | 3254.7 | 3642.4 | 7.0221 |
| 700 | 0.07355 | 3453.0 | 3894.3 | 7.4247 | 0.062850 | 3448.3 | 3888.3 | 7.3487 | 0.054829 | 3443.6 | 3882.2 | 7.2822 |
| 800 | 0.08165 | 3643.2 | 4133.1 | 7.6582 | 0.069856 | 3639.5 | 4128.5 | 7.5836 | 0.061011 | 3635.7 | 4123.8 | 7.5185 |
| 900 | 0.08964 | 3838.8 | 4376.6 | 7.8751 | 0.076750 | 3835.7 | 4373.0 | 7.8014 | 0.067082 | 3832.7 | 4369.3 | 7.7372 |
| 1000 | 0.09756 | 4040.1 | 4625.4 | 8.0786 | 0.083571 | 4037.5 | 4622.5 | 8.0055 | 0.073079 | 4035.0 | 4619.6 | 7.9419 |
| 1100 | 0.10543 | 4247.1 | 4879.7 | 8.2709 | 0.090341 | 4245.0 | 4877.4 | 8.1982 | 0.079025 | 4242.8 | 4875.0 | 8.1350 |
| 1200 | 0.11326 | 4459.8 | 5139.4 | 8.4534 | 0.097075 | 4457.9 | 5137.4 | 8.3810 | 0.084934 | 4456.1 | 5135.5 | 8.3181 |
| 1300 | 0.12107 | 4677.7 | 5404.1 | 8.6273 | 0.103781 | 4676.1 | 5402.6 | 8.5551 | 0.090817 | 4674.5 | 5401.0 | 8.4925 |
| | P = 9.0 MPa (303.35°C) | | | | P = 10.0 MPa (311.00°C) | | | | P = 12.5 MPa (327.81°C) | | | |
| Sat. | 0.020489 | 2558.5 | 2742.9 | 5.6791 | 0.018028 | 2545.2 | 2725.5 | 5.6159 | 0.013496 | 2505.6 | 2674.3 | 5.4638 |
| 325 | 0.023284 | 2647.6 | 2857.1 | 5.8738 | 0.019877 | 2611.6 | 2810.3 | 5.7596 | | | | |
| 350 | 0.025816 | 2725.0 | 2957.3 | 6.0380 | 0.022440 | 2699.6 | 2924.0 | 5.9460 | 0.016138 | 2624.9 | 2826.6 | 5.7130 |
| 400 | 0.029960 | 2849.2 | 3118.8 | 6.2876 | 0.026436 | 2833.1 | 3097.5 | 6.2141 | 0.020030 | 2789.6 | 3040.0 | 6.0433 |
| 450 | 0.033524 | 2956.3 | 3258.0 | 6.4872 | 0.029782 | 2944.5 | 3242.4 | 6.4219 | 0.023019 | 2913.7 | 3201.5 | 6.2749 |
| 500 | 0.036793 | 3056.3 | 3387.4 | 6.6603 | 0.032811 | 3047.0 | 3375.1 | 6.5995 | 0.025630 | 3023.2 | 3343.6 | 6.4651 |
| 550 | 0.039885 | 3153.0 | 3512.0 | 6.8164 | 0.035655 | 3145.4 | 3502.0 | 6.7585 | 0.028033 | 3126.1 | 3476.5 | 6.6317 |
| 600 | 0.042861 | 3248.4 | 3634.1 | 6.9605 | 0.038378 | 3242.0 | 3625.8 | 6.9045 | 0.030306 | 3225.8 | 3604.6 | 6.7828 |
| 650 | 0.045755 | 3343.4 | 3755.2 | 7.0954 | 0.041018 | 3338.0 | 3748.1 | 7.0408 | 0.032491 | 3324.1 | 3730.2 | 6.9227 |
| 700 | 0.048589 | 3438.8 | 3876.1 | 7.2229 | 0.043597 | 3434.0 | 3870.0 | 7.1693 | 0.034612 | 3422.0 | 3854.6 | 7.0540 |
| 800 | 0.054132 | 3632.0 | 4119.2 | 7.4606 | 0.048629 | 3628.2 | 4114.5 | 7.4085 | 0.038724 | 3618.8 | 4102.8 | 7.2967 |
| 900 | 0.059562 | 3829.6 | 4365.7 | 7.6802 | 0.053547 | 3826.5 | 4362.0 | 7.6290 | 0.042720 | 3818.9 | 4352.9 | 7.5195 |
| 1000 | 0.064919 | 4032.4 | 4616.7 | 7.8855 | 0.058391 | 4029.9 | 4613.8 | 7.8349 | 0.046641 | 4023.5 | 4606.5 | 7.7269 |
| 1100 | 0.070224 | 4240.7 | 4872.7 | 8.0791 | 0.063183 | 4238.5 | 4870.3 | 8.0289 | 0.050510 | 4233.1 | 4864.5 | 7.9220 |
| 1200 | 0.075492 | 4454.2 | 5133.6 | 8.2625 | 0.067938 | 4452.4 | 5131.7 | 8.2126 | 0.054342 | 4447.7 | 5127.0 | 8.1065 |
| 1300 | 0.080733 | 4672.9 | 5399.5 | 8.4371 | 0.072667 | 4671.3 | 5398.0 | 8.3874 | 0.058147 | 4667.3 | 5394.1 | 8.2819 |

## TABLE A–6

Superheated water (*Concluded*)

| T °C | v m³/kg | u kJ/kg | h kJ/kg | s kJ/kg·K | v m³/kg | u kJ/kg | h kJ/kg | s kJ/kg·K | v m³/kg | u kJ/kg | h kJ/kg | s kJ/kg·K |
|---|---|---|---|---|---|---|---|---|---|---|---|---|
| | P = 15.0 MPa (342.16°C) | | | | P = 17.5 MPa (354.67°C) | | | | P = 20.0 MPa (365.75°C) | | | |
| Sat. | 0.010341 | 2455.7 | 2610.8 | 5.3108 | 0.007932 | 2390.7 | 2529.5 | 5.1435 | 0.005862 | 2294.8 | 2412.1 | 4.9310 |
| 350 | 0.011481 | 2520.9 | 2693.1 | 5.4438 | | | | | | | | |
| 400 | 0.015671 | 2740.6 | 2975.7 | 5.8819 | 0.012463 | 2684.3 | 2902.4 | 5.7211 | 0.009950 | 2617.9 | 2816.9 | 5.5526 |
| 450 | 0.018477 | 2880.8 | 3157.9 | 6.1434 | 0.015204 | 2845.4 | 3111.4 | 6.0212 | 0.012721 | 2807.3 | 3061.7 | 5.9043 |
| 500 | 0.020828 | 2998.4 | 3310.8 | 6.3480 | 0.017385 | 2972.4 | 3276.7 | 6.2424 | 0.014793 | 2945.3 | 3241.2 | 6.1446 |
| 550 | 0.022945 | 3106.2 | 3450.4 | 6.5230 | 0.019305 | 3085.8 | 3423.6 | 6.4266 | 0.016571 | 3064.7 | 3396.2 | 6.3390 |
| 600 | 0.024921 | 3209.3 | 3583.1 | 6.6796 | 0.021073 | 3192.5 | 3561.3 | 6.5890 | 0.018185 | 3175.3 | 3539.0 | 6.5075 |
| 650 | 0.026804 | 3310.1 | 3712.1 | 6.8233 | 0.022742 | 3295.8 | 3693.8 | 6.7366 | 0.019695 | 3281.4 | 3675.3 | 6.6593 |
| 700 | 0.028621 | 3409.8 | 3839.1 | 6.9573 | 0.024342 | 3397.5 | 3823.5 | 6.8735 | 0.021134 | 3385.1 | 3807.8 | 6.7991 |
| 800 | 0.032121 | 3609.3 | 4091.1 | 7.2037 | 0.027405 | 3599.7 | 4079.3 | 7.1237 | 0.023870 | 3590.1 | 4067.5 | 7.0531 |
| 900 | 0.035503 | 3811.2 | 4343.7 | 7.4288 | 0.030348 | 3803.5 | 4334.6 | 7.3511 | 0.026484 | 3795.7 | 4325.4 | 7.2829 |
| 1000 | 0.038808 | 4017.1 | 4599.2 | 7.6378 | 0.033215 | 4010.7 | 4592.0 | 7.5616 | 0.029020 | 4004.3 | 4584.7 | 7.4950 |
| 1100 | 0.042062 | 4227.7 | 4858.6 | 7.8339 | 0.036029 | 4222.3 | 4852.8 | 7.7588 | 0.031504 | 4216.9 | 4847.0 | 7.6933 |
| 1200 | 0.045279 | 4443.1 | 5122.3 | 8.0192 | 0.038806 | 4438.5 | 5117.6 | 7.9449 | 0.033952 | 4433.8 | 5112.9 | 7.8802 |
| 1300 | 0.048469 | 4663.3 | 5390.3 | 8.1952 | 0.041556 | 4659.2 | 5386.5 | 8.1215 | 0.036371 | 4655.2 | 5382.7 | 8.0574 |
| | P = 25.0 MPa | | | | P = 30.0 MPa | | | | P = 35.0 MPa | | | |
| 375 | 0.001978 | 1799.9 | 1849.4 | 4.0345 | 0.001792 | 1738.1 | 1791.9 | 3.9313 | 0.001701 | 1702.8 | 1762.4 | 3.8724 |
| 400 | 0.006005 | 2428.5 | 2578.7 | 5.1400 | 0.002798 | 2068.9 | 2152.8 | 4.4758 | 0.002105 | 1914.9 | 1988.6 | 4.2144 |
| 425 | 0.007886 | 2607.8 | 2805.0 | 5.4708 | 0.005299 | 2452.9 | 2611.8 | 5.1473 | 0.003434 | 2253.3 | 2373.5 | 4.7751 |
| 450 | 0.009176 | 2721.2 | 2950.6 | 5.6759 | 0.006737 | 2618.9 | 2821.0 | 5.4422 | 0.004957 | 2497.5 | 2671.0 | 5.1946 |
| 500 | 0.011143 | 2887.3 | 3165.9 | 5.9643 | 0.008691 | 2824.0 | 3084.8 | 5.7956 | 0.006933 | 2755.3 | 2997.9 | 5.6331 |
| 550 | 0.012736 | 3020.8 | 3339.2 | 6.1816 | 0.010175 | 2974.5 | 3279.7 | 6.0403 | 0.008348 | 2925.8 | 3218.0 | 5.9093 |
| 600 | 0.014140 | 3140.0 | 3493.5 | 6.3637 | 0.011445 | 3103.4 | 3446.8 | 6.2373 | 0.009523 | 3065.6 | 3399.0 | 6.1229 |
| 650 | 0.015430 | 3251.9 | 3637.7 | 6.5243 | 0.012590 | 3221.7 | 3599.4 | 6.4074 | 0.010565 | 3190.9 | 3560.7 | 6.3030 |
| 700 | 0.016643 | 3359.9 | 3776.0 | 6.6702 | 0.013654 | 3334.3 | 3743.9 | 6.5599 | 0.011523 | 3308.3 | 3711.6 | 6.4623 |
| 800 | 0.018922 | 3570.7 | 4043.8 | 6.9322 | 0.015628 | 3551.2 | 4020.0 | 6.8301 | 0.013278 | 3531.6 | 3996.3 | 6.7409 |
| 900 | 0.021075 | 3780.2 | 4307.1 | 7.1668 | 0.017473 | 3764.6 | 4288.8 | 7.0695 | 0.014904 | 3749.0 | 4270.6 | 6.9853 |
| 1000 | 0.023150 | 3991.5 | 4570.2 | 7.3821 | 0.019240 | 3978.6 | 4555.8 | 7.2880 | 0.016450 | 3965.8 | 4541.5 | 7.2069 |
| 1100 | 0.025172 | 4206.1 | 4835.4 | 7.5825 | 0.020954 | 4195.2 | 4823.9 | 7.4906 | 0.017942 | 4184.4 | 4812.4 | 7.4118 |
| 1200 | 0.027157 | 4424.6 | 5103.5 | 7.7710 | 0.022630 | 4415.3 | 5094.2 | 7.6807 | 0.019398 | 4406.1 | 5085.0 | 7.6034 |
| 1300 | 0.029115 | 4647.2 | 5375.1 | 7.9494 | 0.024279 | 4639.2 | 5367.6 | 7.8602 | 0.020827 | 4631.2 | 5360.2 | 7.7841 |
| | P = 40.0 MPa | | | | P = 50.0 MPa | | | | P = 60.0 MPa | | | |
| 375 | 0.001641 | 1677.0 | 1742.6 | 3.8290 | 0.001560 | 1638.6 | 1716.6 | 3.7642 | 0.001503 | 1609.7 | 1699.9 | 3.7149 |
| 400 | 0.001911 | 1855.0 | 1931.4 | 4.1145 | 0.001731 | 1787.8 | 1874.4 | 4.0029 | 0.001633 | 1745.2 | 1843.2 | 3.9317 |
| 425 | 0.002538 | 2097.5 | 2199.0 | 4.5044 | 0.002009 | 1960.3 | 2060.7 | 4.2746 | 0.001816 | 1892.9 | 2001.8 | 4.1630 |
| 450 | 0.003692 | 2364.2 | 2511.8 | 4.9449 | 0.002487 | 2160.3 | 2284.7 | 4.5896 | 0.002086 | 2055.1 | 2180.2 | 4.4140 |
| 500 | 0.005623 | 2681.6 | 2906.5 | 5.4744 | 0.003890 | 2528.1 | 2722.6 | 5.1762 | 0.002952 | 2393.2 | 2570.3 | 4.9356 |
| 550 | 0.006985 | 2875.1 | 3154.4 | 5.7857 | 0.005118 | 2769.5 | 3025.4 | 5.5563 | 0.003955 | 2664.6 | 2901.9 | 5.3517 |
| 600 | 0.008089 | 3026.8 | 3350.4 | 6.0170 | 0.006108 | 2947.1 | 3252.6 | 5.8245 | 0.004833 | 2866.8 | 3156.8 | 5.6527 |
| 650 | 0.009053 | 3159.5 | 3521.6 | 6.2078 | 0.006957 | 3095.6 | 3443.5 | 6.0373 | 0.005591 | 3031.3 | 3366.8 | 5.8867 |
| 700 | 0.009930 | 3282.0 | 3679.2 | 6.3740 | 0.007717 | 3228.7 | 3614.6 | 6.2179 | 0.006265 | 3175.4 | 3551.3 | 6.0814 |
| 800 | 0.011521 | 3511.8 | 3972.6 | 6.6613 | 0.009073 | 3472.2 | 3925.8 | 6.5225 | 0.007456 | 3432.6 | 3880.0 | 6.4033 |
| 900 | 0.012980 | 3733.3 | 4252.5 | 6.9107 | 0.010296 | 3702.0 | 4216.8 | 6.7819 | 0.008519 | 3670.9 | 4182.1 | 6.6725 |
| 1000 | 0.014360 | 3952.9 | 4527.3 | 7.1355 | 0.011441 | 3927.4 | 4499.4 | 7.0131 | 0.009504 | 3902.0 | 4472.2 | 6.9099 |
| 1100 | 0.015686 | 4173.7 | 4801.1 | 7.3425 | 0.012534 | 4152.2 | 4778.9 | 7.2244 | 0.010439 | 4130.9 | 4757.3 | 7.1255 |
| 1200 | 0.016976 | 4396.9 | 5075.9 | 7.5357 | 0.013590 | 4378.6 | 5058.1 | 7.4207 | 0.011339 | 4360.5 | 5040.8 | 7.3248 |
| 1300 | 0.018239 | 4623.3 | 5352.8 | 7.7175 | 0.014620 | 4607.5 | 5338.5 | 7.6048 | 0.012213 | 4591.8 | 5324.5 | 7.5111 |

## TABLE A–7

Compressed liquid water

| $T$ | $v$ | $u$ | $h$ | $s$ | $v$ | $u$ | $h$ | $s$ | $v$ | $u$ | $h$ | $s$ |
|---|---|---|---|---|---|---|---|---|---|---|---|---|
| °C | m³/kg | kJ/kg | kJ/kg | kJ/kg·K | m³/kg | kJ/kg | kJ/kg | kJ/kg·K | m³/kg | kJ/kg | kJ/kg | kJ/kg·K |
| | P = 5 MPa (263.94°C) | | | | P = 10 MPa (311.00°C) | | | | P = 15 MPa (342.16°C) | | | |
| Sat. | 0.0012862 | 1148.1 | 1154.5 | 2.9207 | 0.0014522 | 1393.3 | 1407.9 | 3.3603 | 0.0016572 | 1585.5 | 1610.3 | 3.6848 |
| 0 | 0.0009977 | 0.04 | 5.03 | 0.0001 | 0.0009952 | 0.12 | 10.07 | 0.0003 | 0.0009928 | 0.18 | 15.07 | 0.0004 |
| 20 | 0.0009996 | 83.61 | 88.61 | 0.2954 | 0.0009973 | 83.31 | 93.28 | 0.2943 | 0.0009951 | 83.01 | 97.93 | 0.2932 |
| 40 | 0.0010057 | 166.92 | 171.95 | 0.5705 | 0.0010035 | 166.33 | 176.37 | 0.5685 | 0.0010013 | 165.75 | 180.77 | 0.5666 |
| 60 | 0.0010149 | 250.29 | 255.36 | 0.8287 | 0.0010127 | 249.43 | 259.55 | 0.8260 | 0.0010105 | 248.58 | 263.74 | 0.8234 |
| 80 | 0.0010267 | 333.82 | 338.96 | 1.0723 | 0.0010244 | 332.69 | 342.94 | 1.0691 | 0.0010221 | 331.59 | 346.92 | 1.0659 |
| 100 | 0.0010410 | 417.65 | 422.85 | 1.3034 | 0.0010385 | 416.23 | 426.62 | 1.2996 | 0.0010361 | 414.85 | 430.39 | 1.2958 |
| 120 | 0.0010576 | 501.91 | 507.19 | 1.5236 | 0.0010549 | 500.18 | 510.73 | 1.5191 | 0.0010522 | 498.50 | 514.28 | 1.5148 |
| 140 | 0.0010769 | 586.80 | 592.18 | 1.7344 | 0.0010738 | 584.72 | 595.45 | 1.7293 | 0.0010708 | 582.69 | 598.75 | 1.7243 |
| 160 | 0.0010988 | 672.55 | 678.04 | 1.9374 | 0.0010954 | 670.06 | 681.01 | 1.9316 | 0.0010920 | 667.63 | 684.01 | 1.9259 |
| 180 | 0.0011240 | 759.47 | 765.09 | 2.1338 | 0.0011200 | 756.48 | 767.68 | 2.1271 | 0.0011160 | 753.58 | 770.32 | 2.1206 |
| 200 | 0.0011531 | 847.92 | 853.68 | 2.3251 | 0.0011482 | 844.32 | 855.80 | 2.3174 | 0.0011435 | 840.84 | 858.00 | 2.3100 |
| 220 | 0.0011868 | 938.39 | 944.32 | 2.5127 | 0.0011809 | 934.01 | 945.82 | 2.5037 | 0.0011752 | 929.81 | 947.43 | 2.4951 |
| 240 | 0.0012268 | 1031.6 | 1037.7 | 2.6983 | 0.0012192 | 1026.2 | 1038.3 | 2.6876 | 0.0012121 | 1021.0 | 1039.2 | 2.6774 |
| 260 | 0.0012755 | 1128.5 | 1134.9 | 2.8841 | 0.0012653 | 1121.6 | 1134.3 | 2.8710 | 0.0012560 | 1115.1 | 1134.0 | 2.8586 |
| 280 | | | | | 0.0013226 | 1221.8 | 1235.0 | 3.0565 | 0.0013096 | 1213.4 | 1233.0 | 3.0410 |
| 300 | | | | | 0.0013980 | 1329.4 | 1343.3 | 3.2488 | 0.0013783 | 1317.6 | 1338.3 | 3.2279 |
| 320 | | | | | | | | | 0.0014733 | 1431.9 | 1454.0 | 3.4263 |
| 340 | | | | | | | | | 0.0016311 | 1567.9 | 1592.4 | 3.6555 |
| | P = 20 MPa (365.75°C) | | | | P = 30 MPa | | | | P = 50 MPa | | | |
| Sat. | 0.0020378 | 1785.8 | 1826.6 | 4.0146 | | | | | | | | |
| 0 | 0.0009904 | 0.23 | 20.03 | 0.0005 | 0.0009857 | 0.29 | 29.86 | 0.0003 | 0.0009767 | 0.29 | 49.13 | −0.0010 |
| 20 | 0.0009929 | 82.71 | 102.57 | 0.2921 | 0.0009886 | 82.11 | 111.77 | 0.2897 | 0.0009805 | 80.93 | 129.95 | 0.2845 |
| 40 | 0.0009992 | 165.17 | 185.16 | 0.5646 | 0.0009951 | 164.05 | 193.90 | 0.5607 | 0.0009872 | 161.90 | 211.25 | 0.5528 |
| 60 | 0.0010084 | 247.75 | 267.92 | 0.8208 | 0.0010042 | 246.14 | 276.26 | 0.8156 | 0.0009962 | 243.08 | 292.88 | 0.8055 |
| 80 | 0.0010199 | 330.50 | 350.90 | 1.0627 | 0.0010155 | 328.40 | 358.86 | 1.0564 | 0.0010072 | 324.42 | 374.78 | 1.0442 |
| 100 | 0.0010337 | 413.50 | 434.17 | 1.2920 | 0.0010290 | 410.87 | 441.74 | 1.2847 | 0.0010201 | 405.94 | 456.94 | 1.2705 |
| 120 | 0.0010496 | 496.85 | 517.84 | 1.5105 | 0.0010445 | 493.66 | 525.00 | 1.5020 | 0.0010349 | 487.69 | 539.43 | 1.4859 |
| 140 | 0.0010679 | 580.71 | 602.07 | 1.7194 | 0.0010623 | 576.90 | 608.76 | 1.7098 | 0.0010517 | 569.77 | 622.36 | 1.6916 |
| 160 | 0.0010886 | 665.28 | 687.05 | 1.9203 | 0.0010823 | 660.74 | 693.21 | 1.9094 | 0.0010704 | 652.33 | 705.85 | 1.8889 |
| 180 | 0.0011122 | 750.78 | 773.02 | 2.1143 | 0.0011049 | 745.40 | 778.55 | 2.1020 | 0.0010914 | 735.49 | 790.06 | 2.0790 |
| 200 | 0.0011390 | 837.49 | 860.27 | 2.3027 | 0.0011304 | 831.11 | 865.02 | 2.2888 | 0.0011149 | 819.45 | 875.19 | 2.2628 |
| 220 | 0.0011697 | 925.77 | 949.16 | 2.4867 | 0.0011595 | 918.15 | 952.93 | 2.4707 | 0.0011412 | 904.39 | 961.45 | 2.4414 |
| 240 | 0.0012053 | 1016.1 | 1040.2 | 2.6676 | 0.0011927 | 1006.9 | 1042.7 | 2.6491 | 0.0011708 | 990.55 | 1049.1 | 2.6156 |
| 260 | 0.0012472 | 1109.0 | 1134.0 | 2.8469 | 0.0012314 | 1097.8 | 1134.7 | 2.8250 | 0.0012044 | 1078.2 | 1138.4 | 2.7864 |
| 280 | 0.0012978 | 1205.6 | 1231.5 | 3.0265 | 0.0012770 | 1191.5 | 1229.8 | 3.0001 | 0.0012430 | 1167.7 | 1229.9 | 2.9547 |
| 300 | 0.0013611 | 1307.2 | 1334.4 | 3.2091 | 0.0013322 | 1288.9 | 1328.9 | 3.1761 | 0.0012879 | 1259.6 | 1324.0 | 3.1218 |
| 320 | 0.0014450 | 1416.6 | 1445.5 | 3.3996 | 0.0014014 | 1391.7 | 1433.7 | 3.3558 | 0.0013409 | 1354.3 | 1421.4 | 3.2888 |
| 340 | 0.0015693 | 1540.2 | 1571.6 | 3.6086 | 0.0014932 | 1502.4 | 1547.1 | 3.5438 | 0.0014049 | 1452.9 | 1523.1 | 3.4575 |
| 360 | 0.0018248 | 1703.6 | 1740.1 | 3.8787 | 0.0016276 | 1626.8 | 1675.6 | 3.7499 | 0.0014848 | 1556.5 | 1630.7 | 3.6301 |
| 380 | | | | | 0.0018729 | 1782.0 | 1838.2 | 4.0026 | 0.0015884 | 1667.1 | 1746.5 | 3.8102 |

## TABLE A–8

Saturated ice–water vapor

| Temp., $T\,°C$ | Sat. press., $P_{sat}$ kPa | Specific volume, m³/kg | | Internal energy, kJ/kg | | | Enthalpy, kJ/kg | | | Entropy, kJ/kg·K | | |
|---|---|---|---|---|---|---|---|---|---|---|---|---|
| | | Sat. ice, $v_i$ | Sat. vapor, $v_g$ | Sat. ice, $u_i$ | Subl., $u_{ig}$ | Sat. vapor, $u_g$ | Sat. ice, $h_i$ | Subl., $h_{ig}$ | Sat. vapor, $h_g$ | Sat. ice, $s_i$ | Subl., $s_{ig}$ | Sat. vapor, $s_g$ |
| 0.01 | 0.61169 | 0.001091 | 205.99 | −333.40 | 2707.9 | 2374.5 | −333.40 | 2833.9 | 2500.5 | −1.2202 | 10.374 | 9.154 |
| 0 | 0.61115 | 0.001091 | 206.17 | −333.43 | 2707.9 | 2374.5 | −333.43 | 2833.9 | 2500.5 | −1.2204 | 10.375 | 9.154 |
| −2 | 0.51772 | 0.001091 | 241.62 | −337.63 | 2709.4 | 2371.8 | −337.63 | 2834.5 | 2496.8 | −1.2358 | 10.453 | 9.218 |
| −4 | 0.43748 | 0.001090 | 283.84 | −341.80 | 2710.8 | 2369.0 | −341.80 | 2835.0 | 2493.2 | −1.2513 | 10.533 | 9.282 |
| −6 | 0.36873 | 0.001090 | 334.27 | −345.94 | 2712.2 | 2366.2 | −345.93 | 2835.4 | 2489.5 | −1.2667 | 10.613 | 9.347 |
| −8 | 0.30998 | 0.001090 | 394.66 | −350.04 | 2713.5 | 2363.5 | −350.04 | 2835.8 | 2485.8 | −1.2821 | 10.695 | 9.413 |
| −10 | 0.25990 | 0.001089 | 467.17 | −354.12 | 2714.8 | 2360.7 | −354.12 | 2836.2 | 2482.1 | −1.2976 | 10.778 | 9.480 |
| −12 | 0.21732 | 0.001089 | 554.47 | −358.17 | 2716.1 | 2357.9 | −358.17 | 2836.6 | 2478.4 | −1.3130 | 10.862 | 9.549 |
| −14 | 0.18121 | 0.001088 | 659.88 | −362.18 | 2717.3 | 2355.2 | −362.18 | 2836.9 | 2474.7 | −1.3284 | 10.947 | 9.618 |
| −16 | 0.15068 | 0.001088 | 787.51 | −366.17 | 2718.6 | 2352.4 | −366.17 | 2837.2 | 2471.0 | −1.3439 | 11.033 | 9.689 |
| −18 | 0.12492 | 0.001088 | 942.51 | −370.13 | 2719.7 | 2349.6 | −370.13 | 2837.5 | 2467.3 | −1.3593 | 11.121 | 9.761 |
| −20 | 0.10326 | 0.001087 | 1131.3 | −374.06 | 2720.9 | 2346.8 | −374.06 | 2837.7 | 2463.6 | −1.3748 | 11.209 | 9.835 |
| −22 | 0.08510 | 0.001087 | 1362.0 | −377.95 | 2722.0 | 2344.1 | −377.95 | 2837.9 | 2459.9 | −1.3903 | 11.300 | 9.909 |
| −24 | 0.06991 | 0.001087 | 1644.7 | −381.82 | 2723.1 | 2341.3 | −381.82 | 2838.1 | 2456.2 | −1.4057 | 11.391 | 9.985 |
| −26 | 0.05725 | 0.001087 | 1992.2 | −385.66 | 2724.2 | 2338.5 | −385.66 | 2838.2 | 2452.5 | −1.4212 | 11.484 | 10.063 |
| −28 | 0.04673 | 0.001086 | 2421.0 | −389.47 | 2725.2 | 2335.7 | −389.47 | 2838.3 | 2448.8 | −1.4367 | 11.578 | 10.141 |
| −30 | 0.03802 | 0.001086 | 2951.7 | −393.25 | 2726.2 | 2332.9 | −393.25 | 2838.4 | 2445.1 | −1.4521 | 11.673 | 10.221 |
| −32 | 0.03082 | 0.001086 | 3610.9 | −397.00 | 2727.2 | 2330.2 | −397.00 | 2838.4 | 2441.4 | −1.4676 | 11.770 | 10.303 |
| −34 | 0.02490 | 0.001085 | 4432.4 | −400.72 | 2728.1 | 2327.4 | −400.72 | 2838.5 | 2437.7 | −1.4831 | 11.869 | 10.386 |
| −36 | 0.02004 | 0.001085 | 5460.1 | −404.40 | 2729.0 | 2324.6 | −404.40 | 2838.4 | 2434.0 | −1.4986 | 11.969 | 10.470 |
| −38 | 0.01608 | 0.001085 | 6750.5 | −408.07 | 2729.9 | 2321.8 | −408.07 | 2838.4 | 2430.3 | −1.5141 | 12.071 | 10.557 |
| −40 | 0.01285 | 0.001084 | 8376.7 | −411.70 | 2730.7 | 2319.0 | −411.70 | 2838.3 | 2426.6 | −1.5296 | 12.174 | 10.644 |

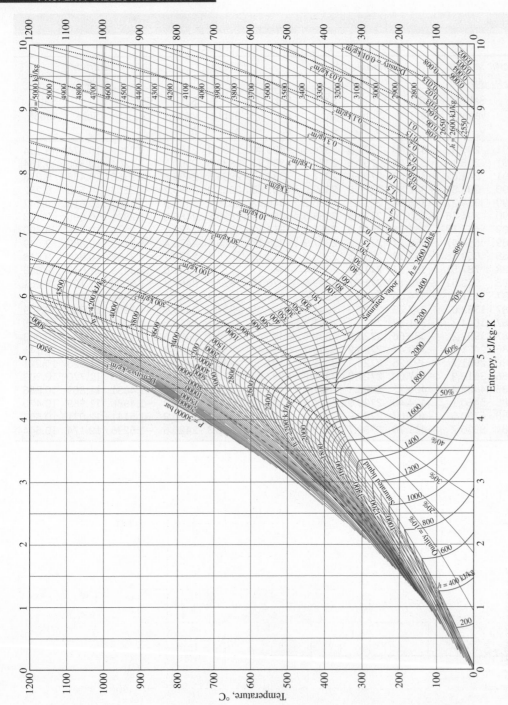

**FIGURE A–9**

*T-s* diagram for water.

*Source of Data: From NBS/NRC Steam Tables/1 by Lester Haar, John S. Gallagher, and George S. Kell. Routledge/Taylor & Francis Books, Inc., 1984.*

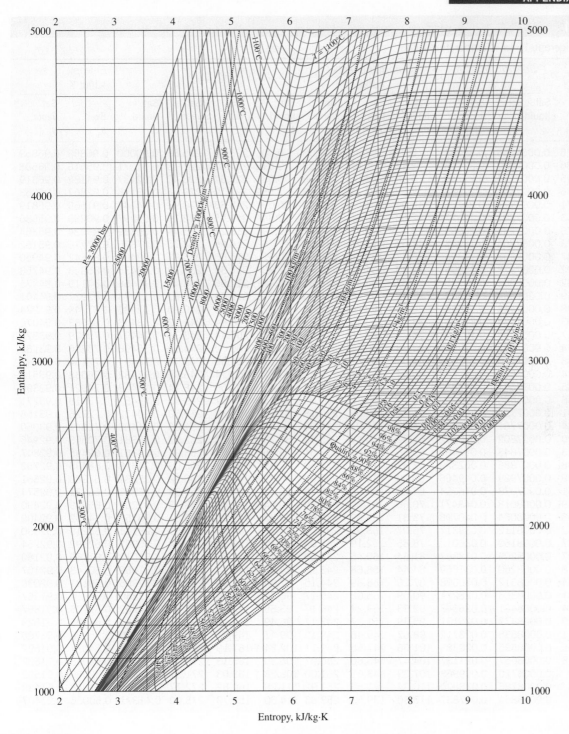

**FIGURE A-10**
Mollier diagram for water.

*Source of Data: From* NBS/NRC Steam Tables/1 *by Lester Haar, John S. Gallagher, and George S. Kell. Routledge/Taylor & Francis Books, Inc., 1984.*

## TABLE A–11

Saturated refrigerant-134a—Temperature table

| Temp., $T$°C | Sat. press., $P_{sat}$ kPa | Specific volume, m³/kg | | Internal energy, kJ/kg | | | Enthalpy, kJ/kg | | | Entropy, kJ/kg·K | | |
|---|---|---|---|---|---|---|---|---|---|---|---|---|
| | | Sat. liquid, $v_f$ | Sat. vapor, $v_g$ | Sat. liquid, $u_f$ | Evap., $u_{fg}$ | Sat. vapor, $u_g$ | Sat. liquid, $h_f$ | Evap., $h_{fg}$ | Sat. vapor, $h_g$ | Sat. liquid, $s_f$ | Evap., $s_{fg}$ | Sat. vapor, $s_g$ |
| −40 | 51.25 | 0.0007053 | 0.36064 | −0.036 | 207.42 | 207.38 | 0.00 | 225.86 | 225.86 | 0.00000 | 0.96869 | 0.96869 |
| −38 | 56.86 | 0.0007082 | 0.32718 | 2.472 | 206.06 | 208.53 | 2.512 | 224.62 | 227.13 | 0.01071 | 0.95516 | 0.96588 |
| −36 | 62.95 | 0.0007111 | 0.29740 | 4.987 | 204.69 | 209.68 | 5.032 | 223.37 | 228.40 | 0.02137 | 0.94182 | 0.96319 |
| −34 | 69.56 | 0.0007141 | 0.27082 | 7.509 | 203.32 | 210.83 | 7.559 | 222.10 | 229.66 | 0.03196 | 0.92867 | 0.96063 |
| −32 | 76.71 | 0.0007171 | 0.24706 | 10.04 | 201.94 | 211.97 | 10.09 | 220.83 | 230.93 | 0.04249 | 0.91569 | 0.95819 |
| −30 | 84.43 | 0.0007201 | 0.22577 | 12.58 | 200.55 | 213.12 | 12.64 | 219.55 | 232.19 | 0.05297 | 0.90289 | 0.95586 |
| −28 | 92.76 | 0.0007232 | 0.20666 | 15.12 | 199.15 | 214.27 | 15.19 | 218.25 | 233.44 | 0.06339 | 0.89024 | 0.95364 |
| −26 | 101.73 | 0.0007264 | 0.18947 | 17.67 | 197.75 | 215.42 | 17.75 | 216.95 | 234.70 | 0.07376 | 0.87776 | 0.95152 |
| −24 | 111.37 | 0.0007296 | 0.17398 | 20.23 | 196.34 | 216.57 | 20.31 | 215.63 | 235.94 | 0.08408 | 0.86542 | 0.94950 |
| −22 | 121.72 | 0.0007328 | 0.15999 | 22.80 | 194.92 | 217.71 | 22.89 | 214.30 | 237.19 | 0.09435 | 0.85323 | 0.94758 |
| −20 | 132.82 | 0.0007361 | 0.14735 | 25.37 | 193.49 | 218.86 | 25.47 | 212.96 | 238.43 | 0.10456 | 0.84119 | 0.94575 |
| −18 | 144.69 | 0.0007394 | 0.13589 | 27.96 | 192.05 | 220.00 | 28.07 | 211.60 | 239.67 | 0.11473 | 0.82927 | 0.94401 |
| −16 | 157.38 | 0.0007428 | 0.12550 | 30.55 | 190.60 | 221.15 | 30.67 | 210.23 | 240.90 | 0.12486 | 0.81749 | 0.94234 |
| −14 | 170.93 | 0.0007463 | 0.11605 | 33.15 | 189.14 | 222.29 | 33.28 | 208.84 | 242.12 | 0.13493 | 0.80583 | 0.94076 |
| −12 | 185.37 | 0.0007498 | 0.10744 | 35.76 | 187.66 | 223.42 | 35.90 | 207.44 | 243.34 | 0.14497 | 0.79429 | 0.93925 |
| −10 | 200.74 | 0.0007533 | 0.099600 | 38.38 | 186.18 | 224.56 | 38.53 | 206.02 | 244.55 | 0.15496 | 0.78286 | 0.93782 |
| −8 | 217.08 | 0.0007570 | 0.092438 | 41.01 | 184.69 | 225.69 | 41.17 | 204.59 | 245.76 | 0.16491 | 0.77154 | 0.93645 |
| −6 | 234.44 | 0.0007607 | 0.085888 | 43.64 | 183.18 | 226.82 | 43.82 | 203.14 | 246.95 | 0.17482 | 0.76033 | 0.93514 |
| −4 | 252.85 | 0.0007644 | 0.079889 | 46.29 | 181.66 | 227.94 | 46.48 | 201.66 | 248.14 | 0.18469 | 0.74921 | 0.93390 |
| −2 | 272.36 | 0.0007683 | 0.074388 | 48.94 | 180.12 | 229.07 | 49.15 | 200.17 | 249.33 | 0.19452 | 0.73819 | 0.93271 |
| 0 | 293.01 | 0.0007722 | 0.069335 | 51.61 | 178.58 | 230.18 | 51.83 | 198.67 | 250.50 | 0.20432 | 0.72726 | 0.93158 |
| 2 | 314.84 | 0.0007761 | 0.064690 | 54.28 | 177.01 | 231.30 | 54.53 | 197.14 | 251.66 | 0.21408 | 0.71641 | 0.93050 |
| 4 | 337.90 | 0.0007802 | 0.060412 | 56.97 | 175.44 | 232.40 | 57.23 | 195.58 | 252.82 | 0.22381 | 0.70565 | 0.92946 |
| 6 | 362.23 | 0.0007843 | 0.056469 | 59.66 | 173.84 | 233.51 | 59.95 | 194.01 | 253.96 | 0.23351 | 0.69496 | 0.92847 |
| 8 | 387.88 | 0.0007886 | 0.052829 | 62.37 | 172.23 | 234.60 | 62.68 | 192.42 | 255.09 | 0.24318 | 0.68435 | 0.92752 |
| 10 | 414.89 | 0.0007929 | 0.049466 | 65.09 | 170.61 | 235.69 | 65.42 | 190.80 | 256.22 | 0.25282 | 0.67380 | 0.92661 |
| 12 | 443.31 | 0.0007973 | 0.046354 | 67.82 | 168.96 | 236.78 | 68.17 | 189.16 | 257.33 | 0.26243 | 0.66331 | 0.92574 |
| 14 | 473.19 | 0.0008018 | 0.043471 | 70.56 | 167.30 | 237.86 | 70.94 | 187.49 | 258.43 | 0.27201 | 0.65289 | 0.92490 |
| 16 | 504.58 | 0.0008064 | 0.040798 | 73.31 | 165.62 | 238.93 | 73.72 | 185.80 | 259.51 | 0.28157 | 0.64252 | 0.92409 |
| 18 | 537.52 | 0.0008112 | 0.038317 | 76.07 | 163.92 | 239.99 | 76.51 | 184.08 | 260.59 | 0.29111 | 0.63219 | 0.92330 |
| 20 | 572.07 | 0.0008160 | 0.036012 | 78.85 | 162.19 | 241.04 | 79.32 | 182.33 | 261.64 | 0.30062 | 0.62192 | 0.92254 |
| 22 | 608.27 | 0.0008209 | 0.033867 | 81.64 | 160.45 | 242.09 | 82.14 | 180.55 | 262.69 | 0.31012 | 0.61168 | 0.92180 |
| 24 | 646.18 | 0.0008260 | 0.031869 | 84.44 | 158.68 | 243.13 | 84.98 | 178.74 | 263.72 | 0.31959 | 0.60148 | 0.92107 |
| 26 | 685.84 | 0.0008312 | 0.030008 | 87.26 | 156.89 | 244.15 | 87.83 | 176.90 | 264.73 | 0.32905 | 0.59131 | 0.92036 |
| 28 | 727.31 | 0.0008366 | 0.028271 | 90.09 | 155.08 | 245.17 | 90.70 | 175.03 | 265.73 | 0.33849 | 0.58117 | 0.91967 |
| 30 | 770.64 | 0.0008421 | 0.026648 | 92.93 | 153.24 | 246.17 | 93.58 | 173.13 | 266.71 | 0.34792 | 0.57105 | 0.91897 |
| 32 | 815.89 | 0.0008477 | 0.025131 | 95.79 | 151.37 | 247.17 | 96.49 | 171.19 | 267.67 | 0.35734 | 0.56095 | 0.91829 |
| 34 | 863.11 | 0.0008535 | 0.023712 | 98.67 | 149.48 | 248.15 | 99.41 | 169.21 | 268.61 | 0.36675 | 0.55086 | 0.91760 |
| 36 | 912.35 | 0.0008595 | 0.022383 | 101.56 | 147.55 | 249.11 | 102.34 | 167.19 | 269.53 | 0.37615 | 0.54077 | 0.91692 |
| 38 | 963.68 | 0.0008657 | 0.021137 | 104.47 | 145.60 | 250.07 | 105.30 | 165.13 | 270.44 | 0.38554 | 0.53068 | 0.91622 |
| 40 | 1017.1 | 0.0008720 | 0.019968 | 107.39 | 143.61 | 251.00 | 108.28 | 163.03 | 271.31 | 0.39493 | 0.52059 | 0.91552 |
| 42 | 1072.8 | 0.0008786 | 0.018870 | 110.34 | 141.59 | 251.92 | 111.28 | 160.89 | 272.17 | 0.40432 | 0.51048 | 0.91480 |
| 44 | 1130.7 | 0.0008854 | 0.017837 | 113.30 | 139.53 | 252.83 | 114.30 | 158.70 | 273.00 | 0.41371 | 0.50036 | 0.91407 |

## TABLE A–11

Saturated refrigerant-134a—Temperature table (*Concluded*)

| Temp., $T$°C | Sat. press., $P_{sat}$ kPa | Specific volume, m³/kg | | Internal energy, kJ/kg | | | Enthalpy, kJ/kg | | | Entropy, kJ/kg·K | | |
|---|---|---|---|---|---|---|---|---|---|---|---|---|
| | | Sat. liquid, $v_f$ | Sat. vapor, $v_g$ | Sat. liquid, $u_f$ | Evap., $u_{fg}$ | Sat. vapor, $u_g$ | Sat. liquid, $h_f$ | Evap., $h_{fg}$ | Sat. vapor, $h_g$ | Sat. liquid, $s_f$ | Evap., $s_{fg}$ | Sat. vapor, $s_g$ |
| 46 | 1191.0 | 0.0008924 | 0.016866 | 116.28 | 137.43 | 253.71 | 117.34 | 156.46 | 273.80 | 0.42311 | 0.49020 | 0.91331 |
| 48 | 1253.6 | 0.0008997 | 0.015951 | 119.28 | 135.30 | 254.58 | 120.41 | 154.17 | 274.57 | 0.43251 | 0.48001 | 0.91252 |
| 52 | 1386.2 | 0.0009151 | 0.014276 | 125.35 | 130.89 | 256.24 | 126.62 | 149.41 | 276.03 | 0.45136 | 0.45948 | 0.91084 |
| 56 | 1529.1 | 0.0009317 | 0.012782 | 131.52 | 126.29 | 257.81 | 132.94 | 144.41 | 277.35 | 0.47028 | 0.43870 | 0.90898 |
| 60 | 1682.8 | 0.0009498 | 0.011434 | 137.79 | 121.45 | 259.23 | 139.38 | 139.09 | 278.47 | 0.48930 | 0.41746 | 0.90676 |
| 65 | 1891.0 | 0.0009751 | 0.009959 | 145.80 | 115.06 | 260.86 | 147.64 | 132.05 | 279.69 | 0.51330 | 0.39048 | 0.90379 |
| 70 | 2118.2 | 0.0010037 | 0.008650 | 154.03 | 108.17 | 262.20 | 156.15 | 124.37 | 280.52 | 0.53763 | 0.36239 | 0.90002 |
| 75 | 2365.8 | 0.0010373 | 0.007486 | 162.55 | 100.62 | 263.17 | 165.01 | 115.87 | 280.88 | 0.56252 | 0.33279 | 0.89531 |
| 80 | 2635.3 | 0.0010774 | 0.006439 | 171.43 | 92.22 | 263.66 | 174.27 | 106.35 | 280.63 | 0.58812 | 0.30113 | 0.88925 |
| 85 | 2928.2 | 0.0011273 | 0.005484 | 180.81 | 82.64 | 263.45 | 184.11 | 95.39 | 279.51 | 0.61487 | 0.26632 | 0.88120 |
| 90 | 3246.9 | 0.0011938 | 0.004591 | 190.94 | 71.19 | 262.13 | 194.82 | 82.22 | 277.04 | 0.64354 | 0.22638 | 0.86991 |
| 95 | 3594.1 | 0.0012945 | 0.003713 | 202.49 | 56.25 | 258.73 | 207.14 | 64.94 | 272.08 | 0.67605 | 0.17638 | 0.85243 |
| 100 | 3975.1 | 0.0015269 | 0.002657 | 218.73 | 29.72 | 248.46 | 224.80 | 34.22 | 259.02 | 0.72224 | 0.09169 | 0.81393 |

*Source of Data:* Tables A–11 through A–13 are generated using the Engineering Equation Solver (EES) software developed by S. A. Klein and F. L. Alvarado. The routine used in calculations is the R134a, which is based on the fundamental equation of state developed by R. Tillner–Roth and H.D. Baehr, "An International Standard Formulation for the Thermodynamic Properties of 1,1,1,2-Tetrafluoroethane (HFC-134a) for temperatures from 170 K to 455 K and pressures up to 70 MPa," *J. Phys. Chem, Ref. Data*, Vol. 23, No. 5, 1994. The enthalpy and entropy values of saturated liquid are set to zero at −40°C (and −40°F).

## TABLE A–12

Saturated refrigerant-134a—Pressure table

| Press., $P$ kPa | Sat. temp., $T_{sat}$ °C | Specific volume, m³/kg | | Internal energy, kJ/kg | | | Enthalpy, kJ/kg | | | Entropy, kJ/kg·K | | |
|---|---|---|---|---|---|---|---|---|---|---|---|---|
| | | Sat. liquid, $v_f$ | Sat. vapor, $v_g$ | Sat. liquid, $u_f$ | Evap., $u_{fg}$ | Sat. vapor, $u_g$ | Sat. liquid, $h_f$ | Evap., $h_{fg}$ | Sat. vapor, $h_g$ | Sat. liquid, $s_f$ | Evap., $s_{fg}$ | Sat. vapor, $s_g$ |
| 60 | −36.95 | 0.0007097 | 0.31108 | 3.795 | 205.34 | 209.13 | 3.837 | 223.96 | 227.80 | 0.01633 | 0.94812 | 0.96445 |
| 70 | −33.87 | 0.0007143 | 0.26921 | 7.672 | 203.23 | 210.90 | 7.722 | 222.02 | 229.74 | 0.03264 | 0.92783 | 0.96047 |
| 80 | −31.13 | 0.0007184 | 0.23749 | 11.14 | 201.33 | 212.48 | 11.20 | 220.27 | 231.47 | 0.04707 | 0.91009 | 0.95716 |
| 90 | −28.65 | 0.0007222 | 0.21261 | 14.30 | 199.60 | 213.90 | 14.36 | 218.67 | 233.04 | 0.06003 | 0.89431 | 0.95434 |
| 100 | −26.37 | 0.0007258 | 0.19255 | 17.19 | 198.01 | 215.21 | 17.27 | 217.19 | 234.46 | 0.07182 | 0.88008 | 0.95191 |
| 120 | −22.32 | 0.0007323 | 0.16216 | 22.38 | 195.15 | 217.53 | 22.47 | 214.52 | 236.99 | 0.09269 | 0.85520 | 0.94789 |
| 140 | −18.77 | 0.0007381 | 0.14020 | 26.96 | 192.60 | 219.56 | 27.06 | 212.13 | 239.19 | 0.11080 | 0.83387 | 0.94467 |
| 160 | −15.60 | 0.0007435 | 0.12355 | 31.06 | 190.31 | 221.37 | 31.18 | 209.96 | 241.14 | 0.12686 | 0.81517 | 0.94202 |
| 180 | −12.73 | 0.0007485 | 0.11049 | 34.81 | 188.20 | 223.01 | 34.94 | 207.95 | 242.90 | 0.14131 | 0.79848 | 0.93979 |
| 200 | −10.09 | 0.0007532 | 0.099951 | 38.26 | 186.25 | 224.51 | 38.41 | 206.09 | 244.50 | 0.15449 | 0.78339 | 0.93788 |
| 240 | −5.38 | 0.0007618 | 0.083983 | 44.46 | 182.71 | 227.17 | 44.64 | 202.68 | 247.32 | 0.17786 | 0.75689 | 0.93475 |
| 280 | −1.25 | 0.0007697 | 0.072434 | 49.95 | 179.54 | 229.49 | 50.16 | 199.61 | 249.77 | 0.19822 | 0.73406 | 0.93228 |
| 320 | 2.46 | 0.0007771 | 0.063681 | 54.90 | 176.65 | 231.55 | 55.14 | 196.78 | 251.93 | 0.21631 | 0.71395 | 0.93026 |
| 360 | 5.82 | 0.0007840 | 0.056809 | 59.42 | 173.99 | 233.41 | 59.70 | 194.15 | 253.86 | 0.23265 | 0.69591 | 0.92856 |
| 400 | 8.91 | 0.0007905 | 0.051266 | 63.61 | 171.49 | 235.10 | 63.92 | 191.68 | 255.61 | 0.24757 | 0.67954 | 0.92711 |
| 450 | 12.46 | 0.0007983 | 0.045677 | 68.44 | 168.58 | 237.03 | 68.80 | 188.78 | 257.58 | 0.26462 | 0.66093 | 0.92555 |
| 500 | 15.71 | 0.0008058 | 0.041168 | 72.92 | 165.86 | 238.77 | 73.32 | 186.04 | 259.36 | 0.28021 | 0.64399 | 0.92420 |
| 550 | 18.73 | 0.0008129 | 0.037452 | 77.09 | 163.29 | 240.38 | 77.54 | 183.44 | 260.98 | 0.29460 | 0.62842 | 0.92302 |
| 600 | 21.55 | 0.0008198 | 0.034335 | 81.01 | 160.84 | 241.86 | 81.50 | 180.95 | 262.46 | 0.30799 | 0.61398 | 0.92196 |
| 650 | 24.20 | 0.0008265 | 0.031680 | 84.72 | 158.51 | 243.23 | 85.26 | 178.56 | 263.82 | 0.32052 | 0.60048 | 0.92100 |
| 700 | 26.69 | 0.0008331 | 0.029392 | 88.24 | 156.27 | 244.51 | 88.82 | 176.26 | 265.08 | 0.33232 | 0.58780 | 0.92012 |
| 750 | 29.06 | 0.0008395 | 0.027398 | 91.59 | 154.11 | 245.70 | 92.22 | 174.03 | 266.25 | 0.34348 | 0.57582 | 0.91930 |
| 800 | 31.31 | 0.0008457 | 0.025645 | 94.80 | 152.02 | 246.82 | 95.48 | 171.86 | 267.34 | 0.35408 | 0.56445 | 0.91853 |
| 850 | 33.45 | 0.0008519 | 0.024091 | 97.88 | 150.00 | 247.88 | 98.61 | 169.75 | 268.36 | 0.36417 | 0.55362 | 0.91779 |
| 900 | 35.51 | 0.0008580 | 0.022703 | 100.84 | 148.03 | 248.88 | 101.62 | 167.69 | 269.31 | 0.37383 | 0.54326 | 0.91709 |
| 950 | 37.48 | 0.0008640 | 0.021456 | 103.70 | 146.11 | 249.82 | 104.52 | 165.68 | 270.20 | 0.38307 | 0.53333 | 0.91641 |
| 1000 | 39.37 | 0.0008700 | 0.020329 | 106.47 | 144.24 | 250.71 | 107.34 | 163.70 | 271.04 | 0.39196 | 0.52378 | 0.91574 |
| 1200 | 46.29 | 0.0008935 | 0.016728 | 116.72 | 137.12 | 253.84 | 117.79 | 156.12 | 273.92 | 0.42449 | 0.48870 | 0.91320 |
| 1400 | 52.40 | 0.0009167 | 0.014119 | 125.96 | 130.44 | 256.40 | 127.25 | 148.92 | 276.17 | 0.45325 | 0.45742 | 0.91067 |
| 1600 | 57.88 | 0.0009400 | 0.012134 | 134.45 | 124.05 | 258.50 | 135.96 | 141.96 | 277.92 | 0.47921 | 0.42881 | 0.90802 |
| 1800 | 62.87 | 0.0009639 | 0.010568 | 142.36 | 117.85 | 260.21 | 144.09 | 135.14 | 279.23 | 0.50304 | 0.40213 | 0.90517 |
| 2000 | 67.45 | 0.0009887 | 0.009297 | 149.81 | 111.75 | 261.56 | 151.78 | 128.36 | 280.15 | 0.52519 | 0.37684 | 0.90204 |
| 2500 | 77.54 | 0.0010567 | 0.006941 | 167.02 | 96.47 | 263.49 | 169.66 | 111.18 | 280.84 | 0.57542 | 0.31701 | 0.89243 |
| 3000 | 86.16 | 0.0011410 | 0.005272 | 183.09 | 80.17 | 263.26 | 186.51 | 92.57 | 279.08 | 0.62133 | 0.25759 | 0.87893 |

## TABLE A-13

Superheated refrigerant-134a

| T °C | v m³/kg | u kJ/kg | h kJ/kg | s kJ/kg·K | v m³/kg | u kJ/kg | h kJ/kg | s kJ/kg·K | v m³/kg | u kJ/kg | h kJ/kg | s kJ/kg·K |
|---|---|---|---|---|---|---|---|---|---|---|---|---|
| | $P = 0.06$ MPa ($T_{sat} = -36.95$°C) | | | | $P = 0.10$ MPa ($T_{sat} = -26.37$°C) | | | | $P = 0.14$ MPa ($T_{sat} = -18.77$°C) | | | |
| Sat. | 0.31108 | 209.13 | 227.80 | 0.9645 | 0.19255 | 215.21 | 234.46 | 0.9519 | 0.14020 | 219.56 | 239.19 | 0.9447 |
| −20 | 0.33608 | 220.62 | 240.78 | 1.0175 | 0.19841 | 219.68 | 239.52 | 0.9721 | | | | |
| −10 | 0.35048 | 227.57 | 248.60 | 1.0478 | 0.20743 | 226.77 | 247.51 | 1.0031 | 0.14605 | 225.93 | 246.37 | 0.9724 |
| 0 | 0.36476 | 234.67 | 256.56 | 1.0775 | 0.21630 | 233.97 | 255.60 | 1.0333 | 0.15263 | 233.25 | 254.61 | 1.0032 |
| 10 | 0.37893 | 241.94 | 264.68 | 1.1067 | 0.22506 | 241.32 | 263.82 | 1.0628 | 0.15908 | 240.68 | 262.95 | 1.0331 |
| 20 | 0.39302 | 249.37 | 272.95 | 1.1354 | 0.23373 | 248.81 | 272.18 | 1.0919 | 0.16544 | 248.24 | 271.40 | 1.0625 |
| 30 | 0.40705 | 256.97 | 281.39 | 1.1637 | 0.24233 | 256.46 | 280.69 | 1.1204 | 0.17172 | 255.95 | 279.99 | 1.0913 |
| 40 | 0.42102 | 264.73 | 289.99 | 1.1916 | 0.25088 | 264.27 | 289.36 | 1.1485 | 0.17794 | 263.80 | 288.72 | 1.1196 |
| 50 | 0.43495 | 272.66 | 298.75 | 1.2192 | 0.25937 | 272.24 | 298.17 | 1.1762 | 0.18412 | 271.81 | 297.59 | 1.1475 |
| 60 | 0.44883 | 280.75 | 307.68 | 1.2464 | 0.26783 | 280.36 | 307.15 | 1.2036 | 0.19025 | 279.97 | 306.61 | 1.1750 |
| 70 | 0.46269 | 289.01 | 316.77 | 1.2732 | 0.27626 | 288.65 | 316.28 | 1.2306 | 0.19635 | 288.29 | 315.78 | 1.2021 |
| 80 | 0.47651 | 297.43 | 326.02 | 1.2998 | 0.28465 | 297.10 | 325.57 | 1.2573 | 0.20242 | 296.77 | 325.11 | 1.2289 |
| 90 | 0.49032 | 306.02 | 335.43 | 1.3261 | 0.29303 | 305.71 | 335.01 | 1.2836 | 0.20847 | 305.40 | 334.59 | 1.2554 |
| 100 | 0.50410 | 314.76 | 345.01 | 1.3521 | 0.30138 | 314.48 | 344.61 | 1.3097 | 0.21449 | 314.19 | 344.22 | 1.2815 |
| | $P = 0.18$ MPa ($T_{sat} = -12.73$°C) | | | | $P = 0.20$ MPa ($T_{sat} = -10.09$°C) | | | | $P = 0.24$ MPa ($T_{sat} = -5.38$°C) | | | |
| Sat. | 0.11049 | 223.01 | 242.90 | 0.9398 | 0.09995 | 224.51 | 244.50 | 0.9379 | 0.08398 | 227.17 | 247.32 | 0.9348 |
| −10 | 0.11189 | 225.04 | 245.18 | 0.9485 | 0.09991 | 224.57 | 244.56 | 0.9381 | | | | |
| 0 | 0.11722 | 232.49 | 253.59 | 0.9799 | 0.10481 | 232.11 | 253.07 | 0.9699 | 0.08617 | 231.30 | 251.98 | 0.9520 |
| 10 | 0.12240 | 240.02 | 262.05 | 1.0103 | 0.10955 | 239.69 | 261.60 | 1.0005 | 0.09026 | 239.00 | 260.66 | 0.9832 |
| 20 | 0.12748 | 247.66 | 270.60 | 1.0400 | 0.11418 | 247.36 | 270.20 | 1.0304 | 0.09423 | 246.76 | 269.38 | 1.0134 |
| 30 | 0.13248 | 255.43 | 279.27 | 1.0691 | 0.11874 | 255.16 | 278.91 | 1.0596 | 0.09812 | 254.63 | 278.17 | 1.0429 |
| 40 | 0.13741 | 263.33 | 288.07 | 1.0976 | 0.12322 | 263.09 | 287.74 | 1.0882 | 0.10193 | 262.61 | 287.07 | 1.0718 |
| 50 | 0.14230 | 271.38 | 297.00 | 1.1257 | 0.12766 | 271.16 | 296.70 | 1.1164 | 0.10570 | 270.73 | 296.09 | 1.1002 |
| 60 | 0.14715 | 279.58 | 306.07 | 1.1533 | 0.13206 | 279.38 | 305.79 | 1.1441 | 0.10942 | 278.98 | 305.24 | 1.1281 |
| 70 | 0.15196 | 287.93 | 315.28 | 1.1806 | 0.13641 | 287.75 | 315.03 | 1.1714 | 0.11310 | 287.38 | 314.53 | 1.1555 |
| 80 | 0.15673 | 296.43 | 324.65 | 1.2075 | 0.14074 | 296.27 | 324.41 | 1.1984 | 0.11675 | 295.93 | 323.95 | 1.1826 |
| 90 | 0.16149 | 305.09 | 334.16 | 1.2340 | 0.14504 | 304.93 | 333.94 | 1.2250 | 0.12038 | 304.62 | 333.51 | 1.2093 |
| 100 | 0.16622 | 313.90 | 343.82 | 1.2603 | 0.14933 | 313.75 | 343.62 | 1.2513 | 0.12398 | 313.46 | 343.22 | 1.2356 |
| | $P = 0.28$ MPa ($T_{sat} = -1.25$°C) | | | | $P = 0.32$ MPa ($T_{sat} = 2.46$°C) | | | | $P = 0.40$ MPa ($T_{sat} = 8.91$°C) | | | |
| Sat. | 0.07243 | 229.49 | 249.77 | 0.9323 | 0.06368 | 231.55 | 251.93 | 0.9303 | 0.051266 | 235.10 | 255.61 | 0.9271 |
| 0 | 0.07282 | 230.46 | 250.85 | 0.9362 | | | | | | | | |
| 10 | 0.07646 | 238.29 | 259.70 | 0.9681 | 0.06609 | 237.56 | 258.70 | 0.9545 | 0.051506 | 235.99 | 256.59 | 0.9306 |
| 20 | 0.07997 | 246.15 | 268.54 | 0.9987 | 0.06925 | 245.51 | 267.67 | 0.9856 | 0.054213 | 244.19 | 265.88 | 0.9628 |
| 30 | 0.08338 | 254.08 | 277.42 | 1.0285 | 0.07231 | 253.52 | 276.66 | 1.0158 | 0.056796 | 252.37 | 275.09 | 0.9937 |
| 40 | 0.08672 | 262.12 | 286.40 | 1.0577 | 0.07530 | 261.62 | 285.72 | 1.0452 | 0.059292 | 260.60 | 284.32 | 1.0237 |
| 50 | 0.09000 | 270.28 | 295.48 | 1.0862 | 0.07823 | 269.83 | 294.87 | 1.0739 | 0.061724 | 268.92 | 293.61 | 1.0529 |
| 60 | 0.09324 | 278.58 | 304.69 | 1.1143 | 0.08111 | 278.17 | 304.12 | 1.1022 | 0.064104 | 277.34 | 302.98 | 1.0814 |
| 70 | 0.09644 | 287.01 | 314.01 | 1.1419 | 0.08395 | 286.64 | 313.50 | 1.1299 | 0.066443 | 285.88 | 312.45 | 1.1095 |
| 80 | 0.09961 | 295.59 | 323.48 | 1.1690 | 0.08675 | 295.24 | 323.00 | 1.1572 | 0.068747 | 294.54 | 322.04 | 1.1370 |
| 90 | 0.10275 | 304.30 | 333.07 | 1.1958 | 0.08953 | 303.99 | 332.64 | 1.1841 | 0.071023 | 303.34 | 331.75 | 1.1641 |
| 100 | 0.10587 | 313.17 | 342.81 | 1.2223 | 0.09229 | 312.87 | 342.41 | 1.2106 | 0.073274 | 312.28 | 341.59 | 1.1908 |
| 110 | 0.10897 | 322.18 | 352.69 | 1.2484 | 0.09503 | 321.91 | 352.31 | 1.2368 | 0.075504 | 321.35 | 351.55 | 1.2172 |
| 120 | 0.11205 | 331.34 | 362.72 | 1.2742 | 0.09775 | 331.08 | 362.36 | 1.2627 | 0.077717 | 330.56 | 361.65 | 1.2432 |
| 130 | 0.11512 | 340.65 | 372.88 | 1.2998 | 0.10045 | 340.41 | 372.55 | 1.2883 | 0.079913 | 339.92 | 371.89 | 1.2689 |
| 140 | 0.11818 | 350.11 | 383.20 | 1.3251 | 0.10314 | 349.88 | 382.89 | 1.3136 | 0.082096 | 349.42 | 382.26 | 1.2943 |

## TABLE A–13

Superheated refrigerant-134a (*Concluded*)

| T °C | v m³/kg | u kJ/kg | h kJ/kg | s kJ/kg·K | v m³/kg | u kJ/kg | h kJ/kg | s kJ/kg·K | v m³/kg | u kJ/kg | h kJ/kg | s kJ/kg·K |
|---|---|---|---|---|---|---|---|---|---|---|---|---|
| | P = 0.50 MPa ($T_{sat}$ = 15.71°C) | | | | P = 0.60 MPa ($T_{sat}$ = 21.55°C) | | | | P = 0.70 MPa ($T_{sat}$ = 26.69°C) | | | |
| Sat. | 0.041168 | 238.77 | 259.36 | 0.9242 | 0.034335 | 241.86 | 262.46 | 0.9220 | 0.029392 | 244.51 | 265.08 | 0.9201 |
| 20 | 0.042115 | 242.42 | 263.48 | 0.9384 | | | | | | | | |
| 30 | 0.044338 | 250.86 | 273.03 | 0.9704 | 0.035984 | 249.24 | 270.83 | 0.9500 | 0.029966 | 247.49 | 268.47 | 0.9314 |
| 40 | 0.046456 | 259.27 | 282.50 | 1.0011 | 0.037865 | 257.88 | 280.60 | 0.9817 | 0.031696 | 256.41 | 278.59 | 0.9642 |
| 50 | 0.048499 | 267.73 | 291.98 | 1.0309 | 0.039659 | 266.50 | 290.30 | 1.0122 | 0.033322 | 265.22 | 288.54 | 0.9955 |
| 60 | 0.050485 | 276.27 | 301.51 | 1.0600 | 0.041389 | 275.17 | 300.00 | 1.0417 | 0.034875 | 274.03 | 298.44 | 1.0257 |
| 70 | 0.052427 | 284.91 | 311.12 | 1.0884 | 0.043069 | 283.91 | 309.75 | 1.0706 | 0.036373 | 282.88 | 308.34 | 1.0550 |
| 80 | 0.054331 | 293.65 | 320.82 | 1.1163 | 0.044710 | 292.74 | 319.57 | 1.0988 | 0.037829 | 291.81 | 318.29 | 1.0835 |
| 90 | 0.056205 | 302.52 | 330.63 | 1.1436 | 0.046318 | 301.69 | 329.48 | 1.1265 | 0.039250 | 300.84 | 328.31 | 1.1115 |
| 100 | 0.058053 | 311.52 | 340.55 | 1.1706 | 0.047900 | 310.75 | 339.49 | 1.1536 | 0.040642 | 309.96 | 338.41 | 1.1389 |
| 110 | 0.059880 | 320.65 | 350.59 | 1.1971 | 0.049458 | 319.93 | 349.61 | 1.1804 | 0.042010 | 319.21 | 348.61 | 1.1659 |
| 120 | 0.061687 | 329.91 | 360.75 | 1.2233 | 0.050997 | 329.24 | 359.84 | 1.2068 | 0.043358 | 328.57 | 358.92 | 1.1925 |
| 130 | 0.063479 | 339.31 | 371.05 | 1.2492 | 0.052519 | 338.69 | 370.20 | 1.2328 | 0.044688 | 338.06 | 369.34 | 1.2186 |
| 140 | 0.065256 | 348.85 | 381.47 | 1.2747 | 0.054027 | 348.26 | 380.68 | 1.2585 | 0.046004 | 347.67 | 379.88 | 1.2445 |
| 150 | 0.067021 | 358.52 | 392.04 | 1.3000 | 0.055522 | 357.98 | 391.29 | 1.2838 | 0.047306 | 357.42 | 390.54 | 1.2700 |
| 160 | 0.068775 | 368.34 | 402.73 | 1.3250 | 0.057006 | 367.83 | 402.03 | 1.3089 | 0.048597 | 367.31 | 401.32 | 1.2952 |
| | P = 0.80 MPa ($T_{sat}$ = 31.31°C) | | | | P = 0.90 MPa ($T_{sat}$ = 35.51°C) | | | | P = 1.00 MPa ($T_{sat}$ = 39.37°C) | | | |
| Sat. | 0.025645 | 246.82 | 267.34 | 0.9185 | 0.022686 | 248.82 | 269.25 | 0.9169 | 0.020319 | 250.71 | 271.04 | 0.9157 |
| 40 | 0.027035 | 254.84 | 276.46 | 0.9481 | 0.023375 | 253.15 | 274.19 | 0.9328 | 0.020406 | 251.32 | 271.73 | 0.9180 |
| 50 | 0.028547 | 263.87 | 286.71 | 0.9803 | 0.024809 | 262.46 | 284.79 | 0.9661 | 0.021796 | 260.96 | 282.76 | 0.9526 |
| 60 | 0.029973 | 272.85 | 296.82 | 1.0111 | 0.026146 | 271.62 | 295.15 | 0.9977 | 0.023068 | 270.33 | 293.40 | 0.9851 |
| 70 | 0.031340 | 281.83 | 306.90 | 1.0409 | 0.027413 | 280.74 | 305.41 | 1.0280 | 0.024261 | 279.61 | 303.87 | 1.0160 |
| 80 | 0.032659 | 290.86 | 316.99 | 1.0699 | 0.028630 | 289.88 | 315.65 | 1.0574 | 0.025398 | 288.87 | 314.27 | 1.0459 |
| 90 | 0.033941 | 299.97 | 327.12 | 1.0982 | 0.029806 | 299.08 | 325.90 | 1.0861 | 0.026492 | 298.17 | 324.66 | 1.0749 |
| 100 | 0.035193 | 309.17 | 337.32 | 1.1259 | 0.030951 | 308.35 | 336.21 | 1.1141 | 0.027552 | 307.52 | 335.08 | 1.1032 |
| 110 | 0.036420 | 318.47 | 347.61 | 1.1531 | 0.032068 | 317.72 | 346.58 | 1.1415 | 0.028584 | 316.96 | 345.54 | 1.1309 |
| 120 | 0.037625 | 327.89 | 357.99 | 1.1798 | 0.033164 | 327.19 | 357.04 | 1.1684 | 0.029592 | 326.49 | 356.08 | 1.1580 |
| 130 | 0.038813 | 337.42 | 368.47 | 1.2062 | 0.034241 | 336.78 | 367.59 | 1.1949 | 0.030581 | 336.12 | 366.70 | 1.1847 |
| 140 | 0.039985 | 347.08 | 379.07 | 1.2321 | 0.035302 | 346.48 | 378.25 | 1.2211 | 0.031554 | 345.87 | 377.42 | 1.2110 |
| 150 | 0.041143 | 356.86 | 389.78 | 1.2577 | 0.036349 | 356.30 | 389.01 | 1.2468 | 0.032512 | 355.73 | 388.24 | 1.2369 |
| 160 | 0.042290 | 366.78 | 400.61 | 1.2830 | 0.037384 | 366.25 | 399.89 | 1.2722 | 0.033457 | 365.71 | 399.17 | 1.2624 |
| 170 | 0.043427 | 376.83 | 411.57 | 1.3081 | 0.038408 | 376.33 | 410.89 | 1.2973 | 0.034392 | 375.82 | 410.22 | 1.2876 |
| 180 | 0.044554 | 387.01 | 422.65 | 1.3328 | 0.039423 | 386.54 | 422.02 | 1.3221 | 0.035317 | 386.06 | 421.38 | 1.3125 |
| | P = 1.20 MPa ($T_{sat}$ = 46.29°C) | | | | P = 1.40 MPa ($T_{sat}$ = 52.40°C) | | | | P = 1.60 MPa ($T_{sat}$ = 57.88°C) | | | |
| Sat. | 0.016728 | 253.84 | 273.92 | 0.9132 | 0.014119 | 256.40 | 276.17 | 0.9107 | 0.012134 | 258.50 | 277.92 | 0.9080 |
| 50 | 0.017201 | 257.64 | 278.28 | 0.9268 | | | | | | | | |
| 60 | 0.018404 | 267.57 | 289.66 | 0.9615 | 0.015005 | 264.46 | 285.47 | 0.9389 | 0.012372 | 260.91 | 280.71 | 0.9164 |
| 70 | 0.019502 | 277.23 | 300.63 | 0.9939 | 0.016060 | 274.62 | 297.10 | 0.9733 | 0.013430 | 271.78 | 293.27 | 0.9536 |
| 80 | 0.020529 | 286.77 | 311.40 | 1.0249 | 0.017023 | 284.51 | 308.34 | 1.0056 | 0.014362 | 282.11 | 305.09 | 0.9875 |
| 90 | 0.021506 | 296.28 | 322.09 | 1.0547 | 0.017923 | 294.28 | 319.37 | 1.0364 | 0.015215 | 292.19 | 316.53 | 1.0195 |
| 100 | 0.022442 | 305.81 | 332.74 | 1.0836 | 0.018778 | 304.01 | 330.30 | 1.0661 | 0.016014 | 302.16 | 327.78 | 1.0501 |
| 110 | 0.023348 | 315.40 | 343.41 | 1.1119 | 0.019597 | 313.76 | 341.19 | 1.0949 | 0.016773 | 312.09 | 338.93 | 1.0795 |
| 120 | 0.024228 | 325.05 | 354.12 | 1.1395 | 0.020388 | 323.55 | 352.09 | 1.1230 | 0.017500 | 322.03 | 350.03 | 1.1081 |
| 130 | 0.025086 | 334.79 | 364.90 | 1.1665 | 0.021155 | 333.41 | 363.02 | 1.1504 | 0.018201 | 332.02 | 361.14 | 1.1360 |
| 140 | 0.025927 | 344.63 | 375.74 | 1.1931 | 0.021904 | 343.34 | 374.01 | 1.1773 | 0.018882 | 342.06 | 372.27 | 1.1633 |
| 150 | 0.026753 | 354.57 | 386.68 | 1.2192 | 0.022636 | 353.37 | 385.07 | 1.2038 | 0.019545 | 352.19 | 383.46 | 1.1901 |
| 160 | 0.027566 | 364.63 | 397.71 | 1.2450 | 0.023355 | 363.51 | 396.20 | 1.2298 | 0.020194 | 362.40 | 394.71 | 1.2164 |
| 170 | 0.028367 | 374.80 | 408.84 | 1.2704 | 0.024061 | 373.75 | 407.43 | 1.2554 | 0.020830 | 372.71 | 406.04 | 1.2422 |
| 180 | 0.029158 | 385.10 | 420.09 | 1.2955 | 0.024757 | 384.12 | 418.78 | 1.2808 | 0.021456 | 383.13 | 417.46 | 1.2677 |

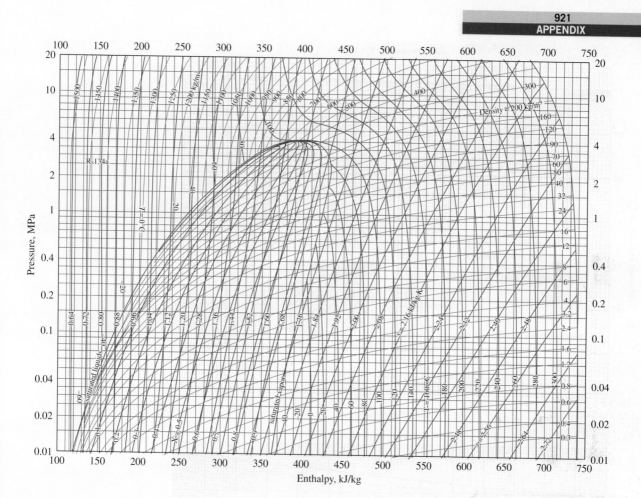

**FIGURE A–14**

*P-h* diagram for refrigerant-134a.

*Note:* The reference point used for the chart is different than that used in the R-134a tables. Therefore, problems should be solved using all property data either from the tables or from the chart, but not from both.

*Source of Data: American Society of Heating, Refrigerating, and Air-Conditioning Engineers, Inc., Atlanta, GA.*

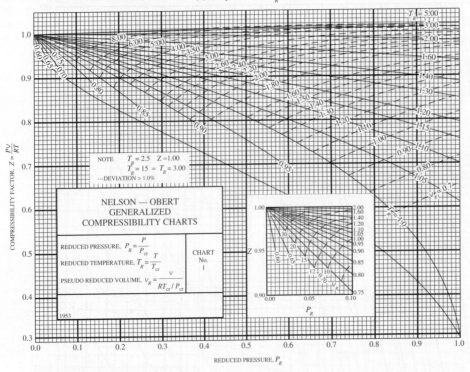

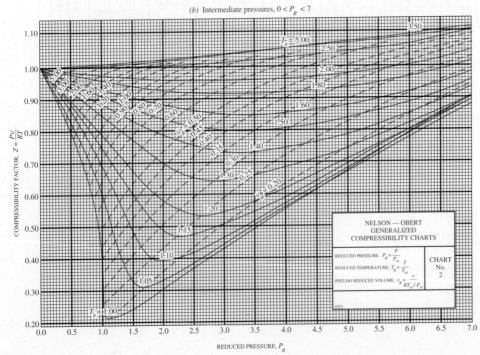

# FIGURE A–15

Nelson–Obert generalized compressibility chart.

*Used with permission of Dr. Edward E. Obert, University of Wisconsin.*

## TABLE A–16

Properties of the atmosphere at high altitude

| Altitude, m | Temperature, °C | Pressure, kPa | Gravity $g$, m/s² | Speed of Sound, m/s | Density, kg/m³ | Viscosity $\mu$, kg/m·s | Thermal Conductivity, W/m·K |
|---|---|---|---|---|---|---|---|
| 0 | 15.00 | 101.33 | 9.807 | 340.3 | 1.225 | $1.789 \times 10^{-5}$ | 0.0253 |
| 200 | 13.70 | 98.95 | 9.806 | 339.5 | 1.202 | $1.783 \times 10^{-5}$ | 0.0252 |
| 400 | 12.40 | 96.61 | 9.805 | 338.8 | 1.179 | $1.777 \times 10^{-5}$ | 0.0252 |
| 600 | 11.10 | 94.32 | 9.805 | 338.0 | 1.156 | $1.771 \times 10^{-5}$ | 0.0251 |
| 800 | 9.80 | 92.08 | 9.804 | 337.2 | 1.134 | $1.764 \times 10^{-5}$ | 0.0250 |
| 1000 | 8.50 | 89.88 | 9.804 | 336.4 | 1.112 | $1.758 \times 10^{-5}$ | 0.0249 |
| 1200 | 7.20 | 87.72 | 9.803 | 335.7 | 1.090 | $1.752 \times 10^{-5}$ | 0.0248 |
| 1400 | 5.90 | 85.60 | 9.802 | 334.9 | 1.069 | $1.745 \times 10^{-5}$ | 0.0247 |
| 1600 | 4.60 | 83.53 | 9.802 | 334.1 | 1.048 | $1.739 \times 10^{-5}$ | 0.0245 |
| 1800 | 3.30 | 81.49 | 9.801 | 333.3 | 1.027 | $1.732 \times 10^{-5}$ | 0.0244 |
| 2000 | 2.00 | 79.50 | 9.800 | 332.5 | 1.007 | $1.726 \times 10^{-5}$ | 0.0243 |
| 2200 | 0.70 | 77.55 | 9.800 | 331.7 | 0.987 | $1.720 \times 10^{-5}$ | 0.0242 |
| 2400 | −0.59 | 75.63 | 9.799 | 331.0 | 0.967 | $1.713 \times 10^{-5}$ | 0.0241 |
| 2600 | −1.89 | 73.76 | 9.799 | 330.2 | 0.947 | $1.707 \times 10^{-5}$ | 0.0240 |
| 2800 | −3.19 | 71.92 | 9.798 | 329.4 | 0.928 | $1.700 \times 10^{-5}$ | 0.0239 |
| 3000 | −4.49 | 70.12 | 9.797 | 328.6 | 0.909 | $1.694 \times 10^{-5}$ | 0.0238 |
| 3200 | −5.79 | 68.36 | 9.797 | 327.8 | 0.891 | $1.687 \times 10^{-5}$ | 0.0237 |
| 3400 | −7.09 | 66.63 | 9.796 | 327.0 | 0.872 | $1.681 \times 10^{-5}$ | 0.0236 |
| 3600 | −8.39 | 64.94 | 9.796 | 326.2 | 0.854 | $1.674 \times 10^{-5}$ | 0.0235 |
| 3800 | −9.69 | 63.28 | 9.795 | 325.4 | 0.837 | $1.668 \times 10^{-5}$ | 0.0234 |
| 4000 | −10.98 | 61.66 | 9.794 | 324.6 | 0.819 | $1.661 \times 10^{-5}$ | 0.0233 |
| 4200 | −12.3 | 60.07 | 9.794 | 323.8 | 0.802 | $1.655 \times 10^{-5}$ | 0.0232 |
| 4400 | −13.6 | 58.52 | 9.793 | 323.0 | 0.785 | $1.648 \times 10^{-5}$ | 0.0231 |
| 4600 | −14.9 | 57.00 | 9.793 | 322.2 | 0.769 | $1.642 \times 10^{-5}$ | 0.0230 |
| 4800 | −16.2 | 55.51 | 9.792 | 321.4 | 0.752 | $1.635 \times 10^{-5}$ | 0.0229 |
| 5000 | −17.5 | 54.05 | 9.791 | 320.5 | 0.736 | $1.628 \times 10^{-5}$ | 0.0228 |
| 5200 | −18.8 | 52.62 | 9.791 | 319.7 | 0.721 | $1.622 \times 10^{-5}$ | 0.0227 |
| 5400 | −20.1 | 51.23 | 9.790 | 318.9 | 0.705 | $1.615 \times 10^{-5}$ | 0.0226 |
| 5600 | −21.4 | 49.86 | 9.789 | 318.1 | 0.690 | $1.608 \times 10^{-5}$ | 0.0224 |
| 5800 | −22.7 | 48.52 | 9.785 | 317.3 | 0.675 | $1.602 \times 10^{-5}$ | 0.0223 |
| 6000 | −24.0 | 47.22 | 9.788 | 316.5 | 0.660 | $1.595 \times 10^{-5}$ | 0.0222 |
| 6200 | −25.3 | 45.94 | 9.788 | 315.6 | 0.646 | $1.588 \times 10^{-5}$ | 0.0221 |
| 6400 | −26.6 | 44.69 | 9.787 | 314.8 | 0.631 | $1.582 \times 10^{-5}$ | 0.0220 |
| 6600 | −27.9 | 43.47 | 9.786 | 314.0 | 0.617 | $1.575 \times 10^{-5}$ | 0.0219 |
| 6800 | −29.2 | 42.27 | 9.785 | 313.1 | 0.604 | $1.568 \times 10^{-5}$ | 0.0218 |
| 7000 | −30.5 | 41.11 | 9.785 | 312.3 | 0.590 | $1.561 \times 10^{-5}$ | 0.0217 |
| 8000 | −36.9 | 35.65 | 9.782 | 308.1 | 0.526 | $1.527 \times 10^{-5}$ | 0.0212 |
| 9000 | −43.4 | 30.80 | 9.779 | 303.8 | 0.467 | $1.493 \times 10^{-5}$ | 0.0206 |
| 10,000 | −49.9 | 26.50 | 9.776 | 299.5 | 0.414 | $1.458 \times 10^{-5}$ | 0.0201 |
| 12,000 | −56.5 | 19.40 | 9.770 | 295.1 | 0.312 | $1.422 \times 10^{-5}$ | 0.0195 |
| 14,000 | −56.5 | 14.17 | 9.764 | 295.1 | 0.228 | $1.422 \times 10^{-5}$ | 0.0195 |
| 16,000 | −56.5 | 10.53 | 9.758 | 295.1 | 0.166 | $1.422 \times 10^{-5}$ | 0.0195 |
| 18,000 | −56.5 | 7.57 | 9.751 | 295.1 | 0.122 | $1.422 \times 10^{-5}$ | 0.0195 |

*Source of Data:* U.S. Standard Atmosphere Supplements, U.S. Government Printing Office, 1966. Based on year-round mean conditions at 45° latitude and varies with the time of the year and the weather patterns. The conditions at sea level ($z = 0$) are taken to be $P = 101.325$ kPa, $T = 15$°C, $\rho = 1.2250$ kg/m³, $g = 9.80665$ m²/s.

## TABLE A–17

Ideal-gas properties of air

| T K | h kJ/kg | $P_r$ | u kJ/kg | $v_r$ | $s°$ kJ/kg·K | T K | h kJ/kg | $P_r$ | u kJ/kg | $v_r$ | $s°$ kJ/kg·K |
|---|---|---|---|---|---|---|---|---|---|---|---|
| 200 | 199.97 | 0.3363 | 142.56 | 1707.0 | 1.29559 | 580 | 586.04 | 14.38 | 419.55 | 115.7 | 2.37348 |
| 210 | 209.97 | 0.3987 | 149.69 | 1512.0 | 1.34444 | 590 | 596.52 | 15.31 | 427.15 | 110.6 | 2.39140 |
| 220 | 219.97 | 0.4690 | 156.82 | 1346.0 | 1.39105 | 600 | 607.02 | 16.28 | 434.78 | 105.8 | 2.40902 |
| 230 | 230.02 | 0.5477 | 164.00 | 1205.0 | 1.43557 | 610 | 617.53 | 17.30 | 442.42 | 101.2 | 2.42644 |
| 240 | 240.02 | 0.6355 | 171.13 | 1084.0 | 1.47824 | 620 | 628.07 | 18.36 | 450.09 | 96.92 | 2.44356 |
| 250 | 250.05 | 0.7329 | 178.28 | 979.0 | 1.51917 | 630 | 638.63 | 19.84 | 457.78 | 92.84 | 2.46048 |
| 260 | 260.09 | 0.8405 | 185.45 | 887.8 | 1.55848 | 640 | 649.22 | 20.64 | 465.50 | 88.99 | 2.47716 |
| 270 | 270.11 | 0.9590 | 192.60 | 808.0 | 1.59634 | 650 | 659.84 | 21.86 | 473.25 | 85.34 | 2.49364 |
| 280 | 280.13 | 1.0889 | 199.75 | 738.0 | 1.63279 | 660 | 670.47 | 23.13 | 481.01 | 81.89 | 2.50985 |
| 285 | 285.14 | 1.1584 | 203.33 | 706.1 | 1.65055 | 670 | 681.14 | 24.46 | 488.81 | 78.61 | 2.52589 |
| 290 | 290.16 | 1.2311 | 206.91 | 676.1 | 1.66802 | 680 | 691.82 | 25.85 | 496.62 | 75.50 | 2.54175 |
| 295 | 295.17 | 1.3068 | 210.49 | 647.9 | 1.68515 | 690 | 702.52 | 27.29 | 504.45 | 72.56 | 2.55731 |
| 298 | 298.18 | 1.3543 | 212.64 | 631.9 | 1.69528 | 700 | 713.27 | 28.80 | 512.33 | 69.76 | 2.57277 |
| 300 | 300.19 | 1.3860 | 214.07 | 621.2 | 1.70203 | 710 | 724.04 | 30.38 | 520.23 | 67.07 | 2.58810 |
| 305 | 305.22 | 1.4686 | 217.67 | 596.0 | 1.71865 | 720 | 734.82 | 32.02 | 528.14 | 64.53 | 2.60319 |
| 310 | 310.24 | 1.5546 | 221.25 | 572.3 | 1.73498 | 730 | 745.62 | 33.72 | 536.07 | 62.13 | 2.61803 |
| 315 | 315.27 | 1.6442 | 224.85 | 549.8 | 1.75106 | 740 | 756.44 | 35.50 | 544.02 | 59.82 | 2.63280 |
| 320 | 320.29 | 1.7375 | 228.42 | 528.6 | 1.76690 | 750 | 767.29 | 37.35 | 551.99 | 57.63 | 2.64737 |
| 325 | 325.31 | 1.8345 | 232.02 | 508.4 | 1.78249 | 760 | 778.18 | 39.27 | 560.01 | 55.54 | 2.66176 |
| 330 | 330.34 | 1.9352 | 235.61 | 489.4 | 1.79783 | 780 | 800.03 | 43.35 | 576.12 | 51.64 | 2.69013 |
| 340 | 340.42 | 2.149 | 242.82 | 454.1 | 1.82790 | 800 | 821.95 | 47.75 | 592.30 | 48.08 | 2.71787 |
| 350 | 350.49 | 2.379 | 250.02 | 422.2 | 1.85708 | 820 | 843.98 | 52.59 | 608.59 | 44.84 | 2.74504 |
| 360 | 360.58 | 2.626 | 257.24 | 393.4 | 1.88543 | 840 | 866.08 | 57.60 | 624.95 | 41.85 | 2.77170 |
| 370 | 370.67 | 2.892 | 264.46 | 367.2 | 1.91313 | 860 | 888.27 | 63.09 | 641.40 | 39.12 | 2.79783 |
| 380 | 380.77 | 3.176 | 271.69 | 343.4 | 1.94001 | 880 | 910.56 | 68.98 | 657.95 | 36.61 | 2.82344 |
| 390 | 390.88 | 3.481 | 278.93 | 321.5 | 1.96633 | 900 | 932.93 | 75.29 | 674.58 | 34.31 | 2.84856 |
| 400 | 400.98 | 3.806 | 286.16 | 301.6 | 1.99194 | 920 | 955.38 | 82.05 | 691.28 | 32.18 | 2.87324 |
| 410 | 411.12 | 4.153 | 293.43 | 283.3 | 2.01699 | 940 | 977.92 | 89.28 | 708.08 | 30.22 | 2.89748 |
| 420 | 421.26 | 4.522 | 300.69 | 266.6 | 2.04142 | 960 | 1000.55 | 97.00 | 725.02 | 28.40 | 2.92128 |
| 430 | 431.43 | 4.915 | 307.99 | 251.1 | 2.06533 | 980 | 1023.25 | 105.2 | 741.98 | 26.73 | 2.94468 |
| 440 | 441.61 | 5.332 | 315.30 | 236.8 | 2.08870 | 1000 | 1046.04 | 114.0 | 758.94 | 25.17 | 2.96770 |
| 450 | 451.80 | 5.775 | 322.62 | 223.6 | 2.11161 | 1020 | 1068.89 | 123.4 | 776.10 | 23.72 | 2.99034 |
| 460 | 462.02 | 6.245 | 329.97 | 211.4 | 2.13407 | 1040 | 1091.85 | 133.3 | 793.36 | 23.29 | 3.01260 |
| 470 | 472.24 | 6.742 | 337.32 | 200.1 | 2.15604 | 1060 | 1114.86 | 143.9 | 810.62 | 21.14 | 3.03449 |
| 480 | 482.49 | 7.268 | 344.70 | 189.5 | 2.17760 | 1080 | 1137.89 | 155.2 | 827.88 | 19.98 | 3.05608 |
| 490 | 492.74 | 7.824 | 352.08 | 179.7 | 2.19876 | 1100 | 1161.07 | 167.1 | 845.33 | 18.896 | 3.07732 |
| 500 | 503.02 | 8.411 | 359.49 | 170.6 | 2.21952 | 1120 | 1184.28 | 179.7 | 862.79 | 17.886 | 3.09825 |
| 510 | 513.32 | 9.031 | 366.92 | 162.1 | 2.23993 | 1140 | 1207.57 | 193.1 | 880.35 | 16.946 | 3.11883 |
| 520 | 523.63 | 9.684 | 374.36 | 154.1 | 2.25997 | 1160 | 1230.92 | 207.2 | 897.91 | 16.064 | 3.13916 |
| 530 | 533.98 | 10.37 | 381.84 | 146.7 | 2.27967 | 1180 | 1254.34 | 222.2 | 915.57 | 15.241 | 3.15916 |
| 540 | 544.35 | 11.10 | 389.34 | 139.7 | 2.29906 | 1200 | 1277.79 | 238.0 | 933.33 | 14.470 | 3.17888 |
| 550 | 555.74 | 11.86 | 396.86 | 133.1 | 2.31809 | 1220 | 1301.31 | 254.7 | 951.09 | 13.747 | 3.19834 |
| 560 | 565.17 | 12.66 | 404.42 | 127.0 | 2.33685 | 1240 | 1324.93 | 272.3 | 968.95 | 13.069 | 3.21751 |
| 570 | 575.59 | 13.50 | 411.97 | 121.2 | 2.35531 | | | | | | |

## TABLE A–17

Ideal-gas properties of air (*Concluded*)

| T K | h kJ/kg | $P_r$ | u kJ/kg | $v_r$ | $s°$ kJ/kg·K | T K | h kJ/kg | $P_r$ | u kJ/kg | $v_r$ | $s°$ kJ/kg·K |
|---|---|---|---|---|---|---|---|---|---|---|---|
| 1260 | 1348.55 | 290.8 | 986.90 | 12.435 | 3.23638 | 1600 | 1757.57 | 791.2 | 1298.30 | 5.804 | 3.52364 |
| 1280 | 1372.24 | 310.4 | 1004.76 | 11.835 | 3.25510 | 1620 | 1782.00 | 834.1 | 1316.96 | 5.574 | 3.53879 |
| 1300 | 1395.97 | 330.9 | 1022.82 | 11.275 | 3.27345 | 1640 | 1806.46 | 878.9 | 1335.72 | 5.355 | 3.55381 |
| 1320 | 1419.76 | 352.5 | 1040.88 | 10.747 | 3.29160 | 1660 | 1830.96 | 925.6 | 1354.48 | 5.147 | 3.56867 |
| 1340 | 1443.60 | 375.3 | 1058.94 | 10.247 | 3.30959 | 1680 | 1855.50 | 974.2 | 1373.24 | 4.949 | 3.58335 |
| 1360 | 1467.49 | 399.1 | 1077.10 | 9.780 | 3.32724 | 1700 | 1880.1 | 1025 | 1392.7 | 4.761 | 3.5979 |
| 1380 | 1491.44 | 424.2 | 1095.26 | 9.337 | 3.34474 | 1750 | 1941.6 | 1161 | 1439.8 | 4.328 | 3.6336 |
| 1400 | 1515.42 | 450.5 | 1113.52 | 8.919 | 3.36200 | 1800 | 2003.3 | 1310 | 1487.2 | 3.994 | 3.6684 |
| 1420 | 1539.44 | 478.0 | 1131.77 | 8.526 | 3.37901 | 1850 | 2065.3 | 1475 | 1534.9 | 3.601 | 3.7023 |
| 1440 | 1563.51 | 506.9 | 1150.13 | 8.153 | 3.39586 | 1900 | 2127.4 | 1655 | 1582.6 | 3.295 | 3.7354 |
| 1460 | 1587.63 | 537.1 | 1168.49 | 7.801 | 3.41247 | 1950 | 2189.7 | 1852 | 1630.6 | 3.022 | 3.7677 |
| 1480 | 1611.79 | 568.8 | 1186.95 | 7.468 | 3.42892 | 2000 | 2252.1 | 2068 | 1678.7 | 2.776 | 3.7994 |
| 1500 | 1635.97 | 601.9 | 1205.41 | 7.152 | 3.44516 | 2050 | 2314.6 | 2303 | 1726.8 | 2.555 | 3.8303 |
| 1520 | 1660.23 | 636.5 | 1223.87 | 6.854 | 3.46120 | 2100 | 2377.7 | 2559 | 1775.3 | 2.356 | 3.8605 |
| 1540 | 1684.51 | 672.8 | 1242.43 | 6.569 | 3.47712 | 2150 | 2440.3 | 2837 | 1823.8 | 2.175 | 3.8901 |
| 1560 | 1708.82 | 710.5 | 1260.99 | 6.301 | 3.49276 | 2200 | 2503.2 | 3138 | 1872.4 | 2.012 | 3.9191 |
| 1580 | 1733.17 | 750.0 | 1279.65 | 6.046 | 3.50829 | 2250 | 2566.4 | 3464 | 1921.3 | 1.864 | 3.9474 |

*Note:* The properties $P_r$ (relative pressure) and $v_r$ (relative specific volume) are dimensionless quantities used in the analysis of isentropic processes, and should not be confused with the properties pressure and specific volume.

*Source of Data:* Kenneth Wark, *Thermodynamics*, 4th ed. (New York: McGraw-Hill, 1983), pp. 785–86, table A–5. Originally published in J. H. Keenan and J. Kaye, *Gas Tables* (New York: John Wiley & Sons, 1948).

## TABLE A–18

Ideal-gas properties of nitrogen, $N_2$

| $T$ K | $\bar{h}$ kJ/kmol | $\bar{u}$ kJ/kmol | $\bar{s}°$ kJ/kmol·K | $T$ K | $\bar{h}$ kJ/kmol | $\bar{u}$ kJ/kmol | $\bar{s}°$ kJ/kmol·K |
|---|---|---|---|---|---|---|---|
| 0 | 0 | 0 | 0 | 600 | 17,563 | 12,574 | 212.066 |
| 220 | 6,391 | 4,562 | 182.639 | 610 | 17,864 | 12,792 | 212.564 |
| 230 | 6,683 | 4,770 | 183.938 | 620 | 18,166 | 13,011 | 213.055 |
| 240 | 6,975 | 4,979 | 185.180 | 630 | 18,468 | 13,230 | 213.541 |
| 250 | 7,266 | 5,188 | 186.370 | 640 | 18,772 | 13,450 | 214.018 |
| 260 | 7,558 | 5,396 | 187.514 | 650 | 19,075 | 13,671 | 214.489 |
| 270 | 7,849 | 5,604 | 188.614 | 660 | 19,380 | 13,892 | 214.954 |
| 280 | 8,141 | 5,813 | 189.673 | 670 | 19,685 | 14,114 | 215.413 |
| 290 | 8,432 | 6,021 | 190.695 | 680 | 19,991 | 14,337 | 215.866 |
| 298 | 8,669 | 6,190 | 191.502 | 690 | 20,297 | 14,560 | 216.314 |
| 300 | 8,723 | 6,229 | 191.682 | 700 | 20,604 | 14,784 | 216.756 |
| 310 | 9,014 | 6,437 | 192.638 | 710 | 20,912 | 15,008 | 217.192 |
| 320 | 9,306 | 6,645 | 193.562 | 720 | 21,220 | 15,234 | 217.624 |
| 330 | 9,597 | 6,853 | 194.459 | 730 | 21,529 | 15,460 | 218.059 |
| 340 | 9,888 | 7,061 | 195.328 | 740 | 21,839 | 15,686 | 218.472 |
| 350 | 10,180 | 7,270 | 196.173 | 750 | 22,149 | 15,913 | 218.889 |
| 360 | 10,471 | 7,478 | 196.995 | 760 | 22,460 | 16,141 | 219.301 |
| 370 | 10,763 | 7,687 | 197.794 | 770 | 22,772 | 16,370 | 219.709 |
| 380 | 11,055 | 7,895 | 198.572 | 780 | 23,085 | 16,599 | 220.113 |
| 390 | 11,347 | 8,104 | 199.331 | 790 | 23,398 | 16,830 | 220.512 |
| 400 | 11,640 | 8,314 | 200.071 | 800 | 23,714 | 17,061 | 220.907 |
| 410 | 11,932 | 8,523 | 200.794 | 810 | 24,027 | 17,292 | 221.298 |
| 420 | 12,225 | 8,733 | 201.499 | 820 | 24,342 | 17,524 | 221.684 |
| 430 | 12,518 | 8,943 | 202.189 | 830 | 24,658 | 17,757 | 222.067 |
| 440 | 12,811 | 9,153 | 202.863 | 840 | 24,974 | 17,990 | 222.447 |
| 450 | 13,105 | 9,363 | 203.523 | 850 | 25,292 | 18,224 | 222.822 |
| 460 | 13,399 | 9,574 | 204.170 | 860 | 25,610 | 18,459 | 223.194 |
| 470 | 13,693 | 9,786 | 204.803 | 870 | 25,928 | 18,695 | 223.562 |
| 480 | 13,988 | 9,997 | 205.424 | 880 | 26,248 | 18,931 | 223.927 |
| 490 | 14,285 | 10,210 | 206.033 | 890 | 26,568 | 19,168 | 224.288 |
| 500 | 14,581 | 10,423 | 206.630 | 900 | 26,890 | 19,407 | 224.647 |
| 510 | 14,876 | 10,635 | 207.216 | 910 | 27,210 | 19,644 | 225.002 |
| 520 | 15,172 | 10,848 | 207.792 | 920 | 27,532 | 19,883 | 225.353 |
| 530 | 15,469 | 11,062 | 208.358 | 930 | 27,854 | 20,122 | 225.701 |
| 540 | 15,766 | 11,277 | 208.914 | 940 | 28,178 | 20,362 | 226.047 |
| 550 | 16,064 | 11,492 | 209.461 | 950 | 28,501 | 20,603 | 226.389 |
| 560 | 16,363 | 11,707 | 209.999 | 960 | 28,826 | 20,844 | 226.728 |
| 570 | 16,662 | 11,923 | 210.528 | 970 | 29,151 | 21,086 | 227.064 |
| 580 | 16,962 | 12,139 | 211.049 | 980 | 29,476 | 21,328 | 227.398 |
| 590 | 17,262 | 12,356 | 211.562 | 990 | 29,803 | 21,571 | 227.728 |

## TABLE A–18

Ideal-gas properties of nitrogen, $N_2$ (*Concluded*)

| T K | $\bar{h}$ kJ/kmol | $\bar{u}$ kJ/kmol | $\bar{s}°$ kJ/kmol·K | T K | $\bar{h}$ kJ/kmol | $\bar{u}$ kJ/kmol | $\bar{s}°$ kJ/kmol·K |
|---|---|---|---|---|---|---|---|
| 1000 | 30,129 | 21,815 | 228.057 | 1760 | 56,227 | 41,594 | 247.396 |
| 1020 | 30,784 | 22,304 | 228.706 | 1780 | 56,938 | 42,139 | 247.798 |
| 1040 | 31,442 | 22,795 | 229.344 | 1800 | 57,651 | 42,685 | 248.195 |
| 1060 | 32,101 | 23,288 | 229.973 | 1820 | 58,363 | 43,231 | 248.589 |
| 1080 | 32,762 | 23,782 | 230.591 | 1840 | 59,075 | 43,777 | 248.979 |
| 1100 | 33,426 | 24,280 | 231.199 | 1860 | 59,790 | 44,324 | 249.365 |
| 1120 | 34,092 | 24,780 | 231.799 | 1880 | 60,504 | 44,873 | 249.748 |
| 1140 | 34,760 | 25,282 | 232.391 | 1900 | 61,220 | 45,423 | 250.128 |
| 1160 | 35,430 | 25,786 | 232.973 | 1920 | 61,936 | 45,973 | 250.502 |
| 1180 | 36,104 | 26,291 | 233.549 | 1940 | 62,654 | 46,524 | 250.874 |
| 1200 | 36,777 | 26,799 | 234.115 | 1960 | 63,381 | 47,075 | 251.242 |
| 1220 | 37,452 | 27,308 | 234.673 | 1980 | 64,090 | 47,627 | 251.607 |
| 1240 | 38,129 | 27,819 | 235.223 | 2000 | 64,810 | 48,181 | 251.969 |
| 1260 | 38,807 | 28,331 | 235.766 | 2050 | 66,612 | 49,567 | 252.858 |
| 1280 | 39,488 | 28,845 | 236.302 | 2100 | 68,417 | 50,957 | 253.726 |
| 1300 | 40,170 | 29,361 | 236.831 | 2150 | 70,226 | 52,351 | 254.578 |
| 1320 | 40,853 | 29,378 | 237.353 | 2200 | 72,040 | 53,749 | 255.412 |
| 1340 | 41,539 | 30,398 | 237.867 | 2250 | 73,856 | 55,149 | 256.227 |
| 1360 | 42,227 | 30,919 | 238.376 | 2300 | 75,676 | 56,553 | 257.027 |
| 1380 | 42,915 | 31,441 | 238.878 | 2350 | 77,496 | 57,958 | 257.810 |
| 1400 | 43,605 | 31,964 | 239.375 | 2400 | 79,320 | 59,366 | 258.580 |
| 1420 | 44,295 | 32,489 | 239.865 | 2450 | 81,149 | 60,779 | 259.332 |
| 1440 | 44,988 | 33,014 | 240.350 | 2500 | 82,981 | 62,195 | 260.073 |
| 1460 | 45,682 | 33,543 | 240.827 | 2550 | 84,814 | 63,613 | 260.799 |
| 1480 | 46,377 | 34,071 | 241.301 | 2600 | 86,650 | 65,033 | 261.512 |
| 1500 | 47,073 | 34,601 | 241.768 | 2650 | 88,488 | 66,455 | 262.213 |
| 1520 | 47,771 | 35,133 | 242.228 | 2700 | 90,328 | 67,880 | 262.902 |
| 1540 | 48,470 | 35,665 | 242.685 | 2750 | 92,171 | 69,306 | 263.577 |
| 1560 | 49,168 | 36,197 | 243.137 | 2800 | 94,014 | 70,734 | 264.241 |
| 1580 | 49,869 | 36,732 | 243.585 | 2850 | 95,859 | 72,163 | 264.895 |
| 1600 | 50,571 | 37,268 | 244.028 | 2900 | 97,705 | 73,593 | 265.538 |
| 1620 | 51,275 | 37,806 | 244.464 | 2950 | 99,556 | 75,028 | 266.170 |
| 1640 | 51,980 | 38,344 | 244.896 | 3000 | 101,407 | 76,464 | 266.793 |
| 1660 | 52,686 | 38,884 | 245.324 | 3050 | 103,260 | 77,902 | 267.404 |
| 1680 | 53,393 | 39,424 | 245.747 | 3100 | 105,115 | 79,341 | 268.007 |
| 1700 | 54,099 | 39,965 | 246.166 | 3150 | 106,972 | 80,782 | 268.601 |
| 1720 | 54,807 | 40,507 | 246.580 | 3200 | 108,830 | 82,224 | 269.186 |
| 1740 | 55,516 | 41,049 | 246.990 | 3250 | 110,690 | 83,668 | 269.763 |

*Source of Data:* Tables A–18 through A–25 are adapted from Kenneth Wark, *Thermodynamics,* 4th ed. (New York: McGraw-Hill, 1983), pp. 787–98. Originally published in JANAF, *Thermochemical Tables,* NSRDS-NBS-37, 1971.

## TABLE A–19

Ideal-gas properties of oxygen, $O_2$

| $T$ K | $\bar{h}$ kJ/kmol | $\bar{u}$ kJ/kmol | $\bar{s}°$ kJ/kmol·K | $T$ K | $\bar{h}$ kJ/kmol | $\bar{u}$ kJ/kmol | $\bar{s}°$ kJ/kmol·K |
|---|---|---|---|---|---|---|---|
| 0 | 0 | 0 | 0 | 600 | 17,929 | 12,940 | 226.346 |
| 220 | 6,404 | 4,575 | 196.171 | 610 | 18,250 | 13,178 | 226.877 |
| 230 | 6,694 | 4,782 | 197.461 | 620 | 18,572 | 13,417 | 227.400 |
| 240 | 6,984 | 4,989 | 198.696 | 630 | 18,895 | 13,657 | 227.918 |
| 250 | 7,275 | 5,197 | 199.885 | 640 | 19,219 | 13,898 | 228.429 |
| 260 | 7,566 | 5,405 | 201.027 | 650 | 19,544 | 14,140 | 228.932 |
| 270 | 7,858 | 5,613 | 202.128 | 660 | 19,870 | 14,383 | 229.430 |
| 280 | 8,150 | 5,822 | 203.191 | 670 | 20,197 | 14,626 | 229.920 |
| 290 | 8,443 | 6,032 | 204.218 | 680 | 20,524 | 14,871 | 230.405 |
| 298 | 8,682 | 6,203 | 205.033 | 690 | 20,854 | 15,116 | 230.885 |
| 300 | 8,736 | 6,242 | 205.213 | 700 | 21,184 | 15,364 | 231.358 |
| 310 | 9,030 | 6,453 | 206.177 | 710 | 21,514 | 15,611 | 231.827 |
| 320 | 9,325 | 6,664 | 207.112 | 720 | 21,845 | 15,859 | 232.291 |
| 330 | 9,620 | 6,877 | 208.020 | 730 | 22,177 | 16,107 | 232.748 |
| 340 | 9,916 | 7,090 | 208.904 | 740 | 22,510 | 16,357 | 233.201 |
| 350 | 10,213 | 7,303 | 209.765 | 750 | 22,844 | 16,607 | 233.649 |
| 360 | 10,511 | 7,518 | 210.604 | 760 | 23,178 | 16,859 | 234.091 |
| 370 | 10,809 | 7,733 | 211.423 | 770 | 23,513 | 17,111 | 234.528 |
| 380 | 11,109 | 7,949 | 212.222 | 780 | 23,850 | 17,364 | 234.960 |
| 390 | 11,409 | 8,166 | 213.002 | 790 | 24,186 | 17,618 | 235.387 |
| 400 | 11,711 | 8,384 | 213.765 | 800 | 24,523 | 17,872 | 235.810 |
| 410 | 12,012 | 8,603 | 214.510 | 810 | 24,861 | 18,126 | 236.230 |
| 420 | 12,314 | 8,822 | 215.241 | 820 | 25,199 | 18,382 | 236.644 |
| 430 | 12,618 | 9,043 | 215.955 | 830 | 25,537 | 18,637 | 237.055 |
| 440 | 12,923 | 9,264 | 216.656 | 840 | 25,877 | 18,893 | 237.462 |
| 450 | 13,228 | 9,487 | 217.342 | 850 | 26,218 | 19,150 | 237.864 |
| 460 | 13,525 | 9,710 | 218.016 | 860 | 26,559 | 19,408 | 238.264 |
| 470 | 13,842 | 9,935 | 218.676 | 870 | 26,899 | 19,666 | 238.660 |
| 480 | 14,151 | 10,160 | 219.326 | 880 | 27,242 | 19,925 | 239.051 |
| 490 | 14,460 | 10,386 | 219.963 | 890 | 27,584 | 20,185 | 239.439 |
| 500 | 14,770 | 10,614 | 220.589 | 900 | 27,928 | 20,445 | 239.823 |
| 510 | 15,082 | 10,842 | 221.206 | 910 | 28,272 | 20,706 | 240.203 |
| 520 | 15,395 | 11,071 | 221.812 | 920 | 28,616 | 20,967 | 240.580 |
| 530 | 15,708 | 11,301 | 222.409 | 930 | 28,960 | 21,228 | 240.953 |
| 540 | 16,022 | 11,533 | 222.997 | 940 | 29,306 | 21,491 | 241.323 |
| 550 | 16,338 | 11,765 | 223.576 | 950 | 29,652 | 21,754 | 241.689 |
| 560 | 16,654 | 11,998 | 224.146 | 960 | 29,999 | 22,017 | 242.052 |
| 570 | 16,971 | 12,232 | 224.708 | 970 | 30,345 | 22,280 | 242.411 |
| 580 | 17,290 | 12,467 | 225.262 | 980 | 30,692 | 22,544 | 242.768 |
| 590 | 17,609 | 12,703 | 225.808 | 990 | 31,041 | 22,809 | 242.120 |

## TABLE A-19

Ideal-gas properties of oxygen, $O_2$ (*Concluded*)

| $T$ K | $\bar{h}$ kJ/kmol | $\bar{u}$ kJ/kmol | $\bar{s}°$ kJ/kmol·K | $T$ K | $\bar{h}$ kJ/kmol | $\bar{u}$ kJ/kmol | $\bar{s}°$ kJ/kmol·K |
|---|---|---|---|---|---|---|---|
| 1000 | 31,389 | 23,075 | 243.471 | 1760 | 58,880 | 44,247 | 263.861 |
| 1020 | 32,088 | 23,607 | 244.164 | 1780 | 59,624 | 44,825 | 264.283 |
| 1040 | 32,789 | 24,142 | 244.844 | 1800 | 60,371 | 45,405 | 264.701 |
| 1060 | 33,490 | 24,677 | 245.513 | 1820 | 61,118 | 45,986 | 265.113 |
| 1080 | 34,194 | 25,214 | 246.171 | 1840 | 61,866 | 46,568 | 265.521 |
| 1100 | 34,899 | 25,753 | 246.818 | 1860 | 62,616 | 47,151 | 265.925 |
| 1120 | 35,606 | 26,294 | 247.454 | 1880 | 63,365 | 47,734 | 266.326 |
| 1140 | 36,314 | 26,836 | 248.081 | 1900 | 64,116 | 48,319 | 266.722 |
| 1160 | 37,023 | 27,379 | 248.698 | 1920 | 64,868 | 48,904 | 267.115 |
| 1180 | 37,734 | 27,923 | 249.307 | 1940 | 65,620 | 49,490 | 267.505 |
| 1200 | 38,447 | 28,469 | 249.906 | 1960 | 66,374 | 50,078 | 267.891 |
| 1220 | 39,162 | 29,018 | 250.497 | 1980 | 67,127 | 50,665 | 268.275 |
| 1240 | 39,877 | 29,568 | 251.079 | 2000 | 67,881 | 51,253 | 268.655 |
| 1260 | 40,594 | 30,118 | 251.653 | 2050 | 69,772 | 52,727 | 269.588 |
| 1280 | 41,312 | 30,670 | 252.219 | 2100 | 71,668 | 54,208 | 270.504 |
| 1300 | 42,033 | 31,224 | 252.776 | 2150 | 73,573 | 55,697 | 271.399 |
| 1320 | 42,753 | 31,778 | 253.325 | 2200 | 75,484 | 57,192 | 272.278 |
| 1340 | 43,475 | 32,334 | 253.868 | 2250 | 77,397 | 58,690 | 273.136 |
| 1360 | 44,198 | 32,891 | 254.404 | 2300 | 79,316 | 60,193 | 273.891 |
| 1380 | 44,923 | 33,449 | 254.932 | 2350 | 81,243 | 61,704 | 274.809 |
| 1400 | 45,648 | 34,008 | 255.454 | 2400 | 83,174 | 63,219 | 275.625 |
| 1420 | 46,374 | 34,567 | 255.968 | 2450 | 85,112 | 64,742 | 276.424 |
| 1440 | 47,102 | 35,129 | 256.475 | 2500 | 87,057 | 66,271 | 277.207 |
| 1460 | 47,831 | 35,692 | 256.978 | 2550 | 89,004 | 67,802 | 277.979 |
| 1480 | 48,561 | 36,256 | 257.474 | 2600 | 90,956 | 69,339 | 278.738 |
| 1500 | 49,292 | 36,821 | 257.965 | 2650 | 92,916 | 70,883 | 279.485 |
| 1520 | 50,024 | 37,387 | 258.450 | 2700 | 94,881 | 72,433 | 280.219 |
| 1540 | 50,756 | 37,952 | 258.928 | 2750 | 96,852 | 73,987 | 280.942 |
| 1560 | 51,490 | 38,520 | 259.402 | 2800 | 98,826 | 75,546 | 281.654 |
| 1580 | 52,224 | 39,088 | 259.870 | 2850 | 100,808 | 77,112 | 282.357 |
| 1600 | 52,961 | 39,658 | 260.333 | 2900 | 102,793 | 78,682 | 283.048 |
| 1620 | 53,696 | 40,227 | 260.791 | 2950 | 104,785 | 80,258 | 283.728 |
| 1640 | 54,434 | 40,799 | 261.242 | 3000 | 106,780 | 81,837 | 284.399 |
| 1660 | 55,172 | 41,370 | 261.690 | 3050 | 108,778 | 83,419 | 285.060 |
| 1680 | 55,912 | 41,944 | 262.132 | 3100 | 110,784 | 85,009 | 285.713 |
| 1700 | 56,652 | 42,517 | 262.571 | 3150 | 112,795 | 86,601 | 286.355 |
| 1720 | 57,394 | 43,093 | 263.005 | 3200 | 114,809 | 88,203 | 286.989 |
| 1740 | 58,136 | 43,669 | 263.435 | 3250 | 116,827 | 89,804 | 287.614 |

## TABLE A–20

Ideal-gas properties of carbon dioxide, $CO_2$

| $T$ K | $\bar{h}$ kJ/kmol | $\bar{u}$ kJ/kmol | $\bar{s}°$ kJ/kmol·K | $T$ K | $\bar{h}$ kJ/kmol | $\bar{u}$ kJ/kmol | $\bar{s}°$ kJ/kmol·K |
|---|---|---|---|---|---|---|---|
| 0 | 0 | 0 | 0 | 600 | 22,280 | 17,291 | 243.199 |
| 220 | 6,601 | 4,772 | 202.966 | 610 | 22,754 | 17,683 | 243.983 |
| 230 | 6,938 | 5,026 | 204.464 | 620 | 23,231 | 18,076 | 244.758 |
| 240 | 7,280 | 5,285 | 205.920 | 630 | 23,709 | 18,471 | 245.524 |
| 250 | 7,627 | 5,548 | 207.337 | 640 | 24,190 | 18,869 | 246.282 |
| 260 | 7,979 | 5,817 | 208.717 | 650 | 24,674 | 19,270 | 247.032 |
| 270 | 8,335 | 6,091 | 210.062 | 660 | 25,160 | 19,672 | 247.773 |
| 280 | 8,697 | 6,369 | 211.376 | 670 | 25,648 | 20,078 | 248.507 |
| 290 | 9,063 | 6,651 | 212.660 | 680 | 26,138 | 20,484 | 249.233 |
| 298 | 9,364 | 6,885 | 213.685 | 690 | 26,631 | 20,894 | 249.952 |
| 300 | 9,431 | 6,939 | 213.915 | 700 | 27,125 | 21,305 | 250.663 |
| 310 | 9,807 | 7,230 | 215.146 | 710 | 27,622 | 21,719 | 251.368 |
| 320 | 10,186 | 7,526 | 216.351 | 720 | 28,121 | 22,134 | 252.065 |
| 330 | 10,570 | 7,826 | 217.534 | 730 | 28,622 | 22,522 | 252.755 |
| 340 | 10,959 | 8,131 | 218.694 | 740 | 29,124 | 22,972 | 253.439 |
| 350 | 11,351 | 8,439 | 219.831 | 750 | 29,629 | 23,393 | 254.117 |
| 360 | 11,748 | 8,752 | 220.948 | 760 | 30,135 | 23,817 | 254.787 |
| 370 | 12,148 | 9,068 | 222.044 | 770 | 30,644 | 24,242 | 255.452 |
| 380 | 12,552 | 9,392 | 223.122 | 780 | 31,154 | 24,669 | 256.110 |
| 390 | 12,960 | 9,718 | 224.182 | 790 | 31,665 | 25,097 | 256.762 |
| 400 | 13,372 | 10,046 | 225.225 | 800 | 32,179 | 25,527 | 257.408 |
| 410 | 13,787 | 10,378 | 226.250 | 810 | 32,694 | 25,959 | 258.048 |
| 420 | 14,206 | 10,714 | 227.258 | 820 | 33,212 | 26,394 | 258.682 |
| 430 | 14,628 | 11,053 | 228.252 | 830 | 33,730 | 26,829 | 259.311 |
| 440 | 15,054 | 11,393 | 229.230 | 840 | 34,251 | 27,267 | 259.934 |
| 450 | 15,483 | 11,742 | 230.194 | 850 | 34,773 | 27,706 | 260.551 |
| 460 | 15,916 | 12,091 | 231.144 | 860 | 35,296 | 28,125 | 261.164 |
| 470 | 16,351 | 12,444 | 232.080 | 870 | 35,821 | 28,588 | 261.770 |
| 480 | 16,791 | 12,800 | 233.004 | 880 | 36,347 | 29,031 | 262.371 |
| 490 | 17,232 | 13,158 | 233.916 | 890 | 36,876 | 29,476 | 262.968 |
| 500 | 17,678 | 13,521 | 234.814 | 900 | 37,405 | 29,922 | 263.559 |
| 510 | 18,126 | 13,885 | 235.700 | 910 | 37,935 | 30,369 | 264.146 |
| 520 | 18,576 | 14,253 | 236.575 | 920 | 38,467 | 30,818 | 264.728 |
| 530 | 19,029 | 14,622 | 237.439 | 930 | 39,000 | 31,268 | 265.304 |
| 540 | 19,485 | 14,996 | 238.292 | 940 | 39,535 | 31,719 | 265.877 |
| 550 | 19,945 | 15,372 | 239.135 | 950 | 40,070 | 32,171 | 266.444 |
| 560 | 20,407 | 15,751 | 239.962 | 960 | 40,607 | 32,625 | 267.007 |
| 570 | 20,870 | 16,131 | 240.789 | 970 | 41,145 | 33,081 | 267.566 |
| 580 | 21,337 | 16,515 | 241.602 | 980 | 41,685 | 33,537 | 268.119 |
| 590 | 21,807 | 16,902 | 242.405 | 990 | 42,226 | 33,995 | 268.670 |

## TABLE A-20

Ideal-gas properties of carbon dioxide, $CO_2$ (Concluded)

| T<br>K | $\bar{h}$<br>kJ/kmol | $\bar{u}$<br>kJ/kmol | $\bar{s}°$<br>kJ/kmol·K | T<br>K | $\bar{h}$<br>kJ/kmol | $\bar{u}$<br>kJ/kmol | $\bar{s}°$<br>kJ/kmol·K |
|---|---|---|---|---|---|---|---|
| 1000 | 42,769 | 34,455 | 269.215 | 1760 | 86,420 | 71,787 | 301.543 |
| 1020 | 43,859 | 35,378 | 270.293 | 1780 | 87,612 | 72,812 | 302.217 |
| 1040 | 44,953 | 36,306 | 271.354 | 1800 | 88,806 | 73,840 | 302.884 |
| 1060 | 46,051 | 37,238 | 272.400 | 1820 | 90,000 | 74,868 | 303.544 |
| 1080 | 47,153 | 38,174 | 273.430 | 1840 | 91,196 | 75,897 | 304.198 |
| 1100 | 48,258 | 39,112 | 274.445 | 1860 | 92,394 | 76,929 | 304.845 |
| 1120 | 49,369 | 40,057 | 275.444 | 1880 | 93,593 | 77,962 | 305.487 |
| 1140 | 50,484 | 41,006 | 276.430 | 1900 | 94,793 | 78,996 | 306.122 |
| 1160 | 51,602 | 41,957 | 277.403 | 1920 | 95,995 | 80,031 | 306.751 |
| 1180 | 52,724 | 42,913 | 278.361 | 1940 | 97,197 | 81,067 | 307.374 |
| 1200 | 53,848 | 43,871 | 297.307 | 1960 | 98,401 | 82,105 | 307.992 |
| 1220 | 54,977 | 44,834 | 280.238 | 1980 | 99,606 | 83,144 | 308.604 |
| 1240 | 56,108 | 45,799 | 281.158 | 2000 | 100,804 | 84,185 | 309.210 |
| 1260 | 57,244 | 46,768 | 282.066 | 2050 | 103,835 | 86,791 | 310.701 |
| 1280 | 58,381 | 47,739 | 282.962 | 2100 | 106,864 | 89,404 | 312.160 |
| 1300 | 59,522 | 48,713 | 283.847 | 2150 | 109,898 | 92,023 | 313.589 |
| 1320 | 60,666 | 49,691 | 284.722 | 2200 | 112,939 | 94,648 | 314.988 |
| 1340 | 61,813 | 50,672 | 285.586 | 2250 | 115,984 | 97,277 | 316.356 |
| 1360 | 62,963 | 51,656 | 286.439 | 2300 | 119,035 | 99,912 | 317.695 |
| 1380 | 64,116 | 52,643 | 287.283 | 2350 | 122,091 | 102,552 | 319.011 |
| 1400 | 65,271 | 53,631 | 288.106 | 2400 | 125,152 | 105,197 | 320.302 |
| 1420 | 66,427 | 54,621 | 288.934 | 2450 | 128,219 | 107,849 | 321.566 |
| 1440 | 67,586 | 55,614 | 289.743 | 2500 | 131,290 | 110,504 | 322.808 |
| 1460 | 68,748 | 56,609 | 290.542 | 2550 | 134,368 | 113,166 | 324.026 |
| 1480 | 66,911 | 57,606 | 291.333 | 2600 | 137,449 | 115,832 | 325.222 |
| 1500 | 71,078 | 58,606 | 292.114 | 2650 | 140,533 | 118,500 | 326.396 |
| 1520 | 72,246 | 59,609 | 292.888 | 2700 | 143,620 | 121,172 | 327.549 |
| 1540 | 73,417 | 60,613 | 292.654 | 2750 | 146,713 | 123,849 | 328.684 |
| 1560 | 74,590 | 61,620 | 294.411 | 2800 | 149,808 | 126,528 | 329.800 |
| 1580 | 76,767 | 62,630 | 295.161 | 2850 | 152,908 | 129,212 | 330.896 |
| 1600 | 76,944 | 63,741 | 295.901 | 2900 | 156,009 | 131,898 | 331.975 |
| 1620 | 78,123 | 64,653 | 296.632 | 2950 | 159,117 | 134,589 | 333.037 |
| 1640 | 79,303 | 65,668 | 297.356 | 3000 | 162,226 | 137,283 | 334.084 |
| 1660 | 80,486 | 66,592 | 298.072 | 3050 | 165,341 | 139,982 | 335.114 |
| 1680 | 81,670 | 67,702 | 298.781 | 3100 | 168,456 | 142,681 | 336.126 |
| 1700 | 82,856 | 68,721 | 299.482 | 3150 | 171,576 | 145,385 | 337.124 |
| 1720 | 84,043 | 69,742 | 300.177 | 3200 | 174,695 | 148,089 | 338.109 |
| 1740 | 85,231 | 70,764 | 300.863 | 3250 | 177,822 | 150,801 | 339.069 |

## TABLE A–21

Ideal-gas properties of carbon monoxide, CO

| $T$ | $\bar{h}$ | $\bar{u}$ | $\bar{s}°$ | $T$ | $\bar{h}$ | $\bar{u}$ | $\bar{s}°$ |
|---|---|---|---|---|---|---|---|
| K | kJ/kmol | kJ/kmol | kJ/kmol·K | K | kJ/kmol | kJ/kmol | kJ/kmol·K |
| 0 | 0 | 0 | 0 | 600 | 17,611 | 12,622 | 218.204 |
| 220 | 6,391 | 4,562 | 188.683 | 610 | 17,915 | 12,843 | 218.708 |
| 230 | 6,683 | 4,771 | 189.980 | 620 | 18,221 | 13,066 | 219.205 |
| 240 | 6,975 | 4,979 | 191.221 | 630 | 18,527 | 13,289 | 219.695 |
| 250 | 7,266 | 5,188 | 192.411 | 640 | 18,833 | 13,512 | 220.179 |
| 260 | 7,558 | 5,396 | 193.554 | 650 | 19,141 | 13,736 | 220.656 |
| 270 | 7,849 | 5,604 | 194.654 | 660 | 19,449 | 13,962 | 221.127 |
| 280 | 8,140 | 5,812 | 195.713 | 670 | 19,758 | 14,187 | 221.592 |
| 290 | 8,432 | 6,020 | 196.735 | 680 | 20,068 | 14,414 | 222.052 |
| 298 | 8,669 | 6,190 | 197.543 | 690 | 20,378 | 14,641 | 222.505 |
| 300 | 8,723 | 6,229 | 197.723 | 700 | 20,690 | 14,870 | 222.953 |
| 310 | 9,014 | 6,437 | 198.678 | 710 | 21,002 | 15,099 | 223.396 |
| 320 | 9,306 | 6,645 | 199.603 | 720 | 21,315 | 15,328 | 223.833 |
| 330 | 9,597 | 6,854 | 200.500 | 730 | 21,628 | 15,558 | 224.265 |
| 340 | 9,889 | 7,062 | 201.371 | 740 | 21,943 | 15,789 | 224.692 |
| 350 | 10,181 | 7,271 | 202.217 | 750 | 22,258 | 16,022 | 225.115 |
| 360 | 10,473 | 7,480 | 203.040 | 760 | 22,573 | 16,255 | 225.533 |
| 370 | 10,765 | 7,689 | 203.842 | 770 | 22,890 | 16,488 | 225.947 |
| 380 | 11,058 | 7,899 | 204.622 | 780 | 23,208 | 16,723 | 226.357 |
| 390 | 11,351 | 8,108 | 205.383 | 790 | 23,526 | 16,957 | 226.762 |
| 400 | 11,644 | 8,319 | 206.125 | 800 | 23,844 | 17,193 | 227.162 |
| 410 | 11,938 | 8,529 | 206.850 | 810 | 24,164 | 17,429 | 227.559 |
| 420 | 12,232 | 8,740 | 207.549 | 820 | 24,483 | 17,665 | 227.952 |
| 430 | 12,526 | 8,951 | 208.252 | 830 | 24,803 | 17,902 | 228.339 |
| 440 | 12,821 | 9,163 | 208.929 | 840 | 25,124 | 18,140 | 228.724 |
| 450 | 13,116 | 9,375 | 209.593 | 850 | 25,446 | 18,379 | 229.106 |
| 460 | 13,412 | 9,587 | 210.243 | 860 | 25,768 | 18,617 | 229.482 |
| 470 | 13,708 | 9,800 | 210.880 | 870 | 26,091 | 18,858 | 229.856 |
| 480 | 14,005 | 10,014 | 211.504 | 880 | 26,415 | 19,099 | 230.227 |
| 490 | 14,302 | 10,228 | 212.117 | 890 | 26,740 | 19,341 | 230.593 |
| 500 | 14,600 | 10,443 | 212.719 | 900 | 27,066 | 19,583 | 230.957 |
| 510 | 14,898 | 10,658 | 213.310 | 910 | 27,392 | 19,826 | 231.317 |
| 520 | 15,197 | 10,874 | 213.890 | 920 | 27,719 | 20,070 | 231.674 |
| 530 | 15,497 | 11,090 | 214.460 | 930 | 28,046 | 20,314 | 232.028 |
| 540 | 15,797 | 11,307 | 215.020 | 940 | 28,375 | 20,559 | 232.379 |
| 550 | 16,097 | 11,524 | 215.572 | 950 | 28,703 | 20,805 | 232.727 |
| 560 | 16,399 | 11,743 | 216.115 | 960 | 29,033 | 21,051 | 233.072 |
| 570 | 16,701 | 11,961 | 216.649 | 970 | 29,362 | 21,298 | 233.413 |
| 580 | 17,003 | 12,181 | 217.175 | 980 | 29,693 | 21,545 | 233.752 |
| 590 | 17,307 | 12,401 | 217.693 | 990 | 30,024 | 21,793 | 234.088 |

## TABLE A–21

Ideal-gas properties of carbon monoxide, CO (*Concluded*)

| $T$ K | $\bar{h}$ kJ/kmol | $\bar{u}$ kJ/kmol | $\bar{s}°$ kJ/kmol·K | $T$ K | $\bar{h}$ kJ/kmol | $\bar{u}$ kJ/kmol | $\bar{s}°$ kJ/kmol·K |
|---|---|---|---|---|---|---|---|
| 1000 | 30,355 | 22,041 | 234.421 | 1760 | 56,756 | 42,123 | 253.991 |
| 1020 | 31,020 | 22,540 | 235.079 | 1780 | 57,473 | 42,673 | 254.398 |
| 1040 | 31,688 | 23,041 | 235.728 | 1800 | 58,191 | 43,225 | 254.797 |
| 1060 | 32,357 | 23,544 | 236.364 | 1820 | 58,910 | 43,778 | 255.194 |
| 1080 | 33,029 | 24,049 | 236.992 | 1840 | 59,629 | 44,331 | 255.587 |
| 1100 | 33,702 | 24,557 | 237.609 | 1860 | 60,351 | 44,886 | 255.976 |
| 1120 | 34,377 | 25,065 | 238.217 | 1880 | 61,072 | 45,441 | 256.361 |
| 1140 | 35,054 | 25,575 | 238.817 | 1900 | 61,794 | 45,997 | 256.743 |
| 1160 | 35,733 | 26,088 | 239.407 | 1920 | 62,516 | 46,552 | 257.122 |
| 1180 | 36,406 | 26,602 | 239.989 | 1940 | 63,238 | 47,108 | 257.497 |
| 1200 | 37,095 | 27,118 | 240.663 | 1960 | 63,961 | 47,665 | 257.868 |
| 1220 | 37,780 | 27,637 | 241.128 | 1980 | 64,684 | 48,221 | 258.236 |
| 1240 | 38,466 | 28,426 | 241.686 | 2000 | 65,408 | 48,780 | 258.600 |
| 1260 | 39,154 | 28,678 | 242.236 | 2050 | 67,224 | 50,179 | 259.494 |
| 1280 | 39,844 | 29,201 | 242.780 | 2100 | 69,044 | 51,584 | 260.370 |
| 1300 | 40,534 | 29,725 | 243.316 | 2150 | 70,864 | 52,988 | 261.226 |
| 1320 | 41,226 | 30,251 | 243.844 | 2200 | 72,688 | 54,396 | 262.065 |
| 1340 | 41,919 | 30,778 | 244.366 | 2250 | 74,516 | 55,809 | 262.887 |
| 1360 | 42,613 | 31,306 | 244.880 | 2300 | 76,345 | 57,222 | 263.692 |
| 1380 | 43,309 | 31,836 | 245.388 | 2350 | 78,178 | 58,640 | 264.480 |
| 1400 | 44,007 | 32,367 | 245.889 | 2400 | 80,015 | 60,060 | 265.253 |
| 1420 | 44,707 | 32,900 | 246.385 | 2450 | 81,852 | 61,482 | 266.012 |
| 1440 | 45,408 | 33,434 | 246.876 | 2500 | 83,692 | 62,906 | 266.755 |
| 1460 | 46,110 | 33,971 | 247.360 | 2550 | 85,537 | 64,335 | 267.485 |
| 1480 | 46,813 | 34,508 | 247.839 | 2600 | 87,383 | 65,766 | 268.202 |
| 1500 | 47,517 | 35,046 | 248.312 | 2650 | 89,230 | 67,197 | 268.905 |
| 1520 | 48,222 | 35,584 | 248.778 | 2700 | 91,077 | 68,628 | 269.596 |
| 1540 | 48,928 | 36,124 | 249.240 | 2750 | 92,930 | 70,066 | 270.285 |
| 1560 | 49,635 | 36,665 | 249.695 | 2800 | 94,784 | 71,504 | 270.943 |
| 1580 | 50,344 | 37,207 | 250.147 | 2850 | 96,639 | 72,945 | 271.602 |
| 1600 | 51,053 | 37,750 | 250.592 | 2900 | 98,495 | 74,383 | 272.249 |
| 1620 | 51,763 | 38,293 | 251.033 | 2950 | 100,352 | 75,825 | 272.884 |
| 1640 | 52,472 | 38,837 | 251.470 | 3000 | 102,210 | 77,267 | 273.508 |
| 1660 | 53,184 | 39,382 | 251.901 | 3050 | 104,073 | 78,715 | 274.123 |
| 1680 | 53,895 | 39,927 | 252.329 | 3100 | 105,939 | 80,164 | 274.730 |
| 1700 | 54,609 | 40,474 | 252.751 | 3150 | 107,802 | 81,612 | 275.326 |
| 1720 | 55,323 | 41,023 | 253.169 | 3200 | 109,667 | 83,061 | 275.914 |
| 1740 | 56,039 | 41,572 | 253.582 | 3250 | 111,534 | 84,513 | 276.494 |

## TABLE A–22

Ideal-gas properties of hydrogen, $H_2$

| T | $\bar{h}$ | $\bar{u}$ | $\bar{s}°$ | T | $\bar{h}$ | $\bar{u}$ | $\bar{s}°$ |
|---|---|---|---|---|---|---|---|
| K | kJ/kmol | kJ/kmol | kJ/kmol·K | K | kJ/kmol | kJ/kmol | kJ/kmol·K |
| 0 | 0 | 0 | 0 | 1440 | 42,808 | 30,835 | 177.410 |
| 260 | 7,370 | 5,209 | 126.636 | 1480 | 44,091 | 31,786 | 178.291 |
| 270 | 7,657 | 5,412 | 127.719 | 1520 | 45,384 | 32,746 | 179.153 |
| 280 | 7,945 | 5,617 | 128.765 | 1560 | 46,683 | 33,713 | 179.995 |
| 290 | 8,233 | 5,822 | 129.775 | 1600 | 47,990 | 34,687 | 180.820 |
| 298 | 8,468 | 5,989 | 130.574 | 1640 | 49,303 | 35,668 | 181.632 |
| 300 | 8,522 | 6,027 | 130.754 | 1680 | 50,622 | 36,654 | 182.428 |
| 320 | 9,100 | 6,440 | 132.621 | 1720 | 51,947 | 37,646 | 183.208 |
| 340 | 9,680 | 6,853 | 134.378 | 1760 | 53,279 | 38,645 | 183.973 |
| 360 | 10,262 | 7,268 | 136.039 | 1800 | 54,618 | 39,652 | 184.724 |
| 380 | 10,843 | 7,684 | 137.612 | 1840 | 55,962 | 40,663 | 185.463 |
| 400 | 11,426 | 8,100 | 139.106 | 1880 | 57,311 | 41,680 | 186.190 |
| 420 | 12,010 | 8,518 | 140.529 | 1920 | 58,668 | 42,705 | 186.904 |
| 440 | 12,594 | 8,936 | 141.888 | 1960 | 60,031 | 43,735 | 187.607 |
| 460 | 13,179 | 9,355 | 143.187 | 2000 | 61,400 | 44,771 | 188.297 |
| 480 | 13,764 | 9,773 | 144.432 | 2050 | 63,119 | 46,074 | 189.148 |
| 500 | 14,350 | 10,193 | 145.628 | 2100 | 64,847 | 47,386 | 189.979 |
| 520 | 14,935 | 10,611 | 146.775 | 2150 | 66,584 | 48,708 | 190.796 |
| 560 | 16,107 | 11,451 | 148.945 | 2200 | 68,328 | 50,037 | 191.598 |
| 600 | 17,280 | 12,291 | 150.968 | 2250 | 70,080 | 51,373 | 192.385 |
| 640 | 18,453 | 13,133 | 152.863 | 2300 | 71,839 | 52,716 | 193.159 |
| 680 | 19,630 | 13,976 | 154.645 | 2350 | 73,608 | 54,069 | 193.921 |
| 720 | 20,807 | 14,821 | 156.328 | 2400 | 75,383 | 55,429 | 194.669 |
| 760 | 21,988 | 15,669 | 157.923 | 2450 | 77,168 | 56,798 | 195.403 |
| 800 | 23,171 | 16,520 | 159.440 | 2500 | 78,960 | 58,175 | 196.125 |
| 840 | 24,359 | 17,375 | 160.891 | 2550 | 80,755 | 59,554 | 196.837 |
| 880 | 25,551 | 18,235 | 162.277 | 2600 | 82,558 | 60,941 | 197.539 |
| 920 | 26,747 | 19,098 | 163.607 | 2650 | 84,368 | 62,335 | 198.229 |
| 960 | 27,948 | 19,966 | 164.884 | 2700 | 86,186 | 63,737 | 198.907 |
| 1000 | 29,154 | 20,839 | 166.114 | 2750 | 88,008 | 65,144 | 199.575 |
| 1040 | 30,364 | 21,717 | 167.300 | 2800 | 89,838 | 66,558 | 200.234 |
| 1080 | 31,580 | 22,601 | 168.449 | 2850 | 91,671 | 67,976 | 200.885 |
| 1120 | 32,802 | 23,490 | 169.560 | 2900 | 93,512 | 69,401 | 201.527 |
| 1160 | 34,028 | 24,384 | 170.636 | 2950 | 95,358 | 70,831 | 202.157 |
| 1200 | 35,262 | 25,284 | 171.682 | 3000 | 97,211 | 72,268 | 202.778 |
| 1240 | 36,502 | 26,192 | 172.698 | 3050 | 99,065 | 73,707 | 203.391 |
| 1280 | 37,749 | 27,106 | 173.687 | 3100 | 100,926 | 75,152 | 203.995 |
| 1320 | 39,002 | 28,027 | 174.652 | 3150 | 102,793 | 76,604 | 204.592 |
| 1360 | 40,263 | 28,955 | 175.593 | 3200 | 104,667 | 78,061 | 205.181 |
| 1400 | 41,530 | 29,889 | 176.510 | 3250 | 106,545 | 79,523 | 205.765 |

## TABLE A–23

Ideal-gas properties of water vapor, $H_2O$

| $T$ K | $\bar{h}$ kJ/kmol | $\bar{u}$ kJ/kmol | $\bar{s}°$ kJ/kmol·K | $T$ K | $\bar{h}$ kJ/kmol | $\bar{u}$ kJ/kmol | $\bar{s}°$ kJ/kmol·K |
|---|---|---|---|---|---|---|---|
| 0 | 0 | 0 | 0 | 600 | 20,402 | 15,413 | 212.920 |
| 220 | 7,295 | 5,466 | 178.576 | 610 | 20,765 | 15,693 | 213.529 |
| 230 | 7,628 | 5,715 | 180.054 | 620 | 21,130 | 15,975 | 214.122 |
| 240 | 7,961 | 5,965 | 181.471 | 630 | 21,495 | 16,257 | 214.707 |
| 250 | 8,294 | 6,215 | 182.831 | 640 | 21,862 | 16,541 | 215.285 |
| 260 | 8,627 | 6,466 | 184.139 | 650 | 22,230 | 16,826 | 215.856 |
| 270 | 8,961 | 6,716 | 185.399 | 660 | 22,600 | 17,112 | 216.419 |
| 280 | 9,296 | 6,968 | 186.616 | 670 | 22,970 | 17,399 | 216.976 |
| 290 | 9,631 | 7,219 | 187.791 | 680 | 23,342 | 17,688 | 217.527 |
| 298 | 9,904 | 7,425 | 188.720 | 690 | 23,714 | 17,978 | 218.071 |
| 300 | 9,966 | 7,472 | 188.928 | 700 | 24,088 | 18,268 | 218.610 |
| 310 | 10,302 | 7,725 | 190.030 | 710 | 24,464 | 18,561 | 219.142 |
| 320 | 10,639 | 7,978 | 191.098 | 720 | 24,840 | 18,854 | 219.668 |
| 330 | 10,976 | 8,232 | 192.136 | 730 | 25,218 | 19,148 | 220.189 |
| 340 | 11,314 | 8,487 | 193.144 | 740 | 25,597 | 19,444 | 220.707 |
| 350 | 11,652 | 8,742 | 194.125 | 750 | 25,977 | 19,741 | 221.215 |
| 360 | 11,992 | 8,998 | 195.081 | 760 | 26,358 | 20,039 | 221.720 |
| 370 | 12,331 | 9,255 | 196.012 | 770 | 26,741 | 20,339 | 222.221 |
| 380 | 12,672 | 9,513 | 196.920 | 780 | 27,125 | 20,639 | 222.717 |
| 390 | 13,014 | 9,771 | 197.807 | 790 | 27,510 | 20,941 | 223.207 |
| 400 | 13,356 | 10,030 | 198.673 | 800 | 27,896 | 21,245 | 223.693 |
| 410 | 13,699 | 10,290 | 199.521 | 810 | 28,284 | 21,549 | 224.174 |
| 420 | 14,043 | 10,551 | 200.350 | 820 | 28,672 | 21,855 | 224.651 |
| 430 | 14,388 | 10,813 | 201.160 | 830 | 29,062 | 22,162 | 225.123 |
| 440 | 14,734 | 11,075 | 201.955 | 840 | 29,454 | 22,470 | 225.592 |
| 450 | 15,080 | 11,339 | 202.734 | 850 | 29,846 | 22,779 | 226.057 |
| 460 | 15,428 | 11,603 | 203.497 | 860 | 30,240 | 23,090 | 226.517 |
| 470 | 15,777 | 11,869 | 204.247 | 870 | 30,635 | 23,402 | 226.973 |
| 480 | 16,126 | 12,135 | 204.982 | 880 | 31,032 | 23,715 | 227.426 |
| 490 | 16,477 | 12,403 | 205.705 | 890 | 31,429 | 24,029 | 227.875 |
| 500 | 16,828 | 12,671 | 206.413 | 900 | 31,828 | 24,345 | 228.321 |
| 510 | 17,181 | 12,940 | 207.112 | 910 | 32,228 | 24,662 | 228.763 |
| 520 | 17,534 | 13,211 | 207.799 | 920 | 32,629 | 24,980 | 229.202 |
| 530 | 17,889 | 13,482 | 208.475 | 930 | 33,032 | 25,300 | 229.637 |
| 540 | 18,245 | 13,755 | 209.139 | 940 | 33,436 | 25,621 | 230.070 |
| 550 | 18,601 | 14,028 | 209.795 | 950 | 33,841 | 25,943 | 230.499 |
| 560 | 18,959 | 14,303 | 210.440 | 960 | 34,247 | 26,265 | 230.924 |
| 570 | 19,318 | 14,579 | 211.075 | 970 | 34,653 | 26,588 | 231.347 |
| 580 | 19,678 | 14,856 | 211.702 | 980 | 35,061 | 26,913 | 231.767 |
| 590 | 20,039 | 15,134 | 212.320 | 990 | 35,472 | 27,240 | 232.184 |

## TABLE A–23

Ideal-gas properties of water vapor, $H_2O$ (*Continued*)

| T | $\bar{h}$ | $\bar{u}$ | $\bar{s}°$ | T | $\bar{h}$ | $\bar{u}$ | $\bar{s}°$ |
|---|---|---|---|---|---|---|---|
| K | kJ/kmol | kJ/kmol | kJ/kmol·K | K | kJ/kmol | kJ/kmol | kJ/kmol·K |
| 1000 | 35,882 | 27,568 | 232.597 | 1760 | 70,535 | 55,902 | 258.151 |
| 1020 | 36,709 | 28,228 | 233.415 | 1780 | 71,523 | 56,723 | 258.708 |
| 1040 | 37,542 | 28,895 | 234.223 | 1800 | 72,513 | 57,547 | 259.262 |
| 1060 | 38,380 | 29,567 | 235.020 | 1820 | 73,507 | 58,375 | 259.811 |
| 1080 | 39,223 | 30,243 | 235.806 | 1840 | 74,506 | 59,207 | 260.357 |
| 1100 | 40,071 | 30,925 | 236.584 | 1860 | 75,506 | 60,042 | 260.898 |
| 1120 | 40,923 | 31,611 | 237.352 | 1880 | 76,511 | 60,880 | 261.436 |
| 1140 | 41,780 | 32,301 | 238.110 | 1900 | 77,517 | 61,720 | 261.969 |
| 1160 | 42,642 | 32,997 | 238.859 | 1920 | 78,527 | 62,564 | 262.497 |
| 1180 | 43,509 | 33,698 | 239.600 | 1940 | 79,540 | 63,411 | 263.022 |
| 1200 | 44,380 | 34,403 | 240.333 | 1960 | 80,555 | 64,259 | 263.542 |
| 1220 | 45,256 | 35,112 | 241.057 | 1980 | 81,573 | 65,111 | 264.059 |
| 1240 | 46,137 | 35,827 | 241.773 | 2000 | 82,593 | 65,965 | 264.571 |
| 1260 | 47,022 | 36,546 | 242.482 | 2050 | 85,156 | 68,111 | 265.838 |
| 1280 | 47,912 | 37,270 | 243.183 | 2100 | 87,735 | 70,275 | 267.081 |
| 1300 | 48,807 | 38,000 | 243.877 | 2150 | 90,330 | 72,454 | 268.301 |
| 1320 | 49,707 | 38,732 | 244.564 | 2200 | 92,940 | 74,649 | 269.500 |
| 1340 | 50,612 | 39,470 | 245.243 | 2250 | 95,562 | 76,855 | 270.679 |
| 1360 | 51,521 | 40,213 | 245.915 | 2300 | 98,199 | 79,076 | 271.839 |
| 1380 | 52,434 | 40,960 | 246.582 | 2350 | 100,846 | 81,308 | 272.978 |
| 1400 | 53,351 | 41,711 | 247.241 | 2400 | 103,508 | 83,553 | 274.098 |
| 1420 | 54,273 | 42,466 | 247.895 | 2450 | 106,183 | 85,811 | 275.201 |
| 1440 | 55,198 | 43,226 | 248.543 | 2500 | 108,868 | 88,082 | 276.286 |
| 1460 | 56,128 | 43,989 | 249.185 | 2550 | 111,565 | 90,364 | 277.354 |
| 1480 | 57,062 | 44,756 | 249.820 | 2600 | 114,273 | 92,656 | 278.407 |
| 1500 | 57,999 | 45,528 | 250.450 | 2650 | 116,991 | 94,958 | 279.441 |
| 1520 | 58,942 | 46,304 | 251.074 | 2700 | 119,717 | 97,269 | 280.462 |
| 1540 | 59,888 | 47,084 | 251.693 | 2750 | 122,453 | 99,588 | 281.464 |
| 1560 | 60,838 | 47,868 | 252.305 | 2800 | 125,198 | 101,917 | 282.453 |
| 1580 | 61,792 | 48,655 | 252.912 | 2850 | 127,952 | 104,256 | 283.429 |
| 1600 | 62,748 | 49,445 | 253.513 | 2900 | 130,717 | 106,605 | 284.390 |
| 1620 | 63,709 | 50,240 | 254.111 | 2950 | 133,486 | 108,959 | 285.338 |
| 1640 | 64,675 | 51,039 | 254.703 | 3000 | 136,264 | 111,321 | 286.273 |
| 1660 | 65,643 | 51,841 | 255.290 | 3050 | 139,051 | 113,692 | 287.194 |
| 1680 | 66,614 | 52,646 | 255.873 | 3100 | 141,846 | 116,072 | 288.102 |
| 1700 | 67,589 | 53,455 | 256.450 | 3150 | 144,648 | 118,458 | 288.999 |
| 1720 | 68,567 | 54,267 | 257.022 | 3200 | 147,457 | 120,851 | 289.884 |
| 1740 | 69,550 | 55,083 | 257.589 | 3250 | 150,272 | 123,250 | 290.756 |

## TABLE A–24

Ideal-gas properties of monatomic oxygen, O

| T<br>K | $\bar{h}$<br>kJ/kmol | $\bar{u}$<br>kJ/kmol | $\bar{s}°$<br>kJ/kmol·K | T<br>K | $\bar{h}$<br>kJ/kmol | $\bar{u}$<br>kJ/kmol | $\bar{s}°$<br>kJ/kmol·K |
|---|---|---|---|---|---|---|---|
| 0 | 0 | 0 | 0 | 2400 | 50,894 | 30,940 | 204.932 |
| 298 | 6,852 | 4,373 | 160.944 | 2450 | 51,936 | 31,566 | 205.362 |
| 300 | 6,892 | 4,398 | 161.079 | 2500 | 52,979 | 32,193 | 205.783 |
| 500 | 11,197 | 7,040 | 172.088 | 2550 | 54,021 | 32,820 | 206.196 |
| 1000 | 21,713 | 13,398 | 186.678 | 2600 | 55,064 | 33,447 | 206.601 |
| 1500 | 32,150 | 19,679 | 195.143 | 2650 | 56,108 | 34,075 | 206.999 |
| 1600 | 34,234 | 20,931 | 196.488 | 2700 | 57,152 | 34,703 | 207.389 |
| 1700 | 36,317 | 22,183 | 197.751 | 2750 | 58,196 | 35,332 | 207.772 |
| 1800 | 38,400 | 23,434 | 198.941 | 2800 | 59,241 | 35,961 | 208.148 |
| 1900 | 40,482 | 24,685 | 200.067 | 2850 | 60,286 | 36,590 | 208.518 |
| 2000 | 42,564 | 25,935 | 201.135 | 2900 | 61,332 | 37,220 | 208.882 |
| 2050 | 43,605 | 26,560 | 201.649 | 2950 | 62,378 | 37,851 | 209.240 |
| 2100 | 44,646 | 27,186 | 202.151 | 3000 | 63,425 | 38,482 | 209.592 |
| 2150 | 45,687 | 27,811 | 202.641 | 3100 | 65,520 | 39,746 | 210.279 |
| 2200 | 46,728 | 28,436 | 203.119 | 3200 | 67,619 | 41,013 | 210.945 |
| 2250 | 47,769 | 29,062 | 203.588 | 3300 | 69,720 | 42,283 | 211.592 |
| 2300 | 48,811 | 29,688 | 204.045 | 3400 | 71,824 | 43,556 | 212.220 |
| 2350 | 49,852 | 30,314 | 204.493 | 3500 | 73,932 | 44,832 | 212.831 |

## TABLE A–25

Ideal-gas properties of hydroxyl, OH

| T<br>K | $\bar{h}$<br>kJ/kmol | $\bar{u}$<br>kJ/kmol | $\bar{s}°$<br>kJ/kmol·K | T<br>K | $\bar{h}$<br>kJ/kmol | $\bar{u}$<br>kJ/kmol | $\bar{s}°$<br>kJ/kmol·K |
|---|---|---|---|---|---|---|---|
| 0 | 0 | 0 | 0 | 2400 | 77,015 | 57,061 | 248.628 |
| 298 | 9,188 | 6,709 | 183.594 | 2450 | 78,801 | 58,431 | 249.364 |
| 300 | 9,244 | 6,749 | 183.779 | 2500 | 80,592 | 59,806 | 250.088 |
| 500 | 15,181 | 11,024 | 198.955 | 2550 | 82,388 | 61,186 | 250.799 |
| 1000 | 30,123 | 21,809 | 219.624 | 2600 | 84,189 | 62,572 | 251.499 |
| 1500 | 46,046 | 33,575 | 232.506 | 2650 | 85,995 | 63,962 | 252.187 |
| 1600 | 49,358 | 36,055 | 234.642 | 2700 | 87,806 | 65,358 | 252.864 |
| 1700 | 52,706 | 38,571 | 236.672 | 2750 | 89,622 | 66,757 | 253.530 |
| 1800 | 56,089 | 41,123 | 238.606 | 2800 | 91,442 | 68,162 | 254.186 |
| 1900 | 59,505 | 43,708 | 240.453 | 2850 | 93,266 | 69,570 | 254.832 |
| 2000 | 62,952 | 46,323 | 242.221 | 2900 | 95,095 | 70,983 | 255.468 |
| 2050 | 64,687 | 47,642 | 243.077 | 2950 | 96,927 | 72,400 | 256.094 |
| 2100 | 66,428 | 48,968 | 243.917 | 3000 | 98,763 | 73,820 | 256.712 |
| 2150 | 68,177 | 50,301 | 244.740 | 3100 | 102,447 | 76,673 | 257.919 |
| 2200 | 69,932 | 51,641 | 245.547 | 3200 | 106,145 | 79,539 | 259.093 |
| 2250 | 71,694 | 52,987 | 246.338 | 3300 | 109,855 | 82,418 | 260.235 |
| 2300 | 73,462 | 54,339 | 247.116 | 3400 | 113,578 | 85,309 | 261.347 |
| 2350 | 75,236 | 55,697 | 247.879 | 3500 | 117,312 | 88,212 | 262.429 |

## TABLE A–26

Enthalpy of formation, Gibbs function of formation, and absolute entropy at 25°C, 1 atm

| Substance | Formula | $\overline{h}_f^\circ$ kJ/kmol | $\overline{g}_f^\circ$ kJ/kmol | $\overline{s}^\circ$ kJ/kmol·K |
|-----------|---------|---------|---------|---------|
| Carbon | C(s) | 0 | 0 | 5.74 |
| Hydrogen | $H_2(g)$ | 0 | 0 | 130.68 |
| Nitrogen | $N_2(g)$ | 0 | 0 | 191.61 |
| Oxygen | $O_2(g)$ | 0 | 0 | 205.04 |
| Carbon monoxide | CO(g) | −110,530 | −137,150 | 197.65 |
| Carbon dioxide | $CO_2(g)$ | −393,520 | −394,360 | 213.80 |
| Water vapor | $H_2O(g)$ | −241,820 | −228,590 | 188.83 |
| Water | $H_2O(\ell)$ | −285,830 | −237,180 | 69.92 |
| Hydrogen peroxide | $H_2O_2(g)$ | −136,310 | −105,600 | 232.63 |
| Ammonia | $NH_3(g)$ | −46,190 | −16,590 | 192.33 |
| Methane | $CH_4(g)$ | −74,850 | −50,790 | 186.16 |
| Acetylene | $C_2H_2(g)$ | +226,730 | +209,170 | 200.85 |
| Ethylene | $C_2H_4(g)$ | +52,280 | +68,120 | 219.83 |
| Ethane | $C_2H_6(g)$ | −84,680 | −32,890 | 229.49 |
| Propylene | $C_3H_6(g)$ | +20,410 | +62,720 | 266.94 |
| Propane | $C_3H_8(g)$ | −103,850 | −23,490 | 269.91 |
| n-Butane | $C_4H_{10}(g)$ | −126,150 | −15,710 | 310.12 |
| n-Octane | $C_8H_{18}(g)$ | −208,450 | +16,530 | 466.73 |
| n-Octane | $C_8H_{18}(\ell)$ | −249,950 | +6,610 | 360.79 |
| n-Dodecane | $C_{12}H_{26}(g)$ | −291,010 | +50,150 | 622.83 |
| Benzene | $C_6H_6(g)$ | +82,930 | +129,660 | 269.20 |
| Methyl alcohol | $CH_3OH(g)$ | −200,670 | −162,000 | 239.70 |
| Methyl alcohol | $CH_3OH(\ell)$ | −238,660 | −166,360 | 126.80 |
| Ethyl alcohol | $C_2H_5OH(g)$ | −235,310 | −168,570 | 282.59 |
| Ethyl alcohol | $C_2H_5OH(\ell)$ | −277,690 | −174,890 | 160.70 |
| Oxygen | O(g) | +249,190 | +231,770 | 161.06 |
| Hydrogen | H(g) | +218,000 | +203,290 | 114.72 |
| Nitrogen | N(g) | +472,650 | +455,510 | 153.30 |
| Hydroxyl | OH(g) | +39,460 | +34,280 | 183.70 |

*Source of Data:* From JANAF, *Thermochemical Tables* (Midland, MI: Dow Chemical Co., 1971); *Selected Values of Chemical Thermodynamic Properties,* NBS Technical Note 270-3, 1968; and *API Research Project 44* (Carnegie Press, 1953).

## TABLE A–27

Properties of some common fuels and hydrocarbons

| Fuel (phase) | Formula | Molar mass, kg/kmol | Density,[1] kg/L | Enthalpy of vaporization,[2] kJ/kg | Specific heat,[1] $c_p$ kJ/kg·K | Higher heating value,[3] kJ/kg | Lower heating value,[3] kJ/kg |
|---|---|---|---|---|---|---|---|
| Carbon (s) | C | 12.011 | 2 | — | 0.708 | 32,800 | 32,800 |
| Hydrogen (g) | $H_2$ | 2.016 | — | — | 14.4 | 141,800 | 120,000 |
| Carbon monoxide (g) | CO | 28.013 | — | — | 1.05 | 10,100 | 10,100 |
| Methane (g) | $CH_4$ | 16.043 | — | 509 | 2.20 | 55,530 | 50,050 |
| Methanol (ℓ) | $CH_4O$ | 32.042 | 0.790 | 1168 | 2.53 | 22,660 | 19,920 |
| Acetylene (g) | $C_2H_2$ | 26.038 | — | — | 1.69 | 49,970 | 48,280 |
| Ethane (g) | $C_2H_6$ | 30.070 | — | 172 | 1.75 | 51,900 | 47,520 |
| Ethanol (ℓ) | $C_2H_6O$ | 46.069 | 0.790 | 919 | 2.44 | 29,670 | 26,810 |
| Propane (ℓ) | $C_3H_8$ | 44.097 | 0.500 | 335 | 2.77 | 50,330 | 46,340 |
| Butane (ℓ) | $C_4H_{10}$ | 58.123 | 0.579 | 362 | 2.42 | 49,150 | 45,370 |
| 1-Pentene (ℓ) | $C_5H_{10}$ | 70.134 | 0.641 | 363 | 2.20 | 47,760 | 44,630 |
| Isopentane (ℓ) | $C_5H_{12}$ | 72.150 | 0.626 | — | 2.32 | 48,570 | 44,910 |
| Benzene (ℓ) | $C_6H_6$ | 78.114 | 0.877 | 433 | 1.72 | 41,800 | 40,100 |
| Hexene (ℓ) | $C_6H_{12}$ | 84.161 | 0.673 | 392 | 1.84 | 47,500 | 44,400 |
| Hexane (ℓ) | $C_6H_{14}$ | 86.177 | 0.660 | 366 | 2.27 | 48,310 | 44,740 |
| Toluene (ℓ) | $C_7H_8$ | 92.141 | 0.867 | 412 | 1.71 | 42,400 | 40,500 |
| Heptane (ℓ) | $C_7H_{16}$ | 100.204 | 0.684 | 365 | 2.24 | 48,100 | 44,600 |
| Octane (ℓ) | $C_8H_{18}$ | 114.231 | 0.703 | 363 | 2.23 | 47,890 | 44,430 |
| Decane (ℓ) | $C_{10}H_{22}$ | 142.285 | 0.730 | 361 | 2.21 | 47,640 | 44,240 |
| Gasoline (ℓ) | $C_nH_{1.87n}$ | 100–110 | 0.72–0.78 | 350 | 2.4 | 47,300 | 44,000 |
| Light diesel (ℓ) | $C_nH_{1.8n}$ | 170 | 0.78–0.84 | 270 | 2.2 | 46,100 | 43,200 |
| Heavy diesel (ℓ) | $C_nH_{1.7n}$ | 200 | 0.82–0.88 | 230 | 1.9 | 45,500 | 42,800 |
| Natural gas (g) | $C_nH_{3.8n}N_{0.1n}$ | 18 | — | — | 2 | 50,000 | 45,000 |

[1] At 1 atm and 20°C.
[2] At 25°C for liquid fuels, and 1 atm and normal boiling temperature for gaseous fuels.
[3] At 25°C. Multiply by molar mass to obtain heating values in kJ/kmol.

## TABLE A–28

Natural logarithms of the equilibrium constant $K_p$

The equilibrium constant $K_p$ for the reaction $\nu_A A + \nu_B B \rightleftharpoons \nu_C C + \nu_D D$ is defined as $K_p \equiv \dfrac{P_C^{\nu_C} P_D^{\nu_D}}{P_A^{\nu_A} P_B^{\nu_B}}$

| Temp., K | $H_2 \rightleftharpoons 2H$ | $O_2 \rightleftharpoons 2O$ | $N_2 \rightleftharpoons 2N$ | $H_2O \rightleftharpoons H_2 + \frac{1}{2}O_2$ | $H_2O \rightleftharpoons \frac{1}{2}H_2 + OH$ | $CO_2 \rightleftharpoons CO + \frac{1}{2}O_2$ | $\frac{1}{2}N_2 + \frac{1}{2}O_2 \rightleftharpoons NO$ |
|---|---|---|---|---|---|---|---|
| 298 | −164.005 | −186.975 | −367.480 | −92.208 | −106.208 | −103.762 | −35.052 |
| 500 | −92.827 | −105.630 | −213.372 | −52.691 | −60.281 | −57.616 | −20.295 |
| 1000 | −39.803 | −45.150 | −99.127 | −23.163 | −26.034 | −23.529 | −9.388 |
| 1200 | −30.874 | −35.005 | −80.011 | −18.182 | −20.283 | −17.871 | −7.569 |
| 1400 | −24.463 | −27.742 | −66.329 | −14.609 | −16.099 | −13.842 | −6.270 |
| 1600 | −19.637 | −22.285 | −56.055 | −11.921 | −13.066 | −10.830 | −5.294 |
| 1800 | −15.866 | −18.030 | −48.051 | −9.826 | −10.657 | −8.497 | −4.536 |
| 2000 | −12.840 | −14.622 | −41.645 | −8.145 | −8.728 | −6.635 | −3.931 |
| 2200 | −10.353 | −11.827 | −36.391 | −6.768 | −7.148 | −5.120 | −3.433 |
| 2400 | −8.276 | −9.497 | −32.011 | −5.619 | −5.832 | −3.860 | −3.019 |
| 2600 | −6.517 | −7.521 | −28.304 | −4.648 | −4.719 | −2.801 | −2.671 |
| 2800 | −5.002 | −5.826 | −25.117 | −3.812 | −3.763 | −1.894 | −2.372 |
| 3000 | −3.685 | −4.357 | −22.359 | −3.086 | −2.937 | −1.111 | −2.114 |
| 3200 | −2.534 | −3.072 | −19.937 | −2.451 | −2.212 | −0.429 | −1.888 |
| 3400 | −1.516 | −1.935 | −17.800 | −1.891 | −1.576 | 0.169 | −1.690 |
| 3600 | −0.609 | −0.926 | −15.898 | −1.392 | −1.088 | 0.701 | −1.513 |
| 3800 | 0.202 | −0.019 | −14.199 | −0.945 | −0.501 | 1.176 | −1.356 |
| 4000 | 0.934 | 0.796 | −12.660 | −0.542 | −0.044 | 1.599 | −1.216 |
| 4500 | 2.486 | 2.513 | −9.414 | 0.312 | 0.920 | 2.490 | −0.921 |
| 5000 | 3.725 | 3.895 | −6.807 | 0.996 | 1.689 | 3.197 | −0.686 |
| 5500 | 4.743 | 5.023 | −4.666 | 1.560 | 2.318 | 3.771 | −0.497 |
| 6000 | 5.590 | 5.963 | −2.865 | 2.032 | 2.843 | 4.245 | −0.341 |

*Source of Data:* Gordon J. Van Wylen and Richard E. Sonntag, *Fundamentals of Classical Thermodynamics,* English/SI Version, 3rd ed. (New York: John Wiley & Sons, 1986), p. 723, table A.14. Based on thermodynamic data given in JANAF, *Thermochemical Tables* (Midland, MI: Thermal Research Laboratory, The Dow Chemical Company, 1971).

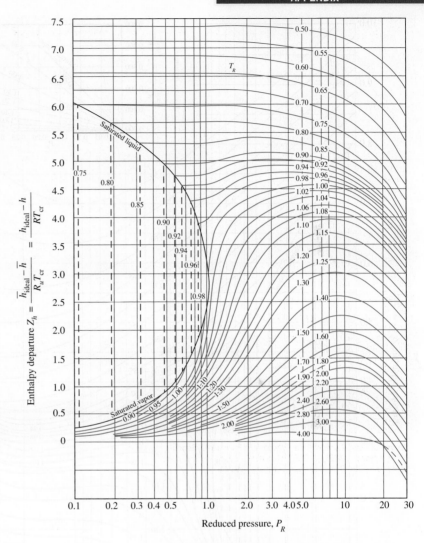

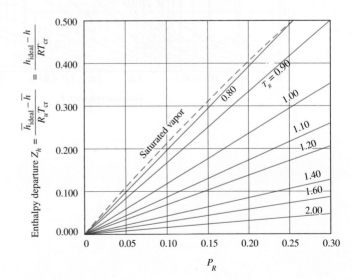

**FIGURE A–29**

Generalized enthalpy departure chart.

*Source of Data: Redrawn from Gordon van Wylen and Richard Sontag,* Fundamentals of Classical Thermodynamics, *(SI version), 2d ed., Wiley, New York, 1976.*

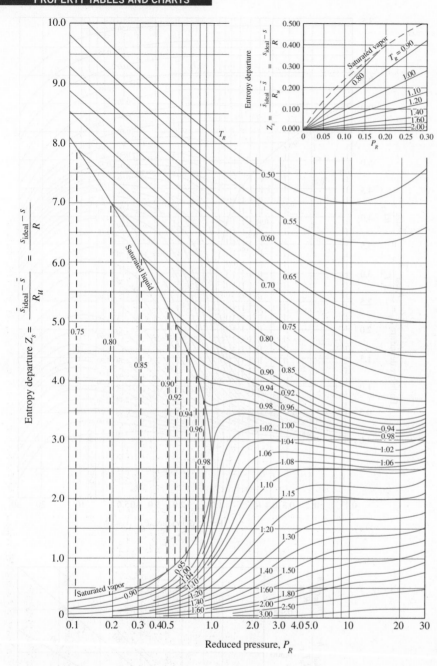

**FIGURE A–30**

Generalized entropy departure chart.

*Source of Data: Redrawn from Gordon van Wylen and Richard Sontag,* Fundamentals of Classical Thermodynamics, *(SI version), 2d ed., Wiley, New York, 1976.*

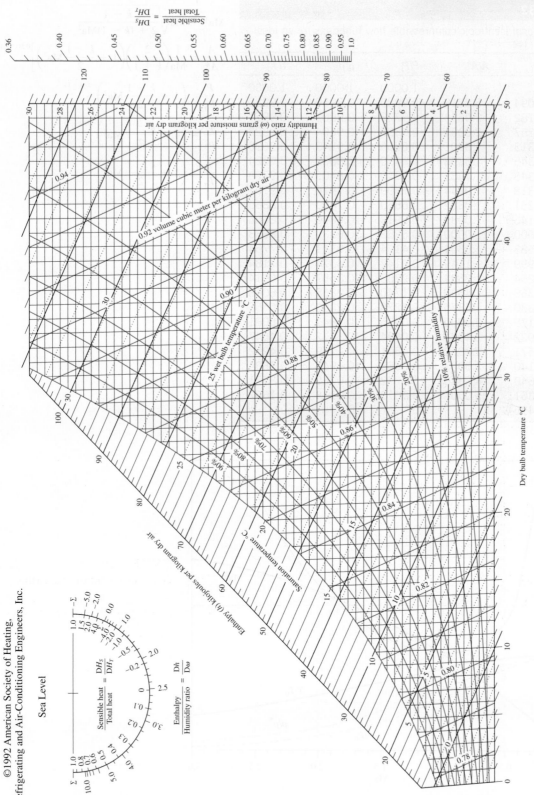

## ASHRAE Psychrometric Chart No. 1
Normal Temperature
Barometric Pressure: 101.325 kPa

©1992 American Society of Heating,
Refrigerating and Air-Conditioning Engineers, Inc.

Sea Level

$$\text{Sensible heat} = \frac{DH_S}{DH_T}$$
$$\text{Total heat}$$

$$\frac{\text{Enthalpy}}{\text{Humidity ratio}} = \frac{Dh}{D\omega}$$

Prepared by Center for Applied Thermodynamic Studies, University of Idaho.

## FIGURE A–31
Psychrometric chart at 1 atm total pressure.

*Reprinted by permission of the American Society of Heating, Refrigerating and Air-Conditioning Engineers, Inc., Atlanta, GA; used with permission.*

## TABLE A–32

One-dimensional isentropic compressible-flow functions for an ideal gas with $k = 1.4$

| Ma | Ma* | $A/A^*$ | $P/P_0$ | $\rho/\rho_0$ | $T/T_0$ |
|---|---|---|---|---|---|
| 0 | 0 | $\infty$ | 1.0000 | 1.0000 | 1.0000 |
| 0.1 | 0.1094 | 5.8218 | 0.9930 | 0.9950 | 0.9980 |
| 0.2 | 0.2182 | 2.9635 | 0.9725 | 0.9803 | 0.9921 |
| 0.3 | 0.3257 | 2.0351 | 0.9395 | 0.9564 | 0.9823 |
| 0.4 | 0.4313 | 1.5901 | 0.8956 | 0.9243 | 0.9690 |
| 0.5 | 0.5345 | 1.3398 | 0.8430 | 0.8852 | 0.9524 |
| 0.6 | 0.6348 | 1.1882 | 0.7840 | 0.8405 | 0.9328 |
| 0.7 | 0.7318 | 1.0944 | 0.7209 | 0.7916 | 0.9107 |
| 0.8 | 0.8251 | 1.0382 | 0.6560 | 0.7400 | 0.8865 |
| 0.9 | 0.9146 | 1.0089 | 0.5913 | 0.6870 | 0.8606 |
| 1.0 | 1.0000 | 1.0000 | 0.5283 | 0.6339 | 0.8333 |
| 1.2 | 1.1583 | 1.0304 | 0.4124 | 0.5311 | 0.7764 |
| 1.4 | 1.2999 | 1.1149 | 0.3142 | 0.4374 | 0.7184 |
| 1.6 | 1.4254 | 1.2502 | 0.2353 | 0.3557 | 0.6614 |
| 1.8 | 1.5360 | 1.4390 | 0.1740 | 0.2868 | 0.6068 |
| 2.0 | 1.6330 | 1.6875 | 0.1278 | 0.2300 | 0.5556 |
| 2.2 | 1.7179 | 2.0050 | 0.0935 | 0.1841 | 0.5081 |
| 2.4 | 1.7922 | 2.4031 | 0.0684 | 0.1472 | 0.4647 |
| 2.6 | 1.8571 | 2.8960 | 0.0501 | 0.1179 | 0.4252 |
| 2.8 | 1.9140 | 3.5001 | 0.0368 | 0.0946 | 0.3894 |
| 3.0 | 1.9640 | 4.2346 | 0.0272 | 0.0760 | 0.3571 |
| 5.0 | 2.2361 | 25.000 | 0.0019 | 0.0113 | 0.1667 |
| $\infty$ | 2.2495 | $\infty$ | 0 | 0 | 0 |

$$\text{Ma}^* = \text{Ma}\sqrt{\frac{k+1}{2 + (k-1)\text{Ma}^2}}$$

$$\frac{A}{A^*} = \frac{1}{\text{Ma}}\left[\left(\frac{2}{k+1}\right)\left(1 + \frac{k-1}{2}\text{Ma}^2\right)\right]^{0.5(k+1)/(k-1)}$$

$$\frac{P}{P_0} = \left(1 + \frac{k-1}{2}\text{Ma}^2\right)^{-k/(k-1)}$$

$$\frac{\rho}{\rho_0} = \left(1 + \frac{k-1}{2}\text{Ma}^2\right)^{-1/(k-1)}$$

$$\frac{T}{T_0} = \left(1 + \frac{k-1}{2}\text{Ma}^2\right)^{-1}$$

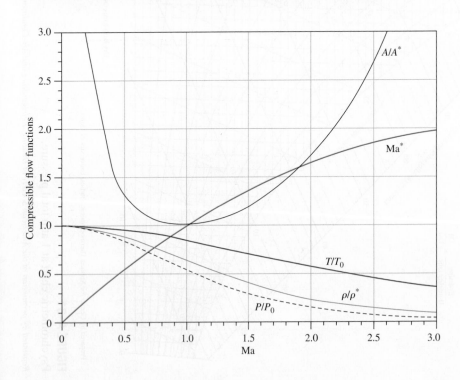

$$T_{01} = T_{02}$$

$$Ma_2 = \sqrt{\frac{(k-1)Ma_1^2 + 2}{2kMa_1^2 - k + 1}}$$

$$\frac{P_2}{P_1} = \frac{1 + kMa_1^2}{1 + kMa_2^2} = \frac{2kMa_1^2 - k + 1}{k + 1}$$

$$\frac{\rho_2}{\rho_1} = \frac{P_2/P_1}{T_2/T_1} = \frac{(k+1)Ma_1^2}{2 + (k-1)Ma_1^2} = \frac{V_1}{V_2}$$

$$\frac{T_2}{T_1} = \frac{2 + Ma_1^2(k-1)}{2 + Ma_2^2(k-1)}$$

$$\frac{P_{02}}{P_{01}} = \frac{Ma_1}{Ma_2}\left[\frac{1 + Ma_2^2(k-1)/2}{1 + Ma_1^2(k-1)/2}\right]^{(k+1)/[2(k-1)]}$$

$$\frac{P_{02}}{P_1} = \frac{(1 + kMa_1^2)[1 + Ma_2^2(k-1)/2]^{k/(k-1)}}{1 + kMa_2^2}$$

### TABLE A–33

One-dimensional normal-shock functions for an ideal gas with $k = 1.4$

| $Ma_1$ | $Ma_2$ | $P_2/P_1$ | $\rho_2/\rho_1$ | $T_2/T_1$ | $P_{02}/P_{01}$ | $P_{02}/P_1$ |
|---|---|---|---|---|---|---|
| 1.0 | 1.0000 | 1.0000 | 1.0000 | 1.0000 | 1.0000 | 1.8929 |
| 1.1 | 0.9118 | 1.2450 | 1.1691 | 1.0649 | 0.9989 | 2.1328 |
| 1.2 | 0.8422 | 1.5133 | 1.3416 | 1.1280 | 0.9928 | 2.4075 |
| 1.3 | 0.7860 | 1.8050 | 1.5157 | 1.1909 | 0.9794 | 2.7136 |
| 1.4 | 0.7397 | 2.1200 | 1.6897 | 1.2547 | 0.9582 | 3.0492 |
| 1.5 | 0.7011 | 2.4583 | 1.8621 | 1.3202 | 0.9298 | 3.4133 |
| 1.6 | 0.6684 | 2.8200 | 2.0317 | 1.3880 | 0.8952 | 3.8050 |
| 1.7 | 0.6405 | 3.2050 | 2.1977 | 1.4583 | 0.8557 | 4.2238 |
| 1.8 | 0.6165 | 3.6133 | 2.3592 | 1.5316 | 0.8127 | 4.6695 |
| 1.9 | 0.5956 | 4.0450 | 2.5157 | 1.6079 | 0.7674 | 5.1418 |
| 2.0 | 0.5774 | 4.5000 | 2.6667 | 1.6875 | 0.7209 | 5.6404 |
| 2.1 | 0.5613 | 4.9783 | 2.8119 | 1.7705 | 0.6742 | 6.1654 |
| 2.2 | 0.5471 | 5.4800 | 2.9512 | 1.8569 | 0.6281 | 6.7165 |
| 2.3 | 0.5344 | 6.0050 | 3.0845 | 1.9468 | 0.5833 | 7.2937 |
| 2.4 | 0.5231 | 6.5533 | 3.2119 | 2.0403 | 0.5401 | 7.8969 |
| 2.5 | 0.5130 | 7.1250 | 3.3333 | 2.1375 | 0.4990 | 8.5261 |
| 2.6 | 0.5039 | 7.7200 | 3.4490 | 2.2383 | 0.4601 | 9.1813 |
| 2.7 | 0.4956 | 8.3383 | 3.5590 | 2.3429 | 0.4236 | 9.8624 |
| 2.8 | 0.4882 | 8.9800 | 3.6636 | 2.4512 | 0.3895 | 10.5694 |
| 2.9 | 0.4814 | 9.6450 | 3.7629 | 2.5632 | 0.3577 | 11.3022 |
| 3.0 | 0.4752 | 10.3333 | 3.8571 | 2.6790 | 0.3283 | 12.0610 |
| 4.0 | 0.4350 | 18.5000 | 4.5714 | 4.0469 | 0.1388 | 21.0681 |
| 5.0 | 0.4152 | 29.000 | 5.0000 | 5.8000 | 0.0617 | 32.6335 |
| $\infty$ | 0.3780 | $\infty$ | 6.0000 | $\infty$ | 0 | $\infty$ |

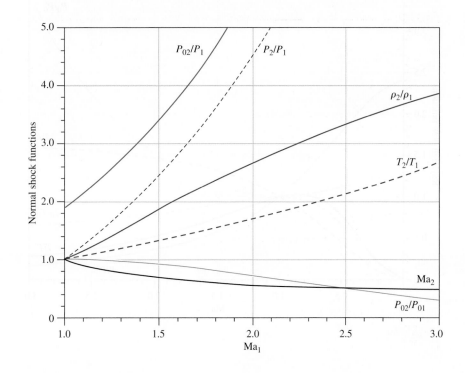

## TABLE A–34

Rayleigh flow functions for an ideal gas with $k = 1.4$

| Ma | $T_0/T_0^*$ | $P_0/P_0^*$ | $T/T^*$ | $P/P^*$ | $V/V^*$ |
|---|---|---|---|---|---|
| 0.0 | 0.0000 | 1.2679 | 0.0000 | 2.4000 | 0.0000 |
| 0.1 | 0.0468 | 1.2591 | 0.0560 | 2.3669 | 0.0237 |
| 0.2 | 0.1736 | 1.2346 | 0.2066 | 2.2727 | 0.0909 |
| 0.3 | 0.3469 | 1.1985 | 0.4089 | 2.1314 | 0.1918 |
| 0.4 | 0.5290 | 1.1566 | 0.6151 | 1.9608 | 0.3137 |
| 0.5 | 0.6914 | 1.1141 | 0.7901 | 1.7778 | 0.4444 |
| 0.6 | 0.8189 | 1.0753 | 0.9167 | 1.5957 | 0.5745 |
| 0.7 | 0.9085 | 1.0431 | 0.9929 | 1.4235 | 0.6975 |
| 0.8 | 0.9639 | 1.0193 | 1.0255 | 1.2658 | 0.8101 |
| 0.9 | 0.9921 | 1.0049 | 1.0245 | 1.1246 | 0.9110 |
| 1.0 | 1.0000 | 1.0000 | 1.0000 | 1.0000 | 1.0000 |
| 1.2 | 0.9787 | 1.0194 | 0.9118 | 0.7958 | 1.1459 |
| 1.4 | 0.9343 | 1.0777 | 0.8054 | 0.6410 | 1.2564 |
| 1.6 | 0.8842 | 1.1756 | 0.7017 | 0.5236 | 1.3403 |
| 1.8 | 0.8363 | 1.3159 | 0.6089 | 0.4335 | 1.4046 |
| 2.0 | 0.7934 | 1.5031 | 0.5289 | 0.3636 | 1.4545 |
| 2.2 | 0.7561 | 1.7434 | 0.4611 | 0.3086 | 1.4938 |
| 2.4 | 0.7242 | 2.0451 | 0.4038 | 0.2648 | 1.5252 |
| 2.6 | 0.6970 | 2.4177 | 0.3556 | 0.2294 | 1.5505 |
| 2.8 | 0.6738 | 2.8731 | 0.3149 | 0.2004 | 1.5711 |
| 3.0 | 0.6540 | 3.4245 | 0.2803 | 0.1765 | 1.5882 |

$$\frac{T_0}{T_0^*} = \frac{(k + 1)\text{Ma}^2[2 + (k - 1)\text{Ma}^2]}{(1 + k\text{Ma}^2)^2}$$

$$\frac{P_0}{P_0^*} = \frac{k + 1}{1 + k\text{Ma}^2}\left(\frac{2 + (k - 1)\text{Ma}^2}{k + 1}\right)^{k/(k - 1)}$$

$$\frac{T}{T^*} = \left(\frac{\text{Ma}(1 + k)}{1 + k\text{Ma}^2}\right)^2$$

$$\frac{P}{P^*} = \frac{1 + k}{1 + k\text{Ma}^2}$$

$$\frac{V}{V^*} = \frac{\rho^*}{\rho} = \frac{(1 + k)\text{Ma}^2}{1 + k\text{Ma}^2}$$

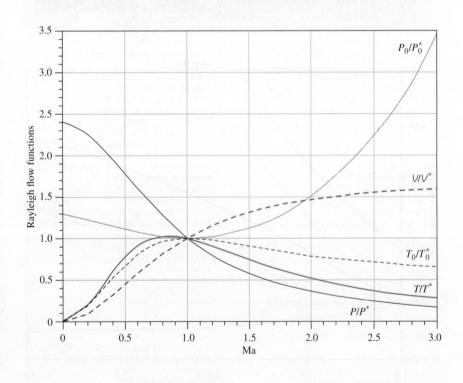

# INDEX

# Nomenclature

| | | | | |
|---|---|---|---|---|
| $a$ | Acceleration, m/s$^2$ | | MEP | Mean effective pressure, kPa |
| $a$ | Specific Helmholtz function, $u - Ts$, kJ/kg | | mf | Mass fraction |
| $A$ | Area, m$^2$ | | $n$ | Polytropic exponent |
| $A$ | Helmholtz function, $U - TS$, kJ | | $N$ | Number of moles, kmol |
| AF | Air–fuel ratio | | $P$ | Pressure, kPa |
| $c$ | Speed of sound, m/s | | $P_{cr}$ | Critical pressure, kPa |
| $c$ | Specific heat, kJ/kg·K | | $P_i$ | Partial pressure, kPa |
| $c_p$ | Constant pressure specific heat, kJ/kg·K | | $P_m$ | Mixture pressure, kPa |
| $c_v$ | Constant volume specific heat, kJ/kg·K | | $P_r$ | Relative pressure |
| COP | Coefficient of performance | | $P_R$ | Reduced pressure |
| COP$_{HP}$ | Coefficient of performance of a heat pump | | $P_v$ | Vapor pressure, kPa |
| COP$_R$ | Coefficient of performance of a refrigerator | | $P_0$ | Surroundings pressure, kPa |
| $d, D$ | Diameter, m | | pe | Specific potential energy, $gz$, kJ/kg |
| $e$ | Specific total energy, kJ/kg | | PE | Total potential energy, $mgz$, kJ |
| $E$ | Total energy, kJ | | $q$ | Heat transfer per unit mass, kJ/kg |
| EER | Energy efficiency rating | | $Q$ | Total heat transfer, kJ |
| $F$ | Force, N | | $\dot{Q}$ | Heat transfer rate, kW |
| FA | Fuel–air ratio | | $Q_H$ | Heat transfer with high-temperature body, kJ |
| $g$ | Gravitational acceleration, m/s$^2$ | | $Q_L$ | Heat transfer with low-temperature body, kJ |
| $g$ | Specific Gibbs function, $h - Ts$, kJ/kg | | $r$ | Compression ratio |
| $G$ | Total Gibbs function, $H - TS$, kJ | | $R$ | Gas constant, kJ/kg·K |
| $h$ | Convection heat transfer coefficient, W/m$^2$·K | | $r_c$ | Cutoff ratio |
| | | | $r_p$ | Pressure ratio |
| $h$ | Specific enthalpy, $u + Pv$, kJ/kg | | $R_u$ | Universal gas constant, kJ/kmol·K |
| $H$ | Total enthalpy, $U + PV$, kJ | | $s$ | Specific entropy, kJ/kg·K |
| $\bar{h}_C$ | Enthalpy of combustion, kJ/kmol fuel | | $S$ | Total entropy, kJ/K |
| $\bar{h}_f$ | Enthalpy of formation, kJ/kmol | | $s_{gen}$ | Specific entropy generation, kJ/kg·K |
| $\bar{h}_R$ | Enthalpy of reaction, kJ/kmol | | $S_{gen}$ | Total entropy generation, kJ/K |
| HHV | Higher heating value, kJ/kg fuel | | SG | Specific gravity or relative density |
| $i$ | Specific irreversibility, kJ/kg | | $t$ | Time, s |
| $I$ | Electric current, A | | $T$ | Temperature, °C or K |
| $I$ | Total irreversibility, kJ | | T | Torque, N·m |
| $k$ | Specific heat ratio, $c_p/c_v$ | | $T_{cr}$ | Critical temperature, K |
| $k_s$ | Spring constant | | $T_{db}$ | Dry-bulb temperature, °C |
| $k_t$ | Thermal conductivity | | $T_{dp}$ | Dew-point temperature, °C |
| $K_p$ | Equilibrium constant | | $T_f$ | Bulk fluid temperature, °C |
| ke | Specific kinetic energy, $V^2/2$, kJ/kg | | $T_H$ | Temperature of high-temperature body, K |
| KE | Total kinetic energy, $mV^2/2$, kJ | | $T_L$ | Temperature of low-temperature body, K |
| LHV | Lower heating value, kJ/kg fuel | | $T_R$ | Reduced temperature |
| $m$ | Mass, kg | | $T_{wb}$ | Wet-bulb temperature, °C |
| $\dot{m}$ | Mass flow rate, kg/s | | $T_0$ | Surroundings temperature, °C or K |
| $M$ | Molar mass, kg/kmol | | $u$ | Specific internal energy, kJ/kg |
| Ma | Mach number | | $U$ | Total internal energy, kJ |

| | | | | |
|---|---|---|---|---|
| $v$ | Specific volume, m³/kg | | $\phi$ | Specific closed system exergy, kJ/kg |

$v$     Specific volume, m³/kg

$v_{cr}$     Critical specific volume, m³/kg

$v_r$     Relative specific volume

$v_R$     Pseudoreduced specific volume

$V$     Total volume, m³

$\dot{V}$     Volume flow rate, m³/s

**V**     Voltage, V

$V$     Velocity, m/s

$V_{avg}$     Average velocity

$w$     Work per unit mass, kJ/kg

$W$     Total work, kJ

$\dot{W}$     Power, kW

$W_{in}$     Work input, kJ

$W_{out}$     Work output, kJ

$W_{rev}$     Reversible work, kJ

$x$     Quality

$x$     Specific exergy, kJ/kg

$X$     Total exergy, kJ

$x_{dest}$     Specific exergy destruction, kJ/kg

$X_{dest}$     Total exergy destruction, kJ

$\dot{X}_{dest}$     Rate of total exergy destruction, kW

$y$     Mole fraction

$z$     Elevation, m

$Z$     Compressibility factor

$Z_h$     Enthalpy departure factor

$Z_s$     Entropy departure factor

## Greek Letters

$\alpha$     Absorptivity

$\alpha$     Isothermal compressibility, 1/kPa

$\beta$     Volume expansivity, 1/K

$\Delta$     Finite change in quantity

$\varepsilon$     Emissivity

$\epsilon$     Effectiveness

$\eta_{th}$     Thermal efficiency

$\eta_{II}$     Second-law efficiency

$\theta$     Total energy of a flowing fluid, kJ/kg

$\mu_{JT}$     Joule-Thomson coefficient, K/kPa

$\mu$     Chemical potential, kJ/kg

$\nu$     Stoichiometric coefficient

$\rho$     Density, kg/m³

$\sigma$     Stefan–Boltzmann constant

$\sigma_n$     Normal stress, N/m²

$\sigma_s$     Surface tension, N/m

$\phi$     Relative humidity

$\phi$     Specific closed system exergy, kJ/kg

$\Phi$     Total closed system exergy, kJ

$\psi$     Stream exergy, kJ/kg

$\gamma_s$     Specific weight, N/m³

$\omega$     Specific or absolute humidity, kg H₂O/kg dry air

## Subscripts

$a$     Air

abs     Absolute

act     Actual

atm     Atmospheric

avg     Average

$c$     Combustion; cross-section

cr     Critical point

CV     Control volume

$e$     Exit conditions

$f$     Saturated liquid

$fg$     Difference in property between saturated liquid and saturated vapor

$g$     Saturated vapor

gen     Generation

$H$     High temperature (as in $T_H$ and $Q_H$)

$i$     Inlet conditions

$i$     $i$th component

$L$     Low temperature (as in $T_L$ and $Q_L$)

$m$     Mixture

$r$     Relative

$R$     Reduced

rev     Reversible

$s$     Isentropic

sat     Saturated

surr     Surroundings

sys     System

$v$     Water vapor

0     Dead state

1     Initial or inlet state

2     Final or exit state

## Superscripts

˙ (over dot)     Quantity per unit time

‾ (over bar)     Quantity per unit mole

° (circle)     Standard reference state

* (asterisk)     Quantity at 1 atm pressure

# Conversion Factors

| DIMENSION | METRIC | METRIC/ENGLISH |
|---|---|---|
| Acceleration | $1\ m/s^2 = 100\ cm/s^2$ | $1\ m/s^2 = 3.2808\ ft/s^2$<br>$1\ ft/s^2 = 0.3048^*\ m/s^2$ |
| Area | $1\ m^2 = 10^4\ cm^2 = 10^6\ mm^2 = 10^{-6}\ km^2$ | $1\ m^2 = 1550\ in^2 = 10.764\ ft^2$<br>$1\ ft^2 = 144\ in^2 = 0.09290304^*\ m^2$ |
| Density | $1\ g/cm^3 = 1\ kg/L = 1000\ kg/m^3$ | $1\ g/cm^3 = 62.428\ lbm/ft^3 = 0.036127\ lbm/in^3$<br>$1\ lbm/in^3 = 1728\ lbm/ft^3$<br>$1\ kg/m^3 = 0.062428\ lbm/ft^3$ |
| Energy, heat, work, internal energy, enthalpy | $1\ kJ = 1000\ J = 1000\ N{\cdot}m = 1\ kPa{\cdot}m^3$<br>$1\ kJ/kg = 1000\ m^2/s^2$<br>$1\ kWh = 3600\ kJ$<br>$1\ cal^\dagger = 4.184\ J$<br>$1\ IT\ cal^\dagger = 4.1868\ J$<br>$1\ Cal^\dagger = 4.1868\ kJ$ | $1\ kJ = 0.94782\ Btu$<br>$1\ Btu = 1.055056\ kJ$<br>$\quad = 5.40395\ psia{\cdot}ft^3 = 778.169\ lbf{\cdot}ft$<br>$1\ Btu/lbm = 25{,}037\ ft^2/s^2 = 2.326^*\ kJ/kg$<br>$1\ kJ/kg = 0.430\ Btu/lbm$<br>$1\ kWh = 3412.14\ Btu$<br>$1\ therm = 10^5\ Btu = 1.055 \times 10^5\ kJ$<br>$\quad$ (natural gas) |
| Force | $1\ N = 1\ kg{\cdot}m/s^2 = 10^5\ dyne$<br>$1\ kgf = 9.80665\ N$ | $1\ N = 0.22481\ lbf$<br>$1\ lbf = 32.174\ lbm{\cdot}ft/s^2 = 4.44822\ N$ |
| Heat flux | $1\ W/cm^2 = 10^4\ W/m^2$ | $1\ W/m^2 = 0.3171\ Btu/h{\cdot}ft^2$ |
| Heat transfer coefficient | $1\ W/m^2{\cdot}°C = 1\ W/m^2{\cdot}K$ | $1\ W/m^2{\cdot}°C = 0.17612\ Btu/h{\cdot}ft^2{\cdot}°F$ |
| Length | $1\ m = 100\ cm = 1000\ mm = 10^6\ \mu m$<br>$1\ km = 1000\ m$ | $1\ m = 39.370\ in = 3.2808\ ft = 1.0926\ yd$<br>$1\ ft = 12\ in = 0.3048^*\ m$<br>$1\ mile = 5280\ ft = 1.6093\ km$<br>$1\ in = 2.54^*\ cm$ |
| Mass | $1\ kg = 1000\ g$<br>$1\ metric\ ton = 1000\ kg$ | $1\ kg = 2.2046226\ lbm$<br>$1\ lbm = 0.45359237^*\ kg$<br>$1\ ounce = 28.3495\ g$<br>$1\ slug = 32.174\ lbm = 14.5939\ kg$<br>$1\ short\ ton = 2000\ lbm = 907.1847\ kg$ |
| Power, heat transfer rate | $1\ W = 1\ J/s$<br>$1\ kW = 1000\ W = 1.341\ hp$<br>$1\ hp^\ddagger = 745.7\ W$ | $1\ kW = 3412.14\ Btu/h$<br>$\quad = 737.56\ lbf{\cdot}ft/s$<br>$1\ hp = 550\ lbf{\cdot}ft/s = 0.7068\ Btu/s$<br>$\quad = 42.41\ Btu/min = 2544.5\ Btu/h$<br>$\quad = 0.74570\ kW$<br>$1\ boiler\ hp = 33{,}475\ Btu/h$<br>$1\ Btu/h = 1.055056\ kJ/h$<br>$1\ ton\ of\ refrigeration = 200\ Btu/min$ |
| Pressure | $1\ Pa = 1\ N/m^2$<br>$1\ kPa = 10^3\ Pa = 10^{-3}\ MPa$<br>$1\ atm = 101.325\ kPa = 1.01325\ bars$<br>$\quad = 760\ mm\ Hg\ at\ 0°C$<br>$\quad = 1.03323\ kgf/cm^2$<br>$1\ mm\ Hg = 0.1333\ kPa$ | $1\ Pa = 1.4504 \times 10^{-4}\ psia$<br>$\quad = 0.020886\ lbf/ft^2$<br>$1\ psi = 144\ lbf/ft^2 = 6.894757\ kPa$<br>$1\ atm = 14.696\ psia = 29.92\ in\ Hg\ at\ 30°F$<br>$1\ in\ Hg = 3.387\ kPa$ |
| Specific heat | $1\ kJ/kg{\cdot}°C = 1\ kJ/kg{\cdot}K = 1\ J/g{\cdot}°C$ | $1\ Btu/lbm{\cdot}°F = 4.1868\ kJ/kg{\cdot}°C$<br>$1\ Btu/lbmol{\cdot}R = 4.1868\ kJ/kmol{\cdot}K$<br>$1\ kJ/kg{\cdot}°C = 0.23885\ Btu/lbm{\cdot}°F$<br>$\quad = 0.23885\ Btu/lbm{\cdot}R$ |

*Exact conversion factor between metric and English units.

†Calorie is originally defined as the amount of heat needed to raise the temperature of 1 g of water by 1°C, but it varies with temperature. The international steam table (IT) calorie (generally preferred by engineers) is exactly 4.1868 J by definition and corresponds to the specific heat of water at 15°C. The thermochemical calorie (generally preferred by physicists) is exactly 4.184 J by definition and corresponds to the specific heat of water at room temperature. The difference between the two is about 0.06 percent, which is negligible. The capitalized Calorie used by nutritionists is actually a kilocalorie (1000 IT calories).

| DIMENSION | METRIC | METRIC/ENGLISH |
|---|---|---|
| Specific volume | $1 \ m^3/kg = 1000 \ L/kg = 1000 \ cm^3/g$ | $1 \ m^3/kg = 16.02 \ ft^3/lbm$ <br> $1 \ ft^3/lbm = 0.062428 \ m^3/kg$ |
| Temperature | $T(K) = T(°C) + 273.15$ <br> $\Delta T(K) = \Delta T(°C)$ | $T(R) = T(°F) + 459.67 = 1.8T(K)$ <br> $T(°F) = 1.8 \ T(°C) + 32$ <br> $\Delta T(°F) = \Delta T(R) = 1.8 \ \Delta T(K)$ |
| Thermal conductivity | $1 \ W/m·°C = 1 \ W/m·K$ | $1 \ W/m·°C = 0.57782 \ Btu/h·ft·°F$ |
| Velocity | $1 \ m/s = 3.60 \ km/h$ | $1 \ m/s = 3.2808 \ ft/s = 2.237 \ mi/h$ <br> $1 \ mi/h = 1.46667 \ ft/s$ <br> $1 \ mi/h = 1.6093 \ km/h$ |
| Volume | $1 \ m^3 = 1000 \ L = 10^6 \ cm^3 \ (cc)$ | $1 \ m^3 = 6.1024 \times 10^4 \ in^3 = 35.315 \ ft^3$ <br> $= 264.17 \ gal \ (U.S.)$ <br> $1 \ U.S. \ gallon = 231 \ in^3 = 3.7854 \ L$ <br> $1 \ fl \ ounce = 29.5735 \ cm^3 = 0.0295735 \ L$ <br> $1 \ U.S. \ gallon = 128 \ fl \ ounces$ |
| Volume flow rate | $1 \ m^3/s = 60,000 \ L/min = 10^6 \ cm^3/s$ | $1 \ m^3/s = 15,850 \ gal/min \ (gpm) = 35.315 \ ft^3/s$ <br> $= 2118.9 \ ft^3/min \ (cfm)$ |

‡Mechanical horsepower. The electrical horsepower is taken to be exactly 746 W.

## Some Physical Constants

| | |
|---|---|
| Universal gas constant | $R_u = 8.31447 \ kJ/kmol·K$ <br> $= 8.31447 \ kPa·m^3/kmol·K$ <br> $= 0.0831447 \ bar·m^3/kmol·K$ <br> $= 82.05 \ L·atm/kmol·K$ <br> $= 1.9858 \ Btu/lbmol·R$ <br> $= 1545.37 \ ft·lbf/lbmol·R$ <br> $= 10.73 \ psia·ft^3/lbmol·R$ |
| Standard acceleration of gravity | $g = 9.80665 \ m/s^2$ <br> $= 32.174 \ ft/s^2$ |
| Standard atmospheric pressure | $1 \ atm = 101.325 \ kPa$ <br> $= 1.01325 \ bar$ <br> $= 14.696 \ psia$ <br> $= 760 \ mm \ Hg \ (0°C)$ <br> $= 29.9213 \ in \ Hg \ (32°F)$ <br> $= 10.3323 \ m \ H_2O \ (4°C)$ |
| Stefan–Boltzmann constant | $\sigma = 5.6704 \times 10^{-8} \ W/m^2·K^4$ <br> $= 0.1714 \times 10^{-8} \ Btu/h·ft^2·R^4$ |
| Boltzmann's constant | $k = 1.380650 \times 10^{-23} \ J/K$ |
| Speed of light in vacuum | $c_o = 2.9979 \times 10^8 \ m/s$ <br> $= 9.836 \times 10^8 \ ft/s$ |
| Speed of sound in dry air at 0°C and 1 atm | $c = 331.36 \ m/s$ <br> $= 1089 \ ft/s$ |
| Heat of fusion of water at 1 atm | $h_{if} = 333.7 \ kJ/kg$ <br> $= 143.5 \ Btu/lbm$ |
| Enthalpy of vaporization of water at 1 atm | $h_{fg} = 2256.5 \ kJ/kg$ <br> $= 970.12 \ Btu/lbm$ |